Dr Johannes Belzner
Inst Organische Chem
Tammannstr 2
37077 GOTTINGEN,
Germany

The chemistry of
organic silicon compounds
Volume 2

THE CHEMISTRY OF FUNCTIONAL GROUPS

A series of advanced treatises under the general editorship of Professors Saul Patai and Zvi Rappoport

The chemistry of alkenes (2 volumes)
The chemistry of the carbonyl group (2 volumes)
The chemistry of the ether linkage
The chemistry of the amino group
The chemistry of the nitro and nitroso groups (2 parts)
The chemistry of carboxylic acids and esters
The chemistry of the carbon–nitrogen double bond
The chemistry of amides
The chemistry of the cyano group
The chemistry of the hydroxyl group (2 parts)
The chemistry of the azido group
The chemistry of acyl halides
The chemistry of the carbon–halogen bond (2 parts)
The chemistry of the quinonoid compounds (2 volumes, 4 parts)
The chemistry of the thiol group (2 parts)
The chemistry of the hydrazo, azo and azoxy groups (2 volumes, 3 parts)
The chemistry of amidines and imidates (2 volumes)
The chemistry of cyanates and their thio derivatives (2 parts)
The chemistry of diazonium and diazo groups (2 parts)
The chemistry of the carbon–carbon triple bond (2 parts)
The chemistry of ketenes, allenes and related compounds (2 parts)
The chemistry of the sulphonium group (2 parts)
Supplement A: The chemistry of double-bonded functional groups (3 volumes, 6 parts)
Supplement B: The chemistry of acid derivatives (2 volumes, 4 parts)
Supplement C: The chemistry of triple-bonded functional groups (2 volumes, 3 parts)
Supplement D: The chemistry of halides, pseudo-halides and azides (2 volumes, 4 parts)
Supplement E: The chemistry of ethers, crown ethers, hydroxyl groups and their sulphur analogues (2 volumes, 3 parts)
Supplement F: The chemistry of amino, nitroso and nitro compounds and their derivatives (2 volumes, 4 parts)
The chemistry of the metal–carbon bond (5 volumes)
The chemistry of peroxides
The chemistry of organic selenium and tellurium compounds (2 volumes)
The chemistry of the cyclopropyl group (2 volumes, 3 parts)
The chemistry of sulphones and sulphoxides
The chemistry of organic silicon compounds (2 volumes, 5 parts)
The chemistry of enones (2 parts)
The chemistry of sulphinic acids, esters and their derivatives
The chemistry of sulphenic acids and their derivatives
The chemistry of enols
The chemistry of organophosphorus compounds (4 volumes)
The chemistry of sulphonic acids, esters and their derivatives
The chemistry of alkanes and cycloalkanes
Supplement S: The chemistry of sulphur-containing functional groups
The chemistry of organic arsenic, antimony and bismuth compounds
The chemistry of enamines (2 parts)
The chemistry of organic germanium, tin and lead compounds

UPDATES

The chemistry of α-haloketones, α-haloaldehydes and α-haloimines
Nitrones, nitronates and nitroxides
Crown ethers and analogs
Cyclopropane derived reactive intermediates
Synthesis of carboxylic acids, esters and their derivatives
The silicon–heteroatom bond
Synthesis of lactones and lactams
Syntheses of sulphones, sulphoxides and cyclic sulphides
Patai's 1992 guide to the chemistry of functional groups—*Saul Patai*

The chemistry of
organic silicon compounds

Volume 2

Part 1

Edited by

ZVI RAPPOPORT

The Hebrew University, Jerusalem

and

YITZHAK APELOIG

Technion–Israel Institute of Technology, Haifa

1998

JOHN WILEY & SONS
CHICHESTER–NEW YORK–WEINHEIM–BRISBANE–SINGAPORE–TORONTO
An Interscience® Publication

Copyright © 1998 John Wiley & Sons Ltd,
Baffins Lane, Chichester,
West Sussex PO19 1UD, England

National 01243 779777
International (+44) 1243 779777
e-mail (for orders and customer service enquiries): cs-books@wiley.co.uk
Visit our Home Page on http://www.wiley.co.uk
or http://www.wiley.com

All Rights Reserved. No part of this publication may be reproduced, stored in a retrieval system, or transmitted, in any form or by any means, electronic, mechanical, photocopying, recording, scanning or otherwise, except under the terms of the Copyright Designs and Patents Act 1988 or under the terms of a licence issued by the Copyright Licensing Agency, 90 Tottenham Court Road, London W1P 9HE, UK, without the permission in writing of the Publisher

Other Wiley Editorial Offices

John Wiley & Sons, Inc., 605 Third Avenue,
New York, NY 10158-0012, USA

WILEY-VCH Verlag GmbH, Pappelallee 3,
D-69469 Weinheim, Germany

Jacaranda Wiley Ltd, 33 Park Road, Milton,
Queensland 4064, Australia

John Wiley & Sons (Asia) Pte Ltd, Clementi Loop #02-01,
Jin Xing Distripark, Singapore 129809

John Wiley & Sons (Canada) Ltd, 22 Worcester Road,
Rexdale, Ontario M9W 1L1, Canada

British Library Cataloguing in Publication Data

A catalogue record for this book is available from the British Library

ISBN 0 471 96757 2

Typeset in 9/10pt Times by Laser Words, Madras, India
Printed and bound in Great Britain by Biddles Ltd, Guildford, Surrey
This book is printed on acid-free paper responsibly manufactured from sustainable forestry, in which at least two trees are planted for each one used for paper production.

To

Sara and Zippi

Contributing authors

Wataru Ando	Department of Chemistry, University of Tsukuba, Tsukuba, Ibaraki 305, Japan
Yitzhak Apeloig	Department of Chemistry and the Lise-Meitner Minerva Center for Computational Quantum Chemistry, Technion–Israel Institute of Technology, Haifa 32000, Israel
D. A. ('Fred') Armitage	Department of Chemistry, King's College London, Strand, London, WC2R 2LS, UK
Norbert Auner	Fachinstitut für Anorganische und Allgemeine Chemie, Humboldt-Universität zu Berlin, Hessische Str. 1–2, D-10115 Berlin, Germany
David Avnir	Institute of Chemistry, The Hebrew University of Jerusalem, Jerusalem 91904, Israel
Alan R. Bassindale	Department of Chemistry, The Open University, Milton Keynes, MK7 6AA, UK
Rosa Becerra	Instituto de Quimica Fisica 'Rocasolano', C/Serrano 119, 28006 Madrid, Spain
Johannes Belzner	Institut für Organische Chemie der Georg-August-Universität Göttingen, Tammannstrasse 2, D-37077 Göttingen, Germany
M. B. Boisen Jr	Department of Materials Science and Engineering, Virginia Tech, Blacksburg, VA 24061, USA
Mark Botoshansky	Department of Chemistry, Technion–Israel Institute of Technology, Haifa 32000, Israel
A. G. Brook	Lash Miller Chemical Laboratories, University of Toronto, Toronto, Ontario M5S 3H6, Canada
C. Chatgilialoglu	I. Co. C. E. A., Consiglio Nazionale delle Ricerche, Via P. Gobetti 101, 40129 Bologna, Italy
Buh-Luen Cheng	Department of Chemistry, Tsing Hua University, Hsinchu, Taiwan 30043, Republic of China
Nami Choi	Department of Chemistry, University of Tsukuba, Tsukuba, Ibaraki 305, Japan
Ernest W. Colvin	Department of Chemistry, University of Glasgow, Glasgow, G12 8QQ, UK

Contributing authors

Uwe Dehnert	Institut für Organische Chemie der Georg-August-Universität Göttingen, Tammannstrasse 2, D-37077 Göttingen, Germany
Robert Drake	Dow Corning Ltd, Cardiff Road, Barry, South Glamorgan, CF63 2YL, UK
Jacques Dubac	Hétérochimie Fondamentale et Appliqué, ESA-CNRS 5069, Université Paul-Sabatier, 118 route de Narbonne, 31062 Toulouse Cedex, France
Moris S. Eisen	Department of Chemistry, Technion–Israel Institute of Technology, Kiryat Hatechnion, Haifa 32000, Israel
C. Ferreri	Departimento di Chimica Organica e Biologica, Università di Napoli 'Federico II', Via Mezzocannone 16, 80134 Napoli, Italy
Toshio Fuchigami	Department of Electrochemistry, Tokyo Institute of Technology, 4259 Nagatsuta, Midori-ku, Yokohama 226, Japan
Peter P. Gaspar	Department of Chemistry, Washington University, St Louis, Missouri 63130-4899, USA
Christian Guérin	Chimie Moléculaire et Organisation du Solide, UMR-CNRS 5637, Université Montpellier II, Place E. Bataillon, 34095 Montpellier Cedex 5, France
G. V. Gibbs	Department of Materials Science and Engineering, Virginia Tech, Blacksburg, VA 24061, USA
T. Gimisis	I. Co. C. E. A., Consiglio Nazionale delle Ricerche, Via P. Gobetti 101, 40129 Bologna, Italy
Simon J. Glynn	Department of Chemistry, The Open University, Milton Keynes, MK7 6AA, UK
Norman Goldberg	Technische Universität Braunschweig, Institut für Organische Chemie, Hagenring 30, D-38106 Braunschweig, Germany
Edwin Hengge	(Deceased)
Reuben Jih-Ru Hwu	Department of Chemistry, Tsing Hua University, Hsinchu, Taiwan 30043, Republic of China
Jörg Jung	Institut für Organische Chemie der Justus-Liebig Universität Giessen, Heinrich-Buff-Ring 58, D-35392 Giessen, Germany
Peter Jutzi	Faculty of Chemistry, University of Bielefeld, Universitätsstr. 25, D-33615 Bielefeld, Germany
Yoshio Kabe	Department of Chemistry, University of Tsukuba, Tsukuba, Ibaraki 305, Japan
Menahem Kaftory	Department of Chemistry, Technion–Israel Institute of Technology, Haifa 32000, Israel
Inna Kalikhman	Department of Chemistry, Ben-Gurion University of the Negev, Beer Sheva 84105, Israel
Moshe Kapon	Department of Chemistry, Technion-Israel Institute of Technology, Haifa 32000, Israel

Contributing authors

Miriam Karni	Department of Chemistry and the Lise-Meitner Minerva Center for Computational Quantum Chemistry, Technion–Israel Institute of Technology, Haifa 32000, Israel
Mitsuo Kira	Department of Chemistry, Graduate School of Science, Tohoku University, Aoba-ku, Sendai 980-77, Japan
Sukhbinder S. Klair	Department of Chemistry, Loughborough University, Loughborough, Leicestershire, LE11 3TU, UK
Lisa C. Klein	Ceramics Department, Rutgers–The State University of New Jersey, Piscataway, New Jersey 08855-0909, USA
Daniel Kost	Department of Chemistry, Ben-Gurion University of the Negev, Beer Sheva 84105, Israel
Takahiro Kusukawa	Department of Chemistry, University of Tsukuba, Tsukuba, Ibaraki 305, Japan
R. M. Laine	Department of Chemistry, University of Michigan, Ann Arbor, Michigan 48109-2136, USA
David Levy	Instituto de Ciencia de Materiales de Madrid, C.S.I.C., Cantoblanco, 28049 Madrid, Spain
Larry N. Lewis	GE Corporate Research and Development Center, Schenectady, NY 12309, USA
Zhaoyang Li	Department of Chemistry, State University of New York at Stony Brook, Stony Brook, New York 11794-3400, USA
Paul D. Lickiss	Department of Chemistry, Imperial College of Science, Technology and Medicine, London, SW7 2AY, UK
Shiuh-Tzung Liu	Department of Chemistry, National Taiwan University, Taipei, Taiwan 106
Tien-Yau Luh	Department of Chemistry, National Taiwan University, Taipei, Taiwan 106
Gerhard Maas	Abteilung Organische Chemie I, Universität Ulm, Albert-Einstein-Allee 11, D-89081 Ulm, Germany
Iain MacKinnon	Dow Corning Ltd, Cardiff Road, Barry, South Glamorgan, CF63 2YL, UK
Svetlana Kirpichenko	Irkutsk Institute of Chemistry, Siberian Branch of the Russian Academy of Sciences, 1 Favorsky St, 664033 Irkutsk, Russia
Christoph Maerker	Laboratoire de Chimie Biophysique, Institut Le Bel, Université Louis Pasteur, 4 rue Blaise Pascal, F-67000 Strasbourg, France
Günther Maier	Institut für Organische Chemie der Justus-Liebig Universität Giessen, Heinrich-Buff-Ring 58, D-35392 Giessen, Germany
Michael J. McKenzie	Department of Chemistry, Loughborough University, Loughborough, Leicestershire, LE11 3TU, UK
Andreas Meudt	Institut für Organische Chemie der Justus-Liebig Universität Giessen, Heinrich-Buff-Ring 58, D-35392 Giessen, Germany

Philippe Meunier	Synthèse et Electrosynthèse Organométalliques, UMR-CNRS 5632, Université de Bourgogne, 6 Boulevard Gabriel, 21004 Dijon Cedex, France
Takashi Miyazawa	Photodynamics Research Center, The Institute of Physical and Chemical Research, 19-1399, Koeji, Nagamachi, Aoba-ku, Sendai 980, Japan
Thomas Müller	Fachinstitut für Anorganische und Allgemeine Chemie, Humboldt-Universität zu Berlin, Hessische Str. 1–2, D-10115 Berlin, Germany
Shigeru Nagase	Department of Chemistry, Faculty of Science, Tokyo Metropolitan University, Hachioji, Tokyo 192-03, Japan
Iwao Ojima	Department of Chemistry, State University of New York at Stony Brook, Stony Brook, New York 11794-3400, USA
Renji Okazaki	Department of Chemistry, School of Science, The University of Tokyo, Bunkyo-ku, Tokyo 113, Japan
Harald Pacl	Institut für Organische Chemie der Justus-Liebig Universität Giessen, Heinrich-Buff-Ring 58, D-35392 Giessen, Germany
Philip C. Bulman Page	Department of Chemistry, Loughborough University, Loughborough, Leicestershire, LE11 3TU, UK
Vadim Pestunovich	Irkutsk Institute of Chemistry, Siberian Branch of the Russian Academy of Sciences, 1 Favorsky St, 664033 Irkutsk, Russia
Stephen Rosenthal	Department of Chemistry, Loughborough University, Loughborough, Leicestershire, LE11 3TU, UK
Hideki Sakurai	Department of Industrial Chemistry, Faculty of Science and Technology, Science University of Tokyo, Yamazaki 2641, Noda, Chiba 278, Japan
Paul von Ragué Schleyer	Center for Computational Quantum Chemistry, The University of Georgia, Athens, Georgia 30602, USA
Ulrich Schubert	Institute for Inorganic Chemistry, The Technical University of Vienna, A-1060 Vienna, Austria
Helmut Schwarz	Institut für Organische Chemie der Technischen Universität Berlin, Straße des 17 Juni 135, D-10623 Berlin, Germany
Akira Sekiguchi	Department of Chemistry, Graduate School of Science, University of Tsukuba, Tsukuba, Ibaraki 305, Japan
A. Sellinger	Sandia National Laboratory, Advanced Materials Laboratory, 1001 University Blvd, University of New Mexico, Albuquerque, New Mexico 87106, USA
Hans-Ullrich Siehl	Abteilung für Organische Chemie I der Universität Ulm, D-86069 Ulm, Germany
Harald Stüger	Institut für Anorganische Chemie, Erzherzog-Johann-Universität Graz, Stremayrgasse 16, A-8010 Graz, Austria
Reinhold Tacke	Institut für Anorganische Chemie, Universität Würzburg, Am Hubland, D-97074 Würzburg, Germany

Toshio Takayama	Department of Applied Chemistry, Faculty of Engineering, Kanagawa University, 3-27-1 Rokkakubashi, Yokohama, Japan 221
Yoshito Takeuchi	Department of Chemistry, Faculty of Science, Kanagawa University, 2946 Tsuchiya, Hiratsuka, Japan 259-12
Peter G. Taylor	Department of Chemistry, The Open University, Milton Keynes, MK7 6AA, UK
Richard Taylor	Dow Corning Ltd, Cardiff Road, Barry, South Glamorgan, CF63 2YL, UK
Norihiro Tokitoh	Department of Chemistry, School of Science, The University of Tokyo, Bunkyo-ku, Tokyo 113, Japan
Shwu-Chen Tsay	Department of Chemistry, Tsing Hua University, Hsinchu, Taiwan 30043, Republic of China
Mikhail Voronkov	Irkutsk Institute of Chemistry, Siberian Branch of the Russian Academy of Sciences, 1 Favorsky St, 664033 Irkutsk, Russia
Stephan A. Wagner	Institut für Anorganische Chemie, Universität Würzburg, Am Hubland, D-97074 Würzburg, Germany
Robin Walsh	The Department of Chemistry, The University of Reading, P O Box 224, Whiteknights, Reading, RG6 6AD, UK
Robert West	Department of Chemistry, University of Wisconsin at Madison, Madison, Wisconsin 53706, USA
Anna B. Wojcik	Ceramics Department, Rutgers–The State University of New Jersey, Piscataway, New Jersey 08855-0909, USA
Jiawang Zhu	Department of Chemistry, State University of New York at Stony Brook, Stony Brook, New York 11794-3400, USA
Wolfgang Ziche	Fachinstitut für Anorganische und Allgemeine Chemie, Humboldt-Universität zu Berlin, Hessische Str. 1–2, D-10115 Berlin, Germany

Foreword

The preceding volume in 'The Chemistry of Functional Groups' series, *The chemistry of organic silicon compounds* (S. Patai and Z. Rappoport, Eds), appeared a decade ago and was followed in 1991 by an update volume, *The silicon–heteroatom bond*. Since then the chemistry of organic silicon compounds has continued its rapid growth, with many important contributions in the synthesis of new and novel types of compounds, in industrial applications, in theory and in understanding the chemical bonds of silicon, as well as in many other directions. The extremely rapid growth of the field and the continued fascination with the chemistry of this unique element, a higher congener of carbon — yet so dramatically different in its chemistry — convinced us that a new authoritative book in the field is highly desired.

Many of the recent developments, as well as topics not covered in the previous volume are reviewed in the present volume, which is the largest in 'The Chemistry of Functional Groups' series. The 43 chapters, written by leading silicon chemists from 12 countries, deal with a wide variety of topics in organosilicon chemistry, including theoretical aspects of several classes of compounds, their structural and spectral properties, their thermochemistry, photochemistry and electrochemistry and the effect of silicon as a substituent. Several chapters review the chemistry of various classes of reactive intermediates, such as silicenium ions, silyl anions, silylenes, and of hypervalent silicon compounds. Multiple-bonded silicon compounds, which have attracted much interest and activity over the last decade, are reviewed in three chapters: one on silicon–carbon and silicon–nitrogen multiple bonds, one on silicon–silicon multiple bonds and one on silicon–hereroatom multiple bonds. Other chapters review the synthesis of several classes of organosilicon compounds and their applications as synthons in organic synthesis. Several chapters deal with practical and industrial aspects of silicon chemistry in which important advances have recently been made, such as silicon polymers, silicon-containing ceramic precursors and the rapidly growing field of organosilica sol–gel chemistry.

The literature covered in the book is mostly up to mid-1997.

Several of the originally planned chapters, on comparison of silicon compounds with their higher group 14 congeners, interplay between theory and experiment in organisilicon chemistry, silyl radicals, recent advances in the chemistry of silicon–phosphorous,–arsenic,–antimony and –bismuth compounds, and the chemistry of polysilanes, regrettably did not materialize. We hope to include these important chapters in a future complementary volume. The current pace of research in silicon chemistry will certainly soon require the publication of an additional updated volume.

We are grateful to the authors for the immense effort they have invested in the 43 chapters and we hope that this book will serve as a major reference in the field of silicon chemistry for years to come.

We will be grateful to readers who will draw our attention to mistakes and who will point out to us topics which should be included in a future volume of this series.

Jerusalem and Haifa
March, 1998

ZVI RAPPOPORT
YITZHAK APELOIG

The Chemistry of Functional Groups
Preface to the series

The series 'The Chemistry of Functional Groups' was originally planned to cover in each volume all aspects of the chemistry of one of the important functional groups in organic chemistry. The emphasis is laid on the preparation, properties and reactions of the functional group treated and on the effects which it exerts both in the immediate vicinity of the group in question and in the whole molecule.

A voluntary restriction on the treatment of the various functional groups in these volumes is that material included in easily and generally available secondary or tertiary sources, such as Chemical Reviews, Quarterly Reviews, Organic Reactions, various 'Advances' and 'Progress' series and in textbooks (i.e. in books which are usually found in the chemical libraries of most universities and research institutes), should not, as a rule, be repeated in detail, unless it is necessary for the balanced treatment of the topic. Therefore each of the authors is asked not to give an encyclopaedic coverage of his subject, but to concentrate on the most important recent developments and mainly on material that has not been adequately covered by reviews or other secondary sources by the time of writing of the chapter, and to address himself to a reader who is assumed to be at a fairly advanced postgraduate level.

It is realized that no plan can be devised for a volume that would give a complete coverage of the field with no overlap between chapters, while at the same time preserving the readability of the text. The Editors set themselves the goal of attaining reasonable coverage with moderate overlap, with a minimum of cross-references between the chapters. In this manner, sufficient freedom is given to the authors to produce readable quasi-monographic chapters.

The general plan of each volume includes the following main sections:

(a) An introductory chapter deals with the general and theoretical aspects of the group.

(b) Chapters discuss the characterization and characteristics of the functional groups, i.e. qualitative and quantitative methods of determination including chemical and physical methods, MS, UV, IR, NMR, ESR and PES — as well as activating and directive effects exerted by the group, and its basicity, acidity and complex-forming ability.

(c) One or more chapters deal with the formation of the functional group in question, either from other groups already present in the molecule or by introducing the new group directly or indirectly. This is usually followed by a description of the synthetic uses of the group, including its reactions, transformations and rearrangements.

(d) Additional chapters deal with special topics such as electrochemistry, photochemistry, radiation chemistry, thermochemistry, syntheses and uses of isotopically labelled compounds, as well as with biochemistry, pharmacology and toxicology. Whenever applicable, unique chapters relevant only to single functional groups are also included (e.g. 'Polyethers', 'Tetraaminoethylenes' or 'Siloxanes').

This plan entails that the breadth, depth and thought-provoking nature of each chapter will differ with the views and inclinations of the authors and the presentation will necessarily be somewhat uneven. Moreover, a serious problem is caused by authors who deliver their manuscript late or not at all. In order to overcome this problem at least to some extent, some volumes may be published without giving consideration to the originally planned logical order of the chapters.

Since the beginning of the Series in 1964, two main developments have occurred. The first of these is the publication of supplementary volumes which contain material relating to several kindred functional groups (Supplements A, B, C, D, E, F and S). The second ramification is the publication of a series of 'Updates', which contain in each volume selected and related chapters, reprinted in the original form in which they were published, together with an extensive updating of the subjects, if possible, by the authors of the original chapters. A complete list of all above mentioned volumes published to date will be found on the page opposite the inner title page of this book. Unfortunately, the publication of the 'Updates' has been discontinued for economic reasons.

Advice or criticism regarding the plan and execution of this series will be welcomed by the Editors.

The publication of this series would never have been started, let alone continued, without the support of many persons in Israel and overseas, including colleagues, friends and family. The efficient and patient co-operation of staff-members of the publisher also rendered us invaluable aid. Our sincere thanks are due to all of them.

The Hebrew University SAUL PATAI
Jerusalem, Israel ZVI RAPPOPORT

Contents

1. Theoretical aspects and quantum mechanical calculations of silaaromatic compounds — 1
   Yitzhak Apeloig and Miriam Karni

2. A molecular modeling of the bonded interactions of crystalline silica — 103
   G. V. Gibbs and M. B. Boisen

3. Polyhedral silicon compounds — 119
   Akira Sekiguchi and Shigeru Nagase

4. Thermochemistry — 153
   Rosa Becerra and Robin Walsh

5. The structural chemistry of organosilicon compounds — 181
   Menahem Kaftory, Moshe Kapon and Mark Botoshansky

6. ^{29}Si NMR spectroscopy of organosilicon compounds — 267
   Yoshito Takeuchi and Toshio Takayama

7. Activating and directive effects of silicon — 355
   Alan R. Bassindale, Simon J. Glynn and Peter G. Taylor

8. Steric effects of silyl groups — 431
   Jih Ru Hwu, Shwu-Chen Tsay and Buh-Luen Cheng

9. Reaction mechanisms of nucleophilic attack at silicon — 495
   Alan R. Bassindale, Simon J. Glynn and Peter G. Taylor

10. Silicenium ions: Quantum chemical computations — 513
    Christoph Maerker and Paul von Ragué Schleyer

11. Silicenium ions — experimental aspects — 557
    Paul D. Lickiss

12. Silyl-substituted carbocations — 595
    Hans-Ullrich Siehl and Thomas Müller

13. Silicon-substituted carbenes — 703
    Gerhard Maas

Contents

14 Alkaline and alkaline earth silyl compounds — preparation and structure ... 779
   Johannes Belzner and Uwe Dehnert

15 Mechanism and structures in alcohol addition reactions of disilenes and silenes ... 827
   Hideki Sakurai

16 Silicon-carbon and silicon-nitrogen multiply bonded compounds ... 857
   Thomas Müller, Wolfgang Ziche and Norbert Auner

17 Recent advances in the chemistry of silicon-heteroatom multiple bonds ... 1063
   Norihiro Tokitoh and Renji Okazaki

18 Gas-phase ion chemistry of silicon-containing molecules ... 1105
   Norman Goldberg and Helmut Schwarz

19 Matrix isolation studies of silicon compounds ... 1143
   Günther Maier, Andreas Meudt, Jörg Jung and Harald Pacl

20 Electrochemistry of organosilicon compounds ... 1187
   Toshio Fuchigami

21 The photochemistry of organosilicon compounds ... 1233
   A. G. Brook

22 Mechanistic aspects of the photochemistry of organosilicon compounds ... 1311
   Mitsuo Kira and Takashi Miyazawa

23 Hypervalent silicon compounds ... 1339
   Daniel Kost and Inna Kalikhman

24 Silatranes and their tricyclic analogs ... 1447
   Vadim Pestunovich, Svetlana Kirpichenko and Mikhail Voronkov

25 Tris(trimethylsilyl)silane in organic synthesis ... 1539
   C. Chatgilialoglu, C. Ferreri and T. Gimisis

26 Recent advances in the direct process ... 1581
   Larry N. Lewis

27 Acyl silanes ... 1599
   Philip C. Bulman Page, Michael J. McKenzie, Sukhbinder S. Klair and Stephen Rosenthal

28 Recent synthetic applications of organosilicon reagents ... 1667
   Ernest W. Colvin

29 Recent advances in the hydrosilylation and related reactions ... 1687
   Iwao Ojima, Zhaoyang Li and Jiawang Zhu

30	Synthetic applications of allylsilanes and vinylsilanes Tien-Yau Luh and Shiuh-Tzung Liu	1793
31	Chemistry of compounds with silicon-sulphur, silicon-selenium and silicon-tellurium bonds D. A. ('Fred') Armitage	1869
32	Cyclic polychalcogenide compounds with silicon Nami Choi and Wataru Ando	1895
33	Organosilicon derivatives of fullerenes Wataru Ando and Takahiro Kusukawa	1929
34	Group 14 metalloles, ionic species and coordination compounds Jacques Dubac, Christian Guérin and Philippe Meunier	1961
35	Transition-metal silyl complexes Moris S. Eisen	2037
36	Cyclopentadienyl silicon compounds Peter Jutzi	2129
37	Recent advances in the chemistry of cyclopolysilanes Edwin Hengge and Harald Stüger	2177
38	Recent advances in the chemistry of siloxane polymers and copolymers Robert Drake, Iain MacKinnon and Richard Taylor	2217
39	Si-containing ceramic precursors R. M. Laine and A. Sellinger	2245
40	Organo-silica sol–gel materials David Avnir, Lisa C. Klein, David Levy, Ulrich Schubert and Anna B. Wojcik	2317
41	Chirality in bioorganosilicon chemistry Reinhold Tacke and Stephan A. Wagner	2363
42	Highly reactive small-ring monosilacycles and medium-ring oligosilacycles Wataru Ando and Yoshio Kabe	2401
43	Silylenes Peter P. Gaspar and Robert West	2463
Author index		2569
Subject index		2721

List of abbreviations used

Ac	acetyl (MeCO)
acac	acetylacetone
Ad	adamantyl
AIBN	azoisobutyronitrile
Alk	alkyl
All	allyl
An	anisyl
Ar	aryl
Bn	benzyl
Bz	benzoyl (C_6H_5CO)
Bu	butyl (also t-Bu or Bu^t)
CD	circular dichroism
CI	chemical ionization
CIDNP	chemically induced dynamic nuclear polarization
CNDO	complete neglect of differential overlap
Cp	η^5-cyclopentadienyl
Cp*	η^5-pentamethylcyclopentadienyl
DABCO	1,4-diazabicyclo[2.2.2]octane
DBN	1,5-diazabicyclo[4.3.0]non-5-ene
DBU	1,8-diazabicyclo[5.4.0]undec-7-ene
DIBAH	diisobutylaluminium hydride
DME	1,2-dimethoxyethane
DMF	N,N-dimethylformamide
DMSO	dimethyl sulphoxide
ee	enantiomeric excess
EI	electron impact
ESCA	electron spectroscopy for chemical analysis
ESR	electron spin resonance
Et	ethyl
eV	electron volt

Fc	ferrocenyl	
FD	field desorption	
FI	field ionization	
FT	Fourier transform	
Fu	furyl(OC_4H_3)	
GLC	gas liquid chromatography	
Hex	hexyl(C_6H_{13})	
c-Hex	cyclohexyl(C_6H_{11})	
HMPA	hexamethylphosphortriamide	
HOMO	highest occupied molecular orbital	
HPLC	high performance liquid chromatography	
i-	iso	
Ip	ionization potential	
IR	infrared	
ICR	ion cyclotron resonance	
LAH	lithium aluminium hydride	
LCAO	linear combination of atomic orbitals	
LDA	lithium diisopropylamide	
LUMO	lowest unoccupied molecular orbital	
M	metal	
M	parent molecule	
MCPBA	m-chloroperbenzoic acid	
Me	methyl	
MNDO	modified neglect of diatomic overlap	
MS	mass spectrum	
n	normal	
Naph	naphthyl	
NBS	N-bromosuccinimide	
NCS	N-chlorosuccinimide	
NMR	nuclear magnetic resonance	
Pc	phthalocyanine	
Pen	pentyl(C_5H_{11})	
Pip	piperidyl($C_5H_{10}N$)	
Ph	phenyl	
ppm	parts per million	
Pr	propyl (also i-Pr or Pr^i)	
PTC	phase transfer catalysis or phase transfer conditions	
Pyr	pyridyl (C_5H_4N)	

R	any radical
RT	room temperature
s-	secondary
SET	single electron transfer
SOMO	singly occupied molecular orbital
t-	tertiary
TCNE	tetracyanoethylene
TFA	trifluoroacetic acid
THF	tetrahydrofuran
Thi	thienyl(SC_4H_3)
TLC	thin layer chromatography
TMEDA	tetramethylethylene diamine
TMS	trimethylsilyl or tetramethylsilane
Tol	tolyl(MeC_6H_4)
Tos or Ts	tosyl(*p*-toluenesulphonyl)
Trityl	triphenylmethyl(Ph_3C)
Xyl	xylyl($Me_2C_6H_3$)

In addition, entries in the 'List of Radical Names' in *IUPAC Nomenclature of Organic Chemistry*, 1979 Edition, Pergamon Press, Oxford, 1979, p. 305–322, will also be used in their unabbreviated forms, both in the text and in formulae instead of explicitly drawn structures.

CHAPTER 1

Theoretical aspects and quantum mechanical calculations of silaaromatic compounds

YITZHAK APELOIG and MIRIAM KARNI

Department of Chemistry and the Lise Meitner-Minerva Center for Computational Quantum Chemistry, Technion—Israel Institute of Technology, Haifa 32000, Israel

I. INTRODUCTION	3
II. COMPUTATIONAL CRITERIA FOR ESTIMATING THE DEGREE OF AROMATICITY	5
A. The Geometric Criterion	5
B. The Energetic Criterion	6
C. The Magnetic Criterion	6
1. Magnetic susceptibility anisotropy and susceptibility exaltation	6
2. Anomalous ^{1}H chemical shifts	7
3. Other magnetic probes of aromaticity (e.g. NICS)	7
D. Interrelations between the Geometric, Energetic and Magnetic Criteria	8
III. SILABENZENOIDS	10
A. Experimental Background	10
B. Monosilabenzenes	11
1. Silabenzene	11
2. Substituted silabenzenes	13
C. Silanaphthalenes	16
D. Disilabenzenes	17
1. Geometry and the degree of aromatic stabilization	17
2. Relation to isomers	19
E. Trisilabenzenes	21
1. Geometry and the degree of aromaticity	21
2. Relation to isomers	22
F. Hexasilabenzenes	22
1. The parent hexasilabenzene	22
a. Geometry and the degree of aromaticity	22
b. Relation to isomers	25

The chemistry of organic silicon compounds, Vol. 2
Edited by Z. Rappoport and Y. Apeloig © 1998 John Wiley & Sons Ltd

		2. Substituted hexasilabenzenes	28	
	G.	Higher Congeners of Hexasilabenzene	29	
	H.	Silabenzenoids Containing Heteroatoms	30	
	I.	Concluding Comments	31	
IV.	CHARGED SILAAROMATIC SYSTEMS		32	
	A.	Systems with Two π-Electrons	33	
		1. The monosilacyclopropenium cation	33	
		2. The trisilacyclopropenium cation and its higher congeners	33	
		3. The tetrasilacyclobutadiene dication	34	
	B.	The Silatropylium Cation and its Isomers	34	
	C.	$C_5H_5Si^+$ Isomers	36	
	D.	Silacyclopentadienyl Anions and Dianions	37	
		1. Experimental background	37	
		2. c-$C_4H_4SiR^-$ silolyl anions	42	
		a. The parent ion c-$C_4H_4SiH^-$ and its lithiated complexes	42	
		b. The silyl substituted silolyl anions, c-$C_4H_4Si(SiH_3)^-$	47	
		3. The parent silole dianion c-$C_4H_4Si^{2-}$ and its lithiated complexes	47	
		4. Higher congeners of the silolyl anion and silole dianion	50	
		5. The pentasilacyclopentadienyl anion	50	
V.	SILYLENES WITH AROMATIC CHARACTER		52	
	A.	Stable 'Arduengo-type' Silylenes	52	
		1. Experimental background	52	
		2. Aromaticity	53	
		a. Structural criteria	55	
		b. Energetic criteria	57	
		c. Magnetic criteria	59	
		d. Charge distribution and ionization potentials	61	
		3. Reactions	62	
		a. 1,2-H rearrangements	62	
		b. Dimerization	63	
	B.	Miscellaneous Potentially Aromatic Silylenes	64	
		1. Cyclic aminosilylenes	64	
		2. Silacyclopropenylidene and silacyclopropynylidene	66	
		3. Trisilapropenylidene and other Si_3H_2 isomers	68	
VI.	SILAFULLERENES		69	
	A.	Are Fullerenes and Silafullerenes Aromatic?	69	
	B.	Si_{60}	70	
		1. Structure	70	
		2. Endohedral complexes of Si_{60}	72	
	C.	Si_{70}	75	
VII.	SILICON COMPOUNDS WITH POTENTIAL 'THREE-DIMENSIONAL AROMATICITY'		77	
	A.	Silicocenes	77	
	B.	Persilaferrocene, $(H_5Si_5)_2Fe$	81	
	C.	Aromaticity in $closo$-Silaboranes, $B_3Si_2H_3$	81	
VIII.	POTENTIAL SILAANTIAROMATIC COMPOUNDS		83	
	A.	Silacyclobutadienes	83	
		1. Monosilacyclobutadiene	83	
		2. Disilacyclobutadienes	86	
		3. Tetrasilacyclobutadiene	87	
	B.	Charged Silaantiaromatic Compounds	88	
		1. The silirenyl anion, $C_2SiH_3^-$	88	
		2. Silacyclopentadienyl cations	89	

a. $C_4H_4SiH^+$	89
b. $Si_5H_5^+$	90
IX. CONCLUSIONS	93
X. ACKNOWLEDGEMENTS	93
XI. LIST OF ABBREVIATIONS	93
XII. REFERENCES	94

I. INTRODUCTION

The last two decades have witnessed increasing interest in the synthesis and chemistry of doubly-bonded silicon compounds[1-3]. The successful synthesis and isolation in 1981 of the first stable silene[4] and disilene[5] stimulated efforts to synthesize other stable doubly-bonded silicon compounds. These efforts led to the synthesis of new stable silenes[1a,2a,3a,b] and disilenes[1a,2b,3c], to a stable 1-sillaallene[6], as well as to stable compounds containing a variety of other double bonds to silicon, i.e. Si=N[2a,3d], Si=P[3e], Si=As[3e], Si=S[2c] and Si=Se[2c,2f]. Stable silanones (i.e. compounds containing a Si=O bond) are still elusive, but there is strong evidence for their existence in a matrix at low temperature[1a,2c,d] and in the gas phase[7].

Aromatic compounds are common in carbon chemistry and in the chemistry of other first-row elements, such as nitrogen and oxygen. In contrast, the first stable benzenoid silaaromatic compound **1**, protected by the large Tbt group (**2**), was isolated by Okazaki and coworkers only in 1997[8]. A marginally stable silaaromatic compound is 1,4-di-*tert*-butyl-2,6-bis(trimethylsilyl)silabenzene (**3**) which is stable in solution below $-100\,°C^9$. All other attempts to produce silaaromatic compounds resulted only in their observation in a matrix or in the gas phase[2d]. The synthesis and characterization of stable benzenoid silaaromatic compounds remains a major challenge in contemporary organosilicon chemistry.

In addition to silaaromatic benzenoid compounds, new exciting families of silaaromatic non-benzenoid compounds, possessing various degrees of aromaticity, have recently attracted much attention. Some examples are: (a) the recently isolated stable silylenes **(4)**[2e]; (b) alkali metal complexes of silole anions **(5a)** and dianions **(5b)**[10]; (c) transition metal complexes of silole (e.g. **6a** and **6b**)[10d]; (d) silafullerenes.

The experimental knowledge on silaaromatic compounds is still very limited, leaving theory as one of the main sources of reliable information about the basic properties of these compounds[1b]. Computational methods have proved to be unusually helpful in the study of organosilicon compounds and in particular of transients and other unstable silicon compounds, for which experimental data are scarce[1b].

In this chapter we will review the available computational studies on silaaromatic and silaantiaromatic compounds. The compounds that are reviewed belong mainly to the categories obeying the Hückel rule[11], and we will refer to compounds having $4n + 2$ π-electrons as being 'aromatic' and to compounds with $4n$ π-electrons as being 'antiaromatic'. However, we stress that by using the 'aromatic–antiaromatic' notation we do not imply that the compound thus referred to actually has the characteristics of an 'aromatic' or of an 'antiaromatic' compound. As discussed in detail in Section II below, the degree of 'aromaticity' of a particular compound is difficult to define and is highly controversial even for simple organic molecules, and the issue is even more problematic for organosilicon compounds. The reader is therefore advised to regard the notation 'aromatic' or 'antiaromatic' only as a guideline and he should consult the specific discussion regarding the 'degree of aromaticity' for each particular compound of interest.

We have made an effort to review all the important theoretical studies on silaaromatic compounds, emphasizing those published in the period January 1990–May 1997, and we include also unpublished data that was brought to our attention. Although we did our best to cover all studies, we might have missed some, and if so we apologize to their authors. Earlier studies theoretical and experimental have been reviewed by Apeloig in the preceding volume of *The Chemistry of Organic Silicon Compounds*[1b]. However, for the sake of completeness we will also discuss older studies which are of importance. We do not review or explain the theoretical methods which were used by the various researchers and the interested reader is referred to the original papers or to the recent extensive encyclopedia of computational chemistry[12]. We use the generally accepted notation to give the level of theory at which the calculations were carried out; e.g. MP2/6-31G*//RHF/6-31G* denotes a single point MP2/6-31G* calculation at the RHF/6-31G* optimized geometry. Relevant experimental studies are mentioned, but no attempt was made to fully review all the experimental studies related to silaaromatic compounds. In some cases we also discuss the analogous compounds of the higher congeners of Si in group 14 of the Periodic Table, i.e. Ge, Sn and Pb. As the reader will realize, this chapter points to many open questions which still require up-to-date calculations to provide reliable data on the still elusive family of silaaromatic compounds.

The list of abbreviations used in this chapter, including those of the theoretical methods, are given in Section XI.

II. COMPUTATIONAL CRITERIA FOR ESTIMATING THE DEGREE OF AROMATICITY

What is aromaticity? This fundamental and widely used chemical concept is nevertheless one of the most controversial concepts in modern chemistry[13–16]. Many definitions and criteria for characterizing and estimating the degree of aromaticity have been considered, and they are discussed extensively and critically in a recently published comprehensive book by Minkin and coworkers[13a]. Schleyer and Jiao have recently emphasized the importance of magnetic properties for estimating aromaticity[16]. We will summarize here briefly the most widely used *computational criteria* for estimating the degree of aromaticity whilst the reader is referred to References 13, 14 and 16 and the papers cited therein for a more comprehensive discussion.

Aromaticity is associated with cyclic arrays of 'mobile electrons' with favourable symmetries. The 'mobile electron' arrays may be π, σ or mixed in character[16]. The $4n+2/4n$ Hückel rule provides a quantum mechanical framework which allows one to relate the stability and structure of π-systems to their π-electron count, and is widely utilized for classifying π-cyclic systems as aromatic or antiaromatic[13].

The main computational criteria used for determining aromaticity are: (a) the geometric criterion, (b) the energetic criterion, (c) the magnetic criterion. These criteria are discussed below.

A. The Geometric Criterion

This widely used criterion is manifested in: (a) bond length equalization due to cyclic delocalization and (b) planarity of the cyclic systems[13–16]. The planarity of cyclic conjugated systems is regarded as a manifestation of aromaticity, while the non-planarity of cyclic conjugated systems is often taken as an indication of antiaromaticity.

The most widely used method for quantifying the equalization of bond lengths is by calculating Julg's parameter A, using equation 1[17]. This parameter defines the degree of aromaticity in terms of the deviation of n individual CC bond lengths (r_i) from the mean CC bond length (r); $A = 1$ for benzene (D_{6h}) and $A = 0$ for a 'Kékule benzene' (D_{3h}).

$$A = 1 - [(225/n)\sum(1 - r_i/r)^2] \qquad (1)$$

Both geometric criteria (i.e. bond equalization and planarity) do not provide definite answers. Thus, bond equalization is found in borazine ($B_3N_3H_6$) and boroxine ($B_3O_3H_3$) which have 6 π-electrons, but these molecules were shown to be non-aromatic according to the magnetic criterion[16,18,19]. Similarly, bond equalization was found in acyclic conjugated non-aromatic compounds, such as $H_2NCH=CH-CH=CHNH_2$[16]. On the other hand, substantial bond-length alternation can be found in compounds which are believed to be highly aromatic such as naphthalene, anthracene, phenanthrene etc.[13a,15a,16]. Cyclic π-electron delocalization stabilizes planar structures, but other effects such as σ-skeleton strain may cause deviation from planarity, diminishing the degree of aromaticity[14]. Deviation from planarity is found in many silicon compounds such as hexasilabenzene and silole anions which, according to other criteria, are believed to be aromatic (see below).

Shaik and coworkers, have argued recently that the bond length equalization in aromatic molecules is *not* due to π-electron delocalization but is forced by the preference of the σ-framework for symmetric structures[15]. According to Shaik and coworkers, the tendency to adopt a symmetric or a distorted geometry results from competition between two opposite driving forces: that of the σ-skeleton which is always symmetrizing, and that of the π-electrons which, in all cases studied, favours localized non-symmetric structures. Thus, according to Shaik bond-length equalization does *not* indicate a high degree of aromaticity[15]. Shaik and coworkers have criticized in particular the use of the $4n + 2/4n$ Hückel rule for predicting the geometries of molecules other than hydrocarbons, and this criticism is therefore particularly relevant to silaaromatic compounds.

B. The Energetic Criterion

This criterion is a measure of the stabilization/(destabilization) of the aromatic/(antiaromatic) compounds due to cyclic electron delocalization, relative to suitable reference systems such as olefins or conjugated polyenes[13,14,16,20,21]. This is demonstrated for benzene in equations 2–5. The resonance stabilization energy (RE) may be estimated by isodesmic equations[21a] (e.g. equation 2 or 3) and the aromatic stabilization energy (ASE) by homodesmotic equations[21b,c] (e.g. equation 4 or 5) or hyperhomodesmotic equations[21d]. The homodesmotic and hyperhomodesmotic equations, and thus the ASEs, are considered to be the more reliable as there is a better match of the number of single and double bonds on both sides of the equation. The major problem with the energetic criterion is that the estimation of the extra stability resulting from aromaticity is dependent on the reference compound chosen, as demonstrated for benzene by equations 2–5[16] (the RE and ASE values reported for equations 2–5 use experimental data[22]). In addition to aromatic stabilization, these equations reflect also the contribution of strain energy and other factors. It is therefore difficult to apply the energy criterion to strained systems, but even for regular systems its use is controversial

$$C_6H_6 + 6CH_4 \longrightarrow 3C_2H_4 + 3C_2H_6 \qquad RE = 64.2 \text{ kcal mol}^{-1} \qquad (2)$$

$$C_6H_6 + 3C_2H_6 \longrightarrow 3C_2H_4 + \text{cyclohexane} \qquad RE = 48.9 \text{ kcal mol}^{-1} \qquad (3)$$

$$C_6H_6 + 3C_2H_4 \longrightarrow 3(\textit{trans}\text{-1, 3-butadiene}) \qquad ASE = 21.7 \text{ kcal mol}^{-1} \qquad (4)$$

$$C_6H_6 + 3(\text{cyclohexane}) \longrightarrow 3(1, 3\text{-cyclohexadiene}) \qquad ASE = 35.2 \text{ kcal mol}^{-1} \qquad (5)$$

C. The Magnetic Criterion

A special magnetic behaviour of aromatic compounds is believed to result from the occurrence of aromatic ring currents. The following magnetic criteria can be used to determine if a molecule is aromatic.

1. Magnetic susceptibility anisotropy and susceptibility exaltation

Magnetic susceptibility anisotropy, $\Delta\chi$, is a characteristic attribute of aromaticity[16,23]. The magnetic tensor which is normal to the aromatic ring is much larger than the average of the other tensors. For aromatic systems $\Delta\chi$ is highly negative, while highly antiaromatic compounds exhibit positive $\Delta\chi$ values. Unfortunately, the absolute $\Delta\chi$ values cannot serve as an aromaticity index because, in addition to ring currents, local π- and σ-bonds as well as local paramagnetic ring currents also contribute to the anisotropy[13a]. These additional effects have to be subtracted from the total anisotropy in order to find the net contribution of the aromaticity to $\Delta\chi$, e.g. by comparing $\Delta\chi$ of benzene with the sum of the $\Delta\chi$ of three ethylenes.

Another characteristic manifestation of aromaticity is the exaltation of the diamagnetic susceptibility, Λ, which is defined as the difference between the diamagnetic susceptibility of a conjugated cyclic system (χ_M) and the diamagnetic susceptibility of a hypothetical reference system without cyclic electron delocalization (χ'_M)[24]. A molecule is considered to be aromatic when $\Lambda < 0$, antiaromatic when $\Lambda > 0$ and non-aromatic when $\Lambda \sim 0$. Thus, the determination of aromaticity from Λ is dependent on the reference system, as is the calculation of RE or ASE. The magnetic susceptibility associated with the ring current (London susceptibility[25]) is dependent on the ring area, and so are the magnetic susceptibility exaltations. This fact must be taken into consideration when systems of different ring sizes are compared, for example, by defining a ring-size-adjusted aromaticity index ρ which is calculated by equation 6, where n is the number of electrons, S is the ring area and k is a scaling factor[13a,19].

$$\rho = k[n\Lambda/S^2] \quad (6)$$

2. Anomalous ^1H chemical shifts

Anomalous ^1H chemical shifts manifest the deshielding of protons of aromatic cyclic molecules, caused by ring currents induced by an external magnetic field. For aromatic compounds, the protons located inside the ring are shifted upfield and those outside the ring are shifted downfield; for antiaromatic compounds the direction is opposite[13,16,26]. Although this criterion is widely used[13a], it is not general and many exceptions are known[16].

3. Other magnetic probes of aromaticity (e.g. NICS)

(i) For aromatic compounds that do not include protons, such as C_{60}, the chemical shift of an encapsulated ^3He is used, both experimentally and computationally, as a probe for aromaticity[27].

(ii) The chemical shift value of a Li^+ ion which complexes the π-face of aromatic systems is shifted upfield due to ring current effects, and thus can serve as an inner-ring current probe.

(iii) An efficient computational probe for diatropic and paratropic ring currents associated with aromaticity and antiaromaticity, respectively, is the recently developed Nucleus-Independent Chemical Shifts (NICS) method, which computes the absolute magnetic shieldings at ring centres[26]. The total NICS values can be divided into the diamagnetic contributions from the π-bonds, NICS(π), the paramagnetic contributions from the ring σ-bonds, NICS(σ), and the contributions of other electrons (i.e. bonds to hydrogens, in-plane lone pairs and core orbitals)[19]. In aromatic delocalized systems the negative NICS(π) values are larger than the positive NICS(σ) values and thus they have significantly negative total NICS values. Positive total NICS values indicate antiaromaticity, and in the absence of significant delocalization the NICS(π) and NICS(σ) values tend to cancel, resulting in total NICS values that are near zero[19]. The total NICS values show only modest dependence on the ring size and they can therefore be used also to evaluate the aromaticity and antiaromaticity contributions of individual rings in polycyclic systems[26].

D. Interrelations between the Geometric, Energetic and Magnetic Criteria

The extensive literature on aromaticity reveals a significant degree of confusion and controversy regarding the assessment of aromaticity using the above-mentioned criteria and regarding the correlation between them. The examples below demonstrate the current state of the issue:

(i) According to Minkin and coworkers, the best way to evaluate aromaticity is by the energetic criterion which is based on the determination of aromatic stabilization by 'making use of various schemes for calculating resonance energy'[13a].

(ii) Haddon proposed the existence of an analytic relationship between ring currents and the resonance energy[28].

(iii) Katritzky and coworkers concluded from a comparison of a variety of heterocycles with different ring sizes that the geometric and energetic criteria of aromaticity on the one hand, and the magnetic criterion on the other, are almost completely orthogonal[29].

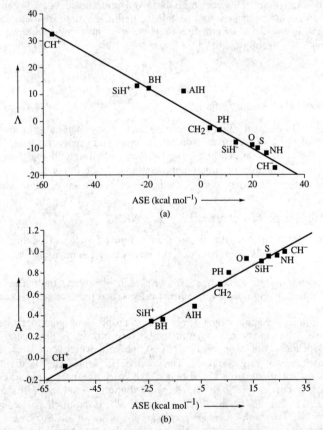

FIGURE 1. Correlation between various criteria for aromaticity for cyclic-C_4H_4X molecules: (a) Magnetic susceptibility exaltations $\Lambda(10^{-6}$ cm^3 mol^{-1}) vs ASE (kcal mol^{-1}); (b) Julg's paprameter A vs ASE (kcal mol^{-1}); (c) Julg's parameter A vs magnetic susceptibility exaltations Λ; (d) NICS (ppm) vs ASE (kcal mol^{-1}). Plots (a)–(c) are reproduced by permission of WILEY-VCH, D-69451 Weinheim, 1995 from Reference 31a. Plot (d) is reprinted with permission from Reference 26. Copyright (1996). American Chemical Society

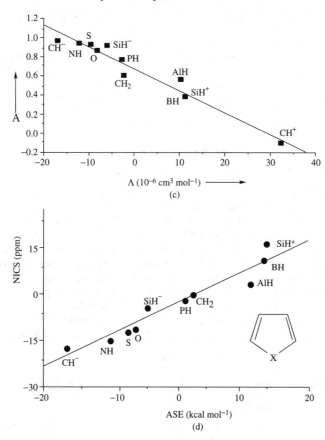

FIGURE 1. *(continued)*

(iv) Jug and Köster concluded that the geometric and energetic criteria are orthogonal to each other while the magnetic criterion correlates with the energetic criteria[30].

(v) Schleyer and coworkers found for a set of cyclic C_4H_4X 5-membered ring systems with 4π- and 6π-electrons[26,31a] that linear relationships exist between the energetic, geometric and magnetic criteria of aromaticity. Diagrams showing the linear correlation between ASEs, Julg's parameter A, the magnetic susceptibility exaltation Λ and the NICS values are shown in Figure 1. Houk and Mendel suggested[31b] that the contradiction between the conclusions of Katritzky[29] and Schleyer[31a] may result from a combination of experimental uncertainties in the magnetic Λ values that Katritzky used, as well as from a comparison of systems with different ring sizes. In a recent study of nearly 50 aromatic and heteroaromatic ring systems, Bird[31c] confirms the conclusions of Schleyer and coworkers[31a] and reports the existence of a good linear relationship between experimental diamagnetic susceptibility enhancements and the corresponding resonance energies as well as other aromaticity indices[31c].

(vi) According to Schleyer and coworkers exaltation of the diamagnetic susceptibility Λ is 'the only uniquely applicable' criterion for aromaticity; other criteria serve only as useful supplementary tools for characterizing aromaticity[16,26]. Schleyer and Jiao recently proposed the following definition of aromaticity: 'Compounds which exhibit significantly exalted diamagnetic susceptibility are aromatic. Cyclic electron delocalization may also result in bond length equalization, abnormal chemical shifts and magnetic anisotropies, as well as chemical and physical properties which reflect energetic stabilization'[16].

In view of the above-mentioned controversy about the definition of aromaticity, it is obvious that there is no unique way to quantify the 'degree of aromaticity' of a particular compound and that any quantification of aromaticity is likely to be highly controversial. Keeping in mind the above discussion on the nature and definition of aromaticity we now examine the available theoretical studies of silaaromatic and silaantiaromatic compounds.

III. SILABENZENOIDS

A. Experimental Background

Barton and Banasink were the first in 1977 to generate and trap a silaaromatic compound, the silatoluene, **7**[32a]. This discovery was soon followed by a matrix isolation and IR, UV and PE spectral characterization of both **7**[32b,c] and silabenzene (**8**)[33]. In 1988 Märkel and Schlosser reported the synthesis of the substituted silabenzene **3** which was stable in solution up to 170 K[9]. Kinetic stabilization by bulky substituents led to the isolation in an argon matrix of **9**, R = Me$_3$Si or i-PrMe$_2$Si by Jutzi, Maier and coworkers[34]; **9**, R = i-PrMe$_2$Si was stable up to 90 K even without an argon cage[34]. In 1991 Maier and coworkers isolated and characterized spectroscopically in the gas phase and in an argon matrix at 12 K the 9-silaanthracenes, **10**, R = H, Ph[35]. Most recently, Okazaki and coworkers reported the synthesis and isolation of 2-silanaphthalene (**1**), the first silabenzenoid compound which is a stable crystalline material even at room temperature[8].

(7)   (8)   (9) R = Me$_3$Si, i-PrMe$_2$Si   (10) R = H, Ph

Several studies reported on silabenzenes in which more than one carbon was substituted by silicon. Maier and coworkers observed 1,4-disilabenzene (**11**) in a matrix and identified it by its IR absorbtion at 1273 cm^{-1} and its electronic absorptions at 408, 340 and 275 nm[36]. Evidence for the formation in the gas phase of 1,3,5-trisilabenzene (**12**) as a ligand in the dehydrogenation reaction of 1,3,5-trisilacyclohexane with [Cp,Fe]$^+$, [Cp,Co]$^+$ and [Cp,Ni]$^+$ was recently reported by Bjarnason and Arnason[37]. The structure of **12** was assigned on the basis of the collision-induced dissociation spectrum and from deuterium-labelling experiments[37].

(11)   (12)

B. Monosilabenzenes

1. Silabenzene

Theory preceded experiment in predicting that silabenzene should be stable with respect to other isomers and to possible decomposition products. Many calculations are available for silabenzene, but most of them are relatively old and thus have used low levels of theory. The available calculations include the following levels of theory: MINDO/3[38], floating spherical Gaussian orbitals[39], MNDO[40], HF/STO-2G[41], HF/STO-3G[40,42], HF/3-21G*[43a] and HF/6-31G*[43b]. The best available calculations are the recent B3LYP/6-311+G** calculations[8], but this study which concentrates on 2-silanaphthalene does not discuss silabenzene in detail.

Is silabenzene aromatic? **8** is calculated to be a planar molecule with a Si−C bond length of 1.771 Å at B3LYP/6-311+G**[8] (1.760 Å at 3-21G*[43a], see Figure 2a for a

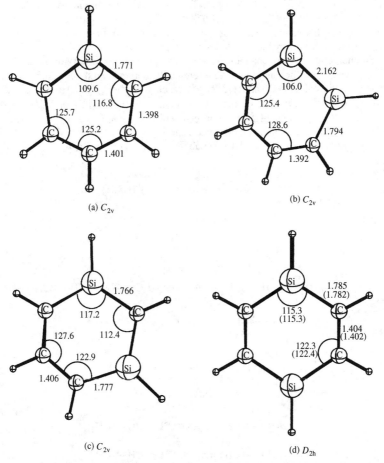

FIGURE 2. (a) B3LYP/6-311+G** optimized structure of silabenzene (**8**); (b−d) B3LYP/3-21G* optimized geometries of *ortho*-, *meta*- and *para*- disilabenzenes, respectively; values in parentheses are at MP2/6-31G**. All data are from Y. Apeloig, M. Bendikov and S. Sklenak, unpublished results

detailed structure). This distance is intermediate between that of a single and a double C—Si bond, displaying almost the same relative shortening relative to a Si—C single bond as found for the C—C bond in benzene. Thus, the calculated structure exhibits the expected partial double-bond character which is indicative of a delocalized system, and all researchers have therefore concluded that silabenzene possesses some degree of aromaticity. Further support for the aromatic character of **8** is provided by the molecular orbital coefficients, which indicate that the π-electrons are delocalized over the entire heavy-atom framework and by the nearly uniform π-electron distribution. The silicon atom has a somewhat smaller electron density than the carbons, i.e. 0.82 electrons on Si vs 1.09, 0.96 and 1.09 electrons on the *ortho*-, *meta*- and *para*-carbons, respectively, exhibiting the charge build-up on the *ortho*- and *para*-carbons, typical of a benzenoid π-system[42]. The σ-electrons are polarized away from the electropositive silicon towards the neighbouring carbon atoms[39,42].

The photoelectron spectra of silabenzene was measured and assigned on the basis of a double-zeta *ab initio* SCF calculation[44]. The lowest ionization energy of **8** is 8.11 eV ($^2B_1$ state, calculated to be 7.8 eV at HF/DZ[44]), by *ca* 1 eV lower than in benzene. The authors conclude that silabenzene is best considered as a symmetry-distorted (due to the presence of the Si atom) cyclic 6π-electron system.

Several energetic comparisons indicate that the degree of aromaticity of **8** is *ca* 70–85% of that of benzene[42,43] (depending on the reference compound and the level of calculation). For example, the calculated (at HF/3-21G*) resonance energy (RE) of **8** (equation 7) of 47.0 kcal mol^{-1} is *ca* 77% of the RE of benzene (equation 2) of 61.0 kcal mol^{-1} (64.2 kcal mol^{-1}, experimental)[43]. Similar results were obtained by using the more reliable homodesmotic equation 8 and hyperhomodesmotic equation 9, according to which the aromatic stabilization energy (ASE) of silabenzene is 70–85% (HF/6-31G*) of that of benzene[43b].

$$8 + 5CH_4 + SiH_4 \longrightarrow 2C_2H_6 + 2C_2H_4 + CH_3SiH_3 + CH_2{=}SiH_2 \qquad (7)$$

$$\underset{X}{\bigcirc} + 3H_2C{=}CH_2 \longrightarrow$$

$$2\ trans\text{-}H_2C{=}CH\text{—}CH{=}CH_2 + trans\text{-}H_2C{=}XH\text{—}CH{=}CH_2 \qquad (8)$$

ASE (HF/6-31G*, kcal mol^{-1}): X = Si, 17.2[43b]; X = C, 24.8[45]

$$\underset{X}{\bigcirc} + 3\ trans\text{-}H_2C{=}CH\text{—}CH{=}CH_2 \longrightarrow$$

$$2\ trans\text{-}H_2C{=}CH\text{—}CH{=}CH\text{—}CH{=}CH_2 +$$

$$trans\text{-}H_2C{=}CH\text{—}XH{=}CH\text{—}CH{=}CH_2 \qquad (9)$$

ASE (HF/6-31G*, kcal mol^{-1}): X = Si, 17.6[43b]; X = C, 23.5[45]

The high degree of aromaticity of silabenzene suggested by the geometric and energetic criteria was supported in a recent paper by its high NICS value of −9.1 (GIAO-HF/6-31+G*//B3LYP/6-311+G**), which is almost as high as that of benzene (−9.7)[8].

Schleyer and coworkers[40] studied in detail the energetic relationships between silabenzene **8** and four of its isomers, **13–16**. However, as this study is quite old, the calculated relative energies of **8** and **13–16**, which have been calculated only at HF/3-21G(*)//HF/STO-3G, are probably only approximate and higher level calculations are required for a more reliable estimate of these relative energies.

(8)   (13)   (14)   (15)   (16)

According to the HF/3-21G(*)//HF/STO-3G calculations the silylene isomers of silabenzene, **14** and **15**, are 20–25 kcal mol^{-1} higher in energy than **8**. The carbene isomer, **16**, which is a ground-state triplet, lies about 60 kcal mol^{-1} above silabenzene. The **8–14** (or **8–15**) energy difference is much smaller than that between the analogous isomers of benzene, reflecting the preference of silicon to adopt divalent structures instead of forming multiple bonds (e.g. H$_3$CS̈iH and H$_2$C=SiH$_2$ have almost the same energy[46a] while ethylene is by *ca* 80 kcal mol^{-1} more stable than CH$_3$C̈H[46b]). The 20–25 kcal mol^{-1} energy difference between **8** and **14** can be taken as yet another measure of the degree of aromaticity of silabenzene. Dewar silabenzene **13** is by 38 kcal mol^{-1} less stable than silabenzene. This value is expected to be reduced at higher levels of theory[40] but even this value is much smaller than the corresponding energy difference of 60 kcal mol^{-1} (experimental) between benzene and Dewar benzene.

In summary, all the calculations find silabenzene to be highly aromatic. However, in view of the relatively basic level of theory which was used in most of the available studies, it is highly desirable to repeat the calculations for this fundamentally important molecule using more reliable levels of theory.

2. Substituted silabenzenes

An extensive systematic study of 48 substituted silabenzenes was carried out by Baldridge and Gordon[41b]. Twelve substituents, namely F, Cl, OH, SH, OCH$_3$, NH$_2$, PH$_2$, CH$_3$, SiH$_3$, CN, NO$_2$ and COOH, were attached to the four unique positions of silabenzene, i.e. at the silicon, and at the *ortho*, *meta* and *para* positions (cf **17**). The analogous substituted benzenes were studied for comparison[41b]. The questions studied included the relative energies of the isomers, the degree of aromaticity, the electron density distributions and the dipole moments. Although the calculations used a relatively simple level of theory (in general HF/3-21G//HF/STO-2G, and in some cases HF/3-21G(*)//HF/STO-2G), the qualitative conclusions will probably hold also when better calculations become available (which is highly desirable). A summary of the major findings is given below.

Of the four possible isomers, substitution at silicon produces the most stable isomer (except for COOH), as shown in Table 1. The authors concluded that 'the degree of stability

ipso
↓
Si
ortho
meta
↑
para
(17)

appears to be a reflection of the electronegativity of the substituent; i.e. the more electronegative the substituent, the more stable the silicon-substituted species relative to the other isomers'[41b]. However, examination of Table 1 seems not to support this conclusion, e.g. for the highly electronegative NO_2 or CN substituents the energy difference between the *ipso* and the other isomers is much smaller than for the significantly less electronegative SH or NH_2 substituents (Table 1). Apparently π-effects also play an important role. A full analysis as well as better calculations of the energy differences between the isomers are still lacking. The relation between the stability of Si- and C-substituted silenes and the electronegativity of the substituent was discussed earlier by Apeloig and Karni[47]. According to Baldridge and Gordon[41b] the relative stability of the *ortho*-, *meta*- and *para*-isomers appears to be controlled by the effect of the substituent on the charge density at silicon, favouring positions in which the substituent directs electron density away from the silicon. Thus, the *meta*-substituted isomer is the most stable (after the *ipso*-substituted isomer) when the substituents are *ortho* and *para* directing (i.e. directing electron density away from the Si). Similarly, the *ortho*-substituted species is the most stable for *meta*-directing substituents (except for NO_2 and SH) (Table 1).

TABLE 1. Relative energies (HF/3-21G//HF/STO-2G, kcal mol^{-1}) of the four isomers of several substituted silabenzenes (see **17**) and the directing effects of the substituents[a]

Substituent	Relative energies position of substitution				Directing effect[b]
	ipso	ortho	meta	para	
F	0.0	45.2	35.2	38.1	*o, p*
Cl	0.0	33.8	31.9	32.6	*o, p*
OH	0.0	38.8	29.1	33.6	*o, p*
SH	0.0	24.0	22.3	24.3	*m*
NH_2	0.0	31.7	27.3	28.6	*o, p*
PH_2	0.0	11.7	13.3	12.9	*m*
CN	0.0	8.6	12.9	10.2	*m*
NO_2	0.0	6.9	10.7	6.4	*m*
CH_3	0.0	14.0	13.0	14.2	*o, p*
SiH_4	0.0	2.3	6.2	5.6	*m*
OCH_3	0.0	35.2	29.4	32.7	*o, p*
COOH	0.0	−7.0	−2.3	−6.4	*m*

[a] From Reference 41b.
[b] As predicted by electron density difference plots (see text). *o,p* = *ortho*- and *para*-directing (relative to silicon); *m* = *meta*-directing (relative to silicon).

The π-directing effects of the substituents were determined from density difference plots, constructed by subtracting the electron density of silabenzene (or benzene) from that of the substituted silabenzene (or benzene). From such density difference plots Baldridge and Gordon concluded that all the substituents studied have the same qualitative directing effects in silabenzene and in benzene (see Table 1)[41b]. It should be pointed out, however, that charge distributions are not always a reliable guide for predicting the substitution site in electrophilic reactions. In principle, the relative energies of the appropriate intermediates should be used for such predictions and such calculations are not yet available. A general feature of all the electron density difference plots is the lack of any negative charge build-up on silicon, regardless of whether the particular substituent is *ortho*-, *para*- or *meta*-directing. Silicon, which is the least electronegative atom in the molecules considered, prefers to remain positive.

The effect of *ipso*-substitution on the stability of the silabenzene was evaluated by equation 10,[41b] which compares the effects of the substituent X in the aromatic system with that in H_3SiX (we think that a comparison with $H_2C=SiHX$ would have been a better choice). The results presented in Table 2[41b] show that F, Cl, SH, NO_2 and CN

TABLE 2. Calculated substituent effects on the thermodynamic stability, ΔE (in kcal mol^{-1}), of *ipso*-substituted silabenzenes and of substituted benzenes (HF/3-21G)[a]

Substituent X	Conformation	$\Delta E^{b,c}$ silabenzenes	$\Delta E^{c,d}$ benzenes
F		−1.2	12.9
Cl		−2.3	3.5
OH	planar[e]	−1.5	12.3
	orthogonal[f]	1.9	10.1
OCH_3	planar[e]	−0.8	12.5
	orthogonal[f]	1.4	11.0
SH	planar[e]	−1.0	2.7
	orthogonal[f]	−0.1	4.4
NH_2	pyramidal (1)[g]	3.0	5.8
	pyramidal (2)[h]	1.3	9.2
	planar[e]	10.4	—
PH_2	pyramidal (1)[g]	1.9	3.3
	pyramidal (2)[h]	3.5	4.0
	planar[e]	−18.5	—
CN		−1.2	1.9
NO_2	planar	−4.6	7.4
CH_3		1.1	2.1
SiH_3		3.4	1.4
COOH	planar	13.1	11.3

[a]From Reference 41b.
[b]Calculated by equation 10.
[c]Positive values indicate stabilization.
[d]Calculated by the equation: $XC_6H_5 + CH_4 \rightarrow C_6H_6 + H_3CX$.
[e]∠ HYSiC (or ∠ HYCC in benzene)$^i = 0°$.
[f]∠ HYSiC (or ∠HYCC in benzene)$^i = 90°$.
[g]∠ HYSiC (or ∠ HYCC in benzene)$^i = 60°$.
[h]∠ HYSiC (or ∠ HYCC in benzene)$^i = 150°$.
[i]Y is the atom in group X directly bonded to the ring.

destabilize the silabenzene (or alternatively, stabilize H_3SiX more than the ring) while the other substituents slightly stabilize the ring. Relatively large stabilizing effects are found for X=COOH and planar NH_2. In general, the energies of equation 10 are similar to those calculated for the analogous substitutions in $H_2C=SiHX$[47]. The same substituents in benzene are always stabilizing but their effects are generally much larger than for the silabenzenes. This reflects the smaller conjugating ability of silabenzene relative to benzene[41b].

$$ipso\text{-}XSiC_5H_5 + SiH_4 \longrightarrow C_5H_6Si \text{ (8)} + H_3SiX \tag{10}$$

C. Silanaphthalenes

The recent synthesis and spectroscopic characterization of the crystalline, stable 2-silanaphthalene **1**, was accompanied by density-functional calculations of the geometries, the NICS values and the NMR chemical shifts of three isomeric silanaphthalenes **18**, **19a** and **20**[8] and two substituted 2-silanaphthalenes **19b** and **19c** (Table 3), of which **19c** is the closest model for **1**. The reliability of the calculations is demonstrated by their ability to reproduce the experimental Raman and NMR spectra of **1**. Thus, the strongest observed Raman shifts for **1** (at 1368 cm^{-1}) and for naphthalene (1382 cm^{-1}) are in good agreement with the calculated (B3LYP/6-31G*, scaled by 0.98) vibrational frequencies for **19a** (1377 cm^{-1}), for **19c** (1378 cm^{-1}) and for naphthalene (1389 cm^{-1}).

(18)

(19a) R = H
(19b) R = Me
(19c) R = Ph

(20)

The computational results point to an aromatic character of the silanaphthalenes **18–20**. The chemical shifts which were observed for **1** are similar to those calculated for **19a–c** (Table 3). The observed (in **1**) and calculated (for **19a–c**) $\delta(^{29}Si)$ and $\delta(^{13}C)$ chemical shifts are similar to those previously reported for other sp^2-hybridized Si and C atoms[3a,8]. All the ^1H NMR signals are, as expected, in the aromatic region (Table 3). The highly negative NICS values calculated for **18**, **19a** and **20** (Figure 3), which are very similar to those of benzene, also indicate that all the calculated silanaphthalenes (**18–20**) are aromatic.

TABLE 3. Calculated ^1H, ^{13}C and ^{29}Si NMR chemical shifts[a] for **19a**, **19b** and **19c** and the experimental values measured for **1**

	$\delta(^1H)$			$\delta(^{13}C)$			$\delta(^{29}Si)$
	$H_{(1)}$	$H_{(3)}$	$H_{(4)}$	$C_{(1)}$	$C_{(3)}$	$C_{(4)}$	$Si_{(2)}$
19a	7.74	7.27	8.64	128.45	125.13	153.38	67.80
19b	6.97	7.03	8.50	120.43	122.68	153.26	100.97
19c	7.32	7.08	8.55	121.63	123.56	152.45	94.32
1	7.40	7.24	8.48	116.01	122.56	148.95	87.35

[a]In ppm, calculated with GIAO-B3LYP/6-311G(3d) for Si and GIAO-B3LYP/6-311G* for C and H, at the B3LYP/6-31G* optimized geometries, the atom numbering is defined in Figure 3; from Reference 8.

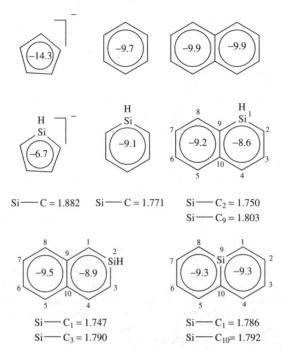

FIGURE 3. Calculated NICS values (ppm) at the ring centres for silanaphthalenes **18**, **19a**, **20** and related aromatic systems, at GIAO-SCF/6-31+G*. The optimized Si–C bond lengths (Å) are at B3LYP/6-311+G**. Reprinted with permission from Reference 8. Copyright (1997). American Chemical Society

The calculated Si–C bond lengths (B3LYP/6-311+G*, Figure 3) also suggest the occurrence of a π-electron delocalization in the three isomeric silanaphthalenes[8]. Thus, for **18**: $r(Si-C_{(2)}) = 1.750$ Å and $r(Si-C_{(9)}) = 1.803$ Å; for **19a**: $r(Si-C_{(1)}) = 1.747$ Å (exp. in **1**: 1.704 Å) and $r(Si-C_{(3)}) = 1.790$ Å (exp. in **1**: 1.765 Å); for **20**: $r(Si-C_{(1)}) = 1.786$ Å and $r(Si-C_{(10)}) = 1.792$ Å, compared to 1.771 Å in silabenzene, 1.708 Å in $H_2Si=CH_2$ and 1.885 Å in H_3SiCH_3. (Note that the reported experimental Si–C bond lengths in **1** are considerably shorter that the calculated values for **19a**. However, the authors comment that the reported structure of **1** is preliminary, as the naphthalene ring in **1** is severely disordered.) The calculated vibrational modes of **19a** and **19c** are similar to those of naphthalene, also suggesting that the 2-silanaphthalene (**1**) possesses a significant aromatic character[8].

Unfortunately, the relative energies of the 3 isomeric silanaphthalenes **18**, **19a** and **20** were not reported[8].

D. Disilabenzenes

1. Geometry and the degree of aromatic stabilization

Of the three possible disilabenzene isomers, *ortho*-**21**, *meta*-**22** and *para*-**11**, only **11** is known experimentally and was observed in a matrix[36]. The published calculations on disilabenzenes are all at relatively low levels of theory (i.e. they used a relatively small basis set and electron correlation was not included), except for unpublished density functional, B3LYP/3-21G*, calculations by Apeloig and coworkers which are reported below[48]. Schaefer and coworkers found, in a study of 1,3,5-trisilabenzene, that upon the addition of a set of d-orbitals to the basis set (especially important for Si) the Si—C bond lengths are contracted dramatically[49] (see Section III.E). This shows that results of calculations which use small basis sets (e.g. HF/STO-2G or HF/STO-3G) should be treated with caution. Unfortunately, there is very little experimental data[36] that can be compared with the calculations for **11**. Experimental UV and IR absorptions of **11** were reported, but unfortunately related computational results which might help to assess the reliability of the computational studies were not reported.

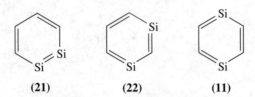

Density functional B3LYP/3-21G*[48] (as well as HF/STO-2G[50a]) calculations show that the three disilabenzene isomers have planar geometries. The fact that *ortho*-disilabenzene is planar whereas disilene is slightly non-planar may indicate some electron delocalization in **21** which stabilizes the planar Si=Si moiety. The calculated geometries of the disilabenzenes at B3LYP/3-21G* (shown in Figure 2b–d) reveal that the bond lengths, including the bonds to silicon, are intermediate between single and double bond lengths, reflecting electron delocalization and aromaticity[48].

meta-Disilabenzene (**22**) is the most stable disilabenzene. The other isomers **21** and **11** are by 6.1 kcal mol^{-1} and 11.1 kcal mol^{-1}, respectively (at B3LYP/3-21G*//B3LYP/3-21G*), less stable. At HF/3-21G//HF/STO-2G the relative energies are: **22** (0.0) > **21** (2.0) > **11** (10.8)[50a]. Baldridge and Gordon suggested that the relative stabilities of the disilabenzenes may be rationalized by considering the contributing ionic Kekulé structures in the three isomers. Thus, only for the *meta*-isomer **22** do all contributing structures (e.g. **22a**) have a formal positive charge on silicon, while for **11** and **21** the undesirable resonance structures with negative charge on silicon are also possible (e.g. **11a** and **21a**)[50a]. We believe that other factors, such as the relative energies of C=Si vs Si=Si bonds, might be more important than these charge effects, and a high-level computational study of these interesting molecules is highly desirable.

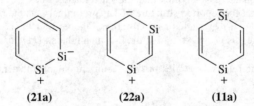

Using the appropriate bond-separation reactions, the HF/3-21G aromatic stabilization energies are calculated to be 47.2, 36.4 and 22.5 kcal mol^{-1} for **22**, **11** and **21**, respectively, compared to 59.0 kcal mol^{-1} for benzene[50a]; thus the *meta*-, *para*- and *ortho*-isomers have 80, 62 and 38% of the aromaticity of benzene. The different orders of the thermodynamic stability of the three isomeric disilabenzene and of their aromatic stabilization energies result from the different reference compounds in the bond separation equations for **21**, **22** and **11**. Thus, the aromatic stabilization of **21** is calculated relative to disilene + ethylene, while the aromatic stabilizations of **22** and **11** are calculated relative to two molecules of silaethylene[50a].

The molecular orbitals (MOs) of the disilabenezenes shown in Figure 4 are also of interest. In all three isomers, the highest lying MO (HOMO) has the largest silicon contribution. In **21** and **11** this orbital is strongly polarized toward the silicons, while in **22** the HOMO is more delocalized[50a].

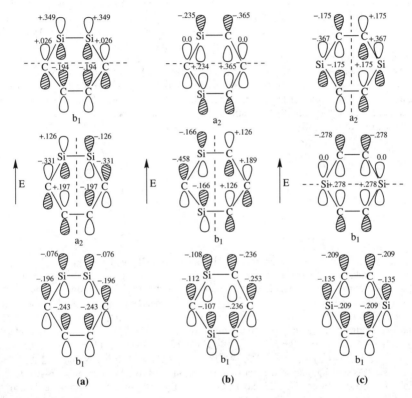

FIGURE 4. The π-molecular orbitals of disilabenzenes: (a) 1,2-disilabenzene (**21**); (b) 1,3-disilabenzene (**22**); (c) 1,4-disilabenzene (**11**). The orbital coefficients given are of the outer valence π-atomic orbitals (at HF/3-21G). Reproduced by permission of Elsevier Sequoia S.A., from Reference 50a

2. Relation to isomers

Comparison of the energy of *para*-disilabenzene (**11**) with that of the isomeric Dewar disilabenzene (**23**) and the silylenes **24** and **25** follows the trends discussed above for the isomers of silabenzene. At HF/3-21G(*)//HF/STO-3G the most stable isomer is the silylene **24**, which is more stable than **11** by 10 kcal mol^{-1}[50b]. The remarkable stability of the silylene **24** relative to **11** results mainly from the fact that C=C bonds are by at least 25 kcal mol^{-1} stronger than C=Si bonds, thus compensating for the loss of aromaticity in **24**, which is estimated to stabilize **11** by only 23 kcal mol^{-1}[50b]. The Dewar isomer **23** is by only 6 kcal mol^{-1} higher in energy than **11**[50b] (2.3 kcal mol^{-1} at HF/6-31G*[51a]). **25** is the highest in energy being by 21 kcal mol^{-1} (at HF/STO-3G, a HF/3-21G* value was not reported) less stable than **11**. When a larger basis set is used and the contribution of electron correlation is taken into account, ΔE (**23**–**11**) increases to 10.0 kcal mol^{-1} at CCSD(T)/6-31G**//MP2/6-31G** +ZPE (13.3 kcal mol^{-1} at MP2/6-31G**//MP2/6-31G** +ZPE and 14.4 kcal mol^{-1} at B3LYP/6-31G**//B3LYP/6-31G** +ZPE)[48]. The **23**–**11** energy difference is considerably smaller than the energy difference between Dewar silabenzene (**13**) and silabenzene (38 kcal mol^{-1} at HF/3-21G*//HF/STO-3G) and dramatically smaller than the calculated Dewar benzene–benzene energy difference of 81.1 kcal mol^{-1} (at MP2/6-31G**//MP2/6-31G**[52]). Two major factors are responsible for the observed energy differences: (1) Ring strain decreases in the order: Dewar benzene > Dewar silabenzene > Dewar disilabenzene, because silicon accommodates better than carbon smaller valence angles. (2) Successive replacement of unsaturated carbon centres by silicon destabilizes the planar 'aromatic' structure. Experimentally, it was observed that irradiation at λ = 405 nm of the matrix-isolated 1,4-disilabenzene caused its disappearance, probably yielding Dewar disilabenzene[36]. Most recently Ando has isolated and characterized spectroscopically a substituted Dewar benzene (having Me groups at Si and SiMe$_3$ groups at the carbon atoms)[51b].

(23)   **(24)**   **(25)**

Ando and coworkers have very recently isolated a bis(silacyclopropene) **26a**, a valence isomer of a substituted *para*-disilabenzene, and have shown that its thermolysis in benzene afforded the corresponding stable crystalline disilabenzvalene **27a**, yet another substituted valence isomer of *para*-disilabenzene[51a]. It was suggested that the isomerization of **26a** to **27a** could proceed via the initial formation of the corresponding Dewar disilabenezene, or *para*-disilabenzene, or the silylenic isomer **28a**. At HF/6-31G* the calculated relative thermodynamic stabilities (kcal mol^{-1}) of the valence isomers of **11** considered here are: **11** (0) > **23** (2.3) > **27b** (8.5) > **26b** (32.7) > **28b** (36.3)[51a]. It was suggested that since large basis sets and inclusion of electron correlation tend to favour tricyclic over bicyclic or monocyclic structures, **27b**, **23** and **11** may actually have similar energies (and even the stability order may be reversed)[51a]. The possible reaction paths connecting these isomers were not studied, and concrete conclusions about the mechanism of the rearrangement of **26a** to **27a** were not given.

(**26a**) R = SiMe$_3$; R' = Ph   (**27a**) R = SiMe$_3$; R' = Ph   (**28a**) R = SiMe$_3$; R' = Ph
(**26b**) R = R' = H   (**27b**) R = R' = H   (**28b**) R = R' = H

E. Trisilabenzenes

1. Geometry and the degree of aromaticity

The D_{3h} 1,3,5-trisilabenzene **12** (TSB) was studied by Gordon and coworkers[53], who used effective core potentials[54] and a polarized basis set for geometry optimizations, and more recently by Schaefer and coworkers at HF/TZ2P and at CCSD(T)/DZP//CISD/DZP[49] and by Apeloig and Hrusàk at MP2/6-31G**[55]. The planar D_{3h} geometry is a minimum on the PES at all these levels of theory. The optimized C=Si bond length of 1.754–1.759 Å (1.755 Å at HF/DZP; 1.754 Å at CISD/DZP[49]; 1.759 Å at MP2/6-31G**[55]), which is intermediate between those of typical C=Si (1.70 Å) and C—Si (1.87 Å) bond lengths, and the CSiC and SiCSi bond angles of 120.5° and 119.5° (CISD/DZP[49]), respectively, are indicative of an aromatic system. It was found that the inclusion of polarization functions in the basis set is crucial for obtaining a reliable geometry[49]. The calculated r(C=Si) is shortened significantly when going from a DZ basis set (1.778 Å) to a DZP basis set (1.755 Å), while the change on going to the larger TZ2P basis set is relatively small (1.750 Å). The HF/DZP Si=C bond length remains almost unchanged upon inclusion of electron correlation (i.e. 1.754 Å at CISD/DZP). From these calculations one can conclude that earlier calculations of various silabenzenes[40,41b,50] which used small non-polarized basis sets (e.g. HF/STO-3G, HF/3-21G) produced unreliable geometries. The DZ basis set also seriously underestimates the spatial extent of electron delocalization in **12**; upon the addition of a set of d-orbitals (especially on Si) the orbital overlap and electron delocalization increase dramatically, contracting the Si—C bond length[49].

(**12**)

The calculated resonance energies (RE) of **12** (equation 11) and of benzene (equation 2) are 53.1 kcal mol^{-1} and 63.8 kcal mol^{-1}, respectively (at CCSD(T)/DZP//CISD/DZP), indicating that **12** has 83% of the aromaticity of benzene[49]. Similar results were obtained with lower levels of theory[53a], but this is probably fortuitous. The addition of polarization functions changes the **12**/benzene RE ratio from 84% at DZ to 99% at DZP, while electron correlation acts in the opposite direction, decreasing the above ratio to 83%[49]. The existence of aromatic stabilization in **12** can also be concluded from the large difference of 27 kcal mol^{-1} (MP2/6-31G**//MP2/6-31G**) in the stability of the Si=C double bonds in the aromatic 1,3,5-trisilabenzene and the non-aromatic 1,3,5-silacyclohexadiene; i.e. $\Delta E = -35.3$ and 8.5 kcal mol^{-1} for equations 12 and 13, respectively[55]. The calculated π-atomic populations and π-bond-orders point to a similar degree of aromaticity in C_6H_6, $C_3Si_3H_6$ and Si_6H_6[53a].

$$\mathbf{12} + 3CH_4 + 3SiH_4 \longrightarrow 3H_3SiCH_3 + 3H_2C=SiH_2 \qquad (11)$$

$$\mathbf{12} + H_3CSiH_3 \longrightarrow H_2C=SiH_2 + c\text{-}C_3Si_3H_8 \qquad (12)$$

$$c\text{-}C_3Si_3H_8 + H_3CSiH_3 \longrightarrow H_2C=SiH_2 + c\text{-}C_3Si_3H_{10} \qquad (13)$$

2. Relation to isomers

Schaefer and coworkers[49] have also studied the relative stability of 1,3,5-trisilabenzene (**12**) and its isomeric 1,3,5-trisilylene **29**. Both isomers are possible products of the triple dehydrogenation reaction of 1,3,5-trisilacyclohexane[37]. Two minima were located for **29**: one having a chair conformation (**29a**) and one having a boat conformation (**29b**), the latter being by 1.5 kcal mol^{-1} higher in energy. The trisilylene **29a**, is by 35.0 kcal mol^{-1} less stable than the aromatic **12** [CCSD(T)/DZP//CISD/DZP][49]. A third isomer, the trisilaprismane **30**, is by 35.3 kcal mol^{-1} higher in energy than **12** (MP2/6-311G**//HF/effective core potentials)[53b]. These computational results may be taken as partial support for the claim by Bjarnason and Arnason that 1,3,5-trisilabenzene is formed in the triple dehydrogenation reaction of 1,3,5-trisilacyclohexane in the gas phase[37], but additional experimental and theoretical studies are required to support more firmly the intermediacy of **12** in this reaction.

(29) (29a) (29b) (30)

F. Hexasilabenzenes

1. The parent hexasilabenzene

a. Geometry and the degree of aromaticity. Hexasilabenzene (**31**), being a complete silicon analogue of benzene, has attracted considerable theoretical interest and as a result our knowledge about this fascinating molecule is quite extensive, although it has not yet been observed experimentally.

(**31**) D_{6h} (**32**) D_{3d}

Is Si_6H_6 a planar D_{6h} molecule analogous to benzene? The answer varies between the D_{6h} structure (**31**) and the D_{3d} structure (**32**), depending on the theoretical method used[19,52,56−61]. The calculated optimized geometries of the D_{6h} and D_{3d} structures of Si_6H_6 at various theoretical levels are shown in Figure 5. While early HF/3-21G calculations found Si_6H_6 to have D_{6h} symmetry with $r(Si-Si)$ equal to 2.220 Å[56], this is not the case at higher levels of theory[19,52,58−61]. Thus when d-functions are added to

HF/6-31G*
HF/DZ+d
MCSCF/DZ
MP2/6-31G**
B3LYP/6-311+G**

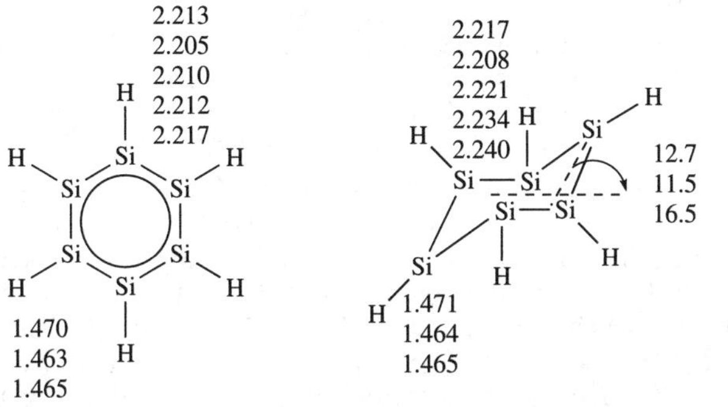

D_{6h} structure Si–Si bond lengths (top to bottom): 2.213, 2.205, 2.210, 2.212, 2.217; Si–H: 1.470, 1.463, 1.465

D_{3d} structure Si–Si bond lengths (top to bottom): 2.217, 2.208, 2.221, 2.234, 2.240; Si–H: 1.471, 1.464, 1.465; puckering angle: 12.7, 11.5, 16.5

FIGURE 5. Optimized structures of D_{6h} and D_{3d} Si$_6$H$_6$ at (from top to bottom): HF/6-31G*, HF/DZ+d, MCSCF/DZ, MP2/6-31G** and B3LYP/6-311+G**. The D_{6h} values and the D_{3d} MCSCF/DZ values are from Reference 60, the D_{3d} HF values are from Reference 59. The MP2/6-31G** and B3LYP/6-311+G** Si–Si bond lengths for both structures are from References 52 and 19, respectively

the basis set the planar D_{6h} structure becomes a maximum on the Si$_6$H$_6$ potential energy surface[59]. Full geometry optimization at HF/6-31G* (or at HF/6-31G** or at HF/DZ+d) leads to a minimum which is a chair-like puckered structure of D_{3d} symmetry (**32**) with equal Si–Si (2.218 Å at HF/6-31G**) and Si–H (1.471 Å) bond lengths[59]. The deviation from planarity is, however, relatively small ($\theta = 13.5°$ at HF/6-31G**[59]), as is the gain in energy (0.6 kcal mol^{-1}[59] at HF/6-31G**). However, the use of more extended basis sets [including double sets of d-functions (HF/6-31++G(2d,p)] or a set of f-functions on Si [HF/6-31G(df,p)] reverses the relative stability of the two structures and the D_{3d} minimum disappears, leaving the D_{6h} structure as a minimum[59]. On the other hand, the addition of electron correlation makes the Si$_6$H$_6$ potential energy surface less flat, and leads again to a greater preference of the D_{3d} puckered structure over the D_{6h} planar structure [$\Delta E = 2.3$ kcal mol^{-1} at B3LYP/6-311+G**//B3LYP/6-311+G**[19]; 4.3 kcal mol^{-1} at MP2/6-31G**//MP2/6-31G**[52]; 4.8 kcal mol^{-1} at MCSCF/DZ[60]; 3.0 kcal mol^{-1} at MP2/6-31G**//HF/ECP(d)[53b]; 7.05 kcal mol^{-1} at MP2/6-31G*//6-31G*[59]] whilst the deviation from planarity increases [$\theta = 16.5°$ and $19.0°$ at MCSCF/ZD[60] and MP2/6-31G**//HF/ECP(d)[53b], respectively]. The most reliable results available to date[19,53b,59-61] conclude that *hexasilabenzene adopts a chair-like puckered structure of D_{3d} symmetry* (with equal Si–Si and Si–H bonds). The preference of the D_{3d} structure over the D_{6h} structure was explained by Nagase in terms of atom localization of the π-electrons onto each Si, generating a radical character at each silicon atom, and as silyl radicals (in contrast to methyl radicals) have strongly pyramidal structures Si$_6$H$_6$ prefers the puckered structure[61a]. On the other hand, a detailed analysis by Janoschek and coworkers shows that the propensity to puckering is due solely to contributions of electron correlation from the σ-framework while correlation of the π-electrons is of little relevance[60]. It is interesting to note that at all computational levels the Si–Si bond distances in the D_{6h}

and D_{3d} structures are quite similar being always intermediate in length between regular Si=Si (2.15 Å) and Si–Si (2.35 Å) bond lengths, indicating the existence of electron delocalization.

Another important question is the degree of aromatic stabilization in the planar hexasilabenzene. Using the appropriate bond-separation reactions, hexasilabenzene, **31** is calculated to possess *ca* 50–80% of the aromatic stabilization energy of benzene[19,53a,56,58,62] depending on the theoretical method or the equations used. The aromatic stabilization energies calculated according to equations 14a and 14b for benzene and Si$_6$H$_6$ (**31**), predict that **31** has 50–60% of the aromaticity of benzene; i.e. the ASE of **31** and benzene are 15.1 and 24.7 kcal mol^{-1}, respectively (at MP2/6-31G**//MP2/6-31G**[52]), according to equation 14a and 15.6 kcal mol^{-1} and 34.1 kcal mol^{-1}, respectively, according to equation 14b (at B3LYP/6-311+G**//B3LYP/6-311+G**)[19]. The ASE of the D_{3d} non-planar structure (**32**) is by 2.3 kcal mol^{-1} larger than for **31** because of the reduced strain in **32** relative to **31**[19]. The resonance energies (RE) calculated for C$_6$H$_6$ and Si$_6$H$_6$ (according to equation 15) predict that Si$_6$H$_6$ possesses 77% of the aromaticity of benzene, i.e. RE = 57.8 and 74.7 kcal mol^{-1}, respectively [at MP2/6-31G**//HF/ECP(d)][53a]. The significant degree of aromaticity of hexasilabenzene suggested by the structural and energetic criteria is supported also by the magnetic criteria. Thus, the calculated (CSGT-B3LYP/6-311+G**//B3LYP/6-311+G**) π-contributions to the chemical shifts [based on NICS(π) values] decrease from -16.8 ppm for benzene to -14.1 ppm for **31**, a result consistent with a relatively large π-electron delocalization in Si$_6$H$_6$ (**31**)[19]. The computed magnetic susceptibility exaltations (Λ) are larger for **31** (and **32**) than for benzene. However, Λ depends on the ring size, which is larger in Si$_6$H$_6$ than in benzene. Using the ring-size-adjusted aromaticity index ρ (see equation 6 in Section II.C.1), it was deduced that **31** possesses 37% of the aromaticity of benzene[19]. A similar degree of aromaticity was deduced for Ge$_6$H$_6$ according to its ASE and magnetic criteria[19].

$$X_6H_6 + 3H_2X=XH_2 \longrightarrow 3 \; trans\text{-}H_2X=XH-XH=XH_2 \qquad X = C; Si \qquad (14a)$$

$$X_6H_6 + 3H_2X=XH_2 \longrightarrow 3 \; cis\text{-}H_2X=XH-XH=XH_2 \qquad X = C; Si \qquad (14b)$$

$$X_6H_6 + 6XH_4 \longrightarrow 3H_2X=XH_2 + 3H_3X-XH_3 \qquad X = C; Si \qquad (15)$$

In the context of the question of the aromaticity of Si$_6$H$_6$, special attention should be paid to the recent studies by Shaik and coworkers[15]. Using VB crossing diagrams and MO calculations that separate the energetic effects of the σ- and π-electrons on the distortion (localization) of a conjugated system from a fully symmetric delocalized structure (i.e. ΔE_σ and ΔE_π, respectively), they reached the following conclusions:

(a) The geometry is *not* a proper measure of aromaticity. The tendency to adopt a symmetric or a distorted geometry is the result of the balance between two opposite driving forces: that of the σ-skeleton which by nature is always symmetrizing, and that of the π-electrons which favour localized structures; i.e. if $\Delta E_\pi + \Delta E_\sigma > 0$, the structure is dictated by the symmetric σ-skeleton and it is delocalized. The localizing propensity of the π-electrons is dependent on the singlet–triplet energy gap of the π-bond and consequently on its strength. For second-row elements both the π-distortion propensity and the σ-resistance to distortion are weaker than for first-row elements, but the former is weakened proportionally more. Thus, for Si$_6$H$_6$, $\Delta E_\sigma = 5.3$ kcal mol^{-1} (16.3 kcal mol^{-1} for C$_6$H$_6$) and $\Delta E_\pi = -2.4$ kcal mol^{-1} (-9.7 kcal mol^{-1} for benzene), resulting in a net positive $\Delta E_\pi + \Delta E_\sigma$. Consequently, second-row atoms generate less distortive π-systems than first-row atoms, and therefore Si$_6$H$_6$ would tend to maintain its D_{6h} symmetry more than benzene. However, the high stability of σ-bonds relative to π-bonds in second-row elements drives Si_6H_6 towards non-planar structures, e.g. the more stable hexasilaprismane (see below)[15c].

(b) The Quantum Molecular Resonance Energy (QMRE), that is the resonance interaction between the Kékule forms at the symmetric geometry, is an energetic measure of the degree of aromaticity of a cyclic π-system. QMRE is related to measured thermochemical properties, e.g. resonance energy (RE). The calculated QMRE of planar Si_6H_6 is about half of that of C_6H_6 (41.6 kcal mol^{-1} vs 85.2 kcal mol^{-1}, respectively, at HF/6-31G)[15b,c] reflecting the difference in the aromaticity between hexasilabenzene and benzene. Hence the degree of aromaticity of Si_6H_6, which according to VB theory is *ca* 50% of that of benzene, is similar to that deduced on the basis of MO theory.

The HOMO of hexasilabenzene is by *ca* 2 eV higher than in benzene, while the HOMO–LUMO gap is much smaller in **31** than in benzene[60]. This suggest that hexasilabenzene should be more reactive than benzene towards both electrophiles and nucleophiles.

b. Relation to isomers. In addition to the structural isomers of hexasilabenzene itself (i.e. **31** and **32**) discussed above, 5 additional Si_6H_6 isomers, **33** –**37**, were located on the PES at various theoretical levels[52,56,58,60]. Of these isomers the D_{6h} hexasilabenzene (**31**), the Dewar hexasilabenzene (**34a**) and the hexasilabenzvalene (**35a**) are not minima on the MP2/6-31G** PES[52] (the highest available level of calculation for this problem); **34a**[52,60] and **35a**[52] are, however, minima on the HF PES. The calculated relative energies of these isomers and of the analogous C_6H_6 isomers are given in Table 4 and are also shown graphically in Figure 6a.

(33a) X = Si
(33b) X = C

(34a) X = Si
(34b) X = C

(35a) X = Si
(35b) X = C

(36a) X = Si
(36b) X = C

(37)

The most interesting conclusion from Table 4 and Figure 6a is that, contrary to the $(CH)_6$ series where benzene is by far the most stable isomer, hexasilabenzene is not the most stable isomer on the PES. The most stable Si_6H_6 isomer is the hexasilaprismane, **33a**. In general, the energy spacing between the X_6H_6 isomers is much larger for X = C (ranging over *ca* 120 kcal mol^{-1}) than for X = Si where it ranges over *ca* 55 kcal mol^{-1} and five Si_6H_6 isomers have energies within 15 kcal mol^{-1} (Table 4, Figure 6a).

TABLE 4. Relative energies and strain energies (kcal mol^{-1}) of Si$_6$H$_6$ and the analogous C$_6$H$_6$ isomers[a]

Compound	Relative energies		Strain energies	
	Si$_6$H$_6$	C$_6$H$_6$	Si$_6$H$_6$	C$_6$H$_6$
31	11.9[b]	0.0[c]	−15.1[d]	−24.7[c,e]
32	7.6	f	f	f
33	0.0	117.5 (127.6[g])	110.0	148.9[h]
34	10.5[i]	81.0 (88.1[g])	26.2	63.6[h]
35	5.9[i]	74.8 (84.5[g])	68.7	81.3[h]
36	56.8	126.4	72.5	107.2[h]
37	43.4[j]	f	f	f

[a] From Reference 52 at MP2/6-31G**//MP2/6-31G**, unless stated otherwise.
[b] Not a minimum at the HF and the MP2 levels of theory.
[c] For benzene.
[d] ASE energies according to equation 14a, X = Si.
[e] ASE energies according to equation 14a, X = C.
[f] Not available.
[g] At HF/6-31G*//HF/3-21G, from Reference 58.
[h] At MP2/6-31G*//6-31G*, from Reference 63.
[i] Not a minimum at the MP2/6-31G** level, a minimum at HF.
[j] At CISDQ/TZP from Reference 60.

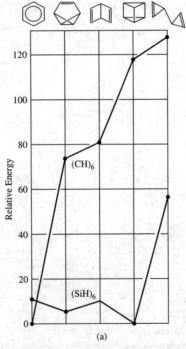

(a)

FIGURE 6. (a) Relative energies of (XH)$_6$ (X = Si, C) isomers; (b) relative energies of (XH)$_6$, X = Si, C, isomers, based on the bond additivity model (centre) and as stabilized by resonance or destabilized by strain (edges). All energies are in kcal mol^{-1} calculated at MP2/6-31G**//MP2/6-31G**. Reprinted with permission from Reference 52. Copyright (1996) American Chemical Society

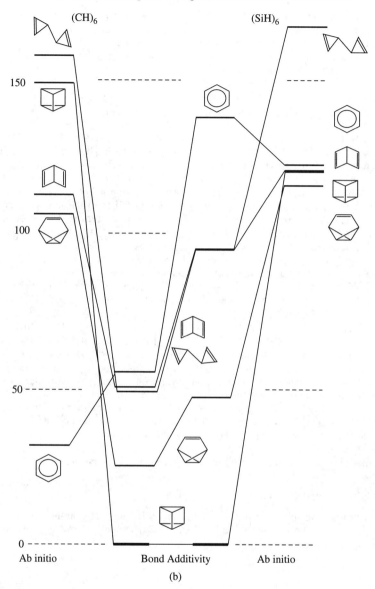

FIGURE 6. *(continued)*

D_{3d} hexasilabenzene **32** (which is by 4.3 kcal mol^{-1} more stable than D_{6h} **31**[52]) is by 7.6 kcal mol^{-1} *less stable* than hexasilaprismane (**33a**)[52]. In contrast, benzene is *more stable* than prismane (**33b**) by 117.5 kcal mol^{-1} (all values are at MP2/6-31G**//MP2/6-31G**)[52]. The aromatic **31** (and **32**) have a similar energy to that of hexasilabenzvalene (**35a**).

The reasons for the dramatic differences between the Si_6H_6 and C_6H_6 systems are already familiar to the reader (see the discussion above regarding silabenzene, and di- and tri-silabenzenes) and they are repeated and extended here. The energy of a Si=X (or C=X) bond is lower than twice the energy of the corresponding Si—X (or C—X) bond. However, these differences are much larger in silicon than in carbon compounds; e.g. the C—C and Si—Si dissociation energies in H_3X-XH_3 and $H_2X=XH_2$ are 86.8 and 176.8 kcal mol^{-1}, respectively, for X = C, and 70.1 and 59.8 kcal mol^{-1}, respectively, for X = Si[64]. This explains the higher preference of Si to be involved in σ-bonds rather than in π-bonds, and it is manifested in the order of stability of the Si_6H_6 isomers (Table 4): **33a** (most stable) > **35a** > **34a** (note that at MP2/6-31G** **34a** and **35a** are not minima on the PES), where the stability increases as the number of Si=Si double bonds decreases. The difference in ring strain between the C_6H_6 and Si_6H_6 isomers also contributes, but to a much smaller degree, i.e. hexasilaprismane (**33a**) is less strained than prismane (**33b**) by 39 kcal mol^{-1} (see Table 4). An interesting insight is provided when the effects of ring strain and resonance energy are subtracted from the relative energies of the Si_6H_6 and C_6H_6 isomers, leaving only the 'inherent differences' in their average single and double bond energies (i.e 'bond additivity model'[52]). The relative energies of the isomers calculated by this procedure are interpreted as given by the value of $m \cdot \Delta$ for a particular system, where m is the number of double bonds in the particular isomer and $\Delta = 2D(X-X) - D(X=X)$, i.e. the energy difference between two X—X single bonds and one X=X double bond. Using this procedure (the reader is referred to the original paper for a more detailed description) the following relative energies of the X_6H_6 isomers (X = C, Si) were calculated (kcal mol^{-1}, in descending stability order): **33** (0, X = C; Si) > **35** [24.9 (X = C); 47.1 (X = Si)] > **34** [48.8 (X = C); 94.3 (X = Si)] > **37** [(50.6 (X = C); 94.3 (X = Si)] > **31** [56.1 (X = C); 137.0 (X = Si)][52]. Figure 6b shows a schematic description of these relative 'inherent' energies and how they are changed when X_6H_6, X = C, Si, are stabilized by resonance or destabilized by strain. These results indicate that when the resonance stabilization, on the one hand, and the destabilization caused by strain, on the other, are removed from the total energies, the relative energies of the various X_6H_6 (X = Si, C) structures increase with the number of double bonds that they possess. The bond energy difference Δ for silicon is almost twice that for carbon, and as a result, according to the 'bond additivity model'[52] the relative energies of the $(SiH)_6$ isomers are spread out compared to the corresponding $(CH)_6$ isomers and hexasilaprismane becomes dramatically more stable than hexasilabenzene.

2. Substituted hexasilabenzenes

The effect of several substituents, i.e. R = F, CH_3, SiH_3 and BH_2, on the structure and the aromatic stabilization energies of R_6Si_6 (**31a** or **32a**) was studied by Nagase and the main results are summarized in Table 5[61a].

(31a) (32a)

TABLE 5. Substituent effects on the structure and aromatic stabilization energy (ASE) of Si_6R_6 (**31a** or **32a**)[a]

R	$r(Si=Si)$ (Å)	$\theta^{\circ b}$	$\Delta E^{c,d}$	$\Delta E_{pyr}^{c,e}$	$ASE^{c,f}$
H	2.231	12.7	−0.4	−2.6	11.8[g]
F	2.278	50.3	−14.4	−7.7	−8.7
CH_3	2.222	10.8	−0.2	−3.6	8.5
SiH_3	2.226	0.0	h	−1.0	21.7
BH_2	2.239	0.0	h	i	15.5

[a] At HF/6-31G*, from Reference 61a.
[b] Defined in **32a**.
[c] In kcal mol^{-1}.
[d] $\Delta E = E$ (non-planar) $-E$ (planar); a minus sign indicates that the non-planar structure (**32a**) is more stable.
[e] $\Delta E_{pyr} = E$ (pyramidal) $-E$ (planar) for the $(H_3Si)_2SiR$ radical. A minus sign indicates that the pyramidal structure is more stable.
[f] For the planar **31a**.
[g] 15.6 and 15.1 kcal mol^{-1} at B3LYP/6-311+G**//B3LYP/6-311+G**[19] and MP2/6-31G**//MP2/6-31G**[52], respectively.
[h] Only the planar structure was identified as a minimum.
[i] A minimum for the pyramidal radical could not be located.

Flourine imposes a very large deviation from planarity ($\theta = 50.3°$) which results also in a long Si−Si bond of 2.278 Å; this is in agreement with the strong preference of the $(H_3Si)_2SiF$ radical to adopt a pyramidal structure [ΔE_{pyr}(pyramidalization) = 7.7 kcal mol^{-1}]. On the other hand, the σ-donor SiH_3 and the π-acceptor BH_2 substituents increase the aromatic stability of the R_6Si_6 skeleton and prevent pyramidalization. Thus, $Si_6(BH_2)_6$ and $Si_6(SiH_3)_6$ are both planar having aromatic stabilization energies (ASEs) of 15.5 and 21.7 kcal mol^{-1}, respectively, compared with 11.8 kcal mol^{-1} for Si_6H_6 and 24.7 kcal mol^{-1} for benzene (HF/6-31G*)[61a]. The substituent effects described above are very similar to the effects that these substituents have on the geometry and stability of disilenes; i.e. F causes a very strong *trans*-bending and elongation of the Si=Si bond, while $(H_3Si)_2Si=Si(SiH_3)_2$ is planar[1c,65]. Based on these results we believe that $Si_6(SiR_3)_6$ with bulky R groups, which are required for kinetic stabilization, may be good candidates for the synthesis and isolation of the first planar hexasilabenzene.

G. Higher Congeners of Hexasilabenzene

Substitution of the Si atoms in Si_6H_6 by heavier group 14 atoms results in an increased deviation from planarity (the planar structures are not minima on the PES) and an increased stabilization upon ring puckering measured by $\Delta E(\mathbf{38-39})$. For $Si_3Ge_3H_6$, $\theta = 36.8°$ and $\Delta E(\mathbf{38-39}) = 4.8$ kcal mol^{-1} [MP2/6-31G**//HF/ECP(d)][53b]; for Ge_6H_6, $\theta = 38.0°$ and $\Delta E(\mathbf{38-39}) = 9.1$ kcal mol^{-1}; for Sn_6H_6, $\theta = 50.8°$ and $\Delta E(\mathbf{38-39}) = 23.1$ kcal mol^{-1}; and for Pb_6H_6, $\theta = 58.0°$ and $\Delta E((\mathbf{38-39}) = 63.3$ kcal mol^{-1} [MP2/DZ(d)//HF/DZ(d)][61b]. The increased deviation from planarity and the larger energy differences between the D_{6h} and D_{3d} structures on moving down along group 14 of the Periodic Table is explained by the increased tendency of the π-electrons to localize, and is consistent with the progressively favoured pyramidal structure on going along the radical series: germyl, stannyl and plumbyl[61]. Similar behaviour was observed for the $H_2X=XH_2$ series, where X = Si[65,66], Ge, Sn and Pb[66,62].

[Structures **(38)**, **(39)** with angle θ, and **(40)** shown at top of page]

Substitution of the Si atoms in Si_6H_6 by its higher congeners has a small effect on their aromatic stabilization energies; i.e. the ASEs calculated according to equation 14a for X_6H_6 (X = Si, Ge, Sn and Pb) are all in the range of 9–12 kcal mol^{-1}[62]. The calculated resonance stabilization energy of **39** (X = Si; Y = Ge) according to equation 15 is 60.2 kcal mol^{-1}, identical to that of 1,3,5-trisilabenzene **(12)** and very similar to that of Si_6H_6 [57.8 kcal mol^{-1} at MP2/ECP(d)//HF/ECP(d)][53a], indicating the small effect of substituting three Si atoms in Si_6H_6 by C or Ge.

The stability of the prismane isomers **(40)** in the $X_3Y_3H_6$ (X, Y = C, Si, Ge, Sn, Pb) series, relative to the corresponding aromatic **38** isomers, decreases in the following order [ΔE, kcal mol^{-1} at MP2/6-311G**//HF/ECP(d)][53b], unless stated otherwise; a negative ΔE indicates that the prismane isomer is more stable]: X = Y = Sn (−67.0, MP2/DZ+d//HF/DZ+d[67]) > X = Y = Pb (−31.3, MP2/DZ+d//HF/DZ+d[67]) > X = Y = Si (−13.6) > X = Y = Ge (−11.1) > X = Si; Y = Ge (−6.4) > X = C; Y = Ge (27.2) > X = C; Y = Si (35.5) > X = Y = C (113.9). The stability of the Dewar benzene **(34)** and the benzvalene **(35)** isomers relative to the corresponding aromatic D_{6h} X_6H_6 isomer also increases along the series X = C → X = Pb, though to a smaller degree than the increase in the stability of the corresponding prismane isomers[62].

H. Silabenzenoids Containing Heteroatoms

The potentially aromatic silabenzenoids, s-1,3,5-trisilatriazine **(41)**[68] and 2-, 3- and 4-silapyridines **(42–44)**[69], were studied by Veszprémi and coworkers. The MP2/6-31G* calculated geometries of **41–44** are given in Figure 7 (almost identical geometries were calculated for **41** also at MP2/6-311G** and CCSD/6-31G*[68]). **41** has D_{3h} symmetry

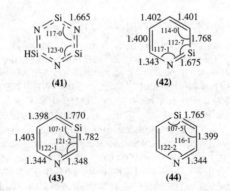

FIGURE 7. MP2/6-31G* optimized geometries of **41–44**. Drawn from data in References 68 and 69

and thus the Si—N bond lengths (1.665 Å) show no alternation, indicating a delocalized aromatic system. The geometries of **42–44** also suggest an aromatic character; the C—C bonds in **42–44** show almost no bond alternation, with bond lengths around 1.40 Å, very similar to those in benzene. The Si—C bond lengths in **42–44** of ca 1.77 Å are intermediate between typical Si—C and Si=C bond lengths, also indicating the existence of electron delocalization.

(41)   **(42)**   **(43)**   **(44)**

The calculated aromatic stabilization energy of **41** is 19.6 kcal mol^{-1} (at MP2/6-31G*//MP2/6-31G* using the homodesmotic equation 16)[68]. A very similar ASE of 19.4 kcal mol^{-1} was calculated for **42** using equation 17 (ASEs for **43** and **44** were not reported)[69]. The ASE calculated for benzene and pyridine (at the same level of theory) are 28.2 and 28.0 kcal mol^{-1}, respectively, indicating a high degree of aromaticity (ca 70% of that of benzene) for both **41** and **42**[69]. This is very similar to the carbon analogue of **41**, s-triazine (c-C$_3$N$_3$H$_3$), which was estimated to have 76% of the aromaticity of benzene according to the calculated relative magnetic susceptibility anisotropies[18] and the relatively high NICS(π) value[19].

$$41 + 3H_2Si=NH \longrightarrow 3H_2Si=N-SiH=NH \qquad (16)$$

$$42 + H_2Si=NH + 2CH_2=CH_2 \longrightarrow \qquad (17)$$

$$CH_2=CH-SiH=NH + CH_2=CH-N=SiH_2 + CH_2=CH-CH=CH_2$$

The relative stability of the three silapyridines **42–44** follows the order (kcal mol^{-1}): **42** (0.0, most stable) > **44** (11.0) > **43** (37.8), but no explanation was given for this stability order[68]. The thermodynamic stability of s-1,3,5-trisilatriazine, **41**, and of the silapyridines **42–44** relative to their isomeric silylenes is discussed in Section V.B.I.

I. Concluding Comments

The silabenzenoids are an exciting family of molecules and much remains to be learned about their properties and chemistry. Table 6 summarizes the degree of aromaticity as calculated from the ASE and NICS values (when available) of the various silabenzenoids that were discussed in this section. Their degree of aromaticity is in the range of ca 40–90% of that of benzene. Unfortunately, as pointed out during the discussion, much of the current data are based on quite old calculations which used relatively simple and therefore unreliable computational methods. Hence a reliable comparison of the aromaticity of the various silabenzenoids is not yet possible. New theoretical studies of these intriguing molecules using higher levels of theory and more sophisticated criteria for evaluating aromaticity (such as the magnetic criteria) are needed for a better understanding of these molecules.

TABLE 6. 'Degree of aromaticity' of silabenzenoids relative to benzene (R) as estimated from their ASEs calculated by various homodesmotic equations

Silabenzenoid	ASE (kcal mol^{-1})	R^a	Equation^b	Method	References
Monosilabenzene (8)	17.2	0.69^c	8	HF/6-31G*//HF/6-31G*	43b,45
meta-Disilabenzene (22)	47.2	0.80	d	HF/3-21G//HF/STO-3G	50a
para-Disilabenzene (11)	36.4	0.62	d	HF/3-21G//HF/STO-3G	50a
ortho-Disilabenzene (21)	22.5	0.38	e	HF/3-21G//HF/STO-3G	50a
Trisilabenzene (12)	53.1^f	0.83	11^g	CCSD(T)/DZP//CISD/DZP	49
Hexasilabenzene (31)	15.1	0.61	14a	MP2/6-31G**//MP2/6-31G**	52
	15.6	0.46^h	14b	B3LYP/6-311G**//B3LYP/6-311G**	19
2-Silapyridine (42)	19.4	0.69	17	MP2/6-31G*//MP2/6-31G*	69
s-Trisilatriazine (41)	19.1	0.68	16	MP2/6-31G*//MP2/6-31G*	68

^{a}ASE(silabenzenoid)/ASE(benzene).
^{b}The number in the text of the homodesmotic equation used to evaluate ASE.
^{c}NICS values are -9.1 ppm and -9.7 ppm for **8** and benzene, respectively[8].
^{d}According to the equation: **22** (or **11**) + 2SiH$_4$ + 4CH$_4$ → 2H$_2$Si=CH$_2$ + C$_2$H$_4$ + C$_2$H$_6$ + 2H$_3$CSiH$_3$.
^{e}According to the equation: **21** + 2SiH$_4$ + 4CH$_4$ → H$_2$Si=SiH$_2$ + C$_2$H$_4$ + C$_2$H$_6$ + 2H$_3$CSiH$_3$.
^{f}Resonance energy (RE); ASE is not reported.
^{g}The RE of benzene was calculated from equation 2.
^{h}NICS(π) values are -14.1 ppm and -16.8 ppm for **31** and benzene, respectively[19].

IV. CHARGED SILAAROMATIC SYSTEMS

In this section we will discuss charged silaaromatic systems which formally possess $4n+2$ π-electrons, such as **45–49**. These species are isoelectronic with the cyclopropenium cation, tropylium cation and cyclopentadienyl anions, respectively, which are all well established aromatic systems[13a,18].

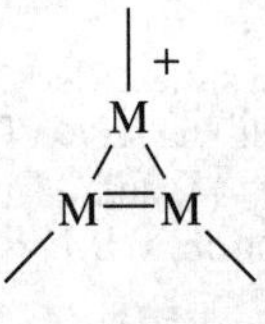

(45) 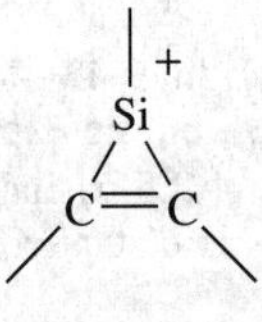 (46) 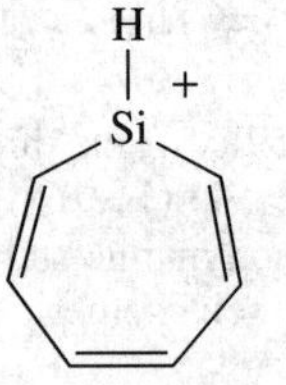 (47)

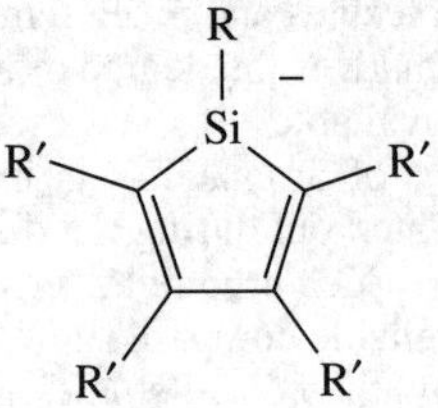

(48)
(48a) R = R' = H

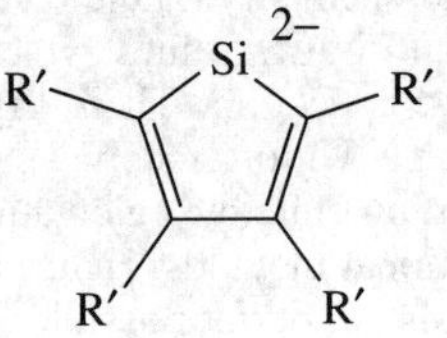

(49)
(49a) R' = H

A. Systems with Two π-Electrons

1. The monosilacyclopropenium cation

The cyclopropenium cation (**45**, M = C) is a well established aromatic system with a high aromatic stabilization energy of 58.7 kcal mol^{-1} [equation 18, M = C, at B3LYP/6-311++G(2d,2p)][70]. What happens when one carbon atom is substituted by silicon? At HF/STO-2G the monosilacyclopropenium cation, **46**, has a rather delocalized structure with r(C−C) = 1.393 Å, r(C−Si) = 1.722 Å and with CSiC and SiCC bond angles of 46.3° and 66.9°, respectively[41a]. However, according to equation 19 which measures the effect of delocalization, **46** is destabilized by 11 kcal mol^{-1} (HF/3-21G//HF/STO-2G) while **45**, M = C is stabilized by 36 kcal mol^{-1}. Unfortunately, this study used a very low computational level and should be repeated using more reliable methods.

$$\begin{array}{c}\text{H}\\\text{M}^+\\/\ \backslash\\\text{HM}=\text{MH}\end{array} + \begin{array}{c}\text{H}_2\\\text{M}\\/\ \backslash\\\text{H}_2\text{M}-\text{MH}_2\end{array} \longrightarrow \begin{array}{c}\text{H}_2\\\text{M}\\/\ \backslash\\\text{HM}=\text{MH}\end{array} + \begin{array}{c}\text{H}\\\text{M}^+\\/\ \backslash\\\text{H}_2\text{M}-\text{MH}_2\end{array} \qquad (18)$$

$$\text{M} = \text{Si, C}$$

$$\begin{array}{c}\text{H}\\\text{M}^+\\/\ \backslash\\\text{HC}=\text{CH}\end{array} + \text{MH}_3^+ + 2\text{CH}_3\text{MH}_3 \longrightarrow \begin{array}{c}\text{H}_2\\\text{M}\\/\ \backslash\\\text{HC}=\text{CH}\end{array} + 2\text{CH}_3\text{MH}_2^+ + \text{MH}_4 \qquad (19)$$

$$\text{M} = \text{Si, C}$$

2. The trisilacyclopropenium cation and its higher congeners

The trisilacyclopropenium cation (**45**, M = Si), the full silicon analogue of the cyclopropenium cation, was calculated to be the global minimum on the Si$_3$H$_3^+$ PES. The aromatic stabilization energy of **45**, M = Si according to equation 18 is 35.6 kcal mol^{-1} [B3LYP/6-311++G(2d,2p)][70], about 60% of the aromatic stabilization energy of the cyclopropenium cation. The silacyclopropenium cation has D_{3h} symmetry with a Si−Si bond length of 2.203 Å [B3LYP/6-311++G(2d,2p)[70]; 2.199 Å at MP2/6-31G*[71]], intermediate between typical Si−Si and Si=Si bond lengths. Based on these facts it can be concluded that **45**, M = Si has a considerable degree of aromaticity. On going down group 14 the π-stabilization of **45** decreases, i.e. ΔE(equation 18) = 58.7, 35.6, 31.9, 26.4 and 24.1 kcal mol^{-1} for M = C, Si, Ge, Sn and Pb[70].

Three non-classical hydrogen-bridged isomers, **50a**, **50b** and **51** (M = Si), which are by about 20 kcal mol^{-1} less stable than **45** (M = Si), were also located as minima on the Si$_3$H$_3^+$ PES[70,71]. On going down group 14 the preference for the hydrogen-bridged isomer **51** increases and for M = Ge, Sn and Pb it becomes the global minimum on the M$_3$H$_3^+$ PES, i.e. ΔE(**51**-**45**) = 23.7, −17.4, −32.4 and −63.3 kcal mol^{-1} for M = Si, Ge, Sn and Pb, respectively[70]. This trend was attributed to the increased stability of divalent arrangements when moving down group 14 of the Periodic Table.[70]

(**50a**) C_1 (**50b**) C_1 (**51**) C_{3v}

3. The tetrasilacyclobutadiene dication

The most stable isomer on the $Si_4H_4^{2+}$ PES is the slightly puckered (D_{2d} symmetry) potentially 2π-aromatic tetrasilacyclobutadiene dication, **52a**, having a SiSiSiSi dihedral angle of 9° (the planar D_{4h} isomer is not a minimum on the PES). The analogous D_{2d} cyclobutadiene dication **52b** is considerably more puckered ($\angle CCCC = 32.5°$)[71]. This is a unique example of a silicon system which is less folded than the carbon analogue. The identical Si—Si bond lengths of 2.268 Å indicate electron delocalization and thus aromaticity. The calculated ASE of **52a** and **52b** according to equation 20 are 32.4 kcal mol^{-1} and 26.8 kcal mol^{-1}, respectively [at MP2(full)/6-31G*]. The unusual higher ASE of **52a** relative to **52b** is attributed, however, to the destabilization of the corresponding reference compounds in equation 20 for M = Si and not to its higher aromatic character[71].

H\ /H
 M—M
 ‖ ‖ 2+
 M—M
/   \
H   H

(**52a**) M = Si
(**52b**) M = C

52a (or **52b**) + $\begin{array}{c} H_2M-MH_2 \\ | \quad | \\ H_2M-MH_2 \end{array}$ ⟶ $\begin{array}{c} H_2M-MH \\ | \quad \| \\ H_2M-MH \end{array}$ + $\begin{array}{c} H_2M-\overset{+}{M}H \\ | \quad | \\ \underset{+}{H}M-MH_2 \end{array}$ (20)

B. The Silatropylium Cation and its Isomers

Gas-phase experiments by Beauchamp and coworkers[72] point to the existence of two distinct $C_6SiH_7^+$ isomers, which are formed by loss of a hydrogen from a silatoluene cation-radical. Based on the observed gas-phase chemistry, the authors proposed that these isomers are the silabenzyl cation **53** and the silatropylium cation **47**. They have also concluded from indirect evidence that the silatropylium ion is thermodynamically more stable than the silabenzyl cation, similar to the stability order in the analogous carbon systems[72]. Based on the reluctance of **47** to react with cycloheptatriene, it was concluded that the hydride affinity of **47** is even lower than that of the tropylium cation, which has one of the lowest hydride affinities known for organic cations[72]. In order to shed more light on these conclusions Nicolaides and Radom studied computationally the $C_6SiH_7^+$ surface[73], using very high computational levels, up to G2 and G2(MP2)[74]. The calculations reveal a fascinating complex $C_6SiH_7^+$ PES with 9 minima, i.e. **47**, **53**–**59**, several of which have 'non-classical'. The most surprising result of the calculations is that *the global minimum on the PES is the pyramidal 'non-classical' structure* **59**, which is by 20–26 kcal mol^{-1} lower in energy than the silabenzyl cation **53** and 29–35 kcal mol^{-1} below the silatropylium cation **47**. The remarkable stability of **59** is also exhibited in its high dissociation energy to Si plus $CH_3C_5H_4^+$ of 145.2 kcal mol^{-1} [at G2(MP2)][73a]. The other $C_6SiH_7^+$ isomers, **47** and **53**–**58**, are clustered in a rather narrow energy range of 14 kcal mol^{-1}, arranged in the following stability order [in kcal mol^{-1} at QCISD(T)/6-311G**//MP2/6-31G*]: **59** (0.0) < **53** (26.3) < **58** (31.8) < **54** (34.0) < **47** (35.4) < **57a** (37.5) < **57b** (38.3) < **55** (39.0) < **56** (40.2)[73].

In contrast to the conclusions of Beauchamp and coworkers[72], the computational results predict that the silatropylium cation (**47**) is *less stable* than the silabenzyl cation, this being the case at all *ab initio* levels, e.g. by 9.1 kcal mol^{-1} at QCISD(T)/6-311G(d,p)//MP2/6-31G(d)+ZPE (in the $C_7H_7^+$ analogue the tropylium cation is by 7.6 kcal mol^{-1} more stable than the benzyl cation)[73]. Moreover, the calculated hydride affinity of **47** is by 11.0 kcal mol^{-1} higher than that of the tropylium cation, in contrast to the experimental suggestion[72]. Based on the calculations it was concluded that, in addition to the silabenzyl cation, the other $C_6SiH_7^+$ isomer that was observed in the gas-phase experiments is most likely the (η^5-methylcyclopentadienyl)silanium cation, **59**. This cation has a calculated hydride affinity 33 kcal mol^{-1} lower than that of the tropylium cation[73a], in agreement with the experimental observation that the hydride affinity of the 'second isomer' is smaller than that of the tropylium cation. However, in a later study by Jarek and Shin[73c] low-energy collision-induced dissociation (CID) of the unreactive $C_6SiH_7^+$ isomer yielded SiH$^+$ and C_6H_6. The authors concluded that the unreactive $C_6SiH_7^+$ isomer is a η^4-$C_6H_6 \cdot SiH^+$ adduct and not **59** as predicted earlier[73a,b] (e.g. **59** is also expected to yield Si$^+$ as a fragment, and this is not observed experimentally)[73c]. *Ab initio* calculations find that the hydride affinity of η^4-$C_6H_6 \cdot SiH^+$ is by 11.9 kcal mol^{-1} at MP2/6-311G** lower than that of a tropylium ion, in agreement with the experimental relative value of -7.0 kcal mol^{-1}. Further experimental and theoretical studies are needed to clarify the identity of the observed $C_6SiH_7^+$ species.

Nicolaides and Radom emphasize the crucial importance of the inclusion of electron correlation, especially for the 'non-classical' isomers, for a reliable calculation of the relative energies, heats of formation and hydride affinities of the various $C_6SiH_7^+$ isomers. For the 'non-classical' structures, even the MP2 computational level does not always describe correctly their relative energies. For cases where the relative stabilities of the isomers oscillate at the MPn levels, the authors recommend using the QCISD(T) method[73b].

The structure of the planar C_{2v} silatropylium cation **47** is quite sensitive to the computational level; whereas at HF/6-31G* the C—C bonds are alternating, being 1.360 Å, 1.440 Å and 1.355 Å; the inclusion of electron correlation tends to narrow their range and at MP2/6-31G* they are 1.387 Å, 1.416 Å and 1.393 Å, consistent with a more delocalized structure. The calculated Si—C bonds are 1.781 Å and 1.782 Å at HF/6-31G* and MP2/6-31G*, respectively. The degree of electron delocalization or aromaticity in the silatropylium ring (and its isomers) was not discussed[73].

The most stable $C_6SiH_7^+$ isomer **59** may be viewed as an η^5-ion-molecule complex involving a Si^{2+} cation and a methylcyclopentadienyl anion. In **59**, the five-membered ring is flat and the distances of the apical Si from the ring carbons are 2.143 Å, 2.147 Å and 2.178 Å at MP2/6-31G*. The cyclopentadienyl C—C bond lengths are almost identical: 1.429 Å, 1.426 Å and 1.422 Å (MP2/6-31G*), indicating a delocalized system.

In this context, it is interesting to note the related $C_6SiH_6^{+\bullet}$ isomers, which are obtained by a gas-phase reaction of $Si^{+\bullet}$ and C_6H_6[75]. MP2/6-31G**//UHF/3-21G* calculations identified 3 minima on the $C_6SiH_6^{+\bullet}$ PES, the π-complex **60**, the C—H insertion isomer **61** and the ionized seven-membered ring structure **62**[75] (these are formally hydrogen abstraction products of **57a**, **53** and **47**, respectively). Their calculated relative energies (kcal mol^{-1}) are: **60** (0.0) > **61** (5.5) > **62** (30.8), the π-complex **60** being the most stable. The potentially aromatic **62** is planar and has C_{2v} symmetry. The C—C bond lengths are alternating (i.e. 1.375 Å, 1.422 Å and 1.382 Å), indicating small π-delocalization. The calculated Si—C bond length is 1.788 Å, i.e. it is slightly shorter than a single Si—C bond. The structure of **62** is very similar to the structure calculated at HF/6-31G* for **47**. However, inclusion of electron correlation in the geometry optimization may result in a more delocalized structure also for **62** as was found for **47** (see above).

(60) **(61)** **(62)**

C. $C_5H_5Si^+$ Isomers

The pyramidal silicon capped cation **63a** is a potential 'three-dimensional' 6π-electron aromatic system[76a], where the formal coordination number of the silicon is five (as in **59**). The interest in **63a** results from the detection of a $C_5SiH_5^+$ fragment in the gas phase[76b]. The crucial question is the stability of the pyramidal ion **63a** relative to other possible $C_5SiH_5^+$ isomers, such as **64a** and **65a**. A comparison with the analogous $C_6H_5^+$ isomers is of interest.

The available, rather low level calculations (HF/STO-3G, HF/STO-3G* and HF/3-21G, all at the HF/STO-3G optimized geometries) show that the pyramidal aromatic structure **63a** is more stable than the two planar structures. In the analogous carbon cations, on the other hand, the planar cations **64b** and **65b** are considerably lower in energy than the

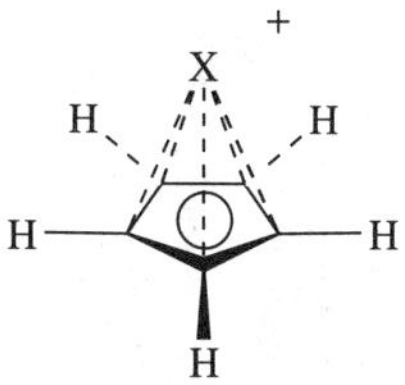

(63a) X = Si
(63b) X = C

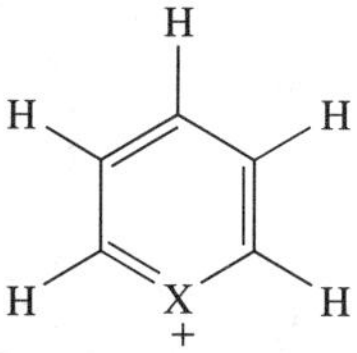

(64a) X = Si
(64b) X = C

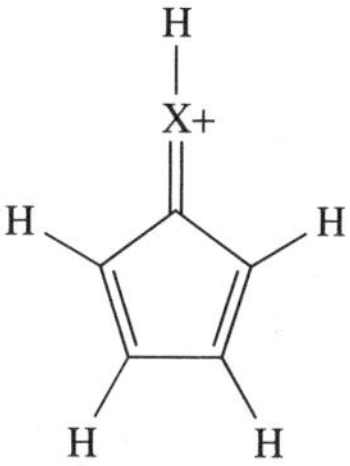

(65a) X = Si
(65b) X = C

pyramidal structure **63b**. Obviously, the reluctance of silicon to participate in multiple bonding destabilizes the planar structures relative to the pyramidal one, while in the all-carbon compounds the situation is reversed. The calculated Si–C distances in **63a** of 1.745 Å are very short, considerably shorter than the value of 2.189 Å (HF/6-31G*, 2.178 Å at MP2/6-31G*)[73a] in the very closely related **59**, and is by only ca 0.04 Å longer than that of a Si=C double bond. This short Si–C bond length is probably an artifact of the small STO-3G basis set used in the geometry optimization. In view of the relatively low level of calculations used so far, we recommend that this interesting system should be reinvestigated.

D. Silacyclopentadienyl Anions and Dianions

1. Experimental background

The search in recent years for silicon compounds with multiple bonds or cyclic π-systems has renewed interest in siloles (**66**)[77] and their mono- and di-anions (**48** and **49**), and led to the successful isolation of stable silole anions coordinated to various metal counter ions (Li^+, Na^+, K^+)[10a–c,78–86] and as complexes with ruthenium (e.g. **6a** and **6b**)[10d].

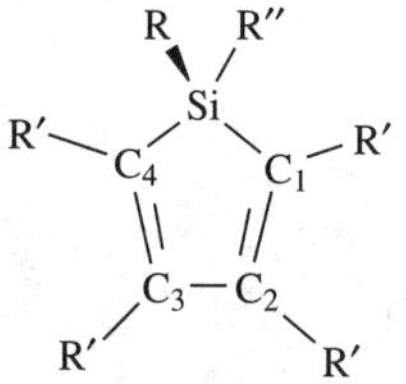

(66)
(66a) R = R' = R'' = H

The conclusions regarding the aromaticity of silole anions and dianions as gathered from the experimental studies of their structural and magnetic properties are summarized in Table 7, which reveals some controversy regarding the degree of aromaticity of silole anions. Hong and Boudjouk[79] found that the chemical shift of the ring Si atom in Li[c-Ph$_4$C$_4$Si(Bu-t)] (in THF) is shifted downfield relative to that of the parent silole, and this was interpreted as indicating charge delocalization in the anion. On the other hand, Tilley and coworkers[10a] find that in Li[c-Me$_4$C$_4$Si(SiMe$_3$)] (in THF) and K(18-crown-6)[c-Me$_4$C$_4$Si(SiMe$_3$)], $\delta(^{29}Si)$ of the ring silicon is shifted considerably upfield with respect to the analogous neutral silole compounds (Table 7). This result was interpreted as indicative of a localized structure[10a]. The latter interpretation is strongly supported

TABLE 7. Experimental structural and magnetic properties of silole anions and dianions and conclusions regarding their degree of aromaticity

Silole anion	Bond length (Å)			$\delta(^{29}Si)$ (ppm)	$\delta(^{13}C)$ (ppm)		Aromaticity	Ref.
	Si–C$_{(1)}$	C$_{(1)}$–C$_{(2)}$	C$_{(2)}$–C$_{(3)}$		C$_{(1)}$	C$_{(3)}$		
a. Monoanions								
Li[c-Ph$_4$C$_4$Si(Bu-*t*)] (in THF)	—	—	—	25.1^a	139.5^b	155.8^b	Magnetic properties indicate charge delocalization	79
[dibenzosilole structure with Si–Li$^+$, Me] (in THF)	—	—	—	−22.1^c	$\Delta T^d = 9.65$		No π-delocalization (NMR)	80
Li[c-Me$_4$C$_4$Si(SiMe$_3$)] (in THF)	—	—	—	−45.4^e	138.7^f	146.4^f	MNR shifts and bond-length alternation suggest non-aromatic character	10a
[K(18-crown-6)][c-Me$_4$C$_4$Si(SiMe$_3$)]	1.880^g	1.360	1.450	−41.5^e	135.8^f	149.6^f		10a
Ruthenium complexes, **6b** (**6a**)	1.830^{h,i}	1.391	1.462	−27.1(−7.35)	74.4(73.1)	108.8(88.5)	Magnetic and structural properties indicate charge delocalization	10d
b. Dianions								10d
Li$^+_2$[c-Me$_4$C$_4$Si]$^{2-}$ (in THF)	—	—	—	29.8^j	139.0^k	120.0^k	NMR shifts indicate some aromatic character	84

Li$^+{}_2$[c-Ph$_4$C$_4$Si]$^{2-}$: (a) Solid^l	1.850	1.448	1.430	87.3 68.5^m	129.7^n	NMR shifts and the structure are consistent with significant charge delocalization	10b
(b) in THF							10c
[K(18-crown-6)$^+$]$_2$[c-Me$_4$C$_4$Si^{2-}]	1.851^o	1.400	1.440	—	151.22^n	The structure indicates significant charge delocalization	82

^{a}Considerably downfield shifted relative to **66**, R = t-Bu; R' = Ph; R'' = [(PhC)$_4$SiBu-t] where $\delta(^{29}\text{Si}) = 3.6$ ppm and relative to the relevant silyl anions (e.g. Ph$_3$SiLi: −30 ppm; Me$_3$SiK: −34.4 ppm).
^{b}Shifted upfield relative to the corresponding precursor, **66** (i.e. C$_{(1)}$ = 144.8 ppm; C$_{(2)}$ = 158.0 ppm).
^{c}Shifted upfield relative to $\delta(^{29}\text{Si})$ of −18.7 ppm of the corresponding precursor, **66**.
$^d\Delta T$ = the total chemical shift change of the phenyl ring carbons relative to the neutral precursor. This small downfield shift relative to the neutral parent was interpreted as indicating that additional π-electron density is not accumulated in the phenyl ring of the anion.
^{e}Considerable upfield shift with respect to analogous neutral siloles, e.g. $\delta(^{29}\text{Si}) = -8.9$ ppm in **66** (R = Me, R' = H, R'' = SiMe$_3$).
^{f}For **66** (R = Me, R' = H, R'' = SiMe$_3$) $\delta(^{13}\text{C}_{(1)}) = 129.5$ ppm and $\delta(^{13}\text{C}_{(2)}) = 150.9$ ppm.
^{g}The silicon is pyramidal.
^{h}The ring Si centre is planar.
^{i}Crystallographic data are for **6b**; **6a** could not be crystallized.
^{j}Downfield shifted relative to $\delta(^{29}\text{Si}) = 8.1$ ppm in the neutral precursor.
^{k}Upfield shifted relative to $\delta(^{13}\text{C}_{(1)}) = 152.3$ ppm and $\delta(^{13}\text{C}_{(2)}) = 124.1$ ppm of its neutral precursor.
^{l}The Li atoms have η^1, η^5-coordination to the five-membered ring; the ring is almost planar, with the Si deviating by only 11 ppm from the plane of the four carbon atoms.
^{m}Shifted downfield relative to the neutral silole precursor [$\delta(^{29}\text{Si}) = 6.8$ ppm].
$^n\delta(^{13}\text{C}_{(1)}) = 132.3$ ppm and $\delta(^{13}\text{C}_{(2)}) = 154.7$ ppm in the relevant silole precursor.
^{o}The K(18-crown-6)$^+$ ions have η^5, η^5-coordination to the silacyclopentadienyl ring.

by the X-ray structure of K(18-crown-6)[c-Me$_4$C$_4$Si(SiMe$_3$)], which shows a difference of ca 0.1 Å between the C$_{(1)}$−C$_{(2)}$ and C$_{(2)}$−C$_{(3)}$ bond lengths (see **66** for the atom notation) and a high degree of pyramidalization at silicon, the angle between the C$_4$Si plane and the Si−Si bond being 99.6° (see Figure 8a for the ORTEP diagram). The long Si···K distance of 3.60 Å suggests a weak Si···K interaction, indicating that the inherent electronic properties of the anion were preserved. This might not be the case for the Li-coordinated silolyl ion in THF, studied by Hong and Boudjouk and whose structure is unknown[79]. A localized structure was also suggested for the analogous germolyl anion, e.g. Li(12-crown-4)$_2$[c-Me$_4$C$_4$Ge(Si(SiMe$_3$)$_3$)], which also does not show any indication for a strong interaction with the counter ion, and which was therefore considered to be a free ion[10a]. The different NMR behaviour of Li[c-Ph$_4$C$_4$Si(Bu-t)][79] and the closely related Li[c-Me$_4$C$_4$Si(SiMe$_3$)][10a] (both in THF) may be attributed to the phenyl vs Me substitution or to the different degrees of interaction between the alkali metal ions and the silolyl ring in the two systems[82], but these points were not further evaluated. Complexation to ruthenium as in **6a** and **6b** reduces the pyramidality at the ring Si (Figure 8b), thus enhancing the electron delocalization within the ring[10d].

The known silole dianions are coordinated to the metallic counter ions either by η^5,η^5-bonding[10a,82] or by η^1,η^5-bonding[10b] (Figures 8c and 8d). Pyramidality at Si is, of course, not a problem for the silole dianions where the Si is dicoordinated, and the known dianions are all planar (Figures 8c and 8d). The structural and magnetic properties of the silole dianions are consistent with charge delocalization and aromatic character of these species, as shown by the data collected in Table 7.

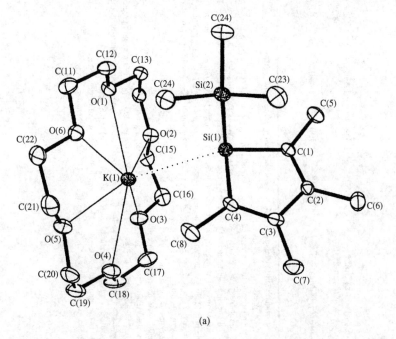

(a)

FIGURE 8. ORTEP drawings of several silolyl anion complexes: (a) K(18-crown-6)[c-Me$_4$C$_4$Si(SiMe$_3$)][10a]; (b) silolyl anion−ruthenium complex **6b**[10d]; (c) η^5,η^5-2[K(18-crown-6)$^+$][c-C$_4$Me$_4$Si^{2-}][10a]; (d) η^1,η^5-2Li$^+$[c-Ph$_4$C$_4$Si^{2-}]·(5THF)[10b]. Reprinted with permission from References 10a, 10b and 10d. Copyright (1994, 1995, 1996) American Chemical Society

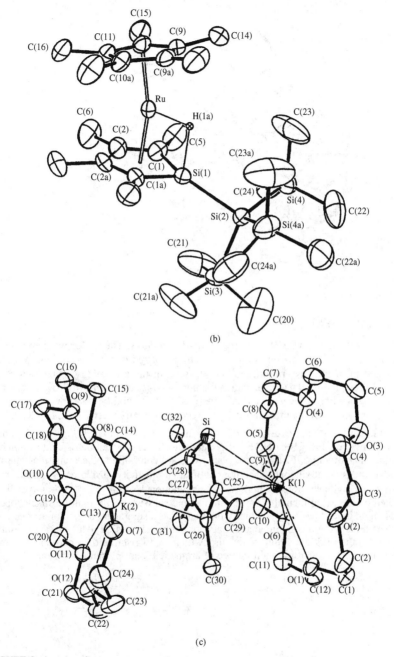

FIGURE 8. *(continued)*

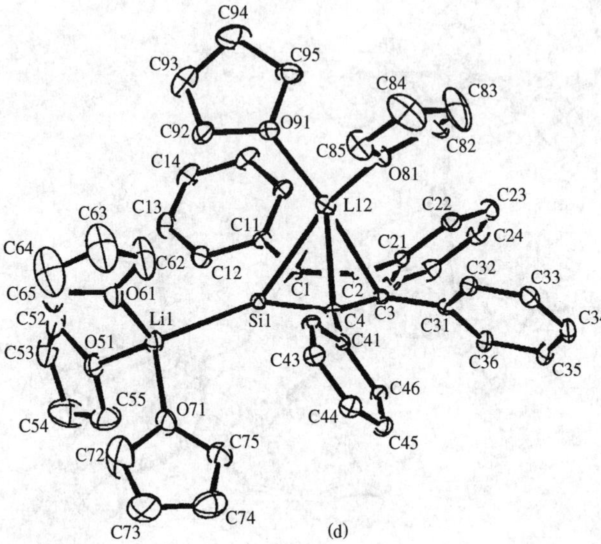

FIGURE 8. *(continued)*

2. c-C$_4$H$_4$SiR$^-$ silolyl anions

a. The parent ion c-C$_4$H$_4$SiH$^-$ and its lithiated complexes. The cyclopentadienyl anion (C$_5$H$_5^-$) is a well characterized aromatic system calculated to have, according to π-ring currents and magnetic susceptibility, about 80–90% of the aromaticity of benzene[18]. In contrast, the analogous silacyclopentadienyl anion, **48a**, is still unknown. Low-level early computational studies of the geometry and thermodynamic stability of **48a** were in conflict regarding the degree of its aromaticity[41a,87]. According to HF/3-21G//HF/STO-2G calculations, **48a** is planar and possesses *ca* 25% of the aromaticity of C$_5$H$_5^-$ [41a]. On the other hand, according to HF/6-31G*//HF/6-31G* calculations, the silacyclopentadienyl anion has a C_s pyramidal structure, **67a**, and the planar C_{2v} structure **48a** is the transition state for pyramidal inversion at silicon[87]. The degree of the aromaticity of **67a** as estimated from its ASE is negligible[87]. Recent high-level computational studies[26,31a,86,88,89] that were stimulated by the experimental progress in this field showed clearly that the silacyclopentadienyl anion indeed has a C_s symmetry but that it has a significant degree of aromaticity. The calculated structural, energetic and magnetic properties of silolyl anions (and of silole dianions) and their lithium silolides, together with a comparison to their carbon analogues and the conclusions regarding their degree of aromaticity, are presented in Table 8.

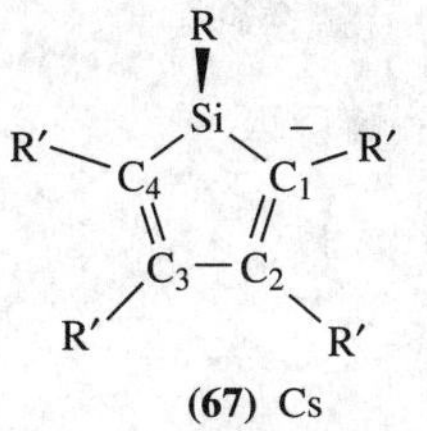

(67) C_s
(67a) R = R' = H

TABLE 8. Calculated structural and magnetic properties of c-$C_4H_4SiH^-$ (**48a** and **67a**) and c-$C_4H_4Si^{2-}$ (**49a**) and their lithium complexes and conclusions regarding the degree of their aromaticity

Compound	ASE or RE[a]	Bond lengths (Å)			A^c	Magnetic properties[b]					Degree of aromaticity	Ref.
		Si–$C_{(1)}$	$C_{(1)}$–$C_{(2)}$	$C_{(2)}$–$C_{(3)}$		$\delta(^{29}Si)$	$\delta(^{13}C_{(1)})$	$\delta(^{13}C_{(2)})$	Λ	NICS		
a. Monoanions												
$C_5H_5^-$ (D_{5h})	28.8[d]; 23.6[e]	—	1.413[f]	1.413[f]	1.0	—	109.2[g]	109.2[g]	−17.3	−14.3[h]	high	31a
C_5H_5Li (C_{5h})	40.2[e]	—	1.423[i]	1.423[i]	1.0	—	112.9[g]	112.9[g]	−14.5	−17.2[j]	high	88,89, 93a
c-$C_4H_4SiH^-$ (**48a**, C_{2v})	23.0[k]	1.821	1.377	1.433	—	—	—	—	—	—	ca 25% of that in $C_5H_5^-$	41a
c-$C_4H_4SiH^-$ (**67a**, C_s)	2.2[l]	1.924	1.341[m]	1.472[m]	—	—	—	—	—	—	negligible	87
c-$C_4H_4SiH^-$ (**48a**, C_{2v})	—	1.794[n]	1.422[n]	1.414[n]	0.998[n]	−6.6[o]	100.5[o]	121.1[o]	−18.5	—	high	88
c-$C_4H_4SiH^-$ (**67a**, C_s)	12.9[e]	1.847[n]	1.399[n]	1.433[n]	0.971[n]	−50.6[o]	153.4[o]	134.6[o]	−10.4	−6.7[h]	ca 50% of that in $C_5H_5^-$	88
η^5-Li[c-C_4H_4SiH] (**68**)	32.0[e]	1.823[n]	1.420[n]	1.424[n]	0.999[n]	−22.4[o]	123.6[o]	130.0[o]	−14.1	—	significant	88
b. Dianions												
c-$C_4H_4Si^{2-}$ (**49a**)	17.9[p]	1.869[n]	1.421[n]	1.419[n]	1.0	51.9[o] (18.5)[q]	145.1[o]	119.2[o]	−30.0	−12.8[r]	high	93a
η^5-Li[+][c-$C_4H_4Si^{2-}$] (**69**)	40.4[d]; 36.4[s]	1.884[t]	1.429[t]	1.425[t]	1.0	83.7[q]	—	—	−20.7	−17.1[r]	high	93a

(continued overleaf)

43

TABLE 8. (continued)

Compound	ASE or RE[a]	Bond lengths (Å)			A^c	$\delta(^{29}Si)$	Magnetic properties[b]			Degree of aromaticity	Ref.
		Si–C$_{(1)}$	C$_{(1)}$–C$_{(2)}$	C$_{(2)}$–C$_{(3)}$			$\delta(^{13}C_{(1)})$	$\delta(^{13}C_{(2)})$	Λ	NICS	
η^5,η^5-2Li$^+$[c-C$_4$H$_4$Si^{2-}] (**70**)	—	1.901^n	1.437^n	1.433^n	1.0	77.7^o (55.7)^q	153.3^{o,u}	125.8^{o,u}	−23.5	−17.6^r	high^v	93a
η^1,η^5-2Li$^+$[c-C$_4$H$_4$Si^{2-}] (**72**)	—	1.855^w	1.420^w	1.426^w	0.999	—	—	—	—	−18.4^r	high	10b

[a] In kcal mol^{-1}, ASE according to equation 21.
[b] δ in ppm; Λ in 10^{-6}cm^3 mol^{-1}; NICS in ppm.
[c] Julg's parameter[17].
[d] At MP2(fc)/6-31G*//MP2(fc)/6-31G*+ZPE.
[e] At MP2(fc)/6-31+G*//MP2(fc)/6-31+G*+ZPE, from Reference 88.
[f] At MP2(fc)/6-31G*.
[g] From Reference 88, calculated by IGLO (basis II) using MP2(fc)6-31+G* geometries.
[h] At GIAO-HF/6-31+G*, from Reference 26.
[i] From Reference 89 at B3LYP/6-31+G*.
[j] From Reference 89 at GIAO-SCF/6-31+G*//B3LYP/6-31G*.
[k] RE at 3-21G, according to the following isodesmic reaction: **48a** (C$_{2v}$) + SiH$_3^-$ + 4CH$_4$ → C$_2$H$_6$ + 2C$_2$H$_4$ + 2CH$_3$SiH$_2^-$; the RE for C$_5$H$_5^-$ (according to an analogous equation) is 87.0 kcal mol^{-1}.
[l] According to the following equation: **48a** + SiH$_4$ → **66a** + SiH$_3^-$; the ASE for the carbon analogue is 73.4 kcal mol^{-1}.
[m] At HF/6-31G*: for **67**, R = SiH$_3$, R' = H, r(C$_{(1)}$–C$_{(2)}$) = 1.367 Å and r(C$_{(2)}$–C$_{(3)}$) = 1.477 Å, indicating a localized ring; the Si–Si bond is strongly bent out of the ring plane[86].
[n] At MP2(fc)/6-31+G**.
[o] At IGLO (basis II).
[p] At MP2(fc)/6-31+G* from Reference 93b.
[q] At GIAO/MP2/tz2p(Si),tzp(C,Li),dz(H)//MP2/6-31+G*, from Reference 93a.
[r] At GIAO/HF/6-31G*//MP2/6-31+G*, from Reference 93b.
[s] At MP2/6-31+G**//MP2/6-31+G*.
[t] At MP2(fc)/6-31+G*.
[u] Better agreement between the experimental C$_{(1)}$/C$_{(2)}$ chemical shifts of 2Li$^+$[c-C$_4$Ph$_4$Si^{2-}] (in THF) and the calculated values for η^5,η^5-Li$_2$[c-C$_4$H$_4$Si^{2-}] is achieved if the experimental C$_{(1)}$(α) and C$_{(2)}$(β) signal assignments are interchanged[93a].
[v] Similar conclusions were drawn when the coordinating metal was Na or K.
[w] At MP2/6-31+G* from Reference 10b.

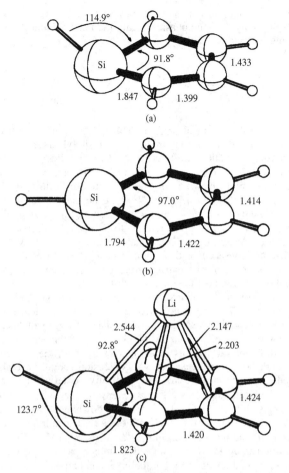

FIGURE 9. Optimized structures of the silolyl anion and its lithium silolide. Bond lengths in Å, bond angles in degrees. (a) The C_s structure of **67a** at MP2/6-31+G**; (b) the C_{2v} structure of **48a** at MP2/6-31+G**; (c) the structure of the lithium silolide **68** at MP2/6-31G**. Reprinted with permission from Reference 88. Copyright (1995) American Chemical Society

Calculations at MP2/6-31+G**[88] find that the parent silolyl anion **67a** has a C_s non-planar structure (Figure 9a), possessing a pyramidal Si centre with an angle sum at the silicon of 321.6°. Yet, the Si atom in **67a** is less pyramidal than in SiH_3^- (angle sum = 289.3°). The difference between $r(C_{(1)}-C_{(2)})$ and $r(C_{(2)}-C_{(3)})$ is relatively small, 0.043 Å compared to 0.124 Å in the neutral **66a**, indicating some degree of electron delocalization in **67a**. The structural evidence for electron delocalization in **67a** is supported by its calculated ASE of 12.9 kcal mol^{-1} compared to 23.6 kcal mol^{-1} for $C_5H_5^-$ (equation 21, X = SiH$^-$ and CH$^-$, respectively) and also by its magnetic properties; **67a** has a negative magnetic susceptibility exaltation (-10.4×10^{-6} cm^3 mol^{-1})[88] and a negative NICS value (-6.7 ppm; Table 8)[26]. Based on these structural, energetic and magnetic properties it was concluded that the silolyl anion, **67a**, has ca 50% of the aromaticity of

$C_5H_5^-$, contrary to the earlier conclusions of Damewood[87] which were based on a significantly lower level of theory. This emphasizes again the need for high level calculations for silicon-containing molecules.

$$\underset{}{\overset{X}{\bigcirc}} + \underset{}{\overset{X}{\bigcirc}} \longrightarrow 2 \underset{}{\overset{X}{\bigcirc}} \quad X = SiH^-, SiHLi, CH^-, CHLi, Si^{2-}, SiLi^- \quad (21)$$

The calculated ^{29}Si NMR chemical shift of **67a** (-50.6 ppm) is shifted upfield by 8.6 ppm relative to **66a**, indicating that in the parent silolyl anion the effect of the negative charge (causing an upfield shift) is larger than the effect of electron delocalization (causing a downfield shift). Upfield shifts of $\delta(^{29}Si)$ were reported when neutral silanes are converted to silyl anions, as in going from $Me_3SiSiMe_3$ (-19.7 ppm) to Me_3SiK [-34.4 ppm, $\Delta\delta(^{29}Si) = 14.7$ ppm][90] where no electron delocalization takes place. A significant upfield shift of 36.5 ppm was measured for the ring silicon in Li[c-$Me_4C_4Si(SiMe_3)$] [$\delta(^{29}Si) = -45.4$ ppm] relative to its neutral precursor (**66**, R = Me, R' = H, R'' = $SiMe_3$). According to Tilley and coworkers, these NMR data as well as the structural data are consistent with a significant localization of the charge on the silicon and with a non-aromatic bond-localized structure[10a] (see above and Table 7). However, this interpretation contradicts the conclusions of Goldfuss and Schleyer (who have also calculated an upfield shift for **67a** relative to **66a**) that **67a** possesses a significant degree of aromaticity[88,89]. This conclusion was based on other evidence, namely ASE, NICS values and the susceptibility exaltation of **67a**. The different conclusions by Tilley and coworkers[10a] on the one hand, and by Goldfuss and Schleyer on the other[88,89], may be due to the different substituents or to interactions with the Li^+ counter ion in Li[c-$Me_4C_4Si(SiMe_3)$]. We believe that Tilley's conclusion should be revised, because chemical shifts and geometry are not the most indicative criteria for aromaticity and, as suggested by Schleyer and Jiao[16], we prefer to rely on magnetic criteria such as Λ and NICS. Interestingly, Hong and Boudjouk found a downfield shift in the ^{29}Si NMR chemical shift in Li[c-$Ph_4C_4Si(Bu$-$t)$] in line with electron delocalization of the negative charge of the silicon into the hydrocarbon moiety[79] (see Section IV.D.1).

The C_{2v} parent silolyl anion **48a**, in which the Si atom is forced to be planar (Figure 9b), is a transition state for the inversion at Si, but the inversion barrier is only 3.8 kcal mol^{-1} (MP2/6-31+G**//MP2/6-31+G**), much smaller than that computed previously at HF/6-31G* (16.2 kcal mol^{-1})[87]. The bond length equalization in **48a** (Figure 9b), its large negative magnetic susceptibility exaltation of -18.4×10^{-6} cm^3 mol^{-1} which is even more negative than that of $C_5H_5^-$ (-17.3×10^{-6} cm^3 mol^{-1}, Table 8) and the downfield shift of its $\delta(^{29}Si)$ relative to **66a** [$\delta(^{29}Si) = -6.6$ and -42 ppm, respectively] are all strong indications of a highly aromatic system[88].

The lower degree of aromaticity of **67a** relative to $C_5H_5^-$ is due mostly to the pyramidal geometry around the silicon. A similar reduction in aromaticity is observed for phosphole c-C_4H_4PH relative to pyrrole[31a]. This implies that reduction or elimination of the pyramidality problem should result in increased charge delocalization. Computationally, this could be demonstrated by η^5-coordination of Li^+ to the silolyl anions, i.e. as in **68**. Such coordination reduces the pyramidality at the Si (sum of angles around the Si is 340.2) and enhances its CC bond equalization (Figure 9c). Furthermore, η^5-Li^+ coordination to **67** increases its ASE to -32 kcal mol^{-1} (equation 21) which is about 80% of the ASE of η^5-C_5H_5Li, causes a considerable downfield shift in $\delta(^{29}Si)$ to -22.4 ppm and increases the diamagnetic susceptibility exaltation to -14.1×10^{-6} cm^3 mol^{-1}, which is nearly as large as that of η^5-C$_5$H$_5$Li (Table 8). All these facts together are best interpreted as indicating that Li$^+$ coordination to **67** results in a strong 'three-dimensional aromaticity'[91] and in stabilization of the π-electrons[92]. Thus, the structural, energetic and magnetic criteria all agree that **68** is highly delocalized and shows a significant aromatic character[88,89,93a]. The recent experimental conclusions regarding the aromaticity of Li[c-Ph$_4$C$_4$Si(Bu-t)][79] are in agreement with this computational conclusion. Similar 'three-dimensional aromaticity' was achieved in **6a** and **6b** by complexation of the silolyl anion to a η^5-Me$_5$C$_5$Ru$^+$ fragment (Figure 8b, Table 7)[10d].

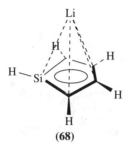

(68)

b. The silyl substituted silolyl anion, c-C$_4$H$_4$Si(SiH$_3$)$^-$. The structure and charge distribution of the silyl substituted metalolyl anions, c-C$_4$H$_4$E(SiH$_3$)$^-$, E = C, Si, Ge and Sn, were studied at HF/6-31G* by Tilley and coworkers[86]. The calculated structure of c-C$_4$H$_4$Si(SiH$_3$)$^-$ exhibits a strong pyramidality at the ring silicon. The angle α between the Si—Si bond and the plane of the ring is 104.5°. This strong pyramidality is accompanied by a localized π-system with C$_{(1)}$—C$_{(2)}$ and C$_{(2)}$—C$_{(3)}$ bond lengths of 1.367 Å and 1.477 Å, respectively[86]. Note, however, that a very similar localized structure was calculated at the same level of theory for **67a**[87], which indicates that the effect of the silyl group is small. Going down along group 14, the pyramidality at E as well as the degree of bond localization increases[86] (similar conclusions were reached by Goldfuss and Schleyer[89] for c-C$_4$H$_4$EH$^-$; see Section IV.D.4)

3. The parent silole dianion c-C$_4$H$_4$Si^{2-} and its lithiated complexes

The parent silole dianion, **49a**, where the pyramidality at the silicon is removed, shows a high degree of aromaticity as concluded from its structure (equalized bond lengths of 1.42 Å; $A = 1$[93a], see Figure 10a) and its magnetic properties, i.e. $\Lambda = -30 \times 10^{-6}$ cm^3 mol^{-1}[93a], NICS = -12.8 ppm[93b], relative to $\Lambda = -17.3 \times 10^{-6}$ cm^3 mol^{-1}[31a] and a NICS value of -14.3 ppm[26] for C$_5$H$_5^-$ (Table 8). The η^5-Li$^+$ coordinated **69** and the η^5,η^5-2Li$^+$ coordinated inverse sandwich structure **70** are also highly aromatic molecules. The optimized structures of **69** and **70** are shown in Figures 10b and 10c, respectively. Their C$_{(1)}$—C$_{(2)}$ and C$_{(2)}$—C$_{(3)}$ bond lengths are nearly equal with a Julg parameter of 1.0, indicating a strong delocalization. Analysis of the nature of the Si—C$_{(1)}$ σ-bonds shows a large p-character on Si, and this may be responsible for the particularly long Si—C$_{(1)}$ bond distances, i.e. 1.884 Å and 1.901 Å in **69** and **70**, respectively[93a]. The diamagnetic susceptibility exaltation calculated for **69** and **70** are highly negative, i.e. -20.7×10^{-6} and -23.5×10^{-6} cm^3 mol^{-1}, respectively (relative to -17.3×10^{-6} cm^3 mol^{-1} for C$_5$H$_5^-$). The NICS values for **69** and **70** of -17.1 ppm

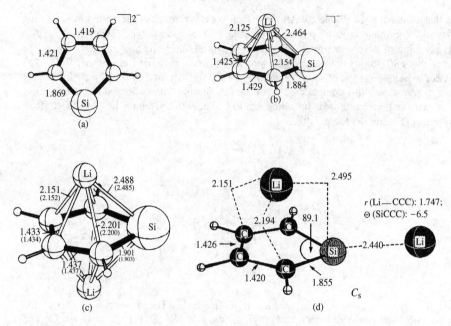

FIGURE 10. Optimized structures of the silole dianion **49a** and its mono- (**69**) and dilithium (**70** and **72**) salts. Bond lengths in Å, bond angles in degrees: (a) **49a** at MP2/6-31+G**[93a]; (b) **69** at MP2/6-31+G*[93a]; (c) η^5,η^5-dilithium silole **70** at MP2/6-31+G** (in parentheses at MP2/6-31G*)[93a]; (d) η^1,η^5-dilithium silole **72** at MP2/6-31+G*[10b]. Reprinted with permission from References 10b and 93a. Copyright (1995, 1996) American Chemical Society

and −17.6 ppm, respectively, are almost identical to that of C_5H_5Li (−17.7 ppm[89]), pointing to large aromatic ring currents in all these molecules. The NICS values for **69** and **70** are considerably more negative than for **49**, and this might point to a remarkable aromaticity of the lithiated complexes or might be caused by diamagnetic contributions from the bonds between Li and the ring atoms[93b]. This point should be further investigated computationally. Note, however, that the diamagnetic susceptibility exaltations of **69** and **70** are less negative than those of the free dianion **49**, which may point to a smaller degree of aromaticity in the lithiated complexes. The aromatic ring currents in **70** are responsible for the strong magnetic shielding of the Li$^+$ ions [$\delta(^7Li)$ = −7.7 ppm[93a]; −6.2 ppm in **68** and −9.1 ppm in $C_5H_5Li^{88}$]. The Na$^+$ and K$^+$ analogues of **69** and **70** show similar characteristics and are therefore also highly aromatic[93a].

West, Apeloig and coworkers reported recently an X-ray structure of an η^1,η^5-dilithium silole (**71**). One Li atom (with its associated two THF solvation molecules) is η^5-bonded to the silole ring and the second Li is η^1-bonded to the Si as well as to three THF molecules (Figure 8d)[10b]. Considering the fact that the solvating THF molecules were not included in the calculations, the MP2/6-31+G* optimized structure of the model system **72** (Figure 10d) is in reasonable agreement with the experimental structure of **71**. The calculated $C_{(1)}-C_{(2)}$ and $C_{(2)}-C_{(3)}$ bond lengths are 1.420 Å and 1.426 Å, respectively, compared to the measured bond lengths of 1.448 Å and 1.430 Å, respectively[10b]. The experimental distances involving the Li atoms are longer in **71** than those calculated for

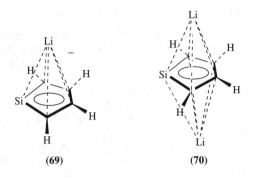

(69)   (70)

72, as expected, because the THF solvation is not present in the calculations. According to the calculations **70** is by 21 kcal mol^{-1} more stable than **72** (MP2/6-31+G*//MP2/6-31+G*), but the authors suggest that solvation by THF may reverse this order. The effect of solvation on the **70–72** energy difference is currently being investigated computationally by Apeloig and coworkers. The structure of 2Li$^+$[c-Ph$_4$C$_4$Si^{2-}] in solution is still uncertain, although some NMR evidence [i.e. an upfield shift of $\delta(^{29}$Si) to 68.5 ppm in solution, from 87.3 ppm in the solid state, and a single signal in the ^{7}Li NMR (at 0.23 ppm) even at $-100\,^\circ$C] might point to a single environment for the two Li atoms as in the therodynamically more stable isomer **70**. The nearly equal, calculated C–C bond lengths in **72** and its NICS value (-18.4 ppm)^{93b} point to the highly aromatic character of **72**, similar to that of **69** and **70**. More recently, the analogous germoles **73a** and **73b** were crystallized from dioxane in two distinct structures: one with η^1,η^5-coordination of the two Li$^+$ cations **(73a)** and the other having a η^5,η^5-coordination **(73b)**. As in the silicon case, also for germanium the calculations (MP2/LANL2DZ) find the η^5,η^5-2Li$^+$[c-H$_4$C$_4$Ge^{2-}] isomer to be by 25 kcal mol^{-1} more stable than the η^1,η^5-2Li$^+$[c-H$_4$C$_4$Ge^{2-}] isomer^{85}.

(71)   (72)

(73a)   (73b)

4. Higher congeners of the silolyl anion and silole dianion

How is the aromaticity of the silolyl anion and dianion affected by substitution of the Si atom by its higher congeners Ge, Sn and Pb? In a comprehensive study Goldfuss and Schleyer[89] evaluated the degree of the aromaticity of mono- and dianions of group 14 metalloles: c-$C_4H_4EH^-$ and c-$C_4H_4E^{2-}$ (E = C, Si, Ge, Sn and Pb) by using a variety of criteria: structural, energetic (ASE) and magnetic (diamagnetic susceptibility exaltations and NICS). Their main conclusions are summarized briefly below.

(a) The aromaticity of the metallolyl monoanions c-$C_4H_4EH^-$ decreases in the order: C (most aromatic) > Si > Ge > Sn > Pb (least aromatic). This is exhibited in an increased bond alternation, stronger pyramidality at E, higher inversion barriers and decreased ASE along the C→Pb series. Similarly, the Λ and NICS values become less negative in going down along group 14, also indicating a decrease in aromaticity.

(b) The aromaticity of the lithium metallolides Li[c-C_4H_4EH] decreases along the C→Pb series similarly to the c-$C_4H_4EH^-$ series. However, Li···H interactions between the η^5-coordinated Li atoms and the E−H hydrogen stabilize significantly the heavier metalloles with E = Sn and Pb relative to their lower congeners.

(c) In contrast to the c-$C_4H_4EH^-$ and Li[c-C_4H_4EH] systems, the degree of aromaticity of the metallole dianions c-$C_4H_4E^{2-}$ and of their dilithium $2Li^+$[c-$C_4H_4E^{2-}$] complexes is remarkably constant for all group 14 elements.

5. The pentasilacyclopentadienyl anion

c-$Si_5H_5^-$ (**74**) has [at MP2(full)/6-31G*] two non-planar minima of C_s and C_2 symmetry which have identical energies[71]. The planar, formally 'fully aromatic' D_{5h} structure (**74**) is not a minimum on the MP2(full)/6-31G* $Si_5H_5^-$ PES and it is by 8.3 kcal mol^{-1} higher in energy than the C_s or C_2 structures. The calculated geometries of the C_s, C_2 and D_{5h} $Si_5H_5^-$ structures are shown in Figure 11. The aromatic stabilization energy of C_s $Si_5H_5^-$ according to equation 22 is 52.8 kcal mol^{-1}, much smaller than for its all-carbon analogue $C_5H_5^-$ (84.7 kcal mol^{-1})[71], pointing to a significantly lower degree of aromaticity in $Si_5H_5^-$. Yet, this ASE is much higher than the ASE of only 2.2 kcal mol^{-1} (HF/6-31G*) calculated for the monosilacyclopentadienyl anion (**67a**) according to an analogous isodesmic equation[87].

(**74**)

$$\text{Si}_5\text{H}_5^- + \text{SiH}_4 \longrightarrow \text{Si}_5\text{H}_6 + \text{SiH}_3^- \qquad (22)$$

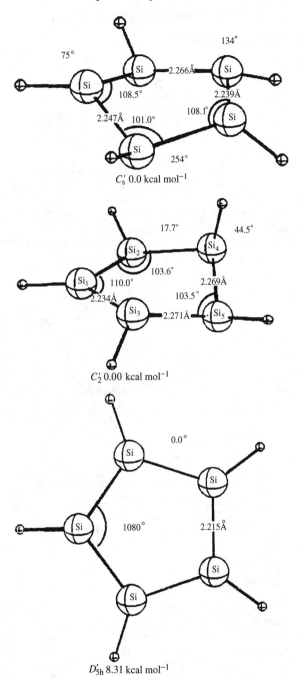

FIGURE 11. Calculated structures and relative energies at MP2(full)/6-31G* of C_s, C_2 and D_{5h} c-$Si_5H_5^-$ (**74**). Bond lengths in Å, bond angles in degrees. The angles given outside the ring are dihedral angles. Reproduced by permission of John Wiley & Sons from Reference 71

V. SILYLENES WITH AROMATIC CHARACTER

A. Stable 'Arduengo-type' Silylenes

1. Experimental background

Divalent group 14 compounds (carbenes, silylenes and germylenes) are generally highly reactive intermediates which, until recently, were directly observed only in matrices at low temperatures[2e,94]. In 1991 Arduengo and coworkers reported the synthesis of the first stable carbenes **75a**[95a,b] and **76a**[95c]. This remarkable achievement stimulated numerous studies which reported physico-chemical measurements[95d-g] and theoretical studies[95d-h,96] of these stable carbenes. Consequently, the isostructural germylenes **75b** and **76b**[97] and silylenes **75c**[98a,c,d] and **76c**[98b,c,d] were synthesized and characterized. The silylene **75c** exhibits remarkable thermal and kinetic stability[98a,c,d]. Thus, it does not dimerize or react with various bases and other known silylene scavengers under conditions in which simple silylenes react instantaneously[98a]. The saturated analogue **76c** is also highly stable, but it slowly dimerizes with a half-life of *ca* 5 days[98b,c,d]. These studies were supplemented by theoretical calculations of the corresponding parent silylenes **75d** and **76d** and the related silanes **77a** and **78**[98].

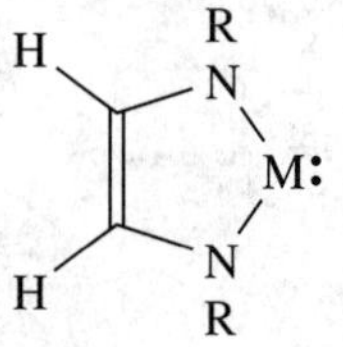

(**75a**) M = C, R = H, *t*-Bu, 1-Ad, Ar
(**75b**) M = Ge, R = *t*-Bu
(**75c**) M = Si, R = *t*-Bu
(**75d**) M = Si, R = H

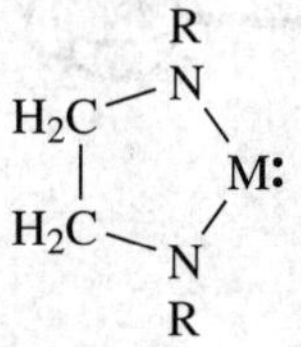

(**76a**) M = C, R = Ar
(**76b**) M = Ge, R = *t*-Bu
(**76c**) M = Si, R = *t*-Bu
(**76d**) M = Si, R = H

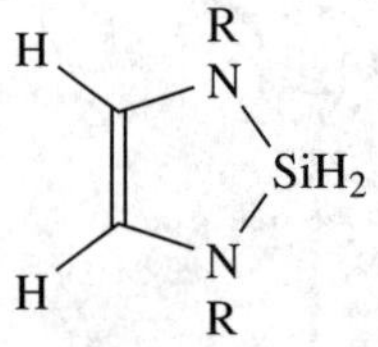

(**77a**) R = H
(**77b**) R = *t*-Bu

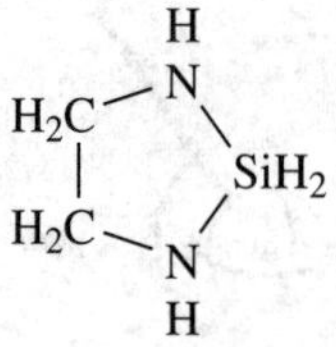

(**78**)

Most recently, stable crystalline 1,3-di(*neo*-pentyl)-2-silabenzimidazol-2-ylidene, **79a**, and its related saturated 2,2-dihydro-1,3-di(*neo*-pentyl)-2-silabenzodiaminosilole, **80a**, were synthesized[99]. **79a** can be stored under nitrogen for more than a year at ambient conditions and it can be distilled or sublimed without significant decomposition[99]. **79a** was characterized by its X-ray and by the UV photoelectron spectrum which was assigned with the help of MP2/6-31G*//MP2/6-31G* calculations for **79b**[99c].

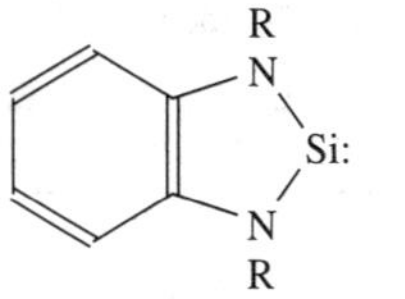

(79a) R = CH₂Bu-t
(79b) R = H

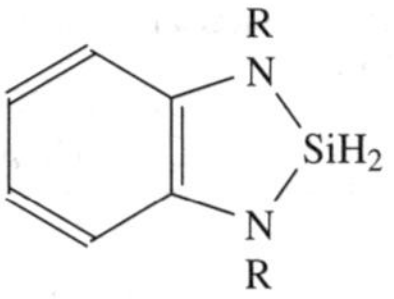

(80a) R = CH₂Bu-t
(80b) R = H

2. Aromaticity

Why are silylenes **75c** or **79a** so highly stable? Is the stabilization due to the presence of the two α-amino groups? Are these silylenes aromatic? The calculations show that $(H_2N)_2Si$ is by 37.2 kcal mol^{-1} more stable than H_2Si (CCSD/6-31G*//MP2/6-31G*)[68], 30.0 kcal mol^{-1} at MP4/6-31G*//6-31G*+ZPE[100]. If this is the only factor contributing to the stability of **75c**, why then is the analogous four-membered ring diamidosilylene (**81**) stable only below 77 K[101]?

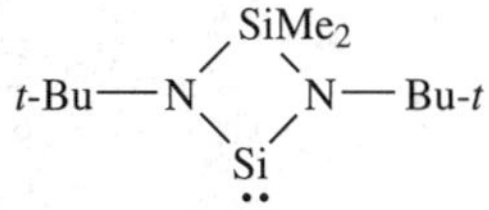

(81)

The high kinetic stability of **75c** and **76c** towards dimerization was explained by Denk and coworkers in terms of their high singlet-triplet energy difference, ΔE_{ST}[98b]. Thus, ΔE_{ST} (**75d**) of 69 kcal mol^{-1} and ΔE_{ST} (**76d**) of 74 kcal mol^{-1} (MP4/6-31G*//6-31G*) are larger than 60 kcal mol^{-1} — a value which was predicted[65] to be the limit above which the resulting dimers (i.e. the disilenes) dissociate spontaneously to the corresponding silylenes. It is therefore expected that **75c** and **76c** would dimerize very slowly, if at all[98b]. However, the lower reactivity of **75c** relative to **76c** towards dimerization and hydrogenation at silicon and its higher thermodynamic stability (see below) point to an additional stabilizing effect which is present in **75c** but not in **76c**[98b]. The additional stabilizing effect in **75c** was attributed to the contribution of the 6π-electron delocalization, i.e. to aromaticity, as shown in resonance structure **82**[96,98,100,102,103].

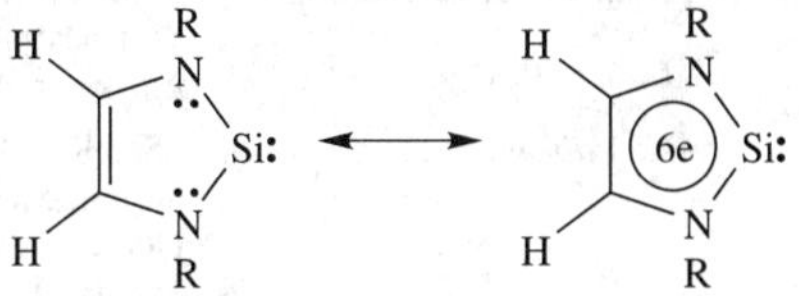

(82)

The occurrence of cyclic 6π-electron delocalization in imidazol-2-ylidene-type systems (**75**) is supported by a variety of structural, energetic, and magnetic criteria, as well as by the charge distributions and low-energy ionization processes which have been carefully analysed theoretically[96,98,100,102,103]. A summery of the conclusions of the theoretical studies regarding the degree of aromaticity in imidazol-2-ylidene-type systems is presented in Table 9.

TABLE 9. Experimental and theoretical studies on imidazol-2-ylidenes (**75a**) and their silicon analogues (**75c**, **75d** and **79a**, **79b**)[a]

Property studied	Method	Conclusions	Ref.
Carbenes			
Electronic structure of imidazol-2-ylidene in the lowest singlet and triplet states; proton affinity	Correlated *ab initio* calculations	Bonding character is carbenic rather than ylidic; π-delocalization is important in the imidazolium cation but not in the carbene	95h
Electronic structure of imidazol-2-ylidene: Bond orders, atomic charges, localized orbitals	Correlated *ab initio* calculations	Stabilization of singlet ground state by σ-back-donation along C-N bonds; π-delocalization plays only a minor role in imidazol-2-ylidene, but a major role in imidazolium cation	95d
Electronic structure of aminocarbenes: Singlet–triplet splittings, Mulliken populations, barriers for 1,2-rearrangements to imines	Correlated *ab initio* calculations	Singlet–triplet splitting in imidazol-2-ylidene is 15 kcal mol^{-1} higher than in imidazolin-2-ylidene, **76a**, R=H; consequently, there is a smaller propensity of the former towards dimerization; imidazol-2-ylidene is kinetically stable towards rearrangement to imidazole	95g
Chemical shielding tensor of a substituted imidazol-2-ylidene	Solid-state NMR; Correlated *ab initio* and density-functional calculations	Dominance in **75a** of carbenic over ylidic resonance structures	95e
Photoelectron spectra of **75a**, R = H and **75d**	Photoelectron spectroscopy; density-functional calculations	Degree of interaction between the π-electrons of the five-membered ring and the divalent group 14 atom is higher in silylenes **75c** and **75d** as compared to carbene **75a**	96
Electron distribution in a substituted imidazol-2-ylidene	X-ray and neutron diffraction; density-functional calculations	π-Delocalization not dominant in imidazol-2-ylidenes; stability of **75a** is kinetic in origin	95f
Electronic structure of stable carbenes, silylenes and germylenes	Correlated *ab initio* calculations	**75a**, R = H has partial aromatic character. π-Delocalization is more extensive in **75a**, R = H than in **76a**, R = H. Similar conclusions were reached for the corresponding silylenes and germylenes	103
Silylenes			
Gas-phase structure and solution-phase NMR of **75c**; theoretical heats of hydrogenation for **75d** and **76d**	Electron diffraction; correlated *ab initio* calculations	**75c** and **75d** benefit from aromatic stabilization	98a

1. Theoretical aspects and quantum mechanical calculations 55

TABLE 9. (continued)

Property studied	Method	Conclusions	Ref.
Photoelectron spectra of **75c** and **76c**; rotational barriers in Si(NH$_2$)$_2$ and C(NH$_2$)$_2$	Photoelectron spectroscopy; correlated *ab initio* calculations	Significant $p_\pi - p_\pi$ interaction between divalent group 14 centres and amino substituents; aromatic resonance structures contribute significantly in **75c** and **75d**	98b
Chemical shifts and anisotropies in aminosilylenes	Correlated *ab initio* calculations	Significant degree of 6π-aromaticity in **75d**	102
Thermodynamic stabilization, structure, chemical shift anisotropies and charge distribution	Correlated *ab initio* calculations, 'atoms-in-molecules' charge distribution analysis	Significant conjugation in **75d** relative to **76d** according to the thermodynamic and magnetic criteria. Small cyclic delocalization according to the topological analysis of charge density. More extensive π-resonance is found in the carbene **75a**	100
Photoelectron spectra, structure and thermodynamic stability of **79a** and **79b**	Photoelectron spectroscopy and correlated *ab initio* calculations	Cyclic delocalization plays an important role in the stability of **79a** and **79b**	99c
Nucleus independent chemical shifts	Correlated *ab initio* calculations	**75c**, **75d**, **79a** and **79b** possess a discernible ring current, which however is *ca* half as large as that in benzene. The ring current in **76c** and **76d** is negligible	107

[a]Based on Reference 100.

The structural, energetic, magnetic and charge distribution data that support the occurrence of 6π-electron delocalization in imidazol-2-ylidene-type silylenes and carbenes are discussed below:

a. Structural criteria. The experimental X-ray and calculated structures of **75c** (at ECP/6-31G*[104]) and **75d** (at 6-31G*[98a,b] and MP2/6-31G*[98a,100,103]) are in very good agreement. The calculated structures of **75c**[104] and **75d**[98a,b,100,103] are very similar. There is also good agreement between the experimental structure of **79a** and the MP2/6-31G*//MP2/6-31G* calculated structure of **79b**[99c]. The MP2/6-31G* optimized structures of **75d**, **76d**, **77a**, **78** and **79b** are shown in Figure 12. There is no significant difference between the experimental structures of **75c** and of the corresponding benzo-derivatives **79a**, or between the calculated structures of the corresponding hydrogen-substituted **75d** and **79b**.

The calculated Si—N bond length in **75d** is 1.774 Å, by 0.034 Å longer than in **76d** (exp. 1.753 Å and 1.719 Å in **75c** and **76c**, respectively)[98b]. The shorter Si—N bonds in **76c** (or **76d**) were interpreted as resulting from the fact that in these molecules the lone-pairs on the nitrogen atoms can conjugate only with the empty 3p(Si) orbital, whereas in **75c** (or **75d**) the N lone-pair electrons are cyclically delocalized as shown in **82**, resulting in a longer Si—N bond in **75** relative to **76**[98b]. The assumption of cyclic delocalization in **75c** (or **75d**) is reinforced by the 0.011 Å longer calculated $r(C=C)$ in **75d** (1.333 Å, MP2/6-31G*) than in the corresponding silane **77a** (where cyclic electron delocalization

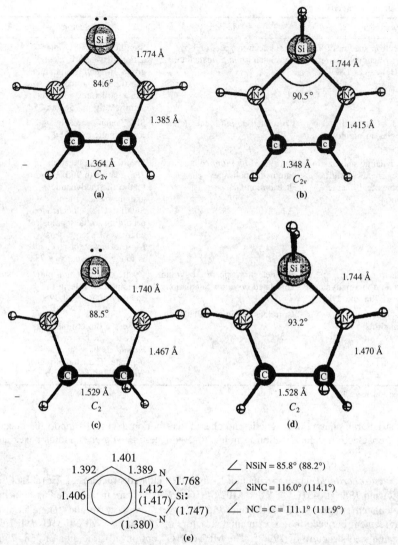

FIGURE 12. MP2/6-31G* optimized geometries of **75d** (a), **76d** (b) and of their corresponding silanes **77a** (c), **78** (d) and of 2-silabenzimidazol-2-ylidene **79b** (e); the experimental structure of **79a** is given in parentheses. Reprinted with permission from References 100 and 99c. Copyright (1996) American Chemical Society

is not possible)[98b,100]. The structures of the 2-silabenzimidazol-2-ylidenes **79a** and **79b** are indicative of a small but significant double-bond character in its Si−N bonds[99c]. Comparison of the cyclosilylenes **75c** and **75d** with the corresponding cyclic carbenes (**75a**, R = *t*-Bu or H) shows that the structural effects in the two types of species are similar, but they are somewhat smaller in the cyclosilylenes, indicating a smaller degree of electron delocalization in the silylenes[100].

b. Energetic criteria. According to the energies of equations 23a, 23b and 23c, the unsaturated cyclic silylene **75d** is by 20.5 and 11.8 kcal mol^{-1} (MP4/6-31G*//6-31G*+ZPE) more stable than (H$_2$N)$_2$Si and **76d**, respectively[100]. Similarly, **75d** is by 25-27 kcal mol^{-1} more stable than **76d** relative to ethane and ethene, respectively (equation 24)[100,103]. Thus, the results of equations 23 and 24 indicate an additional stabilization in **75d** due to the endocyclic C=C double bond. This additional stabilization is not due to a localized conjugative interaction within the HN—CH=CH—NH 'backbone', since 1,2-diaminoethene (in the planar C_{2v} geometry) is stabilized by only 1.4 kcal mol^{-1} (MP4/6-31G*//6-31G*+ZPE) relative to 1,2-diaminoethane[100]. The extra stabilization of **75d** over **76d** was interpreted to indicate cyclic electron delocalization in **75d**, which is absent in the C—C saturated silylene **76d**. However, Heinemann and coworkers pointed out that this interpretation relies on the *a priori* assumption that the π-electrons do in fact undergo cyclic delocalization and that such electron delocalization indeed results in a thermodynamic stabilization[100]. The occurrence of cyclic conjugation in **75d** was consequently studied directly by analysis of the charge distribution in **75d** and **76d** (see below).

$$R_2\ddot{S}i + SiH_4 \longrightarrow R_2SiH_2 + H_2\ddot{S}i \qquad (23)$$

$$R = NH_2 \qquad \Delta E = 30.0 \text{ kcal mol}^{-1[100]} \qquad (23a)$$

$$R_2\ddot{S}i = \textbf{75d} \qquad \Delta E = 50.5 \text{ kcal mol}^{-1[100]} \qquad (23b)$$

$$R_2\ddot{S}i = \textbf{76d} \qquad \Delta E = 38.7 \text{ kcal mol}^{-1[100]} \qquad (23c)$$

$$\text{(75d)} + C_2H_6 \longrightarrow \text{(76d)} + C_2H_4 \qquad (24)$$

$\Delta E = 25.0$ kcal mol^{-1} (MP4/6-31G*//HF/6-31G*[100]);
27.4 kcal mol^{-1} (MP2/6-31G*//MP2/6-31G*[103])

The higher exothermicity, by 13 kcal mol^{-1}, of the hydrogenation energy of the silicon atom in **76d** (equation 25) compared to that of **75d** (equation 26)[98a] and also in the hydrogenation energy of the C=C double bond in **77a** (equation 27) compared to that of **75d** (equation 28)[98b] are also believed to be manifestations of the cyclic 6π-electron delocalization in **75d**[98a,b].

$$\text{(76d)} + H_2 \longrightarrow \text{(78)} \qquad (25)$$

$\Delta E = 15.1$ kcal mol^{-1} (MP2/6-31G*//HF/6-31G*[98a])

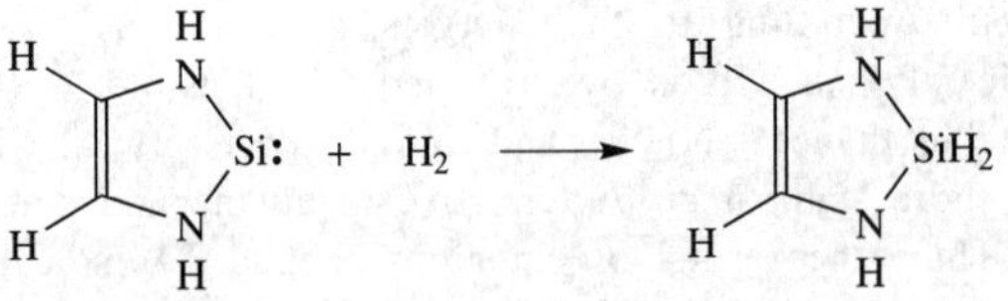

(75d) + H₂ → (77a)   (26)

$\Delta E = -1.7$ kcal mol^{-1} (MP2/6-31G*//HF/6-31G*[98a]);
-9.8 kcal mol^{-1} (MP4/6-311G**//MP2/6-31G*[103])

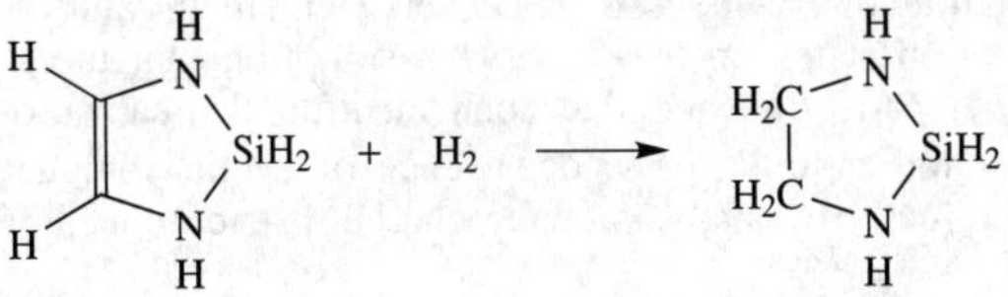

(77a) + H₂ → (78)   (27)

$\Delta E = -19.7$ kcal mol^{-1} (MP4/6-31G*//HF/6-31G*[98b]);

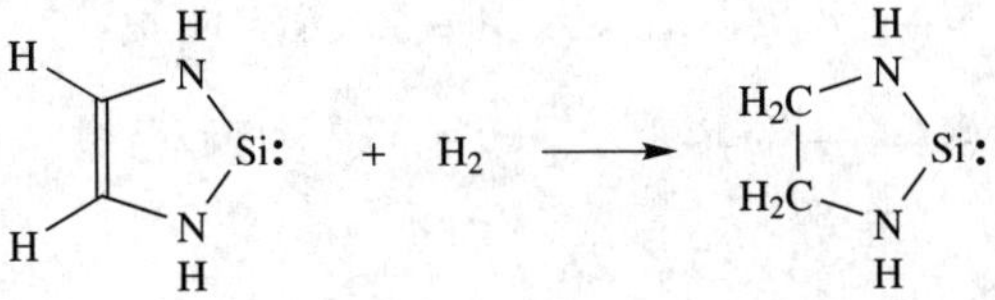

(75d) + H₂ → (76d)   (28)

$\Delta E = -6.7$ kcal mol^{-1} (MP4/6-31G*//HF/6-31G*[98b]);

Cyclic delocalization plays an important role also in the stabilization of the benzimidazol-2-ylidenes, **79**[99c], as concluded from the calculated ring fragmentation energies (equations 29 and 30), i.e. $\Delta E = 40.9$ and 22.9 kcal mol^{-1}, respectively (MP2/6-31G*//MP2/6-31G*). The effect of the five-membered ring is taken into account by keeping the geometry of the N−Si−N units in H₂NSiNH₂ in the same geometry as that in **79** or in **80**. A significant delocalization stabilization of 31.6 kcal mol^{-1} was suggested also for the isomeric silaimine **83** (equation 31)[99c].

$$79b + 2NH_3 + 2CH_4 \longrightarrow H_2N-Si-NH_2 + 2H_2NCH_3 + C_6H_6 \quad (29)$$

$$80b + 2NH_3 + 2CH_4 \longrightarrow H_2N-SiH_2-NH_2 + 2H_2NCH_3 + C_6H_6 \quad (30)$$

$$83 + 2NH_3 + 2CH_4 \longrightarrow H_2N-SiH=NH + 2H_2NCH_3 + C_6H_6 \quad (31)$$

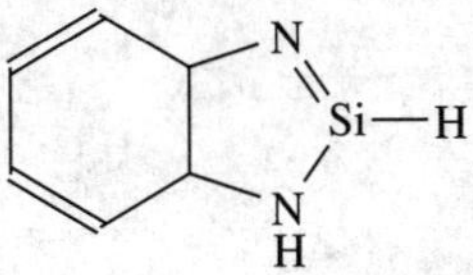

(83)

c. Magnetic criteria. The isolation of **75c** allowed to record the first NMR spectra of a silylene, and the measured chemical shifts[98a] are given in Figure 13.

GIAO[105] [MP2/TZ2P(Si),TZP(C,N),DZ(H)] calculations for the hydrogen substituted silylene **75d** predict chemical shifts which are in very good agreement with the experimental values measured for the substituted **75c** (Figure 13)[102]. In particular, the ^{29}Si chemical shift of **75d** is calculated to be 64 ppm[102], compared with the experimental value for **75c** of 78 ppm[98a]. The experimental–theoretical agreement for the carbon and nitrogen chemical shifts is also good[102]. Similar calculations for the saturated silylene **76d** predict a ^{29}Si chemical shift of 117 ppm[102], again in very good agreement with the experimental value for **76c** of 119 ppm[98d]. This success suggests that GIAO calculations can be used to elucidate the NMR spectra of transient silylenes which cannot yet be studied experimentally. Such studies can provide important fundamental information on the electronic structure of silylenes.

Further insight into the NMR spectrum of **75c** and **76c** was provided by additional GIAO calculations for H$_2$Si and for several diaminosilylenes, carried out by Apeloig, Karni and Müller (Table 10)[102]. The calculated ^{29}Si chemical shift for the parent H$_2$Si is 817 ppm (relative to TMS), and it shows a very large ^{29}Si chemical shift anisotropy (CSA), $\Delta\sigma$, of 1516 ppm (Table 10)[102]. This very high CSA indicates a strong charge anisotropy around the silicon, as is required by its electronic structure having perpendicular empty and filled orbitals, as shown by the magnetic axes of silylenes **75** and **79** drawn in Figure 14. Both the ^{29}Si chemical shift and the CSA of H$_2$Si decrease strongly upon substitution of the hydrogens by two amino groups. In the planar (H$_2$N)$_2$Si $\delta(^{29}$Si$) = 108$ ppm and $\Delta\sigma(^{29}$Si$) = 214$ ppm. This large substituent effect can be attributed to the conjugation between the lone pairs on N and the empty 3p(Si) orbital. This interpretation is supported by the much higher $\delta(^{29}$Si$)$ and CSA in the perpendicular (H$_2$N)$_2$Si, where such 2p(N)–3p(Si) conjugation is not possible (Table 10). The endocyclic C=C π-bond in **75d** causes an additional 53 ppm upfield shift in $\delta(^{29}$Si$)$ relative to **76d**, and to a significant increase in the CSA (from 73 ppm in **76d** to 165 ppm in **75d**)[102]. These trends, in both the ^{29}Si chemical shifts and in the CSAs, seem to support the hypothesis that the 'aromatic' resonance structure **82** contributes significantly to the total wave function of

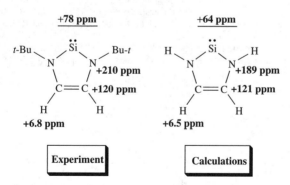

FIGURE 13. Experimental and calculated [GIAO/(MP2/TZ2P(Si)TZP(C)DZ(H))] chemical shifts of **75c** and **75d**, respectively. ^{15}N chemical shifts are relative to NH$_3$. The reported experimental ^{15}N chemical shift (relative to CH$_3$NO$_2$[98a]) was corrected using the experimental difference of 380 ppm between the ^{15}N chemical shifts of CH$_3$NO$_2$ and NH$_3$. Reproduced by permission of VCH Verlagsgesellschaft from Reference 102

75d[102]. The large anisotropies of the magnetic susceptibility, $\Delta \chi$ (calculated using the IGLO method[106]), support the occurrence of ring currents in **75d** ($\Delta \chi = 6.46$ au), as well as in the corresponding potentially aromatic silicenium cation **84** ($\Delta \chi = 5.64$ au), but not in **76d** or **78** — the correspondingly saturated analogues ($\Delta \chi = 1.41$ and 1.96 au, respectively)[100]. The smaller calculated $\Delta \chi$ for the silylene **75d**, relative to that of the analogous carbene **75a**, R = H ($\Delta \chi = 8.17$ au), also indicates a smaller ring current and thus a reduced 'aromaticity' in **75d** relative to that in **75a**, R = H[100].

The NICS values of the silylenes **75c** and **79a**, calculated at a distance of 2.0 Å above the ring centre, are −2.7 ppm and −2.6 ppm, respectively[107]. These values are significantly lower than for benzene (−5.3 ppm) or thiophene (−4.7 ppm), but still indicate that these silylenes possess a discernible diamagnetic ring current of about half of that in benzene. In contrast, the NICS value of the saturated silylene **76c** is negligible (−0.6 ppm), indicating the absence of a significant ring current[107].

TABLE 10. Calculated ^{29}Si chemical shifts [$\delta(^{29}\text{Si})$] and chemical shift anisotropies [CSA, $\Delta\sigma(^{29}\text{Si})$] for several silylenes (in ppm)[a]

Silylene	$\delta(^{29}\text{Si})$[b]	$\Delta\sigma(^{29}\text{Si})$[c]
H$_2$Si	817	1516
(H$_2$N)$_2$Si, per[d]	421	558
(H$_2$N)$_2$Si, pl[e]	108	214
76d	117[f]	73[g]
75d	64[h]	165

[a]From Reference 102 at GIAO[MP2/TZ2P(Si), TZP(C,N), DZ(H)],
[b]Relative to (CH$_3$)$_4$Si,
[c]Calculated from absolute chemical shielding values,
[d]Perpendicular conformation, i.e. ∠HNSiN = 90°C.
[e]Planar conformation, i.e. ∠HNSiN = 0°.
[f] 131.9 ppm at GIAO[B3LYP/6-311+G(2df,p)(Si), HF/6-31G*(C,N,H)//B3LYP/6-31G*[107]], Experimental values for **76c**: 117.0 ppm[98d] and 119.0 ppm[107] in solution and in the solid, respectively.
[g]The CSA values (calculated from chemical shifts) are 447.7 and 363.1 ppm for **76d** and **75d**, respectively [at GIAO(B3LYP/6-311+G(2df,p)(Si), HF/6-31G*(C,N,H)//B3LYP/6-31G*[107]]; the corresponding experimental values for **76c** and **75c** are 354.0 and 314.6 ppm, respectively[107].
[h]84.8 ppm at GIAO[B3LYP/6-311+G(2df,p)(Si), HF/6-31G*(C,N,H)//B3LYP/6-31G*[107]] Experimental values for **75c**: 78 ppm in solution[98d] and 75.2 ppm in the solid[107].

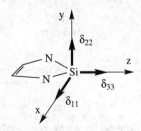

FIGURE 14. Molecular and magnetic axes for silylene **75**. The same axes also apply for **79**. Reprinted with permission from Reference 107. Copyright (1998) American Chemical Society

1. Theoretical aspects and quantum mechanical calculations

$$
\begin{array}{c}
\text{(structure 84)}
\end{array}
$$

(84)

d. Charge distribution and ionization potentials. The degree of π-delocalization in **75**, from the $-$NCH$=$CHN$-$ backbone into the formally empty p atomic orbital of M, can be evaluated by several theoretical approaches. NBO analysis[108] of the charge distribution in **75d** and **76d** (at MP2/6-31G*) shows significant electron occupancy of the 3p(Si) orbital in both silylenes. However, as expected if **75d** is aromatic, the electron occupancy of the 3p(Si) in orbital in **75d** (0.54e) is higher than in the saturated silylene **76d** (0.33e)[103]. The 3p(Si) electron occupation in **75d** is smaller than in the corresponding carbene **75a**, R = H (0.67e) and germylene **75**, M = Ge, R = H (0.63e), as expected from their electronegativites (Ge is more electronegative than Si)[103]. This result contrasts with that of DFT calculations, which predicted that the degree of π-electron delocalization in divalent **75** will decrease in the order M = Ge > Si > C, i.e. delocalization is most efficient for **75**, M = Ge[96]. Arduengo and coworkers suggested a 'chelated-atom' model to account for this trend[96]. Boehme and Frenking pointed out that this model and the DFT trend of the occupation of the 3p-orbital would suggest that the oxidation strength increases in the order Ge > Si > C, which is difficult to understand[103]. Analysis using the 'Atoms in Molecules' theory[109] also provides evidence for some degree of cyclic electron delocalization in the unsaturated Arduengo-type cyclic carbene **75a**, R = H (although to a smaller degree than calculated by other methods) and to an even smaller extent of delocalization in the analogous silylene **75d**[100].

Another argument in favour of the aromaticity of unsaturated cyclic silylenes such as **75c** (or **75d**) comes from the analysis of their ionization potentials. Denk and coworkers argued[98b] that the energy lowering by 0.5 eV (HF/6-31G*) of the HOMO of **75d** shown in **85a** ($E = -7.62$ eV) as compared to the HOMO of the silane **77a** shown in **85b** ($E = -7.1$ eV)[98b], is consistent with a contribution of the 3p(Si) orbital to the stability of the HOMO orbital in **75d** but not in **77a**. The calculated difference in the orbital energies of 0.5 eV is in excellent agreement with the measured vertical ionization potentials of 6.96 eV and 6.56 eV for **75c** and **77b**, respectively[96,98b]. A similar stabilization of the HOMO of **79b**, by 0.33 eV (0.41 eV, experimental), relative to that of the corresponding silane **80b**, was also attributed to the contribution of the empty 3p(Si) orbital to the delocalization of the π-electrons[99c].

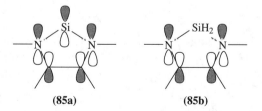

(85a) (85b)

In conclusion, all the criteria discussed above point to the existence of π-electron delocalization and thus to some degree of aromaticity in the unsaturated carbenes, silylenes and germylenes of type **75**. However, the degree of conjugation and aromaticity depends on the criteria that are used to evaluate these effects. The degree of aromaticity in **75** is quite small according to the 'Atoms in Molecules' topological charge analysis[100], but is more significant according to NBO charge analysis[103] or according to the structural, energetic and magnetic properties[100,103]. However, regardless of the criteria used, π-electron delocalization is generally found to be less extensive in the unsaturated silylenes (and germylenes) compared to their carbene analogues, and it is much smaller than in prototypic aromatic systems such as imidazolium cations or benzene[100,103].

3. Reactions

As mentioned above, the cyclic silylenes **75c** and also **76c** (or **79a** and **79b**) are dramatically less reactive than other known silylenes which are all transients. Thus, **75c** has such a low Lewis acidity that, unlike other silylenes, it does not react as an electrophile. However, **75c** reacts as a Lewis base, i.e. as a nucleophilic reagent[98c]. The known reactions of 'Arduengo-type' silylenes are summarized in References 2e, 98c, 98d and 99b and the references cited therein. The theoretical study of the possible reactions of 'Arduengo-type' silylenes is quite limited and awaits the attention of computational chemists.

a. 1,2-H rearrangements. An important factor which contributes to the kinetic stability of silylene **75c** is the fact that 1,2-H rearrangements to give the corresponding tetravalent silaimines do not occur. CCSD(T)/DZP calculations carried out by Heinemann and coworkers[104] showed that the rearrangement of simple aminosilylenes via a 1,2-H shift to produce the corresponding silaimines (equations 32–34) are endothermic, i.e. the silylenes are more stable than the isomeric silaimines. The isomerization energy, ΔE_r, of the cyclic silylene **75d** to the corresponding silaimine, of 31.8 kcal mol^{-1}, is significantly higher than for the rearrangements of the parent aminosilylene **86** to **87** or of the diaminosilylene **88** to **89**. A similar isomerization energy of 33.3 kcal mol^{-1} (MP2/6-31G*//MP2/6-31G*) was calculated for the rearrangement of **79b** to **83**. The high isomerization energy was attributed to the high strain of the five-membered ring in **83**[99c]. Furthermore, the energy barriers for these rearrangements, E_a, are very high, larger than 50 kcal mol^{-1} (above the silylenes). The high barrier of 54.4 kcal mol^{-1} calculated for the **75d**→**90** rearrangement explains why for **75c** such rearrangements were not observed experimentally[2e]. The relatively low barrier of 22.6 kcal mol^{-1} for the reverse **90**→**75d** rearrangement suggests that this type of reaction might be used to synthesize novel cyclic aminosilylenes[104]. The barrier for the analogous isomerization of the carbene **75a**, R = H to the corresponding imidazole of 44.1 kcal mol^{-1} is somewhat lower than for **75d**, but the barrier for the reverse reaction is significantly higher for the carbene (71.1 kcal mol^{-1} at MP4/6-311G**//MP2/6-31G*)[103].

$$\text{(86)} \quad \longrightarrow \quad \text{(87)} \tag{32}$$

$\Delta E_r = 14.2$ kcal mol^{-1}

$E_a = 68.8$ kcal mol^{-1}

1. Theoretical aspects and quantum mechanical calculations

$$\text{(88)} \longrightarrow \text{(89)} \tag{33}$$

$\Delta E_r = 19.2$ kcal mol^{-1}
$E_a = 70.0$ kcal mol^{-1}

$$\text{(75d)} \longrightarrow \text{(90)} \tag{34}$$

$\Delta E_r = 31.8$ kcal mol^{-1}
$E_a = 54.4$ kcal mol^{-1}

b. Dimerization. The report that the saturated silylene **76c** dimerizes slowly to give an unidentified product[98b] stimulated a theoretical study of the structure of this dimer[110]. Silylenes usually dimerize to give disilenes (**91**) as shown schematically in equation 35 for M = Si, and this is the main experimental route for the synthesis of disilenes[1a]. However, Apeloig and coworkers[102,110,111] and Trinquier and coworkers[66d,112] pointed out that the dimerization of two silylenes can in principle lead also to the bridged **92** (equation 36, both *trans* and *cis* arrangements of the R groups are possible). **92** was indeed found to be a minimum on the M_2H_4 surfaces for M = Si and Ge, but not for M = C, and it is even the global minimum for M = Sn and Pb[66d,112]. **92**, M = Si, R = H is by 21.3 kcal mol^{-1} less stable than $H_2Si=SiH_2$ at the G2 level of theory[102,110].

$$\longrightarrow \quad \text{(91)} \tag{35}$$

$$\longrightarrow \quad \text{(92)} \tag{36}$$

Apeloig and Müller found that electronegative substituents having lone-pairs (e.g. R = F, OH and NH$_2$) stabilize considerably the bridged isomer (**92**) relative to the corresponding isomeric disilene (**91**). For example, when the bridging R groups are NH_2 and the exocyclic R groups are H, **92** is more stable than **91** by 10 kcal mol^{-1} at MP4/6-311G**//6-31G**[102,110]. Furthermore, disilenes substituted with four NH_2, F or OH substituents are no longer minima on the Si_2R_4 PES (at HF/6-31G**). On the other hand, the corresponding bridged isomers **92**, R = NH_2, OH and F are local minima on the PES; e.g. **92**, R = NH_2, is by 15.3 kcal mol^{-1} more stable than two isolated diaminosilylenes at MP2(fc)/6-31G*//MP2(fc)/6-31G*[110]. Calculations at HF/6-31G** show that, in analogy to $(H_2N)_2Si=Si(NH_2)_2$, also the disilene **93**, R = H, the formal dimer of **76d**, is not a minimum on the PES. In contrast, the corresponding bridged dimer **94**, R = H is a minimum on the PES, lying 7.8 kcal mol^{-1} below the energy level of two isolated **76d** silylenes. Apeloig and Müller therefore proposed that **76c** does not dimerize to the corresponding disilene **93**, R = t-Bu, but to the bridged cage-like structure **94**, R = t-Bu[110]. This prediction awaits experimental testing.

(93)   (94)

B. Miscellaneous Potentially Aromatic Silylenes

1. Cyclic aminosilylenes

Veszprèmi and coworkers[68,69] suggested that silylenes such as **95** and **96** are also potentially aromatic and that therefore they should be good candidates for synthesis. The formal trisilylene **95** is calculated to be by 30.5 kcal mol^{-1} (CCSD/6-31G*//MP2/6-31G*; 16.5 kcal mol^{-1} at MP2/6-31G*//MP2/6-31G*) more stable than the isomeric aromatic **41**[68] (see Section III.H. for a discussion on **41** and **42**). Similarly, the silylene **96a** is by 8.9 kcal mol^{-1} more stable than 2-silapyridine **42** (MP2/6-311G**//MP2/6-311G**[69]). **96a** and **42** are separated by a high barrier of 56 kcal mol^{-1}, similar to the barrier of 54 kcal mol^{-1} that separates **75d** from **90** (see Section V.A.2.a). The **96–42** stability order is reversed upon methyl substitution; **96b** is by 5.6 kcal mol^{-1} less stable than the methyl-substituted (on Si) **42**.

(95)   (41)   (96a) R = H   (42)
                            (96b) R = Me

The higher stability of **95** and **96a** relative to their silaimine isomers is consistent with the relative energies of other silylimine–aminosilylene pairs (e.g. compare the relative energies of **86** vs **87**, **88** vs **89** and **75d** vs **90**). It is also in harmony with the relative bond energies of Si—H and N—H bonds, N—H being by about 16 kcal mol^{-1} stronger than Si—H[104] (based on the dissociation energies of SiH$_4$[113a] and NH$_3$[113b]). This stability order is in sharp contrast to that for the related phosphinine-2-ylidene (**97**)-phosphinine (**98**) pair, for which the carbene isomer is by 75 kcal mol^{-1} less stable than the phosphinine[114].

(**97**)   (**98**)

Veszprémi and coworkers concluded, on the basis of geometric and energetic criteria, that the silylenes **95** and **96a**, as well as the isomeric silaimines **41** and **42**, exhibit some degree of aromaticity. The geometries of the silaimine isomers, **41** and **42**, were discussed already in Section III.H and they are shown in Figure 7. We will compare here their geometries to those of the silylidene isomers **95** and **96a**. The Si—N bond lengths of 1.756 Å in **95** and 1.665 Å in **41** (MP2/6-31G*) show no bond alternation. The SiNSi and NSiN bond angles in **95** are 129.6° and 110.4°, respectively, while in **41** they are more similar, i.e. 117.0° and 123.0°, respectively. The three C—C bonds in the *cis*-butadiene unit in **42** are almost identical and are similar to those in benzene (i.e. 1.40 Å, Figure 7) while in **96a** these bonds (i.e. 1.382 Å, 1.413 Å and 1.373 Å at MP2/6-31G*[69] show some bond length alternation, similar to that in pyrrole, but smaller than that in *cis*-1,3-butadiene[69]. Based on the C—C bond length alternations, the authors concluded that the delocalization in **42** is somewhat larger than in **96a**[69].

The aromatic stabilization energies (at MP2/6-31G*//MP2/6-31G*) of **95** and **41**, according to the homodesmotic equations 37 and 16, are 6.0 and 19.6 kcal mol^{-1}, respectively, and those of **96a** and **42** (equations 38 and 17) are 14.7 and 19.4 kcal mol^{-1}, respectively. These aromatic stabilization energies, although significant, are considerably smaller than the aromatic stabilization energies of benzene, phosphabenzene, pyridine and silabenzene of 28.2, 27.1, 28.0 and 23.5 kcal mol^{-1}, respectively[69]. Note that these homodesmotic reactions do not provide a pure measure of the aromatic stabilization as they also include the ring strain of 10.0, 5.5 and 14.1 kcal mol^{-1} for **41**, **96a** and **42**, respectively[69] (the ring strain of **95** was not reported), which is released upon ring cleavage.

$$\textbf{95} + 3\text{HSiNH}_2 \longrightarrow 3\text{HSiNHSiNH}_2 \quad (37)$$

$$\textbf{96a} + \text{HSiNH}_2 + 2\text{CH}_2=\text{CH}_2 \longrightarrow \text{CH}_2=\text{CHSiNH}_2 + \text{CH}_2=\text{CHNHSiH} + \text{CH}_2=\text{CH}-\text{CH}=\text{CH}_2 \quad (38)$$

Thus, according to the geometric and energetic criteria the silylenes **95** and **96a** possess some degree of aromaticity, although smaller than that of the corresponding silaimine isomers **41** and **42**. Magnetic properties, which might help in evaluating reliably the degree of aromaticity of these compounds, were not yet reported.

To evaluate the kinetic stability of **96a** its dimerization reaction was studied[69]. The authors find, in analogy to the previous findings of Apeloig and Müller for **76c** (Section V.A.3.b)[102,110], that the head-to-head dimerization of **96a** to give the corresponding disilene **99** (equation 39) is not feasible as **99** is not a minimum on the HF/6-31G* PES. On the other hand, the head-to-tail bridged dimer **100** (equation 40) is a minimum on the PES. However, **100** is by 18.4 kcal mol^{-1} less stable than the two isolated silylenes (**96a**). This contrasts the situation with the bridged dimer **94**, which is by 7.8 kcal mol^{-1} more stable than the corresponding two isolated silylenes **76d**[110]. Thus, **96a** and its substituted derivatives are expected to be highly stable towards dimerization, making silylene **96a** (and especially its sterically protected derivatives) a promising target for synthesis.

(39)

(**96a**)   (**99**)

(40)

(**96a**)   (**100**)

2. Silacyclopropenylidene and silacyclopropynylidene

Kinetic studies of the reaction of Si(^{3}P) with acetylene suggested the formation of stable complexes[115]. The potentially 2π-aromatic silylene, 1-silacyclopropenylidene **101**, was suggested as a possible intermediate, but at that time spectroscopic evidence which supports this suggestion was not available[115]. **101** and its cation-radical were later identified in the gas phase by a neutralization–reionization experiment[116]. Most recently in a beautiful study, Maier and coworkers have generated **101** in a matrix by pulsed flash pyrolysis of HC≡CSiH$_2$Si(CH$_3$)$_3$. Irradiation of **101** produced the non-cyclic isomers **102** and **103** and, surprisingly, due to their high strain, also silacyclopropyne (**104**) and silacyclopropynylidene (**105**)[117a,b]. **104** and **105** are calculated to have strain energies of ca 100 kcal mol^{-1}[117a,b] and are the most strained cycloalkynes ever identified. Isomers **101**–**105** were identified spectroscopically by comparison of their experimental and calculated (MP2/6-31G**[117b]) IR spectra. The structure of 1-silacyclopropenylidene, **101**, was determined by microwave spectroscopy[117c].

Extensive computational studies of **101**[117b,118,119] and also of other singlet[118,119] and triplet[119,120] C$_2$H$_2$Si isomers followed the first prediction of the existence of **101**[115]. The optimized geometries of singlet **101** at various theoretical levels are shown in Figure 15.

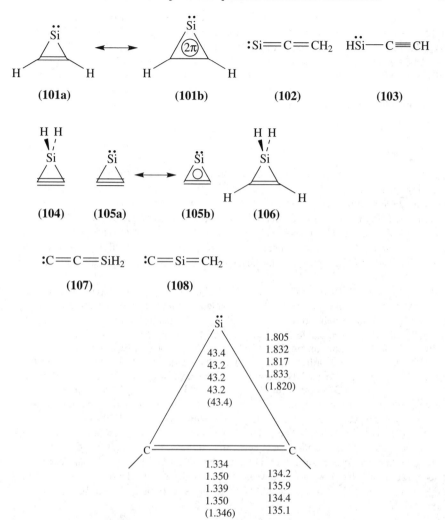

FIGURE 15. Optimized structures of singlet ($^1A_1$) **101**, at (from top to bottom): SCF/TZ2P[118b], MP2/6-311G**[117b], CISD/TZ2P[118b] and CCSD(T)/TZ(2df,2pd)[119]. Experimental values[117c] are given in parentheses, bond lengths in Å, bond angles in degrees

Is **101** aromatic? The C−C and Si−C bond distances in **101** [at CCSD(T)/TZ(2df,2pd), the best available level of theory for this system], of 1.350 Å and 1.833 Å respectively[119], are in good agreement with the experimental values of 1.346 Å and 1.820 Å, respectively[117c]. The C−C bond is somewhat longer than in the analogous cyclopropenylidene (1.328 Å[119]) and the C−Si bond distance is somewhat shorter than a regular single C−Si bond. These structural data and the results of Mulliken[118a,b] and NBO[117b] population analyses indicate some degree of 2π-aromatic delocalization of the C=C π-electrons into the formally empty 3p-orbital of the silicon, as described by resonance structure **101b**. The stabilization energy resulting from π-delocalization in **101**, as calculated according to equation 41a, is very high, 38.1 kcal mol^{-1} (MP2/6-31G**//MP2/6-311G**)[117b]. The authors attributed this large effect to the 'differences in the silanediyl structure'[117b]. A more detailed analysis of this large ASE was not given[117b], but it certainly supports the aromatic character of **101**. The ring strain energy of **101** is estimated to be *ca* 50 kcal mol^{-1} according to equation 41b and assuming that **101** and **106** have similar ring strain energies[117b].

$$101 + H_2Si(CH_3)_2 \longrightarrow 106 + (CH_3)_2Si \qquad (41a)$$

$$106 + C_2H_6 + Si_2H_6 \longrightarrow H_2Si(CH_3)_2 + (Z)\text{-}H_3SiHC=CHSiH_3 \qquad (41b)$$

101 is the global minimum, on both the singlet[117a,118a] and triplet[120] C_2H_2Si PES. The vinylidenesilanediyl **102** and the ethynylsilylene **103** are by 17 and 22 kcal mol^{-1}, respectively, higher in energy than **101** (at CI/DZP+Davidson's correction[118a]; the relative energies at MP2/6-31G** are similar[117a]). The isomeric carbenes **107** and **108** are by 55.8 and 116.7 kcal mol^{-1}, respectively, higher in energy than **101** (at HF/DZ)[118a]. The interesting cyclic silane **104** was calculated to be by 47.4 kcal mol^{-1} higher in energy than **101** at CCSD(T)/TZ(2df,2pd)[119] (54.4 kcal mol^{-1} at MP2/6-31G**[17a,b]; 86.7 kcal mol^{-1} at SCF/DZ[118a]).

Calculations at UMP4/6-31G**//UHF/6-31G* show that the $^2A_1$ cation-radical of **101**, that was detected in the gas phase[116] and which formally is also 2π-aromatic, is by 43.7 kcal mol^{-1} more stable than its Si$^{+\cdot}$ and C_2H_2 fragments, and that it is by only 0.5 and 8.5 kcal mol^{-1} more stable than the isomeric **102**$^{+\cdot}$ and **103**$^{+\cdot}$, respectively[121].

105, one of the irradiation products of **101**, is also a potentially 2π-aromatic silylene (see **105b**). **105** was identified earlier in gas-phase experiments by IR and electronic spectra[122a]. According to an NBO analysis, **105** is best described by a Lewis structure that has a delocalized 3-centre aromatic-type π-orbital and an in-plane CC π-orbital that has essentially a non-bonding character[117b]. The calculated C≡C and C−Si bond lengths in **105** are 1.285 and 1.838 Å, respectively (MP2/6-311G**[117b], 1.294 and 1.838 Å at MBPT(2)/DZ+d[122b]). The C≡C bond length is intermediate between that of a C=C bond in ethylene and a C≡C bond in acetylene, and it was therefore described as a 'weak triple bond'[122c], a description which is supported by NBO analysis[117b]. The calculated delocalization stabilization energy of **105** is very high, 42.0 kcal mol^{-1} (equation 42a), and it is calculated to have an incredibly high ring strain energy of *ca* 100 kcal mol^{-1} (equation 42b)

$$105 + H_2Si(CH_3)_2 \longrightarrow 104 + (CH_3)_2Si \qquad (42a)$$

$$105 + C_2H_6 + Si_2H_6 \longrightarrow H_2Si(CH_3)_2 + H_3SiC\equiv CSiH_3 \qquad (42b)$$

3. Trisilapropenylidene and other Si_3H_2 isomers

The Si_3H_2 PES was studied at CISD/TZP//HF/DZP[123]. The global minimum is the potentially 2π-aromatic trisilapropenylidene (**109**). The degree of aromaticity of **109** was not discussed. The Si=Si bond length is 2.11 Å, typical of a Si=Si double bond, and the Si–̇Si bond distances are 2.28 Å (HF/DZP), very close to that of a typical Si–̇Si single bond of *ca* 2.24 Å. **109** can isomerize with relatively low barriers of 12.5 and 13.6 kcal mol^{-1}, respectively, to the planar hydrogen-bridged structures **110a** and **110b**, which are by only 1.4 and 1.7 kcal mol^{-1}, respectively, less stable than **109**[123]. Another interesting Si_3H_2 isomer is the planar **111a** (C_s symmetry) which is by only

6.6 kcal mol^{-1} less stable than **109**. The conventional 'perpendicular' C_{2v} isomer **111b** is by 47.8 kcal mol^{-1} less stable than **111a**. The remarkable high stability of planar **111a** was rationalized as due to a 3-centre π-interaction in the cyclic Si$_3$ skeleton[123]. Higher level calculations are required to substantiate this interesting PES.

(109)   (110a)   (110b)

(111a)   (111b)

VI. SILAFULLERENES

A. Are Fullerenes and Silafullerenes Aromatic?

The remarkable series of experiments by Kroto and coworkers[124] leading to the discovery of the spheroidal C$_{60}$ Buckminsterfullerene and of larger carbon clusters, followed by the development by Kräschmer and coworkers[125] of methods which enabled the synthesis of macroscopic amounts of C$_{60}$, have stimulated an outburst of interest, both experimental and theoretical, in this class of compounds[126]. The silicon analogues of C$_{60}$ and of larger carbon clusters have not been yet synthesized, but silicon clusters are currently of growing experimental and theoretical interest.

Are fullerenes and silafullerenes, e.g. C$_{60}$, Si$_{60}$, C$_{70}$ and Si$_{70}$, aromatic? The suggestion by Kroto and coworkers[124a] that C$_{60}$ and related molecules are aromatic raised considerable controversy and inspired analysis of this question by a variety of criteria, such as structure, energy, reactivity and magnetic properties[13a,127–133]. Studies of the magnetic properties included the chemical shifts of encapsulated He atoms[128,129] as well as NICS calculations[26,132]. These studies lead to the conclusion that C$_{60}$ is aromatic, but only to a modest extent[127–130]. The degree of aromaticity of C$_{70}$, C$_{60}^{-6}$ and C$_{70}^{-6}$ is considerably larger[129,131,132]. Larger spheroidal carbon clusters such as C$_{76}$, C$_{78}$, C$_{82}$ and C$_{84}$ have a higher degree of aromaticity than C$_{60}$, but smaller than that of C$_{70}$[133]. The interested reader is referred to the above-mentioned references for a full discussion of this interesting topic.

The studies reported to date on silafullerenes have not assessed explicitly their aromatic character. We will assume here that as C$_{60}$ and C$_{70}$ are aromatic, so are their silicon analogues Si$_{60}$ and Si$_{70}$, although they probably possess a smaller degree of aromaticity than their carbon analogues. Consequently, we include a discussion on Si$_{60}$ and Si$_{70}$.

Due to their large size, most theoretical studies on silafullerenes were carried out using relatively simple computational levels. The theoretical studies on Si$_{60}$ conducted up to the end of 1991 were reviewed by Nagase[61], and here we will summarize briefly only the most important conclusions of the earlier studies and add the few new studies which appeared after 1991.

The dramatic developments in computer technology and, in particular, the development of density functional methods will enable one in the near future to repeat and extend the studies discussed below, using more accurate and reliable computational methods. The study of the magnetic properties of silafullerenes and the evaluation of the degree of their aromaticity is also a worthwhile field of exploration for the future.

B. Si_{60}

1. Structure

As mentioned above, Si_{60} is still unknown and theory therefore remains the main tool for establishing its structure. Does Si_{60} possess the buckyball icosohedral structure **112** (Figure 16) that is constructed form 12 five-membered rings (5-MRs) and 20 6-MRs, as in C_{60}? Does Si_{60} adopt an isomeric structure? Several possible Si_{60} isomers **112–116** are shown in Figure 16. In view of the observation that in the gas phase silicon cluster ions lose mainly Si_6 and Si_{10} units[134], it was suggested[135,136] that these clusters are built either as stacked plates of silanaphthalenes (**113**) or as a cylinder of stacked benzenes (**114**), both structures being stabilized by electron conjugation within the silicon plates. However, Nagase and Kobayashi found, using the semiempirical AM1 method[61,137], that the most stable structure of Si_{60} is the icosohedral silafullerene **112**, while the planar stacked **113** and **114** are by 158 and 228 kcal mol^{-1} less stable, respectively. Single-point HF/DZ calculations at the AM1 geometries predict that **113** is by as much as 606 kcal mol^{-1} less stable than **112**, indicating that AM1 strongly underestimates strain energies. This stability order is not surprising in view of the high strain in structures **113** and **114**, which contain many bond angles that deviate significantly from the tetrahedral value of 109.5°. The preference of **112** over **113** and **114** suggests that the gas-phase fragmentation data do not reflect in a simple way the structure of Si_{60}[61,37].

Si_{60} with I_h symmetry (**112**) is by 16.3 and 21.1 kcal mol^{-1} at AM1 and HF/DZ//AM1, respectively, more stable than the C_{2v} isomer of Si_{60} (**115**). The higher stability of **112** ensures its higher population up to 2080 K. Above this temperature the less symmetric **115** is computed to dominate the Si_{60} equilibrium mixture[138]. The finding that the C_{2v} structure is destabilized relative to the I_h structure is consistent with the 'isolated pentagon rule'[139] which states that the fusion of pentagons is energetically unfavourable as a result of increased strain and the resulting induced antiaromatic character[139]. C_{2v} Si_{60} (**115**) has two paris of edge-sharing pentagons, causing its destabilization relative to I_h Si_{60}, which has no edge-sharing pentagons. The number of adjacent pentagons increases as the cluster becomes smaller, causing its destabilization. Thus, at HF/DZ//HF/DZ the smaller spheroidal clusters, Si_{50} (D_{5h}), Si_{30} (C_{2v}) and Si_{24} (D_3), are by 2.0, 5.8 and 9.3 kcal mol^{-1}, respectively, less stable per atom than Si_{60}[137]. This effect is much larger in the analogous carbon clusters, which suffer higher strain and antiaromaticity due to the presence of adjacent pentagons[137]. Larger silicon clusters such as Si_{70}, Si_{78} and Si_{84}, with a larger number of hexagons, are thermodynamically slightly more stable (per Si atom) than Si_{60}[61,137]. The full hydrogenation of Si_{60} leads to the saturated $Si_{60}H_{60}$. The strain energy in $Si_{60}H_{60}$ of 114 kcal mol^{-1} (at AM1, 207 kcal mol^{-1} at HF/DZ//HF/DZ) is considerably smaller than the strain energy in $C_{60}H_{60}$ of 530 kcal mol^{-1} (at AM1). This may also reflect the fact that Si_{60} is significantly less strained than C_{60}[61,137].

Optimization of the structure of Si_{60} without any symmetry restrictions, using a tight binding molecular dynamic technique, resulted in a C_{2h} structure (**116**). This relaxed geometry is probably preferred over the I_h structure because it allows increased

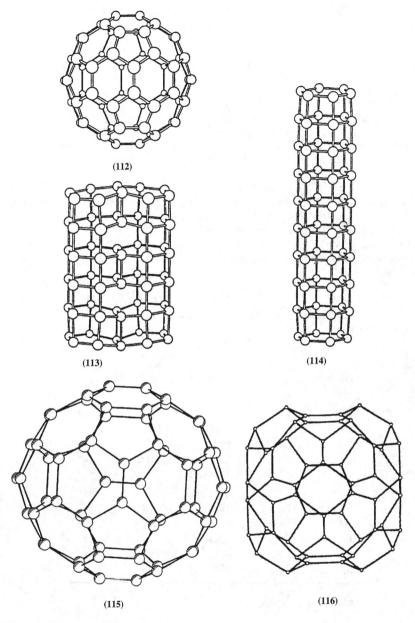

FIGURE 16. Various structures of Si_{60}: fullerene (**112**, I_h)[137]; stacked naphthalene (**113**, D_{2h})[137], stacked benzene (**114**, C_{2v})[137], fullerenes (**115**, C_{2v})[138] and (**166**, C_{2h})[140]. Reproduced from Reference 137, Copyright (1993) by courtesy of Marcel Dekker Inc. and from References 138 and 140, Copyright (1994) with kind permission of Elsevier Science–NL, Sara Burgehartstraat 25, 1055 KV Amsterdam, The Netherlands

tetrahedrality of the atoms, so that out of the 60 vertices, 10 have the ideal tetrahedral angle[140]. Higher level calculations are required to establish whether Si_{60} has I_h or a lower symmetry structure.

I_h Si_{60} has two distinct Si—Si bond lengths, one which fuses two adjacent 6-MRs [designated as $r(6-6)$], and the second which is longer and fuses a 5-MR and a 6-MR [designated as $r(5-6)$]. The calculated bond lengths at various levels of theory are given in Table 11, which also lists the corresponding bond lengths in C_{60}.

The $r(5-6) - r(6-6)$ bond length difference, Δ, is quite large when using the AM1 and PM3 semiempirical levels but is reduced significantly to 0.077–0.088 Å at *ab initio* levels (Table 11). The size of the basis set has a small effect on Δ but, based on the experience with C_{60} (Table 11), it is expected that inclusion of electron correlation will decrease Δ considerably also in Si_{60}. A small Δ may be considered as an indication of aromaticity.

I_h Si_{60} has a five-fold-degenerate HOMO and a three-fold-degenerate LUMO[143,148]. At HF/3-21G the HOMO and LUMO energies are −6.39 and −1.95 eV, respectively[148] (−6.5 and −2.1 eV at HF/DZ+ECP[61,143]), resulting in a small HOMO–LUMO gap of only 5.0 eV, indicating its high polarizability. This HOMO–LUMO gap is only about 65% of that in C_{60} (Table 12). A similar reduction of the HOMO–LUMO gap was calculated also for the hexasilabenzene/benzene pair[59]. The HOMO–LUMO gap is further reduced to 4.3 eV in Ge_{60}[148].

TABLE 11. Optimized X—X bond lengths in I_h X_{60} (X = Si and C) at various levels of calculation

$r(6-6)^a$	$r(5-6)^a$	Δ^b	Method	Reference
		Si_{60}		
2.092	2.297	0.205	AM1	141
1.879	2.247	0.368	PM3	147
2.088	2.169	0.081	STO-3G	142
2.067	2.152	0.085	STO-3G*	142
2.213	2.301	0.088	3-21G	142
2.229	2.307	0.078	6-31G	142
2.189	2.266	0.077	ECP/DZ	143
		C_{60}		
1.385	1.464	0.079	AM1	141
1.400	1.474	0.074	MNDO	144
1.384	1.457	0.073	PM3	147
1.367	1.453	0.086	3-21G	129
1.406	1.446	0.040	MP2/TZP	145
1.401	1.458	0.057	exp.^{c}	146

^{a} In Å.
^{b} Bond length difference, in Å.
^{c} Gas-phase electron diffraction.

2. Endohedral complexes of Si_{60}

One of the most remarkable properties of fullerenes is their ability to encapsulate atoms, ions and small molecules, to form the so-called endohedral fullerene complexes, denoted

1. Theoretical aspects and quantum mechanical calculations

TABLE 12. Encaged ion population (N_{ion}), HOMO and LUMO energies (eV) and stabilization energies (ΔE, kcal mol^{-1}) of endohedral complexes of Y_{60} (Y = C, Si and Ge) with various ions[a,b,c]

Ion	N_{ion}	E(HOMO)	E(LUMO)	Δ^d	ΔE^e
		C_{60}			
none	—	$-8.3\ (-8.0)^f$	$-0.69\ (-0.3)^f$	7.61 (7.3)	—
Ne	—	(−8.0)	(−0.3)	(7.7)	(0.4)
Li$^+$	2.004	−11.78	−4.16	7.62	−9.37
Na$^+$	10.0	−11.79 (−11.4)	−4.17 (−3.8)	7.62 (7.6)	−8.66 (−6.7)
K$^+$	18.05	−11.82	−4.19	7.63	−2.21
Rb$^+$	36.05	−11.83	−4.20	7.63	3.90
F$^-$	9.92	−4.79 (−4.5)	2.81 (3.1)	7.59 (7.6)	−22.54 (−39.8)
Cl$^-$	17.86	−4.82	2.78	7.60	−5.29
Br$^-$	35.78	−4.84	2.74	7.58	7.73
I$^-$	53.72	−4.93	2.67	7.60	39.27
		Si_{60}			
none	—	$-6.39\ (-6.5)^g$	$-1.95\ (-2.1)^g$	4.98 (4.4)	—
Ne	—	(−6.5)	(−2.1)	(4.40)	(0.0)
Li$^+$	2.002	−8.60	−4.14	4.46	−30.27
Na$^+$	9.996	−8.59 (−8.8)	−4.14 (−4.4)	4.45 (4.4)	−30.30 (−20.3)
K$^+$	18.00	−8.58	−4.13	4.45	−30.36
Rb$^+$	36.02	−8.54	−4.13	4.41	−30.09
F$^-$	9.98	−4.16 (−4.2)	0.26 (0.2)	4.42 (4.4)	6.69 (2.2)
Cl$^-$	17.98	−4.18	0.25	4.43	5.49
Br$^-$	35.93	−4.18	0.25	4.43	4.45
I$^-$	53.93	−4.19	0.24	4.43	5.02
		Ge_{60}			
none	—	−5.99	−1.73	4.26	—
Ne	—	-6.1^f	-2.1^f	4.1^f	0.0
Li$^+$	2.002	−8.08	−3.81	4.27	−23.37
Na$^+$	9.997	−8.08	−3.80	4.28	−29.86
K$^+$	18.00	−8.07	−3.80	4.27	−29.83
Rb$^+$	36.01	−8.07	−3.80	4.27	−22.34
F$^-$	9.99	−3.89	0.35	4.24	6.73
Cl$^-$	17.98	−3.90	0.35	4.25	5.97
Br$^-$	35.94	−3.90	0.35	4.25	5.24
I$^-$	53.94	−3.91	0.35	4.26	5.55

[a] Based on data from Reference 148 unless stated otherwise.
[b] At HF/3-21G (3-21+G for the halide ions), the fullerene cages have I_h symmetry with bond distances of 1.370 and 1.450 Å for C_{60}[150], 2.189 and 2.266 Å for Si_{60}[143] and 2.315 and 2.398 Å for Ge_{60}[61] and the ion is placed at the centre of the cage.
[c] Values in parentheses are from Reference 149 for X@C_{60}, at HF/4-31G for the carbons and HF/DZP for the guest atom; from Reference 143 and for X@Si_{60} at HF/DZ+ECP.
[d] HOMO−LUMO energy difference, in eV.
[e] According to equation 43, corrected for basis-set superposition error. Negative values indicate stabilization by the ion.
[f] At AM1 the HOMO and LUMO energies are −9.64 eV and −2.95 eV, respectively[141].
[g] At AM1 the HOMO and LUMO energies are −8.0 eV and −3.38 eV, respectively[141].

as, e.g., X@C_{60}, where X is the encapsulated atom. The diameter of the hollow spherical cage of I_h Si_{60} of 11.1 Å[143] is much larger than that of I_h C_{60} (7.1 Å)[146]. In addition, Si_{60} is also more polarizable than C_{60} and, as a result, it is expected to form endohedral complexes even with larger guest atoms and ions than C_{60}. Nagase and Kobayashi[61,143] have studied the properties of X@Y_{60} complexes where Y = Si, Ge and X = Ne, Na^+ and F^-; Cioslowsky and Fleischmann[149] studied X@C_{60} complexes with the same guests and Geerlings and coworkers[148] studied X@Y_{60} complexes with X = Li^+, Na^+, K^+, Rb^+, F^-, Cl^-, Br^- and I^- and Y = C, Si and Ge. Their main conclusions are summarized below and in Table 12.

(1) Charge transfer from the ion or neutral atom to the Y_{60} cage, or in the opposite direction, is negligible. The interactions in the complexes are electrostatic in origin and are due mainly to polarization[61,143,148].

(2) The ionic radii of the encaged ions are strongly expanded within the fullerene cage, compared to that of the isolated ions, especially in the larger Si_{60} and Ge_{60} cages[148].

(3) Encapsulation of a cation in C_{60}, Si_{60} and Ge_{60} cages conserves the typical orbital structure of the neutral fullerene (i.e. a five-fold-degenerate HOMO and a three-fold-degenerate LUMO). Both the HOMO and LUMO are stabilized by the encapsulated cations, resulting in an increase in the ionization potentials and electron affinities (according to Koopmans' theorem). The HOMO–LUMO gap of Si_{60} is slightly reduced, but is almost unchanged for C_{60} and Ge_{60}. When an anion is placed within the cages the frontier orbitals are destabilized, giving rise even to negative electron affinities of the cages (see Table 12)[148].

(4) The stabilization energies of the X@Y_{60} systems as a result of the encapsulation (ΔE) are calculated by equation 43 (where $E_{(X@Y_{60})}$ is the energy of the endohedral complex whilst E_X and $E_{Y_{60}}$ are the energies of the free ion and of the fullerene, respectively) and are given in Table 12. The stabilization energies, ΔE, depend on the following factors: (i) the endohedral electrostatic potential; (ii) the ion-induced dipole interaction of the central ion with the polarizable cage and (iii) the electrostatic repulsions between the ion and the cage electrons. The stabilization energy sequences within a series of alkali metal ions or halide ions can be explained by using the 'hard'and 'soft' concept[148]. For Si_{60} and Ge_{60} the cations are strongly stabilizing whereas the anions are destabilizing. In the C_{60} cages all ionic species are stabilizing except for Rb^+, Br^- and I^-, which have ionic radii close to the radius of the C_{60} cage so that steric repulsions destabilize the endohedral complex[148]. Note the significant stabilizing effect of an encapsulated F^- in the C_{60} cage (Table 12). Geerlings and coworkers developed an analytical expression (within the framework of the density functional theory) for the endohedral complexation energy of various ions to Y_{60} cages, and the interested reader is referred to their paper for the detailed derivation of this expression[148].

$$\Delta E = E_{(X@Y_{60})} - (E_X + E_{Y_{60}}) \qquad (43)$$

Bühl and Thiel used the chemical shifts of a ^3He atom (see above)[132,133] encaged within a fullerene for evaluating the degree of aromaticity of carbon fullerenes. A similar study is required for evaluating the degree of aromaticity of Si_{60}.

In an attempt to construct a new polymorph of silicon carbide, Osawa and coworkers employed the semi-empirical PM3 method and *ab initio* methods to study a hypothetical double fullerene C_{60}@Si_{60} (**117**, Figure 17) in which a C_{60} fullerene is encapsulated within a Si_{60} cage[147]. However, geometry optimization shows the generation of 60 new Si–C bonds connecting the inner C_{60} shell and the outer Si_{60} shell, forming a new

1. Theoretical aspects and quantum mechanical calculations 75

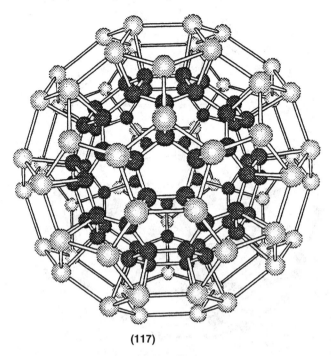

(117)

FIGURE 17. PM3 optimized structure of $C_{60}@Si_{60}$ (117). Reprinted from Reference 147, by courtesy of Marcel Dekker, Inc.

molecule rather than an encapsulated fullerene pair. The C_{60} and Si_{60} radii in the optimized $C_{60}@Si_{60}$ structure have expanded relative to the isolated fullerenes from 3.22 Å to 3.73 Å for C_{60} and from 5.24 Å to 5.75 Å for Si_{60}, forming an inter-shell distance of 2.02 Å. Consequently, the Si—Si and C—C bond lengths became considerably longer as is expected of a sp^2 to sp^3 rehybridization; i.e. the Si—Si $r(6-6)$ and $r(5-6)$ bond distances in 117 are 2.309 and 2.327 Å, respectively, relative to 1.879 and 2.247Å, respectively, in Si_{60} (PM3), whilst the C—C $r(6-6)$ and $r(5-6)$ bond distances are 1.492 and 1.513 Å in 117 relative to 1.384 and 1.457 Å in C_{60}. The calculated bond-order and electron density map also clearly support the formation of new Si—C σ-bonds. As expected, the electron distribution along the Si—C bonds is shifted towards the more electronegative carbon atoms. Hence 117 is electron-deficient on the outer surface and electron-rich in the inner shell. 117 is still a hypothetical molecule, but the authors believe that it may be formed by the reaction of silicon with C_{60} by CVD techniques[147].

C. Si_{70}

An AM1 comparison of the geometry and electronic structure of D_{5h} Si_{70} (118, Figure 18) with that of C_{70}, Si_{60} and C_{60} was reported by Piqueras and coworkers[151]. The calculated bond lengths and bond orders, which are given in Table 13, suggest that the silicon clusters have a more localized structure than the carbon clusters, and that Si_{70}

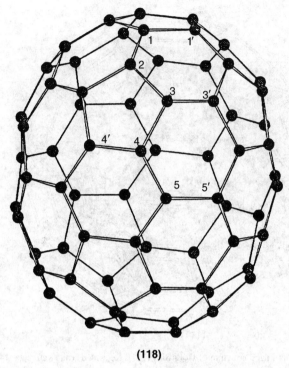

(118)

FIGURE 18. AM1 optimized structure of Si_{70} (**118**). Reprinted from Reference 151, Copyright (1995) by permission of Elsevier Science-NL, Sara Burgerhartstraat 25, 1055 KV Amsterdam, The Netherlands

TABLE 13. AM1 bond lengths and bond orders for Si_{70} and C_{70}[a]

Bond[b]	Bond length (Å)		Bond order	
	Si_{70}	C_{70}	Si_{70}	C_{70}
$X_{(1)}-X'_{(1)}$	2.297	1.464	1.079	1.105
$X_{(1)}-X_{(2)}$	2.094	1.387	1.677	1.480
$X_{(2)}-X_{(3)}$	2.296	1.460	1.079	1.109
$X_{(3)}-X'_{(3)}$	2.067	1.375	1.715	1.526
$X_{(3)}-X_{(4)}$	2.304	1.467	1.049	1.065
$X_{(4)}-X'_{(4)}$	2.218	1.434	1.374	1.302
$X_{(4)}-X_{(5)}$	2.196	1.414	1.369	1.306
$X_{(5)}-X'_{(5)}$	2.296	1.465	1.078	1.094

[a] From Reference 151.
[b] For atom numbering see Figure 18.

is more delocalized than Si_{60}. A similar trend was found for C_{70} vs C_{60} based on their NICS values[132]. Among the four studied clusters, Si_{70} has the lowest ionization potential (7.63 eV), highest electron affinity (3.61 eV) and smallest HOMO–LUMO gap of 4.02 eV. The Koopmans' ionization potential of Si_{70} is by 0.37 eV lower than for Si_{60} and the authors argue that this is due to the more efficient electron delocalization in Si_{70} compared to Si_{60}[151]. The HOMO–LUMO gap for Si_{70} (4.02 eV) is smaller than for Si_{60} (4.62 eV), suggesting that the electrons in Si_{70} are more polarizable than in Si_{60}.

Si_{70} and higher silicon fullerenes, e.g. Si_{78} and Si_{84}, are more stable thermodynamically than Si_{60} because they have a larger number of hexagons and therefore they may be even better candidates for synthesis than Si_{60}[61,137].

VII. SILICON COMPOUNDS WITH POTENTIAL 'THREE-DIMENSIONAL AROMATICITY'

The concept of 'aromaticity' is not restricted only to planar conjugated systems. The '$4n + 2$ interstitial electron rule' extends the concept of aromaticity to three-dimensional delocalized systems[13a,91,152a]. Such systems have marked thermodynamic stability and favour substitution over addition reactions, a property characteristic of planar delocalized aromatic systems. For instance, ferrocene undergoes electrophilic substitution reactions in analogy to benzene[153]. Pyramidal molecules with three-dimensional aromaticity can be divided conceptually into caps (X) and rings (see **119**). The π-electrons of the ring, and the electrons available from the cap that form the cap-ring binding, constitute the interstitial electrons. A system with $4n + 2$ delocalized interstitial electrons is considered to have aromatic properties. Several examples of charged compounds which have potential 'three-dimensional aromaticity' have already been discussed in this review, i.e. **59**, **63a**, and **68–73**.

(**119**)

In this section we will discuss three families of neutral silicon compounds with potential 'three-dimensional aromaticity'.

A. Silicocenes

The synthesis of decamethylsilicocene (**120**), which is regarded as the first stable silylene known, was reported by Jutzi and coworkers in 1986[154a]. This remarkable molecule can be formally described as being formed from 3 units, 2 units of cyclopentadienyl anions (Cp*) and Si^{2+}, which share 14 electrons (12 from the cyclopentadienyl rings and 2 from the Si). In terms of the cap-ring description, 6 interstitial electrons hold each of the rings to the 'sandwiched' silicon atom and the 2 remaining electrons reside in a lone-pair orbital on Si. The molecular orbital picture of **120**, which is discussed below, is analogous to that in ferrocene[152b], and therefore silicocene can be regarded as an aromatic molecule.

Crystal structure analysis of **120** indicated the presence of two structures, **120a** and **120b**, in a ratio of 1 : 2, respectively. In **120a**, the rings are parallel while in **120b** they form a tilt angle θ of 167.4°[154]. According to electron diffraction studies, also in the gas phase **120** has a bent metallocene-type structure with a tilt angle of 169.6°[154b]. Thus, the tilting of the rings in **120b** is not due to crystal packing forces.

R = CH₃ (**120a**) D_{5d} (**120b**) C_s
R = H   (**121a**) D_{5d} (**121b**) C_s (**121c**) C_{2v}

These findings have stimulated computational studies of several possible isomers of the parent silicocene **121**, i.e. **121a–c** and their possible rotamers[154b,155] and the η^1,η^5-isomer (the η^1,η^1-isomer was considered, but was not studied due to computational problems)[154b]. For geometry optimizations these studies used the HF/STO-3G[154,155], HF/STO-3G*[154b,155], HF/DZ+d (in which one set of d orbitals is added only on Si) and HF/DZP[155] (with two sets of d orbitals on the Si and one set of d orbitals on C) methods. Electron correlation was introduced only in single-point energy calculations, using the MP2, CISD and CISD+Q methods[155]. The results of the calculations showed that: (a) With all the methods used the most stable structures are the bent C_s or C_{2v} conformers, **121b** or **121c** (or their C_2 rotamers). All these structures are quite close in energy and their relative energies depend on the level of calculation. **121a** is found to be less stable, i.e. the calculated energy difference between **121a** and **121b** at HF/DZ+d, HF/DZP, MP2/DZ+d and CISD+Q/DZ+d are 15.4, 11.2, 5.7 and 13.8 kcal mol⁻¹, respectively[155]. These findings are in agreement with the experimental electron diffraction results according to which decamethylsilicocene (**120**) has a bent structure[154b]. The η^1,η^5-isomer is much higher in energy; e.g. it is by 52.8 kcal mol⁻¹ (at HF/STO-3G*) higher in energy than **121b**[154b]. (b) Addition of polarization functions to the basis set stabilizes the more symmetric structures, and thus decreases the energy difference between the D_{5d} isomer and the bent isomers. This is due to the participation of the low-lying d-orbitals of silicon in the bonding, a participation which is larger in more symmetric structures[155]. (c) The D_{5d} isomer (**121a**) is not a minimum on the SCF PES, and geometry optimization using correlated levels of theory have not been yet reported.

The most general conclusion from the calculations is that the PES of the parent silicocene **121** is flat with respect to the motion of the cyclopentadienyl rings, and the authors' prediction is that calculations at higher levels of theory will find that **121a**, **121b** and **121c** have similar energies[155]. This conclusion may explain the presence of two structures, **120a** and **120b**, in the solid state of decamethylsilicocene[154].

The calculated (HF/DZP) geometry[155] of **121b** and the X-ray structure of **120b**, which are shown in Figure 19, differ significantly in several parameters: (1) The Si–x2 distance is by as much as 0.4 Å longer than the experimental value. (2) the calculated and experimental C–C bond lengths in one of the cyclopentadienyl rings (x1 in Figure 19) differ by as much as 0.07 Å. Furthermore, the experimental C–C bond lengths in both cyclopentadienyl rings show little bond alternation with a typical average aromatic bond length of *ca* 1.4 Å. In contrast, the calculated C–C bond lengths show equalization in one of the cyclopentadienyl rings (x2) but significant localization in the second ring (x1). (3) The calculated tilt angle θ in **121b** of 152.4° is considerably smaller than the experimental value of 167.4° found in **120b**. This difference is probably due to the steric repulsions associated with the methyl groups, which increase the tilt angle relative to the parent silicocene[154b,155].

How can the bonding in silicocenes be described? The seven highest occupied molecular orbitals of **120a**, in which the 14 relevant electrons reside, are drawn schematically in Figure 20. Most of the bonding between the Si atom and the cyclopentadienyl rings is associated with the two lowest occupied MOs of a_{1g} and a_{2u} symmetry, which are the bonding combinations of the two cyclopentadienyl $2p_\pi$ MOs and the 3s-orbital ($1a_{1g}$) and one of the 3p-orbitals (a_{2u}) of the Si atom. Additional bonding interactions are associated with the e_{1u}-orbitals, which are the bonding linear combinations between the silicon 3p-orbitals and the $2p_\pi$-MOs of the cyclopentadienyl ligands of e_{1u} symmetry. The two highest occupied MOs of e_{1g} symmetry which are localized on the cyclopentadienyl ligands, and the $2a_{1g}$ orbital which is an antibonding combination of the Si 3s-orbital and the $2p_\pi$ MOs on the ligands, contribute little, if at all, to the bonding in the silicocene. This bonding picture, with several bonding orbitals which bind the 3 fragments together, topped by essentially non-bonding orbitals, is similar to the bonding picture of ferrocene[152].

The PE spectrum of **120b** reveals 3 major vertical ionization bands at 6.7, 7.5 and 8.1 eV (very similar ionization energies were calculated at HF/STO-3G* for **121b**, but this good agreement is probably fortuitous[154b]). The first and third ionization potentials were assigned to ionization from the weakly bonding e_{1g} and e_{1u} π-MOs and the ionization energy at 7.5 eV was assigned to ionization from the $2a_{1g}$ orbital[154b].

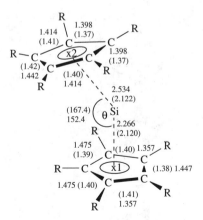

FIGURE 19. The HF/DZP[155] optimized structure of **121b** and the experimental X-ray structure of **120b** (in parentheses)[154]. Bond lengths in Å, bond angles in degrees. x1 and x2 specify the centres of the cyclopentadienyl rings

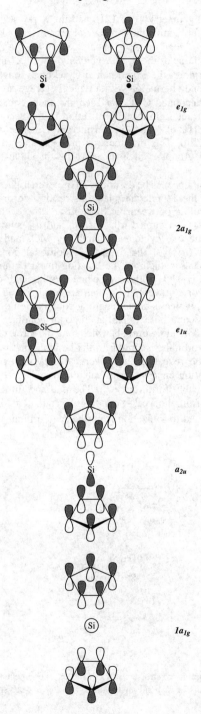

FIGURE 20. Schematic drawing of the seven highest occupied MOs of the parent silicocene, **120a**

B. Persilaferrocene, $(H_5Si_5)_2Fe$

In contrast to the well known ferrocene **122**[153], persilaferrocene $(H_5Si_5)_2Fe$ (**123**) is not known experimentally. However, **123** was studied computationally at the HF and MP2 levels of theory, using DZ and DZP quality basis sets (with ECP for Si and Fe). All-electron calculations were carried out only at the HF level[156].

```
          M
       M-/ \-M
        \ x /
       M-|-M
         |
         |
         Fe
         |
         |
       M-|-M
        / x \
       M-\ /-M
          M
```

(**122**) M = CH
(**123**) M = SiH

The HF calculations show that persilaferrocene adopts the highly symmetric D_{5d} structure in analogy to ferrocene. In **123**, $r(Fe-x)$ (where x is a point at the centre of the cyclopentadienyl rings) ranges from 2.060 to 2.175 Å, depending on the basis set. In **122**, $r(Fe-x)$ is significantly shorter and varies from 1.869 to 1.898 Å. Comparison with experimental data for **122**, where $r(Fe-x)$ is 1.65 Å, shows that the calculated $r(Fe-x)$ distance is significantly overestimated by HF calculations and this conclusion probably holds also for **123**. The Si–Si bond lengths in **123** are all equal and they vary between 2.184 and 2.242 Å as a function of the basis set. This indicates the existence of some Si–Si double-bond character, which is also supported by the calculated Si–Si bond-order of 1.266. The planarity of the Si_5H_5 ligands is noteworthy, since the free cyclic Si_5H_5 radical has a non-planar structure. The calculated vibrational frequencies and relative energies suggest, however, that the D_{5d} structure of **123** lies in a rather shallow minimum and that its transformation to structures of lower symmetry (e.g. C_{2v} or C_{2h}), which involve non-planar Si_5H_5 rings, should therefore be facile[156].

The binding energy of the Fe atom to the two cyclopentadienyl rings is smaller in **123** than in **122**, i.e. 113.6 vs 144.1 kcal mol^{-1}, respectively, at MP2/DZP//HF/DZP. The vibration frequencies for the Fe-ring stretchings, of 103 cm^{-1} for **123** and 236 cm^{-1} for **122**, also indicate that the interaction of the Fe atom with the Si_5H_5 rings is smaller than with the C_5H_5 rings.

The positive charge on the Fe in $(H_5Si_5)_2Fe$ is about half of that in **122**, and the ring Si atoms of **123** are almost neutral. On this basis, it was predicted that the electrophilic reactivity of $(H_5Si_5)_2Fe$ towards polar reagents should be lower than that of $(H_5C_5)_2Fe$, where the cyclopentadienyl rings carry a significant negative charge[156].

C. Aromaticity in *closo*-Silaboranes, $B_3Si_2H_3$

Two bonding alternatives are possible for the *closo*-boranes 1,5-$X_2B_3H_3$ (X = SiH, CH, N, P, BH$^-$), i.e. the classical representation **124** and the non-classical representation

125. Theoretical studies which used Bader's 'Atoms in Molecules' bonding analysis[109] or Mulliken population analysis suggested that these members of the *closo*-borane family favour the classical Lewis representation **124**[157]. In contrast, a more recent study by Schleyer and coworkers[158] concluded that these *closo*-boranes are best represented by the non-classical representation **125** and that these compounds have 'three-dimensional aromaticity'. These conclusions were based on their calculated B−B and B−X distances and the corresponding Wiberg's bond indices (WBI) and natural atomic bond orders (NAO), as well as on their calculated aromatic stabilization energies (ASE) and magnetic properties (susceptibility exaltation and NICS values). Table 14 lists some of these values for the 1,5-$X_2B_2H_3$ *closo*-boranes, with X = SiH, CH and BH$^-$ (for X = P and N see Reference 158), which show clearly their non-classical aromatic character. Natural localized molecular orbital analysis reveals significant electron delocalization on each deltahedral face of the 1,5-$X_2B_3H_3$ *closo*-boranes, also favouring representation **125** over **124**.

(**124**)   X=SiH, CH, N, P, BH$^-$ (**125**)

The degree of the 'three-dimensional aromaticity' and the preference for the non-classical structure **125** over the classical structure **124** follow the electronegativity of X in each row, i.e. the degree of aromaticity follows the order: X = BH$^-$ > SiH > P

TABLE 14. Bonding, energetic and magnetic properties of *closo*-1,5-$X_2B_3H_3$ (**125**, X = SiH, CH and BH$^-$)[a]

Property	X		
	SiH	CH	BH$^-$
$r(B-X)$[b]	1.941	1.554	1.684
$r(B-B)$[b]	2.078	1.844	1.811
WBI(B−X)[c]	0.92	0.933	0.93
WBI(B−B)[c]	0.44	0.202	0.45
NAO(B−X)[d]	0.87	0.89	0.88
NAO(B−B)[d]	0.44	0.35	0.49
ASE[e]	−29.2	−19.8	−34.8
Λ[f]	−36.6	−6.9	−46.1
NICS[g]	−22.4	−17.1	−28.1

[a] From Reference 158.
[b] At MP2/6-31+G*, in Å.
[c] Wiberg's bond index.
[d] Natural atomic bond orders.
[e] In kcal mol^{-1} at MP2/6-31G*+ZPE.
[f] Diamagnetic susceptibility exaltation, calculated at CSGT-HF/6-311+G**//MP2/6-31+G*.
[g] At the cage centre, using GIAO-HF/6-311+G**//MP2/6-31+G*.

> CH > N, being the highest for **125**, X = BH$^-$[158]. According to the calculated ASE, *closo*-disilaborane (**125**, X = SiH) has 84% of the aromaticity of **125**, X = BH$^-$ but it is by *ca* 50% more aromatic than the carbon analogue **125**, X = CH.

VIII. POTENTIAL SILAANTIAROMATIC COMPOUNDS

A. Silacyclobutadienes

Several sila-analogues of cyclobutadiene, the prototype of Hückel 4*n* antiaromatic systems, were studied computationally. The available studies on mono- and disilacyclobutadiene are quite old and therefore use relatively low levels of theory, and their results should therefore be treated with caution.

1. Monosilacyclobutadiene

Monosilacyclobutadiene **126** was first studied by Gordon at the HF/6-31G*//HF/3-21G level of theory[159]. More extensive calculations for the lowest singlet and triplet states of **126** were later carried out by Colvin and Schaefer[160] and a detailed HF/6-31G*//HF/6-31G* study of the C$_3$SiH$_4$ PES was later reported by Gordon and coworkers[161].

$$\begin{array}{c} H \diagdown \quad \diagup H \\ Si - C \\ \| \quad \| \\ \diagup C - C \diagdown \\ H \quad \quad H \end{array}$$

(**126**)

The optimized geometries (at HF/DZ+d and HF/6-31G*) of the singlet and triplet states of **126** are given in Figures 21a and 21b, respectively. Vibrational frequencies of **126** were also reported[160]. The singlet ground state of **126** is predicted to possess a planar geometry with alternating single and double bonds (Figure 21a). At HF/DZ+d, the C—Si bonds are calculated to be 1.866 and 1.688 Å (1.884 and 1.685 Å, respectively, at HF/6-31G*), close to those in methylsilane and silaethylene, respectively. This indicates little π-conjugation, as suggested also by the vibrational frequencies[160]. The structure of the $^3A'$ triplet state of **126** (Figure 21b) is very different. The shortest C—C and C—Si bonds are geminal, and not on opposite sides of the ring as in singlet **126**. Furthermore, the Si—C bond lengths are comparable to those of regular Si—C single bonds, while the C—C bond lengths are closer to those of C=C double bonds. Although silicon substitution rules out a truly square structure, the geometry of triplet **126** is much more 'square-like' than is the singlet monosilacyclobutadiene[160].

The singlet–triplet energy difference is, as expected, strongly dependent on the level of the calculations. At the CISD level the triplet is found to lie only 5 kcal mol^{-1} above the singlet[160]. This singlet–triplet splitting is about 16 kcal mol^{-1} smaller than the calculated splitting of cyclobutadiene at the same level of theory[160].

The 'antiaromatic' destabilization of singlet **126** was estimated (using equation 44) to be 49.1 kcal mol^{-1} (HF/DZ[160]; 53.5 kcal mol^{-1} at HF/6-31G*//HF/3-21G[159]), by *ca* 17 kcal mol^{-1} smaller than the destabilization calculated for cyclobutadiene at

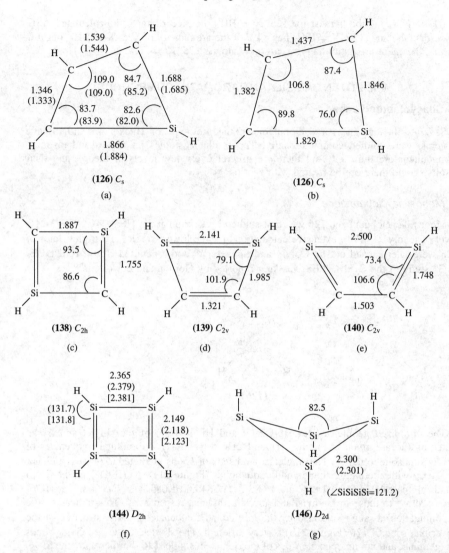

FIGURE 21. (a) Optimized geometries of the lowest $^1A'$ singlet state of monosilacyclobutadiene, **126**; (b) optimized geometries of the lowest $^3A'$ triplet state of **126**, both at HF/DZ+d[160] (HF/6-31G*[161] values in parentheses); (c), (d), (e) optimized geometries of disilacyclobutadienes **138–140** at HF/3-21G[162]; (f) and (g) optimized geometrics of tetrasilacyclobutadienes **144** (D_{2h}) and **146** (D_{2d}), respectively, at MP2/6-31G**[52], HF/DZP[165a] (in round parentheses) and at HF/6-31G*[164] (in square brackets). Bond lengths in Å, bond angles in degrees

a comparable level of theory[160]. However, note that equation 44 cannot distinguish destabilizing conjugation effects from angle strain or 1,3-interactions[160].

$$126 + 3CH_4 + SiH_4 \longrightarrow CH_3SiH_3 + CH_3CH_3 + H_2C=CH_2 + H_2C=SiH_2 \quad (44)$$

1. Theoretical aspects and quantum mechanical calculations

The relative energies at HF/6-31G*//HF/6-31G* of various isomers of monosilacyclobutadiene are given in Figure 22[161]. The global minimum on the C_3SiH_4 PES is silylene **127**, which is stablized by the interaction of the vacant p-orbital on silicon with the C=C π-bond to form a 2π-aromatic system. Four other silylenes **128–131** follow **127**. These silylenes are all lower in energy than the isomeric structures which possess a C=Si double bond or strained rings, such as **132–137**. This stability order contrasts with

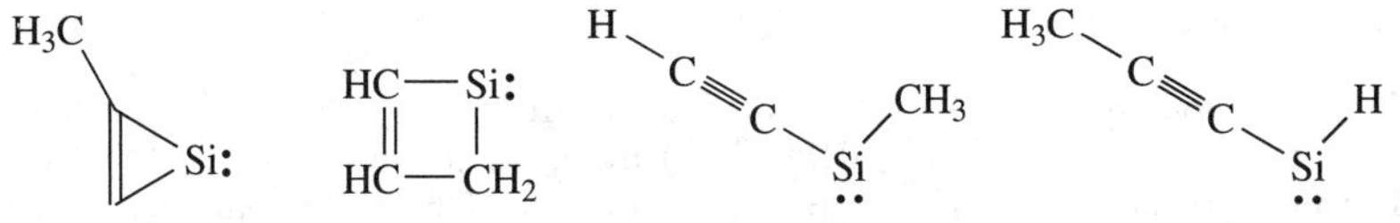

	(127)	(128)	(129)	(130)
RE:	0.0	10.2	18.0	17.8

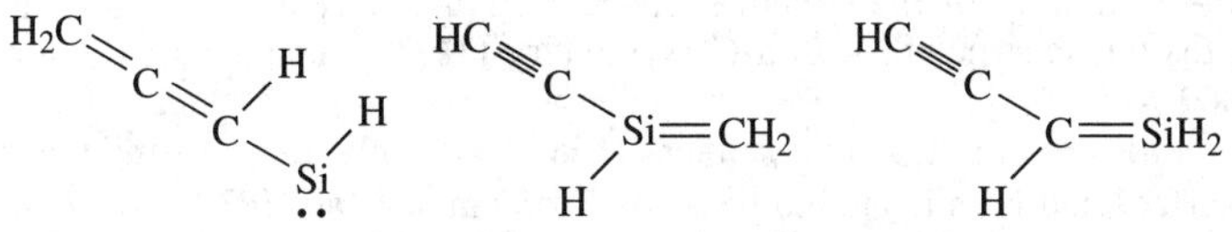

	(131)	(132)	(133)
RE:	26.3	28.2	35.9

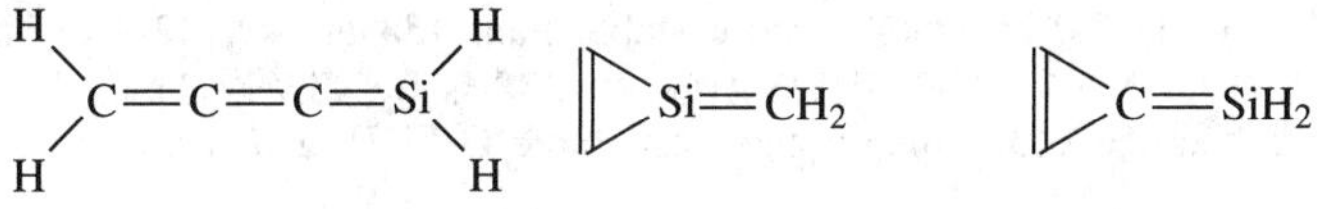

	(134)	(135)	(136)
RE:	36.9	43.6	59.2

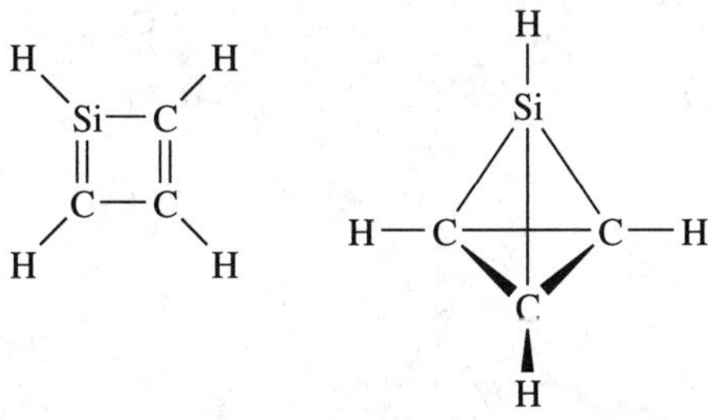

	(126)	(137)
RE:	59.5	91.5

FIGURE 22. Relative energies (RE, in kcal mol^{-1}) calculated at HF/6-31G*//HF/6-31G* of C_3SiH_4 isomers. Data from Reference 161

that in the isoelectronic C_4H_4 system, where structures with multiple bonds are substantially more stable than the corresponding carbenes. Monosilacyclobutadiene (**126**) is by 59.5 kcal mol^{-1} less stable than the silylene, **127**, but it is by 32.0 kcal mol^{-1} more stable than the isomeric, highly strained silatetrahedrane **137**[161]. Due to the high degree of unsaturation or strain in many of the C_3SiH_4 isomers, it is clear that a correlated level of theory is required to obtain more reliable information on the relative energies of these isomers.

2. Disilacyclobutadienes

Holmes, Gordon and coworkers[162] studied at the HF/3-21G//HF/3-21G level (and for some isomers also at MP3/6-31G*//HF/3-21G) the three possible disilacyclobutadienes **138**, **139** and **140**, along with other $C_2Si_2H_4$ isomers. The HF/3-21G optimized structures (within the indicated symmetry constraints) of **138**, **139** and **140** are shown in Figures 21c, 21d and 21e, respectively. All three structures reflect bond localization, as expected for antiaromatic compounds. Of special interest is the Si—Si bond length in **140** of 2.50 Å, which is considerably longer even than a typical Si—Si single bond length of 2.35 Å, indicating the instability of the structure. Indeed, upon full geometry optimization, **140** decays to a disilylene structure with a C_1 symmetry (**141**), which has a very long Si—Si bond of 2.860 Å.

Of the three disilacyclobutadiene isomers (**138–140**), only **138** is a minimum on the HF/3-21G surface and is estimated to be by 49.7 kcal mol^{-1} (at MP3/6-31G*//HF/3-21G) higher in energy than the isomeric aromatic silylene **142**, which is the most stable $C_2Si_2H_4$ isomer (in analogy to **127**), **138** possesses, in analogy to cyclobutadiene, a high diradical character and it therefore cannot be adequately described at the RHF level of theory. The inclusion of the contribution of electron correlation using the MCSCF level or UHF-NO CI methods stabilizes **138**, but **142** is still by ca 20 kcal mol^{-1} more stable[162]. The disilylene **141** is by 7.0 kcal mol^{-1} more stable than **138** and by 42.7 kcal mol^{-1} less stable than **142** (at MP3/6-31G*//HF/3-21G)[162]. **138** is by 37.7 kcal mol^{-1} more stable than disilatetrahedrane **143**, which is the least stable $C_2Si_2H_4$ isomer.

(**138**) C_{2h} (**139**) C_{2v} (**140**) C_{2v} (**141**) C_1 (**142**) C_s

(**143**) C_{2v}

3. Tetrasilacyclobutadiene

The PES of tetrasilacyclobutadiene, **144**, the full silicon analogue of cyclobutadiene, was studied extensively using relatively high levels of theory which include the contributions of polarization functions and of electron correlation[52,163−165]. The calculations revealed a very complex Si_4H_4 PES.

What is the structure of Si_4H_4? Is it a delocalized square or a distorted planar rectangle, or does it possess a three-dimensional structure? The propensity of a planar square Si_4H_4 to distort to the rectangular D_{2h} structure **144** was studied by a valence bond approach (see Section III.F.1). Shaik and coworkers[15b,c] found that the π-distortion energies of square Si_4H_4 are approximately the same as for Si_6H_6, but the resistance of the σ-framework to distortion is smaller for the square than for the hexagon, due to the smaller number of bonds. The total outcome is that Si_4H_4 is more distorted (resulting in alternating bond lengths) than Si_6H_6 and that this distortion is dominated by the σ-framework.

The square and the rectangular **144** (see Figure 21f) are both not local minima on the Si_4H_4 PES at all levels of calculation[52,164,165]. Upon geometry optimization the planar D_{2h} structure (**144**) collapses to a C_{2h} structure, **145**, which lies 1.0 kcal mol^{-1} below **144** (CISD/DZP//HF/DZP[165a]). However, **145** is also not a minimum on the HF PES, and following its a_u imaginary frequency it collapses to a puckered D_{2d} minimum (**146**) which lies 17.4 kcal mol^{-1} below **144** (at CISD/DZP//HF/DZP[165a]; 34.8 kcal mol^{-1} at MP2/6-31G**//MP2/6-31G**[52], 4.7 kcal mol^{-1} at HF/DZP[165a], reflecting a very large electron correlation effect). **146** is thus the most stable $(SiH)_4$ isomer[165a]. The high stability of **146** is due to a large extent to its reduced strain, i.e. the calculated MP2/6-31G** strain energies of **144** and **146** are 38.7 and 8.3 kcal mol^{-1}, respectively[52]. **146** is by 4.5 kcal mol^{-1} less stable than the [1.1.0] bicyclic tetrasilane **147** (MC8SDQCI/TZP)[163b]. The most stable isomer on the Si_4H_4 PES is **148**, which is by 7.5 kcal mol^{-1} more stable than **147** [CISD(Q)/DZP//HF/DZP][165b].

(**144**) D_{2h}

(**145**) C_{2h}

(**146**) D_{2d}

(**147**) C_{2v}

(**148**) C_s

(**149**) T_d

The least stable isomer on the Si_4H_4 PES is tetrasilatetrahedrane (**149**), which is a local minimum at the HF level but has two imaginary vibrational frequencies at the MP2 level[52]. **149** is by only 7.9 kcal mol^{-1} less stable than **144** but by 33.4 kcal mol^{-1} less stable than **146**, at CI/6-31G*//HF/6-31G*[164] (40.9 and 21.9 kcal mol^{-1} at MP2/6-31G**//MP2/6-31G**[52] and CISD/DZP//HF/DZP[165a], respectively). Tetrasilatetrahedrane is also unstable kinetically, and it isomerizes to **146** with a negligible barrier of only 0.6 kcal mol^{-1} [HF/6-31G(2d,p)+ZPE[164]]. However, despite the high predicted reactivity of the parent tetrasilatetrahedrane, a stable derivative of tetrasilatetrahedrane, substituted with 4 bulky t-Bu_3Si groups, was recently synthesized and identified by X-ray crystallography by Wiberg and coworkers[166].

The geometries of **144** and **146** are shown in Figures 21f and 21g, respectively. The planar tetrasilacyclobutadiene (**144**) has a rectangular geometry with typical Si=Si and Si—Si bond lengths of 2.149 and 2.365 Å (MP2/6-31G**[52], 2.123 and 2.381 Å at HF/6-31G*[164], respectively). On the other hand, **146** has 4 equivalent Si—Si bonds of 2.300 Å (MP2/6-31G*; 2.301 Å at HF/DZP[165a] and 2.298 Å at HF/6-31G*[164]).

B. Charged Silaantiaromatic Compounds

1. The silirenyl anion, $C_2SiH_3^-$

The antiaromaticity of three-membered ring anions, **150**, at their ground and excited states was determined by Malar[167] by the bond-order definition of aromaticity suggested by Jug, which relates the lowest bond order in the ring with aromatic ring currents[168]. According to this method the aromaticity index of a ring is defined as being equal to the smallest bond order in the ring[168].

$$\underset{H \quad\quad H}{\overset{X}{C=C}}$$

(**150**) X = SiH$^-$, CH$^-$

The geometries of the ground states and of the first excited singlet and triplet states of **150** with X = SiH$^-$, CH$^-$, NH, O, PH and S were calculated by the CI version of the semi-empirical SINDO1 method[167]. We will discuss here only the silirenyl anion (**150**, X = SiH$^-$) and compare the properties of this anion to those of the analogous cyclopropenyl anion (**150**, X = CH$^-$). The calculated optimized geometries and bond orders for **150**, X = SiH$^-$ and CH$^-$ are given in Table 15.

In the ground state of the silirenyl anion (**150**, X = SiH$^-$) the C=C and C—Si bond lengths are 1.325 and 1.867 Å, respectively, typical values of regular C=C and C—Si bonds. The Si centre in **150**, X = SiH$^-$ is highly pyramidal. The geometries of the excited triplet and singlet states do not differ significantly from that of the ground state, except for the Si centre which becomes less pyramidal upon excitation (Table 15).

The calculated vertical excitation energies to the T_1 and S_1 states are 2.3 and 3.0 eV, respectively (using SINDO1). The C—X bond order in the ground state of **150**, X = SiH$^-$ is 1.21, a typical value for strong single bonds[167] and in the range of non-aromatic systems[168], leading to the conclusion that the ground state of the silirenyl anion is non-aromatic[167]. Upon excitation, the C—Si bond-order is reduced to 0.99 and 1.03 for the S_1 and T_1 states, respectively, leading to the conclusion that the excited states of **150**, X = SiH$^-$ are highly antiaromatic. Generation of the singlet and triplet excited states of

TABLE 15. Calculated geometries and bond orders[a] of the silirenyl anion (**150**, X = SiH$^-$) and the cyclopropenyl anion (**150**, X = CH$^-$) in their ground and first excited singlet and triplet states[a]

State	C–C	C–X	C–H	X–H	CCX	CCH	CXH	HCCX	CCXH
				X = SiH$^-$					
	Geometry								
S_0	1.325	1.867	1.058	1.474	69.2	134.0	103.8	175.0	96.0
S_1	1.336	1.875	1.060	1.484	69.1	131.0	155.4	178.3	146.3
T_1	1.333	1.856	1.067	1.477	68.9	132.5	158.5	182.5	167.6
	Bond order								
S_0	2.05	1.21[b,c]							
S_1	1.96	0.99[b,d]							
T_1	1.95	1.03[b,d]							
				X = CH$^-$					
	Geometry								
S_0	1.313	1.505	1.062	1.138	64.1	140.5	115.0	193.0	101.3
S_1	1.393	1.415	1.089	1.096	60.6	152.4	129.0	235.5	118.2
T_1	1.482	1.427	1.079	1.088	58.7	132.0	126.0	232.0	116.2
	Bond order								
S_0	2.10	1.14[b,e]							
S_1	1.39[b,f]	1.42							
T_1	1.43[b,f]	1.46							

[a] From Reference 167, using the CI/SINDO1 method. Bond lengths in Å, bond angles in degrees.
[b] Aromaticity index determined by the lowest ring bond order.
[c] Non-aromatic.
[d] Highly antiaromatic.
[e] Moderately antiaromatic.
[f] Moderately aromatic.

the silirenyl anion involves a $\sigma(a_1) \rightarrow \pi^*(a_2)$ transition. Both the σ and π^* orbitals are bonding along the C–C bond and antibonding along the C–X bond, but the antibonding along the C–X bonds is more pronounced in the π^* orbital. Thus, excitation weakens the C–X bonds and the S_1 and T_1 states of **150**, X = SiH$^-$ become antiaromatic. In contrast to **150**, X = SiH$^-$, the geometry of **150**, X = CH$^-$ varies from a localized structure in the S_0 ground state (moderately antiaromatic according to its bond order aromaticity index), to a delocalized structure in the S_1 and T_1 states, which are moderately aromatic (Table 15). Excitation of **150**, X = CH$^-$ involves a transition from the $\sigma(a_1)$ orbital to a $\pi^*(b_1)$ orbital. The $\sigma(a_1)$ orbital of **150**, X = CH$^-$ is bonding along the C–C bond and antibonding along the C–X bonds whilst the $\pi^*(b_1)$ orbital is antibonding along the C–C bond and non-bonding along the C–X bonds. Hence excitation weakens, and thus elongates, the C–C bond and strengthens, and thus shortens, the C–X bonds (Table 15)[167].

2. Silacyclopentadienyl cations

a. $C_4H_4SiH^+$. The degree of the antiaromaticity of singlet $C_4H_4SiH^+$, **151** (C_{2v}), a 4π-electron system, in comparison to the aromaticity/antiaromaticity of other five-membered C_4H_4X ring systems was estimated by Schleyer and coworkers using several criteria:

structural (Julg parameter), energetic (ASE, according to equation 21, Section IV.D.2a) and magnetic [i.e. magnetic susceptibility anisotropy (χ_{anis}), diamagnetic susceptibility exaltation (Λ) and NICS values][26,31a]. These calculated properties for **151** and those of the analogous **152**, X = CH^+, CH^- and SiH^- are given in Table 16. A graphical presentation of the correlation between the various criteria for the aromaticity/antiaromaticity of five-membered C_4H_4X ring systems is shown in Figure 1 (Section II.D).

(151)   (152)

$C_4H_4SiH^+$ (C_{2v}) is a minimum on the MP2/6-31G* PES, with a C=C bond length of 1.357 Å, resembling that in polyenes. On the other hand, the conjugated C–C single bond length of 1.518 Å is significantly longer than in polyenes (1.450 Å in *trans*-1,3-butadiene). The Si–C bond lengths of 1.820 Å are very close to those of regular Si–C single bonds. This localized structure results in a small Julg parameter of 0.346. The pronounced bond length alternation points to a significant antiaromatic character of **151**. This conclusion is strongly supported by the relatively high destabilization energy of -24.1 kcal mol^{-1} and by the magnetic properties, i.e. $\chi_{anis} = -0.8 \times 10^{-6}$ cm^3 mol^{-1}, $\Lambda = 13.2 \times 10^{-6}$ cm^3 mol^{-1} and NICS = 12.8 ppm of **151**, which are all characteristic of an antiaromatic compound[26,31e]. Table 16 shows, however, that the antiaromaticity of **151** is smaller than that of $C_5H_5^+$. In analogy, the aromaticity of c-$C_4H_4SiH^-$ (**152**, X = SiH^-) is smaller than that of its carbon analogue **152**, X = CH^-.

b. $Si_5H_5^+$. A detailed study at MP2/6-31G*[169] of the potentially 4π-antiaromatic $Si_5H_5^+$ reveals a complex PES, which is very different from that of $C_5H_5^+$[169,170]. Several important isomers on the PES of $Si_5H_5^+$, **153–160**, arranged according to their relative energies at MP2/6-31G*, are presented in Figure 23 (additional isomers are discussed in Reference 169).

TABLE 16. Calculated structural, energetic and magnetic properties of c-C_4H_4X, **152**[a]

X	Symmetry	A^b	ASE^c	χ_{anis}^d	Λ^d	NICS^e
SiH^+	C_{2v}	0.346	-24.1	-0.8	13.2	12.8
CH^+	C_{2v}	-0.084	-56.7	58.1	32.6	
SiH^-	C_s	0.926	13.8	-41.5	-7.7	-6.7
CH^-	D_{5h}	1.000	28.8	-45.8	-17.2	-14.3

[a]From Reference 31a; all species are minima at MP2/6-31G*.
[b]Julg's parameter.
[c]In kcal mol^{-1}.
[d]In 10^{-6} cm^3 mol^{-1}.
[e]In ppm, at GIAO-SCF/6-31+G*.

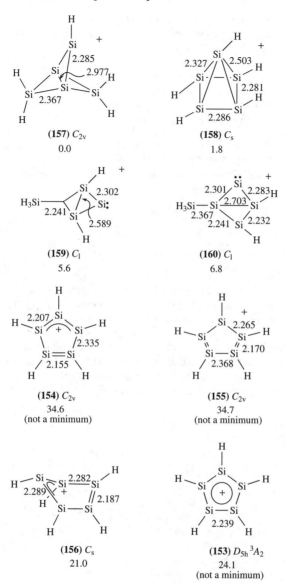

FIGURE 23. Optimized structures (MP2/6-31G*) and relative energies (at MP2/6-31G* + ZPE, in kcal mol^{-1}) of Si$_5$H$_5^+$ isomers. Bond lengths in Å, bond angles in degrees. Data from Reference 169

The singlet and triplet states of the antiaromatic D_{5h} Si$_5$H$_5^+$ (**153**) are not minima on the Si$_5$H$_5^+$ PES. When the D_{5h} symmetry of the singlet is released, it first collapses to the C_{2v} isomers **154** and **155** which are also not minima on the PES, and upon release of the C_{2v} constraints they further collapse to the C_s structure **156**, which is the only minimum on the Si$_5$H$_5^+$ PES. The pentasila-allylic-type cation **156**, which has a structural resemblance to that of the classical antiaromatic D_{5h} $Si_5H_5^+$, is by 21.0 kcal mol^{-1} less stable than the pentasila[1.1.1]propellanyl cation **157**, which is the most stable $Si_5H_5^+$ isomer. Several other $Si_5H_5^+$ isomers, i.e. the C_s pyramidal cation **158** (the pyramidal C_{4v} structure is not a minimum) and the bicyclic cation-silylenes **159** and **160**, lie only a few kcal mol^{-1} above **157** and are more stable than the monocyclic **156**.

The calculated Si—Si bond lengths in **156** are 2.187, 2.282 and 2.289 Å, all in the range intermediate between that of double (2.14 Å) and single (2.35 Å) Si—Si bond lengths. This indicates a certain degree of charge delocalization in the ring, which is also supported by the calculated charge distribution. **156** can be described as being essentially non-aromatic, as the energy change resulting from the presence of the Si=Si double bond is small, only 2 kcal mol^{-1} (equation 45). In contrast, the corresponding antiaromatic carbon analogue is strongly destabilized (35.3 kcal mol^{-1}, according to an equation analogous to equation 45)[169].

$$
\begin{array}{c}
\text{HSi} \overset{\text{H}}{\underset{\text{Si}}{\diagup}} \text{SiH} \\
| \quad + \quad | \\
\text{HSi} = \text{SiH}
\end{array}
+
\begin{array}{c}
\text{H}_2\text{Si} \overset{\text{H}_2}{\underset{\text{Si}}{\diagup}} \text{SiH}_2 \\
| \quad\quad\quad | \\
\text{H}_2\text{Si} - \text{SiH}_2
\end{array}
\longrightarrow
\begin{array}{c}
\text{HSi} \overset{\text{H}}{\underset{\text{Si}}{\diagup}} \text{SiH} \\
| \quad + \quad | \\
\text{H}_2\text{Si} - \text{SiH}_2
\end{array}
+
\begin{array}{c}
\text{H}_2\text{Si} \overset{\text{H}_2}{\underset{\text{Si}}{\diagup}} \text{SiH}_2 \\
| \quad\quad\quad | \\
\text{HSi} = \text{SiH}
\end{array}
\quad (45)
$$

The contrast with the $C_5H_5^+$ surface is interesting. The global minimum on the $C_5H_5^+$ PES (all energies below are at MP2/6-31G*//MP2/6-31G*)[169] is a $^3A_2$ state having D_{5h} symmetry (**161**). The D_{5h} singlet of **161** is not a minimum on the PES and it collapses to the planar singlet C_{2v} cyclopentadienyl cation **162**, which is by 11 kcal mol^{-1} less stable than **161**. The isomeric allylic cation **163** is not a minimum and has nearly the same energy as **162**. The C_{4v} pyramidal structure (**164**) is by 15.6 and 4.6 kcal mol^{-1} less stable than **161** and **162**, respectively. The most stable $C_5H_5^+$ isomer on the singlet surface is the vinylcyclopropeniun ion (**165**), which is by only 1.7 kcal mol^{-1} less stable than **161**.

(**161**) D_{5h}, $^3A_2$ (**162**) C_{2v} (**163**) C_{2v}

(**164**) C_{4v} (**165**) C_s

The relative high stabilities of the pyramidal, bridged and silylenic structures in the case of $Si_5H_5^+$ result from several electronic effects typical of silicon, which were already discussed in detail in this review, such as the reluctance to form multiple bonds, high divalent state stabilization, preference for bridged bonding etc. These effects lead in the case of $Si_5H_5^+$, as in many other examples discussed in this chapter, to stable three-dimensional structures which are not minima on the PES of the analogous carbon compounds.

IX. CONCLUSIONS

Silicon compounds exhibit aromaticity and antiaromaticity in analogy to the corresponding carbon compounds. However, in general, the degree of their aromaticity and antiaromaticity is smaller than that of the analogous carbon compounds.

The reluctance of silicon to form multiple bonds, its high divalent state stabilization, its preference for bridged bonding and other factors result in the existence of compounds which have no stable carbon analogues.

Many important achievements have been reached in the last 20 years in this field, yet the study of aromatic silicon compounds is still in its infancy and much remains to be learned and understood. Quantum chemistry is a particularly powerful tool for studying these intriguing molecules and for directing the difficult and challenging experimental research in this field. The study of aromaticity and antiaromaticity of silicon compounds presents also a major theoretical challenge, as many of their properties are difficult to calculate and the potential energy surfaces on which they reside are extremely complex. As stated throughout the review, many of the currently available theoretical studies are quite old and consequently have not been performed at the appropriate level of theory. New high-level calculations which will lead to more reliable predictions are highly desired.

X. ACKNOWLEDGEMENTS

The authors would like to thank P. v. R. Schleyer, H. Jiao, S. Nagase, R. Kobayashi, W. Thiel, M. Bühl and D. Tilley for preprints of their manuscripts prior to publication. M. Karni thanks her family for their patience during writing this review. The financial support of the German Federal Ministry of Science, Research, Technology and Education and the Minerva Foundation and of the US–Israel Binational Science Foundation (BSF) is gratefully acknowledged.

XI. LIST OF ABBREVIATIONS

Ad	adamantyl
Ar	aryl
ASE	aromatic stabilization energy
B3LYP	Becke's 3-parameter hybrid with Lee, Young and Parr's correlation functional
BSE	bond separation energy
CCSD(T)	coupled cluster with single and double excitations (followed by a perturbation treatment of triple excitations)
CI	configuration interaction
CISD	configuration interaction with single and double excitations
Cp	cyclopentadienyl
CSA	chemical shift anisotropy

CSGT	continuous set of gauge transformations
DFT	density functional theory
DZ	double zeta
DZ+d	double zeta + polarization functions on heavy atoms
DZP	double zeta + polarization functions on all atoms
ECP	electron core potential
GIAO	gauge-independent atomic orbitals
HF	Hartree–Fock
HOMO	highest occupied molecular orbital
IGLO	individual gauge for localized orbitals
IP	ionization potential
LUMO	lowest unoccupied molecular orbital
MBPT	many-body perturbation theory
MCn	multi-configuration; n=number of configurations
MCSCF	multi-configuration self-consistent field
MO	molecular orbital
MP	Møller–Plesset
n-MR	n-membered ring
NAO	natural atomic bond-order
NICS	nucleus-independent chemical shift
NPA	natural population analysis
PES	potential energy surface
QMRE	quantum molecular resonance energy
RE	resonance energy
SCF	self-consistent field
SINDO	symmetrically orthogonalized intermediate neglect of differential overlap
TSB	trisilabenzene
TZP	Triple zeta + polarization
UHF	unrestricted Hartree–Fock
UHF–NO	unrestricted Hartree–Fock –natural orbitals
UMP	unrestricted Møller–Plesset
WBI	Wiberg bond index
ZPE	zero-point vibrational energy

XII. REFERENCES

1. (a) G. Raabe and J. Michl, in *The Chemistry of Organic Silicon Compounds* (Eds. S. Patai and Z. Rappoport), Chap. 17, Wiley, New York, 1989, p. 1015.
   (b) Y. Apeloig, in *The Chemistry of Organic Silicon Compounds* (Eds. S. Patai and Z. Rappoport), Chap. 2, Wiley, New York, 1989, p. 103.
   (c) Y. Apeloig and M. Karni, *Chem. Rev.* (1998), in preparation.

1. Theoretical aspects and quantum mechanical calculations 95

2. Chapters in this book by:
   (a) T. Müller, W. Ziche and N. Auner, Chapter 16.
   (b) H. Sakurai, Chapter 15.
   (c) R. Okazaki and N. Tokitoh, Chapter 17.
   (d) G. Maier, Chapter 19.
   (e) R. West and P. Gaspar, Chapter 43.
3. (a) A. G. Brook and K. M. Baines, *Adv. Organomet. Chem.*, **25**, 1 (1986).
   (b) A. G. Brook and M. A. Brook, *Adv. Organomet. Chem.*, **39**, 71 (1996).
   (c) R. Okazaki and R. West, *Adv. Organomet. Chem.*, **39**, 232 (1996).
   (d) I. Hemme and U. Klingebeil, *Adv. Organomet. Chem.*, **39**, 159 (1996).
   (e) M. Driess, *Adv. Organomet. Chem.*, **39**, 193 (1996).
   (f) N. Auner, G. Fearon and J. Weis in *Organosilicon Chemistry III* (Eds. N. Auner and J. Weis), VCH, Weinheim, 1997.
4. A. G. Brook, F. Abdesaken, B. Gutekunst, G. Gutekunst and R. K. Kallury, *J. Chem. Soc., Chem. Commun.*, 191 (1981).
5. R. West, M. J. Fink and J. Michl, *Science*, **214**, 1343 (1981).
6. G. E. Miracle, J. L. Ball, D. R. Powell and R. West, *J. Am. Chem. Soc.*, **115**, 11598 (1993).
7. (a) S. Bailleaux, M. Bogey, C. Demuynck, J. Destombes and A. Walters, *J. Chem. Phys.*, **101**, 2729 (1994).
   (b) M. Bogey, B. Delcroix, A. Walters and J.-C. Guillemin, *J. Mol. Spectrosc.*, **175**, 421 (1996).
   (c) R. Srinivas, D. K. Böhme, D. Sülzle and H. Schwarz, *J. Phys. Chem.*, **95**, 9836 (1991).
8. N. Tokitoh, K. Wakita, R. Okazaki, S. Nagase and P. v. R. Schleyer, *J. Am. Chem. Soc.*, **119**, 6951 (1997).
9. G. Märkel and W. Schlosser, *Angew. Chem., Int. Ed. Engl.*, **27**, 963 (1988).
10. (a) W. P. Freeman, D. Tilley, L. M. Liable-Sands and A. L. Rheingold, *J. Am. Chem. Soc.*, **118**, 10457 (1996).
    (b) R. West, H. Sohn, U. Bankwitz, J. Calabrese, Y. Apeloig and T. Müller, *J. Am. Chem. Soc.*, **117**, 11608 (1995).
    (c) J.-H. Hong, P. Boudjouk and S. Castellino, *Organometallics*, **13**, 3387 (1994).
    (d) W. P. Freeman, D. Tilley and A. L. Rheingold, *J. Am. Chem. Soc.*, **116**, 8428 (1994). See also references cited in these papers.
11. E. Hückel, *Z. Phys.*, **70**, 204 (1931); E. Hückel, *Z. Phys.*, **72**, 310 (1931).
12. P. v. R. Schleyer, H. F. Schaefer III and N. C. Handy (Eds.), *Encyclopedia of Computational Chemistry*, Wiley, New York, 1998.
13. (a) V. I. Minkin, M. N. Glukhovtsev and B. Ya. Simkin, *Aromaticity and Antiaromaticity*, Wiley, New York, 1994.
    (b) P. J. Garratt, *Aromaticity*, Wiley, New York, 1986.
14. M. N. Glukhovtsev, *J. Chem. Educ.*, **74**, 132 (1997).
15. (a) S. S. Shaik, A. Shurki, D. Danovitch and P. C. Hiberty, *J. Mol. Struct (Theochem)*, **398–399**, 155 (1997) and references cited therein.
    (b) S. S. Shaik, P. C. Hiberty, G. Ohanessian and J.-M. Lefour, *J. Phys. Chem.*, **92**, 5086 (1988).
    (c) G. Ohanessian, P. C. Hiberty, J.-M. Lefour, J.-P. Flament and S. S. Shaik, *Inorg. Chem.*, **27**, 2219 (1988).
16. P. v. R. Schleyer and H. Jiao, *Pure Appl. Chem.*, **68**, 209 (1996).
17. (a) A. Julg and P. Fracois, *Theor. Chim. Acta*, **7**, 249 (1967).
    (b) S. M. Van den Kerk, *J. Organomet. Chem.*, **215**, 315 (1981).
18. P. W. Fowler and E. Steiner, *J. Phys. Chem. A*, **101**, 1409 (1997).
19. P. v. R. Schleyer, H. Jiao, N. v. E. Hommes, V. G. Malkin and O. Malkina, *J. Am. Chem. Soc.*, **119**, 12669 (1992).
20. F. Bernardi, A. Bottoni and A. Venturini, *J. Mol. Struct. (Theochem)*, **163**, 173 (1988) and references cited therein.
21. (a) W. J. Hehre, R. Ditchfield, L. Radom and J. A. Pople, *J. Am. Chem. Soc.*, **92**, 4796 (1970).
    (b) P. George, M. Trachtman, C. W. Bock and A. M. Brett, *Theor. Chim. Acta*, **38**, 121 (1975).
    (c) P. George, M. Trachtman, C. W. Bock and A. M. Brett, *J. Chem. Soc., Perkin Trans. 2*, 1222 (1976).
    (d) B. A. Hass and L. J. Schaad, *J. Am. Chem. Soc.*, **105**, 7500 (1983).
22. Experimental data taken from: H. M. Rosenstock, K. Draxl, B. W. Steiner and J. T. Herron, *J. Phys. Chem. Ref. Data*, **6**, Suppl. 1 (1977); N. Cohen and S. W. Benson, *Chem. Rev.*, **93**, 2419 (1993).

23. (a) D. H. Hutter and W. H. Flygare, *Top. Curr. Chem.*, **63**, 89 (1976) and references cited therein.
    (b) H. Jiao and P. v. R. Schleyer, *J. Chem. Soc., Perkin Trans.* 2, 407 (1994).
24. H. J. Dauben Jr., J. D. Wilson and J. L. Laity, *J. Am. Chem. Soc.*, **90**, 811 (1968); **91**, 1991 (1969); H. J. Dauben Jr., J. D. Wilson and J. L. Laity, in *Non-Benzenoid Aromatics*, Vol. 2 (Ed. J. P. Snyder), Academic Press, New York, 1971.
25. F. London, *J. Phys. Radium*, **8**, 397 (1933).
26. P. v. R. Schleyer, C. Maerker, A. Dransfeld, H. Jiao and N. v. E. Hommes, *J. Am. Chem. Soc.*, **118**, 6317 (1996) and references cited therein.
27. M. Bühl, W. Thiel, H. Jiao, P. v. R. Schleyer, M. Saunders and F. A. L. Anet, *J. Am. Chem. Soc.*, **116**, 7429 (1994); M. Bühl and C. van Wüllen, *Chem. Phys. Lett.*, **247**, 63 (1995).
28. R. C. Haddon, *J. Am. Chem. Soc.*, **101**, 1722 (1979).
29. A. R. Katritzky, P. Barczynski, G. Masummarra, D. Pisano and M. Szafran, *J. Am. Chem. Soc.*, **111**, 7 (1989).
30. K. Jug and A. M. Köster, *J. Phys. Org. Chem.*, **4**, 163 (1991).
31. (a) P. v. R. Schleyer, P. Freeman, H. Jiao and B. Goldfuss, *Angew. Chem., Int. Ed. Engl.*, **34**, 337 (1995).
    (b) K. N. Houk and M. Mendel, *Chemtracts Organic Chemistry*, **9**, 118 (1996).
    (c) C. W. Bird, *Tetrahedron*, **52**, 9945 (1996).
32. Silatoluene: (a) T. J. Barton and D. Banasink, *J. Am. Chem. Soc.*, **99**, 5199 (1977).
    (b) C. L. Kreil, O. L. Chapman, G. T. Burns and T. J. Barton, *J. Am. Chem. Soc.*, **102**, 841 (1980).
    (c) H. Bock, R. A. Bowling, B. Solouki, T. J. Barton and G. T. Burns, *J. Am. Chem. Soc.*, **102**, 429 (1980).
33. Silabenzene: (a) T. J. Barton and G. T. Burns, *J. Am. Chem. Soc.*, **100**, 5246 (1978).
    (b) G. Maier, G. Mihm and H. P. Reisenauer, *Angew. Chem., Int. Ed. Engl.*, **19**, 52 (1980).
    (c) B. Solouki, P. Rosmus, H. Bock and G. Maier, *Angew. Chem., Int. Ed. Engl.*, **19**, 51 (1980).
    (d) G. Maier, G. Nihm and H. P. Reisenauer, *Chem. Ber.*, **115**, 801 (1982).
34. P. Jutzi, M. Meyer, H. P. Reisenauer and G. Maier, *Chem. Ber.*, **122**, 1227 (1989).
35. Y. van den Winkel, B. L. M. van Baar, F. Bickelhaupt, W. Kulik, C. Sierakowski and G. Maier, *Chem. Ber.*, **124**, 185 (1991).
36. G. Maier, K. Schöttler and H. P. Reisenauer, *Tetrahedron Lett.*, **26**, 4079 (1985).
37. A. Bjarnason and I. Arnason, *Angew. Chem., Int. Ed. Engl.*, **31**, 1633 (1992).
38. M. J. S. Dewar, D. H. Lo and C. A. Ramsden, *J. Am. Chem. Soc.*, **97**, 1311 (1975).
39. P. H. Blustin, *J. Organomet. Chem.*, **166**, 21 (1979).
40. J. Chandrasekhar, P. v. R. Schleyer, R. O. W. Baumgartner and M. T. Reets, *J. Org. Chem.*, **48**, 3453 (1983).
41. (a) M. S. Gordon, P. Boudjouk and F. Anwari, *J. Am. Chem. Soc.*, **105**, 4972 (1983).
    (b) K. K. Baldridge and M. S. Gordon, *Organometallics*, **7**, 144 (1988).
42. H. B. Schlegel, B. Coleman and M. Jones Jr., *J. Am. Chem. Soc.*, **100**, 6499 (1978).
43. (a) K. K. Baldridge and M. S. Gordon, *J. Am. Chem. Soc.*, **110**, 4204 (1988).
    (b) P. George, C. W. Bock and M. Trachtman, *Theor. Chim. Acta*, **71**, 289 (1987).
44. H. Bock, P. Rosmus, B. Solouki and G. Maier, *J. Organomet. Chem.*, **271**, 145 (1984).
45. (a) C. W. Bock, P. George and M. Trachtman, *J. Mol. Struct (Theochem)*, **109**, 1 (1984).
    (b) C. W. Bock, P. George and M. Trachtman, *J. Phys. Chem.*, **88**, 1467 (1984).
46. (a) R. S. Grev, G. E. Scuseria,, A. C. Sheiner, H. F. Schaefer III and M. S. Gordon, *J. Am. Chem. Soc.*, **110**, 7337 (1988).
    (b) S. Schroder and W. Thiel, *J. Am. Chem. Soc.*, **107**, 4422 (1985).
47. Y. Apeloig and M. Karni, *J. Am. Chem. Soc.*, **106**, 6676 (1984).
48. Y. Apeloig, M. Bendikov and S. Sklenak, unpublished results.
49. R. A. King, G. Vacek and H. F. Schaefer III, *J. Mol. Struct. (Theochem)*, **358**, 1 (1995).
50. (a) K. K. Baldridge and M. S. Gordon, *J. Organomet. Chem.*, **271**, 369 (1984).
    (b) J. Chandrasekhar and P. v. R. Schleyer, *J. Organomet. Chem.*, **289**, 51 (1985).
51. (a) W. Ando, T. Shiba, T. Hidaka, K. Morihashi and O. Kikuchi, *J. Am. Chem. Soc.*, **119**, 3629 (1997).
    (b) W. Ando, unpublished results.
52. M. Zhao and B. M. Gimarc, *Inorg. Chem.*, **35**, 5378 (1996).

53. (a) N. Matsunaga, T. R. Cundari, M. W. Schmidt and M. S. Gordon, *Theor. Chim. Acta*, **83**, 57 (1992).
    (b) N. Matsunaga and M. S. Gordon, *J. Am. Chem. Soc.*, **116**, 11407 (1994).
54. (a) W. J. Stevens, H. Basch and M. Kraus, *J. Chem. Phys.*, **81**, 6026 (1984).
    (b) W. J. Stevens, M. Kraus and P. Jasian, *Can. J. Chem.*, **70**, 612, (1992).
55. Y. Apeloig and J. Hrusàk, unpublished results.
56. S. Nagase, T. Kudo and M. Aoki, *J. Chem. Soc. Chem. Commun.*, 1121 (1985).
57. D. A. Clabo and H. F. Schaefer III, *J. Chem. Phys.*, **84**, 1664 (1986).
58. A. F. Sax and R. Janoschek, *Angew. Chem., Int. Ed. Engl.*, **25**, 651 (1986).
59. S. Nagase, H. Teramae and T. Kudo, *J. Chem. Phys.*, **86**, 4513 (1987).
60. A. F. Sax, J. Kalcher and R. Janoschek, *J. Comput. Chem.*, **9**, 564 (1988).
61. (a) S. Nagase, *Pure Appl. Chem.*, **65**, 675 (1993).
    (b) S. Nagase, *Polyhedron*, **10**, 1299 (1991).
62. P. v. R. Schleyer, unpublished.
63. J. M. Schulman and R. L. Disch, *J. Am. Chem. Soc.*, **107**, 5059 (1985).
64. H. Jacobson and T. Ziegler, *J. Am. Chem. Soc.*, **116**, 3667 (1994).
65. M. Karni and Y. Apeloig, *J. Am. Chem. Soc.*, **112**, 8589 (1990).
66. (a) R. Grev and H. F. Schaefer III, *J. Chem. Phys.*, **97**, 7990 (1992).
    (b) D. E. Goldberg, P. B. Hitchcock, M. F. Lappert, K. M. Thomas, A. J. Thorne, T. Fjeldberg, A. Haaland and B. E. R. Schilling, *J. Chem. Soc., Dalton Trans.*, 2387 (1986).
    (c) G. Trinquier and J.-P. Malrieu, *J. Phys. Chem.*, **94**, 6184 (1990).
    (d) G. Trinquier, *J. Am. Chem. Soc.*, **112**, 2130 (1990).
    (e) G. Trinquier, J.-P. Malrieu and P. Riviére, *J. Am. Chem. Soc.*, **104**, 4529 (1982).
67. N. Nagase, K. Kobayashi and T. Kudo, *Main Group Metal Chemistry*, **17**, 171 (1994).
68. L. Nyulászi, T. Kárpáti and T. Veszprémi, *J. Am. Chem. Soc.*, **116**, 7239 (1994).
69. T. Veszprémi, L. Nyulászi and T. Kárpáti, *J. Phys. Chem.*, **100**, 6262 (1996).
70. E. D. Jemmis, G. N. Srinivas, J. Leszczynski, J. Kapp, A. Korkin and P. v. R Schleyer, *J. Am. Chem. Soc.*, **117**, 11361 (1995).
71. A. Korkin, M. Glukhovtsev and P. v. R. Schleyer, *Int. J. Quantum Chem.*, **46**, 137 (1993).
72. (a) S. Murthy, Y. Nagano and J. L. Beauchamp, *J. Am. Chem. Soc.*, **114**, 3573, (1992).
    (b) Y. Nagano, S. Murthy and J. L. Beauchamp, *J. Am. Chem. Soc.*, **115**, 10805 (1993).
73. (a) A. Nicolaides and L. Radom, *J. Am. Chem. Soc.*, **118**, 10561 (1996).
    (b) A. Nicolaides and L. Radom, *J. Am. Chem. Soc.*, **116**, 9769 (1994).
    (c) R. L. Jarek and S. K. Shin, *J. Am. Chem. Soc.*, **119**, 6376 (1997).
74. (a) L. A. Curtiss, K. Raghavachari, G. W. Trucks and J. A. Pople, *J. Chem. Phys.*, **94**, 7221 (1991).
    (b) L. A. Curtiss, K. Raghavachari, G. W. Trucks and J. A. Pople, *J. Chem. Phys.*, **98**, 1293 (1993).
    (c) L. A. Curtiss and K. Raghavachari, in *Quantum Mechanical Electronic Structure Calculations with Chemical Accuracy* (Ed. S. R. Langhoff), Kluwer Academic Publishers, Dordrecht, 1995.
    (d) K. Raghavachari and L. A. Curtiss, in *Modern Electronic Structure Theory* (Ed. D. R. Yarkony), World Scientific, Singapore, 1995.
    (e) B. J. Smith and L. Radom, *J. Phys. Chem.*, **99**, 6468 (1995).
    (f) L. A. Curtiss, P. C. Redfern, B. J. Smith and L. Radom, *J. Chem. Phys.*, **104**, 5148 (1996).
75. R. Srinivas, J. Hrusák, D. Sülzle, D. K. Böhme and H. Schwarz, *J. Am. Chem. Soc.*, **114**, 2802 (1992).
76. (a) K. Krogh-Jespersen, J. Chandrasekhar and P. v. R. Schleyer, *J. Org. Chem.*, **45**, 1608 (1980).
    (b) B. J. Aylett and H. M. Colquhoun, *J. Chem. Res. (S)*, **148**, (1977); *J. Chem. Res. (M)*, 1674 (1977).
77. (a) J. Dubac, A. Leporterie and G. Manue, *Chem. Rev.*, **90**, 215 (1990).
    (b) J. Dubac, C. Guerin and P. Meunier, Chapter 34 in this book.
78. K. Tamao and A. Kawachi, *Adv. Organomet. Chem.*, **38**, 1 (1995).
79. J.-H. Hong and P. Boudjouk, *J. Am. Chem. Soc.*, **115**, 5883 (1993).
80. J.-H. Hong, P. Boudjouk and I. Stoenescu, *Organometallics*, **15**, 2179 (1996).
81. W. P. Freeman, D. Tilley, A. L. Rheingold and R. L. Ostrander, *Angew. Chem., Int. Ed. Engl.*, **32**, 1744 (1993).

82. W. P. Freeman, D. Tilley, G. P. A. Yap and A. L. Rheingold, *Angew. Chem., Int. Ed. Engl.*, **35**, 882 (1996).
83. W.-C. Joo, J.-H. Hong, S.-B. Choi, H.-E. Son and C.-H. Kim, *J. Organomet. Chem.*, **391**, 27 (1990).
84. U. Bankwitz, H. Sohn, D. R. Powell and R. West, *J. Organomet. Chem.*, **499**, C7 (1995).
85. R. West, H. Sohn, D. R. Powell, T. Müller and Y. Apeloig, *Angew. Chem., Int. Ed. Engl.*, **35**, 1002 (1996).
86. W. P. Freeman, D. Tilley, F. P. Arnold, A. L. Rheingold and P. K. Gantzel, *Angew, Chem., Int. Ed. Engl.*, **34**, 1887 (1995).
87. J. R. Damewood Jr., *J. Org. Chem.*, **51**, 5028 (1986).
88. B. Goldfuss and P. v. R. Schleyer, *Organometallics*, **14**, 1553 (1995).
89. B. Goldfuss and P. v. R. Schleyer, *Organometallics*, **16**, 1543 (1997).
90. (a) G. A. Olah and R. J. Hundai, *J. Am. Chem. Soc.*, **102**, 6989 (1980).
    (b) For a review on silyl anions see: K. Tamao and A. Kawachi, *Adv. Organomet. Chem.*, **38**, 1 (1995) and J. Beltzner and U. Dehnert, Chapter 14 in this book.
91. E. D. Jemmis and P. v. R. Schleyer, *J. Am. Chem. Soc.*, **104**, 4781 (1982).
92. H. Jiao and P. v. R. Schleyer, *Angew. Chem., Int. Ed. Eng.*, **32**, 1760 (1993).
93. (a) B. Goldfuss, P. v. R. Schleyer and F. Hampel, *Organometallics*, **15**, 1755 (1996).
    (b) T. Müller, Y. Apeloig, H. Son and R. West, in *Organosilicon Chemistry III* (Eds. N. Auner and J. Weis), VCH, Weinheim, 1997.
94. For reviews see:
    (a) Carbenes: O. M. Nefedov, M. P. Egorov, A. I. Joffe, C. G. Menchkov, P. S. Zuev, V. I. Minkin. B. Y. Simkin and M. N. Glukhovtsev, *Pure Appl. Chem.*, **64**, 265 (1992).
    (b) Silylenes: P. P. Gaspar, in *Reactive Intermediates* (Eds. M. Jones and R. A. Moss), Wiley, New York, 1978, Vol. 1, pp. 229–278; 1981, Vol. 2, pp. 335–385; 1985, Vol. 3, pp. 333–427.
    (c) Germylenes: J. Barrau, J. Escudie and J. Satge, *Chem. Rev.*, **90**, 283 (1990); W. P. Neumann, *Chem. Rev.*, **91**, 311 (1991).
95. (a) A. J. Arduengo III, R. L. Harlow and M. Kline, *J. Am. Chem. Soc.*, **113**, 361 (1991).
    (b) A. J. Arduengo III, H. V. Rasika Dias, R. L. Harlow and M. Kline, *J. Am. Chem. Soc.*, **114**, 5530 (1992).
    (c) A. J. Arduengo III, J. R. Goerlich and W. J. Marshall, *J. Am. Chem. Soc.*, **117**, 11027 (1995).
    (d) J. Cioslowski, *Int. Quantum Chem., Quant. Chem. Symp.*, **27**, 309 (1993).
    (e) A. J. Arduengo III, D. A. Dixon, R. L. Harlow, K. K. Kumashiro, C. Lee, W. P. Power and K. W. Zilm, *J. Am. Chem. Soc.*, **116**, 6361 (1994).
    (f) A. J. Arduengo III, H. V. Rasika Dias, D. A. Dixon, R. L. Harlow, W. T. Klooster and T. F. Koetzle, *J. Am. Chem. Soc.*, **116**, 6812 (1994),
    (g) C. Heinemann and W. Thiel, *Chem. Phys. Lett.*, **217**, 11 (1994),
    (h) D. A. Dixon and A. J. Arduengo III, *J. Phys. Chem.*, **95**, 4180 (1991).
96. A. J. Arduengo III, H. Bock, H. Chen, M. Denk, D. A. Dixon, J. C. Green, W. A. Herrmann, N. L. Jones, M. Wagner and R. West, *J. Am. Chem. Soc.*, **116**, 6641 (1994).
97. W. A. Herrmann, M. Denk, J. Behn, W. Scherer, F. R. Klingan, H. Bock, B. Solouki and M. Wagner, *Angew. Chem., Int. Ed. Engl.*, **31**, 1485 (1992).
98. (a) M. Denk, R. Lennon, R. Hayashi, R. West, A. V. Belyakov, H. P. Verne, A. Haaland, M. Wagner and N. Metzler, *J. Am. Chem. Soc.*, **116**, 2691 (1994).
    (b) M. Denk, J. C. Green, N. Metzler and M. Wagner, *J. Chem. Soc., Dalton Trans.*, 2405 (1994).
    (c) R. West and M. Denk, *Pure Appl. Chem.*, **68**, 785 (1996).
    (d) M. Denk, R. West, R. K. Hayashi, Y. Apeloig, R. Paunz and M. Karni in *Organosilicon Chemistry II. From Molecules to Materials* (Eds. N. Auner and J. Weis), VCH, Weinheim, 1996, p. 251.
99. (a) B. Gehrhus, M. Lappert, J. Heinicke, R. Boese and D. Bläser, *J. Chem. Soc., Chem. Commun.*, 1931 (1995).
    (b) B. Gehrhus, P. B. Hitchcock, M. Lappert, J. Heinicke, R. Boese and D. Bläser, *J. Organomet. Chem.*, **521**, 211 (1996).
    (c) P. Blakeman, B. Gehrhus, J. C. Green, J. Heinicke, M. Lappert, M. Kindermann and T. Veszprémi, *J. Chem. Soc., Dalton Trans.*, 1475 (1996).
100. (a) C. Heinemann, T. Müller, Y. Apeloig and H. Schwarz, *J. Am. Chem. Soc.*, **118**, 2023 (1996).
101. M. Veith, E. Werle, R. Lisowsky, R. Köppe and H. Schnöckel, *Chem. Ber.*, **125**, 1375 (1992).

1. Theoretical aspects and quantum mechanical calculations 99

102. Y. Apeloig, M. Karni and T. Müller, in *Organosilicon Chemistry II. From Molecules to Materials*. (Eds N. Auner and J. Weis), VCH, Weinheim, 1996, p. 263.
103. C. Boehme and G. Frenking, *J. Am. Chem. Soc.*, **118**, 2039 (1996).
104. C. Heinemann, W. A. Herrmann and W. Thiel, *J. Organomet. Chem.*, **475**, 73 (1994).
105. (a) R. Ditchfield, *Mol. Phys.*, **27**, 789 (1974).
     (b) J. Gauss, *J. Chem. Phys.*, **99**, 3629 (1993) and references cited therein.
106. (a) W. Kutzelnigg, U. Fleischer and M. Schindler, in *NMR—Basic Principles and Progress*, Vol. 23, Springer, Berlin, 1991.
     (b) W. Kutzelnigg, *Isr. J. Chem.*, **19**, 193 (1980).
     (c) M. Schindler and W. Kutzelnigg, *J. Chem. Phys.*, **76**, 1979 (1982).
107. R. West, J. J. Baffy, M. Haaf, T. Müller, B. Gehrhus, M. F. Lappert and Y. Apeloig, *J. Am. Chem. Soc.*, **120**, 1639 (1998).
108. A. E. Reed, R. B. Weinstock and F. Weinhold, *J. Chem. Phys.*, **83**, 735 (1985).
109. (a) R. W. F. Bader, *Acc. Chem. Res.*, **8**, 34 (1975).
     (b) R. W. F. Bader, in *Atoms in Molecules. A Quantum Theory*, Clarendon Press, Oxford, 1990;
     (c) R. W. F. Bader, *Chem. Rev.*, **91**, 893 (1991); R. W. F. Bader, *Angew. Chem., Int. Ed. Engl.* **33**, 620 (1994).
110. Y. Apeloig and T. Müller, *J. Am. Chem. Soc.*, **117**, 5363 (1995).
111. J. Maxka and Y. Apeloig, *J. Chem. Soc., Chem. Commun.*, 737 (1990).
112. (a) G. Trinquier, and J.-C. Barthelat, *J. Am. Chem. Soc.*, **112**, 9121 (1990).
     (b) G. Trinquier, *J. Am. Chem. Soc.*, **113**, 144 (1991).
     (c) G. Trinquier and J.-P. Malrieu, *J. Am. Chem. Soc.*, **113**, 8634 (1991).
113. (a) R. Walsh, in *The Chemistry of Organic Silicon Compounds* (Eds. S. Patai and Z. Rappoport), Chap. 5, Wiley, Chichester, 1989, p. 371.
     (b) D. R. Lide (Ed.), *CRC Handbook of Chemistry and Physics, 71th edition*, CRC Press, Boca Raton, 1990.
114. L. Nyulászi, D. Szieberth and T. Veszprémi, *J. Org. Chem.*, **60**, 1647 (1995).
115. (a) D. J. Husain and P. E. Norris, *J. Chem. Soc., Faraday Trans. 1* **74**, 106 (1978).
     (b) S. C. Basu and D. Husain, *J. Photochem. Photobiol. A.*, **42**, 1 (1988).
116. R. Srinivas, D. Sulzle, T. Weiske and H. Schwarz, *Int. J. Mass Spectrom. Ion Processes*, **107**, 369 (1991).
117. (a) G. Maier, H. P. Reisenauer and H. Pacl, *Angew. Chem., Int. Ed. Engl.*, **33**, 1248 (1994).
     (b) G. Maier, H. Pacl, H. P. Reisenauer, A. Meudt and R. Janoschek, *J. Am. Chem. Soc.*, **117**, 12712 (1995);
     (c) M. Izuha, S. Yamamoto and S. Saito, *Can. J. Phys.*, **72**, 1206 (1994).
118. (a) G. Frenking, R. B. Remington and H. F. Schaefer III, *J. Am. Chem. Soc.*, **108**, 2169 (1986).
     (b) G. Vacek, B. T. Colegrove and H. F. Schaefer, III, *J. Am. Chem. Sec.*, **113**, 3192 (1991).
     (c) T. J. Lee, A. Bunge and H. F. Schaefer, III, *J. Am. Chem. Soc.*, **107**, 137 (1985).
119. C. D. Sherrill, C. G. Brandow, W. D. Allen and H. F. Schaefer III, *J. Am. Chem. Soc.*, **118**, 7158 (1996).
120. (a) M.-D. Su, R. D. Amos and N. C. Handy, *J. Am. Chem. Soc.*, **112**, 1499 (1990).
     (b) M. S. Gordon and R. D. Koob, *J. Am. Chem. Soc.*, **103**, 2939 (1981).
121. S. Wlodek, A. Fox and D. K. Böhme, *J. Am. Chem. Soc.*, **113**, 4461 (1991).
122. (a) J. D. Presilla-Márquez, W. R. M. Graham and R. A. Shepherd *J. Chem. Phys.*, **93**, 5424 (1990); R. A. Shepherd and W. R. M. Graham, *J. Chem. Phys.*, **88**, 3399 (1988); R. A. Shepherd and W. R. M. Graham, *J. Chem. Phys.*, **82**, 4788 (1985); J. Oddershede, J. R. Sabin, G. H. F. Diercksen and N. E. Grüner, *J. Chem. Phys.*, **83**, 1702 (1985); W. Weltner, Jr. and D. McLeod Jr., *J. Chem. Phys.*, **41**, 235 (1964); B. Kleman, *Astrophys. J.*, **123**, 162 (1956).
     (b) G. Fitzgerald, S. J. Cole and R. J. Bartlett, *J. Chem. Phys.*, **85**, 1701 (1986).
     (c) R. S. Grev and H. F. Schaefer III, *J. Chem. Phys.*, **80**, 3552 (1984).
123. M. C. Ernst, A. F. Sax, J. Kalcher and G. Katzer, *J. Mol. Struct*, **334**, 121 (1995).
124. (a) H. W. Kroto, J. R. Heath, S. C. O'Brien, R. F. Curl and R. E. Smalley, *Nature*, **318**, 162 (1985).
     (b) J. R. Heath, S. C. O'Brien, Q. Zhang, Y. Liu, R. F. Curl, H. W. Kroto, F. K. Tittel and R. E. Smalley, *J. Am. Chem. Soc.*, **107**, 7779 (1985).
     (c) Q. Zhang, S. C. O'Brien, J. R. Heath, Y. Liu, H. W. Kroto and R. E. Smalley, *J. Phys. Chem.*, **90**, 525 (1986).
125. W. Kräschmer, F. Fostiropoulos and D. R. Huffman, *Nature*, **347**, 354 (1990); W. Kräschmer, F. Fostiropoulos and D. R. Huffman, *Chem. Phys. Lett.*, **170**, 167 (1990).

126. M. S. Dresselhaus, G. Dresselhaus and P. C. Eklund, *Science of Fullerenes and Carbon Nanotubes*, Academic Press, San Diego, 1996.
127. R. C. Haddon, *Science*, **261**, 1545 (1993) and references cited therein.
128. R. C. Haddon, *Nature*, **378**, 249 (1995) and references cited therein.
129. M. Bühl, W. Thiel, H. Jiao, P. v. R. Schleyer, M. Saunders and F. A. L. Annet, *J. Am. Chem. Soc.*, **116**, 6005 (1994).
130. R. C. Haddon, L. F. Schneemeyer, J. V. Waszczak, S. H. Glarum, R. Tycko, G. Dabbagh, A. R. Kortan, A. J. Muller, A. M. Mujsce, M. J. Rosseinsky, S. M. Zahurak, A. V. Makhija, F. A. Thiel, K. Raghavachari, E. Cockayne and V. Elser, *Nature*, **350**, 46 (1991).
131. (a) A. Pasquerello, M. Schlüter and R. C. Haddon, *Science*, **257**, 1660 (1992).
    (b) A. Pasquerello, M. Schlüter and R. C. Haddon, *Phys. Rev. A*, **47**, 1783 (1993).
    (c) R. C. Haddon, *Nature*, **367**, 214 (1994).
132. M. Bühl and W. Thiel, *Chem. Eur. J.*, submitted
133. M. Bühl and W. Thiel, *Chem. Phys. Lett.*, **233**, 585 (1995).
134. (a) Q. L. Zhang, T. Liu, R. F. Curl, F. K. Tittle and R. E. Smalley, *J. Phys. Chem.*, **88**, 1670 (1988).
    (b) M. F. Jarrold and E. C. Honea, *J. Phys. Chem.*, **95**, 9181 (1991).
135. D. A. Jelski, Z. C. Wu and T. F. George, *J. Cluster Sci.*, **1**, 143 (1990).
136. C. Zybill, *Angew. Chem., Int. Ed. Engl.*, **31**, 173 (1992).
137. S. Nagase and K. Kobayashi, *Fullerene Science and Technology*, **1**, 299 (1993).
138. Z. Slanina, S.-L. Lee, K. Kobayashi and S. Nagase, *J. Mol. Struct.*, **312**, 175 (1994).
139. (a) H. W. Kroto, *Nature*, **329**, 529 (1987).
    (b) T. G. Schmaltz, W. A. Seitz, D. J. Klein and G. E. Hite, *J. Am. Chem. Soc.*, **110**, 1113 (1988).
    (c) D. E. Manolopoulos, *J. Chem. Soc., Faraday Trans.*, **87**, 2861 (1991).
140. M. Menon and K. R. Subbaswamy, *Chem. Phys. Lett.*, **219**, 219 (1994).
141. M. C. Piqueras, R. Crespo, E. Ortí and F. Tomàs, *Chem. Phys. Lett.*, **213**, 509 (1993).
142. R. Crespo, M. C. Piqueras and F. Tomàs, *Synth. Met.*, **77**, 13 (1996).
143. S. Nagase and K. Kobayashi, *Chem. Phys. Lett.*, **187**, 291 (1991)
144. D. Bakowies and W. Thiel, *J. Am. Chem. Soc.*, **113**, 3704 (1991).
145. M. Häser, J. Almlöf and G. E. Scuseria, *Chem. Phys. Lett.*, **181**, 497 (1991).
146. K. Hedberg, L. Hedberg, D. S. Bethune, C. A. Brown, H. Dorn, R. D. Johnson and M. de Vries, *Science*, **254**, 410 (1991).
147. S. Osawa, M. Harada and E. Osawa, *Fullerene Science and Technology*, **3**, 225 (1995).
148. F. De Proft, C. Van Alesony and P. Geerlings, *J. Phys. Chem.*, **100**, 7440 (1996).
149. J. Cioslowsky and E. D. Fleischmann, *J. Chem. Phys.*, **94**, 3730 (1991).
150. J. Cioslowsky, *J. Am. Chem. Soc.*, **113**, 4139 (1991).
151. M. C. Piqueras. R. Crespo, and F. Tomàs, *J. Mol. Struct. (Theochem)*, **330**, 177 (1995).
152. (a) T. A. Albright, J. K. Burdett and M.-H. Whangbo, *Orbital Interactions in Chemistry*, Wiley, New York, 1984.
    (b) The orbital diagram of ferrocene is discussed on pp. 392–394.
153. J. P. Collman, L. S. Hegedus, J. R. Norton and R. G. Finke, *Principles and Applications of Organotransition Metal Chemistry*, University Science Books, Mill Valley, California, 1987; P. L. Pauson, in *Non-Benzenoid Aromatic Compounds* (Ed. D. Ginsburg), Interscience Publishers, New York, 1959.
154. (a) P. Jutzi, D. Kanne and C. Krüger, *Angew. Chem., Int. Ed. Engl.*, **25**, 164 (1986).
    (b) P. Jutzi, U. Holtmann, D. Kanne, C. Krüger, R. Blom, R. Gleiter and I. Hyla-Krispin, *Chem. Ber.*, **122**, 1629 (1989).
155. T. J. Lee and J. E. Rice, *J. Am. Chem. Soc.*, **111**, 2011 (1989).
156. T. Kudo and S. Nagase, *J. Mol. Struct., (Theochem)*, **311**, 111 (1994).
157. For X = SiH see:
    (a) E. D. Jemmis, G. Subramanian, I. H. Srivastava and S. R. Gadre, *J. Phys. Chem.*, **98**, 6445 (1994).
    (b) E. D. Jemmis and G. Subramanian, *J. Phys. Chem.*, **98**, 9222 (1994).
    (c) J. K. Burdett and O. Eisenstein, *J. Am. Chem. Soc.*, **117**, 11939 (1995).
158. P. v. R. Schleyer, G. Subramanian and A. Dransfeld, *J. Am. Chem. Soc.*, **118**, 9988 (1996).
159. M. S. Gordon, *J. Chem. Soc., Chem. Commun.*, 1131 (1980).
160. M. E. Colvin and H. F. Schaefer III, *Faraday Symp. Chem. Soc.*, **19**, 39 (1984).
161. G. W. Schriver, M. J. Fink and M. S. Gordon, *Organometallics*, **6**, 1977 (1987).

162. T. A. Holmes, M. S. Gordon, S. Yabushita and M. W. Schmidt, *Organometallics*, **3**, 583 (1984).
163. (a) A. F. Sax and J. Kalcher, *J. Chem. Soc., Chem. Commun.*, 809 (1987).
    (b) A. F. Sax and J. Kalcher, *J. Comput. Chem.*, **10**, 309 (1989).
164. S. Nagase and M. Nakano, *Angew. Chem., Int. Ed. Engl.*, **27**, 1081 (1988).
165. (a) B. F. Yates, D. A. Calbo and H. F. Schaefer III, *Chem. Phys. Lett.*, **143**, 421 (1988).
    (b) B. F. Yates and H. F. Schaefer III, *Chem. Phys. Lett.*, **155**, 563 (1989).
166. N. Wiberg, C. M. M. Finger and K. Polborn, *Angew. Chem., Int. Ed. Engl.*, **32**, 1054 (1993).
167. E. J. P. Malar, *Tetrahedron*, **52**, 4709 (1996).
168. K. Jug, *J. Org. Chem.*, **48**, 1344 (1983).
169. A. A. Korkin, V. V. Murashov, J. Leszczynski and P. v. R. Schleyer, *J. Phys. Chem.*, **99**, 17742 (1995).
170. (a) W. T. Borden and E. R. Davidson, *J. Am. Chem. Soc.*, **101**, 3771 (1979).
    (b) E. D. Jemmis and P. v. R. Schleyer, *J. Am. Chem. Soc.*, **104**, 4781 (1982).
    (c) S. Koseki, M. Arai, Y. Fujimura and T. Nakajima, *Chem. Phys.*, **108**, 33 (1986).
    (d) J. Fabian, A. Mehlhorn and N. Tyutyulkov, *J. Mol. Struct. (Theochem)*, **151**, 355 (1987).
    (e) J. Leszczynski, F. Weisman and M. C. Zerner, *Int. J. Quantum Chem., Quantum Chem. Symp.*, **22**, 117 (1988).
    (f) J. Feng, J. Leszczynski, B. Weiner and M. C. Zerner, *J. Am. Chem. Soc.*, **111**, 4648 (1989).
    (g) D. J. Wales and R. G. A. Bone, *J. Am. Chem. Soc.*, **114**, 5399 (1992).
    (h) M. N. Glukhovtsev and P. v. R. Schleyer, *Mendeleev Commun.*, **100** (1993) and references cited therein.

CHAPTER 2

A molecular modeling of the bonded interactions of crystalline silica

G. V. GIBBS and M. B. BOISEN, Jr

Departments of Geological Sciences, Materials Science and Engineering and Mathematics, Virginia Tech, Blacksburg, VA 24061, USA

  I. INTRODUCTION 103
 II. BOND LENGTH-BOND STRENGTH VARIATIONS IN SILICATES AND RELATED MOLECULES 104
III. A POTENTIAL ENERGY SURFACE FOR THE Si-O-Si SKELETON OF THE DISILICIC ACID MOLECULE 109
 IV. BOND CRITICAL POINT PROPERTIES OF THE ELECTRON DENSITY DISTRIBUTION OF THE SKELETAL Si-O-Si UNIT 113
  V. A REPRODUCTION OF THE STRUCTURES AND RELATED PROPERTIES OF THE KNOWN SILICA POLYMORPHS 115
 VI. A GENERATION OF NEW STRUCTURE TYPES FOR SILICA 115
VII. CONCLUDING REMARKS 116
VIII. ACKNOWLEDGMENTS 117
 IX. REFERENCES 117

I. INTRODUCTION

The chemistry of the animal and plant worlds is dominated by C, O, H and N whereas that of the mineral world is dominated by Si and O. Indeed, more than 80% of the volume of the earth's crust is believed to consist of the framework silicates quartz (Figure 1a see Plate 1) and feldspar (Figure 1b) with the remaining rock forming silicates mica, amphibole, pyroxene and olivine (Figures 1c, 1d, 1e and 1f, respectively, see Plates 2 and 3) bringing the grand total volume to more than 96%[1]. Therefore, it is not surprising that a knowledge of the bonded interactions that govern the structures and the properties of silicates is of central importance in attempts to understand inorganic materials in the natural environment and to find new uses for minerals in meeting the ever increasing materials needs of mankind.

The chemistry of organic silicon compounds, Vol. 2
Edited by Z. Rappoport and Y. Apeloig © 1998 John Wiley & Sons Ltd

The bonded interactions in silicates were recently examined in a survey of observed and calculated structures and electron density distributions reported for siloxane and silicic acid molecules and silica polymorph crystals[2]. Employing relatively robust basis sets that include polarization functions, the Si−O bond lengths and Si−O−Si angles of the silica polymorphs were reported to be reproduced in molecular orbital calculations to within a few percent. As observed in earlier studies, polarization functions were found to be a necessary ingredient in the generation of the observed geometries and spectra for silicate crystals. A topological analysis of the observed and calculated electron density distributions of the Si−O bond shows that the valence shell concentration of the oxide ion is locally depleted where the shell crosses the Si−O bond vector, while it is concentrated in the direction of the Si cation. The cross section of the bond was found to become more circular as the Si−O−Si angle widens and the Si−O bond length shortens, a result that can be ascribed to the formation of a weak π-bond involving two of the p-type orbitals on the oxide ion with an increase in the value of the electron density at the bond critical point. The close similarity of the bond length and angle data and the electron density distributions reported for silicate crystals and siloxane molecules were taken to indicate that the bonded interactions that govern the structure of a silicate can be treated as if localized as in a molecule and largely independent of the long-range forces of the ions that comprise the periodic field of a silicate crystal.

One of the goals of this paper is to provide a basis for understanding why the bonded interactions of a silicate crystal can be treated as localized and modeled with a molecular-based potential energy function. Another is to provide a basis for understanding why it is possible to reproduce the crystal structures and the bulk moduli of the known silica polymorphs with such a potential and why it is possible to generate many of the known structures of silica together with a large number of new structure types starting with a random arrangement of Si and O atoms in a unit cell with triclinic $P1$ symmetry, again using a molecular-based potential energy function. In Section II, the bond lengths recorded for silicate and related oxide minerals are compared with those calculated for chemically similar molecules. The Si−O bond lengths and Si−O−Si angles observed for the silica polymorphs are examined in Section III in terms of a potential energy surface calculated for the Si−O−Si skeletal unit of a disilicic acid molecule. Bond critical point properties of an electron density distribution calculated for the Si−O−Si unit is compared with that observed for a silica polymorph in Section IV. A modeling of the structures and the bulk moduli of the known polymorphs, using a potential energy function based on the molecule, is discussed in Section V, while in Section VI, the generation of large numbers of new structure types for silica is examined.

II. BOND LENGTH–BOND STRENGTH VARIATIONS IN SILICATES AND RELATED MOLECULES

More than 60 years ago, the important proposal[3] was made that the strength, s, of an X−O bond in an XO_n-coordinated polyhedron in a crystal like a silicate should depend on the valence, z, and the coordination number, n, of the X^{+z} metal cation such that $s = z/n$. With this simple definition, it was found for instance that the sum of the strengths of each of the t bonds in a crystal reaching a given oxide ion,

$$\zeta = \sum_{i=1}^{t} s_i,$$

often equals 2.0, the valence of the anion with its sign changed. Hence, for a silicate like quartz where each Si^{+4} cation resides in a SiO_4 tetracoordinate polyhedron and each oxide ion is bonded to two Si cations, the strength of the Si−O bond is 1.0 and the sum

PLATE 1

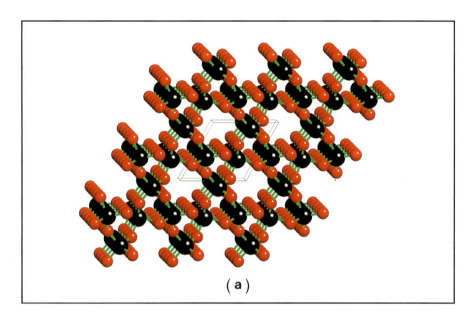

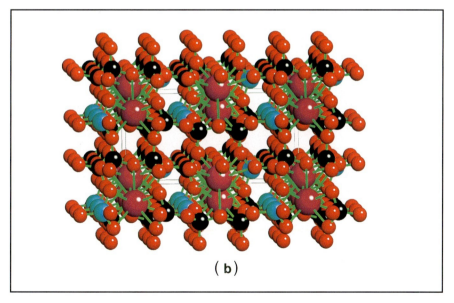

FIGURE 1. Ball and stick model drawings of the crystal structures of (a) quartz, SiO_2; (b) the feldspar, microcline, $KAlSi_3O_8$; (c) the mica, muscovite, $KAl_2(AlSi_3O_{10})(OH)_2$; (d) the amphibole, tremolite, $Ca_2Mg_5(Si_4O_{11})_2(OH)_2$; (e) the pyroxene, diopside, $CaMg(SiO_3)_2$; (f) the olivine, forsterite, Mg_2SiO_4;

PLATE 2

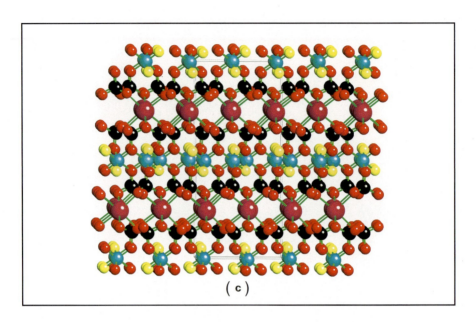

(c)

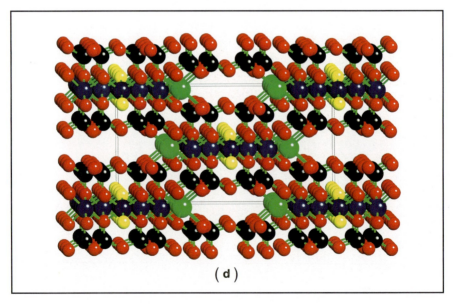

(d)

FIGURE 1 (*continued*):
(g) cristobalite, SiO$_2$; (h) tridymite, SiO$_2$; (i) the high pressure silica polymorph, coesite; (j) wadeite, K$_2$Si$_4$O$_9$ with both 4- and 6-coordinate Si; (k) the very high pressure silica polymorph stishovite with only

PLATE 3

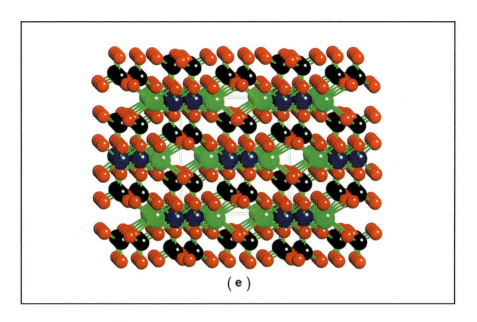

(e)

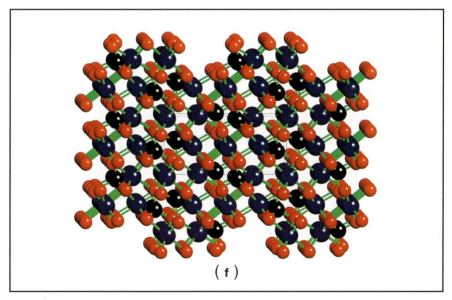

(f)

FIGURE 1 (*continued*):
6-coordinate Si and (l) the zeolite bikitaite, LiAlSi$_2$O$_6$. The structures depicted in (a), (d), (e), (g), (h), (i), (j), (k) and (l) are viewed along [001], those in (c) are viewed along [010] and those in (b) and

PLATE 4

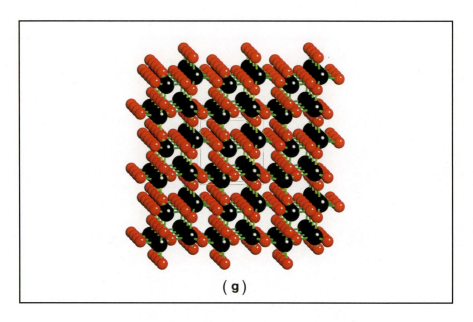

(g)

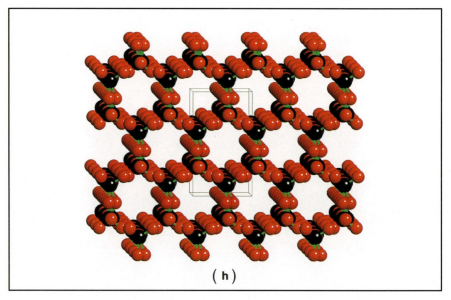

(h)

FIGURE 1 (*continued*):
(f) are viewed along [100]. The red spheres represent oxygen atoms, the black represent 4-coordinate silicon atoms, the violet represent potassium atoms, the light blue represent aluminum atoms in all

PLATE 5

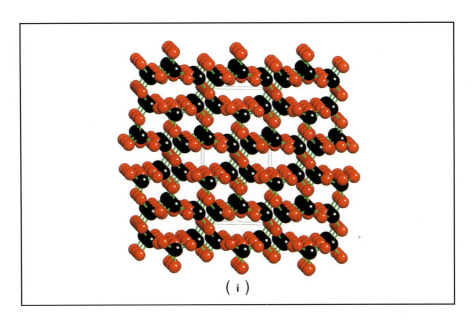

(i)

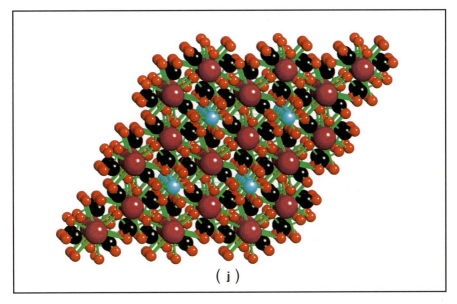

(j)

FIGURE 1 (*continued*):
structures except wadeite where they represent 6-coordinate Si atoms, the dark blue represent magnesium atoms, the green represent calcium atoms, the gray represent lithium atoms and the yellow represent

PLATE 6

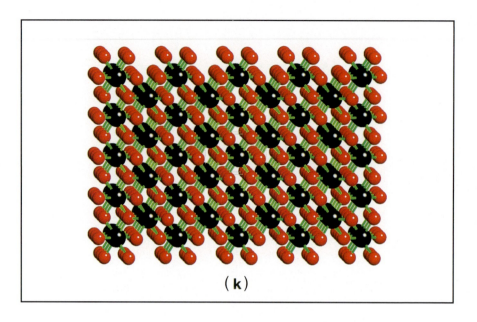

(k)

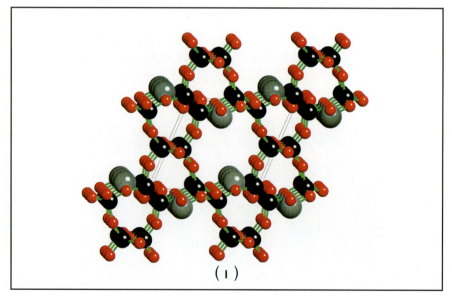

(l)

FIGURE 1 (*continued*):
the hydroxyl (OH) groups. The Al and Si (colored black) atoms in bikitaite are disordered among the tetrahedral coordination polyhedra of the structure. The parallelepiped at the center of each drawing outlines the unit cell

PLATE 7

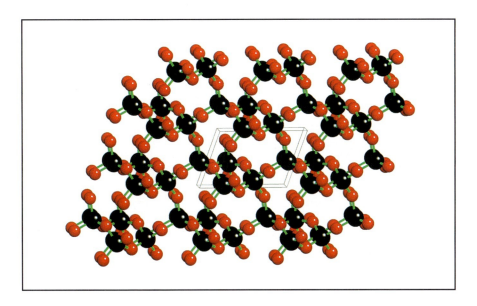

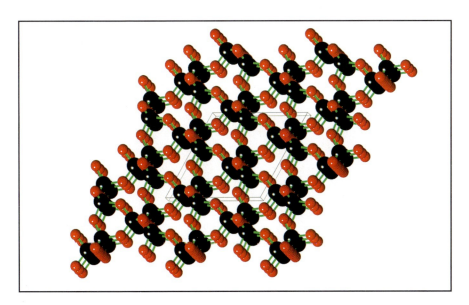

FIGURE 9. Drawings of six silica framework structures generated with simulated annealing strategies starting with a random arrangement of Si and O atoms with the assumption of $P1$ symmetry. The larger black spheres represent silicon and the smaller red spheres represent oxygen. The unit cell of each structure is outlined at the center of each drawing

PLATE 8

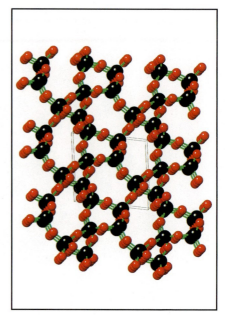
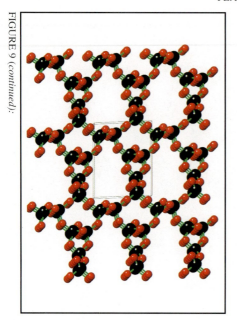
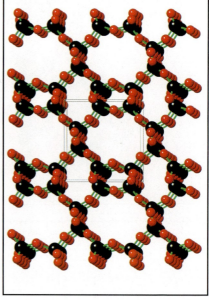
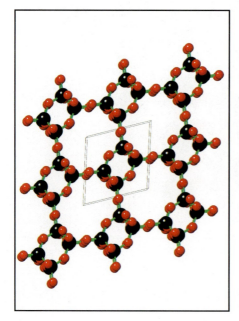

FIGURE 9 (*continued*):

2. A molecular modeling of the bonded interactions of crystalline silica

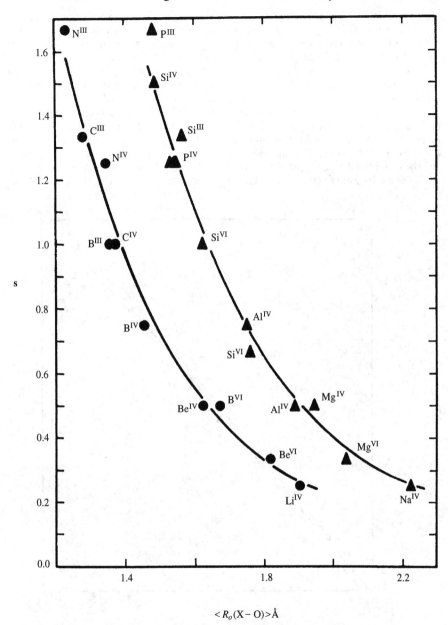

FIGURE 2. A plot of the average bond lengths, $<R_o(X-O)>$, of the XO_n coordination polyhedra observed for silicate and oxide crystals vs the Pauling bond strength, s, of the X–O bond. The Roman numeral superscript denotes the coordination number of the X-cation

of the bond strengths reaching each oxide ion is 2.0, matching exactly the valence of the oxide ion with its sign changed. A plot of s versus the average X—O bond length, $<R_o(X-O)>$, observed for the XO_n-coordination polyhedra in silicate and oxide crystals shows for first-row (Li, Be, B, ...) and second-row (Na, Mg, Al, ...) X-cations that $<R_o(X-O)>$ decreases nonlinearly in two separate but essentially parallel trends from left to right across each row of the periodic table as the strength of the bond increases[4,5] (Figure 2).

In an examination of whether the trends displayed in Figure 2 hold for molecules, molecular orbital calculations were completed[6] on a set of hydroxyacid $H_{2n-m} X^{+m} O_n$ molecules with XO_n-coordination polyhedra containing first and second-row X-cations using a 6-31G* basis set. The energies of the molecules were minimized with the X—O and the O—H bond lengths and the X—O—H angles each treated as single variables. The O—X—O angles were fixed at ideal values (120° for $H_{6-m} X^{+m} O_3$ molecules, 109.47° for

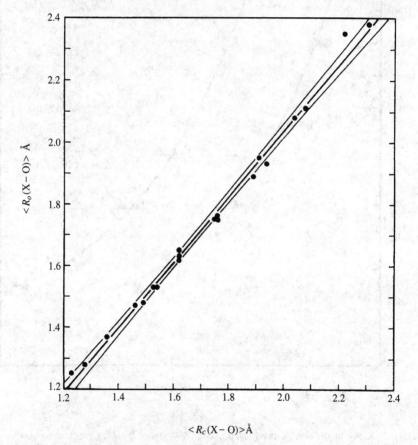

FIGURE 3. A plot of the observed bond length data, $<R_o(X-O)>$, used to prepare Figure 1 vs bond length data, $<R_c(X-O)>$, calculated for hydroxyacid $H_{2n-m} X^{+m}O_n$ molecules with XO_n-coordination polyhedra. The minimum energy X—O bond lengths were calculated with a 6-31G* basis set. A regression analysis of the data yielded the straight line drawn through the data points and its two upper and lower curved 2.7σ-confidence limits

2. A molecular modeling of the bonded interactions of crystalline silica

H_{8-m} X^{+m} O_4 molecules and 90° for H_{12-m} X^{+m} O_6 molecules). The minimum energy X–O bond lengths, $<R_c(X–O)>$, calculated for the molecules were found to reproduce observed $<R_o(X–O)>$ bond length data for silicate and oxide crystals to within ca 0.02 Å, on average (Figure 3). Indeed, a statistical analysis shows that more than 99% of the variation in the experimental $<R_o(X–O)>$ bond lengths can be explained in terms of a linear dependence on $<R_c(X–O)>$. As expected, the s vs $<R_c(X–O)>$ data obtained for the molecules closely parallel the two trends displayed in Figure 2.

Similar calculations have yet to be completed for molecules with the main group X-cations for rows in the periodic table beyond the second. Nonetheless, it was found[6] that the observed bond length data for the main group cations for all six rows of the periodic table correlate with s in six separate but essentially parallel trends similar to those displayed in Figure 2. In a search for a parameter that would rank all of the bond length data in a single trend, a bond order parameter $p = s/r$ was defined where $r = 1, 2, 3, \ldots$ for first-, second-, third-, $\ldots$ row main group X-cations, respectively[6]. When the $<R_c(X–O)>$-values for row one and two X-cations are plotted against this bond order parameter (Figure 4a), it is apparent that p ranks the calculated bond length data in a single trend with shorter bonds involving larger p-values. The data in the figure are observed to conform with a simple power law expression of the form $R(X–O) = \kappa p^{-\beta}$ where $\beta > 0$. It is noteworthy that when $R(X–O)$ is plotted against the value of the electron density at the bond critical point, $\rho(\mathbf{r}_c)$, for each X–O bond the $R(X–O)$ vs $\rho(\mathbf{r}_c)$ data also conform with a power law expression.

A regression analysis of the calculated bond length data used to prepare Figure 4a vs p yields the expression $R(X–O) = 1.39 p^{-\beta}$ where $\beta = ca\, 2/9$. The observed X–O bond length data reported for main-group cations from all six rows of the periodic table are plotted in Figure 4b against p. Not only does the trend parallel that calculated for the molecules, but a regression analysis of the observed data set yields a statistically identical expression with the one obtained for the calculated data displayed in Figure 4a.

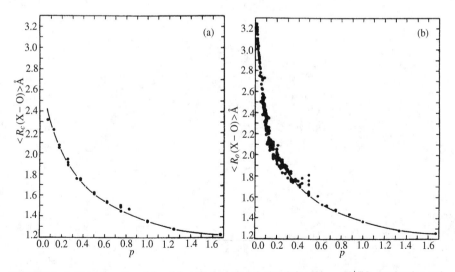

FIGURE 4. Scatter diagrams of bond length data (a) calculated for H_{2n-m} $X^{+m}O_n$ molecules and (b) observed for main group cations vs the bond order parameter p. The expression $R(X–O) = 1.39 p^{-2/9}$ serves to model both sets of data equally well

In addition, a statistical analysis of the observed bond length data shows that more than 99% of the variation of $\ln(<R_o(X-O)>)$ can be explained in terms of a linear dependence upon $\ln(p)$. The observation that the same expression can be used to model the observed and calculated data sets indicates that the bond lengths in oxide molecules and crystals are similar with p playing a similar role in both systems despite the smaller size and density of a molecule.

A graph-theoretic study[7] of the bond length variations observed for 10 different silicate crystals has since yielded a similar expression connecting bond length and resonance bond number, n. A resonance bond number is defined to be the average number of times a bond appears in the family of subgraphs of the connectivity graph that are constrained to have the degree of each node equal to the valence of the atom represented by that node. In the study, resonance bond numbers were calculated for all of the nonequivalent bonds in representative blocks of atoms isolated from the structures of ten different silicate crystals, using an algorithm based on graph theory. A scatter diagram of the resulting n-values versus the individual observed $<R_o(X-O)>$ bond lengths (Figure 5) not only matches the trends discussed above when n is equated with s, but a regression analysis of the data set yields the expression $R(X-O) = 1.39(n/r)^{-\beta}$ where $\beta = ca$ 2/9, in agreement with the form of the expression obtained for the data in Figure 4. A statistical analysis shows that more than 95% of the variation of the bond lengths in the silicate crystals can be explained in terms of the resonance bond numbers calculated for the X—O bonds in the representative fragments of the crystals. Despite the different roles attached to the valence electrons, it is apparent that both models yield similar numbers for the strength of a particular bond.

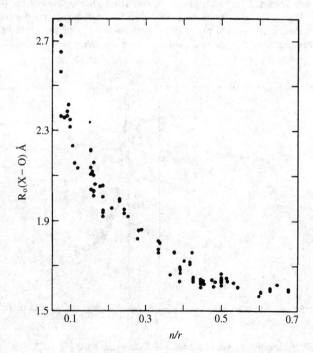

FIGURE 5. A scatter diagram of resonance bond numbers, n, calculated for the individual bonds in ten silicate crystals vs the observed bond lengths where r is the row number of the X-cation

With the discovery that an expression of the form $R(X-O) = \kappa p^{-\beta}$ can be used to model the average bond lengths in oxide molecules and crystals, it was subsequently found that the average bond lengths observed for the coordination polyhedra in sulfide, nitride and fluoride crystals and molecules can also be modeled with a power law expression with essentially the same β-value (ca 2/9) but with κ-values of 1.93, 1.49 and 1.37, respectively[8–10]. In addition, bond length data calculated for cation-containing polyhedra in chemically similar molecules yielded expressions that are statistically identical to those obtained for each set of crystal data. As the relative change of the expression $f(p) = \kappa p^{-\beta}$, as a function of bond order p, is $-\beta/p$ and as β is the same for the bonds in oxide, sulfide, nitride and fluorides molecules and crystals, one can conclude that the relative change in bond length as a function of p, for any given bond order, is the same for all four anions. In short, if a given cation forms a bond of a given bond order in a coordination polyhedron and if it is replaced by another cation, then the relative change in bond length is indicated to be the same, regardless of whether the cation comprises a molecule or a crystal or whether it is bonded to an oxide, nitride, sulfide or a fluoride anion.

III. A POTENTIAL ENERGY SURFACE FOR THE Si−O−Si SKELETON OF THE DISILICIC ACID MOLECULE

The geometry of the disilicic acid $H_6Si_2O_7$ molecule has been partially optimized a number of times ranging from an SCF Hartree–Fock method with an STO-3G basis set to a hybrid density functional Becke3LYP method with a 6-311G(2d,p) basis set. To gain insight into the force field and the energetics of the skeletal Si−O−Si unit, a potential energy surface has been generated with energies calculated for the molecule for more than 70 different combinations of Si−O bridging bond lengths and Si−O−Si angles using a STO-3G basis set and assuming C_{2v} point symmetry[4] (Figure 6). The point of minimum energy (denoted by a '+' sign in the figure) defines an equilibrium Si−O bond length of 1.60 Å and an equilibrium Si−O−Si angle of 142°. Experimental Si−O bond lengths, R(Si−O), and Si−O−Si angle data, observed for the Si−O−Si units of the silica polymorphs quartz, cristobalite (Figure 1g), tridymite (Figure 1h, Plate 4) and

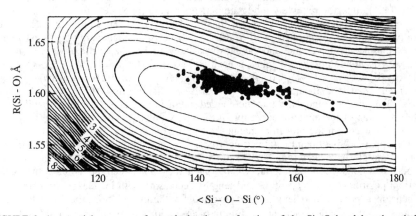

FIGURE 6. A potential energy surface calculated as a function of the Si−O bond length and the Si−O−Si bond angle of the Si−O−Si skeletal unit of the disilicic acid $H_6Si_2O_7$ molecule with a STO-3G basis set. The level contour lines are drawn at intervals of 2.6 kJ mol^{-1}. The bullets represent Si−O bond lengths and Si−O−Si angle data observed for the silica polymorphs. The minimum energy Si−O bond length (1.60 Å) and Si−O−Si angle geometry of the molecule is plotted as a '+' sign

coesite (Figure 1i, Plate 5), are plotted on the surface. The data follow the general trend of the surface, but the observed Si−O bond lengths are, for a given Si−O−Si angle, ca 0.015 Å longer, on average, than generated for the molecule. Also, for a given bond length, the data are shifted to wider angles by ca 5°. The Si−O bond lengths recorded for the silica polymorphs show a relatively small range of values between ca 1.58 Å and ca 1.63 Å while the Si−O−Si angles show a relatively wide range of values between ca 135° and 180°. The shape of the surface conforms with this result with the minimum energy conformation of the molecule lying at the bottom of a relatively narrow, cirque-shaped valley, bounded laterally on both sides by steeply rising energy barriers encountered with a departure of the Si−O bond length from its minimum energy value of 1.60 Å. The valley is blocked on the narrow angle side of the valley floor by a steeply rising headwall while the valley floor extends virtually unimpeded as the Si−O−Si angle widens with the slope along the valley floor, increasing very gradually from the valley bottom with an opening of the angle. The trace of valley floor shows a slight but well-defined curvature which conforms with the curvilinear R(Si−O) vs <Si−O−Si correlation observed for the silica polymorphs[11], a trend that is reproduced by MP2/6-31G* and Becke3LYP/6-311G(2d,p) level calculations on the disilicic acid molecule[5]. The topography of the valley floor indicates that the Si−O−Si energy barrier to linearity is small. If the binding forces that govern the Si−O−Si unit in the disilicic acid molecule and the silica polymorphs are similar, then the force required to deform the angle from its minimum energy value is expected to be small in both systems. Thus, a broad continuum of angles is expected to occur in agreement with the relatively large range of angles observed for the silica polymorphs and observed and calculated for silicic acid and siloxane molecules[12–14]. Also, because of the steeply rising energy headwall at the narrow angle end of the valley, Si−O−Si angles less than ca 120° are indicated to destabilize a structure in agreement with the fact that the Si−O−Si angles reported for a large number of silicate crystals[15] are observed in the range between 120 and 180°.

Like the disilicic acid molecule and silica, the minimum energy Si−O and X−O bond lengths and Si−O−X angles calculated for the skeletal Si−O−X units of a number of $(H_6SiX^{+n}O_7)^{-4+n}$ (X = B, Al, Be) molecules conform with those reported for framework silicate crystals[4,16]. Also, the relatively narrow range of angles observed for the Si−O−B (120-142°) units, the relatively moderate range observed for Si−O−Be (118-152°) units and relatively wide range observed for the Si−O−Al (115-180°) units of each crystal conforms with the shape of the well of the <Si−O−X-potential energy curve calculated for each molecule, the deeper and the narrower the well, the smaller the observed range of angles. The minimum energy angle of each curve (< Si−O−B = 125°, < Si−O−Be = 131°, < Si−O−Al = 139°) also agrees to within a few degrees of each respective average observed value of a silicate (< Si−O−B = 129°, < Si−O−Be = 127°, < Si−O−Al = 138°). This evidence indicates that the bonded interactions that govern the variability and average value of the angle adopted by an Si−O−X unit in a silicate crystal are similar to those that govern the unit in a chemically similar molecule.

The apparent Si−O bond lengths (uncorrected for thermal motion) observed for the silica polymorphs have been found to be shorter than the actual interatomic bond lengths (the bonded and nonbonded interatomic separations that exist between the atoms at their equilibrium positions), with apparent bond lengths decreasing with increasing temperature[17]. When the apparent bond lengths are corrected for thermal motion, the corrected bond lengths for the bonds described in Figure 6 are estimated to be $ca = 0.01$ Å longer than those plotted. As shown in the figure, the bridging Si−O bond length calculated for disilicic acid molecule at absolute zero (without zero-point vibration) with an STO-3G basis set is about 0.01–0.02 Å shorter than the apparent Si−O bond lengths for the silica polymorphs and consequently substantially shorter than the corrected bond lengths.

2. A molecular modeling of the bonded interactions of crystalline silica   111

The geometry of the disilicic acid molecule has since been partially optimized with the relatively more accurate hybrid Becke3LYP method with a 6-311G(2d,p) basis set assuming a staggered structure with C_s point symmetry with its H atoms constrained so as to avoid the formation of bonded interactions between the H atoms of one tetrahedron of the molecule and the O atoms of the other. The calculations yielded a minimum energy Si−O bridging bond length of 1.612 Å and a Si−O−Si angle of 145°. The thermally corrected <Si−O> average bond lengths calculated for the silica polymorphs quartz (1.615 Å), cristobalite (1.615 Å) and coesite (1.612 Å) match the minimum energy bond calculated for the molecule rather well. Also, the minimum energy Si−O−Si angle is only 2.5° narrower than the average value (147.4°) recorded for the silica polymorphs[18]. Thus, if a potential energy surface like the one in Figure 6 were recalculated at the Becke3LYP/6-311G(2d,p) level, then one might expect that the thermally corrected individual <Si−O> bond lengths for the silica polymorphs would lie along the energy valley of the surface rather closely in better agreement than that displayed by the data in the figure.

The two orthogonal curvatures of the energy surface evaluated at the point of minimum energy in Figure 6 indicate that the harmonic stretching force constant of the bridging Si−O bond of the disilicic acid molecule is roughly two orders of magnitude larger than the bending force constant of its Si−O−Si angle, in conformity with the narrow range of Si−O bond lengths and wide range of Si−O−Si angles observed for the silica polymorphs. As is well known, the force constant of a bond can provide useful information about the 'stiffness' of a bond, its 'variability' and the binding forces that exist between a pair of bonded atoms like Si and O. With the proposal that a Morse curve can be used to characterize the electronic energy of a diatomic molecule, it was also proposed that the force constant of the bond can be related to bond length, R, by the power expression $f(R) = \kappa R^{-\alpha}$ where α was taken to be ca 6 and κ is a constant that depends on the identity of the bond[19]. With spectroscopic force constant data determined for the Si−O bonds in a variety of molecules and crystals, together with scaled harmonic force constant data calculated for the molecules H_2SiO_3, H_4SiO_4, $H_6Si_2O_7$ and $H_{12}Si_5O_{16}$ with a 6-31G* basis set, the expression $f(Si-O) = 7.5 \times 10^3 R(Si-O)^{-5.4}$ Newtons/meter (N/m) was calculated that relates the force constant of an Si−O bond, $f(Si-O)$, to its length[20]. When used to generate the force constant for the Si−O bond of quartz, it yielded a value of 593 N/m compared with that observed (597 N/m)[21] and that calculated (600 N/m) for the H_4SiO_4 molecule at the MP2/6-31G** level. It was also reported in the study that the polyhedral compressibilities calculated for oxide, nitride and sulfide coordination polyhedra are similar to those reported for crystals[20].

Of the 3000 known minerals, more than 900 are silicates. Almost all silicates contain the silicate SiO_4 tetrahedra while about 20 are known to contain the SiO_6 octahedra, but none is known to contain an SiO_5 5-coordinate polyhedra. A few exceptions like wadeite (Figure 1j, Plate 5) contain both SiO_4 and SiO_6 polyhedra. Those with SiO_4 tetrahedra either contain monomeric and/or condensed corner sharing SiO_4 tetrahedra with the bulk of silicates containing condensed SiO_4 tetrahedra linked together by Si−O−Si units. The fact that the chemistry of the mineral world is dominated by Si and O certainly must play a role in determining the preponderance and the large number of different kinds of silicates that occur in nature. But the compliant nature of the Si−O−Si unit has also been argued to play a role as well[5,16,22]. Given that the Si−O−Si unit can be easily deformed from its equilibrium value without expending much energy, silicate tetrahedra can be linked together with other such tetrahedra in numerous ways as oligosilicates, cyclosilicates, inosilicates and phyllosilicate anions of various types and as a large variety of tectosilicate structures (see Reference 15 for elegant drawings of many of the condensed anionic units that silicate tetrahedra can adopt) without excessive destabilization unless the structure either requires an Si−O−Si angle that is less than ca 120° or unless the bond length–bond strength requirements of the bonds of the structure are poorly satisfied. The glass-forming tendencies of silica and the ability for silica to adopt a large variety of structure types can also be ascribed in part to the flexible nature of the Si—O—Si unit and the ease with which the Si—O—Si unit can be bent without excessively destabilizing the resulting structure[5,16,22]. The relatively high compressibility and expansion properties of quartz, cristobalite and tridymite can also be related to the compliant nature of the Si—O—Si angle.

As observed earlier, in modeling the Si—O—Si unit of silica and related silicates, the geometry of the $H_6Si_2O_7$ molecule was constrained during its optimization so that the terminating H atoms of each of the silicate groups were directed away from the O atoms of the other group so as to avoid the formation of one or more O···H bonded interactions. If such a constraint is not made, then O···H bonded interactions can be

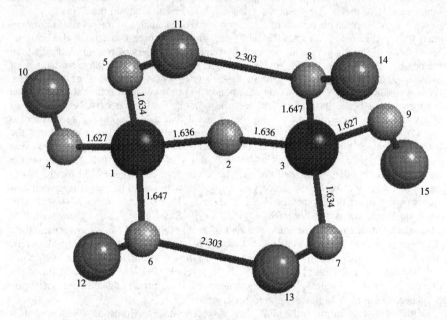

FIGURE 7. A drawing of a ball and stick model of a minimum energy structure of the disilicic acid molecule, $H_6Si_2O_7$. The large dark gray spheres represent Si, the small light gray spheres represent O and the intermediate-size, darker gray spheres represent H. Bond lengths in Å are given next to each of the SiO bonds and the separations between the H and O atoms that define the two O···H interactions are given. The dihedral angles for the molecule, defined in terms of the integers assigned to each atom, are as follows: D(Si3,O2,Si1,O4), −172.38; D(Si3,O2,Si1,O5), 66.12; D(Si3,O2,Si1,O6), −55.44; D(O7,Si3,O2,Si1), 66.12; D(O8,Si3,O2,Si1), −55.43; D(O9,Si3,O2,Si1), −172.37; D(H10,O4,Si1,O2), −84.53; D(H10,O4,Si1,O5), 37.73; D(H10,O4,Si1,O6), 161.5; D(H11,O5,Si1,O2), −29.36; D(H11,O5,Si1,O4), −153.72; D(H11,O5,Si1,O6), 85.76; D(H12,O6,Si1,O2), 179.41; D(H12,O6,Si1,O4), −60.16; D(H12,O6,Si1,O5), 60.47; D(H13,O7,Si3,O2), −29.35; D(H13,O7,Si3,O8), 85.77; D(H13,O7,Si3,O9), −153.71; D(H14,O8,Si3,O2), 179.34; D(H14,O8,Si3,O7), 60.39; D(H14,O8,Si3,O9), −60.24; D(H15,O9,Si3,O2), −84.56; D(H15,O9,Si3,O7), 37.69; D(H15,O9,Si3,O8), 161.48. The predicted energy of the molecule is E(RB + HF -LYP) = −1109.83245 au

expected to form in the calculation with a concomitant lengthening of the Si—O bridging bonds and a narrowing of the Si—O—Si angle[23–25]. Because O···H bonds are absent in the silica polymorphs and in almost all silicates, an unconstrained molecule with O···H bonds is not considered to be a satisfactory moiety for modeling the structure and the elastic properties of crystalline silica.

As a matter of interest, the geometry of the $H_6Si_2O_7$ molecule was fully optimized for this study at the Becke31yp 6-311G(2d,p) level, assuming C_1 point symmetry without any restrictions imposed on the positions of its atoms. A drawing of a ball and stick model of the resulting minimum energy structure is displayed in Figure 7, where the bonded atoms are connected by the sticks. The minimum energy Si—O (br) bridging bond lengths (1.636 Å) of the molecule are ca 0.02 Å longer and the Si—O—Si angle (126.8°) is ca 20° narrower than that calculated for the constrained molecule (see above). The relatively large change in the angle is expected, given its compliant nature. The longer Si—O (br) bond lengths and the narrower Si—O—Si angle can be ascribed to the reduction of the electron density at the bond critical point of the Si—O bond induced by the two O···H bonded interactions, O6···H13 and O8···H11 (Figure 7). The atoms comprising these interactions are at a separation of 2.303 Å. Because a bond critical point exists in the electron density distribution between these atoms at a distance of ca 1.42 Å from each O atom, both O6···H13 and O8···H11 qualify as bonded interactions. But, since the value of the electron density at the critical point, $\rho(\mathbf{r}_c)$, is only 0.08 e Å^{-3}, the bond is indicated to be relatively weak with an $\nabla^2\rho(\mathbf{r}_c)$ value of 1.0 e Å^{-5}. As expected from the $\rho(\mathbf{r}_c)$ values calculated for the molecule, the Si—O bonds involving O6 and O8 are longer (1.647 Å) than either of the two remaining Si—O bonds (1.627, 1.634 Å) of the silicate groups. It is noteworthy that bonds in the molecule with identical environments have identical bond lengths. The dihedral angles and the predicted energy of the molecule are given in Figure 7.

IV. BOND CRITICAL POINT PROPERTIES OF THE ELECTRON DENSITY DISTRIBUTION OF THE SKELETAL Si—O—Si UNIT

If the electron density distribution of a crystal is similar to that of a chemically similar molecule, it can be concluded that the force fields of the two structures are similar. In a mapping of the electron density of coesite[26], deformation electron density distributions, $\Delta\rho(\mathbf{r})$, were generated from experimental X-ray diffraction data to learn whether any of the features in the distribution can be related to the observed bond length and angle variations and the character of the Si—O bond. As the observed Si—O bond lengths in coesite decrease with increasing Si—O—Si angle, it was expected that the heights of the peaks along the bonds in the maps would increase in value as the Si—O bond decreases in length. Although the resulting maps display peaks along each of the Si—O bonds, ranging in height between ca 0.3 and ca 0.5 e Å^{-3}, no statistically significant trend between peak height and bond length could be established.

Because of the nonquantitative nature of $\Delta\rho(\mathbf{r})$ maps[27,28], the X-ray diffraction data recorded for coesite[26] was used to generate a total electron density distribution, $\rho(\mathbf{r})$, for the mineral. In an analysis of the bond critical point properties of the distribution, Downs[29] located the critical points along each of its Si—O bonds, determined the value of the electron density and the Laplacian of $\rho(\mathbf{r})$ at each of these critical points, $\nabla^2\rho(\mathbf{r}_c)$, and mapped $-\nabla^2\rho(\mathbf{r})$ over the domain of each of its Si—O—Si skeletal units. A mapping of the total electron density distribution and its topological properties has a distinct advantage over a mapping of the deformation density in that *'The derivation of a unique and physically meaningful difference (deformation) electron density is a problem that cannot be solved since the choice of the promolecular reference density always implies some arbitrariness'*. This statement[30] seems to be borne out by the following results obtained for coesite: (1) The heights of the peaks in the experimental deformation maps recorded along its Si−O bonds fail to show any obvious trends with the observed bond lengths, and (2) The heights of the peaks in deformation maps calculated for the Si−O bridging bonds of the $H_6Si_2O_7$ molecule actually increase in height as the lengths of the bonds increase (and the Si−O−Si angle decreases) (see Reference 19, Figure 6). But, on the other hand, the value of $\rho(\mathbf{r}_c)$ provided by calculations of the total electron density distribution for a number of silicic acid molecules including the disilicic acid molecule shows that the value of $\rho(\mathbf{r}_c)$ increases in a regular way as the Si−O bond decreases in length (Figure 8).

For purposes of comparison with the results obtained for coesite[29], the electron density distribution of the partially, optimized conformer of $H_6Si_2O_7$ (shown in Figure 8) has been generated using the wave functions obtained in a density functional Becke3LYP/6-311G(2d,p) level calculation. The total electron density distribution measured for coesite attains an average value of 1.05 e Å^{-3} at an average distance of 0.936 Å from the bridging O atom to the critical point, $\mathbf{r}_c$, measured along the Si−O bonds from the oxide ions with an average $\nabla^2\rho(\mathbf{r}_c)$-value at this point of +20.4 e Å^{-5}. These observations together with the relationship between the bonded radius of the oxide ion and the electronegativity of

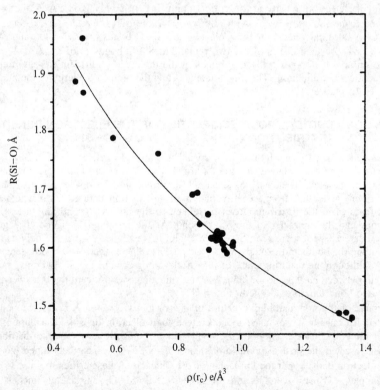

FIGURE 8. A scatter diagram of the minimum energy Si−O bond length, R(Si−O), calculated for a number of hydroxyacid molecules vs the value of the electron density, $\rho(r_c)$, evaluated at $(3,-1)$ critical points. The geometries of the hydroxyacid molecules used to prepare the plot were those of minimum energy at the Hartree−Fock 6-31++G** level

the cation to which it is bonded[2] and the negative value of the local energy density, $H(\mathbf{r}_c)$, at $\mathbf{r}_c$ (ca $- 0.3 \text{H Å}^{-3}$) indicate that the Si—O bonds in coesite have significant covalent character[27,30]. In addition, the curvatures of $\rho(\mathbf{r}_c)$, λ_1 and λ_2, measured at $\mathbf{r}_c$, perpendicular to the Si—O bonds, are -7.2 and -6.8 e Å^{-5}, respectively, on average, while the average curvature of $\rho(\mathbf{r})$ at $\mathbf{r}_c$ along the bond path, λ_3, is observed[29] to be 34.4 e Å^{-5}, resulting in a $\nabla^2(\mathbf{r}_c)$ value of 20.4 e Å^{-5}. The electron density distribution calculated for the molecule has a value of 1.00 e Å^{-3} (*ca* 5% less than that observed on average for coesite) at the critical point along each Si—O bond at a distance of 0.943 Å from the bridging oxide ion (*ca* 1% larger than observed on average for coesite). The curvatures, λ_1, λ_2 and λ_3 evaluated at $\mathbf{r}_c$ are -7.41, -7.35 and 35.6 e Å^{-5}, respectively, resulting in a $\nabla^2 \rho(\mathbf{r}_c)$-value of 20.9 e Å^{-5} (*ca* 2% larger than observed, on average, for coesite). As the curvature of $\rho(\mathbf{r})$ is a very sensitive probe of the topology of $\rho(\mathbf{r})$, as discussed above, it is apparent that a close connection exists between the topological properties of the electron density distribution observed for the crystal and those calculated for the model molecule.

V. A REPRODUCTION OF THE STRUCTURES AND RELATED PROPERTIES OF THE KNOWN SILICA POLYMORPHS

It has been stressed[31] in a study of chemical bonds in crystals that the bonding in molecules differs from that in crystals because bonds are less well defined in crystals and because most crystals are usually denser, much larger and more symmetrical than molecules. Despite these differences, it was acknowledged in the study that a wide range of data *'Shows that nearest neighbor interactions determine most of the properties of a material'*. At the very least, the evidence examined in this review conforms with this statement. For example, the force field that governs the geometry and the electron density distribution of the Si—O—Si skeleton of the disilicic acid molecule can be treated as virtually identical with that of a silica polymorph. The evidence upon which this assertion is based is twofold. First, the Si—O bond length and the Si—O—Si angle variations observed for the silica polymorphs conform in large part with the topographic features of a potential energy surface of the Si$_2$O$_7$ skeleton calculated for the disilicic acid molecule. Second, the bond critical point properties of the electron density distribution observed for the Si—O—Si units of a silica polymorph agrees to within a few percent with that calculated for the molecule. This evidence taken with the observation that spectroscopically determined force constants of a silica polymorph agree well with that calculated for the molecule provides a basis for believing that a silica polymorph like quartz can be viewed as a giant molecule bound together by the same forces that act between the Si and O atoms of the disilicic acid molecule[5]. It also provides a basis for understanding why it has been possible to reproduce the crystal structure of quartz to within *ca* 2% and its bulk modulus to within *ca* 5%, using a potential energy function calculated for the disilicic acid molecule[32]. Since then, a number of other workers have been successful not only in reproducing the structures and volume compressibilities of quartz and cristobalite using a potential energy function based on a molecule, but they have also been successful in reproducing the structures and bulk moduli of coesite and stishovite (Figure 1k, Plate 6) and modeling the observed properties of the α-β transition of quartz and the negative Poisson ratio observed for cristobalite[33–40].

VI. A GENERATION OF NEW STRUCTURE TYPES FOR SILICA

It is one thing to reproduce the known structure of a crystalline material like silica using a molecular potential energy function, but it is quite another to generate the possible crystal structures that silica can adopt with a molecular-based potential energy function, starting with a random arrangement of Si and O atoms in a structure with $P1$ symmetry. To attack this problem, we used simulated annealing. Simulated annealing is a physically initutive strategy that can be used to search for global and local minimum energy crystal structures for silica using a molecular potential energy function based on that calculated for the Si_2O_7 skeleton of the $H_6Si_2O_7$ molecule[37,38]. Simulated annealing strategies accept all downhill steps but, unlike Newton and quasi-Newton methods, allow some uphill steps that increase the value of a potential energy function in a search for viable, low energy structures[37,38].

Calculations have been completed starting with four, six and eight formula units of SiO_2 randomly distributed in a structure with triclinic $P1$ symmetry[38]. Of the several thousand periodic structure types that have been generated, more than two-thirds were found to be framework structures where each Si atom is 4-coordinate and each O atom is 2-coordinate. Further study shows that these framework structures can be classified into a wide variety of distinct structure types. Despite the assumption in the calculations of $P1$ triclinic symmetry, more than two-thirds of the structures possess symmetries higher than $P1$ ranging between the monoclinic space group symmetry Pc and the tetragonal space group symmetry $I\bar{4}2d$. The resulting structures exhibit symmetrically equivalent cell dimensions that agree to within 0.0001 Å and equivalent interaxial angles that agree to within 0.0001°. A large number of low energy structure types match those observed for the left- and right-handed polymorphs (calculated in equal numbers) of quartz and cristobalite, mixed stacking sequences of tridymite and cristobalite and the molecular crystal silica-W. Several exhibit the framework structures of several known aluminosilicates including monoclinic $CaAl_2Si_2O_8$ and the orthorhombic zeolites Li-A(BW) and NaI[42]. Others exhibit structures that are similar to those of cancrinite, sodalite and bikitaite (Figure 11 Plate 6). In addition, the networks defined by the Si atoms of a number of the structures match those enumerated by earlier workers[43,44]. However, many of the remaining structures represent new structure types for silica yet to be synthesized or discovered in nature. Several of the structures generated in the calculations on silica are displayed in Figure 9 (Plates 7 and 8). In the study, two silica structures were considered to be equivalent if their energies are equal and the Schläfli symbols calculated for the 4-connected nets of Si atoms and the coordination sequences of the Si atoms out to the 10th coordination shell for the two are identical[41].

VII. CONCLUDING REMARKS

The notion that molecules can be used to model bonded interactions in crystals is not new. Indeed, it is an idea that has been around for more than 50 years when J. C. Slater[45] concluded, on the basis of the similar types of rigidity and geometry exhibited by several aliphatic molecules and diamond, that *'A diamond is really a molecule of visible dimensions held together by just the same forces acting in small molecules'*. More recently, with the completion of an accurate gas-phase structural determination of disiloxane, it was concluded[43] that the bonding picture of the skeletal Si—O—Si unit of the molecule is similar to that of a silicate crystal because the geometry of the unit $[R(Si-O) = 1.632$ Å; $<$ Si—O—Si $= 142.2°]$ is similar to that observed, on average, for a silicate crystal $[R(Si-O) = 1.626$ Å; $<$ Si—O—Si $= 144°]$. It is noteworthy that a structural analysis of a disiloxane crystal grown at liquid nitrogen temperatures reveals that the geometry of the molecule in the crystal is not significantly different from that in the gas phase, despite differences between the forces acting on the atoms in the gas-phase molecule and those in the crystal[47]. In addition, a survey of the literature shows that the average Si—O bond length (1.634 Å) and Si—O—Si angle (144°) observed for a large variety of siloxane molecules are also similar to those observed for disiloxane and for silicates. Indeed, the structures of the condensed tetrahedral anions of organosiloxanes and silicates are so similar that they can be classified with the well known scheme used by Bragg to classify silicates[48]. Further, as observed above, the calculated geometry and the topological properties of the electron density distribution of the Si_2O_7 skeleton of disilicic acid are strikingly similar to those observed for a silica polymorph. Collectively, these results not only suggest that the force fields that govern the structures and electron density distributions of silicic acid molecules and silicates are not all that different, but they also suggest that the silica polymorphs can be viewed, as asserted earlier[5], as giant molecules bound together by the same forces that bind the Si and O atoms of the Si_2O_7 skeleton of the disilicic molecule, despite the greater size, density and symmetry of the former. It also provides a basis for understanding why the structures and the elastic properties of the silica polymorphs can be modeled with a molecular potential energy function and why a large number of new silica structure types can be generated using simulated annealing strategies and a molecular potential energy function.

Finally, as observed above, the evidence suggests that the forces that bind Si and O ions together in a silicate crystal can be treated as if localized as in a molecule. If the Si—O bond is of intermediate type as asserted by Pauling and others, then it would appear that the effective charges on the ions of the crystal would either be relatively small in conformity with his electroneutrality principle[49] or that the Si cation forms such a strong bonded interaction with its coordinating oxide ions that the field of the crystal has little effect on the geometry and the electron density distribution of a silicate tetrahedral oxyanion. In short, any model that is proposed for the Si—O bond ought to explain why the forces that govern the structure, the elastic properties and the electron density distribution of the Si—O bond of a silicate behave as if in a molecular environment[2].

VIII. ACKNOWLEDGMENTS

This paper was written when GVG was a JSPS Fellow and Invited Research Professor at Kyoto University, Kyoto, Japan and MBB was on Research Leave at UC Berkeley. In particular, GVG wishes to thank Professor Osamu Tamada and Dr. Masanobu Matsumoto for making his stay in Japan a very pleasant and stimulating experience. He also wishes to thank Professors Shegio Sueno, Satoshi Sasaki, Nobuo Ishizwara, Fumikuki Marumo and Yoshi Takeuchi for their kind hospitality. MBB wishes to thank Professor Mark Bukowinski at UC Berkeley for his hospitality and for making his stay an enjoyable and productive one. Dr. F. C. Hill is thanked for reading the manuscript and for her helpful comments. This work was supported by NSF Grant EAR-9627458.

IX. REFERENCES

1. K. H. Wedepohl, *Geochemistry*, Holt, Reinhart and Winston, New York, 1983.
2. G. V. Gibbs, J. W. Downs and M. B. Boisen, Jr., in *SILICA: Physical Behavior, Geochemistry and Materials Applications* (Eds. P. J. Heany, C. T. Prewitt and G. V. Gibbs), Chap. 10, American Mineralogist, Washington, D.C., 1994, p. 331.
3. L. Pauling, *J. Am. Chem. Soc.*, **51**, 1010 (1929).
4. I. D. Brown and R. D. Shannon, *Acta Cryst.*, **A29**, 266 (1973).
5. G. V. Gibbs, *Am. Mineral.*, **67**, 421 (1982).
6. G. V. Gibbs, L. W. Finger and M. B. Boisen, Jr., *Phys. Chem. Min.*, **14**, 327 (1987).
7. M. B. Boisen, Jr., G. V. Gibbs and Z. G. Zhang, *Phys. Chem. Min.*, **15**, 409 (1988).
8. K. L. Bartelmehs, G. V. Gibbs and M. B. Boisen, Jr., *Am. Mineral.*, **74**, 620 (1989).
9. L. A. Buterakos, G. V. Gibbs and M. B. Boisen, Jr., *Phys. Chem. Min.*, **19**, 127 (1992).
10. J. S. Nicoll, G. V. Gibbs, M. B. Boisen, R. T. Downs and K. L. Bartelmehs, *Phys. Chem. Min.*, **20**, 617 (1994).

11. G. V. Gibbs, C. T. Prewitt and K. J. Baldwin, *Z. Kristallogr.*, **145**, 108 (1977).
12. B. C. Chakoumakos, R. J. Hill and G. V. Gibbs, *Am. Mineral.*, **66**, 1237 (1981).
13. I. L. Karle, J. M. Karle and C. J. Nielson, *Acta Cryst.*, **C42**, 64 (1986).
14. H. B. Burği, K. W. Törnroos, G. Calzaferri and H. Bürgy, *Inorg. Chem.*, **32**, 4914 (1993).
15. F. Liebau, *Structural Chemistry of Silicates: Structure, Bonding, and Classification*, Springer-Verlag, Berlin, 1985.
16. K. L. Geisinger, G. V. Gibbs and A. Navrotsky, *Phys. Chem. Min.*, **11**, 266 (1985).
17. R. T. Downs, G. V. Gibbs, and M. B. Boisen, Jr., *Am. Mineral.*, **75**, 1253 (1990).
18. E. P. Meagher, J. A. Tossell and G. V. Gibbs, *Phys. Chem. Min.*, **4**, 11 (1979).
19. P. M. Morse, *Phys. Rev.*, **34**, 57 (1929).
20. F. C. Hill, G. V. Gibbs and M. B. Boisen, Jr., *Struct. Chem.*, **6**, 349 (1994).
21. J. Etchepare, M. Merian, and L. Smetarkine, *J. Chem. Phys.*, **60**, 1873 (1974).
22. K. L. Geisinger and G. V. Gibbs, *Phys. Chem. Min.*, **7**, 204 (1981).
23. D. J. M. Burkhart, B. H. W. S. DeJong, A. J. H. M Meyer and J. H. van Lenthe, *Geochim. Cosmochim. Acta*, **55**, 3453 (1991).
24. J. D. Kubicki and D. Sykes, *Am. Mineral.*, **78**, 253 (1993).
25. B. J. Teppen, D. M. Miller, S. Q. Newton and L. Schafer, *J. Phys. Chem.*, **98**, 12545 (1994).
26. K. L. Geisinger, M. A. Spackman and G. V. Gibbs, *J. Phys. Chem.*, **91**, 3237 (1987).
27. D. Cremer and E. Erakar, *Angew. Chem., Int. Ed. Engl.*, **23**, 627 (1984).
28. M. A. Spackman and E. N. Maslen, *Acta Cryst.*, **A41**, 347 (1985).
29. J. W. Downs, *J. Phys. Chem.*, **99**, 6849 (1995).
30. D. Cremer, in *Modelling of Structure and Properties of Molecules* (Ed. Z. D. Maksić), Chap. 7, Halsted Press, New York, 1987, p. 125.
31. J. C. Phillips, in *Treatise on Solid State Chemistry*, Vol. 1 (Ed. N. B. Hannay), Chap. 1, Plenum Press, New York, 1973, p. 1.
32. A. C. Lasaga and G. V. Gibbs, *Phys. Chem. Min.*, **14**, 107 (1987).
33. L. Stixrude and M. S. T. Bukowinski, *Phys. Chem. Min.*, **15**, 199 (1988).
34. S. Tsuneyuki, H. Aoki, T. Tsukada and M. Matsui, *Phys. Rev. Lett.*, **64**, 776 (1990).
35. G. V. Gibbs, M. B. Boisen, Jr., R. T. Downs and A. C. Lasaga, in *Better Ceramics Through Chemistry, III* (Eds. C. J. Brinker, D. E. Clark and D. R. Ulrich), Mat. Res. Soc. Symp. Proc., Vol. 121, 1988, p. 155.
36. B. W. H. van Beest, G. J. Kramer and R. A. Santen, *Phys. Rev. Lett.*, **64**, 129 (1990).
37. J. R. Chelikowsky, H. E. King, Jr., N. Troullier, J. L. Martins and J. Glinnemann, *Phys. Rev.*, **B44**, 489 (1991).
38. G. J. Kramer and R. A. van Santen, *Phys. Rev. Lett.*, **64**, 1955 (1990).
39. N .R. Keskar and J. R. Chelikowsky, *Phys. Rev.*, **B46**, 1 (1992).
40. M. B. Boisen, Jr. and G. V. Gibbs, *Phys. Chem. Min.*, **20**, 123 (1994).
41. M. B. Boisen, Jr., G. V. Gibbs and M. S. T. Bukowinski, *Phys. Chem. Min.*, **21**, 269 (1994).
42. M. O'Keeffe, *Phys. Chem. Min.*, **22**, 504 (1995).
43. J. V. Smith, *Am. Mineral.*, **62**, 703 (1977).
44. M. O'Keeffe, *Z. Kristallogr.*, **196**, 21 (1991)
45. J. C. Slater, *Introduction to Chemical Physics*, McGraw-Hill, New York, 1939.
46. A. Almennigen, O. Bastiansen, V. Ewing, K. Hedberg and M. Traetteberg, *Acta Chem. Scand.*, **17**, 2455 (1963).
47. M. J. Barrow, E. A. V. Ebsworth and M. M. Harding, *Acta Cryst.*, **B35**, 2093 (1979).
48. W. Noll, *Chemistry and Technology of Silicones*, Academic Press, New York, 1968.
49. L. Pauling, *The Nature of the Chemical Bond*, Cornell University Press, Ithaca, New York, 1960.

CHAPTER 3

Polyhedral silicon compounds

AKIRA SEKIGUCHI

Department of Chemistry, University of Tsukuba, Tsukuba, Ibaraki 305, Japan
Fax: +81-298-53-4314; e-mail: sekiguchi@staff.chem.tsukuba.ac.jp

and

SHIGERU NAGASE

Department of Chemistry, Graduate School of Science, Tokyo Metropolitan University, Hachioji, Tokyo 192-03, Japan
Fax: +81-426-77-2525; e-mail: nagase@SNL70.chem.metro-u.ac.jp

I. INTRODUCTION	120
II. THEORETICAL STUDIES	120
A. Strain Energies	120
B. Bond Lengths and Angles	123
C. Substituent Effects	124
III. SYNTHESIS	125
A. Substituents	125
B. Precursors and Reducing Reagents	125
IV. TETRASILATETRAHEDRANE	125
V. HEXASILAPRISMANE	129
A. Synthesis	129
B. Structure	130
C. Absorption Spectra	133
D. Photochemical Reaction	134
VI. OCTASILACUBANE	136
A. Synthesis	136
B. Structure	137
C. Absorption Spectra	142
D. Reactivity	143
VII. ^{29}Si NMR SPECTRA	146
VIII. SPHERICAL CAGE COMPOUNDS	148
IX. EPILOGUE	150

The chemistry of organic silicon compounds, Vol. 2
Edited by Z. Rappoport and Y. Apeloig © 1998 John Wiley & Sons Ltd

X. ACKNOWLEDGEMENT 150
XI. REFERENCES 150

I. INTRODUCTION

Polyhedral carbon compounds such as tetrahedrane, prismane, and cubane have long fascinated chemists because of their unique properties and aesthetic appeal due to their high symmetry[1,2]. In an effort to enrich silicon chemistry, synthesis of the silicon analogues is a great challenge since it could lead to novel physical and chemical properties unexpected from the carbon compounds. However, such synthesis was believed to be impossible until recently. Triggered by the first successes in synthesizing an octasilacubane derivative[3] and a hexagermaprismane derivative[4], the chemistry of polyhedral compounds of the heavier group 14 atoms has progressed by rapid strides in the last few years and many new derivatives have been prepared[5-11]. At present, only the tin analogues of tetrahedrane and prismane and the entire series of the lead compounds are missing in the series. In this review the successful syntheses, isolation and characterization of polyhedral silicon compounds such as tetrasilatetrahedrane, hexasilaprismane and octasilacubane are summarized together with related theoretical calculations[9,10]. These are also compared with those for the germanium and tin analogues.

II. THEORETICAL STUDIES

A. Strain Energies

Polyhedranes ($C_{2n}H_{2n}$) such as tetrahedrane ($n = 2$), prismane ($n = 3$) and cubane ($n = 4$) have long been interesting synthetic targets. These are highly strained, as is apparent from their carbon bond angles that deviate greatly from the normal tetrahedral value of 109.5°. For instance, the strain energies of tetrahedrane and cubane are evaluated to be as large as 140.0 and 154.7 kcal mol^{-1}, respectively[12]. Since no experimental value is available for the silicon analogues, their strain energies were calculated at the HF/6-31G* level from the appropriate homodesmotic reactions[13]. Table 1 compares the calculated strain energies of the carbon and silicon compounds. The strain energies of 141.4 and 158.6 kcal mol^{-1} calculated for tetrahedrane and cubane are in close agreement with the experimental values[12]. The strain energy of tetrasilatetrahedrane is as large as that of tetrahedrane. However, as the number of four-membered rings increases, the strain of the silicon compounds is significantly decreased while it tends to increase in the carbon compounds. It is noteworthy that hexasilaprismane and octasilacubane are 32 and 65 kcal mol^{-1} less strained than prismane and cubane, respectively.

Also given in Table 1 are the strain energies of the still heavier germanium and tin compounds calculated at the HF/DZ(d) level[14]. Substitution of carbon or silicon by germanium

TABLE 1. Strain energies (kcal mol^{-1}) calculated using homodesmotic reactions[a]

M_nH_n	C	Si[b]	Ge	Sn
Tetrahedrane (M_4H_4, T_d)	141.4	140.9	140.3	128.2
Prismane (M_6H_6, D_{3h})	145.3	113.8	109.4	93.8
Cubane (M_8H_8, O_h)	158.6	93.5	86.0	70.1

[a] $M_nH_n + (3n/2)M_2H_6 \to n(MH_3)_3MH$.
HF/6-31G* for M = C and Si. HF/DZ(d) for M = Ge, Sn and Pb.
[b] The HF/DZ(d) values are 140.3 (Si_4H_4), 118.2 (Si_6H_6) and 99.1 (Si_8H_8) kcal mol^{-1}.

or tin atoms has again only a small effect on the relief of strain in the tetrahedrane system. This is because three-membered germanium and tin rings are as highly strained as three-membered silicon rings, as is apparent from the strain energies of cyclotrigermane (39.4 kcal mol^{-1}) and cyclotristannane (36.6 kcal mol^{-1})[14] which differ only slightly from that of cyclotrisilane (38.9 kcal mol^{-1})[13]; all of these strain energies are significantly larger than that of cyclopropane (28.7 kcal mol^{-1})[13]. However, the strain energies of hexasilaprismane and octasilacubane containing four-membered rings are further decreased upon substitution of the silicon atoms by germanium and tin atoms. Such relief of strain in the prismane and cubane systems reflects the fact that the strain energies of four-membered rings decrease successively in the order: cyclobutane (26.7 kcal mol^{-1})[13] > cyclotetrasilane (16.7 kcal mol^{-1})[13] > cyclotetragermane (15.2 kcal mol^{-1})[14] > cyclotetrastannane (12.2 kcal mol^{-1})[14]. It is a general trend that the strain of polyhedral compounds is progressively relieved as the number of four-membered rings increases and the skeletal atoms become heavier.

The above trend is also applicable to the larger members of the [n]prismane family ($M_{2n}H_{2n}$, $n > 4$). As Figure 1 shows[5,10,15], the strain energies of persila[n]prismanes

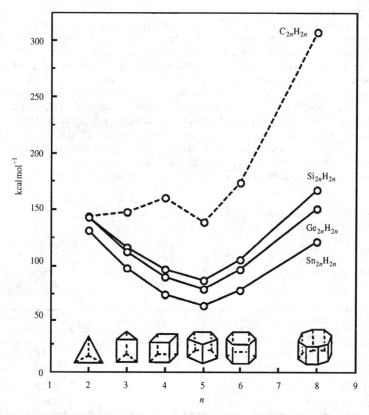

FIGURE 1. The strain energies of the tetrahedrane and [n]prismane system ($M_{2n}H_{2n}$) calculated at the HF/6-31G* level for M = C and Si and the HF/DZ(d) level for M = Ge and Sn. Reprinted with permission from Reference 10. Copyright 1995 American Chemical Society

($Si_{2n}H_{2n}$) are 53.2 ($n = 5$), 70.1 ($n = 6$) and 141.0 ($n = 8$) kcal mol^{-1} smaller than those of the corresponding $C_{2n}H_{2n}$, and even more so as M changes from Si to Ge and to Sn. The strain decreases when n increases from 2 to 5, because the number of four-membered rings increases and the bond angles in the n-membered rings at the top and bottom approach the ideal tetrahedral angle of 109.5°. However, the strain increases sharply with a further increase in n from $n = 5$, despite the increasing number of four-membered rings, because of the increasing deviation of the bond angles (120.0° for $n = 6$ and 135.0° for $n = 8$) in the n-membered rings from the tetrahedral angle. As a result, the strain energy is the smallest when $n = 5$. This suggests that a persila[5]prismane derivative is a reasonable synthetic target. It is interesting that a perstanna[5]prismane derivative was recently synthesized and isolated[16].

The concept of 'hybridization' is most helpful for an intuitive and unified understanding of the important differences in chemical bonding between carbon compounds and their heavier analogues. The size of the valence 2s atomic orbital of a carbon atom is almost equal to that of the 2p atomic orbitals, as measured by the atomic radii (r) of maximal electron density: $r_s = 0.646$ Å and $r_p = 0.644$ Å[17]. However, the valence s and p atomic orbitals differ successively in size for the heavier atoms: $r_p - r_s = 0.203$ (Si), 0.249 (Ge), 0.285 (Sn) and 0.358 (Pb) Å[17]. Therefore, the heavier atoms have a lower tendency to form s–p hybrid orbitals with high p character, and they tend to maintain the $ns^2\ np^2$ electronic valence configuration[5,14,18]. This property of the heavier atoms is favourable for forming bond angles of ca 90° and thus for forming four-membered rings with low strain. In contrast, formation of three-membered rings with bond angles of ca 60° becomes unfavourable since hybrid orbitals with sufficiently high p character are essential for a description of the 'bent-bond' orbitals[19].

This property is also reflected in the relative stability of the M_6H_6 valence isomers. As is well known, benzene (C_6H_6) has a unique stability due to cyclic delocalization of its 6π electrons and it is much more stable than its strained isomers such as Dewar benzene, benzvalene and prismane[1,2]. However, this situation changes drastically in the heavier system. As Table 2 shows, the heavier atoms prefer the isomers with a smaller number of double bonds, since they must hybridize highly to form double bonds[20]. As a result, the saturated prismane structure becomes much more stable than the benzene structure as M

TABLE 2. Relative energies (kcal mol^{-1}) of M_6H_6 valence isomers

M	Benzene D_{6h}	Dewar benzene C_{2v}	Benzvalene C_{2v}	Prismane D_{3h}
C[a]	0.0	81.1	74.9	117.6
Si[b]	0.0 (−0.0)[c]	4.1	−2.0	−8.1
Ge[b]	0.0 (−9.1)[c]	1.8	−1.2	−13.5
Sn[b]	0.0 (−23.1)[c]	−6.5	−11.0	−31.3

[a] MP2/6-31G*//HF/6-31G*.
[b] MP2/DZ(d)//HF/DZ(d) from Reference 15.
[c] Values in parentheses are for chair-like puckered structures of D_{3d} symmetry from Reference 5.

becomes heavier[15,21,22]. This creates another interesting difference between the chemistry of carbon and in heavier congeners.

B. Bond Lengths and Angles

Figure 2 shows the calculated optimized structures of tetrasilatetrahedrane (Si_4H_4), hexasilaprismane (Si_6H_6) and octasilacubane (Si_8H_8) at the HF/6-31G* level[13]. The Si—Si bond lengths in Si_4H_4 are shorter than the single bond length of 2.352 Å calculated for $H_3Si-SiH_3$, and it increases in the order Si_4H_4 (2.314 Å) < Si_6H_6 (2.359 Å and 2.375 Å) < Si_8H_8 (2.396 Å). It is interesting that the Si—Si bond lengths are shorter in the three-membered rings than in the four-membered rings, as is also calculated for the monocyclic rings: cyclotrisilane (2.341 Å) vs cyclotetrasilane (2.373 Å)[13,23]. This trend is enhanced in the heavier compounds.

However, bond lengths are not necessarily correlated with bond strengths; the bonds in three-membered rings are weaker than those in four-membered rings[5]. This is because the heavier atoms are forced to hybridize to a considerable extent in order to achieve and maintain the three-membered skeletons of a given symmetry, at the expense of a large energy loss. To compensate for this energy loss, the bond lengths between skeletal atoms shorten in order to form bonds as effectively as possible. However, the cost for hybridization is too large to be offset just by bond shortening, leading to higher strain and weaker bonds in the three-membered rings.

As a result of the high strain and weak bonds, the heavier polyhedral compounds consisting of only three-membered rings easily undergo bond stretching or bond breaking. As Figure 3 shows, for example, bicyclo[1.1.0]tetrasilane consisting of two fused three-membered rings is subject to 'bond-stretch' isomerism[24,25], unlike bicyclo[2.2.0]hexasilane consisting of four-membered rings which has only one isomer[24d]. The isomer with a longer central bond is more stable than that with a normal short bond length. Since in the 'bond-stretch' isomer, two bridgehead hydrogens approach one another closely to form a H—Si—Si bond angle of 93°, bulky substituents cannot be accommodated at the bridgeheads. It is interesting that a bicyclo[1.1.0]tetrasilane derivative with a normal central bond was synthesized by introducing bulky t-Bu groups at the bridgeheads[26].

Tetrasilatetrahedrane was calculated to correspond to a local minimum on the potential energy surface[27]. However, because of the fusion of four three-membered rings it collapses, almost without a barrier, by breaking two skeletal bonds, to an isomer having one four-membered ring of tetraradical character[28]. As shown schematically in Figure 4 (see also Figure 3), it is general that bond stretching and bond breaking take place so as to decrease the number of three-membered rings and instead increase the number of the less strained four-membered rings.

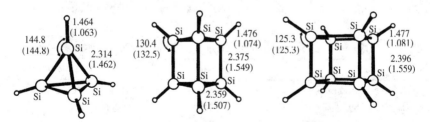

FIGURE 2. The HF/6-31G* optimized geometries of Si_4H_4, Si_6H_6 and Si_8H_8; bond lengths in Å and bond angles in degrees. The values in parentheses are for the corresponding carbon compounds. Reprinted with permission from Reference 13. Copyright 1995 The Chemical Society

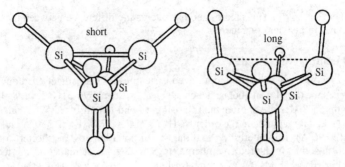

FIGURE 3. Two isomers of Si_4H_6

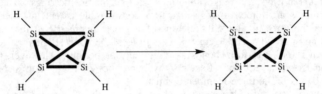

FIGURE 4. Bond stretch in Si_4H_4

C. Substituent Effects

Because of the high strain and weak bonds, the heavier three-membered rings may be regarded as unfavourable as building blocks for polyhedral compounds. In this context, it is important to disclose the role of substituents[8,10,29]. As already mentioned, the strain energy of cyclotrisilane (c-Si_3H_6) is ca 10 kcal mol^{-1} larger than that of cyclopropane (c-C_3H_6). As Table 3 shows, the strain energy decreases only by 1.3 kcal mol^{-1} when the H atoms are substituted by Me groups. In contrast, substitution by SiH_3 groups remarkably decreases the strain energy, c-$Si_3(SiH_3)_6$ being 11 kcal mol^{-1} less strained than c-Si_3H_6, reaching a strain energy as low as that of c-C_3H_6. This suggests that even three-membered rings are not unfavourable as building blocks when they bear suitable substituents.

Accordingly, the strain energies of polyhedral compounds can be also decreased remarkably by substitution, as shown in Table 3; the effect of SiH_3 groups is again larger than that of Me groups. Charge analyses show that the SiH_3 group acts as an electropositive substituent while the Me group is electronegative. The advantage of electron-donating

TABLE 3. Effect of substituents on the strain energies (kcal mol^{-1}) at the HF/6-31G* level

	R = H	R = Me	R = SiH_3
Cyclopropane (c-C_3R_6, D_{3h})	28.7	35.5	34.8
Cyclotrisilane (c-Si_3R_6, D_{3h})	38.9	37.6	28.1
Tetrasilatetrahedrane (Si_4R_4, T_d)	140.9	134.6	114.5
Hexasilaprismane (Si_6R_6, D_{3h})	113.8	105.6	95.7
Octasilacubane (Si_8R_8, O_h)	93.5	88.9	77.9

substituents over electron-accepting ones in the relief of strain is ascribed to the fact that the increased negative charges on the skeletal atoms decrease the size difference between valence s and p atomic orbitals and make s–p hybridization favourable. In addition, the s–p promotion energies are also decreased by the increased negative charges at the skeletal atoms[30].

III. SYNTHESIS

A. Substituents

The choice of substituents is of crucial importance for the successful synthesis and isolation of polyhedral silicon compounds. The Si–Si bonds of the small-ring compounds are readily oxidized because of the existence of high-lying orbitals and their inherent high strain. Therefore, the full protection of the skeleton by bulky substituents is required to suppress the attack by external reagents.

B. Precursors and Reducing Reagents

The most reasonable precursors for the synthesis of polyhedral silicons are halogenated cyclotrisilanes and cyclotetrasilanes. Compounds of the $RSiX_3$ and $RSiX_2-SiX_2R$ types can also serve as precursors of polyhedral silicons through the multi-step reactions when the R group is judiciously selected. The steric bulkiness of the R group determines the ring size and the shape of polyhedral silicons.

The role of metals as the reducing reagent is also crucial. In general, alkali metals such as Li, Na, Na/K, K and lithium naphthalenide (LiNp) are employed as coupling reagents for chlorosilanes. However, these reducing reagents are sometimes such powerful reagents that they cleave the resulting Si–Si bond. In contrast, magnesium metal does not cleave strained Si–Si bonds. $Mg/MgBr_2$ is a particularly useful reagent in this context. The reactive species is presumed to be MgBr, where Mg is at oxidation state $+1$[31]. The first step of the reaction involves one electron transfer from MgBr to the chlorosilane. For the preparation of tetrasilatetrahedrane, t-Bu_3SiNa was used as the electron transfer reagent[32].

IV. TETRASILATETRAHEDRANE

The reductive reaction of 1,2-bis(2,6-diisopropylphenyl)-1,1,2,2-tetrachlorodisilane (**1**) with LiNp led to several products from which **4** was isolated after hydrolytic workup (Scheme 1)[23]. Compound **4** is believed to be formed by the hydrolysis of the intermediate **3**, which arises from the cleavage of an Si–Si bond of the tetrasilatetrahedrane **2**.

The intermediacy of **2** was confirmed by the formation of 1,2,5,6-tetrakis(2,6-diisopropylphenyl)-1,2,5,6-tetrasilatricyclo[3.1.0.0^{2,6}]hexane (**6**) (Scheme 2)[33] in the reductive coupling reaction of 2,6-triisopropylphenyltrichlorosilane (**5**) by the $Mg/MgBr_2$ reagent (which was generated *in situ* by the reaction of Mg and $BrCH_2CH_2Br$). The tricyclic **6** is presumed to be derived from the insertion of ethylene (formed *in situ*, see Scheme 2) into the reactive Si–Si bond of **2**. None of **6** was formed after the complete removal of ethylene from the reaction system.

As pointed out in Section II.B, tetrasilatetrahedrane (Si_4R_4) collapses with no significant barrier to a 'two-bond broken' isomer when R = H, the latter being 37.3 kcal mol^{-1} more stable at the BLYP/6-31G* level[10,28,34]. This energy difference is decreased by 8.9 kcal mol^{-1} with R = Me, but the two bonds still remain broken. However, when R = SiH_3, the 'bond-stretch' isomer is by only 10.4 kcal mol^{-1} more stable than the corresponding tetrasilatetrahedrane, having only one bond stretched. As Figure 5 shows, the structural features of the two isomers, especially the close contact between the substituents

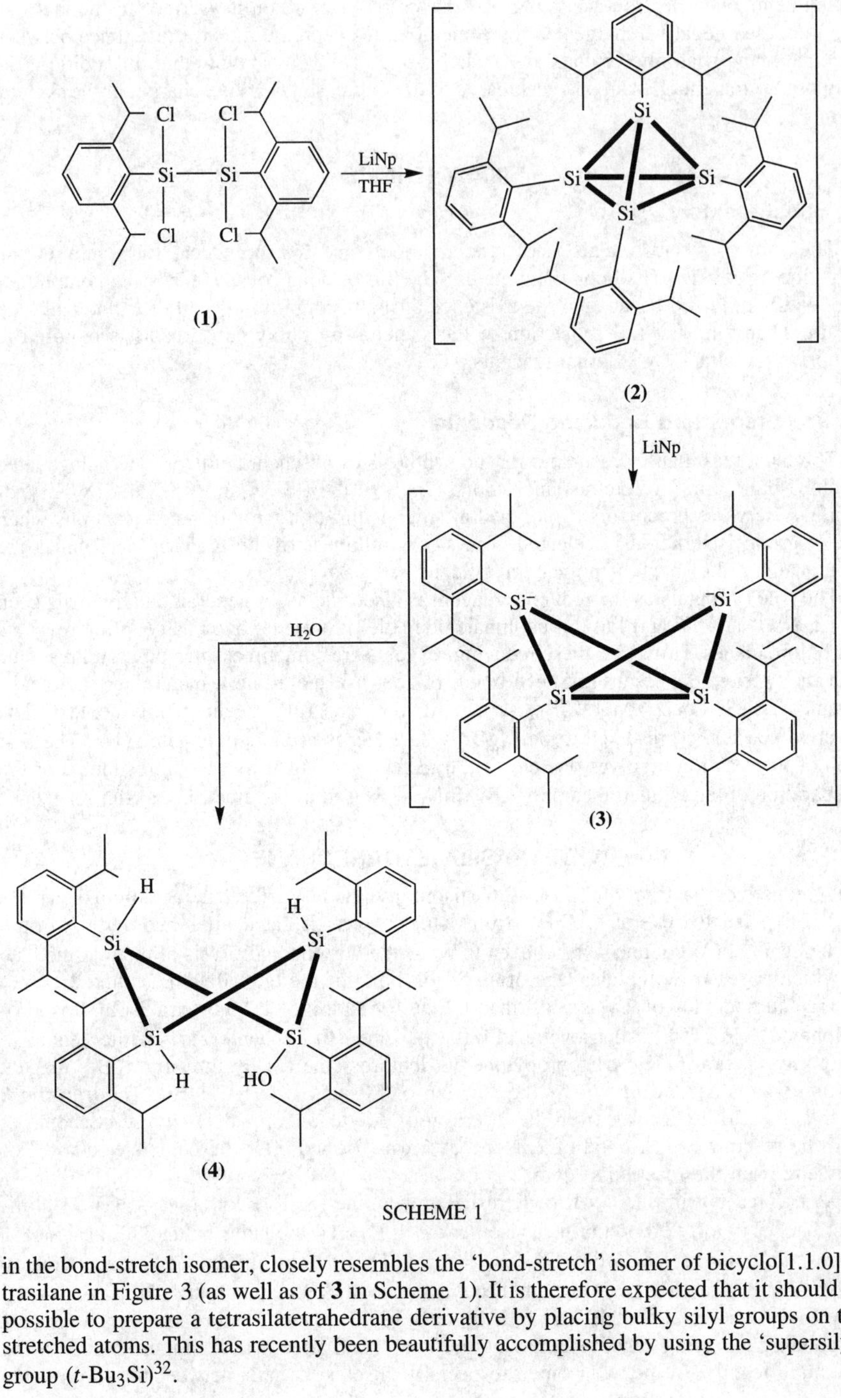

SCHEME 1

in the bond-stretch isomer, closely resembles the 'bond-stretch' isomer of bicyclo[1.1.0]tetrasilane in Figure 3 (as well as of **3** in Scheme 1). It is therefore expected that it should be possible to prepare a tetrasilatetrahedrane derivative by placing bulky silyl groups on the stretched atoms. This has recently been beautifully accomplished by using the 'supersilyl' group (t-Bu$_3$Si)[32].

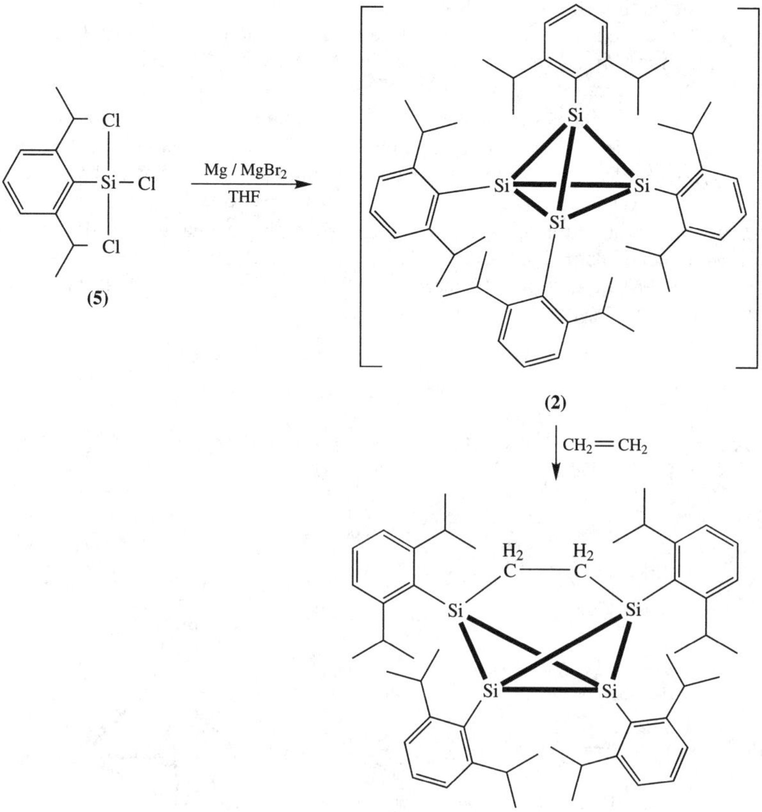

3. Polyhedral silicon compounds

SCHEME 2

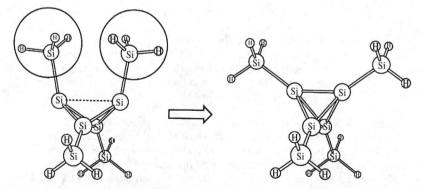

FIGURE 5. Possible conversion of the 'bond-stretch' isomer to a tetrasilatetrahedrane upon replacement of the substituents in circles by more bulky silyl groups. Reproduced with permission from Reference 10. Copyright 1995 American Chemical Society

The dehalogenation of t-Bu$_3$Si−SiCl$_3$ (**7**) with sodium at 80 °C led to the formation of various products such as 1,2-bis(supersilyl)disilane (**8**) and tris(supersilyl)cyclotrisilane (**9**). Bromination of **8** gave the tetrabromodisilane **10**. By the reaction of **10** with the supersilyl anion (t-Bu$_3$SiNa) in THF at −20 °C, tetrakis(tri-t-butylsilyl)tricyclo[1.1.0.0^{2,4}]tetrasilane (**11**) was obtained as yellow-orange crystals (Scheme 3)[32]. While the successful synthesis of **11** may be simply ascribed to the stabilizing effect of silyl groups, it should be emphasized that the large steric size of the t-Bu$_3$Si groups also plays an important role by preventing the collapse of the skeleton to the 'bond-stretch' isomer and by protecting it against reactive reagents[10]. The reaction mechanism that leads to **11** is not clear. However, the reactive disilyne t-Bu$_3$Si−Si≡Si−SiBu$_3$-t which undergoes dimerization to give tetrasilacyclobutadiene, thereby leading to **11**, is a possible intermediate.

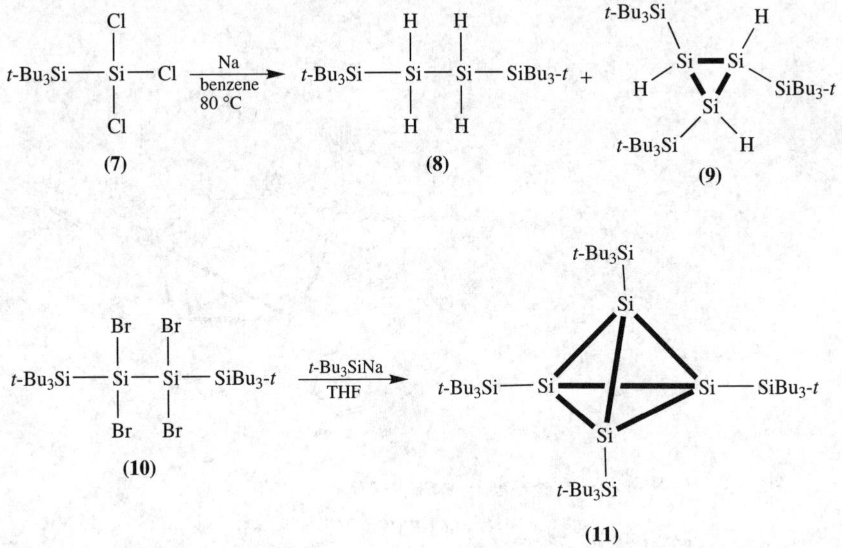

SCHEME 3

Tetrasilatetrahedrane **11** is unexpectedly stable to water, air and light. It cannot be reduced by sodium, but reacts with TCNE and Br$_2$[32]. Unlike t-butyl substituted tetrahedrane of carbon[35], **11** is thermally stable and its crystals do not melt below 350 °C. The bulky t-Bu$_3$Si substituents evidently prevent the collapse of the tetrahedrane skeleton. UV-Vis absorptions were observed at 210 ($\varepsilon = 76\,000$), 235 ($\varepsilon = 71\,000$), 310 ($\varepsilon = 20\,000$) and 451 ($\varepsilon = 3600$) nm.

Recrystallization of a mixture of **11** and hexa-t-butyldisilane from C$_6$D$_6$ leads to $2(t-\text{Bu}_3\text{Si})_4\text{Si}_4 \cdot (t-\text{Bu}_3\text{Si})_2 \cdot \text{C}_6\text{D}_6$ whose structure was established by X-ray crystallography (Figure 6). The cubic unit cell contains two sets of four molecules of **11** (**11A** and **11B**). The skeleton has two different Si−Si distances of 2.320 and 2.315 Å for **11A** and 2.326 and 2.341 Å for **11B**, while the Si−Si−Si bond angles in **11A** and **11B** are 59.9 and 60.4°, respectively. Their Si−Si bond lengths are somewhat longer than those of 2.314 (HF), 2.328 (B3LYP), 2.314 (B3P) and 2.315 (MP2) Å calculated for Si$_4$H$_4$ at several theoretical levels[13,34] with the 6-31G* basis, probably due to steric reasons.

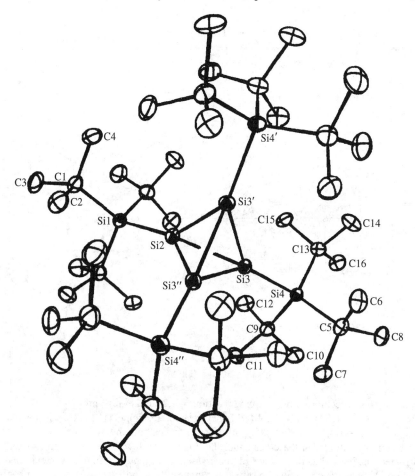

FIGURE 6. ORTEP drawing of tetrasilatetrahedrane (**11**). Reprinted with permission from Reference 32. Copyright 1993 VCH

However, the Si—Si bonds in **11** are shorter than the typical single Si—Si bond (2.34 Å). The exocyclic Si—Si bonds of 2.355 and 2.365 Å for **11A** and 2.371 and 2.356 Å for **11B** are stretched compared with the normal Si—Si bond. A tetragermatetrahedrane derivative was also prepared by using the t-Bu$_3$Si substituent[36].

V. HEXASILAPRISMANE

A. Synthesis

For the synthesis of hexasilaprismane, tetrachlorodisilane (RSiCl$_2$—SiCl$_2$R) and trichlorosilane (RSiCl$_3$) are utilized as the starting compounds. The Mg/MgBr$_2$ reagent is quite useful also for the synthesis of hexasilaprismane. A hexasilaprismane derivative, hexakis(2,6-diisopropylphenyl)tetracyclo[2.2.0.0^{2,6}.0^{3,5}]hexasilane (**12**), was successfully

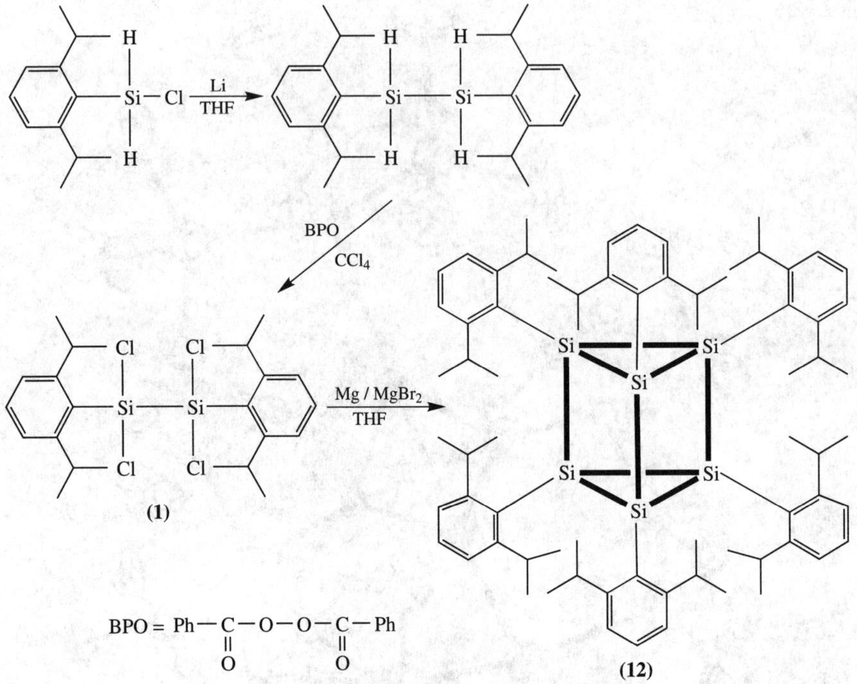

SCHEME 4

prepared by the dechlorinative coupling reaction of 1,2-bis(2,6-diisopropylphenyl)-1,1,2,2-tetrachlorodisilane (1) with the Mg/MgBr$_2$ reagent (Scheme 4)[37]. 12 was isolated as orange crystals (mp > 220 °C), by silica gel chromatography with hexane/toluene of the reaction mixture. On the other hand, the reaction of 1 with LiNp did not give 12[23]. This demonstrates that the choice of the reducing reagent is critical, as mentioned in Section IV. The reaction of (2,6-diisopropylphenyl)trichlorosilane (5) with Mg/MgBr$_2$ also gave 12. The tricyclic derivative 6 was also formed in the presence of ethylene (equation 1)[33]. In the solid state, 12 is thermally and oxidatively fairly stable; no change is observed even after several months in air.

The ^{1}H NMR spectrum of 12 at 25 °C shows that the two isopropyl and the aryl protons are not equivalent due to the restricted rotation of the aryl groups. One set of the methine protons appears at 3.45 ppm along with the two methyl protons at 0.64 and 1.10 ppm. The other set of the methine protons appears at 4.82 ppm with the two methyl protons at 1.01 and 1.62 ppm. The barrier for the rotation of the aryl groups ($\Delta G^{\neq}$) was estimated to be 16.5 kcal mol^{-1}[37].

B. Structure

Figure 7 shows the ORTEP drawing of the hexasilaprismane 12[37]. Its structural parameters are listed in Table 4. The crystal has a two-fold axis of symmetry. The skeleton has a slightly distorted prismane structure with two triangular units [Si–Si = 2.374–2.387 Å (av. 2.380 Å) and ∠Si–Si–Si = 59.8–60.3° (av. 60.0°)] and three rectangular units [Si–Si′ = 2.365–2.389 Å (av. 2.373 Å) and ∠Si–Si–Si′ = 89.6–90.5° (av. 90.0°)]. The exocyclic Si–C$_{ar}$ lengths are 1.901–1.920 Å (av. 1.908 Å). The exocyclic bond angles are significantly expanded: ∠Si–Si–C$_{ar}$ = 126.9–138.7° (av. 133.5°) and ∠Si′–Si–C$_{ar}$ = 124.2–129.2° (av. 126.9°).

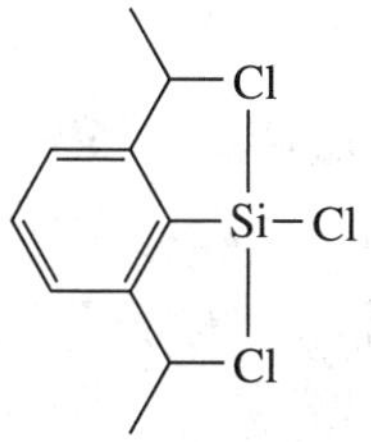

(5) 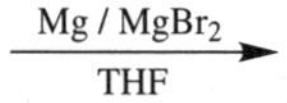 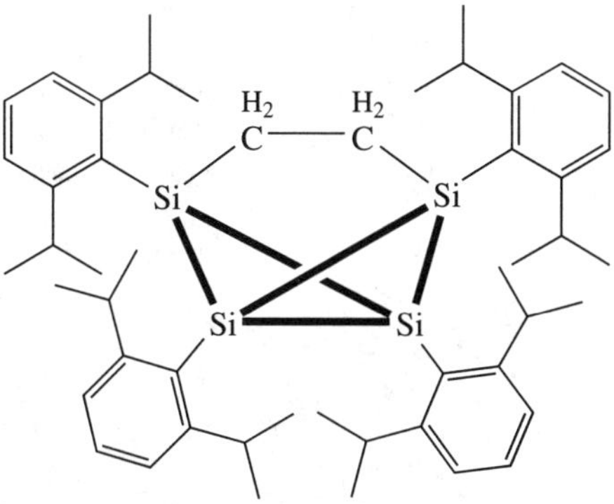 → (6) (1)

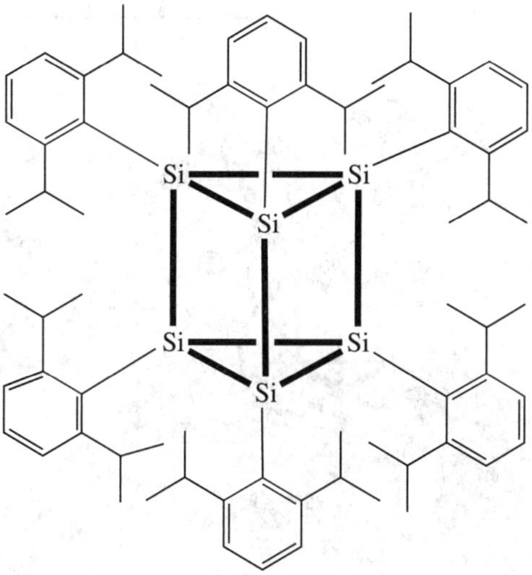

Mg + BrCH$_2$CH$_2$Br ⟶ MgBr$_2$ + CH$_2$=CH$_2$

(12)

All the Si–Si bonds in **12** are elongated from the normal Si–Si bond length (2.34 Å), but are shorter than those in cyclotrisilane (R$_2$Si)$_3$ (R = 2,6-dimethylphenyl: av. 2.407 Å)[38]. The Si–Si bond lengths of **12** are somewhat longer than those calculated with the 6-31G* basis set for Si$_6$H$_6$ [2.359 (HF), 2.369 (B3LYP), 2.354 (B3P) and 2.356 (MP2) Å for the triangular units and 2.375 (HF), 2.376 (B3LYP), 2.361 (B3P), and 2.361 (MP2) Å for the rectangular units][13,34]. The aryl planes are arranged in a screw-shaped

TABLE 4. Selected bond lengths and bond angles of hexasilaprismane **12**

Bond lengths (Å)		Bond angles (deg)	
Si1–Si2	2.374(2)	Si2–Si1–Si3	60.3(0)
Si1–Si3	2.379(2)	Si1–Si2–Si3	60.0(0)
Si2–Si3	2.387(2)	Si1–Si3–Si2	59.8(0)
Si1–Si1′	2.389(2)	Si2–Si1–Si1′	89.6(0)
Si2–Si3′	2.365(2)	Si3–Si1–Si1′	89.9(0)
Si3–Si2′	2.365(2)	Si1–Si2–Si3′	90.5(0)
Si–C$_{ar}$	1.901(7) ≀ 1.920(6)	Si3–Si2–Si3′	90.4(0)
		Si1–Si3–Si2′	90.0(0)
		Si2–Si3–Si2′	89.6(0)
		Si–Si–C$_{ar}$	126.9(2) ≀ 138.7(2)
		Si′–Si–C$_{ar}$	124.2(2) ≀ 129.2(2)

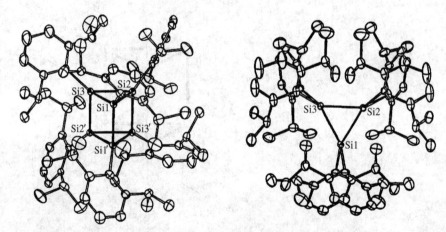

FIGURE 7. ORTEP drawing of hexasilaprismane **12**. Reprinted with permission from Reference 9. Copyright 1995 Academic Press, Inc.

manner around the three-membered rings so that the skeleton is effectively 'covered' by the six 2,6-diisopropylphenyl groups.

Table 5 summarizes the observed structural parameters of prismanes (M_6R_6, M = C, Si, Ge) together with the calculated values for R = H. The M—M bonds within and between the three-membered units are denoted by a and b, respectively. As calculated for R = H[13,14], a is shorter than b in prismane (C_6H_6[39]) and its derivatives C_6R_6 (R = Me[40]

TABLE 5. Structural parameters of prismanes comprising group 14 element

M	R	a (Å)	b (Å)	Method
C	H	1.507	1.549	Calcd.[a]
		1.500	1.585	ED[b]
	Me	1.540	1.551	ED[c]
	SiMe$_3$	1.510	1.582	XRD[d]
Si	H	2.359	2.375	Calcd.[a]
	2,6-i-Pr$_2$C$_6$H$_3$	2.380 (2.374–2.387)	2.373 (2.365–2.389)	XRD[e]
Ge	H	2.502	2.507	Calcd.[f]
	2,6-i-Pr$_2$C$_6$H$_3$	2.503 (2.497–2.507)	2.468 (2.465–2.475)	XRD[e]
	CH(SiMe$_3$)$_2$	2.580 (2.578–2.584)	2.522 (2.516–2.526)	XRD[g]

ED: electron diffraction
XRD: X-ray diffraction

[a] From Reference 13.
[b] From Reference 39.
[c] From Reference 40.
[d] From Reference 41.
[e] From Reference 37.
[f] From Reference 14.
[g] From Reference 4.

and SiMe$_3$[41]). The calculation shows that the length difference between a and b decreases significantly as M becomes heavier. Thus, $|a - b| = 0.042$ (C$_6$H$_6$), 0.016 (Si$_6$H$_6$), 0.005 (Ge$_6$H$_6$) Å. In this context, it is interesting that b is observed to be shorter than a in Si$_6$R$_6$ (**12**: R = 2,6-i-Pr$_2$C$_6$H$_3$) ($a = 2.380$ Å, $b = 2.373$ Å) and Ge$_6$R$_6$ [R = 2,6-i-Pr$_2$C$_6$H$_3$[37], $a = 2.503$ Å, $b = 2.468$ Å; R = CH(SiMe$_3$)$_2$[4], $a = 2.580$ Å, $b = 2.522$ Å].

C. Absorption Spectra

The prismanes with Si and Ge skeletons are yellow to orange. Figure 8 shows the UV-Vis spectra of **12** and its germanium analogue. These prismanes have absorptions tailing into the visible region. For example, **12** has an absorption band with a maximum at 241 ($\epsilon = 78000$) nm tailing to ca 500 nm. The absorption band of Ge$_6$R$_6$ (R = 2,6-i-Pr$_2$C$_6$H$_3$) has a maximum at 261 ($\epsilon = 84000$) nm, which is red-shifted compared to that of **12** because of the higher-lying orbitals of the Ge−Ge bonds.

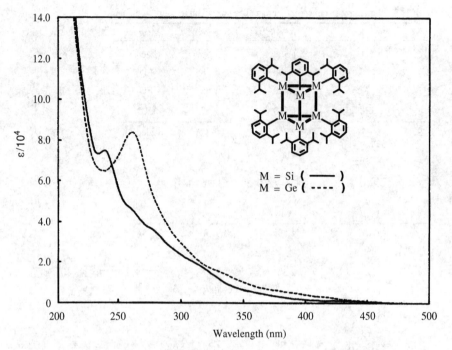

FIGURE 8. Electronic absorption spectra of hexasilaprismane **12** and hexagermaprismane in hexane

D. Photochemical Reaction

Hexasilaprismane **12** is photosensitive. On irradiation in solution at low temperature, with light having wavelengths of 340–380 nm, new absorption bands appeared at 335, 455 and 500 nm assignable to the absorption bands of hexasila-Dewar benzene **13** (equation 2)[37]. Upon excitation of these bands with wavelengths longer than 460 nm, **12** was immediately regenerated. A single chemical species was produced during the photochemical reaction since the bands of **12** and those assigned to **13** appeared and disappeared simultaneously.

The folding angle of the parent hexasila-Dewar benzene (Si_6H_6) with C_{2v} symmetry is 120° at the HF/6-31G* level. The through-space interaction between the Si=Si double bonds splits the π MOs into bonding (π_S) and antibonding (π_A) sets, as shown in Figure 9[42]. Likewise, their π^* MOs split into π^*_S and π^*_A. The allowed transitions from π_S to π^*_S and from π_A to π^*_A correspond to the experimental absorption bands at 455 and 335 nm, respectively. The lowest energy transition from π_A to π^*_S is forbidden. However, it is allowed if the C_{2v} symmetry is lowered. The relatively weak absorption at 500 nm is assigned to the π_A–π^*_S transition.

The hexasila-Dewar benzene **13** is thermally stable at −150 °C, but it gradually reverted to the hexasilaprismane **12**[43]. The half-life is $t_{1/2} = 0.52$ min at 0 °C in 3-methylpentane. The activation parameters for the isomerization of **13** to **12** are $E_a = 13.7$ kcal mol^{-1}, $\Delta H^{\neq} = 13.2$ kcal mol^{-1} and $\Delta S^{\neq} = -17.8$ cal K^{-1} mol^{-1}. The small E_a value is consistent with the high reactivity of Si=Si double bonds. Most probably, the small HOMO–LUMO gap of **13** makes it possible that the Si=Si double bonds undergo a formally symmetry forbidden [2 + 2] thermal reaction. Hexasila-Dewar benzene is a key

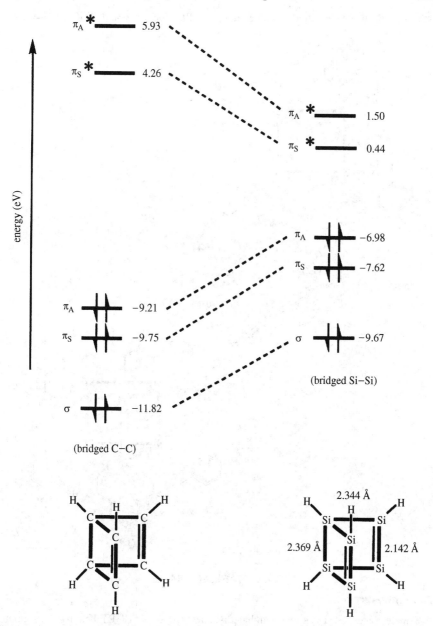

FIGURE 9. Energy diagram of Dewar benzene and of hexasila-Dewar benzene calculated at the HF/6-31G* level

intermediate in the reductive oligomerization of $RCl_2SiSiCl_2R$ and $RSiCl_3$ (R = 2,6-i-$Pr_2C_6H_3R$).

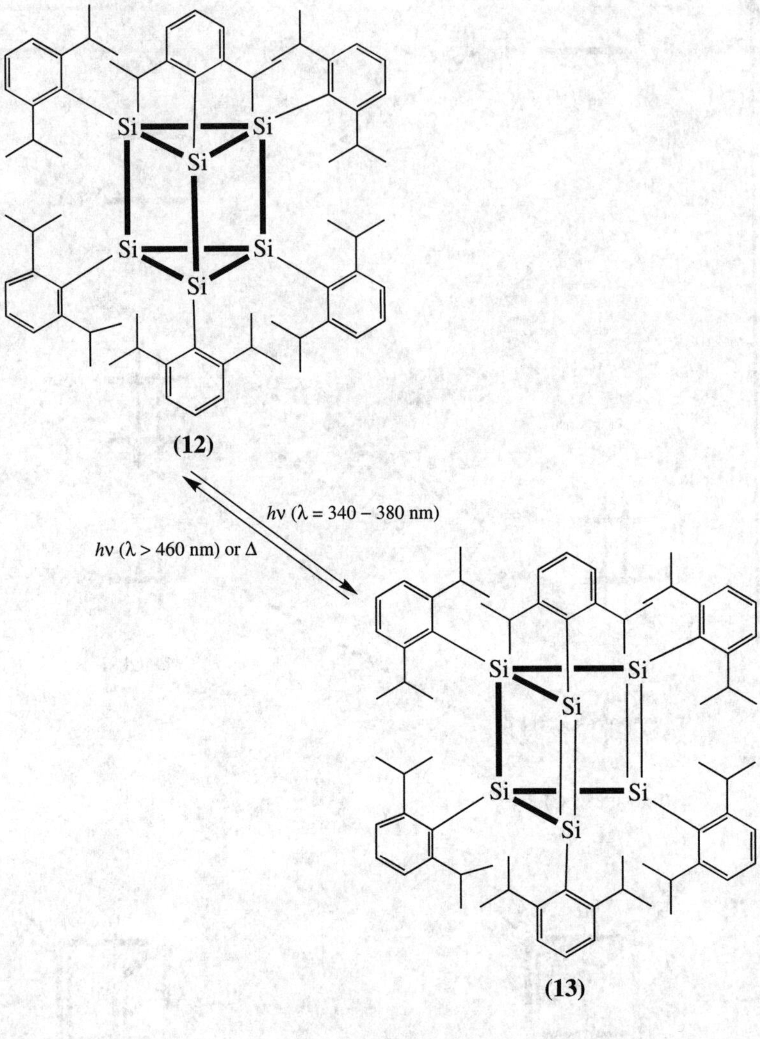

(12)

$h\nu$ (λ > 460 nm) or Δ

$h\nu$ (λ = 340 – 380 nm)

(2)

(13)

VI. OCTASILACUBANE

A. Synthesis

Trihalosilanes ($RSiX_3$) and tetrahalodisilanes ($RSiX_2-SiX_2R$) bearing appropriate substituents R can serve as precursors to octasilacubane by reductive coupling reactions. As pointed out in Section II.C, the strain of octasilacubane is significantly decreased by electropositive silyl groups[8,29]. Accordingly, a silyl-substituted octasilacubane, octakis(t-butyldimethylsilyl)pentacyclo[4.2.0.0^{2,5}.0^{3,8}.0^{4,7}]octasilane (16) was synthesized as bright yellow crystals by condensation of 1,1,1-tribromo-2-t-butyl-2,2-dimethyldisilane (14) and 2,2,3,3-tetrabromo-1,4-di-t-butyl-1,1,4,4-tetramethyltetrasilane (15) with sodium in toluene (equation 3)[3].

$$\text{(14)} \quad t\text{-BuMe}_2\text{Si}-\underset{\underset{\text{Br}}{|}}{\overset{\overset{\text{Br}}{|}}{\text{Si}}}-\text{Br}$$

or

$$\text{(15)} \quad t\text{-BuMe}_2\text{Si}-\underset{\underset{\text{Br}}{|}}{\overset{\overset{\text{Br}}{|}}{\text{Si}}}-\underset{\underset{\text{Br}}{|}}{\overset{\overset{\text{Br}}{|}}{\text{Si}}}-\text{SiMe}_2\text{Bu-}t$$

$$\xrightarrow[\text{90 °C}]{\text{Na, toluene}} \quad \text{(16)} \quad (3)$$

(16) = octakis(SiMe₂Bu-t)-substituted Si₈ cube

Alkyl-substituted octasilacubanes, octakis(t-butyl)octasilacubane (**18a**)[44] and octakis(1,1,2-trimethylpropyl)octasilacubane (**18b**)[45], were also prepared. The condensation of t-butyltrichlorosilane with sodium in the presence of 12-crown-4 produced **18a** as a purple crystalline compound sparingly soluble in organic solvents (equation 4). On the other hand, **18b** was formed as red-orange prisms by the condensation of (1,1,2-trimethylpropyl)trichlorosilane with sodium in toluene (equation 4).

$$R-\underset{\underset{X}{|}}{\overset{\overset{X}{|}}{\text{Si}}}-X \xrightarrow{\text{Na, toluene}} R_8\text{Si}_8 \text{ cube} \quad (4)$$

(**17a**) R = t-Bu (**18a**) R = t-Bu
(**17b**) R = CMe₂CHMe₂ (**18b**) R = CMe₂CHMe₂

Aryl-substituted octasilacubanes **20** and **22** were prepared by the dechlorinative reactions of ArCl₂SiSiCl₂Ar (equation 5) and of ArSiCl₃ (equation 6), respectively. Octakis(2,4,6-trimethylphenyl)octasilacubane (**20**) was isolated as orange crystals by a dechlorinative reaction of 1,2-bis(2,4,6-trimethylphenyl)-1,1,2,2-tetrachlorodisilane with Mg/MgBr₂ reagent (equation 5)[46]. In a similar manner, 2,6-diethylphenyl-substituted octasilacubane **22** was prepared by the reaction of (2,6-diethylphenyl)trichlorosilane (**21**) with Mg/MgBr₂ (equation 6)[47]. The cubane **20** is sparingly soluble in organic solvents, whereas **22** is soluble.

B. Structure

Figure 10 shows the ORTEP drawing of the 1,1,2-trimethylpropyl-substituted octasilacubane **18b**[45]. Its skeleton is slightly distorted from an ideal cubic form by the steric congestion of the eight bulky 1,1,2-trimethylpropyl groups. The Si–Si bond lengths range from 2.398(2) to 2.447(2) Å, and the Si–Si–Si bond angles vary from 87.2(1) to 92.6(1)°. X-ray diffraction analysis revealed that **18a** also has a slightly distorted cubic structure: the Si–Si bond lengths are 2.374–2.400 Å (av. 2.390 Å) and the Si–Si–Si bond angles are 89.3–90.9° (av. 90.0°)[44b].

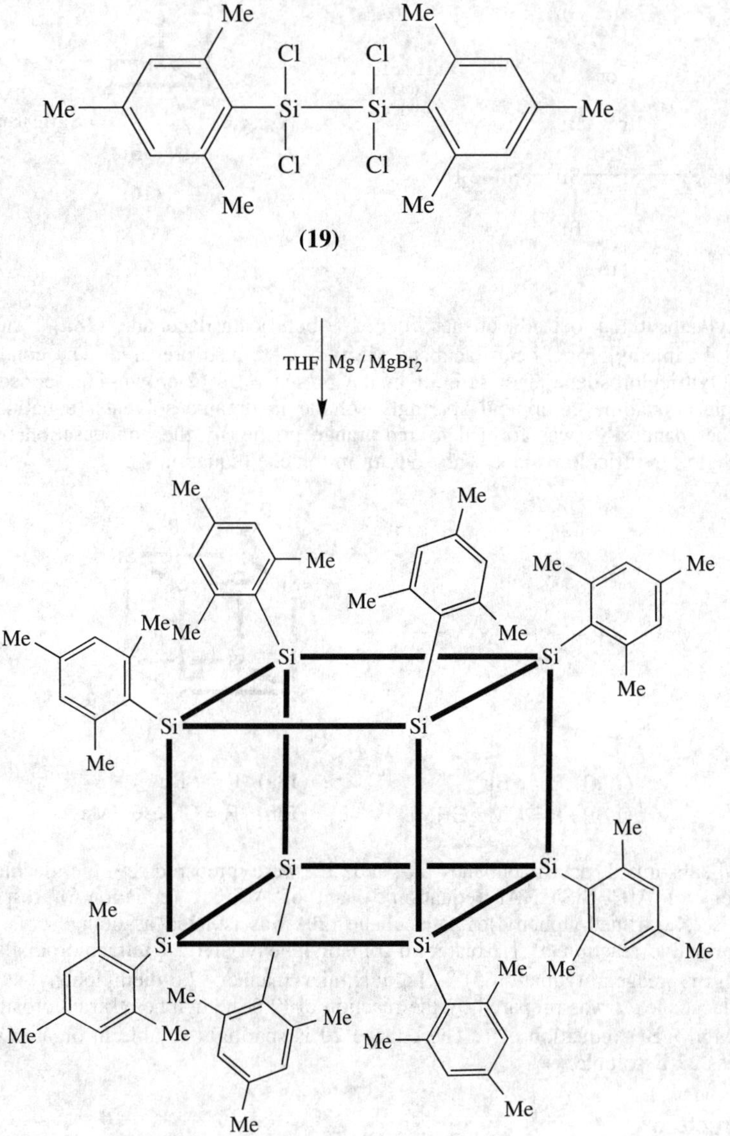

(19)

THF | Mg / MgBr$_2$

(20)   (5)

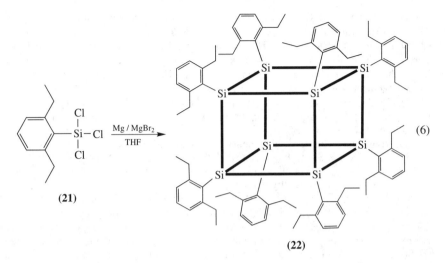

(6)

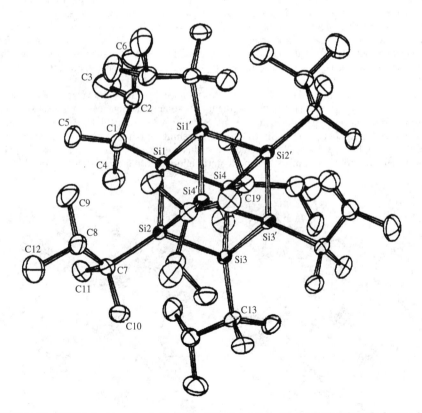

FIGURE 10. ORTEP drawing of octasilacubane **18b**. Reprinted with permission from Reference 45. Copyright 1992 VCH

Figure 11 shows the ORTEP drawing of 2,6-diethylphenyl-substituted octasilacubane **22**[47]. Its structural parameters are listed in Table 6. The X-ray structure indicates that its skeleton is almost perfect cubic. The Si−Si−Si bond angles are 88.9–91.1° (av. 90.0°). The Si−Si bond lengths are in the range of 2.384–2.411 Å (av. 2.399 Å), somewhat longer than the normal Si−Si bond length (2.34 Å). The Si−Si bond lengths of **22** are in close agreement with those calculated for Si_8H_8 of 2.396 (HF/6-31G*), 2.398 (B3LYP/6-31G*), 2.383 (B3P/6-31G*) and 2.385 (MP2/6-31G*) Å[13,14]. The exocyclic Si−C_{ar} bond lengths of 1.911 Å (av) are somewhat longer than the normal Si−C bond length (1.88 Å). The Si−Si−C_{ar} bond angles of 124.4° are significantly expanded due to the endocyclic angular constraint. The aryl substituents form dihedral angles of ca 90° between the benzene ring and the Si−Si bond: ring (Si1)/Si1-Si4′, ring (Si2)/Si2-Si3, ring (Si3)/Si3-Si2′ and (Si4)/Si4-Si1 being all ca 90°. As a result the cubic skeleton of **22** is effectively protected by the eight 2,6-diethylphenyl groups.

Octasilacubanes (Si_8H_8) bearing alkyl, aryl and silyl substituents of various sizes were calculated using the semiempirical AM1 method (Table 7)[10,48]. X-ray structures are available for R = 2,6-$Et_2C_6H_3$[47], CMe_2CHMe_2[45] and t-Bu[44b]. Both the calculated and the X-ray structures show almost perfect cubic skeletons. In addition, the experimental skeletal Si−Si bond lengths are reasonably well reproduced by the calculations, taking into account the overestimation of the Si−Si bond distances by ca 0.05 Å. As Table 7 shows, the O_h symmetry of Si_8H_8 is lowered as the substituents become more bulky. Nevertheless, all of the calculated structures still retain relatively high symmetry (Table 7), in contrast with the available experimental structures in crystals. This suggests that packing forces significantly affect the favourable conformations of bulky substituents around the cubane skeleton, probably because the energy loss due to the conformational changes is very small.

As Table 7 shows, the skeletal bond lengths increase by 0.02–0.05 Å upon substitution by alkyl and aryl groups. This bond lengthening is somewhat enhanced as the substituents become more bulky: e.g. 2.446 Å (R = Ph) vs 2.470 Å (R = 2,6-$Et_2C_6H_3$). There is also a good correlation between the Si−Si bond lengthening and the increased positive charges on the skeletal atoms (Table 7). The skeletal bonds are lengthened even on substitution by the small Me group. It appears that both steric and electronic effects are responsible for

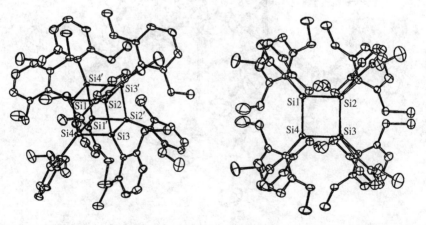

FIGURE 11. ORTEP drawing of octasilacubane **22**. Reprinted with permission from Reference 9. Copyright 1995 Academic Press, Inc.

TABLE 6. Selected bond lengths and bond angles of octasilacubane 22

Bond lengths (Å)		Bond angles (deg)	
Si1−Si2	2.384(2)	Si2−Si1−Si4	90.8(0)
Si2−Si3	2.399(2)	Si2−Si1−Si4'	90.3(0)
Si3−Si4	2.411(2)	Si4−Si1−Si4'	91.1(0)
Si4−Si1	2.400(2)	Si1−Si2−Si3	89.8(0)
Si1−Si4'	2.406(2)	Si1−Si2−Si3'	90.3(0)
Si2−Si3'	2.396(2)	Si3−Si2−Si3'	89.1(0)
Si−C$_{ar}$	1.900(7)	Si2−Si3−Si4	90.2(0)
	⟨	Si2−Si3−Si2'	90.8(0)
	1.924(7)	Si4−Si3−Si2'	89.9(0)
		Si1−Si4−Si3	89.1(0)
		Si1−Si4−Si1'	88.9(0)
		Si3−Si4−Si1'	89.4(0)
		Si−Si−C$_{ar}$	120.7(2)
		⟨	
			128.3(2)

TABLE 7. Symmetries, Si−Si bond lengths, charges and HOMO energies of octasilacubane derivatives (Si$_8$R$_8$) calculated at the AM1 level[a]

R	Symmetry	Si−Si (Å)[b]	Charge[c]	HOMO (eV)
H	O_h	2.421	0.045 (0.068)	−9.70 (−8.13)
Me	O_h	2.437	0.255 (0.343)	−8.68 (−6.65)
t-Bu	D_2	2.445	0.264 (0.481)	−8.68 (−6.71)
CMe$_2$CHMe$_2$	D_2	2.462	0.242 (0.435)	−8.54 (−6.51)
Ph	D_2	2.446	0.366 (0.393)	−8.26 (−6.27)
2,6-Et$_2$C$_6$H$_3$	C_2	2.470	0.343 (0.389)	−8.22 (−6.31)
Si(SiH$_3$)$_3$	D_2	2.414	0.021 (0.054)	−9.22 (−7.99)
SiH$_3$	O_h	2.402	−0.216 (−0.115)	−9.48 (−8.14)
SiMe$_2$CHMe$_2$	D_2	2.408	−0.324 (−0.188)	−8.48 (−7.08)
SiMe$_3$	O_h	2.395	−0.343 (−0.185)	−8.47 (−7.09)
SiF$_3$	O_h	2.391	−0.539 (−0.313)	−9.81 (−10.89)

[a] In parentheses are the HF/3-21G* values calculated on the AM1 optimized structures.
[b] Average skeletal bond lengths.
[c] Average Mulliken charges on the skeletal atoms.

the skeletal bond lengthening. It is noteworthy that substitution by alkyl and aryl groups raises the HOMO levels (localized on the skeletons) and therefore is expected to increase the reactivity. This makes the presence of bulky groups very important since they can fully protect the cubane skeleton, as clearly shown in Figure 12.

Table 7 reveals that the skeletal bond lengths tend to shorten with an increase in the negative charges on the skeletal atoms due to silyl substituents, regardless of their steric bulk. This is also confirmed by the geometry optimization at the HF/3-21G* level of

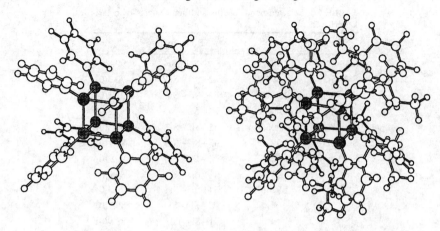

FIGURE 12. Examples of the steric protection of the cubic silicon skeleton by phenyl and 2,6-diethylphenyl groups. The 2,6-diethylphenyl group is apparently more effective for steric protection. Reprinted with permission from Reference 10. Copyright 1995 American Chemical Society

TABLE 8. Calculated data for octasilacubane derivatives (Si_8R_8) optimized at the HF/3-21G* level

R	Symmetry	Si–Si (Å)^b	Charge^c	HOMO (eV)
H	O_h	2.386	0.065	−8.23
Me	O_h	2.391	0.328	−7.05
SiH_3	O_h	2.383	−0.112	−8.31
$SiMe_3$	O_h	2.390	−0.169	−7.45
SiF_3	O_h	2.369	−0.373	−10.42

^{a}Mulliken charges on the skeletal atoms.

several substituted octasilacubanes as shown in Table 8[10,48]. The SiF_3 group which places the largest charge on the skeletal silicon atoms provides the shortest skeletal bond lengths and the lowest HOMO level while the $SiMe_3$-substituted octasilacubane has a relatively high HOMO level.

The geometries of C_8H_8[49] and M_8R_8 (M = Si[47], Ge[47] and Sn[50], R = 2,6-$Et_2C_6H_3$) are compared in Table 9. The M–M–M bond angles in M_8R_8 range from 89–91°, indicating that all these skeletons are almost perfect cubic. The M–M bond lengths of 2.399 Å for Si, 2.490 Å for Ge and 2.854 Å for Sn are in close agreement with those calculated for the corresponding M_8H_8 (2.382 Å for Si, 2.527 Å for Ge and 2.887 Å for Sn)[14]. The range of the M–M–C_{ar} bond angles increases in the order: 121–128° for Si <120–130° for Ge <117–133° for Sn. This implies that the steric congestion between the neighbouring ligands is relaxed by the longer M–C_{ar} bond as M becomes heavier: 1.911 Å for Si <1.982 Å for Ge <2.193 Å for Sn.

C. Absorption Spectra

All cubanes of Si, Ge and Sn are coloured from yellow to purple. The silyl substituted **16** is yellow[3], while the *t*-butyl substituted **18a** is purple[44]. A diffuse reflection absorption

TABLE 9. Structural parameters of cubanes comprising group 14 elements

$M = C, R = H;$
$M = Si, Ge, Sn, R = $ (2,6-diethylphenyl shown in figure)

	C_8H_8[a]	Si_8R_8[b]	Ge_8R_8[b]	Sn_8R_8[c]
M–M (Å)				
X-ray	1.551 (av.)	2.399 (av.)	2.490 (av.)	2.854 (av.)
	(1.549–1.553)	(2.384–2.411)	(2.478–2.503)	(2.839–2.864)
Calculated[d]	1.559	2.382	2.527	2.887
M–C$_{ar}$ (Å)	1.06 (av.)	1.911 (av.)	1.982 (av.)	2.193 (av.)
M–M–M (°)	89.3–90.5	88.9–91.1	88.9–91.1	89.1–91.1
M–M–C$_{ar}$ (°)	123–127	121–128	120–130	117–133

[a] From Reference 49.
[b] From Reference 47.
[c] From Reference 50.
[d] From Reference 14.

spectrum of **18a** shows a broad absorption between 450 and 650 nm[44a]. The colour of **18b** is red-orange[45]. The UV/Vis spectrum of **18b** exhibits absorption bands at 252 ($\epsilon = 30800$), 350 ($\epsilon = 850$) and around 500 ($\epsilon = 70$) nm, as shown in Figure 13[45]. The aryl substituted **20** and **22** are orange. The spectrum depends on the type of the aryl substituent[46,47]. Thus, **20** exhibits three absorption bands at 246 ($\epsilon = 73000$), 280 ($\epsilon = 48500$) and 379 ($\epsilon = 4900$) nm[46], while **22** shows bands at 234 ($\epsilon = 87000$), 284 ($\epsilon = 42000$) and 383 ($\epsilon = 5000$) nm, as shown in Figure 14[47]. The absorption band at around 240 nm is attributed to the $\sigma-\sigma^*$ transition, while the absorption at around 280 nm is caused by transition from a $\sigma-\pi$ mixing between the orbitals of the Si–Si σ bonds and the aromatic π orbitals.

The lowest energy absorption is forbidden when the cubane has high symmetry, but becomes weakly allowed when the symmetry is lowered. CNDO/S calculations show that the spectrum shape in the low energy region depends strongly on the substituents[51]. Among t-butyl-, trimethylsilyl- and phenyl- substituted octasilacubanes, the t-butyl substituted cubane **18a** has the lowest energy absorption band.

Octasilacubanes were used as a model in an attempt to understand the optical properties of porous silicon because both porous silicon and octasilacubane show a broad photoluminescence spectra and large Stokes shifts[52]. **16** for example, shows an absorption edge at ca 3.2 eV and a broad photoluminescence spectrum with a peak at 2.50 eV.

D. Reactivity

The kinetic stability of octasilacubanes depends strongly on the steric bulkiness of the substituents. The silyl-substituted **16** is stable in an inert atmosphere, but is oxidized in air to give colourless solids[3]. The 1,1,2-trimethylpropyl-substituted **18b** is very stable even in air and survives for two weeks in the solid state[45]. The aryl-substituted **20** and **22** are gradually oxidized in the atmospheric air[46,47].

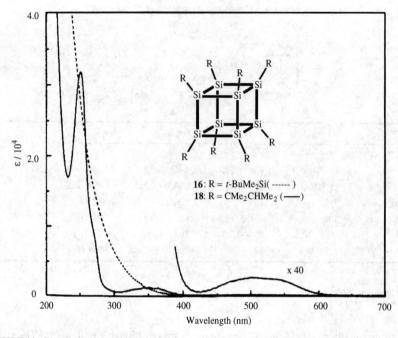

FIGURE 13. Electronic absorption spectra of octasilacubanes **16** and **18b** in hexane. Reproduced with permission from Reference 45. Copyright 1992 VCH

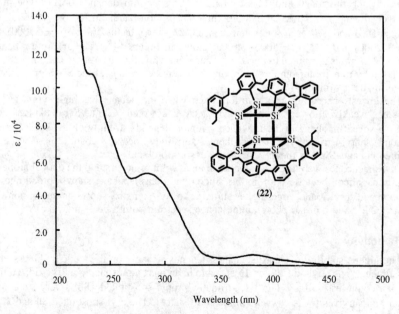

FIGURE 14. Electronic absorption spectrum of octasilacubane **22** in hexane

Despite its low oxidation potential of 0.43 V [vs saturated calomel electrode (SCE) in CH_2Cl_2], **18b** is very air-stable. Upon irradiation of **18b** in the presence of dimethyl sulphoxide with a high pressure mercury lamp, 9-oxaoctasilahomocubane (**23**) and 5,10-dioxaoctasilabishomocubane (**24**) were formed (equation 7)[53]. Without irradiation, neither **23** nor **24** was formed even at 115 °C.

The chlorination of **18b** with PCl_5 results in skeletal rearrangement to give three stereoisomers of 4,8-dichlorooctakis(1,1,2-trimethylpropyl)tetracyclo[3.3.0.0^{2,7}.0^{3,6}]octasilanes (**25**) (equation 8)[54]. Three stereoisomers (*endo–exo, exo–exo, endo–endo*) were isolated in a pure form and characterized by X-ray diffractions. The mechanism by which **25** is formed is not clear, but the first step involves the electrophilic attack by PCl_5 on the strained Si—Si bond, followed by an intramolecular skeletal rearrangement. Bromo and iodo derivatives of **25** are also formed by the reactions of **18b** with Br_2 and with I_2[55].

Reductive dehalogenation of **25** with a large excess of sodium metal at 110–120 °C in toluene reverted to **18b** with formation of a reduced product **26** (equation 9)[55]. Interestingly, all of the three isomers could be used as precursors to **18b**.

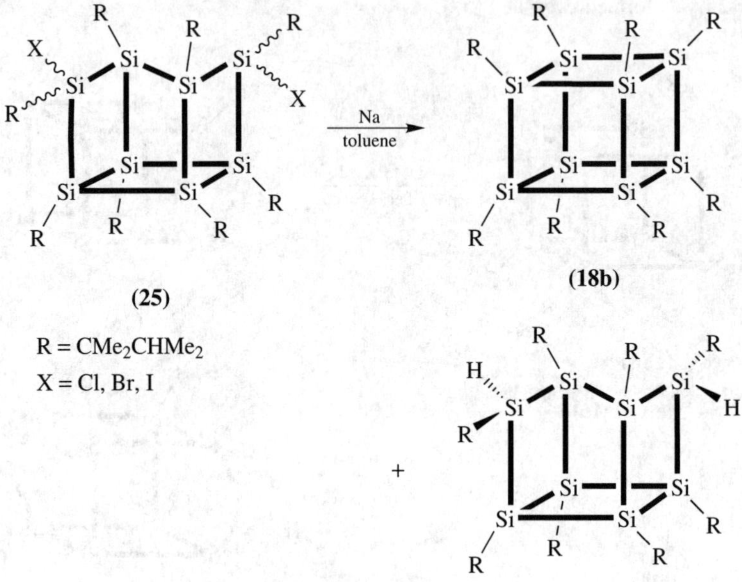

(25) R = CMe$_2$CHMe$_2$; X = Cl, Br, I

(18b)

(26)

(9)

VII. ^{29}Si NMR SPECTRA

NMR spectroscopy is a powerful tool for structural analysis. The chemical shifts of polyhedral silicons range from −22 to 39 ppm. The ^{29}Si chemical shifts of tetrasilatetrahedrane **11**[32], hexasilaprismane **12**[37] and octasilacubanes (**16**[3], **18a**[44b], **18b**[45], **20**[46] and **22**[47]) are listed in Table 10.

^{29}Si NMR spectroscopy of **12** in solution demonstrated that the six skeletal silicons are equivalent, with a single resonance appearing at −22.3 ppm. However, cross polarization magic-angle spinning (CPMAS) ^{29}Si NMR in the solid state shows two signals at −22.2 and −30.8 ppm with relative intensity of 2 : 1[37]. Similarly, CPMAS ^{29}Si NMR of **18a**

TABLE 10. ^{29}Si NMR chemical shifts of poyhedral silicon compounds

Polyhedral silicon	^{29}Si (ppm)	Solvent
[(t-Bu)$_3$Si]$_4$Si$_4$ **11**	38.89 (53.07 for Si(t-Bu)$_3$)	C$_6$D$_6$
(2,6-i-Pr$_2$C$_6$H$_3$)$_6$Si$_6$ **12**	−22.3	C$_6$D$_6$
(t-BuMe$_2$Si)$_8$Si$_8$ **16**	−35.03 (5.60 for t-Bu Me$_2$Si)	C$_6$D$_6$
t-Bu$_8$Si$_8$ **18a**	13.04	o-xylene-d$_{10}$
(Me$_2$CHCMe$_2$)$_8$Si$_8$ **18b**	22.24	C$_6$D$_6$
(2,4,6-Me$_3$C$_6$H$_2$)$_8$Si$_8$ **20**	0.99	CDCl$_3$
(2,6-Et$_2$C$_6$H$_3$)$_8$Si$_8$ **22**	0.36	C$_6$D$_6$

3. Polyhedral silicon compounds

gives two peaks at 6.6 and 10.6 ppm with relative intensity of 1 : 2[44a]. The solid state ^{29}Si NMR spectra of **12** and **18a** indicate that in distorted skeletons the silicon atoms are non-equivalent.

The static solid-state NMR can produce a broad pattern to give the three principal values of the shielding tensors for polyhedral silicons. Figures 15 and 16 show the static powder spectra of octasilacubane **22** and of the hexasilaprismane **12**, respectively[56]. The values of the σ_{11}, σ_{22} and σ_{33} tensors are given in Table 11 together with the calculated values for the methyl substituted compounds[57].

The principal axis of **22**, the least shielded principal value ($\sigma_{11} = 80$ ppm), corresponds to the diagonal of the cube in the direction parallel to the substituent. The other two axes corresponding to σ_{22} and σ_{33} are perpendicular to this axis and must have the same value. However, due to the different orientations of the 2,6-Et$_2$C$_6$H$_4$ groups, σ_{22} and σ_{23} are not equivalent. This is readily understood from the crystal structure of **22**, as shown in Figure 11. The isotropic chemical shift (σ_{iso}) of -0.3 ppm is consistent with the ^{29}Si chemical shift in solution of 0.36 ppm. The remarkable deshielding of σ_{11} is probably attributable to the paramagnetic term which is mainly caused by the low excitation energy, as depicted in Figure 14. As shown in Table 11, **22** shows an anisotropy spread ($\sigma_{11} - \sigma_{33}$) of 134 ppm. This is somewhat larger than those for ordinary cyclotetrasilanes[58].

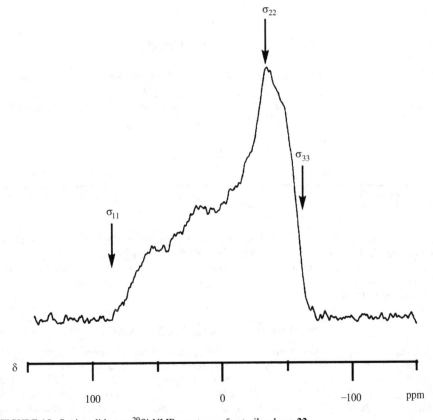

FIGURE 15. Static solid state ^{29}Si NMR spectrum of octasilacubane **22**

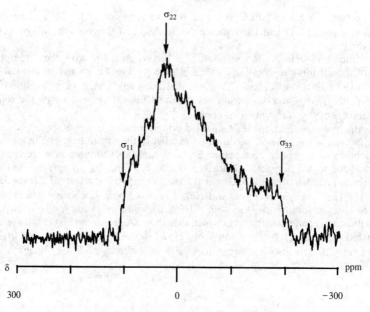

FIGURE 16. Static solid state ^{29}Si NMR spectrum of hexasilaprismane **12**

TABLE 11. Experimental and calculated principal values of shielding tensors for hexasilaprismane and octasilacubane

Tensors	(2,6-i-Pr$_2$C$_6$H$_3$)$_6$Si$_6$ **12**	Me$_6$Si$_6$ (D_{3h})[a]	(2,6-Et$_2$C$_6$H$_3$)$_8$Si$_8$ **22**	Me$_8$Si$_8$ (O_h)[a]
σ_{11}	98	101	80	86
σ_{22}	18	92	−27	−35
σ_{33}	−182	−221	−54	−35
σ_{iso}	−22	−29	−0.3	4.8
$\sigma_{11} - \sigma_{33}$	280	322	134	121

[a]GIAO-B3LYP calculations at B3LYP/6-31G (d) optimized geometries. The basis sets are 6-311G(3d) for Si and 6-311(d) for C and H.

In **12**, the principal axis corresponding to σ_{33}, the most shielded principal value (−182 ppm), is perpendicular to the three-membered rings. High shielding in this direction can be found in all types of cyclotrisilanes[58]. The hexasilaprismane **12** has an unprecedented spread anisotropy of 280 ppm; a high value of 322 ppm is also found theoretically for Me$_6$Si$_6$ (Table 11).

VIII. SPHERICAL CAGE COMPOUNDS

According to Figure 1, the strain energies of persila[n]prismanes increase highly with an increase in n because the bond angles in the n-membered rings at the top and bottom of the persila[n]prismanes deviates greatly from the ideal tetrahedral angle. In persila[10]prismane (**27**, Si$_{20}$H$_{20}$), the bond angles are as large as 144.0°. This great deviation from tetrahedral angles is reflected in the large strain energy of 252.1 kcal mol^{-1}, although this value is much smaller than that of 492.1 kcal mol^{-1} for C$_{20}$H$_{20}$[10,15]. For Si$_{20}$H$_{20}$, the less strained

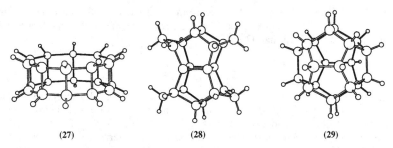

(27)   (28)   (29)

FIGURE 17. Three isomers of $Si_{20}H_{20}$ having the [10]prismane, pagodane, and dodecahedrane structures. Reprinted with permission from References 10. Copyright 1995 American Chemical Society

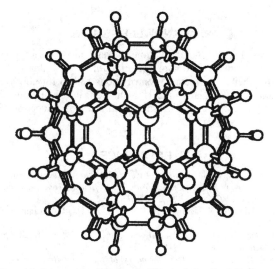

FIGURE 18. $Si_{60}H_{60}$ with I_h symmetry. Reprinted with permission from Reference 8. Copyright 1993 IUPAC

[1.1.1.1]pagodane (**28**) and dodecahedrane (**29**) structures shown in Figure 17 are 200 and 220 kcal mol^{-1} more stable than **27**[10,15]. In the I_h cage structure of **29**, skeletal silicons can form bond angles close to the ideal tetrahedral value and consequently the strain energy become as small as 32.3 kcal mol^{-1}. This is considerably smaller than the value of 43.6 kcal mol^{-1} for dodecahedrane. It may be an interesting challenge to chemists to synthesize persiladodecahedrane[10,15], as it was once to synthesize dodecahedrane[59].

In view of the great progress in the chemistry of fullerenes (spherical carbon clusters consisting of pentagonal and hexagonal rings, C_n), the silicon analogues are also interesting. It has been suggested that silicon clusters (Si_n) have a tendency to take on fullerene-like cage structures as the cluster size increases, as do carbon clusters[8,60]. The diameter (11.1 Å[60a]) of the hollow spherical cage of Si_{60} is much larger than that of C_{60} (7.1 Å[61]). In addition, the electrons of Si_{60} are more polarizable, as expected from the higher HOMO and lower LUMO levels: -6.5 and -2.1 eV for Si_{60}[60a] vs -8.0 and -0.3 eV for C_{60}[62]. The complete hydrogenation of Si_{60} leads to the saturated $Si_{60}H_{60}$ molecule, as shown in Figure 18. The strain energies of $Si_{60}H_{60}$ equal to 114 (AM1) and

207 (HF/DZ/HF/DZ) kcal mol^{-1} calculated from the homodesmotic reactions are much smaller than that of 530 (AM1) kcal mol^{-1} for $C_{60}H_{60}$. This may reflect the fact that Si_{60} is significantly less strained than C_{60}.

IX. EPILOGUE

Polyhedral silicon compounds were once thought to be synthetically inaccessible. However, this view has been drastically changed in the last few years through a close interplay of theoretical predictions and experimental tests. Many new types of polyhedral compounds including Si, Ge and Sn have been synthesized, characterized and isolated, showing novel structures and properties which are often unexpected from the carbon counterparts. These accomplishments have greatly enhanced our understanding of the bonding nature in the heavier systems.

X. ACKNOWLEDGEMENT

We thank Professors Hideki Sakurai, Robert West and Zvi Rappoport for helpful discussions and useful advise. We are also grateful to C. Kabuto, T. Yatabe, S. Doi and H. Kamatani for their experimental contributions and to K. Kobayashi, M. Nakano, T. Kudo, M. Souma and M. Nagashima for theoretical calculations. This work was supported in part by a Grant-in-Aid from the Ministry of Education, Science, Sports, and Culture of Japan.

XI. REFERENCES

1. A. Greenberg and J. F. Liebman, *Strained Organic Molecules*, Academic Press, New York, 1978.
2. A. T. Balaban, M. Banciu and V. Ciorba, *Annulenes, Benzo-, Hetero-, Homo-Derivatives, and Their Valence Isomers*, CRC Press, Florida, 1987.
3. H. Matsumoto, K. Higuchi, Y. Hoshino, H. Koike, Y. Naoi and Y. Nagai, *J. Chem. Soc., Chem. Commun.*, 1083 (1988).
4. A. Sekiguchi, C. Kabuto and H. Sakurai, *Angew. Chem., Int. Ed. Engl.*, **28**, 55 (1989).
5. S. Nagase, *Polyhedron*, **10**, 1299 (1991).
6. A. Sekiguchi and H. Sakurai, in *The Chemistry of Inorganic Ring Systems* (Ed. R. Steudel), Chap. 7, Elsevier, New York, 1992.
7. H. Sakurai and A. Sekiguchi, in *Frontiers of Organogermanium, -Tin and -Lead Chemistry*. (Eds. E. Lukevics and L. Ignatovich), Latvian Institute of Organic Synthesis, Riga, 1993.
8. S. Nagase, *Pure Appl. Chem.*, **65**, 675 (1993).
9. A. Sekiguchi and H. Sakurai, *Adv. Organomet. Chem.*, **37**, 1 (1995).
10. S. Nagase, *Acc. Chem. Res.*, **28**, 469 (1995).
11. L. R. Sita, *Acc. Chem. Res.*, **27**, 191 (1994).
12. K. B. Wiberg, *Angew. Chem., Int. Ed. Engl.*, **25**, 312 (1986).
13. S. Nagase, M. Nakano and T. Kudo, *J. Chem. Soc., Chem. Commun.*, 60 (1987).
14. S. Nagase, *Angew. Chem., Int. Ed. Engl.*, **28**, 329 (1989).
15. S. Nagase, K. Kobayashi and T. Kudo, *Main Group Metal Chem.*, **17**, 171 (1994).
16. L. R. Sita and I. Kinoshita, *J. Am. Chem. Soc.*, **113**, 1856 (1991).
17. J. P. Desclaux, *At. Data Nucl. Data Tables*, **12**, 311 (1973).
18. W. Kutzelnigg, *Angew. Chem., Int. Ed. Engl.*, **23**, 272 (1984). For a similar trend for group 15 elements, see: S. Nagase, in *The Chemistry of Organic Arsenic, Antimony and Bismuth Compounds* (Ed. S. Patai), Chap. 1, Wiley, New York, 1994.
19. D. Cremer, J. Gauss and E. J. Cremer, *J. Mol. Struct. (Theochem)*, **169**, 531 (1988).
20. For the successful synthesis of silicon–silicon double bonds, see:
    (a) R. West, *Pure Appl. Chem.*, **56**, 163 (1984).
    (b) G. Raabe and J. Michl, *Chem. Rev.*, **85**, 419 (1985).
    (c) R. West, *Angew. Chem., Int. Ed. Engl.*, **26**, 1201 (1987).

3. Polyhedral silicon compounds

(d) G. Raabe and J. Michl, in *The Chemistry of Organic Silicon Compounds* (Eds. S. Patai and Z. Rappoport), Chap. 17, Wiley, New York, 1989.
(e) R. Okazaki and R. West, in *Adv. in Organomet. Chem.*, **39**, 232 (1996).
21. (a) S. Nagase, T. Kudo and M. Aoki, *J. Chem. Soc., Chem. Commun.*, 1121 (1985).
(b) A. Sax and R. Janoschek, *Angew. Chem., Int. Ed. Engl.*, **25**, 651 (1986).
(c) A. Sax and R. Janoschek, *Phosphorus Sulfur*, **28**, 151 (1986).
(d) S. Nagase, H. Teramae and T. Kudo, *J. Chem. Phys.*, **86**, 4513 (1987).
(e) N. Matsunaga and M. S. Gordon, *J. Am. Chem. Soc.*, **116**, 11407 (1994).
22. For a similar trend for group 15 elements, see: K. Kobayashi, H. Miura and S. Nagase, *J. Mol. Struct. (Theochem)*, **311**, 69 (1994).
23. For several experimental and calculated data, see: T. Tsumuraya, S. A. Batcheller and S. Masamune, *Angew. Chem., Int. Ed. Engl.*, **30**, 902 (1991).
24. (a) P. v. R. Schleyer, A. F. Sax, J. Kalcher and R. Janoschek, *Angew. Chem., Int. Ed. Engl.*, **26**, 364 (1987).
(b) T. Dabisch and W. W. Schoeller, *J. Chem. Soc., Chem. Commun.*, 896 (1986).
(c) W. W. Schoeller, T. Dabisch and T. Busch, *Inorg. Chem.*, **26**, 4383 (1987).
(d) S. Nagase and T. Kudo, *J. Chem. Soc., Chem. Commun.*, 54 (1988).
(e) J. A. Boatz and M. S. Gordon, *J. Phys. Chem.*, **93**, 2888 (1989).
(f) D. B. Kitchen, J. E. Jackson and L. C. Allen, *J. Am. Chem. Soc.*, **112**, 3408 (1990).
(g) J. A. Boatz and M. S. Gordon, *Organometallics*, **15**, 2118 (1996).
25. For calculations of the Ge, Sn and Pb analogues, see:
(a) S. Nagase and M. Nakano, *J. Chem. Soc., Chem. Commun.*, 1077 (1988).
(b) T. Kudo and S. Nagase, *J. Phys. Chem.*, **96**, 9189 (1992).
26. (a) S. Masamune, Y. Kabe, S. Collins, D. J. Williams and R. Jones, *J. Am. Chem. Soc.*, **107**, 5552 (1985).
(b) R. Jones, D. J. Williams, Y. Kabe and S. Masamune, *Angew. Chem., Int. Ed. Engl.*, **25**, 173 (1986).
27. (a) D. A. Clabo, Jr. and H. F. Schafer III, *J. Am. Chem. Soc.*, **108**, 4344 (1986).
(b) A. F. Sax and J. Kalcher, *J. Chem. Soc., Chem. Commun.*, 809 (1987).
(c) A. F. Sax and J. Kalcher, *J. Comput. Chem.*, **10**, 309 (1989).
28. S. Nagase and M. Nakano, *Angew. Chem., Int. Ed. Engl.*, **27**, 1081 (1988).
29. S. Nagase, K. Kobayashi and M. Nagashima, *J. Chem. Soc., Chem. Commun.*, 1302 (1992).
30. M. Kaupp and P. v. R. Schleyer, *J. Am. Chem. Soc.*, **115**, 1061 (1993).
31. (a) M. Gomberg, *Recl. Trav. Chim. Pays-Bas*, **48**, 847 (1929).
(b) M. Gomberg and W. E. Bachmann, *J. Am. Chem. Soc.*, **49**, 236 (1927).
32. (a) N. Wiberg, C. M. M. Finger and K. Polborn, *Angew. Chem., Int. Ed. Engl.*, **32**, 1054 (1993).
(b) N. Wiberg, C. M. M. Finger, H. Auer and K. Polborn, *J. Organomet. Chem.*, **521**, 377 (1996).
33. A. Sekiguchi, S. Doi and H. Sakurai, to appear.
34. S. Nagase and K. Kobayashi, to appear.
35. G. Maier, *Angew. Chem., Int. Ed. Engl.*, **27**, 309 (1988).
36. N. Wiberg, W. Hochmuth, H. Nöth, A. Appel and M. Schmidt-Amelunxen, *Angew. Chem., Int. Ed. Engl.*, **35**, 1333 (1996).
37. A. Sekiguchi, T. Yatabe, C. Kabuto and H. Sakurai, *J. Am. Chem. Soc.*, **115**, 5853 (1993).
38. S. Masamune, Y. Hanzawa, S. Murakami, T. Bally and J. F. Blount, *J. Am. Chem. Soc.*, **146**, 1150 (1982).
39. R. R. Karl, K. L. Gallaher, Y. C. Wang and S. H. Bauer, unpublished results cited in *J. Am. Chem. Soc.*, **96**, 17 (1974).
40. R. R. Karl, Y. C. Wang and S. H. Bauer, *J. Mol. Struct.*, **25**, 17 (1975).
41. A. Sekiguchi, K. Ebata, C. Kabuto and H. Sakurai, to appear.
42. S. Nagase, A. Sekiguchi, S. Doi and H. Sakurai, to appear.
43. A. Sekiguchi, T. Yatabe, S. Doi and H. Sakurai, *Phosphorus, Sulfur, and Silicon and the Related Elements*, **93 & 94**, 193 (1994).
44. (a) K. Furukawa, M. Fujino and N. Matsumoto, *Appl. Phys. Lett.*, **60**, 2744 (1992).
(b) K. Furukawa, M. Fujino and N. Matsumoto, *J. Organomet. Chem.*, **515**, 37 (1996).
(c) H. Tachibana, M. Goto, M. Matsumoto, H. Kishida and Y. Tokuda, *Appl. Phys. Lett.*, **64**, 2509 (1994). However, the crystal structure reported by Tachibana and coworker was questioned by K. Furukawa, M. Fujino and N. Matsumoto, *Appl. Phys. Lett.*, **66**, 1291 (1995).
45. H. Matsumoto, K. Higuchi, S. Kyushin and M. Goto, *Angew. Chem., Int. Ed. Engl.*, **31**, 1354 (1992).

46. A. Sekiguchi, S. Doi and H. Sakurai, to appear.
47. A. Sekiguchi, T. Yatabe, H. Kamatani, C. Kabuto and H. Sakurai, *J. Am. Chem. Soc.*, **114**, 6260 (1992).
48. K. Kobayashi and S. Nagase, to appear.
49. E. B. Fleischer, *J. Am. Chem. Soc.*, **86**, 3889 (1964).
50. L. R. Sita and I. Kinoshita, *Organometallics*, **9**, 2865 (1990).
51. K. Furukawa, H. Teramae and N. Matsumoto, *65th Annual Meeting of Japan Chemical Society*, Tokyo, March 1993, Abstract I 4F 342 (1993).
52. (a) Y. Kanemitsu, K. Suzuki, H. Uto, Y. Masumoto, T. Matsumoto, S. Kyushin, K. Higuchi and H. Matsumoto, *Appl. Phys. Lett.*, **61**, 2446 (1992).

    (b) Y. Kanemitsu, K. Suzuki, H. Uto, Y. Masumoto, K. Higuchi, S. Kyushin and H. Matsumoto, *Jpn. J. Appl. Phys.*, **32**, 408 (1993).

    (c) S. Kyushin, H. Matsumoto, Y. Kanemitsu and M. Goto, *J. Phys. Soc. Jpn.*, **63**, 46 (1994).

    (d) Y. Kanemitsu, K. Suzuki, M. Kondo, S. Kyushin and H. Matsumoto, *Phys. Rev. B*, **51**, 10666 (1995).
53. M. Unno, T. Yokota and H. Matsumoto, *J. Organomet. Chem.*, **521**, 409 (1996).
54. M. Unno, K. Higuchi, M. Ida, H. Shioyama, S. Kyushin and H. Matsumoto, *Organometallics*, **13**, 4633 (1994).
55. M. Unno, H. Shioyama, M. Ida and H. Matsumoto, *Organometallics*, **14**, 4004 (1995).
56. A. Sekiguchi, S. Doi, H. Sakurai and R. West, to appear.
57. S. Nagase, to appear.
58. (a) R. West, J. D. Cavalieri, J. Duchamp and K. W. Zilm, *Phosphorus, Sulfur, and Silicon and the Related Elements*, **93 & 94**, 213 (1994).

    (b) J. D. Cavalieri, R. West, J. C. Duchamp and K. W. Zilm, *J. Am. Chem. Soc.*, **115**, 3770 (1993).
59. L. A. Paquette, *Chem. Rev.*, **89**, 1051 (1989).
60. (a) S. Nagase and K. Kobayashi, *Chem. Phys. Lett.*, **187**, 291 (1991).

    (b) S. Nagase and K. Kobayashi, *Fullerene Sci. Technol.*, **1**, 299 (1993).

    (c) K. Kobayashi and S. Nagase, *Bull. Chem. Soc. Jpn.*, **66**, 3334 (1993).

    (d) Z. Slanina, S.-L. Lee, K. Kobayashi and S. Nagase, *J. Mol. Struct. (Theochem)*, **312**, 175 (1994).
61. K. Hedberg, L. Hedberg, D. S. Bethune, C. A. Brown, H. C. Dorn, R. D. Johnson and M. de Vries, *Science*, **254**, 410 (1991).
62. J. Cioslowski and E. D. Fleischmann, *J. Chem. Phys.*, **94**, 3730 (1991).

CHAPTER 4

Thermochemistry

ROSA BECERRA

Instituto de Quimica Fisica 'Rocasolano', CSIC, C/Serrano, 119, 28006 Madrid, Spain

and

ROBIN WALSH

Department of Chemistry, University of Reading, Whiteknights, PO Box 224, Reading RG6 6AD, UK

I. INTRODUCTION	154
II. COMPOUNDS OF TETRAVALENT SILICON	155
A. General Considerations, Additivity Rules and Electronegativity Correlations	155
1. Bond additivity	155
2. Group additivity	156
3. Enthalpy/electronegativity correlations	157
B. Experimental Data and Preferred $\Delta H_f°$ Values	158
1. Silicon hydrides (Si/H)	158
2. Alkyl and related organosilanes (Si/C/H)	159
3. Halogen-containing organosilanes (Si/C/H/X)	164
4. Oxygen-containing compounds (Si/C/H/O)	164
5. Nitrogen-containing compounds (Si/C/H/N)	166
6. Other organosilicon compounds	166
III. FREE RADICALS AND BOND DISSOCIATION ENTHALPIES	166
A. General Comments	166
B. Experimental Data: Measured Dissociation Energies and Radical Enthalpies of Formation	167
C. Derived Bond Dissociation Energies	169
IV. OTHER SILICON-CONTAINING SPECIES	171
A. Silylenes	171
1. SiH_2	171
2. MeSiH	172
3. Me_2Si	172

The chemistry of organic silicon compounds, Vol. 2
Edited by Z. Rappoport and Y. Apeloig   © 1998 John Wiley & Sons Ltd

    4. H$_3$SiSiH 173
    5. Silicon dihalides 173
  B. π-Bonded Species 174
    1. Sila-alkenes 174
    2. Disilene 175
    3. Silanone 175
V. APPENDIX 176
VI. REFERENCES 177

I. INTRODUCTION

A knowledge of molecular heats of formation and chemical bond dissociation energies has always been regarded as fundamental to the understanding of chemical structure and reactivity. This chapter deals with the extent and reliability of our knowledge of these quantities for silicon-containing compounds. The information and material presented here represent an update of an earlier review[1] in this series on the same subject. Whilst we were surveying the work published in the intervening seven years since the previous review, we were initially struck by how few experimental papers there were in this area. Indeed we even wondered whether, with the shortage of material, a review was justified. There is no doubt that the traditional science of calorimetry has virtually died out. Just a few laboratories in the world are left in this area. *Equilibrium studies* also have dropped out of fashion, in the direct sense, but if this is taken to include *kinetic studies* of matched pairs of forward and reverse processes, then the view is not so bleak. In this article, as in the previous one[1] and other reviews[2-6], we have always taken the view that thermochemistry should be inclusive of information derived from as wide a range of techniques and sources as possible. As anticipated, the impact of theoretical calculations through the development and implementation of molecular orbital theory *ab initio* methods has been substantial. Calculations of heats of formation through widespread use of computerized packages have become sufficiently routine that many experimental groups use them as a supplement to interpretation of their results. The quality of these calculations appears to have reached a point where they can seriously challenge experimental numbers in some cases (but see caveat below). Thus, in sum, there appears to be sufficient new information, including revisions of earlier results, to make this enterprise worthwhile. But if current trends continue we foresee both a benefit and a danger. The benefit is the increasing ease of calculation of once difficult-to-obtain thermochemical quantities. The danger will be the shortage of a sufficiently broad foundation of reliable experimental values with which to secure the theoretical edifice.

This chapter concentrates on results rather than the details of experimental techniques. These can be found either in the reviews[1-6] already mentioned or in original articles. Nevertheless it is worth recording that the application of oxygen-bomb calorimetry to organosilicon compounds continues to pose problems. The difficulties have been much discussed[1,7] and led to exclusion of most of the older (pre-1970) data on heats of formation[8] from earlier data compilations[7,9] (with one exception[10]). In the last decade or so the group of Voronkov has developed and exploited a method designed to overcome earlier problems of incomplete combustion, whereby the sample is initially vaporized prior to ignition. The initial results of these studies were flagged in our earlier review[1], and since then a series of seven papers has appeared covering the thermochemistry of polyalkylsilanes[11], polyalkoxysilanes[11], organylsilatranes[12], S-containing alkoxysilanes and silatranes[13], silacyclobutanes (siletanes)[14], oligocyclosiloxanes[15], silylamines[16] and cyclosilazanes[17]. These are an impressive series of studies and almost all of the data looks extremely self-consistent. The difficulty is that in some selected cases there are differences with either published data from other sources or reasonable chemical expectations (*vide infra*). This leads to the suspicion that in spite of the claims, there may still be some unassessed source of systematic error in the method. Unfortunately most of the papers do not give sufficient experimental details to enable such an evaluation. This is one of the continuing problems of data republished in western journals from largely unread Russian original sources. Data from other laboratories are relatively sparse and we reiterate that there is still a crying need for more, and reliable, experimentation in this field.

Amongst the theoretical calculations on organosilicon thermochemistry the work of Melius' group is noteworthy[18-25]. For each species a total atomization enthalpy is calculated by means of fourth-order Moller–Plesset (MP4) perturbation theory using the 6-31G** basis set at the HF/6-31G* calculated geometry. Bond additivity corrections (BAC) are then applied to these enthalpies to overcome the deficiencies of truncated wave functions and incomplete basis sets. These corrections require the use of experimental data on reference compounds. Thus the theory is not 'pure' in the sense of totally independent of experiment. Nevertheless the range of species and variety of compounds explored is impressive as is the general consistency and agreement with experiment. Melius and coworkers have studied compounds of Si/H[18,19], Si/Cl[18], Si/H/Cl[18], Si/F[20], Si/H/F[20], Si/H/N/F[21], Si/C/H[22], Si/C/Cl/H[23], Si/O/H/C[24] and Si/O/H[25].

In addition to our previous reviews of this subject, there is a recent review of theoretical investigations of the thermochemistry of organosilicon compounds[26]. An article on three methods to measure RH bond energies by Berkowitz, Ellison and Gutman[27] is a valuable up-to-date source of many bond dissociation energies and Chatgilialoglu's article[28] on structural and chemical properties of silyl radicals contains a useful discussion of silyl radical thermochemistry. In this review we have tried to concentrate on the recent work, but inevitably there is some overlap with the earlier review[1]. This is necessary to bring out a number of data comparisons. If the coverage in some parts is a little thin, due to shortage of new data, readers are urged to refer to the previous article[1]. We have structured this review similarly to the previous one for ease of back reference. As previously, all standard enthalpies of reaction or formation refer to the gas phase at 298.2 K.

II. COMPOUNDS OF TETRAVALENT SILICON

A. General Considerations, Additivity Rules and Electronegativity Correlations

It is now widely recognized that enthalpies of formation for each compound do not exist in isolation from one another. They can be judged by how well they fit into additivity schemes. Most authors of both experimental and theoretical papers recognize this. There are many additivity schemes, but most of them are variants on the same theme. We have used the laws of bond and group additivity devised by Benson and colleagues[29-31] to assess organosilicon compounds[1] and we again take this approach here.

1. Bond additivity

The law of bond additivity[29] states that for a bond redistribution (or disproportionation) reaction (these days called an isodesmic reaction) such as:

$$2SiX_nY_{4-n} \rightleftharpoons SiX_{n+1}Y_{3-n} + SiX_{n-1}Y_{5-n}$$

overall thermodynamic changes (such as $\Delta H°$) should be zero. Another way of stating this (for enthalpy changes) is to say that X- for -Y replacement enthalpies in a sequence of SiX_nY_{4-n} compounds should be constant. For X = Me, Y = H, i.e. the methylsilanes,

TABLE 1. Standard enthalpies of formation (ΔH_f°/kJ mol^{-1}) of fluoro- and chlorosilanes

Compound	Experiment[a]	Theory
SiH$_3$F	-377 ± 42	-359^b, $-358^{c,d}$
SiH$_2$F$_2$	-791 ± 33	-779^b, -780^c, -777^d
SiHF$_3$	-1200 ± 21	-1206^b, -1208^c, -1204^d
SiF$_4$	-1615 ± 1	
SiH$_3$Cl	-136 ± 10	-134^e, -134^c
SiH$_2$Cl$_2$	-315 ± 8	-310^e, -311^c
SiHCl$_3$	-499 ± 6	-489^e, -490^c
SiCl$_4$	-663 ± 5	

[a] Reference 3. [b] Reference 32. [c] Reference 20. [d] Reference 33. [e] Reference 34.

this is known to work very well[1,4]. This is not altogether surprising since the Si—C and Si—H bonds in these compounds are not very polar. A more demanding test is to examine silyl halides where X = F, Y = H or X = Cl, Y = H.

Enthalpies of formation of these compounds are listed in Table 1. Experimental values are taken from one of our earlier reviews[3]: values listed by JANAF[35] are in essential agreement (within experimental error). For the SiH$_n$F$_{4-n}$ series, apart from ΔH_f° (SiF$_4$), there is very little experimental information and the recommended values have been obtained by interpolation. Theoretical values are, however, all in close agreement with one another. The disproportionation enthalpies (kJ mol^{-1}) are -27 (SiH$_3$F), -6 (SiH$_2$F$_2$) and $+16$ (SiHF$_3$) based on the values of Ignacio and Schlegel[32]. Although not zero (as they would be if bond additivity were obeyed), they are relatively small and follow the same trend as for the analogous CH$_n$F$_{4-n}$ series[3]. Thus the theoretical values for ΔH_f° (SiH$_n$F$_{4-n}$) (uncertainties quoted at ± 8 kJ mol^{-1}) are certainly more reliable than experiment at the present time. For the SiH$_n$Cl$_{4-n}$ series, experimental and theoretical values are in reasonable agreement (except for SiHCl$_3$, see below). The disproportionation enthalpies (kJ mol^{-1}) are: for SiH$_3$Cl, -9.6 (expt), -8 (theory); for SiH$_2$Cl$_2$, -4.6 (expt), -1.7 (theory); for SiHCl$_3$, $+21$ (expt), $+4$ (theory). Here the values follow the same trends as for both SiH$_n$F$_{4-n}$ and CH$_n$Cl$_{4-n}$[31], whether experimental or theoretical values are used. Our expectation would be that the disproportionation values should be less for SiH$_n$Cl$_{4-n}$ than for SiH$_n$F$_{4-n}$. This seems to be the case except for SiHCl$_3$. This suggests that the experimental value for ΔH_f° (SiHCl$_3$) may be slightly too low, although it has been determined by several groups[3]. For redistribution reactions involving other groups on silicon, we would thus expect bond additivity to work quite well, with deviations not exceeding the values just discussed.

2. Group additivity

The effectiveness of group additivity as a means to calculate unknown ΔH_f° values depends on an extensive data base. The data base has been substantially enlarged in recent years by the work of Voronkov and coworkers[11-17a] who have shown good consistency using the Tatevsky[17b] as well as a version of Benson's additivity schemes[29]. Because we have reservations about some of the data from Voronkov's work (*vide infra*) and because there are not enough independent checks, we do not feel that substantial extension of the existing set of group contributions[1] is a worthwhile exercise. However, in the Appendix we provide a table showing the current values of the group contributions. These are slightly modified from previously[1] to provide a best fit to the data recommended in this review.

3. Enthalpy/electronegativity correlations

Starting from the classic work of Pauling[36], there have been many attempts to link quantitative measurements of bond energies (and reaction enthalpy changes) to fundamental properties of the bonded atoms, such as electronegativity. One of the most recent, by Luo and Benson[37,38], employs a new scale of electronegativity called 'unshielded core potential' or more simply 'covalent potential', V_x. In a series of papers[37-45], Luo and Benson have applied the covalent potential to the correlation of enthalpies of formation and argued that it is more successful in this exercise than other scales of electronegativity[45]. Some of these papers have been devoted to organosilicon compounds[41-44]. The correlations are generally good and have been exploited by Luo and Benson[43] to obtain new $\Delta H_f°$ values for Me$_3$SiF, Me$_3$SiNH$_2$ and Me$_3$SiSH and a revised $\Delta H_f°$ value for SiH$_3$I. In addition, they have suggested a revision of $\Delta H_f°$ (Si$_2$H$_6$)[44]. These claims led us to examine the correlations more closely[46]. Our conclusion is that, while this scale may be used *generally* to correlate the thermochemistry of silicon and organosilicon compounds, it has to be exercised with caution, particularly where the availability of data is very limited.

We can illustrate this with the example of the correlation of $\Delta_1 = [\Delta H_f°(SiH_3X) - \Delta H_f°(HX)]$ versus V_x. This is shown in Figure 1. The original plot of Luo and Benson[42] featured X = H, I, Br, Cl, F. We have added X = SiH$_3$ and CH$_3$. The 'best fit' line ignores the points we have added. For X = I the fit is clearly not good and this supports revision of $\Delta H_f°$ (SiH$_3$I). For X = SiH$_3$, the situation is more complicated. Luo and

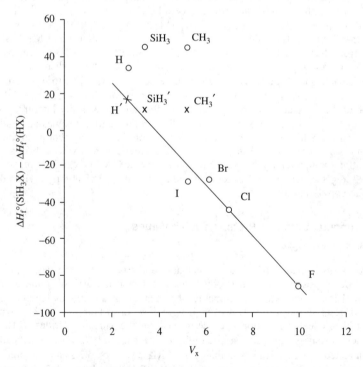

FIGURE 1. Correlation of selected $\Delta H_f°$ differences (Δ_1, kJ mol^{-1}) with covalent potential, V_x(Å^{-1}); O, direct differences, Δ_1; ×, scaled differences, Δ_1/p (see text for definition of p). Primed substituents (H′,SiH$_3'$,CH$_3'$) indicate compounds with scaled differences

TABLE 2. Standard enthalpies of formation ($\Delta H_f°$/kJ mol^{-1}) for selected silicon compounds obtained by correlations with the covalent potential

Compound	Luo and Benson[a]	Re-evaluated[b]
Me$_3$SiF	−572	−568
Me$_3$SiNH$_2$	−289	−291
Me$_3$SiSH	−269	−273
SiH$_3$I	+9.2	+8.6

[a]Reference 43. [b]Reference 46, see text.

Benson[39,42] argue that the true correlation test requires the use of a parameter p (= number of H atoms in the HX molecule), such that the real correlation should be between Δ_1/p and V_x, not Δ_1 and V_x. Thus $\Delta_1(X = H) = [\Delta H_f°(SiH_4) - \Delta H_f°(H_2)]$ is divided by 2 and $\Delta_1(X = SiH_3) = [\Delta H_f°(Si_2H_6) - \Delta H_f°(SiH_4)]$ is divided by 4. If the deviation of the X = SiH$_3$ point is taken as significant, then to fit perfectly requires the suggested alteration[44] in $\Delta H_f°(Si_2H_6)$. We have added the further point $\Delta_1(X = CH_3) = [\Delta H_f°(SiH_3CH_3) - \Delta H_f°(CH_4)]$ and we note that (even after division by 4) the departure from the plot is significant and well beyond any experimental error. Our conclusion is thus that these correlations have to be treated with caution. There may be deep-seated factors causing them to break down for too wide a range of examples. For species of a similar kind (e.g. the halides) they seem to be all right. The revision of $\Delta H_f°(SiH_3I)$ is supported by the correlation[43] of $\Delta_2 = [\Delta H_f°(Me_3SiX) - \Delta H_f°(SiH_3X)]$ versus V_x for X = I, Br, Cl and F. The values of $\Delta H_f°$ for Me$_3$SiF, Me$_3$SiNH$_2$ and Me$_3$SiSH were obtained from the correlation[43] of $\Delta_3 = [\Delta H_f°(Me_3SiX) - \Delta H_f°(CH_3X)]$ versus V_x for X = I, Br, Cl and OH. For both these correlations, the examples where X = H, SiH$_3$ and CH$_3$ deviate substantially, attributed in the case of Δ_3 vs V_x[43], to p$_\pi$-d$_\pi$ bonding. We have tried[46] to extend these ideas to $\Delta_4 = [\Delta H_f°(Me_3SiX) - \Delta H_f°(HX)]$ versus V_x and find the correlation neither so good, nor the deviations easily interpretable. Luo and Benson[44] extended the use of V_x to calculate group additivity values and enthalpies of formation for SiMe$_m$H$_{3-m}$X compounds. This would be a very worthwhile exercise if the data base were more substantial, but with so few concrete values this exercise is dependent on too few reliable figures. The values obtained by Luo and Benson are shown in Table 2, together with minor revisions found by us[46] by use of slightly revised original data in the correlation plots.

B. Experimental Data and Preferred $\Delta H_f°$ Values

1. Silicon hydrides (Si/H)

The heats of formation of silane[47], disilane[47] and trisilane[48] were determined in an explosive decomposition calorimeter by Gunn and Green in the 1960s and, apart from a small revision of the reference enthalpy of formation of amorphous silicon[35] are not thought to be in serious error. However, it is interesting to note that in the most state-of-the-art theoretical calculations yet performed on the silicon hydrides, SiH$_n$, Grev and Schaefer[49] have suggested the possibility of error in $\Delta H_f°(SiH_4)$. Specifically they have calculated the atomization enthalpy of SiH$_4$ to be ca 6.2 kJ mol^{-1} higher than experiment[35]. Thus the difference $\Delta H_f°(Si) - \Delta H_f°(SiH_4)$ needs to be raised to match the theory. The combined experimental uncertainties of these quantities is ±8.2 kJ mol^{-1} and so the discrepancy is within experimental error. Nevertheless, this points to the fact that

4. Thermochemistry

TABLE 3. Standard enthalpies of formation (ΔH_f°/kJ mol^{-1}) of silanes

Compound	$\Delta H_f^{\circ a}$
SiH$_4$	34.3 ± 1.2
Si$_2$H$_6$	80 ± 1.5
Si$_3$H$_8$	121 ± 4.4

[a]Reference 1.

refinement of uncertainties in ΔH_f°(SiH$_4$) and ΔH_f°(Si) would be a worthwhile experimental exercise. The current values for SiH$_4$, Si$_2$H$_6$ and Si$_3$H$_8$ are shown in Table 3. They form a consistent set and are reasonably matched by the theoretical calculations of Ho and coworkers[19], Sax and Kalcher[50] and Leroy and coworkers[51]. As mentioned already Luo and Benson[44] have questioned the value of ΔH_f°(Si$_2$H$_6$) in Table 3 and suggested instead a figure less than 64.9 kJ mol^{-1} based on empirical correlations. There is no other evidence in favour of this figure, and it is clearly inconsistent with the other silicon hydrides.

2. Alkyl and related organosilanes (Si/C/H)

There are no new experimental data for the methylsilanes since the earlier review[1]. The original data are listed in Table 4. It is worth re-emphasizing that the listed values are

TABLE 4. Standard enthalpies of formation (ΔH_f°/kJ mol^{-1}) of methyl, ethyl and propyl silanes

Compound	Experiment	Additivity[b,c,d]	Ab initio
SiH$_4$	34.3 ± 2.0^a		
MeSiH$_3$	−29.1 ± 4.0^b	−29.1^i	−28.5^j, −30.5^k, −26.8^l
Me$_2$SiH$_2$	−94.7 ± 4.0^b	−94.7^i	−94.6^j, −97.1^k, −96.2^l
Me$_3$SiH	−163.4 ± 4.0^b	−163.4^i	−163.6^j, −164.8^k
Me$_4$Si	−233.2 ± 3.2^e	−233.2^i	
	−229 ± 3.0^f		
EtSiH$_3$	−143^g	−46.1^c	−37.7^c, −34.3^l
PrSiH$_3$		−66.8^c	−61.5^c
EtSiH$_2$Me		−111.7^c	−106.3^c
Et$_2$SiH$_2$	−182 ± 6^h	−128.7^c	−117.2^c
i-PrSiH$_3$		−73.3	−59.0^l
t-BuSiH$_3$			−86.6^l
Et$_3$SiH	−201 ± 15^h	214.4	
	−217.5 ± 5^f		
Pr$_3$SiH	−280 ± 6^f	−276.5	
MeSiHEt$_2$	−200 ± 5^f	−197.4	
MeSiHPr$_2$	−240 ± 6^f	−238.8	
MeSiHPr$_2$-i	−255 ± 6^f	−251.8^i	
EtSiHPr$_2$	−259 ± 6^f	−255.8	
EtSiHPr$_2$-i	−270 ± 6^f	−268.8^i	

[a]Reference 35. [b]Reference 4. [c]Reference 52. [d]Reference 1. [e]Reference 53. [f]Reference 11. [g]Reference 54. [h]Reference 9. [i]Used to fix additivity groups. [j]Reference 26. [k]Reference 22. [l]Reference 51.

related to, and therefore dependent upon, Steele's value[53] for $\Delta H_f°$(Me$_4$Si) of -233.2 ± 3.2 kJ mol^{-1}. This is supported by a more recent value from Voronkov and coworkers[11] of -229 ± 3 kJ mol^{-1}, virtually within experimental error. Theoretical calculations using isodesmic reactions[26] and BAC-MP4 calculations by Allendorf and Melius[22] are in essential agreement, although just as with the experimental data for MeSiH$_3$, Me$_2$SiH$_2$ and Me$_3$SiH, they are pinned to the value for $\Delta H_f°$(Me$_4$Si). Semi-empirical[55] and empirical correlations[44] are also in reasonable agreement.

For higher alkylsilanes there have been recent combustion measurements by Voronkov and coworkers[11] of a variety of trialkyl and tetraalkylsilanes. These are shown in Tables 4 and 5. Included also are some of the older data listed in earlier compilations[54]. We have pointed out in the previous review[1] and elsewhere[52] the unreliability of these data on ethylsilanes on the basis of lack of fit to additivity rules[1] and also to *ab initio* values combined with homodesmic reactions. The data of Voronkov are truly astonishing. Tables 4 and 5 include only organosilane molecules with alkyl groups up to C$_3$. This publication[11] contains 30 triorganyl and 14 tetraorganyl silanes. Within the cited data the fits to group additivity[1] show very small deviations. Indeed, if the Si—(C)$_3$(H) group value were lowered by 2.6 kJ mol^{-1} and the Si—(C)$_4$ group value raised by 4 kJ mol^{-1} all compounds would fit to better than ± 2 kJ mol^{-1} (see Table 27 in the Appendix for group values). Within the *total* data set, almost none of the evaluated experimental enthalpies of atomization deviates by more than 2 kJ mol^{-1} from values calculated on the authors' own additivity scheme (similar to groups increments). It is somewhat frustrating that the authors do not provide more details in their papers. The authors have seemingly republished their results in the *Journal of Organometallic Chemistry* after the original publications in *Izvestiya Akademii Nauk SSSR*[56]. The data, however, are not *identical* between the two publications (see Table 5), and many more compounds appear in the second publication. What is a pity is that there is no reference to the earlier publication[56] in the second[11], nor any mention of the changes (albeit small) in the values listed or error limits obtained. The biggest change ($+7$ kJ mol^{-1}) in Table 5 is the value for $\Delta H_f°$(Pr$_3$SiMe). This value was the one previously furthest out of line with the authors' additivity scheme.

Apart from the importance of the individual data, increments which reveal aspects of bonding are particularly significant. We previously[1] emphasized the importance of the methyl to ethyl increment of -17 kJ mol^{-1}, which was based on an earlier analysis[4] of electronegativity trends and found to be consistent with the data of Voronkov and coworkers[11,56]. Luo and Benson[44,41] suggest -13 kJ mol^{-1} from their own analysis on

TABLE 5. Standard enthalpies of formation ($\Delta H_f°$/kJ mol^{-1}) of tetraalkylsilanes

Compound	Voronkov and coworkers[a]	Voronkov and coworkers[b]	Additivity[c]
Me$_4$Si	-230 ± 6	-229 ± 3	-233
Et$_2$SiMe$_2$	-264 ± 6	-263 ± 5	-267
Et$_3$SiMe	-287 ± 5	-281 ± 5	-284
Et$_4$Si	-297 ± 6	-297 ± 5	-301
Pr$_2$SiMe$_2$	-305 ± 7	-306 ± 5	-309
Pr$_3$SiMe	-350 ± 7	-343 ± 5	-346
PrSiEt$_3$		-319 ± 6	-322
Pr$_2$SiEt$_2$	-342 ± 6	-341 ± 5	-343
Pr$_3$SiEt		-360 ± 6	-363
Pr$_4$Si	-379 ± 6	-378 ± 5	-384

[a]Reference 56. [b]Reference 11. [c]Reference 1.

metal alkyl enthalpies of formation. An even less negative value of -5.4 (± 10.5) kJ mol^{-1} has been recently suggested by Jardine and coworkers[57]. The *ab initio* calculations of Boatz, Gordon and Walsh[52] give -8.8 kJ mol^{-1} for this quantity and those of Leroy and coworkers[51] -7.5 kJ mol^{-1}. To be able to use the additivity rules with confidence there is a considerable need to confirm the value of this quantity, which corresponds to the group value C−(Si)(C)(H)$_2$. Another group which comes from fitting the Voronkov data is C−(Si)(C)$_2$(H) which has the value -4.2 kJ mol^{-1}, but for which Luo and Benson[44] recommend $+7.1$ kJ mol^{-1}. The theoretical work of Leroy and coworkers[51] implies $+12.9$ kJ mol^{-1}.

We have commented previously[1] on $\Delta H_f°(Me_3SiC_2H_3)$ and $\Delta H_f°((C_2H_3)_4Si)$, determined by Voronkov and coworkers[11,56]. The interest here concerns whether there is any thermochemical evidence for a d_π-p_π type interaction between the Si atom and the π-bond of the vinyl group. The combustion values for these compounds suggest that it should be *ca* 65 kJ mol^{-1} per vinyl group which we previously questioned[1]. Our argument was based on enthalpies of hydrogenation. There are no new thermochemical data on these compounds since the last review[1]. However, two sets of theoretical calculations by Allendorf and Melius[22], and Ketvirtis and coworkers[58] bear on this question. The relevant data are shown in Table 6. This shows that there is relatively little difference between estimates based on enthalpies of hydrogenation (or isodesmic reactions) and the *ab initio* calculations for the model compounds, vinyl- and ethynylsilane. Thus it would be surprising if the experimental thermochemical data were correct here. Electron diffraction studies of Me$_3$SiC$_2$H$_3$[63] reveal no structural (or geometrical) evidence of any special interaction. It is also worth adding that there is no evidence for any strong π-type interaction between the phenyl ring and the Si atom in tetraphenylsilane[1].

One very useful experimental number to have been obtained recently is that for $\Delta H_f°(Me_3SiSiMe_3) = 303.7 \pm 5.5$ kJ mol^{-1} by reaction solution calorimetry[64]. This involved measuring $\Delta H°$ for the reaction Br$_2$ + Me$_3$SiSiMe$_3$ → 2Me$_3$SiBr. The reaction stoichiometry was verified and the calorimetry carried out under conditions with each reagent independently in excess, and furthermore $\Delta H_f°(Me_3SiBr)$ was checked by measurement of the hydrolysis enthalpy of Me$_3$SiBr [to give (Me$_3$Si)$_2$O] which was in agreement with the literature value. It is gratifying to see this important quantity now reliably established, after our plea in the previous review[1]. This compound provides one of the reference points in bond dissociation energy discussions and also for establishing silylene thermochemistry. Prior to this measurement, values for $\Delta H_f°(Me_3SiSiMe_3)$ ranged from -295 to -359 kJ mol^{-1}. One of us[65] was responsible for one of the less accurate earlier values (-347 kJ mol^{-1}), while the better estimates, from kinetic

TABLE 6. Standard enthalpies of formation ($\Delta H_f°$/kJ mol^{-1}) of vinyl and ethynyl substituted silanes

Compound	Experiment[a]	Additivity[b]	Ab initio
(C$_2$H$_3$)$_4$Si	-81.6 ± 6.8	$+199$	
Me$_3$SiC$_2$H$_3$	-190 ± 5	-125	
H$_2$C=CHSiH$_3$		79 ± 6[c]	86.6 ± 4.2[e]
			84.5[f], 108 ± 15[g]
HC≡CSiH$_3$		243 ± 8[d]	221.8 ± 4.6[e]
			215[f], 253 ± 13[h]

[a]References 11 and 56. [b]Based on $\Delta H°$ (hydrogenation) and isodesmic reactions. [c]Reference 59. [d]Reference 60. [e]Reference 22. [f]Reference 58. [g]Reference 61. [h]Reference 62.

TABLE 7. Standard enthalpies of formation (ΔH_f°/kJ mol^{-1}) of methyl substituted disilanes

Compound	Experiment[a]	Allen scheme[a]	Kinetics[b]	Ab initio[c]
Si$_2$H$_6$	80.3 ± 1.5	80.3		
MeH$_2$SiSiH$_3$		18.4	20.9	18.0
Me$_2$HSiSiH$_3$		−45.8	−46.0	−48.5
MeH$_2$SiSiH$_2$Me		−43.6	−37.7	−43.9
Me$_3$SiSiH$_3$		−111.8	−112.5	−116.3
Me$_2$HSiSiH$_2$Me		−107.7	−104.2	−110.0
Me$_3$SiSiH$_2$Me		−173.7	−172.8	−176.6
Me$_2$HSiSiHMe$_2$		−171.8	−171.1	−174.9
Me$_3$SiSiHMe$_2$		−237.9	−240.6	−244.8
Me$_3$SiSiMe$_3$	−303.7 ± 5.5	−303.9	−313.8	−313.8

[a]Reference 64. [b]Reference 55. [c]Reference 66.

studies, were provided by O'Neal and coworkers[55]. The details of the discussion of ΔH_f°(Me$_3$SiSiMe$_3$) may be found in the original references[1,55,64,65] and are not reproduced here (they may provide aficionados with some entertainment on the subject of how scientists try to cope in the absence of concrete information!). The values have been used to derive ΔH_f° values for the partially methylated disilanes. These values are reproduced in Table 7 together with the theoretical (*ab initio*) values of Boatz and Gordon[66]. Other theoretical values[51] for some of these disilanes are less accurate.

The situation for strained silicon-containing ring compounds has become if anything, more confusing since the previous review[1]. The only new experimental data, are those for a series of siletanes[14] (silacyclobutanes), shown in Table 8. From these, strain enthalpies in the range of 109–120 kJ mol^{-1} can be calculated. These are slightly larger than an

TABLE 8. Standard enthalpies of formation (ΔH_f°/kJ mol^{-1}) of some siletanes and their strain enthalpies

Compound	Experiment[a]	Strain enthalpy[b]
silacyclobutane (H$_2$Si-ring)	−23.0	109.4
methylsilacyclobutane	−85.0	117.2
1-methyl-1-(iso-substituted)silacyclobutane	−110.0	122.2
1,1-dimethylsilacyclobutane	−113.0	119.2
silacyclobutene (vinylidene)	−8.0	—

[a]Reference 14. [b]Based on Group Additivity; see Table 27 and Reference 31.

4. Thermochemistry

TABLE 9. Standard enthalpies of formation ($\Delta H_f°$/kJ mol^{-1}) and strain enthalpies (kJ mol^{-1}) of some silacycloalkanes[a] and their 1,1-dimethyl derivatives

Compound	$\Delta H_f°$	Strain enthalpy	$\Delta H_f°$ (1,1-dimethylcompound)	
			additivity	experiment
Siletane	39	103	−99	−108.7 ± 6[b]
				−138 ± 11[c]
				−82.8 ± 5.9[d]
				−85.0[e]
Silacyclopentane	−66	19	−204	−182 ± 12[b]
Silacyclohexane	−88	17	−227	

[a]For silirane see Table 10. [b]Reference 68. [c]Reference 9. [d]Reference 69. [e]Reference 14.

ab initio estimate (at the MP2/6-31G* //3-21G* level) by Gordon and coworkers[52,67] of 103 kJ mol^{-1}, and also those corresponding to a number of other experimentally based $\Delta H_f°$ values shown in Table 9. For 1,1-dimethylsiletane, the experimental values for $\Delta H_f°$ correspond to strain enthalpies in the range 64–120 kJ mol^{-1}. The most recent analyses[70,71] of the kinetics of the thermal decomposition of this compound require a value of the strain enthalpy of 92 ± 12 kJ mol^{-1} corresponding to a $\Delta H_f°$ value of −110 ± 12 kJ mol^{-1}. For this reason we are still inclined to prefer Steele's (unpublished) value[68] of −108.7 ± 6 kJ mol^{-1}.

The need for thermochemistry for the three membered rings, silirane and silirene, is also driven, in part, by the desire to understand the complexities of gas-phase kinetics, in this case of addition reactions of SiH$_2$ with C$_2$H$_4$[59] and C$_2$H$_2$[60] (these reactions are more experimentally accessible than the reverse reactions of the thermal decompositions of silirane and silirene). Theoretical modelling of the pressure dependencies of rate constants has been used to obtain approximate (±12 kJ mol^{-1}) estimates of $\Delta H_f°$ values for these rings. These values are very consistent with those of the *ab initio* calculations of Gordon and coworkers[52,67,72] shown in Table 10. These values correspond to ring strain enthalpies of *ca* 167 kJ mol^{-1} (silirane) and 207 kJ mol^{-1} (silirene).

It has long been assumed that ring strain enthalpies are functions of the ring alone and independent of substituents. However, recent evidence suggests that for methyl-substituted siliranes this may not be so. Berry[74] has found that hexamethylsilirane has a strain enthalpy of *ca* 237 kJ mol^{-1} from a study of the kinetics of its decomposition: studies in our labs[75] also suggest increased ring strain in 2-methyl, 2,2-dimethyl and 1,1-dimethyl siliranes, compared with silirane itself.

TABLE 10. Standard enthalpies of formation ($\Delta H_f°$/kJ mol^{-1}) of three-membered silicon heterocycles

Compound	Ab initio	Experiment
Silirane	126[a]	124 ± 12[f]
	144[b]	
	153 ± 6[c]	
Silirene	289[d], 302[e]	289 ± 12[g]

[a]References 52 and 67. [b]Reference 73. [c]Reference 61. [d]Reference 72. [e]Reference 62. [f]Reference 59. [g]Reference 60.

TABLE 11. Standard enthalpies of formation ($\Delta H_f°$/kJ mol^{-1}) of methylchlorosilanes and methyl (hydrido) chlorosilanes

Compound	Recommended[a]	Gadzhiev + Agarunov[b]	Ab initio[c]
Me$_3$SiCl	-354 ± 3	-354 ± 3	-361 ± 4
Me$_2$SiCl$_2$	-466 ± 10	-464 ± 2	-476 ± 4
MeSiCl$_3$	-569 ± 11	-585 ± 4	-577 ± 4
MeSiH$_2$Cl	-210 ± 7	—	-210 ± 4
MeSiHCl$_2$	-393 ± 9	-414 ± 3	-394 ± 4
Me$_2$SiHCl	-282 ± 3	-305 ± 3	-285 ± 4

[a]Reference 1. [b]Reference 76. [c]Reference 23.

3. Halogen-containing organosilanes (Si/C/H/X)

There are virtually no new experimental data on this class of compounds. Earlier experimental data were discussed in detail in the previous review[1]. The only new information is the BAC-MP4 calculated data of Allendorf and Melius[23] for the chloromethylsilanes. These are compared with our earlier recommended values[1] in Table 11, which shows agreement within the stated uncertainties.

A new enthalpy of hydrolysis measurement[64] gives $\Delta H_f°$ (Me$_3$SiBr) = -298 ± 4 kJ mol^{-1}, which differs very little from the previous value[9,77] of -293 ± 4 kJ mol^{-1}. However we recommend the new value because it is slightly more consistent with the relative thermochemistry of ion breakdown from Me$_3$SiBr and Me$_4$Si[1,78]. This small change also implies modification[79] of $\Delta H_f°$ (Me$_3$SiI). The revised value is -222 ± 4 kJ mol^{-1}, compared to -217 ± 4 kJ mol^{-1} previously[1].

4. Oxygen-containing compounds (Si/C/H/O)

This review is concerned with organosilanes and therefore purely inorganic Si/O/H species are not considered. Nevertheless, because of the importance of silane oxidation and of oxidized silicon coatings there is considerable interest and activity in this area. Four theoretical papers[25,80-82] have been published concerning stabilities of the potentially important molecules in these processes. Needless to say there are very few experimental data in the area under consideration.

There have been two experimental studies by Voronkov and colleagues[11,15], another by van der Vis and coworkers[83,84], and some theoretical calculations by Ho and Melius[24]. Previously we noted[1] (as with Si/C/H compounds reviewed here) the consistency of the Voronkov data from combustion calorimetry of silyl alkyl ethers. Once again the newer publication[11] gives figures which are slightly different from the old (without comment). The consistency is just as good. The authors[11] claim to check the reliability of their procedure using hexamethyldisiloxane, (Me$_3$Si)$_2$O, for which several independent determinations exist, but the data are missing from the new publication[11], although present in the old[56]. However, there is now an independent check possible on the compound tetraethoxysilane (commonly called TEOS) investigated by van der Vis and Cordfunke[83] by means of calorimetric measurements on the aqueous solution reaction:

$$Si(OEt)_4 + 6HF \longrightarrow H_2SiF_6 + 4EtOH$$

The data are shown in Table 12. This latter study obtained a value some 41 kJ mol^{-1} lower than the combustion value[11], well outside the quoted error limits of both studies.

TABLE 12. Standard enthalpies of formation (ΔH_f°/kJ mol^{-1}) of O-containing organosilanes

Compound	Experimental	BAC-MP4(MP2)[a]
Me$_3$SiOH	-500 ± 3[b]	-500
(Me$_3$Si)$_2$O	-777 ± 6[c]	(-785)[g]
Si(OMe)$_4$	-1180 ± 5[d]	-1190
Si(OEt)$_4$	-1315 ± 6[d], -1356 ± 6[e]	-1328
HSi(OEt)$_3$	-912 ± 8[f]	-975

[a]Reference 24. [b]References 7 and 9. [c]References 1 and 7. [d]Reference 11. [e]References 83 and 84. [f]Reference 10. [g]Estimated in Reference 24.

It is instructive to compare the implications of each value for the other's measurements. If the combustion value is correct, it implies that the ΔH_f° value for the aqueous solution reaction is -245.9 kJ mol^{-1} instead of -200.1 kJ mol^{-1} as measured, an error of ca 23%, well outside the quoted $\pm 0.7\%$. If the -245.9 kJ mol^{-1} value is correct, then ΔH_f° (combustion) $= -5538$ kJ mol^{-1} instead of -5583 kJ mol^{-1} as measured, an error of 0.8% compared to the quoted $\pm 0.07\%$. To the impartial assessor, the second scenario seems more likely than the first. A more detailed discussion of potential sources of uncertainty is contained in the paper of van der Vis and Cordfunke[83], which also lists earlier measurements. The BAC-MP4 calculation[24], while nearer to combustion value, is probably not definitive, since there are too few reliable reference compounds on which to base the Bond Additivity corrections.

The consequences of this are to undermine confidence in the previous analysis[1] of the thermochemistry of siloxanes and alkyl silyl ethers. Thus another study by Voronkov and coworkers[15] of perorganyloligocyclosiloxanes is therefore not easy to assess. This consists of combustion of 12 cyclosiloxanes with various differing ring sizes and substituents (methyl or phenyl groups). We may illustrate these data with hexamethylcyclotrisiloxane, c-(Me$_2$SiO)$_3$, for which $\Delta H_f^\circ = -1568$ kJ mol^{-1} in the gas phase has been obtained. According to the author's own additivity scheme this compound has a ring strain enthalpy of 80 kJ mol^{-1}. Our own estimate, based on our earlier group additivity scheme[1], is 108 kJ mol^{-1}. Since these figures depend on groups derived from the silylalkyl ethers, there must be a question as to their validity. The data for the three simple permethylated cyclic siloxanes (sometimes known as D$_3$, D$_4$ and D$_5$) are shown in Table 13. Other lower estimates are obtained for the strain enthalpies if the van der Vis and coworkers value[83,84] for ΔH_f° (TEOS) is used as the basis of an additivity scheme (see Appendix). Whatever the values of these strain enthalpies it seems that they decrease in magnitude with increasing ring size.

TABLE 13. Standard enthalpies of formation (ΔH_f°/kJ mol^{-1}) and strain enthalpies (kJ mol^{-1}) of selected permethylcyclosiloxanes

Compound	ΔH_f° [a]	Strain enthalpy		
		Voronkov[a]	additivity[b,c]	
c-(Me$_2$SiO)$_3$	-1568 ± 11	80	108[b]	65[c]
c-(Me$_2$SiO)$_4$	-2138 ± 11	60	97[b]	40[c]
c-(Me$_2$SiO)$_5$	-2708 ± 13	40	86[b]	14[c]

[a]Reference 15. [b]Reference 1. [c]This review (see text).

TABLE 14. Standard enthalpies of formation (ΔH_f°/kJ mol^{-1}) of methylsilylamines

Compound	Baldwin and coworkers[77]	Voronkov and coworkers[16]
Me$_3$SiNHMe	-227 ± 4	-238 ± 7
Me$_3$SiNMe$_2$	-248 ± 4	-246 ± 8
(Me$_3$Si)$_2$NH	-477 ± 5	-451 ± 10
(Me$_3$Si)$_2$NMe	-449 ± 8	-456 ± 10
(Me$_3$Si)$_3$N	-671 ± 12	-656 ± 11

5. Nitrogen-containing compounds (Si/C/H/N)

Only organo silylamines are considered here. There are again new combustion data from Voronkov and coworkers[16]. These may be compared with those previously obtained by Pedley's group[25] using solution calorimetry, as shown in Table 14. Only a selection of Voronkov and coworkers' data[16] is shown. There is some agreement here (within experimental error), although not in all cases. Previously[1] we evaluated group redistribution energies for these compounds and noted the variations. With Voronkov's data the variations are less. However, in view of the smaller enthalpy changes associated with solution calorimetry compared to combustion (as discussed in the previous section), it would be rash to prefer the combustion values. A deficiency in Voronkov's paper[16] is the failure to specify the final oxidized form of nitrogen in the combustion calorimetry.

Because of previously[1] and presently expressed doubts about earlier work[10,56] we do not discuss ΔH_f° values for other nitrogen containing compounds[12]. There are as yet no theoretical calculations for this class of compounds, although the purely inorganic silyl amines(Si/N/H system) have been investigated by Melius and Ho[21].

6. Other organosilicon compounds

Amongst Si/C/H/S compounds only Me$_3$SiSBu has been studied[77] ($\Delta H_f^\circ = -381 \pm 3$ kJ mol^{-1}). However, Voronkov and coworkers[13] have investigated a series of (organylthioalkyl) trialkoxysilanes and 1-(organylthioalkyl) silatranes, viz compounds in the Si/C/H/S/O and Si/C/H/S/O/N classes. In view of the complexity of these molecules and the lack of comparisons with any related substances, these results are not reviewed here.

III. FREE RADICALS AND BOND DISSOCIATION ENTHALPIES

A. General Comments

Since our last review of values of Si—H bond dissociation energies[1] there have been small but significant changes in values. This has come about because of the experimental removal of one of the assumptions underlying earlier values[2,5]. The basis of determination of these values was the study of the kinetics of the reversible reaction:

$$X + RH \underset{r}{\overset{f}{\rightleftharpoons}} R + HX$$

where X was usually I, an iodine atom. For this reaction we may write the exact enthalpic equation

$$DH^\circ(R-H) = DH^\circ(H-X) + E_f - E_r$$

where E_f and E_r represent forward and reverse activation energies; E_f was measured experimentally[2], but E_r was only estimated to lie in the range 4–8 kJ mol^{-1}. From measurements with carbon-centred (i.e. alkyl) radicals[85] it has emerged that E_r has small but negative values for many radicals, R, with both HI[85] and HBr[86,87] (X = I, Br). This is also true of silicon centred radicals, although thus far only R = SiH_3, and $SiMe_3$ have been investigated fully[88,89]. Because chemical bond energies are one of the most fundamental quantities of chemistry, this problem has been widely discussed in the literature in the past six years[5,6,27,28,90], and so no further general discussion is given here. However, this does imply that all Si—H bond dissociation energies previously listed[1] require an increase in value, and also that $\Delta H_f°$ values for radicals are also higher. As a consequence, previously derived bond dissociation energies evaluated from thermochemical cycles[1] also need revision. This is carried out in this review.

B. Experimental Data: Measured Dissociation Energies and Radical Enthalpies of Formation

Table 15 shows the new values of bond dissociation energies and radical enthalpies of formation derived using the relationship

$$\Delta H_f°(R^\bullet) = \Delta H_f°(RH) - \Delta H_f°(H^\bullet) + D(R-H)$$

where $\Delta H_f°(RH)$ values used are those recommended in this review. For comparison, the earlier values[1] of $D(R-H)$ are also included in Table 15. The value listed for $\Delta H_f°(SiH_3)$ is that of Seetula and coworkers[88], which is the most precise, although other recent experimental determinations exist[95]. Numerous theoretical calculations[6,20,22,32,49,50,96–98] are well within the experimental error of the listed value. The values listed for $\Delta H_f°(Me_3Si^\bullet)$[89,92] indicate a substantial increase (18 kJ mol^{-1}) over that found previously[1]. The increased value of $D(Me_3Si-H)$ implies an activation energy for the reaction of Me_3Si + HI of -12 kJ mol^{-1}, a value more negative than that for any radical + HI reaction yet studied. The values are, however, consistent with theoretical estimates by Marshall[90,99] and also by Allendorf and Melius[22] who obtained $\Delta H_f°(MeSiH_2^\bullet) = 138 \pm 4$ kJ mol^{-1}, $\Delta H_f°(Me_2SiH^\bullet) = 77 \pm 4$ kJ mol^{-1} and $\Delta H_f°(Me_3Si^\bullet) = 13 \pm 4$ kJ mol^{-1} using the BAC-MP4 method. In addition to the new measured values, we may reasonably assume that the other, as yet uninvestigated, radical + HI reactions will all have negative activation energies. Using estimated values we have obtained revised values for other

TABLE 15. Measured Si—H bond dissociation energies for silanes (kJ mol^{-1}) and derived enthalpies of associated radicals ($\Delta H_f°$/kJ mol^{-1})

Bond	D(old)^a	D(new)	$\Delta H_f°(R^\bullet)^i$
H_3Si-H	378 ± 5	384 ± 2^b	200.5 ± 2
$MeSiH_2-H$	375 ± 8	388 ± 5^c	141 ± 6
Me_2SiH-H	374 ± 8	391 ± 5^c	78 ± 6
Me_3Si-H	378 ± 5	397 ± 2^d, 395 + 9^e	15 + 7
H_3SiSiH_2-H	361 ± 8	372 ± 5^f, 374^g	234 ± 6
$C_6H_5SiH_2-H$	369 ± 5	(382 ± 5)^h	(ca 274)^j
Me_3SiCH_2-H	415 ± 5	(419 ± 5)^h	(−32 ± 6)
$Me_3SiCMe_2CH_2-H$	405 ± 5	(409 ± 5)^h	(ca −121)^j
Cl_3Si-H	382 ± 5	(395 ± 5)^h	(−322 ± 8)
F_3Si-H	419 ± 5	(432 ± 5)^h	(−987 ± 20)

^{a}Reference 1. ^{b}Reference 88. ^{c}Reference 91. ^{d}Reference 89. ^{e}Reference 92. ^{f}Reference 93. ^{g}Reference 94. ^{h}Based on revised assumptions (see text) ^{i}Calculated using $\Delta H_f°(RH)$ from this review. ^{j}Based on bond additivity estimate for $\Delta H_f°(RH)$.

bond dissociation energies previously listed. The values for $\Delta H_f°(\text{R}^•)$ corresponding to these are in reasonable agreement with recent theoretical values [$\Delta H_f°(\text{Me}_3\text{SiCH}_2^•)$ = -29 ± 6 kJ mol^{-1}[122], $\Delta H_f°(\text{SiCl}_3^•) = -318 \pm 7$ kJ mol^{-1}[20,100], $\Delta H_f°(\text{SiF}_3^•)$ values in the range -976 to -1007 kJ mol^{-1}[126]], and also a recent experimental value of Armentrout and coworkers[101] [$\Delta H_f°(\text{SiF}_3^•) = -997 \pm 5$ kJ mol^{-1}].

What emerges from these revised values is the fact that methyl substitution at silicon leads to a strengthening of Si–H bonds, whereas the old data showed no trend. Marshall has pointed out[89,90] that this is completely in accord with the relative electronegativities of carbon and silicon, a point invoked by one of us[2] to account for trends in other cases. Relative values remain unaffected so that, for instance, the SiH$_3$ group weakening effect manifests itself in the reduced Si–H dissociation energy of Si$_2$H$_6$ compared with SiH$_4$. This is again supported by other experimental[94] and theoretical calculations[20,50,90,93,102–104].

Studies of electron affinities of organosilyl radicals combined with gas-phase acidities have been used by Brauman and coworkers[105] to obtain bond dissociation energies. The values obtained are generally in excellent agreement with those of Table 15, although uncertainties tend to be greater (± 8–12 kJ mol^{-1}). In addition to data in Table 15, other values for Si–H bond dissociation energies have been obtained by means of photoacoustic calorimetry[106–108]. These are shown in Table 16. We have increased the values by 19 kJ mol^{-1} on the grounds that the value for $D(\text{Et}_3\text{Si}–\text{H})$ should be the same as that for $D(\text{Me}_3\text{Si}–\text{H})$. Chatgilialoglu[28] has reviewed uncertainties in this method and concluded that while absolute values may be subject to systematic errors, relative values should be reliable. Thus these data show clearly that silyl substitution at the silicon centre systematically weakens Si–H bonds (by ca 12–20 kJ mol^{-1}) per silyl group. This is borne out by the theoretical calculations of Sax and Kalcher[50] who have obtained values for $D(\text{H}_3\text{Si}–\text{H})$, $D(\text{H}_3\text{SiSiH}_2–\text{H})$ and $D((\text{H}_3\text{Si})_2\text{SiH}–\text{H})$ of 383, 374 and 361 kJ mol^{-1}, respectively. The explanation for this effect is not clear. Stabilization by silyl groups is even more marked in silyl anions[109]. Calculations suggest that there is an increase in p character in the Si–Si bonds upon anion formation with the anion electron pair located on silicon in an orbital with substantial s character[109]. This type of explanation has been used by us to explain the stabilities of silylenes[5,110], but for the radicals a simple inductive effect from the silyl substituent to the silicon bearing the odd electron[2] seems equally appealing.

The data also suggest that phenyl substitution weakens Si–H bonds, but only by a small amount (ca 4–8 kJ mol^{-1}) per phenyl group. The difficulty here is that since methyl

TABLE 16. Bond dissociation energies (kJ mol^{-1}) for silanes obtained by photoacoustic calorimetry

Bond	D(old)	D(new)^d
Et$_3$Si–H	377 ± 4^a	396 ± 4
Me$_3$SiSiMe$_2$–H	357 ± 4^a	376 ± 4
(Me$_3$Si)$_3$Si–H	331 ± 4^a	350 ± 4
(Me$_3$Si)$_3$Si–H	345 ± 3^b	364 ± 3
PhMe$_2$Si–H	358 ± 7^c	377 ± 7
Ph$_2$SiH–H	360 ± 6^c	379 ± 6
Ph$_2$MeSi–H	342 ± 10^c	361 ± 10
Ph$_3$Si–H	352 ± 2^c	371 ± 2

^{a}Reference 106. ^{b}Reference 107. ^{c}Reference 108. ^{d}See text for discussion of these values.

groups strengthen the Si−H bonds, the effects of methyl and phenyl substitution tend to cancel out. This is also supported by D(PhMeSiH−H) = 382 ± 13 kJ mol^{-1} from the electron affinity measurements[105]. Despite the revision of these values, the small effect of phenyl substitution remains indicative of very little π-type interaction between the aromatic ring and the odd-electron orbital on the silicon[2]. Silyl stabilization of a carbon-centred radical is very small. New values of bond dissociation energies do not affect these quantities, which were determined by us by relative means. Thus values for α-silyl and β-silyl stabilization of primary alkyl radicals Me$_3$SiCH$_2$• and Me$_3$SiCMe$_2$CH$_2$• are 2 and 12 kJ mol^{-1} respectively[1,111,112]. *Ab initio* calculations give 5 kJ mol^{-1} for the α-silyl stabilization in SiH$_3$CH$_2$•[103]. Higher values (11 kJ mol^{-1}) for α-stabilization[113] came from a different comparative basis which includes other effects, and cannot be strictly compared with these numbers[1].

The determination of an Si−N bond dissociation energy has been carried out by Krasnoperov and coworkers[114] who studied directly the gas-phase equilibrium:

$$\text{Me}_3\text{Si} + \text{NO} \rightleftharpoons \text{Me}_3\text{SiNO}$$

The value for $\Delta H°$ gives directly D(Me$_3$Si−NO). A value of 190.2 ± 3.6 kJ mol^{-1} was obtained using the third law method, and is in excellent agreement with the *ab initio* (BAC-MP4) value of 191.4 kJ mol^{-1}[114]. The same calculation[114] gives a value for D(H$_3$Si−NO) of 149.9 kJ mol^{-1}, slightly higher than the previous value[115]. This provides a more dramatic example of bond strengthening of Si−X bonds by methyl group. These Si−NO bonds are significantly weaker than those of the more typical Si−N bonds in silylamines obtainable only indirectly and discussed in the next section.

C. Derived Bond Dissociation Energies

A set of bond dissociation energies for representative Si−C, Si−Si, Si−halogen, Si−O and Si−N bonds are derived from the molecular and radical heats of formation of the previous sections and, where necessary, those in the Appendix. These are shown in Tables 17−19. The selection of molecules is identical to that of the earlier review[1], but the values are of course revised to take account of the new results. The most significant change is the general increase in values arising from the higher values for the enthalpies of formation of the silicon-centred radicals, particularly Me$_3$Si•. The changes do not affect significantly the trends in values discussed in earlier reviews[1,2]. The bond strengthening effect of methyl groups (at silicon centres), noted already here for Si−H bonds, is also clearly apparent for Si−C bonds, and also evident for Si−Si bonds. The effect was also found previously[1,2] for Si−halogen bonds, but it is even more marked than thought then. The bond weakening effect of silyl groups (at the silicon centres), noted already for Si−H bonds, is also evident for Si−Si bonds. A recent experimental study[92] of the pyrolysis of Me$_3$SiSiMe$_3$ yielded an estimate of D(Me$_3$SiSiMe$_2$−Me) = 372 kJ mol^{-1}. When compared with the values of Table 17 it can be seen that the trimethylsilyl group is exerting a bond weakening effect on the Si−C bond.

The halogen substituent effects, as noted earlier[1], are more complex. Fluorine (as in F$_3$SiX) strengthens the Si−X bond (relative to H and Me). Chlorine and bromine are in between, D(H$_3$Si−X) and D(X$_3$Si−X) being equal within experimental error, and iodine weakens Si−X bonds. Si−OR and Si−NR$_2$ bonds, just like Si−halogen bonds, are very strong in comparison with analogous C−O, C−N and C−X bonds. The figures in parentheses in Table 19 were obtained indirectly by use of estimates of the radical heats of formation, $\Delta H_f°$(Me$_3$SiO) = −223 kJ mol^{-1} and $\Delta H_f°$((Me$_3$Si)$_2$N) = −231 kJ mol^{-1}.

TABLE 17. Derived silicon–carbon and silicon–silicon bond dissociation energies (kJ mol^{-1})

Bond	D	Bond	D
H_3Si-CH_3	375 ± 5	$H_3Si-SiH_3$	321 ± 4
$MeSiH_2-CH_3$	381 ± 7	$H_3Si-Si_2H_5$	313 ± 8
$Me_2SiH-CH_3$	387 ± 7	$H_3SiH_2Si-SiH_2SiH_3$	306 ± 10
Me_3Si-CH_3	394 ± 8	$Me_3Si-SiMe_3$	332 ± 12[a]

[a]Based on the directly measured data of Reference 92.

TABLE 18. Derived silicon–halogen bond dissociation energies (kJ mol^{-1})[a]

Halogen(X)	$D(H_3Si-X)$	$D(Me_3Si-X)$	$D(X_3Si-X)$
F	638 ± 5	662 ± 11	697 ± 6
Cl	458 ± 7	490 ± 8	462 ± 9
Br	376 ± 9	425 ± 8	376 ± 22
I	299 ± 8	344 ± 8	284 ± 26

[a]For $\Delta H_f°$ values for SiH_3X and SiX_4 see Table 29.

TABLE 19. Derived silicon–oxygen and silicon–nitrogen bond dissociation energies (kJ mol^{-1})[a]

Bond	D	Bond	D
Me_3Si-OH	555 ± 8	Me_3SiO-H	(495)
$Me_3Si-OMe$[b]	513 ± 11	$Me_3SiO-Me$[b]	(403)
$Me_3Si-OEt$[b]	512 ± 11	$Me_3SiO-Et$[b]	(412)
$Me_3Si-OSiMe_3$	(569)		
$Me_3Si-NHMe$	419 ± 8	$(Me_3Si)_2N-H$	(464)
$Me_3Si-NMe_2$	408 ± 8	$(Me_3Si)_2N-Me$	(364)
$Me_3Si-N(SiMe_3)_2$	(455)		

[a]Figures in parentheses have been estimated (see text).
[b]$\Delta H_f°$ values for Me_3SiOMe (−480 ± 8 kJ mol^{-1}) and Me_3SiOEt (−514 ± 8 kJ mol^{-1}) were estimated using additivity.

These were obtained, as previously[1], by assuming that sequential Si–O and Si–N dissociation energies in $(Me_3Si)_2O$ and $(Me_3Si)_3N$ follow the same proportionate relationships as O–H and N–H dissociation energies in H_2O and NH_3. It is interesting to note that the derived values for O–H and N–H dissociation energies in Me_3SiOH and $(Me_3Si)_2NH$ are very similar to $D(HO-H) = 499$ kJ mol^{-1} and $D(H_2N-H) = 453$ kJ mol^{-1}. As a substituent on these electronegative atoms, clearly the Me_3Si group is similar to the H atom itself in its electronic effects. Although there are no theoretically calculated (ab initio) values for the bond dissociation energies of Table 19, values of Si–O and Si–N dissociation energies have been obtained by the Melius group[21,24,25], Darling and Schlegel[81] and Zachariah and Tsang[82]. For the prototype molecules H_3SiOH and H_3SiNH_2, the following values (kJ mol^{-1}) have been found: $D(H_3Si-OH) = 505$[81], 514[25]; $D(H_3SiO-H) = 522$[81], 517[82], 513[25]; $D(H_3Si-NH_2) = 439$[21]; $D(H_3SiNH-H) = 480$[21].

Many other bond dissociation energies have been derived from theoretically calculated enthalpies of formation of radicals in recent years. These include Si–H, Si–C, Si–Si, Si–halogen, Si–O and Si–N. These are too many examples to consider in the context of this (mainly) experimentally orientated account. Where comparisons exist, as indicated, there has been a convergence in recent years, so that agreement between experiment and theory is generally good.

IV. OTHER SILICON CONTAINING SPECIES

There is great interest in silicon-containing transient species. This is because they are invariably involved in the thermal and photochemical breakdown mechanisms of more stable organosilicon compounds. Apart from SiX_3 free radical species dealt with in the previous section, these include the silylenes, SiX_2, and the π-bonded analogues of organic alkenes and carbonyl compounds containing Si=C, Si=Si and Si=O double bonds. An excellent account of the structure and energetics of these compounds has been given by Grev[116].

A. Silylenes

There has been a great deal of discussion about the thermochemistry of silylenes, particularly SiH_2, MeSiH and $SiMe_2$.

1. SiH₂

Together with Jasinski, we have recently reviewed[6] $\Delta H_f°(SiH_2)$. Even more recently we have listed further published values[117] as part of a study of the kinetics and mechanism of the reaction

$$SiH_2 + SiH_4 \rightleftharpoons Si_2H_6$$

The latest data are shown in Table 20. In the previous article[1] we reported $\Delta H_f°(SiH_2) = 273 \pm 6$ kJ mol^{-1} and the current experimentally preferred value[117] is the same but with a reduced uncertainty of ± 2 kJ mol^{-1}. This refinement has come about because the new kinetic data are extensive enough in their temperature and pressure ranges to be used to obtain the equilibrium constant for the above reaction without the need for extrapolation. A number of previous analyses of the kinetics of this system and also that of $SiH_2 + H_2 \rightleftharpoons SiH_4$ have given very similar values[118–122], although the most recent study[123] seems to have given two different values (288 and 275 kJ mol^{-1}). The higher

TABLE 20. Recent values for $\Delta H_f°(SiH_2)$/kJ mol^{-1}

	Experimental values			Theoretical values	
year	value	Reference	year	value	Reference
1986	273 ± 6	118	1985	265	128
1986	289 ± 13	125	1985	285	18
1987	273 ± 3	127	1986	273	129
1987	287 ± 6	95	1986	287	19
1987	269 ± 1	119	1988	272	96
1987	268–287	120	1988	266	97
1988	274 ± 7	126	1990	271 ± 9	20
1989	267 ± 8	55	1991	275 ± 8	32
1991	274 ± 4	121	1991	277	50
1991	266 ± 6	64	1992	271 ± 9	22
1992	273 ± 2	5	1992	273	49
1992	269 ± 4	122	1992	267	49
1994	285 ± 13	124			
1995	273 ± 2	117			
1996	288, 275	123			

value is associated with an RRKM modelling analysis of the high temperature, shock tube pyrolysis of SiH_4[124]. This could have a number of uncertainties associated with energy transfer and pressure dependence and is therefore not definitive. Other experimental values of $\Delta H_f°(SiH_2)$ from ion cyclotron resonance mass spectrometry[125], Si^+ ion beam reaction thresholds[95], electronic excitation spectroscopy[126] and photoionization mass spectrometry[127] are in reasonable agreement with the preferred value[117] of 273 kJ mol^{-1}. On the theoretical side, advances have reflected the increasing use of electron-correlated wavefunctions and large basis sets. The highest level calculations by Grev and Schaefer[49] are in precise agreement with experiment[117] assuming $\Delta H_f°(SiH_4)$ is correct. However, as indicated earlier, the discrepancy between $\Delta H_f°(SiH_4)$ and $\Delta H_f°(Si)$ means that a lower value of 267 kJ mol^{-1} is obtained for $\Delta H_f°(SiH_2)$ starting from $\Delta H_f°(Si)$.

2. MeSiH

These data are shown in Table 21. In 1988 Walsh published a value[130] for $\Delta H_f°(MeSiH)$ of 184 ± 13 kJ mol^{-1}. This was based on an analysis of the decomposition reactions of certain methyl substituted disilanes and was included in the previous review[1]. At the same time ion cyclotron resonance studies[131] gave a value of 222 ± 17 kJ mol^{-1}. The next year we extended our analysis[65] but with no different outcome, although O'Neal and coworkers[55], using the same basic kinetic data, obtained 201 ± 8 kJ mol^{-1}. The essential difference lay in the choice of values for $\Delta H_f°$ for methyl substituted disilanes which were at that time rather uncertain. This problem was resolved by the calorimetric measurement of $\Delta H_f°(Si_2Me_6)$ by Pilcher and coworkers[64] from which the securely based value of $\Delta H_f°(MeSiH)$ of 201 ± 6 kJ mol^{-1} was obtained. Theoretical calculations[22,132,144] give values in reasonable agreement with this. More recently Becerra and coworkers[133] have obtained a value of 202 ± 6 kJ mol^{-1} from analysis of the same kinetic systems but incorporating new and directly measured kinetic data for MeSiH insertion reactions. Although the analysis is now more complete, there remains enough experimental uncertainty that it is not realistic to reduce error limits below ± 6 kJ mol^{-1}.

3. Me$_2$Si

The data are shown in Table 22. Just as with $\Delta H_f°(MeSiH)$, Walsh published[130] a value of $\Delta H_f°(SiMe_2)$ equal to 109 ± 8 kJ mol^{-1} in 1988 based on an analysis including

TABLE 21. Recent values for $\Delta H_f°(MeSiH)$/kJ mol^{-1}

Year	Value	Reference
1988	184 ± 13	130[a]
1988	222 ± 17	131[a]
1989	206	144[b]
1989	184 ± 13	65[a]
1989	201 ± 8	55[a]
1990	212	132[b]
1991	201 ± 6	64[a]
1992	204 ± 10	22[b]
1993	201.8 ± 6	133[a]

[a]Experimental value. [b]Theoretical value.

TABLE 22. Recent values for $\Delta H_f°(Me_2Si)/kJ\,mol^{-1}$

Year	Value	Reference
1988	109 ± 8	130[a]
1988	155 ± 25	131[a]
1989	136 ± 13	144[b]
1989	108 ± 13	65[a]
1989	134 ± 8	55[a]
1990	134–138	132[b]
1991	140 ± 6	64[a]
1992	135 ± 10	22[b]
1995	135 ± 8	134[a]

[a]Experimental value. [b]Theoretical value.

erroneous values for $\Delta H_f°$ for methyl substituted disilanes. This was the value cited in our earlier review[1]. Once again, O'Neal and coworkers[55] were more foresighted and obtained a value of 134 ± 8 kJ mol^{-1}. Ion cyclotron resonance studies[131] gave an even higher value. The problem was resolved, as in the case of MeSiH, with the calorimetric determination of $\Delta H_f°(Si_2Me_6)$[64]. Theory[22,132,144] is in excellent agreement with our current best estimate of 135 ± 8 kJ mol^{-1}.

4. H_3SiSiH

The only experimental value for $\Delta H_f°(H_3SiSiH)$ is 312 ± 8 kJ mol^{-1} [135]. However, theoretical calculations have yielded values (kJ mol^{-1}) of 349[96], 313 ± 11[20], 305[132], 317[50] and 300[102]. Agreement is thus tolerable if not perfect.

5. Silicon dihalides

The thermochemistry of these species was reviewed by Walsh[3] some years ago. Since then new data have appeared for SiF_2 and $SiCl_2$. This has been recently reviewed again by Gordon and coworkers[26]. Theoretical values for $\Delta H_f°(SiF_2)$, all in good agreement[20,32,136], have placed its value some 42 kJ mol^{-1} lower than the earlier experimental values[3,35]. Recent experimental ion-beam studies[101] have confirmed the theoretical values. $\Delta H_f°(SiCl_2)$ seems to be experimentally well established[3,35] at 169 ± 3 kJ mol^{-1}. A more recent value[137] is in reasonable agreement, as are theoretical calculations[20,100]. $\Delta H_f°$ values for $SiBr_2$ and SiI_2 have not changed since earlier[3,35]. No theoretical calculations have been carried out on these species. The data are shown in Table 23.

TABLE 23. Standard enthalpies of formation ($\Delta H_f°/kJ\,mol^{-1}$) of silicon dihalides

Silylene	Experiment	Theory
SiF_2	−588 ± 13[a], −638 ± 6[b]	−627 ± 17[c], −640 ± 8[d], −642 (−638)[e]
$SiCl_2$	−169 ± 3[a], −165 ± 14[f]	−151 ± 16[c], −163[g]
$SiBr_2$	−52 ± 17[a], −46 ± 8[h]	
SiI_2	93 ± 8[a], 92 ± 8[h]	

[a]Reference 35. [b]Reference 101. [c]Reference 20. [d]Reference 32. [e]Reference 136. [f]Reference 137. [g]Reference 100. [h]Reference 3.

TABLE 24. Recommended standard enthalpies of formation (ΔH_f°/kJ mol^{-1}) for silylenes together with DSSE values

Silylene	$\Delta H_f^{\circ a}$	DSSE[b]
SiH$_2$	273 ± 2	94 ± 4
MeSiH	202 ± 6	113 ± 11
SiMe$_2$	135 ± 8	128 ± 11
H$_3$SiSiH	312 ± 8	76 ± 10
SiF$_2$	−638 ± 6	259 ± 8
SiCl$_2$	−169 ± 3	188 ± 10
SiBr$_2$	−46 ± 8	161 ± 27
SiI$_2$	92 ± 8	152 ± 27

[a] See text and Tables 20–23 for source values.
[b] See text for definition.

Values for the important index of reactivity for silylenes, DSSE (= divalent state stabilization energy) defined as follows:

$$\text{DSSE}(\text{SiR}_2) = D(\text{R}_3\text{Si}-\text{R}) - D(\text{R}_2\text{Si}-\text{R})$$

where R is the substituent, are shown in Table 24. This updates earlier estimates[3,5,110,134]. It can be generally seen that DSSE increases with the electronegativity of the substituent, as noted previously[5,110].

B. π-Bonded Species

There has been considerable attention paid to the thermochemistry of these species by theoreticians, but very little by experimentalists, since the earlier review[1].

1. Sila-alkenes

Experimental data for these Si=C double bonded species come from ion-cyclotron resonance measurements[131] or thermochemical analyses of siletane (silacyclobutanes) decomposition[71]. Our own previous estimate[78] was based on the latter approach. The value of $\Delta H_f^\circ(\text{Me}_2\text{Si}=\text{CH}_2) = 36 \pm 7$ kJ mol^{-1} determined by Brix and coworkers[71] is the current best available value for any of the sila-alkenes, having reduced considerably the uncertainties and assumptions of our earlier analyses[78]. It is still, however, dependent on the value[68] for ΔH_f° for 1,1-dimethylsiletane (−108.7 ± 6 kJ mol^{-1}) discussed elsewhere in this review (Section II.B.2). If the value of the latter is taken as −85 kJ mol^{-1} as obtained by Voronkov and coworkers[14], then $\Delta H_f^\circ(\text{Me}_2\text{Si}=\text{CH}_2)$ becomes 61 kJ mol^{-1} as suggested recently by Ahmed and coworkers[138]. Theoretical values of ΔH_f° have been calculated for all the methylsila-alkenes (H$_2$Si=CH$_2$, MeSiH=CH$_2$, Me$_2$Si=CH$_2$) by the BAC-MP4 method[22] and these show a Me-for-H replacement enthalpy of −59 kJ mol^{-1}, a rather smaller value than might be expected from examination of other methyl-substituted silicon species (e.g. −67 kJ mol^{-1} for the methyl silanes, −69 kJ mol^{-1} for the methylsilylenes). Most theoretical activity has focused on $\Delta H_f^\circ(\text{H}_2\text{Si}=\text{CH}_2)$ for which the highest level calculation[139] has yielded a value 15 kJ mol^{-1} lower than for $\Delta H_f^\circ(\text{MeSiH})$. From

TABLE 25. Standard enthalpies of formation ($\Delta H_f°$/kJ mol^{-1}) of sila-alkenes

Sila-alkene	Experiment	Theory
$H_2Si=CH_2$	155 ± 20^a, 180 ± 13^b 187 ± 6^c	170 ± 10^d, 194^e, 190^f
MeSiH=CH$_2$	88 ± 20^a	110 ± 8^d
Me$_2$Si=CH$_2$	21 ± 20^a, 36 ± 7^g $(61 \pm 7)^h$	51 ± 6^d

[a]Reference 1. [b]Reference 131. [c]Estimated from $\Delta H_f°$(MeSiH); see text.
[d]Reference 22. [e]Reference 132. [f]Reference 58. [g]Reference 71. [h]Reference 138.

our own experimental value for the latter, this would mean $\Delta H_f°(H_2Si=CH_2) = 187 \pm 6$ kJ mol^{-1}. At the G-1 level of theory, Boatz and Gordon[132] have obtained a value of 194 kJ mol^{-1}. These are somewhat higher than the Allendorf and Melius (BAC-MP4) value[22] of 170 ± 10 kJ mol^{-1}. On the other hand, comparison of the highest of these values with $\Delta H_f°(Me_2Si=CH_2)$ discussed above implies a rather high Me-for-H replacement (79 kJ mol^{-1}). There is thus room for further refinement of the enthalpies of formation for these species. The values are listed in Table 25. Earlier theoretical values are discussed by Gordon, Francisco and Schlegel[26].

2. Disilene

Our value[135] for $\Delta H_f°(H_2Si=SiH_2)$ of 261 ± 8 kJ mol^{-1} is in conflict with the photoionization studies of Ruscic and Berkowitz[94] who obtained a value of 275 ± 4 kJ mol^{-1}. The discrepancy is not large but tends to be supported by theoretical values (kJ mol^{-1}) of 281[50], 272[132] and 270[102]. Ho and Melius[20] with 263 ± 10 are in between, and Horowitz and Goddard's value[96] of 322 looks to be too high. Our experimental value[135] is based on a mechanistic interpretation and is less direct than the photoionization value[94] which should therefore be the best available experimental estimate.

3. Silanone

An experimental value for $\Delta H_f°(H_2Si=O)$ is urgently needed. None currently exists. In the previous review[1] we estimated a value of -90 ± 30 kJ mol^{-1} which we have revised to -92 ± 20[140] based on subsequent theoretical values. Hartman and coworkers[141] have estimated a value of -115 kJ mol^{-1} from bond energy arguments and recent theoretical values[25,81,82] are in agreement at -98 kJ mol^{-1}.

Values for the important index of reactivity, the π-bond energy, $D_\pi(Si=X)$, may be obtained from:

$$D_\pi(Si=X) = D_\sigma(Si-H) + D_\sigma(X-H) - D_\sigma(H-H) + \Delta H_{hyd}$$

where ΔH_{hyd} is the enthalpy of hydrogenation of the species $H_2Si=X$. $D_\pi(Si=X)$ values for several Si=X bonds are shown in Table 26.

There are many other important reactive silicon-containing species, for example the ground state 'butterfly' Si_2H_2 molecule, $Si(H_2)Si$. There is virtually no experimental thermochemistry on such species. However, theoretical values of $\Delta H_f°$ for many other species have been calculated in the articles on theory cited in this review. Since the

TABLE 26. Recommended standard enthalpies of formation ($\Delta H_f°$/kJ mol^{-1}) of unsaturated silicon-containing molecules and π-bond energies[a]

Species	$\Delta H_f°$	π-Bond energy (D_π)[b]
$H_2Si=CH_2$	187 ± 6	146 ± 9
$MeHSi=CH_2$	116 ± 10	154 ± 12
$Me_2Si=CH_2$	46 ± 10	161 ± 12
$H_2Si=SiH_2$	275 ± 4	113 ± 8
$H_2Si=O$	-92 ± 20	256 ± 26

[a] See text and Table 25 for references.
[b] See text for definition.

emphasis is on experimental values we do not extend this article to include them. Interested readers are referred to the original articles.

V. APPENDIX

Group increments for silicon compounds based on the data reviewed here, and derived in accordance with the group additivity scheme[29–31], are shown in Table 27. Radical and atomic heats of formation for non-silicon-containing species used in this chapter to derive bond dissociation energies are shown in Table 28. Miscellaneous inorganic silane heats of formation are included in Table 29.

TABLE 27. Group increment contributions to standard enthalpies of formation ($\Delta H_f°$/kJ mol^{-1}) for organosilicon compounds

Group	$\Delta H_f°$
$Si-(C)(H)_3$	13.7
$Si-(C)_2(H)_2$	-9.1
$Si-(C)_3(H)$	-35.0
$Si-(C)_4$	-62.0
$C-(Si)(H)_3$	-42.8[a]
$C-(Si)(C)(H)_2$	-17.0
$C-(Si)(C)_2(H)$	-4.2
$O-(Si)(H)$	-309.2
$O-(Si)_2$	-396.6
$O-(Si)(C)$	-246.8
$Si-(C)_3(O)$	-62.0[b]
$Si-(C)_2(O)_2$	-62.2
$Si-(C)(O)_3$	-61.4
$Si-(O)_4$	-61.6
$C-(C)(H)_3$	-42.8
$C-(C)_2(H)_2$	-20.7
$C-(O)(C)(H)_2$	-33.9
$C-(O)(C)_2(H)$	-30.1

[a] Arbitrary value: $C-(Si)(H)_3$ set equal to $C-(C)(H)_3$.
[b] Arbitrary value: $Si-(C)_3(O)$ set equal to $Si-(C)_4$.

TABLE 28. Standard enthalpies of formation (ΔH_f°/kJ mol^{-1}) for various atoms and free radicals

Species	ΔH_f°	Species	ΔH_f°
H	218.0^a	C	716.7^a
CH$_3$	145.6^b	Si	450.6^a
C$_2$H$_5$	120.9^b	F	79.4^a
OH	39.0^a	Cl	121.3^a
OMe	17.2^b	Br	111.9^a
OEt	−15.5^b	I	106.8^a
NH$_2$	185^c		
NHMe	177^c		
NMe$_2$	145^c		

^{a}Reference 35. ^{b}Reference 27. ^{c}Reference 142.

TABLE 29. Standard enthalpies of formation (ΔH_f°/kJ mol^{-1}) of silicon hydrides and halides

Molecule	$\Delta H_f^{\circ a}$	Molecule	ΔH_f°	Molecule	$\Delta H_f^{\circ d}$
SiH$_4$	34.3 ± 1.2	SiH$_3$F	−359 ± 8^c	SiF$_4$	−1615 ± 1
Si$_2$H$_6$	80 ± 1.5	SiH$_3$Cl	−134 ± 6^c	SiCl$_4$	−663 ± 5
Si$_3$H$_8$	121 ± 4.4	SiH$_3$Br	−64 ± 9^d	SiBr$_4$	−415 ± 8
Si$_4$H$_{10}$	(162)^b	SiH$_3$I	+8.6 ± 8^e	SiI$_4$	−110 ± 16

^{a}References 1 and 143. ^{b}Additivity estimate. ^{c}See Table 1. ^{d}Reference 3. ^{e}See Table 2.

VI. REFERENCES

1. R. Walsh, in *The Chemistry of Organic Silicon Compounds* (Eds. S. Patai and Z. Rappoport), Chap. 5, Wiley, Chichester, 1989, p. 371.
2. R. Walsh, *Acc. Chem. Res.*, **14**, 246 (1981).
3. R. Walsh, *J. Chem. Soc., Faraday Trans. 1*, **79**, 2233 (1983).
4. A. M. Doncaster and R. Walsh, *J. Chem. Soc., Faraday Trans. 2*, **82**, 707 (1986).
5. R. Walsh, in *Energetics of Organometallic Species* (Ed. J. A. Martinho Simões), NATO-ASI Series C, Vol. 367, Chap. 11, Kluwer, Dordrecht, 1992, p. 171.
6. J. M. Jasinski, R. Becerra and R. Walsh, *Chem. Rev.*, **95**, 1203 (1995).
7. J. D. Cox and G. Pilcher, *Thermochemistry of Organic and Organometallic Compounds*, Academic Press, London, 1970.
8. S. Tannenbaum, *J. Am. Chem. Soc.*, **76**, 1027 (1954).
9. J. B. Pedley and J. Rylance, *Sussex-NPL Computer Analysed Thermochemical Data: Organic and Organometallic Compounds*, University of Sussex, 1977.
10. V. I. Tel'noi and I. B. Rabinovitch, *Russ. Chem. Rev.*, **49**, 603 (1980).
11. M. G. Voronkov, V. P. Baryshok, V. A. Klyuchnikov, T. F. Danilova, V. I. Pepikin, A. N. Korchagina and Yu. I. Khudobin, *J. Organomet. Chem.*, **345**, 27 (1988).
12. M. G. Voronkov, V. P. Baryshok, V. A. Klyuchnikov, A. N. Korchagina and V. I. Pepikin, *J. Organomet. Chem.*, **359**, 169 (1989).
13. M. G. Voronkov, M. S. Sorokin, V. A. Klyuchnikov, G. N. Shvets and V. I. Pepikin, *J. Organomet. Chem.*, **359**, 301 (1989).
14. M. G. Voronkov, V. A. Klyuchnikov, E. V. Sokolova, T. F. Danilova, G. N. Shvets, A. N. Korchagina, L. E. Gusel'nikov and V. V. Volkova, *J. Organomet. Chem.*, **401**, 245 (1991).
15. M. G. Voronkov, V. A. Klyuchnikov, E. V. Mironenko, G. N. Shvets, T. F. Danilova and Yu. I. Khudobin, *J. Organomet. Chem.*, **406**, 91 (1991).
16. M. G. Voronkov, V. A. Klyuchnikov, L. I. Marenkova, T. F. Danilova, G. N. Shvets, S. I. Tsvetnitskaya and Yu. I. Khudobin, *J. Organomet. Chem.*, **406**, 99 (1991).

17. (a) M. G. Voronkov, V. A. Klyuchnikov and L. I. Marenkova, *J. Organomet. Chem.*, **510**, 263 (1996).
    (b) V. M. Tatevskii, *The Structure of Molecules*, Khimaya Publishers, Moscow, 1978, p. 51.
18. P. Ho, M. E. Coltrin, J. S. Binkley and C. F. Melius, *J. Phys. Chem.*, **89**, 4647 (1985).
19. P. Ho, M. E. Coltrin, J. S. Binkley and C. F. Melius, *J. Phys. Chem.*, **90**, 3399 (1986).
20. P. Ho and C. F. Melius, *J. Phys. Chem.*, **94**, 5120 (1990).
21. C. F. Melius and P. Ho, *J. Phys. Chem.*, **95**, 1410 (1991).
22. M. D. Allendorf and C. F. Melius, *J. Phys. Chem.*, **96**, 428 (1992).
23. M. D. Allendorf and C. F. Melius, *J. Phys. Chem.*, **97**, 720 (1993).
24. P. Ho and C. F. Melius, *J. Phys. Chem.*, **99**, 2166 (1995).
25. M. D. Allendorf, C. F. Melius, P. Ho and M. R. Zachariah, *J. Phys. Chem.*, **99**, 15285 (1995).
26. M. S. Gordon, J. S. Francisco and H. B. Schlegel, *Advances in Silicon Chemistry*, **2**, 137 (1993).
27. J. Berkowitz, G. B. Ellison and D. Gutman, *J. Phys. Chem.*, **98**, 2744 (1994).
28. C. Chatgilialoglu, *Chem. Rev.*, **95**, 1229 (1995).
29. S. W. Benson and J. H. Buss, *J. Chem. Phys.*, **29**, 546 (1958).
30. S. W. Benson, F. R. Cruickshank, D. M. Golden, G. R. Haugen, H. E. O'Neal, A. S. Rodgers, R. Shaw and R. Walsh, *Chem. Rev.*, **69**, 279 (1969).
31. S. W. Benson, *Thermochemical Kinetics*, 2nd edn., Wiley-Interscience, New York, 1976.
32. E. W. Ignacio and H. B. Schlegel, *J. Chem. Phys.*, **92**, 5404 (1990).
33. D. A. Dixon, *J. Phys. Chem.*, **92**, 86 (1988).
34. M-D. Su and H. B. Schlegel, *J. Phys. Chem.*, **97**, 8732 (1993).
35. M. W. Chase, C. A. Davies, J. R. Downey, D. J. Frurip, R. A. McDonald and A. N. Syverud, *JANAF Thermochemical Tables 3rd Edn.*, *J. Phys. Chem. Ref. Data*, **14** (1985), supplement 1.
36. L. Pauling, *The Nature of the Chemical Bond*, 3rd edn., Cornell University Press, Ithaca, New York, 1960.
37. Y-R. Luo and S. W. Benson, *Acc. Chem. Res.*, **25**, 375 (1992).
38. Y-R. Luo and S. W. Benson, *J. Phys. Chem.*, **92**, 5255 (1988).
39. Y-R. Luo and S. W. Benson, *J. Am. Chem. Soc.*, **111**, 2480 (1989).
40. Y-R. Luo and S. W. Benson, *J. Phys. Chem.*, **93**, 3304 (1989).
41. Y-R. Luo and S. W. Benson, *J. Phys. Chem.*, **93**, 3306 (1989).
42. Y-R. Luo and S. W. Benson, *J. Phys. Chem.*, **93**, 1674 (1989).
43. Y-R. Luo and S. W. Benson, *J. Phys. Chem.*, **93**, 4643 (1989).
44. Y-R. Luo and S. W. Benson, *J. Phys. Chem.*, **93**, 3791 (1989).
45. Y-R. Luo and S. W. Benson, *J. Phys. Chem.*, **94**, 914 (1990).
46. R. Becerra and R. Walsh, unpublished calculations.
47. S. R. Gunn and L. G. Green, *J. Phys. Chem.*, **65**, 779 (1961).
48. S. R. Gunn and L. G. Green, *J. Phys. Chem.*, **68**, 946 (1964).
49. R. S. Grev and H. F. Schaefer III, *J. Chem. Phys.*, **97**, 8389 (1992).
50. A. F. Sax and J. Kalcher, *J. Phys. Chem.*, **95**, 1768 (1991).
51. G. Leroy, M. Sana, C. Wilante and D. R. Temsamani, *J. Mol. Struct. (Theochem)*, **259**, 369 (1992).
52. M. S. Gordon, J. A. Boatz and R. Walsh, *J. Phys. Chem.*, **93**, 1584 (1989).
53. W. V. Steele, *J. Chem. Thermodyn.*, **15**, 595 (1983).
54. D. D. Wagman, W. H. Evans, V. B. Parker, R. H. Shumm, I. Halow, S. M. Bailey, K. L. Churney and R. L. Nuttall, *J. Phys. Chem. Ref. Data, 1982 Supplement*, **11**, 2 (1982).
55. H. E. O'Neal, M. A. Ring, W. H. Richardson and G. F. Licciardi, *Organometallics*, **8**, 1968 (1989).
56. M. G. Voronkov, V. A. Klyuchnikov, T. F. Danilova, A. N. Korchagina, V. P. Baryshok and L. M. Landa, *Bull. Acad. Sci. USSR, Div. Chem. Sci.*, 1790, 1795 (1986).
57. R. E. Jardine, H. E. O'Neal, M. A. Ring and M. E. Beatie, *J. Phys. Chem.*, **99**, 12507 (1995).
58. A. E. Ketvirtis, D. K. Bohme and A. C. Hopkinson, *J. Phys. Chem.*, **99**, 16121 (1995).
59. N. Al-Rubaiey and R. Walsh, *J. Phys. Chem.*, **98**, 5303 (1994).
60. R. Becerra and R. Walsh, *Int. J. Chem. Kinet.*, **26**, 45 (1994).
61. D. Sengupta and M. T. Nguyen, *Mol. Phys.*, **89**, 1567 (1996).
62. M. T. Nguyen, D. Sengupta and L. G. Vanquickenborne, *Chem. Phys. Lett.*, **240**, 513 (1995).
63. E. M. Page, K. Hagen, D. A. Rice and R. Walsh, *J. Mol. Struct.*, in press (1997).
64. G. Pilcher, M. L. P. Leitão, Y. Meng-Yan and R. Walsh, *J. Chem. Soc., Faraday Trans.*, **87**, 841 (1991).

4. Thermochemistry

65. R. Walsh, *Organometallics*, **8**, 1973 (1989).
66. J. A. Boatz and M. S. Gordon, *J. Phys. Chem.*, **94**, 3874 (1990).
67. J. A. Boatz, M. S. Gordon and R. L. Hildebrandt, *J. Am. Chem. Soc.*, **110**, 352 (1988).
68. W. V. Steele, unpublished results (private communication).
69. V. G. Genchel, N. V. Demidova, N. S. Nametkin, L. E. Gusel'nikov, E. A. Volnina, E. N. Burdasov and V. N. Vdovin, *Izv. Akad. Nauk SSSR, Ser. Khim.*, **10**, 2337 (1976).
70. R. T. Conlin, M. Namavari, J. S. Chickos and R. Walsh, *Organometallics*, **8**, 168 (1989).
71. Th. Brix, N. L. Arthur and P. Potzinger, *J. Phys. Chem.*, **93**, 8193 (1989).
72. J. A. Boatz, M. S. Gordon and L. R. Sita, *J. Phys. Chem.*, **94**, 5488 (1990).
73. D. A. Horner, R. S. Grev and H. F. Schaefer III, *J. Am. Chem. Soc.*, **114**, 2093 (1992).
74. D. H. Berry, unpublished results (private communication).
75. N. Al-Rubaiey, I. W. Carpenter and R. Walsh, unpublished results.
76. S. N. Gadzhiev and M. J. Agarunov, *J. Organomet. Chem.*, **11**, 415 (1968); **22**, 305 (1970).
77. J. C. Baldwin, M. F. Lappert, J. B. Pedley and J. A. Treverton, *J. Chem. Soc. (A)*, 1980 (1967).
78. R. Walsh, *J. Phys. Chem.*, **90**, 389 (1986).
79. A. M. Doncaster and R. Walsh, *J. Phys. Chem.*, **83**, 3037 (1979).
80. D. J. Lucas, L. A. Curtiss and J. A. Pople, *J. Chem. Phys.*, **99**, 6697 (1993).
81. C. L. Darling and H. B. Schlegel, *J. Phys. Chem.*, **97**, 8207 (1993).
82. M. R. Zachariah and W. Tsang, *J. Phys. Chem.*, **99**, 5308 (1995).
83. M. G. M. van der Vis and E. P. H. Cordfunke, *J. Chem. Thermodyn.*, **25**, 1205 (1993).
84. M. G. M. van der Vis, E. P. H. Cordfunke and R. J. M. Konings, *J. Phys. IV*, **C3**, 75 (1993).
85. J. A. Seetula, J. J. Russell and D. Gutman, *J. Am. Chem. Soc.*, **112**, 1347 (1990).
86. J. J. Russell, J. A. Seetula, R. S. Timonen, D. Gutman and D. F. Nava, *J. Am. Chem. Soc.*, **110**, 3084 (1988).
87. J. J. Russell, J. A. Seetula and D. Gutman, *J. Am. Chem. Soc.*, **110**, 3092 (1988).
88. J. A. Seetula, Y. Feng, D. Gutman, P. W. Seakins and M. J. Pilling, *J. Phys. Chem.*, **95**, 1658 (1991).
89. I. J. Kalinovski, D. Gutman, L. Krasnoperov, A. Goumri, W-J. Yuan and P. Marshall, *J. Phys. Chem.*, **98**, 9551 (1994).
90. P. Marshall, *J. Mol. Struct. (Theochem)*, **313**, 19 (1994).
91. L. Ding and P. Marshall, *J. Chem. Soc., Faraday Trans.*, **89**, 419 (1993).
92. W. J. Bullock, R. Walsh and K. King, *J. Phys. Chem.*, **98**, 2595 (1994).
93. A. Goumri, W-J. Yuan, L. Ding and P. Marshall, *Chem. Phys. Lett.*, **204**, 296 (1993).
94. B. Ruscic and J. Berkowitz, *J. Chem. Phys.*, **95**, 2416 (1991).
95. B. H. Boo and P. B. Armentrout, *J. Am. Chem. Soc.*, **109**, 3549 (1987).
96. D. S. Horowitz and W. A. Goddard III, *J. Mol. Struct. (Theochem)*, **163**, 207 (1988)
97. L. A. Curtiss and J. A. Pople, *Chem. Phys. Lett.*, **144**, 38 (1988).
98. L. A. Curtiss, K. Raghavachari, G. W. Trucks and J. A. Pople, *J. Chem. Phys.*, **94**, 7221 (1991).
99. L. Ding and P. Marshall, *J. Am. Chem. Soc.*, **114**, 5754 (1992).
100. C. L. Darling and H. B. Schlegel, *J. Phys. Chem.*, **97**, 1368 (1993).
101. E. R. Fischer, B. L. Kickel and P. B. Armentrout, *J. Phys. Chem.*, **97**, 10204 (1993).
102. L. A. Curtiss, K. Raghavachari, P. W. Deutsch and J. A. Pople, *J. Chem. Phys.*, **95**, 2432 (1991).
103. M. B. Coolidge and W. T. Borden, *J. Am. Chem. Soc.*, **110**, 2298 (1988).
104. Y-D. Wu and C-L. Wong, *J. Org. Chem.*, **60**, 821 (1995).
105. D. M. Wetzel, K. E. Salomon, S. Berger and J. I. Brauman, *J. Am. Chem. Soc.*, **111**, 3835 (1989).
106. J. M. Kanabus-Kaminska, J. A. Hawari, D. Griller and C. Chatgilialoglu, *J. Am. Chem. Soc.*, **109**, 5267 (1987).
107. C. Chatgilialoglu, M. Guerra, A. Guerrini, A. Seconi, K. B. Clark, D. Griller, J. M. Kanabus-Kaminska and J. A. Martinho Simões, *J. Org. Chem.*, **57**, 2427 (1992).
108. A. R. Dias, H. P. Diogo, D. Griller, M. E. Pinas de Piedade and J. A. Martinho Simões, *Bonding Energetics in Organometallic Compounds*, ACS Symposium Series No. 428 (Ed. T. Marks), Chap. 14 *(Metal Bond Dissociation Enthalpies from Classical and Nonclassical Calorimetric Studies)*, 1990, p. 205.
109. E. A. Brinkman, S. Berger and J. I. Brauman, *J. Am. Chem. Soc.*, **116**, 8304 (1994).
110. R. Walsh, *Pure Appl. Chem.*, **59**, 69 (1987).
111. A. M. Doncaster and R. Walsh, *J. Chem. Soc., Faraday Trans. 1*, **72**, 2908 (1976).
112. N. Auner, R. Walsh and J. Westrup, *J. Chem. Soc., Chem. Commun.*, 207 (1986).
113. I. M. Davidson, T. J. Barton, K. J. Hughes, S. Ijadi-Maghsoodi, A. Revis and G. C. Paul, *Organometallics*, **6**, 644 (1987).

114. L. N. Krasnoperov, J. T. Niiranen, D. Gutman, C. F. Melius and M. D. Allendorf, *J. Phys. Chem.*, **99**, 14347 (1995).
115. P. Marshall, *Chem. Phys. Lett.*, **201**, 493 (1993).
116. R. S. Grev, *Adv. Organomet. Chem.*, **33**, 125 (1991).
117. R. Becerra, H. M. Frey, B. P. Mason, R. Walsh and M. Gordon, *J. Chem. Soc., Faraday Trans.*, **91**, 2723 (1995).
118. H. M. Frey, R. Walsh and I. M. Watts, *J. Chem. Soc., Chem. Commun.*, 1189 (1986).
119. J. G. Martin, M. A. Ring and H. E. O'Neal, *Int. J. Chem. Kinet.*, **19**, 715 (1987).
120. K. F. Roenijk, K. F. Jensen and R. W. Carr, *J. Phys. Chem.*, **91**, 5732 (1987).
121. H. K. Moffat, K. F. Jensen and R. W. Carr, *J. Phys. Chem.*, **95**, 145 (1991).
122. H. K. Moffat, K. F. Jensen and R. W. Carr, *J. Phys. Chem.*, **96**, 7683 (1992).
123. H. J. Mick, P. Roth and V. N. Smirnov, *Kinet. Katal.*, **37**, 5 (1996).
124. H. J. Mick, P. Roth and V. N. Smirnov and I. S. Zaslonko, *Kinet. Katal.*, **35**, 485 (1994).
125. S. K. Shin and J. L. Beauchamp, *J. Phys. Chem.*, **90**, 1507 (1986).
126. C. M. Van Zoeren, J. W. Thoman, J. I. Steinfeld and N. Rainbird, *J. Phys. Chem.*, **92**, 9 (1988).
127. J. Berkowitz, J. P. Greene, H. Cho and B. Ruscic, *J. Chem. Phys.*, **86**, 1235 (1987).
128. J. A. Pople, B. T. Luke, M. J. Frisch and J. S. Binkley, *J. Phys. Chem.*, **89**, 2198 (1985).
129. M. S. Gordon, D. R. Gano, J. S. Binkley and M. J. Frisch, *J. Am. Chem. Soc.*, **108**, 2191 (1986).
130. R. Walsh, *Organometallics*, **7**, 75 (1988).
131. S. K. Shin, K. K. Irikura, J. L. Beauchamp and W. A. Goddard III, *J. Am. Chem. Soc.*, **110**, 24 (1988).
132. J. A. Boatz and M. S. Gordon, *J. Phys. Chem.*, **94**, 7331 (1990).
133. R. Becerra, H. M. Frey, B. P. Mason and R. Walsh, *J. Chem. Soc., Faraday Trans.*, **89**, 411 (1993).
134. R. Becerra and R. Walsh, *Kinetics and mechanisms of silylene reactions: a prototype for gas-phase acid/base chemistry*, in *Research in Chemical Kinetics* (Eds. R. G. Compton and G. Hancock), Vol. 3, Elsevier, Amsterdam, 1995, p. 263.
135. R. Becerra and R. Walsh, *J. Phys. Chem.*, **91**, 5765 (1987).
136. H. H. Michels and R. H. Hobbs, *Chem. Phys. Lett.*, **207**, 389 (1993).
137. E. R. Fischer and P. B. Armentrout, *J. Phys. Chem.*, **95**, 4765 (1991).
138. M. Ahmed, P. Potzinger and H. Gg. Wagner, *J. Photochem. Photobiol. A*, **86**, 33 (1995).
139. R. S. Grev, G. E. Scuseria, A. C. Scheiner, H. F. Schaefer III and M. S. Gordon, *J. Am. Chem. Soc.*, **110**, 7337 (1988).
140. R. Becerra, H. M. Frey, B. P. Mason and R. Walsh, *Chem. Phys. Lett.*, **185**, 415 (1991).
141. J. R. Hartman, J. Famil-Ghiria, M. A. Ring and H. E. O'Neal, *Combust. Flame*, **68**, 43 (1987).
142. D. F. McMillan and D. M. Golden, *Annu. Rev. Phys. Chem.*, **33**, 493 (1982).
143. J. B. Pedley and B. S. Iseard, *CATCH Tables for Silicon Compounds*, University of Sussex, 1972.
144. M. S. Gordon and J. A. Boatz, *Organometallics*, **8**, 1978 (1989).

CHAPTER 5

The structural chemistry of organosilicon compounds

MENAHEM KAFTORY, MOSHE KAPON and MARK BOTOSHANSKY

Department of Chemistry, Technion–Israel Institute of Technology, Haifa 32000, Israel

I. INTRODUCTION	182
II. SINGLE BONDS TO SILICON	185
A. Structural Chemistry of the Si—C Bond	185
1. Si—C(sp) bonds	185
2. Si—C(sp^2) bonds	189
3. Si—C(aryl) bonds	191
4. Si—C(sp^3) bonds	192
5. Si—C bonds in silacyclopropanes and silacyclopropenes	194
6. Structural chemistry of Si bonds with special functional groups	195
B. Structural Chemistry of the Si—Si Bond	197
1. Si—Si bonds in compounds with tetracoordinate silicon atoms	197
C. Structural Chemistry of the Si—N Bond	198
1. Si—N bonds in compounds with tetracoordinate silicon and dicoordinate nitrogen atoms	203
2. Si—N bonds in compounds with tetracoordinate silicon and tricoordinate nitrogen atoms	206
3. Si—N bonds in compounds with pentacoordinate silicon and tricoordinate nitrogen atoms	210
4. Si—N bonds in compounds with hexacoordinate silicon and tricoordinate nitrogen atoms	211
D. Structural Chemistry of the Si—P Bond	211
1. Si—P bonds in compounds with tetracoordinate silicon and dicoordinate phosphorus atoms	211
2. Si—P bonds in compounds with tetracoordinate silicon and tri- or tetracoordinate phosphorus atoms	211
E. Structural Chemistry of the Si—O Bond	213
1. Si—O bonds in compounds with tetracoordinate silicon and dicoordinate oxygen atoms	218

The chemistry of organic silicon compounds, Vol. 2
Edited by Z. Rappoport and Y. Apeloig   © 1998 John Wiley & Sons Ltd

    2. Si—O bonds in silanols 221
    3. Si—O bonds in disiloxanes 222
    4. Si—O bonds in compounds with pentacoordinate silicon and
       dicoordinate oxygen atoms 225
    5. Si—O bonds in compounds with hexacoordinate silicon and
       dicoordinate oxygen atoms 227
  F. Structural Chemistry of the Si—S Bond 228
    1. Si—S bonds in compounds containing the Si—S—Si group 229
    2. Si—S bonds in compounds containing the $(t\text{-BuO})_3$—Si—S—X
       group 229
    3. Si—S bonds in compounds containing the $(Ph)_3$—Si—S—X
       group 229
    4. Si—S bonds in compounds containing the Si—S—X group 233
  G. Structural Chemistry of the Si—Halogen Bond (Si—F, Cl, Br, I) 233
    1. Si—F bond in compounds with tetracoordinate silicon 234
    2. Si—F bond in compounds with pentacoordinate silicon 235
    3. Si—F bond in compounds with hexacoordinate silicon 239
    4. Si—Cl bond in compounds with tetracoordinate silicon 239
    5. Si—Cl bond in compounds with pentacoordinate silicon 242
    6. Si—Cl bond in compounds with hexacoordinate silicon 245
    7. Si—Br and Si—I bonds 245
III. DOUBLE BONDS TO SILICON 247
  A. Disilenes (>Si=Si<) 247
  B. Silenes (>Si=C<) 250
  C. Silanimines (>Si=N—) 251
  D. Silanephosphimines (>Si=P—) 252
  E. Silanones (>Si=O) and Silanethiones (>Si=S) 252
IV. SUMMARY 254
V. REFERENCES 257

I. INTRODUCTION

The question of how closely do the properties of organic compounds of silicon resemble those of derivatives of its congener carbon atom was raised more than a century ago. A question that follows is how close are the structural chemistry of the two atoms in their analogue compounds. The major properties that distinguish silicon from carbon and have a direct effect on the structural chemistry of the compounds of the two atoms are compared in Table 1. From the differences between these properties it is seen that, since the silicon atom is about 50% larger than the carbon atom, the bond lengths involving silicon are expected to be longer than the lengths of similar bonds involving carbon. The longer distance between silicon and its bonded element will cause a weakening of the π-bonding and lowering the barrier to rotation about the bonds will influence the out-of-plane bending, which will in turn affect the planarity of the silicon atom. The smaller electronegativity of silicon will affect the bond polarity and ionicity, which will in turn have a major influence on the bond lengths involving silicon. Steric effects caused by bulky substituents will, on the one hand be less important than in the carbon analogues because of the longer bond lengths, but, on the other hand, will have a greater effect because of the lower barrier to distortion. One of the striking differences between silicon and carbon atoms is the ease of formation of hypervalent species with silicon. Five- and six-coordinate silicon compounds are stable and the role of d orbitals in the bonding of silicon in these compounds is a subject of continuing debate[7,8].

5. The structural chemistry of organosilicon compounds

TABLE 1. A comparison of some properties of carbon and silicon

Property	C	Si
Atomic radius (Bragg–Slater)[1] (Å)	0.7	1.1
Covalent radius[2] (Å)	0.772	1.169
Van der Waals radius[3] (Å)	1.70	2.17
Ionization energies (eV)[4]		
E_1	11.26	8.15
E_2	24.38	16.34
E_3	47.89	34.49
Electron affinity[5] (eV)	1.12	1.39
Electronegativity[2]	2.746	2.138
Dipole polarizability[6] (a.u.)	11.8	36.3

This review is not the first to discuss the structural chemistry of organosilicon compounds. There have been a few earlier reviews discussing various aspects of the structural chemistry of organosilicon compounds. The stereochemistry of elements of Group 15 and 16 bonded to silicon was reviewed in 1973[9,10]. In 1985 the geometry of silatranes was reviewed[11] and simulation of the reaction pathway for S_N2 substitution reactions at tetrahedral silicon using structural data was published[12]. Later on, in 1986, the structural chemistry of tricoordinate silicon was reviewed[13], and the X-ray and NMR studies on penta- and hexacoordinate silicon compounds were summarized[14]. The most comprehensive review on the structural chemistry of organosilicon compounds was published in 1989 by W. S. Sheldrick[15].

In the last decade we have been witnessing an explosion of information on the geometry of organosilicon compounds extracted from X-ray crystal structure determination (later referred to as XRD). The histogram given in Figure 1 shows the growth in the number of organosilicon compounds whose structures have been determined by XRD. Until 1980 the structure of only 525 such compounds were known, and most of the structural data were obtained by other methods such as electron diffraction (ED) or microwave spectroscopy (MW). Both methods were restricted to small and symmetric compounds. The crystal structures of 1446 compounds were known at the time that Sheldrick was writing his review. At the time of writing the present review the crystal structures of more than 6300 organosilicon compounds are known.

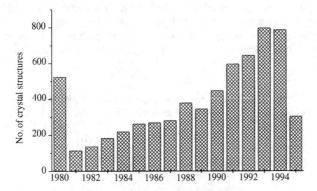

FIGURE 1. Annual growth in X-ray crystal structure determination of organosilicon compounds (the value for 1980 is the number of structures available in that year; not all compounds of 1995 are included)

Since most of the geometric parameters discussed in this review are obtained from X-ray crystal structure determination, it is important that the reader be familiar with some facts regarding the method. X-ray diffractions obtained from crystalline materials are being used to produce a model structure. The accuracy of the model and the refined structure is achieved by a least-squares procedure that minimizes the difference between the calculated and experimental structure factors (derived from the diffraction intensities). The accuracy of the refined structure is determined by the agreement factor (R); the smaller the R value, the better the experimentally determined structure. The reader should keep in mind that the position of the atom is determined by the centre of its electron density. However, comparison with the results of neutron diffraction, that provides the position of an atom as its centre of mass, shows that the shift of the centre of electron density from the centre of mass is in most cases of the order of only 0.001 Å. It should also be noted that in the presence of 'heavy' atoms (atoms with many electrons) in the molecule, the positions of the 'light' atoms are less accurate. The main disadvantage of the X-ray diffraction method for the structural determination of molecules and crystals is that the compound should be in its crystalline state. Consequently, the structures of unstable compounds and of small gaseous compounds are usually missing from XRD data. It means that calculated geometry by *ab initio* methods that are regularly conducted on small model molecules should be compared with caution with the experimental XRD molecular structures. The advantage of the method is evident from its extensive use and the data it provides form the base for every discussion related to molecular geometry. It should also be mentioned that, unlike isolated molecules in the gaseous state, molecules in the crystal are imposed on intermolecular forces that, when they are strong, they might affect the molecular geometry. The vast amount of XRD structural data available today, the ease of extracting the relevant parameters from computerized data banks and the possibility to perform statistical analysis of the data, all determine the character of the current review. The structural chemistry of organosilicon compounds reviewed here is mainly based on XRD structural data gathered from the Cambridge Structural Database (CSD)[16]. Comparison of the experimental geometry with optimized calculated geometry is given as well[17-29]. Statistical analysis was executed in most cases for the relevant geometric parameter. The review covers Si—X structural chemistry where X = C, Si, N, P, O, S, Hal. Silicon may be either tri-, tetra-, penta- or hexacoordinate and X may have different coordination numbers. For each type of Si—X bond, typical and exceptional compounds are shown and the relevant geometric parameters are listed in tables. Histograms are provided whenever statistical analysis was performed. Each compound that appears in the tables is designated by a six-letter code called Refcode, which is used in the CSD system. A complete list of Refcodes, IUPAC names and chemical formulae of all compounds appearing in this review can be obtained on request from the authors. The statistical analysis was executed with the Origin Program[30], the average value of a geometric parameter, its standard deviation (s.d.) and the standard deviation of the mean (s.m.) values were calculated according to equations 1–3:

$$d(\text{mean}) = (1/N)\Sigma d_i \quad (1)$$

where $d(\text{mean})$ is the calculated average, N is the number of data points; the sum is taken over all data points; d_i is the experimental value;

$$\text{s.d.} = \{[1/(N-1)]\Sigma[d_i - d(\text{mean})]^2\}^{1/2} \quad (2)$$

where s.d. is the standard deviation;

$$\text{s.m.} = (\text{s.d.})/(N)^{1/2} \quad (3)$$

where s.m. is the standard deviation of the mean.

II. SINGLE BONDS TO SILICON

A. Structural Chemistry of the Si—C Bond

Most of the organosilicon compounds contain bonds between the silicon and carbon atom. In the following paragraph the structural chemistry of the Si—C single bond is discussed, mostly in compounds with tetracoordinate silicon and tetracoordinate carbon atoms. The structural chemistry of the Si—C bond in compounds where the carbon coordination state is different, is also discussed. The Si—C bond is markedly polarized and the increase of the bond ionicity by attaching different substituents to either the silicon or the carbon atoms may affect its length. The electronic and steric effects are discussed later.

1. Si—C(sp) bonds

A single bond between a silicon and dicoordinate carbon atom involves a carbon atom in sp hybridization. XRD crystal structure analysis provided 238 individual Si—C(sp) bond lengths and Si—C(sp)—X bond angles. The average Si—C(sp) bond length was calculated from 226 values to be 1.839 Å (s.d. 0.02 Å, s.m. 0.001 Å). There were few exceptionally longer or shorter bonds; however, in almost all cases they resulted from disorder in the crystal or inaccurate crystal structure refinement. The histogram of Si—C(sp) bond lengths is shown in Figure 2. A list of Si—C(sp) bond lengths and Si—C(sp)—X bond angles for selected acyclic and cyclic compounds is given in Table 2. The histogram for the bond angles (Figure 2) shows that in most compounds the Si—C(sp)—X moiety is practically linear. The bonds that deviate from linearity are those in small cyclic molecular systems: Figure 3 shows the correlation between the Si—C(sp)—X bond angle and the ring size. Linearity is reached at a ring size of 10–12 atoms. No correlation was found between bond lengths and bond angles.

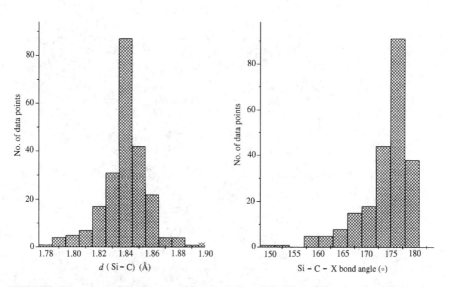

FIGURE 2. Histograms of Si—C(sp) bond lengths (left) and Si—C(sp)—X bond angles (right)

TABLE 2. Geometry of the Si—C(sp) bond in acyclic and cyclic compounds

Refcode	Compound	d(Si—C) (Å)	Si—C—X (°)	Reference
Acyclic				
HACLAY[a]	1	1.828	173.4	31
HEGGUV[a]	2	1.837	175.8	32
JUGBIW	3	1.840	179.0	33
JUGBOC[a]	4	1.855	174.3	33
SULMIV[a]	5	1.837	175.8	34
SUMTUP	6	1.807	177.6	35
VOBREJ[a]	7	1.837	177.1	36
WAWRER[a]	8	1.845	173.9	37
Cyclic				
KUPLAI[a]	9	1.843	148.7	38
TACKEN[a]	10	1.822	160.9	39
KUCVEJ[a]	11	1.842	166.7	40
KUCVIN[a]	12	1.832	164.3	40
BUXJUZ[a]	13	1.844	166.0	41
DUWCAZ[a]	14	1.842	177.0	42
DUWCED[a]	15	1.844	173.6	42
LENHUH[a]	16	1.840	173.1	43
DUWCIH[a]	17	1.850	179.4	42
DUWCON[a]	18	1.844	178.2	42
JALGOS[a]	19	1.843	177.5	44
SAYJIL[a]	20	1.821	176.3	45
PEZFOP[a]	21	1.839	177.9	46
JOHJUL	22	1.834	178.5	47

[a] Average values are given.

5. The structural chemistry of organosilicon compounds

(11) (12) (13) (14) (15) (16) (17) (18) (19) (20)

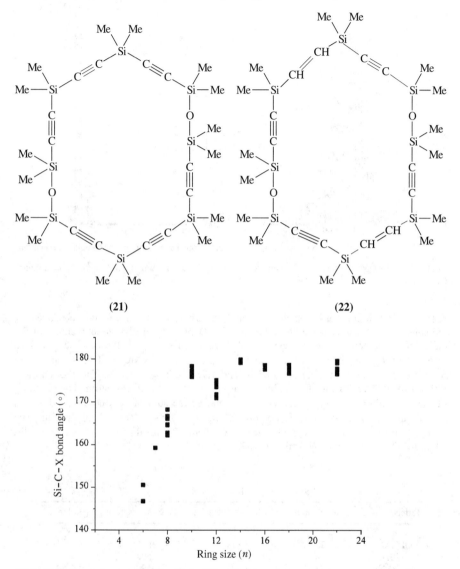

FIGURE 3. Correlation between Si–C(sp)–X bond angle and ring size

2. Si–C(sp²) bonds

Replacing an sp carbon atom by an sp² carbon affects the lengthening of the Si–C bond by only 0.04 Å. The average length of the bond was calculated from 633 individual values obtained from XRD to be 1.878 Å (s.d. 0.03 Å, s.m. 0.001 Å). The histogram is shown in Figure 4. A list of Si–C (sp²) bond angles and the sum of valence bond angles at the carbon atom are given in Table 3. Exceptionally longer bond lengths have

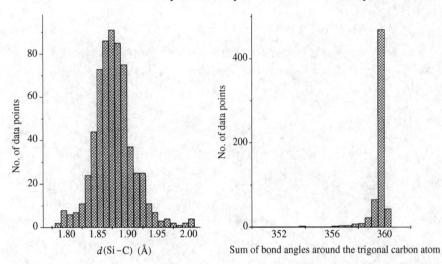

FIGURE 4. Histograms of Si−C(sp^2) bond lengths d (left) and the sum of Si−C(sp^2)−X bond angles (right)

been detected in compounds such as **29** and **30**, where the bond ionicity is reduced by the substitution on the carbon atom. Compound **31** provides a very good example of the effect of attachment of electronegative substituents on silicon on the bond length. Two different groups are bonded to the same sp^2 carbon atom, SiMe$_3$ and SiCl$_2$ Bu-*t*. The Si−C bond length in the first is 1.867 Å, and 1.784 Å in the second. The shortening is a result of the attachment of two chlorine atoms to the same silicon atom. The planarity of the sp^2 carbon atom was checked by calculating the sum of its valence bond angles (see Figure 4). In almost all the compounds the sp^2 carbon is practically planar. The exceptional pyramidality of the sp^2 carbon found in **32** with the sum of valence angles equal to 323.8° is suspicious and might be in error.

TABLE 3. Geometry of Si−C(sp^2) bonds

Refcode	Compound	d (Si−C) (Å)	Σ Si−C−X (°)	Reference
Typical				
VOLZAX	23	1.879	360.0	48
WEGBAL	24	1.880	360.0	49
SUCVAN[a]	25	1.840	359.6	50
YIMYEY[a]	26	1.873	358.0	51
YIMYIC[a]	27	1.878	357.6	51
YOSSUU	28	1.906	360.0	52
Exceptions				
FERMIY	29	2.006	360.0	53
PESICR	30	2.003	360.0	54
LEBPIR	31	1.784	359.7	55
LEBPIR	31	1.867	359.7	55
KIZDUS	32	1.909	323.8	56

[a] Average values are given.

3. Si–C(aryl) bonds

The average Si–C(aryl) bond length was calculated from 3371 individual values to be 1.879 Å (s.d. 0.03 Å, s.m. 0.0004 Å). The histogram is shown in Figure 5. The experimental Si–C(aryl) bond length is in very good agreement with the calculated value.

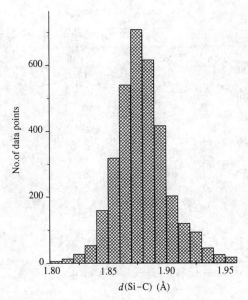

FIGURE 5. Histogram of Si—C(aryl) bond lengths

Ab initio computations at the Hartree–Fock level were carried out on two conformations of silylbenzene[17] and the Si—C(phenyl) bond length was calculated to be 1.8729 Å. Longer Si—C(aryl) bonds have been found in only a few compounds, such as **33**[57] (1.941, 1.954 and 1.972 Å) and **34**[58] (1.910 and 1.993 Å). In both cases the lengthening may be attributed to steric congestion.

$$\begin{array}{cc}
\text{Ph—Sn(Ph)(Ph)—O—Si(Ph)(Ph)—Ph} & \text{OC—Fe—Au←P(Tol-}p\text{)(Ph)} \\
(33) & (34)
\end{array}$$

4. Si—C(sp³) bonds

Perhaps the most abundant organosilicon fragment is trimethylsilyl. The average Si—C(sp^3) bond length was calculated from 19169 individual XRD experimental values to be 1.860 Å (s.d. 0.02 Å, s.m. 0.0002 Å). The histogram is shown in Figure 6a. The strong effect of electronegative groups attached to the silicon atom is shown in Figure 6b. The average Si—CH$_3$ bond is significantly shorter by 0.014 Å, upon attachment of at least one fluorine atom to the silicon.

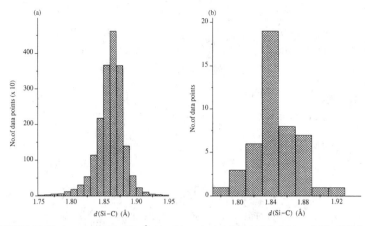

FIGURE 6. Histograms of Si−C(sp^3) bond lengths in trimethylsilyl (a) and Si−CH$_3$ bond lengths in fluorosilanes (b)

Replacement of a methyl by t-Bu causes a marked lengthening of the Si−C(sp^3) bond length (by 0.047 Å) to 1.907 Å (calculated from 889 XRD bond lengths, s.d. 0.03 Å, s.m. 0.001 Å). The histogram is shown in Figure 7. The effect of electronegative substituents is demonstrated by three compounds, **35**[59], **36**[60] and **37**[61]. In **35**, where a fluorine atom is attached to silicon, increasing the Si−C bond ionicity, the bond length is 'short' (1.888 Å); additional shortening is observed (1.801, 1.823 Å) when three fluorine atoms are attached to the silicon atom, as in **36**. In **37**, where a fluorine atom is bonded to the carbon atom, the Si−C bond length is 'long' (1.923 Å).

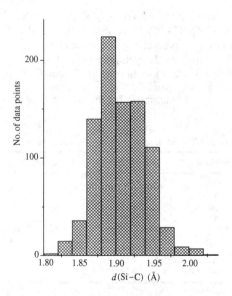

FIGURE 7. Histogram of Si−C(t−Bu) bond lengths

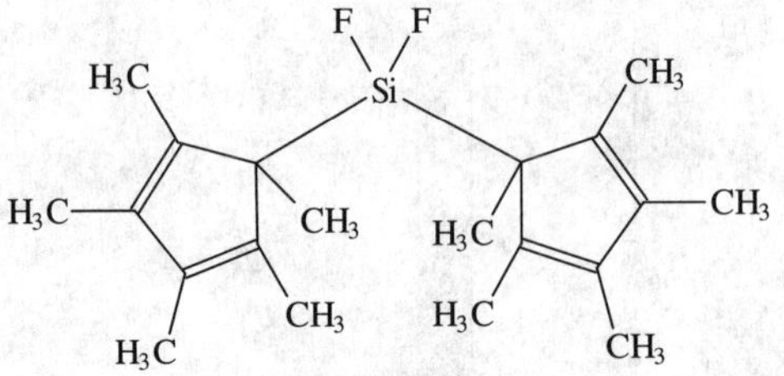

(35) 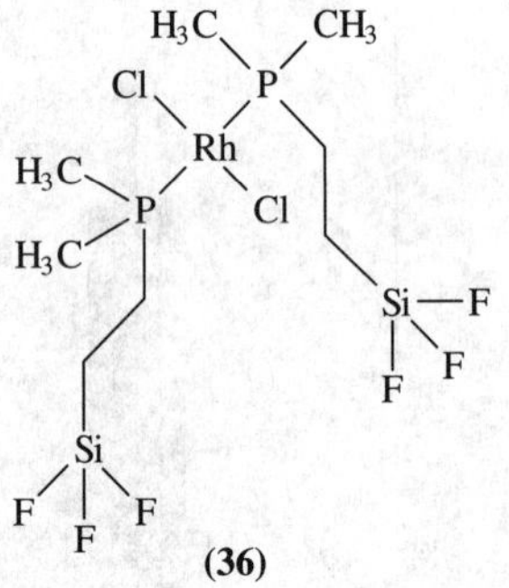   (36)

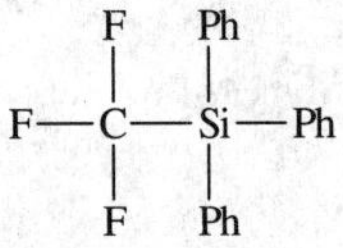

(37) 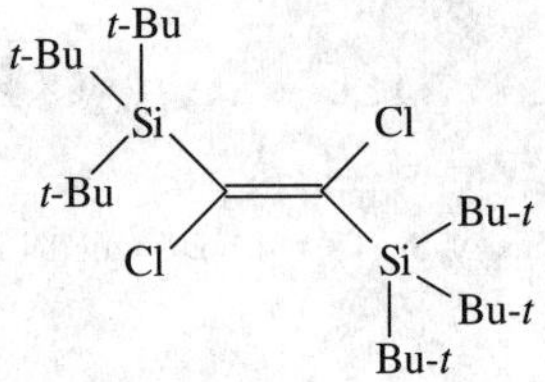   (38)

The lengthening of the Si–C(sp^3) bond length can also be a result of steric congestion, which is demonstrated by **38**[62] containing tri-*tert*-butylsilyl moieties. The average Si–C bond length in **38** is 1.953 Å, compared to 1.907 Å in compounds with a single *t*-Bu bonded to silicon.

5. Si–C bonds in silacyclopropanes and silacyclopropenes

Si–C bond lengths in silacyclopropanes are only slightly shorter (1.84–1.87 Å; see Table 4) than in any other Si–C(sp^3) bonds, such as with methyl (1.860 Å). However, the same bond in silacyclopropenes is significantly shorter (1.81–1.82 Å) than that found for Si–C(sp^2) bonds (1.879 Å). In both silacyclopropanes and silacyclopropenes C–Si–C bond angles are smaller than 60° due to the longer Si–C than C–C bond length. Since the C=C bond lengths in silacyclopropenes are even shorter than the C–C bond lengths in silacyclopropanes, the C–Si–C bond angle in the former (42–44°) is smaller than in the latter (47–52°).

TABLE 4. Geometry of Si–C bond in silacyclopropanes and in silacyclopropenes

Refcode	Compound	d(Si–C) (Å)	C–Si–C (°)	Reference
Cyclopropane				
GIMNUL[a]	39	1.869	50.2	63
WARREM[a]	40	1.843	46.9	64
KIPDES[a]	41	1.868	49.5	65
Cyclopropene				
FIBJUV[a]	42	1.811	43.0	66
YOSZEL[a]	43	1.817	43.7	67
SUJGOT	44	1.827[b]	42.7	68

[a] Average values are given.
[b] The second bond length seems to be in error.

6. Structural chemistry of Si bonds with special functional groups

The structural chemistry of silicon bonded to special functional groups such as acetyl group, cyanide and isocyanide deserves special attention. Acylsilanes having the general formula R_3SiCOR constitute an interesting class of chemical compounds. They are sensitive to light and rather unstable, particularly in a basic environment, where they react to give aldehydes and more complex rearrangement products. Furthermore, their electronic and vibrational spectra display some unusual features. The enhanced reactivity of substituents attached to the carbon atom in the α-position to silicon is referred to as the α-silicon effect. The experimental Si−C(sp^2) bond length in these compounds containing acetyl groups is 1.925 Å, significantly longer (by 0.05 Å) than the regular Si−C(sp^2) bonds of 1.878 Å. A list of selected compounds is given in Table 5. Optimization[24] of the geometry of SiH$_3$CHO at the Hartree−Fock 6-31++G** level reveals a Si−CO bond length of 1.930 Å, slightly longer than the experimental value. The lengthening is attributed to the effect of attachment of the electronegative oxygen atom to the carbon atom, thus lowering the Si−C bond ionicity.

The crystal structures of only three silyl cyanides **51−53** are known. The Si−CN bond length in **51**[75] is 1.823 Å, the CN triple bond length is 1.211 Å and the Si−C−N bond angle is 178.7°. In the dicyanide **52**[76] the Si−CN bond length increases to 1.862 and 1.873 Å while CN triple bonds decrease to 1.121 and 1.148 Å. The Si−C−N bond angle is unchanged (178.0, 178.1°). When the silicon is pentacoordinate, as in **53**[77], the Si−CN bond length is elongated to 2.051 Å, the CN triple bond is unchanged (1.144 Å) and the Si−C−N bond angle is 175.0°. The silyl cyanide[78] H$_3$SiCN was used as a model for calculating the geometry of the Si−C and the CN triple bond lengths which resulted in values of 1.872 Å and 1.141 Å, respectively.

Agreement between the calculated[78] geometry and experimental geometry of silyl isocyanide is poor. The calculated Si−N bond length in H$_3$SiNC is 1.745 Å, and the NC triple bond length is 1.165 Å. The experimental values are 1.819 and 1.785 Å for the

TABLE 5. Geometry of Si−C bonds

Refcode	Compound	d(Si−C) (Å)	Σ Si−C−X (°)	Reference
ACTPSI	45	1.925	360.0	69
GIFPUG	46	1.925	359.8	70
JEMXII	47	1.925	360.0	71
JOPDUN	48	1.925	360.0	72
PSITET	49	1.921	360.0	73
VIGZOA[a]	50	1.907	360.0	74

[a] Average values are given.

Si−N bond, 1.087 and 1.152 Å for the N−C triple bond, and 175.1 and 169.9° for the Si−N−C bond angles in **54**[79] and **55**[80], respectively.

(Structures 51, 52, 53, 54, 55)

B. Structural Chemistry of the Si−Si Bond

1. Si−Si bonds in compounds with tetracoordinate silicon atoms

The sum of covalent radii for the Si−Si bond length in disilanes is 2.341 Å. The average bond length in acyclic disilanes was calculated from 541 individual XRD values to be 2.358 Å (s.d. 0.02 Å, s.m. 0.0009 Å). The histogram is shown in Figure 8. Bond lengths

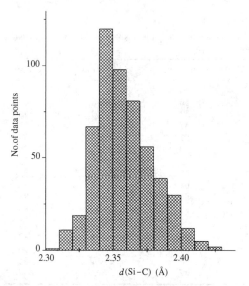

FIGURE 8. Histogram of Si−Si bond lengths in acyclic compounds

for selected compounds whose XRD data were published between 1990 and 1995 are listed in Table 6. Compounds with exceptionally long Si—Si bond lengths are also listed in Table 6. The significant lengthening (0.23 Å on average) of the Si—Si bond in **64**–**67** is attributed to the steric effect. In all four compounds the silicon atoms are substituted by more than one bulky group such as t-Bu or iodine atoms.

Slightly longer Si—Si bond lengths are found in cyclic disilane compounds. The average of this bond was calculated from 472 XRD values to be 2.372 Å (s.d. 0.04 Å, s.m. 0.002 Å). The histogram is shown in Figure 9. The cyclopolysilanes show a range of unique electronic and spectroscopic properties, which arise from electron delocalization in the Si—Si σ framework. Examples of cyclopolysilanes with ring sizes of 3, 4, 5, 6, 7, 13 and 16 are shown in **68**–**74**. Si—Si bond lengths and Si—Si—Si bond angles for these compounds are given in Table 7. Figure 10 shows the dependence of the Si—Si—Si bond angles on the ring size. The ring strain is relieved when the ring size is larger than 8.

Exceptionally long Si—Si bonds in cyclopolysilanes are found in compounds such as **75**–**79** which are bearing bulky substituents such as t-Bu.

C. Structural Chemistry of the Si—N Bond

More than 700 individual Si—N bond lengths were known at the time Sheldrick[15] wrote his review (1989). Although he did not make any remark on the source of the data, it seems that most were available from X-ray crystal structure analysis. At the time of writing the present review (July 1996) more than 3000 individual experimental Si—N bond lengths were obtained from the CSD bank for crystal structure data. It is not possible to present a comprehensive list in this review and therefore we provide a statistical analysis of the data and mention a few typical and exceptional compounds together with their Si—N structural features.

The Schomaker-Stevenson corrected sum of covalent radii of silicon and nitrogen is 1.80 Å and, in general, it was shown[15] that shorter bonds have been found for compounds in which the ionicity of the bond is enhanced by the attachment of either a very

TABLE 6. Typical and exceptional Si—Si bond lengths in acyclic compounds

Refcode	Compound	d(Si—Si) (Å)	Reference
Typical			
PERPAD	**56**	2.346	81
JUKYIX[a]	**57**	2.388	82
PAFNOZ[a]	**57**	2.363	83
PERPOR[a]	**57**	2.375	83
LINKAU[a]	**58**	2.362	84
YASYEW[a]	**59**	2.395	85
YASYIA[a]	**60**	2.373	85
JUXPIB	**61**	2.368	86
WAJZUC	**62**	2.335	87
WIGNUV	**63**	2.391	88
Exceptional			
TAGNOE	**64**	2.453	89
DIVKIC	**65**	2.697	90
FAPBON[a]	**66**	2.614	91
GERSEB	**67**	2.593	92

[a] Average values are given.

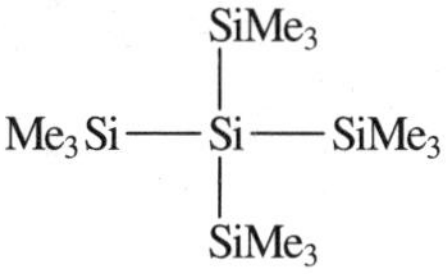

5. The structural chemistry of organosilicon compounds

$Me_3Si-Si(SiMe_3)(SiMe_3)-SiMe_3$

(56)

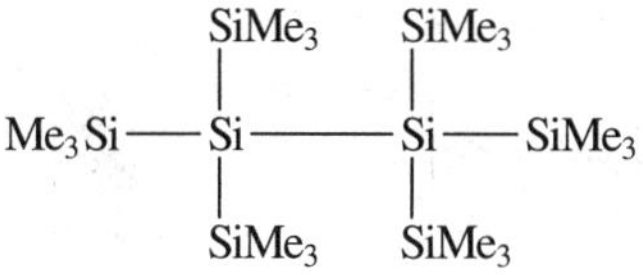

$Me_3Si-Si(SiMe_3)(SiMe_3)-Si(SiMe_3)(SiMe_3)-SiMe_3$

(57)

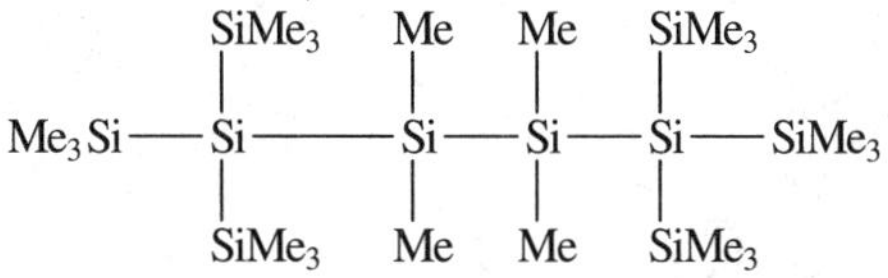

$Me_3Si-Si(SiMe_3)(SiMe_3)-Si(Me)(Me)-Si(Me)(Me)-Si(SiMe_3)(SiMe_3)-SiMe_3$

(58)

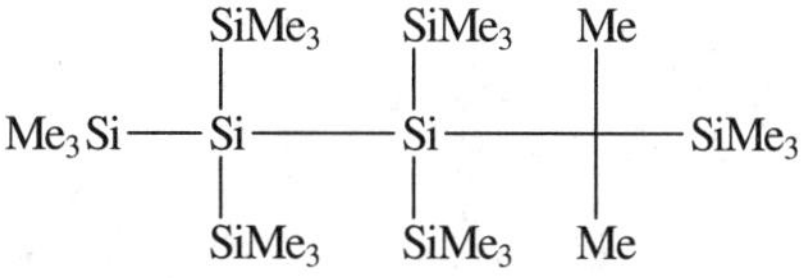

$Me_3Si-Si(SiMe_3)(SiMe_3)-Si(SiMe_3)(SiMe_3)-C(Me)(Me)-SiMe_3$

(59)

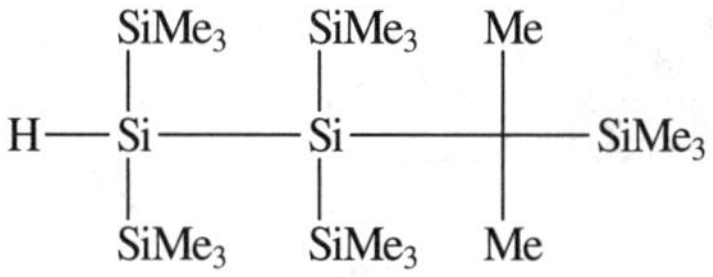

$H-Si(SiMe_3)(SiMe_3)-Si(SiMe_3)(SiMe_3)-C(Me)(Me)-SiMe_3$

(60)

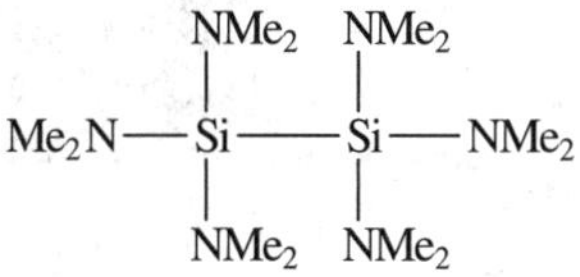

$Me_2N-Si(NMe_2)(NMe_2)-Si(NMe_2)(NMe_2)-NMe_2$

(61)

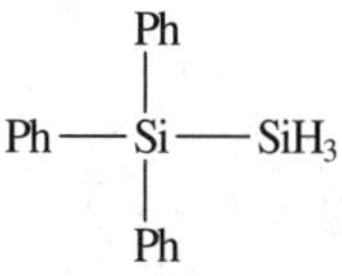

$Ph-Si(Ph)(Ph)-SiH_3$

(62)

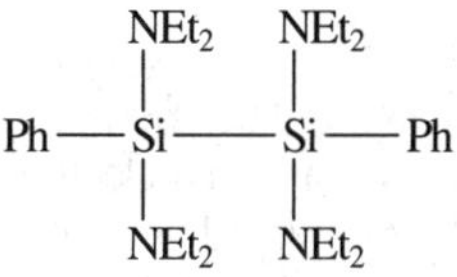

$Ph-Si(NEt_2)(NEt_2)-Si(NEt_2)(NEt_2)-Ph$

(63)

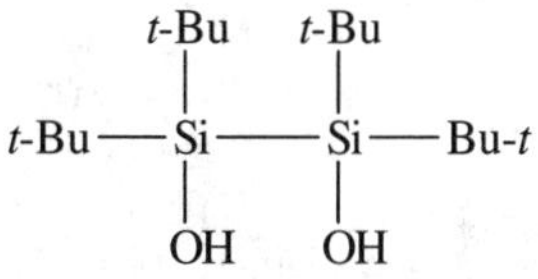

$t\text{-}Bu-Si(t\text{-}Bu)(OH)-Si(t\text{-}Bu)(OH)-Bu\text{-}t$

(64)

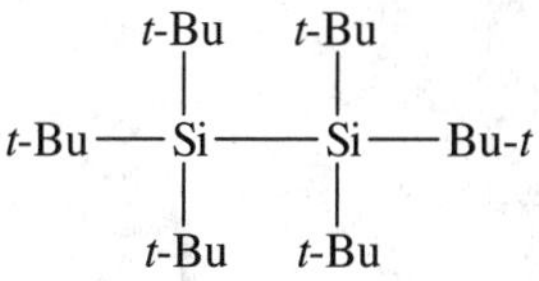

$t\text{-}Bu-Si(t\text{-}Bu)(t\text{-}Bu)-Si(t\text{-}Bu)(t\text{-}Bu)-Bu\text{-}t$

(65)

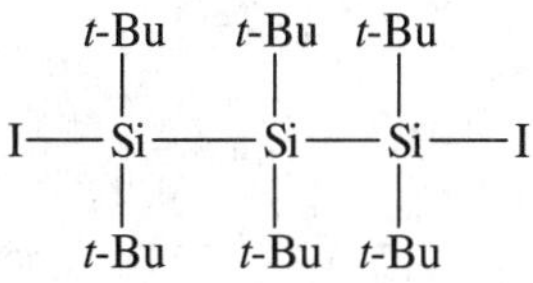

$I-Si(t\text{-}Bu)(t\text{-}Bu)-Si(t\text{-}Bu)(t\text{-}Bu)-Si(t\text{-}Bu)(t\text{-}Bu)-I$

(66)

$t\text{-}Bu-Si(t\text{-}Bu)(t\text{-}Bu)-Si(t\text{-}Bu)(t\text{-}Bu)-Si(t\text{-}Bu)(t\text{-}Bu)-Bu\text{-}t$

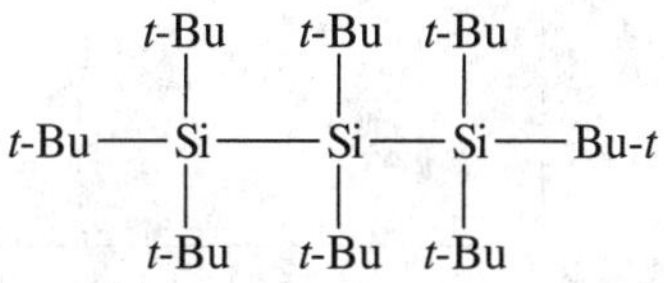

(67)

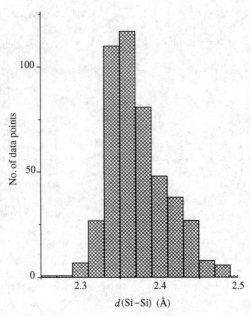

FIGURE 9. Histogram of Si—Si bond lengths in cyclic compounds

electronegative substituent to silicon or an electropositive metal to nitrogen. Long Si—N bonds are due to the reduction of the ionicity by the attachment of a substituent in the opposite manner, namely attachment of either an electronegative substituent to nitrogen or an electropositive substituent to silicon. Long bonds may also be a result of steric effects. In cases where the nitrogen atom is involved in a very strong hydrogen bond, the Si—N bond may lengthen.

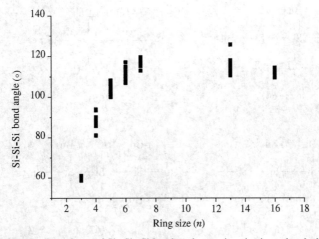

FIGURE 10. The dependence of Si—Si—Si bond angles on ring size in cyclopolysilanes

5. The structural chemistry of organosilicon compounds

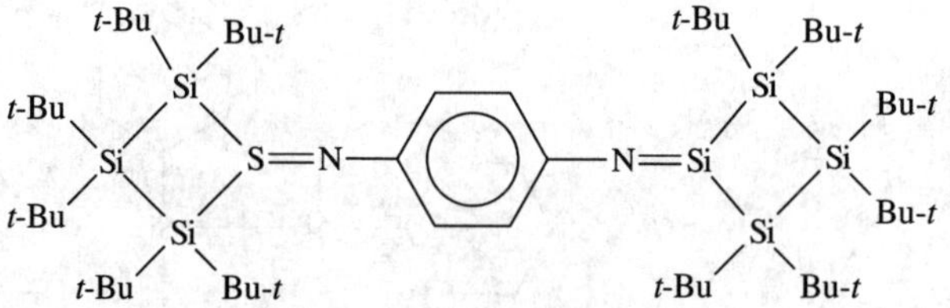

(77)

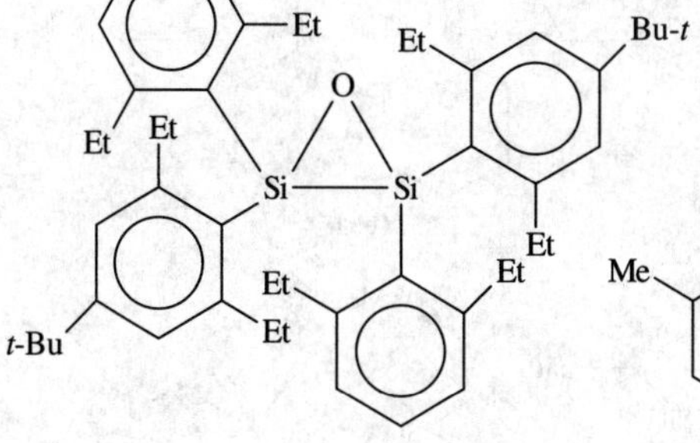

(78)

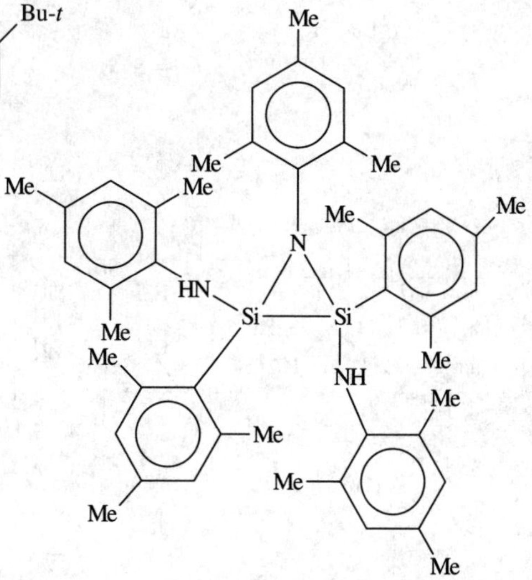

(79)

TABLE 7. Typical and exceptional Si—Si bond lengths in cyclic compounds with tetra-coordinate silicon atoms

Refcode	Compound	d(Si—Si) (Å)	Si—Si—Si(°)	Reference
Typical				
COHBIC[a]	68	2.385	59.1	93
SOWBEL[a]	69	2.391	88.4	94
DPHPSI[a]	70	2.394	104.5	95
HASKAN[a]	71	2.332	117.0	96
DIDJOP[a]	72	2.341	116.2	97
FALSEQ[a]	73	2.355	115.7	98
FALSIU[a]	74	2.355	111.9	98
Exceptional				
CIHRAM	75	2.511	60.0	99
YISJUF[a]	76	2.494	90.1	100
SUDJEG[a]	77	2.500	78.8	101
LANLAN	78	2.214	—	102
LAWZIS	79	2.228	—	103

[a]Average values are given.

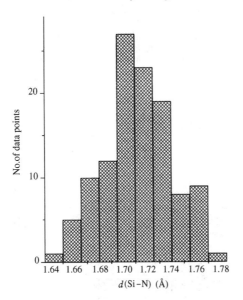

FIGURE 11. Histogram of Si—N bond lengths in compounds with tetracoordinate silicon and dicoordinate nitrogen atoms

1. Si—N bonds in compounds with tetracoordinate silicon and dicoordinate nitrogen atoms

The average Si—N bond length in this class of compounds was calculated from 115 individual experimental values obtained by XRD, and was found to be 1.713 Å (s.d. 0.03 Å and s.m. 0.003 Å). The histogram is shown in Figure 11. Selected typical and outlying Si—N bond lengths and Si—N—X bond angles are listed in Tables 8 and 9, respectively. The calculated[25] (3-21G*) Si—N(sp^2) bond length in H_3SiNCH_2 of 1.755 Å is significantly longer than the experimental value. In **80-84** Si=N double bonds are also present, but such bonds are discussed later in this review.

TABLE 8. Geometry of Si—N bonds in compounds with tetracoordinate silicon and dicoordinate nitrogen atoms

Refcode	Compound	d(Si—N) (Å)	Si—N—X (°)	Reference
DOKWUV	**80**	1.696	177.8	104
DOKXEG[a]	**81**	1.660	161.2	104
JIWGAX	**82**	1.661	174.2	105
GEVRII	**83**	1.678	169.4	104
LENVUV	**84**	1.662	149.8	106
POHXEP[a]	**85**	1.689	158.9	107
WECWIK	**86**	1.725	119.9	108
YEFJUO[a]	**87**	1.758	129.8	109
WEDDOY[a]	**88**	1.728	149.2	110
YOVLEA	**89**	1.739	155.7	111
LECDIG[a]	**90**	1.707	155.0	112

[a] Average values are given.

TABLE 9. Exceptionally long and short Si—N bonds in compounds with tetracoordinate silicon and dicoordinate nitrogen atoms

Refcode	Compound	d(Si—N) (Å)	Si—N—X (°)	Reference
Long bonds				
JALSUK	91	1.798	109.7	113
KEGTIZ[a]	92	1.808	168.0	114
KETMON	93	1.822	178.3	115
TMSIDM	94	1.807	120.0	116
Short bonds				
YEBVAC	95	1.673	136.0	117
YEBVAC	95	1.791	130.4	117
SETYEV	96	1.649	151.1	118

[a] Average values are given.

Si—N=X bond angles are wider than in the carbon analogues. Upon bending, the HOMO (which effectively contains the lone pair of nitrogen) and a σ^* orbital of higher energy can interact, leading to the stabilization of a bent geometry provided that the gap in energy between the two orbitals is small. Upon increasing the electronegativity on going from carbon to silicon, the energy gap will increase and the Si—N=X bond angle will widen. A similar effect will occur by the attachment of more electropositive substituent to nitrogen. Examples of the two extremes are seen in **80** where, by attachment of an electropositive Si atom to the nitrogen, the Si—N=Si bond angle widens to 177.8°, while in **86** where the electronegative substituent, the sulphur atom, is attached to nitrogen, the Si—N=S bond angle closes to 119.9°. It is also evident from the Si—N bond lengths listed in Table 8 that the Si—N bond length shortens upon linearization of the Si—N=X bond.

91–96 are compounds with exceptionally long or short Si–N bond lengths. Compound **95** provides an interesting example in which both long and short Si–N bonds exist in the same molecule. The attachment of the more electropositive germanium atom to nitrogen imposes an increase of ionicity and Si–N bond shortening (1.673 Å). The attachment of the azide to silicon causes a lengthening of the bond (1.791 Å). The shortening of the Si–N bond length caused by the attachment of germanium to the nitrogen atom is also seen in **96** (1.649 Å).

2. Si–N bonds in compounds with tetracoordinate silicon and tricoordinate nitrogen atoms

There are about 3000 individual experimental Si–N bond lengths obtained by XRD. Histograms of the Si–N bonds and of the sum of valence angles around the nitrogen atom are given in Figure 12. Selected typical and exceptional Si–N bond lengths and the sum of valence bond angles are listed in Tables 10 and 11, respectively. The average Si–N bond length was calculated from 2908 individual values to be 1.739 Å (s.d. 0.03 Å, and s.m. 0.0006 Å). Lengthening of the Si–N bond was observed in many compounds such as **112–120**, where the substituent at the nitrogen atom has a comparable electronegative character to that of nitrogen. Boron and phosphorus are most frequently found to be the substituents on nitrogen that cause the lengthening. On the other hand, the attachment of a strong electropositive atom to nitrogen increases the ionicity of the bond and the extent of its shortening. The effect is exemplified by compounds **124,125** and **126** where Li or Ca are the atoms attached to nitrogen. The same effect may also be achieved by the attachment of a strong electronegative atom or group to the silicon atom. Thus, in **121** and **122** where chlorine atoms are attached to silicon, the Si–N bonds are short (1.678 and 1.679 Å, respectively), while in a similar compound lacking the chlorine atoms **123** the Si–N bond is longer (1.703 Å).

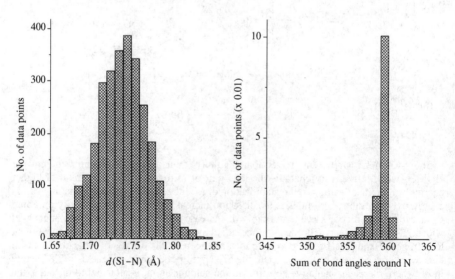

FIGURE 12. Histograms of Si–N bonds (left) and sum of valence bond angles around nitrogen atoms (right)

5. The structural chemistry of organosilicon compounds

TABLE 10. Geometry of the Si−N bond in compounds with tetracoordinate silicon and tricoordinate nitrogen atoms

Refcode	Compound	d(Si−N) (Å)	Σ (X−N−X) (°)	Reference
PECBUU	97	1.710	351.9	119
LEHHEL[a]	98	1.700	360.0	120
LINTIL[a]	99	1.733	359.3	121
YEPTUI[a]	100	1.724	360.0	122
JUXPIB[a]	101	1.716	359.1	123
WIGNUV[a]	102	1.726	359.4	124
YEPTOC	103	1.718	358.8	122
PEWDAW[a]	104	1.738	360.0	125
WAHGOB[a]	105	1.728	358.7	126
WAJLOI[a]	106	1.729	359.0	127
WAJLUO[a]	107	1.729	359.8	127
PECCEF[a]	108	1.711	350.3	119
WEHMAX[a]	109	1.736	359.1	128
WEHLOK[a]	110	1.740	359.9	128
YUDLAK[a]	111	1.717	360.0	129

[a] Average values are given.

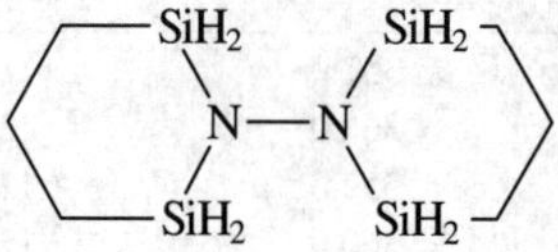

(106) 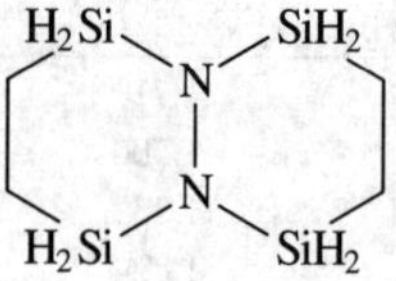 (107) 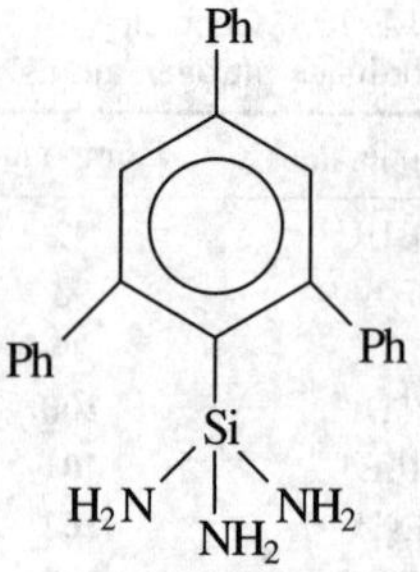 (108)

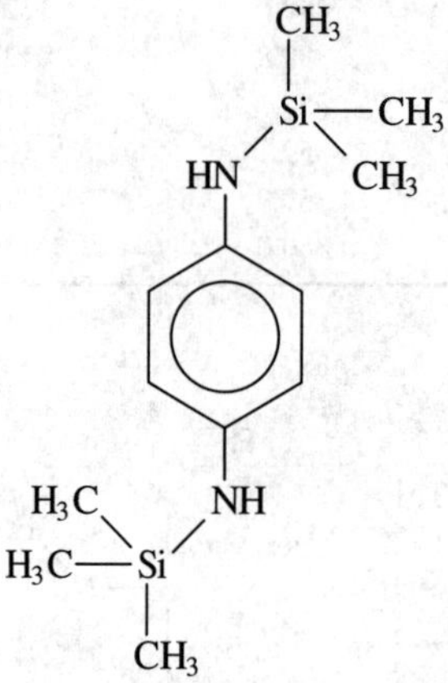

(109) 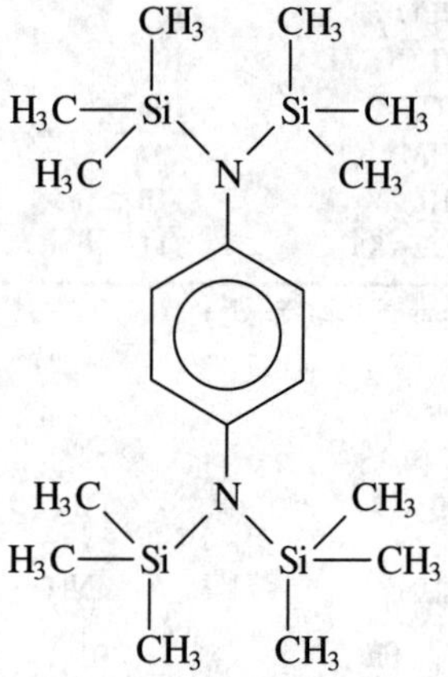 (110)

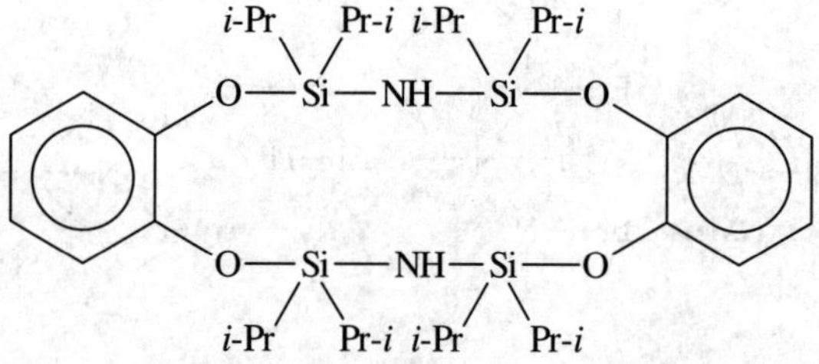

(111)

It is interesting to note that, with no exception, the tricoordinate nitrogen atom that is bonded to the tetracoordinate silicon atom is not pyramidal. The theoretical sum of valence angles in tetrahedral geometry (sp^3 for nitrogen atom) is 329.5°, and 360° for trigonal geometry (sp^2 for nitrogen atom). The histogram shown, in Figure 12 (right) shows very clearly that the nitrogen atom is practically planar. The average value is

TABLE 11. Geometry of exceptionally long and short Si—N bonds in compounds with tetracoordinate silicon and tricoordinate nitrogen atoms

Refcode	Compound	d(Si—N) (Å)	Σ (X—N—X) (°)	Reference
Long bonds				
LAWNAY[a]	112	1.804	359.9	130
LEMMIZ[a]	113	1.791	357.6	131
PEFPEV[a]	114	1.768	360.0	132
PEFPOF[a]	115	1.780	360.0	132
PEFRAT	116	1.756	358.2	132
PEPNON	117	1.795	357.1	133
PODSUW[a]	118	1.770	359.0	134
YERWOH[a]	119	1.816	358.2	135
YOVYEN	120	1.825	360.0	136
Short bonds				
HEZBUJ	121	1.679	359.8	137
HEZCEU[a]	122	1.678	354.7	137
HEZCIY[a]	123	1.703	356.1	137
LEZREN[a]	124	1.678	360.0	138
SUPSUR	125	1.675	358.8	139
WAKGIY	126	1.650	359.9	140

[a] Average values are given.

calculated to be 358.4° (s.d. 9.3°, s.m. 0.2°). The planarity of the tricoordinate nitrogen atom bonded to the silicon atom was established more than 40 years ago. The interaction of nitrogen 2p-orbitals normal to the NSi_3 plane in $(H_3Si)_3N$ with silicon 3d-orbitals of appropriate symmetry provided a simple explanation for both the planarity of nitrogen and the shortening of the bond compared with the corrected sum of the covalent radii (1.80 Å).

3. Si–N bonds in compounds with pentacoordinate silicon and tricoordinate nitrogen atoms

Pentacoordinate silicon forms two types of bonds with tricoordinate nitrogen atoms, a pure covalent bond and a N → Si dative bond. The first is significantly shorter than the second. The average covalent Si–N bond length in compounds where pentacoordinate silicon atom is bonded to tricoordinate nitrogen atom was calculated from 48 XRD experimental values to be 1.761 Å (s.d. 0.06 Å, s.m. 0.009 Å). An example of the difference in bond length is shown in **127**[141] where the covalent Si–N bond lengths are 1.766 and 1.770 Å and the dative bond is 2.333 Å.

5. The structural chemistry of organosilicon compounds

(127)

4. Si–N bonds in compounds with hexacoordinate silicon and tricoordinate nitrogen atoms

All Si–N bonds in compounds where hexacoordinate silicon atom is bonded to tricoordinate nitrogen atoms are dative bonds, where the nitrogen atom provides its lone-pair electrons to the bond. The average Si → N bond was calculated from 31 individual values to be 1.969 Å (s.d. 0.05 Å and s.m. 0.008 Å).

D. Structural Chemistry of the Si–P Bond

1. Si–P bonds in compounds with tetracoordinate silicon and dicoordinate phosphorus atoms

There are only five known X-ray crystal structures of compounds containing a tetracoordinate silicon atom that is bonded to a dicoordinate phosphorus atom. In three, **128–130**, the phosphorus atom is singly bonded to silicon and doubly bonded to either carbon or phosphorus atoms while in the remaining two, **131** and **132**, the phosphorus atom is singly bonded to both substituents. The Si–P bond lengths and the Si–P–X bond angles are given in Table 12. On the average the Si–P bond length in the first three compounds is longer by 0.1 Å than the average in the last two compounds. The double bond of phosphorus has the same effect as two single bonds in regards to the Si–P bond length and can be compared to compounds with tricoordinate phosphorus atoms.

2. Si–P bonds in compounds with tetracoordinate silicon and tri- or tetracoordinate phosphorus atoms

The sum of the covalent radii of silicon and phosphorus is 2.27 Å and, since the ionic character of the Si–P bond is relatively limited, no significant difference of the Si–P bond length is expected. The average of the experimental (X-ray diffraction) Si–P bond

TABLE 12. Geometry of the Si–P bond in compounds with tetracoordinate silicon and dicoordinate phosphorus atoms

Refcode	Compound	d(Si–P) (Å)	Si–P–X (°)	Reference
DARPOB	**128**	2.237	106.2	142
KAXZOY	**129**	2.269	98.6	143
PINZIV	**130**	2.263	109.3	144
KITKEV[a]	**131**	2.161	103.5	145
VEJNED	**132**	2.167	105.4	146

[a] Average values are given.

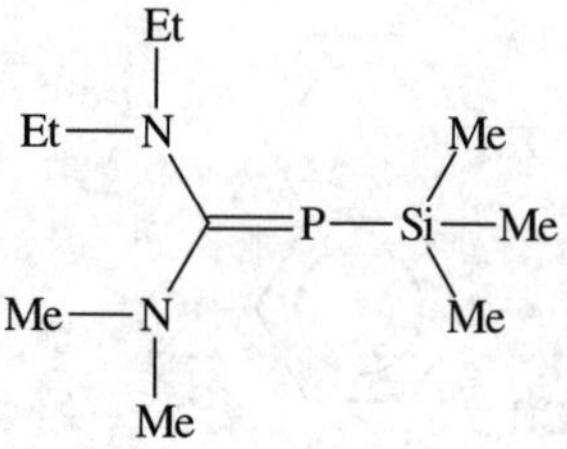

(128) 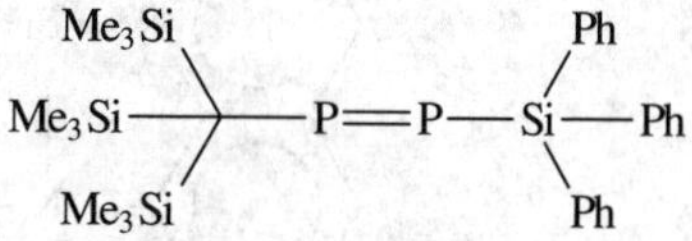 **(129)**

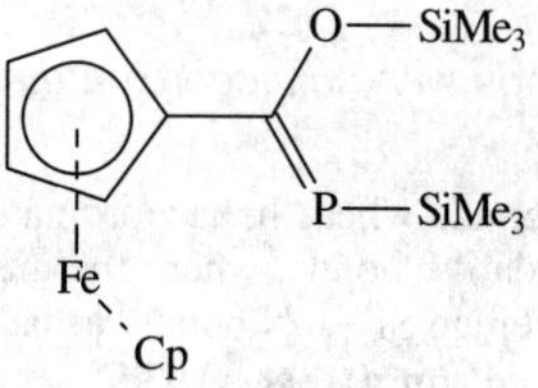

(130)

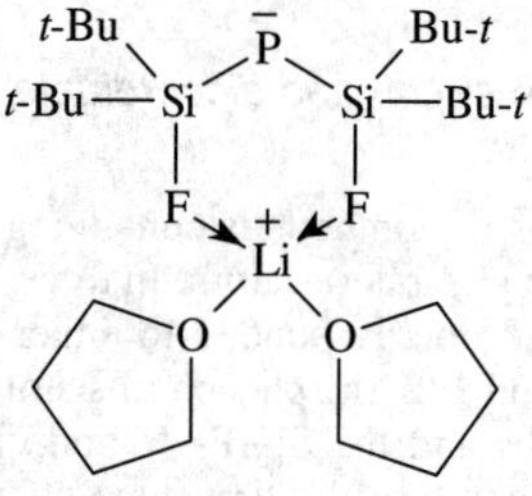

(131) 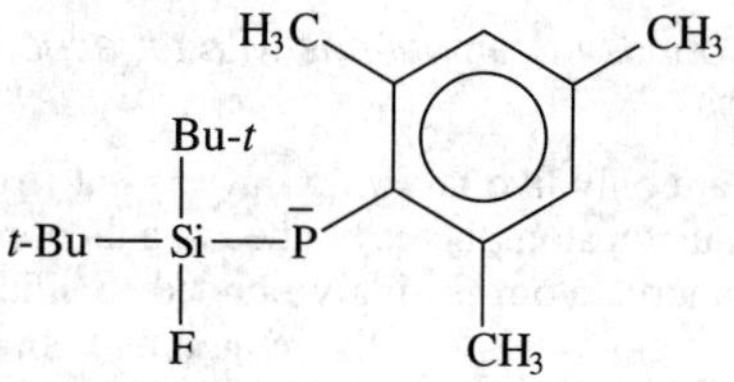 **(132)**

length in silylphosphines containing a tricoordinate phosphorus atom is calculated for 165 values to be 2.265 Å (s.d. 0.02 Å, s.m. 0.002 Å). The histogram for the Si—P bond lengths is shown in Figure 13. The geometry of typical compounds **133–142** is given in Table 13. In contrast to nitrogen, the tricoordinate phosphorus atom is pyramidal as shown by the sum of the bond angles it forms (Tables 13 and 14). The planar tricoordinate phosphorus atom has not been observed. The closest to planarity are observed whenever the phosphorus atom is bonded to a phenyl ring such as in **137** or in **148**, as indicated by the sum of bond angles involving the phosphorus atom, 345.2° and 351.7°, respectively.

The geometry of exceptionally long or short Si—P bond lengths is given in Table 14 for compounds **143–145** and **146–148**, respectively. The lengthening may be attributed to strain and steric congestion in compounds **144** and **145** and to the reduction of polarity in **143**. An electropositive substituent on the phosphorus atom will reduce the Si—P bond length as found in **147** and **148**.

The average Si—P bond length in compounds containing a tetracoordinate phosphorus atom was calculated with 144 individual bond lengths obtained experimentally by X-ray crystal structures. The Si—P bond length is 2.265 Å (s.d. 0.02 Å and s.m. 0.002 Å). The histogram is shown in Figure 14. Selected Si—P bond lengths for typical compounds **149–153** are listed in Table 15.

Exceptionally long and short Si—P bond lengths are found in **154–156** and in **157–163**, respectively, and the X-ray experimental values are summarized in Table 16. Most of the

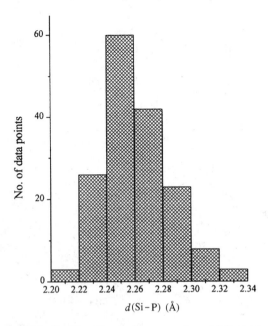

FIGURE 13. Histogram of the Si—P bond lengths in compounds containing tetracoordinate silicon and tricoordinate phosphorus atoms

TABLE 13. Geometry of the Si—P bond in compounds with tetracoordinate silicon and tricoordinate phosphorus atoms

Refcode	Compound	d(Si—P) (Å)	Σ (Si—P—X) (°)	Reference
HAZKUO	133	2.246	328.1	147
HAZLAV	134	2.298	301.4	147
JIXZAR[a]	135	2.228	201.6	148
SENLAY[a]	136	2.280	231.1	149
SIFMEZ[a]	137	2.264	345.2	150
SIYXUT[a]	138	2.278	333.0	151
VIXVAZ	139	2.256	290.4	152
VOBSOU[a]	140	2.245	294.3	153
WAPLOO[a]	141	2.293	304.6	154
WAPLUU[a]	142	2.257	321.3	154

[a] Average values are given.

short Si—P bonds are detected in compounds where the phosphorus is bonded to a very electropositive atom such as lithium, thus increasing the ionicity of the Si—P bond and its shortening.

E. Structural Chemistry of the Si—O Bond

In 1976, Si—O structural data were available for only 51 organosilicon compounds. The structural features of Si—O bonds have been reviewed by Voronkov and colleagues[172] in

(133) (134) (135) (136) (137) (138) (139) (140) (141) (142)

5. The structural chemistry of organosilicon compounds

TABLE 14. Exceptionally long and short Si–P bonds in compounds with tetracoordinate silicon and tricoordinate phosphorus atoms

Refcode	Compound	d(Si–P) (Å)	Σ (P–Si–X) (°)	Reference
Long bonds				
DABLAT	143	2.306	301.3	155
JUWJAM[a]	144	2.318	331.4	156
VENKUU[a]	145	2.300	260.7	157
Short bonds				
DETTAX	146	2.227	307.7	158
FOWLOS[a]	147	2.218	325.4	159
WAPLEE	148	2.223	351.7	154

[a] Average values are given.

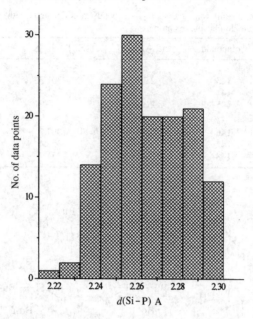

FIGURE 14. Histogram of the Si−P bond lengths in compounds containing tetracoordinate silicon and tetracoordinate phosphorus atoms

(149)   (150)   (151)

TABLE 15. Geometry of the Si−P bond in compounds with tetracoordinate silicon and tetracoordinate phosphorus atoms

Refcode	Compound	d(Si−P) (Å)	Reference
KAKFAD	**149**	2.242	160
KAKFEH	**150**	2.248	160
KUDXAI	**151**	2.254	161
TADFIN[a]	**152**	2.259	162
TADFOT[a]	**153**	2.265	162

[a] Average values are given.

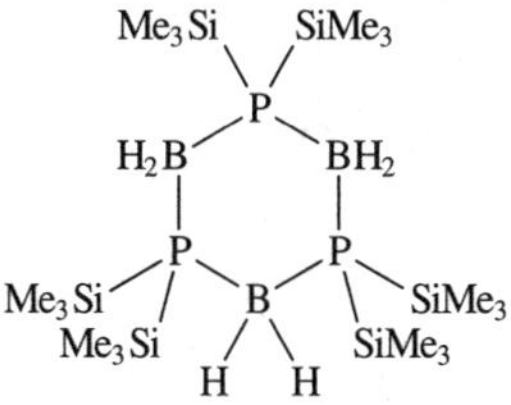

5. The structural chemistry of organosilicon compounds

(152) 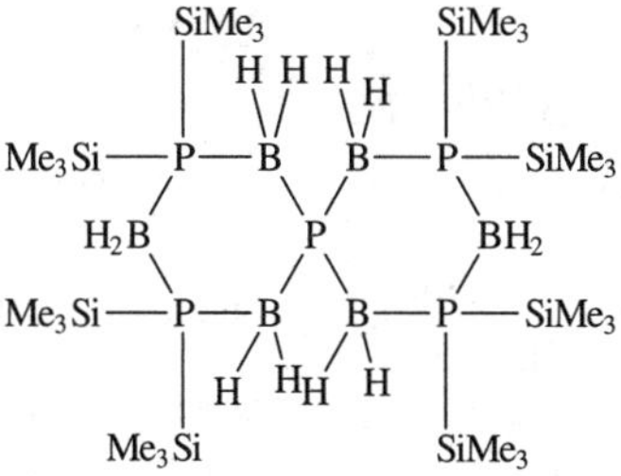 **(153)**

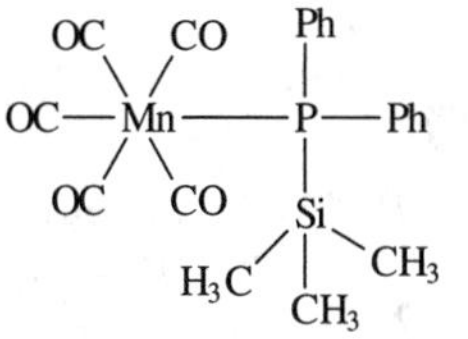

(154) 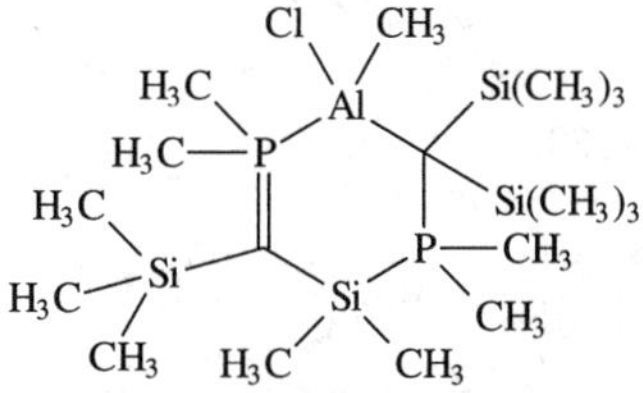 **(155)**

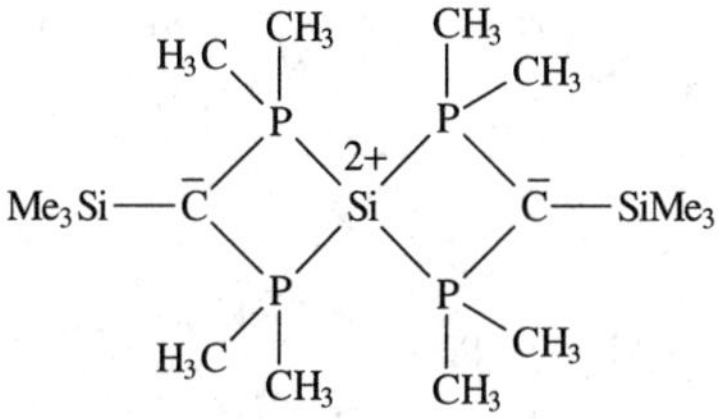

(156) 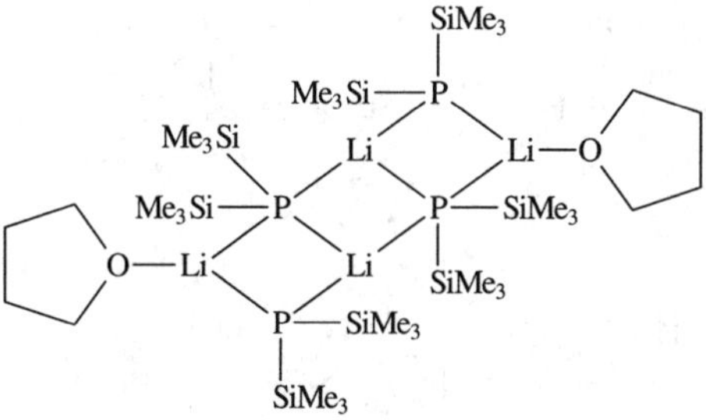 **(157)**

TABLE 16. Exceptionally long and short Si−P bonds in compounds with tetracoordinate silicon and tetracoordinate phosphorus atoms

Refcode	Compound	d(Si−P) (Å)	Reference
Long bonds			
GANRAO	154	2.328	163
KIBSAP	155	2.329	164
SEVRIE[a]	156	2.416	165
Short bonds			
FOFFUB[a]	157	2.211	166
FOFGAI[a]	158	2.194	166
SEHHOC	159	2.205	167
PADDED[a]	160	2.211	168
VAFWEE[a]	161	2.200	169
LAJLAJ[a]	162	2.220	170
YEZGOZ[a]	163	2.214	171

[a] Average values are given.

(158) (159) (160) (161) (162) (163)

1978. At that time the calculated Si−O bond length was 1.64(3) Å. In 1989, Sheldrick[15] indicated in his review that the additional data available at the time he wrote his review confirmed this finding. This value is smaller than the Stevenson-Schomaker corrected sum of the covalent radii of silicon and oxygen (1.76 Å). The large amount of XRD structural data available today enables more accurate statistical analysis and the ability to verify it by theoretical calculation.

1. Si−O bonds in compounds with tetracoordinate silicon and dicoordinate oxygen atoms

The 3276 bond lengths obtained by XRD have been used to calculate the average Si−O bond length in compounds containing tetracoordinate silicon atom bonded to a dicoordinate oxygen atom. The average Si−O bond length is 1.629 Å (s.d. 0.03 Å, s.m.

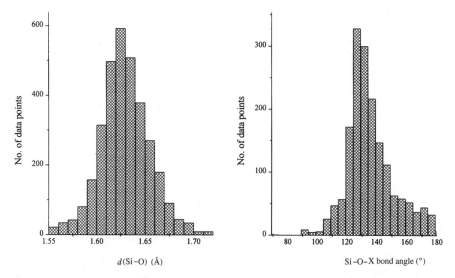

FIGURE 15. Histograms of Si—O bond lengths (left) and Si—O—X (X ≠ Si) (right) in compounds with tetracoordinate silicon and dicoordinate oxygen atoms

0.0005 Å). An average value for the Si—O—X (X = silicon was excluded) bond angle was calculated from 1722 individual values to be 135.4° (s.d. 15.8°, s.m. 0.4°). The histograms of the bond lengths and bond angles are given in Figure 15. A list of Si—O bond lengths and bond angles for typical compounds that have been published in the scientific literature since 1993, and their crystal structures, were very accurately refined ($R < 0.05$) are given in Table 17.

Most of the exceptional bond lengths were obtained for compounds that were found to be disordered in their crystalline state. Shorter bond lengths have been found to belong to similar systems as presented by **174–176**. The average Si—O bond lengths in these compounds are 1.552, 1.551 and 1.503 Å in **174**[183], **175**[184] and **176**[185], respectively. The attachment of an electronegative sulphur atom to the silicon atom increases the Si—O bond ionicity and therefore the bond tends to become shorter.

TABLE 17. Si—O bond lengths and Si—O—X bond angles in compounds with tetracoordinate silicon and dicoordinate oxygen atoms

Refcode	Compound	d(Si—O) (Å)	Si—O—X (°)	Reference
HMDSIX[a]	**164**	1.631	148.2	173
PODJIB	**165**	1.613	180.0	174
WEHJOI	**166**	1.635	148.7	175
WEHJOI[a]	**167**	1.616	141.2	176
PITHOP	**168**	1.642	121.7	177
YECTEF[a]	**169**	1.666	126.4	178
WEKFEX[a]	**170**	1.597	138.7	179
WEKTOV	**171**	1.627	117.9	180
PIJJIB[a]	**172**	1.621	147.5	181
LIJGUG[a]	**173**	1.606	147.4	182

[a] Average values are given.

(164) (165) (166) (167) (168) (169) (170) (171) (172) (173)

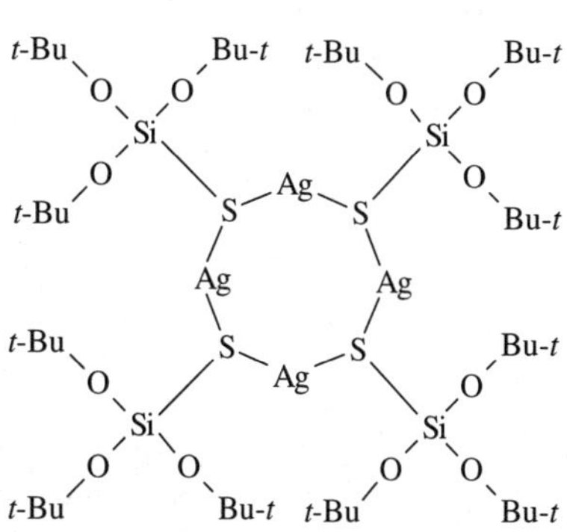

(174) 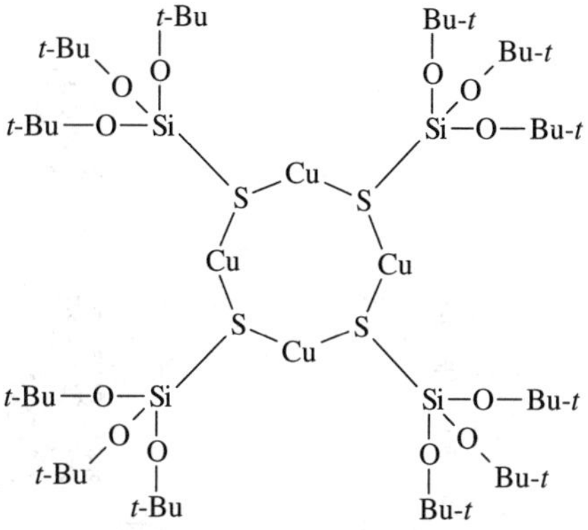   **(175)**

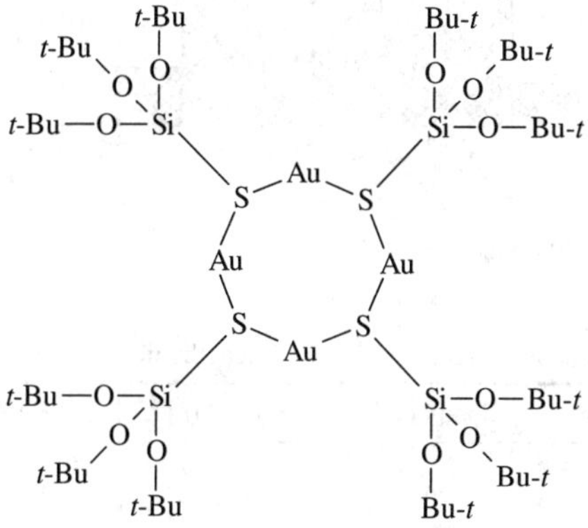

(176)

2. Si—O bonds in silanols

The geometry of silanols was the subject of many theoretical investigations. The Si—OH bond length in methylsilanol was calculated[18] by use of the 3-21G basis set to be 1.679 Å, significantly longer than the experimental value (1.636 Å). Shorter bond lengths were found[19] by optimization of the geometry using *ab initio* calculation at the 3-21G* (modified) level, 1.653 and 1.657 Å in H_3Si-OH and CH_3SiH_2-OH, respectively. Geometry optimization of silanol and methanol at the self-consistent field (SCF) level was carried out[20] and revealed a Si—OH bond length of 1.646 Å. Optimization of the geometry of silanol at the restricted Hartree–Fock (RHF) level with a series of basis sets of increasing quality was undertaken to determine the dependence of the structure on the basis set[21]. Starting with the STO-3G basis set the Si—OH bond length was calculated to be 1.686 Å. Further optimization using the double-zeta and triple-zeta split valence basis sets, both with and without polarization functions, reduces the bond length to a value of 1.631 Å. The histogram of the experimental (XRD) Si—OH bond lengths in silanols is given in Figure 16. The average bond length was calculated from 116 individual parameters to be 1.636 Å (s.d. 0.01 Å, s.m. 0.001 Å). The bond lengths for selected compounds are listed in Table 18. The Si—O bond lengths in silanols are only slightly longer than the average bond length found for alkoxysilanes (1.629 Å).

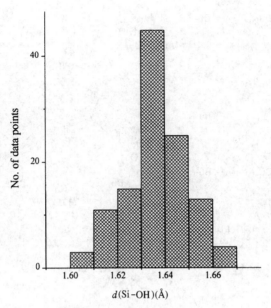

FIGURE 16. Histogram of Si—O bond lengths in silanols

TABLE 18. Geometry of Si—O bond lengths in silanoles

Refcode	Compound	d(Si—O) (Å)	Reference
JODYAC[a]	**177**	1.637	186
SELFIY	**177**	1.632	187
SITKEL[a]	**177**	1.645	188
VUFKEM[a]	**178**	1.627	189
JODYEG	**179**	1.636	186
VOZJOJ	**179**	1.638	190
VOZJUP	**179**	1.637	190
VOZKAW	**179**	1.642	190
VOZKEA	**179**	1.642	190
JODFAJ	**180**	1.646	191
KOMROT	**181**	1.631	192
KOSFUT	**182**	1.617	193
VAFWII[a]	**183**	1.632	194
WEKTUB	**184**	1.622	195
WEKTOV	**184**	1.627	195

[a] Average values are given.

3. Si—O bonds in disiloxanes

The geometry of disiloxane was a subject of many theoretical investigations. The molecular structure of $H_3Si-O-SiH_3$ was calculated at the 3-21G* level[19] and it was shown that the Si—O bond length is 1.645 Å and the Si—O—Si bond angle is 149.5°. Analyses of the Cambridge Structural Database and results of *ab initio* molecular orbital calculations[22] provide insights into the bond-angle widening at oxygen as compared to the alkyl ethers.

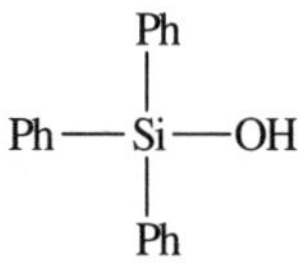

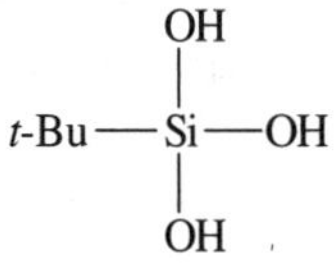

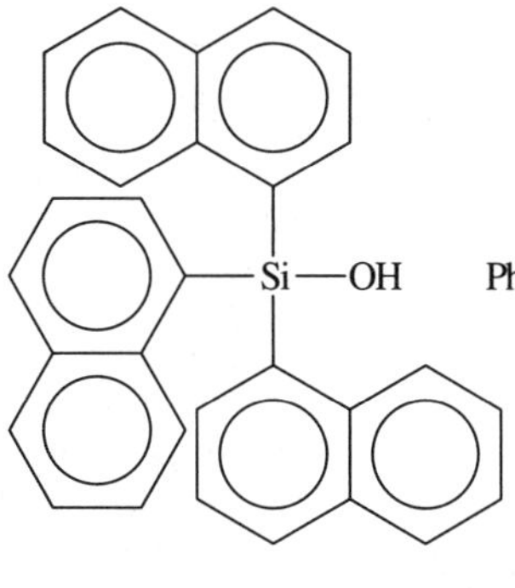

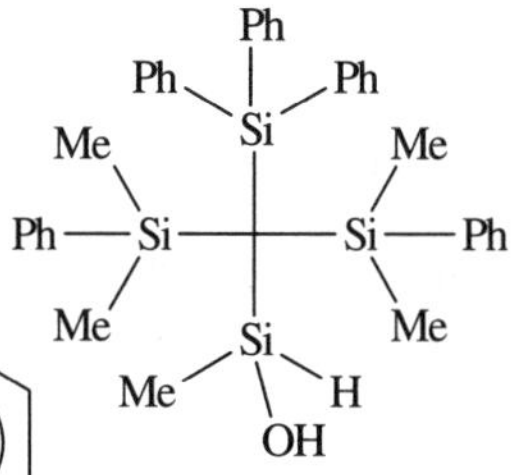

(177) **(178)** **(179)** **(180)**

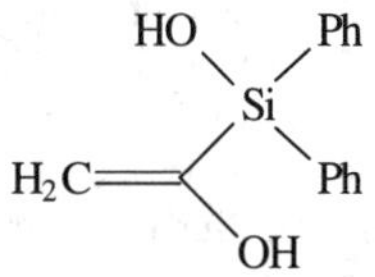

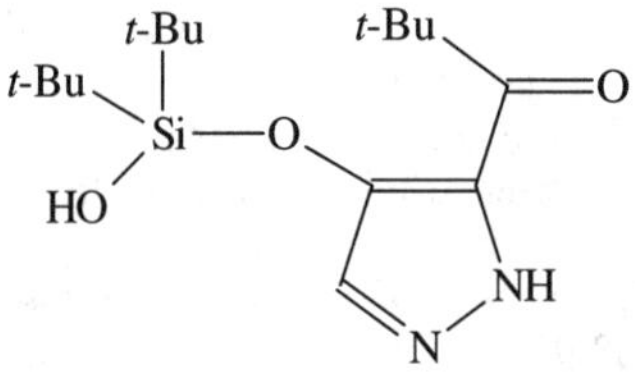

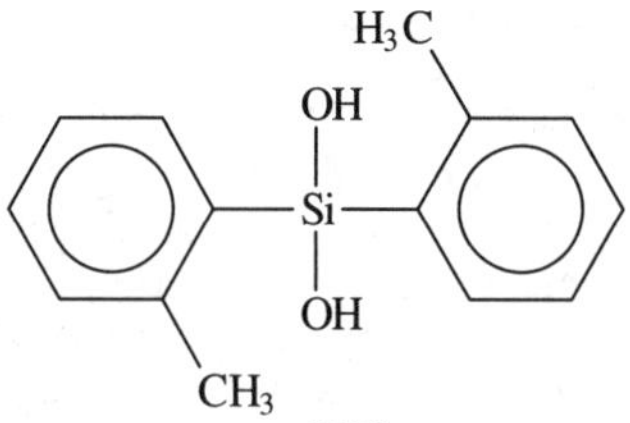

(181) **(182)** **(183)**

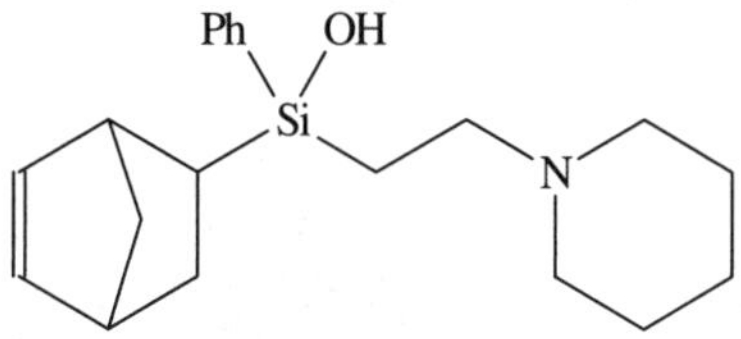

(184)

The XRD data of the acyclic Si−O−C(sp^3) fragments provided the basis for calculating the average bond angle of 134.2° and average Si−C bond distance of 1.62 Å. A trend of decreasing bond lengths with increasing bond angle was also shown. Optimization of the geometry of methoxysilane and disiloxane by calculation (at 6-31G) show that the Si−O−C and Si−O−Si bond angles are 125.0° and 170.1°, respectively, and the Si−O bond lengths are 1.640 and 1.626 Å, respectively. For both molecules the bond angles differ significantly from the experimental values. The calculated Si−O bond length in disiloxane is in very good agreement with the XRD experimental value (1.621 Å).

Optimization of the geometry of disiloxane at the restricted Hartree–Fock (RHF) level with a series of basis sets of increasing quality was conducted to determine the dependence of the structure on the basis set[21]. Starting with the STO-3G basis set, the Si−O bond length and Si−O−Si bond angle were calculated to be 1.658 Å and 124.0°; further optimization using the double-zeta and triple-zeta split valence basis sets, both with and without polarization functions, reduces the bond length and angle to values of 1.620 Å and 147.4°. The bond length is in very good agreement with the XRD experimental average (1.621 Å). No comparison is made with the XRD experimental bond angle because of the large scattering of the data (for more details see below). Linear disiloxanes and cyclosiloxanes of the general formula (R^{1}R^{2}Si)$_2$O and (R^{1}R^{2}SiO)$_n$ were the subject of intensive XRD structural investigations. Karle and coworkers[196] found that the Si−O bond length in linear disiloxanes tends to shorten upon opening of the Si−O−Si bond angle. A linear correlation between the two parameters was found by applying a least-squares procedure to eight disiloxanes. XRD of such compounds provide more than 140 individual values for the Si−O−Si bond angles in disiloxanes. The trend of shortening the bond upon linearization of the bond angle can still be seen, but it is not as clear as that shown by Karle and coworkers. The average bond length for disiloxanes was calculated from 147 bond lengths to be 1.621 Å (s.d. 0.02 Å, s.m. 0.002 Å).

The average Si−O bond length in cyclosiloxanes was calculated from 161 individual values to be 1.618 Å. Both values are not significantly different from the average Si−O bond length calculated for the whole set of data mentioned above. In his review, Sheldrick[15] summarizes briefly the development in the study of structural chemistry since an earlier review[197] on organocyclosiloxanes published in 1980. The main structural features that are mentioned concern the dependence of the Si−O bond length and Si−O−Si bond angle on ring size. We have used geometric data selected on the basis of simplicity of the ring system, with $n = 3, 4, 6, 7, 8$. It is interesting to note that while the scattering of the O−Si−O XRD experimental bond angles is very small (105-114°, see also the histogram in Figure 17), the scattering of the Si−O−Si bond angle is very large (129-177°). It seems that the barrier to linearization of the Si−O−Si bond may be reduced by the attachment of different substituents on silicon, and therefore linearization of various degrees are detected. The average of the O−Si−O bond angle was calculated for 161 individual values to be 108.9° (s.d. 1.7°, s.m. 0.1°). Since the O−Si−O bond angle remains unchanged upon changing the ring size, the Si−O−Si bond angle must be varied with the ring size. The average Si−O−Si bond angles for $n = 3, 4, 6, 7$ and

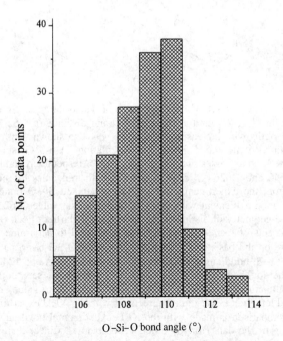

FIGURE 17. Histogram of O−Si−O bond angles in selected cyclosiloxanes

8 are 132.8, 148.9, 152.9, 164.9 and 153.3°, respectively. It should be emphasized that although it seems that the trend is real, a direct proof for it and for the correlation will be available only when cyclosiloxanes of different ring sizes with identical substituents are prepared and analysed structurally.

4. Si—O bonds in compounds with pentacoordinate silicon and dicoordinate oxygen atoms

The pentacoordinate silicon atom appears either as a neutral complex with Lewis base or as an anion. The average Si—O bond length for this coordination was calculated from 481 individual values to be 1.699 Å (s.d. 0.05 Å, s.m. 0.002 Å). The histogram is shown in Figure 18 and a list of selected compounds with Si—O bond lengths is given in Table 19. Compounds **185–190** are examples of neutral Lewis base complexes with N → Si dative bonds. In these compounds the Si—O bond length range is 1.650–1.666 Å. Compounds **191–194** are representative of anionic compounds where the silicon is negatively charged, and therefore the Si—O bond length is significantly longer. Upon increasing the number of oxygen atoms attached to the same silicon atom, the Si—O bond length shortens as a result of increasing bond ionicity. Thus, when the silicon atom is bonded to two oxygen atoms as in **191**, the Si—O bond length is 1.834 Å; when two additional oxygen atoms are attached to silicon, such as in **192** and **193**, the Si—O bond length shortens to 1.752 and 1.738 Å, respectively. The bond length is further shortened to 1.716 Å when the silicon atom is attached to five oxygen atoms as in **194**.

Exceptionally long Si—O bond lengths are found in compounds **195–198**. The longer bond lengths are the dative O → Si bonds such as in **195** (2.040 Å). In **196–198** two

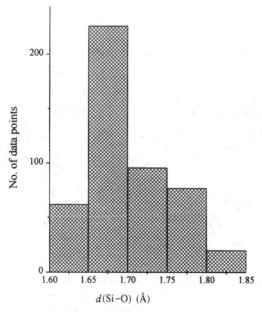

FIGURE 18. Histogram of Si—O bond lengths in compounds with pentacoordinate silicon and dicoordinate oxygen atoms

TABLE 19. Si—O bond lengths in compounds with pentacoordinate silicon and dicoordinate oxygen atoms

Refcode	Compound	d(Si—O) (Å)	Reference
MSILTR[a]	185	1.666	198
SITBEC[a]	186	1.663	199
TALVUX[a]	187	1.650	200
VIFVAH[a]	188	1.657	201
WABDIM[a]	189	1.661	202
WABDIM[a]	190	1.661	202
KELSAV[a]	191	1.834	203
KELSEZ[a]	192	1.752	203
KUXPIC[a]	193	1.738	204
YEVWOL[a]	194	1.716	205
PENDOB	195	2.050	206
PENDUH	196	2.242	206
WENBOG	197	1.778	207
WENBOG	197	2.227	207
WENCAT	198	1.787	207
WENCAT	198	2.077	207

[a]Average values are given.

types, i.e. covalent and dative bonds, are found and the distinction between the two is easily conducted on the basis of the differences in the Si—O bond length. In **197** the covalent Si—O bond length is 1.778 Å and the dative O → Si bond is 2.227 Å. In **198** these bonds are 1.787 and 2.077 Å respectively.

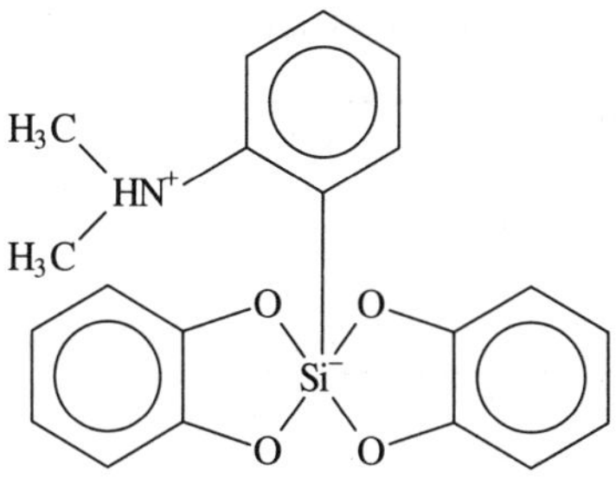

(193) 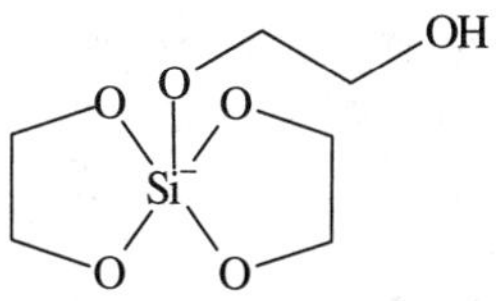 **(194)**

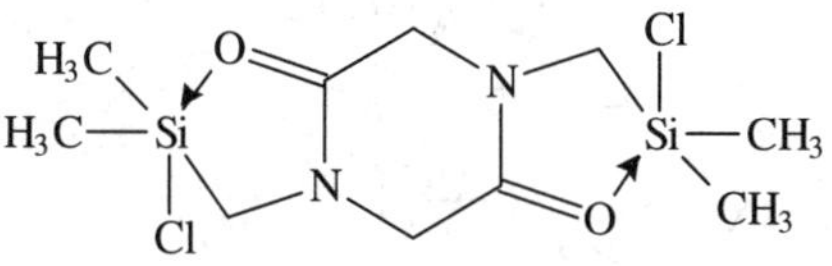

(195) 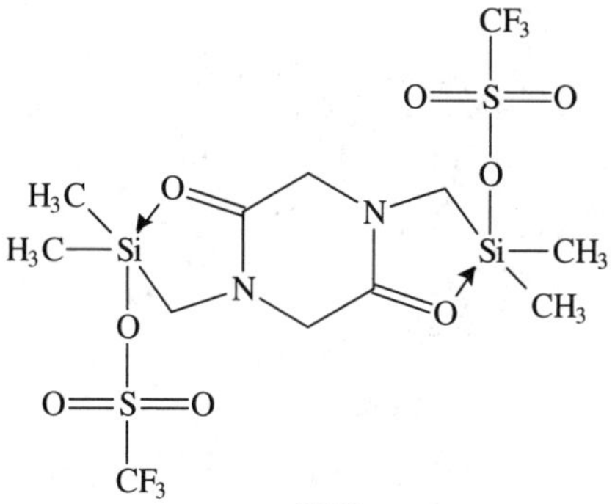 **(196)**

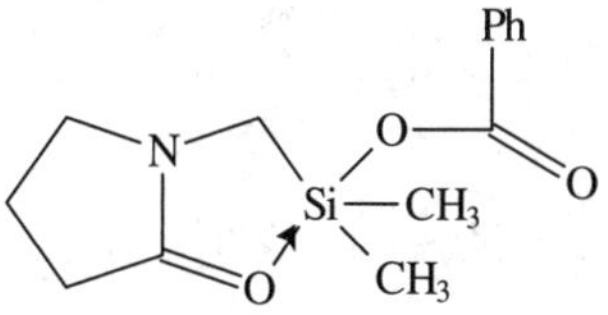

(197) 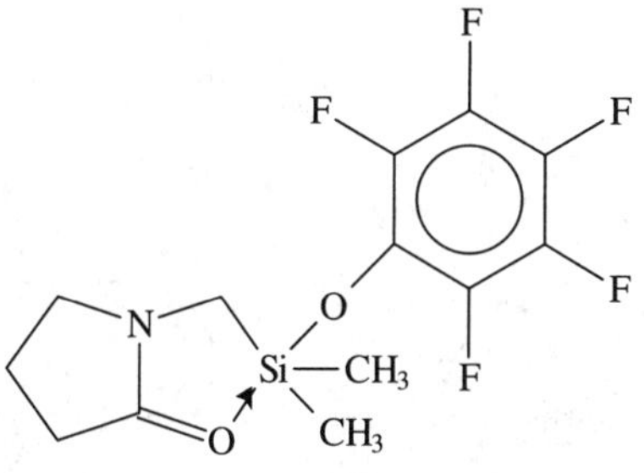 **(198)**

5. Si–O bonds in compounds with hexacoordinate silicon and dicoordinate oxygen atoms

Three types of compounds with hexacoordinate silicon atom bonded to dicoordinate oxygen atom are found; neutral Lewis acid–base complexes, such as **199**, compounds with positively charged silicon atom, such as **200**, and compounds with negatively charged silicon atoms, such as **201** and **202**. A list of Si–O bond lengths in these compounds is given in Table 20. It is clear that the Si–O bond length increases as one goes from positively charged silicon **200** to negatively charged silicon **201** and **202**, in accordance with the increase of ionicity of the bond.

TABLE 20. Si—O bond lengths in compounds with tetracoordinate silicon and hexacoordinate oxygen atoms

Refcode	Compound	d(Si—O) (Å)	Reference
YOMCEA[a]	199	1.724	208
BIPSTI[a]	200	1.643	209
PYPHSI[a]	201	1.784	210
JUVJEP[a]	202	1.785	211

[a] Average values are given.

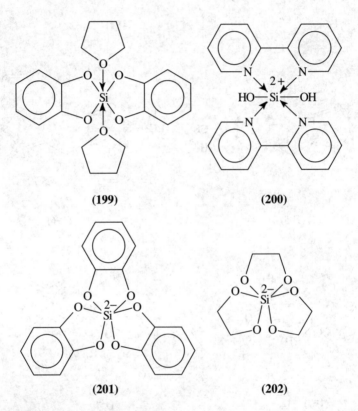

(199) (200)

(201) (202)

F. Structural Chemistry of the Si—S Bond

The sum of the covalent radii[3] for silicon and sulphur is 2.21 Å. The electronegativity of the sulphur atom is smaller than that of its congener oxygen atom and therefore the expected polarization and ionicity of the Si—S bond is smaller than that of the Si—O bond. As a result it is expected that the shortening of the Si—S bond length from the sum of covalent radii will be significantly smaller than that observed for the Si—O bond.

The Si—S—X bond angle is highly dependent on the HOMO-LUMO gap and it is expected that the barrier to linearization will be greater for sulphur than for oxygen.

The crystal structures of over 70 different compounds possessing Si—S bonds have been determined. These structures provide more than 90 bond lengths and bond angles. The complexity and diversity of the compounds do not allow fruitful discussion of all of them.

5. The structural chemistry of organosilicon compounds

We therefore discuss three different classes of compounds and also analyse statistically the whole set of data.

1. Si–S bonds in compounds containing the Si–S–Si group

The fragment Si–S–Si was found in compounds **203–216** and contributed 39 Si–S bond lengths and bond angles. The geometric parameters are summarized in Table 21. Statistical analysis in the form of a histogram is given in Figure 19. The range of bond lengths is 2.07–2.19 Å with an average of 2.141 Å (s.d. 0.02 Å, s.m. 0.003 Å). We shall not compare and discuss Si–S–Si bond angles here because in some cases the fragment is confined within a cyclic compound, and therefore the Si–S–Si bond angles are determined by the ring size. We discuss these bond angles below.

2. Si–S bonds in compounds containing the (t-BuO)$_3$–Si–S–X group

Compounds containing the $(t\text{-BuO})_3$–Si–S–X group may be regarded as a special group because they clearly show shortening of the Si–S bond length. The geometric data are given in Table 22. The average Si–S bond length is 2.100 Å (s.d. 0.03 Å, s.m. 0.01 Å). The remarkable shortening of the bonds is mainly observed for compounds where X is a metal such as Cu, Cd, V, Co as well as Pb and Tl[15]. Shortening of the Si–S bond length may be attributed to the increase of ionicity and polarization of the bond caused by the electron-withdrawing substituents at the silicon atom and the presence of a metal bonded to the sulphur atom.

3. Si–S bonds in compounds containing the (Ph)$_3$–Si–S–X group

There are only a few compounds containing triphenylsilanethiol derivatives of type **224a–e** that can be used to examine the effect of substitution at the sulphur atom on the Si–S–X geometry. It was found that the Si–S bond lengths range between 2.138 Å (for **224a**) to 2.161 Å (for **224e**) in accordance with the increase of the electronegative properties of the substituent on sulphur.

TABLE 21. Geometric parameters in compounds containing the Si–S–Si fragment

Refcode	Compound	d(Si–S) (Å)	Si–S–Si (°)	Reference
CADLOI	**203**	2.152	82.4	212
CUKFUJ	**204**	2.131	82.1	213
DEBYAK[a]	**205**	2.151	111.9	214
DIXJUP	**206**	2.169	94.9	215
DOBCEC[a]	**207**	2.118	110.5	216
FAPYAW[a]	**208**	2.148	104.0	217
FEKGAD[a]	**209**	2.147	108.8	218
FIGJIO[a]	**210**	2.139	105.4	219
FIGJOU[a]	**211**	2.109	108.0	219
FIGJUA[a]	**212**	2.136	107.7	219
GINCEL[a]	**213**	2.148	108.7	220
JITKIG[a]	**214**	2.172	82.9	221
MSISUL[a]	**215**	2.130	104.6	222
TMTPSS[a]	**216**	2.143	106.2	223

[a]Average values are given.

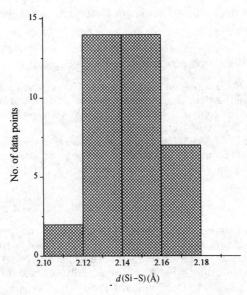

FIGURE 19. Histogram of Si—S bond lengths in compounds containing Si—S—Si fragments

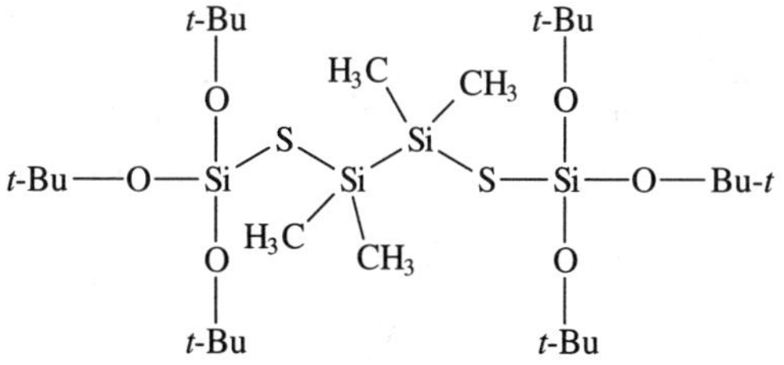

5. The structural chemistry of organosilicon compounds

(210)

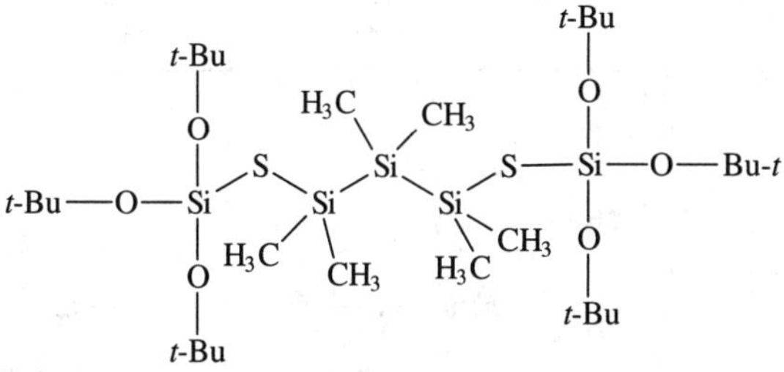

(211)

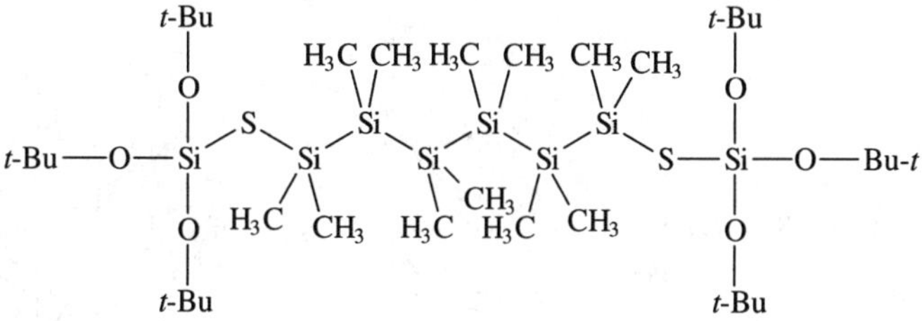

(212)

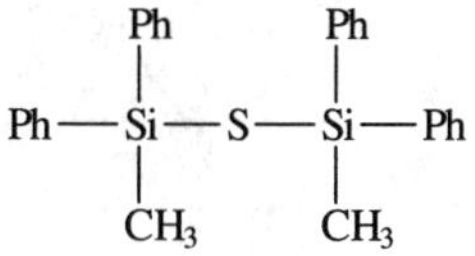

(213)

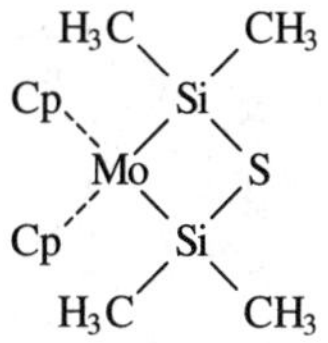

(214)

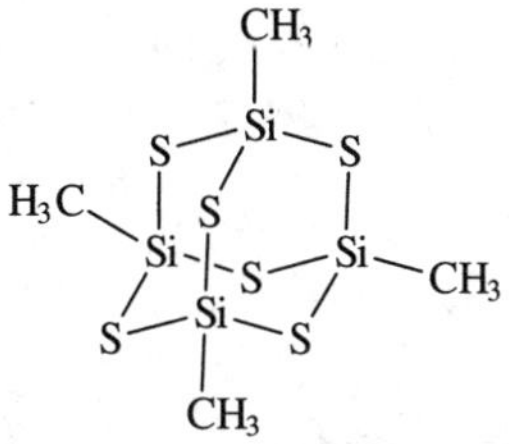

(215)

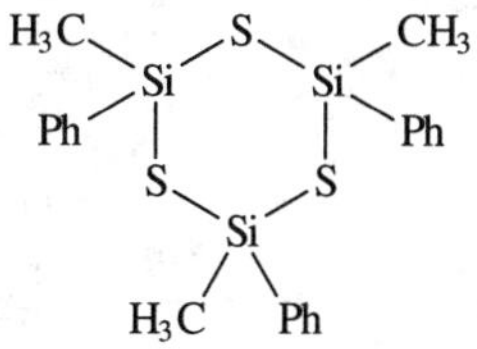

(216)

TABLE 22. Geometric parameters in compounds containing the $(t\text{-BuO})_3-\text{Si}-\text{S}-\text{X}$ fragment

Refcode	Compound	$d(\text{Si}-\text{S})$ (Å)	Si–S–Si (°)	Reference
GAJBAU	217	2.111	110.9	224
KACBIZ[a]	218	2.131	100.8	225
KIJLUK[a]	175	2.132	98.3	184
LEBZIB	219	2.075	103.7	226
PAFZUR	220	2.085	102.1	227
PAGBAA[a]	221	2.075	91.7	228
POHXOZ	176	2.169	101.9	185
VIHPIL[a]	222	2.058	82.8	229
YOYNIJ	223	2.068	85.2	230

[a] Average values are given.

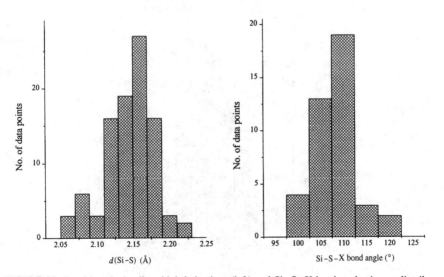

4. Si–S bonds in compounds containing the Si–S–X group

The average Si–S bond length obtained by statistical analysis of 95 data points is 2.144 Å (s.d. 0.03 Å, s.m. 0.004 Å). The results are shown in Figure 20. The average of 41 Si–S–X bond angles in acyclic derivatives is 108.3° (s.d. 4.7°, s.m. 0.7°) and the histogram is also shown in Figure 20.

G. Structural Chemistry of the Si–Halogen Bond (Si–F, Cl, Br, I)

Most of the available X-ray experimental geometry data for compounds containing silicon–halogen bonds are those with fluorine and chlorine atoms. The X-ray crystal structure of only a few compounds containing Si–Br and Si–I bonds are known.

FIGURE 20. Bond lengths in silanethiol derivatives (left) and Si–S–X bond angles in acyclic silanethiol derivatives (right)

1. Si−F bond in compounds with tetracoordinate silicon

The geometries of fluorosilanes were fully optimized[23] at the Hartree−Fock level with the 3-21G and 6-31G* basis sets using analytical gradient methods. The latter basis set was shown to provide better agreement with experimental geometry. The calculated Si−F bond length in SiH_3F is 1.5941 Å, and 1.605 Å when the 6-31++G** basis set is being used for geometry optimization[24]. The Si−F bond length becomes shorter on increasing substitution by fluorine atoms to 1.5570 Å in SiF_4. The sum of the covalent radii for silicon and fluorine atoms is 1.69 Å, which is significantly longer than the experimental value. Sheldrick in his review[15] pointed out that the addition of fluorine atoms on the same silicon atom causes a progressive shortening of the Si−F bond length. The experimental data obtained by microwave spectroscopy shows that the Si−F bond length is 1.593 Å in SiH_3F[234], 1.577 Å in SiH_2F_2[235], 1.562 Å in $SiHF_3$[236] and 1.556 Å in SiF_4[237] (obtained by electron diffraction). The average Si−F bond length was calculated using 168 values obtained by X-ray crystal structures with no distinction between the different numbers of fluorine atoms attached to the same silicon atom. The results are shown in Figure 21. The average Si−F bond length is 1.594 Å (s.d. 0.02, s.m. 0.002 Å). Bond lengths for typical compounds are given in Table 23. The shortening of the bond caused by the increase in the number of fluorine substituents on the same silicon atom can be seen by comparing the Si−F bond length in **226** (1.607 Å), where a single fluorine atom is attached to silicon, with that in **225, 227** and **228** (1.590, 1.587 and 1.581 Å, respectively), where two fluorine atoms are bonded to the same silicon atom.

Exceptionally long Si−F bond lengths in the tetracoordinate silicon atom are found in compounds such as **232**[241] (1.656 Å), **233**[242] (1.684 Å on the average) and **234**[243] (1.667 Å). The first two are compounds having bulky substituents and the lengthening might be attributed to a steric effect. The third is an example of compounds where the Si−F bond ionicity decreases due to the negatively charged nitrogen atom attached to the same silicon atom.

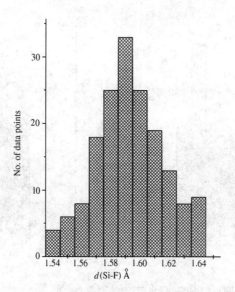

FIGURE 21. Histogram of Si−F bond lengths in compounds with a tetracoordinate silicon atom

5. The structural chemistry of organosilicon compounds

TABLE 23. Si—F bond lengths in compounds with a tetracoordinate silicon atom

Refcode	Compound	d(Si—F) (Å)	Reference
JESLOI[a]	**225**	1.586	238
JESLUO	**226**	1.607	238
JESMAV[a]	**227**	1.593	238
JESMEZ[a]	**228**	1.584	238
JIXXIX	**229**	1.601	239
JIYNUA[a]	**230**	1.575	240
JIYPAI[a]	**231**	1.599	240

[a] Average values are given.

2. Si—F bond in compounds with pentacoordinate silicon

Three types of halogenosilanes (**235, 236** and **237**) with pentacoordinate silicon atom are considered. The first comprises anions of type **235**, the second is a neutral species containing a Lewis base (D) (**236**) and the last is a cation formed and stabilized by a chelate ligand of a Lewis base character (**237**).

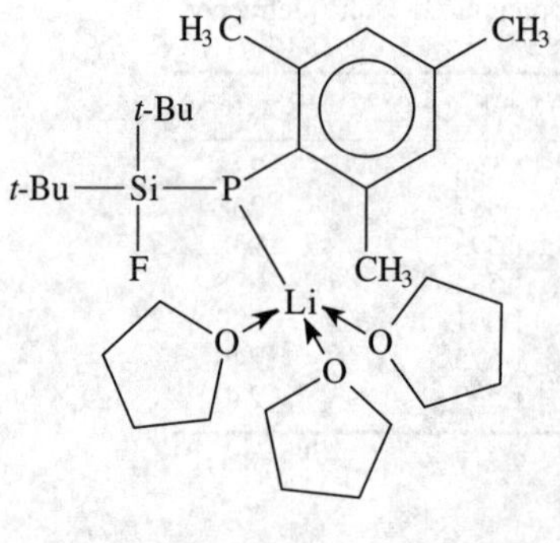

(232)

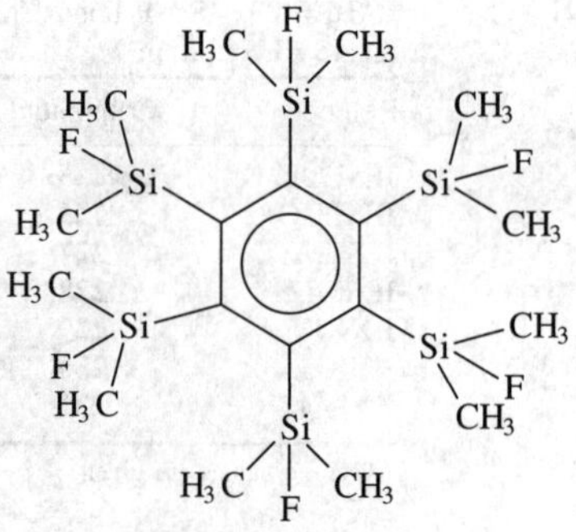

(233)

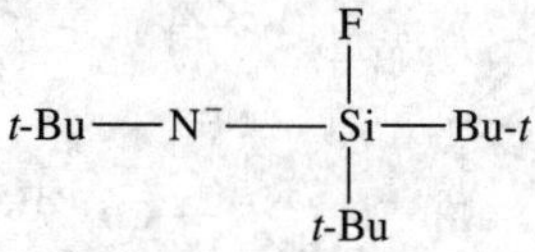

(234)

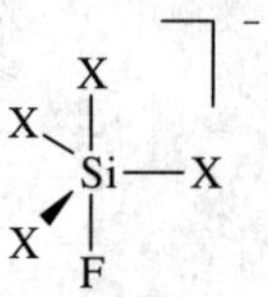

(235) 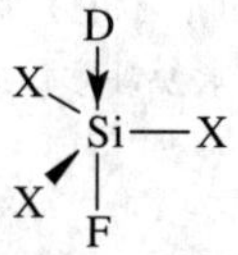 (236) 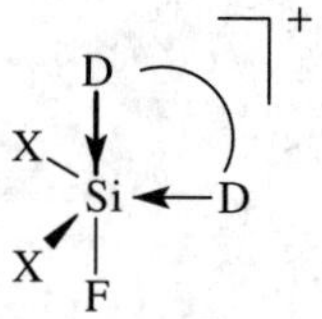 (237)

The average Si–F bond length in pentacoordinate silicon compounds was calculated from 153 XRD experimental bond lengths to be 1.631 Å (s.d. 0.04 Å, s.m. 0.004 Å) from 60 values obtained by X-ray crystal structure determination and its histogram is shown in Figure 22. Some typical values for compounds of type **236** are given in Table 24. All compounds **238–247** show much shorter Si–F bond lengths, resulting from the attachment of three fluorine atoms to the same silicon.

Si–F bond lengths in the anions of type **235** show significant lengthening, as can be seen in Table 25 for compounds **248–254**.

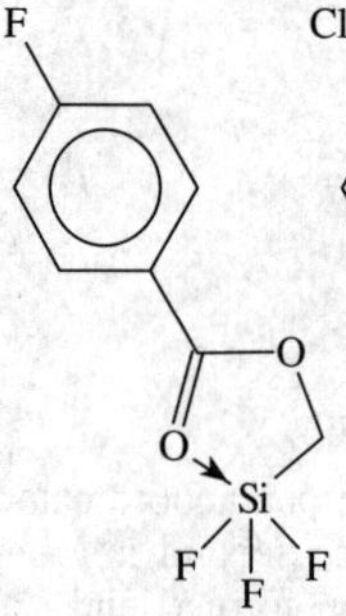

(238) 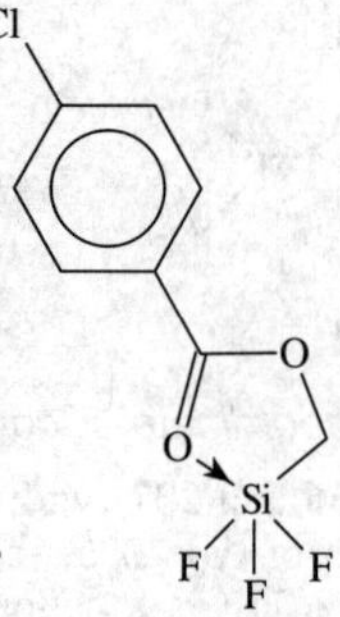 (239) 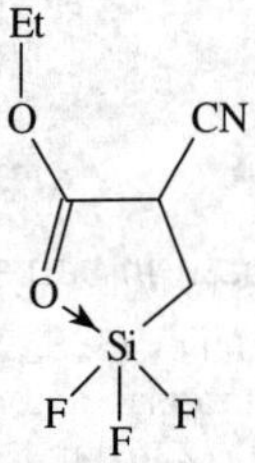 (240) 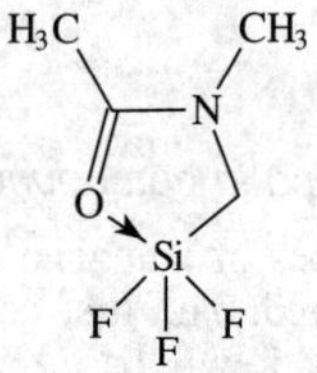 (241)

5. The structural chemistry of organosilicon compounds

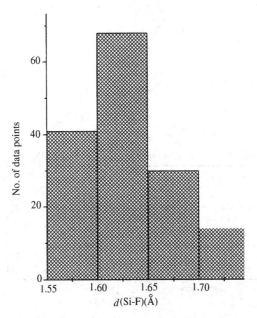

FIGURE 22. Histogram of Si–F bond lengths in compounds with a pentacoordinate silicon atom

TABLE 24. Si–F bond lengths in compounds with a pentacoordinate silicon atom

Refcode	Compound	d(Si–F) (Å)	Reference
BABZAF[a]	**238**	1.591	244
BABZEJ[a]	**239**	1.595	244
HEPCUA[a]	**240**	1.571	245
HEPDEL[a]	**241**	1.605	245
HEPDIP[a]	**242**	1.595	245
KEZROW[a]	**243**	1.591	246
VILJUV[a]	**244**	1.601	247
WEDKOF[a]	**245**	1.605	248
WEDKUL[a]	**246**	1.604	248
WIHYER[a]	**247**	1.591	249

[a] Average values are given.

(245) (246) (247)

TABLE 25. Si−F bond lengths in anions of type 235

Refcode	Compound	d(Si−F) (Å)	Reference
BEPCAA	248	1.670	250
FEDDUN	249	1.707	251
FESCOV[a]	250	1.671	252
FESCUB[a]	251	1.721	252
PERROT[a]	252	1.685	253
PERRUZ[a]	253	1.677	253
PILKEA[a]	254	1.757	251

[a]Average values are given.

(248) (249) (250) (251)

(252) (253) (254)

(255) (256) (257)

3. Si—F bond in compounds with hexacoordinate silicon

X-ray crystal structures of 24 compounds containing a hexacoordinate silicon atom bonded to a fluorine atom are known in the literature. Most of them contain the commonly used $[SiF_6]^{2-}$ anion **255** and similar anions with various numbers of fluorine atoms replaced by other substituents. The second type is the neutral species based on **237**. An average Si—F bond length of 1.677 Å (s.d. 0.02 Å, s.m. 0.004 Å) was calculated from 26 available bond lengths. The shortest Si—F bond length, within this family of compounds, was detected in **256**[254] (1.632 Å), and the longest was observed in **257**[255] (1.717 and 1.740 Å). Such a lengthening may be attributed to the involvement of fluorine in strong hydrogen bonding.

The average of all Si—F bond lengths for compounds with hexacoordinate silicon was calculated from 100 experimental values to be 1.675 Å (s.d. 0.02 Å, s.m. 0.002 Å) and the histogram is given in Figure 23.

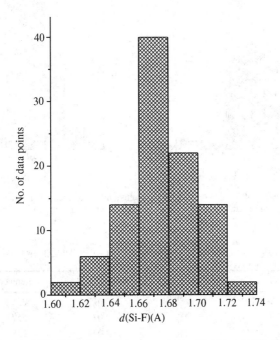

FIGURE 23. Histogram of Si—F bond lengths in compounds with a hexacoordinate silicon atom

4. Si—Cl bond in compounds with tetracoordinate silicon

The geometries of chlorosilanes were fully optimized[23] at the Hartree–Fock level with the 3-21G and 6-31G* basis sets using analytical gradient methods. The latter basis set was shown to provide better agreement with experimental geometry. The calculated Si—Cl bond length in SiH_3Cl is 2.0666 Å; the Si—Cl bond length becomes shorter when the number of chlorine atoms attached to Si is increasing such as in $SiCl_4$ (2.0290 Å). The sum of the covalent radii of chlorine and silicon is 2.05 Å. The X-ray crystal structure of compounds with a tetracoordinate silicon atom containing Si—Cl bonds provides 423 bond lengths that have been used for the statistical analysis. The average of the experimental Si—Cl bond lengths is 2.050 Å (s.d. 0.03 Å, s.m. 0.002 Å), in very good agreement with the sum of covalent radii. A histogram of the bond lengths is given in Figure 24. Bond lengths for the simplest representative compounds (**258–261**) are given in Table 26. As was pointed out for the Si—F bond lengths, Si—Cl bond lengths also become shorter upon increasing the number of chlorine atoms bonded to the same silicon atom. Therefore, the bond length in **260** is shorter than in the other representatives.

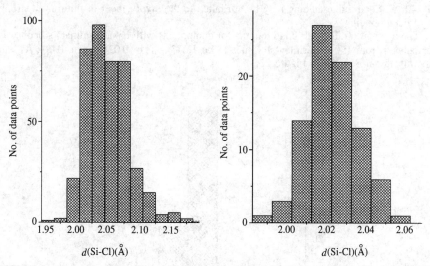

FIGURE 24. Histogram of Si—Cl bond lengths in compounds with a tetracoordinate silicon atom: (left) all 423 bond lengths are used, (right) data from compounds with multiple Si—Cl bonds

TABLE 26. Si—Cl bond lengths in compounds with a tetracoordinate silicon atom

Refcode	Compound	d(Si—Cl) (Å)	Reference
BARNUD[a]	**258**	2.077	256
COJKIV	**259**	2.075	257
JIRMAY[a]	**260**	2.020	258
LEFTEV	**261**	2.087	259

[a]The average value is given.

5. The structural chemistry of organosilicon compounds

TABLE 27. Si—Cl bond lengths in compounds with multiple Si—Cl bonds

Refcode	Compound	d(Si—Cl) (Å)	Reference
LAYBES[a]	262	2.018	260
LAYBIW[a]	263	2.023	260
LEGKEN[a]	264	2.018	261
LEGKIR[a]	265	2.001	261
WIHJAY[a]	266	2.032	262
WIHJEC[a]	267	2.043	262
WIHJIG[a]	268	2.044	262
WILHUJ	269	2.005	263
WILXEU[a]	270	2.016	263
WILXIY[a]	271	2.020	263

[a]The average value is given.

Statistics on Si—Cl bonds extracted from the X-ray crystal structure of compounds **262–271** that contain several groups of silicon atoms, each bonded to at least two chlorine atoms, show that the average Si—Cl bond length is 2.022 Å (s.d. 0.01 Å, s.m. 0.001 Å), indeed shorter than found in compounds with a single Si—Cl bond. Bond lengths for **262–271** are given in Table 27 and the statistics is shown in Figure 24.

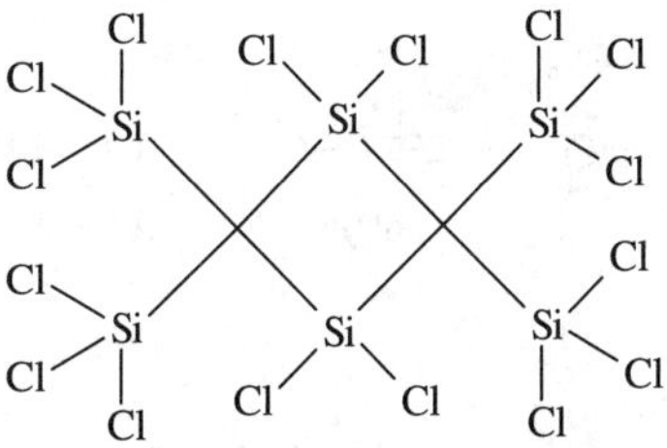

(262)

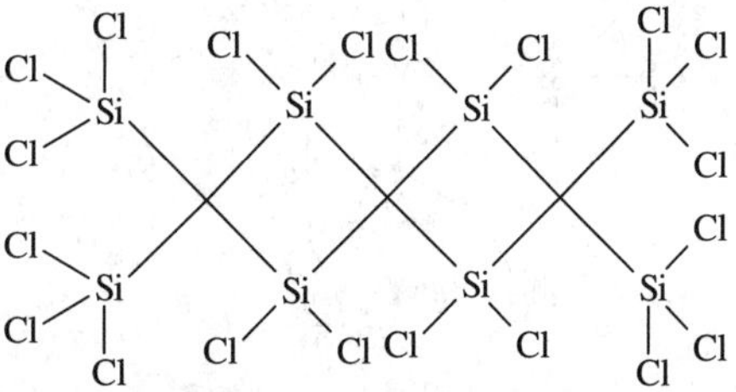

(263)

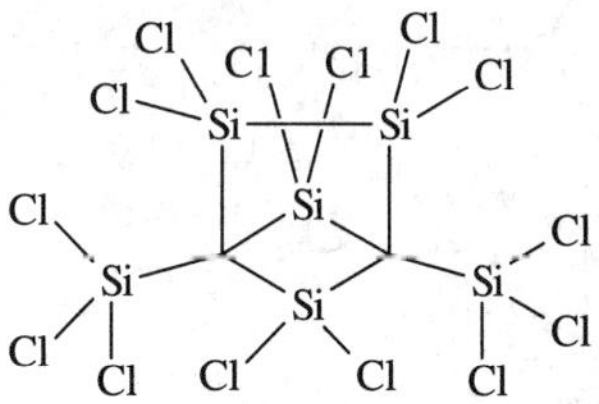

(264)

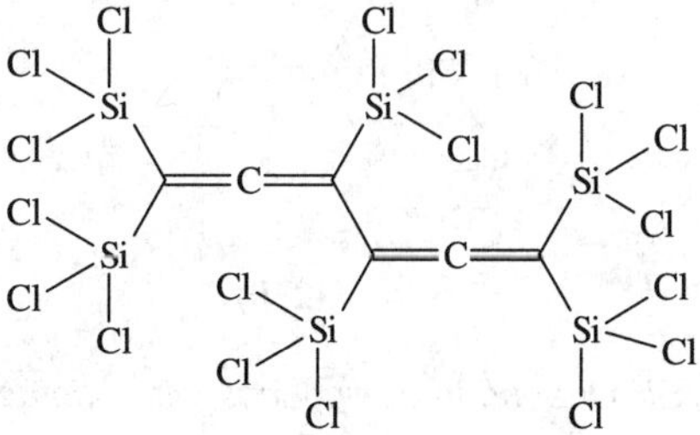

(265)

Longer Si—Cl bonds are found in compounds such as **272–275**[264–267], where the bond lengths are 2.137, 2.146, 2.124 and 2.112 Å, respectively.

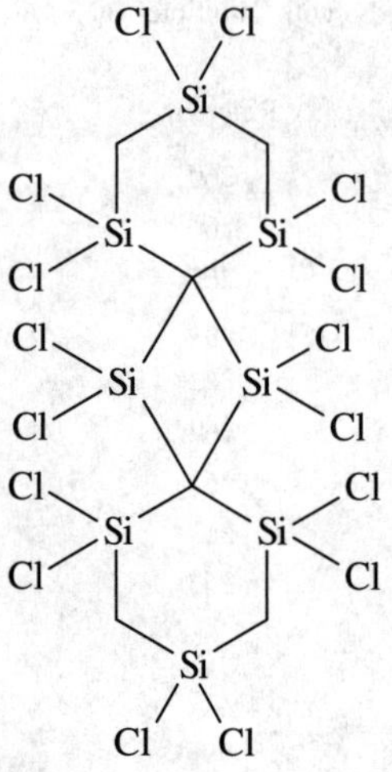

(266)

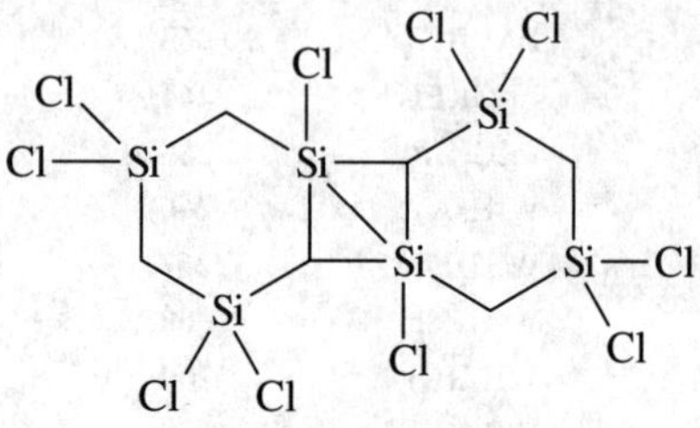

(267)

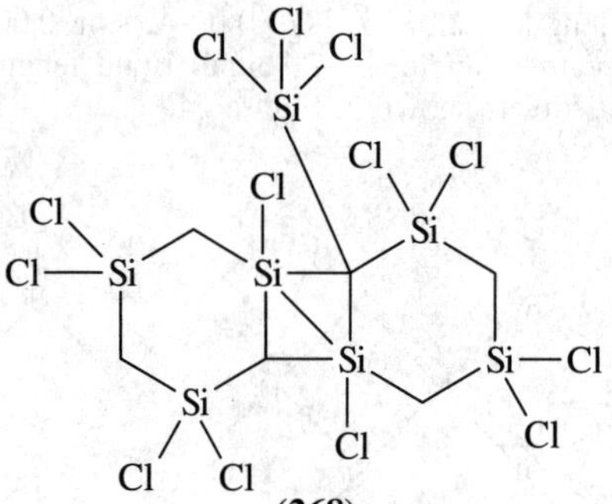

(268)

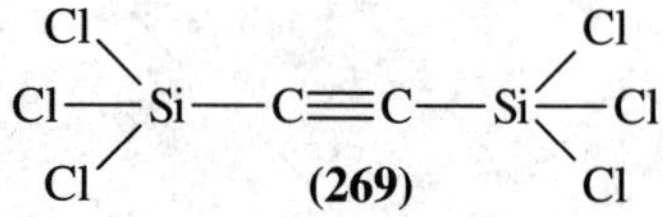

(269)

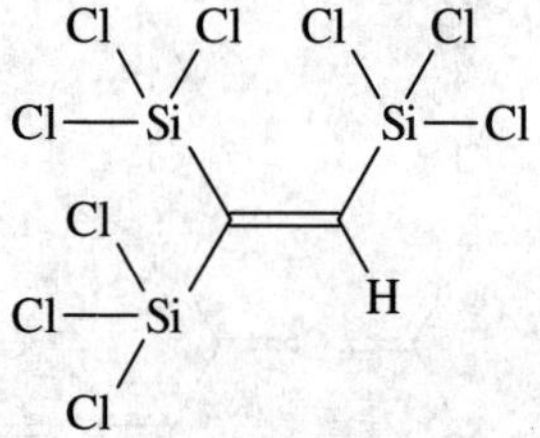

(270)

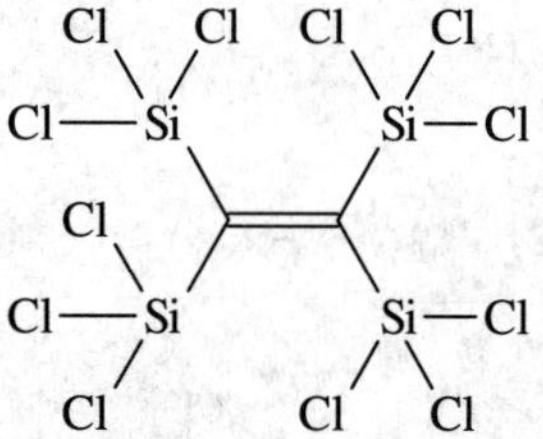

(271)

5. Si–Cl bond in compounds with pentacoordinate silicon

In contrast to fluorosilanes with pentacoordinate silicon systems such as **235–237**, it is very difficult to find similar compounds of chlorosilanes to establish the Si–Cl bond length in similar cases. X-ray crystal structure is available for only few of these compounds and only typical cases of the shortest and longest Si–Cl bonds are given. The shortest bond lengths are found in compounds where the silicon is bonded to a metal such as

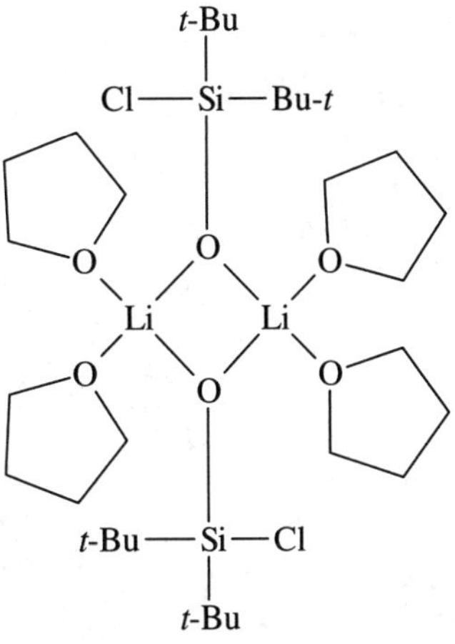

5. The structural chemistry of organosilicon compounds

(272)

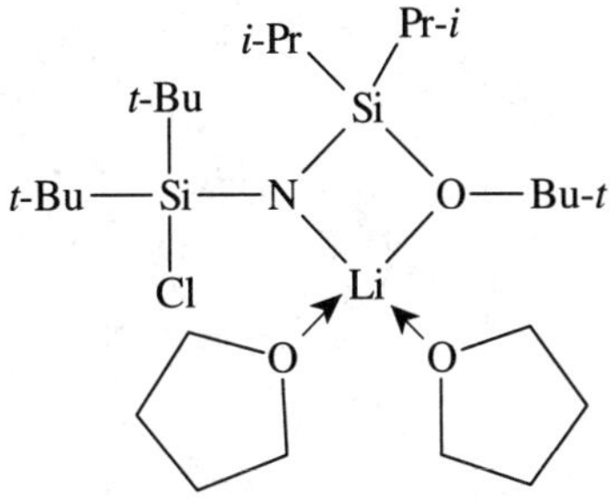

(273)

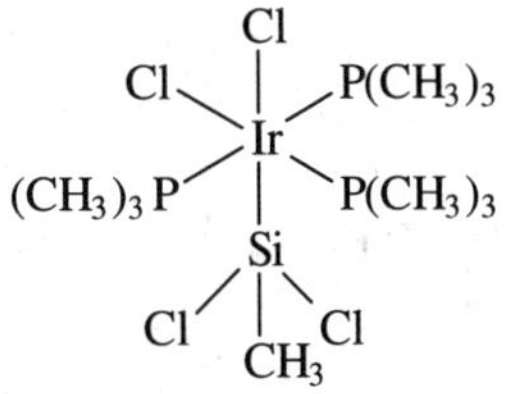

(274)

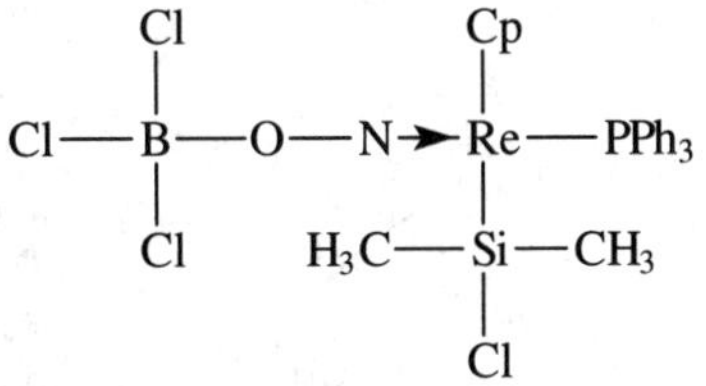

(275)

Mn **276**[268] or Cr **277**[269] with an average bond length of 2.062 and 2.072 Å, respectively. The longest bond lengths are found in compounds of the chloro analogue of **236**. On average, the Si—Cl bond length is 0.5 Å longer in **278–286** with the exception of **284** in which there are three chlorine atoms on the same silicon. The geometrical parameters for **278–286** are given in Table 28.

TABLE 28. Si—Cl bond lengths in compounds with pentacoordinate silicon atom

Refcode	Compound	d(Si–Cl) (Å)	Reference
FUPBOH	278	2.432	270
FUSYIB	279	2.307	271
GEGDAX	280	2.315	272
GEGDEB[a]	281	2.310	273
GILYIJ	282	2.624	274
VEKROS	283	2.609	275
VEKRUY	284	2.129	275
YAMSOU	285	2.678	276
YIJWOD	286	2.423	277

[a]The average value is given.

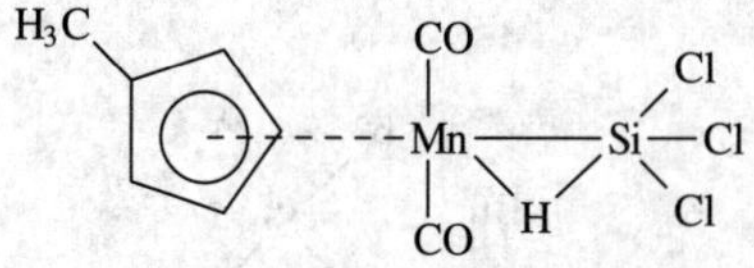

(276)

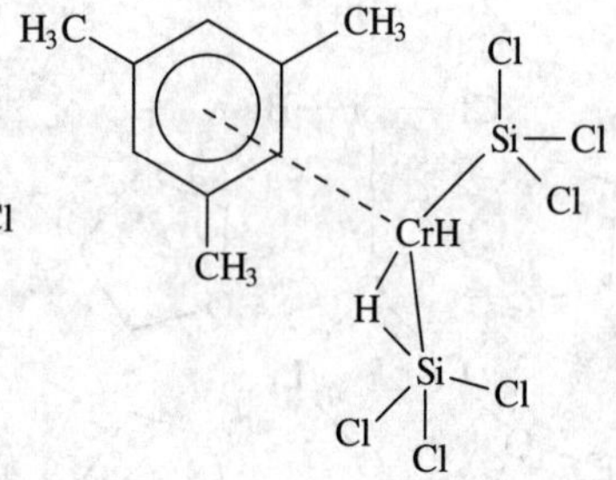

(277)

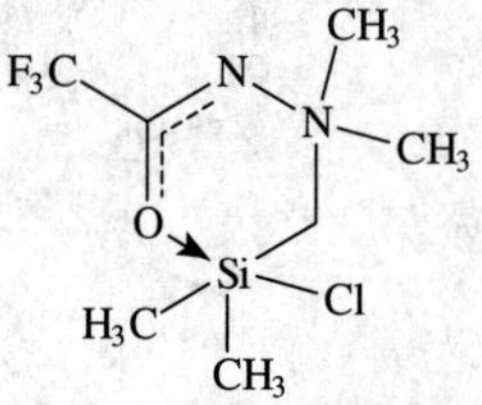

(278)

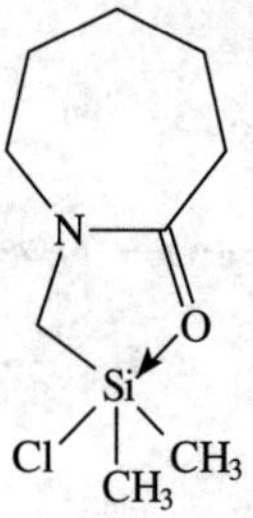

(279) 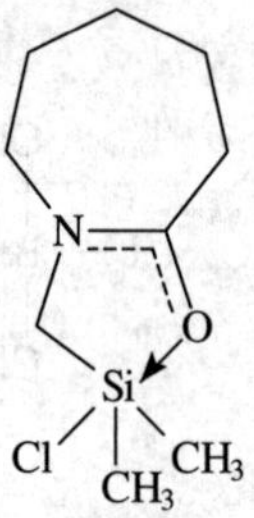 (280) 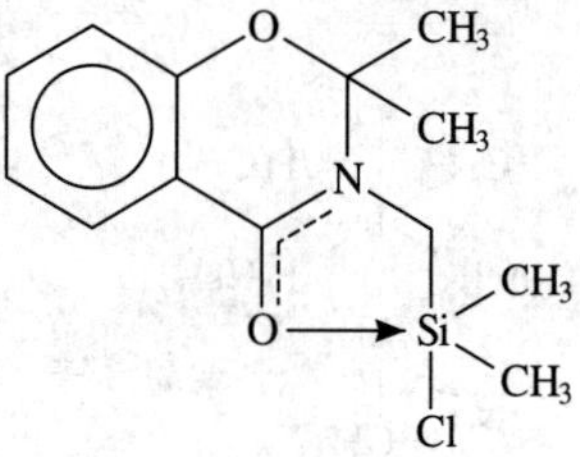 (281)

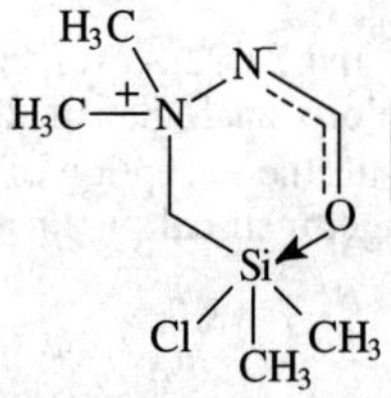

(282) 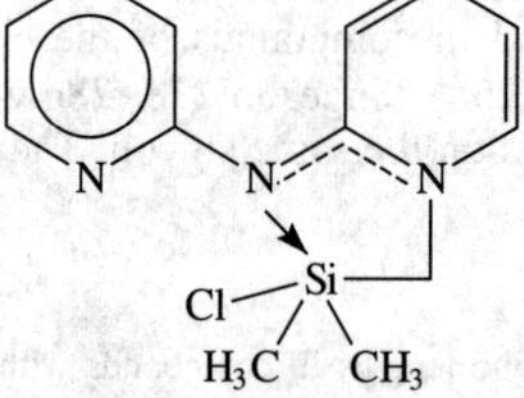 (283) 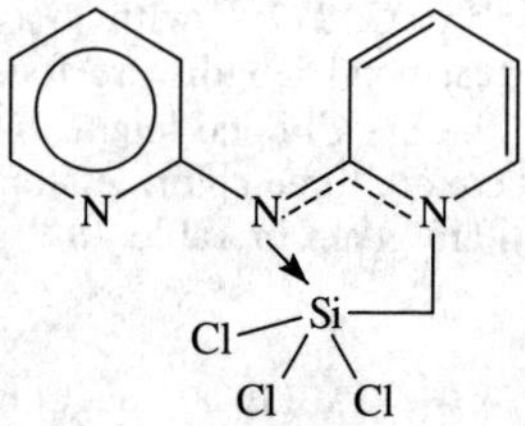 (284)

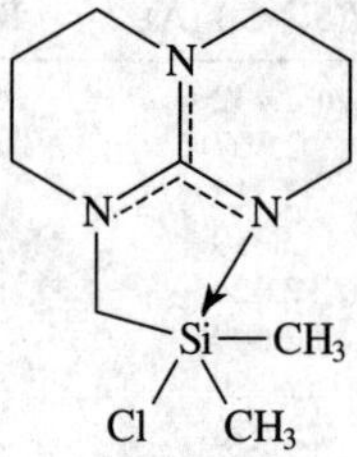

(285) 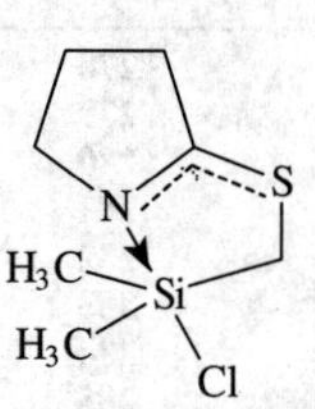 (286)

6. Si–Cl bond in compounds with hexacoordinate silicon

The average Si–Cl bond length in compounds with a hexacoordinate silicon atom was determined from 17 experimental values to be 2.184 Å (s.d. 0.02 Å, s.m. 0.006 Å). The results are shown in Figure 25 and Table 29.

7. Si–Br and Si–I bonds

The crystal structures of only three compounds containing Si–Br bonds and only two with Si–I bonds are known. Si–Br bond length in tetracoordinate silicon is 2.197 and 2.284 Å in **295**[284] and **296**[285], respectively. The bond is significantly longer (3.122 Å) in **297**[286] when the silicon is pentacoordinate (analogous to **235**). Si–I bond lengths are 2.527 and 2.574 Å in **298**[287], and 2.487 Å in **299**[288].

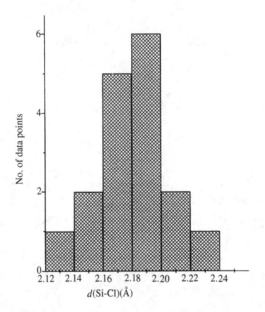

FIGURE 25. Si–Cl bond lengths in compounds with a hexacoordinate silicon atom

TABLE 29. Si–Cl bond lengths in compounds with a hexacoordinate silicon atom

Refcode	Compound	d(Si–Cl) (Å)	Reference
JEPCOW	**287**	2.184	278
CELFUU	**288**	2.198	279
FUMFAU[a]	**289**	2.169	280
SINBEW[a]	**290**	2.205	281
KENHUG[a]	**291**	2.196	282
KENJAO[a]	**292**	2.175	282
KENJES[a]	**293**	2.205	282
KUJFAW	**294**	2.140	283

[a]The average value is given.

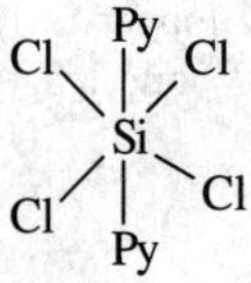

(287) 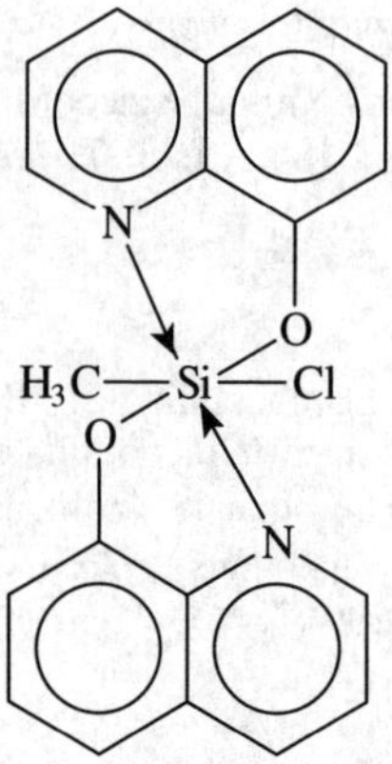 (288) 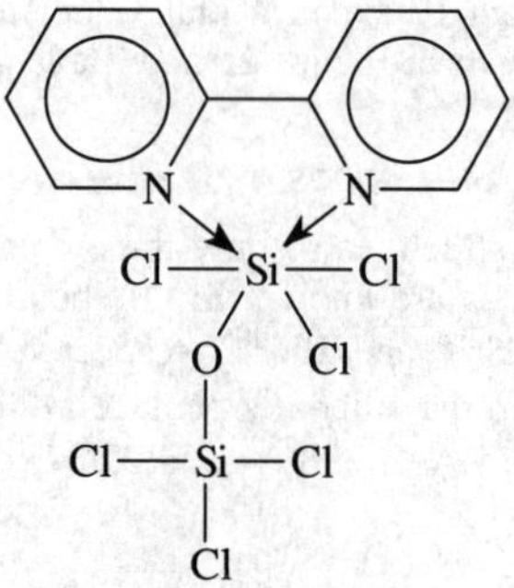 (289)

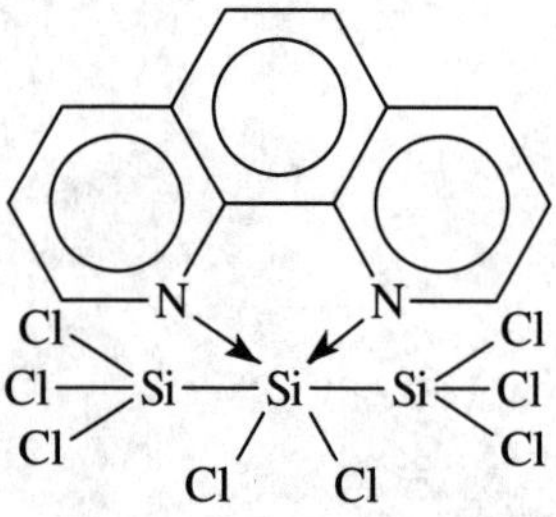

(290) 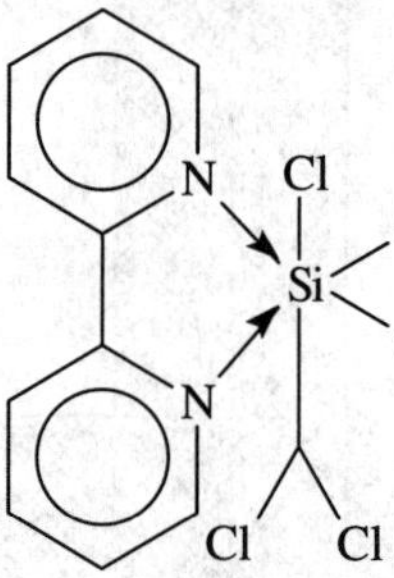 (291) 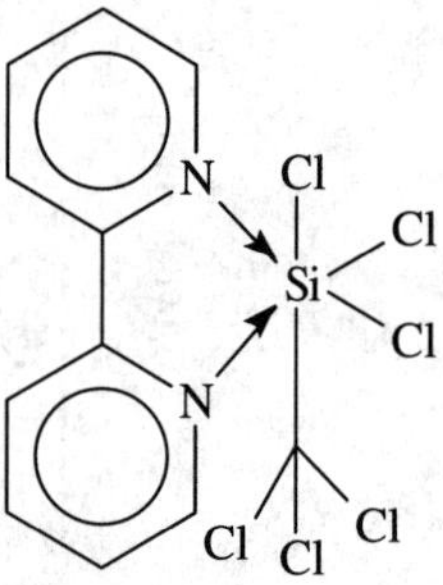 (292)

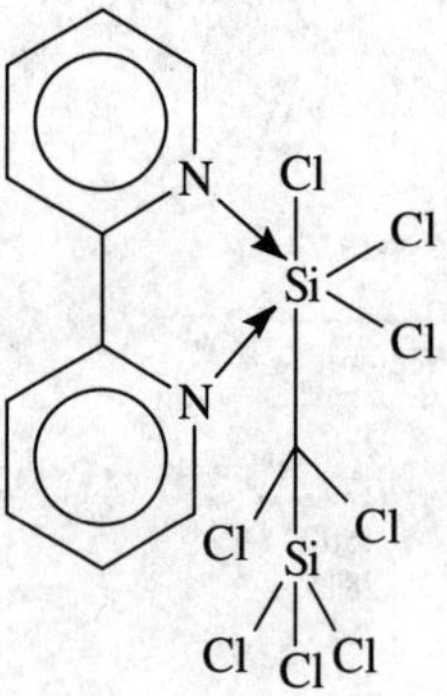

(293) 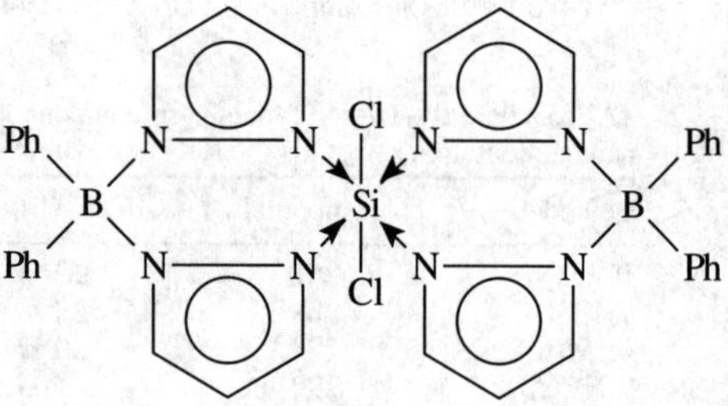 (294)

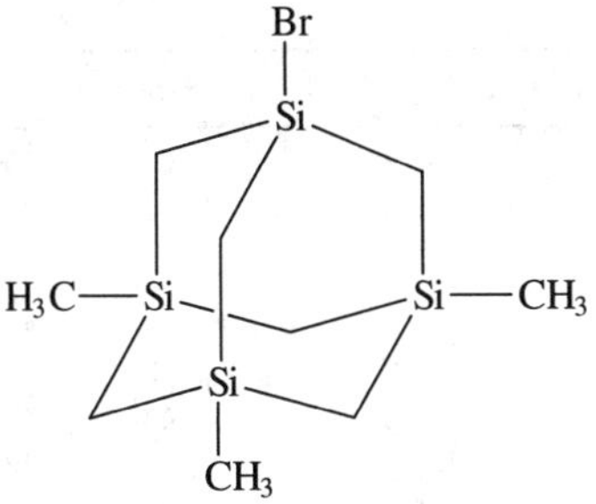

(295) 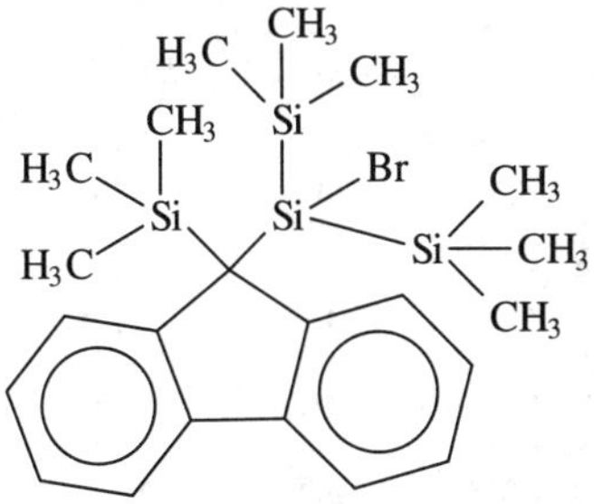   (296) 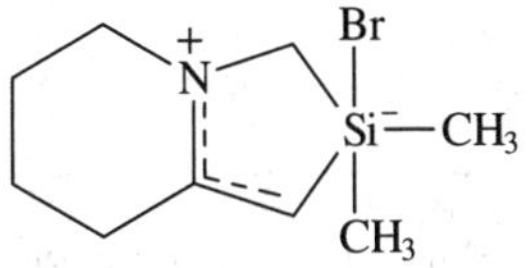   (297)

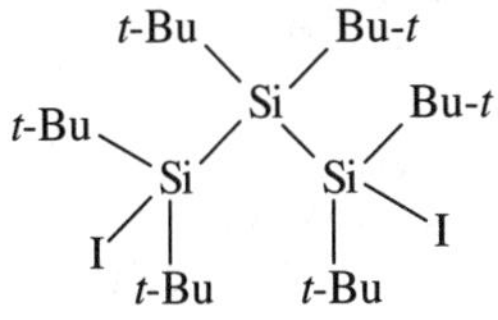

(298) 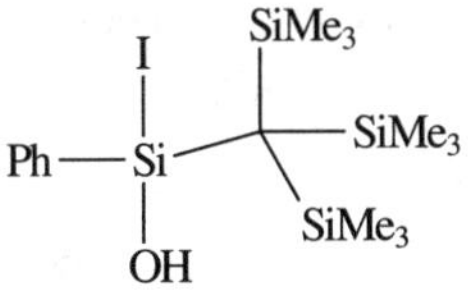   (299)

III. DOUBLE BONDS TO SILICON

A. Disilenes (>Si=Si<)

Stabilization of disilenes is achieved by the use of bulky substituents. Only a few disilenes have been prepared and X-ray diffraction studies were available for only three compounds when Sheldrick[15] published his review. A few other relevant reviews were published at the same time by West[289-291], Michl[292], Cowley[293,294] and their coworkers. There are crystallographic data for 14 crystal structures of disilenes in the CSD[16] updated on January 1996. Those 14 crystal structures provide geometric data for only 11 unique compounds. The most interesting geometric parameters are the bond lengths between the silicon atoms, the pyramidalization of the silicon atoms expressed by the sum of bond angles involving those atoms and the twist angle τ, signifying the bending at the double bond. The last two parameters are related to the steric congestion imposed by the substituents. The structural parameters are summarized in Table 30.

Disilenes are considered to be planar, but are very floppy in the sense that pyramidalization on silicon up to quite large angles (20°) lead to structures which are essentially equal in energy[295], within 4 or 8 kJ mol^{-1}. Raabe and Michl[292] suggested that such floppy molecules are likely to explore a large part of the nuclear configuration space at elevated temperatures, and since the $\pi\pi^*$ excitation energy is likely to decrease with pyramidalization and twisting, they can be expected to be thermochromic. Indeed, tetramesityldisilene[296] is bright orange at room temperature and pale yellow at $-100\,°C$. The bond length between the silicon atoms is calculated[297] to be 2.125 Å,

TABLE 30. Structural parameters for disilenes

Refcode	Compound	d(Si=Si) (Å)	$\Sigma^a(Si_1)$ (°)^a	$\Sigma(Si_2)$ (°)	τ (°)^b	Sym.^{g}	Reference
BUYYOJ^{c,d}	300	2.160	356.3	356.3	6.5	Ci	291
CELSER	301	2.140	360.0	360.0	9.6	Ci	298
CIJRAO	302	2.143	359.9	359.9	0.0	Ci	291
GICSAM	303	2.145	360.1	359.9	1.6	none	299
JAXSAC	304	2.138	359.9	359.9	0.0	Ci	300
JUMSEP	305	2.153	360.0	360.0	0.0	Ci	301
JUMSIT	306	2.156	360.0	360.0	0.0	Ci	301
SESSUE^{c,e}	300	2.142	358.7	358.4	3.1	none	302
SESSUE01^{e,f}	300	2.142	358.4	358.0	2.9	none	302
WICVEJ	307	2.201	360.0	360.0	8.9	Ci	303
WICVAF	308	2.226	359.7	359.7	0.0	Ci	303
WICVIN	309	2.252	358.9	358.9	0.0	Ci	303
HATDOV	310	2.229	358.8	357.3	8.7	none	304

$^a\Sigma$(Si) = sum of bond angle at Si.
^{b}The average torsion angle in the general structure **311** (R^2SiSiR^3 and R^1SiSiR^4).
^{c}Structural data from tetramesityldisilene obtained from different crystals.
^{d}The crystal contains toluene.
^{e}Crystal modification without solvent; data collected at room temperature.
^{f}The same crystal modification as in footnote e; data collected at $-100\,°$C.
^{g}Ci = inversion centre.

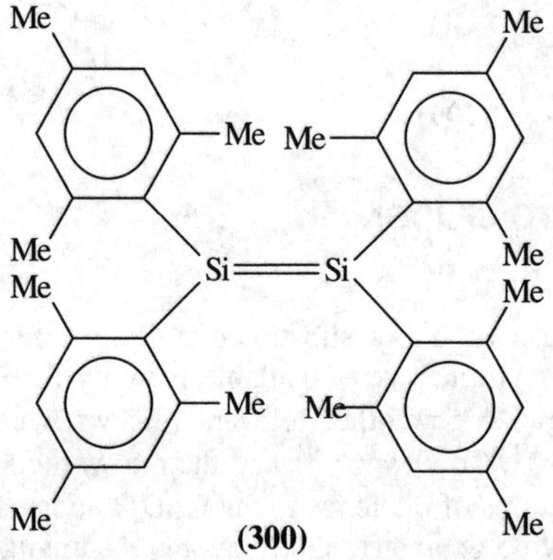

(300)

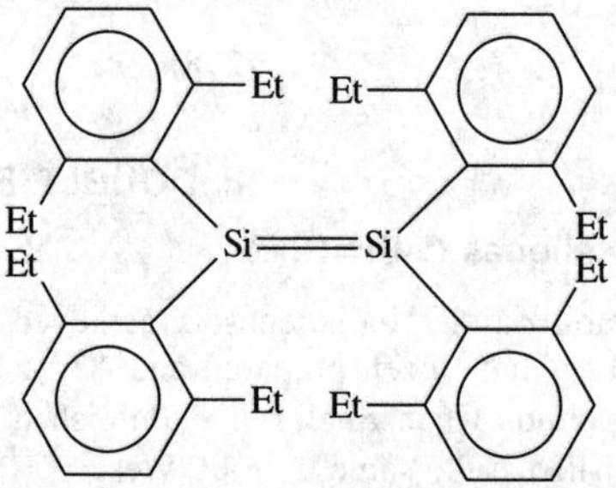

(301)

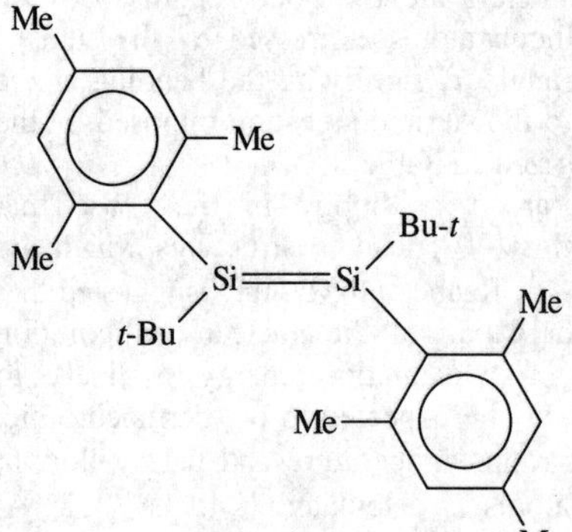

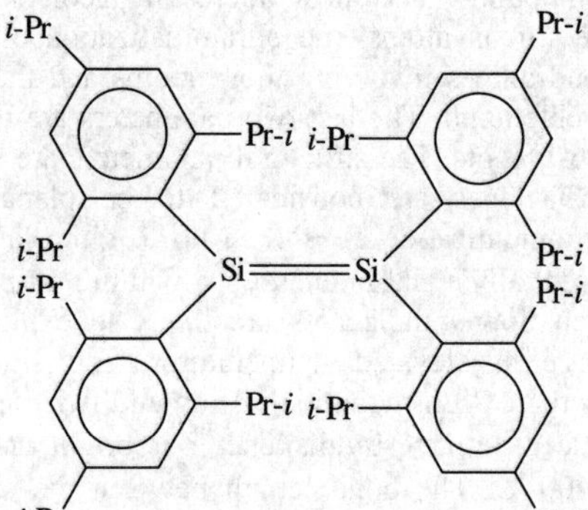

(302)

(303)

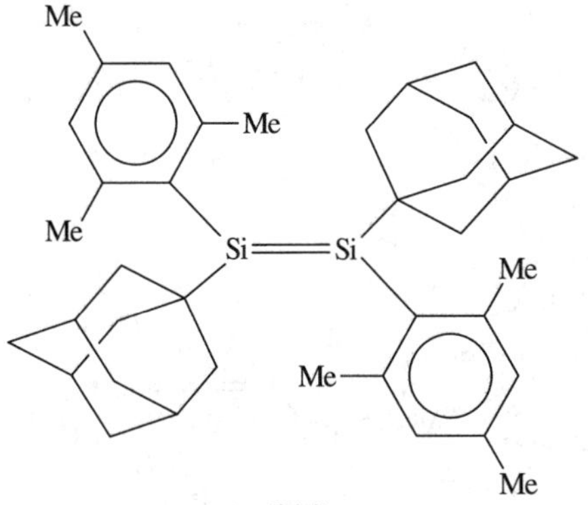

(304) 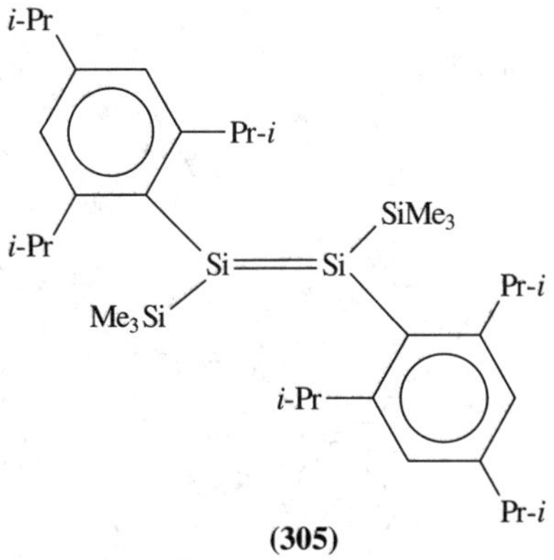 **(305)**

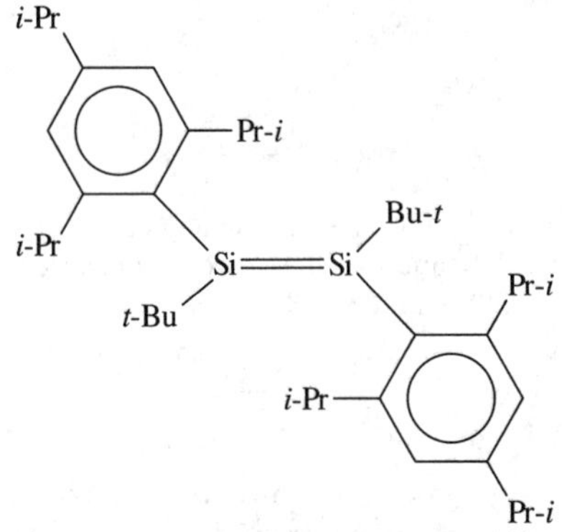

(306) 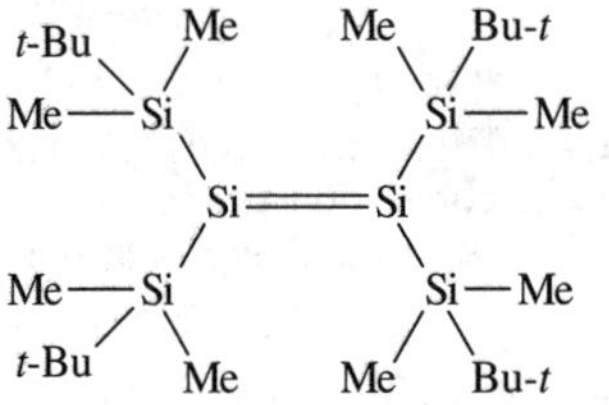 **(307)**

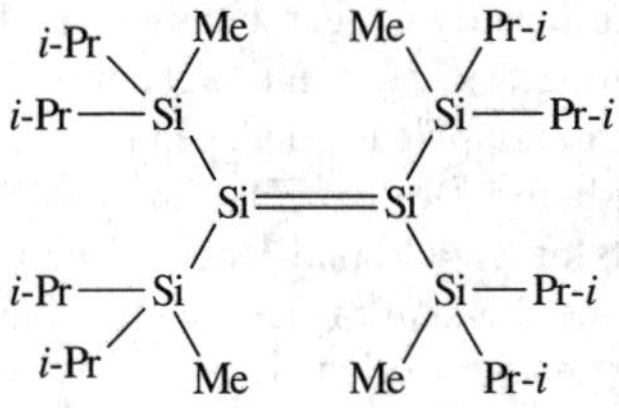

(308) 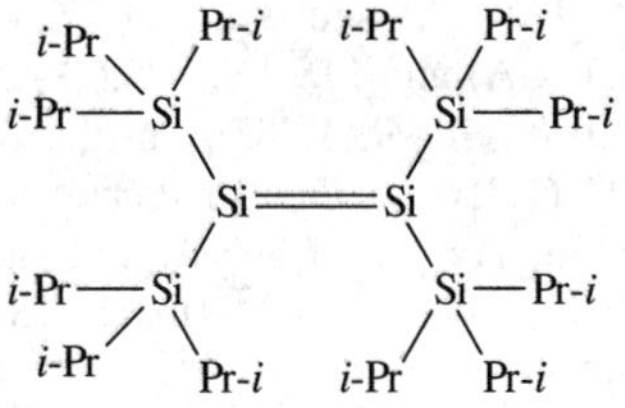 **(309)**

and recently[26] a Si=Si bond length of 2.150 Å was calculated, based on approximate density functional theory within the local density approximation and augmented by non-local exchange and correlation corrections. The experimental bond lengths obtained by X-ray crystal structure show two distinct ranges of values: 2.138–2.160 Å and 2.201–2.261 Å. The somewhat longer bond lengths were observed in compounds where all the substituents are alkylsilyls. The lengthening is attributed to an electronic effect. The same lengthening was also found where the substituents are extremely bulky, such

(310)

(311)

as 2,4,6-tris(bis(trimethylsilyl)methylphenyl) (HATDOV). The lengthening can be a result of steric congestion as manifested in the pyramidalization of the silicon atoms, deduced by the sum of bond angles at the silicon atoms (358.8 and 357.3°).

B. Silenes (>Si=C<)

The first stable silene was reported in 1981 by Brook and coworkers[305]. The X-ray crystal structures of four stable silenes have been published up to 1987[306–311]. In the last decade the X-ray crystal structures of only two more compounds possessing double bonds between silicon and carbon were published as well as two organometallic compounds containing Si double-bonded to carbon. The shorter Si=C bond length was found in **312**[309] (1.703 Å) and **313**[312] (1.704 Å). Significantly longer bonds were found in **314**[307] (1.764 Å) and **315**[313] (1.758 Å) and the longest ones in the organometallic compounds **316**[314] (1.799, 1.783 Å in the two crystallographically independent molecules) and **317**[315] (1.750 Å). The predicted bond length for $Me_2Si=CH_2$ based on *ab initio* MO calculation[316] was 1.692 Å, and later results of Si=C bond length obtained by *ab initio* calculations are 1.703[27] and 1.718[28] Å, while the sum of the double-bond radii of the participant atoms is 1.74 Å[3]. There are two possible explanations for the significant differences in the Si=C bond lengths. The first suggests that the differences between the bond lengths are the result of steric congestion, while the second attributes the differences to electronic effects[317–319]. According to the latter, the polarization of the Si=C bond is the dominant effect and substitution that alters the degree of polarization (and hence the degree of ionicity) will also cause variation of the bond lengths. Therefore, the presence of a Me_3SiO group that includes the very electronegative oxygen atom as the substituent on the carbon atom in **314** results in the withdrawal of charge from the carbon atom, thus reducing the polarity of the Si=C bond. Consequently, the bond lengthens compared to that found in **312**. The shortening of the Si=C bond in the latter is a result of the increase of polarity compared with the calculated value for the model molecule $Me_2Si=CH_2$.

Of particular interest is the lengthening of the Si=C bonds in **316** (1.799 Å) and in **317** (1.783 and 1.790 Å). These values lie between typical Si=C double and Si—C single bond lengths, which can presumably be attributed to a partially double-bond character. The shortening of the similar bond in silaallene **313** (1.704 Å) exhibits a partial 'reversed polarity' of the Si=C double bond, comparable in effect to that of an electron-donating oxygen substituent on carbon, as in **314**, without bond elongation. This 'reversed polarity' of the Si=C π-bond is believed to be 'the most important single electronic factor that reduces the reactivity of silenes'[318]. Another group of silenes that should be mentioned are those including substituents with increasing Lewis basicity which leads to a Si=C bond lengthening, such as in **315** (1.758 Å).

C. Silanimines (>Si=N—)

Theoretical calculations[320] of the geometry in $H_2Si=NH$ were carried out at the SCF (3-21G* and 6-31G**), SCF-MP4 (6-31G**) and MCSCF+CI (6-31G*) levels and suggest that the molecule is planar with a Si=N bond length of 1.576 Å and a large SiNH valence angle of 129.5°. Further calculation[321] using a 6-31G* basis set with d-functions on Si and N atoms was carried out for $H_2Si=NSiH_3$ as a model molecule. The bond lengths were calculated to be 1.549 and 1.688 Å for the double and the single SiN bonds, with a Si=N−Si bond angle of 175.6°. The widening of the bond angle was attributed to an electronic effect. The potential energy surface for widening the bond angle is flat and the barrier for linearization of the bond is only 13.8 kJ mol^{-1}, ten times smaller than for the carbon analogue. The difference stems from the larger electronegativity differences between silicon and nitrogen compared with carbon and nitrogen. Experimental geometry obtained by the X-ray crystal structure of the first stable silanimine[322] was published in 1986[323]. Structural parameters in silanimines are given in Table 31.

It is interesting to compare the relevant geometries of compounds **318–322** as they all consist of identical Si=N−Si skeleton. The bond lengths in **319–321**, where the silicon is also complexed to THF, lie in a very small range: the Si=N bond length in the range 1.593 to 1.597 Å, and the Si−N bond length in the range 1.653 to 1.667 Å. In **318**, on the other hand, where the silicon is purely tricoordinated, the double bond decreases to 1.568 Å and the single bond increases to 1.696 Å. It should also be noted that in **318** the Si=N−Si bond angle is 177.8° while in the other compounds it is in a range between 160.9 and 174.2°. Upon linearization, the Si=N double bond is strengthened due to the increase of the s contribution to nitrogen hybridization. In compounds of the type Si=N−X, where X = sulphur (**325** and **326**) the Si=N bond length is significantly longer (1.714 and 1.704 Å, respectively).

TABLE 31. Structural parameters for silanimines

Refcode	Compound	d(Si=N) (Å)	d(Si−N) (Å)	Si=N−X^a (°)	Reference
DOKWUV	**318**	1.568	1.696	177.8	324
DOKXAC	**319**	1.588	1.654	161.5	323
DOKXAC	**319**	1.574	1.667	161.0	323
DOKXEG	**320**	1.589	1.653	161.5	324
DOKXEG	**320**	1.573	1.666	160.9	324
JIWGAX	**321**	1.597	1.661	174.2	325
GEVRII	**322**	1.601	1.678	169.4	324
KIDRUK	**323**	1.605	—	—	326
FEWYAH	**323**^{a}	1.619	—	172.1	327
YETNVG	**324**	1.599	—	134.6	328
KUCCIV	**325**	1.714	—	135.6	329
KUCCOA	**326**	1.704	—	137.1	329

^{a}Similar to **323**, but at the silicon the CH_3 and t-Bu are replaced by i-Pr.

D. Silanephosphimines (>Si=P−)

There is a theoretical investigation of silanephosphimine $H_2Si=PH$ and of its isomer $HSiPH_2$[330]. Experimental X-ray crystal structure are known for only two silanephosphimines: **327**[331] and **328**[332]. The Si=P bond lengths are 2.094 and 2.063 Å in **327** and **328**, respectively. The trigonal silicon is planar in **328** [Σ(Si) is 359.9°] while it is slightly pyramidal in **327** [Σ(Si) is 356.7°].

E. Silanones (>Si=O) and Silanethiones (>Si=S)

The silanone molecule is calculated[333] to be planar and to contain the very polar and strong Si=O double bond. The bond length in silanone is calculated[334] to be 1.545 Å,

5. The structural chemistry of organosilicon compounds

(318) $t\text{-Bu}_2\text{Si}=\text{N}-\text{Si}(t\text{-Bu})_2-\text{Bu-}t$

(319) THF→Si(t-Bu)_2=N−Si(t-Bu)_2−Bu-t

(320) THF→Si(Me)(H_3C)=N−Si(t-Bu)_2−Bu-t

(321) THF→Si(t-Bu)_2=N−Si(t-Bu)(t-Bu)−CH_3

(322) Ph_2C=O→Si(t-Bu)_2=N−Si(t-Bu)_2−Bu-t

(323) F→Si(CH_3)(t-Bu)=N−Ar (Ar = 2,4,6-tri-t-Bu-phenyl)

(324) diazasilene with THF coordination, =N−CPh_3

(325) Ph_2(t-Bu)Si=N−S(=O)(CH_3)−CH(i-Pr)(Ph)−NH−Ph

(326) Ph_2(t-Bu)Si=N−S(=O)(CH_3)−CH(Ph)−CH(OH)(Et)

(327) bis(aryl)phosphino-silicon compound with P=Si and PPh_2 substituents

(328)

and later it was calculated[29] using gradient techniques with 3-21G to be 1.599 Å; the addition of the DZ + P basis set causes a shortening of the calculated bond to 1.499 Å. Few XRD structures of compounds possessing Si=O double bonds are known, but in all of them the silicon atom is tetracoordinate, such as in **329–332**. The average Si=O bond length in **329**[335] is 1.664 Å, and it is 1.687 Å in **330**[336], 1.622 Å in **331**[337] and 1.615 Å in **332**[337]. Those bond lengths are significantly longer than the calculated ones, which signify that these are not purely double bonds but dative bonds.

(329)   **(330)**

(331)   **(332)**

Unfortunately there are no structural data for silanethiones.

IV. SUMMARY

All Si—X bond lengths obtained by X-ray crystal structures of the organosilicon compounds discussed in this review are summarized in Table 32.

TABLE 32. Summary of bond lengths in organosilicon compounds

X	Si–X	d(Si–X) (Å)
C(sp)	$>$Si—C≡	1.839
C(sp^2)	$>$Si—C$<$	1.878

5. The structural chemistry of organosilicon compounds

TABLE 32. (continued)

Bond type	Structure	Length (Å)
C(C=O)	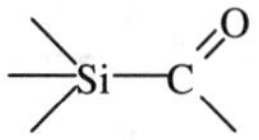 >Si—C(=O)—	1.925
C(aryl)	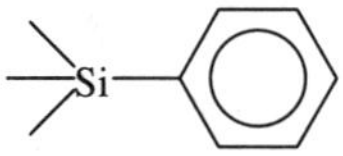 >Si—C₆H₅	1.879
C(methyl)	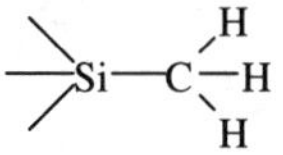 >Si—CH₃	1.860
C(t-Bu)	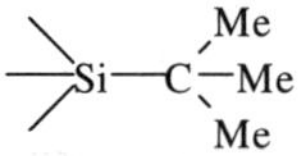 >Si—C(Me)₃	1.907
C	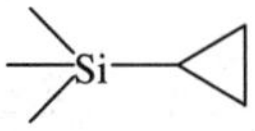 >Si—△ (cyclopropyl)	1.855
C	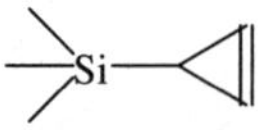 >Si—△∥ (cyclopropenyl)	1.815
C(sp^2)	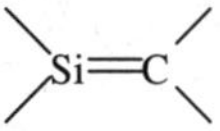 >Si=C<	1.703–1.764
Si(acyclic)	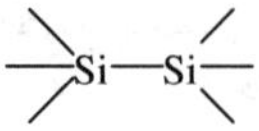 >Si—Si<	2.358
Si(cyclic)	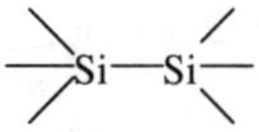 >Si—Si<	2.372
Si	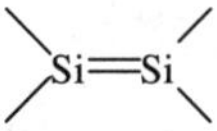 >Si=Si<	2.138–2.261
N(sp^2)	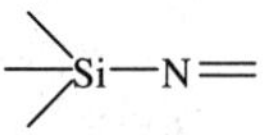 >Si—N=	1.713
N(sp^3)	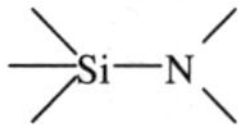 >Si—N<	1.739
N(sp^3)	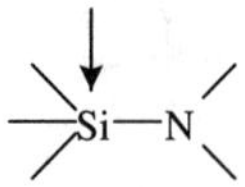 >Si←N<	1.761
N(sp^3)	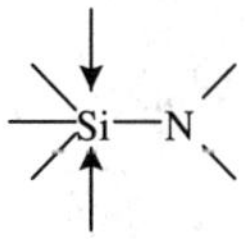 >Si↔N<	1.969
N(sp^2)	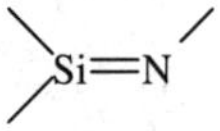 >Si=N—	1.568–1.619
P	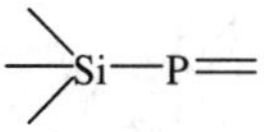 >Si—P=	2.161–2.269

(*continued overleaf*)

TABLE 32. (*continued*)

X	Si–X	d(Si–X) (Å)
P	>Si–P<	2.265
P	>Si=P<	2.063–2.094
O	>Si–O–X	1.629
O	>Si–O–H	1.636
O(acyclic)	>Si–O–Si<	1.621
O(cyclic)	>Si–O–Si<	1.618
O	↓>Si–O–X	1.699
S	>Si–S–Si<	2.141
S	>Si–S–	2.144
F	>Si–F	1.594
F	O↓>Si–F	1.631
F	O↓>Si–F↑O	1.677
Cl	>Si–Cl	2.050
Cl	Cl\\>Si–Cl	2.022
Cl	O↓>Si–Cl↑O	2.184

TABLE 32. *(continued)*

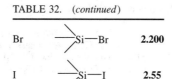

V. REFERENCES

1. J. C. Slater, *J. Chem. Phys.*, **41**, 3199 (1964).
2. R. T. Sanderson, *J. Am. Chem. Soc.*, **105**, 2259 (1983).
3. L. Pauling, *The Nature of the Chemical Bond*, 3rd ed., Cornell University Press, Ithaca, 1960.
4. C. E. Roore, *Ionization Potentials and Ionization Limits Derived from the Analysis of Optical Spectra*, NSRDS-NBS34, National Bureau of Standards, Washington, DC, 1970.
5. M. C. Day, Jr. and J. Selbin, *Theoretical Inorganic Chemistry*, Reinhold, New York, 1969.
6. M. J. S. Dewar, D. H. Lo and C. A. Ramsden, *J. Am. Chem. Soc.*, **97**, 1311 (1975).
7. H. Kwart and K. G. King, *d-Orbitals in the Chemistry of Silicon, Phosphorus and Sulfur*, Springer-Verlag, Berlin, 1977.
8. W. Kutzelnigg, *Angew. Chem., Int. Ed. Engl.*, **23**, 272 (1984).
9. C. Glidewell, *Inorg. Chim. Acta Rev.*, **7**, 69 (1973).
10. H. Bürger, *Angew. Chem., Int. Ed. Engl.*, **12**, 474 (1973).
11. P. Hencsei and L. Parkanyi, *Rev. Silicon, Germanium, Tin, Lead Compd.*, **8**, 191 (1985).
12. G. Klebe, *J. Organomet. Chem.*, **293**, 147 (1985).
13. G. Müller, *Nachr. Chem. Tech. Lab.*, **34**, 778 (1986).
14. S. N. Tandura, M. G. Voronkov and N. V. Alekseev, *Top. Curr. Chem.*, **131**, 99 (1986).
15. W. S. Sheldrick, in *The Chemistry of Organic Silicon Compounds* (Eds. S. Patai and Z. Rappoport), Wiley, Chichester, 1989
16. CSD, Cambridge Structural Database System, version April 1996, Cambridge Crystallographic Data Centre, Cambridge, England.
17. D. C. McKean, I. Torto, J. E. Boggs and K. Fan, *J. Mol. Spectrosc.*, **152**, 389 (1992).
18. M. S. Gordon and C. George, *J. Am. Chem. Soc.*, **106**, 609 (1984).
19. S. Grigoras and T. H. Lane, *J. Comput. Chem.*, **9**, 25 (1987).
20. P. Ugliengo, A. Bleiber, E. Garone and A. M. Ferrari, *Chem. Phys. Lett.*, **191**, 537 (1992).
21. J. B. Nicholas, R. E. Winans, R. J. Harrison, L. E. Iton, L. A. Curtiss and A. J. Hopfinger, *J. Phys. Chem.*, **96**, 10247 (1992).
22. S. Shambayati, J. F.Blake, S. G. Wierschke, W. L. Jorgensen and S. L. Schreiber, *J. Am. Chem. Soc.*, **112**, 697 (1990).
23. E. W. Ignacio and H. B. Schlegel, *J. Phys. Chem.*, **96**, 5830 (1992).
24. H. B. Schlegel and P. N. Skancke, *J. Am. Chem. Soc.*, **115**, 10916 (1993).
25. C. Guimon and G. Pfister-Guillouzo, *Organometallics*, **6**, 1387 (1987).
26. H. Jacobsen and T. Ziegler, *J.Am. Chem. Soc.*, **116**, 3667 (1994).
27. M. S. Gordon, *J. Am. Chem. Soc.*, **104**, 4352 (1982).
28. M. E. Colvin, J. Kobayashi, J. Bicerano and H. F. Schaefer III, *J. Chem. Phys.*, **85**, 4563 (1986).
29. D. A. Dixon and J. L. Gole, *Chem. Phys. Lett.*, **125**, 179 (1986).
30. ORIGIN, software for technical graphics and data analysis, Microcal Software Inc.
31. A. S. Dreiding, J. H. Bieri, R. Prewo, A. Linden and E. Gesing, private communication.
32. Li Guo, J. D. Bradshaw, C. A. Tessier and W. J. Youngs, *J. Chem. Soc., Chem. Commun.*, 243 (1994).
33. A. M. Boldi, J. Anthony, C. B. Knobler and F. Diederich, *Angew. Chem., Int. Ed. Engl.*, **31**, 1240 (1992).
34. K. S. Feldman, C. K. Weinreb, W. J. Youngs and J. D. Bradshaw, *J. Am. Chem. Soc.*, **116**, 9019 (1994).
35. K. S. Feldman, C. K. Weinreb, W. J. Youngs and J. D. Bradshaw, *J. Org. Chem.*, **59**, 1213 (1994).
36. Y. Rubin, C. B. Knobler and F. Diederich, *Angew. Chem., Int. Ed. Engl.*, **30**, 698 (1991).
37. J.-D. van Loon, P. Seiler and F. Diederich, *Angew. Chem., Int. Ed. Engl.*, **32**, 1187 (1993).

38. Y. Pang, A. Schneider, T. J. Barton, M. S. Gordon and M. R. Carroll, *J. Am. Chem. Soc.*, **114**, 4920 (1992).
39. W. Ando, N. Nakayama, Y. Kabe and T. Shimizu, *Tetrahedron Lett.*, **31**, 3597 (1990).
40. M. Ishikawa, T. Hatano, Y. Hasegawa, T. Horio, A. Kunai, A. Miyai, T. Ishida, T. Tsukihara, T. Yamanaka, T. Koike and J. Shioya, *Organometallics*, **11**, 1604 (1992).
41. H. Sakurai, Y. Nakadaira, A. Hosomi, Y. Eriyama and C. Kabuto, *J. Am. Chem. Soc.*, **105**, 3360 (1983).
42. G. A. Eliassen, E. Kloster-Jensen and C. Romming, *Acta Chem. Scand., Ser. B*, **40**, 574 (1986).
43. A. Baumegger, E. Hengge, S. Gamper, E. Herdtweck and R. Janoschek, private communication.
44. M. G. Voronkov, Yu. E. Ovchinnikov, V. E. Shklover, Yu. T. Struchkov, I. A. Zamaev, O. G. Yarosh, G. Yu. Turkina and V. Yu. Vitkovskii, *Dokl. Akad. Nauk SSSR*, **296**, 130 (1987); *Chem. Abstr.*, **109**, 93118b (1988).
45. R. Bortolin, S. S. D. Brown and B. Parbhoo, *Inorg. Chim. Acta*, **158**, 137 (1989).
46. Yu. E. Ovchinnikov, Yu. T. Struchkov, M. G. Voronkov, O. G. Yarosh, G. Yu. Turkina and T. M. Orlova, *Metalloorg. Khim.*, **5**, 1280 (1992); *Chem. Abstr.*, **119**, 226033a (1993).
47. I. A. Zamaev, Yu. E. Ovchinnikov, V. E. Shklover, Yu. T. Struchkov, O. G. Yarosh, M. G. Voronkov, G. Yu. Turkina and T. M. Orlova, *Metalloorg. Khim.*, **1**, 1265 (1988); *Chem. Abstr.*, **110**, 163900x (1989).
48. L. N. Lewis, K. G. Sy, G. L. Bryant Jr. and P. E. Donahue, *Organometallics*, **10**, 3750 (1991).
49. P. Jutzi, J. Kleimeier, R. Krallmann, H.-G. Stammler and B. Neumann, *J. Organomet. Chem.*, **462**, 57 (1993).
50. Yu. E. Ovchinnikov, V. A. Igonin, I. A. Zamaev, V. E. Shklover, Yu. T. Struchkov, O. G. Yarosh, M. G. Voronkov and G. Y. Turkina, *J. Struct. Chem.*, **32**, 250 (1991).
51. K. Tamao, S. Yamaguchi and M. Shiro, *J. Am. Chem. Soc.*, **116**, 11715 (1994).
52. A. Sekiguchi, M. Ichinohe, C. Kabuto and H. Sakurai, *Organometallics*, **14**, 1092 (1995).
53. M. Weidenbruch, A. Schaefer, K. Peters and H. G. von Schnering, *J. Organomet. Chem.*, **314**, 25 (1986).
54. E. O. Fischer, H. Hollfelder, P. Friedrich, F. R. Kreissl and G. Huttner, *Chem. Ber.*, **110**, 3467 (1977).
55. H. H. Karsch, R. Richter and A. Schier, *Z. Naturforsch., Teil B*, **48**, 1533 (1993).
56. M. Fritz, J. Hiermeier, N. Hertkorn, F. H. Kohler, G. Muller, G. Reber and O. Steigelmann, *Chem. Ber.*, **124**, 1531 (1991).
57. B. Morosin and L. A. Harrah, *Acta Crystallogr., Sect. B*, **37**, 579 (1981).
58. G. Reinhard, B. Hirle and U. Schubert, *J. Organomet. Chem.*, **427**, 173 (1992).
59. P. Jutzi, D. Kanne, M. Hursthouse and A. J. Howes, *Chem. Ber.*, **121**, 1299 (1988).
60. J. Grobe, R. Martin, G. Huttner and L. Zolnai, *Z. Anorg. Allg. Chem.*, **607**, 79 (1992).
61. H. Beckers, D. J. Brauer, H. Burger and C. J. Wilke, *J. Organomet. Chem.*, **356**, 31 (1988).
62. M. Weidenbruch, H. Flott, B. Ralle, K. Peters and H. G. von Schnering, *Z. Naturforsch., Teil B*, **38**, 1062 (1983).
63. W. Ando, M. Fujita, H. Toshida and A. Sekiguchi, *J. Am. Chem. Soc.*, **110**, 3310 (1988).
64. Y. Yammamoto, Y. Kebe and W. Ando, *Organometallics*, **12**, 1996 (1993).
65. Dong Ho Pae, M. Xiao, M. Y. Chiang and P. P. Gaspar, *J. Am. Chem. Soc.*, **113**, 1281 (1991).
66. M. Yu. Antipin, A. V. Polyakov, Yu. T. Struchkov, M. P. Egorov, A. L. Gal'minas, S. P. Golensnikov and O. M. Nefedov, *Metalloorg. Khim.*, **2**, 593 (1989); *Chem. Abstr.* **112**, 118996p (1990).
67. A. Kunai, Y. Matsuo, J. Ohshita, M. Ishikawa, Y. Aso, T. Otsubo and F. Ogura, *Organometallics*, **14**, 1204 (1995).
68. F. Hojo, S. Sekigawa, N. Nakayama, T. Shimizu and W. Ando, *Organometallics*, **12**, 803 (1993).
69. P. C. Chien and J. Trotter, *J. Chem. Soc. (A)*, 1778 (1969).
70. S. J. Rettig and J. Trotter, *Acta Crystallogr., Sect. C*, **44**, 1850 (1988).
71. Y. E. Ovchinnikov, V. E. Shklover, Yu. T. Struchkov, T. M. Frunze, V. V. Dement'ev, B. A. Antipova and T. M. Ezhova, *J. Struct. Chem.*, **30**, 281 (1989).
72. R. Tacke, H. Hengelsberg, E. Klingner and H. Henke, *Chem. Ber.*, **125**, 607 (1992).
73. J. P. Vidal, J. L. Galigne and J. Falgueirettes, *Acta Crystallogr., Sect. B*, **28**, 3130 (1972).
74. S. S. Al-Juaid, Y. Derouiche, P. B. Hitchcock, P. D. Lickiss and A. G. Brook, *J. Organomet. Chem.*, **403**, 293 (1991).
75. M. J. Barrow, *Acta Crystallogr., Sect. B*, **38**, 150 (1982).
76. J. Konnert, D. Britton and Y. M. Chow, *Acta Crystallogr., Sect. B*, **28**, 180 (1972).

77. D. A. Dixon, W. R. Hertler, D. B. Chase, W. B. Farnham and F. Davidson, *Inorg. Chem.*, **27**, 4012 (1988).
78. W. R. Hertler, D. A. Dixon, E. W. Matthews, F. Davidson and F. G. Kitson, *J. Am. Chem. Soc.*, **109**, 6532 (1987).
79. Z. Xie, D. J. Liston, T. Jelinek, V. Mitro, R. Bau and C. A. Reed, *J. Chem. Soc., Chem. Commun.*, 384 (1993).
80. G. Rajca, W. Schwarz and J. Weidlein, *Z. Naturforsch., Teil B*, **39**, 1219 (1984).
81. A. Heine, R. Herbst-Irmer, G. M. Sheldrick and D. Stalke, *Inorg. Chem.*, **32**, 2694 (1993).
82. F. R. Fronczek and P. D. Lickiss, *Acta Crystallogr., Sect. C*, **49**, 331 (1993).
83. S. P. Mallela, I. Bernal and R. A. Geanangel, *Inorg. Chem.*, **31**, 1626 (1992).
84. J. B. Lambert, J. L. Pflug, A. M. Allgeier, D. J. Campbell, T. B. Higgins, E. T. Singewald and C. L. Stern, *Acta Crystallogr., Sect. C*, **51**, 713 (1995).
85. H. Oehme, R. Wustrack, A. Heine, G. M. Sheldrick and D. Stalke, *J. Organomet. Chem.*, **452**, 33 (1993).
86. J. Wan and J. G. Verkade, *Inorg. Chem.*, **32**, 341 (1993).
87. A. Haas, R. Sullentrup and C. Kruger, *Z. Anorg. Allg. Chem.*, **619**, 819 (1993).
88. K. Tamao, A. Kawachi, Y. Nakagawa and Y. Ito, *J. Organomet. Chem.*, **473**, 29 (1994).
89. R. West and E. K. Pham, *J. Organomet. Chem.*, **403**, 43 (1991).
90. N. Wiberg, H. Schuster, A. Simon and K. Peters, *Angew. Chem., Int. Ed. Engl.*, **25**, 79 (1986).
91. M. Weidenbruch, B. Flintjer, K. Peters and H. G. von Schnering, *Angew. Chem., Int. Ed. Engl.*, **25**, 1129 (1986).
92. M. Weidenbruch, B. Flintjer, K. Kramer, K. Peters and H. G. von Schnering, *J. Organomet. Chem.*, **340**, 13 (1988).
93. H. Watanabe, M. Kato, T. Okawa, Y. Nagai and M. Goto, *J. Organomet. Chem.*, **271**, 225 (1984).
94. M. Weidenbruch, K.-L. Thom, S. Pohl and W. Saak, private communication.
95. L. Parkanyi, K. Sasvari, J. P. Declercq and G. Germain, *Acta Crystallogr., Sect. B*, **34**, 3678 (1978).
96. H. Li, I. S. Butler and J. F. Harrod, *Organometallics*, **12**, 4553 (1993).
97. F. Shafiee, J. R. Damewood Jr., K. J. Haller and R. West, *J. Am. Chem. Soc.*, **107**, 6950 (1985).
98. F. Shafiee, K. J. Haller and R. West, *J. Am. Chem. Soc.*, **108**, 5478 (1986).
99. A. Schafer, M. Weidenbruch, K. Peters and H. G. von Schnering, *Angew. Chem., Int. Ed. Engl.*, **23**, 302 (1984).
100. H. Tachibana, M. Goto, M. Matsumoto, H. Kishida and Y. Tokura, *Appl. Phys. Lett.*, **64**, 2509 (1994).
101. M. Weidenbruch, J. Hamann, S. Pohl and W. Saak, *Chem. Ber.*, **125**, 1043 (1992).
102. W. Ando, M. Kako, T. Akasaka and S. Nagase, *Organometallics*, **12**, 1514 (1993).
103. J. L. Shibley, R. West, C. A. Tessier and R. K. Hayashi, *Organometallics*, **12**, 3480 (1993).
104. G. Reber, J. Riede, N. Wiberg, K. Schurz and G. Muller, *Z. Naturforsch., Teil B*, **44**, 786 (1989).
105. S. Walter, U. Klingebiel and D. Schmidt-Base, *J. Organomet. Chem.*, **412**, 319 (1991).
106. R. Hasselbring, H. W. Roesky, M. Rietzel and M. Noltmeyer, private communication.
107. M. Jansen and H. Jungermann, *Z. Kristallogr.*, **209**, 779 (1994).
108. S. Freitag, W. Kolodziejski, F. Pauer and D. Stalke, *J. Chem. Soc., Dalton Trans.*, 3479 (1993).
109. M. Herberhold, S. Gerstmann, B. Wrackmeyer and H. Bormann, *J. Chem. Soc., Dalton Trans.*, 633 (1994).
110. M. Herberhold, S. Gerstmann, W. Milius and B. Wrackmeyer, *Z. Naturforsch., Teil B*, **48**, 1041 (1993).
111. S. S. Al-Juaid, A. A. Al-Nasr, G. A. Ayoko, C. Eaborn and P. Hitchcock, *J. Organomet. Chem.*, **488**, 155 (1995).
112. A. Kienzle, A. Obermeyer, R. Riedel, F. Aldinger and A. Simon, *Chem. Ber.*, **126**, 2569 (1993).
113. M. Weidenbruch, B. Flintjer, S. Pohl and W. Saak, *Angew. Chem., Int. Ed. Engl.*, **28**, 95 (1989).
114. J. D. Lichtenhan, S. C. Critchlow and N. M. Doherty, *Inorg. Chem.*, **29**, 439 (1990).
115. C. J. Burns, W. H. Smith, J. C. Huffman and A. P. Sattelberger, *J. Am. Chem. Soc.*, **112**, 3237 (1990).
116. M. Veith and H. Barnighausen, *Acta Crystallogr., Sect. B*, **30**, 1806 (1974).
117. W. Ando, T. Ohtaki and Y. Kabe, *Organometallics*, **13**, 434 (1994).
118. M. Veith, S. Becker and V. Huch, *Angew. Chem., Int. Ed. Engl.*, **29**, 216 (1990).
119. K. Ruhlandt-Senge, R. A. Bartlett, M. Olmstead and P. P. Power, *Angew. Chem., Int. Ed. Engl.*, **32**, 425 (1993).

120. J. He, Hua Qin Liu, J. F. Harrod and R. Hynes, *Organometallics*, **13**, 336 (1994).
121. N. W. Mitzel, J. Riede, A. Schier and H. Schmidbaur, *Acta Crystallogr., Sect. C*, **51**, 756 (1995).
122. F. Huppmann, M. Noltmeyer and A. Meller, *J. Organomet. Chem.*, **483**, 217 (1994).
123. Y. Wan and J. G. Verkade, *Inorg. Chem.*, **32**, 341 (1993).
124. K. Tamao, A. Kawachi, Y. Nakagawa and Y. Ito, *J. Organomet. Chem.*, **473**, 29 (1994).
125. N. W. Mitzel, J. Riede, A. Schier, M. Paul and H. Schmidbaur, *Chem. Ber.*, **126**, 2027 (1993).
126. N. W. Mitzel, P. Bissinger and H. Schmidbaur, *Chem. Ber.*, **126**, 345 (1993).
127. N. W. Mitzel, P. Bissinger, J. Riede, K.-H. Dreihaupl and H. Schmidbaur, *Organometallics*, **12**, 413 (1993).
128. H. Bock, J. Neuret, C. Nather and U. Krynitz, *Tetrahedron Lett.*, **34**, 7553 (1993).
129. L. Marcus, U. Klingebiel and M. Noltmeyer, *Z. Naturforsch., Teil B*, **50**, 687 (1995).
130. R. Hasselbring, I. Leichtweis, M. Noltmeyer, H. W. Roesky, H.-G. Schmidt and A. Herzog, *Z. Anorg. Allg. Chem.*, **619**, 1543 (1993).
131. V. Sum, C. A. Baird, T. P. Kee and M. Thornton-Pett, *J. Chem. Soc., Perkin Trans. 1*, 3183 (1994).
132. D. Dou, M. Westerhausen, G. L. Wood, G. Linti, E. N. Duesler, H. Noth and R. T. Paine, *Chem. Ber.*, **126**, 379 (1993).
133. H. Braunschweig, P. Paetzold and T. P. Spaniol, *Chem. Ber.*, **126**, 1565 (1993).
134. M. Yu. Antipin, A. N. Chernega and Yu. T. Struchkov, *Phosphorus, Sulfur & Silicon*, **78**, 289 (1993).
135. R. Oberdorfer, M. Nieger and E. Niecke, *Chem. Ber.*, **127**, 2397 (1994).
136. B. Wrackmeyer, C. Kohler, W. Milius and M. Herberhold, *Phosphorus, Sulfur & Silicon*, **89**, 151 (1994).
137. Th. Schlosser, A. Sladek, W. Hiller and H. Schmidbaur, *Z. Naturforsch., Teil B*, **49**, 1247 (1994).
138. D. J. Burkey, E. K. Alexander and T. P. Hanusa, *Organometallics*, **13**, 2773 (1994).
139. A. Steiner and D. Stalke, *Angew. Chem., Int. Ed. Engl.*, **34**, 1752 (1995).
140. K. Dippel, U. Klingebiel and D. Schmidt-Base, *Z. Anorg. Allg. Chem.*, **619**, 836 (1993).
141. L. M. Englehardt, P. C. Junk, W. C. Patalinghug, R. E. Sue, C. L. Raston, B. W. Skelton and A. H. White, *J. Chem. Soc., Chem. Commun.*, 930 (1991).
142. A. N. Chernega, M. Yu. Antipin, Yu. T. Struchkov, I. E. Boldeskul, T. V. Sarina and V. D. Romanenko, *Dokl. Akad. Nauk SSSR*, **278**, 1146 (1984); *Chem. Abstr.*, **102**, 149379n (1984).
143. A. H. Cowley, P. C. Knuppel and C. M. Nunn, *Organometallics*, **8**, 2490 (1989).
144. A. Grunhagen, U. Pieper, T. Kottke and H. W. Roesky, *Z. Anorg. Allg. Chem.*, **620**, 716 (1994).
145. U. Klingebiel, M. Meyer, U. Pieper and D. Stalke, *J. Organomet. Chem.*, **408**, 19 (1991).
146. M. Andrianarison, D. Stalke and U. Klingebiel, *J. Organomet. Chem.*, **381**, C38 (1990).
147. H. R. G. Bender, M. Nieger and E. Niecke, *Z. Naturforsch., Teil B*, **48**, 1742 (1993).
148. M. Driess, H. Pritzkow and M. Reisgys, *Chem. Ber.*, **124**, 1923 (1991).
149. K.-F. Tebbe and M. Feher, *Acta Crystallogr., Sect. C*, **46**, 1071 (1990).
150. M. Andrianarison, U. Klingebiel, D. Stalke and G. M. Sheldrick, *Phosphorus, Sulfur & Silicon*, **46**, 183 (1989).
151. H. Westermann and M. Nieger, *Inorg. Chim. Acta*, **177**, 11 (1990).
152. R. Appel, D. Gudat, E. Niecke, M. Nieger, C. Porz and H. Westermann, *Z. Naturforsch., Teil B*, **46**, 865 (1991).
153. M. Baudler, W. Oehlert and K.-F. Tebbe, *Z. Anorg. Allg. Chem.*, **598/599**, 9 (1991).
154. M. A. Petrie and P. P. Power, *J. Chem. Soc., Dalton Trans.*, 1737 (1993).
155. G. Becker, W. Massa, R. E. Schmidt and G. Uhl, *Z. Anorg. Allg. Chem.*, **520**, 139 (1985).
156. I. Kovacs, G. Baum, G. Fritz, D. Fenske, N. Wiberg, H. Schuster and K. Karaghiosoff, *Z. Anorg. Allg. Chem.*, **619**, 453 (1993).
157. M. Baudler, G. Schlotz and K.-F. Tebbe, *Z. Anorg. Allg. Chem.*, **581**, 111 (1991).
158. L. Weber, K. Reizig and R. Boese, *Organometallics*, **4**, 2097 (1985).
159. H. Schafer, D. Binder, B. Deppisch and G. Mattern, *Z. Anorg. Allg. Chem.*, **546**, 79 (1987).
160. R. Koster, G. Seidel, G. Muller, R. Boese and B. Wrackmeyer, *Chem. Ber.*, **121**, 1381 (1988).
161. P. Frankhauser, M. Driess, H. Pritzkow and W. Siebert, *Chem. Ber.*, **125**, 1341 (1992).
162. G. L. Wood, D. Dou, D. K. Narula, E. N. Duesler, R. T. Paine and H. Noth, *Chem. Ber.*, **123**, 1455 (1990).
163. G. Effinger, W. Hiller and I.-P. Lorenz, *Z. Naturforsch., Teil B.*, **42**, 1315 (1987).
164. H. H. Karsch, K. Zellner and G. Muller, *Organometallics.*, **10**, 2884 (1991).

5. The structural chemistry of organosilicon compounds

165. H. H. Karsch, U. Keller, S. Gamper and G. Muller, *Angew. Chem., Int. Ed. Engl.*, **29**, 295 (1990).
166. E. Hey, P. B. Hitchcock, M. F. Lappert and A. K. Rai, *J. Organomet. Chem.*, **325**, 1 (1987).
167. G. Becker, H.-M. Hartmann and W. Schwartz, *Z. Anorg. Allg. Chem.*, **577**, 9 (1989).
168. E. Hey-Hawkins and E. Sattler, *J. Chem. Soc., Chem. Commun.*, 725 (1992).
169. E. Hey, C. L. Raston, B. W. Skelton and A. H. White, *J. Organomet. Chem.*, **362**, 1 (1989).
170. D. A. Atwood, A. H. Cowley, R. A. Jones and M. A. Mardones, *J. Organomet. Chem.*, **449**, C1 (1993).
171. M. Westerhausen, *J. Organomet. Chem.*, **479**, 141 (1994).
172. M. G. Voronkov, V. P. Mileshkevich and Yu. A. Yuzhelevskii, *The Siloxane Bond*, Consultants Bureau, New York, London, 1978.
173. A. N. Chernega, M. Yu. Antipin, Yu. T. Struchkov and D. F. Nikson, *Ukr. Khim. Zh.*, **59**, 196 (1993); *Chem. Abstr.*, **118**, 95614 (1993).
174. J. He, J. F. Harrod and R. Hynes, *Organometallics*, **13**, 2496 (1994).
175. D. Murphy, J. P. Sheehan, T. R. Spalding, G. Ferguson, A. J. Lough and J. F. Gallagher, *J. Mater. Chem.*, **3**, 1275 (1993).
176. S. Dielkus, D. Grosskopf, R. Herbst-Irmer and U. Klingebiel, *Z. Naturforsch., Teil B*, **50**, 844 (1995).
177. T. R. Prout, M. L. Thompson, R. C. Haltiwanger, R. Schaeffer and A. D. Norman, *Inorg. Chem.*, **33**, 1778 (1994).
178. H. Bock, J. Meuret, J. W. Bats and Z. Havlas, *Z. Naturforsch., Teil B*, **49**, 288 (1994).
179. W. Wojnowski, J. Pikies, K. Peters, E.-M. Peters, D. Thiery and H. G. von Schnering, *Z. Anorg. Allg. Chem.*, **620**, 377 (1994).
180. R. Tacke, J. Pikies, F. Wiesenberger, L. Ernst, D. Schomburg, M. Waelbroeck, J. Christophe, G. Lambecht, J. Gross and E. Mutchler, *J. Organomet. Chem.*, **466**, 15 (1994).
181. J. J. Edema, R. Libbers, A. Ridder, R. M. Kellog and A. L. Spek, *J. Organomet. Chem.*, **464**, 127 (1994).
182. G. Galzaferri, R. Imhof and K. W. Tornroos, *J. Chem. Soc., Dalton Trans.*, 3123 (1994).
183. W. Wojnowski, M. Wojnowski, K. Peters, E.-M. Peters and H. G. von Schnering, *Z. Anorg. Allg. Chem.*, **530**, 79 (1985).
184. B. Becker, W. Wojnowski, K. Peters, E.-M. Peters and H. G. von Schnering *Polyhedron*, **9**, 1659 (1990).
185. W. Wojnowski, B. Becker, J. Sassmannshausen, E.-M. Peters, K. Peters and H. G. von Schnering, *Z. Anorg. Allg. Chem.*, **620**, 1417 (1994).
186. S. A. Bourne, L. Johnson, C. Marais, L. R. Nassimbeni, E. Weber, K. Skobridis and F. Toda, *J. Chem. Soc., Perkin Trans. 2*, 1707 (1991).
187. E. A. Babaian, M. Huff, F. A. Tibbals and D. C. Hrncir, *J. Chem. Soc., Chem. Commun.*, 306 (1990).
188. S. A. Bourne, L. R. Nassimbeni, K. Skobridis and E. Weber, *J. Chem. Soc., Chem. Commun.*, 282 (1991).
189. N. Winkhofer, H. W. Roesky, M. Noltemeyer and W. T. Robinson, *Angew. Chem., Int. Ed. Engl.*, **31**, 599 (1992).
190. S. A. Bourne, L. R. Nassimbeni, E. Weber and K. Skobridis, *J. Org. Chem.*, **57**, 2438 (1992).
191. S. S. Al-Juaid, A. K. A. Al-Nasr, C. Eaborn and P. B. Hitchcock, *J. Chem. Soc., Chem. Commun.*, 1482 (1991).
192. J. D. Buynak, J. B. Strickland, G. W. Lamb, D. Khasnis, S. Modi, D. Williams and H. Zhang, *J. Org. Chem.*, **56**, 7076 (1991).
193. A. Fronda, F. Krebs, B. Daucher, T. Werle and G. Maas, *J. Organomet. Chem.*, **424**, 253 (1992).
194. S. S. Al-Juaid, C. Eabron, P. B. Hitchcock and P. D. Lickiss, *J. Organomet. Chem.*, **362**, 17 (1989).
195. R. Tacke, J. Pikies, F. Eiesenberger, L. Ernst, D. Schomburg, M. Waelbroeck, J. Christophe, G. Lambrecht, J. Gross and E. Mutschler, *J. Organomet. Chem.*, **466**, 15 (1994).
196. I. L. Karle, J. M. Karle and C. J. Nielsen, *Acta Crystallogr., Sect. C*, **42**, 64 (1986).
197. V. E. Shklover and Yu. T. Struchkov, *Usp. Khim.*, **49**, 518 (1980); *Chem. Abstr.*, **92**, 198795d (1980).
198. W.-J. Lay, M.-S. Hong, M.-S. Huang and S.-J. Hu, *Jiegou Huaxue O. Struct. Chem.*, **10**, 258 (1991); *Chem. Abstr.*, **116**, 266136f (1992).
199. L. Parkanyi, V. Fulop, P. Hencsei and I. Kovacs, *J. Organomet. Chem.*, **418**, 173 (1991).
200. R. J. Garant, L. M. Daniels, S. K. Das, M. N. Janakiraman, R. A. Jacobson and J. G. Verkade, *J. Am. Chem. Soc.*, **113**, 5728 (1991).

201. Y. Yang, C. Yin, G. Chen and C. He, *Gaodeng Xuexiao Huaxue Xuebao (Chem. J. Chin. Uni.)*, **11**, 102 (1990).
202. M. Nasim, V. S. Petrosyan, G. S. Zeitseva, L. J. Lorbeth, S. Wocadlo and W. Massa, *J. Organomet. Chem.*, **441**, 27 (1992).
203. K. C. K. Swamy, V. Chandrasekhar, J. J. Harland, J. M. Holmes, R. O. Day and R. R. Holmes, *J. Am. Chem. Soc.*, **112**, 2341 (1990).
204. R. Tacke, F. Wiesenberger, A. Lopez-Mras, J. Sperlich and G. Mattern, *Z. Naturforsch., Teil B*, **47**, 1370 (1992).
205. K. Y. Blohowiak, D. R. Traedwell, B. L. Mueller, M. L. Hoppe, S. Jouppi, P. Kansal, K. W. Chew, C. L. S. Scotto, F. Babonneau, J. Kampf and R. M. Laine, *Chemistry of Materials*, **6**, 2177 (1994).
206. A. O. Mozzhurkhin, M. Y. Antipin, Y. T. Struchkov, A. G. Shipov, E. P. Kramerova and Y. T. Baukov, *Metalloorg. Khim.*, **5**, 906 (1992); *Chem. Abstr.*, **118**, 102109q (1993).
207. Yu. E. Ovchinnikov, A. A. Macharashvili, Yu. T. Struchkov, A. G. Shipov and Yu. T. Baukov, *J. Struct. Chem.*, **35**, 91 (1994).
208. F. E. Hahn, M. Keck and K. N. Raymond, *Inorg. Chem.*, **34**, 1402 (1995).
209. G. Sawitzki, H. G. von Schnering, D. Kummer and T. Seshadri, *Chem. Ber.*, **111**, 3705 (1978).
210. J. J. Flynn and F. P. Boer, *J. Am. Chem. Soc.*, **91**, 5756 (1969).
211. M. L. Hoppe, R. M. Laine, J. Kampf, M. S. Gordon and L. W. Burggraf, *Angew. Chem., Int. Ed. Engl.*, **32**, 287 (1993).
212. W. E. Schklower, Y. T. Struchkov, L. E. Guselnikow, W. W. Wolkowa and W. G. Awakyan, *Z. Anorg. Allg. Chem.*, **501**, 153 (1983).
213. W. Wojnowski, K. Peters, D. Weber and H. G. von Schnering, *Z. Anorg. Allg. Chem.*, **519**, 134 (1984).
214. W. Wojnowski, K. Peters, E.-M. Peters and H. G. von Schnering, *Z. Anorg. Allg. Chem.*, **525**, 121 (1985).
215. W. Wojnowski, B. Dreczewski, A. Herman, K. Peters, E.-M. Peters and H. G. von Schnering, *Angew Chem., Int. Ed. Engl.*, **24**, 992 (1985).
216. W. Wojnowski, W. Bochenska, K. Peters, E.-M. Peters and H. G. von Schnering, *Z. Anorg. Allg. Chem.*, **533**, 165 (1986).
217. C. Habben, A. Meller, M. Noltmeyer and G. M. Sheldrick, *Angew. Chem., Int. Ed. Engl.*, **25**, 741 (1986).
218. B. Becker, R. J. P. Corriu, B. J. L. Henner, W. Wojnowski, K. Peters and H. G. von Schnering, *J. Organomet. Chem.*, **312**, 305 (1986).
219. W. Wojnowski, B. Dreczewski, K. Peters, E.-M. Peters and H. G. von Schnering, *Z. Anorg. Allg. Chem.*, **540**, 271 (1986).
220. W. Wojnowski, B. Becker, K. Peters, E.-M. Peters and H. G. von Schnering, *Z. Anorg. Allg. Chem.*, **563**, 48 (1988).
221. D. H. Berry, J. Chey, H. S. Zipin and P. J. Carroll, *Polyhedron*, **10**, 1189 (1991).
222. J. C. J. Bart and J. J. Daly, *J. Chem. Soc., Dalton Trans.*, 2063 (1975).
223. L. Pazdernik, F. Brisse and R. Rivest, *Acta Crystallogr., Sect. B*, **33**, 1780 (1977).
224. W. Wojnowski, K. Przyjemska, K. Peters and H. G. von Schnering, *Z. Anorg. Allg. Chem.*, **556**, 92 (1988).
225. W. Wojnowski, M. Wojnowska, B. Becker and M. Noltmeyer, *Z. Anorg. Allg. Chem.*, **561**, 167 (1988).
226. B. Becker, W. Wojnowski, K. Peters, E.-M. Peters and H. G. von Schnering, *Inorg. Chim. Acta*, **214**, 9 (1993).
227. B. Becker, W. Wojnowski, K. Peters, E.-M. Peters and H. G. von Schnering, *Polyhedron*, **11**, 613 (1992).
228. W. Wojnowski, B. Becker, L. Waltz, K. Peters, E.-M. Peters and H. G. von Schering, *Polyhedron*, **11**, 607 (1992).
229. F. Preuss, M. Steidel and R. Exner, *Z. Naturforsch., Teil B*, **45**, 1618 (1990).
230. B. Becker, K. Radacki, A. Konitz and W. Wojnowski, *Z. Anorg. Allg. Chem.*, **621**, 904 (1995).
231. G. D. Andreetti, G. Calestani and P. Sgarabotto, *J. Organomet. Chem.*, **273**, 31 (1984).
232. R. Minkwitz, A. Kornath and H. Preut, *Z. Anorg. Allg. Chem.*, **619**, 877 (1993).
233. C. R. Lucas, M. J. Newlands, E. J. Gabe and F. L. Lee, *Can. J. Chem.*, **65**, 898 (1987).
234. R. Kewley, P. M. McKinney and A. G. Robiette, *J. Mol. Spectrosc.*, **34**, 309 (1970).
235. V. W. Laurie, *J. Chem. Phys.*, **26**, 1359 (1973).

236. A. R. Hoy, M. Bertram and I. M. Mills, *J. Mol. Spectrosc.*, **46**, 429 (1973).
237. B. Beagley, D. P. Brown and J. M. Freeman, *J. Mol. Struct.*, **18**, 337 (1973).
238. K. Dippel, U. Klingebiel, T. Kottke, F. Pauer, G. M. Sheldrick and D. Stalke, *Z. Anorg. Allg. Chem.*, **584**, 87 (1990).
239. T. Kottke, U. Klingebiel, M. Noltemeyer, U. Pieper, S. Walter and D. Stalke, *Chem. Ber.*, **124**, 1941 (1991).
240. S. Walter, U. Klingebiel, M. Noltemeyer and D. Schmidt-Base, *Z. Naturforsch., Teil B*, **46**, 1149 (1991).
241. M. Andrianarison, U. Klingebiel, D. Stalke and G. M. Sheldrick, *Phosphorus, Sulfur & Silicon*, **46**, 183 (1989).
242. K. Ebata, T. Inada, C. Kabuto and H. Sakurai, *J. Am. Chem. Soc.*, **116**, 3595 (1994).
243. U. Pieper, D. Stalke, S. Vollbrecht and U. Klingebiel, *Chem. Ber.*, **123**, 1039 (1990).
244. E. A. Zelbst, V. E. Shklover, Yu. T. Struchkov, Yu. L. Frolov, A. A. Kashaev, L. I. Gubanova, V. M. D'yakov and M. G. Voronkov, *J. Struct. Chem.*, **22**, 377 (1981).
245. Yu. E. Ovchinnikov, A. O. Mozzhukhin, M. Yu. Antipin, Yu. T. Struchkov, V. P. Baryshok, N. F. Lazareva and M. G. Voronkov, *J. Struct. Chem.*, **34**, 888 (1993).
246. E. A. Zelbst, V. S. Fundamenskii, A. A. Kashaev, L. I. Gubanova, and M. G. Voronkov, *Dokl. Akad. Nauk SSSR*, **312**, 612 (1990); *Chem. Abstr.*, **113**, 174207q (1989).
247. M. G. Voronkov, E. A. Zelbst, V. S. Fundamenskii, A. A. Kashaev and L. I. Gubanova, *Dokl. Akad. Nauk SSSR*, **305**, 1124 (1989); *Chem. Abstr.*, **111**, 174207y (1989).
248. Yu. E. Ovchinnikov, Yu. T. Struchkov, N. F. Chernov, O. M. Trofimova and M. G. Voronkov, *J. Organomet. Chem.*, **461**, 27 (1993).
249. Yu. E. Ovchinnikov, M. S. Sorokon, Yu. T. Struchkov and M. G. Voronkov, *Dokl. Akad. Nauk SSSR*, **330**, 337 (1993); *Chem. Abstr.*, **119**, 282716n (1993).
250. D. Schomburg, *J. Organomet. Chem.*, **221**, 137 (1981).
251. D. A. Dixon, W. B. Farnham, W. Heilemann, R. Mews and M. Noltemeyer, *Heteroatom Chemistry*, **4**, 287 (1993).
252. J. J. Harland, J. S. Payne, R. O. Day and R. R. Holmes, *Inorg. Chem.*, **26**, 760 (1987).
253. R. Tacke, J. Becht, A. Lopez-Mras, W. S. Sheldrick and A. Sebald, *Inorg. Chem.*, **32**, 2761 (1993).
254. A. D. Adley, P. H. Bird, A. R. Fraser and M. Onyszchuk, *Inorg. Chem.*, **11**, 1402 (1972).
255. R. Tacke and M. Muhleisen, *Angew. Chem., Int. Ed. Engl.*, **33**, 1359 (1994).
256. E. B. Lobkovski, V. N. Fokin and K. N. Semeneko, *J. Struct. Chem.*, **22**, 603 (1981).
257. S. N. Gurkova, A. J. Gusev, N. V. Alexeev and Yu. M. Varezhkin, *J. Struct. Chem.*, **25**, 655 (1984).
258. N. A. Avdyukhina, E. B. Chuklanova, I. A. Abronin, A. I. Gusev, V. I. Zhun and V. D. Sheludyakov, *Metalloorg. Khim.*, **1**, 878 (1988); *Chem. Abstr.*, **111**, 153893d (1989).
259. A. Sekiguchi, I. Maruki and H. Sakurai, *J. Am. Chem. Soc.*, **115**, 11460 (1993).
260. G. Fritz, S. Lauble, R. Befurt, K. Peters, E.-M. Peters and H. G. von Schnering, *Z. Anorg. Allg. Chem.*, **619**, 1494 (1993).
261. G. Fritz, A. G. Beetz, E. Matern, K. Peters, E.-M. Peters and H. G. von Schnering, *Z. Anorg. Allg. Chem.*, **620**, 136 (1994).
262. G. Fritz, P. Fusik, E. Matern, K. Peters, E.-M. Peters and H. G. von Schnering, *Z. Anorg. Allg. Chem.*, **620**, 1253 (1994).
263. C. Rudinger, H. Beruda and H. Schmidbaur, *Z. Naturforsch., Teil B*, **49**, 1348 (1994).
264. D. Schmidt-Base and U. Klingebiel, *Chem. Ber.*, **123**, 449 (1990).
265. K. Dippel, U. Klingebiel and D. Schmidt-Base, *Z. Anorg. Allg. Chem.*, **619**, 836 (1993).
266. A. A. Zlota, F. Frolow and D. Milstein, *J. Chem. Soc., Chem. Commun.*, 1826 (1989).
267. K. E. Lee, A. M. Arif and J. A. Gladysz, *Chem. Ber.*, **124**, 309 (1991).
268. U. Schubert, K. Ackermann, G. Kraft and B. Worle, *Z. Naturforsch., Teil B*, **38**, 1488 (1983).
269. B. R. Jagirdar, R. Palmer, K. J. Klabunde and L. J. Radonovich, *Inorg. Chem.*, **34**, 278 (1995).
270. M. Yu. Antipin, A. A. Macharashvili, Yu. T. Struchkov and V. E. Shklover, *Metalloorg. Khim.*, **3**, 998 (1990); *Chem. Abstr.*, **114**, 102172 (1991).
271. A. A. Macharashvili, Y. I. Baukov, E. P. Kramarova, G. I. Olenova, V. A. Pestunovich, Y. T. Struchkov and V. E. Shklover, *J. Struct. Chem.*, **28**, 730 (1987).
272. A. A. Macharashvili, V. E. Shklover, Yu. T. Struchkov, V. A. Pestunovich, Yu. I. Baukov, E. P. Kramarova and G. I. Olenova, *J. Struct. Chem.*, **29**, 759 (1988).
273. A. A. Macharashvili, V. E. Shklover, Yu. T. Struchkov, G. I. Olenova, E. P. Kramarova, A. G. Shipov and Yu. I. Baukov, *J. Chem. Soc., Chem. Commun.*, 683 (1988).

274. A. A. Macharashvili, V. E. Shklover, Yu. T. Struchkov, B. A. Gostevskii, I. D. Kalikhman, O. B. Bannikova, M. G. Voronkov and V. A. Pestunovich, *J. Organomet. Chem.*, **356**, 23 (1988).
275. D. Kummer, S. C. Chaudhry, J. Seifert, B. Deppisch and G. Mattern, *J. Organomet. Chem.*, **382**, 345 (1990).
276. D. Kummer, S. H. A. Halim, W. Kuhs and G. Mattern, *J. Organomet. Chem.*, **446**, 51 (1993).
277. A. A. Macharashvili, Yu. E. Ovchinnikov, Yu. T. Struchkov, V. N. Sergeev, S. V. Pestunovich and Yu. I. Baukov, *Izv. Akad. Nauk SSSR, Ser. Khim.*, 189 (1993).
278. O. Bechstein, B. Ziemer, D. Hass, S. I. Trojanov, V. B. Rybakov and G. N. Maso, *Z. Anorg. Allg. Chem.*, **582**, 211 (1990).
279. G. Klebe and D. Tranqui, *Acta Crystallogr., Sect. C*, **40**, 476 (1984).
280. D. Kummer, S. C. Chaudhry, U. Thewalt and T. Debaerdemaeker, *Z. Anorg. Allg. Chem.*, **553**, 147 (1987).
281. D. Kummer, S. C. Chaudhry, W. Dempeier and G. Mattern, *Chem. Ber.*, **123**, 2241 (1990).
282. D. Kummer, S. C. Chaudhry, T. Debaerdemaeker and U. Thewalt, *Chem. Ber.*, **123**, 945 (1990).
283. V. O. Atwood, D. A. Atwood, A. H. Cowley and S. Trofimenko, *Polyhedron*, **11**, 711 (1992).
284. S. N. Gurkova, A. I. Gusev, V. A. Sharapov, T. K. Gar and N. V. Alexeev, *J. Struct. Chem.*, **20**, 302 (1979).
285. U. Schubert and C. Steib, *J. Organomet. Chem.*, **238**, C1 (1982).
286. A. A. Macharashvili, Yu. I. Baukov, E. P. Kramarova, G. I. Olenova, V. A. Shestunovich, Yu. T. Struchkov and V. E. Shklover, *J. Struct. Chem.*, **28**, 552 (1987).
287. M. Weidenbruch, B. Blintjer, K. Peters and H. G. von Schnering, *Angew. Chem., Int. Ed. Engl.*, **25**, 1129 (1986).
288. S. S. Al-Juaid, A. K. A. Al-Nasr, C. Eaborn and P. B. Hitchcock, *J. Organomet. Chem.*, **429**, C9 (1992).
289. R. West, *Pure Appl. Chem.*, **56**, 163 (1984).
290. R. West, *Science*, **225**, 1109 (1984).
291. R. West, M. J. Fink, M. J. Michaczyk and D. J. De Young, in *Organisilicon and Bioorganosilicon Chemistry* (Ed. H. Sakurai), Ellis Horwood, Chichester, 1985, p. 3.
292. G. R. Raabe and J. Michl, *Chem. Rev.*, **85**, 419 (1985).
293. A. H. Cowley, *Acc. Chem. Res.*, **17**, 386 (1984).
294. A. H. Cowley and N. C. Norman, *Prog. Inorg. Chem.*, **34**, 1 (1986).
295. K. Krogh-Jespersen, *J. Am. Chem. Soc.*, **107**, 537 (1985).
296. B. D. Shepherd, C. F. Campana and R. West, *Heteroatom Chemistry*, **1**, 1 (1990).
297. R. S. Grev, H. F. Schaefer and K. M. Baines, *J. Am. Chem. Soc.*, **112**, 9458 (1990).
298. S. Masamune, S. Murakami, J. T. Snow, H. Tobita and D. J. Williams, *Organometallics*, **3**, 333 (1984).
299. H. Watanabe, K. Takeuchi, N. Fukawa, M. Kato, M. Goto and Y. Nagai, *Chem. Lett.*, 1341 (1987).
300. B. D. Shepherd, D. R. Powell and R. West, *Organometallics*, **8**, 2664 (1989).
301. R. S. Archibald, Y. van der Winkel, A. J. Millevolte, J. M. Desper and R. West, *Organometallics*, **11**, 3276 (1992).
302. B. D. Shepherd, C. F. Campana and R. West, *Heteroatom Chemistry*, **1**, 1 (1990).
303. M. Kira, T. Maruyama, C. Kabuto, K. Ebata and H. Sakurai, *Angew. Chem., Int. Ed. Engl.*, **33**, 1489 (1994).
304. N. Tokitoh, H. Suzuki and R. Okazaki, *J. Am. Chem. Soc.*, **115**, 10428 (1993).
305. A. G. Brook, F. Abdesaken, B. Gutekunst, G. Gutekunst and R. K. Kallury, *J. Chem. Soc., Chem. Commun.*, 191 (1981).
306. A. G. Brook, S. C. Nyburg, F. Abdesaken, B. Gutekunst, G. Gutekunst, R. Krishna, M. R. Kallury, Y. C. Poon, Y.-M. Chang and W. Wong-Ng, *J. Am. Chem. Soc.*, **104**, 5667 (1982).
307. S. C. Nyburg, A. G. Brook, F. Abdesaken, G. Gutekunst and W. Wong-Ng, *Acta Crystallogs.*, **C41**, 1632 (1985).
308. N. Wiberg, G. Wagner and G. Müller, *Angew. Chem.*, **97**, 220 (1985).
309. N. Wiberg, G. Wagner, J. Riede and G. Müller, *Organometallics*, **6**, 32 (1987).
310. N. Wiberg, G. Wagner, J. Riede and G. Müller, *J. Organomet. Chem.*, **271**, 381 (1984).
311. N. Wiberg, G. Wagner, G. Reber, J. Riede and G. Müller, *Organometallics*, **6**, 35 (1987).
312. G. E. Miracle, J. L. Ball, D. R. Powell and R. West, *J. Am. Chem. Soc.*, **115**, 11598 (1993).
313. N. Wiberg, K.-S. Joo and K. Polborn, *Chem. Ber.*, **126**, 67 (1993).
314. T. S. Koloski, P. J. Carrol and D. H. Berry, *J. Am. Chem. Soc.*, **112**, 6405 (1990).

315. B. K. Campion, R. H. Heyn, T. D. Tilley and A. L. Rheingold, *J. Am. Chem. Soc.*, **115**, 5527 (1993).
316. H. F. Schaefer, *Acc. Chem. Res.*, **15**, 283 (1982).
317. Y. Apeloig and M. Karni, *J. Chem. Soc., Chem. Commun.*, 768 (1984).
318. Y. Apeloig and M. Karni, *J. Am. Chem. Soc.*, **106**, 6676 (1984).
319. B. T. Luke, J. A. Pople, M.-B. Krogh-Jespersen, Y. Apeloig, M. Karni, J. Chandrasekhar and P. v. R. Schleyer, *J. Am. Chem. Soc.*, **108**, 270 (1986).
320. M. S. Gordon and T. N. Truong, *Chem. Phys. Lett.*, **142**, 110 (1987).
321. P. v. R. Schleyer and P. D. Stout, *J. Chem. Soc., Chem. Commun.*, 1373 (1986).
322. N. Wiberg, K. Schurz and G. Fischer, *Angew. Chem.*, **97**, 1058 (1985).
323. N. Wiberg, K. Schurz, G. Reber and G. Müller, *J. Chem. Soc., Chem. Commun.*, 591 (1986).
324. G. Reber, J. Riede, N. Wiberg, K. Schurz and G. Müller, *Z. Naturforsch., Teil B*, **44**, 786 (1989).
325. S. Walter, U. Klingebiel and D. Schmidt-Base, *J. Organomet. Chem.*, **412**, 319 (1991).
326. R. Boese and U. Klingebiel, *J. Organomet. Chem.*, **315**, C17 (1986).
327. D. Stalke, U. Pieper, S. Vollbrecht and U. Klingebiel, *Z. Naturforsch., Teil B*, **45**, 1513 (1990).
328. M. Deak, R. K. Hayashi and R. West, *J. Am. Chem. Soc.*, **116**, 10813 (1994).
329. S. G. Pyne, B. Dikic, B. W. Skelton and A. H. White, *Aust. J. Chem.*, **45**, 807 (1992).
330. K. J. Dykema, T. N. Truong and M. S. Gordon, *J. Am. Chem. Soc.*, **107**, 4535 (1985).
331. H. R. G. Bender, E. Niecke and M. Nieger, *J. Am. Chem. Soc.*, **115**, 3314 (1993).
332. M. Driess, S. Rell and H. Pritzkow, *J. Chem. Soc., Chem. Commun.*, 253 (1995).
333. R. Jaquet, W. Kutzelnigg and V. Staemmler, *Theor. Chim. Acta*, **54**, 205 (1980).
334. T. Kudo and S. Nagase, *J. Phys. Chem.*, **88**, 2833 (1984).
335. A. P. Polishchuk, M. Y. Antipin, T. V. Timofeeva, N. N. Makarova, N. A. Golovina and Yu. T. Struchkov, *Soriet Phys. Crystallogr.*, **36**, 50 (1991).
336. T. Rubenstahl, D. W. von Grudenberg, F. Weller, K. Dehnicke and H. Goesmann, *Z. Naturforsch., Teil B*, **49**, 15 (1994).
337. D. W. von Grudenberg, H.-C. Kan, W. Massa, K. Dehnicke, C. Maichle-Mossmer and J. Strahle, *Z. Anorg. Allg. Chem.*, **620**, 1719 (1994).

CHAPTER 6

^{29}Si NMR spectroscopy of organosilicon compounds

YOSHITO TAKEUCHI

Department of Chemistry, Faculty of Science, Kanagawa University, 2946 Tsuchiya, Hiratsuka, Japan 259-12
Fax: +81 463 58 9684;   e-mail: yoshito@info.kanagawa-u.ac.jp

and

TOSHIO TAKAYAMA

Department of Applied Chemistry, Faculty of Engineering, Kanagawa University, 3-27-1 Rokkakubashi, Yokohama, Japan 221
Fax: +81 45 413 9770;   e-mail: takayama@cc.kanagawa-u.ac.jp

I. INTRODUCTION	268
II. NEW TECHNIQUES	269
A. Modern Pulse Sequences	269
B. Combination of NMR and Other Techniques	275
III. NEW COMPOUNDS	277
A. Some Interesting Compounds	277
B. Silicenium Ions	280
C. Compounds with Hypervalent Si Nuclei	284
D. Organically Modified Silicates	290
IV. NMR PARAMETERS	293
A. Theory of ^{29}Si NMR Parameters	293
B. Coupling Constants	296
C. Relaxation Times and Exchange Phenomena	306
V. SOLID-STATE ^{29}Si NMR	309
A. Introduction	309
B. New Techniques	311
C. NMR and X-ray Studies	315
D. Relaxation Times	325
E. Dynamics	329

The chemistry of organic silicon compounds, Vol. 2
Edited by Z. Rappoport and Y. Apeloig   © 1998 John Wiley & Sons Ltd

F. Hypervalent Silicon 334
G. Miscellaneous 343
VI. REFERENCES 350

I. INTRODUCTION

Apart from ^{1}H and ^{13}C NMR ^{29}Si is one of the most frequently used method for NMR spectroscopy. The natural abundance is 4.7%, and its gyromagnetic ratio is -5.314 (10^7 radian T^{-1} s^{-1}), which is about one-fifth of that of the proton. Hence, in the advent of NMR spectroscopy, ^{29}Si NMR spectroscopy has been difficult if not esoteric. The increase in the magnetic field produced by the NMR instruments together with rapid development in computer hardware and software has gradually made ^{29}Si NMR spectroscopy more feasible.

Excellent reviews have occasionally been published on this topic. In the previous volume of this book the topic was also reviewed[1]. Hence in the present review, advances made mostly after that review, i.e. literature for the last ten years (ca 1985–1995), will be covered.

The advances in NMR spectroscopy in the last ten years were enormous. Thus, almost all laboratories which produce papers are equipped with SCM (Super Conducting Magnet) instruments with 400–500 MHz magnets, and 2D or multidimensional NMR experiments are now routinely employed. In addition, solid-state NMR and NMR imaging (MRI) have widened their scope to a considerable extent. ^{29}Si NMR has enjoyed this general progress.

A very noticeable change was found in the methodology associated with NMR. Previously NMR spectroscopies were used for structure elucidation with other spectroscopic means such as UV or IR. During the last decade, the partner of NMR was changed from conventional spectrometry to X-ray analysis. As expected, this combination is very much more powerful and often produces results which cannot be expected from the use of other combinations of methodologies.

At the same time, advances in the field of organosilicon chemistry have also been incredibly large[2]. Characterization of unstable species such as divalent silylenes or compounds with silicon-containing double bonds were successfully achieved for many compounds. Advances in the field of siloxanes and other polymeric materials are also remarkable.

Under such circumstances it would be difficult to cover all the literature of the last ten years. Rather, topics which, from the viewpoint of the authors of this review, demonstrate this fruitful period will be selected and discussed. Thus, firstly, new developments in NMR techniques will be discussed in relation to ^{29}Si NMR. Discussion of the ^{29}Si NMR spectra of novel and hypervalent organosilicon compounds will follow. New data on ^{29}Si NMR parameters are preceded by the recent development in the theory of ^{29}Si NMR. In most of the reviews on NMR, this section will come first. We have inverted this order because the purpose of this review is not to cover all the literature, but to bring to the attention of organosilicon chemists the most crucial points of recent ^{29}Si NMR spectroscopy. The development of solid-state ^{29}Si NMR will be described in a separate section. We deliberately make this section independent since the use of solid-state ^{29}Si NMR is increasing particularly rapidly. This tendency will no doubt be accelerated owing to the relevance of silicon chemistry to materials science.

A compilation of chemical shifts and other spectroscopic data was not intended, since the amount of newly reported data is much too large for a review of reasonable length. In addition such data can be found in previously published reviews and references cited therein[1,3]. Readers who are interested in a compilation of chemical shifts and coupling constants can consult Williams's review in Vol. 1 of this book.

6. ^{29}Si NMR spectroscopy of organosilicon compounds

II. NEW TECHNIQUES

A. Modern Pulse Sequences

Recent improvements in high resolution NMR of solutions have added a new dimension to the structural study of silicon compounds. Many new methods in measuring ^{29}Si NMR have been reported. The low sensitivity of ^{29}Si can be circumvented by isotopic enrichment and/or the use of very high fields. For aqueous ^{29}Si-enriched (99%) sodium silicate solutions, experiments at high field (11.7 T) have been reported by Harris and coworkers[4]. Furthermore, a large number of species have been identified[5] using two-dimensional (2D) experiments such as COSY for observation of ^{29}Si–^{29}Si correlations in chemical exchange studies.

For an effective measurement of ^nJ(SiH) ($1 < n < 4$) values in silylated silyl enol ethers, a selective population transfer experiment was modified so that selective decoupling was applied during acquisition. ^{1}H–^{29}Si heteronuclear COSY confirmed the presence of silylated groups in the molecule[6].

Another way to improve sensitivity is to use NMR methods such as INEPT and DEPT[7–9] which, through scalar spin–spin coupling, will allow polarization to be transferred from the abundant high-γ protons to the rare low-γ ^{29}Si nuclei. Some ten years ago, measurements of ^{29}Si NMR spectra of trimethylsilyl derivatives by employing the INEPT technique were reported[10]. The authors suggested that the use of this technique will substantially shorten the measuring time and widen the scope of the analytical applications of ^{29}Si NMR spectroscopy.

Until now the double-quantum coherence spectroscopy has not often been used for silicon frameworks. In double-quantum coherence spectroscopy the silicon nucleus is almost ten times as sensitive as the ^{13}C nucleus because of the higher natural abundance of ^{29}Si (4.7% compared with 1.1% ^{13}C). Disadvantages of the ^{29}Si nucleus are the long relaxation times (T_1) and the negative nuclear Overhauser effect (NOE). In practice, this means that a very long recording time is necessary unless a relaxation reagent such as Cr(acac)$_3$ is used.

The usefulness of ^{29}Si-INADEQUATE spectroscopy has been demonstrated by Hengge and Schrank[11] for the four known cyclosilanes: trimethylsilylnonamethylcyclopentasilane (Si$_5$Me$_9$–SiMe$_3$, **1**), dimethylsilylnonamethylcyclopentasilane (Si$_5$Me$_9$–SiMe$_2$H, **2**), bis(undecamethylcyclohexasilanyl) ((Si$_6$Me$_{11}$)$_2$, **3**) and bis(undecamethylcyclohexasilanyl) dimethylsilane ((Si$_6$Me$_{11}$)$_2$- SiMe$_2$, **4**).

The chemical shifts and the ^{29}Si–^{29}Si coupling constants (determined from the 1D-INADEQUATE spectra) are summarized in Table 1. Figure 1 shows the 1D-INADEQUATE spectrum of **4**. If all the ^{29}Si–^{29}Si couplings can be determined, the structure of the compound can be easily derived.

When resonance frequencies are very close, overlapping or extinguishing (both satellite signals are in antiphase) of signals might occur. In such cases the 2D-INADEQUATE experiment provides the desired information about the structure, though the long recording time is sometimes a serious disadvantage. The applicability of the 2D-INADEQUATE technique to organosilicon compounds is demonstrated by the spectrum of **1** shown in Figure 2. Note, however, that in this case all the required information can be obtained from the 1D experiment[10].

A second means of enhancing sensitivity is to increase the concentration of the sample in solution. Thus, 2D-INADEQUATE studies of ^{29}Si–^{29}Si correlations on polysilanes at natural ^{29}Si abundance at high concentrations (1–2 M) was reported[12]. Sophisticated solid-state NMR studies (see Section V) have also been performed, including two-dimensional

Structures (1)–(4):

(1) Me–Si¹(Me)–Si¹(Me)–Me core with Si²(Me)(Me) branches and Si³(Me)(Si⁴Me₃)

(2) Analogous to (1) but with Si⁴Me₂H in place of Si⁴Me₃

(3) Two Si⁴ centers linked, each bearing Si¹ with two Si²–Si³ branches

(4) As (3) but with central Si⁵ inserted between the two Si⁴ groups

TABLE 1. ^{29}Si NMR data of compounds **1–4**

	Chemical shifts (ppm)				^1J (^{29}Si–^{29}Si) (Hz)		
1[a]	1 : −40.9	2 : −36.1	3 : −83.2	4 : −10.2	1,2 : 61.1	2,3 : 51.1	3,4 : 62.5
2[b]	1 : −41.0	2 : −36.2	3 : −83.3	4 : −34.2	1,2 : 61.6	2,3 : 51.8	3,4 : 60.5
3[c]	1 : −42.6	2 : −39.6	3 : −36.3	4 : −68.2	1,2 : 61.4	2,3 : 59.6	3,4 : 51.1
4[d]	1 : −42.5	2 : −40.3	3 : −36.6	4 : −73.5	1,2 : 61.4	2,3 : 59.8	3,4 : 50.7
	5 : −23.4				4,5 : 46.3		

[a] D2 40 μs, 900 mg, $T = 20\,°C$.
[b] D2 60 μs, 950 mg, $T = 20\,°C$.
[c] D2 50 μs, 650 mg, $T = 60\,°C$.
[d] D2 60 μs, 500 mg, $T = 60\,°C$.
Reproduced by permission of Elsevier Science from Reference 11.

COSY magic angle spinning NMR experiments on silicate glasses[13] and COSY and 2D-INADEQUATE measurements on Zeolites[14–16].

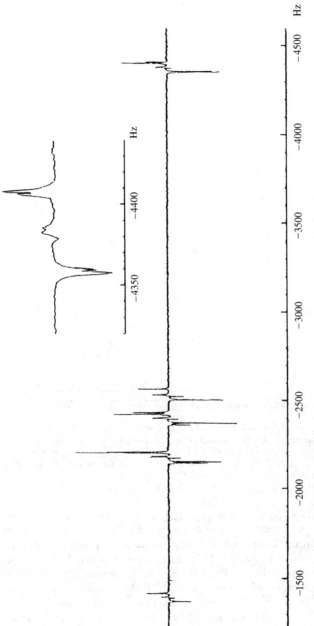

FIGURE 1. ID-INADEQUATE spectrum of $(Si_6Me_{11})_2SiMe_2$ (**4**): 500 mg silane in 2 ml C_6D_6, 30 mg $Cr(acac)_3$ added; $T = 333$ K; 6400 FIDs accumulated; gentle exponential multiplication; insert shows the couplings at Si^4 in an expanded scale. Reproduced by permission of Elsevier Science from Reference 11

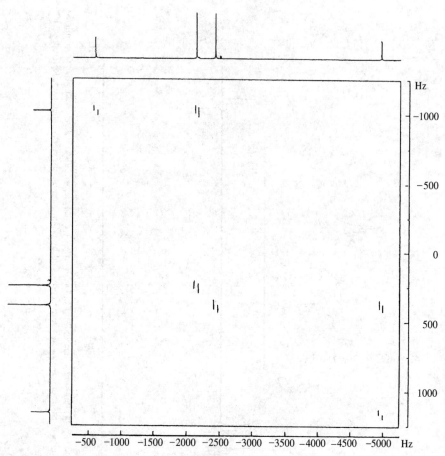

FIGURE 2. 2D-INADEQUATE spectrum of $(Si_5Me_9-SiMe_3)$ (1); 900 mg silane in 2 ml C_6D_6 containing 30 mg $Cr(acac)_3$; $^{29}Si-^{1}H$ couplings refocused; 256 rows recorded, each 96 FIDs accumulated; 4×4 K data matrix, spectral width in F1-dimension is 6000 Hz (2.9 Hz digital resolution); mild gaussian multiplication in both dimensions; 24 h performance time. Reproduced by permission of Elsevier Science from Reference 11

^{13}C 2D-INADEQUATE spectroscopy can conveniently disclose the connectivity of each carbon atom embedded in a molecule and has found widespread use in structural elucidation of organic compounds[17]. This technique is extremely powerful when the size of $^{13}C-^{13}C$ one-bond coupling (^1J) is much larger than that of long-range couplings (^nJ, $n \geqq 2$), which is usually the case in the acyclic and unstrained alicyclic systems so far investigated[18]. Using the INEPT INADEQUATE technique specifically modified for ^{29}Si nuclei[19], Masamune and coworkers[12] examined $^{29}Si-^{29}Si$ couplings in strained polycyclic compounds such as tetracyclo[3.3.0.0^{2,7}0^{3,5}]octasilane (5) and tricyclo[2.2.0.0^{2,5}]hexasilane (6).

The 2D NMR spectrum of the tetracycle 5a (Figure 3a) reveals the presence of 15 $^{29}Si-^{29}Si$ couplings in the range of ca 20 to 40 Hz. Since this system is constructed

6. ^{29}Si NMR spectroscopy of organosilicon compounds

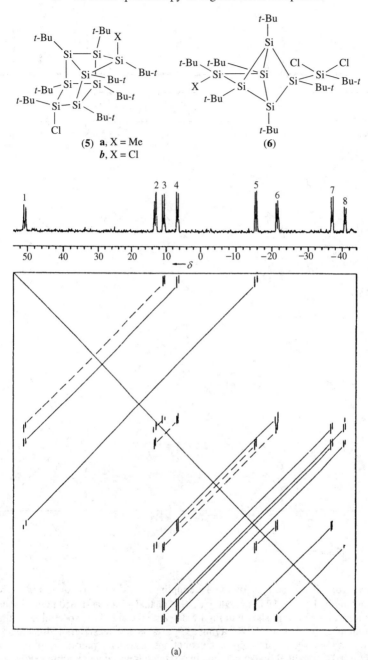

FIGURE 3. 53.7-MHz ^{29}Si–^{29}Si 2D-INEPT-INADEQUATE spectrum (a) Tetracycle **5a** in C$_6$D$_6$. (b) Tricycle **6** in CDCl$_3$ (40 °C). and projection of the 2D spectrum on the F2 axis (top). Signals connected by solid lines correspond to one-bond interactions (^1J); those connected by dotted lines denote long-range couplings (^nJ). Reproduced by permission of VCH Verlagsgesellschaft from Reference 12

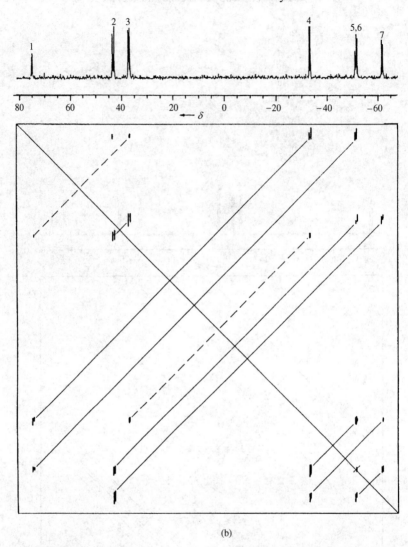

FIGURE 3. *(continued)*

from eleven Si—Si bonds, four of the 15 couplings must be long-range (^nJ). Through somewhat lengthy but logical arguments[20], ^{29}Si signals 1–8 are assigned to the Si nuclei of compound **5a** in the manner shown in Figure 3a. With the aid of the spectral assignment of **5a**, the assignment of signals of **5b** (having a C_2 axis) is now feasible. In **5b** there are three pairs of observable, significant long-range couplings of which a pair of couplings is rather small (5.8 Hz). Compound **6** also exhibits two sizable long-range couplings demonstrated by its spectrum (Figure 3b) as there are eleven observed *J*s of approximately 15 to 40 Hz while only nine Si—Si bonds build the framework of **6**. The ^{29}Si–^{29}Si coupling constants for **5** and **6** are tabulated in Table 2.

6. ^{29}Si NMR spectroscopy of organosilicon compounds

TABLE 2. ^{29}Si–^{29}Si coupling constants for the polycyclic polysilanes **5a**, **5b** and **6**[a]

Compound	^1J(Hz)		^2J ($n \geq 2$) (Hz)
5a	42.2(2–6)	41.8(1–5)	24.1 (1–3)
	41.0(2–3)	33.4(1–4)	23.1(4–5)
	30.8(4–7)	28.9(3–8)	22.5(3–6)
	28.3(3–7)	25.7(4–8)	19.9(2–4)
	25.7(5–6)	18.0(5–7)	
	17.3(6–8)		
5b	41.0(1–3, 1'–3')		23.5(1–2', 1'–2)
	35.1(1–2, 1'–2')		23.5(2–3, 2'–3')
	29.3(2–4, 2'–4, or 2'–4')		5.8(1–4, 1'–4')
	17.5(3–4, 3'–4')		
6	43.0(2–3)	36.1(1–4)	13.7(1–3)
	34.1(2–6)	32.2(2–7)	16.6(3–4)
	29.3(1–5)	27.4(5–7)	
	24.5(4–6)	23.4(4–7)	
	not determined(5–6)[b]		

[a]The silicon nuclei of each molecule are numbered according to the order of their chemical shifts as shown in the structures below. [b]The chemical shifts of these ^{29}Si nuclei were too close to permit accurate measurement of the coupling constant between them. Reproduced by permission of VCH Verlagsgesellschaft from Reference 12.

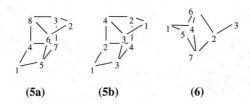

(5a)   (5b)   (6)

B. Combination of NMR and Other Techniques

It is usual for NMR spectroscopy to be employed together with other physicochemical methods in a complementary manner. Combined use of NMR and X-ray crystallography is already a main stream among the methodologies of physical organic chemistry. It must be pointed out that not only the combination of solution NMR and X-ray analysis but also that of solid-state NMR and X-ray analysis is possible.

Thus, Brook and coworkers[21] prepared a series of tetrakis(trimethylsilyl)ethynyl derivatives of Si, Ge, Sn and Pb and studied their structure by means of NMR, Mössbauer and X-ray methods. The combined results clearly indicated that there exists an electronic interaction between the central metal atom and distal Me$_3$Si groups, leading to a strong shielding of the central atom which was clearly demonstrated by chemical shifts of respective nuclei (Figure 4). The presence of the four distal Me$_3$Si groups serves to increase substantially the upfield shifts of the central atom. The chemical shifts reported here are -101.6, -188.5, -384.5 [-356.3 for (HC≡C)$_4$Sn], and -760.7 ppm for ^{29}Si, ^{73}Ge, ^{119}Sn and ^{207}Pb respectively.

Many other examples of this combination of techniques will be found in various parts of this review. Probably the most persuasive example of this combination are the criteria applied in attempts to establish the structure of silicenium ions. This topic will be discussed in the next section.

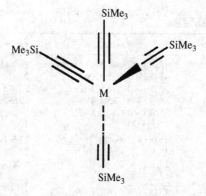

(a) M = Si, Ge, Sn, Pb

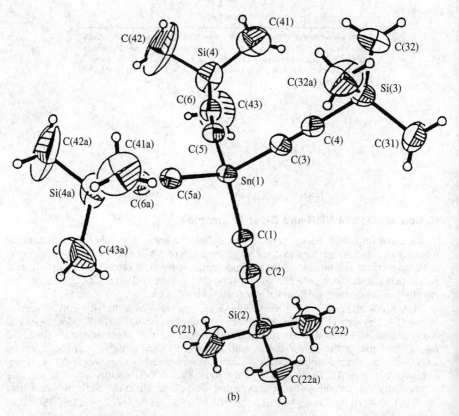

(b)

FIGURE 4. (a) The structure of the compounds investigated; (b) ORTEP plot from the X-ray crystal structure of $(Me_3SiC\equiv C)_4 Sn$. Reproduced by permission of the Canadian Chemical Society from Reference 21

III. NEW COMPOUNDS

A. Some Interesting Compounds

The last ten years have experienced a great advance in the synthesis of novel and unusual organosilicon compounds which are interesting in view of their structure or relevance to theory. ^{29}Si NMR study of such interesting and unusual compounds will be treated in this section.

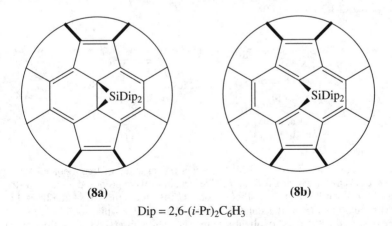

(7)

The first [2.2]paracyclophane bridged by silicon, tetrasila[2.2]paracyclophane (**7**) was prepared and characterized by Sakurai and coworkers[22] with the aid of UV, IR, ^{1}H, ^{13}C and ^{29}Si NMR spectra ($\delta = 6.45$) together with X-ray crystallographic data. Cyclophane **7** displays a strong $\sigma-\pi$ mixing between Si—Si bonds and aromatic rings as evidenced by a large red shift in the UV spectra.

Ando and coworkers[23] reacted C_{60} with bis(2,6-diisopropylphenyl)silylene generated *in situ* by the photolysis of the corresponding trisilane. The dark brown powder obtained (58%) exhibits ^{29}Si signal at $\delta - 72.74$ which, together with other experimental evidence, supports the silirane structure (**8a**) rather than the bridged annulene structure (**8b**).

(8a) (8b)

Dip = 2,6-(*i*-Pr)$_2$C$_6$H$_3$

A photochemical 1 : 1 addition of bis(alkylidene)silacyclobutanes to C_{60} was reported (equation 1)[24]. The structure of adducts **9a** and **9b** were identified by various means including ^{29}Si NMR spectra (**9a**; $\delta = -14.30, 4.20$; **9b**; $-11.47, 8.12$). A Si—H HMBC heteronuclear shift correlation NMR spectra of **9a** was also taken. A photochemical addition of octaarylcyclotetrasilane to C_{60} was also reported, together with full ^{29}Si NMR data, by Ando and coworkers[25].

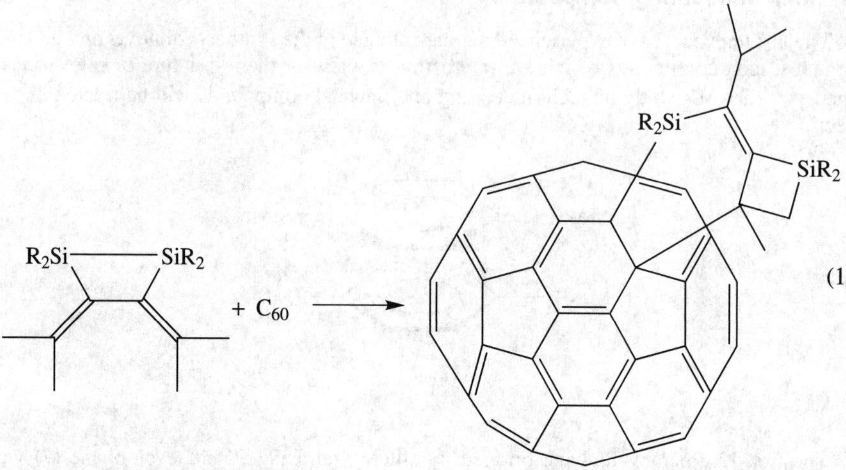

$$R_2Si=C(Me)-C(Me)=SiR_2 + C_{60} \longrightarrow \text{[C}_{60}\text{ adduct]} \tag{1}$$

9a: R = n-Pr
9b: R = Et

Another interesting class of compounds reported by Ando and coworkers is tetrasilacyclohexynes and polysilabridged allenes. Thus, 1,1,2,2,3,3,4,4-octaalkyl-1,2,3,4-tetrasilacyclohexynes (**10a** and **10b**) were prepared and their ^{29}Si NMR spectral data were reported (**10a**; δ = −30.8, −19.2; **10b**: −17.7, −8.3)[26]. ^{29}Si NMR spectral data (−32.2, −7.9) were also reported for 1,3,5,5-tetraphenyl-4,4,6,6-tetramethyl-4,5,6-trisilacyclohexa-1,2-diene (**11**)[27].

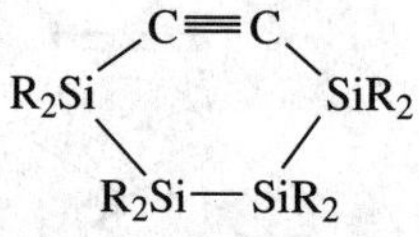

(**10**)

a, R = M
b, R = Et

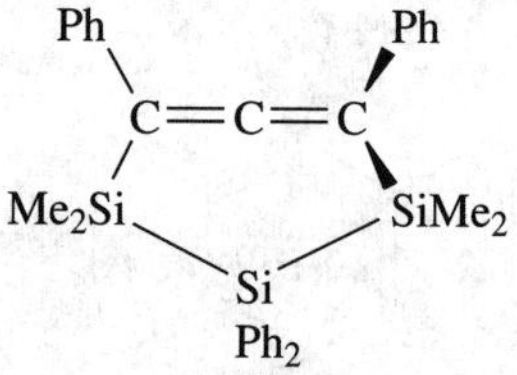

(**11**)

West and coworkers[28] used J(SiSi) to discuss the structure of long-debated 1,3-cyclodisiloxanes (**12a**) for which a few alternative structures were proposed. One such structure retains the σ bond between two silicon atoms (**12b**). In a structure such **12a** as **12b**, the J(SiSi) should be close to the standard values of 80–90 Hz while for **12a** ^2J(Si−O−Si) will be observed with values of about 4 ppm. The values observed for a variety of 1,1,3,3-tetraaryl-1,3-cyclodisiloxanes are in the range of 3.85–4.02 Hz, which support structure **12a**. There is another proposal for the structure of 1,3-cyclodisiloxanes (dibridged π complex) and the authors suggested this structure is also consistent with the observed coupling constants.

The ^1J(Si=Si) values of a variety of unsymmetrically substituted disilenes were first reported in the same paper. The values are in the range 155–158 Hz, which is ca 1.8

6. ^{29}Si NMR spectroscopy of organosilicon compounds 279

times the standard $^1J(SiSi)$ values for organodisilanes. This is to be compared with the corresponding values in the carbon series (1.98 times; 67.6 Hz for ethylene and 34.6 Hz for ethane).

Watanabe and coworkers[29] reported a ^{29}Si chemical shift of a cyclotrisilane, i.e, hexaneopentylcyclotrisilane ($\delta - 81.68$), which is substantially upfield as compared with the values for larger rings.

Hengge and coworkers[30] assigned the structure and ^{29}Si δ values of halodimethylsilylnonamethylcyclopentasilanes and halononamethylcyclopentasilanes based on $^1J(SiSi)$ and $^2J(SiSi)$ derived from ^{29}Si INADEQUATE and ^{29}Si INEPT INADEQUATE spectra. The compounds exhibit good correlations between $^1J(SiSi)$ chemical shifts and Pauling electronegativities. Some ^{29}Si NMR spectral data were also reported for a series of cyclopolysilanes $(R^1R^2Si)_n (n = 3-6)$[31].

Watanabe and coworkers reported the synthesis and structure of a variety of polysilacycloalkanes in which other heteronuclei are also incorporated in the ring. An example is a series of peralkyl-1-germa-2,3,4-trisilacyclobutanes[32]. ^{29}Si NMR spectral data are as follows: 1,1-bis(trimethylsilyl)methyl-1-germa-2,2,3,3,4,4-hexaisopropyl-2,3,4-trisilacyclobutane, δ 1.13, 2.23. 5.78; 1,1-bis(trimethylsilyl)methyl-1-germa-2,2,3,3,4,4-hexaneopentyl-2,3,4-silacyclobutane, -23.45, -11.82, 1.83. A germanium analogue of the above compounds, 1,1-diphenyl-1-germa-2,2,3,3,4,4,5,5- octaisopropyl-2,3,4,5-tetrasilacyclopentane, was also reported by Watanabe and coworkers[33]. The ^{29}Si δ values for the compound are -13.76 and -12.16, respectively.

Stable disilenes have long been an intriguing target for organosilicon chemists. The first successful isolation of a stable disilene derivative, tetramesityldisilene was reported by West and coworkers[34] as early as 1981. Since then several investigators worked toward this target. Watanabe and coworkers[35] reported an air-stable disilene, tetrakis(2,4,6-triisopropylphenyl)disilene. ^{29}Si chemical shift of this compound was reported as $\delta+53.4$.

Okazaki and coworkers[36] reported some extremely hindered and stable disilenes, (E)- and (Z)-Tbt(Mes)Si=Si(Mes)Tbt where Tbt = 2, 4, 6-tris[bis(trimethylsilyl)methyl]-phenyl and Mes = 2, 4, 6-trimethylphenyl. ^{29}Si chemical shifts of the (E)-isomer showed only one signal at δ 66.49 while that of the (Z)-isomer exhibited four peaks with roughly equal intensity at 56.16, 56.74, 57.12 and 58.12, for which the authors claimed the possible existence of two or more conformational isomers on the NMR time scale.

Silylenes, the carbene equivalent of organosilicon compounds, has been another intriguing target. Okazaki and coworkers carried out thermolysis of the hindered disilenes mentioned above with the expectation that a silylene, Tbt(Mes)Si:, will be formed. The thermolysis was monitored by means of ^{29}Si NMR spectra, but the generation of silylene was not proved by NMR, although trapping experiments established the generation of silylene[37].

West and coworkers[38] isolated a silylene (13) stable enough to be distilled at 85 °C/0.1 Torr and reported ^{29}Si (and other nuclei) NMR chemical shift ($\delta + 78.3$). The structure was also confirmed by X-ray crystallographic analysis (Figure 5) and quantum chemical calculations. It was suggested that the compound has an aromatic ground state.

(12a) (12b) (13)

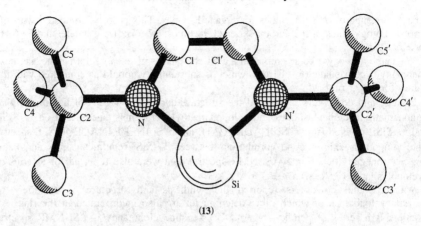

(13)

FIGURE 5. Molecular model of **13**. Hydrogen atoms have been omitted for clarity. Reprinted with permission from Reference 38. Copyright 1994 American Chemical Society

B. Silicenium Ions

There has been much dispute on the question whether triorganosilicenium cations, (R_3Si^+), can be generated in solution and whether some of the features found in the rich carbocation chemistry can also be found in silyl cation chemistry[39] (See Schleyer, Chapter 44, and Lickiss, Chapter 11 in this book).

In a recent publication, Lambert and Zhang[40] have presented a promising new procedure for isolating silicenium cations (equation 2)

$$Ph_3C^+X^- + R_3Si-H \longrightarrow Ph_3C-H + R_3Si^+X^- \qquad (2)$$

by using as a counterion the weakly coordinating anion tetrakis(pentafluorophenyl)borate, $(C_6F_5)_4B^-$ (TPFPB$^-$), in aromatic solvents such as benzene. From measured ^{29}Si NMR chemical shifts, they concluded that they had obtained R_3Si^+ (R = Me, Et, Pr, Me$_3$Si) ions which do not have interaction with anions but have weak interaction with the aromatic solvent, and therefore with a nearly free cationic structure. Lambert and coworkers[41] added further support to the postulated existence of silicenium ions by investigating the crystal structure of Et$_3$Si$^+$ TPFPB$^-$ (**14a**). The unit cell of the crystal contains four molecules of **14a** and eight molecules of the aromatic solvent toluene, which was used for recrystallizing **14a**. Each molecule of **14a** has one associated toluene molecule with the closest contact between Si and a toluene C atom being 2.18 Å. The geometry at the Si atom is not planar, as expected for a silicenium cation, but pyramidal, while the toluene ring is essentially planar and does not indicate any stronger interactions with the triethylsilyl moiety[41].

Furthermore, Lambert and coworkers[42] obtained a white solid from the reaction of silanes and Ph$_3$C$^+$ TPFPB$^-$, which was examined as a solid or in solution. ^{29}Si chemical shift is highly dependent on the nucleophilicity of the solvent, with large downfield shifts. The shift in toluene is *ca* 100 ppm, far short of the expectation for a fully trigonal silicenium ion. For (Me$_3$Si)$_3$Si$^+$ TPFPB$^-$ (**14c**) the shift is 111 ppm, much closer to the trigonal ideal. ^{29}Si shifts for the solid are almost identical to those in benzene solution. Based on the X-ray crystallographic data of **14a** which is coordinated with toluene in the solid state, they proposed that the ions are best termed silicenium cations with weak $\eta^1-\pi$ coordination to toluene.

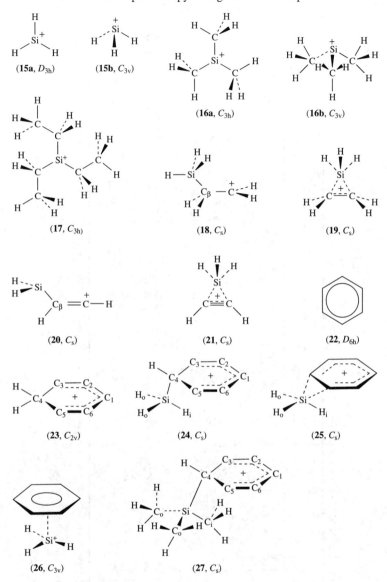

FIGURE 6. Structure and equilibrium conformation of compounds **15–27**. Reproduced by permission of Elsevier Science from Reference 43

A thorough theoretical study was carried out to cast light on this problem[43]. Olsson and Cremer calculated IGLO (individual gauge for localized orbitals) NMR chemical shifts (^{13}C and ^{29}Si) for molecules **15–27** shown in Figure 6, namely silicenium cations **15–17**, reference compounds **18–23** and silicenium–benzene interaction complexes **24–27** employing the [7s6p2d/5s4p1d/3s1p] basis set recommended by Kutzelnigg and Schindler[44].

TABLE 3. IGLO/[7s6p2d/5s4p1d/3s1p] NMR chemical shifts[a]

Molecule		Symmetry	Chemical shifts	
			δ^{29}Si	δ^{13}C
SiH_3^+	15a	D_{3h}	270.2	
	15b	C_{3v}	363.1	
$(CH_3)_3Si^+$	16a	C_{3h}	355.9	9.0
	16b	C_{3v}	397.1	23.5
$(CH_3CH_2)_3Si^+$	17	C_s	376.9	16.5(Cα), 7.5(Cα)
$H_3SiCH_2CH_2^+$	18	C_s	−29.8	80.9(Cα), 262.8(C$^+$)
	19	C_s	−54.1	139.5
H_3Si CHCH$^+$	20	C_s	−51.7	63.6(Cα), 288.3(C$^+$)
	21	C_{2v}	−31.9	89.1
C_6H_6	22	C_{6h}		135.8
$C_6H_7^+$	23	C_{2v}		205.2(C2), 134.2(C2), 199.9(C3), 43.8(C4)
$H_3Si\ C_6H_6^+$	24	C_s	−23.8	176.7(C1), 136.5(C3), 180.8(C3), 77.3(C4)
	25	C_s	10.5	152.1(Cγ), 145.0(Cβ), 126.3(Cα)
	26	C_{3v}	201.9	138.3
$(CH_3)_3SiC_6H_6^+$	27	C_s	83.1 (83.6)[b]	166.2(C1), 137.1(C2), 163.1(C3), 91.7(C4) −4.4(Me$_i$), 4.5(Me$_o$)

[a] All shifts in ppm relative to TMS. For the numbering of atoms, see structures. The IGLO calculations for **26** have been conducted with the 6-31 G (d) basis set.
[b] Experimental δ^{29}Si value.
Reproduced by permission of Elsevier Science from Reference 43.

6. ^{29}Si NMR spectroscopy of organosilicon compounds 283

The calculated chemical shifts are tabulated in Table 3. They imply the three conclusions given below.

(1) The postulated silicenium cations are covalently bonded C-centered π-complexes between SiR$_3$$^+$ and aromatic solvents. Formally, these may be described as Wheland σ complexes, in which the silicenium cation character is totally lost.

(2) Nevertheless, the use of the TPFPB$^-$ in connection with aromatic solvents opens an intriguing route to nearly free silicenium ions in the form of ring-centered SiR$_3$$^+$-arene van der Waals complexes with δ^{29}Si values of about 310–320 ppm. These complexes can be best generated if both the Si and the benzene ring are substituted by sterically demanding groups.

(3) Compound **14c** investigated by Lambert and Zhang realizes by 70% the situation of a free tricovalent silicenium cation in a condensed phase. These authors further suggested that experiments with properly substituted aromatic solvents and **14c** will generate a free silicenium cation in solution. Another interesting theoretical study was reported by Frenking and coworkers[45] for a series of silaguanidium Si(NH$_2$)$_3$$^+$ ions and derivatives in order to clarify the effective stabilization by the heteroatoms.

Reed and coworkers[46] employed hexahalocarboranes (X$_6$B$_{11}$H$_6$; X = Cl, Br, I) as counterions to stabilize and characterize i-Pr$_3$Si$^+$. They employed, as others did, two experimental criteria of developing silicenium ion character: (i) downfield ^{29}Si chemical shifts and (ii) the geometrical approach of silicon toward planarity. They reported that the chemical shift of i-Pr$_3$Si(Cl$_6$B$_{11}$H$_6$) is 115 ppm while the upper limit of the expectation value for i-Pr$_3$Si$^+$ is around 220 ppm, and concluded that their compound may have >50% silicenium ion character. The C–Si–C angle as determined by X-ray crystallographic analysis of these carborane derivatives is in the range of 111–120°, in good agreement with the ideal angle for tricoordination.

Lambert and Zhao[47] chose a new compound, Mes$_3$Si$^+$ TPFPB$^-$ (**14d**) for preparing a free silyl cation. According to their molecular mechanics calculations, the *ortho* methyl groups shield the silicon center from attack by the large nucleophile but are prohibited by their geometry from interacting with the silicon. The ^{29}Si chemical shift of **14d** is $\delta = 225.5$ in C$_6$D$_6$. This is the highest value ever observed for species with silicenium ion character. The authors remarked that though the possibility of existence of interaction between the silicenium ion and TPFTB anion still remains to some extent, the chemical shift strongly favors a nearly free, tricoordinate silicenium ion.

(**14d**)

It must be added that a new theoretical calculation of triarylsilicenium moieties nicely reproduced the experimental results of **14d**[48]. The calculated ^{29}Si chemical shift is 251.4. The 25.0 ppm difference may be the result of the applied computational method. Thus, the long debated controversy of the silicenium cation problem seems to be solved[49].

TABLE 4. ^{29}Si NMR data for silylnitrilium species

Species generated	^{29}Si $(\delta)^a$
[(t-Bu)$_2$(s-Bu)$_2$Si(NCPr)]$^+$ TFPB$^-$	30.39
[t-BuMe$_2$Si(NCPr)]$^+$ TFPB$^-$	36.50
[Et$_3$Si(NCPr)]$^+$ TFPB$^-$	37.01
[Ph$_2$MeSi (NCCD$_3$)]$^+$ TFPB$^-$	4.23
[(t-Bu)$_2$ SiH(NCCD$_3$)$^+$ TFPB$^-$	19.3 (d, J_{SiH} = 242 Hz)
[(PrCN)Si(OCH$_2$CH$_2$)$_3$N]$^+$ TFPB$^-$	−94.9

^{a}TFPB = [bis(3,5-trifluoromethyl)phenyl] borate.
^{b}All ^{29}Si NMR spectra were taken in butyronitrile except for Ph$_2$MeSiH and t-Bu$_2$SiH$_2$, which were taken in CD$_3$CN.
Reproduced by permission of the American Chemical Society from Reference 50.

^{29}Si chemical shift of a nitrile-stabilized silicenium ion (silylnitrilium ion), R$_3$Si(NCC$_3$H$_7$)$^+$B[C$_6$H$_3$(CF$_3$)$_2$-3,5]$^{4-}$ is reported[50]. The compound was prepared by the reaction between trityl tetrakis [bis(3,5-trifluoromethyl)phenyl] borate and trialkylsilanes in butyronitrile. ^{29}Si chemical shifts are summarized in Table 4. All of the ^{29}Si NMR shifts for the alkylated or phenylated silicenium cations are consistently 12–30 ppm downfield of that of the starting hydrosilanes. Considering that only one nitrile group is coordinating with the bulky (t-Bu)$_2$(s-Bu) Si moiety but that two or three nitrile groups may coordinate with the less bulky silyl moieties, the shifts for the five- or six-coordinate silicon species would be expected to be further upfield than the values observed. The authors suggested that the silicenium cation generated in their study was four-coordinate, with stabilization resulting from only one nitrile group.

C. Compounds with Hypervalent Si Nuclei

Penta- and hexacoordinate silicon compounds have recently attracted a great deal of interest from structural and mechanistic points of view[51]. In particular, pentacoordinate anionic siliconates have long been recognized as the reaction intermediates in nucleophilic substitution at silicon atoms[52]. The geometry of pentacoordinate anionic siliconate was first confirmed in 1981, by Schomburg[53], who performed an X-ray structural analysis of [PhSiF$_4$][n-Pr$_4$N] and found that the geometry about the silicon atom was trigonal bipyramidal with two fluorine nuclei preferentially occupying the apical positions.

Since then a large number of reports have dealt with the structure and stability of anionic pentacoordinate mono(siliconates), which contain only one silicon atom in a molecule. Damrauer and his coworkers first reported nonhygroscopic fluorosiliconates as the K$^+$. 18-crown-6 salts[54]. Subsequently, Holmes and his coworkers have reported the isolation and structural analysis of a series of [R$_n$SiF$_{5-n}$]$^-$ species[55,56], especially sterically crowed fluorosiliconates such as [Mes$_2$SiF$_3$]$^-$ and [(TTBP)SiF$_4$]$^-$, where TTBP stands for the 2,4,6-tri-t-butylphenyl group. These studies demonstrated that the bond parameters around the silicon atom depend highly on the steric hindrance of the organic groups R and on the number of fluorine ligands[55]. A number of pentacoordinate mono(siliconates) containing anionic chelate ligands such as catecholate, pinacolate, and Martin ligands[57] which involve the first examples of cyanosiliconates[58], silylsiliconates[59] and pentaalkoxysiliconate[60] have also been studied.

One interesting example of pentacoordinate silicon is a 1,2-oxasiletanide, reported by Okazaki and coworkers[61], which can be regarded as the intermediate of the Peterson reaction. Thus, treatment of a β-hydroxysilane (**28**) with butyllithium afforded oxasiletanide

(29). A large upfield shift of ^{29}Si chemical shift from **28** (δ 10.66) to **29** (δ − 72.45) strongly supports the structure of a pentacoordinate silicate.

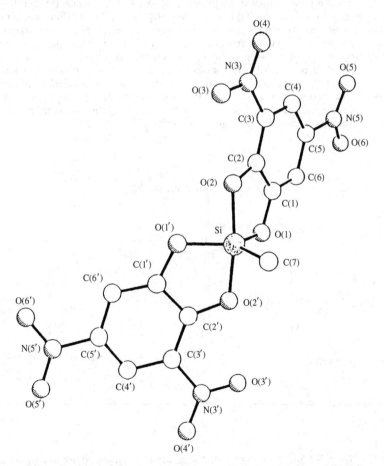

Evans and coworkers[62] described the synthesis and structural study of a series of pentacoordinate bis(catecholate) complexes of silicon(IV). The crystal structure of the anion of [Et$_3$NH][SiMe(3,5-dncat)$_2$] [H$_2$(3, 5-dncat) = 3, 5-dinitrocatechol] is reported (Figure 7).

FIGURE 7. The crystal structure of the anion in [Et$_3$NH][SiMe(3,5-dncat)$_2$]. Reproduced by permission of the Royal Society of Chemistry from Reference 62

Further analysis of the NMR data of these compounds reveals that the 'stronger', i.e. more electronegative, catechols give rise to complexes with more square pyramidal character. Conversely, the effect of the R group may be steric, i.e. the phenyl moiety twists the catechols out of the square plane, thus increasing the trigonal bipyramidal character. Two mixtures were prepared by dissolution of equivalents of each of two complexes in Me$_2$SO for ^{29}Si NMR analysis. Mixture A {K[SiPh(cat)$_2$] and K[SiPh(tccat)$_2$]}. (cat: catechol, tccat: tetrachlorocatechol) contained two complexes with symmetric catechols and each complex gave its own characteristic ^{29}Si NMR (δ − 87.5 and −83.9, respectively). The equilibrated mixture contained a high, nonstatistical proportion of the [SiPh(cat)(tccat)]$^-$ species. This was characterized by a ^{29}Si NMR resonance appearing between those of the individual complexes at δ − 85.3 (Figure 8). Equilibrium was achieved at ambient temperature over a period of *ca* 24 h.

Mixture B {K[SiPh(3-fcat)$_2$ and K[SiPh(dbcat)$_2$]} (3-fcat: 2,3-dihydroxybenzaldehyde, dbcat: 3,5-di-*t*-butylcatechol) contained two complexes with asymmetric catechols. Each complex showed the presence of two resonances due to the isomerism described above. The equilibrated mixtures showed the presence of two further species (Figure 9). These are attributed to isomers of the [SiPh(3-fcat) (dbcat)]$^-$ anion. Equilibrium was not established even after 8 weeks, whereupon decomposition prevented a more quantitative kinetic analysis. However, it is apparent from the two experiments described that the kinetics of redistribution of ligands between complexes varies dramatically according to the catecholate involved. It is reasonable to conclude that the rate of redistribution decreases as the 'strength' of the catecholate derivative increases. The nonstatistical distribution of complexes in a mixture indicates a thermodynamic stability of the complexes in Me$_2$SO. The likely explanation lies in the electronic rather than the steric effects in the complex, since the five-coordination imposes little steric constraint.

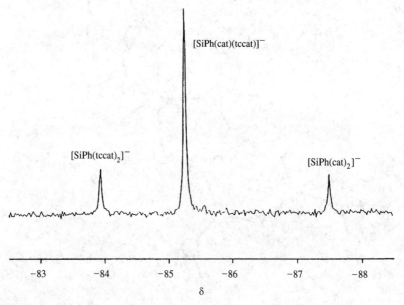

FIGURE 8. The ^{29}Si NMR spectrum of a mixture of K[SiPh(cat$_2$)] and K[SiPh(tccat)$_2$] [53.7 MHz, Me$_2$SO, internal SiMe$_4$, 0.08 mol dm^{-3} Cr(acac)$_3$ relaxant]. Reproduced by permission of the Royal Society of Chemistry from Reference 62]

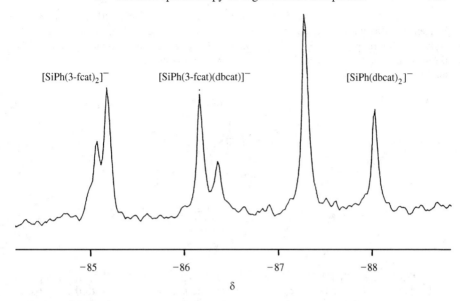

FIGURE 9. The ^{29}Si NMR spectrum of a mixture of K[SiPh(3-fcat)$_2$] and K[SiPh(dbcat)$_2$]. Details as in Figure 8. Reproduced by permission of the Royal Society of Chemistry from Reference 62

The silicon–hydrogen bond in pentacoordinate complexes has an enhanced reactivity compared with the equivalent bond in the related tetracoordinate compounds, as has been convincingly demonstrated by many investigators (e.g. Chopra and Martin[63]). It is established that the active hydrogen is normally expected to be in an equatorial position in trigonal bipyramidal complexes[64]. Bassindale and Jiang[65] were interested in pentacoordinate silicon hydrides as intermediates, or models for intermediates, in nucleophilic substitutions at silicon, and, with varying ligands around silicon, as potentially 'tunable' reducing agents. Most of the species in which they were interested were only available in solution and therefore were not susceptible to definitive structure determination by X-ray crystallography. Hence they employed ^{29}Si NMR extensively, and established the effect of coordination on ^1J(SiH) and δ (^{29}Si) for some reactions of the type shown in equation 3.

$$\text{Nu}^n + \text{R}_3\text{SiH} \rightleftharpoons \text{H}-\underset{\underset{\text{R}}{|}}{\overset{\overset{\text{Nu}^{n+}}{|}}{\text{Si}}}\begin{matrix}\nearrow \text{R} \\ \searrow \text{R}\end{matrix} \qquad (3)$$

Thus, the formation of adducts R$_3$SiH/Nu produces changes in the ^{29}Si NMR chemical shift and ^1J(SiH) that follow similar trends. Coordination of highly electronegative, soft ligands is accompanied by the strongest low frequency shifts and the greatest increase in ^1J(SiH). Formal coordination of hard electropositive ligands such as methyl can result in high frequency shifts and a decrease in ^1J(SiH). The ^{29}Si chemical shift range for the pentacoordinate neutral, anionic and cationic adducts is about $\delta - 110$ to -47 ppm, whereas the range for the related tetracoordinate silanes is greater at $\delta + 24$ to -76 ppm.

As a part of a study on persilylated π-electron systems, particularly on persilylated benzenes[66], Sakurai and coworkers[67] were interested in hexakis(fluorodimethylsilyl)benzene (**30**) for its possible dynamic properties as a gear-meshed structure[68] and for the possible presence of nonclassical neutral pentacoordinate silicon atoms[52]. The X-ray analysis of **30** was also reported (Figure 10).

(**30**)

At 273 K, the ^{29}Si NMR of **30** shows a triplet [J(SiF) = 127 Hz], indicating that each of the silicon nuclei interacts with two fluorine nuclei in accord with the solid-state structure. This triplet did not change to a doublet of doublets at the low-temperature limit; however, these signals transformed at higher temperatures to a septet [J(SiF) = 43 Hz at 328 K] as shown in Figure 11. Correspondingly, all the ^{1}H and ^{13}C NMR signals which are triplets at 273 K become septets at 328 K. Noteworthy is the fact that chemical shifts in these NMR spectra did not change at all; only the coupling pattern changed. This unusual dynamic behavior can be explained by a mechanism where, at the low temperature, rotation of the silyl groups is frozen but fluorine atom transfer between vicinal silyl groups is rapid. This process corresponds to a cyclic network of

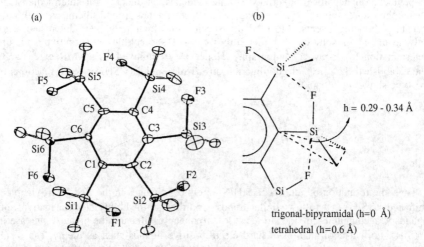

FIGURE 10. Molecular structure of hexakis(fluorodimethylsilyl)benzene **30**: (a) top view and (b) geometry around silicon. Reprinted with permission from Reference 67, Copyright 1994 American Chemical Society

intramolecular consecutive $S_N2(Si)$-type Walden inversions which are very rapid because each silicon atom already forms a quasipentacoordinate structure. As a consequence, triplet signals are observed. From a symmetry perspective, Si−F bond alternation in this study is analogous to the inversion mechanism studied by Mislow and coworkers[68], for hexakis(dimethylamino)benzene, although the molecular mechanism of the exchange is quite different.

Septet signals mean that silicon and other nuclei interact equally with six fluorine nuclei at the higher temperature. At high temperatures, rotation of the silyl groups is allowed and all the fluorine nuclei migrate throughout the ring by a combination of Si−F bond alternation and rotation. Although not proven in a strict sense, the gear-meshed motion is highly likely as the mechanism of rotation, since the molecule already takes the gear-meshed structure in the solid state. Fluorine nuclei thus move like in a merry-go-round.

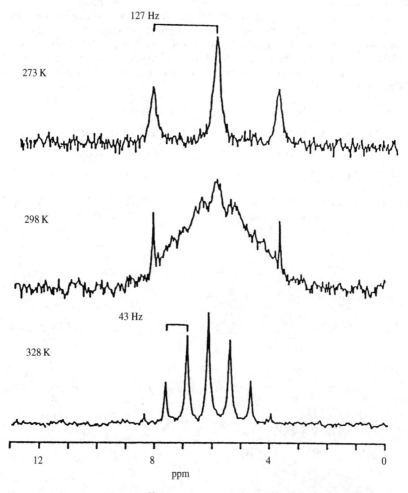

FIGURE 11. Temperature-dependent ^{29}Si NMR sepctra of hexakis(fluorodimethylsilyl)benzene in toluene-d_8. Reprinted with permission from Reference 67. Copyright 1994 American Chemical Society

D. Organically Modified Silicates

Organically modified silicates(ormosils) constitute an important new family of amorphous solids. Since the successful preparation of these new ormosils using the sol–gel method[69,70], there has been increasing interest in making new organic/inorganic hybrid materials. These materials are synthesized by chemically incorporating organic polymers into inorganic networks, resulting in improved mechanical properties such as ductility and toughness. Recently, structure-related rubber-like ormosils of the polydimethylsiloxane (PDMS)/tetraethoxysilane (TEOS) system have been reported and some characterization studies of these ormosils have been made[71–73]. The sol-gel reaction mechanisms of these ormosils are still unclear, though ^{29}Si NMR spectroscopy has proven to be a powerful tool for the structural characterization of organic and inorganic silicon compounds. Iwamoto and coworkers[74] investigated the reaction mechanisms leading to the formation of the ormosils of the PDMS/TEOS system by liquid-state ^{29}Si NMR spectroscopy.

In order to assign the ^{29}Si chemical shift relevant to the bonding between PDMS and TEOS relative to TMS, the liquid-state ^{29}Si NMR spectrum of the solution of the dimethyldiethoxysilane (DMDES)/TEOS system was recorded. DMDES is a "monomer" of PDMS and can form chains (equations 4 and 5), rings (especially cyclic D_4 tetramers in this acid catalyzed system[75]) (equation 6) and copolymerized species with condensed TEOS (equation 7).

$$SiMe_2(OR)_2 + 2H_2O \longrightarrow SiMe_2(OH)_2, \qquad (4)$$

$$-O-SiMe_2-OH + HO-SiMe_2-O- \longrightarrow -O-SiMe_2-O-SiMe_2-O- + H_2O \qquad (5)$$

$$4SiMe_2(OH)_2 \longrightarrow [SiMe_2-O-]_4(cyclic) + 4H_2O \qquad (6)$$

$$-O-SiMe_2-OH + HO-Si(-O-)_3 \longrightarrow -O-SiMe_2-O-Si(-O-)_3 + H_2O \qquad (7)$$

For clarity, the various silicate structures are shown in Figure 12. The ^{29}Si NMR spectra of the solutions whose starting compositions were TEOS: PDMS: H_2O: HCl = 1 : 0.082 : 2 : 0.1 at 70 °C are shown in Figure 13.

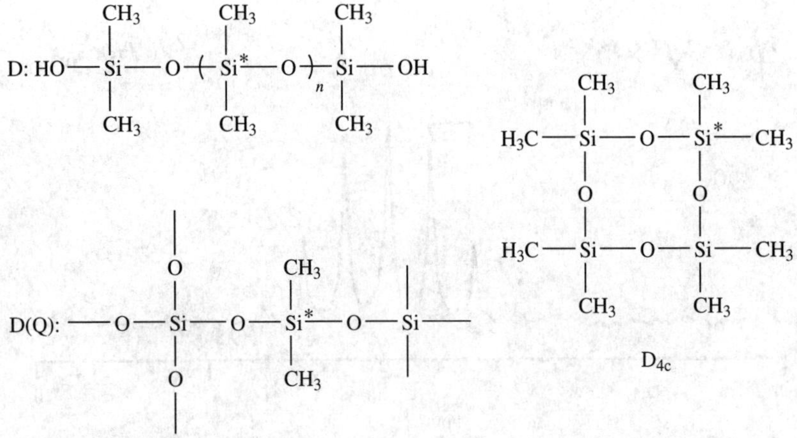

FIGURE 12. Various silicate structures. The asterisks denote the peaks for silicon. Reproduced by permission of Elsevier Science from Reference 74

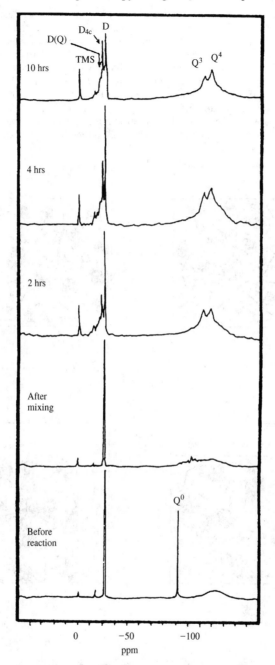

FIGURE 13. ^{29}Si NMR spectra of solutions with a TEOS: PDMS: H$_2$O: HCl = 1 : 0.082 : 2 : 0.1 composition at varying reaction times at 70 °C. Reproduced by permission of Elsevier Science from Reference 74

The self-condensation reaction of TEOS was predominant rather than copolymerization between PDMS and TEOS immediately after mixing. As the reaction proceeded, D decreased and D(Q) increased. Furthermore, the presence of D_{4c} was observed. These observations indicate that bonds between PDMS and TEOS are formed in the ormosils of the PDMS/TEOS system and that the PDMS chains which contained $-O-SiMe_2-O-$ units were broken into shorter chains and/or cyclic D_{4c} tetramers. These facts indicate that hydrolyzed TEOS reacted with $-O-SiMe_2-O-$ in the middle of PDMS chains as well with the silanol end groups, $HO-SiMe_2-O-$, of PDMS. D and D_{4c} remained after 10 h at 70 °C (Figure 13). The peak intensity of D after 10 h at 70 °C was about one-tenth of the peak intensity before reaction, and the peak intensity of D_{4c} after 10 h was also

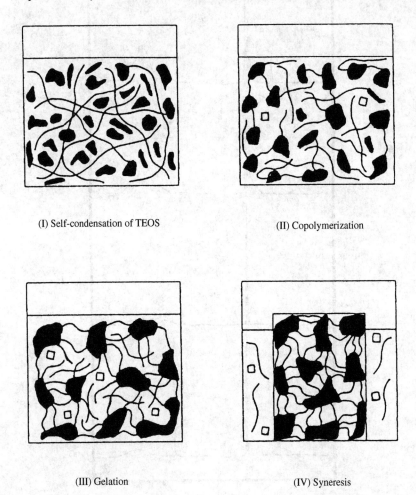

(I) Self-condensation of TEOS   (II) Copolymerization

(III) Gelation   (IV) Syneresis

(● : condensed TEOS, ~ : PDMS, □ : cyclic D_4 tetramer)

FIGURE 14. Structural models of the sol-gel reaction products of ormosils. Reproduced by permission of Elsevier Science from Reference 74

about one-tenth of the peak intensity of D before reaction. Three peaks, D, D_{4c} and D(Q), appeared in the region of -14.5 to -22.0 ppm in Figure 13. Therefore, the intensity of the broad D(Q) peak was about 8/10 of the peak intensity of D before reaction. Since more PDMS would be copolymerized with TEOS until the solution gelled, the amount of PDMS that was copolymerized with TEOS in this solution can be estimated to be more than 80%.

Condensation of TEOS could be controlled by the reaction rate and/or the diffusion of water, while copolymerization could be controlled solely by the diffusion rate of PDMS. Proposed structural models of ormosils based on the reaction mechanisms before gelation are shown in Figure 14. The TEOS/PDMS ratio of the ormosils was 1/0.082. Immediately after mixing, the self-condensation of TEOS(I) was predominant over copolymerization between PDMS and TEOS. As the reaction time increased, copolymerization between PDMS and TEOS(II) was promoted. At this time, the PDMS chains were broken into shorter chains and/or cyclic D_{4c} tetramers. As copolymerization and condensation reactions of TEOS proceeded, the solution gelled (III). After gelation, syneresis (IV) occurred and nonbridging PDMS chains and cyclic D_{4c} tetramers were released from the gel.

IV. NMR PARAMETERS

A. Theory of ^{29}Si NMR Parameters

The widespread interest in ^{13}C NMR spectroscopy has ensured that the chemical shift of this nucleus has received much attention at both the semiempirical MO- and ab initio MO-level approaches, and comparison of the calculated and experimental ^{13}C chemical shifts has provided useful information on the electronic distribution and molecular structure[76,77]. On the other hand, in spite of the fact that ^{29}Si NMR spectroscopy is widely used for investigating molecular structures and electronic distributions in organosilicon compounds, theoretical investigation has been carried out only by using a rough approximate theory such as the averaged excitation energy (ΔE) method. This method has an ambiguity in the estimation of the value of ΔE as a parameter in the paramagnetic term by which ^{29}Si chemical shift is predominantly governed. According to our best knowledge, there have been few ^{29}Si chemical shift calculations using sophisticated methods. In this chapter, therefore, we show calculations of the ^{29}Si chemical shifts of several organic compounds containing a silicon atom by the use of semiempirical MO and ab initio MO methods.

Takayama and Ando[78] discussed the relationship between the ^{29}Si chemical shift and the electronic structure through a composition of the calculated finite perturbation theory (FPT) within the CNDO/2 framework, which successfully reproduced the experimental trend for the ^{13}C chemical-shift values. The diamagnetic, paramagnetic and total contributions for the ^{29}Si chemical shift calculated by using the 'new' value of $\beta_{Si} = -12$ eV), together with the experimental values, are listed in Table 5 for several silicon derivatives.

As may be seen from Table 5, the diamagnetic term, σ^d, etc. moves upfield by about 2 ppm in going from SiH_4 to $SiHEt_3$. This variation is quite small, and the chemical shift displacement is in the opposite direction compared with the experimental one. On the other hand, the paramagnetic term, σ^p, moves downfield by up to 54 ppm. This means that the ^{29}Si chemical shift is predominantly governed by the paramagnetic term. Next, the authors examined the ^{29}Si chemical shift behavior of organic silicon compounds containing fluorine atoms. The electronegative fluorine nuclei are responsible for a higher electron unbalance in the Si–F bond and may lead to a wider range of ^{29}Si chemical shifts as compared with the case of silicon–hydride compounds. In the ^{29}Si chemical shift calculation, the value of β_{Si} determined above in silicon–hydride compounds was used; also, the authors adopted as the β_F parameter for the fluorine atom a value of -20 eV, by

TABLE 5. Calculated ^{29}Si chemical shifts of silicon hydrides[a,b]

Compound	d	σ^d	σ^p	σ^{Total}	δ_{calcd}[c,d]	δ_{exp}[d]
SiH$_4$	3.4578	66.78	−321.64	−254.86	−63.87	−92.5
SiH$_3$Me	3.5050	67.43	−336.78	−269.35	−49.38	−65.2
SiH$_3$Ph	3.5236	67.69	−333.55	−265.87	−52.86	−60.0
SiH$_3$Bz	3.5175	67.60	−341.22	−273.62	−45.11	−56.0
SiH$_2$Me$_2$	3.5376	67.88	−348.38	−280.50	−38.23	−40.0
SiH$_2$MePh	3.5546	68.11	−346.13	−278.02	−40.71	−36.9
SiH$_2$Ph$_2$	3.5874	68.56	−341.51	−272.95	−45.78	−33.6
SiHMePh$_2$	3.6058	68.80	−358.63	−289.82	−28.91	−19.5
SiHPh$_3$	3.6286	69.11	−352.99	−283.88	−34.85	−17.8
SiHMe$_2$Ph	3.5715	68.34	−367.09	−298.75	−19.98	−17.6
SiHMe$_3$	3.5577	68.15	−360.83	−292.67	−26.06	−16.3
SiHPr$_3$	3.5918	68.62	−375.43	−306.82	−11.91	−8.5
SiHEt$_3$	3.5798	68.45	−372.51	−304.05	−14.68	0.2

[a] Data taken from E. A. Williams and J. D. Cargioli, in *Annual Reports on NMR Spectroscopy*, Vol. 9, Academic Press, New York, p. 287, 1979.
[b] The bonding parameters used are $\beta_H = -13$ eV, $\beta_C = -15$ eV and $\beta_{Si} = -12$ eV.
[c] Values given in ppm. Chemical shifts calculated with respect to TMS.
[d] The negative sign means an upfield shift from TMS.
Reproduced by permission of the Chemical Society of Japan from Reference 78.

the use of which the ^{19}F chemical shift calculation reproduced the experiment reasonably well. The diamagnetic, paramagnetic and total contributions calculated by using these β parameters are listed, together with the experimental chemical shift values, in Table 6. It may be seen that the chemical shift range is much expanded relative to the case of silicon–hydride compounds.

TABLE 6. Calculated ^{29}Si chemical shifts of fluorosilanes[a,b]

Compound	d	σ^d	σ^p	σ^{Total}	δ_{calcd}[c,d]	δ_{exp}[d]
SiF$_4$	2.1098	45.17	−254.38	−209.20	−109.53	−111.0
SiF$_3$Ph	2.5888	53.50	−296.91	−243.41	−75.32	−73.7
SiF$_3$C$_6$H$_4$Me-p	2.5952	53.61	−297.43	−243.83	−74.90	−72.0
SiF$_3$CH$_4$OMe-p	2.5892	53.50	−297.29	−243.78	−74.95	−71.4
SiF$_3$Bz	2.5805	53.36	−303.18	−249.82	−68.91	−64.2
SiF$_3$Me	2.5345	52.59	−296.31	−243.72	−75.01	−51.8
SiF$_2$Ph$_2$	2.9720	59.64	−331.09	−271.45	−47.28	−30.5
SiF$_2$MePh	2.9314	59.01	−337.96	−278.95	−39.78	−12.4
SiFPh$_3$	3.3213	64.85	−351.88	−278.03	−31.70	−4.7
SiF$_2$Et$_2$	2.9258	58.93	−335.45	−276.52	−42.21	0.5
SiF$_2$Me$_2$	2.9006	58.53	−336.55	−278.02	−40.71	6.2
SiFMePh$_2$	3.2980	64.51	−358.52	−294.01	−24.72	7.7
SiFMe$_2$Ph	3.2633	64.01	−367.00	−302.99	−15.74	19.8
SiFPr$_3$	3.2819	64.28	−375.19	−310.91	−7.82	28.8
SiFMe$_3$	3.2426	63.71	−359.88	−296.17	−22.56	31.9

[a] Data taken from E. A. Williams and J. D. Cargioli, in *Annual Reports on NMR Spectroscopy*, Vol. 9, Academic Press, New York, p. 287 1979.
[b] The bonding parameters used are $\beta_H = -13$ eV, $\beta_C = -15$ eV, $\beta_{Si} = -12$ eV and $\beta_F = -20$ eV.
[c] Values given in ppm. Chemical shifts calculated with respect to TMS.
[d] The negative sign means an upfield shift from TMS.
Reproduced by permission of the Chemical Society of Japan from Reference 78.

It can also be seen that the relative chemical shift is predominantly governed by the paramagnetic term, although the diamagnetic term is much more varied by the degree of fluorine substitution than the value in the case of silicon–hydride compounds. A plot of the experimental chemical shifts vs the calculated values is shown in Figure 15. It can be seen that data points deviate slightly from the theoretical line with the slope of 1.0, but the overall trend of the calculation reasonably reproduce the experimental data. In particular, it can be said that it has been established that the ^{29}Si chemical shift in SiR$_n$F$_{4-n}$ moves upfield as the degree of fluorine substitution is increased. (It is notable that the upfield shift with an increase in the degree of fluorine substitution is opposite to the case of the ^{13}C chemical shift in CR$_n$F$_{4-n}$.)

Tossell and Lazzeretti[79] have recently applied *ab initio* coupled Hartree–Fock perturbation theory (CHFPT) to the calculation of ^{29}Si NMR shifts in SiH$_4$, SiF$_4$ and other molecules. They also carried out CHFPT calculations of ^{29}Si chemical shielding tensors for SiH$_4$, Si$_2$H$_6$, Si$_2$H$_4$ and H$_2$SiO. Experimental geometries were employed for SiH$_4$ and Si$_2$H$_6$ (D_{3h} symmetry was utilized for Si$_2$H$_6$ in order to shorten the calculation time) and optimized geometries from high level *ab initio* SCF calculations were employed for Si$_2$H$_4$[80] and H$_2$SiO.

Total energies and atomization energies refereed to Hartree–Fock nuclei are shown in Table 7. For SiH$_4$ and Si$_2$H$_6$ a comparison can be made with experiment using tabulated bond dissociation energies. Calculated values were roughly 90% of experiment. It is evident that the calculated atomization energy for Si$_2$H$_4$ is only slightly greater than that for

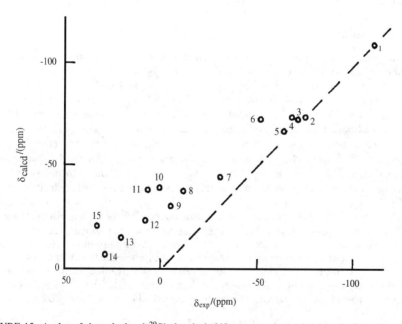

FIGURE 15. A plot of the calculated ^{29}Si chemical shifts vs experimental values in fluorosilanes. β bonding parameters are: ($\beta_H = -13$ eV, $\beta_C = -15$ eV, $\beta_{Si} = -12$ eV and $\beta_F = -20$ eV). The negative sign means upfield shift relative to TMS. The numbers refer to the following compounds: 1, SiF$_4$; 2, SiF$_3$Ph; 3, F$_3$SiC$_6$H$_4$Me-p; 4, SiF$_3$ C$_6$H$_4$OMe-p; 5, SiF$_3$Bz; 6, SiF$_3$Me; 7, SiF$_2$Ph$_2$; 8, SiF$_2$MePh; 9, SiFPh$_3$; 10, SiF$_2$Et$_2$; 11, SiF$_2$Me$_2$; 12, SiFMePh$_2$; 13, SiFMe$_2$Ph; 14, SiFPr$_3$; 15, SiFMe$_3$. Reproduced by permission of the Chemical Society of Japan from Reference 78

TABLE 7. Calculated total energies E (hartree) and heats of atomization (kcal mol^{-1}) for several silicon derivatives

	SiH$_4$	Si$_2$H$_6$	Si$_2$H$_4$	H$_2$SiO
E	−291.260	−581.365	−580.1355	−364.987
heat of atomization, calc.	279	461	317	222
exp.	304	509	—	—

Reproduced by permission of Elsevier Science from Reference 79.

SiH$_4$, indicative of the weakness of the Si=Si bond. (Note that the calculated minimum energy geometry rather than the experimental one was used for Si$_2$H$_4$; this should stabilize with respect to SiH$_4$ and Si$_2$H$_6$, for which the experimental geometries were used.) Indeed, the reaction Si$_2$H$_4$ + H$_2$ → Si$_2$H$_6$ is exothermic by about 34 kcal mol^{-1} (assuming a calculated H$_2$ atomization energy of about 100 kcal mol^{-1}). The calculated Si=O atomization energy in H$_2$SiO (assuming the calculated Si−H atomization energy value of 70 kcal mol^{-1} from SiH$_4$) is only 82 kcal mol^{-1}, less than the experimental Si−O atomization energy of 108 kcal mol^{-1}.

Calculated chemical shielding tensors $\sigma_{\alpha\beta}$ are given in Table 8 for SiH$_4$, Si$_2$H$_6$, Si$_2$H$_4$ and H$_2$SiO. Comparison can be made with experimental chemical shift values for SiH$_4$ and Si$_2$H$_6$ and for (Mes)$_2$Si=Si(Mes)$_2$. For each molecule, σ and its components are given with the Si as gauge origin.

An example of the use of IGLO calculations was already mentioned[43].

B. Coupling Constants

One-bond nuclear spin–spin coupling constants (^1J) are a valuable source of information on the nature of chemical bonding[81]. The sign and magnitude of $^1J(^{29}\text{Si}^{15}\text{N})$ were first reported for ^{15}N-enriched (H$_3$Si)$_3$N (+6 Hz) in 1973[82]. Very few studies of these couplings appeared during the next decade, mainly due to a low natural abundance of both ^{15}N (0.36%) and ^{29}Si (4.7%) isotopes[83]. As a result, only scattered values of ^1J(SiN) had been reported in preliminary communications[84,85]. This is in contrast with the current upsurge of interest in Si−N bonding explored by both theoretical[86−88] and experimental[89,90] methods. Such attention is largely accounted for by the controversial assumption of d$_\pi$-p$_\pi$ interaction along the Si−N bond[91]. Attempts to invoke ^{15}N and ^{29}Si chemical shifts in order to explore this phenomenon have met with little success so far[89].

^{1}H, ^{13}C and ^{29}Si NMR spectral analyses carried out earlier for bis(trimethylsilyl)ethylenes[92] and some types of vinylsilanes[93] indicate an interaction between the vacant d-orbitals on the silicon atom and the π-electron system of the vinyl group. In an effort to broaden these notions and in continuation of previous studies concerned with the influence of electronic effects of silicon and vinyl substituents on the chemical shifts as well as the coupling constants, Lukevics and coworkers[94] determined the ^{1}H, ^{13}C, ^{29}Si and ^{17}O NMR spectra for the following chlorinated silylethylenes (**31–34**) and 1,2-disilylethylenes (**35**):

$$\text{Me}_{3-n}\text{X}_n\text{SiCH=CHCl} \quad (\mathbf{31} = trans, \mathbf{32} = cis)$$

$$\text{Me}_{3-n}\text{X}_n\text{SiCH=CCl}_2 \quad (\mathbf{33})$$

$$\text{Me}_{3-n}\text{X}_n\text{SiCCl=CCl}_2 \quad (\mathbf{34})$$

$$\text{Me}_{3-n}\text{X}_n\text{SiCH=CHSiX}_n\text{Me}_{3-n} \quad (n = 0-3, \text{X} = \text{Cl, OEt, OMe}) \quad (\mathbf{35})$$

TABLE 8. Calculated diamagnetic, paramagnetic and total ^{29}Si chemical shielding tensors σ (in ppm, gauge origin at Si) compared with experimental chemical shifts, τ

	SiH$_4$	Si$_2$H$_6$			Si$_2$H$_4$				H$_2$SiO			
	xx, yy, zz	xx, yy	zz	av[a]	xx	yy	zz	av[a]	xx	yy	zz	av[a]
calc.												
σ^d	899.8	985.4	907.4	959.4	990.5	977.6	895.9	954.6	967.3	952.4	889.9	936.5
σ^p	−420.0	−476.6	−421.3	−458.2	−523.5	−801.6	−501.4	−608.8	−593.3	−819.6	−414.4	−609.1
σ^{Total}	479.8	508.7	486.2	501.2	467.0	176.0	394.5	345.8	374.0	132.8	475.5	327.4
exp.												
τ	−93		−104.8			+64						

[a] av = $(xx + yy + zz)/3$
Reproduced by permission of Elsevier Science from Reference 79.

TABLE 9. ^{29}Si–^{13}C and ^{29}Si–^{1}H spin–spin coupling constants (Hz) in silylethylenes

Compound	X	n	^{29}Si–^{13}CH$_3$	^{29}Si–^{13}C$_\alpha$	^{29}Si–^{13}C$_\beta$	^{29}Si–C–^{1}H	^{29}Si–C$_\alpha$–^{1}H	^{29}Si–C$_\alpha$=C$_\beta$–^{1}H
31a	Cl	0	53.33	59.32	5.76	6.81	4.25	2.34
31b	Cl	1	60.69	69.87	7.42	7.14	5.71	4.76
31c	Cl	2	73.00	86.11	9.03	7.93	6.75	6.81
31d	Cl	3	—	111.86	11.54	—	8.22	8.50
31e	OEt	1	62.64	68.20	7.06	6.81	2.83	4.75
31f	OEt	2	78.38	86.90	7.01	7.40	3.02	4.30
31g	OEt	3	—	114.56	7.79	—	3.22	5.06
31h	OMe	3	—	115.68	8.05	3.76(^3J(SiOCH))	3.03	5.13
32a	Cl	0	53.49	61.82	—	6.85	—	—
32b	Cl	1	60.49	73.02	—	7.31	2.17	13.40
32c	Cl	2	72.28	89.62	—	8.10	3.25	17.23
32d	Cl	3	—	115.58	—	—	7.69	21.76
32e	OEt	1	62.64	71.24	—	6.92	1.48	11.76
32f	OEt	2	78.54	89.68	—	7.51	1.20	13.25
32g	OEt	3	—	117.08	—	—	0.4	15.32
32h	OMe	3	—	116.74	—	3.95(^3J(SiOCH))	1.03	15.53
33a	Cl	0	54.24	58.60	—	6.85	1.46	—
33b	Cl	1	61.77	70.19	—	7.25	1.10	—
33c	Cl	2	74.37	87.81	—	8.17	1.60	—
33d	Cl	3	—	114.62	—	—	5.13	—
33f	OEt	2	81.19	87.35	—	7.58	1.13	—
33g	OEt	3	—	115.67	—	—	0.81	—
34a	Cl	0	55.49	60.81	6.07	6.89	—	—
34b	Cl	1	63.82	74.80	8.08	7.36	—	—

Compound	X	n					
34c	Cl	2	77.58	96.77	10.72	8.39	—	—
34d	Cl	3	—	130.77	13.42	—	—	—
34f	OEt	2	84.34	91.30	8.64	7.73	—	—
34g	OEt	3	—	127.28	9.75	—	—	—
35a	Cl	0	51.89	63.34		6.63	—	7.55
35b	Cl	1	59.30	71.11		7.07	—	7.66
35c	Cl	2	71.01	84.42		7.74	—	7.90
35d	Cl	3	—	107.46		—	—	8.10
35g	OEt		—	115.96		—	—	—
Me$_3$SiCH=CHBr		trans	53.05	57.10	4.91	6.77	0.70	10.93
		cis	53.48	62.19	—	6.85	4.30	6.00
Et$_3$SiCH=CHCOOEt			53.48(SiCH$_2$)	57.86	8.28	8.00(SiCH$_2$)		
						6.60(SiCH$_2$)		
Me$_{3-n}$X$_n$Si–CR^1=C(H^2)(H^3)	Cl	0	52.25	64.16	—	—	6.19	15.12(SiH$_2$)
								8.17(SiH$_2$)
	Cl	1	—	—	—	—	6.42	15.26
	Cl	1	—	—	—	—		8.59
	Cl	2	70.40	89.70	—	—	11.80	23.81(SiH$_2$)
								12.53(SiH$_2$)
	Cl	3	—	112.40	—	—	15.44	30.61(SiH$_2$)
								15.54(SiH$_2$)

Reproduced by permission of Elsevier Science from Reference 94.

The $^1J(^{29}Si^{13}C)$ coupling constants of **31–34** increase with increasing number of substituents X in the molecules. According to Bent's hypothesis[95], the value of this coupling depends on the s-character of both silicon and carbon. This is supported by the fact that $^1J(^{29}SiC_\alpha) > {}^1J(^{29}SiMe)$ (Table 9). Nearly linear correlations were found between the $^1J(^{29}SiC_\alpha)$ coupling constants and the sum of the electronegativities of substituents attached to the silicon atom ($\sum E$) (equations 8–11).

$$^1J(^{29}SiC_\alpha) = -71.0 + 20.2 \sum E \qquad n = 8;\ r = 0.964\ (\mathbf{31}) \qquad (8)$$

$$^1J(^{29}SiC_\alpha) = -69.6 + 20.4 \sum E \qquad n = 8;\ r = 0.969\ (\mathbf{32}) \qquad (9)$$

$$^1J(^{29}SiC_\alpha) = -62.2 + 19.0 \sum E \qquad n = 6;\ r = 0.984\ (\mathbf{33}) \qquad (10)$$

$$^1J(^{29}SiC_\alpha) = -89.7 + 23.5 \sum E \qquad n = 6;\ r = 0.978\ (\mathbf{34}) \qquad (11)$$

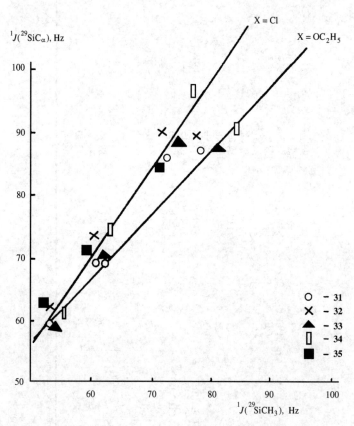

FIGURE 16. $^1J(^{29}SiC_\alpha)$ as a function of the $^1J(^{29}SiCH_3)$ coupling constants in silylethylenes; X = Cl, $^1J(^{29}SiC_\alpha) = -13.7 + 1.39\ ^1J(^{29}SiCH_3)$ ($r = 0.980$); X = OEt, $^1J(^{29}SiC_\alpha) = 5.1 + 1.02\ ^1J(^{29}SiCH_3)$ ($r = 0.979$). Reproduced by permission of Elsevier Science from Reference 94

6. ^{29}Si NMR spectroscopy of organosilicon compounds

These equations indicate the predominant role of positive charge in determining $^1J(^{29}SiC_\alpha)$ coupling constants. The long-range coupling constants $^nJ(SiH)$ in vinylsilanes have been studied[96,97]. Based on this argument, the authors measured these coupling constants in compounds **31–35** (Table 9). There is a certain decrease in the geminal $^2J(^{29}SiC_\alpha H)$ coupling constants in compounds **31–33** as compared to $Me_3SiC_\alpha H = CH_2$ [$^2J(^{29}SiC_\alpha H)$ 6.42 Hz[98]] and $N(CH_2CH_2O)_3SiC_\alpha H = CH_2$ [$^2J(^{29}SiC_\alpha H)$ 5.74 Hz[21]]. This can possibly be attributed not only to the influence of electronic charge on chlorine, but also to changes in the Si—C—H valence angle. The role of the latter factor can be deduced from NMR data obtained for organotin compounds[99].

As was shown above, SiOR and SiCl substituents exert a completely different effect on the chemical shifts. The same difference clearly comes out in the case of coupling constants. Thus, comparing the two sets of $^1J(^{29}SiC_\alpha)$ and $^1J(^{29}SiCH_3)$ values measured for the same SiCl-substituted molecule, one can find a linear correlation between these quantities with a slope coefficient close to unity (Figure 16). However, compounds bearing SiOR substituents form another straight line, showing additional electronic and/or steric effects. The same picture is found when comparing $^1J(^{29}Si^{13}CH_3)$ and $^2J(^{29}Si^{13}CH)$ coupling constants. Here, too, SiOR-substituted compounds account for a separate correlation line (Figure 17).

These findings can possibly be explained by a stronger $(p-d)_\pi$ conjugation in the Si—O bond in comparison with the Si—Cl bond. It is, however, necessary to take into account also the steric differences of the SiOR and SiCl substituents. The importance of the steric

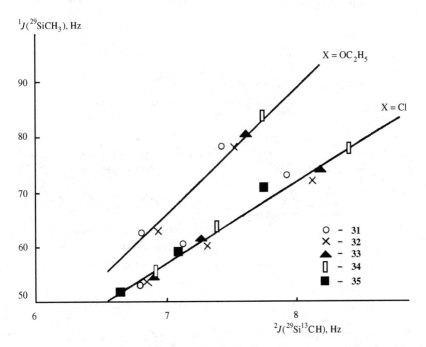

FIGURE 17. Correlation between $^1J(^{29}Si^{13}CH_3)$ and $^2J(^{29}Si^{13}CH)$ coupling constants in silylethylenes; X = Cl, $^1J(^{29}Si^{13}CH_3) = -49.5 + 15.2\ ^2J(^{29}Si^{13}CH)$ ($r = 0.989$); X = OC_2H_5, $^1J(^{29}Si^{13}CH_3) = -110.5 + 25.2\ ^2J(^{29}Si^{13}CH)$ C,H) ($r = 0.990$). Reproduced by permission of Elsevier Science from Reference 94

contribution can be inferred from the lack of correlation between the $^1J(^{29}SiC_\alpha)$ and $^2J(^{29}SiC_\alpha H)$ couplings.

Measurements of spin–spin coupling constants make it possible to obtain information on electron distribution in molecules. $^{29}Si-^{15}N$ spin–spin coupling constants were, as mentioned previously, first measured in 1973[78]. The development of polarization transfer methods (INEPT, DEPT etc.) allowed their measurements under natural isotope abundance. As a result, systematic investigation of these spin–spin coupling constants and their application for structural studies became important[100–103]. Kupce, Lukevics and coworkers conducted an extensive study on $^{29}Si-^{15}N$ coupling constants in silazanes **36–51**, along with studies on ^{29}Si and ^{15}N chemical shifts, $^{15}N-H$, $^{29}Si-^{13}C$ and $^{29}Si-^{29}Si$ coupling constants and $^{15/14}N$ isotope effects on ^{29}Si chemical shifts[104]. This enabled the authors to attain a more reliable interpretation of $^{29}Si-^{15}N$ coupling constants scarcely studied so far and to characterize in detail the structure of the investigated compounds.

$(R_3Si)_2NH$ $(Me_3Si)_2NNa$ $Me_3SiNHOSiMe_3$

(**36**) $R_3 = Me_3$
(**37**) $R_3 = Me_2Ph$ (**41**)   (**42**)
(**38**) $R_3 = MePh_2$
(**39**) $R_3 = Me_2Cl$ $Ph_2MeSiNH_2$ $Ph_2MeSiNHBu-t$
(**40**) $R_3 = Me_2O(SiMe_2O)_5SiMe_3$
                              (**43**)   (**44**)

$RMe_2Si-N(\underset{Si Me_2}{\overset{Si Me_2}{}})N-SiMe_2R$

(**45**) R = Cl
(**46**) R = NH_2 (**48**) R = Me
(**47**) R = OEt   (**49**) R = Ph

(**50**)   (**51**)

The spin–spin coupling constants and chemical shifts of the compounds studied are presented in Table 10. An increase in $^1J(SiN)$ with the electronegativities of the substituents on the Si atom in the SiR series Me < Ph < Cl < OR (compounds **36–40** and **46–48**) is indicative of positive $^1J(SiN)$ values [$^1J(SiN) > 0$]. Since the variation in the electronegativities of the substituents on the N atom is insignificant, it can be

TABLE 10. ^{29}Si NMR data of silazanes **36–51**[a]

Compound	Solvent	Position	δ (^{15}N)	δ (^{29}Si)	^1J(NH)	^1J(SiC)[b]	^1J(SiN)	$^1\Delta$ ($^{15/14}$N)[c]	Other data
36	acetone–d$_6$		−354.2	2.1	69.6	56.2	13.5	10.7	
37	CDCl$_3$		−357.8	−3.5	66.1	58.2	13.8	10.5	^1J(SiC) = 72.2 (Ph)
38	CDCl$_3$		−361.5	−9.4	66.3	60.0	14.2	8.5	^1J(SiC) = 74.6 (Ph)
39	CDCl$_3$		−334.3	13.3	68.1	68.1	14.4	8.2	
40	CDCl$_3$		−335.7	−11.9	nm[d]	69.4	17.4	5.9	
41	C$_6$D$_6$		nm[d]	−14.6		51.9	7.8	11.6	
42	C$_6$D$_6$		−255.6	10.9	64.8	56.7	6.5	14.3	δ (^{29}Si) 20.9 (OSiMe$_3$)
43	C$_6$D$_6$		−373.8	−10.2	74.0	59.8	17.5	10.8	^1J(SiC) = 74.4 (Ph)
44	acetone-d$_6$		−327.3	−15.4	75.0	nm	19.8	nm[d]	
45	CDCl$_3$	endo	−322.1	7.3		61.9	7.1	10.5	^2J(SiSi) = 2.2 (exo–endo)
		exo		6.2		67.2	16.9	4.0	
46	C$_6$D$_6$	endo	−323.8	3.0		60.9	7.4	10.0	^2J(SiSi) = 2.3 (exo–endo)
		exo	−363.1	−6.7	72.8	63.9	17.8	nm[d]	^1J(SiN) = 18.4 (NH$_2$)
									^3J(SiN) = 3.1
47	CDCl$_3$	endo	−325.4	3.8		61.2	7.3	9.5	^2J(SiSi) = 2.0 (exo–endo)
		exo		−8.9		68.1	18.3	4.5	
48	CDCl$_3$		−347.3	−4.6	69.4	63.0	15.4	nm[d]	
49	CDCl$_3$	(SiMe$_2$)	−347.6	−3.1	69.2	63.1	16.0	nm[d]	^1J(SiN) = 15.3 (SiMe$_2$)
		(SiPh$_2$)	−351.0	−21.7	70.1	83.8	18.1	nm[d]	^3J(SiN) = 2.8
50	CDCl$_3$	(NSi$_3$)	−341.7	−8.2	67.0	64.0	16.9	8.2	
51	CDCl$_3$	(NSi$_3$)	nm[d]	−6.8	nm[d]	64.1	9.8	8.7	
							14.3		

[a]Chemical shifts (δ) in ppm relative to TMS(^{29}Si) and MeNO$_2$(^{15}N).
[b]For the SiMe group.
[c]Isotope shift in ppb.
[d]nm = not measured.
Reproduced by permission of Elsevier Science from Reference 104.

suggested that the increase of $^1J(\text{SiN})$ in this series is related to enhanced s character of the Si–N bond according to Bent's law[105]. A parallel increase in the $^{29}\text{Si}-^{13}\text{C}$ spin-spin coupling constants supports this suggestion. No general correlation exists between these coupling constants although, as in the case of aminosilanes[102], such correlation can be found for compounds with identical substituents at the N atom [e.g. disilazanes **36–40** (equation 12)]. Deviation from this relation occurs in the case of compound **39**.

$$^1J(\text{SiN}) = 0.31\,^1J(\text{SiC}) - 4.0 \qquad r = 0.993 \tag{12}$$

The less steep slope of the correlation line, as compared with the slope found for the analogous relation in aminosilanes $[^1J(\text{SiN}) = 0.41\,^1J(\text{SiC}) - 7.3]^{103}$, results from the lower electronegativities of substituents on the N atom in the silazanes.

Much attention has been devoted in recent years to studies of isotope shifts in NMR spectra caused by the replacement of ^{12}C isotopes with ^{13}C in the molecule. As for the ethynylsilanes, the $^{13/12}\text{C}$ isotope shifts in the ^{29}Si NMR spectra have so far been determined only for three derivatives. Therefore, in order to study this phenomenon more thoroughly Lukevics and coworkers[104] synthesized a wide range of ethynylsilanes: $\text{Me}_3\text{SiC}{\equiv}\text{CX}$, where X = H, Br, I, SMe, SEt, SC_6F_5, Me, CH_2Cl, CH_2OEt, CH_2NEt_2, CH_2SiMe_3, $\text{CH}_2\text{Sn(Bu-}t)_3$, CF_3, CN, Ph, C_6F_5, SiMe_3, GeMe_3, GeEt_3, SnMe_3, $\text{Sn(Bu-}t)_3$, and examined their ^{29}Si NMR spectra.

The isotope shifts $^n\Delta^{29}\text{Si}\,(^{13/12}\text{C})$ measured in the ^{29}Si NMR spectra of ethynylsilanes are listed in Table 11. For some of these ethynylsilanes the coupling constants $^nJ(\text{SiC})$ have been reported[106] but not discussed. Lukevics and coworkers[107] showed that the values of the $^1J(\text{SiC}_\alpha)$ coupling constants largely depend on the electronic properties of substituent X according to equation 13.

$$^1J(\text{SiC}_\alpha) = 82.9 - 21.7\sigma^p \qquad n = 12;\ r = 0.98 \tag{13}$$

Coupling constants over two bonds $^2J\,(^{29}\text{Si}^{13}\text{C}_\beta)$ are influenced similarly by the substituent X. This is demonstrated by the correlation between $^1J(^{29}\text{Si}^{13}\text{C}_\alpha)$ and $^2J(^{29}\text{Si}^{13}\text{C}_\beta)$:

$$^2J(^{29}\text{Si}^{13}\text{C}_\beta) = -3.44 + 0.23\,^1J(^{29}\text{Si}^{13}\text{C}_\alpha) \qquad n = 15; r = 0.94 \tag{14}$$

Thus, an increase in the electron-accepting ability of the substituent X leads to a decrease in the coupling constant. A correlation similar to that in equation 14 has been reported for the coupling constants $^nJ\,(^{119}\text{Sn}^{13}\text{C})$ in the ethynylstannanes $\text{Me}_3\text{SnC}{\equiv}\text{CX}$. The slope of this correlation for the tin derivatives (0.28) is very close to that found for the ethynylsilanes (0.23), which indicates that the transmission of electronic effects of the substituent X through the triple bond is very similar for both classes of compounds.

It is noteworthy that the points obtained for ethynylsilanes with X = SiMe_3, GeMe_3, SnMe_3 deviate strongly from correlations 13 and 14. In the case of equation 13, one can speculate that the σ^p values for X = MMe_3 (M = Si, Ge, Sn) substituents inadequately describe the electronic effects in ethynylsilanes; however, the analogous deviations in the case of correlation 14 make this explanation questionable. These effects may be connected both with the violation of the nonlinearity of $\text{R}_3\text{MC}{\equiv}\text{CM}'\text{R}_3$ acetylenides[108] and/or with the existing additional concurrent hyperconjugation effect $\text{H}_3{\equiv}\text{C}-\text{M}-\text{C}{\equiv}$ in these molecules[109]. The latter effect must affect the state of π- and σ-electrons in the triple bond, which are involved in the transmission of spin information between the various nuclei.

In contrast to the $^1J(^{29}\text{Si}^{13}\text{C}_\alpha)$ and the $^2J(^{29}\text{Si}^{13}\text{C}_\beta)$, the $^1J(^{29}\text{Si}^{13}\text{CH}_3)$ coupling constants increase with increasing acceptor properties of substituent X. Such changes can be

TABLE 11. ^nJ (^{29}Si–^{13}C) coupling constants (Hz) and $^{13/12}$C isotope shifts (ppb) in the ^{29}Si NMR spectra of ethynylsilanes Me$_3$SiC$_\alpha$≡C$_\beta$X

No.	X	$^1\Delta^{29}$Si ($^{13/12}$C$_\alpha$)	^1J (^{29}Si:^{13}C$_\alpha$)	$^1\Delta^{29}$Si ($^{13/12}$C$_\beta$)	^1J(^{29}Si:^{13}C$_\beta$)	$^1\Delta^{29}$Si ($^{13/12}$C$_{Me}$)	^1J(^{29}Si:^{13}C$_{Me}$)
1	H	−16.4	81.47	−5.2	15.41	−1.2	56.28
2	Me	−14.8	85.75	−4.2	15.74	−1.3	56.28
3	CH$_2$SiMe$_3$	−13.3	89.54	−3.1	17.86	−1.8	56.06
4	CH$_2$GeMe$_3$	−12.9	88.31	−3.0	17.69	−2.0	56.01
5	CH$_2$Sn(Bu-t)$_3$	−13.5	90.80	−3.1	18.02	−1.8	55.91
6	CH$_2$NEt$_2$	−15.0	93.84	−4.3	15.75	−1.1	56.24
7	CH$_2$OEt	−15.3	80.57	−4.0	15.49	−1.1	56.41
8	CH$_2$Cl	−16.7	80.77	−5.0	15.42	−1.1	56.51
9	SMe	−16.2	82.63	−4.8	15.54	−1.0	56.44
10	SC$_6$F$_5$	−17.2	78.50	−4.9	14.61	−0.1	56.70
11	SEt	−16.2	82.73	−4.5	16.67	−0.9	56.39
12	Ph	−15.6	83.12	−4.7	16.01	−1.1	56.28
13	C$_6$F$_5$	−19.0	77.62	−6.2	14.00	−0.8	56.57
14	Br	−16.7	80.22	−5.5	15.05	−1.0	56.68
15	I	−17.9	77.73	−5.5	13.15	−0.6	56.49
16	CN	−21.8	68.10	−7.6	13.14	−0.5	57.27
17	CF$_3$	−20.9	71.77	a	a	−0.2	57.33
18	SiMe$_3$	−16.1	76.75	−4.2	12.37	−1.4	56.13
19	GeMe$_3$	−15.4	78.25	−4.1	12.20	−1.3	56.09
20	GeEt$_3$	−15.0	78.30	−4.3	12.30	−1.3	55.99
21	SnMe$_3$	−15.3	78.83	a	a	−1.3	55.92
22	Sn(Bu-t)$_3$	−14.3	78.66	a	a	−1.3	55.79

^{a}Not recorded.
Reproduced by permission of Elsevier Science from Reference 107.

attributed to the occurrence of the same hyperconjugation effect. This shows that both types of Si—C bond are interrelated: a change in the properties of substituent X strengthens the Si—C_α and weakens the Si—Me bond. However, the data in Table 11 show that the $^1J(^{29}Si^{13}CH_3)$ values are subject to minor variations, thus making a more detailed discussion impossible.

Isotope shifts $^1\Delta^{29}Si(^{13/12}C_\alpha)$ in the ^{29}Si NMR spectra of ethynylsilanes reveal the same tendencies as the $^1\Delta^{119}Sn(^{13/12}C_\alpha)$ values in the ^{119}Sn NMR spectra of ethynylstannanes (equation 15 and Figure 18).

$$^1\Delta^{119}Sn(^{13/12}C_\alpha) = 0.59 + 3.6\,^1\Delta^{29}Si(^{13/12}C_\alpha) \qquad n = 11; \; r = 0.966 \qquad (15)$$

The correlation 15 suggests that the changes in the isotope shifts are uniform regardless of the type of the central atom in the group 14 acetylenides.

C. Relaxation Times and Exchange Phenomena

The high structural sensitivity of chemical shifts makes ^{29}Si NMR a powerful tool for determination of the structure of oligomeric and polymeric siloxanes. Detailed information can be obtained for the characterization of the different structural units, for the

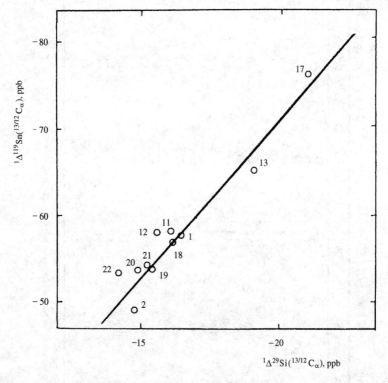

FIGURE 18. The correlation between one-bond $^1\Delta$ M$(^{13/12}C)$ isotope shifts in the ^{29}Si and ^{119}Sn NMR spectra of isostructural acetylenes (CH$_3$)$_3$ MC$_\alpha\equiv$C$_\beta$ X (M = ^{29}Si, ^{119}Sn). The numbering of the compounds corresponds to that in Table 11. Reproduced by permission of Elsevier Science from Reference 107

6. ^{29}Si NMR spectroscopy of organosilicon compounds

determination of average chain lengths or for the degree of condensation of the siloxane framework. Both the ^{29}Si chemical shifts as well as the spin–lattice relaxation times (T_1) of the silicon nuclei in low molecular weight linear polydimethylsiloxane, Me$_3$SiO[Me$_2$SiO]$_n$SiMe$_3$(MD$_n$M), have been determined[110]. In the ^{29}Si NMR spectra of the oligomers MD$_n$M, $n = 1$–8, individual resonance lines can be found for each distinct silicon nucleus. This degree of chemical shift resolution surpasses that observed in the ^{13}C or ^{1}H NMR of the same materials. Spin–lattice relaxation times give information about the mobility at different points in polymer chains. The relatively high values of the ^{29}Si T_1 values found in MD$_n$M are consistent with high mobility of these polymer chains[111].

In the case of polymethylhydrosiloxanes, Me$_3$SiO[MeHSiO]$_n$SiMe$_3$ (MD$_n^H$M), some of the long-range substituent chemical shifts are obscured[112]. These complications result from the asymmetry of the MeHSiO(D^H) unit. In the ^{29}Si NMR spectra the terminal trimethylsilyl (M) groups and the D^H moieties exhibit resonances in substantially different regions. Tacticity effects are essential in understanding the stereochemical features of substituted vinyl polymer chains. Tacticity and end group effects were used to interpret the observed fine structure in the proton decoupled spectrum of MD$_5^H$M, i.e. the appearance of a triplet and a doublet. Analogous features in the ^{29}Si NMR of MD$_{50}^H$M were explained as resulting from either complete atacticity or from the presence of equal amounts of oligomers of different tacticities.

Pai and coworkers[113] investigated in detail both the ^{29}Si chemical shifts and the first spin–lattice relaxation times reported of these systems (Table 12). The spin–lattice relaxation time (T_1) for all ^{29}Si nuclei were measured simultaneously by the

TABLE 12. ^{29}Si chemical shifts and T_1 values for polymethylhydrosiloxanes [δ^a, (T_1) ^{b}]

Compound^c	M	D$_a^H$	D$_b^H$	D$_c^H$	D$_x^H$
MD$_3^H$M	10.00(50)	−35.84(42)	−35.67		
		−35.87(42)	−35.70(30)		
			−35.72		
MD$_4^H$M	10.22(50)	−35.60(44)	−35.11(32)		
		−35.65(38)	−35.15(36)		
			−35.19(41)		
MD$_5^H$M	10.21(52)	−35.61(43)	−35.06(44)	−34.69(42)	
		−35.65(43)	−35.10(39)	−34.76(42)	
			−35.20(38)	−34.82(41)	
MD$_6^H$M	10.15(47)	−35.64(44)	−35.08	−34.72	
		−35.69(46)	−35.12(42)	−34.76(43)	
			−35.16	−34.81	
MD$_7^H$M	10.21(48)	−35.60(48)	−35.02	−34.65(35)	
		−35.64(43)	−35.06(41)	−34.66(35)	
			−35.10	−34.69(32)	
MD$_8^H$M	10.26(45)	−35.58(46)	−34.96	−34.56	
		−35.61(48)	−34.99(42)	−34.58(36)	
			−35.04	−34.60	
MD$_{35}^H$M	10.07(59)	−35.84	−35.15		−34.66(37)
		−35.87	−35.18		−34.70(37)
			−35.23		−34.75(37)

^{a}chemical shifts (ppm).
^{b}In seconds; experimentally determined T_1 values are given in parentheses.
^{c}For the definition of the symbols, see the text. The D units are identified as follows: MD$_a^H$D$_b^H$D$_c^H$....
Reproduced by permission of Elsevier Science from Reference 113.

inversion-recovery pulse method $[(180° - t - 90° - T)_n]$ under conditions of proton noise decoupling. The delay between the pulse sequence (T) was set at 300 s. The ^{29}Si spectrum of MD_n^HM ($n = 5-8, 35$) oligomers shows four distinct regions of absorption, M at 10.00 to 10.26 ppm, D_a^H from -35.58 to -35.84 ppm, D_b^H from -34.96 to -35.72 ppm and D_x^H ($x = 3$ to $n-2$) from -34.56 to -34.82. The predominant feature, a triplet at -34.66 to -34.75 ppm, can be assigned to the D^H groups which are three or more D^H groups away from the terminal M group of the oligomer. These coalesce to a triplet structure. Hence there is no detectable change in these chemical shifts and no new stereochemical effects for oligomers MD_n^HM beyond $n = 5$ (Figure 19).

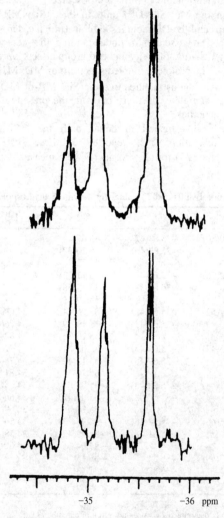

FIGURE 19. ^{29}Si {^{1}H} NMR of the D^H region of MD_5^HM (upper) and MD_7^HM (lower) oligomers. Reproduced by permission of Elsevier Science from Reference 113

The ^{29}Si spin–lattice relaxation times of these polymethylhydrosiloxanes were determined in order to provide information about their molecular motion. The general trends in T_1 values can be seen in Table 12. The T_1 for the M end groups are longest. The D_α^H units have longer relaxation times than those of more internal D^H units. It is reasonable that M has the longest T_1 because M units are able to spin freely at the ends of the chain while D^H units may only rotate through a restricted angle which obviously decreases their T_1 values. These values decrease gradually as the D^H units approach the middle of the chain. All D^H units have comparable relaxation times. This implies that the motional processes along the chains are similar. The relatively long values of T_1 make it clear that these systems are quite mobile. This is to be expected in the absence of crosslinking of linear MD_n^HM chains. Finally, the T_1 values of the different stereoisomers do not change significantly in oligomers and polymer systems. This implies that tacticity and chain length have little effect on the motion occurring in these systems. The ^{29}Si T_1 measurements on the series of MD_n^HM oligomers demonstrate that localized motions along segments of the oligomer backbone result in ^{29}Si relaxation that rapidly becomes independent of chain length.

For comparison, the T_1 values for the D^H groups in the MD_n^HM oligomers are consistently shorter than the T_1 values for the D groups in MD_nM oligomers. The Si–H groups present in the MD_n^HM oligomers provide a dipolar contribution to the relaxation mechanism which shortens the ^{29}Si T_1 relaxation times.

V. SOLID-STATE ^{29}Si NMR

A. Introduction

In recent years the literature has furnished a wealth of information concerning chemical structures from solid-state ^{29}Si NMR studies of crystalline and noncrystalline silicates and aluminosilicate, polysiloxanes, polysilanes and other organosilicon compounds.

Most spectroscopic techniques (e.g. infrared and Raman spectroscopy) provide a 'snapshot' view of the structure of a liquid because the timescale of the techniques is of the order of lattice vibration. However, NMR can probe much lower frequency motions, motions which are important in the glass transition and the viscosity of a silicate liquid. In addition, the timescale of the NMR experiment may be varied (by changing the magnetic field, or the type of experiment, T_1 or $T_{1\rho}$, or observing quadrupolar effects) from a few hertz to several hundred megahertz.

The advent of pulse Fourier Transform (FT) NMR techniques in the middle 1970s set the stage for the use of ^{29}Si NMR for qualitative and quantitative analysis in liquids. However, for solid samples the effects of ^{1}H–^{29}Si magnetic dipole–dipole interactions and ^{29}Si chemical shift anisotropies and the time bottleneck of long ^{29}Si spin–lattice relaxation times render the direct application of the liquid-state ^{29}Si NMR technique essentially useless yielding broad, featureless spectra of low intensity. For understanding the aspects for solid-state ^{29}Si NMR, those for solid-state NMR will be briefly presented.

Pines, Gibby and Waugh introduced the technique of high-power ^{1}H decoupling for eliminating the broadening effect of ^{1}H–^{13}C dipolar interactions, with ^{13}C–^{1}H cross-polarization (CP) to circumvent the ^{13}C T_1 bottleneck[114]. Schaefer and Stejskal[115] then introduced the use of magic angle spinning (MAS) to average out the ^{13}C chemical shift anisotropy[116,117] and demonstrated that the CPMAS combination provides an approach that is capable of yielding high-resolution ^{13}C NMR spectra of solid samples. Line widths of the order of 1 ppm or less can be achieved by this method on crystalline samples, often providing a higher order of structural discrimination than one can achieve in a corresponding liquid (e.g. because of motional averaging of different conformations in the liquid state)[118].

In solid-state high-resolution NMR spectra with MAS, line broadening due to chemical shift anisotropy and dipolar coupling is removed by magic angle spinning and rf irradiation, respectively, for attaining high resolution. Such anisotropy parameters, however, give information about static and dynamic molecular structures in more detail than isotropic chemical shifts do. In a static powder sample, an anisotropic interaction yields a peculiar line shape well known as a powder pattern[119], whose singularities give the principal values of the tensorial interaction. In general, however, the existence of some inequivalent nuclei results in an extensive overlap of their powder patterns, making interpretation almost impossible.

Alternatively, the spectral resolution can be increased at the expense of the angular resolution by MAS in particular in ^{29}Si NMR spectroscopy to allow discrimination between the different carbon residues in the sample. In order to retain the information about the anisotropic interactions, the spinning angular velocity ω_R must be low enough to produce a sufficient number of spinning sidebands[119]. The use of sideband patterns for studying molecular motions was demonstrated by Maricq and Waugh[120]. Not until 1986 was it realized how spinning sidebands can be exploited to study the degree of molecular alignment in partially ordered samples by applying two-dimensional NMR spectroscopy[121,122].

In fact, two-dimensional solid-state NMR is far superior to one-dimensional techniques for studying structure and dynamics[123]. In particular, the two-dimensional exchange NMR spectrum[124] of a static sample is identical with a two-time distribution function[125]. Thus a two-dimensional NMR spectrum, which is detected for a fixed mixing time, is an image of the state of the dynamic process under study at that time.

Experimental examples were first provided through ^{2}H NMR in powder samples of molecular crystals[126,127] and in polymers at their glass transition[128,129]. Switching-Angle Sample Spinning(SASS)[130,131] has some advantages: it can be applied to complex systems contrary to spinning side band analysis[132] or off-magic-angle spinning[133]; it requires no critical adjustments for the experimental parameters, contrary to the rotation-synchronized pulse methods, and, moreover, provides reliable principal values from nondistorted powder patterns.

Solid-state NMR methods allow the investigation of local orientation, dynamics and conformational order of polymer chain segments. ^{29}Si chemical shifts contain information about molecular moieties and about the conformation of chain segments. The line width of a ^{1}H wideline spectrum characterizes the strength of the dipolar couplings among protons and, therefore, the molecular mobility. ^{1}H spin diffusion, mediated by the homonuclear dipolar couplings, is a powerful technique to obtain information about the spatial proximity of molecular moieties. These concepts have recently been combined in a two-dimensional ^{1}H–^{13}C Wideline Separation Experiment (WISE-NMR spectroscopy) by Spiess and coworkers,[134]. They present the results of applying WISE-NMR and proton spin diffusion experiments to stiff macromolecules with flexible side chains. The WISE-NMR spectra demonstrate the existence of rigid and mobile side-chain domains and characterize them with respect to the predominant chain conformations and the local chain mobility. Proton spin diffusion experiments with ^{13}C detection are used to obtain morphological data, like the typical sizes of such domains. Under the normal conditions, to obtain a solid-state ^{13}C spectrum one might observe ^{29}Si signals. The other methods of obtaining ^{13}C NMR spectra are practical techniques for molecules containing multiple silicon atoms. These techniques would be applicable to a wide range of solid-state ^{29}Si NMR methods.

B. New Techniques

^{1}H–^{29}Si dipolar-dephasing experiments (Figures 20 and 21) indicate that various hydroxyl groups of silanols in silica gel undergo rapid ^{1}H spin exchange and that the most strongly coupled protons provide the dominant source of cross-polarization to geminal-silanol silicons. The ^{1}H–^{1}H dipolar dephasing prior to ^{1}H → ^{29}Si cross-polarization shows a rapid ^{1}H spin exchange between ^{1}H reservoirs of single-silanol groups and of geminal-silanol groups; however, the ^{1}H spin exchange rate is too fast to be measured by this strategy. A slower ^{1}H–^{1}H dipolar-dephasing decay due to ^{1}H spin exchanges is found for the ^{1}H spin reservoir of single-silanol groups[135].

Much of the power of MAS NMR has come from the ability of exploiting cross-polarization for facile signal detection and spectral editing for insensitive and rare-spin nuclei. To date, protons have almost invariably constituted the abundant-spin reservoir for cross-polarization experiments, although, most recently, ^{19}F has found increasing use.

There has been only a single CP study using any other nucleus (^{31}P) as the abundant-spin magnetization reservoir, but even that study benefited from the presence of highly abundant ^{1}H spins(which were detected). Thus, to date solid-state NMR spectroscopy of inorganic materials that are devoid of protons has been mostly limited to Bloch decay and spin echo studies. The spin–lattice relaxation times encountered in such systems are often excessively long, resulting in poor signal-to-noise ratios and long measurement times. Furthermore, the line shapes of disordered inorganic solids are frequently rather broad, poorly resolved and thus hard to interpret in the absence of spectral editing experiments.

Recently, various heteronuclear X−Y double resonance approaches have helped to increase the informational content of static and MAS NMR spectra of such systems. Frank and coworkers[136] demonstrate, for the first time, that cross-polarization from ^{31}P to insensitive ^{29}Si nuclei is possible in conjunction with MAS results in significant sensitivity enhancements, and can provide important insights into the structure of inorganic

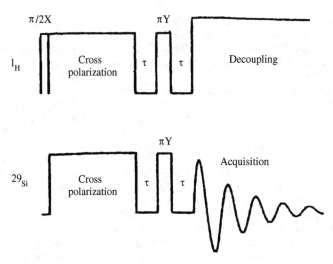

FIGURE 20. A ^{1}H–^{29}Si dipolar-dephasing ^{29}Si CPMAS NMR experiment. Reprinted with permission from Reference 135. Copyright 1992 American Chemical Society

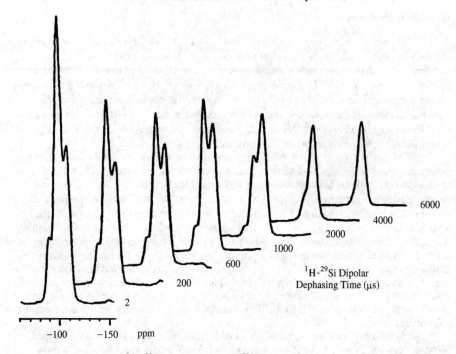

FIGURE 21. 39.75-MHz ^{1}H-^{29}Si dipolar dephasing ^{29}Si CPMAS NMR spectra of Fisher S-679 silica gel, with ^{1}H-^{29}Si dipolar-dephasing times shown. Cross-polarization contact times, 5 ms; magic angle spinning speed, 2.0 kHz. Each spectrum is the result of 3000 accumulations. Reprinted with permission from Reference 135. Copyright 1992 American Chemical Society

semiconductors. Figure 22 compares single-pulse ^{29}Si CPMAS NMR spectra obtained in crystalline $CdSiP_2$ with the corresponding ^{31}P-^{29}Si CPMAS NMR spectra, illustrating the expected sensitivity advantage of CPMAS NMR. This advantage is especially critical for the detection of the ^{29}Si nuclei which have excessively long spin–lattice relaxation times.

Silica gel has frequently been used as a convenient, inert, high-surface-area and non-swelling support that can be easily removed by filtration from the reaction medium if necessary. In these systems, the bonding of the reactant functionality, R, to the surface is via a bridging silicon attached to the silica surface by an Si—O—Si linkage. These functionalized silica gels are usually made by activating the silica surface to produce hydroxyl groups and then reacting it with a trichloro (or trialkoxy) silane. The reaction is often represented as in equation 16. However, the trifunctional silane can also be considered as a trifunctional monomer molecule that would polymerize to give a highly cross-linked polysiloxane polymer as in equation 17. Reaction 16 may be favored by making the system as anhydrous as possible, but this is difficult to do because of the hydrophilic nature of the silica surface, and in practice there may well be an unknown amount of self-condensation of the organosilane occurring, giving oligomeric polysiloxane species bonded to the surface, making the reproducibility of these reactions somewhat variable. Fyfe and coworkers[137] outline the use of an alternative preparation of functionalized materials of this general type and characterization of the systems formed by two-dimensional

6. ^{29}Si NMR spectroscopy of organosilicon compounds

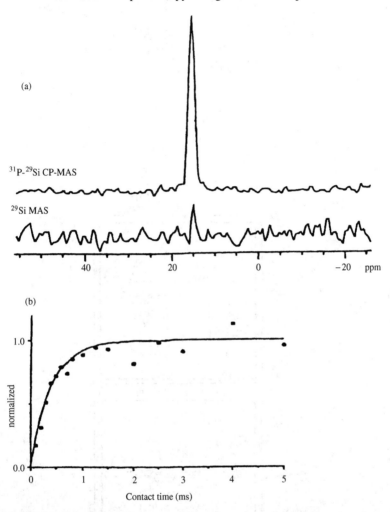

FIGURE 22. ^{31}P–^{29}Si CPMAS in crystalline CdSiP$_2$. (a) Comparison of single-pulse spectrum (bottom trace) and CPMAS spectrum (top trace). The spectrum is recorded by 16 scans with a 2-min relaxation delay and a single-pulse ^{29}Si spectrum (two scans, 15-min relaxation delay). The CPMAS contact times was 5.5 ms for the ^{29}Si detection. (b) Variable contact time ^{31}P–^{29}Si (MAS at 3.0 kHz) CPMAS experiment. The solid curve is a fit to an exponential cross-relaxation process time of 0.4 ms (^{31}P–^{29}Si). Reprinted with permission from Reference 136. Copyright 1992 American Chemical Society

solid-state NMR techniques.

$$\text{Si}-\text{OH} + (\text{EtO})_3\text{Si}-\text{R} \xrightarrow[\text{H}_2\text{O}]{\text{H}^+} \text{Si}\begin{matrix}-\text{O}\\-\text{O}\\-\text{O}\end{matrix}\text{Si}-\text{R} + 3\text{EtOH} \quad (16)$$

$$(\text{EtO})_3\text{Si}—\text{R} \xrightarrow[\text{H}_2\text{O}]{\text{H}^+} \left[\text{R}—\underset{\underset{|}{\text{O}}}{\overset{\overset{|}{\text{O}}}{\text{Si}}}—\text{O} \right]_n \quad (17)$$

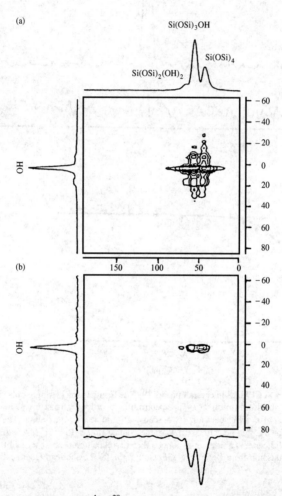

FIGURE 23. Contour plots of the 2D ^1H–^{29}Si correlation experiments on silica gel obtained with a 22.0 ms contact time, a 3.0 s repetition time and a 4.0 kHz sample spinning rate. The vertical axis represents the proton chemical-shift scale and the horizontal axis the ^{29}Si chemical-shift scale. The spectra above and at the side of the figures are the one-dimensional projections. The 2D spectrum was obtained from 64 individual experiments: (a) unwashed silica gel, 80 scans for each individual experiment; (b) D_2O washed sample, 200 scans for each experiment. Reprinted with permission from Reference 137. Copyright 1988 American Chemical Society

6. ^{29}Si NMR spectroscopy of organosilicon compounds

The viability of an experiment of this type between ^{1}H and ^{29}Si has previously been demonstrated by Vega[138]. At first sight, it might seem that this two-dimensional experiment would be limited by the same factors as the one-dimensional CP experiments. However, because the proton spins are relatively isolated MAS alone gives enough resolution to clearly distinguish the OH and Me signals, thus making it possible to identify the source of the polarization transfers. The isolation of the proton spins also limits ^{1}H–^{1}H spin diffusion. In addition, because there are two related frequency scales in the experiment, the chemical shift resolution is better than in the simple one-dimensional experiment. Figure 23 shows the results of the ^{1}H/^{29}Si connectivity experiments carried out on unfunctionalized silica gel using the experimental conditions detailed in the figure caption. The one-dimension spectra shown in the figure are projections of the data onto the F1 and F2 axes and can be used to establish the connectivities. However, the intensities of the signals do not reflect those of the corresponding 1D spectra.

In Figure 23a, the gel has not been deuterium-exchanged and an intense signal is observed due to OH protons in the ^{1}H projection. There is, as expected, a series of intense connectivities to the three silicon environments in the gel.

In Figure 23b, the gel has been exchanged twice with D_2O and dried at 100 °C, with considerable care being taken to avoid subsequent water adsorption. As can be seen from the figure, there is a very marked decrease in the intensity of the heteronuclear connectivity, but there is still a residual interaction arising from the trace amounts of hydroxy protons indicated in the projection of F1. It is this situation that leads to the possible ambiguities in the one-dimensional experiments described above.

Figure 24 shows experiments carried out on a simple physical mixture of unfunctionalized silica gel and polymethylsiloxane whose one-dimensional ^{29}Si MAS NMR spectrum is shown in Figure 25. In this case, there should be no connectivity between the methyl protons and any of the silicon nuclei in the gel. Figure 24a shows the experimental results when no deuterium exchange has been carried out. There are clear connectivities to both groups of silicon. In the case of the lower-field methyl-substituted silicon signal, the connectivity is mainly from the methyl protons, and there is substantial intensity from spinning sidebands consistent with this. In the case of the three high-field signals due to the silica gel, the connectivities must be to hydroxyl protons, as is borne out by the limited sideband pattern. Most importantly, however, the sources of polarization can be unambiguously identified from the chemical shifts. Thus, as indicated in the figure, the two different sets of silicon nuclei are polarized from two distinct proton sources: methyl groups for the polymethylsiloxane and hydroxyl groups for the silica gel. The projection on F1 identifies their proton chemical shifts as $\delta = 0.45$ ppm and $\delta = 1.9$ ppm, respectively. Consistent with this interpretation, D_2O exchange of the mixture removes the connectivities to the silica gel, while a very strong connectivity to the polymethylsiloxane silicons with its associated sideband pattern remains (Figure 24b).

The layered alkali metal silicate magadiite (Figure 26) is analyzed by solid-state 2D MAS NMR techniques (Figure 27). The ^{1}H peak at $\delta 3.5$ can be attributed to water molecules, whereas that at $\delta 14.9$ must arise from strongly hydrogen-bonded protons that presumably involve $-Si-OH$ groups in unusual environments. The ^{29}Si projection spectrum consists of four peaks at $\delta -99.2$, -109.5, -111.2 and -113.6. These can confidently be assigned to one Q3 site and three Q4 sites respectively. It has been proposed that the three Q4 sites may differ by their average Si–O–Si bond angles.

C. NMR and X-ray Studies

^{29}Si shielding tensor information may reflect the structure of organosilicon compounds while the structure might have been determined by X-ray crystallographic analysis. Thus,

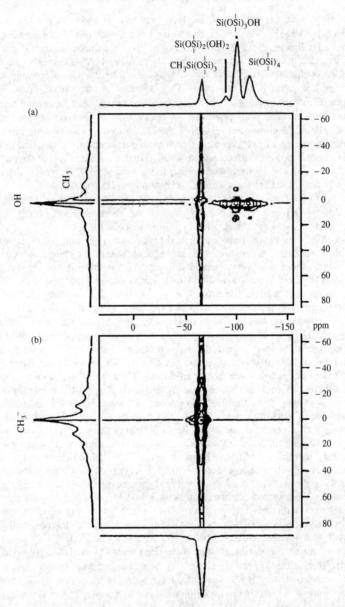

FIGURE 24. Contour plots of the 2D $^1H-^{29}Si$ heteronuclear correlation experiments on a mixture of silica gel and polymethylsiloxane. The spectra were obtained under the same conditions described in Figure 23: (a) untreated sample, 120 scans for each individual experiment: (b) D$_2$O washed sample, 200 scans for each individual experiment. The spinning rate used (4.3 kHz) was high enough to prevent overlap of spinning sidebands and isotropic peaks between the two sets of resonances. Reprinted with permission from Reference 137. Copyright 1988 American Chemical Society

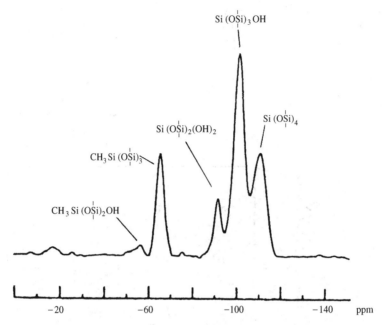

FIGURE 25. ^{29}Si CPMAS NMR spectrum of mechanical mixture of silica gel and polymethylsiloxane. Reprinted with permission from Reference 137, Copyright 1988 American Chemical Society

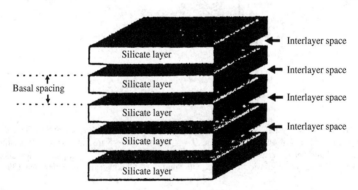

FIGURE 26. A schematic diagram of the layered alkali metal silicate magadiite. Reproduced by permission of the Royal Society of Chemistry from Reference 139

much work has been concerned with the measurement of isotropic chemical shift, and the correlation of these with various structural parameters. One difficulty with such an approach is that, in rapid magic-angle spinning experiments, only the isotropic part of the second-rank shielding tensor (σ) is generally obtained for ^{29}Si since shielding anisotropies are relatively small (less than 50 ppm in magnitude, which is smaller than typical spinning rates). Information concerning the principal compounds is lost. Harris and coworkers[140] have used slow MAS (i.e. at rates of ca 200 Hz) in order to obtain extensive spinning sideband manifolds from which shielding tensor components can be derived. The

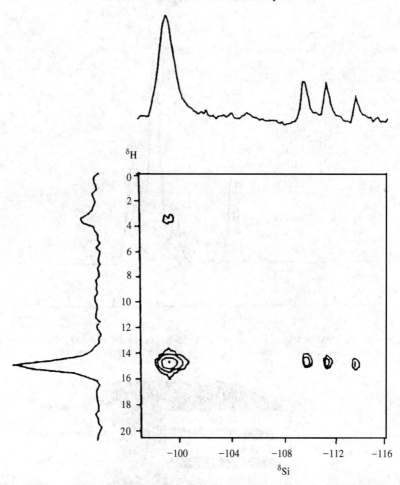

FIGURE 27. Two-dimensional ^{1}H–^{29}Si correlation spectrum (contour mode) of carefully-dried magadiite. Experimental conditions: recycle time 0.5 s, contact time 8 ms, number of experiments 128, number of transients per t_1 point 432, spinning rate 4.87 kHz. ^{1}H and ^{29}Si projections are shown along the axes. Reproduced by permission of the Royal Society of Chemistry from Reference 139

determination of the principal components of the tensor σ is important for several reasons. Firstly, it may help to provide a better explanation for and physical description of the trends observed in ^{29}Si chemical shifts. Secondly, the tensor components provide data which may be used to test theories of shielding. Finally, shielding tensor information may give evidence of molecular motion. They presented results on the isotropic chemical shifts and ^{29}Si shielding tensor components for a number of organosilicon compounds, in an attempt to investigate the various factors which influence the shielding anisotropy observed.

It is assumed that shielding is the only magnetic influence on spinning sideband intensities. The isotropic chemical shift $\sigma_{\text{iso}} = -\delta$ Si/ppm is defined as equal to one-third of the trace of the shielding tensor (Trσ/3). The principal values of the shielding tensor (σ_{11},

6. ^{29}Si NMR spectroscopy of organosilicon compounds

TABLE 13. ^{29}Si shielding tensor data for organosilicon compounds[a]

Compound	δ (^{29}Si) (ppm) solution	solid	Δσ (ppm)	$σ_{11}$ (ppm)	$σ_{22}$ (ppm)	$σ_{33}$ (ppm)	$η$[b]
Ph$_3$Si—SiMe$_3$	−18.4(23)	−18.9	21.9	11.6	11.6	33.5	0.00
Ph$_3$Si—SiMe$_3$	−21.0(23)	−21.0	−33.6	32.2	32.2	−1.4	0.00
Ph$_3$Si—SiPh$_3$	−26.61(24)	−25.4	−25.4	37.2	37.2	1.8	0.00
Ph$_3$Si—O—SiPh$_3$	−18.5	−17.0	27.9	1.5	13.9	35.6	0.67
Ph$_3$SiH	−21.1(25)	−21.4	45.0	−0.4	13.2	51.4	0.45
Ph$_8$Si$_4$	−20.93(26)	−24.2	−44.8	51.7	26.6	−5.7	0.84
(PhCH$_2$)$_3$SiH	−3.83(27)	−4.5	11.6	−2.8	4.1	12.2	0.90

[a] The errors in $σ_{11}$ values are typically ca ± 2 ppm.
[b] A typical error in $η$ is ±0.1.
Reproduced by permission of the Royal Society of Chemistry from Reference 140.

$σ_{22}$ and $σ_{33}$) are designated by equation 18.

$$|σ_{33} − σ_{iso}| \geq |σ_{11} − σ_{iso}| \geq |σ_{22} − σ_{iso}| \qquad (18)$$

The shielding anisotropy (Δσ) is defined by equation 19

$$Δσ = σ_{33} − (σ_{11} + σ_{22})/2 \qquad (19)$$

and the asymmetry parameter ($η$) by equation 20

$$η = (σ_{22} − σ_{11})/(σ_{33} − σ_{iso}) \qquad (20)$$

The results of these studies are summarized in Table 13. Some typical ^{29}Si CPMAS NMR spectra are displayed in Figures 28–30.

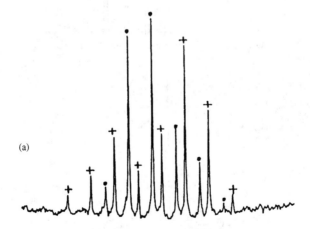

FIGURE 28. 39.758-MHz CPMAS ^{29}Si NMR spectrum of solid 1,1,1-trimethyltriphenyldisilane. (a) Slow spinning (206 Hz). (b) Fast spinning (3.2 kHz). The two spinning sideband manifolds in (a) are indicated by different symbols. Experimental conditions: contact time 8 ms, recycle time 10 s, number of transients 2090 (a) and 405 (b). Reproduced by permission of the Royal Society of Chemistry from Reference 140

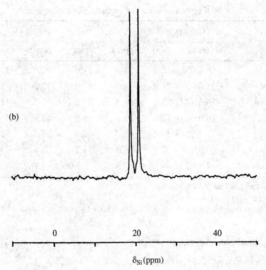

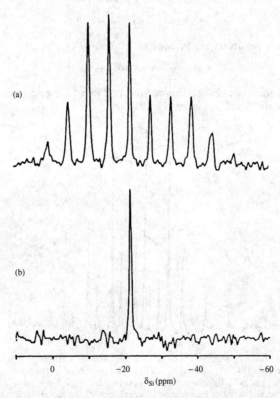

FIGURE 29. 39.758-MHz CPMAS ^{29}Si NMR spectrum of solid triphenylsilane. (a) Slow spinning (228 Hz). (b) Fast spinning (3.1 kHz). Experimental conditions: contact time 3 ms, recycle time 30 s, number of transients 1782 (a) and 24 (b). Reproduced by permission of the Royal Society of Chemistry from Reference 140

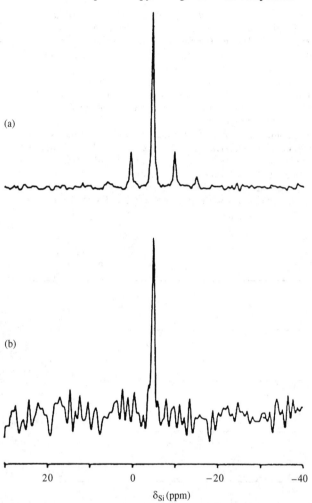

FIGURE 30. 39.758-MHz CPMAS ^{29}Si NMR spectrum of solid tribenzylsilane. (a) Slow spinning (196 Hz). (b) Fast spinning (3.2 kHz). Experimental conditions: contact time 1 ms, recycle time 30 s, number of transients 1480 (a) and 170 (b). Reproduced by permission of The Chemical Society from Reference 140

It is noticeable that the isotropic chemical shifts of the molecules do not change appreciably from the solution to the solid state. Therefore, effects caused by intermolecular interactions may be considered as negligible, and there are no major structural changes with phase. The ^{29}Si CPMAS spectra displayed two resonances at -18.9 and -21.0 ppm, which were assigned to silicon nuclei present within the Me$_3$Si and Ph$_3$Si moieties, respectively. The crystal structure of this compound has been published. It has space group P$\overline{3}$ and the asymmetric unit has been shown to consist of one molecule. Each of the silicon nuclei lies on an axis of C$_{3v}$ symmetry. Hence slow-spinning ^{29}Si CPMAS NMR experiments reveal spinning sideband manifolds resulting from axially symmetric shielding tensors. The results (Table 13) obtained in this work are a positive anisotropy being observed for the Me$_3$Si while a negative anisotropy is seen for the silicon atom of the Ph$_3$Si group. Figure 31 summarizes the data on shielding tensor components for the six compounds studied. It is remarkable that for the four results involving Ph$_3$Si groups, the value of the lowest shielding component varies only to a small extent (within the range -1.5 to $+1.8$ ppm) whereas the other components vary widely. The relatively constant component is known to be directed along the Ph$_3$Si axis in two of the cases. However, the atom to which this group is bonded differs, being Si for the two disilanes, O for the disiloxane and H for triphenylsilane.

The low-temperature reaction (Figure 32) of (E)-1,2-dimesityl-1,2-di-t-butyldisilene (**52a**) with dioxgen gives (E)-1,2-dimesityl-1,2-di-t-butyl-1,2-disiladioxethane (**53a**), the structure of which has been established by X-ray crystallographic analysis (Figure 33).

This result establishes that both oxidation of **52a** and rearrangement of **53a** to 1,3-cyclodisiloxane **55a** take place with retention of configuration at silicon. Disiladioxethane **53a** forms monoclinic crystals with space group C2/c. Extra electron density found above

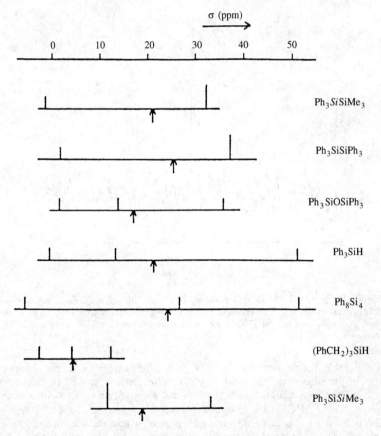

FIGURE 31. Summary chart of the principal components for the ^{29}Si shielding tensors of the compounds studied. Axial symmetry is indicated by the double intensity of lines at $\sigma_\perp$. The arrows indicate the isotropic values. The shielding scale is relative to the isotropic value for tetramethylsilane. Reproduced by permission of the Royal Society of Chemistry from Reference 140

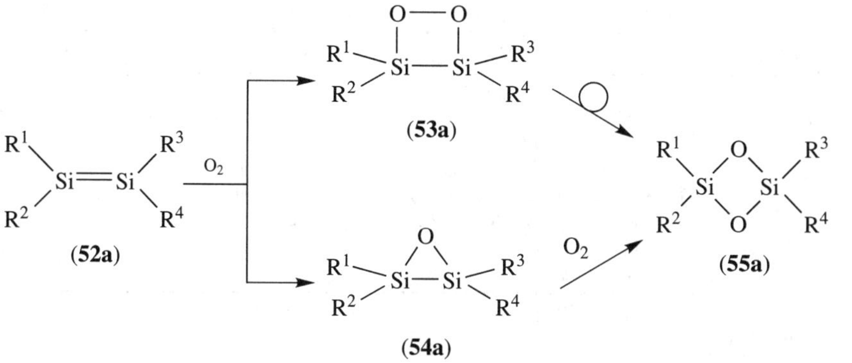

6. ^{29}Si NMR spectroscopy of organosilicon compounds

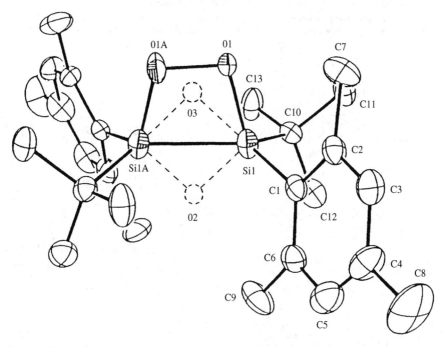

FIGURE 32. Oxidation of disilenes with $^3O_2$. Reprinted with permission from Reference 141. Copyright 1992 American Chemical Society

FIGURE 33. Thermal ellipsoid drawing of the crystal structure of 1,2-dimesityl-1,2-di-*t*-butyldisiladioxethane (**53a**) with hydrogen atoms omitted for clarify. Dotted lines indicate the 5.6% **55a** present. Selected bond distances (pm) and angles (deg) are as follows: Si(1)−Si(1A), 233.1(2); O(1)−O(1A), 148.2(5); Si(1)−O(1), 173.2(3); Si(1)−C(10), 189.4(3); Si(1)−C(1), 188.7(4); Si(1)−O(3), 170.2(2); Si(1)−O(2), 170.2(2); Si(1A)−Si(1)−O(1), 75.5(1); Si(1)−O(1)−O(1A), 103.7(1); Si(1)−O(3)−Si(1A), 86.5(1); Si(1)−O(2)−Si(1A), 86.4(1); O(3)−Si(1)−O(2), 93.6(1); C(10)−Si(1)−Si(1A), 123.0(1); C(1)−Si(1)−C(10), 112.9(2); C(1)−Si(1)−Si(1A), 119.5(1). Reprinted with permission from Reference 141. Copyright 1992 American Chemical Society

TABLE 14. ^{29}Si NMR chemical shifts δ (ppm) for **53** and **55**

System	Substitution[a]	53	55
a	$R^1 = R^4 = $ Mes; $R^2 = R^3 = t$-Bu	+54.80	+13.22
b	$R^1 = R^2 = R^3 = R^4 = $ Mes	+41.37	−3.35
c	$R^1 = R^2 = R^3 = R^4 = $ Xyl	+40.96	−3.30
d	$R^1 = R^2 = R^3 = R^4 = $ Dmt	+41.19	−2.46
e[b]	$R^1 = R^2 = $ Mes; $R^3 = R^4 = $ Xyl	+41.46, +40.67	−3.63, −2.33
f[c]	$R^1 = R^2 = R^3 = $ Xyl; $R^4 = $ Mes	+41.26, +40.77	−3.49, −2.75
g	$R^1 = R^3 = $ Mes; $R^2 = R^4 = $ Xyl	+41.16, +41.15	
h	$R^1 = R^3 = $ Ad; $R^2 = R^4 = $ Mes	+49.10	+8.1

[a] Mes = 2,4,6-trimethylphenyl, Xyl = 2,6-dimethylphenyl, Dmt = 2,6-dimethyl-4-(t-butyl)phenyl, Ad = 1-adamantyl.
[b] ^1J(SiSi) = 98 Hz for **52e**.
[c] ^1J(SiSi) = 94 Hz for **52f**.
Reproduced by permission of the American Chemical Society from Reference 141.

and below the Si−Si bond was shown by ^{1}H and solid-state ^{29}Si NMR to be due to **55a** (Table 14), arising from rearrangement of **53a** in the solid[141]. In the presence of phospines or sulfides, **53a** is partly deoxygenated to form disilaoxirane **54a**, which was characterized by means of ^{29}Si NMR.

The zwitterion λ^5-spirosilicate bis[2,3-naphthalendiolato][2-(dimethylammonio)phenyl]silicate (**56**; isolated as **57** = **56**·1/2 MeCN) was synthesized by reaction of [2-(dimethylamino)phenyl]dimethoxyorganosilanes with 2,3-dihydroxynaphthalene in acetonitrile at room temperature. Reaction of **57** or of [2-(dimethylamino)phenyl]trimethoxysilane with water in acetonitrile yielded the cage-like silasesquioxane **58** ($R = 2 - $ Me$_2$NC$_6$H$_4$). The crystal structures of **57** and **58** were studied by X-ray diffraction. In addition, **57** and **58** were characterized by solid-state ^{29}Si CPMAS NMR[142] (Figure 34).

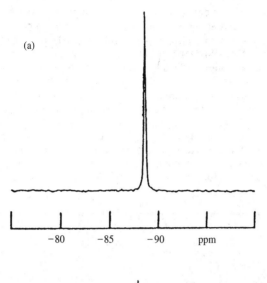

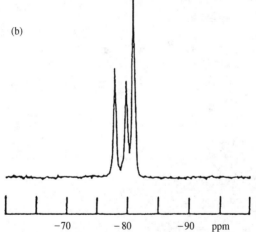

FIGURE 34. ^{29}Si CPMAS NMR spectra of **57** (a) and **58** (b). (a) δ^{29}Si $= -88.6$, $\nu_{1/2} = 12$ Hz, (b) δ^{29}Si $= -77.9, -79.7, -80.9$. Reproduced by permission of Johann Ambrosius Barth from Reference 142

D. Relaxation Times

Understanding of the physical chemistry of silicate liquids is important in both earth sciences and materials science: the chemical and physical behavior of magmas dominates many geological processes, and most technological glasses and glass ceramics start off in the molten state. The ability of NMR measurements to give information about the dynamics of systems (as well as about their structure) through the measurement of relaxation times and line shapes over a range of temperature is well known. Farnan and Stebbins[143] have applied a variable-temperature study over a temperature range of 1200 °C to silicate systems with different SiO_2 contents. Samples were prepared with SiO_2 which was enriched to 95% in ^{29}Si and high-purity Na_2CO_3, Al_2O_3, K_2CO_3, Li_2CO_3. These samples were albite ($NaAlSi_3O_8$) and potassium tetrasilicate ($K_2Si_4O_9$) glasses, and a mixture of crystalline lithium orthosilicate and metasilicate (Li_4SiO_4/Li_2SiO_3) with a near-eutectic composition. Each represents a different bridging and nonbridging oxygen distribution, from framework Q^4 (albite) to Q^2 and Q^0 (lithium orthosilicate/metasilicate). In the Q^n nomenclature, n refers to the number of oxygens per SiO_4 unit (Q) which are shared with another SiO_4 unit. The glass samples were made by heating stoichiometric amounts of their components to about 1200 °C above the melting point for several hours; they were then quenched, reground and remelted until homogeneous. The crystalline sample was made by slow cooling from the melt. In each preparation the weight loss at each stage was carefully monitored to ensure that there was only a negligible loss

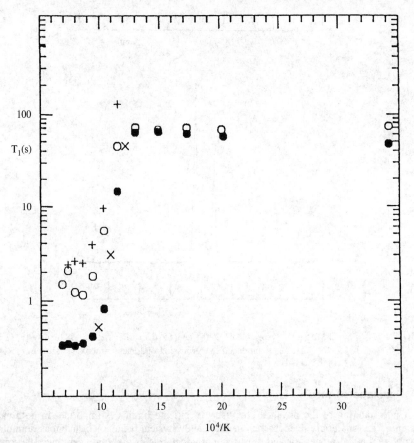

FIGURE 35. ^{29}Si T_1 versus inverse temperature for potassium tetrasilicate glass: undoped sample measured with increasing temperature (O), undoped sample measured with decreasing temperature (+); sample doped with 500 ppm Fe_2O_3 measured with increasing temperature (●); measured with decreasing temperature (×). Reprinted with permission from Reference 143. Copyright 1990 American Chemical Society

6. ^{29}Si NMR spectroscopy of organosilicon compounds

of alkalis. High-temperature (up to 1250 °C) ^{29}Si NMR T_1 and line-shape measurements were conducted on silicate samples with varying SiO_2 contents. The samples represented a range of bridging and nonbridging oxygen distributions from Q^4 in albite ($NaAlSi_3O_8$) to Q^2 and Q^0 in a mixture of lithium orthosilicate and metasilicate (Li_4SiO_4/Li_2SiO_3). T_1 relaxation data as a function of temperature for glassy samples showed a dramatic increase in efficiency at the glass transition (Figure 35). This was ascribed to relaxation by paramagnetic impurities becoming more efficient as silicon nuclei and impurity ions begin to diffuse through the material at temperatures above T_g.

In each sample chemical exchange was observed at high temperature, indicating that the lifetime of a silicate tetrahedron in the melt is short on the NMR timescale, i.e. a few microseconds. Variable-temperature line-shape data for potassium tetrasilicate ($K_2Si_4O_9$) allowed the exchange process to be modeled and spectra to be simulated (Figures 36 and 37) yielding an activation energy for the process. This was in good agreement with the activation energy for viscous flow derived from viscosity measurements. It appears that

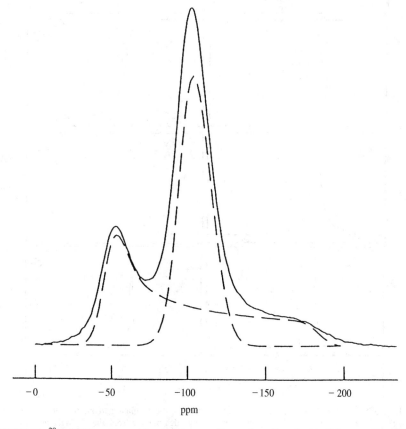

FIGURE 36. ^{29}Si NMR spectrum of potassium tetrasilicate glass (doped with 500 ppm of Fe_2O_3) at 20 °C: 128 pulses ($\pi/2$) with a delay of 300 s. The dashed lines show the two resonances (Q^3 and Q^4) which compose the total line shape. Reprinted with permission from Reference 143. Copyright 1990 American Chemical Society

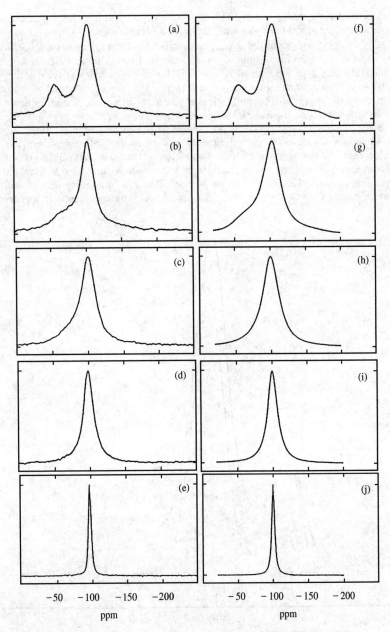

FIGURE 37. ^{29}Si NMR spectra (a-e) of potassium tetrasilicate glass and simulations (f-j). Spectra were recorded with 64 pulses ($< \pi/6$) and delays determined from the T_1 data at (a) 697 °C, (b) 774 °C, (c) 800 °C, (d) 847 °C, (e) 997 °C. Simulations of these spectra were judged 'by eye' to be the best match with exchange rates of (f) 2,000 Hz, (g) 10,000 Hz, (h) 25,000 Hz, (i) 50,000 Hz, (j) 500,000 Hz. Reprinted with permission from Reference 143, Copyright 1990 American Chemical Society

^{29}Si NMR comes close to detecting the fundamental step in viscous flow in silicates with good agreement between the time constant of the exchange process determined by NMR and the shear relaxation time of the $K_2Si_4O_9$ liquid.

High-resolution ^{29}Si NMR spectra and ^{29}Si NMR relaxation times of a polysiloxane containing rigid groups, namely copoly(tetramethyl-*p*-silaphenylenesiloxane/dimethylsiloxane) (CPTMPS/DMS) (equation 21) in the solid state, were measured over a wide range of temperature in order to obtain information about the conformation and dynamics[144]. The ^{29}Si T_1 values of CPTMPS/DMS in the solid state using the inversion recovery method at −90, −60, −30, 27 and 80 °C were determined. Figure 38 shows the ^{29}Si stack spectrum of CPTMPS/DMS at 80 °C. The plots of ^{29}Si T_1 against 1/T (K^{-1}) for the TMPS (tetramethyl-*p*-silaphenylene siloxane) and DMS (dimethylsiloxane) moieties from −90 to 80 °C are shown in Figure 39.

$$\text{(21)}$$

The T_1 value for the TMPS moiety decreased as the temperature increased from −90 to 80 °C. This indicates that the molecular motion is in the slow-motion region in this temperature range. However, the T_1 value for the DMS moiety is increased as the temperature is increased from −67 to 80 °C, becomes a minimum at −67 °C and is increased again as the temperature is decreased from −67 to −90 °C. This indicates that the molecular motion is in the extreme narrow region above −67 °C. The copolymer retains both rigid and soft parts in the main chain.

E. Dynamics

Solid-state NMR experiments showed the dynamic nature of the disordering. It has been possible to monitor the type and the changes of segmental motions within the mesomorphic phase and at the corresponding phase transitions for typical examples. The thermal

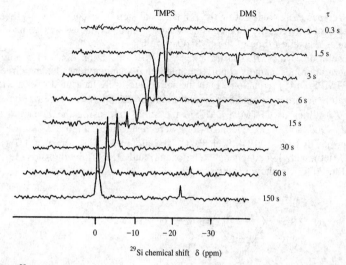

FIGURE 38. ^{29}Si GHDMAS NMR spectra of CPTMPS/DMS in the solid state at 80 °C using the inversion recovery method (180°·– τ – 90° – 150 s) as a function of τ. Reproduced by permission of Elsevier Science from Reference 144

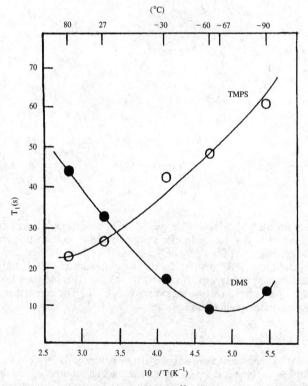

FIGURE 39. Temperature dependence of the observed ^{29}Si T_1 in CPTMPS/DMS in the solid state. Reproduced by permission of Elsevier Science from Reference 144

6. ^{29}Si NMR spectroscopy of organosilicon compounds

behavior of poly(diethylsiloxane) (PDES)^{145} is summarized schematically in Figure 40. High-resolution solid-state ^{29}Si NMR spectra of PDES are shown in Figure 41.

The temperature-dependent MAS spectra were obtained by heating pure β-PDES (a) and also by heating a mixture of the α- and β-modification (b). The β-polymorph gave one sharp signal at -17.9 ppm for the fully ordered crystalline state at 190 K, which did not change on raising the temperature until the $\beta_1-\beta_2$ transition was reached. At the $\beta_1-\beta_2$ transition the resonance was shifted upfield by $\Delta\nu = 3.3$ ppm to -21.2 ppm. Increasing the temperature further resulted in a gradual upfield shift from -21.2 ppm at 220 K to -22.2 ppm at 280 K, which is directly below the $\beta_2-\mu$ transition. On passing the $\beta_2-\mu$ transition, the isotropic chemical shift of the silicon nuclei was shifted to -23.6 ppm. Further heating, even above the melt transition (isotropization), did not result in further variation of the chemical shift. The chemical shift remained the same at 298, 315 and 325 K.

The observed upfield shifts indicate changes in the molecular packing and the bond conformation as the sample is converted from a highly ordered crystal to the isotropic melt. In the melt, the ^{29}Si resonance gives the fast exchange-averaged chemical shift for a dynamic equilibrium between different rotational isomeric states of the Si–O and Si–C bonds. The fact that the ^{29}Si chemical shift is identical for the melt and the μ-phase demonstrates a dynamically disordered conformational state also below the isotropic transition. This is confirmed by ^{29}Si spin–lattice relaxation experiments. The T_1 time was 23 s for the melt at 330 K and 25 s for the μ-phase at 300 K. Thus, the motional state and the conformational equilibrium of the molecular segments remain very much the

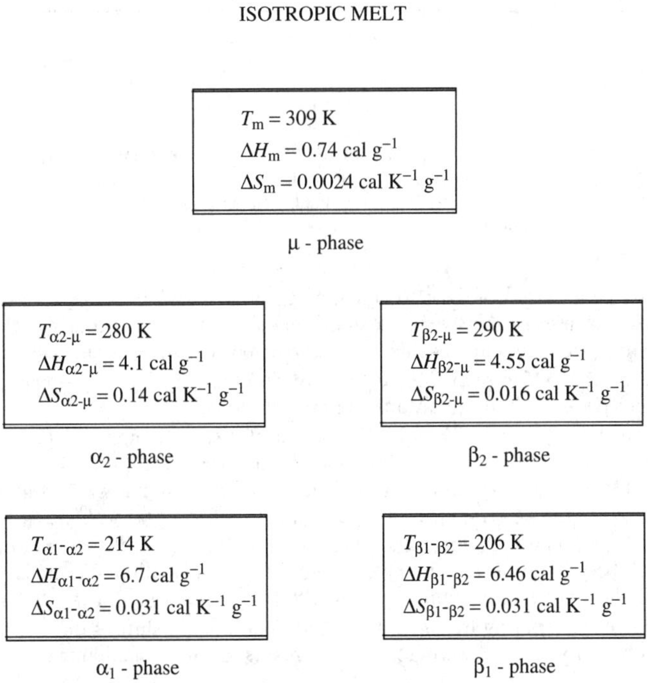

ISOTROPIC MELT

$T_m = 309$ K
$\Delta H_m = 0.74$ cal g^{-1}
$\Delta S_m = 0.0024$ cal K^{-1} g^{-1}

μ - phase

$T_{\alpha2-\mu} = 280$ K
$\Delta H_{\alpha2-\mu} = 4.1$ cal g^{-1}
$\Delta S_{\alpha2-\mu} = 0.14$ cal K^{-1} g^{-1}

α_2 - phase

$T_{\beta2-\mu} = 290$ K
$\Delta H_{\beta2-\mu} = 4.55$ cal g^{-1}
$\Delta S_{\beta2-\mu} = 0.016$ cal K^{-1} g^{-1}

β_2 - phase

$T_{\alpha1-\alpha2} = 214$ K
$\Delta H_{\alpha1-\alpha2} = 6.7$ cal g^{-1}
$\Delta S_{\alpha1-\alpha2} = 0.031$ cal K^{-1} g^{-1}

α_1 - phase

$T_{\beta1-\beta2} = 206$ K
$\Delta H_{\beta1-\beta2} = 6.46$ cal g^{-1}
$\Delta S_{\beta1-\beta2} = 0.031$ cal K^{-1} g^{-1}

β_1 - phase

FIGURE 40. Polymorphism and melting of poly(diethylsiloxane). Reprinted with permission from Reference 145. Copyright 1989 American Chemical Society

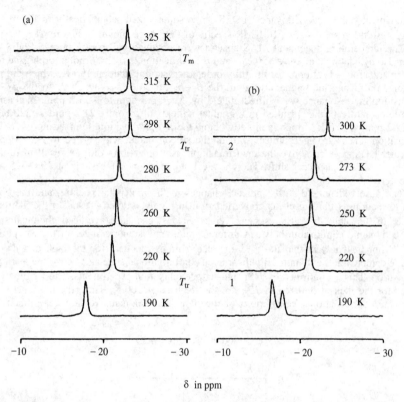

FIGURE 41. Temperature variation of the MAS ^{29}Si NMR spectrum of poly(diethylsiloxane): (a) spectra of the pure β-modification; (b) spectra of a mixture of α and β-modification. Spectra were recorded by stepwise heating of the samples. Reprinted with permission from Reference 145. Copyright 1989 American Chemical Society

same. The downfield shift of the ^{29}Si signal on cooling below the μ–β_2 transition may be explained by a change of the segmental conformations and the molecular packing. Below the β_1–β_2 transition, the silicon nuclei appear to be locked in a single state, which is represented by the −17.6 ppm ^{29}Si resonance. Thus, finally in the β_1-phase, the PDES molecules are packed rigidly in an ordered crystal lattice.

Figure 41b shows the ^{29}Si MAS NMR spectra of a sample that was quenched rapidly from the μ-phase and thus contains both the α- and β-modifications. Two signals could be resolved for the rigid crystal at 190 K. The −16.6 ppm signal was assigned to the α-polymorph while the −17.6 ppm signal is identical with the 190 K ^{29}Si NMR resonance in Figure 41b and represents the β-modification. The fraction of the α_1-modification was calculated to be 64% from the signal intensities as the relaxation and cross-polarization behavior of the two modifications did not differ significantly. Raising the temperature above the lower disordering transitions resulted in an upfield shift of the ^{29}Si resonances. Up to 220 K, two signals were resolved. This is demonstrated more clearly in the enlargement in Figure 42a.

The −20.9 ppm signal can be assigned to the α_2- and the resonance at −21.2 ppm to the β_2-modification by comparison with the corresponding spectrum of Figure 42b. Thus,

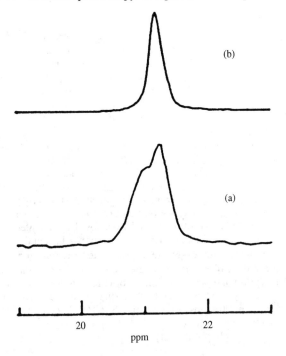

FIGURE 42. Enlargement of the MAS ^{29}Si NMR resonance of pure β-PDES (b) and a mixture of α- and β-PDES (a), $T = 220$ K. Reprinted with permission from Reference 145. Copyright 1989 American Chemical Society

the α_2- and β_2-modifications are clearly distinguished by the ^{29}Si NMR chemical shift. The deconvolution of the 220 K resonances yielded only 60% α_2-PDES, which is only little less than the value obtained from the spectrum at 190 K. When the temperature was raised further, the shoulder disappeared, and only one signal could be resolved. The slow rates for the conversion of α_2- to β_2-PDES as demonstrated by the calorimetric experiments cannot explain this phenomenon.

High-resolution solid-state ^{29}Si and ^{13}C NMR spectroscopy have been used to investigate the dynamic properties of tetrakis(trimethylsilyl)silane [Si(SiMe$_3$)$_4$:TTMSS][146]. It is clear from the temperature dependence of the spectra as a function of temperature that TTMSS shows the spectral changes associated with the phase transition. ^{29}Si chemical shifts are given relative to SiMe$_4$ as an external standard. The spectrum recorded at 153 K contains two peaks for the SiMe$_3$ environments ($\delta - 8.7$ and -9.6) with 1 : 3 intensity ratio; coalescence of these peaks occurs at 206 K ($\delta - 9.5$), and there are further changes of chemical shift on heating to 304 K ($\delta - 9.8$). The spectrum contains a single peak for the central Si environment at $\delta - 140.8$ (153 K), -138.8 (206 K) and -135.3 (304 K). As a result, the following represent two plausible types of motion that may be occurring in the low-temperature phase of solid TTMSS: (a) rotation of the whole molecule around a space-fixed crystallographic axis parallel to one of the Si—Si bonds, and (b) rotation of each SiMe$_3$ group about the local C_3 symmetry axis coincident with the relevant Si—Si bond. Rotation of each Me group around the relevant Si—C bond is assumed to be rapid on the experimental timescale at all temperatures studied here.

F. Hypervalent Silicon

An interesting question is whether hypervalent silicon nuclei can be monitored by ^{29}Si NMR spectroscopy. A systematic ^{29}Si MAS NMR measurement of nine silicates containing hexacoordinate silicon nuclei shows that a decrease in the mean Si—O bond distance $d(\text{Si}^{VI}-\text{O})$ corresponds to an increase in the magnetic shielding of the Si^{VI} nucleus. All isotropic chemical shifts lie within the range -142 to -220 ppm. The spectrum of Si$_5$O(PO$_4$)$_6$ (Figure 43) also shows a broad (ca 400 Hz) signal centered at -112 ppm and a narrow signal (ca 60 Hz) at -119.9 ppm; these lines are characteristic of tetrahedrally coordinated silicon, Si^{VI}.

The high-field signals between -142 and -220 ppm reflect the presence of octahedral silicon[147]. ^{29}Si MAS NMR is capable of detecting hexacoordinated silicon in heterogeneous phase in catalysts and other materials. Compounds **59** and **60** (R = Me, Ph) are zwitterionic (ammonioalkyl)organotrifluorosilicates. The zwitterionic organofluorosilicates contain a tetracoordinate nitrogen atom (formally positively charged). In the crystal, the coordination polyhedrons around the Si nuclei can be described as distorted trigonal bipyramids: two of the F nuclei occupy the axial sites, whereas the third F atom and two C nuclei are in the equatorial positions. In solution, the zwitterionic species display a rapid ligand exchange at room temperature (one ^{19}F resonance). Solid-state ^{15}N and ^{29}Si NMR studies on authentic tetrafluoro(pyrrolidiniomethyl)silicate [F$_4$SiCH$_2$N(H)C$_4$H$_8$] (**61**) and on (3-ammoniopropyl)tetrafluorosilicate [F$_4$Si(CH$_2$)$_3$NH$_3$] (**62**) revealed evidence that **62** is indeed the earlier postulated product formed in the reaction of (3-aminopropyl)triethoxysilane [EtO)$_3$Si(CH$_2$)$_3$)NH$_2$] with HF in ethanol/HF [**61**, $\delta(^{15}\text{N}) = -318.5$ ppm, $\delta(^{29}\text{Si}) = -121.1$ ppm; **62**, $\delta(^{15}\text{N}) = -345.3$ ppm, $\delta(^{29}\text{Si}) = -112.4$ ppm][148,149] (Figure 44).

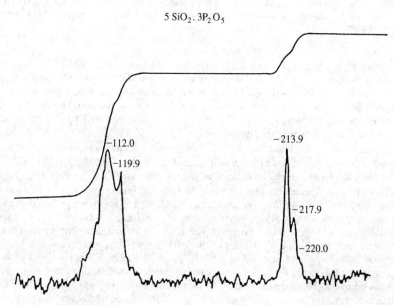

FIGURE 43. ^{29}Si NMR MAS spectrum of Si$_5$O(PO$_4$)$_6$. The chemical shift is referenced to the M signal of the Q$_8$M$_8$ external standard (+11.5 ppm). Delay between 30° pulses is 20 s, number of scans is 2242 and the frequency is 39.74 MHz. Reproduced by permission of Elsevier Science from Reference 147

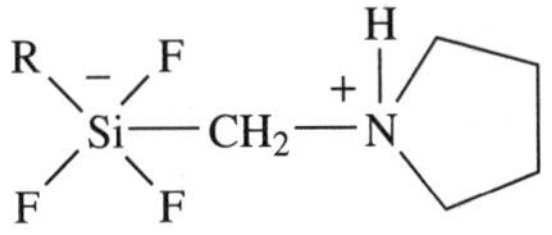

6. ^{29}Si NMR spectroscopy of organosilicon compounds

(59) R = Me
(60) R = Ph

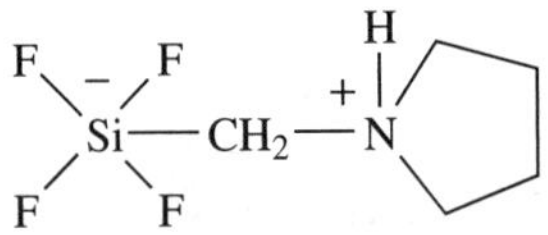

(61)

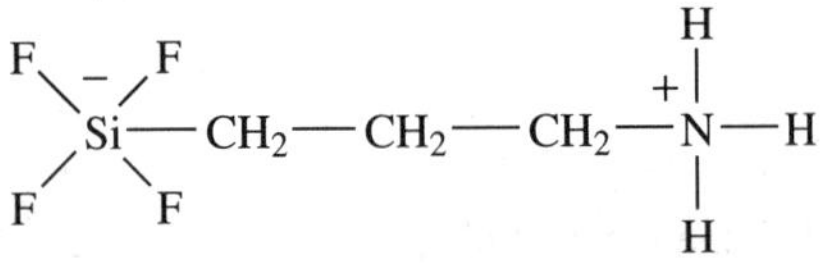

(62)

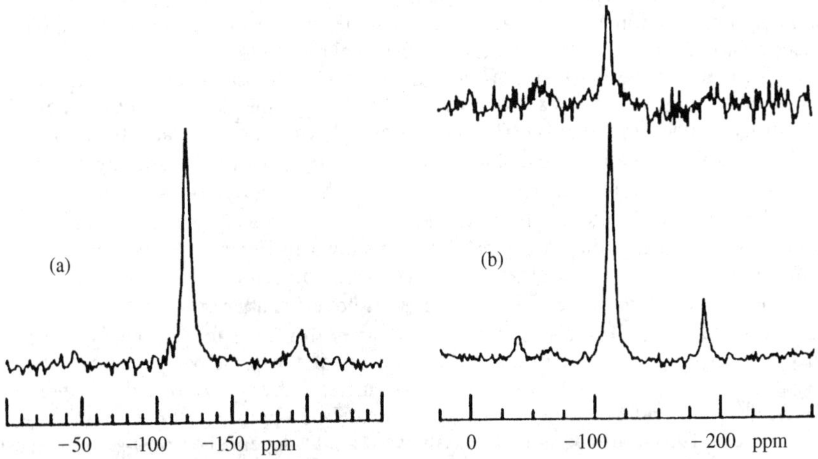

FIGURE 44. ^{29}Si MAS spectrum of **61** (a) and ^{29}Si MAS and ^{19}F $\rightarrow$ ^{29}Si CP/MAS spectra of **62** (b) Key: (a) ^{19}F high-power decoupled ^{29}Si MAS spectrum of **61**, spinning rate = 4.5 kHz, 30° ^{29}Si pulse, recycle delay time = 30 s, 2271 transients, exponential line broadening of 25 Hz, $\delta(^{29}$Si) = -121.1 ppm, $v_{1/2}$ = 350 Hz; (b) ^{19}F $\rightarrow$ ^{29}Si CP/MAS spectra of **62** (top), with spinning rate = 4.5 kHz, contact time = 10 ms, recycle delay time = 15 s, 1167 transients, exponential line broadening of 25 Hz and ^{19}F high-power decoupled ^{29}Si MAS spectrum of **62** (bottom) with spinning rate = 4.5 kHz, 30° ^{29}Si pulse, recycle delay time = 30 s, 2165 transients, exponential line broadening of 25 Hz, $\delta(^{29}$Si) = -112.4 ppm, $v_{1/2}$ = 300 Hz. Reproduced by permission of VCH Weinheim from Tacke *et al.*, *Chem. Ber.*, **126**, 851 (1993)

The isotropic chemical shift is the average value of the diagonal elements of the chemical shift tensor. Advances in solid state NMR spectroscopy allow one to determine the orientation dependence, or anisotropy, of the chemical shift interaction. It is now possible to determine the principal elements of a chemical shift powder pattern conveniently, and the orientation of the principal axes with more effort. Hence, instead of settling for just the average value of the chemical shift powder pattern, one can now aim for values of the three principal elements and the corresponding orientations in a molecular axis system.

```
          X
          |
      O  Si  O
      /  O:  \
   H₂C  H₂C   CH₂
      \  N    /
    H₂C / \ CH₂
        CH₂
```

(63)

The chemical shift parameters of the ^{13}C, ^{15}N and ^{29}Si resonances in a set of 2,8,9-trioxa-5-aza-1-silabicyclo[3.3.3]undecanes (silatranes) **(63)** were determined from powdered, crystalline samples with and without magic angle spinning. Silatranes are a class of organosilicon compounds that feature a silicon atom that can be discussed as nominally pentacoordinate. The interest in silatranes is due to their intriguing molecular structure, biological activity and patterns of chemical reactivity. The most intriguing aspect of this structure is the existence of and influence of a 'transannular bond' between the silicon and nitrogen atoms, as indicated by the dashed line drawn between the silicon and nitrogen nuclei in structure **63**. Not surprisingly, because the substituent is directly attached to silicon, the chemical shift tensor for ^{29}Si yields a more striking dependence on substituent variation than that for ^{15}N. Although the correlations are not as 'clean' as for the ^{15}N chemical shift, the observed variation in the ^{29}Si chemical shift is much larger. ^{29}Si chemical shift powder patterns are shown in Figure 45. They correspond to chemical shift tensors that are nearly axially symmetric and exhibit wider variation in their anisotropies than do the ^{15}N chemical shift powder patterns.

For ^{29}Si, $\delta_{33} - \delta_{11} = \Delta\delta$ (the anisotropy) varies from 112 to 32 ppm for the chloro and ethyl derivatives of **63**, respectively, with all of the principal elements exhibiting a dependence on the substituent. δ_{11} and δ_{22} increase by approximately 40 ppm with increasing r_{Si-N}, and δ_{33} decreases by approximately 20 ppm with increasing r_{Si-N}.

The result of these trends is that δ_{iso} shows a slight dependence on r_{Si-N}, increasing by approximately 30 ppm as r_{Si-N} increases. The ^{29}Si chemical shift powder patterns exhibit a trend that is the reverse of what is seen with the ^{15}N chemical shift powder patterns; the more electron-withdrawing substituents give the widest ^{29}Si chemical shift powder patterns and the least electron-withdrawing substituents give the narrowest ^{29}Si chemical shift powder patterns. As in the case of the ^{15}N chemical shift parameters, the ^{29}Si chemical shift parameters show general trends with respect to r_{Si-N}. Figure 46 displays a set of chemical shift correlation curves for δ_{iso}, $\delta_{33} - \delta_{11}$, δ_{33}, δ_{22} and δ_{11} vs r_{Si-N}. The correlations show some scatter as those discussed above for the ^{15}N data. The added complication is the presence and the local effect of the directly-attached substituent. Thus, the substituent effect on the ^{29}Si chemical shift can be viewed as having three types of origins:

(1) the geometry-structure effect associated with variations in the $\angle$OSiO bond angle;
(2) the substituent effect due to variations in the N·Si transannular interaction;
(3) the direct substituent effect that would be induced even in the absence of the first two, e.g. in an analogous system without nitrogen but with the substituent attached to silicon.

This third type of effect is, of course, absent in the ^{15}N case, so one can expect that interpretation of chemical shift data in terms of the transannular bond should be more

6. ^{29}Si NMR spectroscopy of organosilicon compounds

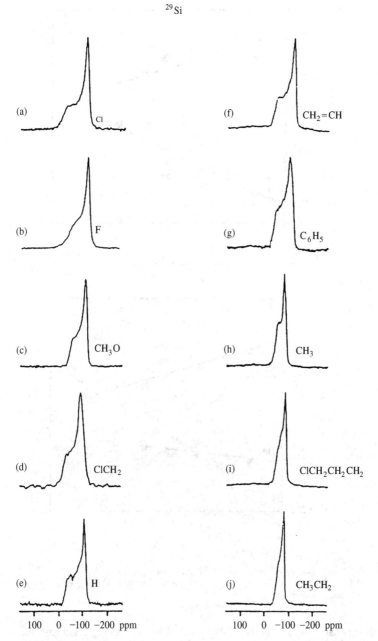

FIGURE 45. ^{29}Si chemical shift powder patterns collected under cross-polarization conditions at 39.8 MHz on X-substituted silatranes **63** with X indicated. The fluorosilatrane chemical shift powder pattern was collected under high-power ^{1}H and ^{19}F decoupling conditions at 200 MHz. Reprinted with permission from Reference 150. Copyright 1993 American Chemical Society

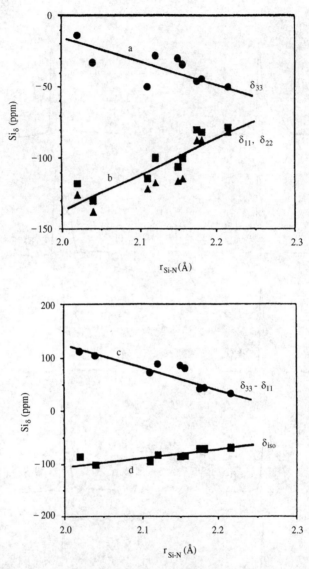

FIGURE 46. Correlation curves for some ^{29}Si chemical shift parameters vs r_{Si-N}: (a) δ_{33}, (b) δ_{11} (▲), δ_{22}(■), (c) $\delta_{33} - \delta_{11}$ and (d) δ_{iso}. Reprinted with permission from Reference 150. Copyright 1993 American Chemical Society

difficult for ^{29}Si than for ^{15}N. Nevertheless, the plots in Figure 46 indicate that as r_{Si-N} decreases in length, $\delta_{33} - \delta_{11}$ increases, the values of δ_{11} and δ_{22} decrease and δ_{33} increases, with the overall result that δ_{iso} decreases. In contrast to the ^{15}N data, each of the principal elements of the ^{29}Si chemical shift tensor shows a rough dependence on r_{Si-N}, with the δ_{11} and δ_{22} elements being more sensitive to changes in r_{Si-N} than the δ_{33} element[150].

6. ^{29}Si NMR spectroscopy of organosilicon compounds 339

Tamao and coworkers[151] presented the first example of a bis(siliconate) containing an Si−F−Si bond, [o-C$_6$H$_4$(SiPhF$_2$)$_2$F]$^-$, K$^+$· 18-crown-6 (64). This report describes the full details of the solid-state structures of 64 and the unsymmetrical analogues [o-C$_6$H$_4$(SiF$_3$)(SiPh$_2$F)F]$^-$, K$^+$· 18-crown-6 (65) and [o-C$_6$H$_4$(SiPhF$_2$)(SiPh$_2$F)F]$^-$, K$^+$· 18-crown-6 (66). For clarity, these bis(siliconates) 64−66 may be abbreviated to [F2{F}F2]$^-$, [F{F}F3]$^-$ and [F{F}F2]$^-$, respectively, in which the central {F} represents the bridging fluorine atom while the left and right sides represent the number of fluorine nuclei on the two silicon atoms. An X-ray structural analysis of the three bis(siliconates) has afforded new significant information about bond lengths and bond angles (Figures 47−49).

K$^+$ - 18-crown-6   K$^+$ - 18-crown-6   K$^+$ - 18-crown-6
(64) [F2{F}F2]$^-$ (65) [F{F}F3]$^-$ (66) [F{F}F2]$^-$

(67)   (68)

The ^{29}Si chemical shift and coupling constant, ^1J(SiF), for [F2{F}F2]$^-$ (64) are compared with data for other pertinent compounds in Table 15.

Three significant features should be mentioned.

(1) At 20 °C, the ^{29}Si NMR peak appears as a sextet for two silicon atoms, consistent with the fast exchange of all five fluorine nuclei.

(2) The ^{29}Si chemical shift (δ − 90.0 ppm) of 64 is intermediate between those of the tetracoordinate precursors o-bis(difluorphenylsilyl)benzene (67; δ − 30.21 ppm) and Ph$_2$SiF$_2$ (δ − 29.0 ppm) and that of the pentacoordinate mono(siliconate) [Ph$_2$SiF$_3$]$^-$ (δ − 109.5 ppm).

(3) The Si−F coupling constant [^1J(SiF) = 134.7 Hz] is smaller than those of Ph$_2$SiF$_2$ [^1J(SiF) = 291.2 Hz] and [Ph$_2$SiF$_3$]$^-$ [^1J(SiF) = 238.0 Hz] and is comparable with the calculated average value (130 Hz) on the assumption of ^1J(SiF) = 291 Hz (× 2F), ^1J(SiF) = 238 Hz (× 3F) and ^4J (SiF) = 0 (×5F). Thus, these data strongly support fast fluoride transfer between tetracoordinate and pentacoordinate silicon atoms. The unsymmetrical bis(siliconates) 65 and 66 showed no detectable peaks of the silicon nuclei in solution.

Solid-state MAS ^{29}Si NMR data for the pentacoordinate bis(siliconates) 64 and 65 and mono(siliconate) [PhSiF$_4$]$^-$, K$^+$·18-crown-6 are also summarized in Table 15 for comparison. The solid-state spectrum of 64 shows an uncharacterizable multiplet, possibly due to the slightly unsymmetrical structure. In contrast to the absence of peaks in solution,

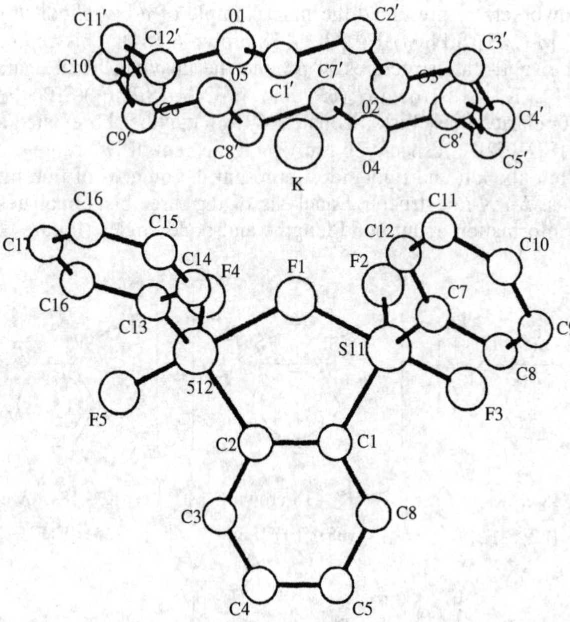

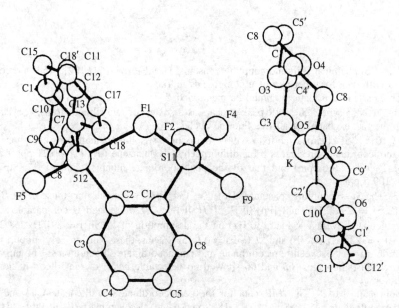

FIGURE 47. X-ray crystal structure of $[o\text{-}C_6H_4(SiPh\ F_2)_2F]^-$, $K^+ \cdot$18-crown-6 (64). Reprinted with permission from Reference 151. Copyright 1990 American Chemical Society

FIGURE 48. X-ray crystal structure of $[o\text{-}C_6H_4(SiF_3)(SiPh_2F)\ F]^-$, $K^+ \cdot$18-crown-6 (65). Reprinted with permission from Reference 151. Copyright 1990 American Chemical Society

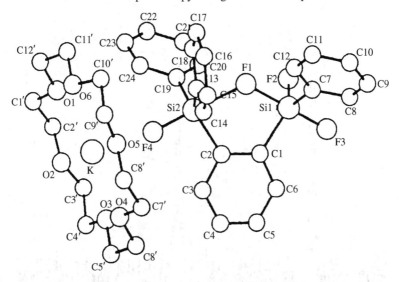

FIGURE 49. X-ray crystal structure of [o-C$_6$H$_4$(SiPh F$_2$)(SiPh$_2$F) F]$^-$, K$^+$•18-crown-6 (**66**). Reprinted with permission from Reference 151. Copyright 1990 American Chemical Society

TABLE 15. ^{29}Si NMR spectroscopic data for bis(siliconates) and mono(siliconates) and their precursors, fluorosilanes, in solution and/or in the solid state[a]

Siliconate and precursor[b]	Chemical shift (ppm)	Multiplicity	J(SiF) coupling constant (Hz)
anion of **64**	−90.03	sextet	134.74
	[−75 to −105]	[multiplet]	
anion of **65**	[−129.3(Si1)]	[triplet]	207.9
	[−31.3 (Si2)]	[doublet]	256.5
67	−30.21	triplet	293.9
68	−73.46(Si1)	quintet	265.4
	−3.63(Si2)	doublet	282.1
[Ph$_2$SiF$_3$]$^-$	−109.55	quartet	238.06
	−128.28	triplet	200.2
[PhSiF$_4$]$^-$	−125.90	quintet	210
	[−129.28]	[triplet]	200.2
Ph$_2$SiF$_2$	−29.00	triplet	291.20
PhSiF$_3$	−72.42	quintet	266.8

[a] Spectra were recorded at 20 °C in acetone-d$_6$ unless otherwise stated. Solid-state spectral data are given in brackets.
[b] The counterion of the siliconate is K$^+$•18-crown-6.
[c] No signal in solution.
Reproduced by permission of the American Chemical Society from Reference 151.

the solid-state spectrum of [F{F}F3]$^-$ (**65**) shows two distinct silicon peaks at −31.3 ppm as a doublet [^{1}J(SiF) = 256.5 Hz] for Si^2 and at −129.3 ppm as a triplet [^{1}J(SiF) = 207.9 Hz] for Si^1, as shown in Table 15 and in Figure 50. While Si^2 couples with only one fluorine, the Si^1 couples with only two of the four fluorine nuclei on Si^1 to appear as a triplet. It may be noted here that the silicon atom in the mono(siliconate) [PhSiF$_4$]$^-$ also appears as a triplet in the solid state (Table 15), although it cannot readily be deduced

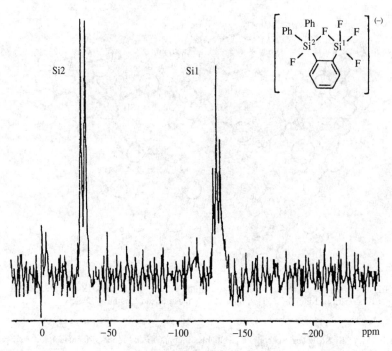

FIGURE 50. Solid-state ^{29}Si NMR spectrum for **65** (79.46 MHz, external standard DSS at δ 0 ppm). Reprinted with permission from Reference 151. Copyright 1990 American Chemical Society

whether the triplet arises from coupling with two apical or two equatorial fluorines. The Si^1 signal of **65** appears *ca* 56 ppm upfield from that of the $-$SiF$_3$ group in the precursor *o*-(fluorodiphenylsilyl)(trifluorosilyl)benzene (**68**). This difference is nearly equal to that (56.8 ppm) between [PhSiF$_4$]$^-$ in the solid state and PhSiF$_3$ in acetone-d$_6$, as shown in Table 15. Similarly, the chemical shift of Si^2 in **65** also shows a 27.7 ppm upfield shift from that of the $-$SiPh$_2$F group in **68**, but the chemical shift difference is less than half of that for Si^1. Thus, it may be deduced that the Si^1 side is almost completely pentacoordinated, as for the trigonal-bipyramidal mono(siliconate) [PhSiF$_4$]$^-$, while the Si^2 side is of low pentacoordination character, in accordance with the results obtained by X-ray structural analysis. The ^{29}Si$-^{19}$F spin–spin coupling constant observed in the solid-state ^{29}Si NMR spectrum also provides useful information on the degree of interaction of the bridging fluorine atom with silicon atoms, i.e. on the pentacoordination character. ^1J (SiF) in the bis(siliconate) **65** and ^1J (SiF) in the precursor **68** are 207.9 and 265.4 Hz, respectively. Thus, the coupling constant in **65** is 57.5 Hz smaller than that in **68**. This suggests that Si^1 in **65** is highly pentacoordinated, since the difference in ^1J (SiF) between [PhSiF$_4$]$^-$ and PhSiF$_3$ is 66.5 Hz. In contrast, the coupling constant of ^1J (Si$_2$F) in **65** is 256.5 Hz, which is 25.6 Hz smaller than ^1J(Si$_2$F) = 282.1 Hz in **68**, suggesting lower pentacoordination character of Si^2 in **65**. On the basis of the X-ray structural analysis of **65**, pentacoordination characters of Si^1 and Si^2 have been estimated to be 93% and 50%, respectively. Thus, the geometries about the silicon nuclei estimated from the ^{29}Si chemical shifts and ^{29}Si–^{19}F coupling constants by solid-state ^{29}Si NMR spectroscopy are consistent with the geometries found by the X-ray structural analysis (Figure 50).

G. Miscellaneous

In many applications of reversed-phase high-performance liquid chromatography (RP-HPLC), stationary-phase degradation is a major drawback when using alkylsilane-modified silica surfaces because of the hydrolytic instability of siloxane bonds. The lifetime of one column packing may not even be sufficient to perform adequate experimental designs for optimizing separation efficiencies. Consequently, much research was conducted in order to identify the most important factors involved in phase deterioration and to design new stationary phases with improved stability. The bulky substituents in the silanizing reagent (for example, diisobutyl-n-octadecylsilane instead of dimethyl-n-octadecylsilane) would result in a more efficient steric protection of the silica surface and, in particular, the ligand siloxane bond. These so-called stable bond phases indeed exhibit superior hydrolytic stability at low pH. However, the improved steric protection was not observed as such; it was postulated using the increased chromatographic stability as a criterion. Concerning the physicochemical methods used to investigate chromatographic silica surfaces, solid-state NMR has proven to be a powerful tool that enables identification of different chemical surface structures. The goal of much research has been to relate NMR characteristics of the detected chemical surface species to the observed chromatographic behavior of silica surfaces.

Scholten and coworkers[152] present ^{29}Si CPMAS NMR evidence for a decreased contribution of hydrogen bonding groups to the ligand silane signal in diisobutyl-n-octadecylsilane-modified silica gel, compared to the dimethyl-n-octadecylsilane analogue. This lower extent of hydrogen bonding is brought about by steric protection of the ligand siloxane bond by the bulky isobutyl substituents. Figure 51 displays the ^{29}Si CPMAS NMR spectra of the Aerosil samples in the silane ligand region for four different degrees of trimethylsilylation. Clearly, the chemical shift of the maximum of the ligand signal decreases with increasing surface coverage. Also, the asymmetry of the signals due to a shoulder at the left of the peak maximum is evident. Figure 52 displays the ^{29}Si CP MAS NMR spectra of the two Zorbax C_{18} phases. Before considering the asymmetry of the silane ligand NMR signals, it should be noted that the peak maximum of the SB-C_{18} ligand signal is shifted 2 ppm upfield from the maximum of the Rx-C_{18} ligand signal. This is due to the β effect on ^{29}Si upon substitution of two hydrogen nuclei for two isopropyl groups. This chemical shift difference is, however, irrelevant in the following discussion. The attention is focused on the degree of asymmetry of both signals. It appears that the shoulder in the SB-C_{18} spectrum is much less pronounced, indicating that the ligand siloxane bond is involved in hydrogen bonding only to a small extent. In the Rx-C_{18} spectrum, on the other hand, the shoulder is clearly discernible. It should be noted that the surface coverage by the diisobutyl-n-octadecylsilane ligands is much lower than that of the dimethyl-n-octadecylsilane ligands.

Bearing in mind the result of the trimethylsilylated Aerosil surfaces, where increasing surface coverage is accompanied by a decreasing hydrogen bonding contribution to the NMR signal, the slight asymmetry of the SB-C_{18} ligand NMR signal strongly suggests the superior steric shielding properties of the isobutyl groups. This is schematically illustrated in Figure 53.

Contact time dependences of signal intensities in ^{29}Si and ^{13}C CP MAS NMR spectrum (Figure 54) were measured for kaolinite and kaolinite-DMSO and kaolinite-DMSO-d_6

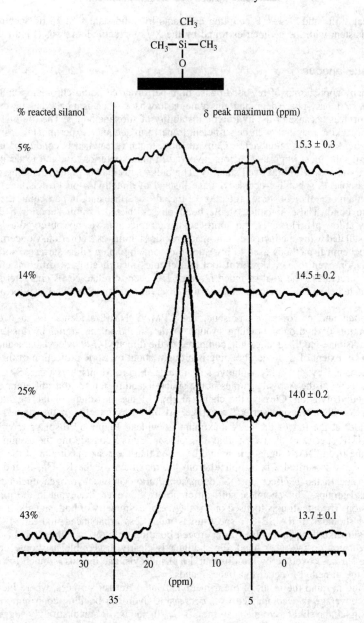

FIGURE 51. ^{29}Si CPMAS NMR spectra of Aerosil A-200 with trimethylsiloxane surface coverage and chemical shifts of the peak maxima (± maximum error) as indicated. All spectra are on the same intensity scale. Reproduced by permission of Elsevier Science from Reference 152

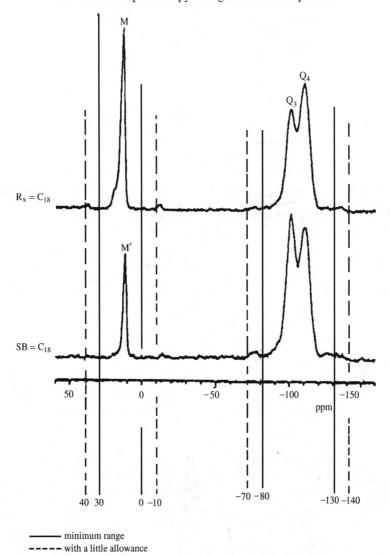

FIGURE 52. ^{29}Si CPMAS NMR spectra of the Zorbax octadecyl RP-HPLC phases. Both spectra are on the same intensity scale. M = dimethyl-n-octadecylsiloxane, M′ = diisobutyl-n-octadecylsiloxane, Q_3 = single silanol, Q_4 = siloxane. Reproduced by permission of Elsevier Science from Reference 152

intercalation compounds. Cross-relaxation times between ^{1}H and ^{29}Si and between ^{1}H and ^{13}C were estimated experimentally, which reflect the internuclear distances. Relaxation times were calculated theoretically using reported crystal structure models (Figure 55), and the validity of those models is discussed[153].

In 1997, West and coworkers[154] reported the solid-state NMR study of the ^{29}Si chemical shift tensors for a series of disilenes with different substitution at the Si=Si double bond

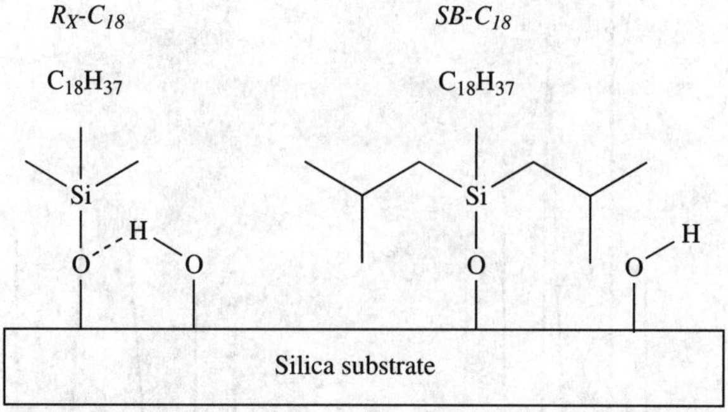

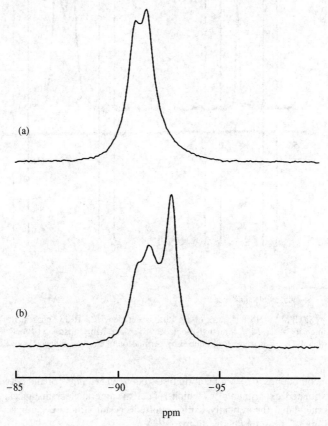

FIGURE 53. Schematic drawing of the dimethyl-n-octadecylsiloxane and the diisobutyl-n-octadecylsiloxane surface structures, illustrating the increased steric protection of the ligand siloxane bond by the bulky side groups in the latter. Reproduced by permission of Elsevier Science from Reference 152

FIGURE 54. ^{29}Si CPMAS NMR spectra of (a) kaolinite and (b) kaolinite-DMSO. The spinning rate of the sample was 3.00 kHz. Reproduced by permission of Elsevier Science from Reference 153

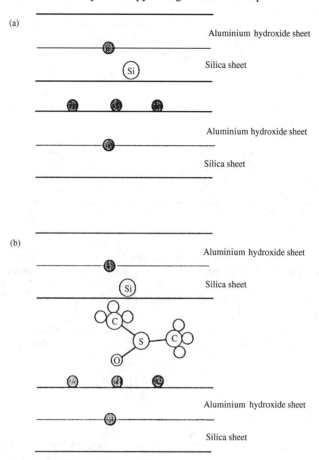

FIGURE 55. Schematic representation of layered structures in (a) kaolinite and (b) kaolinite-DMSO. Dark circles indicate hydrogen atoms. Reproduced by permission of Elsevier Science from Reference 153

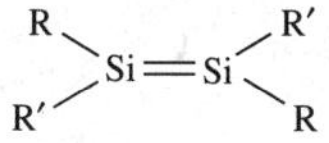

(69) R = R' = mesityl
(70) R = R' = 2,4,6-triisopropylphenyl
(71) R = mesityl, R' = *t*-butyl
(72) R = R' = (Me$_3$Si)$_2$CH
(73) R = Me$_3$Si, R' = 2,4,6-triisopropylphenyl
(74) R = R' = *i*-Pr$_2$MeSi
(75) R = R' = *i*-Pr$_3$Si

FIGURE 56. Structures of disilenes studied by ^{29}Si NMR. Reprinted with permission from Reference 154. Copyright 1997 American Chemical Society

TABLE 16. Chemical shift tensors and structural parameters for disilenes

Disilene	δ_{11}	δ_{22}	δ_{33}	δ_{iso} solid	δ_{iso} soln	$\Delta\delta$	CSA[a]	Si=Si[b] (pm)
69	181	31	−22	63.2	63.3	203	176	214
69·C₇H₈	185	34	−22	65.0	63.3	207	179	216
69·THF	165	40	−25	59.6	63.0	190	157	215
70	155	30	−31	50.8 53.2	53.4	186	155	214
71	178	77	3	86.1	90.3	175	138	214
72	182	55	21	86.1	90.4	161	144	
	199	54	9	87.4		190		
73	296	46	−59	94.5	94.4	355	168	215
74	414	114	−100	143	144.5	514	408	228
75	412	149	−69	164	154.5	481	372	225

[a]CSA = $\delta_{11} - (\delta_{22} + \delta_{33})/2$.
[b]Data from Reference 159.

(Figure 56)[155]. Compounds investigated included three forms of tetramesityldisilene: the solvent-free form **69**[156], the toluene adduct **69·C₇H₈**[157] and the tetrahydrofuran solvate **69·THF**[158]. Also studied were a second tetraaryldisilene (**70**)[159] dialkyldiaryl-substituted disilene, **71**[160], and the only tetraalkyldisilene known to be stable as a solid, **72**[161]. Three silyl-substituted disilenes, **73**, **74** and **75**, were also investigared[162,163]. To assist in the interpretation of the experimental results, *ab initio* molecular orbital calculations of the ^{29}Si chemical shift tensors were carried out for model disilene molecules.

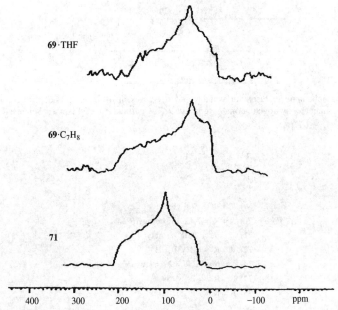

FIGURE 57. Static ^{29}Si NMR spectra of disilenes. Reprinted with permission from Reference 154. Copyright 1997 American Chemical Society

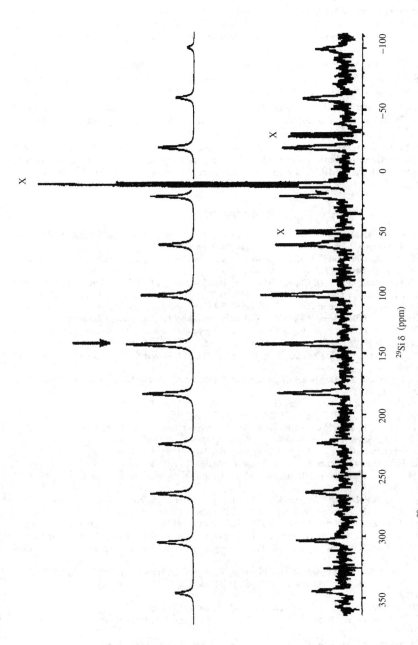

FIGURE 58. Slow-spinning MAS ^{29}Si NMR spectrum for $(i\text{-PrMe}_2)_2\text{Si}_2\text{Si}=\text{Si}(\text{SiMe}_2\text{Pr-}i)_2$ (74): below, recorded spectrum; above, computer simulation for the doubly-bonded silicons. The isotropic peak for the disilene silicons is marked with an arrow. Peaks marked x are due to the i-Pr–Si silicons; these isotropic peaks are shown at reduced gain. Reprinted with permission from Reference 154. Copyright 1997 American Chemical Society

Chemical shift tensors for the doubly-bonded silicon atoms in the nine disilene samples are given in Table 16. Typical powder patterns are shown in Figure 57, and the slow-spinning spectrum for **75** is illustrated in Figure 58. All of the disilenes show significant deshielding along one axis (δ_{11}). For tetrasilyldisilenes this deshielding is extreme, +414 ppm for **75**. Together with greater shielding in the δ_{33} direction, these lead to $\Delta\delta$ values ($\Delta\delta = \delta_{11} - \delta_{33}$) more than a factor of 2 greater than any previously reported for ^{29}Si[164]. The $\Delta\delta$ values for the doubly-bonded silicon atoms in all of the disilenes are significantly larger than those for singly-bonded silicon compounds, which typically have $\Delta\delta$ values of 0–60 ppm[165]. For disilenes **74** and **75**, tensors for the four-coordinate silicon atoms were also determined: for **74**, $\delta_{11} = 36.4$, $\delta_{22} = -0.9$ and $\delta_{33} = -4.2$ ppm; and for **75**, the corresponding values are 46.0, 21.1 and 17.8 ppm, giving $\Delta\delta$ values of about 30 ppm, as expected for sp^3-type silicons.

The MO calculations on model disilenes indicate that the deshielding results from a paramagnetic contribution along the in-plane axis perpendicular to the Si—Si vector. Implications of the NMR data and the theoretical computations for Si=Si bonding are discussed.

VI. REFERENCES

1. E. A. Williams, in *The Chemistry of Organic Silicon Compounds*, Vol. 1 (Eds. S. Patai and Z. Rappoport), Chap. 8, Wiley, New York, 1989, pp. 511–544.
2. A. R. Bassindale and P. P. Gasper (Eds.) *Frontiers of Organosilicon Chemistry*, The Royal Society of Chemistry, 1991.
3. H. Marsmann, in *NMR Basic Principles and Progress*, **17** (Eds. P. Diehl, E. Fluck and R. Kosfeld), Springer-Verlag, Berlin, 1981, p. 235; H. C. Marsmann, 'Silicon-29 NMR', and G. Engelhardt, 'Silicon-29 NMR of Solid Silicates', in *Encyclopedia of Nuclear Magnetic Resonance*, Wiley, New York, 1996; *Specialist Periodical Report, Nuclear Magnetic Resonance*, Vols 1–24, The Royal Society of Chemistry, These volumes are useful since most of them contain some ^{29}Si NMR results.
4. R. K. Harris, N. J. O'Connor, E. H. Curzon and O. W. Howarth, *J. Magn. Reson.*, **57**, 115 (1984).
5. G. Engelhardt and D. Michel, *High Resolution Solid State NMR of Silicates and Zeolites*, Wiley, New York, 1987.
6. V. Baudrillard, D. Davoust and G. Ple, *Magn. Reson. Chem.*, **32**, 40 (1994).
7. G. A. Morris and R. Freeman, *J. Am. Chem. Soc.*, **101**, 760 (1979).
8. Th. Blinka, B. J. Helmer and R. West, *Adv. Organomet. Chem.*, **23**, 193 (1984).
9. O. W. Sorensen, *Prog. Nuclear Magn. Reson. Spectrosc.*, **21**, 503 (1989).
10. J. Schraml, *Collect. Czech. Chem. Commun.*, **48**, 3402 (1983).
11. E. Hengge and F. Schrank, *J. Organomet. Chem.*, **362**, 11 (1989).
12. M. Kuroda, Y. Kabe, M. Hashimoto and S. Masamune, *Angew. Chem., Int. Ed. Engl.*, **27**, 1727 (1988).
13. C. T. G. Knight, R. J. Fitzpatrick and E. Oldfield, *J. Non-Cryst. Solids*, **116**, 140 (1990).
14. C. A. Fyfe, H. Grondey, Y. Feng and G. T. Kokotailo, *Chem. Phys. Lett.*, **173**, 211 (1990).
15. C. A. Fyfe, H. Grondey, Y. Feng and G. T. Kokotailo, *J. Am. Chem. Soc.*, **112**, 8812 (1990).
16. C. A. Fyfe, H. Grondey, Y. Feng and G. T. Kokotailo, *Nature (London)*, **341**, 223 (1989).
17. M. R. Churchill, J. W. Ziller, J. H. Freudenberger and R. R. Scrock, *Organometallics*, **3**, 1554 (1984).
18. J. Okuda, R. C. Murray, J. C. Dewan and R. R. Schrock, *Organometallics*, **5**, 1681 (1986).
19. R. R. Schrock, S. F. Pedersen, M. R. Churchill and J. W. Ziller, *Organometallics*, **3**, 1574 (1984).
20. T. M. Sivavec and T. J. Katz, *Tetrahedron Lett.*, **26**, 2159 (1985).
21. C. Dallaire, M. A. Brook, A. D. Bain, C. S. Frampton and J. F. Britten, *Can. J. Chem.*, **71**, 1676 (1993).
22. H. Sakurai, S. Hoshi, A. Kamiya and A. Hosomi, *Chem. Lett.*, 1781 (1986).
23. K. Akasaka, W. Ando, K. Kobayashi and S. Nagase, *J. Am. Chem. Soc.*, **115**, 1605 (1993).
24. T. Kusakawa, Y. Kabe, T. Erata, B. Nestler and W. Ando, *Organometallics*, **13**, 4186 (1994).

6. ^{29}Si NMR spectroscopy of organosilicon compounds   351

25. T. Kusakawa, Y. Kabe and W. Ando, *Organometallics*, **14**, 2142 (1995).
26. W. Ando, F. Hojo, S. Sakigawa, N. Nakayama and T. Shimizu, *J. Am. Chem. Soc.*, **115**, 3111 (1993).
27. T. Shimizu, F. Hojo and W. Ando, *Organometallics*, **11**, 1009 (1992).
28. H. B. Yokelson, A. J. Millevolte, B. R. Adams and R. West, *J. Am. Chem. Soc.*, **109**, 4116 (1987).
29. H. Watanabe, T. Okawa, M. Kato and Y. Nagai, *J. Chem. Soc., Chem. Commun.*, 781 (1983).
30. P. K. Jenkner, A. Spielberger, M. Eibl and E. Hengge, *Spectrochim. Acta*, **49A**, 161 (1993).
31. H. Watanabe, M. Kato, T. Okawa, Y. Kougo, Y. Nagai and M. Goto, *Appl. Organometal. Chem.*, **1**, 157 (1987).
32. H. Suzuki, K. Okabe, R. Kato, N. Sato, Y. Fukuda, H. Watanabe and M. Goto, *Organometallics*, **12**, 4833 (1993).
33. H. Suzuki, N. Kenmotu, K. Tanaka, H. Watanabe and M. Goto, *Chem. Lett.*, 811 (1995).
34. For a review see: R. West, *Angew. Chem., Int. Ed. Engl.*, **26**, 1201 (1987).
35. H. Watanabe, K. Takeuchi, N. Fukawa, M. Kato, M. Goto and Y. Nagai, *Chem. Lett.*, 1341 (1987).
36. H. Suzuki, N. Tokitoh, R. Okazaki, J. Harada, K. Ogawa, S. Tomoda and M. Goto, *Organometallics*, **14**, 1016 (1995).
37. N. Tokitoh, H. Suzuki, R. Okazaki and K. Ogawa, *J. Am. Chem. Soc.*, **115**, 10428 (1993).
38. M. Denk, R. Lennon, R. Hayashi, R. West, A. V. Belyakov, H. P. Verne, A. Haaland, M. Wagner and N. Metxler, *J. Am. Chem. Soc.*, **116**, 2691 (1994).
39. J. B. Lambert, L. Kania and S. Zhang, *Chem. Rev.*, **95**, 1191 (1995); P. D. Lickiss, *J. Chem. Soc., Dalton Trans.*, 1333 (1992).
40. J. B. Lambert and S. Zhang, *J. Chem. Soc., Chem. Commun.*, 383 (1993).
41. J. B. Lambert, S. Zhang, C. L. Stern and J. C. Huffman, *Science*, **260**, 1917 (1993).
42. J. B. Lambert, S. Zhang and S. M. Ciro, *Organometallics*, **13**, 2430 (1994).
43. L. Olsson and D. Cremer, *Chem. Phys. Lett.*, **215**, 433 (1993).
44. W. Kutzelnigg, M. Schindler and U. Fleischer, in *NMR Basic Principles and Progress*, Vol. 23, Springer, Berlin, 1989.
45. U. Pidum, M. Stahl and G. Frenking, *Chem. Eur. J.*, **2**, 1996 (1997).
46. Z. Xie, J. Manning, R. W. Reed, R. Mathur, P. D. W. Boyd, A. Benesi and C. A. Reed, *J. Am. Chem. Soc.*, **118**, 2922 (1996).
47. J. B. Lambert and Y. Zhao, *Angew. Chem., Int. Ed. Engl.*, **36**, 400 (1997).
48. T. Müller, Y. Zhao and J. Lambert, private communication.
49. P. v. R. Schleyer, *Science*, **275**, 39 (1997).
50. S. R. Bahr and P. Boudjouk, *J. Am. Chem. Soc.*, **115**, 4514 (1993).
51. For a review, see: R. J. P. Corriu and J. C. Young, in *The Chemistry of Organic Silicon Compounds*, Vol. 1 Part 2 (Eds. S. Patai and Z. Rappoport), Chap. 20, Wiley, Chichester, 1989; St. N. Tandura, N. V. Alekseev and M. G. Voronkov, *Top. Curr. Chem.*, **131**, 99 (1986).
52. R. R. Holmes, *Chem. Rev.*, **90**, 17 (1990); R. J. P. Corriu, *J. Organomet. Chem.*, **400**, 81 (1990); R. J. P. Corriu, C. Guerin and J. J. E. Moreau, in *The Chemistry of Organic Silicon Compounds*, Vol. 1, Part 1 (Eds. S. Patai and Z. Rappoport), Chap. 4, Wiley, Chichester, 1989; A.R. Bassindale and P. G. Taylor, in *The Chemistry of Organic Silicon Compounds*, Part 1 (Eds. S. Patai and Z. Rappoport), Chap. 13, Wiley, Chichester, 1989; C. E. DePuy, R. Damrauer, J. H. Bowie and J. C. Sheldon, *Acc. Chem. Res.*, **20**, 127 (1987).
53. D. J. Schomburg, *J. Organomet. Chem.*, **221**, 137 (1981).
54. R. Damrauer and S. E. Danahey, *Organometallics*, **5**, 1490 (1986); R. Damrauer, B. O'Connell, S. E. Danahey and R. Simon, *Organometallics*, **8**, 1167 (1989).
55. J. Harland, J. S. Payne, R. O. Day and R. R. Holmes, *Inorg. Chem.*, **26**, 760 (1987).
56. S. E. Johnson, J. A. Deiters, R. O. Day and R. R. Holmes, *J. Am. Chem. Soc.*, **111**, 3250 (1989); S. E. Johnson, R. O. Day and R. R. Holmes, *Inorg. Chem.*, **28**, 3182 (1989); S. E. Johnson, J. S. Payne, R. O. Day, J. M. Holmes and R. R. Holmes, *Inorg. Chem.*, **28**, 3190 (1989).
57. W. H. Stevenson, III, S. Wilson, J. C. Martin and W. B. Farnham, *J. Am. Chem. Soc.*, **107**, 6340 (1985); W. H. Stevenson, III and J. C. Martin, *J. Am. Chem. Soc.*, **107**, 6352 (1985).
58. D. A. Dixon, W. R. Hertler, D. B. Chaseand, W. G. Farnham and F. Davidson, *Inorg. Chem.*, **27**, 4012 (1988).
59. M. Kira, K. Sato, C. Kabuto and H. Sakurai, *J. Am. Chem. Soc.*, **111**, 3747 (1989).
60. S. K. C. Kumara, V. Chandraskhar, J. J. Harland, J. M. Holmes, R. O. Day and R. R. Holmes, *J. Am. Chem. Soc.*, **112**, 2341 (1990).

61. T. Kawashima, N. Iwama and R. Okazaki, *J. Am. Chem. Soc.*, **114**, 7599 (1992).
62. D. F. Evans, A. M. Z. Slawin, D. J. Williams, C. Y. Wong and J. D. Woollins, *J. Chem. Soc., Dalton Trans.*, 2383 (1992).
63. S. K. Chopra and J. C. Martin, *J. Am. Chem. Soc.*, **112**, 5342 (1990).
64. C. Breliere, F. Care, R. J. P. Corriu, M. Poirier and G. Royo, *Organometallics*, **5**, 388 (1986).
65. A. R. Bassindale and J. Jiang, *J. Organomet. Chem.*, **446**, C3 (1993).
66. H. Sakurai, K. Ebata, C. Kabuto and A. Sekiguti, *J. Am. Chem. Soc.*, **112**, 1799 (1990); A. Sekiguti, K. Ebata, C. Kabuto and H. Sakurai, *J. Am. Chem. Soc.*, **113**, 1464 (1991); A. Sekiguti, K. Ebata, C. Kabuto and H. Sakurai, *J. Am. Chem. Soc.*, **113**, 7081 (1991); A. Sekiguti, K. Ebata, Y. Terui and H. Sakurai, *Chem. Lett.*, 1417 (1991).
67. K. Ebata, T. Inada, C. Kabuto and H. Sakurai, *J. Am. Chem. Soc.*, **116**, 3595 (1994).
68. J. M. Chance, B. Kahr, A. B. Buda, J. P. Toscano and K. Mislow, *J. Org. Chem.*, **53**, 3226 (1988) and references cited therein.
69. G. Philipp and H. Schmidt, *J. Non-Cryst. Solids*, **63**, 283 (1984).
70. G. L. Wilkesg, B. Orler and H. Huang, *Polym. Prepr.*, **26**, 300 (1985).
71. H. Huang, B. Orler and G. L. Wilkesg, *Polym. Bull.*, **14**, 557 (1985).
72. H. Huang, B. Orler and G. L. Wilkesg, *Macromolecules*, **20**, 1322 (1987).
73. H. Huang, R. H. Glaser and G. L. Wilkesg, *Polym. Prepr.*, **28**, 434 (1987).
74. T. Iwamoto, K. Morita and J. D. Mackenzie, *J. Non-Cryst. Solids*, **159**, 65 (1993).
75. T. W. Zerda, I. Artaki and J. Jonas, *J. Non-Cryst. Solids*, **81**, 365 (1986).
76. K. A. K. Ebraheen and W. A. Webb, *Prog. Nucl. Magn. Reson. Spectrosc.*, **11**, 149 (1977).
77. I. Ando and W. A. Webb, *Theory of NMR Parameters*, Academic Press, London, 1983.
78. T. Takayama and I. Ando, *Bull. Chem. Soc. Jpn.*, **60**, 3125 (1987).
79. For a more extensive paper see: J. A. Tossell and P. Lazzeretti, *Chem. Phys. Lett.*, **128**, 420 (1986).
80. H. Lischka and H. -J. Kohler, *Chem. Phys. Lett.*, **85**, 467 (1982).
81. J. Kowalewski, *Annu. Rep. NMR Spectrosc.*, **12**, 81 (1982); A. Laaksonen, Specialist Periodical Report, *Nucl. Magn. Reson.*, **14**, 62 (1985); **13**, 64 (1984); I. Ando and G. A. Webb, *Theory of NMR Parameters*, Academic Press, London, 1983, p. 83.
82. D. W. N. Anderson, J. E. Bentham and D. W. H. Rankin, *J. Chem. Soc., Dalton Trans.*, 1215 (1973).
83. B. Wrackmeyer, *J. Magn. Reson.*, **61**, 536 (1985).
84. E. Kupce, E. Liepins, O. Pudova and E. Lukevics, *J. Chem. Soc., Chem. Commun.*, 581 (1984).
85. E. Kupce, E. Liepins and E. Lukevics, *Angew. Chem.*, **97**, 588 (1985); *Angew. Chem., Int. Ed. Engl.*, **24**, 568 (1985).
86. T. N. Truong and M. S. Gordon, *J. Am. Chem. Soc.*, **108**, 1775 (1986).
87. P. v. R. Schleyer and P. D. Stout, *J. Chem. Soc., Chem. Commun.*, 1373 (1986).
88. M. S. Gordon, *Chem. Phys. Lett.*, **126**, 451 (1986).
89. R. H. Cragg and R. D. Lane, *J. Organomet. Chem.*, **294**, 7 (1985).
90. B. Wrackmeyer, S. Kerschl, C. Stader and K. Horchler, *Spectrochim. Acta, Part A*, **42**, 1113 (1986).
91. B. Coleman, in *NMR of Newly Accessible Nuclei*, Vol. 2 (Ed. P. Laszlo), Academic Press, New York, 1983, p. 197.
92. E. Lippmaa, M. Magi, V. Chvalovsky and J. Schraml, *Collect. Czech. Chem. Commun.*, **42**, 318 (1977).
93. H. Schmidbaur, J. Ebenhoch and G. Muller, *Z. Naturforsch.*, **42**, 142 (1987).
94. E. Liepins, I. Birgele, E. Lukevics, V. D. Sheludyakov and V. G. Lahtin, *J. Organometal. Chem.*, **385**, 185 (1990).
95. H. A. Bent, *Chem. Rev.*, **61**, 275 (1961).
96. E. Liepins, I. Birgele, P. Tomsons and E. Lukevics, *Magn. Reson. Chem.*, **23**, 485 (1985).
97. M. Grignon-Dubois and M. Laquerre, *Organometallics*, **7**, 1443 (1988).
98. H. J. Jacobsen, P. J. Kanyha and W. S. Brey, *J. Magn. Reson.*, **54**, 134 (1983).
99. B. De Poorter, *J. Organomet. Chem.*, **128**, 361 (1977).
100. E. Kupce, E. Liepins and E. Lukevics, *Angew. Chem.*, **97**, 588 (1985); *Angew. Chem., Int. Ed. Engl.*, **24**, 568 (1985).
101. E. Kupce, E. Liepins, E. Lukevics and B. Astapov, *J. Chem. Soc., Dalton Trans.*, 1593 (1987).
102. B. Wrackmeyer, S. Kerschl, C. Stader and K. Horchler, *Spectrochim. Acta, Part A*, **42A**, 1113 (1986).

103. E. Kupce and E. Lukevics, *J. Magn. Reson.*, **76**, 63 (1988).
104. E. Kupce, E. Lukevics, Y. M. Varezhkin, A. N. Mikhailova and V. D. Sheludyakov, *Organometallics*, **7**, 1649 (1988).
105. V. A. Chertkov and N. M. Sergeyev, *J. Magn. Reson.*, **52**, 400 (1983).
106. K. Kamenska-Trela, Z. Biedrzicka, R. Machinek, B. Knieren and W. Luttke, *Org. Magn. Reson.*, **22**, 317 (1984).
107. E. Liepins, I. Birgele, E. Lukevics, E. T. Bogoradovsky and V. S. Zavgovodny, *J. Organomet. Chem.*, **393**, 11 (1990).
108. K. Kamenska-Trela, H. Ilcewicz, M. Rospenk, M. Pajdowska and L. Sobczyk, *J. Chem. Res.(S)*, 122 (1987).
109. G. A. Ranzuvaev, A. N. Egorochkin, S. E. Skobeleva, V. A. Kuzenetsov, V. S. Zavgorodny and E. T. Bogoradovsky, *J. Organomet. Chem.*, **222**, 55 (1981).
110. G. C. Levy, J. D. Cargioli, P. C. Juliano and T. D. Mitchell, *J. Magn. Reson.*, **8**, 399 (1972).
111. C. G. Levy, J. D. Cargioli, P. C. Juliano and T. D. Mitchell, *J. Am. Chem. Soc.*, **95**, 3445 (1973).
112. R. K. Harris and B. J. Kimber, *J. Organomet. Chem.*, **70**, 43 (1974).
113. Y. M. Pai, W. P. Weber and K. L. Servis, *J. Organomet. Chem.*, **288**, 269 (1985).
114. A. Pines, M. G. Gibby and J. S. Waugh, *J. Chem. Phys.*, **59**, 569 (1973).
115. J. Schaefer and E. O. Stejskal, *J. Am. Chem. Soc.*, **98**, 1031 (1976).
116. E. R. Andrew, *Prog. Nucl. Magn. Reson. Spectrosc.*, **8**, 1 (1971).
117. J. Schaefer and E. O. Stejskal, in *Topics in Carbon-13 NMR Spectroscopy*, Vol. 3 (Ed. G.C. Levy), Wiley-Interscience, New York, 1979, p. 283.
118. M. J. Sullivan and G. E. Maciel, *Anal. Chem.*, **54**, 1606 (1982).
119. M. Mehring, *High Resolution NMR in Solids*, Springer-Verlag, Berlin, 1983.
120. M. M. Maricq and J. S. Waugh, *J. Chem. Phys.*, **70**, 3300 (1979).
121. G. S. Harbison and H. W. Spiess, *Chem. Phys. Lett.*, **124**, 128 (1986).
122. G. S. Harbison, V. -D. Vogt and H. W. Spiess, *J. Chem. Phys.*, **86**, 1206 (1987).
123. B. Blumich and H. W. Spiess, *Angew. Chem., Int. Ed. Engl.*, **27**, 1655 (1988).
124. R. R. Ernst, G. Bodenhausen and A. Wokaun, *Principles of Nuclear Magnetic Resonance in One and Two Dimensions*, Oxford Univ. Press (Clarendon), London and New York, 1987.
125. S. Wefing and H. W. Spiess, *J. Chem. Phys.*, **89**, 1219 (1988).
126. C. Schmidt, S. Wefing, B. Blumich and H. W. Spiess, *Chem. Phys. Lett.*, **130**, 84 (1986).
127. C. Schmidt, B. Blumich and H. W. Spiess, *J. Magn. Reson.*, **79**, 269 (1988).
128. S. Wefing, S. Kaufmann and H. W. Spiess, *J. Chem. Phys.*, **89**, 1234 (1988).
129. A. Hagemeyer, K. Schmidt-Rohr and H. W. Spiess, *Adv. Magn. Reson.*, **13**, 85 (1989).
130. T. Terao, T. Fujii, T. Onodera and A. Saika, *Chem. Phys. Lett.*, **107**, 145 (1984); J. Ashida, T. Nakai and T. Terao, *Chem. Phys. Lett.*, **168**, 523 (1990).
131. A. Bax, N. M. Szeverenyi and G. E. Maciel, *J. Magn. Reson.*, **55**, 494 (1983).
132. M. M. Maricq and J. S. Waugh, *J. Chem. Phys.*, **70**, 3300 (1979); J. Herzfeld and A. Berger, *J. Chem. Phys.*, **73**, 6021 (1989); W. P. Aue, D. J. Ruben and R. G. Griffin, *J. Magn. Reson.*, **43**, 472 (1981); A. C. Kolbert, H. J. M. de Groot and R. G. Griffin, *J. Magn. Reson.*, **85**, 60 (1989).
133. E. O. Stejskal, J. Schafer and R. A. McKay, *J. Magn. Reson.*, **25**, 569 (1977).
134. J. Clauss, K. Schmidt-Rohr, A. Adam, C. Boeffel and H. W. Spiess, *Macromolecules*, **25**, 5208 (1992).
135. S. Chuang, D. R. Kinney, C. E. Bronnimann, R. Zeigler and G. E. Maciel, *J. Phys. Chem.*, **96**, 4027 (1992).
136. D. Frank, C. Hudalla, R. Maxwell and H. Eckert, *J. Phys. Chem.*, **96**, 7506 (1992).
137. C. A. Fyfe, Y. Zhang and P. Aroca, *J. Am. Chem. Soc.*, **114**, 3252 (1992).
138. A. J. Vega, *J. Am. Chem. Soc.*, **110**, 1049 (1988).
139. G. G. Almond, R. K. Harris and P. Graham, *J. Chem. Soc., Chem. Commun.*, 851 (1994).
140. R. K. Harris, T. N. Pritchard and E. G. Smith, *J. Chem. Soc., Faraday Trans. 1*, **85**, 1853 (1989).
141. K. L. McKillop, G. R. Gillette, D. R. Powell and R. West, *J. Am. Chem. Soc.*, **114**, 5203 (1992).
142. R. Tacke, A. Lopez-Mras, W. S. Sheldrick and A. Sebald, *Z. Anorg. Allg. Chem.*, **619**, 347 (1993).
143. I. Farnan and J. F. Stebbins, *J. Am. Chem. Soc.*, **112**, 32 (1990).
144. T. Takayama and I. Ando, *J. Mol. Struct.*, **271**, 75 (1992).
145. G. Koegler, A. Hasenhindl and M. Moeller, *Macromolecules*, **22**, 4190 (1989).
146. A. E. Aliev, K. D. M. Harris and D. C. Apperley, *J. Chem. Soc., Chem. Commun.*, 251 (1993).
147. A. R. Grimmer, F. Von-Lampe and M. Magi, *Chem. Phys. Lett.*, **132**, 549 (1986).

148. R. Tacke, J. Becht, A. Lopez-Mras, W. S. Sheldrick and A. Sebald, *Inorg. Chem.*, **32**, 2761 (1993).
149. R. Tacke, A. Lopez-Mras, J. Sperlich, C. Strohmann, W. F. Kuhs, G. Mattern and A. Sebald, *Chem. Ber.*, **126**, 851 (1993).
150. J. H. Iwamiya and G. E. Maciel, *J. Am. Chem. Soc.*, **115**, 6835 (1993).
151. K. Tamao, T. Hayashi, Y. Ito and M. Shiro, *Organometallics*, **11**, 2099 (1992).
152. A. B. Scholten, J. W. de Haan, H. A. Claessens, L. J. M. von de Ven and C. A. Cramers, *J. Chromatogr. A*, **688**, 25 (1994).
153. S. Hayashi and E. Akiba, *Chem. Phys. Lett.*, **226**, 495 (1994).
154. R. West, J. D. Cavalieri, J. J. Buffy, C. Fry, K. W. Zilm, J. C. Duchamp, M. Kira, T. Iwamoto, T. Mullerand and Y. Apeloig, *J. Am. Chem. Soc.*, **119**, 4972 (1997).
155. Experimental data for a few disilenes were reported earlier in a communication. See: J. D. Cavalieri, R. West, J. C. Duchamp and K. W. Zilm, *Phosphorus, Sulfur, Silicon*, **93-94**, 213 (1994).
156. B. D. Shepherd, C. F. Campana and R. West, *Heteroat. Chem.*, **1**, 1 (1990).
157. M. J. Fink, M. J. Michalczyk, K. J. Haller, R. West and J. Michl, *J. Chem. Soc., Chem. Commun.*, 1010 (1983).
158. M. Wind, D. J. Powell and R. West, *Organometallics*, **15**, 5772 (1996).
159. H. Watanabe, K. Takeuchi, N. Fukawa, M. Kato, M. Goto and Y. Nagai, *Chem. Soc. Jpn., Chem. Lett.*, 1341 (1987).
160. M. J. Fink, M. J. Michalczyk, K. J. Haller, R. West and J. Michl, *Organometallics*, **3**, 793 (1984).
161. S. Masamune, Y. Eriyama and T. Kawase, *Angew. Chem., Int. Ed. Engl.*, **26**, 584 (1987).
162. R. S. Archibald, Y. van den Winkel, A. Millevolte, D. R. Powell and R. West, *Organometallics*, **11**, 3276 (1992).
163. M. Kira, T. Maruyama, C. Kabuto, K. Ebata and H. Sakurai, *Angew. Chem., Int. Ed. Engl.*, **33**, 1489 (1994).
164. T. M. Duncan, *A Compliation of Chemical Shift Anisotropies*, Farragut Press, Chicago, 1990.
165. R. Okazaki and R. West, *Adv. Organomet. Chem.*, **39**, 232 (1995).

CHAPTER 7

Activating and directive effects of silicon

ALAN R. BASSINDALE, SIMON J. GLYNN and PETER G. TAYLOR

Department of Chemistry, The Open University, Milton Keynes, MK7 6AA, UK

I. INTRODUCTION	356
II. ELECTRONIC EFFECTS OF R_3Si	356
A. Introduction	356
B. Inductive Effects	356
C. Field Effects	357
D. (p–d) π Bonding	357
E. Hyperconjugation	358
III. THE MEASUREMENT AND INTERPRETATION OF R_3Si ACTIVATING AND DIRECTIVE PARAMETERS	359
A. R_3Si as a Substituent in Aromatic Compounds	359
B. R_3SiCH_2 as a Substituent in Aromatic Compounds	361
C. The Effect of R_3Si on α-Silylcarbocations	362
D. The Effect of R_3SiCH_2 on Adjacent Carbocations	364
E. The Effect of $R_3SiCH_2CH_2$ on Adjacent Carbocations	375
F. The Effect of $R_3SiCH_2CH_2CH_2$ on Adjacent Carbocations	380
G. The Effect of R_3Si on α-Silylcarbanions	381
H. The Effect of R_3SiCH_2 on Adjacent Carbanions	382
IV. ACTIVATING AND DIRECTIVE EFFECTS OF SILICON IN ELECTROPHILIC AROMATIC SUBSTITUTION	382
A. Reactions of Compounds with SiR_3 Directly Bonded to the Aromatic Ring	382
B. Electrophilic Aromatic Substitution in $ArCH_2SiR_3$ Compounds	388
V. ACTIVATING AND DIRECTIVE EFFECTS IN ALIPHATIC ELECTROPHILIC REACTIONS	388
A. Reactions of Vinylsilanes	388
1. Introduction	388
2. Addition to vinylsilanes	390
3. Substitution in vinylsilanes	392
B. Reactions of Alkynylsilanes	397
C. Reactions of Allylsilanes	398
D. Reactions of Propargylsilanes	411

The chemistry of organic silicon compounds, Vol. 2
Edited by Z. Rappoport and Y. Apeloig © 1998 John Wiley & Sons Ltd

E. Reactions of Allenylsilanes 412
F. Miscellaneous Reactions Controlled by the Formation of a
   β-Silylcarbocation 413
G. Reactions Involving γ-Silylcarbocations 414
VI. ACTIVATING AND DIRECTIVE EFFECTS OF SILICON IN
    CARBANIONIC REACTIONS 415
  A. Methods of Formation of α-Silylcarbanions 415
    1. Proton abstraction 415
    2. Metal-halogen exchange 416
    3. Transmetallation 416
    4. Organometallic addition to vinylsilanes 417
  B. The Peterson Reaction 417
  C. Variations of the Peterson Reaction 421
VII. STABILIZATION OF DEVELOPING NEGATIVE CHARGE BY
     SILICON 423
  A. α-Silylepoxides 423
VIII. REFERENCES 426

I. INTRODUCTION

This chapter extends and updates our previous review in *The Chemistry of Organic Silicon Compounds*[1]. Each section starts by summarizing some of the key conclusions of the previous review, then goes on to discuss the work that has been published since 1989.

In Section II we begin by surveying the factors that lead to activation and direction. In Section III these concepts are then applied to the various systems in which silicon exerts a stabilizing or destabilizing effect. Where possible we have tried to give quantitative information on the size of the effect. In the final sections we survey reactions which exemplify the activating and directing effects of silicon.

II. ELECTRONIC EFFECTS OF R_3Si

A. Introduction

Under appropriate conditions a proximate silicon atom may exhibit electron-donating or electron-accepting properties, stabilizing positive or negative charge.

The electronic effects of an R_3Si group can be divided into four components: (i) inductive effects, (ii) field effects, (iii) (p-d) π bonding and (iv) hyperconjugative effects. The total electronic effect of an R_3Si group in a molecule or intermediate will be a combination of these effects, and much effort has been made in the last ten years to quantify the contributions of the various effects to certain properties of silicon compounds, particularly as regards the importance of (p-d) π bonding and hyperconjugative effects.

The four electronic effects are defined in the following sections.

B. Inductive Effects

Inductive effects are generally considered to act through the σ-framework of a molecule, and the electronegativity of an element is taken as a measure of the tendency of the element to attract σ-electrons. Ebsworth[2] cited twelve electronegativity scales, all of which agree that carbon is more electronegative than silicon. The most commonly referenced values are those of Pauling[3]; C, 2.5; Si, 1.8; H, 2.1, and Allred and Rochow[4]; C, 2.55; Si, 1.93. The electronegativity of hydrogen is intermediate between that of silicon and carbon.

7. Activating and directive effects of silicon

Therefore the polarity of the silicon–carbon bond is Si^+C^-, and silicon–carbon bonds are generally observed to cleave in this direction, either via electrophilic attack at carbon, or nucleophilic attack at silicon.

Trialkylsilyl and triarylsilyl groups are electron-supplying through inductive effects. The inductive effect is short-range in character, having the greatest effect on atoms directly bonded to silicon, and falling off rapidly after two or three atoms.

C. Field Effects

Field effects describe the response of a neighbouring π-system to the dipole moment of the R_3Si group. There are two types of π-inductive effects[5,6]. The first, known as π_s, is a consequence of charge differences in the σ-system brought about by inductive effects. For example, a substitutent Y which is more electronegative than H will induce a fractional positive charge at C-1 in a benzene ring, and this may induce a redistribution of the π-system that can be predicted by consideration of the resonance canonicals shown in equation 1. Thus, through the π_s effect, the electron density is diminished at C-2 and C-4, and enhanced at C-1, by an electron-withdrawing substituent Y.

$$(1)$$

The second π-inductive effect is the field effect, π_F, which arises through polarization of the whole π-system due to the electric dipole of $(CH_2)_n$ Y, as shown in **1**.

(1)

Both these effects contribute to the overall π-electron density, but are difficult to separate, particularly as they both operate mainly at C-1.

D. (p–d) π Bonding

The chemical and physical properties of the R_3Si group show that it can act as a π-electron-withdrawing group; however, the mechanism by which this electron withdrawal takes place has been open to dispute. The first explanation is that the relatively low-lying, unoccupied silicon d orbitals can participate in (p–d) π bonding. In this way electron density from the p orbital on X can be partially devolved onto silicon through a donor–acceptor interaction with the vacant silicon 3d-orbitals. The conceptually simple (p–d) π bonding model was introduced by Pauling[7] to account for the unexpected shortness of silicon–oxygen and silicon–halogen bonds, and is most easily applied to systems in which electron density, in a p-type orbital adjacent to silicon, is transferred partially onto silicon.

An important feature of the (p–d) π-bonding model is the lack of conformational requirements. The degree of (p–d) π overlap is constant, regardless of rotation around the Si—X bond, as a consequence of the symmetry of the five 3d orbitals on silicon.

E. Hyperconjugation

Hyperconjugation[8], also known as vertical stabilization[9], has been advanced as an alternative to (p–d) π bonding[8]. If two adjacent molecular orbitals are relatively close in energy and have appropriate symmetry, they can undergo interaction so that the energy of one is lowered and that of the other is raised. For example, the hyperconjugative interaction of a π orbital with a Si—C σ^* orbital is shown in Figure 1.

The magnitude of the hyperconjugative interaction is dependent on the energy difference between the orbitals and the orbital coefficients. To evaluate fully the effect of hyperconjugation it is necessary to also consider the $\sigma^*-\pi^*$, $\sigma-\pi$ and $\sigma^*-\pi^*$ interactions. However, in many cases the overall effect of hyperconjugation can be predicted simply by the use of resonance canonicals and knowledge of the polarity of bonds to silicon. The hyperconjugative effect of a methyl group can be represented as shown in equation 2.

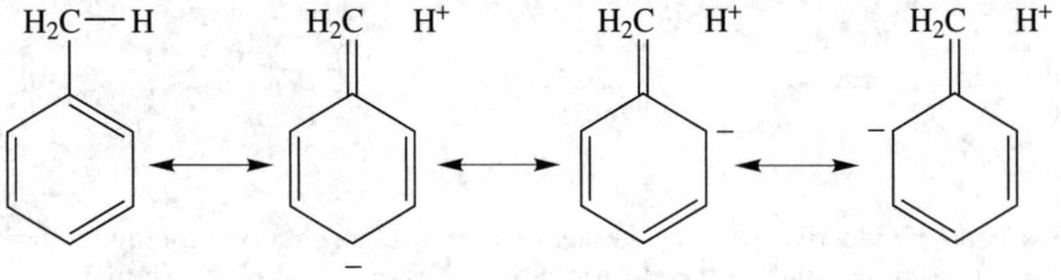

$$\text{(2)}$$

Based on the polarity of the C—H bond being C^-H^+, the resonance canonicals predict that the methyl group activates the *ortho* and *para* carbons in an aromatic ring to electrophilic substitution. For SiH_3 or $SiMe_3$ the polarities are Si^+H^- and Si^+C^-, which gives the resonance canonicals, shown in equation 3.

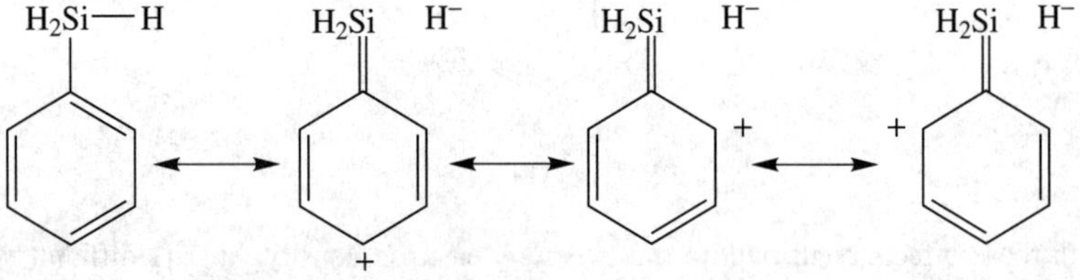

$$\text{(3)}$$

These suggest that, in contrast to a methyl group, the hyperconjugative effect of a SiH_3 or a $SiMe_3$ group will act to deactivate the *ortho* and *para* positions, although the electron-supplying inductive effect will complicate this.

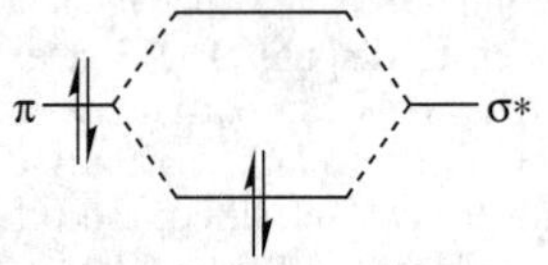

FIGURE 1. The hyperconjugation interaction of a π and σ^* orbital

Hyperconjugation also predicts that the R_3SiCH_2 group will be electron releasing at the *ortho* and *para* positions of an aromatic ring, and activating to electrophilic aromatic substitution, as shown in equation 4.

$$\text{[Resonance structures showing } H_2C\text{—}SiR_3 \text{ substituted benzene with hyperconjugation to ortho and para positions]} \tag{4}$$

The simple hyperconjugation approach also predicts the stabilization of carbocations $R_3SiCH_2CH_2^+$, and of carbanions $R_3SiCH_2^-$, as shown in equations 5 and 6, respectively.

$$\overset{+}{H_2C}-\underset{\underset{\displaystyle SiR_3}{|}}{CH_2} \longleftrightarrow H_2C=CH_2 \text{ with } \overset{+}{SiR_3} \tag{5}$$

$$\overset{-}{H_2C}-\underset{\underset{\displaystyle SiR_2}{|}}{\overset{\overset{\displaystyle R}{|}}{}} \longleftrightarrow H_2C=SiR_2 \text{ with } R^- \tag{6}$$

No hyperconjugative stabilization is predicted, on the basis of bond polarities, for α-silylcarbocations or for β-silylcarbanions, as shown in equations 7 and 8, respectively.

$$\overset{+}{H_2C}-\underset{\underset{\displaystyle SiR_2}{|}}{\overset{\overset{\displaystyle R}{|}}{}} \longleftrightarrow H_2C=SiR_2 \text{ with } R^+ \tag{7}$$

$$\overset{-}{H_2C}-\underset{\underset{\displaystyle CH_2}{|}}{\overset{\overset{\displaystyle SiR_3}{|}}{}} \longleftrightarrow H_2C=CH_2 \text{ with } {}^-SiR_3 \tag{8}$$

Unlike (p–d) π bonding, hyperconjugation has strict conformational requirements. For hyperconjugation to be at a maximum the C—Si bond must be coplanar with the p orbital with which it is interacting[9]. When the C—Si bond is orthogonal to the p orbital, hyperconjugation is necessarily zero. This marked difference in conformational dependence between hyperconjugation and (p–d) π bonding has been utilized in attempts to determine which is the more important effect.

III. THE MEASUREMENT AND INTERPRETATION OF R_3Si ACTIVATING AND DIRECTIVE PARAMETERS

A. R_3Si as a Substituent in Aromatic Compounds

The first quantitative measures of the electronic effect of R_3Si groups were obtained through application of the Hammett equation[10] which is usually expressed in the forms:

$$\log k = \log k_0 + \rho\sigma$$

$$\log K = \log K_0 + \rho\sigma$$

where k or K is the rate or equilibrium constant for a side-chain reaction of a *meta-* or *para-*substituted benzene derivative. The quantities k_0 and K_0 approximate to k or K for the unsubstituted or parent compound. The substituent constant σ measures, for a *meta* or *para* substituent, the polar electronic effect relative to hydrogen. The σ-constant is, in principle, independent of the nature of the reaction provided there is no conjugation between the reaction centre and the substituent. Electron-withdrawing substituents have positive σ-values, and electron-supplying substituents have negative σ-values. The reaction constant ρ is a measure of the susceptibility of the reaction to polar effects and varies from reaction to reaction.

The σ_m and σ_p values for R_3Si groups have been measured in a variety of ways for a variety of reactions. Hansch, Leo and Taft[11] have surveyed Hammett substituent constants, resonance and field parameters and a selection of values of σ_m and σ_p for R_3Si groups are presented in Table 1.

The values of $\sigma_m = -0.04$ and $\sigma_p = -0.07$ for the trimethylsilyl substituent compare with values of -0.07 and -0.17 for the methyl substituent and values of -0.10 and -0.20 for the *t*-butyl substituent[11]. Hence compared with alkyl groups the Me_3Si group has a modest electronic effect in both the *meta* and *para* positions. The σ_m and σ_p constants suggest that Me_3Si will be a weakly activating substituent, but will not have a significant directing effect in aromatic substitutions.

It is clear from Table 1 that the nature of R is important in determining the electronic effect of R_3Si. Unlike $SiMe_3$, SiH_3 is found to be electron-withdrawing in both *meta*

TABLE 1. Selected Hammett substituent constants, resonance and field parameters for silicon-based substituents

R	σ_m	σ_p	F	R	σ_p^+	σ_p^-
SiH_3	0.05	0.10	0.06	0.04	0.14	
$SiMe_3$	−0.04	−0.07	0.01	−0.08	0.02	0.11
$SiHMe_2$	0.01	0.04	0.03	0.01	−0.04	
$SiPhMe_2$	0.04	0.07	0.06	0.01	0.08	
$SiMePh_2$	0.10	0.13	0.11	0.02	−0.04	
$SiPh_3$	−0.03	0.10	−0.04	0.14	0.12	0.29
$SiMe_2(OMe)$	0.04	−0.02	0.09	−0.11	−0.02	
$SiMe(OMe)_2$	0.04	−0.10	0.05	0.05	0.01	
$Si(OMe_3)$	0.09	0.13	0.10	0.03	0.13	
$Si(OEt)_3$	0.02	0.08	0.03	0.05	0.17	
$SiMe_2(OSiMe_3)$	0.00	−0.01	0.04	−0.05		
$SiMe(OSiMe_3)_2$	−0.02	−0.01	0.01	−0.02		
$Si(OSiMe_3)_3$	−0.09	−0.01	−0.08	0.07		
$Si(NMe_2)_3$	−0.04	−0.04	0.00	−0.04		
$SiMe_2F$	0.12	0.17	0.12	0.04	0.17	
$SiMe_2Cl$	0.16	0.21	0.16	0.05	0.02	
$SiMeF_2$	0.29	0.23	0.32	−0.09	0.23	
$SiMeCl_2$	0.31	0.39	0.29	0.10	0.08	
SiF_3	0.54	0.69	0.47	0.22		
$SiCl_3$	0.48	0.56	0.44	0.12	0.57	
$SiBr_3$	0.48	0.57	0.44	0.13	0.41	

and *para* positions. The triphenylsilyl group appears to be slightly activating in the *meta* position, but deactivating in the *para* position. Substitution of halogen atoms for methyl groups causes incremental increases in both σ_m and σ_p. This is mainly due to σ-electron withdrawal by the halogen atom, although enhanced (p–d) π bonding or hyperconjugation will also make a contribution. Replacement of methyl groups by methoxy groups on the silicon has a much lesser deactivating effect, whereas replacement by trimethylsilyloxy appears to show no simple trend.

The Hammett σ_p constants are inadequate for aromatic reactions in which cross-conjugation between substituents is possible. For example, when the substituent X is electron-supplying by resonance and Y is electron-accepting, cross-conjugation is possible as shown in equation 9. This direct cross-conjugation is not possible for *meta* substituents. To allow for these interactions the σ-constants σ_p^+ and σ_p^- have been defined[12] for resonance electron-withdrawing and resonance electron-supplying substituents respectively, and some values for these constants are shown in Table 1.

$$\text{(9)}$$

Unlike alkyl groups, the Me$_3$Si group does not show a much increased σ_p^+ value compared to σ_p. In alkyl groups C–H or C–C hyperconjugation is usually invoked to explain the enhanced electron supply. For the trimethysilyl group hyperconjugation in this direction is not favoured.

By contrast, σ_p^- for Me$_3$Si has a value of +0.11, compared to −0.07 for σ_p. This is attributed to the Me$_3$Si group being electron-withdrawing by resonance, as predicted by (p–d) π bonding or hyperconjugation. The SiPh$_3$ group also shows a greatly enhanced σ_p^- value.

The Hammett σ-values contain contributions from both inductive/field effects and the resonance effect. The σ-constant can be separated quantitatively into a resonance component R, which operates mainly in the *para* position, and an inductive component F, which is assumed to be equal in the *meta* and *para* positions. Hansch, Leo and Taft[11] have calculated the F and R values of Me$_3$Si to be 0.01 and −0.08, respectively, as quoted in Table 1. These values seem somewhat at odds with experimentally determined values[12–18] for the inductive and resonance parameters, which give mean values of −0.08 and 0.06, respectively. These values confirm the generally accepted view that Me$_3$Si is electron-supplying by inductive effects and electron-withdrawing by resonance effects.

B. R$_3$SiCH$_2$ as a Substituent in Aromatic Compounds

In contrast to the R$_3$Si group, R$_3$SiCH$_2$ is quite strongly electron-supplying to the aromatic ring; Hammett σ-values of $\sigma_m = -0.16$ and $\sigma_p = -0.21$ are quoted by Hansch, Leo and Taft[11] for Me$_3$SiCH$_2$. By comparison values of $\sigma_m = -0.05$ and $\sigma_p = -0.17$ are quoted for Me$_3$CCH$_2$. Initially it was thought that the trimethylsilylmethyl substituent was electron-releasing through inductive effects only, but the measurement of a σ_p^+ value of −0.62 indicated a large degree of resonance contribution due to hyperconjugative release[19].

Schaefer and coworkers[20] have used long-range NMR coupling constants to investigate rotational barriers about the $C(sp^2)-C(sp^3)$ bonds in benzyl compounds. The barrier for benzylsilane was found to be 1.77 kcal mol^{-1}, compared to 1.2 kcal mol^{-1} for ethylbenzene. The increased barrier for benzylsilane is attributed to increased stabilization of the stable conformer, in which the C−Si bond lies in a plane perpendicular to the benzene plane, by a hyperconjugative interaction between the C−Si bond and the π-system.

On this evidence the Me$_3$SiCH$_2$ group attached to an aromatic ring should be quite strongly activating and *ortho/para* directing. Crestoni and Fornarini[21] have studied the gas-phase reactivity of benzyltrimethylsilane and have confirmed that it is highly activated towards electrophilic attack directed to the *ortho/para* positions.

C. The Effect of R$_3$Si on α-Silylcarbocations

Despite the electron-releasing inductive effect of trialkylsilyl groups, silicon does not have a large stabilizing effect on R$_3$SiCH$_2{}^+$. An often cited observation is that Me$_3$SiCMe$_2$Br has been found to solvolyse 38,000 times more slowly than Me$_3$CCMe$_2$Br[22]. Eaborn and coworkers[23] found that there was little rate increase in the solvolyses of Me$_3$SiCH$_2$Cl, (Me$_3$Si)$_2$CHCl and (Me$_3$Si)$_3$CCl, compared to the large rate increases observed in the carbon substituted series ethyl, propyl and *t*-butyl. In the carbon series, the rate increase is attributed to C−H hyperconjugation, $\sigma-\pi$ delocalization leading to a carbon-carbon double bond and a positively charged hydrogen. The lack of a rate increase on silyl substitution is therefore attributable to the poor ability of the Si−C bond to hyperconjugate, as shown in equation 10.

$$\underset{Me_2Si-\overset{+}{C}H_2}{\overset{\overset{CH_3}{|}}{}} \longleftrightarrow \underset{Me_2Si=CH_2}{\overset{\overset{+}{C}H_3}{}} \qquad (10)$$

α-Silyl carbocations generated in the gas phase by chloride elimination were observed to rearrange to silicenium ions with migratory aptitudes Ph>H≫Me[24] (equation 11).

$$RMe_2SiCH_2Cl \longrightarrow \underset{Me_2Si-CH_2{}^+}{\overset{\overset{R}{|}}{}} \longrightarrow Me_2\overset{+}{Si}-CH_2R \qquad (11)$$

Cho has calculated[25], using *ab initio* methods (MP2/6-31G*), that all three of the above groups migrate from silicon to the adjacent α-carbon with no energy barrier. However, Hartee−Fock calculations (HF/6-31G*) predict small energy barriers of 1.4 kcal mol^{-1} and 1.5 kcal mol^{-1} for hydrogen and methyl migrations respectively.

The solvolysis of Me$_3$SiCH$_2$Br is observed to be enhanced relative to Me$_3$CCH$_2$Br, with a rate ratio of 2600 : 1 in basic ethanol[26]. However, further studies[27] showed that the reaction took place without formation of the carbocation. The solvolysis of Me$_3$CCH$_2$X (X = triflate or tosylate) took place with almost total rearrangement, whereas no rearrangement was observed in the silicon case, although calculations indicated that this should take place on formation of the α-silyl carbocation. Together with kinetic information, these observations led to the conclusion that the α-silyl compound undergoes solvolysis by a S$_N$2 mechanism whilst the corresponding carbon compound involves considerable neighbouring group participation.

Calculations at the 6-31G* level[27] predicted CH$_3$CH$_2{}^+$ to be more stable than H$_3$SiCH$_2{}^+$ by 13.2 kcal mol^{-1}, and Me$_3$SiCH$_2{}^+$ to be less stable than Me$_3$CCH$_2{}^+$ by

11 kcal mol^{-1}. However, $H_3SiCH_2^+$ was more stable than CH_3^+ by 16.1 kcal mol^{-1}[27]. High level calculations using polarization functions and electron correlation at the MP3/6-31G* level have been performed for the isodesmic reaction shown in equation 12a.

$$XCH_2^+ + CH_4 \longrightarrow XCH_3 + CH_3^+ \quad (12a)$$

From the calculations it was found that when $X = CH_3$ the cation is 34 kcal mol^{-1} more stable than when $X = H$. However, when $X = SiH_3$ the cation is only 17.8 kcal mol^{-1} more stable than when $X = H$[28].

For the vinyl cation, using the isodesmic reaction shown in equation 12b, it was found that CH_3 and SiH_3 were 27 and 24 kcal mol^{-1} more stabilizing than H, respectively[28]. In this case the alpha effect of silicon is comparable to that of carbon.

$$H_2C=CX^+ + H_2C=CH_2 \longrightarrow H_2C=CHX + H_2C=CH^+ \quad (12b)$$

This has been supported by the mass spectroscopic experiments of McGibbon, Brook and Terlouw[29], who determined the stabilization of the trimethylsilylvinyl cation to be 29.5 kcal mol^{-1} relative to the vinyl cation. The trimethylsilyl group was stabilizing by 2 kcal mol^{-1} relative to a methyl group. These values are in good agreement with the theoretical figures above.

Bausch and Gong[30] have experimentally derived values for the free energies of the heterolytic C—H bond cleavage in the substituted fluorenes **2**. The α-SiMe$_3$ group and the α-CMe$_3$ groups have near-equal effects on the free energies of C—H heterolysis. The α-Me substituent is found to stabilize the fluorenium cation by 6 kcal mol^{-1} compared to the α-SiMe$_3$, whereas the α-SiMe$_3$ substituent gives a stabilization of 2 kcal mol^{-1} relative to hydrogen.

(2) (3) (4)

R = H, Me, SiMe$_3$, CMe$_3$

2,2-Dimethyl-2-sila-1-indanyl bromide **3** was found to solvolyse 4.98×10^2 times less rapidly than the carbon reference 2,2-dimethyl-1-indanyl bromide **4**[31]. This corresponds to the solvolytic generation of the α-silylated benzylic cation being some 4 kcal mol^{-1} less favourable than the α-alkylated benzylic cation.

Apeloig and coworkers studied solvolysis rates of Me$_3$CX and (Me$_3$Si)Me$_2$CX in 97% trifluoroethanol and derived a value of 4.8 kcal mol^{-1} for the stability of the *t*-butyl cation over the 2-trimethylsilylpropyl cation[32a].

Apeloig and coworkers have also studied the effect of silyl substituents on keto-enol equilibria (equation 13)[32b,32c].

$$\underset{(5a)}{\overset{H}{\underset{H}{>}}C=C\overset{R}{\underset{OH}{<}}} \rightleftharpoons \underset{(5b)}{H_3C-C\overset{R}{\underset{O}{\lessgtr}}} \longleftrightarrow \underset{(5c)}{H_3C-\overset{+}{C}\overset{R}{\underset{O^-}{<}}} \qquad (13)$$

The equilibrium between the enol (5a) and the keto (5b) forms depends upon the nature of R, electropositive substituents destabilizing the keto form 5b. This destabilization is due to the adjacent carbon having some carbocationic character, as can be seen in the resonance form 5c.

Calculations at the 6-31G*//3-21G level showed ΔE (keto–enol) for R = SiH$_3$ to be 12.9 kcal mol^{-1}, which is 6.1 kcal mol^{-1} lower than ΔE for R = Me. The silyl group was calculated to destabilize the keto form by 3.8 kcal mol^{-1} relative to R = H. The theoretical predictions were confirmed experimentally.

The gas-phase basicities of α-trimethylsilylstyrenes were determined by Mishima and coworkers by measurement of proton transfer equilibrium constants[33]. The basicity of α-trimethylsilylstyrene was found to be comparable to that of α-alkyl styrenes, which was taken to suggest that an α-trimethylsilyl group stabilizes a carbocation.

D. The Effect of R$_3$SiCH$_2$ on Adjacent Carbocations

In this section we concentrate mainly on mechanistic aspects of reactions involving β-silylcarbocations. The structure and properties of β-silylcarbocations are discussed in detail in Chapter 12 on Silicon-substituted Carbocations in this volume. The strong stabilization of β-silylcarbocations is of particular importance in relation to activating and directing effects in organic syntheses using silicon compounds, and is still an important area for mechanistic, theoretical and synthetic studies.

The β-effect was first noted by Ushakov and Itenberg in 1937[34], and in 1946 Sommer, Whitmore and coworkers reported the high reactivity of β-chlorosilyl systems to elimination (equation 14), compared to the corresponding α- and γ-systems[35,36].

$$R_3Si-CH_2CH_2-X \longrightarrow CH_2=CH_2 + R_3SiX \qquad (14)$$

The current evidence supports a carbocationic (E1) mechanism for this reaction. However, the nature of the carbocationic intermediate has caused much interest. There are two possibilities:

(i) Interaction of silicon by a purely hyperconjugative mechanism. In this case there is no significant movement of the silicon atom, and the process is termed 'vertical' participation (of 6a).

$$\underset{(6a)}{\overset{R_3Si}{\underset{H_2C-CH_2^+}{\diagdown}}} \qquad \underset{(6b)}{\overset{\overset{+}{\underset{Si}{R_3}}}{\underset{H_2C-CH_2}{\diagup\diagdown}}}$$

(ii) Interaction of silicon by internal neighbouring group participation, to form a three-membered ring siliconium ion 6. This is termed 'non-vertical' participation.

Studies by Jarvie and coworkers[37] showed that solvolysis of *erythro*- Me$_3$SiCHBrCHBr-CH$_3$ led predominantly to *cis*-1-bromopropene, via an antiperiplanar elimination. This

7. Activating and directive effects of silicon

was taken to be evidence in favour of the cyclic siliconium intermediate, although the open-chain cation could also maintain the antiperiplanar geometry if hyperconjugation disfavours rotation about the C—C bond.

A further study[38] found that the treatment of $Me_3SiCH_2CD_2OH$ with PBr_3 gave the products $Me_3SiCH_2CD_2Br$ and $Me_3SiCD_2CH_2Br$ in equal amounts. The deuterium scrambling again appears to support the cyclic siliconium intermediate, although again the open cation would also be consistent provided that it undergoes a rapid 1,2 migration of the trimethylsilyl group, as shown in equation 15.

$$Me_3Si\diagdown_{H_2C-CD_2^+} \rightleftharpoons {}^{SiMe_3}\diagup_{H_2C^+-CD_2} \quad (15)$$

Calculations by Jorgensen and co-workers[28] on primary systems assessed the size of the β-effect by calculation of the energy change for the isodesmic reaction shown in equation 16.

$$CH_4 + H_3SiCH_2CH_2^+ \longrightarrow H_3SiCH_2CH_3 + CH_3^+ \quad (16)$$

The energies at the MP3/6-31G* level of the three geometries cyclic (**7**), parallel open (**8**) and orthogonal open (**9**) were calculated.

$$\underset{\substack{\text{cyclic} \\ (\mathbf{7})}}{\overset{H_3}{\underset{H_2C-CH_2}{{}^+Si}}} \qquad \underset{\substack{\text{parallel} \\ (\mathbf{8})}}{\text{parallel}} \qquad \underset{\substack{\text{orthogonal} \\ (\mathbf{9})}}{\text{orthogonal}}$$

The cyclic form **7** had a stabilization of 74.4 kcal mol^{-1} relative to CH_3^+, the parallel form **8**, 72.0 kcal mol^{-1}, and the orthogonal form **9**, 42.4 kcal mol^{-1}. The lesser stability of the orthogonal form is attributable to the lack of hyperconjugation in this geometry.

Compared to the ion $HCH_2CH_2^+$, $H_3SiCH_2CH_2^+$ in the parallel geometry is stabilized by 38 kcal mol^{-1}, this stabilization arising from a combination of hyperconjugative and inductive effects. In the orthogonal geometry, the cation is stabilized by 8.9 kcal mol^{-1} which must necessarily arise from angle-independent inductive effects. The hyperconjugative stabilization in the parallel geometry is therefore around 29 kcal mol^{-1}.

Further calculations[39] on secondary and tertiary systems at the MP2/6-31G(d) level showed that for the 2-propyl cation, a β-silyl group in the bisected orientation provides 22.1 kcal mol^{-1} of stabilization compared to a β-hydrogen, as opposed to 6.6 kcal mol^{-1} provided by a β-methyl group. For the secondary system, the bridged carbocation was found to have a stabilization of 18.4 kcal mol^{-1}, i.e. less than the open form.

The lesser magnitude of the β-effect in the secondary cation (22.1 kcal mol^{-1} compared to 38 kcal mol^{-1} for the primary cation) is attributed to reduced electron demand in the secondary cation, as a consequence of the stabilizing effect of alkyl substituents.

β-Effects for the tertiary butyl carbocation were also calculated and, as would be expected, the value for β-silicon stabilization was smaller than in the secondary system, having a value of 15.9 kcal mol^{-1}. For comparison, the β-methyl stabilization was 5.0 kcal mol^{-1}.

In all these calculations SiH_3 has been used as a model for $SiMe_3$, which is the more common group used in practice. Recently, calculations by Adcock and coworkers have shown that the two silyl groups can have markedly different stabilizing effects[40].

Lambert and coworkers[41] studied the solvolysis reactions of the two cyclohexyl systems **10** and **11**.

(10)   (11)

X = CF$_3$CO$_2$

It was found that the rate of solvolysis of the trifluoroacetate of the *trans* isomer **10** was 10^{12} times faster than that of cyclohexyl trifluoroacetate, whereas the rate acceleration of the trifluoroacetate of the *cis* isomer **11** was only 10^4. In the *trans* isomer, the Me$_3$Si group is frozen into the antiperiplanar (diaxial) relationship with respect to the leaving group, due to the *t*-butyl group. This is the ideal conformation for maximal hyperconjugation, whereas in the *cis* isomer, the dihedral angle between the two substituents is around 60°, and a much smaller contribution from hyperconjugation would be expected. Analysis of the various contributing effects gave rate accelerations of a factor of 10^2 in both cases arising from inductive effects. Hyperconjugative contributions were shown to be 10^{10} for the *trans* isomer and 10^2 for the *cis* isomer.

Norbornyl systems were used to study the β-effect in the synperiplanar geometry[42]. The solvolyses of the mesylates (X = MeSO$_2$O) of *endo*-3-(trimethylsilyl)-*endo*-2-norborneol **12** and *endo*-2-norborneol **13** were shown to take place via carbocationic mechanisms, and the trimethylsilyl-substituted system showed a rate acceleration of 10^5.

(12)   (13)   (14)

A direct comparison of *exo*-3-trimethylsilyl-*endo*-2-norbornyl mesylate **14** with **12** and **13** could not be obtained because of solubility problems. However, measurements of the acid catalysed elimination of the corresponding alcohols suggested that **14**, which has an anticlinal stereochemistry (dihedral angle = 120°), has a β-effect of a similar or slightly larger magnitude to that in **12**.

These studies imply that the synperiplanar β-effect is considerably less than the antiperiplanar β-effect. Although the synperiplanar β-effect was measured in a norbornyl system and the antiperiplanar β-effect measured in a cyclohexyl system, it is unlikely that the skeletal differences between the two systems could cause such a large difference. The β-effect of $10^4 - 10^5$ observed for the synperiplanar, synclinal and anticlinal geometries is of too great a magnitude to be attributed solely to inductive effects. Non-vertical participation cannot contribute at dihedral angles of 0° and 60°, so the β-effect at these geometries necessarily arises from hyperconjugation. However, from this study it cannot be determined whether the large β-effect in the antiperiplanar geometry contains a contribution from non-vertical participation, or whether the β-effect in the synperiplanar geometry is attenuated by reduced orbital overlap arising from steric and/or electronic factors.

To ascertain the origin of the large β-effect in the antiperiplanar geometry, α-secondary deuterium isotope effects were measured for the solvolysis of **15**[43].

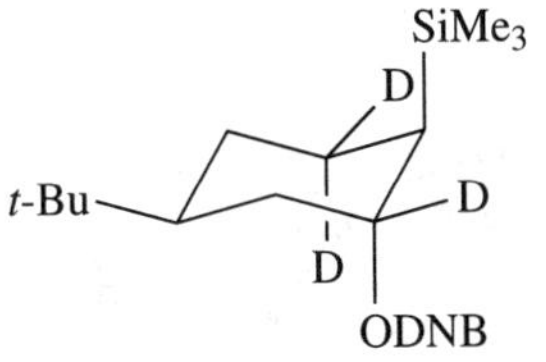

(**15**)

ODNB = 3,5-$(O_2N)_2C_6H_3CO_2$

For a vertical mechanism, the hybridization of the β-carbon atom changes from sp^3 to sp^2, for which the α-hydrogen/deuterium kinetic isotope effect is normally in the range 1.15–1.25.

For the non-vertical mechanism, the transition state is analogous to a S_N2 reaction, for which the isotope effect is negligible or inverse, in the range 0.95–1.05.

Solvolysis of **15** in 97% trifluoroethanol gave a secondary isotope effect of 1.17, which indicates a vertically stabilized transition state. Thus the highly unsymmetrical dihedral dependence of silicon participation can almost entirely be attributed to the hyperconjugation model with little non-vertical involvement of the silicon nucleophile.

A recent investigation has involved a system in which the vacant p orbital of the carbocation has an orthogonal relationship with the Si—C bond (equation 17)[44].

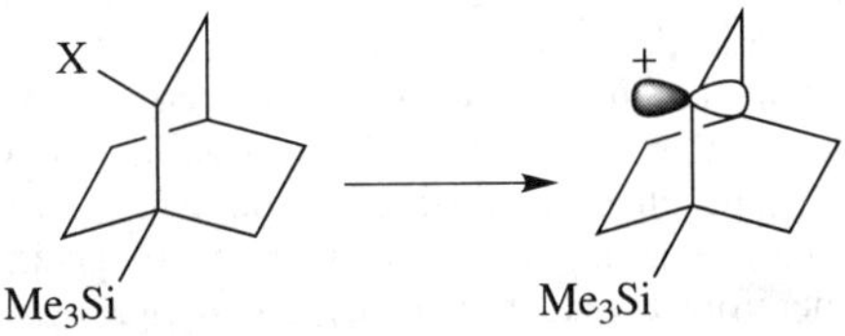

(17)

(**16**)

In the ion **16** neither bridging nor hyperconjugation is possible and any stabilization must result from inductive effects.

Solvolysis of the tosylates **17** and **18** in 97% trifluoroethanol gave a rate ratio β-silyl/β-H of 1.31.

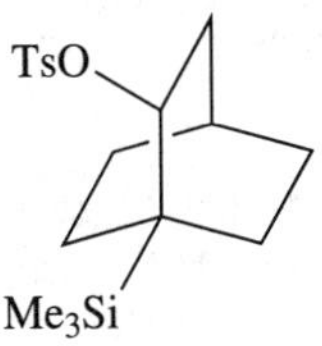

(**17**) 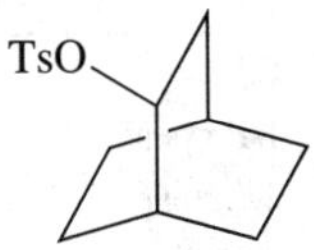   (**18**)

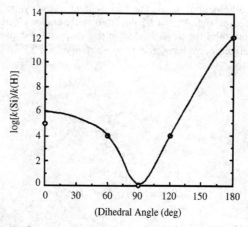

FIGURE 2. The dihedral angle dependence of the β effect of silicon

In the absence of hyperconjugative and internal participation, there is essentially no silicon β-effect. The inductive effect of the Me$_3$Si group is essentially zero. This agrees with the calculations of Ibrahim and Jorgensen[39], which predict no inductive stabilization for secondary systems.

From these data the dependence of the β-silicon effect on the dihedral angle can be represented graphically as shown in Figure 2, and resembles a cosine-squared curve with a highly flattened left side.

A variety of methods have been used to determine the energy of stabilization of a carbocation by a β-silicon substituent. Li and Stone[45] studied the association of the trimethylsilicenium ion with alkenes in a mass spectrometer and have calculated the β-silyl stabilization energies for the carbocations produced as shown in Table 2.

These data show a decrease in the extent of β-silyl stabilization with successive methyl substitution. The methyl (and phenyl) substituents stabilize the carbocation by polarization and inductive effects, resulting in a delocalization of positive charge away from the carbocation, and therefore a reduction in hyperconjugative interaction with the β-substituent bond.

Mass spectral studies have also shown that the vinyl cation Me$_3$SiCH=CH$^+$ is stabilized by 43.5 kcal mol^{-1} relative to CH$_2$=CH^{+}[29]. The association of trimethylsilicenium ion

TABLE 2. Thermodynamic data for the reaction Me$_3$SiX$^+$ + XH$_2$ = Me$_3$SiXH + XH$^+$

X	ΔH° (kcal mol^{-1})
Ethene	48.2
1-Propene	38.4
2-Butene	38.2
2-Methylpropene	28.1
2-Methyl-2-butene	28.8
2,3-Dimethyl-2-butene	25.8
Styrene	21.8

TABLE 3. Stabilization of vinyl cations by β-trimethylsilyl groups

Cation	β-Silicon stabilization (kcal mol^{-1})
$\underset{H}{\overset{Me_3Si}{>}}\!\!=\!\!\overset{+}{-}C_4H_9\text{-}n$ (19)	11
$\underset{Me}{\overset{Me_3Si}{>}}\!\!=\!\!\overset{+}{-}C_3H_7\text{-}n$ (20)	12
$\underset{Me}{\overset{Me_3Si}{>}}\!\!=\!\!\overset{+}{-}Ph$ (21)	9

with alkynes gave stabilization energies (kcal mol^{-1}) as shown in Table 3 for some further vinyl cations **19–21**[46].

The smaller stabilizations of the vinyl cations **19–21** compared to the alkyl cations in Table 2 is attributed to the α-aryl or alkyl substituents having a greater stabilizing effect on the vinylic systems.

Kresge and Tobin[47] found that the protonation of phenyl(trimethylsilyl)acetylene in superacid media was only 300 times faster than that of phenylacetylene, corresponding to a stabilization of 3.4 kcal mol^{-1} for the β-trimethysilylvinyl cation. This small stabilization was attributed to a reduction of the hyperconjugative interaction. Although the dihedral angle in the system has the optimum value of 0°, the angle φ between the C—SiMe$_3$ bond and the adjacent C—C bond varies from 180° in the reactant **22** to 120° in the product **23** (equation 18). Hyperconjugative interaction would be expected to be maximum when $\varphi = 90°$. Thus, the small β-effect in this system arises from a large φ value in the transition state.

$$Me_3Si\text{—}C\!\!\equiv\!\!\!\equiv\text{—}Ph \quad \longrightarrow \quad \underset{Me_3Si}{\overset{H}{>}}C\!\!=\!\!\overset{+}{=}\text{—}Ph \qquad (18)$$

$\varphi = 180°$ $\varphi = 120°$
(**22**)   (**23**)

Kresge and Tobin[48] also studied the hydrolysis of vinyl ethers and found a rate ratio of 130 between methyl vinyl ether and ethyl cis-trimethylsilylvinyl ether, corresponding to a stabilization of the β-silyl carbocation of 2.9 kcal mol^{-1}. In this case the small rate acceleration (compared to the cyclohexyl systems studied by Lambert) can be attributed to the unfavourable dihedral angle. The dihedral angle in the vinyl ether is 90° (**24**), and on protonation it drops to 60° (**25**), whereas maximum hyperconjugative interaction requires a dihedral angle of 0°.

The rate accelerations $k(X = SiMe_3)/k(X = H)$ and corresponding free energy differences produced by trimethylsilyl substituents in alkene and alkyne protonation reactions are shown in Table 4[49].

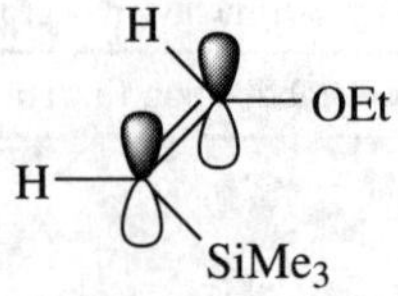

(24) 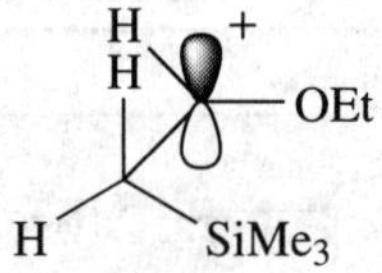 (25)

TABLE 4. Rate accelerations and free energy differences in the protonation of alkenes and alkynes

Substrate, X = H, SiMe$_3$	$k(X = SiMe_3)/k(X = H)$	$\delta(\Delta G)$ (kcal mol^{-1})
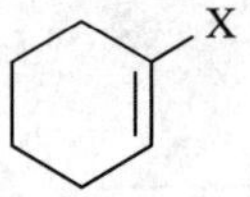 cyclohexenyl-X	15,300	5.7
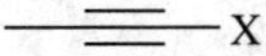 HC≡C–X	54,300	6.5
 n-Bu–C≡C–X	56,800	6.5
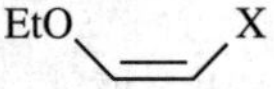 EtO–CH=CH–X	129	2.9
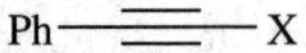 Ph–C≡C–X	312	3.4

The β-silyl effects are much greater for the purely aliphatic systems than the vinyl ether or phenylacetylene, indicating that the stabilization in these latter systems is attenuated by the carbocation-stabilizing ability of the ethoxy and phenyl groups, respectively.

However, the β-effects in the aliphatic systems are still much smaller than the 10^{12} rate acceleration observed in the cyclohexyl system. As mentioned earlier, this is due to the unfavourable dihedral angle θ in the alkene protonations and the angle φ in the acetylene protonations.

A variety of other methods have been used to measure the β-silicon stabilization of carbocations. From gas-phase studies, Hajdasz and Squires[50] derived a value of 39 kcal mol^{-1} for the stabilization of the cation Me$_3$SiCH$_2$CH$_2^+$ relative to the ethyl cation. This is in agreement with calculations by Ibrahim and Jorgenson[39]. Siehl and Kaufmann[51] have used carbon-13 NMR spectroscopic data to give an indication of the β-silyl stabilizing effect in some aryl vinyl cations.

Shimizu and coworkers have extensively studied the solvolyses of β-silyl benzyl systems. They observed a rate acceleration of 3×10^5 for solvolysis of 1-phenyl-2-(trimethylsilyl) ethyl trifluoroacetate **26** compared to the corresponding β-t-butyl system **27**. This indicates that the solvolytic generation of the β-silyl carbocation is about 7.5 kcal mol^{-1} more favourable[52].

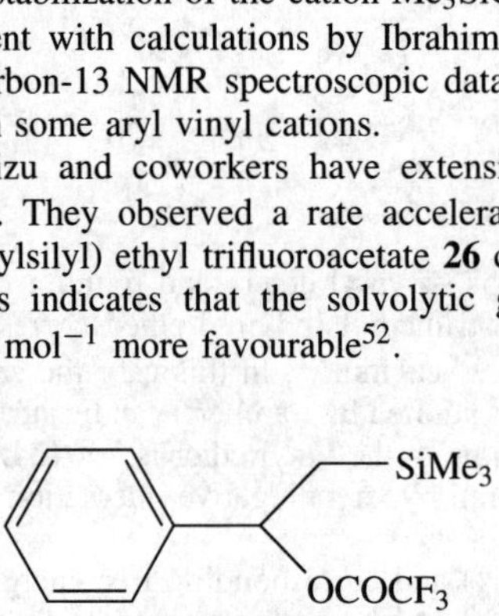

(26) 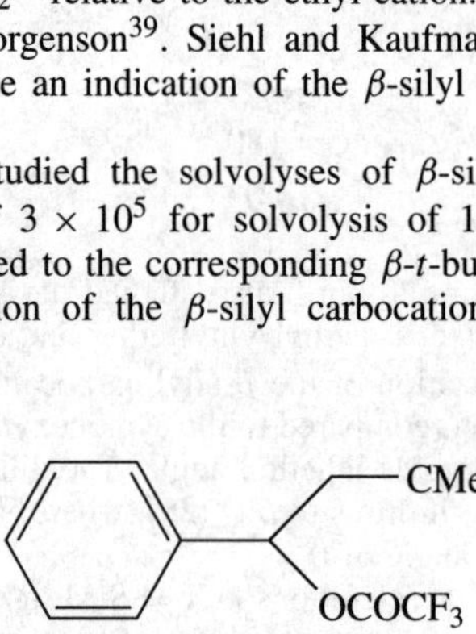 (27)

7. Activating and directive effects of silicon

The rate of hydrolysis of the disilanyl benzyl bromide **28** is about 1×10^5 larger than that of **29**[53]. Thus the β-effect of **28** is similar in magnitude to **26**.

PhCH(Br)–SiMe$_2$–SiMe$_3$ (**28**)

PhCH(Br)–SiMe$_2$–CMe$_3$ (**29**)

The effect of a methoxy substituent at the *para* position in **30** was found to be different from that in **31**[54].

(30) MeO–C$_6$H$_4$–CH(Br)–CH$_2$–SiMe$_3$
k(Y=MeO/Y=H) = 269

(31) MeO–C$_6$H$_4$–CH(SiMe$_3$)–CH$_2$–Br
k(Y=MeO/Y=H) = 2.1

This was taken to indicate that solvolysis of the two substrates does not give the same cyclic siliconium ion intermediate **32**, but that one (if not both) has an open β-carbocation form.

However, the studies of Fujiyama and Munechika[55] on the solvolyses of 2-(aryldimethylsilylethyl) chlorides **33** revealed aryl substituent effects which were consistent with the formation of a cyclic siliconium ion **34**.

(32) MeO–C$_6$H$_4$–CH–CH$_2$ (cyclic with SiMe$_3^+$)

(33) X–C$_6$H$_4$–Si(Me)$_2$–CH$_2$CH$_2$Cl

(34) X–C$_6$H$_4$–Si(Me)$_2^+$(CH$_2$CH$_2$) (cyclic)

Shimizu and coworkers also compared the rates of solvolysis of β-silyl benzyl trifluoroacetates **35**, and chlorides **36**, with varying substituents at silicon[56]. The relative rates of solvolysis observed are given in Table 5 and 6.

(35) PhCH(OCOCF$_3$)–CH$_2$–R

(36) PhCH(Cl)–CH(SiMe$_3$)–R

TABLE 5. Relative rates of solvolysis for compounds having structure **35**

R	k_{rel}
Me$_3$Si	1.0
Me$_3$SiMe$_2$Si	5.57
PhMe$_2$Si	0.31

TABLE 6. Relative rates of solvolysis for compounds having structure **36**

R	k_{rel}
Me$_3$Si	1.0
Me$_3$SiMe$_2$Si	7.65
(*i*-PrO)Me$_2$Si	0.50
(CH$_3$OCH$_2$)Me$_2$Si	0.29

The γ-substituent data in Tables 5 and 6 show that phenyl and alkoxy substituents at silicon are less effective than methyl at stabilizing the partial positive charge build-up on β-silicon by hyperconjugation. However, the γ-trimethylsilyl group does stabilize this build-up of positive charge.

The effect of the substituents at silicon to stabilize a β carbocation has also been investigated by Brook and Neuy[57], who studied the degree of *syn* addition to (*E*)-β-silylstyrenes. Addition of bromine to vinyltrimethylsilanes normally proceeds in an *anti* sense as shown in Scheme 1, the *trans* vinylsilane **37** giving the *cis* product **38**.

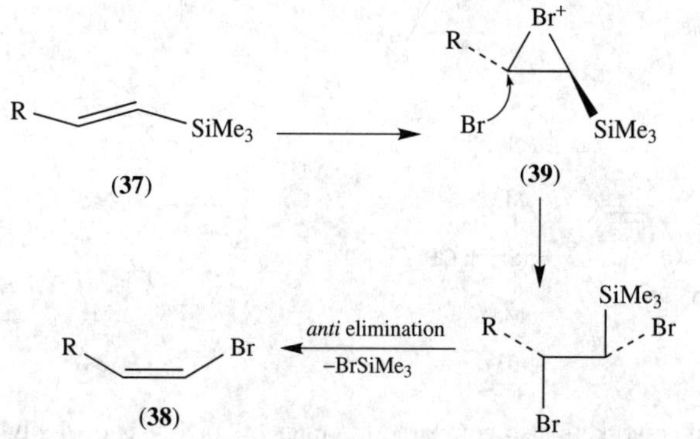

SCHEME 1

However, when R is phenyl, the bromonium ion **39** is found to open to give a β-silyl carbocation **40**. Overall *syn* addition can then occur, followed by *anti* elimination to give the *trans* alkene **41**, as seen in Scheme 2.

Table 7 shows *syn/anti* ratios obtained for a variety of styrylsilanes, together with group electronegativities.

It was concluded that for halogen and methyl substituted silyl groups, the magnitude of the β-effect is directly related to the electron-withdrawing ability of the ligands.

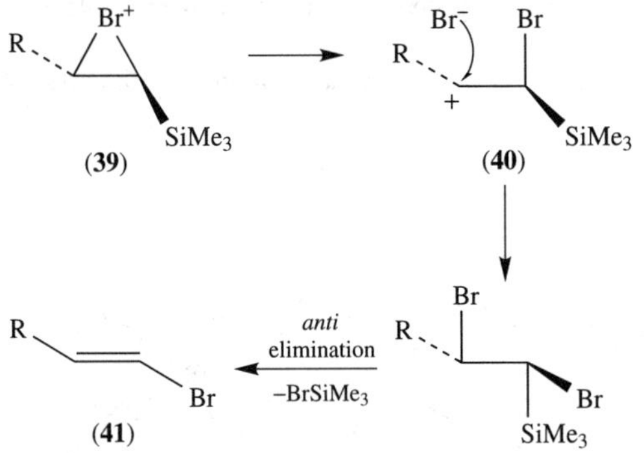

7. Activating and directive effects of silicon

SCHEME 2

TABLE 7. Syn/anti ratio in the bromination of styrylsilanes

Ligands on Si	syn/anti	Group electronegativity
Me$_3$	100/0	2.06
Me$_2$Cl	100/0	2.12
Me$_2$F	85/15	2.18
MeCl$_2$	75/25	2.19
Cl$_3$	55/45	2.26
MeF$_2$	40/60	2.32
F$_3$	15/85	2.46

In addition to the stabilization of β-carbocations, the β-effect of silicon can also be observed in the ground states of neutral molecules. Lambert and Singer[58] studied compounds of the type **42** where hyperconjugation should be enhanced by increasing the electron-accepting properties of the substituent X (MeO < Me < H < CN). σ–π overlap in this system gives the resonance structure **43** shown in equation 19.

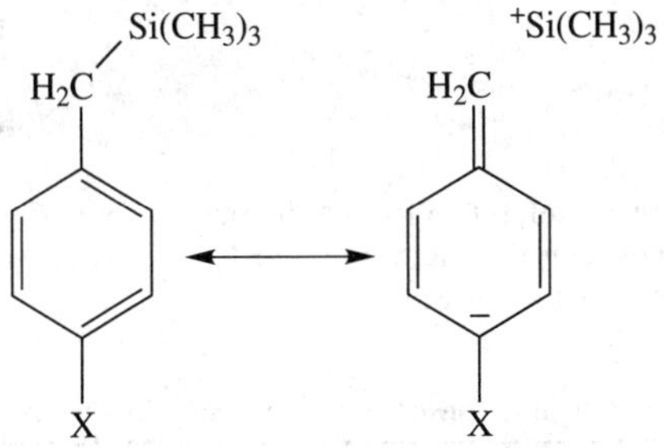

(19)

Hyperconjugation should raise the bond order between the *ipso* and benzylic carbons, and lower the bond order between the benzylic carbon and the silicon atom. As the *ipso*-benzylic bond length decreases, the ^{13}C–^{13}C coupling constant should increase, and as the benzylic carbon–silicon bond length increases, the ^{13}C–^{29}Si coupling constant should decrease. The observed coupling constants are shown in Table 8, and demonstrate the expected variations.

Kirby and coworkers[59,60] obtained crystal structures of the esters **44** and **45**.

(44) **(45)**

In 2-trimethylsilylethyl 4-phenylbenzoate **44**, the C—Si and C—O ester bonds are in an antiperiplanar conformation, and the C—O ester bond is significantly longer than expected for an alkyl ester bond at a primary centre. This is consistent with a $\sigma_{C-Si}-\sigma^*_{C-O}$ interaction in the ground state. In *trans*-2-(dimethylphenylsilyl)cyclohexyl-3,5-dinitrobenzoate **45**, the dihedral angle between the C—Si and C—O ester bonds is found to be just over 60°, and there appears to be little lengthening of the C—O bond.

White and coworkers have made further studies in this area[61–64]. The crystal structures of seven β-trimethylsilyl-substituted cyclohexylnitrobenzyl esters were obtained, together with two silicon-free model compounds[62]. In the molecules where Si—C and C—O bonds were antiperiplanar, the C—O bond lengths were found to be increased by an average of 0.014 Å. For the molecules where the Si—C and C—O bonds were *gauche*, no such systematic lengthening of the C—O bonds is observed.

NMR analysis of the conformational change in the ester **46** (equation 20) revealed the equilibrium data given in Table 9[63].

(46a) K_{eq} **(46b)** (20)

All the substrates show a preference for the diaxial conformation, **(46b)**. It appears that there is a relationship between the increasing electronegativity of the ester function and the increasing preference for the diaxial conformation.

TABLE 8. Coupling constants in trimethylsilylbenzyl systems **42**

1X	^1J ($^{13}C_{ipso}-^{13}CH_2$) (Hz)	^1J ($^{13}CH_2-^{29}Si$) (Hz)
MeO	41.5	46.9
Me	41.1	46.5
H	40.9	46.1
CN	40.4	43.8

7. Activating and directive effects of silicon

TABLE 9. Ratio of diequatorial (**46a**) and diaxial (**46b**) conformers of 2-trimethysilyl cyclohexyl esters

R	Diaxial (%)	Diequatorial (%)	log K_{eq}
Me	58.0	42.0	0.140
ClCH$_2$	70.0	30.0	0.367
Cl$_2$CH	80.8	19.2	0.624
Cl$_3$C	87.0	13.0	0.825
Ph	68.5	31.5	0.337
3-MeOC$_6$H$_4$	70.0	30.0	0.367
4-MeOC$_6$H$_4$	67.0	33.0	0.307
3-O$_2$NC$_6$H$_4$	70.0	30.0	0.367
4-O$_2$NC$_6$H$_4$	72.9	27.1	0.429
2,4-(O$_2$N)$_2$C$_6$H$_3$	81.8	18.2	0.647
3,4-(O$_2$N)$_2$C$_6$H$_3$	74.5	25.5	0.466
3,5-(O$_2$N)$_2$C$_6$H$_3$	73.0	27.0	0.431

The results demonstrate the presence of significant $\sigma-\sigma^*$ interactions between the σ_{C-Si} orbital and the σ^*_{C-O} orbital, which stabilizes the diaxial conformation, in a similar fashion to the anomeric effect between an oxygen p-type lone pair and a σ^*_{C-X} orbital. The effect of increasing the electronegativity of the ester function is to decrease the energy of the σ^*_{C-O} orbital, resulting in a closer energy match with the σ_{C-Si} orbital, and therefore a greater interaction.

Analysis of the crystal structure of the *p*-nitrobenzoate and 2,4-dinitrobenzoate derivatives of *endo*-3-(trimethylsilyl)-*endo*-2-norborneol **47**, in which the Si—C and C—O bonds are in the synperiplanar geometry, showed that there was no significant lengthening of the C—O bond in these cases[64]. By comparison, the cyclohexyl ester **48**, in which the Si—C and C—O bonds are in the antiperiplanar conformation, did show a significant lengthening of the C—O bond compared to model compounds[64].

(**47**)   (**48**)

The small ground state effect of a β-silicon in the synperiplanar geometry compared to an antiperiplanar geometry is presumably the result of much poorer overlap between the σ_{C-Si} orbital and the neighbouring σ^*_{C-O} orbital in the synperiplanar geometry (**49**) compared to the antiperiplanar geometry (**50**).

E. The Effect of R$_3$SiCH$_2$CH$_2$ on Adjacent Carbocations

Sommer, Whitmore and their coworkers were the first to recognize, in 1946, that a γ silicon atom could interact with a positive charge. They found that ClCH$_2$CH$_2$CH$_2$SiCl$_3$

underwent hydrolysis faster than the corresponding α isomer, though slower than the β isomer[35].

The magnitude of the γ-effect was reported by Shiner and coworkers, who studied the solvolyses of the three cyclohexyl brosylates **51–53**[65].

The *cis* isomer **52** was found to react about 450 times faster than its unsilylated analogue **51** in 97% trifluoroethanol. However, the *trans* isomer **53** showed essentially no acceleration. The secondary deuterium isotope effect observed with the β tetradeuterated analogue of the *cis* isomer **52** confirmed that the molecule reacts via the diequatorial conformation. In this conformation the back lobes of the Si—C bond at the 3-position can interact with the developing p orbital at the 1-position. This through-space interaction is often referred to as homohyperconjugation.

This interaction can also be modelled by the simple hyperconjugation approach, as shown in equation 21.

$$\tag{21}$$

Solvolysis of the optically active $Me_3SiCH_2CH_2CH(CH_3)OBs$ had a rate acceleration of 130 relative to the carbon analogue, and gave a racemic substitution product, which indicates that the cationic intermediate is attacked equally from both sides[66].

Grob and Sawlewicz examined the effect of γ-trimethylsilyl groups using the adamantyl systems **54**[67].

Compared with hydrogen (**54**; R = R′ = H), one trimethylsilyl group (**54**; R = H, R′ = $SiMe_3$) accelerated the ethanolysis by a factor of 8.6 and two trimethylsilyl groups gave a further acceleration by a factor of 3.8, giving a total acceleration of 33, and forming the doubly γ-silyl stabilized ion **55**.

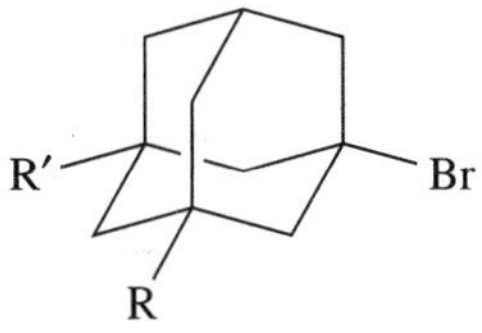

7. Activating and directive effects of silicon

(54) 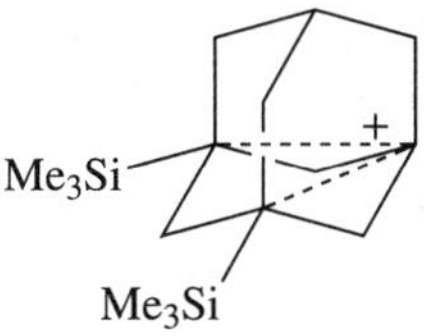 (55)

Further studies by Shiner and coworkers on the diastereomeric 4-(trimethylsilyl)-3-methyl-2-butyl brosylates **56** showed that solvolysis of one diastereomer gave mainly substitution products with a retained configuration[68]. A minor product is a cyclopropane resulting from nucleophilic attack at silicon in the ionic intermediate. From the stereochemistry of the cyclopropanes obtained it was deduced that the 'W' conformation is favoured over the '*endo*-sickle', as shown in Figures 3a and 3b.

Kirmse and Sollenbohmer[69] studied the trifluoroethanolysis of the norbornyl *p*-nitrobenzoate **57**, and observed the formation of the intermediate **58**, followed by a 6,2 migration of silicon and Wagner–Meerwein shifts to give a mixture of products. The solvolysis of the norbornyl system **57** is accelerated by a factor of 3×10^4 compared to unsubstituted 2-norbornyl *p*-nitrobenzoate.

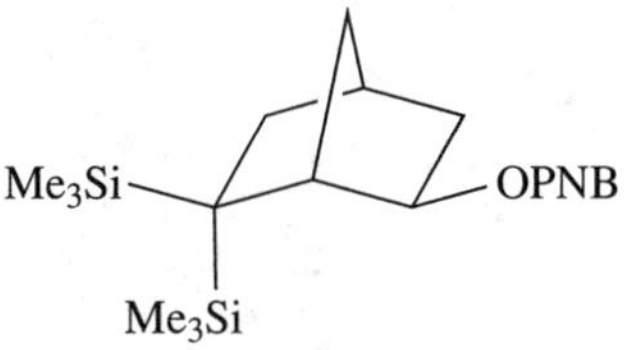

(57) 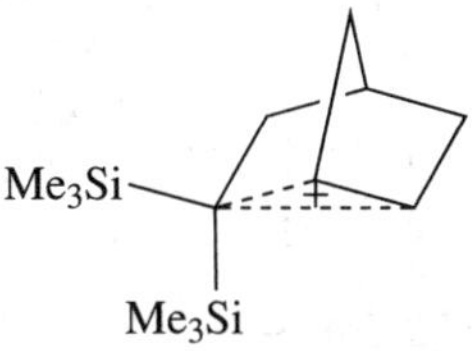 (58)

Ground state γ-effects of silicon may be responsible for the elongated C(alkyl)-O(ester) bond in *cis*-3-trimethylsilylcyclohexyl *p*-nitrobenzoate **59** relative to the silicon-free derivative[61]. It is suggested that the ground state γ-effect could be due either to homohyperconjugation, **60**, or to inductively enhanced C–C hyperconjugation where the trimethylsilyl substituent increases the importance of the resonance form **61** relative to the silicon-free derivative.

(59)

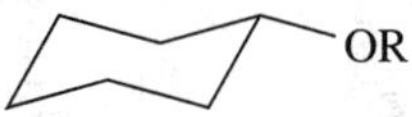

(60)  (61)

OR = $OCOC_6H_4NO_2$-*p*

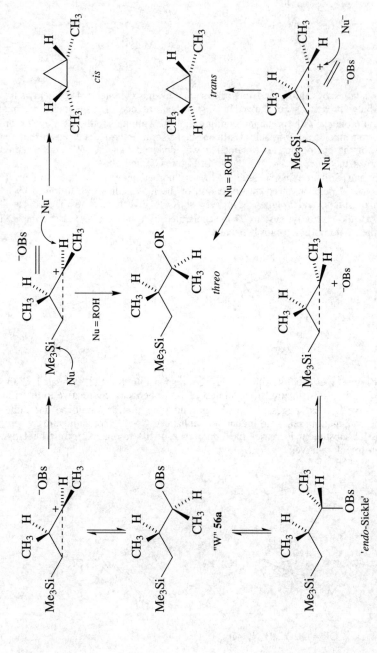

FIGURE 3a. Mechanism for the solvolysis of *threo*-4-(trimethylsilyl)-3-methyl-2-butyl brosylate, **56a**

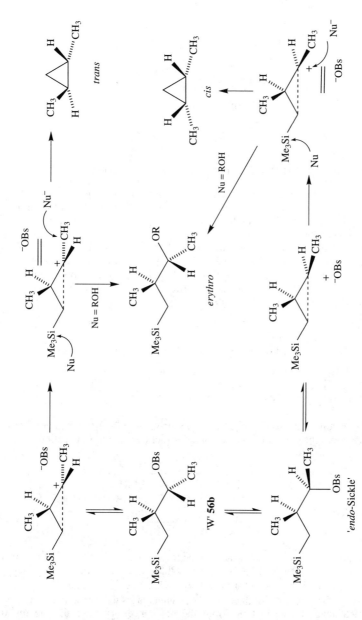

FIGURE 3b. Mechanism for the solvolysis of *erythro*-4-(trimethylsilyl)-3-methyl-2-butyl brosylate, **56b**

Adcock and Kok have analysed the ^{19}F chemical shifts of a range of 3-substituted 1-fluoroadamantane derivatives **62**. The substituent chemical shift (^{19}F SCS) can be factorized into polar field ($\rho_F \sigma_F$) and residual contributions (^{19}F SCS - $\rho_F \sigma_F$). A significant solvent-independent residual contribution is observed when R = SiMe$_3$, and is attributed to homohyperconjugation[70].

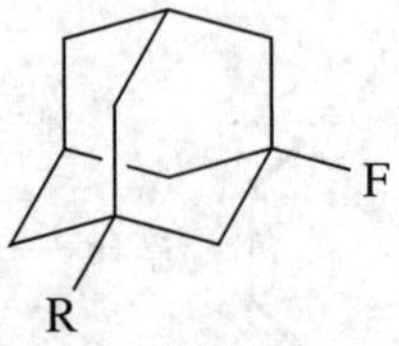

(62)

F. The Effect of R$_3$SiCH$_2$CH$_2$CH$_2$ on Adjacent Carbocations

Until recently it was thought that silicon substituents beyond the γ-position have no effect on reactivity[71]. Fessenden and coworkers[72] found that the rates of ethanolysis of *cis*- and *trans*-4-(trimethylsilyl)cyclohexyl tosylates were almost identical to those of the *cis*- and *trans*-4-*tert*-butylcyclohexyl tosylates.

However, Adcock, Shiner and coworkers found increased rates of solvolysis for 4-metalloidal-substituted bicyclo[2.2.2]oct-1-yl *p*-nitrobenzenesulphonates and methanesulphonates, in the order tin>germanium>silicon[73], as also observed for the β-effect. The relative rate of solvolysis of the 4-trimethylsilyl mesylate **63** compared to the non-substituted parent compound was 49 : 1.

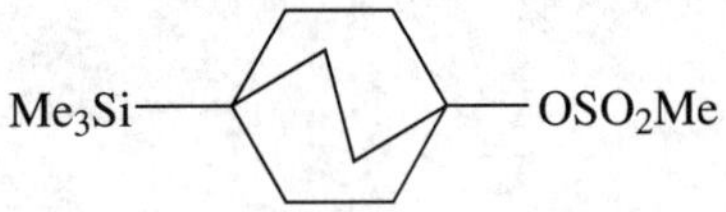

(63)

The stabilization of a positive charge by δ-silicon is believed to be through double hyperconjugation, represented by the resonance structures shown in equation 22.

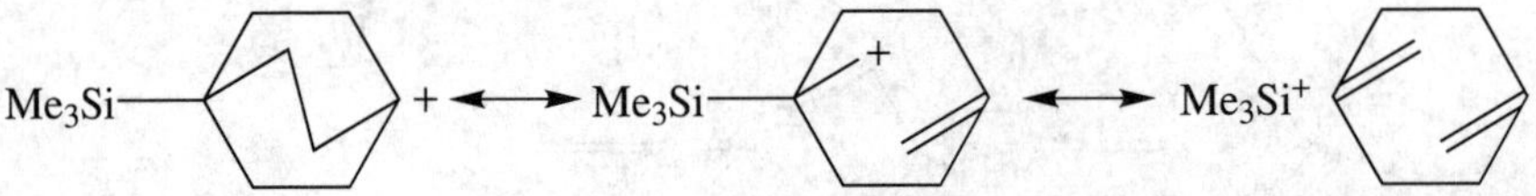

(22)

This mechanism also accounts for the very large δ-deuterium isotope effect observed for the solvolysis of (4-D)bicyclo[2.2.2]oct-1-yl mesylate. This mechanism is also supported by theoretical calculations of 4-substituted bicyclo[2.2.2]oct-1-yl cations[74].

^{19}F NMR studies on 5-substituted 2-fluoroadamantanes **64** showed residual contributions which suggest contributions from the resonance structures **65** and **66** to the ground-state structure (equation 23)[75].

7. Activating and directive effects of silicon

(23)

(64) (65) (66)

Similar conclusions are obtained from the ^{19}F NMR of 4-substituted bicyclo[2.2.2]oct-1-yl fluorides[76].

An X-ray crystal structure analysis of *trans*-4-trimethylsilylcyclohexyl *p*-nitrobenzene-sulphonate **67** reveals an elongated C(alkyl)-O(ester) bond, compared to the non-silicon substituted derivative[61].

(67)

G. The Effect of R$_3$Si on α-Silylcarbanions

It is a general observation in organosilicon chemistry that R$_3$Si tends to stabilize an adjacent negative charge. In the series (Me$_3$Si)$_n$H$_{3-n}$CH the ease of metallation increases as n increases from 0 to 3[77].

The stabilization of α-carbanions by silicon has been the subject of several theoretical studies. Using *ab initio* calculations, Hopkinson and Lien[78] calculated the proton affinities of C$_2$H$_5^-$ and H$_3$SiCH$_2^-$. They found a difference of 31.5 kcal mol^{-1}, which represents the stabilization of the carbanion by the SiH$_3$ group relative to methyl. The α-methyl group was found to be weakly destabilizing, by 2.2 kcal mol^{-1}, relative to hydrogen, such that for the isodesmic proton-transfer reaction in equation 24, the proton affinity relative to methane is 29.3 kcal mol^{-1}.

$$CH_4 + H_3SiCH_2^- \longrightarrow CH_3^- + H_3SiCH_3 \qquad (24)$$

Glidewell and Thomson[79] calculated the proton affinities for a variety of silyl substituted carbanions. In the series H$_3$SiCH$_2^-$, (H$_3$Si)$_2$CH$^-$ and (H$_3$Si)$_3$C$^-$, stability increased by about 20 kcal mol^{-1} per silyl substituent.

The proton affinity of the trimethylsilylmethyl anion was also calculated from measurements of the electron affinity of the trimethylsilylmethyl radical[80]. This gave a proton affinity of 25.7 kcal mol^{-1} for Me$_3$SiCH$_2^-$, relative to methane. This compares well with the calculated value for H$_3$SiCH$_2^-$ given earlier.

Brinkman and coworkers[81] obtained a value of around 30 kcal mol^{-1} from a similar gas-phase study for the same proton affinity relative to methane. Calculations at the 6-311+G(d,p) level gave a value of 25 kcal mol^{-1} (for H$_3$SiCH$_2^-$). The proton affinities of (H$_3$Si)$_2$CH$^-$ and (H$_3$Si)$_3$C$^-$ relative to methane were calculated to be 47 and 66 kcal mol^{-1}, respectively. The experimental proton affinity of (Me$_3$Si)$_2$CH$^-$ was found to be approximately 43 kcal mol^{-1}.

The computed geometries of the anionic and neutral species reveal a decrease in carbon–silicon bond length by approximately 0.1 Å on deprotonation. This is consistent with the important role played by hyperconjugation in stabilizing the carbanion. In addition, Mulliken population analysis indicates that in the neutral molecule, the central atom has close to a full negative charge. Upon deprotonation, most of the charge is picked up throughout the rest of the molecule. Previous studies have shown that the inclusion of silicon d-functions in calculations of the energies of α-silyl carbanions has little effect on the results[78–80], indicating that the d orbitals on silicon are not involved in the stabilization of anions by α-silicon. This suggests that the stabilization arises solely through hyperconjugation, as shown in equation 25.

$$\overset{-}{H_2C}-\underset{R}{\overset{}{SiR_2}} \longleftrightarrow H_2C=\underset{R^-}{\overset{}{SiR_2}} \qquad (25)$$

Sieburth and Somers[82] studied the product ratio of metallation of *t*-butyl-*N*-phenylmethyl-*N*-trimethylsilylmethyl carbamate **68** (equation 26). This ratio reflects the relative abilities of an α-phenyl and α-trimethylsilyl group to stabilize an adjacent carbanion. It was found that metallation α to silicon (**70**) was slightly favoured kinetically, but at equilibrium the α-phenyl anion **69** was strongly favoured.

(26)

(**68**)   (**69**)   (**70**)

Bordwell and coworkers[83] investigated the deprotonation of 9-trimethylsilylfluorene and trimethylsilylmethyl phenyl sulphones. The presence of an α-trimethylsilyl group increased the acidity by 2 and 4.2 kcal mol^{-1}, respectively. Introduction of a Ph$_3$Si group increased the acidity by 5.9 and 10.5 kcal mol^{-1}, respectively.

H. The Effect of R$_3$SiCH$_2$ on Adjacent Carbanions

Very little experimental or theoretical work has been reported on the stability of R$_3$SiCH$_2$CH$_2^-$. All available evidence points to the formation of R$_3$SiCH$_2^-$ instead of R$_3$SiCH$_2$CH$_2^-$, where there is a choice.

IV. ACTIVATING AND DIRECTIVE EFFECTS OF SILICON IN ELECTROPHILIC AROMATIC SUBSTITUTION

A. Reactions of Compounds with SiR$_3$ Directly Bonded to the Aromatic Ring

Electrophilic aromatic substitution normally proceeds via a positively charged intermediate **71** (known as a Wheland intermediate or σ-complex) (equation 27)[84].

7. Activating and directive effects of silicon

$$\text{ArY} \xrightleftharpoons{E^+} \mathbf{(71)} \xrightleftharpoons{-Y^+} \text{ArE} \quad (27)$$

The transition states in both steps of the reaction are not likely to be far removed in energy or structure from the intermediate, which may be used as a model to rationalize variations in the rates and products of such reactions. If silicon is in a position such that it is β to the positive charge in one of the resonance forms, this might be expected to lower the energy and increase the rate, provided the carbon–silicon bond can overlap with the vacant π-orbital.

Substitution of a trimethylsilyl group directly onto an aromatic ring does lead to a rate increase relative to hydrogen, however the reaction usually occurs via *ipso* substitution (equation 28).

$$\text{ArSiMe}_3 \xrightleftharpoons{E^+} \text{intermediate} \xrightleftharpoons{} \text{ArE} \quad (28)$$

Cleavage of the C–Si bond is in the direction $C^-SiR_3^+$, in the same sense as aryl–H bonds are broken C^-H^+. Activation to electrophilic attack arises from β-stabilization of the carbocation in one of the resonance forms **(72)** of the intermediate[85,86].

(72)

For each *ipso* substitution there is a competing electrophilic substitution of hydrogen (equation 29).

$$\quad (29)$$

For most electrophiles substitution of the silyl group is faster than replacement of hydrogen.

As would be expected, substituents on the silicon have a profound effect on the reactivity of the arylsilicon compound. Increased electron supply from R in R_3Si increases the rate of reaction. Aryl–SiX_3 bonds are cleaved much less readily than aryl–$SiMe_3$ bonds when X is a more electronegative element than carbon. When X is halogen, desilylation is deactivated to such an extent that ring substitution occurs without loss of silicon. In the presence of other substituents the degree of Si–C cleavage compared to C–H bond cleavage depends upon the relative ability of the silyl group and the other directing group to stabilize the intermediate. The hydroxy, amino, methoxy and dimethylamino groups usually have a more powerful directing effect than the $-SiMe_3$ group (equation 30)[87,88], whereas the directing effect of the methyl group is often less (equation 31)[89].

The ability of the silyl group to direct substitution in competition with other directing groups is also found to depend on the nature of the electrophile (equation 32).

7. Activating and directive effects of silicon

As expected, electronegative substituents on silicon decrease the preference for *ipso* substitution and silicon substituents in normally activated sites are not substituted, as shown in equation 33[92].

$$\text{MeO-C}_6\text{H}_4\text{-SiCl}_3 \xrightarrow{\text{Br}_2} \text{3-Br-4-MeO-C}_6\text{H}_3\text{-SiCl}_3 \tag{33}$$

Silicon-substituted arenes have particular usefulness in organic synthesis. *Ipso* substitution always gives a single isomer, and desilylation can be carried out on deactivated systems, or systems that would undergo side-reactions under normal conditions (equations 34–36). *Ipso* substitution also provides a useful route to labelled compounds (equation 37)[95–97].

$$\text{2-SiMe}_3\text{-C}_6\text{H}_4\text{-COOH} \xrightarrow{\text{Br}_2} \text{2-Br-C}_6\text{H}_4\text{-COOH} \quad \text{(Ref. 93)} \tag{34}$$

$$\text{(4-MeO, 3-OMe, 2-SiMe}_3\text{)-C}_6\text{H}_2\text{-CONEt}_2 \xrightarrow{\text{Br}_2} \text{(4-MeO, 3-OMe, 2-Br)-C}_6\text{H}_2\text{-CONEt}_2 \quad \text{(Ref. 94)} \tag{35}$$

$$\text{2-Cl-C}_6\text{H}_4\text{-OH} \xrightarrow[\text{(ii) Na/Me}_3\text{SiCl}]{\text{(i) Me}_3\text{SiCl/Et}_3\text{N}} \text{2-SiMe}_3\text{-C}_6\text{H}_4\text{-OSiMe}_3 \xrightarrow[\text{(ii) H}_2\text{O}]{\text{(i) ClSO}_3\text{SiMe}_3} \text{2-SO}_3\text{H-C}_6\text{H}_4\text{-OH} \tag{36}$$

(Ref. 95)

$$\text{4-SiMe}_3\text{-C}_6\text{H}_4\text{-OMe} \xrightarrow[\text{CH}_3\text{CO}_2{}^{18}\text{F/AcOH, }-25\,°\text{C}]{{}^{18}\text{F}_2/\text{CFCl}_3,\ -78\,°\text{C}} \text{4-}{}^{18}\text{F-C}_6\text{H}_4\text{-OMe} \quad \text{(Ref. 96)} \tag{37}$$

Felix, Dunogues and Calas[98-101] have extended this strategy such that a range of disubstituted benzene derivatives **74** can be regiospecifically synthesized starting from the readily accessible *o*-, *m*- or *p*-bis(trimethylsilyl)benzenes **73** (equation 38).

$$(38)$$

Since trimethylsilylarenes can be prepared by metallation of the arene followed by treatment with chlorotrimethylsilane, this provides an alternative route into a range of difficult substitution patterns. For example, the *ortho/para* directing effects of the methoxy groups in 1,3-dimethoxybenzene **75** direct the electrophile to the 4-position. However, lithiation of 1,3-dimethoxybenzene takes place at the 2-position. Reaction with chlorotrimethylsilane then gives the 2-trimethylsilyl compound **76**, which undergoes *ipso* substitution with the electrophile to give the 1,2,3-trisubstituted product **77** (equation 39)[101,102].

$$(39)$$

Mills, Taylor and Snieckus[103] describe a methodology based on the preferred *o*-metallation of benzamides (equation 40). In this case the trimethylsilyl group is used to block one of the *ortho* positions, directing the electrophile to the other.

In the absence of desilylation the —SiMe$_3$ group is very slightly activating with no discernable directing effect on electrophilic aromatic substitution. Although electrophilic attack at the *meta* position generates a β-carbocation **78**, the carbon–silicon bond and the vacant p orbital are orthogonal, which precludes any stabilization by hyperconjugation.

A recent study by Ishibashi and coworkers[104] found that the reaction of trimethylphenylsilane with methyl chloro(methylthio)acetate in the presence of tin(IV) chloride gave no *ipso* substitution (equation 41). This was attributed to steric factors, since when the primary chloride ClCH$_2$SCH$_2$CO$_2$Et was used, a 20% yield of *ipso*-substituted product was obtained.

TABLE 10. Partial rate factors for the reaction of arenes with methyl chloro (methylthio)acetate

Ph-X	$k_{Ar}/k_{benzene}$	% ortho	% meta	% para	f_{ortho}	f_{meta}	f_{para}
X = Me$_3$Si	3.81	4.6	65.8	30.0	0.53	7.53	6.87
X = Me	152.4	11.7	0.3	88.0	52.4	1.5	804

The rate of electrophilic addition relative to the rate of reaction of benzene and of toluene was measured in this study, and the partial rate factors f calculated, as shown in Table 10. These data demonstrate the weak activating effect of Me$_3$Si compared to methyl.

Substitution of R in R$_3$Si by more electronegative groups both decreases the extent of desilylation and increases the proportion of *meta* substitution. The overall rate of electrophilic substitution is also decreased as the R$_3$Si group becomes more electron-withdrawing.

B. Electrophilic Aromatic Substitution in ArCH$_2$SiR$_3$ Compounds

The p-CH$_2$SiMe$_3$ group has a large activating effect in protiodetritiation and protiodesilylation, the —CH$_2$SiMe$_3$ substituted compound reacting up to 180 times faster than the corresponding methyl compound[105,106]. Presumably this is due to β-stabilization of the resonance form **79**.

(79)

The —CH$_2$SiMe$_3$ group is *ortho/para* directing, as mentioned earlier.

V. ACTIVATING AND DIRECTIVE EFFECTS IN ALIPHATIC ELECTROPHILIC REACTIONS

A. Reactions of Vinylsilanes

1. Introduction

Electrophilic addition to carbon–carbon double bonds normally occurs through cationic intermediates. As discussed previously, β-silylcarbocations are strongly stabilized, whereas α-silylcarbocations are not stabilized. Thus, if the R$_3$Si group is arranged so that it can be β to the carbocation, then hyperconjugative stabilization may be possible, and this will affect the regio- and stereochemical outcome of the reaction.

Vinylsilanes react readily with a variety of electrophiles either by addition to the double bond or substitution of the R$_3$Si group. The addition product **80** often undergoes subsequent elimination to give the substitution product **81**, as shown in equation 42.

The first step in substitution or addition is normally addition of the electrophile to the carbon bearing the silicon to give a β-silylcarbocation (equation 43).

7. Activating and directive effects of silicon

$$H_2C=CHSiMe_3 \xrightarrow{H^+} H_2\overset{+}{C}-CH_2SiMe_3 \quad (43)$$

However, the directing influence of silicon can be overcome if the vinylsilane contains another substituent that can stabilize a carbocation more strongly than silicon. For example, when the silyl group is attached to C-2 of a terminal alkene, reaction occurs to give the more substituted carbocation **82** (equation 44)[107]. Similarly, if the silicon is bound to the same carbon atom as a phenyl group, reaction occurs via the benzyl cation to give the product shown in equation 45[108].

2. Addition to vinylsilanes

As with alkenes, in general, *anti*-addition is often the course of reaction, especially when halonium ions are involved[109-112]. However, as mentioned earlier, *syn* addition can take place in the bromination of β-silylstyrenes. This stereochemistry is explained by stabilization of the open-chain carbocation by the aromatic group, compared to the cyclic bromonium ion. In this case the conformer **83** has the maximum hyperconjugative stabilization, and is formed by the least motion rotation about the carbon–carbon bond.

Attack of bromide then takes place on the less hindered side of the cation, *anti* to the β-silyl group, to give overall *syn* addition (equation 46).

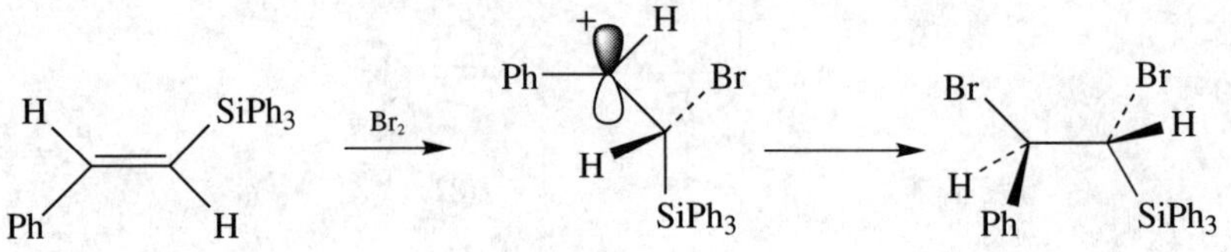

$$\text{(46)}$$

(83)

Miura and coworkers have studied the intramolecular addition of alcohols to vinylsilanes[113]. 5-Silyl-4-penten-1-ols **84** are readily transformed into 2-substituted tetrahydrofurans **(85)** in the presence of a catalytic amount of *p*-toluenesulphonic acid or titanium tetrachloride (equation 47).

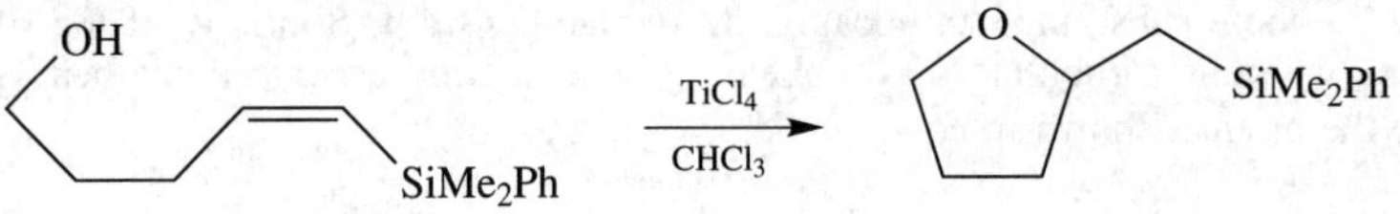

$$\text{(47)}$$

(84)   **(85)**

Such cyclization does not take place in the absence of the silyl group.

The *syn* stereochemistry of the addition is revealed when 5-deuteriated vinylsilanes **86** are employed, as shown in equation 48.

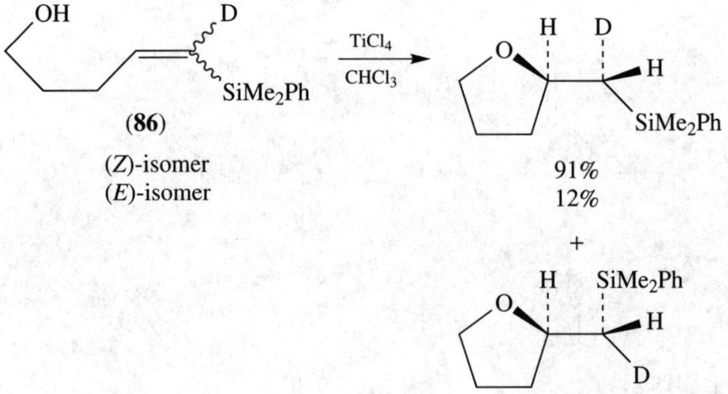

(86)
(Z)-isomer
(E)-isomer

91%
12%

+

9%
88%

$$\text{(48)}$$

Under identical conditions, 6-silyl-5-penten-1-ols **87** cyclize to give 2-substituted tetrahydropyrans **88** and **89**. Again *cis* addition is strongly favoured, as shown in equation 49.

[Structure of (87) Z-isomer and E-isomer with OH, D, SiMe₂Ph groups] → [Structure (88) 92% (Z), 16% (E)] + [Structure (89) 8% (Z), 84% (E)]   (49)

The cyclization of (Z)-1-substituted-5-silyl-4-penten-1-ols **90**, shown in equation 50, gave 2,5-disubstituted tetrahydrofurans **91** and **92** with a high *trans*-selectivity (% *trans* > 86 for R = Ph, *i*-Pr, C$_6$H$_{13}$)[114].

[Structure of (90) with R¹, OH, SiMe₂R²] —TiCl₄→ [Structure (91) trans] + [Structure (92) cis]   (50)

Addition to vinylsilanes is favoured over substitution by (i) the presence of bulky spectator ligands which hinder nucleophilic attack at silicon, and (ii) when the presence of electron-withdrawing groups on silicon lowers its leaving group ability.

Brook and coworkers have studied the effects of electronegative groups on silicon on the reactivities of vinylsilanes[115]. Unlike the substitution reaction of 1-trimethylsilylprop-1-ene with acetyl chloride (equation 51a) the major product of the reaction of β-trichlorosilylstyrene with phenylacetyl chloride arises from addition, as shown in equation 51b.

[Reaction scheme showing CH₃CH=CHSiMe₃ + MeCOCl/AlCl₃ → intermediate → product]   (51a)

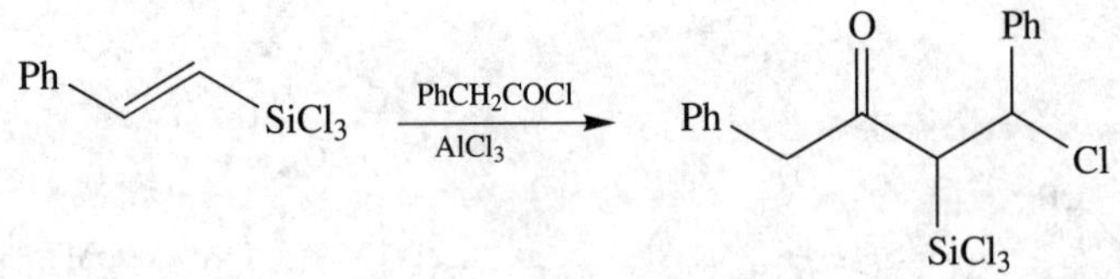

[Scheme showing reaction: Ph-CH=CH-SiCl₃ + PhCH₂COCl / AlCl₃ → Ph-C(=O)-CH₂-CH(Ph)-CHCl with SiCl₃ substituent] (51b)

The reduced leaving ability of electron-poor silicon is demonstrated by the reactions of styrene derivatives with trifluoromethanesulphonic acid, as shown in Scheme 3. When the silicon has one or zero electronegative substituents, protiodesilylation takes place. However, when two chloro substituents are present, the loss of the silyl group is disfavoured such that an intramolecular cyclization of the intermediate β-carbocation **93** takes place[116,117].

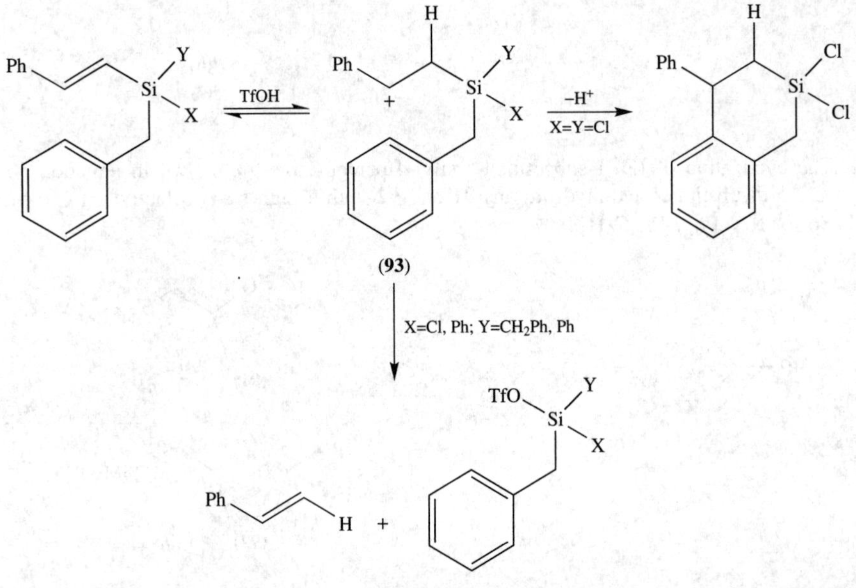

[Scheme 3: reaction sequence showing vinylsilane with TfOH giving β-carbocation (93), then −H⁺ (X=Y=Cl) giving cyclized product; alternative path X=Cl, Ph; Y=CH₂Ph, Ph giving styrene + silyl triflate]

SCHEME 3

3. Substitution in vinylsilanes

The electrophilic substitution of vinylsilanes has been reviewed in detail by Fleming, Dunogues and Smithers[118]. In most cases, substitution of vinylsilanes take place with retention of configuration. This can be rationalized as follows. If the electrophile attacks on the top face of the π-bond, as shown in Scheme 4, the silyl group will then rotate into a conformation in which there is a maximum hyperconjugative interaction between the C−Si bond and the vacant p orbital[119]. There are two ways the silyl group can rotate into the plane of the vacant p orbital — either via a 120° counterclockwise rotation, which would give inversion of configuration in the product **94**, or via rotation through 60° clockwise, which gives the observed retention product **95**. This shorter path would be expected to be the more likely, especially when the effects of hyperconjugation are taken into account. In the 60° rotation, hyperconjugation increases as the silyl group moves into the plane of the p orbital. However, in the 120° rotation, hyperconjugative stabilization will initially decrease to zero as the silyl group passes through an orientation perpendicular to the p orbital.

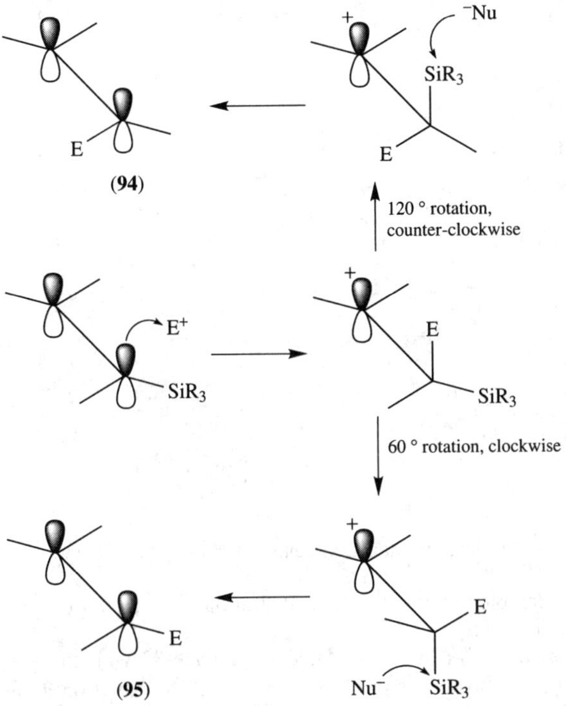

SCHEME 4

With halogen electrophiles both retention and inversion of stereochemistry have been observed. In this case the addition of the electrophile may lead to the β-silicon cation, or a cyclic halonium ion. Scheme 5 shows a generalized mechanism for the reaction of vinylsilanes with electrophilic reagents[120].

The position of the equilibrium A depends upon which provides the greater stabilization, hyperconjugation or onium ion bridging.

When the electrophile E^+ is a proton, the equilibrium A lies to the left, and the routes B (addition of nucleophile X at carbon to give overall *syn* addition, followed by *anti* elimination of Me$_3$SiX in D) and C (attack of nucleophile X at silicon) are followed. Both these routes give overall retention of configuration.

When R is phenyl, the stabilizing effect of the phenyl group on an α-cation drives the equilibrium A to the left, and substitution with retention of configuration dominates, whatever the electrophile.

For other vinylsilanes, the reaction with chlorine or bromine usually proceeds with inversion of configuration, i.e. the equilibrium A lies to the right.

The situation is more complicated with electrophiles such as iodine[121], cyanogen bromide and cyanogen chloride[122] in the presence of aluminium chloride. In these cases products with retained stereochemistry are obtained, despite the equilibrium A lying to the right. This is because the addition step F does not take place, and reaction occurs via the minor β-silylcarbocation species **96**. When E^+ is I^+, changing the nature of X^-

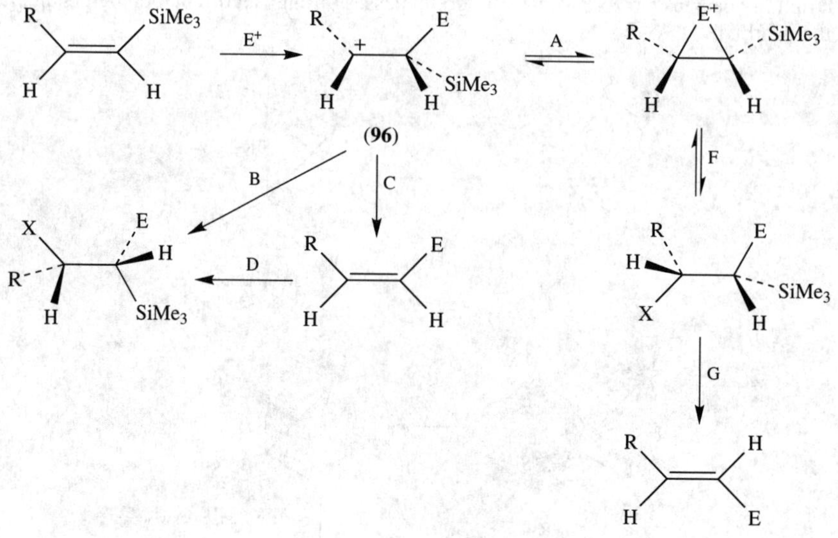

SCHEME 5

to F$^-$[123] or Cl$^-$[111] favours the addition reaction F and the expected inverted product is observed.

Some recent examples of vinylsilane substitutions demonstrating the synthetic utility of such reactions follow.

The reaction of the (E)-α,β-disubstituted vinylsilane **97** with the glyoxalate **98** gives exclusively the (E)-trisubstituted product with the 2S configuration **99** (equation 52)[124].

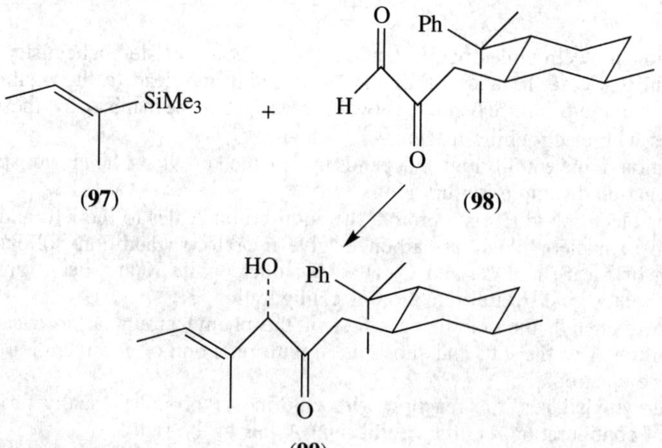

(52)

Kishi, Mikami and Nakai[125] have studied the intramolecular acylation of trimethylsilyl alkenoyl chlorides, using aluminium trichloride. They found two possible outcomes. 5-Trimethylsilylhept-5-enoyl chloride **100** cyclized in the expected fashion (α-cyclization) via the β-silyl carbocation **101**, to give the cyclopentanone **102** (equation 53).

$$(100) \longrightarrow (101) \longrightarrow (102) \quad (53)$$

However, 4-methyl-5-trimethylsilylhex-5-enoyl chloride **103** gave a cyclohexenone product **104**, as shown in equation 54.

$$(103) \longrightarrow (104) \quad (54)$$

This product must arise from β-cyclization followed by migration of the methyl group to give a β-silyl carbocation intermediate **106**, as shown in equation 55.

$$(103) \longrightarrow [(105) \longrightarrow (106) \longrightarrow] \longrightarrow (104) \quad (55)$$

These results suggest that the tertiary α-silyl carbocation **105** has a higher stability than the primary β-silyl cation **107**, formed by α-cyclization. This shows that the β-effect does not always dominate the regioselectivity of reactions of vinylsilanes.

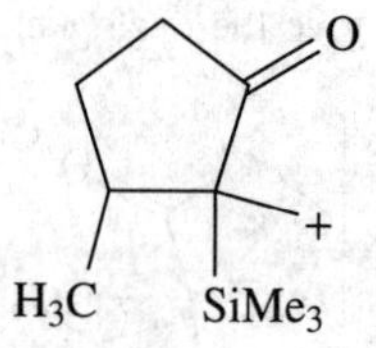

(**107**)

Another example of an important cyclization is the intramolecular addition of iminium ions to vinylsilanes, as shown in equation 56. This synthetically important cyclization has been studied in detail by Overman and coworkers[126–130].

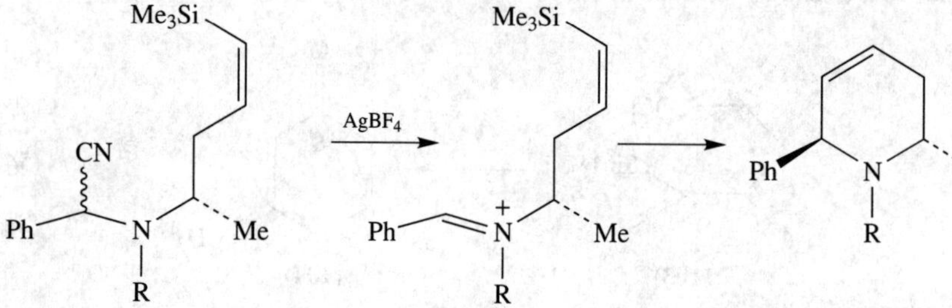

(56)

Brook and Henry[131] studied the reactivity of aryldimethylsilyl styrenes towards acetyl chloride. They found that when the aryl group was phenyl (**108**), the expected vinylsilane substitution reaction took place to give the ketone **109** (equation 57). However, when a mesityl group was used (**110**), *ipso* substitution on the aromatic ring occurred (equation 58). Although the reactivity of the aryl group would be expected to increase with increasing alkylation, it would not be expected to surpass the reactivity of the styryl group. The cleavage of the silicon–aryl bond is facilitated in this case by the relief of steric congestion.

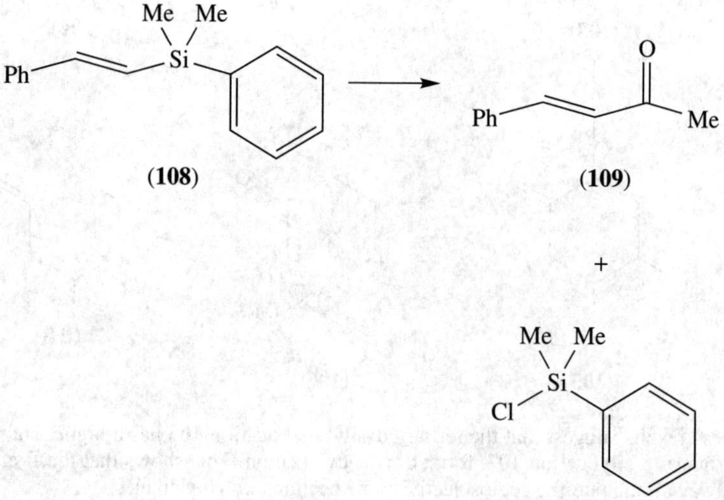

(**108**)   (**109**)

(57)

$$\text{(equation diagram 58)}$$

(58)

B. Reactions of Alkynylsilanes

Alkynylsilanes have been studied less than vinylsilanes. Electrophilic addition to an alkynylsilane can give a α- or β-silyl-substituted vinyl cation, in the first step (equation 59).

$$R-C\equiv C-SiMe_3 \xrightarrow{E^+} R-\overset{+}{C}=C\begin{smallmatrix}SiMe_3\\E\end{smallmatrix} \quad \text{or} \quad \begin{smallmatrix}R\\E\end{smallmatrix}C=\overset{+}{C}-SiMe_3 \qquad (59)$$

Provided that the silicon–carbon bond can be coplanar with the vacant p orbital, the β-silyl substituted carbocation should be stabilized by hyperconjugation, and this has been demonstrated by Kresge and coworkers[47,49].

Overall electrophilic addition to alkynes is rare, although bromination of alkynylsilanes has been described and takes place in the *trans* sense (equation 60)[132].

$$R-C\equiv C-SiMe_3 \xrightarrow{Br_2} \begin{smallmatrix}R\\Br\end{smallmatrix}C=C\begin{smallmatrix}Br\\SiMe_3\end{smallmatrix} \qquad (60)$$

Substitution reactions normally predominate, as shown in equations 61[132] and 62[133].

$$Me_3Si-C\equiv C-SiMe_3 \xrightarrow{ICl} Me_3Si-C\equiv C-I \qquad (61)$$

$$R-C\equiv C-SiMe_3 \xrightarrow[AlCl_3]{R^1COCl} R-C\equiv C-\overset{O}{\underset{\|}{C}}-R^1 \qquad (62)$$

The titanium tetrachloride catalysed addition of alkenes to the alkynylsilane **111** selectively produces cyclopentene products via a β-silylcarbocation **112** (equation 63)[134].

C. Reactions of Allylsilanes

Electrophilic substitution reactions of allylsilanes have been reviewed in detail by Fleming, Dunogues and Smithers[118].

In general, allylsilanes undergo electrophilic addition or substitution via an intermediate β-silylcarbocation (equation 64).

If a substituent is present that can stabilize the cationic intermediate more effectively than the silicon, this may alter the course of the reaction. For example, deuterionation of the allylsilane **113** leads to a mixture of the expected product **114** and an anomalous product **115** (equation 65)[135].

in ratio of 2:3

This can be explained if the initial attack on **113** forms a carbocation α to the phenyl group (**116**) (Markovnikov attack), followed by a 1,2 shift of hydride or deuteride to give the β-silyl carbocations **117** and **118** respectively, and hence the alkenes **114** and **115** (Scheme 6).

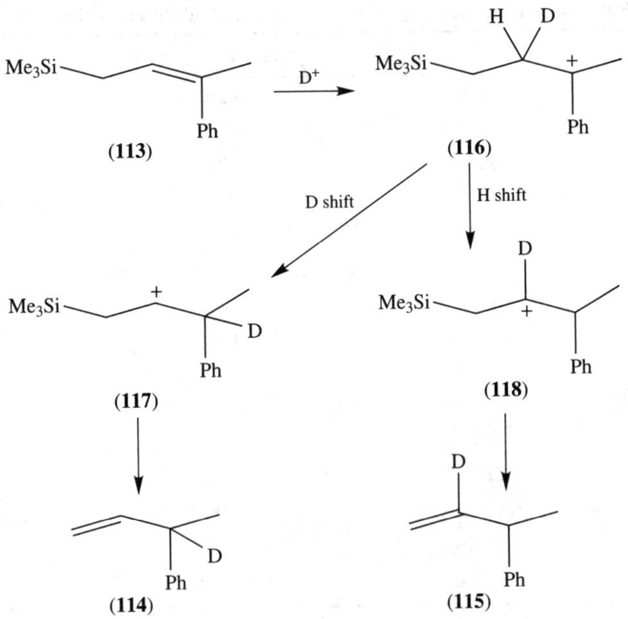

7. Activating and directive effects of silicon

SCHEME 6

As with vinylsilanes and alkynylsilanes, substitution is favoured over addition for allylsilanes. However, this can be affected by the steric and electronic effects of the silicon substituents. Mayr and Hagen have studied the reactivities of allylsilanes towards the *p*-methoxy substituted diphenylcarbocation (Scheme 7)[136,137]. Relative rate data and observed products are summarized in Table 11 for allylsilanes of the structures **119–122**, with various silicon substituents.

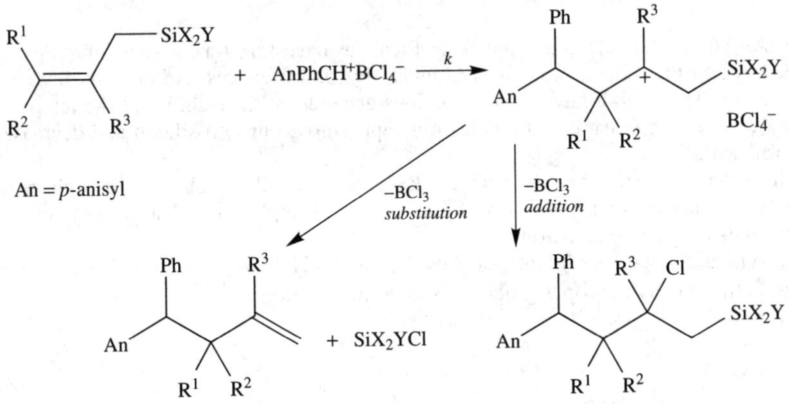

SCHEME 7

From these data it seems that steric hindrance at the silicon atom is responsible for the course of the reaction, whilst the electronegativity of the substituents affects the reaction

TABLE 11. Rate data and reaction process for the reaction of allylsilanes with AnPhCH$^+$

Allylsilane	X	k (1 mol^{-1} s^{-1})[a]	Process
119	SiCl$_3$	no reaction	—
119	SiClMe$_2$	0.276	substitution
119	SiPh$_3$	3.21	add:subst 1.5 : 1
119	SiMe$_2$Ph	38.7	substitution
119	SiMe$_3$	197	substitution
119	SiMe$_2$Bu-t	204	addition
119	SiEt$_3$	313	substitution
119	Si(Pr-i)$_3$	439	addition
119	Si(Bu-n)$_3$	507	substitution
119	Si(Hex-n)$_3$	542	substitution
120	SiCl$_3$	0.066	substitution
120	SiPh$_3$	1.91×10^4	substitution
120	SiMe$_3$	$> 10^5$	substitution
121	SiMe$_3$	4.15×10^3	substitution
122	SiMe$_3$	1.56×10^3	substitution

[a] Reference rate constant 23.8 l mol^{-1} s^{-1} for 2-methylpropene.

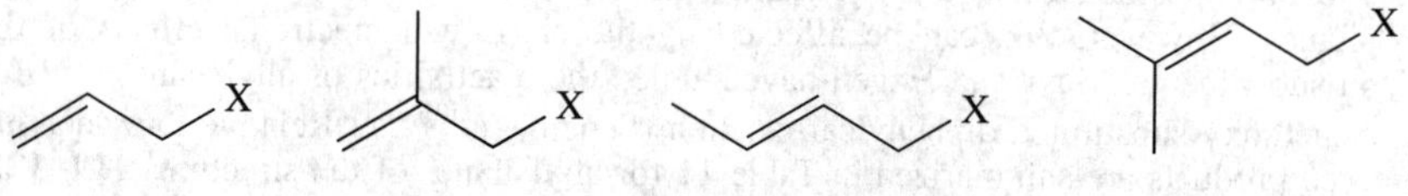

(119) (120) (121) (122)

rates. As expected, nucleophilic attack at silicon, as necessary for the substitution pathway, is slow when the silicon atom is surrounded by bulky groups. Addition is favoured in these cases. The overall rate of both reactions depends on the stability of the intermediate carbocation which is increased by electron-supplying groups on silicon and decreased by electron withdrawal.

Allylsilanes undergo reactions with a large range of electrophiles, although catalysis by Lewis acids is often necessary. Some recent examples of substitution reactions of allylsilanes are discussed below.

Hosomi and coworkers found that β-cyclopropyl allylsilanes undergo substitution without affecting the cyclopropyl group, as shown in equation 66[138].

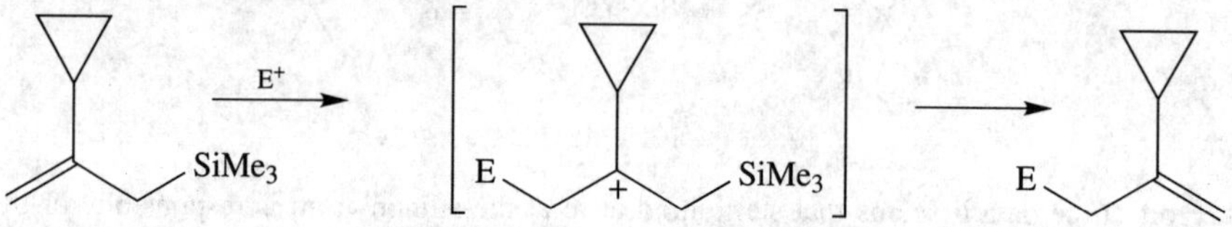

(66)

7. Activating and directive effects of silicon

Generally, systems in which a carbocation is formed next to a cyclopropyl group are avoided, as this leads to isomerization of the cyclopropyl group (equation 67).

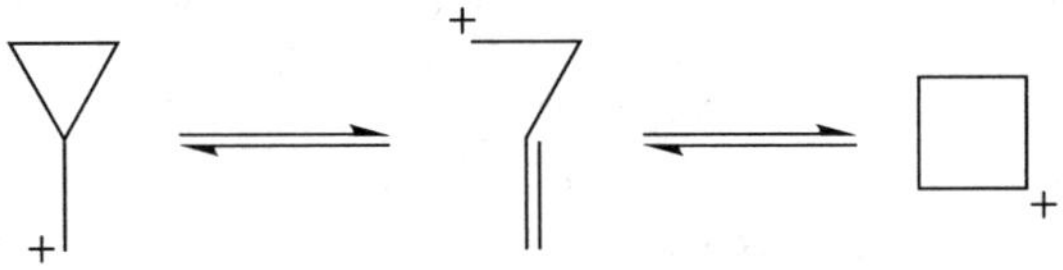

$$\text{(67)}$$

The increased stabilization by the β-silicon of the carbocation α to the cyclopropyl ring presumably disfavours the isomerizations in which the charge is located on other carbon atoms in the molecule.

Aluminium chloride is often used as a Lewis acid catalyst (equations 68 and 69), although there are many other suitable catalysts (equations 70 and 71). Strong electrophiles such as chlorosulphonyl isocyanate or an aluminium salt do not require catalysis (equations 72 and 73).

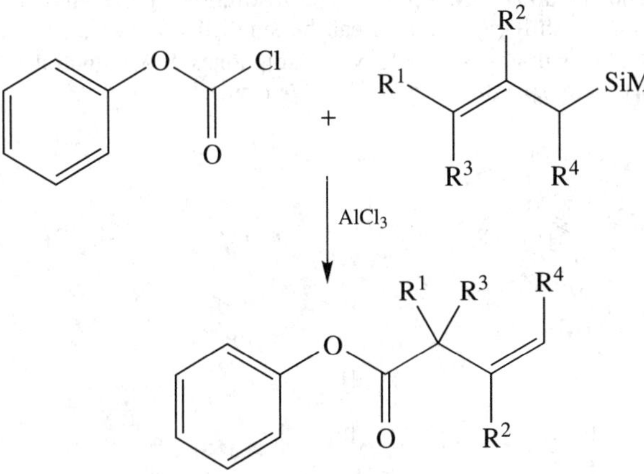

(Ref. 139) (68)

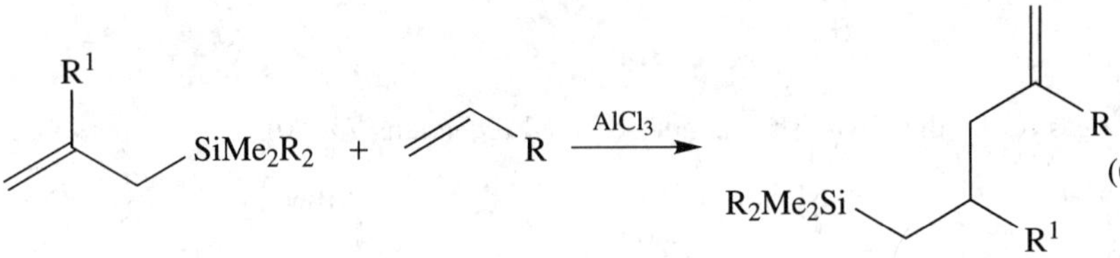

(Ref. 140) (69)

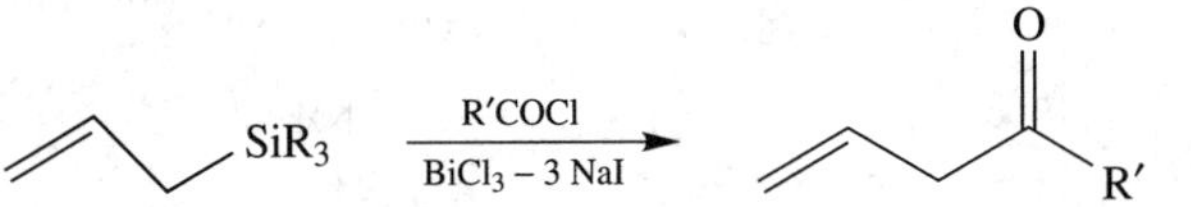

(Ref. 141) (70)

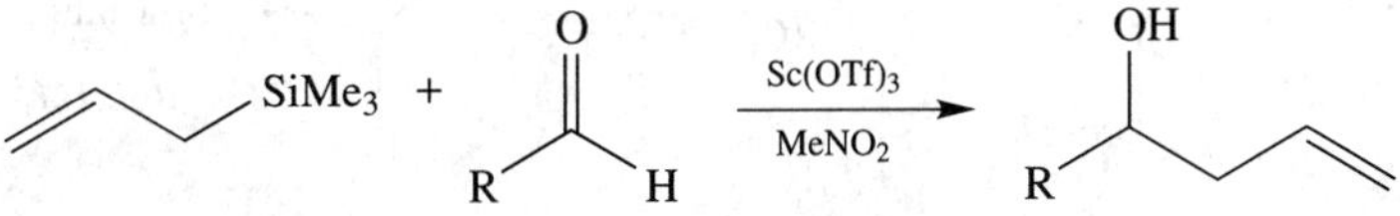

(Ref. 142) (71)

$$\text{CH}_2=\text{CHCH}_2\text{SiMe}_3 \xrightarrow{\text{ClSO}_2\text{NCO}} \underset{\text{NSO}_2\text{Cl}}{\overset{\text{OSiMe}_3}{\text{CH}_2=\text{CHCH}_2\text{C}}} \quad \text{(Ref. 143)} \quad (72)$$

$$\text{H}_2\text{C}=\text{C}(\text{CH}_2\text{SiMe}_3)_2 + \text{Me}_2\text{HC}\overset{+}{-}\text{N}(\text{H})=\text{CH}_2 \xrightarrow{\text{H}^+ \ (\text{CH}_2\text{O})_n} \text{H}_2\text{C}=\overset{\frown}{\underset{\smile}{\text{C}}}\text{N}-\text{CHMe}_2 \quad \text{(Ref. 144)} \quad (73)$$

As reactions 68 and 69 above exemplify, the substitution of the allylsilane usually takes place with an allylic shift (S_E2'). This can be synthetically useful, for example in the isomerization of allyl sulphones **123** to vinyl sulphones **124** (equation 74)[145]. The reaction is also highly stereospecific (>90% E) (*vide infra*).

$$\underset{(123)}{\text{PhSO}_2\text{CH}_2\text{C}(R^1)=\text{C}(R^2)(R^3)} \xrightarrow[\text{Me}_3\text{SiCl}]{\text{BuLi}} \text{PhSO}_2\text{CH}(\text{SiMe}_3)\text{C}(R^1)=\text{C}(R^2)(R^3) \xrightarrow{\text{HX}} \underset{(124)}{\text{PhSO}_2\text{CH}=\text{C}(R^1)\text{CH}(R^2)(R^3)} \quad (74)$$

Occasionally the allylic shift is not observed (equations 75–80).

$$\text{Me}_3\text{Si}-\text{CH}_2-\text{CH}=\text{CH}-\text{CH}(\text{NHBoc})\text{Me} \xrightarrow{\text{RCOCl}} \text{CH}_2=\text{CH}-\text{CH}(\text{COR})-\text{CH}(\text{NHBoc})\text{Me} \quad \text{allylic shift} \quad \text{(Ref. 146)} \quad (75)$$

$$\text{Me}_3\text{Si}-\text{CH}_2-\text{CH}=\text{CH}-\text{CH}(\text{NHBoc})\text{CHMe}_2 \xrightarrow{\text{RCOCl}} \text{RCO}-\text{CH}_2-\text{CH}=\text{CH}-\text{CH}(\text{NHBoc})\text{CHMe}_2 \quad \alpha\text{-substitution} \quad \text{(Ref. 146)} \quad (76)$$

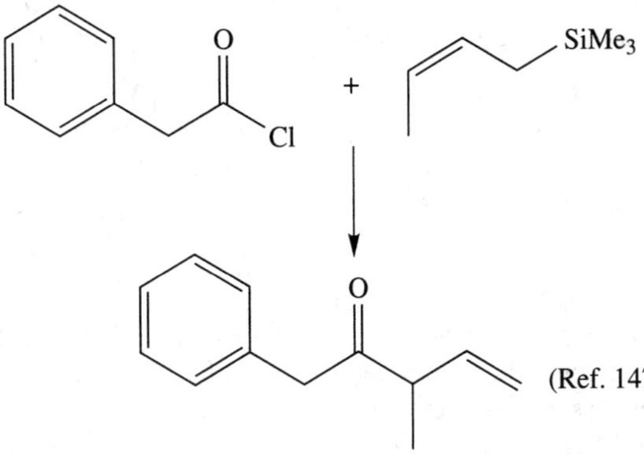

7. Activating and directive effects of silicon

(77) (Ref. 147)

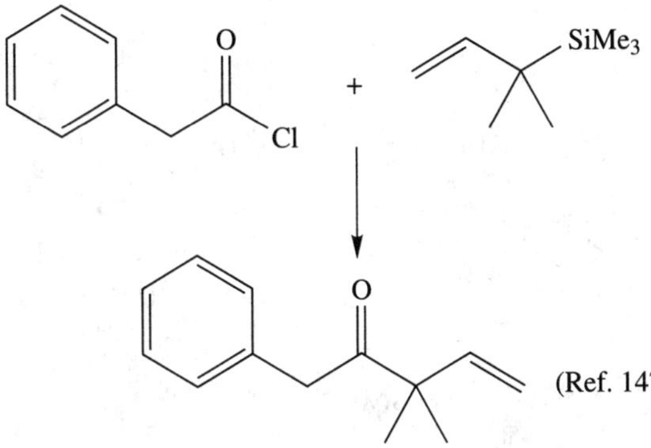

(78) (Ref. 147)

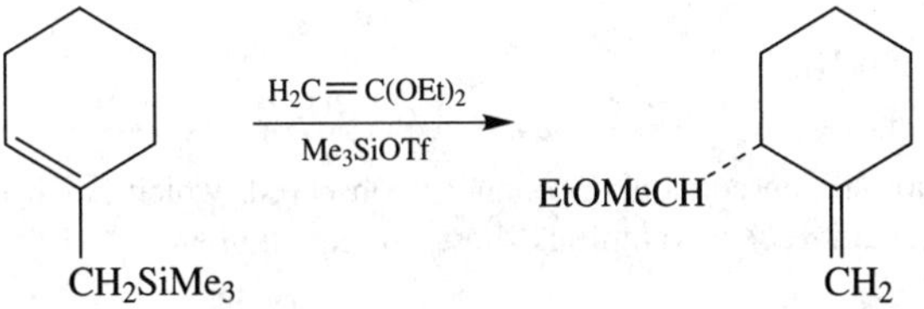

(79) (Ref. 148)

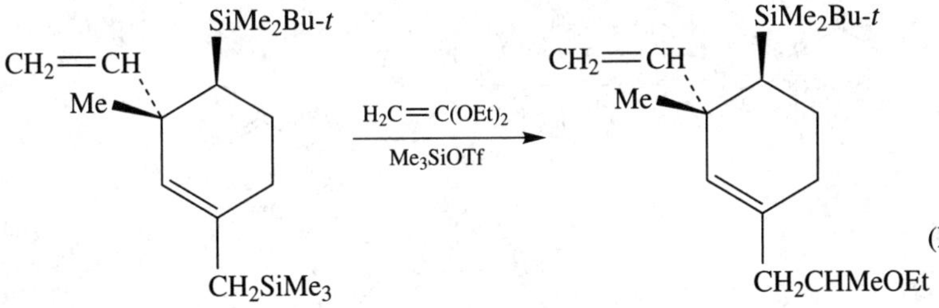

(80) (Ref. 148)

An increase in steric hindrance at the γ-position appears to favour α-substitution in equations 76 and 80, compared to γ-substitution in equations 75 and 79. However, this does not explain the α-substitution in equation 78. Polla and Frejd[148] attribute the α-substitution shown in equation 80 to protiodesilylation followed by electrophilic alkylation of the resulting olefin.

With α, β-unsaturated carbonyl compounds as the electrophile, 1,4-addition of the allylsilane to the α,β-unsaturated carbonyl is often observed, as shown in equation 81.

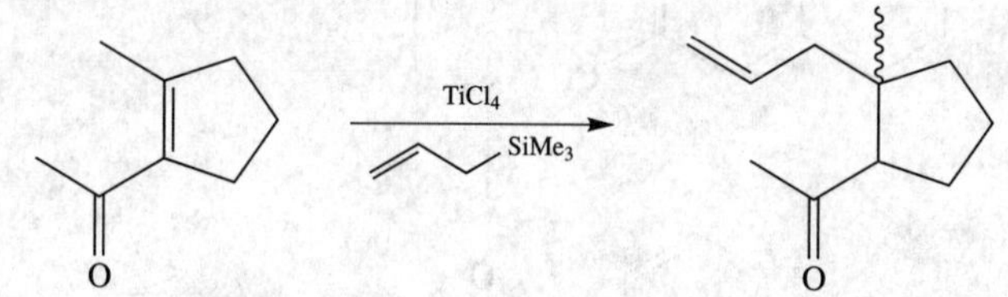

$$(81)$$

This synthetically important reaction is known as the Sakurai or Hosomi–Sakurai reaction[149]. The intramolecular Sakurai reaction is useful for the synthesis of spiro **(125)** and fused **(126)** ring systems, as shown in equations 82 and 83[150].

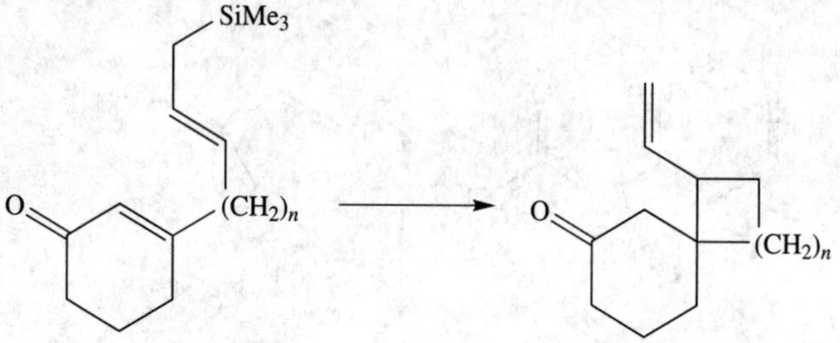

$$(82)$$

(125)

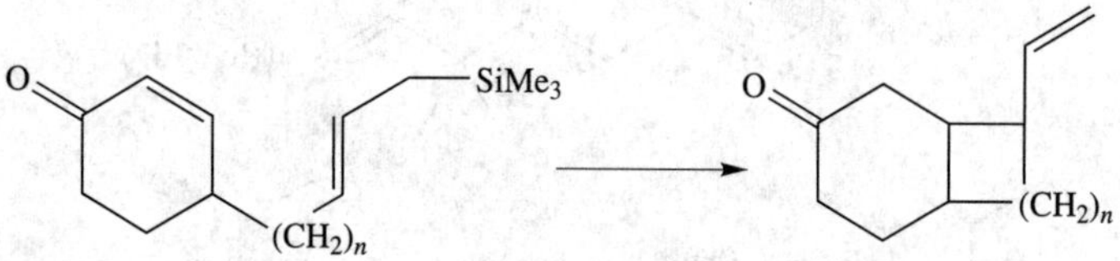

$$(83)$$

(126)

In some cases, a minor silicon-containing product is observed, which has been shown to be a cyclopentane derivative. An example is shown in equation 84[151].

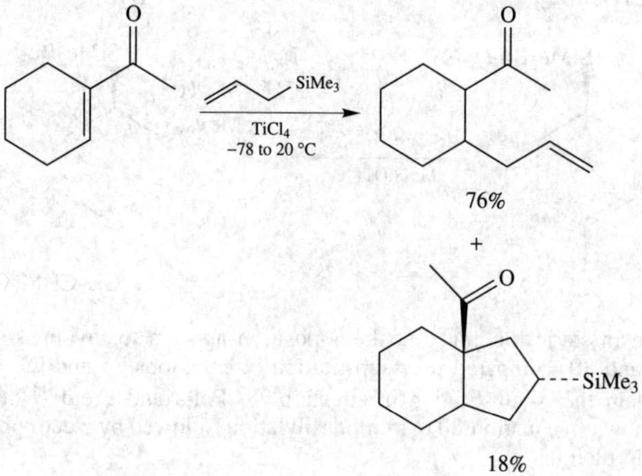

$$(84)$$

76%

+

18%

7. Activating and directive effects of silicon

When bulky groups are present on silicon, this reaction pathway dominates (equations 85 and 86).

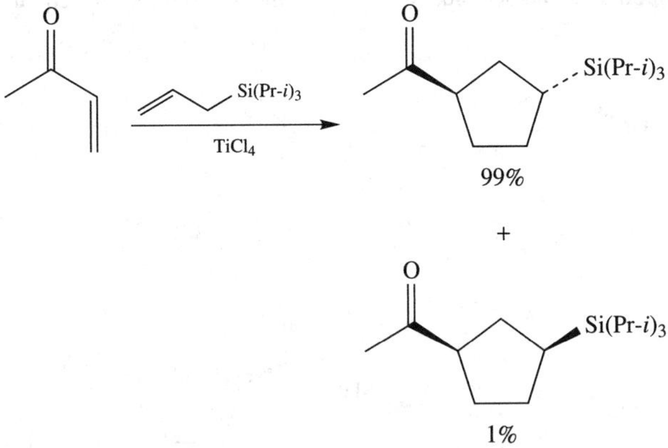

(85) (Ref. 152)

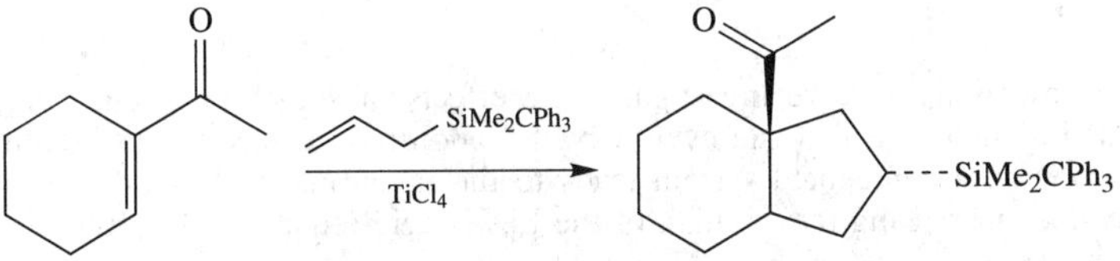

(86) (Ref. 153)

The formation of these products can be explained by consideration of the reaction mechanism, which is most usefully pictured as proceeding via a cyclic siliconium ion **128** (equation 87).

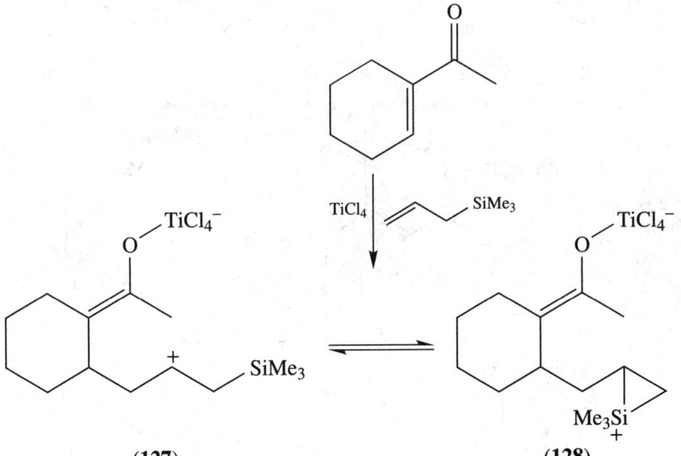

(87)

(**127**) (**128**)

Attack of a nucleophile on the β-silyl carbocation **127** or the cyclic siliconium ion **128** leads to desilylation and formation of the Sakurai product. When nucleophilic attack is disfavoured by steric hindrance at the silicon, competing intramolecular attack by the enolate becomes important. This 5-*exo-tet* cyclization gives the trimethylsilylcyclopentane product with high stereospecificity, the trimethylsilyl group having undergone a 1,2 shift.

It has recently been found that replacement of ketones by esters in this reaction leads to the production of cyclobutanes, sometimes as the major product, arising from intramolecular attack of the enolate at the secondary carbon of the siliconium ion (equation 88)[154].

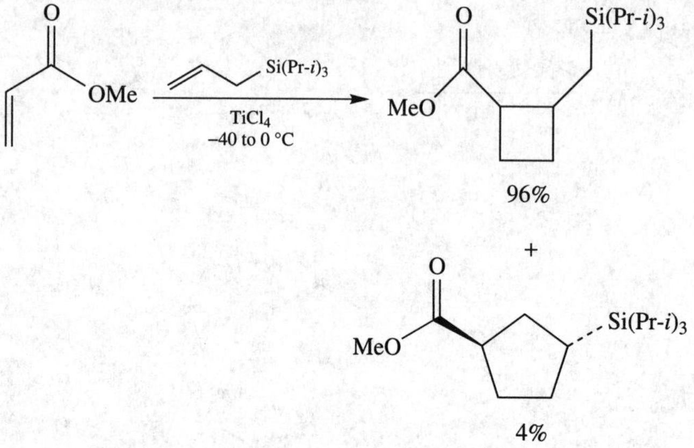

(88)

Majetich and coworkers[155] have investigated the effects of substitution and catalyst on the conjugated dienones, which can cyclize by 1,6-addition. Increased substitution at the terminal carbon of a conjugated system leads to the favouring of 1,2-addition over 1,6-addition, as does increasing the strength of the Lewis acid (equations 89–91).

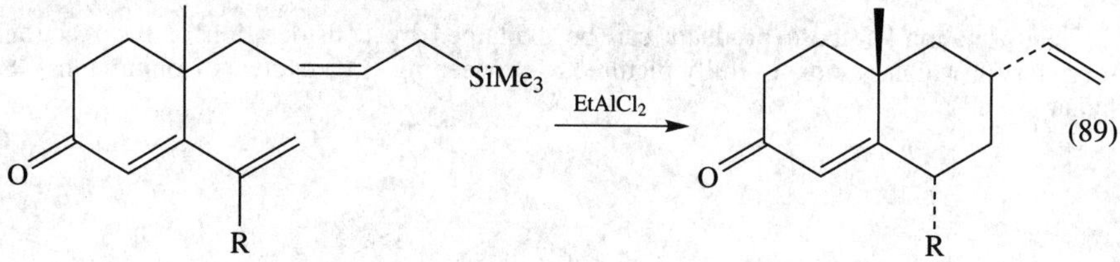

(89)

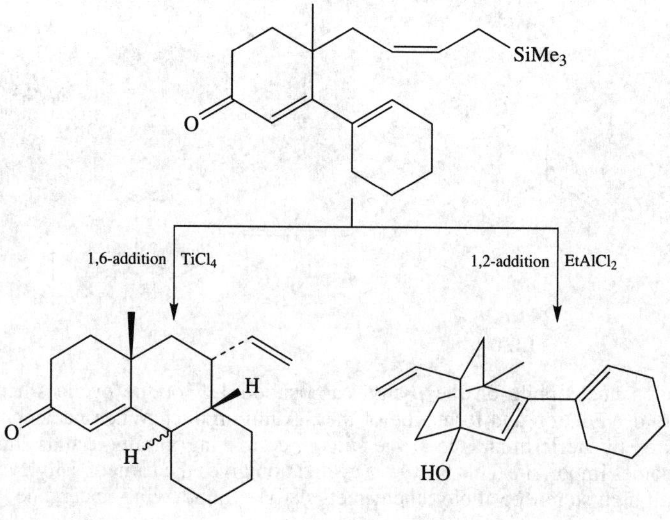

(90)

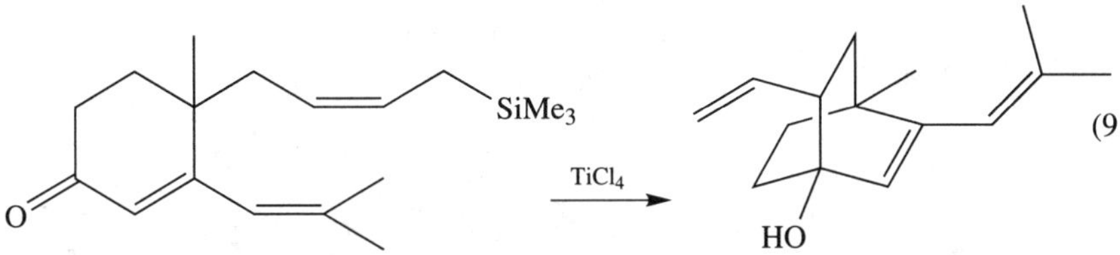

7. Activating and directive effects of silicon

[Scheme showing reaction of silyl compound with TiCl₄ to give bicyclic product] (91)

In most cases, open-chain allylsilanes react with electrophiles with *anti* stereoselectivity[156–162]. The simple explanation for this observation follows from the probable conformation of the allylsilane. The preferred conformation **129** will have the small substituent H eclipsing the double bond.

The large and electropositive silyl group encourages attack of the electrophile on the lower surface as shown in equation 92, and only a 30° rotation is necessary to give the β-silyl cation **130** with the optimum geometry for hyperconjugative stabilization. This stabilization is probably large enough that the configuration is maintained until the silyl group is lost in the second step, to give a product **131** containing a *trans* double bond, with the overall reaction being stereoselectively *anti*.

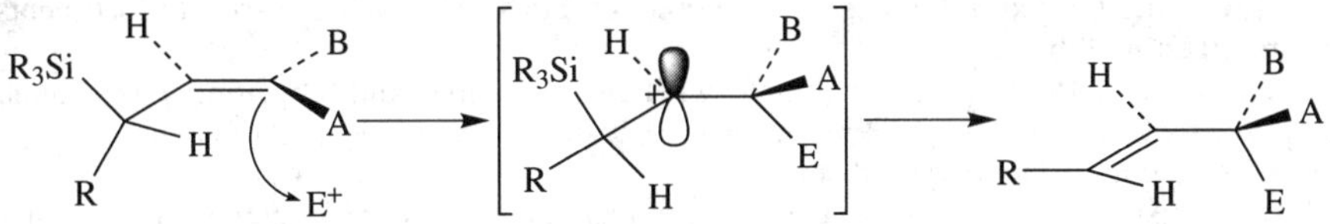

[Scheme 92: structures (129) → (130) → (131)]

However, several factors can result in this simple pattern of stereospecific *anti* attack not being followed. The difference in size between the substituents H and R may not be sufficient to prevent reaction through the alternative conformation **132** (equation 93). This is most common when the substituent A is a proton.

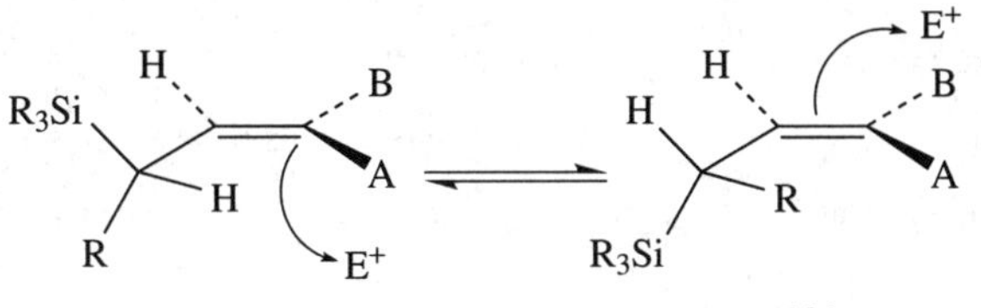

[Scheme 93: structures (129) ⇌ (132)]

The alternative conformation **132** may also be favoured when the electrophile is a very bulky species, in which case interactions between the electrophile and R are minimized. Increasing the size of the R group might also be expected to change the direction of electrophilic attack as its hindrance to the incoming electrophile becomes similar to that of the silyl group. This also acts to disfavour the conformation **132** due to increased 1,3 interactions.

Fleming and Higgins[163] studied the protiodesilylation of the two cyclohexene derivatives **133** and **134**. If protonation takes place as described in the simple picture above, then the protiodesilylation products should be **135** and **136**, respectively, as shown in equations 94a and 94b.

$$\text{(133)} \xrightarrow{H^+} \text{(135)} \tag{94a}$$

$$\text{(134)} \xrightarrow{H^+} \text{(136)} \tag{94b}$$

The protonations of the two cyclohexenes were not completely stereospecific, both **135** and **136** being formed in both cases. The product ratios observed for various substituents R are given in Table 12.

Selectivity is found to be negligible when R is methyl, and surprisingly not much improved when R is phenyl. However, the isopropyl group gives a substantial degree of stereoselectivity in the expected sense.

Another study by Fleming and coworkers examined the stereoselectivity of the titanium tetrachloride catalysed reaction of the allylsilane **137** with 1-adamantyl chloride (equation 95)[164].

$$E\text{-}(137) \xrightarrow[\text{TiCl}_4]{\text{AdCl}} \begin{array}{c} anti\ 90\% \quad syn\ 10\% \quad E\ 40\% \\ anti > 99\% \quad syn < 1\% \quad Z\ 40\% \end{array} \tag{95}$$

Ad = 1-Adamantyl

TABLE 12. Product ratios in the protiodesilylations of allylsilanes **133** and **134**

Allylsilane	R	Product ratio **135 : 136**
133	Me	57 : 43
134	Me	38 : 62
133	Ph	69 : 31
134	Ph	38 : 62
133	i-Pr	91 : 9
134	i-Pr	8 : 92

7. Activating and directive effects of silicon 409

The Z product is enantiomerically pure, but the E isomer is formed in a 90 : 10 ratio of enantiomers. One reason for this was revealed by considering the relevant conformations of the allylsilane, shown in equation 96.

$$(138) \rightleftharpoons (139) \qquad (96)$$

$$(138) \rightarrow E \qquad (139) \rightarrow Z$$

Attack in the conformation **139** which leads to the Z-isomer takes place from the upper surface of the molecule, *anti* to the trimethylsilyl group. This face is occupied by a hydrogen group. In the conformation **138** which leads to the E-isomer, this surface is occupied by a methyl group. The greater steric effect of this group leads to the small amount of attack on the other surface.

In order to determine whether attack *anti* to silicon was due to steric effects or electronic effects, Fleming and coworkers studied the attack of electrophiles on the pentadienylsilanes **140**, in which the electrophile attacks at the second double bond, as shown in equation 97[165].

$$(140) \longrightarrow \qquad (97)$$

The electrophiles isobutyraldehyde and its dimethyl acetal gave surprisingly high ratios of *syn* : *anti* attack, around 10 : 90. Since the silicon centre is more removed from the reactive site, this could be taken to indicate that electronic effects predominate. However, the intramolecular cyclization of **141** (equation 98) was found to take place with an enantiomeric excess of only 20%.

$$(141) \xrightarrow{TiCl_4} \qquad (98)$$

The corresponding allyl system reacts to give a product with an enantiomeric excess of 90%. The small electrophilic centre in this reaction is very unlikely to experience any steric interaction with the silyl group, and the small *anti* stereoselectivity (60 : 40) in this reaction was taken to be an indication of the size of the electronic effect. The high stereoselectivities observed for the reactions of isobutyraldehyde and isobutyraldehyde dimethyl acetal were thought to be due to steric interactions. Following these and other studies Fleming and coworkers concluded: 'whether the high levels of *anti* stereoselectivity seen with allylsilanes are largely steric or electronic in origin remains unknown'[165].

The reactions of chiral allylsilanes with electrophiles to give diastereoselective products has recently been extensively reviewed by Masse and Panek[166]. An example is shown in equation 99. The reaction of the chiral allylsilanes **142** with phenylsulphenyl chloride (PSC) and chlorosulphonyl isocyanate (CSI) takes place with a diastereoselectivity which increases with the increasing steric bulk of the substituent R[167].

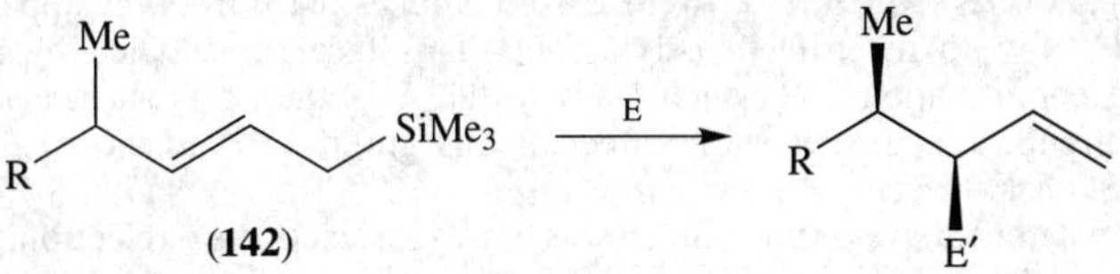

$$E = PhSCl; E' = PhS$$
$$E = ClSO_2NCO;$$
$$E' = CN$$

(99)

The product diastereomeric excesses (de) obtained with various R substituents in **142** are summarized in Table 13.

The increasing diastereoselectivity, as the steric requirements of R increase, can be rationalized by considering the conformation of the allylsilane. Again the smallest group (H) is positioned in the plane of the double bond **(143)**, and the steric influence of R determines the ratio of attack on the two faces of the allylsilane.

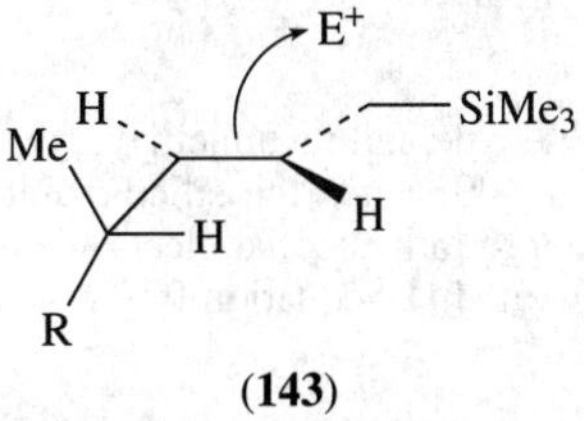

Allylsilanes are more reactive than vinylsilanes. Firstly, hyperconjugative overlap of the C—Si bond with the π bond raises the energy of the HOMO, making it more

TABLE 13. Diastereomeric excesses (de) in the products of the reactions of the allylsilane **142** with electrophiles E

R	E	de
Et	PSC	20%
Ph	PSC	90%
t-Bu	PSC	98%
Et	CSI	24%
Ph	CSI	90%
t-Bu	CSI	98%

7. Activating and directive effects of silicon

reactive towards electrophiles, and secondly, the C—Si bond can stabilize the positive charge buildup throughout electrophilic addition to allylsilanes. This is in contrast with vinylsilanes, where full hyperconjugative stabilization is only possible after rotation of the C—C bond through 90°. For these reasons, systems which are both an allylsilane and a vinylsilane generally react as an allylsilane. An example is the cyclohexene **144**, which reacts with acetyl chloride as an allylsilane (equation 100)[168].

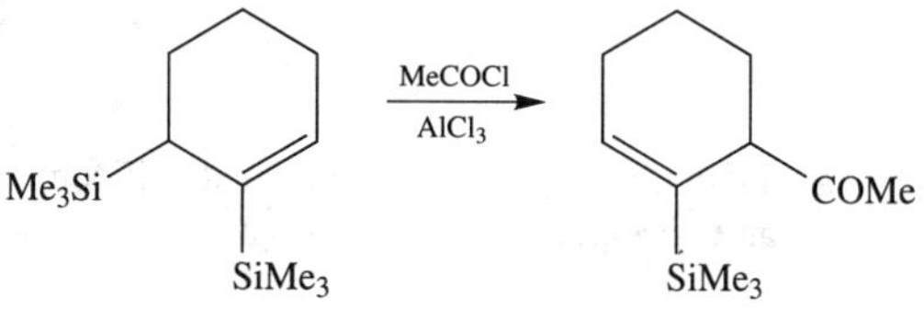

(100)

D. Reactions of Propargylsilanes

Propargylsilanes undergo addition to give a β-silylcarbocation, which can then react further to give either addition or substitution, as shown in equation 101.

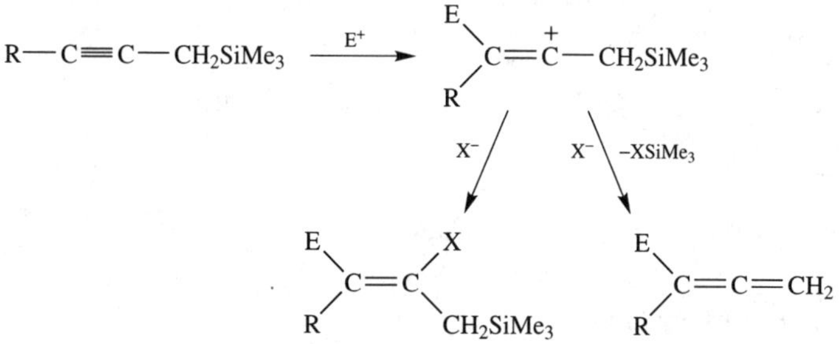

(101)

As with allylsilanes, substitution is found to be the dominant reaction pathway (equation 102)[169].

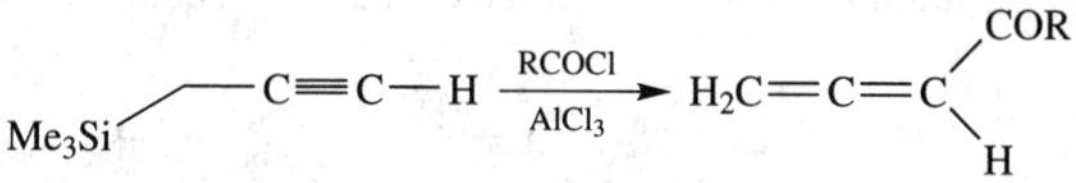

(102)

Electrophilic substitution of propargylsilanes is an important route to allenes. Cyclization gives allene-substituted cyclic compounds as shown in equations 103 and 104.

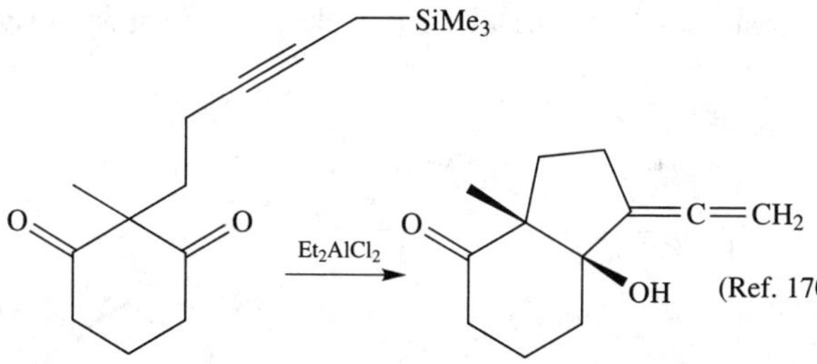

(103)

(Ref. 170)

As with allylsilanes, increased steric hindrance at silicon favours annulation (equation 105)[172] rather than the Sakurai reaction (equations 103 and 104).

E. Reactions of Allenylsilanes

Allenylsilanes may behave as allyl or vinylsilanes. In general, allylsilane behaviour dominates over vinylsilane behaviour. The reactions of allenylsilanes with electrophiles have been comprehensively reviewed by Fleming, Dunogues and Smithers[118].

Electrophilic substitution of allenylsilanes gives alkynes (equation 106)[173].

Allenylsilanes with α-alkyl substituents undergo annulation reactions, as in equation 107[174].

Buckle and Fleming studied the reactions of the allenylsilane **145** with 1-adamantyl chloride (equation 108) and with isobutyraldehyde and found these S_E2 reactions to be stereospecifically *anti* to a very high degree[175].

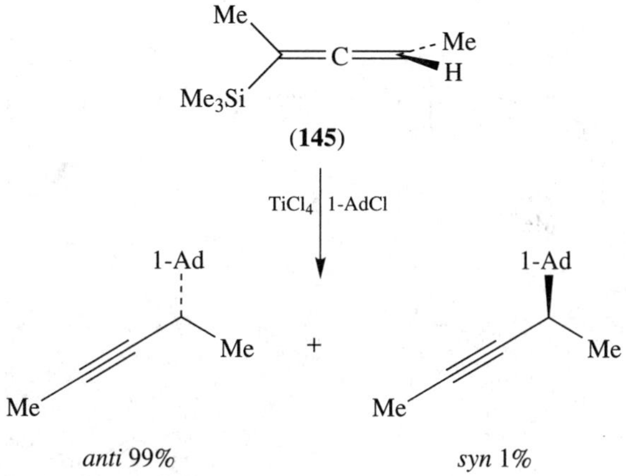

(108)

anti 99% syn 1%

F. Miscellaneous Reactions Controlled by the Formation of a β-Silylcarbocation

There are a number of reactions in which the chemical and stereochemical outcome is controlled by the formation of a β-silylcarbocation.

One example is the Baeyer–Villiger reaction. The migratory aptitude of groups in the Baeyer–Villiger reaction depends on their ability to bear a positive charge[176]. Hudrlik and coworkers have shown that the presence of a silicon substituent β to a ketone stabilizes the incipient carbocation leading to a regiospecific oxygen insertion, as shown in equation 109[177].

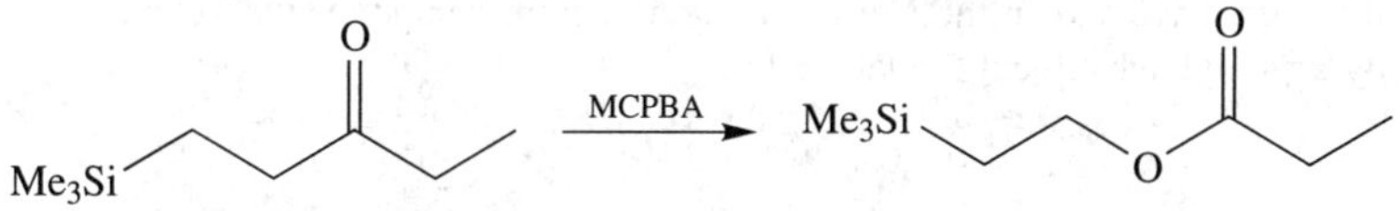

(109)

This regioselectivity has been useful in ring expansions (equation 110)[178].

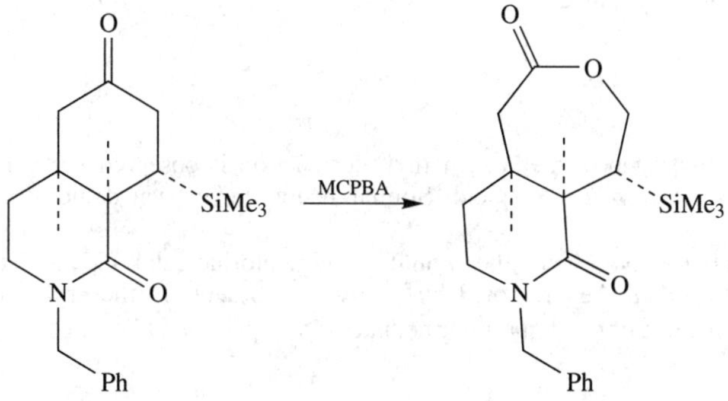

(110)

The ring contraction shown in equation 111 is also promoted by formation of a β-silyl carbocation in the intermediate **146**[179].

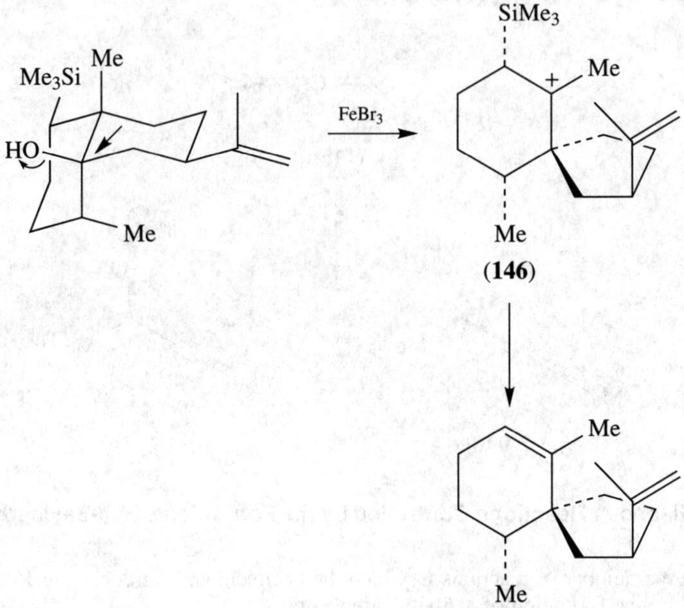

(111)

G. Reactions Involving γ-Silylcarbocations

The γ-effect of silicon has so far found little application in organic synthesis, apart from the solvolysis studies described previously.

Hwu and Gilbert have shown the Nef conversion of nitro compounds to ketones to be promoted by a γ-silyl substituent (equation 112)[180].

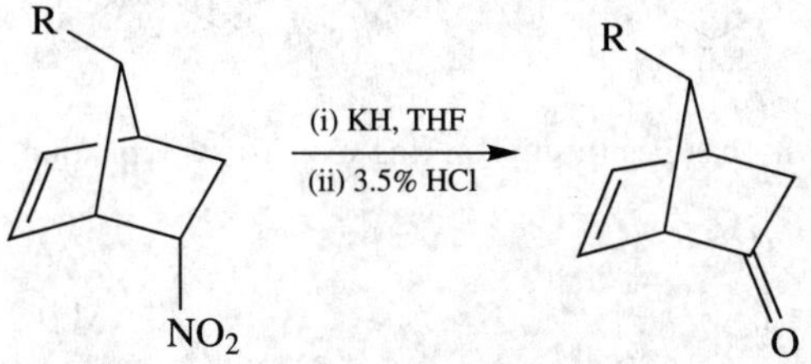

(112)

When R is hydrogen or methoxymethyl, no reaction is observed, but for R = SiMe$_3$, reaction takes place in a 64% yield. Similar results have been demonstrated for other systems[180].

Engler and Reddy found the titanium tetrachloride-catalysed arylation of 1,4-benzoquinones **147** to be promoted by γ-silyl substituents in the aryl component, via stabilization of the carbocationic intermediate **148** (equation 113)[181].

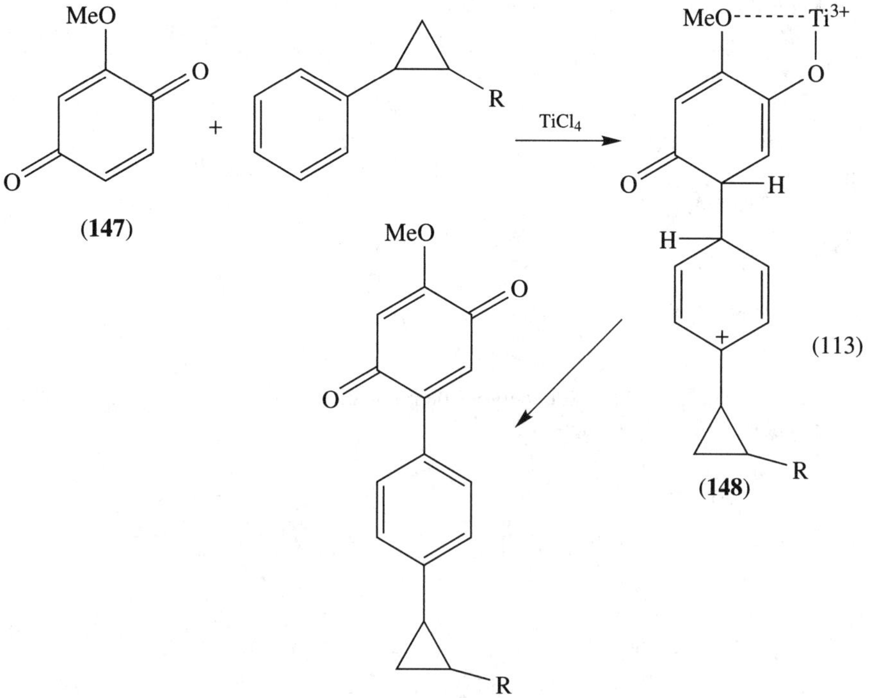

The yield in this reaction increased from 46% (R = H) to 69% (R = SiMe$_3$).

VI. ACTIVATING AND DIRECTIVE EFFECTS OF SILICON IN CARBANIONIC REACTIONS

A. Methods of Formation of α-Silylcarbanions

Silicon stabilizes an adjacent C–metal bond despite being more electropositive than hydrogen or carbon. Stabilization of the free anion is often discussed, although it should be noted that the free ion is rarely involved. The generation of α-silylcarbanions is discussed in detail in Ager's review of the Peterson olefination reaction[182]. A selection of methods for generating α-silylcarbanions are illustrated below.

1. Proton abstraction (equation 114)[183]

$$\text{Me}_3\text{SiCH}_2\text{R} \xrightarrow{\text{BuLi/TMEDA}} \text{Me}_3\text{Si}\bar{\text{C}}\text{HR} + \text{Li}^+ \quad (114)$$

Generally, the direct metallation of unactivated alkyl groups is not a synthetically useful reaction. However, under certain circumstances unactivated α-silylcarbanion formation has been reported. Treatment of *ortho*-silylated benzamides **149** with LDA gave an α-silyl carbanion **150**, stabilized by a complex-induced proximity effect, which then underwent intramolecular attack on the amide group (equation 115)[184].

$$\text{(149)} \xrightarrow{\text{LDA}} \text{(150)} \longrightarrow \text{product} \tag{115}$$

The TBDMS ether **151** also undergoes deprotonation of the silyl methyl group, again due to chelation control (equation 116)[185].

$$\text{(151)} \xrightarrow{\text{LDA}} \text{product} \tag{116}$$

2. Metal–halogen exchange (equation 117)[186,187]

$$\text{Me}_3\text{SiCHIR} \xrightarrow{\text{BuLi}} \text{Me}_3\text{Si}\bar{\text{C}}\text{HR} + \text{Li}^+ \tag{117}$$

3. Transmetallation (equation 118)[188]

$$\underset{\underset{\text{SeR}}{|}}{\text{Me}_3\text{SiCHR}} \xrightarrow{\text{BuLi}} \text{Me}_3\text{Si}\bar{\text{C}}\text{HR} + \text{Li}^+ \tag{118}$$

Included in this category are desilylations of bis(silyl) compounds with fluoride or alkoxide ions which is a particularly clean method of anion production (equation 119)[189].

$$\underset{\underset{R}{|}}{\text{Me}_3\text{Si}\text{—CHSiMe}_3} \xrightarrow{\text{F}^-} \underset{\underset{R}{|}}{\bar{\text{C}}\text{HSiMe}_3} + \text{Me}_3\text{SiF} \tag{119}$$

4. Organometallic addition to vinylsilanes

Organolithium[190-192] and Grignard reagents[193] react with vinylsilanes to give α-silylcarbanions (equation 120). Such reactions are difficult with unsubstituted vinylsilanes.

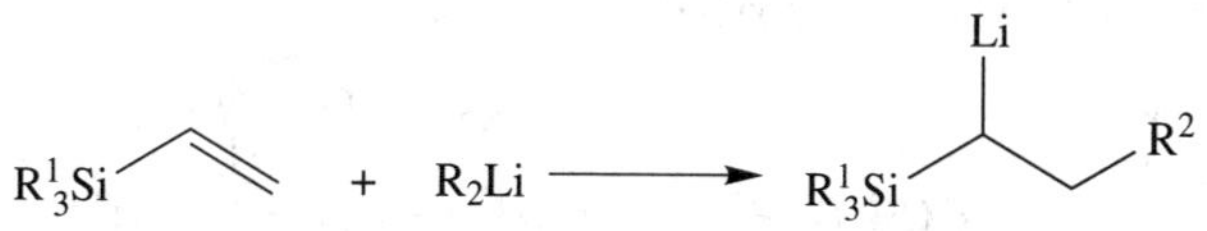

$$R^1_3Si-CH=CH_2 + R_2Li \longrightarrow R^1_3Si-CH(Li)-CH_2-R^2 \quad (120)$$

The SiMe$_3$ group does not stabilize an α-carbanion to the same extent as a nitro group, as shown by the regioselectivity of the reaction shown in equation 121[194].

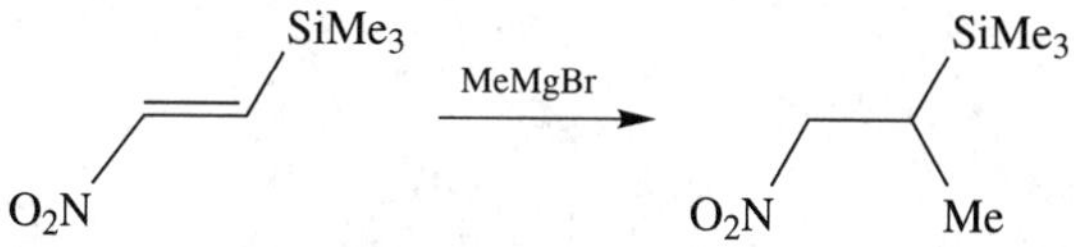

$$\text{(equation 121)}$$

The stability of an α-silyl carbanion is responsible for the improved synthetic utility of the Stork annulation over other annulations[195,196]. These reactions involve the Michael addition of an enolate ion to an enone, and in the absence of a α-silyl substituent suffer drawbacks due to the reversibility of the Michael reaction. However, the addition of enolate ions to α-trimethylsilylvinyl ketones is not reversible, owing to α-silicon stabilization of the canonical form **152** shown in equation 122.

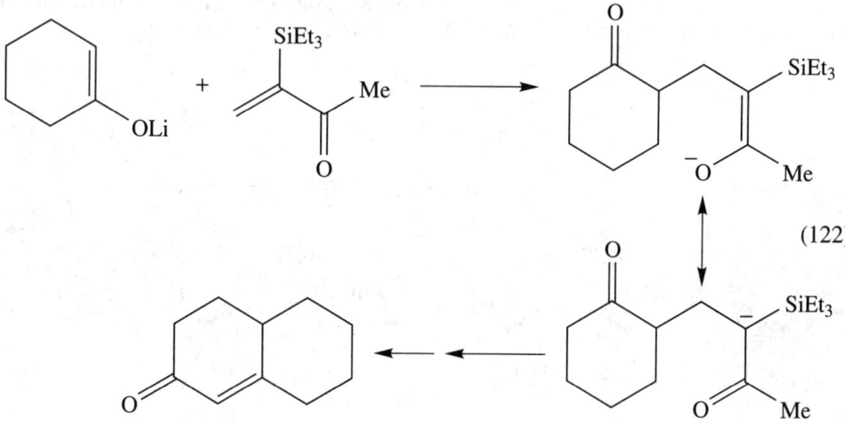

(122)

(**152**)

B. The Peterson Reaction

α-Silylcarbanions undergo the normal range of carbanion reactions. Of particular importance is their reaction with carbonyl compounds (equation 123). Depending on the counterion, the β-silyl alkoxide **153** can undergo protonation to give the corresponding β-silyl alcohol **154**, or can eliminate R$_3$SiO$^-$ to give the alkene **155**. This latter route is the Peterson olefination[183], which has great synthetic utility and is often used in situations where the Wittig reaction has failed. The carbonyl compound can be an aldehyde or a ketone and a range of carbanions have been used. The Peterson reaction has two general advantages over the Wittig reaction: (i) the silanolate ion is water-soluble, which makes the reaction cleaner than the Wittig; and (ii) the silyl carbanion is less sterically hindered and more reactive than the Wittig reagent. The Peterson reaction has recently been comprehensively reviewed by Ager[182].

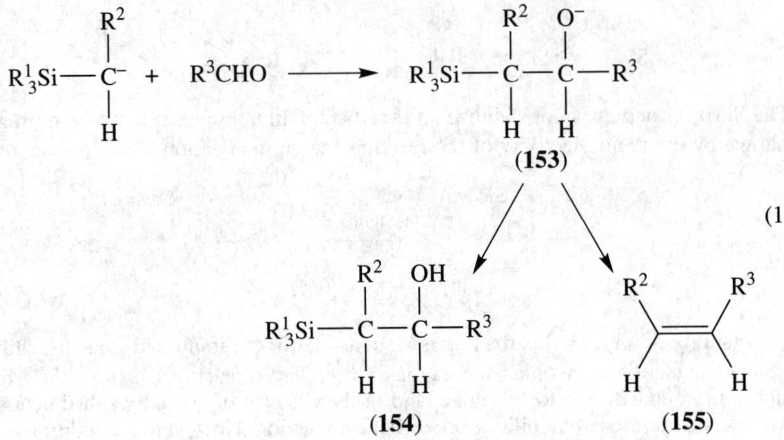

$$R_3^1Si-\underset{\underset{H}{|}}{\overset{\overset{R^2}{|}}{C^-}} + R^3CHO \longrightarrow R_3^1Si-\underset{\underset{H}{|}}{\overset{\overset{R^2}{|}}{C}}-\underset{\underset{H}{|}}{\overset{\overset{O^-}{|}}{C}}-R^3 \quad (123)$$

(153)

$$R_3^1Si-\underset{\underset{H}{|}}{\overset{\overset{R^2}{|}}{C}}-\underset{\underset{H}{|}}{\overset{\overset{OH}{|}}{C}}-R^3 \qquad \underset{H}{\overset{R^2}{\diagdown}}C=C\underset{H}{\overset{R^3}{\diagup}}$$

(154)   (155)

The detailed mechanism of the Peterson reaction has not yet been revealed. When only alkyl, hydrogen or electron-donating substituents are present on the carbon atom bonded to silicon, the β-hydroxysilane **156** can be isolated, usually as a diastereomeric mixture (equation 124), which can be separated using the usual physical methods.

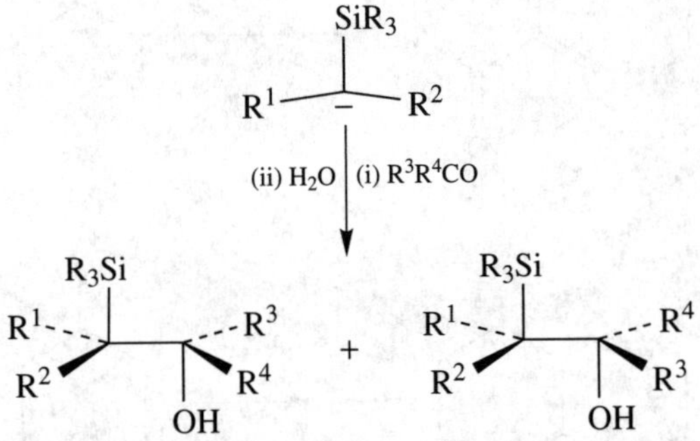

(124)

(156)

Treatment of the β-hydroxysilane **156** with acid gives an *anti* elimination, as shown in equation 125.

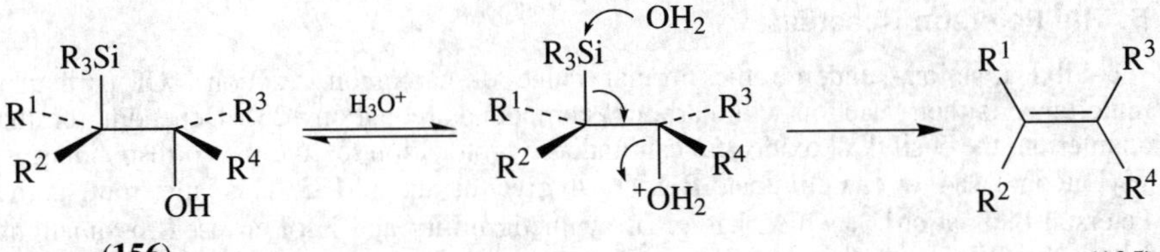

(156)   (125)

Treatment of the β-hydroxysilane **156** with base gives *syn* elimination[197], producing the isomeric alkene (equation 126).

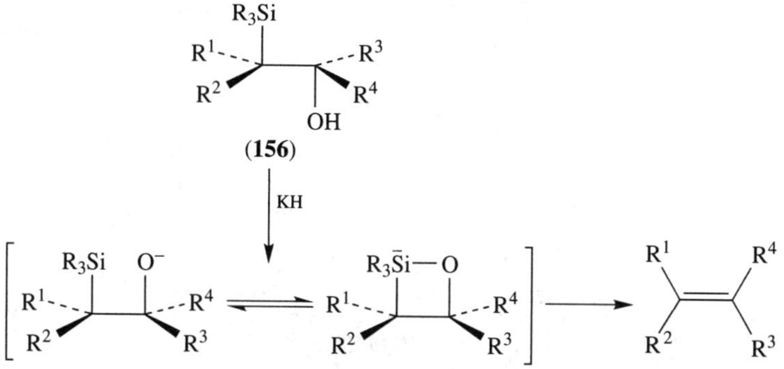

$$\underset{(156)}{\overset{R_3Si}{\underset{R^2}{R^1\!\!-\!\!\overset{|}{\underset{|}{C}}\!\!-\!\!\overset{R^3}{\underset{R^4}{C}}}}}\;\xrightarrow{KH}\;\left[\underset{R^2}{\overset{R_3Si\;\;O^-}{R^1\!\!-\!\!\overset{|}{C}\!\!-\!\!\overset{R^4}{\underset{R^3}{C}}}}\;\rightleftharpoons\;\underset{R^2}{\overset{R_3\bar{S}i\!-\!O}{R^1\!\!-\!\!\overset{|}{C}\!\!-\!\!\overset{R^4}{\underset{R^3}{C}}}}\right]\;\longrightarrow\;\underset{R^2}{\overset{R^1}{\diagdown}}C\!=\!C\underset{R^3}{\overset{R^4}{\diagup}}$$

(126)

Using these routes either alkene is available from each diastereoisomer of the β-hydroxysilane[197].

To avoid the need for separation of the diastereoisomers of the β-hydroxysilane, various stereoselective routes to β-hydroxysilanes have been developed.

When an electron-withdrawing, carbanion-stabilizing group (Z) is present on the carbon-bearing silicon, the olefin is generally isolated directly from the reaction mixture, usually as a mixture of isomers (equation 127).

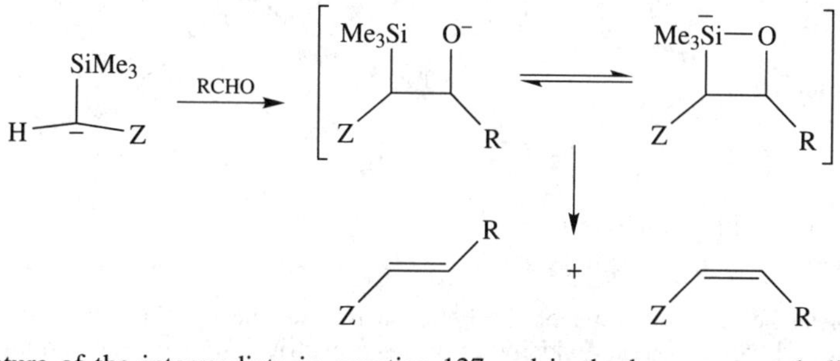

(127)

The nature of the intermediate in equation 127 and in the base-promoted elimination of β-hydroxysilanes is still uncertain. Hudrlik and co-workers have shown that the β-oxidosilane may not be an intermediate[198]. When the β-oxidosilane **158** was generated by deprotonation of the β-hydroxysilane **157** or by hydride opening of the epoxide **159**, the product arising from elimination from the β-oxidosilane was exclusively the *trans* alkene **160** (Scheme 8).

However, the reaction of benzaldehyde with bis(trimethylsilyl)methyllithium gave a mixture of *trans* and *cis* isomers in a ratio of 1.4 : 1 (equation 128). If this reaction involved the β-oxidosilane intermediate **158**, the same stereochemical outcome would be expected. This was taken to suggest that, in this particular Peterson olefination reaction at least, the β-oxidosilane **158** is not a major intermediate, and that the oxasiletane anion is formed directly by simultaneous formation of C—C and Si—O bonds.

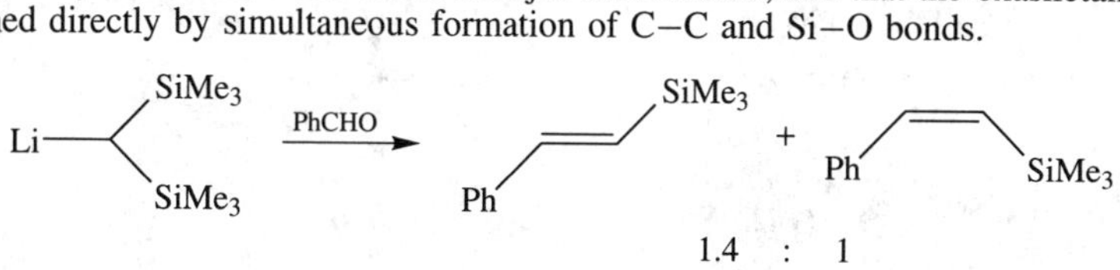

(128)

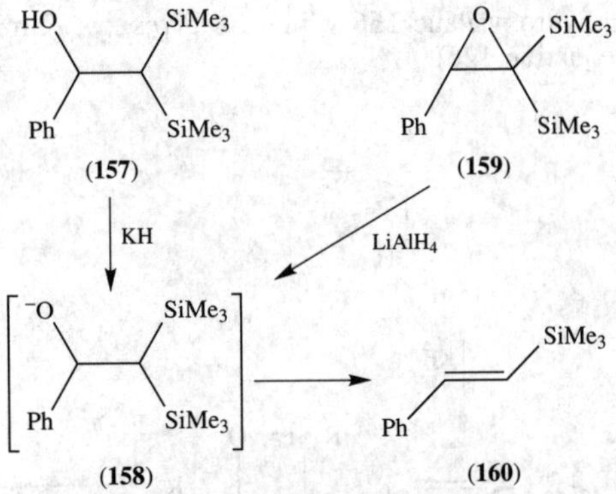

SCHEME 8

1,2-Oxasiletanide **161a** has been synthesized and its structure determined by X-ray crystallographic analysis[199]. On slight heating a similar oxasiletanide **161b** was observed to decompose to the alkene and lithium silanoxide, as shown in equation 129.

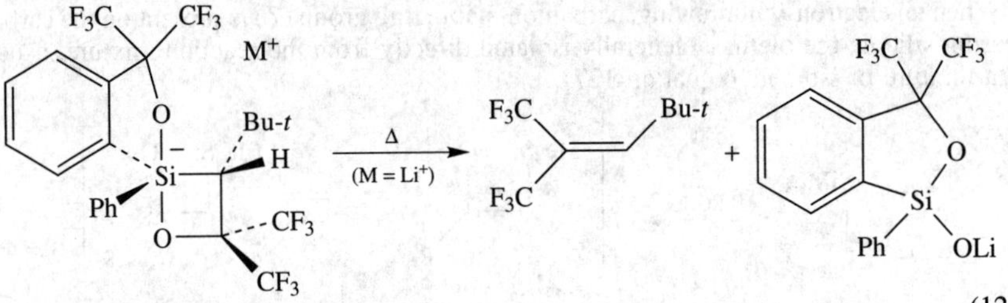

(129)

(**161a**) M = K (18-crown-6)
(**161b**) M = Li

The stereochemical outcome of the Peterson reaction between unsymmetrically substituted α-silyl carbanions and aldehydes or unsymmetrical ketones is determined by the relative rates of formation of the *threo* and *erythro* β-oxidosilanes. Often the rates are similar, to give a product alkene $E : Z$ ratio of 1 : 1, although some workers report a predominance of *cis* olefins in the reactions of aldehydes.

The $E : Z$ ratio is found to vary as the steric bulk of R_3Si increases, the Z isomer becoming more favoured. This is easily rationalized by consideration of the attack of the carbanion on the carbonyl compound[200,201]. The two possible attacks of the α-silylbenzyl anion on benzaldehyde are shown in **162a** and **162b**.

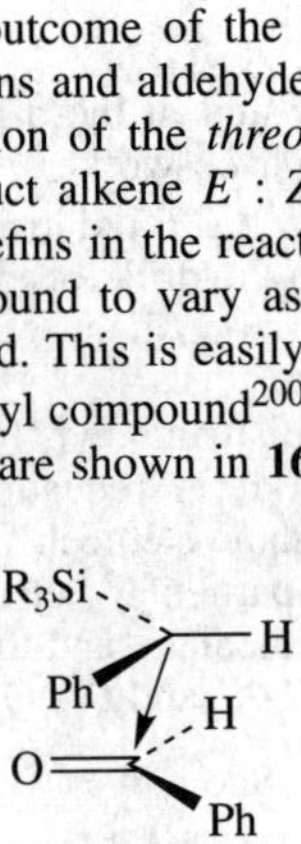

(*erythro*)
(**162a**)

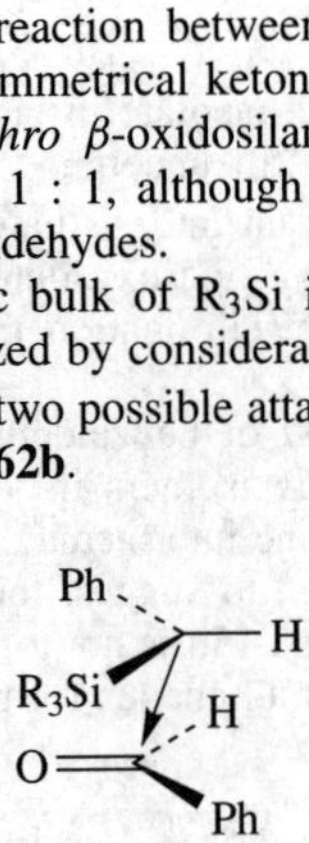

(*threo*)
(**162b**)

7. Activating and directive effects of silicon   421

TABLE 14. Effect on the Z/E product ratio of electron-withdrawing aryl substituents on silicon in the Peterson reaction

Ar	Z : E ratio
Ph	1 : 0.55
m-C$_6$H$_4$F	1 : 0.88
p-C$_6$H$_4$CF$_3$	1 : 1.22

The smallest group on the anion is placed between the substituents on the aldehyde to minimize steric repulsion. The other two groups are placed so that the largest group is placed opposite the bulky phenyl group on the aldehyde. Thus as the steric bulk at silicon is increased, the transition state **162a** leading to the *erythro* isomer and the Z alkene becomes increasingly favoured.

However, electronic factors can also have an effect. It was found that placing electron-withdrawing groups on the aryl substituents at silicon led to an increase in the proportion of E-alkene formed, as shown by the data in Table 14[202].

The slight increase in steric bulk on going from phenyl to fluorophenyl to trifluoromethylphenyl would be expected to increase the proportion of Z isomer. However, this is not observed. It was proposed that the increased electrophilicity of silicon when electron-withdrawing substituents are present causes an increase in the degree of silicon–oxygen interaction in the transition state and a concomitant modification of the geometry of the transition state to allow this. If a decrease in the dihedral angle between silicon and oxygen is considered in the two transition states **162a** and **162b**, the steric repulsion will increase in the *erythro* transition state **162a** as the two phenyl groups become closer to being eclipsed, whereas in the *threo* transition state **162b**, steric repulsion will decrease. Therefore, the proportion of E-alkene product increases accordingly.

C. Variations of the Peterson Reaction

An intramolecular Peterson reaction led to cycloalkene products (equation 130)[203].

(130)

α-Silylcarbanions with α-halo groups **163** may give epoxides **164**, as shown in equation 131[204].

$$\text{(163)} \xrightarrow{R^1R^2CO} \text{intermediate} \longrightarrow \text{(164)} \quad (131)$$

α-Silylallyl anions **165** have been used in the Peterson reaction and often give γ-attack (equation 132)[205].

$$\text{(165)} \xrightarrow{RCOR} \text{product} \quad (132)$$

Under certain conditions α-attack is favoured (equation 133)[206].

$$\text{(165)} \xrightarrow[\text{HMPT/MgBr}_2]{RCOR} \text{product} \longrightarrow \text{diene} \quad (133)$$

Reaction of α-silylcarbanions with epoxides has led to observation of the homo-Peterson reaction (equation 134)[207].

$$(Me_3Si)_3C^- + \text{epoxide} \longrightarrow \text{intermediate} \xrightarrow{-OSiMe_3} \text{cyclopropane} \quad (134)$$

A vinylogous Peterson reaction has also been described, and the elimination has been found to be stereospecifically *syn*, as shown in equations 135 and 136[208].

$$\xrightarrow{KH} \quad (135)$$

7. Activating and directive effects of silicon

[Structure: cyclohexyl-CH(SiMe₂Ph)-CH=CH-CH(OH)-CH₃ → KH → cyclohexyl-CH=CH-CH=CH-CH₃] (136)

The *cis* double bond is found to be selectively placed adjacent to the carbon that originally carried the hydroxy group.

A recent modification of the Peterson reaction involves the use of fluoride ion catalysts. Reaction of bis(trimethylsilyl)methyl derivatives **166** and carbonyl compounds gives the expected alkenes **167** (as shown in equation 137) in high yields, especially for non-enolisable carbonyl compounds, and in some cases with high stereoselectivity[209].

$$\underset{(166)}{\underset{Me_3Si}{\overset{Me_3Si}{>}}C\underset{R^4}{\overset{R^3}{<}}} + \underset{R^1}{\overset{O}{\underset{\|}{C}}}R^2 \xrightarrow{TBAF} \underset{(167)}{\underset{R^2}{\overset{R^1}{>}}=\underset{R^4}{\overset{R^3}{<}}} \quad (137)$$

In a similar catalysed reaction alkenes are produced by the addition of monosilylated derivatives **168** to aldehydes in the presence of a catalytic amount of caesium fluoride in dimethyl sulphoxide (equation 138)[210].

$$\underset{(168)}{Me_3Si\diagup\diagdown CO_2Et} + RCHO \xrightarrow[DMSO]{CsF} \underset{R}{\overset{OSiMe_3}{\diagup}}\diagdown CO_2Et \quad (138)$$

$$\xrightarrow{\Delta} R\diagup\diagdown CO_2Et$$

VII. STABILIZATION OF DEVELOPING NEGATIVE CHARGE BY SILICON

A. α-Silylepoxides

In addition to stabilizing an α-carbanion, silicon also stabilizes negative charge build-up on the α-carbon in a transition state. This situation is found in nucleophilic substitutions where substantial Nu—C bond formation precedes C—X bond cleavage. One example of this phenomenon is the ring-opening reactions of α,β-epoxysilanes. In the absence of stronger activating effects, α,β-epoxysilanes undergo α C—O bond cleavage in both nucleophilic and electrophilic ring opening (equation 139).

[Structure: epoxide with SiR₃ → NuE → EO-C-C(SiR₃)(Nu)] (139)

It might be predicted that electrophilic attack would take place by a S_N1 - type process, to give a β-silyl carbocation on cleavage of the β C—O bond. However, the relative orientations of the C—Si bond and the developing positive charge are such that hyperconjugative overlap is minimal (169).

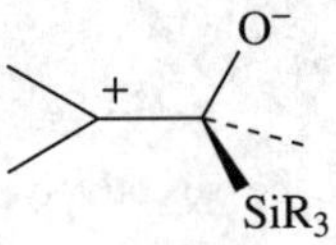

(169)

Consequently, the predominant reaction of α-silylepoxides is α-cleavage, due to the stabilization of the transition state for the S_N2 reaction α to silicon. Some recent examples are shown in equations 140[211] and 141[212].

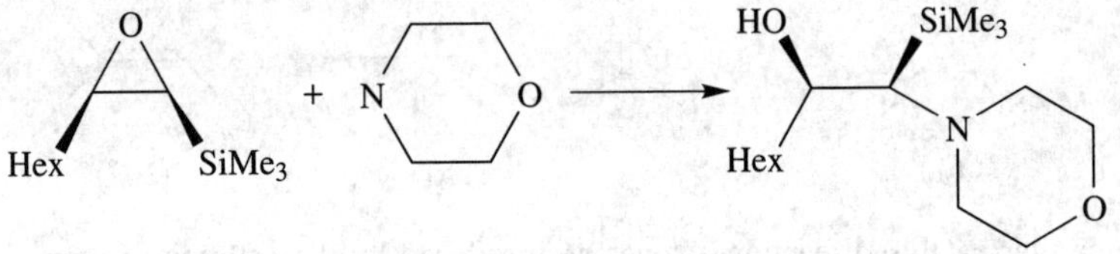

(140)

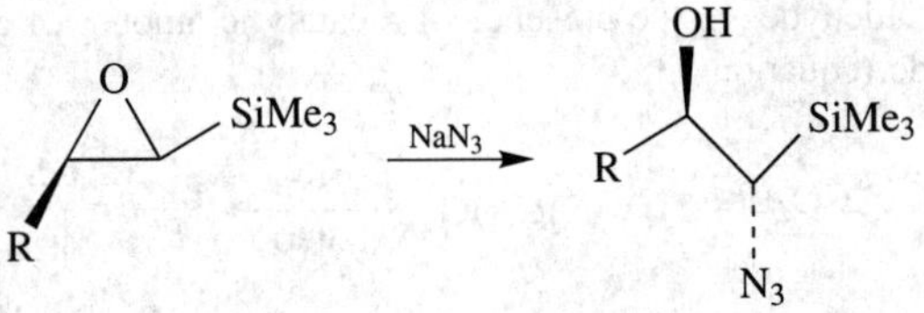

(141)

However, β-cleavage is observed in some cases, especially when steric factors are important. The regioselectivity of the reaction of epoxysilane 170 with α-sulphonyl anions derived from 171 (Scheme 9) was found to depend upon the substitution in the anion[213].

The ratios of products 172 and 173 arising from α-attack and β-attack are shown in Table 15 for sulphones with varying R^1 and R^2 groups.

Due to steric hindrance by the trimethylsilyl group, attack at the α-carbon becomes increasingly disfavoured as the steric bulk of the carbanion increases. With cyclohexyl phenyl sulphone the hindrance is such that exclusive β-attack is observed.

Replacement of the trimethylsilyl group by a triphenylsilyl group gave exclusive addition in the β-position for all sulphones studied[214].

TABLE 15. Products formed in the reaction of trimethylsilylepoxide 170 with α-sulphonyl anions

R^1, R^2	% Yield 172 (α-attack)	% Yield 173 (β-attack)
n-C_5H_{11}, H	57	—
i-Pr, H	35	40
Me, Me	31	45
—$(CH_2)_5$—	—	73

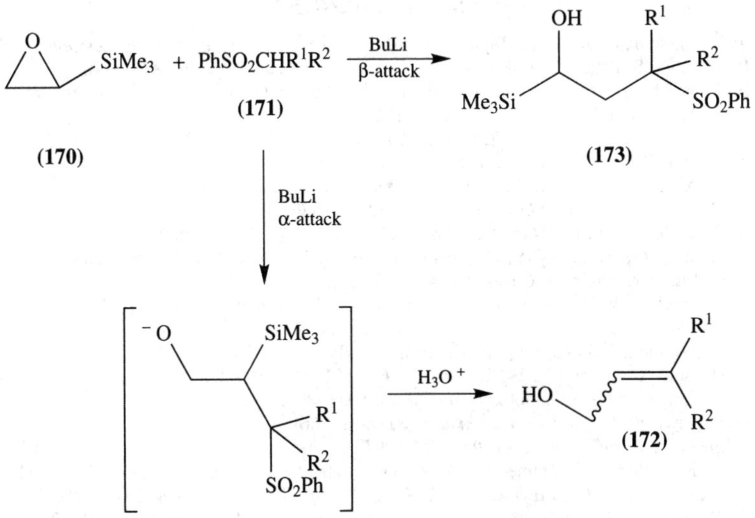

7. Activating and directive effects of silicon

SCHEME 9

The reaction of triphenylsilylepoxide **174** with Grignard reagents gives the unexpected product **175** (equation 142)[215].

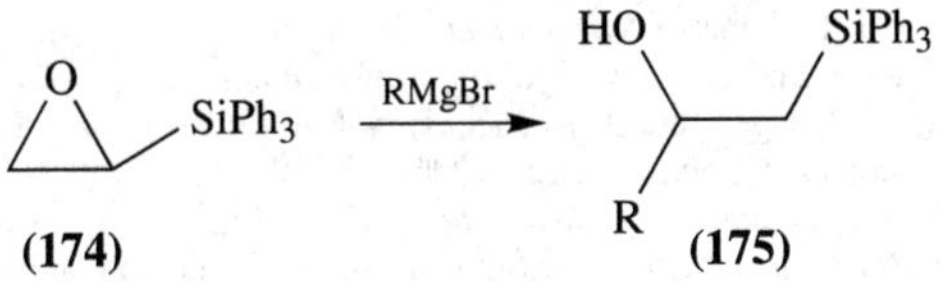

(142)

Magnesium halides are known to promote rearrangement of α, β-epoxysilanes **176** to β-silyl carbonyl compounds **177**, as shown in equation 143[216,217].

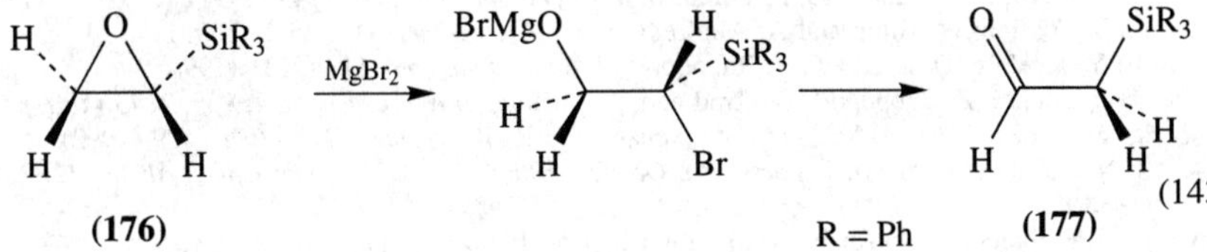

(143)

In the reaction in equation 142, the aldehyde **177**, R = Ph, produced by the magnesium halide promoted rearrangement, is then trapped by the Grignard reagent to give the observed product **175** (equation 144).

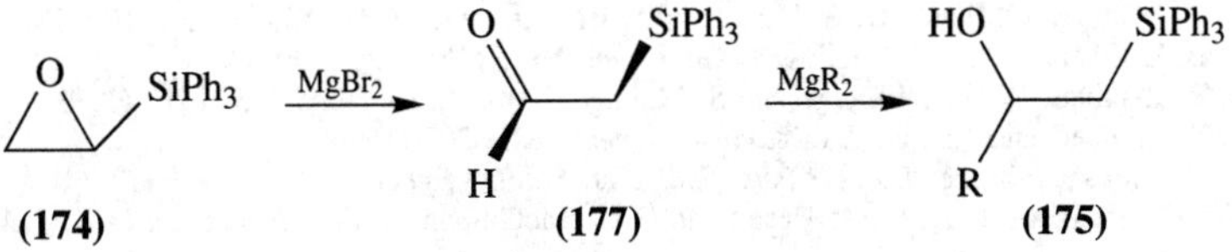

(144)

VIII. REFERENCES

1. A. R. Bassindale and P. G. Taylor, in *The Chemistry of Organic Silicon Compounds*, Vol. 1, Part 2 (Eds. S. Patai and Z. Rappoport), Wiley, Chichester, 1989, pp. 893–964.
2. E. A. V. Ebsworth, in *Organometallic Compounds of the Group IV Elements*, Vol. 1 (Ed. A. G. MacDiarmid), Dekker, New York, 1968, pp. 1–104.
3. L. Pauling, *The Nature of the Chemical Bond and the Structure of Molecules and Crystals*, 3rd edn., Cornell Univ. Press, Ithaca, New York, 1960, p. 93.
4. A. L. Allred and E. G. Rochow, *J. Inorg. Nucl. Chem.*, **5**, 264 (1958).
5. R. D. Topsom, in *Prog. Phys. Org. Chem.*, **12**, 1 (1976).
6. M. J. S. Dewar and P. J. Grisdale, *J. Am. Chem. Soc.*, **84**, 3539 (1962).
7. L. Pauling, *The Nature of the Chemical Bond*, 2nd edn., Cornell Univ. Press, New York, 1950.
8. C. G. Pitt, *J. Organomet. Chem.*, **61**, 49 (1973).
9. T. G. Traylor, W. Hanstein, H. J. Berwin, N. A. Clinton and R. S. Brown, *J. Am. Chem. Soc.*, **93**, 5715 (1971).
10. L. P. Hammett, *Physical Organic Chemistry*, 2nd edn., McGraw-Hill, New York, 1970.
11. C. Hansch, A. Leo and R. W. Taft, *Chem. Rev.*, **91**, 165 (1991).
12. H. C. Brown and Y. Okamoto, *J. Am. Chem. Soc.*, **80**, 4979 (1958).
13. O. Exner, *Coll. Czech. Chem. Commun.*, **31**, 65 (1966).
14. J. Lipowitz, *J. Am. Chem. Soc.*, **94**, 1582 (1972).
15. W. F. Reynolds, G. K. Hamer and A. R. Bassindale, *J. Chem. Soc., Perkin Trans. 2*, 971 (1977).
16. S. Ehrenson, R. T. C. Brownlee and R. W. Taft, *Prog. Phys. Org. Chem.*, **10**, 1 (1973).
17. W. Adcock, J. Alste, S. Q. A. Rizvi and M. Auranagzeb, *J. Am. Chem. Soc.*, **98**, 1701 (1976).
18. A. R. Katritzky, R. F. Pinelli, M. V. Sinnott and R. D. Topsom, *J. Am. Chem. Soc.*, **92**, 6801 (1970).
19. A. R. Bassindale, C. Eaborn, D. R. M. Walton and D. J. Young, *J. Organomet. Chem.*, **20**, 49 (1969).
20. T. Schaefer, R. Sebastian and G. H. Penner, *Can. J. Chem.*, **69**, 496 (1991).
21. M. E. Crestoni and S. Fornarini, *J. Organomet. Chem.*, **465**, 109 (1994).
22. F. R. Cartledge and J. P. Jones, *J. Organomet. Chem.*, **67**, 379 (1974).
23. M. A. Cook, C. Eaborn and D. R. M. Walton, *J. Organomet. Chem.*, **29**, 389 (1971).
24. R. Bakhtiar, C. M. Holznagel and D. B. Jacobson, *J. Am. Chem. Soc.*, **114**, 3227 (1992).
25. S. G. Cho, *J. Organomet. Chem.*, **510**, 25 (1996).
26. I. Dostrovsky and E. D. Hughes, *J. Chem. Soc.*, 157 (1946).
27. P. J. Stang, M. Ladika, Y. Apeloig, A. Stanger, M. D. Schiavelli and M. R. Hughey, *J. Am. Chem. Soc.*, **104**, 6852 (1982).
28. S. G. Wierschke, J. Chandrasekhar and W. L. Jorgensen, *J. Am. Chem. Soc.*, **107**, 1496 (1985).
29. G. A. McGibbon, M. A. Brook and J. K. Terlouw, *J. Chem. Soc., Chem. Commun.*, 360 (1992).
30. M. J. Bausch and Y. Gong, *J. Am. Chem. Soc.*, **116**, 5963 (1994).
31. N. Shimizu, E. Osijama and Y. Tsuno, *Bull. Chem. Soc. Jpn.*, **64**, 1145 (1991).
32. (a) Y. Apeloig, R. Biton and A. Abufreih, *J. Am. Chem. Soc.*, **115**, 2522 (1993).
    (b) Y. Apeloig, D. Arad and Z. Rappoport, *J. Am. Chem. Soc.*, **112**, 9131 (1990).
    (c) E. Nadler, Z. Rappoport, D. Arad and Y. Apeloig, *J. Am. Chem. Soc.*, **109**, 7873 (1987).
33. M. Mishima, T. Ariga, Y. Tsuno, K. Ikenaga and K. Kikukawa, *Chem. Lett.*, 489 (1992).
34. S. N. Ushakov and I. M. Itenberg, *Zh. Obshch. Khim.*, **7**, 2495 (1937); *Chem. Abstr.*, **32**, 2083 (1938).
35. L. H. Sommer, E. Dorfman, G. M. Goldberg and F. C. Whitmore, *J. Am. Chem. Soc.*, **68**, 488 (1946).
36. L. H. Sommer, D. L. Bailey and F. C. Whitmore, *J. Am. Chem. Soc.*, **70**, 2869 (1948).
37. A. W. P. Jarvie, *Organometal. Chem. Rev. A*, **6**, 153 (1970).
38. A. W. P. Jarvie, A. Holt and J. Thompson, *J. Chem. Soc. (B)*, 746 (1970).
39. M. R. Ibrahim and W. L. Jorgensen, *J. Am. Chem. Soc.*, **111**, 819 (1989).
40. W. Adcock, C. I. Clark and C. H. Schiesser, *J. Am. Chem. Soc.*, **118**, 11541 (1996).
41. J. B. Lambert, G. Wang, R. B. Finzel and D. H. Teramura, *J. Am. Chem. Soc.*, **109**, 7838 (1987).
42. J. B. Lambert and E. C. Chelius, *J. Am. Chem. Soc.*, **112**, 8120 (1990).
43. J. B. Lambert, R. W. Emblidge and S. Malony, *J. Am. Chem. Soc.*, **115**, 1317 (1993).
44. J. B. Lambert and X. Liu, *J. Organomet. Chem.*, **521**, 203 (1996).
45. X. Li and J. A. Stone, *J. Am. Chem. Soc.*, **111**, 5586 (1989).
46. W. Zhang, J. A. Stone, M. A. Brook and G. A. McGibbon, *J. Am. Chem. Soc.*, **118**, 5764 (1996).

47. A. J. Kresge and J. B. Tobin, *Angew. Chem., Int. Ed. Engl.*, **32**, 721 (1993).
48. A. J. Kresge and J. B. Tobin, *J. Phys. Org. Chem.*, **4**, 587 (1991).
49. V. Gabelica and A. J. Kresge, *J. Am. Chem. Soc.*, **118**, 3838 (1996).
50. D. Hajdasz and R. Squires, *J. Chem. Soc., Chem. Commun.*, 1212 (1988).
51. H.-U. Siehl and F.-P. Kaufmann, *J. Am. Chem. Soc.*, **114**, 4937 (1992).
52. N. Shimizu, S. Watanabe and Y. Tsuno, *Bull. Chem. Soc. Jpn.*, **64**, 2249 (1991).
53. N. Shimizu, C. Kinoshita, E. Osajina, F. Hayakawa and Y. Tsuno, *Bull. Chem. Soc. Jpn.*, **64**, 3280 (1991).
54. F. Hayakawa, S. Watanabe, N. Shimizu and Y. Tsuno, *Bull. Chem. Soc. Jpn.*, **66**, 153 (1993).
55. R. Fujiyama and T. Munechika, *Tetrahedron Lett.*, **34**, 5907 (1993).
56. N. Shimizu, S. Watanabe, F. Hayakawa, S. Yasuhara, Y. Tsuno and T. Inazu, *Bull. Chem. Soc. Jpn.*, **67**, 500 (1994).
57. M. A. Brook and A. Neuy, *J. Org. Chem.*, **55**, 3609 (1990).
58. J. B. Lambert and R. A. Singer, *J. Am. Chem. Soc.*, **114**, 10246 (1992).
59. M. J. Doyle, A. J. Kirby, J. M. Percy and P. R. Raithby, *Acta Crystallogr., Sect. C*, **48**, 866 (1992).
60. P. G. Jones, A. J. Kirby and J. K. Parker, *Acta Crystallogr., Sect. C*, **48**, 868 (1992).
61. J. M. White, *Aust. J. Chem.*, **48**, 1227 (1995).
62. J. M. White and G. B. Robertson, *J. Org. Chem.*, **57**, 4638 (1992).
63. Y. Kuan and J. M. White, *J. Chem. Soc., Chem. Commun.*, 1195 (1994).
64. A. J. Green, Y. Kuan and J. M. White, *J. Org. Chem.*, **60**, 2734 (1995).
65. V. J. Shiner, Jr., M. W. Ensinger, G. S. Kriz and K. A. Halley, *J. Org. Chem.*, **55**, 653 (1990).
66. V. J. Shiner, Jr., M. W. Ensinger and R. D. Rutkowske, *J. Am. Chem. Soc.*, **109**, 804 (1987).
67. C. A. Grob and P. Sawlewicz, *Tetrahedron Lett.*, **28**, 951 (1987).
68. V. J. Shiner, Jr., M. W. Ensinger and J. C. Huffman, *J. Am. Chem. Soc.*, **111**, 7199 (1989).
69. W. Kirmse and F. Sollenbohmer, *J. Am. Chem. Soc.*, **111**, 4127 (1989).
70. W. Adcock and G. B. Kok, *J. Org. Chem.*, **52**, 356 (1987).
71. J. B. Lambert, *Tetrahedron*, **46**, 2677 (1990).
72. R. J. Fessenden, K. Seiles and M. Dagani, *J. Org. Chem.*, **31**, 2433 (1966).
73. W. Adcock, A. R. Krstic, P. J. Duggan, V. J. Shiner, Jr., J. Coope and M. W. Ensinger, *J. Am. Chem. Soc.*, **112**, 3140 (1990).
74. D. Hrovat and W. T. Borden, *J. Org. Chem.*, **57**, 2519 (1992).
75. W. Adcock and N. A. Trout, *J. Org. Chem.*, **56**, 3229 (1991).
76. W. Adcock, H. Gangodawila, G. B. Kok, V. S. Iyer, W. Kitching, G. M. Drew and D. Young, *Organometallics*, **6**, 156 (1987).
77. I. Fleming, in *Comprehensive Organic Chemistry*, Vol. 3 (Eds. D. H. R. Barton and W. D. Ollis), Chap. 13, Pergamon Press, Oxford, 1979.
78. A. C. Hopkinson and M. H. Lien, *J. Org. Chem.*, **46**, 998 (1981).
79. C. Glidewell and C. Thomson, *J. Comput. Chem.*, **3**, 495 (1982).
80. D. M. Wetzel and J. I. Brauman, *J. Am. Chem. Soc.*, **110**, 8333 (1988).
81. E. A. Brinkman, S. Berger and J. I. Brauman, *J. Am. Chem. Soc.*, **116**, 8304 (1994).
82. S. McN. Sieburth and J. J. Somers, *Tetrahedron*, **52**, 5683 (1996).
83. S. Zhang, X-M. Zhang and F. G. Bordwell, *J. Am. Chem. Soc.*, **117**, 602 (1995).
84. E. Berliner, *Prog. Phys. Org. Chem.*, **2**, 253 (1964).
85. C. Eaborn, *J. Chem. Soc., Chem. Commun.*, 1255 (1972).
86. R. W. Bott, C. Eaborn and P. M. Greasley, *J. Chem. Soc.*, 4804 (1964).
87. T. Hashimoto and M. Seki, *Yakugaku Zasshi*, **81**, 204 (1961); *Chem. Abstr.*, **55**, 14340 (1961).
88. Y. Sakata and T. Hashimoto, *Yakugaku Zasshi*, **80**, 730 (1960); *Chem. Abstr.*, **54**, 24480 (1960).
89. C. Eaborn and D. E. Webster, *J. Chem. Soc.*, 4449 (1957).
90. C. Eaborn and D. E. Webster, *J. Chem. Soc.*, 179 (1960).
91. G. V. Motsarev, V. T. Inshhakova, V. I. Kolbasov and V. R. Rosenberg, *Zh. Obshch. Khim.*, **44**, 1053 (1974); *Chem. Abstr.*, **81**, 105611 (1974).
92. T. Hashimoto, *Yakugaku Zasshi*, **87**, 528 (1967); *Chem. Abstr.*, **61**, 54206 (1967).
93. R. J. Mills and V. Snieckus, *Tetrahedron Lett.*, **25**, 483 (1984).
94. P. Babin, B. Bennetau, P. Bourgeois, F. Rajarison and J. Dunogues, *Bull. Soc. Chim. Fr.*, **129**, 25 (1992).
95. G. W. Kalbalka and R. S. Varma, *Tetrahedron*, **45**, 6601 (1989).
96. S. M. Moerlein, W. Beyer amd G. Stocklin, *J. Chem. Soc., Perkin Trans. 1*, 779 (1988).

97. M. Speranza, C-Y. Shiue, A. P. Wolf, D. S. Wilbur and J. Angelini, *J. Fluorine Chem.*, **30**, 97 (1985).
98. G. Felix, J. Dunogues and R. Calas, *Angew. Chem., Int. Ed. Engl.*, **18**, 402 (1979).
99. G. Felix, M. Laguerre, J. Dunogues and R. Calas, *J. Chem. Res.*, 236 (1980).
100. R. Calas, *Compt. Rend. Acad. Sci., Ser. 2*, **301**, 1289 (1985).
101. B. Bennetau and J. Dunogues, *Synlett*, 171 (1993).
102. B. Bennetau, F. Rajarison, J. Dunogues and P. Babin, *Tetrahedron*, **49**, 10843 (1993).
103. R. J. Mills, N. J. Taylor and V. Snieckus, *J. Org. Chem.*, **54**, 4372 (1989).
104. H. Ishibashi, H. Sakashita and M. Ikeda, *J. Chem. Soc., Perkin Trans. 1*, 1953 (1992).
105. C. Eaborn, T. A. Emokpae, V. I. Sidorov and R. Taylor, *J. Chem. Soc., Perkin Trans. 2*, 1454 (1974).
106. R. W. Bott, C. Eaborn and R. Taylor, *J. Chem. Soc.*, 4927 (1961).
107. K. Mikami, K. Nishi and T. Nakai, *Tetrahedron Lett.*, **24**, 795 (1983).
108. I. Fleming and A. Pearce, *J. Chem. Soc., Perkin Trans. 1*, 2485 (1980).
109. F. Duboudin, *J. Organomet. Chem.*, **156**, C25 (1978).
110. R. B. Miller and T. Reichenbach, *Tetrahedron Lett.*, 543 (1974).
111. R. B. Miller and G. McGarvey, *Synth. Commun.*, **8**, 291 (1978).
112. R. B. Miller and G. McGarvey, *Synth. Commun.*, **7**, 475 (1977).
113. K. Miura, S. Okajima, T. Hondo and A. Hosomi, *Tetrahedron Lett.*, **36**, 1483 (1995).
114. K. Miura, T. Hondo, S. Okajima and A. Hosomi, *Tetrahedron Lett.*, **37**, 387 (1996).
115. M. A. Brook, C. Henry, R. Jueschke and P. Modi, *Synlett*, 97 (1993).
116. C. Henry, R. Juschke and M. A. Brook, *Inorg. Chim. Acta*, **220**, 145 (1994).
117. C. Henry and M. A. Brook, *Tetrahedron*, **50**, 11379 (1994).
118. I. Fleming, J. Dunogues and R. Smithers, *Org. React.*, **37**, 57 (1989).
119. K. E. Koenig and W. P. Weber, *J. Am. Chem. Soc.*, **95**, 3416 (1973).
120. T. H. Chan and T. Fleming, *Synthesis*, 761 (1979).
121. T. H. Chan and K. Koumaglo, *Tetrahedron Lett.*, **27**, 883 (1986).
122. T. H. Chan, P. W. K. Lau and W. Mychajlowskij, *Tetrahedron Lett.*, 3317 (1977).
123. T. H. Chan and K. Koumaglo, *J. Organomet. Chem.*, **285**, 109 (1985).
124. K. Mikami, H. Wakabayashi and T. Nakai, *J. Org. Chem.*, **56**, 4337 (1991).
125. N. Kishi, K. Mikami and T. Nakai, *Tetrahedron*, **47**, 8111 (1991).
126. L. E. Overman and A. T. Robichard, *J. Am. Chem. Soc.*, **111**, 300 (1989).
127. G. W. Daub, D. A. Heerding and L. E. Overman, *Tetrahedron*, **44**, 3919 (1988).
128. R. M. Lett, L. E. Overman and J. Zablocki, *Tetrahedron Lett.*, **29**, 6541 (1988).
129. R. M. Burk and L. E. Overman, *Heterocycles*, **35**, 205 (1993).
130. P. Castro, L. E. Overman, X. M. Zhang and P. S. Mariano, *Tetrahedron Lett.*, **34**, 5243 (1993).
131. M. A. Brook and C. Henry, *Tetrahedron*, **52**, 861 (1996).
132. M. I. Alhassan, *J. Organomet. Chem.*, **372**, 183 (1989).
133. D. R. M. Walton and F. Waugh, *J. Organomet. Chem.*, **37**, 45 (1972).
134. H. Mayr, E. Bauml, G. Cibura and R. Koschinsky, *J. Org. Chem.*, **57**, 768 (1992).
135. I. Fleming and S. K. Patel, *Tetrahedron Lett.*, **22**, 2321 (1981).
136. H. Mayr and G. Hagen, *J. Chem. Soc., Chem. Commun.*, 91 (1989).
137. G. Hagen and H. Mayr, *J. Am. Chem. Soc.*, **113**, 4954 (1991).
138. M. Hojo, K. Ohsumi and A. Hosomi, *Tetrahedron Lett.*, **33**, 5981 (1992).
139. G. A. Olah, D. S. Van Vliet, Q. Wang and G. S. Prakash, *Synthesis*, 159 (1995).
140. S. H. Yeon, B. W. Lee, B. R. Yoo, M-Y. Suk and I. N. Jung, *Organometallics*, **14**, 2361 (1995).
141. C. LeRoux and J. Dubac, *Organometallics*, **15**, 4646 (1996).
142. V. K. Aggarwal and G. P. Vennall, *Tetrahedron Lett.*, **37**, 3745 (1996).
143. E. W. Colvin and M. Monteith, *J. Chem. Soc., Chem. Commun.*, 1230 (1990).
144. B. Guyot, J. Pornet and L. Miginiac, *Tetrahedron*, **47**, 3981 (1991).
145. R. L. Funk, J. Umstead-Daggett and K. M. Brummond, *Tetrahedron Lett.*, **34**, 2867 (1993).
146. M. Franciotti, A. Mann, A. Mordini and M. Taddei, *Tetrahedron Lett.*, **34**, 1355 (1993).
147. H. Mayr, A. O. Gabriel and R. Schumacher, *Justus Liebigs Ann. Chem.*, 1583 (1995).
148. M. Polla and T. Frejd, *Acta Chem. Scand.*, **47**, 716 (1993).
149. A. Hosomi and H. Sakurai, *J. Am. Chem. Soc.*, **99**, 1673 (1977).
150. D. Schinzer, *Synthesis*, 263 (1988).
151. H. J. Knolker, N. Foitzik, R. Graf and J-B. Pannek, *Tetrahedron*, **49**, 9955 (1993).
152. R. L. Danheiser, B. R. Dixon and R. W. Gleason, *J. Org. Chem.*, **57**, 6094 (1992).

153. G. P. Brengal and A. I. Meyers, *J. Org. Chem.*, **61**, 3230 (1996).
154. H. J. Knolker, G. Baum and R. Graf, *Angew. Chem., Int. Ed. Engl.*, **33**, 1612 (1994).
155. G. Majetich, K. Hull, A. M. Casares and V. Khetani, *J. Org. Chem.*, **56**, 3958 (1991).
156. H. Wetter and P. Scherer, *Helv. Chim. Acta*, **66**, 118 (1983).
157. T. Hayashi, M. Konishi, H. Ito and M. Kumada, *J. Am. Chem. Soc.*, **104**, 4962 (1982).
158. T. Hayashi, M. Konishi and M. Kumada, *J. Am. Chem. Soc.*, **104**, 4963 (1982).
159. T. Hayashi, H. Ito and M. Kumada, *Tetrahedron Lett.*, **23**, 4605 (1982).
160. G. Wickham and W. J. Kitching, *J. Org. Chem.*, **48**, 612 (1983).
161. I. Fleming and N. K. Terrett, *J. Organomet. Chem.*, **264**, 99 (1984).
162. S. D. Khan, C. F. Pau, A. R. Chamberlin and W. J. Hehre, *J. Am. Chem. Soc.*, **109**, 650 (1987).
163. I. Fleming and D. Higgins, *J. Chem. Soc., Perkin Trans. 1*, 3327 (1992).
164. M. J. C. Buckle, I. Fleming and S. Gil, *Tetrahedron Lett.*, **33**, 4481 (1992).
165. I. Fleming, G. R. Jones, N. D. Kindon, Y. Landais, C. P. Leslie, I. T. Morgan, S. Peukert and A. K. Sarkar, *J. Chem. Soc., Perkin Trans. 1*, 1171 (1996).
166. C. E. Masse and J. S. Panek, *Chem. Rev.*, **95**, 1293 (1995).
167. C. Nativi, G. Palio and M. Taddei, *Tetrahedron Lett.*, **32**, 1583 (1991).
168. M. Laguerre, M. Grignon-Dubois and J. Dunogues, *Tetrahedron*, **37**, 1161 (1981).
169. J. P. Pillot, B. Bennetau, J. Dunogues and R. Calas, *Tetrahedron Lett.*, **22**, 3401 (1981).
170. D. Schinzer and G. Panke, *J. Org. Chem.*, **61**, 4496 (1996).
171. D. Schinzer, J. Kabbara and K. Ringe, *Tetrahedron Lett.*, **33**, 8017 (1992).
172. R. L. Danheiser, B. R. Dixon and R. W. Gleason, *J. Org. Chem.*, **57**, 6094 (1992).
173. R. L. Danheiser and D. J. Carini, *J. Org. Chem.*, **45**, 3925 (1980).
174. R. L. Danheiser, D. J. Carini and A. Basak, *J. Am. Chem. Soc.*, **103**, 1604 (1981).
175. M. J. C. Buckle and I. Fleming, *Tetrahedron Lett.*, **34**, 2383 (1993).
176. P. A. Smith, in *Molecular Rearrangements*, Part 1 (Ed. P. de Mayo), Wiley-Interscience, New York, 1963, pp. 577–589.
177. P. F. Hudrlik, A. M. Hudrlik, G. Nagandrappa, T. Yimenu, E. T. Zellers and E. Chin, *J. Am. Chem. Soc.*, **102**, 6894 (1980).
178. S. R. Wilson and M. J. diGrandi, *J. Org. Chem.*, **56**, 4766 (1991).
179. J. R. Hwu and J. M. Wetzel, *J. Org. Chem.*, **57**, 922 (1992).
180. J. R. Hwu and B. A. Gilbert, *J. Am. Chem. Soc.*, **113**, 5917 (1991).
181. T. A. Engler and J. P. Reddy, *J. Org. Chem.*, **56**, 6491 (1991).
182. D. J. Ager, *Org. React.*, **38**, 1 (1990).
183. D. J. Peterson, *J. Org. Chem.*, **33**, 780 (1968).
184. P. A. Brough, S. Fisher, B. Zhao, R. C. Thomas and V. Snieckus, *Tetrahedron Lett.*, **37**, 2915 (1996).
185. H. Imanieh, P. Quayle, M. Voaden, J. Conway and S. D. A. Street, *Tetrahedron Lett.*, **33**, 543 (1992).
186. A. G. Brook, J. M. Duff and D. G. Anderson, *Can. J. Chem.*, **48**, 561 (1970).
187. A. G. M. Barrett and J. A. Flygare, *J. Org. Chem.*, **56**, 638 (1991).
188. W. Dumont and A. Krief, *Angew. Chem., Int. Ed. Engl.*, **15**, 161 (1976).
189. A. R. Bassindale, R. J. Ellis and P. G. Taylor, *Tetrahedron Lett.*, **25**, 2705 (1984).
190. D. Seyferth, T. Wada and G. Raab, *Tetrahedron Lett.*, 20 (1960).
191. D. Seyferth and T. Wada, *Inorg. Chem.*, **1**, 78 (1962).
192. M. R. Stober, K. W. Michael and O. L. Speier, *J. Org. Chem.*, **32**, 2740 (1967).
193. K. Tomao, R. Kanatani and M. Kumada, *Tetrahedron Lett.*, **25**, 1905 (1984).
194. T. Hayama, S. Tomoda, Y. Takeuchi and Y. Nomura, *Tetrahedron Lett.*, **24**, 2795 (1983).
195. G. Stork and J. Singh, *J. Am. Chem. Soc.*, **96**, 6181 (1974).
196. R. K. Boeckman Jr., *J. Am. Chem. Soc.*, **96**, 6179 (1974).
197. H. O. House, *Acc. Chem. Res.*, **9**, 59 (1976).
198. P. F. Hudrlik, E. L. O. Agwaramgbo and A. M. Hudrlik, *J. Org. Chem.*, **54**, 5613 (1989).
199. T. Kawashima, N. Iwama and R. Okazaki, *J. Am. Chem. Soc.*, **114**, 7598 (1992).
200. A. R. Bassindale, R. J. Ellis, J. C-Y. Lau and P. G. Taylor, *J. Chem. Soc., Chem. Commun.*, 98 (1986).
201. H. B. Burgi and J. D. Dunitz, *Acc. Chem. Res.*, **16**, 153 (1983).
202. A. R. Bassindale, R. J. Ellis and P. G. Taylor, *J. Chem. Res (S)*, 34 (1996).
203. A. Couture, H. Cornet and P. Grandclaudon, *J. Organomet. Chem.*, **440**, 7 (1992).
204. C. Burford, F. Cooke, E. Ehlinger and P. Magnus, *J. Am. Chem. Soc.*, **99**, 4536 (1977).

205. E. Ehlinger and P. Magnus, *J. Am. Chem. Soc.*, **102**, 5004 (1976).
206. P. W. K. Lau and T. H. Chan, *Tetrahedron Lett.*, 1137 (1978).
207. I. Fleming and C. D. Lloyd, *J. Chem. Soc., Perkin Trans. 1*, 969 (1981).
208. I. Fleming, I. T. Morgan and A. K. Sarkar, *J. Chem. Soc., Chem. Commun.*, 1575 (1990).
209. C. Palomo, J. M. Aizpurua, J. M. Garcia, I. Ganboa, F. P. Cossio, B. Lecea and C. Lopez, *J. Org. Chem.*, **55**, 2498 (1990).
210. M. Bellassoued and N. Ozanne, *J. Org. Chem.*, **60**, 6582 (1995).
211. P. F. Hudrlik, A. M. Hudrlik and A. K. Kulkarni, *Tetrahedron Lett.*, **26**, 139 (1985).
212. A. R. Bassindale, P. G. Taylor and Y. Xu, *Tetrahedron Lett.*, **37**, 555 (1996).
213. P. Jankowski, S. Marczak, M. Masnyk and J. Wicha, *J. Chem. Soc., Chem. Commun.*, 297 (1991).
214. P. Jankowski and J. Wicha, *J. Chem. Soc., Chem. Commun.*, 802 (1992).
215. P. F. Hudrlik, M. E. Ahmed, R. R. Roberts and A. M. Hudrlik, *J. Org. Chem.*, **61**, 4395 (1996).
216. A. G. Brook, D. McRae and A. R. Bassindale, *J. Organomet. Chem.*, **86**, 185 (1975).
217. P. F. Hudrlik, R. N. Misra, G. P. Withers, A. M. Hudrlik, R. J. Rona and J. P. Arcoleo, *Tetrahedron Lett.*, 1453 (1976).

CHAPTER 8

Steric effects of silyl groups

R. JIH-RU HWU, SHWU-CHEN TSAY and BUH-LUEN CHENG

Organosilicon and Synthesis Laboratory, Institute of Chemistry, Academia Sinica, Nankang, Taipei, Taiwan 11529, Republic of China and Department of Chemistry, National Tsing Hua University, Hsinchu, Taiwan 30043, Republic of China
Fax: 886-2-7881337; e-mail: JRHWU@chem.nthu.edu.tw

I. INTRODUCTION	433
II. AN ORDER IN INCREASING SIZE OF SILYL GROUPS	433
III. COMPARISON OF ORGANOSILYL GROUPS IN SUBSTRATES	438
A. General Considerations	438
B. Organic Reactions of Various Types	438
1. Acylation	438
2. Addition	438
3. Aldol condensation	441
4. Alkylation	442
5. Allylation	443
6. Cyclization	443
7. [2 + 2] Cycloaddition	445
8. [4 + 2] Cycloaddition	446
9. Decomposition	446
10. Deprotonation	447
11. Desilylation	447
12. Elimination	448
13. Ene reaction	449
14. Epoxidation	449
15. Hydroboration	450
16. Hydroethoxycarbonylation	451
17. Hydroformylation	451
18. Hydrosilylation	451
19. Isomerization	452
20. Migration and rearrangement	452
21. Osmylation	454
22. Oxymercuration	455
23. Ozonolysis	455
24. Peterson olefination	456
25. Phosphonylation	456

The chemistry of organic silicon compounds, Vol. 2
Edited by Z. Rappoport and Y. Apeloig © 1998 John Wiley & Sons Ltd

26. Polymerization	457
27. Reduction	457
28. Simmons–Smith reaction	457
29. Substitution	458

IV. COMPARISON OF THE INFLUENCE RESULTING FROM HYDROGEN ATOM AND ALKYL GROUPS VERSUS SILYL GROUPS ... 459
  A. General Considerations ... 459
  B. Organic Reactions of Various Types ... 459
    1. Addition ... 459
    2. Aldol reaction ... 460
    3. Alkylation ... 460
    4. Allylboration ... 461
    5. Allylsilylation ... 461
    6. Carbenoid rearrangement ... 462
    7. Cyclization ... 462
    8. [2 + 2] Cycloaddition ... 465
    9. [3 + 2] Cycloaddition ... 465
    10. [4 + 2] Cycloaddition ... 466
    11. Cyclopropanation ... 467
    12. Dehydration ... 469
    13. Ene reaction ... 469
    14. Epoxidation ... 470
    15. Hydride reduction ... 470
    16. Hydroboration ... 472
    17. Hydrogenation ... 473
    18. Lithiation ... 473
    19. Oxidation ... 473
    20. Photocyclization ... 473
    21. Reduction ... 474
    22. [2,3]-Sigmatropic rearrangement ... 474
    23. Substitution ... 474

V. REACTIVITY OF SILICON-CONTAINING REAGENTS ... 475
  A. General Considerations ... 475
  B. Organic Reactions of Various Types ... 475
    1. Addition and substitution ... 475
    2. Condensation ... 477
    3. [3 + 2] Cycloaddition ... 477
    4. [4 + 2] Cycloaddition ... 478
    5. Hydrosilylation ... 479
    6. Insertion reaction ... 479
    7. Nitrone formation ... 479
    8. Olefination ... 480
    9. Oxygen–oxygen bond cleavage ... 481
    10. Silylation ... 481
    11. Silylformylation ... 481
    12. Silylstannation ... 482

VI. SOLVOLYSIS OF VARIOUS ORGANOSILANES ... 482
  A. Adamantane p-Nitrobenzoates ... 482
  B. Benzylic p-Toluenesulfonates ... 482
  C. Silyl Ethers ... 482
  D. Triorganosilyl Chlorides and Fluorides ... 483
  E. Triorganosilyl Hydrides ... 483

VII. CHANGE OF PHYSICAL, CHEMICAL AND SPECTROSCOPIC PROPERTIES BY INTRODUCTION OF SILYL GROUPS 483
A. Change of Physical and Chemical Properties 484
1. Conformation, bond angles and bond lengths 484
2. Stability, free energy of activation and reaction rates 485
3. Migratory aptitudes of organic groups 487
B. Change of Spectroscopic Properties 487
1. ^{1}H, ^{13}C, ^{19}F and ^{29}Si NMR spectroscopy 487
2. X-ray crystallography 487
VIII. CONCLUSION 488
IX. ACKNOWLEDGMENT 488
X. REFERENCES 488

I. INTRODUCTION

Triorganosilyl groups are frequently used in organic reactions. Appropriate placement of those groups in substrates, in reagents or in catalysts can control the stereochemical outcome of the reactions because of their steric influence. The most popular silyl group is the Me$_3$Si, which is referred to as the 'bulky proton'[1-3]. In 1989, a review was published on the steric effect of the Me$_3$Si group[2].

In this chapter, the influence resulting from many different silyl groups on organic compounds and reactions will be discussed. Topics are selected from papers published mainly between mid-1988 and mid-1996. During this period of time, several informative reviews and articles on organosilicon chemistry have been published by Chan and Wang[4,5], Fleming[6], Page and coworkers[7], Cirillo and Panek[8], Rücker[9] and Steinmetz[10]. The chemistry discussed in those papers will not be emphasized in this chapter although some crucial information is cited in the text or listed in Table 1.

II. AN ORDER IN INCREASING SIZE OF SILYL GROUPS

A comparison of steric influence resulting from more than 80 different silyl groups is summarized in Table 1. The trend is established on the basis of reaction yields, rates and selectivity, as well as on reactivity and physical properties of organosilanes. Two different effects may be involved: (a) the effect of a silyl group on reactions taking place at the neighboring centers, and (b) the effect of the groups attached to silicon on the nucleophilic attack at the silicon atom.

The results from analyses of many research groups' reports published in the past decade indicate that often both effects display a parallel trend. An order of the bulk of 44 silyl groups is thus established as follows:

smaller ⟶ bulkier
(less influence)   (greater influence)

Me$_2$HSi < Me$_3$Si < PhMe$_2$Si < Ph$_2$MeSi < EtMe$_2$Si < (PhCH$_2$)Me$_2$Si < (n-Pr)Me$_2$Si ∼ Ph$_3$Si ∼ (t-Bu)$_2$HSi < Et$_2$MeSi ∼ (n-Bu)Me$_2$Si ∼ (n-C$_6$H$_{13}$)Me$_2$Si < (Me$_3$Si)Me$_2$Si < (i-Bu)Me$_2$Si < (cyclopentyl)Me$_2$Si ∼ (n-Pr)$_2$MeSi < (n-Bu)$_2$MeSi < Et$_3$Si < (i-Pr)Me$_2$Si < (cyclohexyl)Me$_2$Si < (s-Bu)Me$_2$Si < (n-Bu)(i-Pr)MeSi ∼ (n-Pr)$_3$Si < (i-Pr)(−CH$_2$CH$_2$CH$_2$CH$_2$−)Si < [(t-Bu)$_2$CH]Me$_2$Si ∼ (t-Bu)Me$_2$Si ∼ (i-Pr)$_2$MeSi < (EtMe$_2$C)Me$_2$Si < (n-PrMe$_2$C)Me$_2$Si < (i-PrMe$_2$C)Me$_2$Si < (n-Bu)$_3$Si < (n-hexyl)$_3$Si < (i-Bu)$_3$Si < [(t-Bu)CH$_2$]$_3$Si < [(t-Bu)EtMeC]Me$_2$Si < (i-Pr)$_3$Si < (t-Bu)Ph$_2$Si < (t-Bu)(−CH$_2$CH$_2$CH$_2$CH$_2$−)Si < (t-Bu)(i-Pr)EtSi < (t-Bu)(i-Pr)$_2$Si < (t-Bu)$_2$MeSi < (t-Bu)$_3$Si < (t-Bu)(cyclohexyl)$_2$Si < (cyclohexyl)$_3$Si

TABLE 1. Order of the size of organosilyl groups on the basis of their influence on various reactions

Entry	Reaction or property studied	Selectivity basis	Order of the size	References
1	1,2-addition	%ee and yield	$(i\text{-Pr})_3\text{Si} < (t\text{-Bu})\text{Ph}_2\text{Si}$	11
2	1,2-addition	rate and diastereo-selectivity	$\text{Me}_3\text{Si} < \text{Et}_3\text{Si} < (t\text{-Bu})\text{Me}_2\text{Si} < (t\text{-Bu})\text{Ph}_2\text{Si} < (i\text{-Pr})_3\text{Si}$	12,13
3	1,2-addition	regioselectivity	$\text{Ph}_3\text{Si} < (t\text{-Bu})\text{Ph}_2\text{Si} < \text{Et}_3\text{Si} < (i\text{-Pr})_3\text{Si}$	14
4	Mukaiyama addition	stereoselectivity	$\text{Me}_3\text{Si} < \text{Et}_3\text{Si} < (t\text{-Bu})\text{Me}_2\text{Si} < (i\text{-Pr})_3\text{Si}$	15
5	Mukaiyama addition	stereoselectivity	$\text{Me}_3\text{Si} < \text{Et}_3\text{Si} < (t\text{-Bu})\text{Me}_2\text{Si} < \text{PhMe}_2\text{Si} < \text{Ph}_2\text{MeSi}$	16
6	addition	cone angle	$\text{Me}_3\text{Si} < \text{ClMe}_2\text{Si} < \text{PhMe}_2\text{Si} < \text{Et}_3\text{Si} < (n\text{-Bu})_3\text{Si} \sim (n\text{-hexyl})_3\text{Si} < (n\text{-Bu})\text{Me}_2\text{Si} < \text{Ph}_3\text{Si} < (i\text{-Pr})_3\text{Si}$	17
7	alkylation	yield	$\text{Me}_3\text{Si} < \text{PhMe}_2\text{Si} < \text{Ph}_2\text{MeSi} < \text{Ph}_3\text{Si}$	18
8	alkylation	γ/α site of alkylation	$\text{Me}_3\text{Si} < \text{Ph}_3\text{Si} < \text{Et}_3\text{Si} < (n\text{-Pr})_3\text{Si}$	19
9	allylation vs dihydrofuranation	ratio	$\text{Me}_3\text{Si} < \text{PhMe}_2\text{Si} < (i\text{-Pr})_3\text{Si}$	20
10	allylation	regio-selectivity	$(t\text{-Bu})\text{Me}_2\text{Si} < (i\text{-Pr})_3\text{Si}$	21
11	dihydrofuranation	yield	$\text{Me}_3\text{Si} < \text{PhMe}_2\text{Si} < (i\text{-Pr})_3\text{Si}$	20
12	cyclization	stereoselectivity	$\text{Me}_3\text{Si} < \text{Et}_3\text{Si}$	22
13	iodolactonization	diastereoselectivity	$(t\text{-Bu})\text{Me}_2\text{Si} < (t\text{-Bu})\text{Ph}_2\text{Si} < (i\text{-Pr})_3\text{Si}$	23
14	iodolactonization	diastereoselectivity	$(t\text{-Bu})\text{Me}_2\text{Si} < (t\text{-Bu})\text{Ph}_2\text{Si} < (i\text{-Pr})_3\text{Si}$	24
15	iodolactonization	diastereoselectivity	$(t\text{-Bu})\text{Me}_2\text{Si} < (i\text{-Pr})_3\text{Si}$	25
16	cyclization	cis/trans ratio	$\text{Me}_3\text{Si} < \text{Ph}_2\text{MeSi} < \text{Ph}_3\text{Si}$	26
17	cyclization	product distribution	$\text{PhMe}_2\text{SiMe}_2\text{Si} < \text{PhMe}_2\text{SiEt}_2\text{Si} < \text{PhMe}_2\text{SiPh}_2\text{Si} < \text{PhMe}_2\text{Si}(i\text{-Bu})_2\text{Si} < \text{PhMe}_2\text{Si}(i\text{-Pr})_2\text{Si}$	27
18	decomposition	product distribution	$\text{Me}_3\text{Si} < \text{Et}_3\text{Si} < (t\text{-Bu})\text{Me}_2\text{Si} < (i\text{-Pr})_3\text{Si}$	28
19	desilylation	rate	$\text{Et}_3\text{Si} < (t\text{-Bu})\text{Me}_2\text{Si} < (i\text{-PrMe}_2\text{C})\text{Me}_2\text{Si} \sim (t\text{-Bu})\text{Ph}_2\text{Si}$	29
20	desilylation	rate	$(t\text{-Bu})\text{Me}_2\text{Si} < (i\text{-PrMe}_2\text{C})\text{Me}_2\text{Si} < (i\text{-Pr})_3\text{Si} < (t\text{-Bu})\text{Ph}_2\text{Si}$	30
21	desilylation	rate	$\text{Me}_3\text{Si} < (t\text{-Bu})\text{Me}_2\text{Si} \sim (i\text{-Pr})_3\text{Si} < (t\text{-Bu})\text{Ph}_2\text{Si}$	31
22	dehydrogenolysis	rate	$(t\text{-Bu})\text{Me}_2\text{Si} < (i\text{-Pr})_3\text{Si} < (t\text{-Bu})\text{Ph}_2\text{Si}$	32
23	desilylation	chemoselectivity	$(t\text{-Bu})\text{Me}_2\text{Si} < (i\text{-Pr})_3\text{Si}$	33
24	desilylation	chemoselectivity	$(t\text{-Bu})\text{Me}_2\text{Si} < (t\text{-Bu})\text{Ph}_2\text{Si}$	34,35
25	elimination	rate	$\text{Me}_3\text{Si} < \text{PhMe}_2\text{Si} < \text{Ph}_2\text{MeSi} < \text{Ph}_3\text{Si}$	36
26	ene reaction	stereoselectivity	$(t\text{-Bu})\text{Me}_2\text{Si} < (t\text{-Bu})\text{Ph}_2\text{Si}$	37

TABLE 1. (continued)

Entry	Reaction or property studied	Selectivity basis	Order of the size	References
27	epoxidation	stereoselectivity	Me_3Si < $(Me_3Si)Me_2Si$	38
28	hydroboration	stereoselectivity	Me_3Si < Et_3Si < $(t\text{-}Bu)Me_2Si$ < $(i\text{-}Pr)_3Si$	39
29	hydroformylation	regioselectivity	Me_3Si < $(t\text{-}Bu)Me_2Si$ < Ph_3Si < $(t\text{-}Bu)Ph_2Si$	40
30	hydrosilylation	reactivity	Ph_2HSi << $PhMe_2Si$ < Ph_2MeSi < Ph_3Si	41
31	isomerization	ratio	$(t\text{-}Bu)Ph_2Si$ < $(t\text{-}Bu)Me_2Si$ < $(i\text{-}Pr)_3Si$	42
32	Brook vs. Wittig rearrangements	chemoselectivity	Me_3Si < $(t\text{-}Bu)Ph_2Si$ < Et_3Si < $(i\text{-}PrMe_2C)Me_2Si$ and $(i\text{-}Pr)_3Si$	43
33	silyl migration	yield	Me_3Si < Et_3Si < $(n\text{-}Pr)_3Si$ < $(i\text{-}Pr)_3Si$	44
34	intermolecular migration	product distribution	Me_3Si < Et_3Si < $(i\text{-}Pr)Me_2Si$ < $(t\text{-}Bu)Me_2Si$ < $(i\text{-}Pr)_3Si$	45
35	osmylation	stereoselectivity	Me_3Si < $PhMe_2Si$ < Et_3Si < $(t\text{-}Bu)Me_2Si$ < Ph_2MeSi	46
36	oxymercuration	diastereoselectivity	$(t\text{-}Bu)Me_2Si$ < $(t\text{-}Bu)Ph_2Si$	47
37	ozonolysis	stereoselectivity	Me_3Si < $(i\text{-}Pr)_3Si$ < $(t\text{-}Bu)Me_2Si$	48
38	ozonolysis	rate	$(n\text{-}Bu)_2HSi$ < $(i\text{-}Pr)_2HSi$ < Et_2MeSi < $(i\text{-}Pr)_2MeSi$ < $(n\text{-}Bu)_2MeSi$ < Et_3Si < $(n\text{-}Pr)_3Si$ < $(n\text{-}Bu)_3Si$ < $(t\text{-}Bu)(cyclohexyl)_2Si$ < $(cyclohexyl)_3Si$	49,50
39	Peterson olefination	diastereomeric ratio	Me_3Si < Et_3Si < $(cyclohexyl)_2MeSi$ < $(t\text{-}Bu)Me_2Si$ < $(t\text{-}Bu)Ph_2Si$ < Ph_3Si	51
40	Peterson olefination	diastereomeric ratio and yield	Me_3Si < Et_3Si < $(t\text{-}Bu)Me_2Si$ < $(t\text{-}Bu)Ph_2Si$	52,53
41	Peterson olefination	E/Z ratio	Me_3Si < Et_3Si < $(i\text{-}Pr)_3Si$	54
42	phosphonylation	diastereoselectivity	Et_3Si < $(t\text{-}Bu)Me_2Si$ < $(t\text{-}Bu)Ph_2Si$ < $(i\text{-}Pr)_3Si$	55
43	reduction	product distribution	Me_3Si < Et_3Si	56
44	substitution	Taft's E_s value	Me_3Si < $EtMe_2Si$ < $(n\text{-}Pr)Me_2Si$ < $(n\text{-}Bu)Me_2Si$ < $(i\text{-}Bu)Me_2Si$ < $(cyclopentyl)Me_2Si$ < $(i\text{-}Pr)Me_2Si$ < $(cyclohexyl)Me_2Si$ < $(s\text{-}Bu)Me_2Si$	57
45	epoxidation	stereoselectivity	Ph_3Si < $(t\text{-}Bu)Me_2Si$	58
46	1,4-addition	stereoselectivity	Me_3Si < $(Me_3Si)Me_2Si$	38
47	[3 + 2] cycloaddition	yield	Me_3Si < $PhMe_2Si$ < $(t\text{-}Bu)Me_2Si$	59
48	hydrosilylation	product distribution	$(n\text{-}C_5H_{11})H_2Si$ < Ph_2HSi < $(n\text{-}Pr)_2HSi$ < $(n\text{-}Bu)MeHSi$ < Et_2MeSi < Et_3Si	60

(continued overleaf)

TABLE 1. (continued)

Entry	Reaction or property studied	Selectivity basis	Order of the size	References
49	hydrosilylation	rate	$EtMe_2Si$ < Et_2MeSi < Et_3Si < $(n\text{-}Pr)_2MeSi$ < $(i\text{-}Pr)_3Si$ < $(i\text{-}Pr)_2MeSi$	61
50	[2 + 1] insertion	rate and yield	Ph_3Si < Et_3Si	62
51	insertion	regioselectivity	$(t\text{-}Bu)Me_2Si$ < $(i\text{-}Pr)_3Si$	63
52	olefination	regioselectivity	Me_3Si < Et_3Si < $(t\text{-}Bu)Me_2Si$ < Ph_3Si	64
53	silylation	heat of reaction and rate	Ph_2MeSi < Me_3Si < $(t\text{-}Bu)Ph_2Si$ < $(i\text{-}Pr)_3Si$	65
54	solvolysis of $PhCH(SiR_3)OTs$	rate	Me_3Si < Et_3Si < $(t\text{-}Bu)Me_2Si$	66
55	solvolysis of R_3SiOPh	rate (H^+, aq EtOH)	Me_3Si < $(t\text{-}Bu)Me_2Si$ < Et_3Si < $(n\text{-}Pr)_3Si$ < $(n\text{-}Bu)_3Si$	67,68
56	solvolysis of R_3SiOPh	rate (OH^-, aq EtOH)	Me_3Si < Et_3Si < $(t\text{-}Bu)Me_2Si$ < $(n\text{-}Pr)_3Si$ < $(n\text{-}Bu)_3Si$	67,68
57	solvolysis of triorgano-chlorosilanes	rate (aq dioxane)	Me_2HSi < $Me_2(Cl_2HC)Si$ < $(CF_3CH_2O)Me_2Si$ < $(ClH_2C)Me_2Si$ < $[MeO(CH_2)_2O]Me_2Si$ < $(p\text{-}O_2NC_6H_4)Me_2Si$ < $[NC(CH_2)_3]Me_2Si$ < Me_3Si < $(MeO)Me_2Si$ < $[MeO(CH_2)_3]Me_2Si$ < $EtMe_2Si$ < $(EtO)Me_2Si$ < $(PhCH_2)Me_2Si$ < $(n\text{-}Pr)Me_2Si$ < $(i\text{-}PrO)Me_2Si$ < $(n\text{-}Bu)Me_2Si$ < $(Me_3Si)Me_2Si$ < $(i\text{-}Bu)Me_2Si$ < $PhMe_2Si$ < $(n\text{-}C_{18}H_{37})Me_2Si$ < $(t\text{-}BuCH_2O)Me_2Si$ < $(p\text{-}MeC_6H_4)Me_2Si$ < $(Me_3SiCH_2)Me_2Si$ < $(t\text{-}BuCH_2)Me_2Si$ < $(i\text{-}Pr)Me_2Si$ < $(s\text{-}Bu)Me_2Si$ < $(n\text{-}Bu)_2MeSi$ < $(cyclohexyl)Me_2Si$ < $(t\text{-}BuO)Me_2Si$ < $(EtO)_3Si$ < $(Et_2CH)Me_2Si$ < Et_3Si < Ph_2MeSi < $(n\text{-}Bu)_3Si$ < $(i\text{-}Pr)_2MeSi$ < Ph_3Si < $(t\text{-}Bu)Me_2Si$ < $(i\text{-}PrMe_2C)Me_2Si$ < $(i\text{-}Pr)_3Si$ < $(t\text{-}Bu)Ph_2Si$	69
58	solvolysis of triorgano-chlorosilanes	reactivity (aq dioxane)	Me_2HSi < $(CF_3CH_2O)Me_2Si$ < $(p\text{-}O_2NC_6H_4)Me_2Si$ < Me_3Si < $(MeO)Me_2Si$ < $(EtO)Me_2Si$ < $EtMe_2Si$ < $(PhCH_2)Me_2Si$ < $(PhO)Me_2Si$ < $(n\text{-}PrO)Me_2Si$ < $(n\text{-}Pr)Me_2Si$ < $(n\text{-}Bu)Me_2Si$ ~ $(n\text{-}C_6H_{13})Me_2Si$ < $(Me_3Si)Me_2Si$ < $(i\text{-}PrO)Me_2Si$ < $(n\text{-}C_{18}H_{37})Me_2Si$ < $PhMe_2Si$ < $(i\text{-}Bu)Me_2Si$ ~ $(Me_3SiO)Me_2Si$ < $(t\text{-}BuCH_2O)Me_2Si$ < $(p\text{-}MeC_6H_4)Me_2Si$ < $(Me_3SiCH_2)Me_2Si$ < $(t\text{-}BuCH_2)Me_2Si$ < $(i\text{-}Pr)Me_2Si$ < $(n\text{-}Bu)_2MeSi$ < $(s\text{-}Bu)Me_2Si$ < $(cyclohexyl)Me_2Si$	70

8. Steric effects of silyl groups

TABLE 1. (continued)

Entry	Reaction or property studied	Selectivity basis	Order of the size	References
			< (Et_2CH)Me_2Si < [(i-Pr)MeCH]Me_2Si < (t-BuO)Me_2Si < Ph_2MeSi < [(i-Pr)Me_2CO]Me_2Si < (EtO)$_3$Si < Et_3Si < [(t-Bu)MeCH]Me_2Si < [(i-Pr)$_2$CH]Me_2Si < (n-Bu)$_3$Si < Ph_3Si < (i-Pr)$_2$MeSi < [(t-Bu)$_2$CH]Me_2Si < (t-Bu)Me_2Si < (EtMe_2C)Me_2Si < (i-PrMe_2C)Me_2Si < (Et_3C)Me_2Si < [(t-Bu)Me_2C]Me_2Si < [(t-Bu)CH_2]$_3$Si < [(t-Bu)EtMeC]Me_2Si < (i-Pr)$_3$Si < (t-Bu)Ph_2Si < (t-Bu)(i-Pr)EtSi < (t-BuO)$_3$Si < (t-Bu)(i-Pr)$_2$Si < (t-Bu)$_2$MeSi < (t-Bu)$_3$Si	
59	solvolysis of R_3SiF	rate (aq acetone)	Et_2MeSi < Et_3Si < (n-Bu)(i-Pr)MeSi < (n-Bu)$_3$Si < (i-Pr)$_3$Si	71
60	solvolysis of R_3SiH	rate (H^+, aq EtOH)	(cyclohexyl)H_2Si < (n-Pr)$_2$HSi < (ClH_2C)Me_2Si < (n-Pr)Me_2Si < (n-Pr)$_2$MeSi < Et_3Si < (n-Pr)$_3$Si < (n-Bu)$_3$Si < (i-Bu)$_3$Si < (i-Pr)$_3$Si	72,73
61	solvolysis of R_3SiH	rate (OH^-, aq EtOH)	$EtMe_2Si$ < (n-Pr)Me_2Si < Et_2MeSi < (n-Pr)$_2$MeSi < Et_3Si < (i-Pr)Me_2Si < (n-Pr)$_3$Si < (i-Pr)$_2$MeSi < (t-Bu)Me_2Si < (i-Pr)$_3$Si	74
62	distortion of geometry	degree of distortion	Me_3Si < Et_3Si < Ph_3Si ~ (t-Bu)$_2$HSi	75
63	complexation	bond length	Me_2HSi < Me_3Si < (t-Bu)$_2$HSi	76
64	complexation	physical data	$PhMe_2Si$ ~ Ph_2MeSi < Me_3Si < Ph_3Si	77
65	complexation	bond length	Me_3Si < (Me_3Si)Me_2Si < (Me_3Si)$_2$MeSi	78
66	ring flip	A value	Me_3Si < (Me_3Si)$_3$Si < (Me_3Si)$_3$C	79
67	decay	rate	Me_3Si < Et_3Si < (i-Pr)$_3$Si	80
68	^{1}H NMR spectroscopy	peak broadening	Me_3Si < Et_3Si < (t-Bu)Me_2Si	81
69	^{1}H NMR spectroscopy	coupling constant	Me_3Si < $PhMe_2Si$ < (t-Bu)Ph_2Si	82
70	1,2-addition	diastereoselectivity	(i-Pr)Me_2Si < (t-Bu)Me_2Si < (i-Pr)$_3$Si	83
71	1,2-addition	selectivity	Me_3Si < Et_3Si < (t-Bu)Me_2Si < (i-Pr)$_3$Si < (t-Bu)Ph_2Si	84
72	Diels–Alder reaction	selectivity	Me_3Si < (t-Bu)Me_2Si < (i-Pr)$_3$Si	85
73	Diels–Alder reaction	chemoselectivity	Me_3Si < (t-Bu)Me_2Si < (i-Pr)$_3$Si	86
74	ene reactions	selectivity	(i-Pr)Me_2Si < (i-PrMe_2C)Me_2Si < (i-Pr)$_3$Si < (t-Bu)Ph_2Si	87
75	rearrangement	regioselectivity	Me_3Si < (t-Bu)Me_2Si < (i-Pr)$_3$Si	88

(continued overleaf)

TABLE 1. (continued)

Entry	Reaction or property studied	Selectivity basis	Order of the size	References
76	[3 + 2] cycloaddition	selectivity	$(t\text{-Bu})\text{Me}_2\text{Si} < (i\text{-Pr})_3\text{Si} < (t\text{-Bu})\text{Ph}_2\text{Si}$	89
77	deoxygenation	diastereoselectivity	$(t\text{-Bu})\text{Me}_2\text{Si} < (i\text{-Pr})_3\text{Si}$	90
78	epoxide opening	regioselectivity	$\text{Et}_3\text{Si} \sim (i\text{-Pr})\text{Me}_2\text{Si} < \text{PhMe}_2\text{Si} < (t\text{-Bu})\text{Me}_2\text{Si} < \text{Ph}_3\text{Si} < (i\text{-Pr})_3\text{Si}$	91
79	Ireland–Claisen rearrangement	selectivity	$\text{Me}_3\text{Si} \sim \text{Et}_3\text{Si} < (n\text{-PrMe}_2\text{C})\text{Me}_2\text{Si} < (t\text{-Bu})\text{Me}_2\text{Si} < (i\text{-Pr})_3\text{Si}$	92
80	oxidation	enantioselectivity	$(t\text{-Bu})\text{Me}_2\text{Si} < (i\text{-Pr})_3\text{Si}$	93
81	photolysis	quantum yield	$\text{Me}_3\text{Si} < \text{Et}_3\text{Si} < (i\text{-Pr})_3\text{Si}$	94

The trend of increasing the bulk agrees with most of the information listed in Table 1, as well as with the order established in 1989[2]. The current order is, however, much more informative than that (including 23 silyl groups) published earlier.

III. COMPARISON OF ORGANOSILYL GROUPS IN SUBSTRATES

A. General Considerations

Different silyl groups in substrates exert steric influence on organic reactions in various degrees. Evidence for this exists in the following 29 types of important organic reactions, which can be controlled by bulky silyl groups. Their steric effect often dominates the outcome of those reactions. Nevertheless, the electronic or the stereoelectronic effect may simultaneously also play a minor role.

B. Organic Reactions of Various Types

1. Acylation

With an unsubstituted phenyl group on silicon, *trans*-PhCH=CHSiMe₂Ph is converted to *trans*-PhCH=CHC(=O)R upon reaction with RC(=O)Cl in the presence of TiCl₄[95]. This reaction involves transfer of a styryl group to acyl chlorides. Introduction of methyl groups into the *ortho* positions of the phenyl group attached to silicon results in synergistic activation through relief of steric strain. Thus dearylation becomes preferable.

2. Addition[83,84]

Diethylzinc adds to α-silyloxyaldehydes **1** in the presence of a catalytic amount of (1*R*,2*R*)-bis(trifluoromethanesulfonamido)cyclohexane (**2**) and Ti(OPr-*i*)₄ (2 equiv) in toluene at −20 °C (equation 1)[11]. The yields of the 1,2-adducts **3** and the enantioselectivity of the transformation depend upon the steric environment resulting from the silyl group (see entry 1 of Table 1). The asymmetric addition involving the substrate bearing an (*i*-Pr)₃Si group gives 89% yield of the corresponding alcohol with 85 %ee. The enantioselectivity can be increased to 92 %ee, although the yield drops to 78%, by use of a bulky (*t*-Bu)Ph₂Si group to shield the aldehydic oxygen.

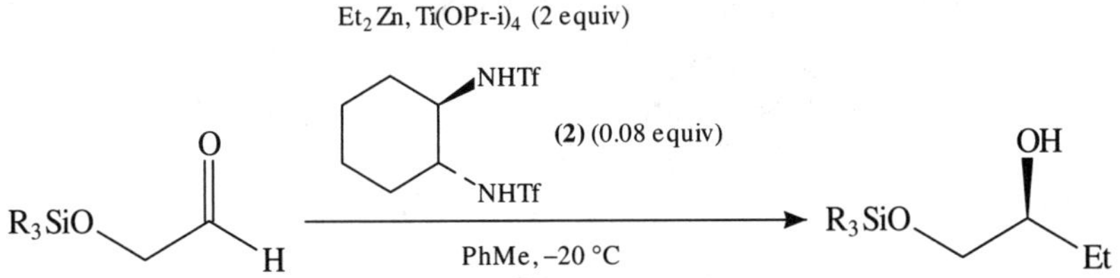

8. Steric effects of silyl groups

$$R_3SiO\text{-CH}_2\text{-CHO} \ (1) \xrightarrow[\text{PhMe, }-20\,°\text{C}]{\text{Et}_2\text{Zn, Ti(OPr-}i)_4 \text{ (2 equiv)}, \ (2)\ (0.08\ \text{equiv})} R_3SiO\text{-CH}_2\text{-CH(OH)Et} \ (3) \quad (1)$$

where (2) is the bis-triflamide cyclohexane ligand (NHTf, NHTf).

R_3Si	Chemical yield (%)	%ee
$(i\text{-Pr})_3Si$	89	85
$(t\text{-Bu})Ph_2Si$	78	92

On the other hand, specific rates of 1,2-additions of Me_2Mg to ketones $PhCOCH(OSiR_3)Me$ **(4)** parallel the diastereoselectivity of the reactions, as predicted by Cram's rule[12]. Experimental results from complexation and kinetics (equation 2) indicate a steady decrease in chelating ability upon an increase in size of the silyl groups[13,96]. The order is shown in entry 2 of Table 1.

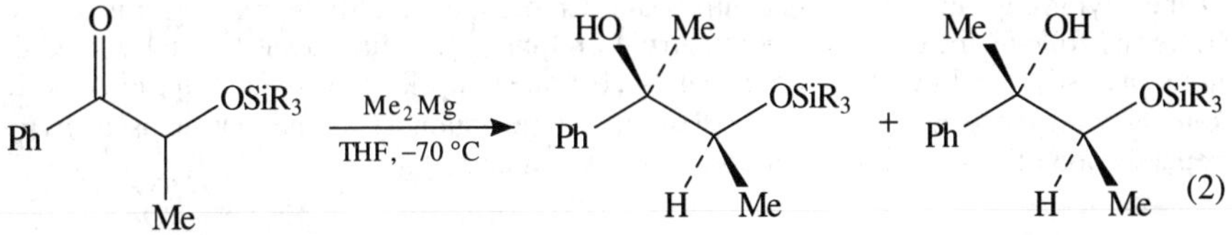

$$Ph\text{-CO-CH(Me)(OSiR}_3) \ \textbf{(4)} \xrightarrow[\text{THF, }-70\,°\text{C}]{Me_2Mg} Ph\text{-C(OH)(Me)-CH(Me)(OSiR}_3) \ (RS/SR) + Ph\text{-C(Me)(OH)-CH(Me)(OSiR}_3) \ (RR/SS) \quad (2)$$

'Cram's rule product'

(4) (excess)

R_3Si	$k_2(\times 10^2\ M^{-1}\ s^{-1})$	$(RS/SR)/(RR/SS)$
Me_3Si	100 ± 30	99/1
Et_3Si	8 ± 1	96/4
$(t\text{-Bu})Me_2Si$	2.5 ± 0.3	88/12
$(t\text{-Bu})Ph_2Si$	0.82 ± 0.06	63/37
$(i\text{-Pr})_3Si$	0.45 ± 0.04	42/58

Furthermore, organobarium reagents, prepared from anhydrous BaI_2 and (silyloxy)allyllithium, react with carbonyl compounds at the least sterically hindered terminus (i.e. the γ position)[14]. The ratio of γ- to α-products depends upon the size of the silyl group and follows the order shown in entry 3 of Table 1.

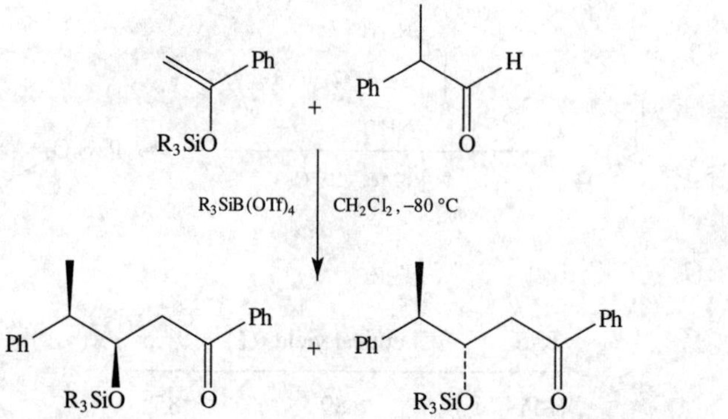

$$
\text{(3)}
$$

(RS/SR) R_3Si	Overall yield (%)	Selectivity (RS/SR) : (RS/RS)
Me_3Si	44	8 : 1
Et_3Si	60	18 : 1
$(t\text{-}Bu)Me_2Si$	82	25 : 1
$(i\text{-}Pr)_3Si$	71	97 : 1

The Mukaiyama addition of a silyl enol ether to an aldehyde can be catalyzed by $TiCl_4$, $TiCl_3(OPr\text{-}i)$[97], and 'supersilylating agents' $R_3SiB(OTf)_4$ (equation 3)[15]. The steric bulk of the silyl groups in enol ethers controls the 'Cram-type' selectivity, ranging from 8:1 to 97:1, and follows the order shown in entry 4 of Table 1. Similarly, $SnCl_4$-catalyzed additions of a silyl enol ether to various silyl-substituted acyclic acetals give diastereomeric ketones with ratios ranging from 5.0:1 to 13.0:1 (equation 4)[16]. The diastereoselectivity depends upon the size of silyl groups, as shown in entry 5.

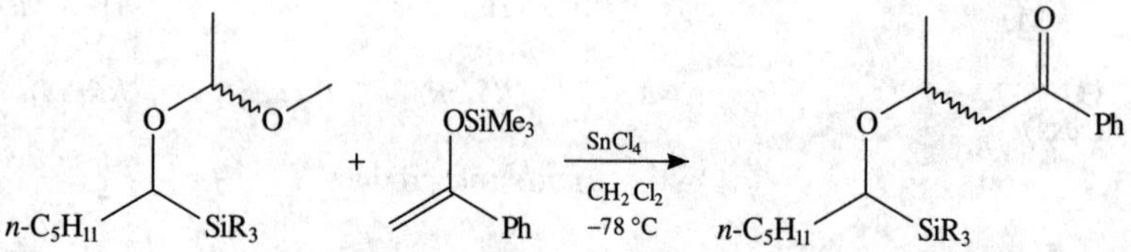

$$
\text{(4)}
$$

R_3Si	Yield (%)	Selectivity
Me_3Si	71	5.0 : 1
Et_3Si	73	5.5 : 1
$(t\text{-}Bu)Me_2Si$	66	6.6 : 1
$PhMe_2Si$	83	11.6 : 1
Ph_2MeSi	59	13.0 : 1

Addition followed by silylation of O-silylcyanohydrins **5**, **7** and **9** gives disilyl enamines **6**, **8** and **10**, respectively (equations 5–7)[98]. In all instances, the C-Me_3Si group of the starting cyanohydrins is transferred to the nitrogen atom in the products. The original O-silyl groups are also transferred to the nitrogen atom except for the highly hindered

8. Steric effects of silyl groups

$(i\text{-Pr})_3\text{Si}$ group[99].

$$\text{(5)} \xrightarrow[\text{2. Et}_3\text{SiCl}]{\text{1. MeLi}} \text{(6)} \quad (88\%) \tag{5}$$

$$\text{(7)} \xrightarrow[\text{2. Me}_3\text{SiCl}]{\text{1. MeLi}} \text{(8)} \quad (80\%) \tag{6}$$

$$\text{(9)} \xrightarrow[\text{2. Me}_3\text{SiCl}]{\text{1. MeLi}} \text{(10)} \quad (80\%) \tag{7}$$

A different way to determine the steric size of various silyl groups is to use the values of cone angle θ[100], which are listed in equation 8[17]. It illustrates the mechanism of addition of the (p-anisyl)phenyl methyl ion (11) to allylsilanes[101]. The rate of addition depends on the size of the silyl groups. Bulky silyl groups, the size of which is reflected by their θ values can accelerate the addition process. Given the θ values, the size of the silyl groups follows the order shown in entry 6 of Table 1.

$$\text{An} = p\text{-MeOC}_6\text{H}_4 \tag{8}$$

R_3Si	θ	R_3Si	θ	R_3Si	θ
Me_3Si	118	Et_3Si	132	$(n\text{-Bu})Me_2Si$	139
$ClMe_2Si$	120	$(n\text{-Bu})_3Si$	136	Ph_3Si	145
$PhMe_2Si$	122	$(\text{hexyl})_3Si$	136	$(i\text{-Pr})_3Si$	160

3. Aldol condensation[102]

Reaction of lithium enolates of α-silyloxyketones with benzaldehyde may yield opposite diastereoselection depending upon the size of the silyl group[103–105]. The corresponding titanium enolates, however, lead to high diastereofacial selectivities[106]. Replacement of the Me_3Si group in acyl silanes by $(t\text{-}Bu)Me_2Si$, Et_3Si and $(n\text{-}Pr)_3Si$ groups leads to diastereoselectivity increased from 1 : 1 to > 20 : 1 in the aldol condensation with aldehydes[107].

4. Alkylation

The size of silyl groups influences the carbon–carbon bond formation between silylalkenes and ketones. Treatment of allylsilanes **12** or vinylsilanes **14** with ketones in the presence of MnO_2 and acetic acid at elevated temperature gives α-alkylation products **13** and **15**, respectively (equations 9 and 10)[18]. The steric effect resulting from the silyl groups plays an essential role on the exclusive C—C bond formation at the terminal sp^2 carbon of silylalkenes **12** and **14**. The yield of the alkylation is inverse to the size of silyl groups and follows the order listed in entry 7 of Table 1.

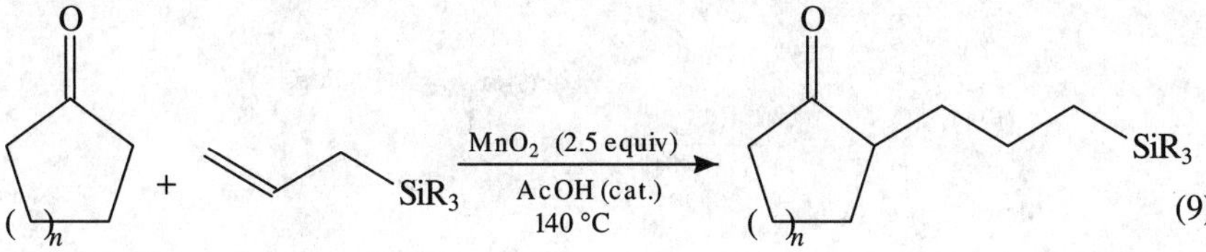

$$n = 1-8$$

(**12**)

($R_3Si = Me_3Si, PhMe_2Si, Ph_2MeSi, Ph_3Si$)

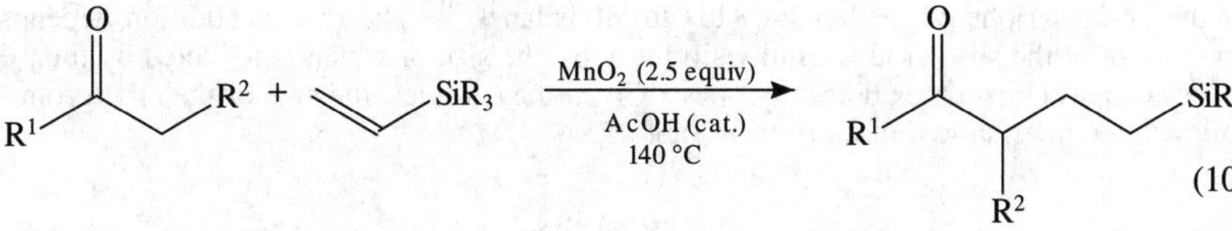

(**13**) (54–85%)   (9)

(**14**)   (**15**) (59–75%)   (10)

R^1, R^2 = Alkyl; $R_3Si = Me_3Si, Et_2MeSi$

On the other hand, methylaluminum bis(4-bromo-2,6-di-*tert*-butylphenoxide) can effect the α-alkylation of enol silyl ethers of a variety of ketones, esters and some aldehydes[108]. Use of the $(t\text{-}Bu)Me_2Si$ group is recommended in the ketene silyl acetal substrates. Use of a less bulky Me_3Si or Et_3Si group leads to a mixture of monoalkylation products and rearranged α-silyl esters.

Reaction of silylallyl anions **16** with alkyl halides gives a mixture of α- and γ-alkylated products **17** and **18**, respectively (equation 11)[19]. The ratio of γ/α alkylation increases along with the bulk of the silyl groups by following the order shown in entry 8 of Table 1.

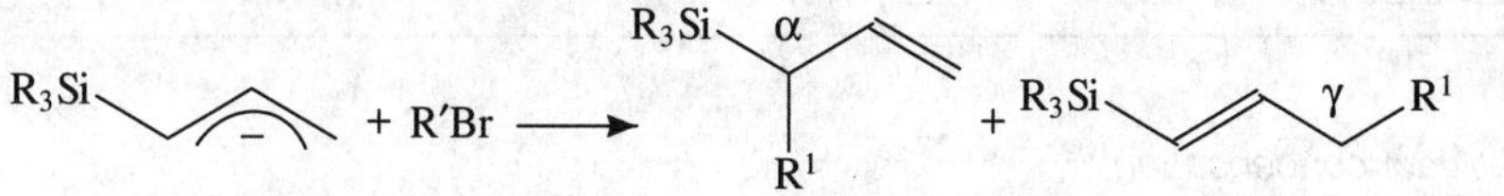

(**16**)   R' = Alkyl   (**17**)   (**18**)   (11)

R_3Si	$\alpha : \gamma$
Me_3Si	1 : 5.5
Ph_3Si	1 : 16
Et_3Si	1 : 18
$(n\text{-}Pr)_3Si$	1 : 46

5. Allylation

Allylsilanes **12** react with β-ketoester **19** in the presence of ceric ammonium nitrate (**20**) in methanol to give allylated ketoesters **21** or silylated dihydrofurans **22** or both in 76–84% overall yields (equation 12)[20]. The ratio of **21/22** depends upon the steric bulk of the silyl groups in the order listed in entry 9 of Table 1. A larger silyl group offers a greater chance to form a dihydrofuran through an intramolecular cyclization.

R_3Si	Overall yield (%)	Ratio		
		21	:	22
Me_3Si	84	100	:	0
$PhMe_2Si$	80	16	:	1
$(i\text{-}Pr)_3Si$	76	0	:	100

In the ruthenium-catalyzed allylation of terminal propargylic silyl ethers, a regioisomeric mixture of 1,4-dienes is produced[21]. Their ratio can reflect the size of the silyl groups as shown in entry 10 of Table 1.

Furthermore, regioselective (α versus γ) allylation and propargylation by use of Grignard, organozinc and tin reagents can be accomplished by use of acylsilanes as electrophiles[109–111]. Because of the bulkiness of the $(i\text{-}Pr)_3Si$ group, its use results in a greater extent of regiocontrol than that involving the Me_3Si group.

6. Cyclization

Allylsilanes **12** react with β-ketoester **19** in the presence of $Mn(OAc)_3 \cdot 2H_2O$ and acetic acid to give dihydrofurans **22** in 76–85% yields (equation 13)[20]. Efficiency of the dihydrofuran formation depends upon the steric bulk of the silyl groups shown in entry 11 of Table 1.

$$\text{(12)} \quad + \quad \text{(19)} \xrightarrow[\text{HOAc, 80 °C}]{\text{Mn(OAc)}_3 \cdot 2H_2O} \quad \text{(22)} \quad (13)$$

R_3Si	Yield (%)
Me_3Si	85
$PhMe_2Si$	82
$(i\text{-Pr})_3Si$	76

An example involving intramolecular cyclization is the treatment of silyl ethers **23** with I_2 and triethylamine in acetonitrile at 0 °C to give diastereomeric tetrahydrofurans **24** and **25** (equation 14)[22]. The Me_3Si and Et_3Si groups are expelled in the reaction and do not appear in the final cyclization products. In comparison with the Me_3Si group, cyclization of the substrate carrying a Et_3Si group gives a much higher selectivity (entry 12 of Table 1). Moreover, iodolactonizations of silyloxy-containing alkenylcarboxylic acids also give diastereomeric mixtures[23-25]. Their ratio often depends upon the size of the silyl groups as shown in entries 13–15.

$$\text{(23)} \xrightarrow[\text{MeCN, 0 °C}]{I_2, Et_3N} \text{(24)} + \text{(25)} \quad (14)$$

R_3Si	Yield (%)	Ratio 24 : 25
Me_3Si	93	6 : 1
Et_3Si	91	75 : 1

N-Bromosuccinimide can also initiate an intramolecular cyclization. Thus vinylsilane alcohols **26** are converted to a mixture of 2,5-*cis*- and *trans*-furans **27** in a ratio ranging

8. Steric effects of silyl groups

from 1 : 1 to 1 : 2.3 (equation 15)[26]. The thermodynamically favored 2,5-*trans*-isomer can be generated in a higher ratio by increasing the bulk of the silyl group in the order shown in entry 16 of Table 1.

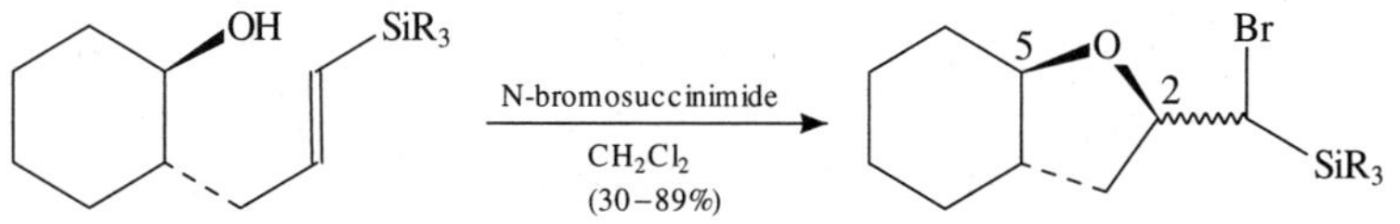

$$\text{(26)} \xrightarrow[\substack{CH_2Cl_2 \\ (30-89\%)}]{\text{N-bromosuccinimide}} \text{(27)}$$

(26) R = Me or Ph

(27) *cis*/*trans* 1:1–1:2.3

(15)

Furthermore, intramolecular bis-silylative cyclization of dienes **28** with face selectivity gives a diastereomeric mixture of **29** and **30** (equation 16). The size of the substituents on the nonterminal silicon influences the ratio of **29/30**, which follows the order Me < Et < Ph < *i*-Bu < *i*-Pr (cf. entry 17 of Table 1)[27].

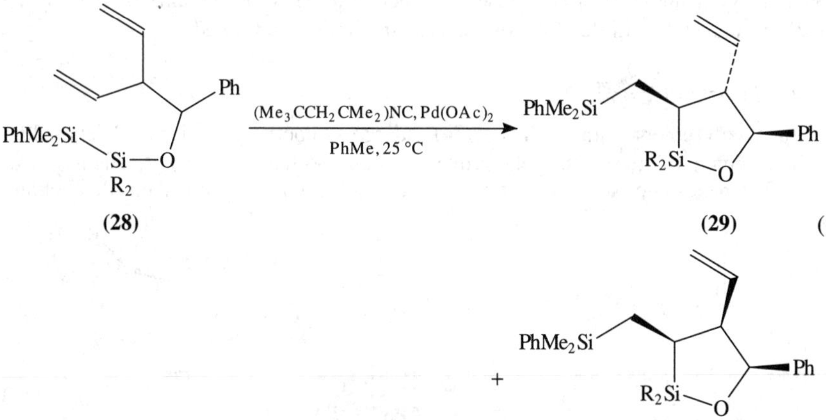

(28) → (29) + (30)   (16)

Reagents: $(Me_3CCH_2CMe_2)NC$, $Pd(OAc)_2$, PhMe, 25 °C

R	Yield (%)	Ratio		
		29	:	30
Me	98	59	:	41
Et	82	75	:	25
Ph	87	83	:	17
i-Bu	90	88	:	12
i-Pr	27	92	:	8

7. [2 + 2] Cycloaddition

In certain reactions, introduction of a bulkier silyl group in substrates may decrease the stereoselectivity. For example, photolysis of a mixture of benzaldehyde and silyl vinyl ethers **31** affords a mixture of diastereomeric [2 + 2] cycloadducts **32** and **33** (equation 17)[112]. The stereoisomeric ratio of 10 : 1 obtained by use of **31a** containing a Me_3Si group drops to 5.7 : 1 by use of **31b**, which contains a $(t\text{-}Bu)Me_2Si$ group.

$$Ph\text{-}CHO + t\text{-}Bu\text{-}C(=CH_2)\text{-}OSiR_3 \xrightarrow[PhH]{h\nu} \mathbf{(32)} + \mathbf{(33)} \quad (17)$$

(31)   (32)   (33)

	R₃Si	Yield (%)	Ratio 32	:	33
(a)	Me_3Si	65	10	:	1
(b)	$(t\text{-}Bu)Me_2Si$	61	5.7	:	1

On the other hand, upon reaction with methyl glyoxalate imines, an increase in bulk of the silyl moiety [i.e., $PhMe_2Si$, Ph_2MeSi, Ph_3Si and $(t\text{-}Bu)Ph_2Si$] in β-(triorganosilyl)alkanoyl chlorides causes a moderate increase in the *anti/syn* ratio of the [2 + 2] *cis*-adducts[113]. Little effect on the *cis/trans* ratio is observed.

8. [4 + 2] Cycloaddition[85,86]

Diorganylsilyl ethers of sorbyl alcohol (**34**) undergo a smooth intramolecular Diels–Alder reaction to give a mixture of bicyclic products **35** and **36** (equation 18)[114]. The *endo/exo* selectivity depends upon the steric influence resulting from the substituent on silicon atom.

(34)   + SiR_2Cl $\xrightarrow{\Delta}$ (35)   +   (36)   (18)

R	Ratio 35	:	36
Me	2	:	1
Ph	1	:	1
t-Bu	1	:	4

9. Decomposition

The size of the silyl groups in compounds $R_3SiC(N_2)COOMe$ influences their decomposition by copper(I) triflate, dirhodium tetraacetate and dirhodium tetrakis(perfluorobutyrate)[28]. These reactions may lead to $R_2(MeO)SiC(R)=C=O$, carbene dimers, azines and 5,5-dimethoxy-3,4-bis(organosilyl)-2(5H)-furanones. As the silyl groups become larger (see entry 18 of Table 1), the azine formation becomes preferable.

10. Deprotonation

Steric bulk of various silyl groups, including Me_3Si, $(i\text{-}Pr)Me_2Si$, $(t\text{-}Bu)Me_2Si$, $(t\text{-}Bu)Ph_2Si$ and $(i\text{-}Pr)_3Si$, at the C-2 position can effectively block a base to coordinate the oxygen atom of the furan ring[115]. Thus, coordination occurs between a base and a hydroxymethyl group at the C-3 position and consequently results in C-4 deprotonation.

11. Desilylation

Tetrafluorosilane can cleave various silyl ethers in CH_2Cl_2 or MeCN at 23 °C to give the corresponding alcohols via trifluorosilyl ethers (equation 19)[29]. The steric bulk of the series of silyl groups, shown in entry 19 of Table 1, is opposite to the relative rates of desilylation of the silyl ethers.

Very recently, the bulk of various silyl groups[116] was also determined by the time required for desilylation of 5'-O-silylthymidine **37** with a complex of Bu_4NF and $BF_3 \cdot OEt_2$ in MeCN at room temperature (equation 20)[30]. The size follows the order listed in entry 20 of Table 1. Furthermore, the steric bulk of silyl groups is determined independently, as shown in entry 21, by utilization of neutral alumina for selective cleavage of various primary and secondary silyl ethers[31]. Catalytic transfer dehydrogenolysis by Pd(II)O has also been used to cleave silyl protecting groups selectively[32]. The rate of deprotection appears to be sensitive to the steric bulk of the silyl groups by following the order listed in entry 22.

Selective removal of a $(t\text{-Bu})Me_2Si$ group in the presence of the $(i\text{-Pr})_3Si$ group in disilylated p-xylene-α,α'-diol **38** can be accomplished by use of a stoichiometric amount of fluorosilicic acid (H_2SiF_6) in *tert*-butanol (equation 21). Alcohol **39** is thus obtained in 91% yield[33,117]. The results indicate that the $(i\text{-Pr})_3Si$ group is sterically bulkier than the $(t\text{-Bu})Me_2Si$ group (entry 23 of Table 1).

$$(i\text{-Pr})_3SiO-\text{C}_6H_4-CH_2OSiMe_2(Bu\text{-}t) + H_2SiF_6 \xrightarrow[23\,°C\;(91\%)]{t\text{-BuOH}} (i\text{-Pr})_3SiO-\text{C}_6H_4-CH_2OH \quad (21)$$

(38) → **(39)**

In addition to fluorosilicic acid[34], pyridinium p-toluenesulfonate can selectively remove the less bulky $(t\text{-Bu})Me_2Si$ group in the presence of the $(t\text{-Bu})Ph_2Si$ group from the corresponding silyl ethers (entry 24 of Table 1)[35]. In contrast, under basic conditions involving sodium hydride in HMPA, the rate of cleavage of *t*-butyldiphenylsilyl ethers is significantly faster than that of the corresponding *t*-butyldimethylsilyl ethers[118]. Furthermore, sodium azide in DMF effects the cleavage of the methyldiphenylsilyl ether bond, yet *t*-butyldimethylsilyl ethers and *t*-butyldiphenylsilyl ethers survive[119].

12. Elimination

Upon thermolysis, *erythro*-silyl ethers **40** undergo 1,2-*syn*-elimination. The ratio of the possible products **41** and **42** depends upon the bulk of the silyl groups (equation 22). These reactions involve four-membered ring transition states[81].

$$Mes_2B\text{-}CH(Me)\text{-}CH(Ph)(OSiR_3) \xrightarrow[\text{solvent}]{130-170\,°C} \underset{H\;\;\;\;H}{\overset{Me\;\;\;\;Ph}{C=C}} + \underset{H\;\;\;\;Ph}{\overset{Me\;\;\;\;OSiR_3}{C=C}} \quad (22)$$

(40) → **(41)** **(42)**

Mes = 2,4,6-trimethylphenyl

R_3Si	Solvent	Ratio 41	:	42
Me_3Si	C_6D_6	100	:	0
Et_3Si	C_6D_6	32	:	68
$(t\text{-Bu})Me_2Si$	toluene-d_8	<5	:	95

In a different study, the rates (k) are measured at 600 K for a thermal elimination of R_3SiCH_2COOEt to give R_3SiOEt and $CH_2=C=O$[36]. The k values are 1.98, 2.12, 1.86 and 1.42×10^{-2} s^{-1} for Me_3SiCH_2COOEt, $PhMe_2SiCH_2COOEt$, $Ph_2MeSiCH_2COOEt$ and Ph_3SiCH_2COOEt, respectively. Increase in the electrophilicity of silicon through replacement of one methyl group by a phenyl group causes a minor increase in reactivity[36]. Furthermore, the similar replacements result in a rate decrease, which is attributed to steric hindrance as shown in entry 25 in Table 1.

13. Ene reaction[87,120]

Intramolecular ene reaction takes place in activated 1,6-dienes **43** to give diastereomeric cyclopentanes **44** and **45** (equation 23)[37]. The stereoselectivity of ene adducts increases with the steric bulk of the silyl groups on oxygen atom by following the order shown in entry 26 of Table 1.

$$\text{(43)} \xrightarrow[235\,°C]{PhMe} \text{(44)} + \text{(45)} \quad (23)$$

R_3Si	Yield (%)	Ratio 44 : 45
(t-Bu)Me$_2$Si	83	79 : 21
(t-Bu)Ph$_2$Si	79	88 : 12

A profound steric effect resulting from the Me$_3$Si group also exists in the regio- and diastereoselective ene reaction between 4-methyl-1,2,4-triazoline-3,5-dione and electronically activated vinylsilanes[121]. The steric effect resulting from t-Bu, Me$_3$Si, and Me$_3$Sn groups attached to a 1,2-dimethylvinyl moiety also influences the regioselectivity in photooxygenation with singlet oxygen[122]. The influence increases in the order Me$_3$Sn < Me$_3$Si < t-Bu; a longer carbon–metal bond would decrease the effective size of the substituents[123].

14. Epoxidation

Epoxidation of diene **46a** with MCPBA occurs at the more electron-rich trisubstituted C$_4$–C$_5$ double bond, instead of at the isopropenyl unit (equation 24)[38,124]. The Me$_3$Si group with α configuration prevails over the β-angular methyl group in directing epoxidation. Thus β-epoxide **47a** is obtained as the major product along with the corresponding α-epoxide as the by-product (ratio = 35 : 1). The bulkier (Me$_3$Si)Me$_2$Si group in **46b**, however, can fully dominate the epoxidizing orientation (cf. entry 27 of Table 1). Accordingly, β-epoxide **47b** is obtained exclusively from **46b** under the same epoxidation conditions.

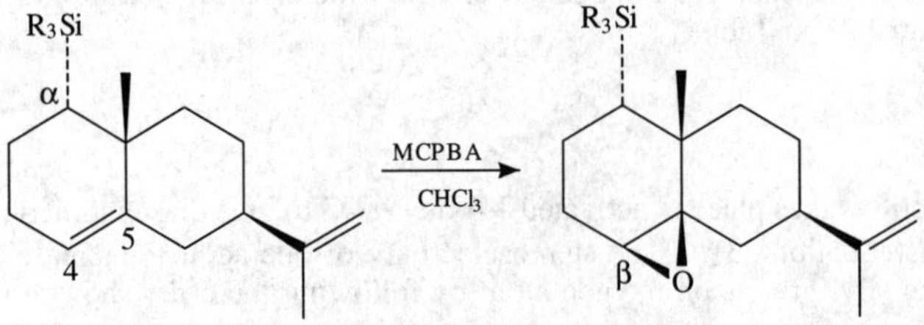

$$\text{(46)} \xrightarrow{\text{MCPBA}}_{\text{CHCl}_3} \text{(47)} \quad (24)$$

	R₃Si	Yield (%)	Selectivity
(a)	Me₃Si	64	β/α = 35 : 1
(b)	Me₃SiMe₂Si	59	β exclusively

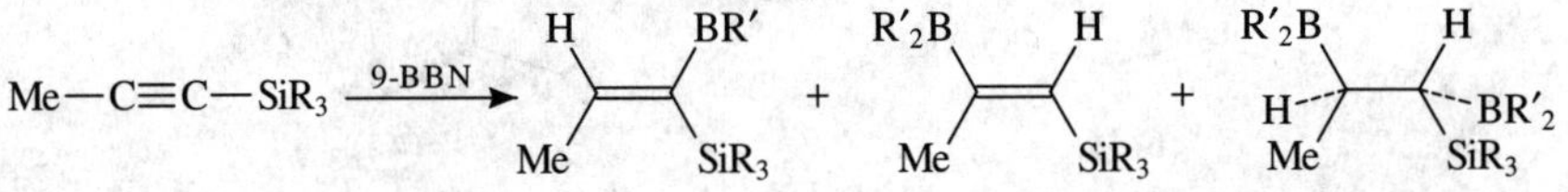

$$\text{Me}-\text{C}\equiv\text{C}-\text{SiR}_3 \xrightarrow{\text{9-BBN}} \text{(49)} + \text{(50)} + \text{(51)}$$

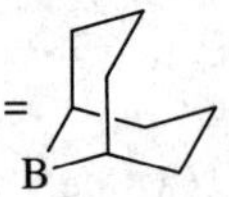

(48)

BR'₂ = 9-BBN group

R₃Si	Ratio 49 : 50 : 51
Me₃Si	100 : 0
	54 : 23
Et₃Si	97 : 3
	70 : 15
(t-Bu)Me₂Si	95 : 5
	71 : 14
(i-Pr)₃Si	0 : 100
	100 : 0

(25)

15. Hydroboration

The size of the silyl groups in alkynylsilanes **48** effects their regio- and chemoselectivity in hydroboration with 9-borabicyclo[3.3.1]nonane[125] (9-BBN). Alkynylsilanes

8. Steric effects of silyl groups 451

48 bearing a silyl group smaller than $(i\text{-Pr})_3\text{Si}$ (see entry 28 of Table 1) are led to (α-borylvinyl)silanes **49** in high regioisomeric purity (equation 25)[39]. The use of the $(i\text{-Pr})_3\text{Si}$ substituent not only provides the (β-borylvinyl)silane **50** exclusively, but also completely suppresses the formation of 1,2-diboryl adduct **51**.

16. Hydroethoxycarbonylation

The steric hindrance created by the silyl group, including Me_3Si and $(t\text{-Bu})\text{Me}_2\text{Si}$, influences the efficiency of stereo-defined hydroethoxycarbonylation of silylacetylenes[126]. In the presence of $\text{PdCl}_2[1,1'\text{-bis(diphenylphosphino)ferrocene}]$ and $\text{SnCl}_2 \cdot \text{H}_2\text{O}$, silylacetylenes react with CO and ethanol to give (E)-β-ethoxycarbonylvinylsilanes in good to excellent yields.

17. Hydroformylation

Regiochemistry of hydroformylation of alkenes can be controlled by use of bulky silyl groups directly attached to the sp^2 carbon. Treatment of silylalkenes **52** with an initial pressure of 400 psi of H_2 and CO (1 : 1 molar ratio) in the presence of $\text{HRhCO(PPh}_3)_3$ and PPh_3 provides a mixture of aldehydes **53** and **54**[40,127]. The regioselectivity increases dramatically as the bulk of the silyl groups increases in the order shown in entry 29 of Table 1 and equation 26. Moreover, sterically demanding silyl substituents can efficiently control the regioselectivity of hydrocyanation of alkynes[128].

$R_3\text{Si}$	Yield (%)	Ratio		
		53	:	54
Me_3Si	81	50	:	50
$(t\text{-Bu})\text{Me}_2\text{Si}$	87	70	:	30
Ph_3Si	69	90	:	10
$(t\text{-Bu})\text{Ph}_2\text{Si}$	80	96	:	4

18. Hydrosilylation

The fluoride ion-catalyzed reduction of 2-methylcyclohexanone by various hydrosilanes gives the corresponding silyl ethers in 40–99% yields (equation 27)[41]. The reactivity, markedly influenced by the steric bulk of hydrosilanes (see entry 30 of Table 1), decreases in the order of $\text{Ph}_2\text{SiH}_2 >> \text{PhMe}_2\text{SiH} > \text{Ph}_2\text{MeSiH} > \text{Ph}_3\text{SiH}$.

R_3Si	Temp (°C)	Time (h)	Yield (%)
Ph_2HSi	0	5	74
$PhMe_2Si$	0	24	99
Ph_2MeSi	0	24	81
Ph_3Si	25	12	40

19. Isomerization

(1-Diazo-2-oxoalkyl)silanes **55** react with cyclopropene **56** to give a mixture of diazabicyclohexenes **57** and 1,4-dihydropyridazines **58** (equation 28)[42]. Steric interactions between the vicinal acyl and silyl groups exist in **58**. Therefore the monocyclic isomer **58** is increasingly destabilized and isomerized to give **57** as the more bulky these silyl groups are (see entry 31 of Table 1).

(28)

R_3Si		Ratio	
	57	:	**58**
$(t\text{-Bu})Ph_2Si$	30	:	70
$(t\text{-Bu})Me_2Si$	52	:	48
$(i\text{-Pr})_3Si$	>97	:	3

20. Migration and rearrangement[88]

Palladium(0)-catalyzed rearrangement of 2-silyl-3-vinyloxiranes **59** depends upon the size of the silyl groups[129]. The substrates containing a bulky $(t\text{-Bu})Me_2Si$ or $(i\text{-Pr})_3Si$ group give α-silyl-β,γ-unsaturated aldehydes **60** (equation 29). In contrast, substrates bearing a less steric congested group, such as Me_3Si, Et_3Si and $PhMe_2Si$, are led to silyldienol ethers **61** through the Brook rearrangement[130].

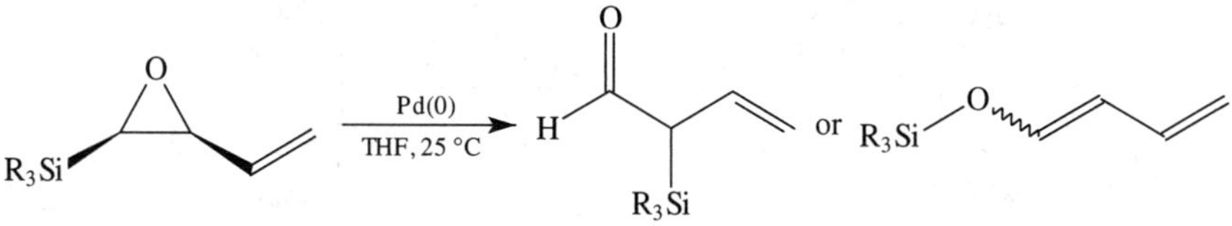

(59) **(60)** **(61)**

(29)

Upon treatment with lithium naphthalenide, silyloxy *O,S*-acetals **62** may lead to α-silyl ethers **63** through a retro [1,4]-Brook rearrangement or to 1,3-diol **64** through a [2,3]-Wittig rearrangement (equation 30)[43]. With a small (e.g. Me_3Si) or a medium-sized silyl group [e.g. $(t\text{-Bu})Ph_2Si$], the retro Brook rearrangement prevails. Contrarily, protection of **62** with a bulkier $(i\text{-PrMe}_2C)Me_2Si$ or $(i\text{-Pr})_3Si$ group gives Wittig products. Use of the Et_3Si group affords a mixture of **63** and **64** in a *ca* 4 : 1 ratio (see entry 32 of Table 1).

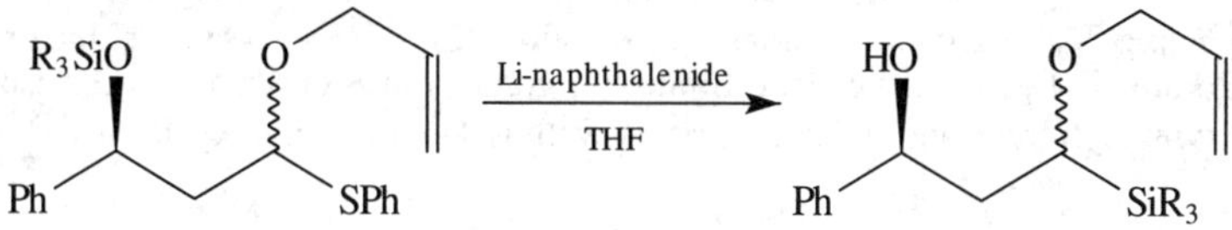

(62) **(63)**

retro [1,4]-Brook rearrangement

(30)

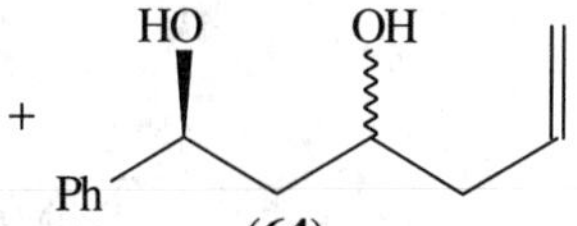

(64)

[2,3]-Wittig rearrangement

R_3Si	Yield (%) **63**	**64**
Me_3Si	81	0
$(t\text{-Bu})Ph_2Si$	77	0
Et_3Si	49	12
$(i\text{-PrMe}_2C)Me_2Si$	0	66
$(i\text{-Pr})_3Si$	0	59

Migratory aptitude, however, highly depends on the bulk of the substituents on silicon[131,132], and is opposite to the size of the silyl groups in the order listed in entry 33 of Table 1[44]. As shown in equation 31, homoallylic silyl ethers **65** react with lithium di-*tert*-butylbiphenylide to give allylsilanes **66**. The trend of migratory aptitude of silyl groups parallels the corresponding yields of **66**.

$$\text{(65)} \xrightarrow[\text{THF, -78 °C}]{\text{lithium di-}tert\text{-butylbiphenylide}} \text{(66)} \quad (31)$$

R$_3$Si	Yield (%)
Me$_3$Si	83
Et$_3$Si	70
(n-Pr)$_3$Si	49
(i-Pr)$_3$Si	0

Furthermore, upon treatment with lithium diisopropylamide (LDA) and HMPA in THF at −78 °C, 3-(silyloxycarbonyl)furans **67** undergo 1,4 O→C silyl migration to give the corresponding 2-silyl-3-carboxyfurans (**68**, equation 32)[133–135]. The yields, ranging from 43–72%, are silyl group dependent. Similar migrations also occur to the corresponding 3-silyloxymethylfurans and thiophenes[134] as well as 1-iodo-2-silyloxyalkenes[135].

$$\text{(67)} \xrightarrow[\text{THF, -78 °C}]{\text{LDA, HMPA}} \text{(68)} \quad (32)$$

R$_3$Si	Yield (%)
(t-Bu)Me$_2$Si	72
Et$_3$Si	57
(i-Pr)$_3$Si	56
(n-Bu)$_3$Si	51
(t-Bu)Ph$_2$Si	43

Under basic conditions, 1,3-migration of triorganosilyl groups (i.e., Me$_3$Si, PhMe$_2$Si and Ph$_2$MeSi) from carbon to oxygen takes place in highly crowded organosilanols[136]. These migrations $(\text{R}_3\text{Si})(\text{R}'_3\text{Si})_2\text{C}-\text{SiR}''_2-\text{OH} \rightarrow \text{H}(\text{R}'_3\text{Si})_2\text{C}-\text{SiR}''_2-\text{O}-\text{SiR}_3$ are facilitated, in part, by relief of steric strain created by silyl groups.

On the other hand, an intermolecular migration of silyl groups also takes place among 5-methyl-2-(trialkylsilyl)thiophenetricarbonylchromium(0) complexes under basic conditions[45]. Increased steric hindrance from the silyl groups (see entry 34 of Table 1) results in preferential deprotonation.

21. Osmylation

Osmium-catalyzed vicinal hydroxylations of various allylic silanes **69** give a mixture of *anti*- and *syn*-diols **70** and **71** (equation 33)[46]. The silyl groups participate in vicinal

8. Steric effects of silyl groups

hydroxylation reactions with useful levels of selectivity[137-140]. The *anti* selectivity (i.e., **70/71**) improves as the size of the silyl group increases as shown in entry 35 of Table 1.

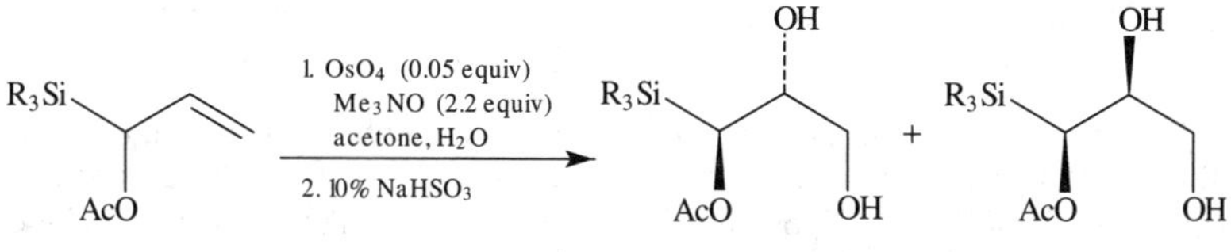

(33)

R_3Si	Yield (%)	Ratio 70 : 71
Me_3Si	57	6.5 : 1
$PhMe_2Si$	70	7.0 : 1
Et_3Si	67	7.5 : 1
$(t\text{-}Bu)Me_2Si$	70	11.3 : 1
Ph_2MeSi	58	11.5 : 1

22. Oxymercuration

Use of remote allylic silyl ethers, such as **72b,c**, rather than of allylic alcohol **72a** in intramolecular oxymercurations leads to a higher 1,3-*syn*-diastereoselectivity (equation 34)[47]. The overall yields for *syn*-**73** and *anti*-**74** range between 85–93%. The best selectivity reaches 7 : 1 involving the $(t\text{-}Bu)Ph_2Si$ group (entry 36 of Table 1).

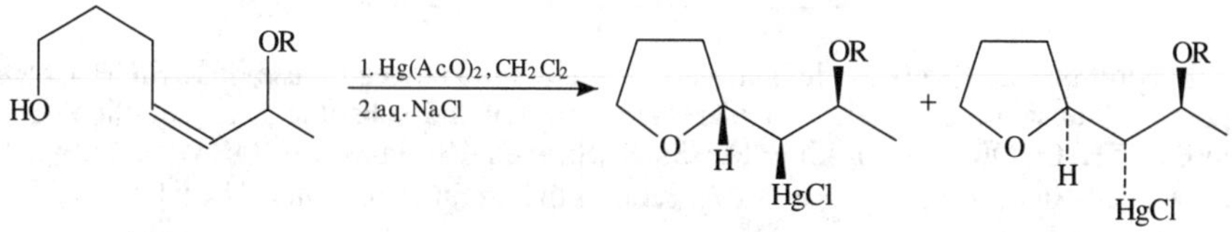

(34)

R	Yield (%)	Ratio 73 : 74
(a) H	85	2.5 : 1
(b) $(t\text{-}Bu)Me_2Si$	89	6 : 1
(c) $(t\text{-}Bu)Ph_2Si$	93	7 : 1

23. Ozonolysis

Ozonolysis of silyl ethers of 1-cyclopenten-3-ol gives a high selectivity in favor of the *exo*-substituted ozonide[48]. The bulkier silyl blocking groups, as shown in entry 37 of Table 1, exhibit a somewhat higher degree of stereoselection. Furthermore, the size of the silyl groups in R_3SiH influences their ozonolysis rates in CCl_4[49,50]. The order is listed in entry 38 of Table 1.

24. Peterson olefination[141]

The diastereomeric ratio of *cis*- to *trans*-stilbenes formed in the Peterson olefination of PhC̄HSiR₃ with benzaldehyde increases significantly as the bulk of silyl groups increases (see entry 39 of Table 1)[51]. Upon reaction with *n*-decanal, the steric bulk of the silyl groups in reagent, **75**[142,143] also affects the E/Z ratio and the yields of the product **76**[52,53]. Increased steric congestion by the order shown in entry 40 results in higher diastereoselectivity, but at the expense of the yield (equation 35). Those outcomes are in accord with those of the addition of prochiral carbanions, including α-silyl carbanions, to aldehydes[144]. Furthermore, the Peterson olefination between a ketone and bis-heterocycles bearing various silyl groups [e.g. Me_3Si, Ph_3Si and $(t\text{-}Bu)Me_2Si$] is applied as a key step in the synthesis of BRL 49467, an anti-bacterial compound[145].

$$n\text{-}C_6H_{13}\overset{\overset{Li}{|}}{\underset{\underset{SiR_3}{|}}{CH}} + n\text{-}C_9H_{19}CHO \xrightarrow{TFA} n\text{-}C_6H_{13}CH=CH(C_9H_{19}\text{-}n) \quad (35)$$

(75)   (76)

R_3Si	Yield (%)	E/Z
Me_3Si	76	1 : 2
Et_3Si	60	1 : 2
$(t\text{-}Bu)Me_2Si$	51	6 : 1
$(t\text{-}Bu)Ph_2Si$	47	6 : 1

In addition, the Peterson olefination of 3-phenylthioketones by use of R_3SiCH_2COOR' gives α,β-unsaturated esters. An increase in size of the substituent at the silicon atom in R_3SiCH_2COOR' [e.g., $R_3Si = Me_3Si$, Et_3Si, $(i\text{-}Pr)_3Si$, $PhMe_2Si$, Ph_2MeSi and Ph_3Si] only results in a slight decrease in Z/E ratio of the products (cf entry 41 of Table 1)[54,146].

25. Phosphonylation

Nucleophilic phosphonylations of α-silyloxy aldehydes (e.g. **77**) with phosphite **78** give a mixture of **79** and **80** with moderate to good diastereoselectivity (equation 36)[55]. The ratio of **79/80** depends greatly upon the size of the silyl groups in **77**, according to the order shown in entry 42 of Table 1.

$$\underset{(77)}{R_3SiO\text{-}CH(CH_3)\text{-}CHO} + \underset{(78)}{(MeO)_3Si\text{-}O\text{-}P(OEt)_2} \xrightarrow[25\,°C]{CH_2Cl_2} \underset{(79)}{R_3SiO\text{-}CH(OSiMe_3)\text{-}P(O)(OEt)_2} + \underset{(80)}{R_3SiO\text{-}CH(OSiMe_3)\text{-}P(O)(OEt)_2} \quad (36)$$

R₃Si	Yield (%)	Ratio		
		79	:	80
Et₃Si	55	63	:	37
(t-Bu)Me₂Si	71	71	:	29
(t-Bu)Ph₂Si	66	90	:	10
(i-Pr)₃Si	67	92	:	8

26. Polymerization

A (MeO)₃Si or (i-PrO)₃Si group attached at the C-2 position of 1,3-butadienes enables anionic polymerization to take place under various conditions and to give a stable propagating end[147]. The geometry of the monomer units in the polymers is influenced by the steric hindrance on the silyl group; 1,4-E geometry is the predominant micro-structure of the resultant polymers.

For co-polymerization, various silyl methacrylates (CH₂=C(Me)CO₂SiR₃) can react with methyl methacrylate in bulk at 60 °C[148]. Their reactivity depends upon, in part, the steric effect resulting from the silyl groups including (t-Bu)Me₂Si, (t-Bu)Ph₂Si, Ph₂MeSi and Ph₃Si.

27. Reduction

The steric bulk of the silyl groups in acylsilanes influences their asymmetric reduction to give chiral secondary alcohols by borane complexed with (S)-(−)-2-amino-3-methyl-1,1-diphenylbutan-1-ol[149]. The enantiomeric excess increases from 50% to 94% by replacement of the PhMe₂Si group with the Ph₃Si group.

Palladium(0)-catalyzed reduction of trialkylsilylallyl esters **81** in the presence of n-BuZnCl leads to a mixture of allylsilanes **82**, vinylsilanes **83** and dienylsilanes **84** (equation 37). Because the Et₃Si group exerts a greater steric influence than the Me₃Si group (cf entry 43 of Table 1), the yields for allylsilanes are higher for Et₃Si-containing products[56].

R₃Si		Ratio		
	82	: 83	:	84
Me₃Si	37	: 16	:	47
Et₃Si	67	: 17	:	16

28. Simmons–Smith reaction

Reaction of γ-silyl allylic alcohols **85** with Et₂Zn and CH₂I₂ in the presence of (+)-diethyl tartrate as a chiral auxiliary affords the corresponding cyclopropylmethyl

TABLE 2. Taft's E_s values for alkyl and cycloalkyl groups R' and relative rate constants $k(R^1Me_2SiCl)/k(Me_3SiCl)$ for the reaction of the two chlorides with lithium silanolates and lithium isopropylate in Et_2O at 20 °C

R^1	E_s (R^1)	Relative rate constants for their reactions with		
		Me_3SiOLi	$PhMe_2SiOLi$	Me_2CHOLi
Me	0.0	1.00	1.00	1.00
Et	−0.28	0.40	0.25	0.46
n-Pr	−0.36	0.30	0.25	0.38
n-Bu	−0.39	0.29	0.22	0.33
i-Bu	−0.55	0.08	0.17	0.25
Cyclopentyl	−0.70	0.10	—	0.14
i-Pr	−0.76	0.10	0.07	0.11
Cyclohexyl	−0.79	0.08	0.07	0.09
s-Bu	−0.87	0.09	0.06	0.07

alcohols **86** with high stereoselectivity (equation 38)[150]. The effective silyl groups for these Simmons–Smith reactions include the Me_3Si, $PhMe_2Si$ and Ph_3Si groups.

$$R_3Si-C(Me)=CH-CH_2OH \quad \xrightarrow[\text{3. } CH_2I_2, CH_2Cl_2]{\begin{array}{l}1.\ Et_2Zn\\ 2.\ (+)\text{-diethyl tartrate}\end{array}}\quad R_3Si-\underset{Me}{\triangle}-CH_2OH$$

(85) (86)

(38)

R_3Si	Yield (%)	%ee
Me_3Si	53	87
$PhMe_2Si$	88	92
Ph_3Si	82	90

29. Substitution

Data listed in Table 2 include the substituent constants R' of trialkylchlorosilanes and the relative rate constants $k(R^1Me_2SiCl)/k(Me_3SiCl)$ for the reactions of the two chlorides with lithium silanolates and isopropylate (equation 39)[57]. The reaction rates of silanes are influenced almost exclusively by the steric effects of the alkyl groups attached to the silicon atom. The $\log(k_{rel})$ values of the compounds with various R^1 groups give a satisfactory correlation with Taft's E_s values[151]. Thus the steric hindrance of silyl groups follows the order listed in entry 44[57] of Table 1.

$$R^1Me_2SiCl + R^2OLi \longrightarrow R^1Me_2SiOR^2 + LiCl \qquad (39)$$

IV. COMPARISON OF THE INFLUENCE RESULTING FROM HYDROGEN ATOM AND ALKYL GROUPS VERSUS SILYL GROUPS

A. General Considerations

Replacement of a hydrogen atom or an alkyl group, such as Me, Et, i-Pr, t-Bu and Ph, in organic compounds with a silyl group selected with deliberation often gives improved stereoselectivity. The following reactions of 23 types represent different ways to use a silyl group in stereochemical control. The silyl groups include Me_3Si, $(t\text{-Bu})Me_2Si$, $(i\text{-Pr})_3Si$, $PhMe_2Si$, Ph_3Si, $(Me_3Si)Me_2Si$ and $(i\text{-PrO})Me_2Si$.

B. Organic Reactions of Various Types

1. Addition

Replacement of the aldehydic hydrogen with a Me_3Si group in **87** increases the diastereofacial selectivity in nucleophilic addition[152-154]. An example is shown in equation 40, in which the ratio of **88** to **89** is increased from 5 : 1 to >100 : 1. Furthermore, the γ-Me_3Si group in α-methyl-α,β-cyclopentenone provides stereocontrol of the Michael addition by use of $PhCH_2OCH_2CH_2CMe_2MgCl$ in the presence of $CuBr \cdot SMe_2$, Me_3SiCl and HMPA[155]. The desired β,γ-*trans* adduct is generated in *ca* 63% yield. In addition, the Me_3Si group at the C-5 position of cyclohexenone **90** influences 1,4-addition by organometallic reagents to give *trans* adducts **91** as the major or even the exclusive products (equation 41)[156].

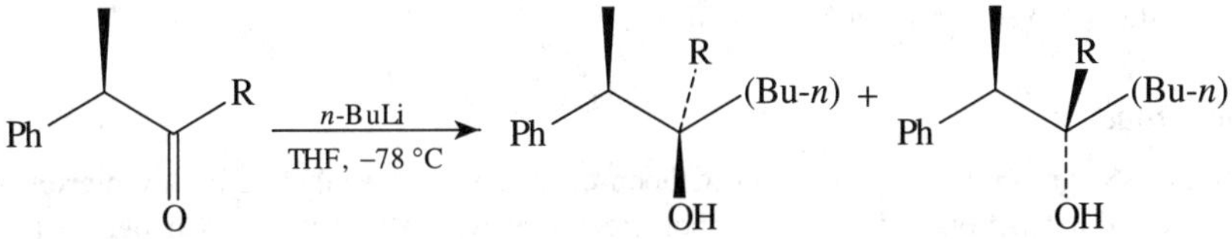

R	Yield (%)	Ratio 88 : 89
H	91	5 : 1
Me_3Si	92	>100 : 1

In asymmetric syntheses, the bulky $(i\text{-Pr})_3Si$ group can direct 1,2-addition of phosphite **92** to (S)-triisopropylsilyloxy lactaldehyde (**93**) to afford adduct **94** preferentially along with the by-product **95** (equation 42)[157]. The stereochemical outcome results from the bulk of the $(i\text{-Pr})_3Si$ protecting group[9]. Furthermore, the steric congestion in allylic cyanohydrin trimethylsilyl and *tert*-butyldimethylsilyl ethers influences their regioselective 1,2- and 1,4-additions to carbonyl compounds under basic conditions[158].

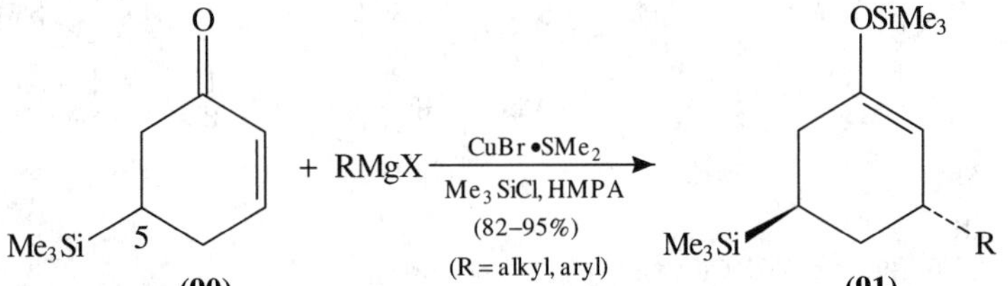

(42)

[Scheme showing reaction: (92) PhCH₂O-P(OSiMe₃)-OCH₂Ph + (93) aldehyde with OSi(Pr-i)₃ → in CH₂Cl₂ at −78 °C → (94) and (95), ratio 9 : 1]

The asymmetric addition of ethyl azidoformate (N_3CO_2Et) to an optically active enamine, prepared from cyclohexanone and (S)-2-pyrrolidinemethyl methyl ether, followed by photolysis produces 2-(ethoxycarbonylamino)cyclohexanone with modest enantiomeric excess (18 %ee) in 40% yield[159]. Use of (S)-2-pyrrolidinemethyl trimethylsilyl ether as a chiral auxiliary increases the steric hindrance and provides the same product with the highest value of %ee (35%) in 51% yield.

2. Aldol reaction

A $PhMe_2Si$ group at the β position of open-chain enolates enables a highly diastereoselective aldol reaction to take place upon treatment with aldehydes[160]. Change of the silyl group results in a small but not always consistent effect on selectivity of related methylation by MeI[161].

3. Alkylation

Regioselective functionalization of metallated 2-aza-1,1-disilylimine allyl anion **96** with an electrophile depends upon competition between the steric and the electronic effects offered by the C-1 Me_3Si groups (equation 43)[162]. Reaction of **96** with alkyl iodides occurs at the C-1 position to give **97** (52–63%); this is because the two Me_3Si groups can stabilize the α-carbanion. In contrast, with bulkier electrophiles such as Me_3SiCl and ClCOOEt, functionalization of **96** takes place at the C-3 position to give **98** in 59–60% yields. Thus the steric congestion created by the Me_3Si groups prevails over the electronic effect. Moreover, a higher level of stereocontrol on alkylation of enol acetates can be accomplished by introduction of a Me_3Si group at the α position[163].

(43)

(96) Ph–N=C(SiMe₃)₂ allyl anion, M⁺ → EX → **(97)** + **(98)**

$M^+ = Li^+, K^+$
EX = MeI, EtI, BuI, Me_3SiCl, ClCOOEt

4. Allylboration

Homochiral borolanes **99–102** can asymmetrically allylborate aldehydes[164]. The better enantioselectivity exhibited by borolane **102** than others is due to the steric origin offered by the Me_3Si group.

	R^1	R^2	%ee
(99)	Me	Me	27
(100)	i-Pr	H	24
(101)	t-Bu	H	72
(102)	Me_3Si	H	81

5. Allylsilylation

In the $AlCl_3$-catalyzed allylsilylation, the silyl group of allyltrimethylsilane adds regioselectively to the terminal carbon of phenylacetylene (equation 44)[165]. The allyl group adds to the inner carbon to form a 1,4-diene, in which the allyl moiety is *cis* to the silyl group. In contrast, allyltrimethylsilane adds to diphenylacetylene to give the corresponding *trans* adduct. Steric interaction is greater between two phenyl groups than between the phenyl and the Me_3Si groups in a *cis* configuration.

(44)

6. Carbenoid rearrangement

Reaction of 1,1-dihalo-2-t-butyldimethylsilyloxyalkanes with lithium diisopropylamide in ether gives (Z)-1-halo-2-t-butyldimethylsilyloxy-1-alkenes regio- and stereoselectively via carbenoids through α-elimination[166]. Replacement of the (t-Bu)Me$_2$Si group with a Me$_3$Si or Et group results in the formation of a by-product through β-elimination.

7. Cyclization

Various types of cyclizations can be controlled by the silyl group present in the substrates or reagents. Thus the desired products with a four-, five-, six- or seven-membered ring can be produced stereoselectively. An example shown in equation 45 includes cyclization of (2-pyridylthio)glycosidic t-butyldimethylsilyl enol ethers **103**, initiated by silver trifluoromethanesulfonate, to give bicyclic ketooxetanes **104**[167]. The steric crowding resulting from the (t-Bu)Me$_2$Si group in the transition state dominates the stereoselectivity during the four-membered ring formation.

The t-Bu group has a condensed bulk, yet it is spatially smaller than the Me$_3$Si group. The example shown in equation 46 indicates this phenomenon. Induced by dibenzoyl peroxide, the free radical cyclization takes place on iodo t-butylacetylenic ester **105a** to give a mixture of (E)- and (Z)-iodoalkylidene lactones **106a** (51%) and **107a** (ca 5%)[168]. The Me$_3$Si-containing iodo acetylenic ester **105b**, however, undergoes cyclization to generate a high yield of (E)-iodoalkylidene lactone **106b** (92%) exclusively.

R	E-(**106**)	Z-(**107**)
(a) t-Bu	51%	(ca 5%)
(b) SiMe$_3$	92%	—

In a total synthesis of (−)-β-vetivone, the key step involves intramolecular alkylation of the bromo ketone **108** to give spiro ketone **111** (equation 47)[169]. The 5-Exo-Tet

8. Steric effects of silyl groups

cyclization taking place preferentially on intermediate **110** over **109** is due to the steric effect of the methyl group, which is fixed in the pseudoaxial position by the bulky Me$_3$Si group. Moreover, 5-*Exo*-Trig cyclization occurs with silyloxyalkene **112** to give diol **113** (equation 48)[170]. This radical cyclization with steric and stereochemical control provides an efficient avenue to introduce an angular hydroxymethyl group in organic compounds.

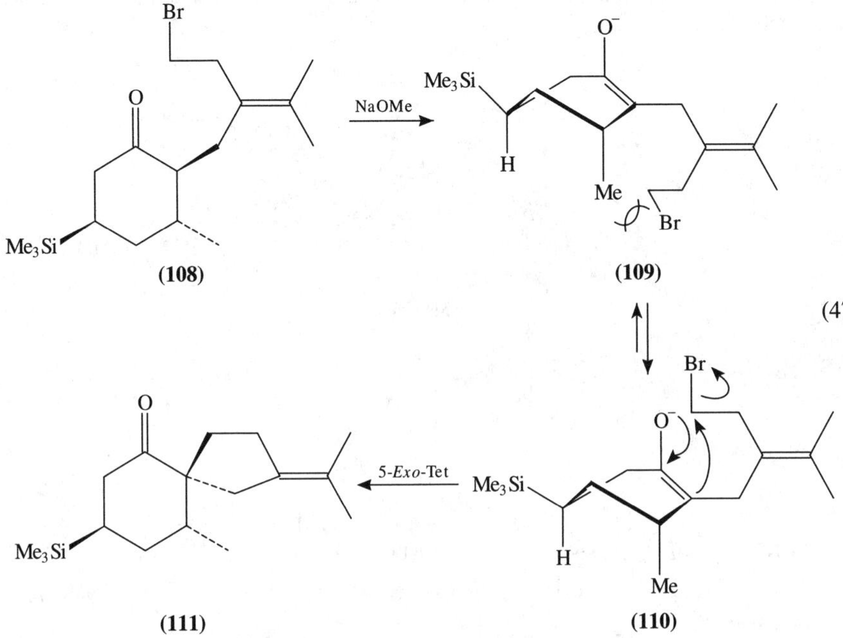

(47)

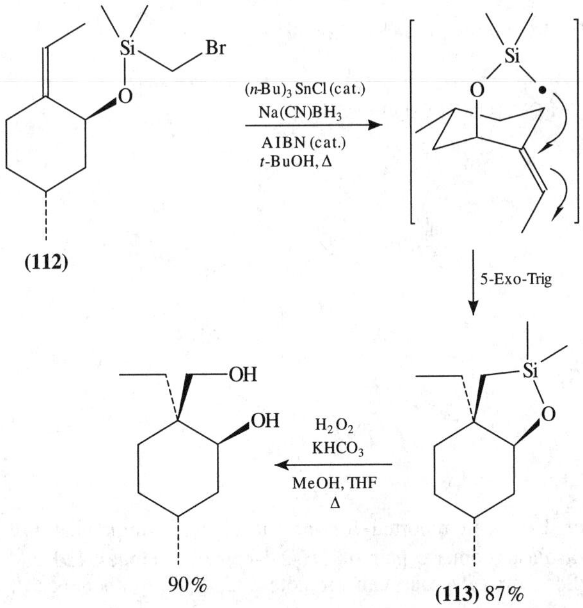

(48)

Polyene cyclization of epoxy trienylsilanes **114** by use of $TiCl_4$ in CH_2Cl_2 containing 2,6-di-*tert*-butylpyridine at $-78\,°C$ gives a mixture of 9,10-*syn* and 9,10-*anti* bicycles **117** and **118** in comparable amounts (equation 49)[171]. The formation of six-membered ring products proceeds through a chair/chair or a chair/boat transition state (i.e. **115** and **116**) with similar steric interactions. These results contrast sharply with the consistent bias favoring the chair/chair orientation in the cyclization of polyene epoxides lacking the Me_3Si substituent[172–174].

$$R^1 = CH_2CH_2CMe\!=\!CHCH_2O\!-\!\overset{\overset{\displaystyle O}{\|}}{C}Me \text{ or}$$

$$CH_2CH_2CMe\!=\!CHCH_2O\!-\!\overset{\overset{\displaystyle O}{\|}}{C}Ph$$

Recently a [1,7] cyclization has been reported for the diazostyryl compounds **119** (equation 50)[175]. It leads to a diastereomeric pair of 1*H*-2,3-benzodiazepines **120** and **121**. Modification of the allylic hydroxyl group can promote attack from either face of the C—C double bond. Its lithium alcoholate gives a selectivity of 85 : 15 for one face and its *tert*-butyldimethylsilyl ether gives a selectivity of 9 : 91 for the other face.

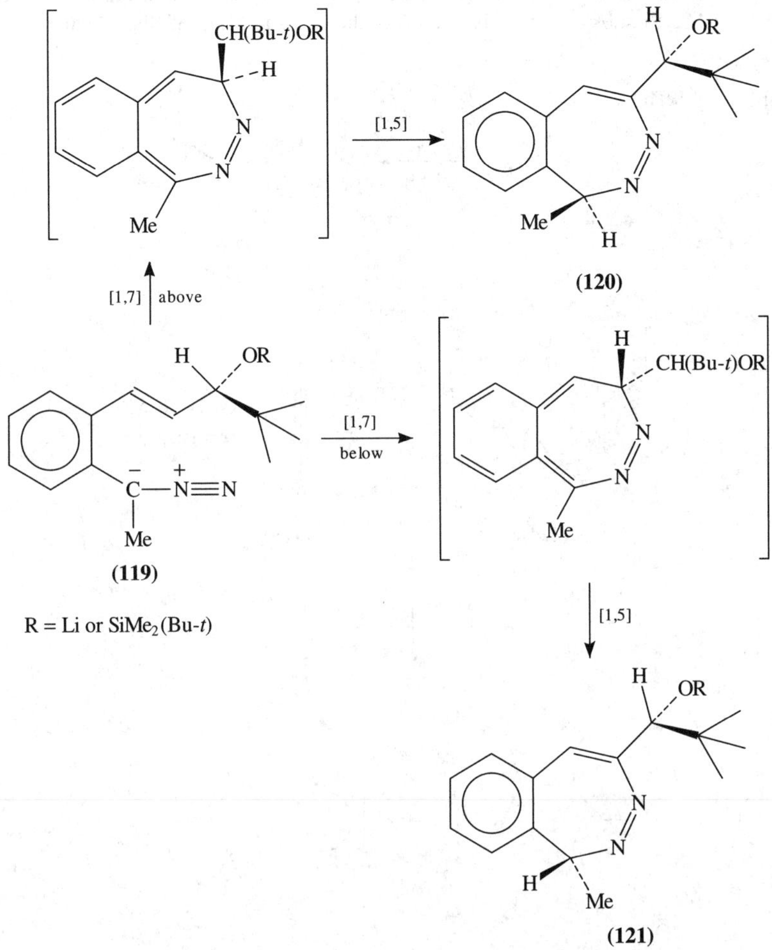

(119)

R = Li or SiMe₂(Bu-*t*)

(120)

(121)

(50)

8. [2 + 2] Cycloaddition

Reaction of (trimethylsilyl)vinyl selenide **122** with methyl vinyl ketone catalyzed by SnCl₄ gives the [2 + 2] adduct **123** in 66% yield (equation 51)[176]. Replacement of the Me₃Si group with a hydrogen atom or a methyl group produces a complex mixture. Thus, the Me₃Si group can suppress side reactions by its steric effect.

9. [3 + 2] Cycloaddition[89]

The intermolecular cycloaddition of *N*-(azidomethyl)benzisothiazolone **124** with various electron-deficient acetylenes **125** generates potential inhibitors of human leukocyte elastases 1,2,3-triazoles **126** and their regioisomers **127** in 72–99% overall yields (equation 52)[177]. As the steric effect increases due to the increase size of R in **125**, the adducts **127** prevail over adducts **126** and become the predominant products. The substituent rank order obtained on the basis of the STERIMOL program (the Verloop steric parameters) is H < Ph < SO$_2$Ph < t-Bu ~ Me$_3$Si[178]. A similar type of dominating effect by a Me$_3$Si substituent is also noted in the cycloaddition of diazoalkanes[179].

$$\text{(122)} + \text{Me} \xrightarrow[\text{CH}_2\text{Cl}_2]{\text{SnCl}_4} \text{(123)} \quad 66\% \tag{51}$$

$$\text{(124)} + \text{(125)} \quad R = \text{H, Ph, } t\text{-Bu, Me}_3\text{Si}; \quad Ar = \text{Ph, } p\text{-Tol}$$

$$\xrightarrow[\Delta]{\text{PhH or H}_2\text{O}} \text{(126)} + \text{(127)} \tag{52}$$

In a different class, palladium(0)-catalyzed diastereo-controlled [3 + 2] cycloaddition of vinyl sulfone **128a** with allylsilyl acetate **129** gives a mixture of spiranes **130a** and **131a** in a ratio of 4.2 : 1 (equation 53)[180]. The diastereomeric ratio increases to >100 : 1 for **130b/131b** by introduction of a (t-Bu)Me$_2$SiO group in the vinyl sulfone (i.e. **128b**).

10. [4 + 2] Cycloaddition

Introduction of a silyl group, such as Me$_3$Si and (t-Bu)Me$_2$Si, on the oxygen of a propargylic alcohol[181,182] or on vinylallenes[183] affects the feasibility and the stereochemical course of intramolecular Diels–Alder reactions with an alkynyl diene, maleic anhydride or N-methylmaleimide. A possible explanation involves steric effects.

8. Steric effects of silyl groups

(53)

R	Yield (%)	Ratio 130 : 131
(a) H	67	4.2 : 1
(b) OSiMe$_2$(Bu-t)	76	>100 : 1

Moreover, enone **132** reacts with (t-Bu)Me$_2$SiOTf and Et$_3$N to give the Diels–Alder adduct **134** in 82% yield via intermediate **133** (equation 54)[184]. Lactam **134** with a *cis* fused ring is more stable than the corresponding *trans* isomer, in which the Me$_3$Si group must rotate into an axial position. The secondary orbital overlap and the steric requirements of the Me$_3$Si group on the dienophile moiety in **133** also appear to be critical to the observed stereoselectivity.

In addition, protection of the hydroxyl group of chiral naphthyl alcohols with a Me$_3$Si group increases the steric crowding. Consequently, the rate decreases for their [4 + 2] cycloaddition with singlet oxygen[185]. [4 + 2] Cycloaddition also takes place between *N*-phenylmaleimide and trimethylsilyl enol ether of *N*-methyl-2-acetylpyrrole in toluene at 115 °C to give the corresponding adduct in 34% yield[186]. Nevertheless, the same reaction does not proceed by use of *N*-trimethylsilyl-2-acetylpyrrole as the substrate. Results from a model study indicate that the bulk of the Me$_3$Si group suppresses a requisite cisoid conformation.

For substrates containing a larger silyl group, the stereocontrol in intramolecular [4+2] cycloaddition of **135** is achieved by use of a sterically demanding PhMe$_2$Si group to affect *endo*-selectivity in the production of **136** as the major product (equation 55)[187]. Absence of this group leads to a 1 : 8 selectivity in favor of the *exo*-product **137**.

11. Cyclopropanation

Attachment of a Me$_3$Si group to the terminal sp^2 carbon of allylic alcohols can influence diastereoselective cyclopropanation involving Sm and CH$_2$I$_2$[188].

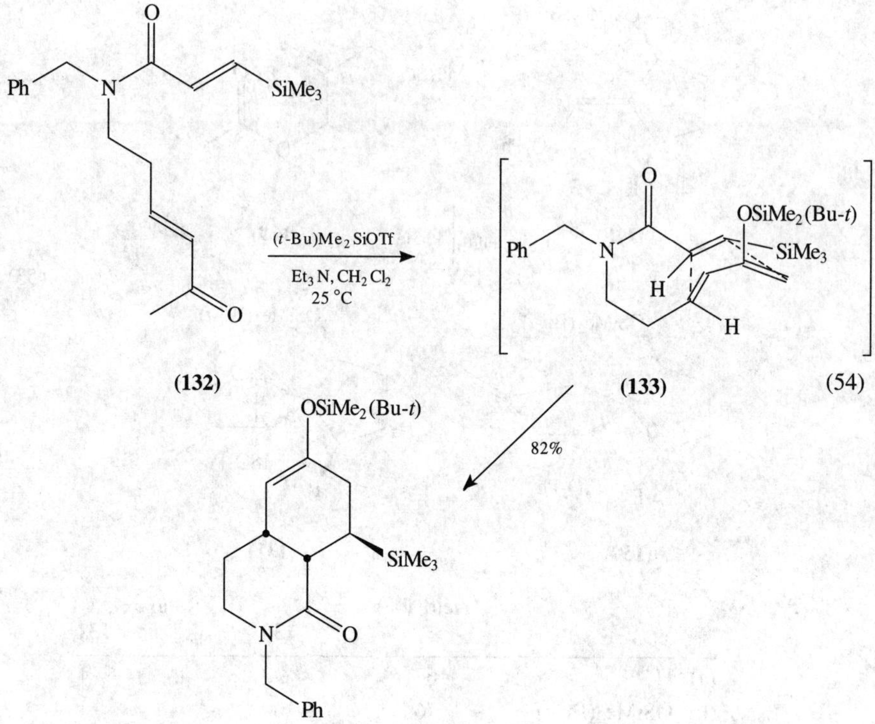

(132) → [(133)] → (134)

(t-Bu)Me$_2$SiOTf, Et$_3$N, CH$_2$Cl$_2$, 25 °C; 82%

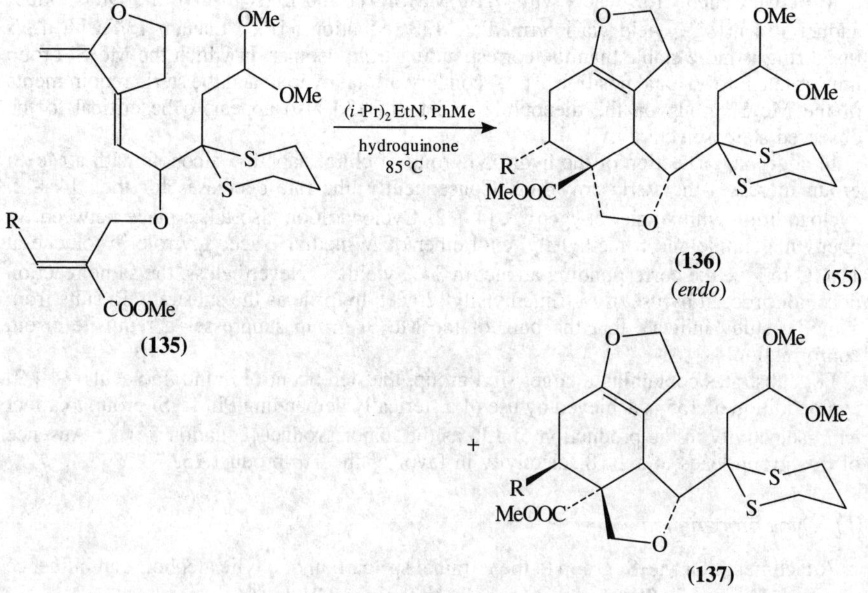

(135) → (136) (*endo*) + (137) (*exo*)

(*i*-Pr)$_2$EtN, PhMe, hydroquinone, 85°C

(54)

(55)

R	136	Ratio	137
PhMe$_2$Si	5	:	2
H	1	:	8

12. Dehydration

Dehydration of α-hydroxy-γ-oxoalkyltrimethylsilanes under acidic conditions gives a mixture of (E)- and (Z)-γ-oxoalkenyltrimethylsilanes[189]. The thermodynamically more stable (Z)-isomer, according to *ab initio* calculation, is isolated as the major component. The ratio of (Z)/(E) isomers ranges from 4.5 : 1 to 27 : 1. The Me$_3$Si group stabilization of the (Z)-isomer is due to steric along with electronic influence.

13. Ene reaction

Carbonyl–ene reactions involving vinylsilanes promoted by a Lewis acid provide an avenue for regio- and stereocontrolled introduction of a vinyl functionality. The selectivity changes dramatically by placement of a Me$_3$Si group into the substrates as shown in equations 56 and 57[190]. Similarly, upon reaction with aldehydes in the presence of Me$_2$AlCl, the (*t*-Bu)Me$_2$Si group in optically pure **138** assists the chirality transfer to give adducts **139** in 80–99 %ee (equation 58)[191,192].

R	Ratio	
	(Z)	(E)
H	15 :	85
Me$_3$Si	98 :	2

R	Ratio	
	(SR, SR)	(RS,SR)
H	72 :	28
Me$_3$Si	7 :	93

R^1CHO + (138) [SR, OSiMe$_2$(Bu-t)] $\xrightarrow[\text{solvent}]{\text{Me}_2\text{AlCl}}$ (139) [OH, SR, OSiMe$_2$(Bu-t), R^1] (58)

−78 to −23 °C

(138) R = Et, Ph; R′ = Alkyl, Aryl, 2-Furyl
(70–98%)
(139) (80–99 %ee)

14. Epoxidation[193,194]

Epoxidation of 4-(hydroxymethyl)cyclopent-1-ene (**140a**) with *m*-CPBA gives a mixture of *syn*- and *anti*-epoxides in 1 : 1.1 ratio (equation 59)[58]. This ratio is slightly raised to 1 : 2.3 in the epoxidation of the corresponding benzoate **140b**. Protection of the hydroxyl group in **140a** with a (*t*-Bu)Me$_2$Si group increases the steric hindrance significantly. Consequently, conversion of **140c** to a mixture of *syn*- and *anti*-epoxides affords a ratio of 1 : 8.2 (entry 45 of Table 1). The selectivity drops to 1 : 4 when the less sterically congested Ph$_3$Si group replaces the (*t*-Bu)Me$_2$Si group (i.e. **140d**).

(140) $\xrightarrow[\text{CH}_2\text{Cl}_2, \text{THF}, \Delta]{\text{MCPBA}}$ syn + anti (59)

R		Ratio	
	syn	:	anti
(a) H	1	:	1.1
(b) COPh	1	:	2.3
(c) (*t*-Bu)Me$_2$Si	1	:	8.2
(d) Ph$_3$Si	1	:	4.0

In a related reaction, stereochemical control during epoxidation of Δ^4-*cis*-1,2-disubstituted cyclohexenes **141** with *m*-CPBA depends upon the hydroxyl functionality[195]. A free alcohol (**141a**) or an acetate (**141b**) affords *syn*-epoxides exclusively. Epoxidation of a *t*-butyldimethylsilyl ether (**141c**) gives predominantly the *anti*-epoxide (54%).

15. Hydride reduction

The Me$_3$Si group can direct stereoselective synthesis of 2-vinyl-1,3-diols. Reduction of 3-hydroxy-2-vinylketone **142a** with LiBEt$_3$H produces a mixture of diastereomers **143a** and **144a** in a ratio of 1 : 49 (equation 60)[196]. Introduction of a Me$_3$Si group to the vinyl moiety of the substrate completely reverses the selectivity. This is evidenced by reduction of **142b** to give a mixture of **143b** and **144b** in a ratio of >99 : 1.

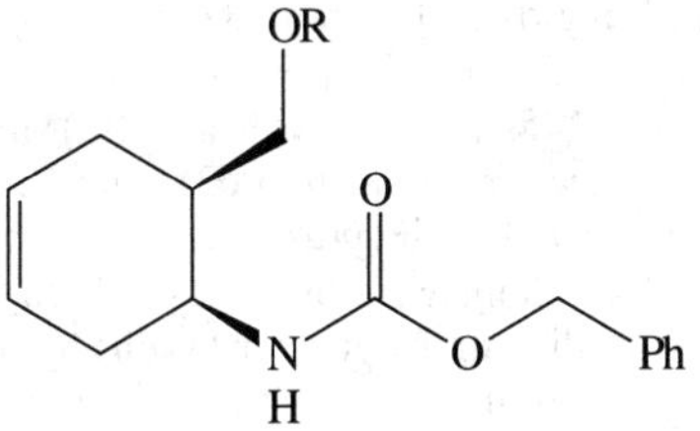

8. Steric effects of silyl groups

(141)
R =
(a) H
(b) Ac
(c) SiMe$_2$(Bu-t)

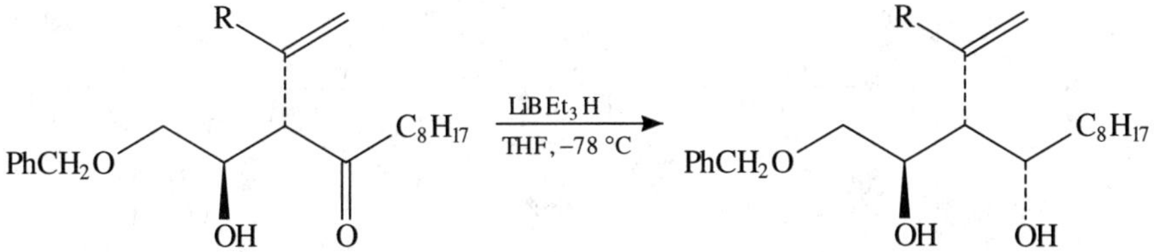

(142) → **(143)** 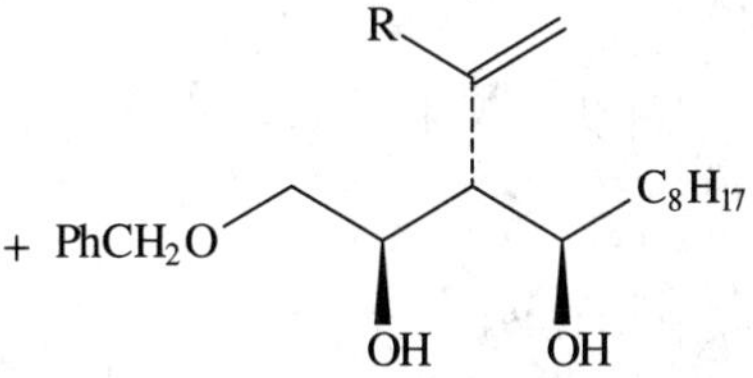 + **(144)** (60)

LiBEt$_3$H, THF, −78 °C

R	Ratio 143 : 144
(a) H	2 : 98
(b) SiMe$_3$	>99 : <1

Furthermore, simple reduction of Me$_3$Si-containing norcamphor **145a** with LiAlH$_4$ gives the corresponding *endo* alcohol **146a** in 98% yield (equation 61)[197]. Under the same conditions, reduction of substrate **145b** possessing geminal di-Me$_3$Si groups produces a mixture of *endo* alcohol **146b** (83%) and silyl ether **147** (15%). Conversion of **146b** to **147** can also be achieved in 80–82% yields by use of MeLi (2 equiv) in ether or t-BuOK (0.01 equiv) in THF. The diverse results of reduction indicate that the unusual cleavage of an Si—Me bond is assisted by steric compression resulting from the Me$_3$Si groups.

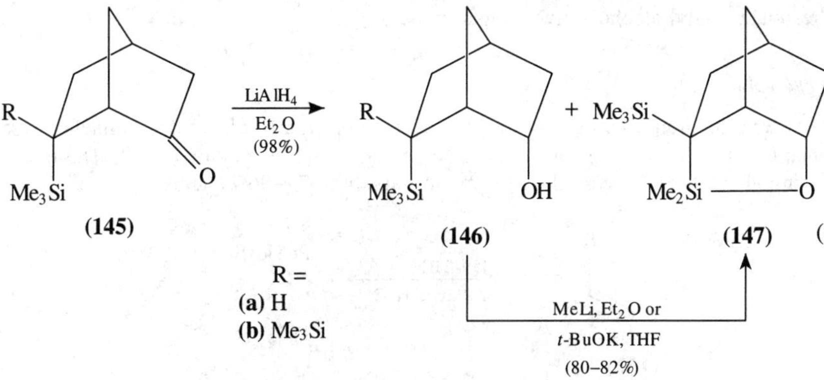

(145) → **(146)** + **(147)** (61)

LiAlH$_4$, Et$_2$O (98%)

R =
(a) H
(b) Me$_3$Si

MeLi, Et$_2$O or t-BuOK, THF (80–82%)

A very recent report discloses that regio- and stereoselective deuteriation of allylic chlorides can be controlled by a neighboring hydroxyl, acetoxyl or silyl ether group[198]. Protection of a free hydroxyl group in **148**, R = H with a (t-Bu)Me$_2$Si group gives, however, a completely different regioselectivity (equation 62). Similarly, reductive ring opening of bicyclic ether **149a** with DIBALH in the presence of MeLi produces homoallylic alcohol **150a** as the major product along with by-product **151a** (equation 63)[199]. Reduction of the corresponding *tert*-butyldimethylsilyl ether (**149b**) by the same reducing agent in the absence of MeLi gives homoallylic alcohol **151b** as the major product along with by-product **150b**. The opposite regioselectivity results from the bulky (t-Bu)Me$_2$Si protecting group.

R	Ratio		
	150	:	151
(a) H	9.5	:	1
(b) (t-Bu)Me$_2$Si	1	:	6.4

Moreover, introduction of the (i-PrO)Me$_2$Si group to ketones at an α position assists their selective reduction by L-*Selectride*[200]. Upon treatment with KF/KHCO$_3$ and H$_2$O$_2$, the resultant α-silyl alcohols give vicinal diols.

16. Hydroboration

The steric hindrance caused by the PhMe$_2$Si group in allylsilanes **152** influences the orientation of hydroboration by 9-BBN, NaOH and H$_2$O$_2$ (equation 64)[139]. The corresponding alcohols **153** are produced stereoselectively in 79–96% yields.

	R¹	R²	R³	R⁴	
(a)	H	Me	(a) OH	H	(96%)
(b)	Me	H	(b) H	OH	(79%)

17. Hydrogenation

Hydroxy-directed hydrogenation of (phenyldimethylsilyl)allyl alcohols with a cationic rhodium complex provides a highly diastereoselective route to β-silyl alcohols[201,202].

18. Lithiation

Introduction of a bulky $(t\text{-Bu})Me_2Si$ group at the C-2 position of 3-furoic acid allows regioselective lithiation at the C-4 position with 2.5 equiv of BuLi[203]. In comparison, lithiation of 3-furoic acid without the $(t\text{-Bu})Me_2Si$ group takes place at the C-2 position with 2.0 equiv of LDA[204].

19. Oxidation

The $(t\text{-Bu})Me_2Si$ group in ketones **154** directs their diastereoselective oxidation to the corresponding α-silyloxy ketones **155** with m-CPBA or 3-phenyl-2-(phenylsulfonyl)oxaziridine (equation 65)[205]. The oxidizing agents approach *anti* to the $(t\text{-Bu})Me_2Si$ group. This process compares well with other silicon-directed electrophilic reactions at an alkene C−C double bond[206].

$$\underset{(154)}{\underset{R^1 = Me, Et, PhCH_2}{\underset{R^2 = Me, Et}{(t\text{-Bu})Me_2Si\diagdown\overset{O}{\underset{}{\diagup}}\diagdown R^2}}} \xrightarrow[\substack{\text{2. MCPBA, hexane, 0 °C} \\ \text{or} \\ \text{oxaziridine, CHCl}_3, \Delta}]{1.\text{LDA, THF, }-78\,°\text{C, Me}_3\text{SiCl}} \underset{(155)}{(t\text{-Bu})Me_2Si\diagdown\overset{O}{\underset{OSiMe_3}{\diagup}}\diagdown R^2}$$

(65)

20. Photocyclization

Photoinduced cyclizations of phenyl triphenylsilyl thioketone **(156)** with *cis*- and *trans*-1,2-dichloroethene give a mixture of silyl thietanes **157** and **158** in a regio- and stereoselective manner (equation 66)[207]. These reactions involving biradical intermediates and ring closure are governed by the steric effect. The steric hindrance between the Ph_3Si group and the vicinal chlorine atom ensures their *trans* relationship in the thietane products.

For a different class of compounds, cyclization of *ortho*- and *meta*-substituted 1-benzyl-1-pyrrolinium perchlorates **159** and **160** under photolytic conditions gives benzopyrrolizidines or benzindolizidines or both[208]. The product distribution is highly influenced by the steric effect resulting from the silyl groups at the benzylic position[209].

(66)

21. Reduction

A sterically demanding PhMe$_2$Si group at the C-3 position of 1-cyclohexanone derivatives affects a highly stereoselective reduction of the C-1 carbonyl group by NaBH$_4$ in methanol[187]. The 3-silyl-1-cyclohexanol product holds a *cis* configuration.

22. [2,3]-Sigmatropic rearrangement

A stereoselective [2,3]-sigmatropic rearrangement can be achieved by introduction of a Me$_3$Si or (*i*-Pr)$_3$Si group at the terminal *sp*-carbon in a prop-2-ynyl-2-(trimethylsilyl)allyl ether[210]. It generates vinylsilanes in the *E* form.

23. Substitution

A sterically demanding group at one terminus of the allyl moiety blocks the incoming nucleophile in palladium-catalyzed allylic substitutions. The Me$_3$Si group in **161** can fulfil such a purpose and prevails over the phenyl group as a controlling element in the preparation of optically active γ-silylallylamines (**162**, equation 67)[211].

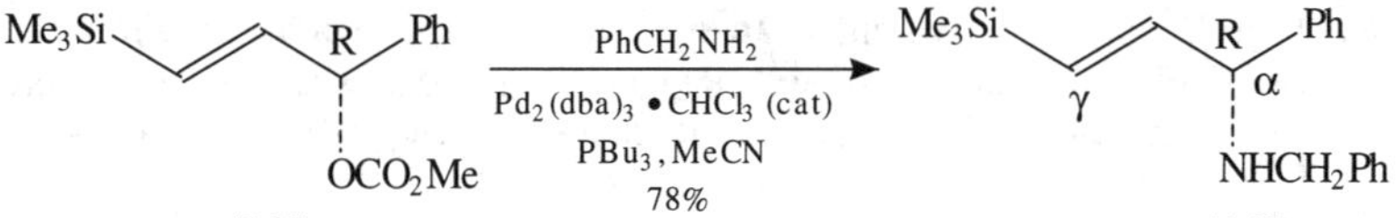

$$\text{Me}_3\text{Si}\diagup\diagdown\underset{\underset{(\textbf{161})}{\text{OCO}_2\text{Me}}}{\text{R}}\diagdown\text{Ph}\xrightarrow[\substack{\text{PBu}_3,\text{MeCN}\\78\%}]{\substack{\text{PhCH}_2\text{NH}_2\\\text{Pd}_2(\text{dba})_3\bullet\text{CHCl}_3\text{ (cat)}}}\text{Me}_3\text{Si}\underset{\gamma}{\diagup}\diagdown\underset{\underset{(\textbf{162})}{\text{NHCH}_2\text{Ph}}}{\overset{\alpha}{\text{R}}}\diagdown\text{Ph} \quad (67)$$

(98% ee)
α/γ > 95:5

V. REACTIVITY OF SILICON-CONTAINING REAGENTS

A. General Considerations

The size of a silyl group in silicon-containing reagents greatly influences their reactivity[212] and selectivity. Some prominent examples are discussed in the following reactions of 12 classes.

B. Organic Reactions of Various Types

1. Addition and substitution

A complete stereocontrol is achieved by addition of the bulky silyl radical $(\text{Me}_3\text{Si})_3\text{Si}^\bullet$ to a chiral and conformationally flexible electron-deficient olefin **163**, as shown in equation 68[213]. Replacement of $(\text{Me}_3\text{Si})_3\text{Si}^\bullet$ with a less sterically hindered $(n\text{-Bu})_3\text{Sn}^\bullet$ gives a mixture of *syn* and *anti* diastereomeric adducts **164** and **165** in a ratio of 7 : 3. The A values (kcal mol^{-1}) of the tin, carbon and silicon species follow the order $(n\text{-Bu})_3\text{Sn}$ (1.1) < Me (1.7) < Me$_3$Si (2.5), and the bond length for C—Sn (2.2 Å) is longer than that for C—Si (1.85 Å)[214,215].

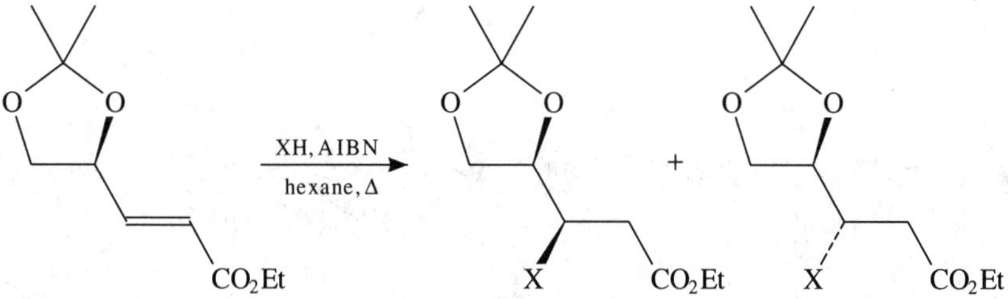

X	Overall yield (%)	Ratio 164 : 165
(Me$_3$Si)$_3$Si	>95	100 : 0
n-Bu$_3$Sn	93	7 : 3

Recent reports indicate that the size of silyl groups, including Ph$_3$Si and Ph$_2$MeSi, influences the 1,2-addition of an imine by α-silyl organocopper reagent[216] and the S_N2 displacement of a styrene oxide by α-silyl organolithium reagents[217]. On the other hand,

1,4-addition of a silyl anion to an α,β-unsaturated enone moiety provides an efficient way to introduce the silicon element onto organic molecules. Reaction of Me$_3$SiLi with dienone **166** at the C$_1$ position gives enone **167a** as the major product along with its α-epimer in a ratio of 19 : 1 (equation 69)[38,124]. Reagent (Me$_3$Si)Me$_2$SiLi, sterically more hindered than Me$_3$SiLi, reacts with **166** to afford exclusively enone **167b** in 63% yield (cf entry 46 of Table 1).

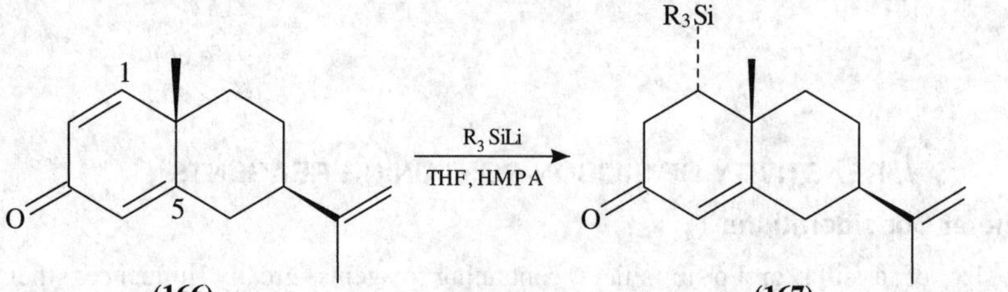

$$\text{(166)} \xrightarrow{\text{R}_3\text{SiLi}, \text{THF, HMPA}} \text{(167)} \tag{69}$$

R$_3$Si =
(a) Me$_3$Si (66%; 19:1)
(b) Me$_3$SiMe$_2$Si (63%)

Silylcuprate reagents add diastereoselectively to various types of functional groups[91,218], including α-alkylidenelactones[219], N-acryloyl lactams[220], N-enoyl sultams[221], 2-phenylselenocyclopent-2-en-1-one[222] as well as α,β-unsaturated enones, amides and esters[218,223,224]. The level of diastereoselection of the addition to N-enoyl sultams **168** giving a mixture of **169** and **170** (equation 70) greatly depends on the nature of silyl cuprates and invariably increases when the higher-order silylcyanocuprates are used. The stereofacial discrimination ability follows the sequence: (R$_3$Si)$_2$CuLi $\ll$ R$_3$SiCu $<$ (R$_3$Si)$_2$Cu(CN)Li$_2$[225].

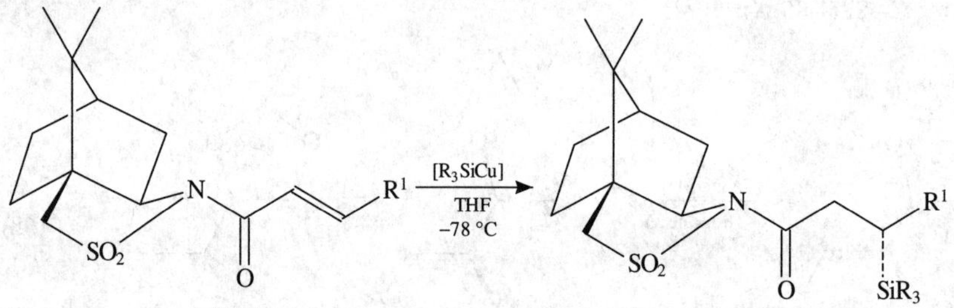

$$\text{(168)} \xrightarrow[\text{THF, } -78\,°\text{C}]{[\text{R}_3\text{SiCu}]} \text{(169)} + \text{(170)} \tag{70}$$

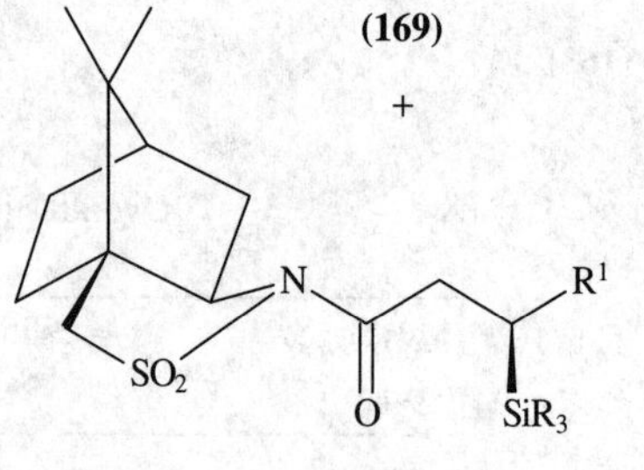

[R$_3$SiCu] =
R$_3$SiCu : R$_3$SiLi + CuI + PBu$_3$ (1 : 1 : 1)
(R$_3$Si)$_2$CuLi : R$_3$SiLi + CuI + PBu$_3$ (2 : 1 : 1)
(R$_3$Si)$_2$Cu(CN)Li$_2$: R$_3$SiLi + CuCN (2 : 1)
R = Me, t-Bu, Ph
R^1 = Me, 4-XC$_6$H$_4$ (X = H, Me, MeO, Cl), 2-Furyl

The diastereoselection levels are uniformly appealing (93–99%) for crotonates and cinnamates when higher (dimethylphenylsilyl)cyano cuprates are used as reagents. Other more hindered silyl groups, such as Ph_2MeSi and $(t-Bu)Ph_2Si$ groups, show somewhat lower diastereoselection levels.

Depending upon the steric nature of the alkyl groups attached to tin and silicon in $R_3Sn-SiR'_2R''$, these silylstannanes undergo ligand exchange upon treatment with a higher-order cuprate $(n-Bu)_2Cu(CN)Li_2$ to afford trialkylsilyl mixed cuprates (equation 71)[226]. The resultant cuprates $RMe_2Si(n-Bu)Cu(CN)Li_2$ (R = t-Bu or thexyl) can participate in substitution and addition reactions.

$$RMe_2SiSnMe_3 + (n-Bu)_2Cu(CN)Li_2 \xrightarrow[\text{THF}]{25\,°C} RMe_2Si(n-Bu)Cu(CN)Li_2 \quad (71)$$

R = t-Bu or thexyl; R^1 = H or Me

2. Condensation

The chiral reagents [(1R,2S)-ephedrine]$POSiR_3$ (**172**) are prepared through the reaction of [(1R,2S)-ephedrine]PCl (**171**) with R_3SiOH in the presence of NEt_3 (equation 72)[227]. The epimers **172** and **173** exist as an equilibrium mixture with a diastereoselectivity (86–94%) that is dependent upon the nature of the substituents on silicon.

(72)

R_3Si	Yield (%)	Diastereoselectivity (%) of 172/173
Et_3Si	82	86
$(t-Bu)Me_2Si$	78	88
Ph_3Si	91	94

3. [3 + 2] Cycloaddition

Allylsilanes **12** react with α-ketoesters **174** to give [3+2] cycloadducts **175** in 50–85% yields through 1,2-silyl migration (equation 73)[59,228]. Use of an allylic trimethylsilane produces an allyl alcohol by-product through a competing silyl elimination process. It can be circumvented by use of a bulkier silicon-containing reagent, such as the

$CH_2=CHCH_2SiMe_2(Bu-t)$ (cf entry 47 of Table 1). Given a similar concern, a new sterically demanding allylsilane, $CH_2=CHCH_2SiMe_2(CPh_3)$, has been developed recently as a mild and efficient reagent for hydroxypropyl annulation of electron-deficient alkenes[229].

$$Ph-C(=O)-C(=O)-OEt \quad + \quad CH_2=CHCH_2-SiR_3 \quad \xrightarrow{SnCl_4, CH_2Cl_2} \quad (175) \quad (73)$$

(174)   (12)   (175)

R_3Si	Yield (%)
Me_3Si	50
$PhMe_2Si$	54
$(t\text{-}Bu)Me_2Si$	85

4. [4 + 2] Cycloaddition

The Me_3Si group in (E)-1,3-bis(trimethylsilyloxy)buta-1,3-diene (**177**) provides the necessary steric requirement to react with 10-amino-9-hydroxy-1,4-anthraquinone derivatives **176** in a regio- and stereospecific manner (equation 74)[230]. Thus a single anthracyclinone derivative (**178**) is obtained in an excellent yield through the Diels–Alder process.

(176) + (177) $\xrightarrow{\Delta, PhH}$ (178)   (74)

$R = Cl_3CCH_2OC$, Boc, CO_2Et; $R^1 = H$, MeO

5. Hydrosilylation

Various organosilanes, including $(n\text{-}C_5H_{11})SiH_3$, $(n\text{-}Bu)MeSiH_2$, $(n\text{-}Pr)_2SiH_2$, Ph_2SiH_2, Et_3SiH and Et_2MeSiH, are used in hydrosilylation of 2,2,4,4-tetramethyl-1,3-cyclobutanedione in the presence of the Wilkinson catalyst [i.e., $Rh(PPh_3)_3Cl$]. These reactions lead to a mixture of the corresponding *cis* and *trans* diols as well as 3-hydroxylcyclobutanone[60]. The selectivity of the reductions depends greatly upon the bulk of silane reagents. The order of the silyl groups in these reagents is listed in entry 48 of Table 1.

For hydrosilylation of 1-hexyne with R_3SiH in the presence of H_2PtCl_6, the trend of steric influence of silyl groups is shown in entry 49 of Table 1[61]. The steric hindrance resulting from various substituents on the silicon of vinylsilanes also influences their hydrosilylation by Et_3SiH in the presence of $Ni(acac)_2$ as a catalyst[231].

6. Insertion reaction

Alkenyl Fischer carbene complexes (e.g. **179**) undergo a [2 + 1] insertion reaction with Et_3SiH or Ph_3SiH to give allylsilanes (e.g. **180**) in 68–87% yields (equation 75)[62]. In comparison with Et_3SiH, use of Ph_3SiH shows a significant (*ca* 10-fold) rate enhancement and a yield improvement of *ca* 20%. The Ph_3Si group, although 'bigger' than the Et_3Si group, is less hindered as a result of the 'propeller effect' of the phenyl rings (entry 50 of Table 1).

$$\text{Ph} \diagdown \text{C=C(Cr(CO)}_5\text{)(OMe)} + R_3SiH \xrightarrow[60\,°C]{\text{hexane}} \text{Ph} \diagdown \text{CH-C(OMe)=CH-SiR}_3 \quad (75)$$

(179) → **(180)**

R_3Si	Yield (%)
Et_3Si	68
Ph_3Si	87

On the other hand, rhodium(II) can catalyze intramolecular C—H insertion of 1-methyl-3-silyloxyl-1-(diazoacetyl)cyclohexanes[63]. The ratio of the products through C_5—H versus C_3—H insertions increases from 2.2 : 1 to 6 : 1 when the bulkier $(i\text{-}Pr)_3Si$ group is used to replace the $(t\text{-}Bu)Me_2Si$ group (entry 51 of Table 1).

7. Nitrone formation

'Bulky proton'-containing reagent $Me_3SiN(Me)OSiMe_3$ (**181**) reacts with aldehydes and ketones to give the corresponding nitrones **182** in good to excellent yields (equation 76)[232,233]. This reagent is sensitive towards steric hindrance as exemplified by reacting it with dicarbonyl compound 5α-pregnan-3,20-dione (**183**, equation 77). Treatment of a 1 : 1 ratio of $Me_3SiN(Me)OSiMe_3$ (**181**) with **183** in benzene at reflux affords mono-nitrone **184** in 71% yield. In comparison with the C-20 carbonyl group, the

C-3 carbonyl moiety in **183** is less sterically congested[3].

$$\underset{(181)}{\underset{R^1}{\overset{R}{>}}\!\!=\!\!O + MeN\!\!<\!\!\overset{OSiMe_3}{\underset{SiMe_3}{}}} \longrightarrow \left[\underset{}{RR^1C\!\!<\!\!\overset{\overset{OSiMe_3}{N-Me}}{\underset{OSiMe_3}{}}}\right] \longrightarrow \underset{(182)}{\underset{R^1}{\overset{R}{>}}\!\!=\!\!\overset{O^-}{\underset{Me}{N+}}} + Me_3SiOSiMe_3 \quad (76)$$

R = H, Me; R[1] = Me, CHMe$_2$, CMe$_3$, (CH$_2$)$_3$CH=CH$_2$, p-XC$_6$H$_4$ (X = H, NH$_2$, NO$_2$), 2-Furyl

(77)

(183) + **(181)** $\xrightarrow[PhH]{\Delta}$ **(184)** (71%)

8. Olefination

Various silyl groups in Grignard reagents **185** mediate their addition to aldehydes to generate mainly Z-enynes **186** (equation 78)[64]. The ratios of **186** to its E-isomer **187** depends upon the size of silyl groups according to the order listed in entry 52 of Table 1.

$$RCHO + \underset{(185)}{\underset{R'_3Si}{\overset{H}{>}}\!\!=\!\!\underset{SiMe_3}{\overset{MgBr}{<}}} \xrightarrow{THF} \underset{(186)}{R\!\!-\!\!\equiv\!\!-SiMe_3} + \underset{(187)}{\underset{SiMe_3}{\overset{R}{>}}} \quad (78)$$

R′$_3$Si = Me$_3$Si, Et$_3$Si, (t-Bu)Me$_2$Si, Ph$_3$Si

9. Oxygen–oxygen bond cleavage

Phenylethynyllithium (**188**) reacts with silylperoxides **189a–d** to give C-silylated products **190a–d** or silyl ethers **191a–d** or both in 48–90% overall yields (equation 79)[234]. The distribution of the products and the yields depend upon the steric hindrance created by the substituents attached to the silicon atom.

$$PhC\equiv C-Li + (R_3Si-O)_2 \longrightarrow PhC\equiv C-SiR_3 + PhC\equiv C-OSiR_3$$

(**188**)   (**189**)   (**190**)   (**191**)

(79)

	R_3Si	Overall yield (%)	Ratio		
			190	:	**191**
(a)	Me_3Si	86	100	:	0
(b)	$(n\text{-}Pr)_3Si$	67	3	:	2
(c)	$(n\text{-}Bu)_3Si$	90	3	:	2
(d)	Ph_2MeSi	48	>9	:	1

10. Silylation

The steric effect resulting from trisubstituted silyl chlorides influences their capability in silylation of alkali phenolates[65]. As the size of the silyl groups increases with the trend shown in entry 53 of Table 1, the activity of the silyl chlorides decreases.

Substitution with a *t*-Bu functionality in the *ortho* position of the phenolate ring retards the reactivity through steric crowding of the reaction site. Moreover, the feasibility of reactions of azasilatranes with CF_3SO_3Me can be rationalized on steric grounds[235].

11. Silylformylation

Dirhodium(II) perfluorobutyrate [i.e. $Rh_2(pfb)_4$] catalyzes the silylformylation of terminal alkynes with carbon monoxide and organosilanes to afford β-silylacrylaldehydes **192** in moderate to excellent yields (equation 80)[236]. Use of organosilanes Et_3SiH and $PhMe_2SiH$ provides appealing stereocontrol for the (Z)-isomer (>10 : 1). Hydrosilylation, however, competes with the silylformylation. The steric effect between the organosilanes and alkynes plays a role in determining the relative rates of these two competitive processes.

$$R^1C\equiv CH + R_3^2SiH + CO \xrightarrow[\text{CH}_2\text{Cl}_2]{Rh_2(pfb)_4} \text{(192)}$$

(24–95%)

(**192**)
(Z/E > 10:1)

(80)

R^1 = Ph, *p*-Tol, *p*-*i*(-Bu)C_6H_4, 6-MeO-Naph-2, *n*-Hex, $AcOCH_2$, $MeOCH_2$, $Me_2C(OH)$
R_3^2Si = Et_3Si, $PhMe_2Si$

12. Silylstannation

Silylstannation of alkenes with organosilylstannanes in the presence of bis(dibenzylideneacetone)palladium [Pd(dba)$_2$] and trialkylphosphine as catalysts affords 1,2-adducts in 36–97% yields (equation 81)[237]. The organosilylstannane reagents include Me$_3$SiSnMe$_3$, Me$_3$SiSn(Bu-n)$_3$, (MeO)Me$_2$SiSn(Bu-n)$_3$, (t-Bu)Me$_2$SiSnMe$_3$, and (t-Bu)Me$_2$SiSn(Bu-n)$_3$. The steric congestion seems to affect the reaction significantly.

$$R^1Me_2SiSnR^2{}_3 + \underset{R^5\quad R^6}{\overset{R^3\quad R^4}{\searrow\!\!=\!\!\swarrow}} \xrightarrow[\text{PR}_3,\,130\,°\text{C}]{\text{Pd(dba)}_2} \underset{R^5\quad R^6}{\overset{R^1Me_2Si\quad SnR^2{}_3}{R^3\!-\!\!-\!\!R^4}} \quad (36\text{–}97\%) \tag{81}$$

R^1 = Me, t-Bu, OMe
R^2 = Me, n-Bu

$R^3R^5C\!\!=\!\!CR^4R^6$ = ethylene, norbornene

VI. SOLVOLYSIS OF VARIOUS ORGANOSILANES

A. Adamantane p-Nitrobenzoates

The Me$_3$Si and the Me groups exhibit similar effective sizes in 2-adamant-2-yl p-nitrobenzoate[238]. The conclusion is drawn on the basis of solvolysis of its derivatives bearing a Me$_3$Si and a Me group as well as on the results of *ab initio* and force-field calculations. The solvolysis rate of the adamant-2-yl derivatives is not sterically accelerated by a 2-Me$_3$Si group, yet it is greatly accelerated by a 2-t-Bu group[239].

B. Benzylic p-Toluenesulfonates

A pronounced rate-retardation of 1.65×10^4-fold by an α-Me$_3$Si group relative to Me in the solvolysis of benzylic p-toluenesulfonates is due to a steric effect[66]. The rate increment in ethanol decreases with increasing steric size of the α-silyl group attached to the benzylic position and follow the order listed in entry 54 of Table 1.

C. Silyl Ethers

Complexation of silyltetralol **193** with Cr(CO)$_6$ gives a mixture of *syn*- and *anti*-complexes (**194/195** = 45 : 55; equation 82)[240]. The *anti*-isomer is favored because the steric effect of the (t-Bu)Me$_2$SiO group outweighs any chelation effect. A greater desilylation rate for **195** than for **194** by use of Bu$_4$NF·3H$_2$O is due to the steric bulk around the silicon in **194** caused by the Cr(CO)$_3$ moiety at the *syn*-position.

The k_2 values (M^{-1} s^{-1}) of the solvolysis of various R$_3$SiOPh in aqueous ethanol at 25 °C were determined under acidic and basic conditions[67,68]. The size of the silyl groups influences the rate and follows the order listed in entries 55 and 56 of Table 1. The correlation between the Si–O–C angle and the oxygen basicities may be completely different, or even reversed, in comparison with trimethylsilyl and *tert*-butyldimethylsilyl ethers[241]. The relative hydrolysis rates may be simply controlled by steric hindrance to solvent assistance for the Si–O bond cleavage following the protonation step.

                                8. Steric effects of silyl groups

                                           (193) OSiMe₂(Bu-t)

                                                Cr(CO)₆ ↓   (82)

        (194) OSiMe₂(Bu-t)   +   (195) OSiMe₂(Bu-t)
        Cr(CO)₃   Cr(CO)₃
        (syn)   (anti)
         45   :   55

D. Triorganosilyl Chlorides and Fluorides

A quantitative scale for the structural effect of various silyl groups is established, as shown in entry 57 of Table 1, by the rates of solvolysis of 40 triorganosilyl chlorides in aqueous dioxane under neutral conditions[69]. The structural effect involves the steric effect and, in some examples, the electronic effect. Because little difference exists in the electronic effect among alkyl groups, their steric effect at silicon follows the order primary < secondary < tertiary substituents.

Moreover, reactivities for 54 triorganosilyl groups toward nucleophilic displacement at silicon have been predicted on the basis of their solvolysis rates in aqueous dioxane (see entry 58)[70]. In addition, results from solvolysis of R_3SiF in aqueous acetone are listed in entry 59[71] of Table 1.

E. Triorganosilyl Hydrides

Two similar orders listed in entries 60 and 61 of Table 1 are established by solvolysis of R_3SiH. Those reactions are performed in aqueous ethanol at 34.5 °C under acidic[72,73] or basic conditions[74].

VII. CHANGE OF PHYSICAL, CHEMICAL AND SPECTROSCOPIC PROPERTIES BY INTRODUCTION OF SILYL GROUPS

Replacement of an H, Me, Et, i-Pr, or t-Bu unit with a bulky silyl group in organic compounds may change their physical or spectroscopic properties to a great extent. The changes in the following systems clearly come from the steric congestion created by the silyl groups.

A. Change of Physical and Chemical Properties

1. Conformation, bond angles and bond lengths

An extreme example reflecting the steric congestion is the replacement of all hydrogens in benzene by six Me$_3$Si groups to cause a large ring distortion into a chair form[242]. A strong σ(C−Si)−π conjugation becomes important. At elevated temperature (e.g. 200 °C), the relative amount of the boat form increases to 70%[243]. Thermolysis of the highly distorted hexakis(trimethylsilyl)benzene results in rupture of the aromatic ring[244]. The internal benzene ring bond angles also reflect the distortion of molecule **196**, resulting from the bulk of the substituents and the size of the Me$_3$Si groups at the C-4 and the C-6 positions[245].

(196)

(197)
(eclipsed)
R$_3$Si =
(a) Me$_3$Si
(b) Et$_3$Si

(198)
(staggered)
R$_3$Si =
Ph$_3$Si =
(i-Pr)$_3$Si

Another example includes replacement of two or four methyl groups in N,N,N',N'-tetramethyl-p-phenylenediamine (i.e. Wurster's Blue) with Me$_3$Si groups. It results in distortion of the SiHN−C$_6$H$_4$−NHSi skeleton[246]. The nitrogen lone pairs of the planar Me$_3$SiNH units in N,N'-disilylated derivatives are co-axial with the π-vector perpendicular to the benzene ring. The p-type electron pairs of the flattened (Me$_3$Si)$_2$N groups in the overcrowded N,N,N',N'-tetrakis(trimethylsilyl)-p-phenylenediamine are twisted into the plane of the benzene ring. Thus it provides the possibility for one-electron reduction to a Wurster's Blue radical anion.

In addition, trimethylsilyl ether **197a** and triethylsilyl ether **197b** have conformations different from those of compounds **198a** and **198b**[247], which bear a bulky Ph$_3$Si or a (i-Pr)$_3$Si group, respectively. The change of conformation is due to the steric hindrance from these two bulky silyl groups.

Furthermore, reactions of R$_3$SiNH$_2$ with R$'_3$Al in hexane at reflux afford dimeric aluminum silylamides [R$'_2$AlNHSiR$_3$]$_2$ [R$_3$ = Me$_3$, Et$_3$, Ph$_3$ or (t-Bu)$_2$H; R$'$ = Me or (i-Bu)][75]. A distorted-tetrahedral geometry at nitrogen of the resultant complexes places the silyl groups nearly in the plane of the Al$_2$N$_2$ ring. The degree of distortion depends on the steric bulk of the silyl groups by following the order shown in entry 62 in Table 1.

In metal complexes, good correlation exists between the size of the silyl ligand and the metal−Si bond distance. For example, the Mo−Si bond distances are 2.538(2), 2.560(1) and 2.604(1) Å for (Cp)$_2$Mo(H)(SiMe$_2$H), (MeCp)$_2$Mo(H)(SiMe$_3$) and (MeCp)$_2$Mo(H)[Si(Bu-t)$_2$H], respectively (cf. entry 63 in Table 1)[76]. Complexes containing a bulky silyl ligand [i.e. (t-Bu)$_2$HSi] exhibit hindered rotation around the metal−Si bond in solution as determined by variable temperature ^{1}H NMR spectroscopy. Moreover, the steric bulk

8. Steric effects of silyl groups

of the silyl groups in (ferrocenylacyl)silanes, $(\eta^5\text{-}C_5H_5)$ Fe $(\eta^5\text{-}C_5H_4)$ COSiR$_3$, is in the order of PhMe$_2$Si = Ph$_2$MeSi < Me$_3$Si < Ph$_3$Si (entry 64)[77].

For the η^5-indenyl complexes $(\eta^5\text{-}C_9H_7)$Fe(CO)$_2$(Si$_n$Me$_{(2n+1)}$), the Fe—Si bond length increases regularly in the order of increasing silyl group bulk as the sequence listed in entry 65 of Table 1[78]. Moreover, the readily accessible ligands —N(SiMe$_2$Ph)$_2$ and —N(SiMePh$_2$)$_2$ are capable of stabilizing two coordinations in the solid state for the metals Mn, Fe and Co[248].

Comparison of two extremely bulky groups[249,250], [i.e. (Me$_3$Si)$_3$Si and (Me$_3$Si)$_3$C], is made on the basis of their A values. The structures of enols Mes$_2$C=C(OH)R [Mes = 2, 4, 6-trimethylphenyl; R = Me$_3$Si, (Me$_3$Si)$_3$Si, (Me$_3$Si)$_3$C and *tert*-Bu] and two-ring flip barriers have been calculated by the MM2* force field[79]. The A values are estimated as 2.5, 4.89, 13.3 and 4.9 kcal mol^{-1} for the Me$_3$Si, (Me$_3$Si)$_3$Si, (Me$_3$Si)$_3$C and *tert*-Bu groups, respectively. Thus the (Me$_3$Si)$_3$Si and the *tert*-Bu groups have similar effective size; the (Me$_3$Si)$_3$C group is significantly larger (entry 66 of Table 1).

In related studies, characteristic structural parameters including dihedral angles of X$_3$—Y—Y—X$_3$ as well as bond distances between Y—Y and Y—X are determined for molecules containing bulky half-shell substituent groups. The molecules include (Me$_3$Si)$_3$Si—Si(SiMe$_3$)$_3$[251,252], (Me$_3$C)$_3$Si—Si(CMe$_3$)$_3$[253], *p*-(Me$_3$Si)$_3$C—C$_6$H$_4$—C(SiMe$_3$)$_3$[254] and (Me$_3$Si)$_3$C—C≡C—C(SiMe$_3$)$_3$[254]. They often adopt staggered conformations of their sterically congested (Me$_3$X)$_3$Y-half shells along substituent axes of C_3-symmetry at different distances Y—Y.

2. Stability, free energy of activation and reaction rates

Steric crowding among the Me$_3$Si groups in boraazatrane **199**, in contrast to the methyl analog **200**, leads to enantiomers sufficiently long-lived to be observed at room temperature on the NMR time scale[255]. The enantiomers racemize by a concerted mechanism rather than by the stepwise process found for silatranes[256]. Moreover, by increasing the bulk of R in compounds **201** to Me$_3$Si (when E = Si and Z = Me), the transannular bond is essentially broken[257].

(**199**) R = Me$_3$Si
(**200**) R = Me

(**201**) E = Si, Ge, Sn, Ti, V
Z = NMe$_2$, N-(Bu-*t*)

On the other hand, an intramolecular reversible rotation of the two mesityl rings in 2,2-dimesityl-1-(trimethylsilyl)ethenol (**202**) occurs to give **203** (equation 83)[258]. The free energy of activation for the rotation ($\Delta G_c^{\ddagger}$) leading to the enantiomerization is

11.1 kcal mol^{-1}, which is intermediate between the $\Delta G_c^{\ddagger}$ values for the corresponding 1-isopropyl enol (10.4 kcal mol^{-1}) and 1-t-butyl enol (11.7 kcal mol^{-1}). Charton's υ values, proportional to the van der Waals radii, are 0.76, 1.24 and 1.40 for i-Pr, t-Bu and Me$_3$Si groups, respectively[259]. According to Charton's values, the Me$_3$Si group occupies more space than the t-Bu group. The lower rotational barrier for 2,2-dimesityl-1-(trimethylsilyl)ethenol than for the corresponding 1-t-butyl enol may reflect small differences in the ground state geometry.

R	$\Delta G_c^{\ddagger}$ (kcal mol^{-1})	Charton's υ value
i-Pr	10.4	0.76
Me$_3$Si	11.1	1.40
t-Bu	11.7	1.24

Rates of decay for (p-methoxybenzyl)trialkylsilane cation radicals, which bear a Me$_3$Si, Et$_3$Si or (i-Pr)$_3$Si group, are determined in acetonitrile as 2.3×10^6, 1.3×10^6 and $<4 \times 10^3$ s^{-1}, respectively[80]. Thus the lifetimes of silane cation radicals increase as the size of the silyl group increases, as shown in entry 67 of Table 1.

Furthermore, the relative steric bulk of the hindered bases 2,4,6-tris[bis(trimethylsilyl)methyl]phenyllithium and 2,4,6-tri-$tert$-butylphenyllithium is determined by investigating the kinetic versus thermodynamic enolate formation from benzyl methyl ketones[260]. The results indicate that the 2,4,6-[(Me$_3$Si)$_2$CH]$_3$C$_6$H$_2$ (Tb) group is bulkier than the 2,4,6-(t-Bu)$_3$C$_6$H$_2$ (Bp) group[261]. The high steric demand of the Tb group is used in the synthesis of extremely hindered and kinetically stable cis- and $trans$-disilenes

Tb(Mes)Si=Si(Mes)Tb (Mes = mesityl) by reductive coupling of Tb(Mes)SiBr$_2$ with lithium naphthalenide[262].

The steric size of the substituent R on silenes (Me$_3$Si)RSi=C(OSiMe$_3$)R' influences their photochemical reactivity[263]. When R = Me$_3$Si or Me, head-to-head dimerization occurs to give 1,2-disilacyclobutanes. When R = Ph or bulky *t*-Bu, the silenes are stable. Although the Me$_3$Si group is also bulky, the long Si−C single bond reduces markedly the effective steric hindrance.

3. Migratory aptitudes of organic groups

A relative ranking of the migratory aptitudes of the σ-bound ligands is established in the study of the regioselective insertions of CO into the W−C σ bonds of complexes (η^5-C$_5$Me$_5$)W(NO)(R)(R') (R,R' = alkyl, aryl). The trend, Me < Ph < *o*-toluenesulfonyl < CH$_2$SiMe$_3$ < CH$_2$CMe$_3$, primarily reflects steric effects[264]. This trend agrees with the results found from kinetic studies on carbonylation of Cp$_3$ThR complexes to form η^2-acyl products. The rate increases in the order PhCH$_2$ < Me < CH$_2$SiMe$_3$ ≪ *n*-Bu < CH$_2$CMe$_3$ ≪ *s*-Bu < *i*-Pr[265]. A similar trend is also noted for the insertion rates of CpFe(CO)$_2$R in DMSO solution to form η^1-acyls. The trend follows the order Me < Et < *n*-Bu < CH$_2$SiMe$_3$ < *i*-Pr < *s*-Bu < CH$_2$CMe$_3$ ≪ CH(SiMe$_3$)$_2$[266].

B. Change of Spectroscopic Properties

1. ^1H, ^{13}C, ^{19}F and ^{29}Si NMR spectroscopy

In the ^{1}H NMR spectra of compound **40**, the signals are broad for the *ortho*-CH$_3$ of two mesityl groups. This is because of the hindered rotation of the mesityl groups caused by the silyl groups[81]. The degree of broadening increases in the order **40a** < **40b** < **40c** with an increase in the bulk of the silyl groups by following the order shown in entry 68 of Table 1.

As the silyl group becomes larger (see entry 69 of Table 1) in 2-methoxy-3-silylhex-4-en-1-oic acids, the coupling constant $^3J_{2,3}$ for the 2,3-*syn* diastereomer increases (7.57, 8.00 and 9.80 Hz)[82]. Yet the $^3J_{2,3}$ values of 2,3-*anti* diastereomers decrease (3.17, 3.10 and 2.40 Hz).

In a different system, 1-(trimethylsilyl)cycloocta-1,5-diene forms complexes with Ag(I), Rh(I), Pd(II) and Pt(II), in which the metals are pushed away from the Me$_3$Si group[267]. The distortion in their structures, in comparison with the near-symmetrical structures of the corresponding 1,5-cyclooctadiene complexes, changes the characteristics of their ^{1}H and ^{13}C NMR spectra.

Substituent contributions to the exchange barrier are determined by ^{19}F NMR for silicates RR'SiF$_3^-$, which are prepared as their K$^+$-18-crown-6 salts[268]. Their free energies of activation range from 9.3−14 kcal mol^{-1}. The steric comparison can be made in two series: Ph$_2$SiF$_3^-$ < Ph(2,6-Me$_2$C$_6$H$_3$)SiF$_3^-$ < (2,6-Me$_2$C$_6$H$_3$)$_2$SiF$_3^-$ and Ph(2,6-Me$_2$C$_6$H$_3$)SiF$_3^-$ < (2-MeC$_6$H$_4$)(2,6-Me$_2$C$_6$H$_3$)SiF$_3^-$ < (2,6-Me$_2$C$_6$H$_3$)$_2$SiF$_3^-$. Moreover, the steric effect resulting from a silyl group can be used for structure determination of hydroxy steroids by means of ^{29}Si NMR spectroscopy[269].

2. X-ray crystallography

Results from X-ray crystallography analyses indicate that the Si−Si bond distance is 2.539 Å in di-thexyl-substituted disilane **204** and 2.476 Å in di-*tert*-butyl disilane

205^{270}. These values are much larger than the typical Si—Si bond length of 2.34 Å^{271}. These distances result from mutual repulsions between the bulky thexyl, *tert*-butyl, and diethylamino groups. In addition, the bulk of the (*t*-Bu)Me$_2$Si group severely restricts its bending back in icosahedral *closo*-dicarbaborane 1-phenyl-2-(*t*-butyldimethylsilyl)-1,2-dicarba-*closo*-dodecaborane (12), C$_{14}$H$_{30}$B$_{10}$Si. In consequence, some deformation of the cage has resulted^{272}.

$$\text{Et}_2\text{N}-\underset{\underset{\text{NEt}_2}{|}}{\overset{\overset{R}{|}}{\text{Si}}}-\underset{\underset{\text{NEt}_2}{|}}{\overset{\overset{R}{|}}{\text{Si}}}-\text{NEt}_2 \quad \begin{array}{l} R = \\ \textbf{(204)} \ \text{CMe}_2(\text{Pr-}i) \\ \textbf{(205)} \ t\text{-Bu} \end{array}$$

On the other hand, a series of tetrasilylethylenes **206–210** is prepared and their structures are studied spectroscopically. Nonplanar distortion (i.e., twisting) is the main factor for relieving the severely crowded environment for **207**^{273} and **208**^{274}, but slight pyramidalization starts to occur for **209**^{275} in addition to twisting. Compound **210**^{276} is the most twisted and congested olefin and **206** is an untwisted tetrasilylethylene.

$$\underset{R^2}{\overset{R^1}{\diagdown}}\!\!=\!\!\underset{R^4}{\overset{R^3}{\diagup}}$$

(**206**) R^1 = R^2 = R^3 = R^4 = HMe$_2$Si
(**207**) R^1 = R^2 = R^3 = R^4 = Me$_3$Si
(**208**) R^1 = R^2 = Me$_3$Si; R^3 = R^4 = (*t*-Bu)Me$_2$Si
(**209**) R^1 = R^3 = Me$_3$Si; R^2 = R^4 = (*t*-Bu)Me$_2$Si
(**210**) R^1 = Me$_3$Si; R^2 = R^3 = R^4 = (*t*-Bu)Me$_2$Si

VIII. CONCLUSION

Triorganosilyl groups can control stereochemistry in organic reactions through a steric, an electronic or a stereoelectronic effect. More than one of these effects may exist simultaneously in some chemical processes. The trend listed in Section II could provide chemists with a clear guideline, yet definitive order, to choose a silyl group with appropriate size in control of reactions based on the steric effect.

IX. ACKNOWLEDGMENT

The authors wish to thank Mr. Chun-Chieh Lin and Miss Shu-Mei Liao for their assistance in the preparation of the manuscript.

X. REFERENCES

1. J. R. Hwu and J. M. Wetzel, *J. Org. Chem.*, **50**, 3946 (1985).
2. J. R. Hwu and N. Wang, *Chem. Rev.*, **89**, 1599 (1989).
3. J. R. Hwu, K. P. Khoudary and S.-C. Tsay, *J. Organomet. Chem.*, **399**, C-13 (1990).
4. T. H. Chan and D. Wang, *Chem. Rev.*, **92**, 995 (1992).
5. T. H. Chan and D. Wang, *Chem. Rev.*, **95**, 1279 (1995).
6. I. Fleming, *Chemtracts: Organic Chemistry*, **9**, 1 (1996).
7. P. C. B. Page, S. S. Klair and S. Rosenthal, *Chem. Soc. Rev.*, **19**, 147 (1990).
8. P. F. Cirillo and J. S. Panek, *Org. Prep. Proc. Int.*, **24**, 555 (1992).
9. C. Rücker, *Chem. Rev.*, **95**, 1009 (1995).
10. M. G. Steinmetz, *Chem. Rev.*, **95**, 1527 (1995).
11. C. Eisenberg and P. Knochel, *J. Org. Chem.*, **59**, 3760 (1994).
12. X. Chen, E. R. Hortelano, E. L. Eliel and S. V. Frye, *J. Am. Chem. Soc.*, **114**, 1778 (1992).
13. X. Chen, E. R. Hortelano, E. L. Eliel and S. V. Frye, *J. Am. Chem. Soc.*, **112**, 6130 (1990).

14. A. Yanagisawa, K. Yasue and H. Yamamoto, *Synlett*, 686 (1993).
15. A. P. Davis and S. J. Plunkett, *J. Chem. Soc., Chem. Commun.*, 2173 (1995).
16. R. J. Linderman and T. V. Anklekar, *J. Org. Chem.*, **57**, 5078 (1992).
17. J. S. Panek, A. Prock, K. Eriks and W. P. Giering, *Organometallics*, **9**, 2175 (1990).
18. J. R. Hwu, B.-L. Chen and S.-S. Shiao, *J. Org. Chem.*, **60**, 2448 (1995).
19. T. H. Chan and K. Koumaglo, *J. Organomet. Chem.*, **285**, 109 (1985).
20. J. R. Hwu, C. N. Chen and S.-S. Shiao, *J. Org. Chem.*, **60**, 856 (1995).
21. B. M. Trost and A. Indolese, *J. Am. Chem. Soc.*, **115**, 4361 (1993).
22. S. H. Kang and S. B. Lee, *Tetrahedron Lett.*, **34**, 7579 (1993).
23. F. Bennett and D. W. Knight, *Tetrahedron Lett.*, **29**, 4865 (1988).
24. F. Bennett, D. W. Knight and G. Fenton, *J. Chem. Soc., Perkin Trans. 1*, 133 (1991).
25. F. Bennett, D. W. Knight and G. Fenton, *J. Chem. Soc., Perkin Trans. 1*, 1543 (1991).
26. G. Adiwidjaja, H. Flörke, A. Kirschning and E. Schaumann, *Justus Liebigs Ann. Chem.*, 501 (1995).
27. M. Suginome, Y. Yamamoto, K. Fujii and Y. Ito, *J. Am. Chem. Soc.*, **117**, 9608 (1995).
28. G. Maas, M. Gimmy and M. Alt, *Organometallics*, **11**, 3813 (1992).
29. E. J. Corey and K. Y. Yi, *Tetrahedron Lett.*, **33**, 2289 (1992).
30. S.-I. Kawahara, T. Wada and M. Sekine, *Tetrahedron Lett.*, **37**, 509 (1996).
31. J. Feixas, A. Capdevila and A. Guerrero, *Tetrahedron*, **50**, 8539 (1994).
32. J. F. Cormier, M. B. Isaac and L.-F. Chen, *Tetrahedron Lett.*, **34**, 243 (1993).
33. A. S. Pilcher and P. DeShong, *J. Org. Chem.*, **58**, 5130 (1993).
34. A. S. Pilcher, D. K. Hill, S. J. Shimshock, R. E. Waltermire and P. DeShong, *J. Org. Chem.*, **57**, 2492 (1992).
35. C. Prakash, S. Saleh and I. A. Blair, *Tetrahedron Lett.*, **30**, 19 (1989).
36. S. E. Chapman and R. Taylor, *J. Chem. Soc., Perkin Trans. 2*, 1119 (1991).
37. T. K. Sarkar, B. K. Ghorai, S. K. Nandy and B. Mukherjee, *Tetrahedron Lett.*, **35**, 6903 (1994).
38. J. R. Hwu and J. M. Wetzel, *J. Org. Chem.*, **57**, 922 (1992).
39. J. A. Soderquist, J. C. Colberg and L. D. Valle, *J. Am. Chem. Soc.*, **111**, 4873 (1989).
40. M. M. Doyle, W. R. Jackson and P. Perlmutter, *Aust. J. Chem.*, **42**, 1907 (1989).
41. M. Fujita and T. Hiyama, *J. Org. Chem.*, **53**, 5405 (1988).
42. R. Munschauer and G. Maas, *Chem. Ber.*, **125**, 1227 (1992).
43. R. Hoffmann and R. Brückner, *Chem. Ber.*, **125**, 1471 (1992).
44. S. Marumoto and I. Kuwajima, *J. Am. Chem. Soc.*, **115**, 9021 (1993).
45. M. S. Loft, D. A. Widdowson and T. J. Mowlem, *Synlett*, 135 (1992).
46. J. S. Panek and P. F. Cirillo, *J. Am. Chem. Soc.*, **112**, 4873 (1990).
47. K. Bratt, A. Garavelas, P. Perlmutter and G. Westman, *J. Org. Chem.*, **61**, 2109 (1996).
48. W. H. Bunnelle and T. A. Isbell, *J. Org. Chem.*, **57**, 729 (1992).
49. L. Spialter, L. Pazdernik, S. Bernstein, W. A. Swansiger, G. R. Buell and M. E. Freeburger, *J. Am. Chem. Soc.*, **93**, 5682 (1971).
50. Y. A. Aleksandrov and B. I. Tarunin, *Dokl. Akad. Nauk SSSR, Ser. Khim.*, **212**, 869 (1973); *Chem. Abstr.*, **80**, 2930d (1974).
51. A. R. Bassindale, R. J. Ellis and P. G. Taylor, *Tetrahedron Lett.*, **25**, 2705 (1984).
52. A. G. M. Barrett, J. M. Hill, E. M. Wallace and J. A. Flygare, *Synlett*, 764 (1991).
53. A. G. M. Barrett and J. A. Flygare, *J. Org. Chem.*, **56**, 638 (1991).
54. Cf B. Santiago, C. Lopez and J. A. Soderquist, *Tetrahedron Lett.*, **32**, 3457 (1991).
55. A. Bongini, M. Panunzio, E. Bandini, G. Martelli and G. Spunta, *Synlett*, 461 (1995).
56. J. Ollivier and J. Salaün, *Synlett*, 949 (1994).
57. K. Käppler, U. Scheim, K. Rühlmann and K. Porzel, *J. Organomet. Chem.*, **441**, 15 (1992).
58. L. Agrofoglio, R. Condom and R. Guedj, *Tetrahedron Lett.*, **33**, 5503 (1992).
59. T. Akiyama, K. Ishikawa and S. Ozaki, *Chem. Lett.*, 627 (1994).
60. B. Török, K. Felföldi, Á. Molnár and M. Bartók, *J. Organomet. Chem.*, **460**, 111 (1993).
61. M. G. Voronkov, V. B. Pukhnarevich, L. I. Kopylova, V. A. Nestunovich, E. O. Tsetlina, B. A. Trofimov, I. Pola and V. Khvalovskii, *Dokl. Akad. Nauk SSSR, Ser. Khim.*, **227**, 91 (1976); *Chem. Abstr.*, **85**, 5012v (1976).
62. C. C. Mak and K. S. Chan, *J. Chem. Soc., Perkin Trans. 1*, 2143 (1993).
63. P. Wang and J. Adams, *J. Am. Chem. Soc.*, **116**, 3296 (1994).
64. Y. Yamakado, M. Ishiguro, N. Ikeda and H. Yamamoto, *J. Am. Chem. Soc.*, **103**, 5568 (1981).
65. J. C. Ellington Jr. and E. M. Arnett, *J. Am. Chem. Soc.*, **110**, 7778 (1988).

66. N. Shimizu, E. Osajima and Y. Tsuno, *Bull. Chem. Soc. Jpn.*, **64**, 1145 (1991).
67. E. Akerman, *Acta Chem. Scand.*, **10**, 298 (1956).
68. E. Akerman, *Acta Chem. Scand.*, **11**, 373 (1957).
69. N. Shimizu, N. Takesue, A. Yamamoto, T. Tsutsumi, S. Yasuhara and Y. Tsuno, *Chem. Lett.*, 1263 (1992).
70. N. Shimizu, N. Takesue, S. Yasuhara and T. Inazu, *Chem. Lett.*, 1807 (1993).
71. L. H. Sommer, *Stereochemistry, Mechanism, and Silicon*, McGraw-Hill, New York, 1965, p. 142.
72. J. E. Baines and C. Eaborn, *J. Chem. Soc.*, 1436 (1956).
73. O. W. Steward and O. R. Pierce, *J. Am. Chem. Soc.*, **83**, 4932 (1961).
74. W. P. Barie Jr., Ph.D. Dissertation, Pennsylvania State University, 1954.
75. D. M. Choquette, M. J. Timm, J. L. Hobbs, M. M. Rahim, K. J. Ahmed and R. P. Planalp, *Organometallics*, **11**, 529 (1992).
76. T. S. Koloski, D. C. Pestana, P. J. Carroll and D. H. Berry, *Organometallics*, **13**, 489 (1994).
77. H. K. Sharma, S. P. Vincenti, R. Vicari, F. Cervantes and K. H. Pannell, *Organometallics*, **9**, 2109 (1990).
78. K. H. Pannell, S.-H. Lin, R. N. Kapoor, F. Cervantes-Lee, M. Pinon and L. Parkanyi, *Organometallics*, **9**, 2454 (1990).
79. J. Frey, E. Schottland, Z. Rappoport, O. Bravo-Zhivotovskii, M. Nakash, M. Botoshansky, M. Kaftory and Y. Apeloig, *J. Chem. Soc., Perkin Trans. 2*, 2555 (1994).
80. J. P. Dinnocenzo, S. Farid, J. L. Goodman, I. R. Gould, W. P. Todd and S. L. Mattes, *J. Am. Chem. Soc.*, **111**, 8973 (1989).
81. T. Kawashima, N. Yamashita and R. Okazaki, *Chem. Lett.*, 1107 (1995).
82. M. A. Sparks and J. S. Panek, *J. Org. Chem.*, **56**, 3431 (1991).
83. H. Yoda, K. Shirakawa and K. Takabe, *Tetrahedron Lett.*, **32**, 3401 (1991).
84. M. Carda, F. González, S. Rodríguez and J. A. Marco, *Tetrahedron: Asymmetry*, **3**, 1511 (1992).
85. P. A. Jacobi and G. Cai, *Tetrahedron Lett.*, **32**, 1765 (1991).
86. P. A. Jacobi and G. Cai, *Heterocycles*, **35**, 1103 (1993).
87. M. Shimizu and K. Mikami, *J. Org. Chem.*, **57**, 6105 (1992).
88. N. Iwasawa and M. Iwamoto, *Chem Lett.*, 1257 (1993).
89. Y. Horiguchi, I. Suehiro, A. Sasaki and I. Kuwajima, *Tetrahedron Lett.*, **34**, 6077 (1993).
90. H. Yoda, H. Kitayama, W. Yamada, T. Katagiri and K. Takabe, *Tetrahedron: Asymmetry*, **4**, 1451 (1993).
91. D. C. Chauret and J. M. Chong, *Tetrahedron Lett.*, **34**, 3695 (1993).
92. K. Araki, and J. T. Welch, *Tetrahedron Lett.*, **34**, 2251 (1993).
93. T. Nakamura, N. Waizumi, Y. Horiguchi and I. Kuwajima, *Tetrahedron Lett.*, **35**, 7813 (1994).
94. M. Okabe, R.-C. Sun and S. Wolff, *Tetrahedron Lett.*, **35**, 2865 (1994).
95. M. A. Brook and C. Henry, *Tetrahedron*, **52**, 861 (1996).
96. E. L. Eliel and H. Satici, *J. Org. Chem.*, **59**, 688 (1994).
97. K. Ishihara, H. Yamamoto and C. H. Heathcock, *Tetrahedron Lett.*, **30**, 1825 (1989).
98. R. F. Cunico and C. P. Kuan, *J. Org. Chem.*, **55**, 4634 (1990).
99. F. K. Cartledge, *Organometallics*, **2**, 425 (1983).
100. For the definition of the cone angle sec C. A. Tolman, *Chem. Rev.*, **77**, 313 (1977).
101. H. Mayr and G. Hagen, *J. Chem. Soc., Chem. Commun.*, 91 (1989).
102. A. Vulpetti, A. Bernardi, C. Gennari, J. M. Goodman and I. Paterson, *Tetrahedron*, **49**, 685 (1993).
103. C. Panyachotipun and E. R. Thornton, *Tetrahedron Lett.*, **31**, 6001 (1990).
104. A. Choudhury and E. R. Thornton, *Tetrahedron Lett.*, **34**, 2221 (1993).
105. C. Mukai, S. Hashizume, K. Nagami and M. Hanaoka, *Chem. Pharm. Bull.*, **38**, 1509 (1990).
106. C. Siegel and E. R. Thornton, *J. Am. Chem. Soc.*, **111**, 5722 (1989).
107. D. Schinzer, *Synthesis*, 179 (1989).
108. K. Maruoka, J. Sato and H. Yamamoto, *J. Am. Chem. Soc.*, **114**, 4422 (1992).
109. A. Yanagisawa, S. Habaue and H. Yamamoto, *J. Org. Chem.*, **54**, 5198 (1989).
110. M. Suzuki, Y. Morita and R. Noyori, *J. Org. Chem.*, **55**, 441 (1990).
111. A. Yanagisawa, S. Habaue and H. Yamamoto, *Tetrahedron*, **48**, 1969 (1992).
112. T. Bach and K. Jödicke, *Chem. Ber.*, **126**, 2457 (1993).
113. C. Palomo, J. M. Aizpurua, R. Urchegui and M. Iturburu, *J. Org. Chem.*, **57**, 1571 (1992).
114. S. M. Sieburth and L. Fensterbank, *J. Org. Chem.*, **57**, 5279 (1992).
115. E. J. Bures and B. A. Keay, *Tetrahedron Lett.*, **29**, 1247 (1988).

116. I. Fleming, J. Dunoguès and R. Smithers, in *Organic Reactions*, Vol. 37 (Ed. A. S. Kende), Chap. 2, Wiley, New York, 1989.
117. S. J. Shimshock, R. E. Waltermire and P. DeShong, *J. Am. Chem. Soc.*, **113**, 8791 (1991).
118. M. S. Shekhani, K. M. Khan, K. Mahmood, P. M. Shah and S. Malik, *Tetrahedron Lett.*, **31**, 1669 (1990).
119. S. J. Monger, D. M. Parry and S. M. Roberts, *J. Chem. Soc., Chem. Commun.*, 381 (1989).
120. H. Shoda, T. Nakamura, K. Tanino and I. Kuwajima, *Tetrahedron Lett.*, **34**, 6281 (1993).
121. W. Adam and M. Richter, *Chem. Ber.*, **125**, 243 (1992).
122. W. Adam and P. Klug, *J. Org. Chem.*, **58**, 3416 (1993).
123. W. Kitching, D. Doddrell and J. B. Grutzner, *J. Organomet. Chem.*, **107**, C-5 (1976).
124. J. R. Hwu, J. M. Wetzel, J. S. Lee and R. J. Butcher, *J. Organomet. Chem.*, **453**, 21 (1993).
125. K. K. Wang, Y. G. Gu and C. Liu, *J. Am. Chem. Soc.*, **112**, 4424 (1990).
126. R. Takeuchi and M. Sugiura, *J. Chem. Soc., Perkin Trans. 1*, 1031 (1993).
127. M. M. Doyle, W. R. Jackson and P. Perlmutter, *Tetrahedron Lett.*, **30**, 233 (1989).
128. N. Fitzmaurice, W. R. Jackson and P. Perlmutter, *J. Organomet. Chem.*, **285**, 375 (1985).
129. F. Le Bideau, F. Gilloir, Y. Nilsson, C. Aubert and M. Malacria, *Tetrahedron Lett.*, **36**, 1641 (1995).
130. R. Hoffmann and R. Brückner, *Chem. Ber.* **125**, 2731 (1992).
131. M. Lautens, P. H. M. Delanghe, J. B. Goh and C. H. Zhang, *J. Org. Chem.*, **60**, 4213 (1995).
132. M. Lautens, P. H. M. Delanghe, J. B. Goh and C. H. Zhang, *J. Org. Chem.*, **57**, 3270 (1992).
133. G. Beese and B. A. Keay, *Synlett*, 33 (1991).
134. Cf. E. J. Bures and B. A. Keay, *Tetrahedron Lett.*, **28**, 5965 (1987).
135. Cf. K. D. Kim and P. A. Magriotis, *Tetrahedron Lett.*, **31**, 6137 (1990).
136. A. I. Al-Mansour, M. A. M. R. Al-Gurashi, C. Eaborn, F. A. Fattah and P. D. Lickiss, *J. Organomet. Chem.*, **393**, 27 (1990).
137. E. Vedejs and C. K. McClure, *J. Am. Chem. Soc.*, **108**, 1094 (1986).
138. I. Fleming, A. K. Sarkar and A. P. Thomas, *J. Chem. Soc., Chem. Commun.*, 157 (1987).
139. I. Fleming and N. J. Lawrence, *Tetrahedron Lett.*, **29**, 2077 (1988).
140. D. A. Evans and S. W. Kaldor, *J. Org. Chem.*, **55**, 1698 (1990).
141. P. F. Hudrlik and A. M. Hudrlik, in *Advances in Silicon Chemistry*, Vol. 2 (Ed. G. L. Larson), JAI Press, London, 1993, pp. 1–89.
142. T. Cohen, J. P. Sherbine, J. R. Matz, R. R. Hutchins, B. M. McHenry and P. R. Willey, *J. Am. Chem. Soc.*, **106**, 3245 (1984).
143. D. J. Ager, *J. Org. Chem.*, **49**, 168 (1984).
144. A. R. Bassindale, R. J. Ellis, J. C.-Y. Lau and P. G. Taylor, *J. Chem. Soc., Chem. Commun.*, 98 (1986).
145. D. Bell, E. A. Crowe, N. J. Dixon, G. R. Geen, I. S. Mann and M. R. Shipton, *Tetrahedron*, **50**, 6643 (1994).
146. N. Y. Grigorieva, O. A. Pinsker and A. M. Moiseenkov, *Mendeleev Commun.*, 129 (1994).
147. K. Takenaka, T. Hattori, A. Hirao and S. Nakahama, *Macromolecules*, **22**, 1563 (1989).
148. P. Durand, A. Margaillan, M. Camail and J. L. Vernet, *Polymer*, **35**, 4392 (1994).
149. J. D. Buynak, J. B. Strickland, T. Hurd and A. Phan, *J. Chem. Soc., Chem. Commun.*, 89 (1989).
150. Y. Ukaji, K. Sada and K. Inomata, *Chem. Lett.*, 1227 (1993).
151. K. Käppler, A. Porzel, U. Scheim and K. Rühlmann, *J. Organomet. Chem.*, **402**, 155 (1991).
152. M. Nakada, Y. Urano, S. Kobayashi and M. Ohno, *J. Am. Chem. Soc.*, **110**, 4826 (1988).
153. P. F. Cirillo and J. S. Panek, *J. Org. Chem.*, **55**, 6071 (1990).
154. G. L. Larson, J. A. Soderquist and M. R. Claudio, *Synth. Commun.*, **20**, 1095 (1990).
155. M. Asaoka, K. Obuchi and H. Takei, *Tetrahedron*, **50**, 655 (1994).
156. M. Asaoka, K. Shima, N. Fujii and H. Takei, *Tetrahedron*, **44**, 4757 (1988).
157. E. Bandini, G. Martelli, G. Spunta and M. Panunzio, *Tetrahedron: Asymmetry*, **6**, 2127 (1995).
158. S. Hünig and M. Schäfer, *Chem. Ber.*, **126**, 177 (1993).
159. S. Fioravanti, M. A. Loreto, L. Pellacani and P. A. Tardella, *Tetrahedron: Asymmetry*, **1**, 931 (1990).
160. I. Fleming and J. D. Kilburn, *J. Chem. Soc., Perkin Trans. 1*, 3295 (1992).
161. I. Fleming, *J. Chem. Soc., Perkin Trans. 1*, 3363 (1992).
162. A. Ricci, A. Guerrini, G. Seconi, A. Mordini, T. Constantieux, J.-P. Picard, J.-M. Aizpurua and C. Palomo, *Synlett*, 955 (1994).
163. J. P. Gilday, J. C. Gallucci and L. A. Paquette, *J. Org. Chem.*, **54**, 1399 (1989).

164. R. P. Short and S. Masamune, *J. Am. Chem. Soc.*, **111**, 1892 (1989).
165. S. H. Yeon, J. S. Han, E. Hong, Y. Do and I. N. Jung, *J. Organomet. Chem.*, **499**, 159 (1995).
166. H. Shinokubo, K. Oshima and K. Utimoto, *Tetrahedron Lett.*, **34**, 4985 (1993).
167. D. Craig and V. R. N. Munasinghe, *J. Chem. Soc., Chem. Commun.*, 901 (1993).
168. G. Haaima, M.-J. Lynch, A. Routledge and R. T. Weavers, *Tetrahedron*, **49**, 4229 (1993).
169. M. Asaoka, K. Takenouchi and H. Takei, *Chem. Lett.*, 1225 (1988).
170. M. Koreeda and D. C. Visger, *Tetrahedron Lett.*, **33**, 6603 (1992).
171. N. K. N. Yee and R. M. Coates, *J. Org. Chem.*, **57**, 4598 (1992).
172. E. E. van Tamelen, *Acc. Chem. Res.*, **8**, 152 (1975).
173. E. E. van Tamelen and J. R. Hwu, *J. Am. Chem. Soc.*, **105**, 2490 (1983).
174. J. R. Hwu and E. J. Leopold, *J. Chem. Soc., Chem. Commun.*, 721 (1984).
175. A. J. Blake, M. Harding and J. T. Sharp, *J. Chem. Soc., Perkin Trans. 1*, 3149 (1994).
176. S. Yamazaki, H. Fujitsuka and S. Yamabe, *J. Org. Chem.*, **57**, 5610 (1992).
177. D. J. Hlasta and J. H. Ackerman, *J. Org. Chem.*, **59**, 6184 (1994).
178. A. Verloop, W. Hoogenstraaten and J. Tipker, in *Drug Design*, Vol. 7 (Ed. E. J. Ariens), Chap. 4, Academic Press, New York, 1976.
179. A. Padwa and M. W. Wannamaker, *Tetrahedron*, **46**, 1145 (1990).
180. B. M. Trost and M. Acemoglu, *Tetrahedron Lett.*, **30**, 1495 (1989).
181. B. M. Trost and R. C. Holcomb, *Tetrahedron Lett.*, **30**, 7157 (1989).
182. For a related intermolecular reaction, see L. Strekowski, S. Kong and M. A. Battiste, *J. Org. Chem.*, **53**, 901 (1988).
183. H. J. Reich, E. K. Eisenhart, W. L. Whipple and M. J. Kelly, *J. Am. Chem. Soc.*, **110**, 6432 (1988).
184. S. R. Wilson and M. J. Di Grandi, *J. Org. Chem.*, **56**, 4766 (1991).
185. W. Adam and M. Prein, *Tetrahedron Lett.*, **35**, 4331 (1994).
186. M. Ohno, S. Shimizu and S. Eguchi, *Tetrahedron Lett.*, **31**, 4613 (1990).
187. H. C. Kolb, S. V. Ley, A. M. Z. Slawin and D. J. Williams, *J. Chem. Soc., Perkin Trans. 1*, 2735 (1992).
188. M. Lautens and P. H. M. Delanghe, *J. Org. Chem.*, **57**, 798 (1992).
189. K. Nakatani, T. Izawa and S. Isoe, *J. Org. Chem.*, **59**, 5961 (1994).
190. K. Mikami, T.-P. Loh and T. Nakai, *J. Am. Chem. Soc.*, **112**, 6737 (1990).
191. K. Tanino, H. Shoda, T. Nakamura and I. Kuwajima, *Tetrahedron Lett.*, **33**, 1337 (1992).
192. K. Tanino, T. Nakamura and I. Kuwajima, *Tetrahedron Lett.*, **31**, 2165 (1990).
193. R. F. W. Jackson, S. P. Standen, W. Clegg and A. McCamley, *Tetrahedron Lett.*, **33**, 6197 (1992).
194. A. B. Bueno, M. C. Carreño and J. L. G. Ruano, *Tetrahedron Lett.*, **34**, 5007 (1993).
195. D. P. Rotella, *Tetrahedron Lett.*, **30**, 1913 (1989).
196. K. Suzuki, M. Miyazawa, M. Shimazaki and G.-I. Tsuchihashi, *Tetrahedron*, **44**, 4061 (1988).
197. W. Kirmse and F. Söllenböhmer, *J. Chem. Soc., Chem. Commun.*, 774 (1989).
198. P. G. Andersson, *J. Org. Chem.*, **61**, 4154 (1996).
199. M. Lautens, *Pure Appl. Chem.*, **64**, 1873 (1992).
200. D. Enders and S. Nakai, *Helv. Chim. Acta*, **73**, 1833 (1990).
201. M. Lautens, C. H. Zhang and C. M. Crudden, *Angew. Chem., Int. Ed. Engl.*, **31**, 232 (1992).
202. M. Lautens, C. H. Zhang, B. J. Goh, C. M. Crudden and M. J. A. Johnson, *J. Org. Chem.*, **59**, 6208 (1994).
203. S. Yu and B. A. Keay, *J. Chem. Soc., Perkin Trans. 1*, 2600 (1991).
204. D. W. Knight and A. P. Nott, *J. Chem. Soc., Perkin Trans. 1*, 1125 (1981).
205. B. B. Lohray and D. Enders, *Helv. Chim. Acta*, **72**, 980 (1989).
206. I. Fleming, *Pure Appl. Chem.*, **60**, 71 (1988).
207. B. F. Bonini, M. C. Franchini, M. Fochi, G. Mazzanti, A. Ricci, P. Zani and B. Zwanenburg, *J. Chem. Soc., Perkin Trans. 1*, 2039 (1995).
208. A. J. Y. Lan, R. O. Heucheroth and P. S. Mariano, *J. Am. Chem. Soc.*, **109**, 2738 (1987).
209. I.-S. Cho and P. S. Mariano, *J. Org. Chem.*, **53**, 1590 (1988).
210. J. E. Crawley, A. D. Kaye, G. Pattenden and S. M. Roberts, *J. Chem. Soc., Perkin Trans. 1*, 2001 (1993).
211. H. Inami, T. Ito, H. Urabe and F. Sato, *Tetrahedron Lett.*, **34**, 5919 (1993).
212. D. W. H. Rankin, in *Frontiers of Organosilicon Chemistry* (Eds. A. R. Bassindale and P. P. Gaspar), Royal Society of Chemistry, Cambridge, 1991, pp. 253–262.

213. W. Smadja, M. Zahouily and M. Malacria, *Tetrahedron Lett.*, **33**, 5511 (1992).
214. C. Chatgilialoglou, *Acc. Chem. Res.*, **25**, 180 (1992) and references cited therein.
215. M. Pereyre, J. P. Quintard and A. Rahm, *Tin in Organic Synthesis*, Butterworths, London, 1987.
216. F. L. van Delft, M. de Kort, G. A. van der Marel and J. H. van Boom, *J. Org. Chem.*, **61**, 1883 (1996).
217. E. J. Corey and Z. Chen, *Tetrahedron Lett.*, **35**, 8731 (1994).
218. R. A. N. C. Crump, I. Fleming, J. H. M. Hill, D. Parker, N. L. Reddy and D. Waterson, *J. Chem. Soc., Perkin Trans. 1*, 3277 (1992).
219. W. Amberg and D. Seebach, *Chem. Ber.*, **123**, 2439 (1990).
220. I. Fleming and N. J. Lawrence, *Tetrahedron Lett.*, **31**, 3645 (1990).
221. W. Oppolzer, R. J. Mills, W. Pachinger and T. Stevenson, *Helv. Chim. Acta*, **69**, 1542 (1986).
222. S. Kusuda, Y. Ueno, T. Hagiwara and T. Toru, *J. Chem. Soc., Perkin Trans. 1*, 1981 (1993).
223. I. Fleming and N. D. Kindon, *J. Chem. Soc., Chem. Commun.*, 1177 (1987).
224. M. R. Hale and A. H. Hoveyda, *J. Org. Chem.*, **59**, 4370 (1994).
225. C. Palomo, J. M. Aizpurua, M. Iturburu and R. Urchegui, *J. Org. Chem.*, **59**, 240 (1994).
226. B. H. Lipshutz, D. C. Reuter and E. L. Ellsworth, *J. Org. Chem.*, **54**, 4975 (1989).
227. V. Sum and T. P. Kee, *J. Chem. Soc., Perkin Trans. 1*, 2701 (1993).
228. J. S. Panek and M. Yang, *J. Am. Chem. Soc.*, **113**, 9868 (1991).
229. G. P. Brengel and A. I. Meyers, *J. Org. Chem.*, **61**, 3230 (1996).
230. F. Fariña, P. Noheda and M. C. Paredes, *J. Org. Chem.*, **58**, 7406 (1993).
231. B. Marciniec and H. Maciejewski, *J. Organomet. Chem.*, **454**, 45 (1993).
232. J. A. Robl and J. R. Hwu, *J. Org. Chem.*, **50**, 5913 (1985).
233. J. R. Hwu, J. A. Robl, N. Wang, D. A. Anderson, J. Ku and E. Chen, *J. Chem. Soc., Perkin Trans. 1*, 1823 (1989).
234. S. Florio and L. Troisi, *Tetrahedron Lett.*, **34**, 3141 (1993).
235. J. Woning and J. G. Verkade, *J. Am. Chem. Soc.*, **113**, 944 (1991).
236. M. P. Doyle and M. S. Shanklin, *Organometallics*, **13**, 1081 (1994).
237. Y. Obora, Y. Tsuji, M. Asayama and T. Kawamura, *Organometallics*, **12**, 4697 (1993).
238. Y. Apeloig and A. Stanger, *J. Am. Chem. Soc.*, **107**, 2806 (1985).
239. Y. Apeloig, R. Biton and A. Abu-Freih, *J. Am. Chem. Soc.*, **115**, 2522 (1993).
240. S. G. Davies and C. L. Goodfellow, *J. Organomet. Chem.*, **340**, 195 (1988).
241. S. Shambayati, J. F. Blake, S. G. Wierschke, W. L. Jorgensen and S. L. Schreiber, *J. Am. Chem. Soc.*, **112**, 697 (1990).
242. A. Sekiguchi, K. Ebata, C. Kabuto and H. Sakurai, *J. Am. Chem. Soc.*, **113**, 1464 (1991).
243. H. Sakurai, K. Ebata, C. Kabuto and A. Sekiguchi, *J. Am. Chem. Soc.*, **112**, 1799 (1990).
244. A. Sekiguchi, K. Ebata, Y. Terui and H. Sakurai, *Chem. Lett.*, 1417 (1991).
245. R. J. Mills, N. J. Taylor and V. Snieckus, *J. Org. Chem.*, **54**, 4372 (1989).
246. H. Bock, J. Meuret, C. Näther and U. Krynitz, *Tetrahedron Lett.*, **34**, 7553 (1993).
247. B. W. Gung, J. P. Melnick, M. A. Wolf and A. King, *J. Org. Chem.*, **60**, 1947 (1995).
248. H. Chen, R. A. Bartlett, H. V. R. Dias, M. M. Olmstead and P. P. Power, *J. Am. Chem. Soc.*, **111**, 4338 (1989).
249. Cf N. Wiberg, in *Frontiers of Organosilicon Chemistry* (Eds. A. R. Bassindale and P. P. Gaspar), Royal Society of Chemistry, Cambridge, 1991, pp. 263–270.
250. Cf. H. Bock, J. Meuret, C. Näther and K. Ruppert, in *Organosilicon Chemistry: From Molecules to Materials* (Eds. N. Auner and J. Weis), VCH, New York, 1994, pp. 11–19.
251. W. S. Sheldrick, in *The Chemistry of Organic Silicon Compounds* (Eds. S. Patai and Z. Rappoport), Wiley, New York, 1989, pp. 227–304.
252. E. Lukevics, O. Pudowa and R. Sturkovich, *Molecular Structure of Organosilicon Compounds*, Wiley, New York, 1989.
253. G. Raabe and J. Michl, in *The Chemistry of Organic Silicon Compounds* (Eds. S. Patai and Z. Rappoport), Wiley, New York, 1989, pp. 1015–1142.
254. H. Bock, J. Meuret and K. Ruppert, *J. Organomet. Chem.*, **462**, 31 (1993).
255. J. Pinkas, B. Gaul and J. G. Verkade, *J. Am. Chem. Soc.*, **115**, 3925 (1993).
256. K.-J. Lee, P. D. Livant, M. L. McKee and S. D. Sorley *J. Am. Chem. Soc.*, **107**, 9901 (1985).
257. D. Gudat, L. M. Daniels and J. G. Verkade, *J. Am. Chem. Soc.*, **111**, 8520 (1989).
258. E. B. Nadler and Z. Rappoport, *Tetrahedron Lett.*, **32**, 1233 (1991).
259. M. Charton, *J. Am. Chem. Soc.*, **97**, 1552 (1975).
260. M. Yoshifuji, T. Nakamura and N. Inamoto, *Tetrahedron Lett.*, **28**, 6325 (1987).

261. R. Okazaki, M. Unno and N. Inamoto, *Chem. Lett.*, 791 (1989).
262. N. Tokitoh, H. Suzuki, R. Okazaki and K. Ogawa, *J. Am. Chem. Soc.*, **115**, 10428 (1993).
263. K. M. Baines, A. G. Brook, R. R. Ford, P. D. Lickiss, A. K. Saxena, W. J. Chatterton, J. F. Sawyer and B. A. Behnam, *Organometallics*, **8**, 693 (1989).
264. J. D. Debad, P. Legzdins, R. J. Batchelor and F. W. B. Einstein, *Organometallics*, **12**, 2094 (1993).
265. T. J. Marks, E. A. Mintz and D. C. Sonnenberger, *J. Am. Chem. Soc.*, **106**, 3484 (1984).
266. J. D. Cotton, G. T. Crisp and L. Latif, *Inorg. Chim. Acta*, **47**, 171 (1981).
267. B. S. Bandodakar and G. Nagendrappa, *J. Organomet. Chem.*, **430**, 373 (1992).
268. R. Damrauer, B. O'Connell, S. E. Danahey and R. Simon, *Organometallics*, **8**, 1167 (1989).
269. J. Schraml, J. Čermák, V. Chvalovsky, A. Kasal, C. Bliefert and E. Krahé, *J. Organomet. Chem.*, **341**, C-6 (1988).
270. M. Unno, M. Saito and H. Matsumoto, *J. Organomet. Chem.*, **499**, 221 (1995).
271. W. S. Sheldrick, in *The Chemistry of Organic Silicon Compounds* (Eds. S. Patai and Z. Rappoport), Chap. 3, Wiley, New York, 1989.
272. T. D. McGrath and A. J. Welch, *Acta Crystallogr.*, **C51**, 654 (1995).
273. H. Sakurai, Y. Nakadaira, H. Tobita, T. Ito, K. Toriumi and H. Ito, *J. Am. Chem. Soc.*, **104**, 300 (1982).
274. H. Sakurai, H. Tobita, Y. Nakadaira and C. Kabuto, *J. Am. Chem. Soc.*, **104**, 4288 (1982).
275. H. Sakurai, K. Ebata, Y. Nakadaira and C. Kabuto, *Chem. Lett.*, 301 (1987).
276. H. Sakurai, K. Ebata, K. Sakamoto, Y. Nakadaira and C. Kabuto, *Chem. Lett.*, 965 (1988).

CHAPTER 9

Reaction mechanisms of nucleophilic attack at silicon

ALAN R. BASSINDALE, SIMON J. GLYNN and PETER G. TAYLOR

Department of Chemistry, The Open University, Milton Keynes, MK7 6AA, UK

I. INTRODUCTION	495
II. DIRECT SUBSTITUTION OF A LEAVING GROUP BY A NUCLEOPHILE	496
A. Stereochemistry of Substitution	496
1. The effect of nucleophile and leaving group	496
2. The effect of substrate structure	497
B. Kinetics of Substitution	498
C. The Geometry of the Intermediate/Transition State	498
1. Inversion	498
2. Retention	502
D. Pseudorotation in Pentacoordinate Silicon Species	503
III. NUCLEOPHILE-CATALYSED SUBSTITUTION AT SILICON AND RACEMIZATION	506
IV. REFERENCES	510

I. INTRODUCTION

The mechanism of reactions involving silicon have proved to be a most fruitful area of research for almost fifty years. Silicon engages in nucleophilic and radical reactions covering almost all known reaction mechanisms.

In this review we have chosen to limit the scope to nucleophilic substitution at silicon. A short overview is given of material covered in detail in our previous review in 'The Chemistry of Organic Silicon Compounds'[1], with recent advances covered in greater depth.

Recently, the stereochemistry of nucleophilic substitution at silicon has been reviewed by Holmes[2], and the role of pentacoordinate silicon compounds as reaction intermediates has been reviewed by Corriu and coworkers[3].

The chemistry of organic silicon compounds, Vol. 2
Edited by Z. Rappoport and Y. Apeloig © 1998 John Wiley & Sons Ltd

II. DIRECT SUBSTITUTION OF A LEAVING GROUP BY A NUCLEOPHILE
A. Stereochemistry of Substitution

We begin by examining the direct exchange of a leaving group X by a nucleophile Y. Much of the discussion in this area has centred on whether the five-coordinate species **1** is an activated complex, as in the elementary mechanism A, or an intermediate, as in mechanism B (equation 1).

$$X-\underset{|}{\overset{\diagdown \diagup}{Si}}-Y$$

(**1**)

$$Y^- + \overset{\diagdown}{\underset{\diagup}{Si}}-X \longrightarrow \left[Y\cdots\underset{|}{\overset{\diagdown \diagup}{Si}}\cdots X\right]^{-\ddagger} \longrightarrow Y-\overset{\diagup}{\underset{\diagdown}{Si}} + X^-$$

Mechanism A

(1)

$$Y^- + \overset{\diagdown}{\underset{\diagup}{Si}}-X \longrightarrow \left[Y-\underset{|}{\overset{\diagdown \diagup}{Si}}-X\right]^- \longrightarrow Y-\overset{\diagup}{\underset{\diagdown}{Si}} + X^-$$

Mechanism B

As in substitution at carbon, stereochemical and kinetic data provide the means of differentiating between these two possibilities. The stereochemical data are examined first. Substitution at silicon leads to both retention and inversion of configuration and the stereochemical outcome depends upon the nature of the leaving group, the nucleophile, solvent, complexing agents and whether or not the silicon is part of a ring.

1. The effect of nucleophile and leaving group

When the leaving group at a chiral silicon is chlorine, inversion of configuration invariably occurs. Fluorine may lead to inversion, retention or a mixture of products. Si—H bond cleavage occurs primarily with retention of configuration. Sommer[4] and Corriu and Guerin[5,6] have both suggested that bond polarization and ease of Si—X stretching during nucleophilic attack may be important factors in the stereochemical outcome. An empirical relationship between the observed stereochemistry and the ability of the leaving group to be displaced has been suggested by Corriu and Guerin[5,6], to give a series Cl, Br, OAc > F, SR > OMe, H where leaving group ability decreases and degree of retention increases towards the right.

It was suggested by Sommer and coworkers[7,8] that retention occurs when the counterion is able to assist the departure of the leaving group. For poor leaving groups such as RO⁻, it is difficult to displace the leaving group as the free ion, and Sommer proposed a cyclic mechanism involving electrophilic assistance by the E of E-Y (cf. **2**)[4].

This electrophilic assistance could account for the observed solvent and counterion effects. The stereochemistry of substitution of alkoxy, mercapto and fluorosilanes with alkoxide ion depends dramatically on the composition of the solvent[4]. As the percentage

$$R_3Si \overset{X}{\underset{Y}{\diamond}} E$$

(2)

of alcohol in the benzene solution increases, there is a move from retention of stereochemistry to inversion. Sommer suggested that in benzene, electrophilic assistance from the counterion is required to enable the leaving group to be expelled. However, as the alcoholic content of the solution increases, hydrogen bonding to the solvent provides sufficient electrophilic assistance for removal of the poor leaving group, thus allowing inversion to proceed.

When the counterion is varied from lithium to sodium to potassium, the proportion of inversion increases. The relatively covalent lithium–oxygen bond favours a retentive mechanism. However, as the metal–oxygen bond becomes more ionic, the components may function more independently, allowing attack of RO^- on the back face of the silicon tetrahedron while electrophilic assistance by M^+ of the leaving group aids inversion of configuration.

The stereochemistry of substitution is also dependent on the nucleophile. Corriu and coworkers have related the preference for retention to the hardness of the nucleophile[5,6,9]. A hard nucleophile such as CH_3CH_2Li, where the negative charge is concentrated on the carbon, gives retention whereas the soft nucleophile $PhCH_2Li$, where the charge is delocalized, leads to inversion. These ideas have received a theoretical justification from the work of Anh and Minot[10], who carried out perturbational studies on substitution at silicon with either retention or inversion. More recently, Deiters and Holmes[11] have performed molecular orbital calculations on the $[SiH_3XY]^-$ species. The experimentally observed tendency of leaving groups towards retention ($H^- > OH^- > SH^- \sim F^- > Cl^-$) is found to correlate with a decrease in the overlap population of the silicon–leaving group bond. The nucleophile was found to influence the apicophilicity of the leaving group, soft nucleophiles increasing the leaving group apicophilicity and favouring inversion.

Corriu has also used the hard/soft nucleophile concept to explain the solvent and countercation effects[5,6,9]. He suggests that in benzene the $RO^- M^+$ species have a localized negative charge on the oxygen atom and thus are hard and react with retention. On increasing the alcohol content the charge on the anion is dispersed by hydrogen bonding producing a softer species which reacts with inversion.

2. The effect of substrate structure

The structure of the alkyl or aryl groups on the silicon has been found to have little effect on the stereochemical outcome of substitution[4]. Changes in both electronic and steric factors leave the stereochemical outcome unchanged[4]. However, there is a significant effect when the silicon is part of a strained-ring structure. When the leaving group is exocyclic there is a move to retention as the angle strain at silicon increases[12,13]. The opposite stereochemical trend is observed with endocyclic leaving groups. In this case angle strain favours inversion[6,14]. These observations can be rationalized by considering the requirements of small rings to be axial/equatorial with respect to a pentacoordinate silicon. With exocyclic leaving groups, axial entry of the nucleophile give the intermediate **3** which results in retention. When the leaving group is endocyclic, the small ring provides extra incentive for the leaving group to be axial. With axial entry this leads to preferred inversion as shown in **4**.

Anh and Minot[10] have forwarded an alternative explanation involving considerations of orbital overlap as the s character of the Si—X bond varies with the bond angle.

B. Kinetics of Substitution

The rates of hydrolysis of *p*-methoxyphenyltriphenylsilane and methoxytriphenylsilane were found by Eaborn and coworkers[15] to be first-order with respect to base concentration. This is consistent with (i) a concerted S_N2 mechanism, (ii) a mechanism involving a pentacoordinate intermediate where loss of alkoxide is not catalysed by HO^- or (iii) a mechanism where such a step is not kinetically significant. Detailed kinetic studies on the solvolyses of silyl ethers have been interpreted by Dietze and coworkers[16-19] as not supporting a mechanism in which there is a pre-equilibrium formation of a pentacoordinate intermediate which then collapses to produce products. They favour a mechanism in which solvent attack occurs on a tetravalent silicon atom with simultaneous breaking of the silicon leaving group bond (S_N2). They suggested that the observation of general acid catalysis, general base catalysis and bifunctional catalysis all arise from a common mechanism involving the transition state **5**, in which A = acid, B = base, OR^1 is the leaving group and HOR^2 is the nucleofuge.

The dependence of the rate of reaction on leaving group has been measured by a number of workers. In solvolysis reactions, Sommer found that the rate of reaction decreases as the basicity of the leaving group increases[4]. However, using organolithium and Grignard nucleophiles, Corriu and coworkers found only a very small variation of rate with leaving group[20,21]. This was taken to suggest that the rate-determining step in this case does not involve Si—X bond cleavage; thus, a mechanism involving a single elementary step could be eliminated.

C. The Geometry of the Intermediate/Transition State

1. Inversion

The simplest model for the transition state and/or intermediate leading to inversion is a distorted trigonal bipyramid with the entering and leaving groups both apical.

Recent theoretical studies of the S_N2(Si) reaction and pentacoordinate silicon intermediates or transition states show that d-orbitals are unlikely to have any significant involvement in the bonding in such species. Using valence bond computations of curve-crossing diagrams, Sini and coworkers[22] have shown that the stability of SiH_5^- relative to CH_5^- originates from the ability of silicon to utilize its Si—H σ^* orbitals for bonding to a much greater degree than carbon does. Consequently, pentavalent silicon has two resonating H—Si—H axial bonds, one arising from the axial p-orbital of the silicon and the other from the equatorial $\sigma^*(SiH_3)$ orbital of the central SiH_3 fragment. Due to the bonding nature of this σ^* orbital SiH_5^- can delocalize the fifth valence-electron pair into the equatorial Si—H bonds. For carbon this delocalization is prohibited by the high p-σ^* promotion energy, and by poor overlap of the σ^* orbital with the axial hydrogens. Thus, the fifth valence electron pair is localized in the H—C—H axial bonds, which are elongated and raise the energy of the system.

Shi and Boyd[23] have compared the reaction pathway of CH_3X with SiH_3X by calculating the Laplacian of the charge density, which gives an indication of differences in bonding and electron distributions. The bonding in CH_3X is essentially completely covalent. In S_N2 reactions, the uncharged carbon atom is always attacked from the area of greatest charge depletion, which is found opposite the C—X bond (see contour map, Figure 1) — hence the reaction takes place with inversion. The pentacoordinate species is a high energy transition state. In the case of SiH_3X, it is found that the bonding has more ionic character with a considerable positive charge on silicon. This positive charge makes nucleophilic attack easier, hence the increased reactivity of silicon with nucleophiles. Additionally, the large positively charged centre can attract nucleophiles from all directions. The stereochemistry is then dependent on the extent of charge depletion in different regions of space (Figure 2) which varies with the substrate. The significant ionic bonding increases the stability of the pentacoordinate species, which is therefore found to lie at the bottom of the potential energy surface.

A similar conclusion was reached by Streitwieser and coworkers[24], who used integrated population analyses to indicate that bonds between Si and H, C, O and F are extensively polarized in the sense Si^+ X^- with significant charge transfer. As a result Si—O and Si—F

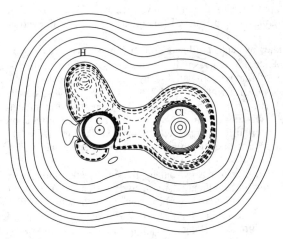

FIGURE 1. Laplacian of the charge density contour map of CH_3Cl in the HCCl plane. Reprinted with permission from Z. Shi and R. J. Boyd, *J. Phys. Chem.*, **95**, 4698. Copyright (1991) American Chemical Society

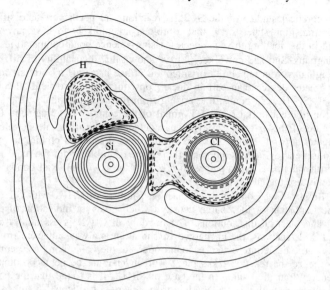

FIGURE 2. Laplacian of the charge density contour map of SiH_3Cl in the HSiCl plane. Reprinted with permission from Z. Shi and R. J. Boyd, *J. Phys. Chem.*, **95**, 4698. Copyright (1991) American Chemical Society

bonds are dominated by ionic interactions, and Si—H and Si—C bonds have an important ionic character. The ionic nature of these bonds allows silicon to expand its coordination sphere to form stable pentacoordinate species.

The increased ionic character of the bonding in SiH_5^- was also described by Carroll, Gordon and Windus[25]. They studied the addition of H^- to MH_4 to form MH_5^-. When M = Si, stabilization of the incoming hydride was dominant over the destabilization of the original MH_4, and ΔE for the addition was negative. When M = C, the destabilization of the atoms originally comprising CH_4 is greater than the stabilization of the incoming hydride and ΔE is positive.

Structural studies of pentacoordinate silicon compounds give valuable insight into the S_N2 process at silicon. Macharashvili and coworkers[26] obtained the X-ray crystal structures of a series of (halodimethylsilylmethyl) lactams. The Si—O and Si—X distances were found to vary over a wide range (1.749–2.461 and 3.734–1.630 Å, respectively) in the series X = I, Br, Cl, F. The compounds can be regarded as models of intermediate stages of the S_N2 substitution process at silicon. The silicon atom coordination changes from a slightly distorted tetrahedron with a long Si—X bond (X = I), through a trigonal bipyramid, and back to a distorted tetrahedron, inverted, with a long Si—O distance (X = F).

Similar studies have been carried out by Bassindale and Borbaruah[27] in the solution phase for a series of N-(halodimethylsilylmethyl) pyridones (**6**) (equation 2). The structures of the pyridones can be envisaged as points on the reaction profile of a nucleophilic substitution of X^- by the carbonyl oxygen (**6, 7, 8**). Analysis of the pyridone ring carbon-13 NMR chemical shifts can, using a mapping method, give a measure of the extent of silicon–oxygen bond formation in each pyridone.

The extent of Si—O bond formation is found to increase as X becomes a better leaving group (F<Cl<Br<OSO_2CF_3). Additionally, substituents Y on the pyridone ring also affect the extent of Si−O bond formation. With a strongly electron-withdrawing substituent, 3-NO$_2$, the extent of Si−O bond formation for each leaving group was significantly less than for the unsubstituted series.

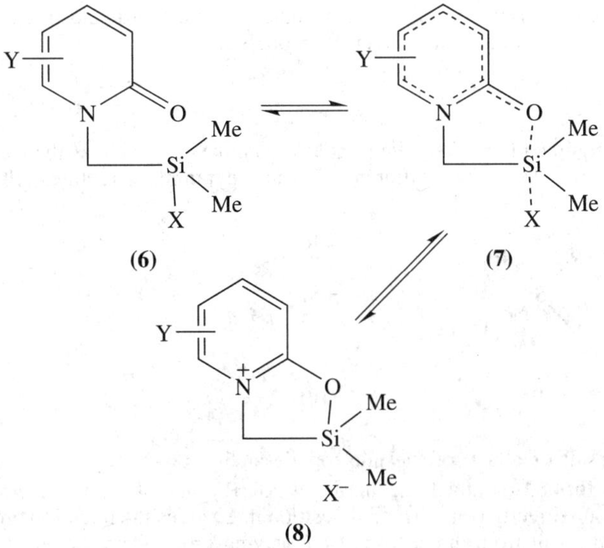

(2)

Sidorkin and coworkers[28] have used quantum chemical (MNDO) methods to model the pathway of intramolecular S$_N$2 substitution at silicon, as shown in equation 3.

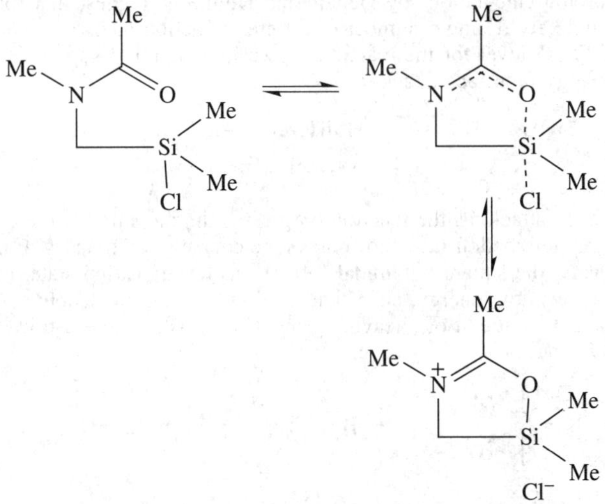

(3)

The changes in the geometric parameters of the ClSiC$_3$O coordination centre are typical of a bimolecular nucleophilic substitution reaction at a tetrahedral atom and consistent with the results obtained using the Burgi−Dunitz structure correlation method with the experimental results of Macharashvili and coworkers[26].

These studies show the simultaneous lengthening of the silicon–leaving group bond as the silicon–nucleophile interaction increases. Molecular orbital calculations at the STO-3G level by Deiters and Holmes[29] also show that as the nucleophile approaches from an apical position, the *trans* axial bond to the leaving group increases in length. However, there is little conclusive evidence as to whether a pentacoordinate intermediate with a finite lifetime is involved in such invertive processes.

2. Retention

Retention requires the nucleophile to attack from the same side as the leaving group in **9** and this can give rise to two different trigonal bipyramidal structures, **10** and **11**.

$$\underset{(9)}{\overset{R^1 \quad R^2}{\underset{X \quad R^3}{\diagdown\text{Si}\diagup}}} \qquad \underset{(10)}{\overset{R^1}{\underset{X}{\text{Nu}-\text{Si}\overset{R^2}{\underset{R^3}{\diagdown}}}}} \qquad \underset{(11)}{\overset{R^1}{\underset{\text{Nu}}{X-\text{Si}\overset{R^2}{\underset{R^3}{\diagdown}}}}}$$

10 is the result of attack of the nucleophile at the R^1–X edge of the tetrahedron **9** whereas **11** is formed by attack on the lower XR^2R^3 face of the tetrahedron **9**. Loss of the leaving group directly from the axial position in **10** gives the retention product. For **11**, retention could result from direct loss of the leaving group from the equatorial position, or by pseudorotation to give an intermediate in which the leaving group is axial and the nucleophile equatorial, followed by loss of the leaving group from the axial position. This latter mechanism is generally accepted for substitution with retention at phosphorus[30].

Molecular-orbital calculations by Deiters and Holmes[29] suggest that for silicon, edge attack is very unlikely. Using a number of different reaction pathways, *ab initio* calculations at the STO-3G level for the reaction shown in equation 4 suggest that edge attack has a higher energy than face attack.

$$SiH_3X + H^- \longrightarrow (HSiH_3X)^- \longrightarrow SiH_3H + X^- \qquad (4)$$
$$X = Cl \text{ or } F$$

In fact, for edge attack, as the reaction progresses the incoming nucleophile slips over to a lower-energy tetrahedral face and follows the course of axial entry. Calculations also suggest that, using the square pyramidal form **12** as the transition state, the pseudorotation has a relatively low energy, such that axial entry, of a nucleophile A followed by pseudorotation and axial exit of a leaving group C or D is a viable pathway (equation 5).

$$\underset{}{\overset{A}{\underset{E}{B-\text{Si}\overset{C}{\underset{D}{\diagdown}}}}} \longrightarrow \underset{(12)}{\overset{A}{\underset{E}{B-\text{Si}\overset{C}{\underset{D}{\diagdown}}}}} \longrightarrow \underset{}{\overset{C}{\underset{D}{B-\text{Si}\overset{E}{\underset{A}{\diagdown}}}}} \qquad (5)$$

The calculations also suggest that face attack leading to inversion and face attack leading to retention were similar in energy for fluorine, but for chlorine inversion was preferred. This agrees with the experimental stereochemical behaviour of fluorine and chlorine. The Si–Cl bond stretched under the influence of the nucleophile to a larger extent than the Si—F bond. The authors proposed that with a good leaving group such as chlorine, the Si—Cl bond breaks as the Si—Nu bond forms, leading to inversion. However, with poor leaving groups such as fluorine, a tightly bound intermediate is formed allowing ligand rearrangement, so that the inversion and retention routes have similar energies. It was also found that the calculated apicophilicities of the leaving groups correlated with the experimentally determined tendency for inversion, with groups of low apicophilicity tending towards retention. Furthermore, the apicophilicities of the leaving groups were found to be dependent on the nucleophile, such that soft nucleophiles increase the apicophilicity of the leaving group and hence favour inversion.

By definition, the retention mechanism involving axial attack and axial exit must involve a pentacoordinated intermediate of sufficient lifetime to allow pseudorotation. The evidence for pseudorotation in the retentive process is now reviewed.

D. Pseudorotation in Pentacoordinate Silicon Species

The possibility of pseudorotation at silicon was first suggested by Klanberg and Muetterties[31] in their study of pentacoordinate fluorosilicates SiF_5^-, $RSiF_4^-$ and $R_2SiF_3^-$. Recently there have been a number of studies on the pseudorotation of pentacoordinate silicon species.

Corriu and coworkers[32] have used compounds of the type **13** to distinguish between two possible routes of ligand exchange — pseudorotation and a dissociative process involving cleavage of the Si—N bond.

The two N-methyl groups are diastereotopic owing to the chirality of the benzyl group. A pseudorotation process will cause equivalence of the fluorines but not of the two N-methyl groups. However, a dissociative pathway as shown in equation 6 causes the fluorines and the methyl groups to be made equivalent.

(6)

(13)

Variable-temperature studies using fluorine-19 NMR show an equivalence mechanism with a $\Delta G^{\ddagger}$ of 13.1 kcal mol^{-1}, whereas proton DNMR shows that the NMe$_2$ groups are made equivalent via a process with a $\Delta G^{\ddagger}$ of 15.8 kcal mol^{-1}. The authors propose that the lower-energy pathway shown by fluorine-19 NMR arises from pseudorotation, whereas the higher-energy process shown by proton NMR is due to a dissociative process. Silicon–fluorine coupling is retained in all cases, showing that intermolecular fluorine exchange does not occur.

The interconversion of diastereomeric pentacoordinate organosilanes **14** was found to have an activation energy which had only a small dependence on the nature of the equatorial X substituent, despite large changes in the electronic effect of the ligand (X = H, F, Cl, OCOR, OPh, Ph)[33].

It is observed that on heating, the NMe$_2$ methyl signals become equivalent with the same $\Delta G^{\ddagger}$ values as the methyl groups of the pinacoxy moiety. This implies a symmetrical geometry in the transition state. The only pseudorotation mechanism which involves the degree of symmetry required for the equivalence of the methyl groups is shown in equation 7.

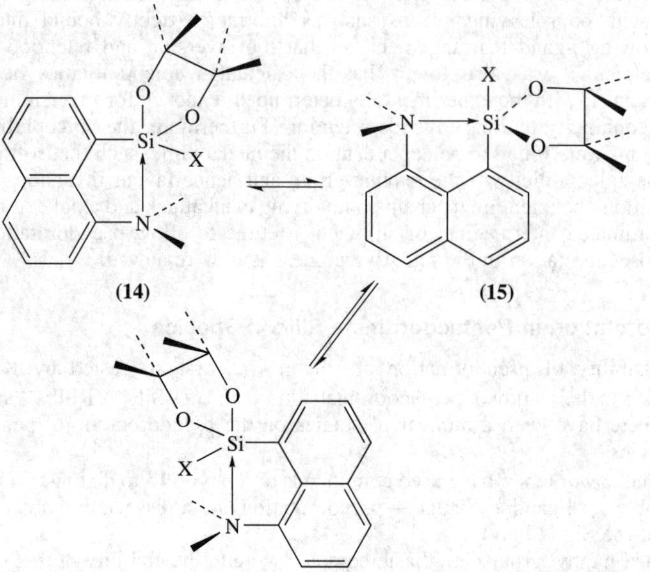

(14) (15) (7)

In conformation **15**, X is apical (which can be favourable) but the pinacoxy moiety is in the highly unfavourable diequatorial position. The authors suggest that this discourages pseudorotation such that the interconversion of diastereomers with $\Delta G^{\ddagger}$ values between 15 and 19 kcal mol^{-1} is thought to result from dissociation and recoordination of the Si—N bond.

Kalikhman, Kost and coworkers[34] have studied ligand exchange reactions in the system **16**.

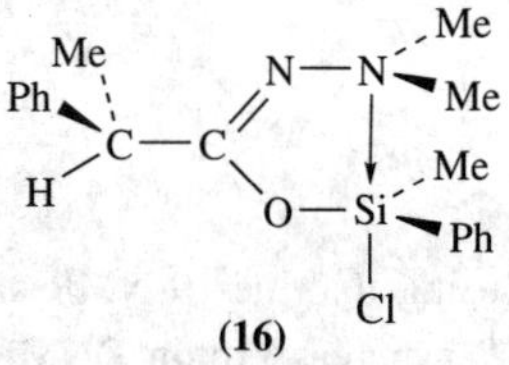

(16)

Two diastereomers are observed at the slow-exchange limit, due to the different relative configurations at the two chiral centres. As the temperature increases the four N-methyl singlets first coalesce into two singlets. This corresponds to a N-methyl interchange process without interconversion of diastereomers. This is attributed to exchange of N-methyls via Si—N bond cleavage, followed by rotation around the N—N bond and reclosure of the chelate ring, without concomitant epimerization at the silicon. $\Delta G^{\ddagger}$ for this process is 11.4 kcal mol^{-1}. On raising the temperature, a second exchange process is observed in which all the signal pairs of the diastereomers coalesce, i.e. epimerization at the silicon centre becomes rapid relative to the NMR time scale. This epimerization process, presumably occuring via a pseudorotation reaction, has a remarkably high activation barrier of 18.7 kcal mol^{-1}. This large activation energy is attributed to two factors: (i) in the pseudorotation process, the electronegative Cl ligand is forced out of the apical position and replaced by a less electronegative carbon ligand, to give a much less stable arrangement,

9. Reaction mechanisms of nucleophilic attack at silicon 505

and (ii) the chelate ring may be forced into a diequatorial position, generating considerable ring strain.

Corriu and coworkers have found that the barrier to pseudorotation seems to depend on the number of electronegative groups present[35]. For species with one electronegative substituent, $\Delta G^{\ddag} > 20$ kcal mol^{-1}. With two electronegative groups, the barrier decreases to ca 9–12 kcal mol^{-1}, and when three electronegative substituents are present, the pseudorotation barrier has $\Delta G^{\ddag} < 7$ kcal mol^{-1}.

Gordon and coworkers have studied the pseudorotation of SiH$_5^-$ using ab initio and semiempirical calculations[36]. Both AM1 and MP2/6-31++G(d,p) calculations predict a pseudorotation barrier of 2.4 kcal mol^{-1}. In a further study[37], detailed ab initio investigations of the pseudorotation of the pentacoordinated silicon anions SiH$_{5-n}$X$_n^-$ (X = F, Cl; $n = 0-5$) were made. The barriers to pseudorotation of SiF$_5^-$ and SiCl$_5^-$ were also found to be less than or equal to 3 kcal mol^{-1}. In fact, pseudorotation is less energetically demanding than loss of X$^-$ for SiH$_5^-$, SiH$_3$F$_2^-$, SiX$_2$X$_3^-$, SiHX$_4^-$ and SiX$_5^-$. Cramer and Squires[38] have described the conformational potential energy surface for the dihydroxysiliconate ion H$_3$Si(OH)$_2^-$. They found it to be remarkably flat with six minima lying within 2.7 kcal mol^{-1} of one another. Interconversions of the minima have pseudorotation barriers of less than 6 kcal mol^{-1}, suggesting that H$_3$Si(OH)$_2^-$ may be considered to be essentially without structure as regards ligand location.

These results show that pseudorotation is to be expected with silicon, and Martin and coworkers[39] have shown that it has a lower barrier than pseudorotation involving phosphorus. Thus axial entry, pseudorotation and axial exit, which is generally accepted for phosphorus, is a viable mechanism for silicon.

Myers and coworkers[40,41] have shown evidence for a pseudorotational mechanism in the reaction of the O-silyl ketene N,O-acetal **17** with aromatic aldehydes (equation 8).

(8)

Kinetic studies were consistent with an associative mechanism involving pentacoordinate silicon. X-ray crystallographic data for the product **20** suggests the trigonal bipyramid **19** as a reasonable precursor. Axial attack of benzaldehyde on the starting material **17** gives the trigonal bipyramid **18**, which can be converted to **19** by a pseudorotation.

Gung and coworkers[42] have located (at the MP2/6-31 G* level of theory) the transition state for the silicon-directed aldol reaction between a silyl enol ether and formaldehyde. They found it to be a boatlike six-membered ring **21** with a pentacoordinate silicon (equation 9).

$$(9)$$

The formaldehyde oxygen occupies the apical position with a Si−O bond length of 2.01 Å. The enol silane oxygen assumes the equatorial orientation with a Si−O bond length of 1.81 Å.

A similar transition state **22** is calculated for the reaction of allylsilanes with aldehydes (equation 10)[43]. In this case, however, the calculations show that the oxygen attacks at an apical site of the silicon centre, while the allyl group departs directly from an equatorial position without causing a pseudorotation, in contrast to the mechanisms previously discussed.

$$(10)$$

In these reactions the pentacoordinate species is a high-energy transition state, not an intermediate.

III. NUCLEOPHILE-CATALYSED SUBSTITUTION AT SILICON AND RACEMIZATION

Nucleophilic substitution at silicon has frequently been found to be catalysed by the presence of other nucleophiles. Corriu and coworkers have found that the hydrolysis of a range of chlorosilanes is first order in nucleophilic catalysts such as pyridines, HMPA, DMSO and DMF[44]. The stereochemistry of this type of reaction has been shown to be retentive, even with chlorosilanes where direct nucleophilic substitution normally gives inversion[45].

A reaction related to the nucleophile-catalysed hydrolysis of silanes is the nucleophile-catalysed racemization of silanes. The racemization of a range of halosilanes was found to have an order with respect to nucleophile varying from 1 to 3[46,47]. The reactions have entropies of activation which are large and negative, and enthalpies of activation which are small and sometimes negative. Both racemization and hydrolysis are slowed by increasing steric crowding at silicon.[48]

It was proposed by Sommer and Bauman that the methanol-catalysed racemization of 1-naphthylphenylmethylfluorosilane involves formation of a pentacoordinate intermediate followed by pseudorotation[49]. Three successive pseudorotations are required for racemization, followed by axial loss of the solvent. However, this mechanism is not consistent with the reaction being subsequently reported to be third order with respect to nucleophile[46,47].

Corriu and coworkers have proposed[50] an alternative mechanism involving extension of the coordination of the silicon atom. The first step involves rapid and reversible attack by a molecule of the nucleophilic catalyst, Nu^-, to give the pentacoordinate species **23** (equation 11).

$$R^1\text{-Si}(X)(R^2)(R^3) + Nu^- \rightleftharpoons R^1\text{-Si}^-(X)(R^2)(R^3)(Nu) \quad (23) \tag{11}$$

The slow step in the process is then the attack by a second nucleophile at an R−X edge to give a hexacoordinate intermediate or transition state. The second nucleophile is the alcohol in the case of alcoholysis or the nucleophilic catalyst in the case of racemization (equations 12 and 13).

racemization

$$(23) \rightleftharpoons (24) \rightleftharpoons (25) \tag{12}$$

alcoholysis

$$(23) \longrightarrow (26) \longrightarrow (27) \longrightarrow (28) \tag{13}$$

For alcoholysis, collapse of **23** via **26** and **27** then leads to the substitution product **28** with retention. For racemization, since both **24** and **25** have a plane of symmetry, the reversible formation of such species leads to retention and inversion. It is not certain whether racemization involves solely the intermediate **24** or whether this is a transition state leading to **25** as the only intermediate.

In addition to explaining the stereochemistry, this mechanism is also in agreement with the kinetics. Corriu has found[48,51] that racemization is second order with respect to the nucleophile, whereas alcoholysis is first order in the alcohol and nucleophile. Both of these rate equations are in agreement with the pre-equilibrium formation of an intermediate followed by rate-limiting attack by a second nucleophile. The small values obtained for $\Delta H^{\ddagger}$ and the large negative values for $\Delta S^{\ddagger}$ show the reaction to be entropy controlled with a highly organized transition state, as would be expected for a hexacoordinate transition state/intermediate.

Corriu's proposed mechanisms were criticised by Frye and coworkers[52], who considered the observed rate enhancements to be irreconcilable with intermediates of increased coordination number. Frye maintained that steric effects would retard attack on a pentacoordinate silicon compared to a tetracoordinate silicon, and nucleophilic attack would also be disfavoured on pentacoordinate silicon as the silicon bears a formal negative charge due to the formation of an essentially dative bond with the nucleophile.

However, recent studies have shown that pentacoordinate silicon centres are indeed more reactive than tetracoordinate silicon centres. Theoretical studies have shown that the positive charge on the central silicon atom is at least maintained[53] and may well be increased[54] by coordination of an additional ligand, even when the ligand is anionic. Calculations by Deiters and Holmes[53] also show a lengthening of bonds, particularly axial bonds, in pentacoordinate silicon species compared to tetracoordinate silicon species. As there is no decrease in positive charge at the central silicon atom on coordination of an anionic ligand, the negative charge on the ligands is increased in the pentacoordinate silicon species compared to the corresponding tetracoordinate species. Thus the leaving group ability of the ligands in the pentacoordinate system is increased.

Numerous experimental observations of the increased reactivity of pentacoordinate silicon over tetracoordinate silicon have been reported. Corriu and coworkers have found that the relative reactivity of $PhMeSiF_3^-$ and $PhMeSiF_2$ towards t-BuMgBr was $>1000:1$[55]. The pentacoordinate species **29** reacts with alcohols and acids to give mono- or disubstitution products whereas the corresponding four-coordinate species **30** does not react at all[56].

(29)   **(30)**   **(31)**

Bassindale and Borbaruah have studied the reactivities of bis-halo N,N-bisdimethylsilylmethylacetamides **31** towards nucleophiles[57]. In these systems there are two silicon centres that, when $X = Y$, differ only in the coordination at silicon. Investigations were carried out at both centres at the same time and under the same conditions.

It was found that silicon–bromine and silicon–chlorine bonds were thermodynamically activated towards substitution by coordination of an oxygen nucleophile, whereas coordination was found to deactivate the silicon–fluorine bond.

A further mechanism for nucleophile-catalysed racemization and substitution has been proposed by Chojnowski and coworkers[58]. They proposed that the HMPA catalysed substitution of halogen at silicon may involve the transient formation of a phosphonium cation (**32**) containing a four-coordinate silicon. The retention of the configuration observed would then be the result of two consecutive inversions (equation 14).

$$R_3SiCl \underset{\text{inversion}}{\overset{\text{HMPA}}{\rightleftharpoons}} R_3SiO-\overset{+}{P}(NMe_2)_3Cl^- \underset{\text{inversion}}{\overset{R'OH}{\longrightarrow}} R_3SiOR' \quad (14)$$

$$(32)$$

The HMPA adducts of trimethylchlorosilane (**32**, R = Me) and trimethylbromosilane were isolated and found to be ionic as shown. If the first step shown in equation 14 is a pre-equilibrium, the observed order for substitution is first order as expected. For racemization, the rate-limiting step is invertive attack of the second HMPA molecule on **32**, such that the reaction is second-order overall with respect to nucleophile.

A wide range of four-coordinate adducts have been formed from the reaction of nucleophiles with Me_3SiX (X = I, Br, OSO_2CF_3)[59,60]. As expected, these salts are highly susceptible to nucleophilic attack[61].

Corriu and coworkers suggest that Chojnowski's mechanism may only operate with very labile Si−Br and Si−I bonds and good nucleophiles[62]. They also criticize the mechanism as it cannot account for the nucleophile-induced epimerization of chlorocyclobutanes. As mentioned earlier, in small rings with exocyclic leaving groups, retentive substitution predominates, so that two consecutive substitutions by HMPA would only lead to retention.

Bassindale and coworkers[63,64] have studied the nucleophile-assisted racemizations of $PhCHMeSiMe_2X$ where X = Cl or Br. Thermodynamic and kinetic studies led to the conclusion that intermediates involving extracoordinate silicon were not being formed. The results were interpreted in terms of two competing mechanisms for racemization: (a) halide–halide exchange, with inversion of configuration (equation 15) and (b) displacement of halide by nucleophile followed by nucleophile–nucleophile exchange, followed by displacement of nucleophile by halide, each step proceeding by inversion of configuration (equation 16).

$$\underset{X}{\overset{R^2}{\underset{|}{R^1\!\!\diagdown\!\!\underset{Si}{\diagup}\!\!R^3}}} \underset{X^-}{\overset{X^-}{\rightleftharpoons}} \underset{R^3}{\overset{X}{\underset{|}{R^1\!\!\diagdown\!\!\underset{Si}{\diagup}\!\!R^2}}} \quad (15)$$

$$\underset{X}{\overset{R^2}{\underset{|}{R^1\!\!\diagdown\!\!\underset{Si}{\diagup}\!\!R^3}}} \underset{X^-}{\overset{Nu}{\rightleftharpoons}} \left[\underset{R^3}{\overset{Nu}{\underset{|}{R^1\!\!\diagdown\!\!\underset{Si}{\diagup}\!\!R^2}}}\right]^+ \underset{Nu}{\overset{Nu}{\rightleftharpoons}} \left[\underset{Nu}{\overset{R^2}{\underset{|}{R^1\!\!\diagdown\!\!\underset{Si}{\diagup}\!\!R^3}}}\right]^+ \underset{Nu}{\overset{X^-}{\rightleftharpoons}} \underset{R^3}{\overset{X}{\underset{|}{R^1\!\!\diagdown\!\!\underset{Si}{\diagup}\!\!R^2}}}$$

$$(16)$$

Substitution with inversion and retention of the pentacoordinate species **33** has been interpreted in terms of two competing dissociative mechanisms: Si−O cleavage and Si−NMI cleavage[65]. A hexacoordinate intermediate is not implicated.

However, Corriu and coworkers[66] postulate a hexacoordinate intermediate (or transition state) in the hydrolysis of organic silicates, with the rate-determining step in the reaction being the coordination of water to a pentacoordinate intermediate formed by initial nucleophilic attack.

(33)   NMI = *N*-methyl imidazole

A hexacoordinate transition state **34** is also postulated in the reaction of allyltrichlorosilanes with aldehydes in the presence of DMF (equation 17)[67].

(17)

(34)

IV. REFERENCES

1. A. R. Bassindale and P. G. Taylor, in *The Chemistry of Organic Silicon Compounds*, Part 1 (Eds. S. Patai and Z. Rappoport), Wiley, Chichester, 1989, pp. 839–892.
2. R. R. Holmes, *Chem. Rev.*, **90**, 17 (1990).
3. C. Chuit, R. J. P. Corriu, C. Reye and J. C. Young, *Chem. Rev.*, **93**, 1371 (1993).
4. L. H. Sommer, *Intra-Sci. Chem. Rep.*, **7**, 1 (1973).
5. R. J. P. Corriu and C. Guerin, *J. Organomet. Chem.*, **74**, 1 (1980).
6. R. J. P. Corriu and C. Guerin, *Adv. Organomet. Chem.*, **20**, 265 (1982).
7. L. H. Sommer and W. D. Korte, *J. Am. Chem. Soc.*, **89**, 5802 (1967).
8. L. H. Sommer, J. McLick and C. M. Golino, *J. Am. Chem. Soc.*, **94**, 669 (1972).
9. R. J. P. Corriu, C. Guerin and J. J. E. Moreau, *Top. Stereochem.*, **15**, 43 (1984).
10. N. T. Anh and C. Minot, *J. Am. Chem. Soc.*, **102**, 103 (1980).
11. J. A. Deiters and R. R. Holmes, *J. Am. Chem. Soc.*, **109**, 1692 (1987).
12. B. G. McKinnie, N. S. Bhacca, F. K. Cartledge and J. Fayssoux, *J. Org. Chem.*, **41**, 1534 (1976).
13. F. K. Cartledge, J. M. Wolcott, J. Dubac, P. Mazerolles and M. Joly, *J. Organomet. Chem.*, **154**, 203 (1978).
14. R. Corriu, C. Guerin and J. Masse, *J. Chem. Res. (M)*, 1877 (1977).
15. C. Eaborn, R. Eidenschink and D. R. M. Walton, *J. Chem. Soc., Chem. Commun.*, 388 (1975).
16. P. E. Dietze, *J. Org. Chem.*, **57**, 6843 (1992).
17. P. E. Dietze, *J. Org. Chem.*, **58**, 5653 (1993).
18. Y. Xu and P. E. Dietze, *J. Am. Chem. Soc.*, **115**, 10722 (1993).
19. P. E. Dietze, C. Foerster and Y. Xu, *J. Org. Chem.*, **59**, 2523 (1994).
20. R. J. P. Corriu and B. J. L. Henner, *J. Organomet. Chem.*, **102**, 407 (1975).
21. G. Chauviere, R. J. P. Corriu and B. J. L. Henner, *J. Organomet. Chem.*, **86**, C1 (1975).
22. G. Sini, G. Ohanessian, P. C. Hiberty and S. S. Shaik, *J. Am. Chem. Soc.*, **112**, 1407 (1990).
23. Z. Shi and R. J. Boyd, *J. Phys. Chem.*, **95**, 4698 (1991).
24. S. Gronert, R. Glaser and A. Streitwieser, *J. Am. Chem. Soc.*, **111**, 3111 (1989).
25. M. T. Carroll, M. S. Gordon and T. L. Windus, *Inorg. Chem.*, **31**, 825 (1992).
26. A. A. Macharashvili, V. E. Shklover, Yu. T. Struchkov, G. I. Oleneva, E. P. Kramarova, A. G. Shipov and Yu. I Baukov, *J. Chem. Soc., Chem. Commun.*, 683 (1988).
27. A. R. Bassindale and M. Borbaruah, *J. Chem. Soc., Chem. Commun.*, 1501 (1991).
28. V. F. Sidorkin, V. V. Vladimirov, M. G. Voronkov and V. A. Pestunovich, *J. Mol. Struct.*, **228**, 1 (1991).
29. J. A. Deiters and R. R. Holmes, *J. Am. Chem. Soc.*, **109**, 1686 (1987).

30. R. Luckenbach, *Dynamic Stereochemistry of Pentacoordinate Phosphorus and Related Elements*, George Thieme Publ., Stuttgart, 1973.
31. F. Klanberg and E. L. Muetterties, *Inorg. Chem.*, **7**, 155 (1968).
32. R. J. P. Corriu, A. Kpoton, M. Poirier, G. Royo and J. Y. Corey, *J. Organomet. Chem.*, **277**, C5 (1986).
33. F. H. Carre, R. J. P. Corriu, G. F. Lanneau and Z. Yu, *Organometallics*, **10**, 1236 (1991).
34. I. Kalikhman, S. Krivinos, A. Ellern and D. Kost, *Organometallics*, **15**, 5073 (1996).
35. J. Boyer, R. J. P. Corriu, A. Kpoton, M. Mazhar, M. Poirier and G. Royo, *J. Organomet. Chem.*, **301**, 131 (1986).
36. M. S. Gordon, T. L. Windus, L. W. Burggraf and L. P. Davis, *J. Am. Chem. Soc.*, **112**, 7167 (1990).
37. T. L. Windus, M. S. Gordon, L. P. Davis and L. W. Burggraf, *J. Am. Chem. Soc.*, **116**, 3568 (1994).
38. C. J. Cramer and R. R. Squires, *J. Am. Chem. Soc.*, **117**, 9285 (1995).
39. W. H. Stevenson III, S. Wilson, J. C. Martin and W. B. Farnham, *J. Am. Chem. Soc.*, **107**, 6340 (1985).
40. A. G. Myers, K. L. Widdowson and P. J. Kukkola, *J. Am. Chem. Soc.*, **114**, 2765 (1992).
41. A. G. Myers, S. E. Kephart and H. Chen, *J. Am. Chem. Soc.*, **114**, 7922 (1992).
42. B. W. Gung, Z. Zhou and R. A. Fouch, *J. Org. Chem.*, **60**, 2860 (1995).
43. K. Omoto, Y. Sawada and H. Fujimoto, *J. Am. Chem. Soc.*, **118**, 1750 (1996).
44. R. J. P. Corriu, G. Dabosi and M. Martineau, *J. Organomet. Chem.*, **150**, 27 (1978).
45. R. J. P. Corriu, G. Dabosi and M. Martineau, *J. Organomet. Chem.*, **154**, 33 (1978).
46. R. J. P. Corriu and M. Henner-Leard, *J. Organomet. Chem.*, **64**, 351 (1974).
47. F. K. Cartledge, B. G. McKinnie and J. M. Wolcott, *J. Organomet. Chem.*, **118**, 7 (1976).
48. R. J. P. Corriu, F. Larcher and G. Royo, *J. Organomet. Chem.*, **104**, 293 (1976).
49. L. H. Sommer and D. L. Bauman, *J. Am. Chem. Soc.*, **91**, 7045 (1969).
50. R. J. P. Corriu, F. Larcher and G. Royo, *J. Organomet. Chem.*, **129**, 299 (1977).
51. R. J. P. Corriu and M. Henner-Leard, *J. Organomet. Chem.*, **65**, C39 (1974).
52. H. K. Chu, M. D. Johnson and C. L. Frye, *J. Organomet. Chem.*, **271**, 327 (1984).
53. J. A. Deiters and R. R. Holmes, *J. Am. Chem. Soc.*, **112**, 7197 (1990).
54. M. S. Gordon, M. T. Carroll, L. P. Davis and L. W. Burggraf, *J. Phys. Chem.*, **94**, 8125 (1990).
55. J. L. Brefort, R. J. P. Corriu, C. Guerin, B. J. L. Henner and W. W. C. Wong Chi Man, *Organometallics*, **9**, 2080 (1990).
56. B. J. Helmer, R. West, R. J. P. Corriu, M. Poirier, G. Royo and A. De Saxce, *J. Organomet. Chem.*, **251**, 295 (1983).
57. A. R. Bassindale and M. Borbaruah, *J. Chem. Soc., Chem. Commun.*, 352 (1993).
58. J. Chojnowski, M. Cypryk and M. Michalski, *J. Organomet. Chem.*, **161**, C31 (1978).
59. A. R. Bassindale and T. Stout, *J. Chem. Soc., Perkin Trans. 2*, 221 (1986).
60. A. R. Bassindale and T. Stout, *J. Organomet. Chem.*, **238**, C41 (1982).
61. A. R. Bassindale, J. C-Y. Lau, T. Stout and P. G. Taylor, *J. Chem. Soc., Perkin Trans. 2*, 227 (1986).
62. R. J. P. Corriu, G. Dabosi and M. Martineau, *J. Organomet. Chem*, **186**, 25 (1980).
63. A. R. Bassindale, J. C-Y. Lau and P. G. Taylor, *J. Organomet. Chem.*, **490**, 75 (1994).
64. A. R. Bassindale, J. C-Y. Lau and P. G. Taylor, *J. Organomet. Chem.*, **499**, 137 (1995).
65. A. R. Bassindale, S. J. Glynn, J. Jiang, D. J. Parker, R. Turtle, P. G. Taylor and S. S. D. Brown, in *Organosilicon Chemistry II. From Molecules to Materials* (Eds. N. Auner and J. Weiss), VCH, Weinheim, 1996, pp. 411–425.
66. R. J. P. Corriu, C. Guerin, B. J. L. Henner and Q. Wang, *Organometallics*, **10**, 3200 (1991).
67. S. Kobayashi and K. Nishio, *J. Org. Chem.*, **59**, 6620 (1994).

CHAPTER 10

Silicenium ions: Quantum chemical computations

CHRISTOPH MAERKER and PAUL VON RAGUÉ SCHLEYER[†]

Computer Chemistry Center of the Institute of Organic Chemistry, The University of Erlangen-Nürnberg, Henkestrasse 42, 91054 Erlangen, Germany
Fax: 49-9131-85-9132,   e-mail: pvrs@ccc.uni-erlangen.de

I. ABBREVIATIONS	514
II. INTRODUCTION	514
III. COMPUTATIONAL METHODS	515
A. 'Traditional' *ab initio* Molecular Orbital (MO) Techniques	515
B. Density Functional Theory (DFT) Based Methods	516
C. Computations of Magnetic Properties	516
IV. QUANTUM MECHANICAL STUDIES OF SILICENIUM CATIONS	517
A. Thermodynamic Stability of Silicenium Cations	517
B. Towards Free Silicenium Ions in Solution: How Can They be Prepared?	519
C. Trialkyl-substituted Silicenium Ions	520
1. Energetic stabilization by multiple alkyl substitution	520
2. The first X-ray structures: The first stable tricoordinate silicenium ions?	520
3. Interactions with Lewis bases: Binary and ternary complexes	524
D. ^{29}Si NMR Chemical Shifts of Silicenium Ions	532
E. π-Donor Stabilized Silicenium Ions	538
1. The silaguanidinium ion	538
2. Resonance stabilization by aromatic substituents	540
3. Intramolecular π-stabilization	543
F. Stabilization by Organoboryl Groups	547
V. CONCLUDING REMARKS AND OUTLOOK	550
VI. REFERENCES	551

[†] New postal address: Center for Computational Quantum Chemistry The University of Georgia, Athens, GA 30602 USA. Fax: +1-706-542-0406

The chemistry of organic silicon compounds, Vol. 2
Edited by Z. Rappoport and Y. Apeloig   © 1998 John Wiley & Sons Ltd

I. ABBREVIATIONS

B	Becke's exchange functional
B3LYP	three-parameter hybrid exchange-correlation-density functional
BII	standard basis set of triple-zeta quality, including polarization functions for all elements, often employed in IGLO calculations
BII'	same basis set as BII for non-hydrogen atoms, combined with a double-zeta quality sp-basis set for hydrogen atoms
9-BBN	9-borabicyclo[3.3.1]nonyl group
CCSD	coupled-cluster theory with single and double excitation terms
CCSD(T)	same as CCSD plus a perturbative estimate of triple excitation terms
DFT	density functional theory
DFPT	density functional perturbation theory
ΔZPE	differential zero-point vibrational energy
E_h	computed total energy in Hartree
ϵ	dielectric constant
fc	frozen-core approximation
GIAO	gauge-including atomic orbitals
HF	Hartree–Fock
http	hypertext transfer protocol
IGLO	individual gauge for localized orbitals
KS	Kohn–Sham
LYP	gradient-corrected correlation functional by Lee, Yang and Parr
MBPT	many-body perturbation theory
Me	methyl group
Mes	mesityl group
MP2	second-order Møller–Plesset perturbation theory
Ng	noble gas
NPA	natural population analysis
PG	(symmetry) point group
π-SE	π-electron stabilization energy
$\rho(\mathbf{r})$	electron density distribution
$\nabla^2 \rho(\mathbf{r})$	Laplacian of the electron density distribution
TMS	tetramethyl silane, Me$_4$Si
URL	uniform resource locator
WBI	Wiberg bond index
WWW	world wide web
ZPE	zero-point vibrational energy

II. INTRODUCTION

Since 1989, when the first volume of *The Chemistry of Organosilicon Compounds* appeared[1], large strides have been made in preparing silicenium cations. The first X-ray structures[2,3] of solvent stabilized silicenium cations are among the most notable successes. Most recently, a free silicenium cation has been prepared in solution[4,5]. In addition to experimental techniques for structure elucidation, e.g. X-ray single-crystal structure analysis, and nuclear magnetic resonance (NMR) spectroscopy, *quantum theory* is now a widely accepted *practical tool* for the exploration of chemical entities and their reaction behaviour. In particular, quantum chemistry has played a major role in exploring and predicting the structure and energetics of organosilicon compounds[6], especially of silenes[7], which very often behave quite differently from their lighter carbon analogues[7,8]. Apeloig[9,10], Gordon[11], Nagase[12] and Morokuma[13], to name but a few, have recently reviewed quantum chemical studies on neutral and ionic silicon compounds.

This chapter focuses almost exclusively on the most recent theoretical investigations of silicenium ions, i.e. of trivalent, positively charged silicon ions, which have not been covered in earlier reviews[6,9]. When appropriate, experimental work on silicenium ions and studies on closely related carbenium ions will be mentioned. This chapter is organized as follows: Section III summarizes briefly recent developments in quantum mechanical methodology pertinent to organosilicon chemistry. Section IV highlights recent computational studies, and Section V presents conclusions and an outlook of future work. The reader is also referred to the accompanying review by Lickiss which describes recent experimental work on silicenium ions.

III. COMPUTATIONAL METHODS

In the past decade quantum chemical investigations of organosilicon compounds have progressed substantially due to the benefits from hardware speed-up and new software developments. Modern RISC processors and new parallel machines facilitate computations on large molecular systems which one could not have imagined even five years ago. Today, routine high-level computational 'experiments' can be carried out on 'real' systems, i.e. the actual system carrying the experimental substituents, rather than on simple model species. Very recently, a massively parallel Intel™ computer reached unprecedented 2.5 Teraflops per second, which is 1000 times faster than in currently used supercomputers. In addition to these hardware achievements, new software and methodology developments are equally important. Density functional theory based methods have become more widely used[14,15]; see the discussion below. Faster algorithms, such as the Fast Multipole Method (FMM)[16], facilitate an efficient linear-scaling computation of the Coulomb problem, both in quantum mechanics[17] (QM) and in molecular[18] mechanics (MM) computations. Combined QM/MM methods[19] are currently under development, mainly in the field of biomedical research[20]. In QM/MM, the active part of the molecule (e.g. the reaction site) is treated quantum mechanically, whereas the outer spheres are described by well-parametrized force fields. Similar computational techniques in principle can be applied to silicon compounds, e.g. in the simulation of semiconductor surface–gas phase interactions. In summary, it is not too far fetched to say that a computational revolution is taking off with ever-increasing speed.

A. 'Traditional' *ab initio* Molecular Orbital (MO) Techniques

A wealth of texts, e.g. those by Clark[21] and by Pople and coworkers[22], and the series *Reviews in Computational Chemistry*[23] provide comprehensive overviews on the whole field of computational chemistry, in particular on quantum chemistry. The *Encyclopedia of Computational Chemistry*[24], to appear in 1998, will provide a comprehensive review on state-of-the-art computational chemistry, written by world-leading experts in the field. The Internet also provides an online forum of computational chemistry related sites, e.g. the Fourth Electronic Computational Chemistry Conference (ECCC4)[25] and the *Journal of Molecular Modeling*[26]. Thus, it suffices here to comment shortly on more recent developments such as quantum mechanical calculations using density functional theory[14,15] (DFT) and *ab initio* computations of magnetic properties[27].

B. Density Functional Theory (DFT) Based Methods

'The physicists have been trying to persuade us for years that we ought to study density functional theory. We ought have listened more carefully, especially since it was one of our own, J. C. Slater who pushed them in that direction with his 1951 contribution[28] ... So although I think we have all worked hard and made great progress ... I rather wonder if we are up against it and the physicists were right after all[29]'. This very emphatic statement by Handy, one of the leading proponents of traditional *ab initio* MO theory, gives the reader an impression to what extent the DFT-based methods have influenced the quantum chemists' thinking since they have plunged into the quantum theory of molecules some years ago.

The key theorems of modern density functional theory were introduced by Hohenberg, Sham and Kohn in the mid-1960s[30]. DFT theory is based on the fact that the total energy E of a chemical species (either in its electronic ground or in its excited states) can in principle be expressed via functionals of the electron density ρ. However, the exact form of these functionals, in particular that of the exchange-correlation functional, $E_{XC}[\rho]$, is unknown[14,15]. Nevertheless, currently-employed DFT-based methods often rival high-level molecular orbital procedures in accuracy, and their computational efficiency (with a formal scaling in computer time of N^3, where N is the number of electrons) provides a superior accuracy per computational cost ratio[14,15]. Excellent monographs on recent implementations and applications of DFT methods are available, where the interested reader can obtain further information[14,15]. Fast multipole methods[16], as mentioned earlier, are now being incorporated into quantum chemical programs[17] and will provide further speed-ups.

C. Computations of Magnetic Properties

NMR spectroscopy is an ubiquitous and indispensable tool for modern structure elucidation. During the last two decades quantum mechanical procedures for the computation of nuclear magnetic shielding constants and the derived chemical shifts have become available. Questionable experimental assignments can now be verified by means of computed data[27]. In addition, high-level computations even allow reliable predictions for yet-to-be prepared species. The recently developed computations of spin–spin coupling constants[31] have analytical and predictive potential, as do computations of chemical shifts.

Despite the importance of theoretical NMR data, magnetic property computations have only become possible recently. The gauge dependence problem of finite basis set calculations is best handled via 'local gauge origin' methods. The most widely employed approaches are the 'individual gauge for localized orbitals' (IGLO)[32], the 'localized orbital/local origin' (LORG)[33] and the 'gauge-including atomic orbitals' (GIAO)[34] methods. The IGLO method (and program), developed by Kutzelnigg and coworkers[32] at the beginning of the 1980s, was the first practical tool for *ab initio* computations of the magnetic properties of medium-sized molecules.

Schleyer and coworkers have shown[35] that in order to use effectively the IGLO/NMR approach to structure elucidation, it is necessary to employ high-level optimized geometries (e.g. at the correlated MP2/6-31G* or DFT levels of theory) in order to obtain best agreement between computed and experimental chemical shifts. Early GIAO-SCF calculations some 20 years ago[36] were hampered by an inefficient implementation which prohibited computations on larger molecules. Improved gradient algorithms in the GIAO scheme[37] enabled GIAO/NMR calculations on large organic molecules at a speed comparable with that of IGLO calculations.

It soon became apparent that the inclusion of electron correlation in the computation of chemical shielding tensors would improve the results considerably[38]. The contributions by Oddershede[39], Galasso[40], Kutzelnigg[41], Gauss[42,43] and their coworkers are noteworthy. In particular the reader is referred to the impressive series of benchmark calculations by Gauss and coworkers[43] of chemical shielding tensors at hierarchically increased electron correlated levels employing the GIAO approach (GIAO-MBPT[n], $n = 2-4$; GIAO-CCSD and GIAO-CCSD(T)), which have also been reviewed recently[27]. However, the inherent scaling problem of these highly correlated levels prevents their application even to medium-sized organic species. Thus, the burgeoning field of DFT-based methods for the computation of magnetic properties might be a valuable alternative for larger systems, and implementations of density functional theory based computer codes for the calculation of magnetic properties have already appeared recently[44].

IV. QUANTUM MECHANICAL STUDIES OF SILICENIUM CATIONS

A. Thermodynamic Stability of Silicenium Cations

Unlike carbocations[45–47], for which the first quantum mechanical studies were reported in the 1950s[48,49], computational studies on silicenium ions only date back some 20 years[50–54]. Schleyer, Apeloig and coworkers were the first to report comprehensive investigations on the structure and stability of α-substituted silicenium cations[51,54]. Although those early computations only employed modest SCF levels, the authors were able to analyse the importance of the various properties of α-substituents, such as σ-inductive effects, lone pair π-donation, conjugation and hyperconjugation, in stabilizing silicenium ions[51,54]. The relative stabilities of silicenium ions were explored by means of isodesmic reactions 1 to 3. Isodesmic reactions, where the same type and number of bonds appear on both sides of the equation, were introduced by Pople and coworkers[55] in order to reduce errors, particularly when using lower theoretical levels.

$$R_n SiH_{3-n}^+ + SiH_4 \longrightarrow R_n SiH_{4-n} + SiH_3^+ \qquad n = 0-3 \qquad (1)$$

$$R_n CH_{3-n}^+ + CH_4 \longrightarrow R_n CH_{4-n} + CH_3^+ \qquad n = 0-3 \qquad (2)$$

$$R_n SiH_{3-n}^+ + R_n CH_{4-n} \longrightarrow R_n SiH_{4-n} + R_n CH_{3-n}^+ \qquad n = 0-3 \qquad (3)$$

Equation 1 compares the stability of substituted silicenium ions versus the parent SiH_3^+ (**1**). Equation 2 compares the stabilities of the correspondingly substituted carbocations versus CH_3^+. Equation 3 compares the stabilities of silicenium ions with those of the corresponding carbocations. The calculated energies of equations 1–3 are presented in Table 1, and the following conclusions can be drawn[51]:

(1) Relative to hydrogen, all the first-row substituents, with the exception of fluorine, stabilize the silicenium ion. The greater stability of α-substituted silicenium cations versus their corresponding carbocations can be attributed in part to the lower electronegativity of silicon (1.7) compared to carbon (2.5). Equation 3 also takes into account the relative stability of alkanes versus silanes, i.e. the stability of C–H versus Si–H bonds. The Si–H bond is weaker than the C–H bond by *ca* 13 kcal mol^{-1} in the gas phase, e.g. the bond dissociation energy DH_{298}^0(C–H) in CH_4 is 104.9 kcal mol^{-1}[56], compared

TABLE 1. Computed stabilization energies of α-substituents on carbenium RCH_2^+ and silicenium $RSiH_2^+$ cations (kcal mol^{-1})

R	Equation 1		Equation 2		Equation 3	
	this group[a]	Ref. 9[b]	this group[a]	Ref. 9[b]	this group[a]	Ref. 9[b]
H	0.0	0.0	0.0	0.0	57.4	54.9
Li	57.8		75.3		39.8	
BeH	14.3		15.1		56.5	
BH$_2$, planar		8.3		2.7		60.4
BH$_2$, perp.	13.7	13.9	25.9	25.8	45.1	43.0
CH$_3$	15.1	15.1	40.6	34.1	31.8	35.9
NH$_2$, planar	37.5	36.8	100.5	97.8	−5.6	−6.1
NH$_2$, perp.		11.9		15.9		50.9
OH, planar	19.1	17.9	66.3	62.7	10.1	10.1
F	−1.1	−2.3	25.3	21.5	31.0	31.1
SiH$_3$	12.3	12.6	16.5	17.7	53.2	49.8
PH$_2$, planar	17.6	13.5	63.4	60.0	11.6	8.4
SH, planar	18.5	18.4	63.6	60.9	12.2	12.4
SH, perp.		0.4		8.1		47.3
Cl	2.2	2.0	29.2	26.6	30.3	30.3

[a]MP2(fc)/6-31G*//MP2(fc)/6-31G* level of theory, Reference 6.
[b]MP3/6-31G*//HF/6-31G*.

to DH_{298}^0(Si−H) of 91.9 kcal mol^{-1} in SiH$_4$[57]. Trimethyl substitution decreases the difference in dissociation energies DH_{298}^0(E−H) slightly: 96.4 kcal mol^{-1} in Me$_3$C−H[58], compared to 90.3 kcal mol^{-1} in Me$_3$Si−H[59]. Thus, hydride transfer from silicon to carbon in the gas phase is favoured by the weaker silicon–hydrogen bonds. Hydride transfer from organosilicon compounds to trityl salts in solution was first reported and utilized by Corey and coworkers[60] in an attempted preparation of silicenium ions. In this context it is important to note that commonly employed leaving groups other than hydrogen, e.g. halides, do not lead to silicenium ions, and only Lewis acid–organosilane adducts are formed[61].

(2) First-row α-substituents are more effective in stabilizing carbocations than in stabilizing silicenium ions. In particular, lone pair π-electron donation, e.g. by a planar amino group in SiH$_2$NH$_2^+$, is *ca.* 60 kcal mol^{-1} less effective than in CH$_2$NH$_2^+$. This dramatic decrease in the stabilizing power of π-donors was attributed to the unfavorable interaction (reduced overlap) between the 2p lone pair donor and the 3p(Si$^+$) acceptor orbitals[62]. One might therefore assume that second-row substituents are more effective in stabilizing silicenium ions, either due to a better 3p lone pair to 3p(Si$^+$) overlap or due to a reduced inductive destabilization caused by the less electronegative second-row substituents. In particular, the SH group even competes in its stabilizing power for silicenium ions with the OH group, whereas chlorine substitution gives a stabilizing effect contrary to the destabilization found for fluorine. Noteworthy is also the remarkable stabilization of silicenium and carbenium ions by electropositive substituents, e.g. by Li, BeH and by BH$_2$.

The *intrinsic* π-donation capabilities of heavy element atoms (group 15 to group 17), compared to first-row atoms, has been investigated recently by Schleyer and coworkers[63]. The superior ability of nitrogen to act as a π-donor in α-substituted carbocations

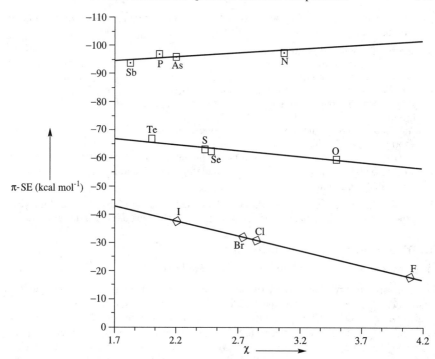

FIGURE 1. $CH_2-XH_n^+$ ($n = 0, 1, 2$) π-stabilization energies (π-SE) [kcal mol^{-1}; at QCISD(T)/-DZ++PP//MP2/DZ+P+ZPE] versus the electronegativity χ of X. Reproduced by permission of Wiley-VCH from Reference 63

$H_2C-XH_n^+$ relative to heavier group 15 substituents, e.g. comparing $-NH_2$ with $-PH_2$, was found to be due to the small planarization energy of the amino group relative to that of the PH_2 group. Thus, the *inherent* π-donor stabilization energies of the heavy element group 15 atoms is *not* reduced compared to first-row substituents, as long as a planarized, sp^2-hybridized $-XH_2$ group is taken into account, although such idealized groups might be of little practical relevance. Group 16 $-XH$ and group 17 $-X$ heavy element atoms (X = S, Se, Te; and Cl, Br, I, respectively) even were found to stabilize carbocations better than the parent first-row substituents $-OH$ and $-F$, respectively (Figure 1). The authors concluded that: 'In contrast to the still common misconception that 2p/3p/4p/5p overlap is ineffective, our comparison emphasizes that the *inherent* π-donor capabilities of the heavier elements are as large as or even larger than their first-row counterparts'[63].

B. Towards Free Silicenium Ions in Solution: How Can They be Prepared?

While appropriate substitution can enhance the thermodynamic stability of silicenium ions, this does not suffice to produce stable silicenium ions in solution. The propensity of silicon to form intermediates with higher coordination numbers[64], in combination with the large positive charge on silicon, makes any free silicenium ion an inviting target for Lewis bases, e.g. solvent molecules or counterions. Hence, both electronic stabilization and steric shielding of the cationic silicon centre, e.g. by bulky groups, are necessary simultaneously. The following approaches towards the generation of silicenium ions in solution have been proposed, and most of them have already been examined:

- substitution with alkyl groups R (e.g. R = methyl, ethyl and especially the sterically demanding isopropyl and t-butyl), which provide electronic stabilization via both inductive and hyperconjugative (C−H and C−C hyperconjugation) effects.
- electronic stabilization by lone pair donor atoms (e.g. by N, O or S),
- substitution by bulky boraorganyl substituents, e.g. the 9-borabicyclo[3.3.1]nonyl group,
- stabilization via conjugation with aryl groups, which might provide resonance stabilization and steric shielding simultaneously,
- incorporation of the silicenium ion function into an aromatic ring, a strategy which has worked well in many other cases, e.g. in preparing silanol anions[65], isolable carbenes[66] and stable nitrenium ions[67].

Let us first have a closer look at trialkyl-substituted silicenium ions[2,3] since, four years ago, the X-ray structures of these species started a lively debate on silicenium ions.

C. Trialkyl-substituted Silicenium Ions

1. Energetic stabilization by multiple alkyl substitution

Successive methyl substitution of SiH_3^+ was investigated in a 1989 ion cyclotron resonance (ICR) study[68]. The first, second and third methyl substitution decreases the experimental hydride affinity of the corresponding silicenium ions by 15.5, 15.8 and 9.6 kcal mol^{-1}, respectively, relative to SiH_3^+ (equation 1)[68]. Apeloig reported computed values of 15.1, 12.9 and 10.6 kcal mol^{-1} (MP3/6-31G*//6-31G*)[9], respectively, which fit the experimental numbers quite well. Thus, the incremental stabilization by the methyl substituent decreases with increasing degree of substitution (Table 2). Note that a similar trend of decreasing incremental methyl stabilization is also known for carbenium ions[68] (Table 2). According to equation 3, the trimethyl silicenium cation is only 21.0 kcal mol^{-1} more stable than the corresponding t-butyl cation, compared to 57.4 kcal mol^{-1} for the parent species SiH_3^+ vs CH_3^+ [at MP2(fc)/6-31G*]. Nevertheless, hydride transfer from trialkyl silanes to trialkyl carbenium ions still is favoured thermodynamically.

2. The first X-ray structures: The first stable tricoordinate silicenium ions?

X-ray crystal structures of 'silicenium ion candidates' have been reported as early as in 1983[69], but in all cases it turned out that the silicon atom formed a covalent bond to a neighbouring solvent molecule or to the counterion[70,71]. In 1993, both Lambert's[2] and

TABLE 2. Effect of multiple methyl substitution on the stability of carbenium $R_nCH_{3-n}^+$ and silicenium $R_nSiH_{3-n}^+$ ions (kcal mol^{-1})

Species	Equation 1		Equation 2		Equation 3	
	calc^a	exp^b	calc^a	exp^b	calc^a	exp^b
R = Me						
$n = 1$	15.1 (15.1)^c	15.5	40.6	43.9	31.8 (27.4)^c	24.8
$n = 2$	27.9 (28.0)^c	31.3	58.7	62.9	26.6 (17.4)^c	21.8
$n = 3$	38.4 (38.6)^c	40.9	74.8	80.8	21.0 (12.9)^c	12.9

^{a}At MP2(fc)/6-31G*//MP2(fc)/6-31G*, from Reference 6.
^{b}Data taken from Reference 68.
^{c}MP3/6-31G*//6-31G* from Reference 9.

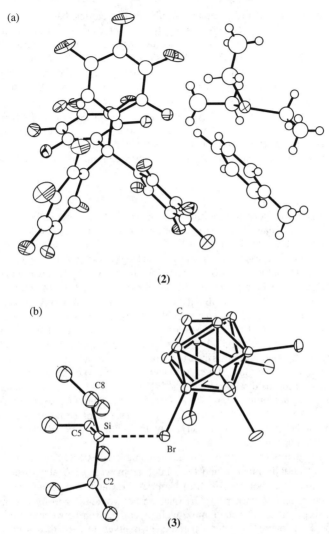

FIGURE 2. (a) Experimental molecular structure (ORTEP) of Et_3Si^+ $(C_6F_5)_4B^-$ (**2**). Silicenium ion upper right, anion left; the toluene solvent molecule, lower right. Reproduced by permission of the American Association for the Advancement of Science from Reference 2a. (b) Experimental molecular structure (ORTEP) of $i\text{-}Pr_3Si^+(Br_6CB_{11}H_6)^-$ (**3**). Bond distances (Å): Si−Br = 2.479(9) (dashed); Si−C(2) = 1.860(27); Si−C(5) = 1.908(27); and Si−C(8) = 1.799(35). Bond angles (deg): C(2)−Si−C(5) = 102.2(12); C(2)−Si−C(8) = 111.2(14); and C(5)−Si−C(8) = 119.6(13). Reproduced by permission of the American Association for the Advancement of Science from Reference 3a

Reed's[3] groups reported single-crystal X-ray structures of alkyl-substituted 'silicenium ion candidates', namely $Et_3Si^+B[C_6F_5]_4^-$ (**2**) and $i\text{-}Pr_3Si^+$ $Br_6CB_{11}H_6^-$ (**3**), as shown in Figures 2a and 2b, respectively. Their claims to have prepared tricoordinate positively charged silicon ions in the solid state[2,3] opened a lively discussion[72−78]. Note that

Lambert and his coworkers have made similar claims earlier, based on insufficient evidence in solution[79,80,81]. Lambert's 1993 study[2a], for example, showed an organosilyl moiety with no coordination to the anion, but with a toluene molecule close nearby (R(Si−C) = 2.18 Å). Lambert and coworkers claimed that in **2** "covalent bonding is weak or absent"[2a] i.e. there is no fourth coordination between the silicon atom and toluene. The elongated Si−C_{ipso} distance in **2** led Lambert and his coworkers to the stable silicenium ion description[2a]. The X-ray structure of **3**, reported by Reed and coworkers, showed a $Si^+ \cdots Br$ distance of 2.48 Å, 0.24 Å longer than a normal Si−Br single bond[3a].

Pauling, in one of his last scientific contributions, pointed out that, based on the observed Si−C bond length, there is a covalent silicon−carbon (of the toluene) bonding in the triethylsilicenium−toluene complex **2**[72]. The bond number, n, is 0.35, according to equation 4,

$$D(n) = D(1) - 0.60 \log n \tag{4}$$

where $D(n)$ is the bond length in question and $D(1)$ is the bond length of a 'normal' covalent single bond (i.e. n equals 1), whilst n is the bond number of the bond length in question. Pauling further noted that 'It has been [his] experience that calculated bond numbers as small as 0.10 (which represent an increase of bond length as great as 0.60 Å) need to be taken into account'[72]. Thus, the 2.18 Å Si−C bond length observed by Lambert and coworkers[2a] can *not* be regarded as being 'well outside the range (1.9 to 2.0 Å) of "long" Si−C bonds that unusually involve Si−*t*-butyl' bonds[2], as stated by Lambert and coworkers.

Similar conclusions were drawn by Schleyer, Apeloig, Siehl and coworkers, whose calculated structures for the [Me$_3$Si−toluene]$^+$ complex (**4**) and the [H$_3$Si−benzene]$^+$ complex (**5**) are shown in Figure 3[73]. These authors reported a Wiberg bond index[82] of 0.44 for the Si−C bond and described the silicenium ion−toluene complexes **4** and **5** as silylated arenium ions (Figure 3), i.e. σ-bonded Wheland-type species, with considerable charge delocalization to the aromatic ring[73]. The long Si−C_{ipso} bond distance is due to extensive Si−C hyperconjugation, as represented by the stabilizing resonance forms shown in Figure 4[73]. Reed's group acknowledged the presence of some covalent interactions between the arene and the silyl moiety, e.g. in **2**, but interpreted the structure of the adduct as reflecting a weak π-complex[76].

The X-ray structure of **2** (Figure 2a) showed an average CSiC bond angle of 114° (the toluene was essentially planar), and the silicon atom was 0.4 Å above the plane of the three adjacent ethyl carbons[2a]. *Ab initio* Hartree−Fock and DFT geometry optimizations of **4** by Schleyer and coworkers[6,73], by Olah and coworkers[77] and by Cremer's group[83] reproduced these structural details quite well. A comparison between the calculated and some of the experimental structural parameters is given in parentheses in **4** (Figure 3). Lambert and coworkers attributed the significant non-planarity of the silicenium moiety in **2** to 'a combination of steric effects, long-range orbital interaction, and crystal packing forces. ... the silyl cation herein easily distorts from the plane to relieve external sources of strain'. However, even these arguments were refuted by *ab initio* computations[77,83]. Pyramidalization towards the observed CSiC bond angle of 114° *raises* the energy of the silicenium ion by 9.3 kcal mol^{-1} (MP2/6-31G*)[77]. Moreover, IGLO NMR computations showed that pyramidalization leads to a *deshielded* silyl resonance (see below)[77,83] and not to an increased shielding as surmised by Lambert and coworkers[2a].

Olah and coworkers[77] further characterized Reed's silicenium ion **3** as a silylated bromonium ion[6] [exp. R(Si−Br) = 2.479 Å]. Despite all the criticism and theoretical evidence, both Lambert and Reed defended their own interpretation of structures **2** and **3** as being chemical entities with predominant silicenium ion character[75,76].

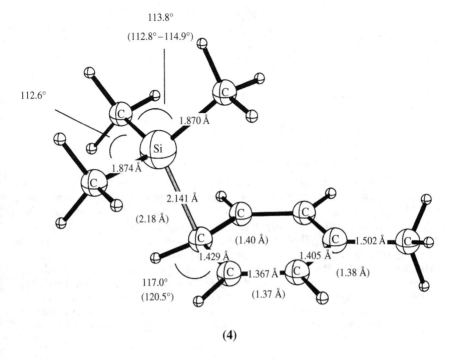

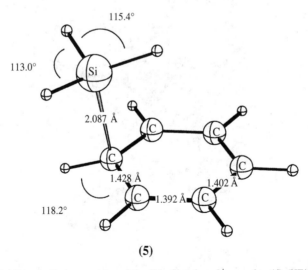

FIGURE 3. *Ab initio* optimized geometries of the [Me₃Si–toluene]⁺ complex (**4**) (HF/6-31G*, top) and of the [H₃Si–benzene]⁺ complex (**5**) [MP2(fc)/6-31G*, bottom]. Values in parentheses correspond to experimental X-ray data of Et₃Si⁺ (C₆F₅)₄B⁻ (**2**). Reproduced by permission of Wiley-VCH from Reference 73

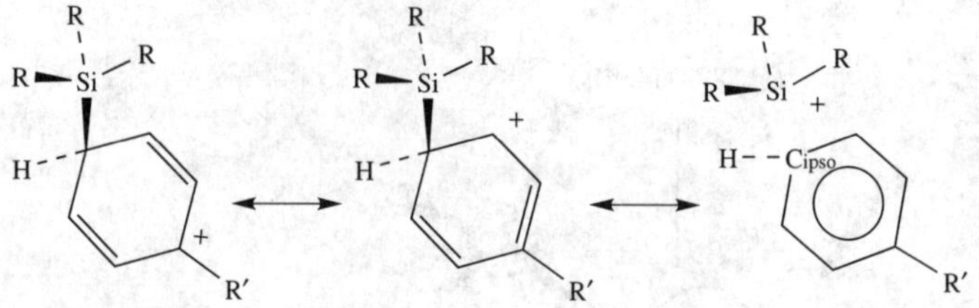

FIGURE 4. Resonance structures of silyl-substituted arenium ions

3. Interactions with Lewis bases: Binary and ternary complexes

Previously attempted preparations of silicenium ions in condensed phases gave structures with silicon covalently bonded to perchlorate, e.g. in triphenylsilyl perchlorate (**6**)[70], to pyridine, e.g. in [Me$_3$Si–pyridine]$^+$-I$^-$ (**7**)[69] (Figures 5 and 6) and to acetonitrile, e.g. in [Me$_3$Si–NCMe]$^+$ (**8**)[3b]. Recently, Reed and coworkers reported on the X-ray structure of the 1 : 1 (t-Bu)$_3$Si$^+$–water complex **9** (Figure 7), formed by protonation of (t-Bu)$_3$SiOH[71]. The latter complex featured a Si–O bond length of 1.779 Å, and a ^{29}Si NMR chemical shift of 46.7 ppm versus TMS[71] (see below).

Quantum theory sheds more light on the character of such silicenium ion–Lewis base complexes, e.g. on the nature of the bonds formed. Are they covalent? Do d-orbitals contribute? How large are the complexation energies and how do they change upon substitution? Can free tricoordinate silicenium ions be formed at all in π-donor or aromatic solvents?

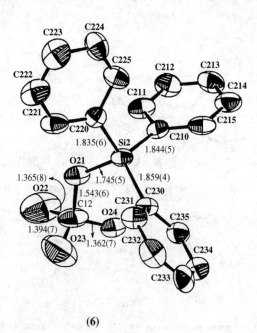

(**6**)

FIGURE 5. ORTEP drawing of the experimental structure of triphenylsilyl perchlorate (**6**). Hydrogen atoms are omitted for clarity. Bond lengths in Å. Reprinted with permission from Reference 70. Copyright (1987) American Chemical Society

10. Silicenium ions: Quantum chemical computations

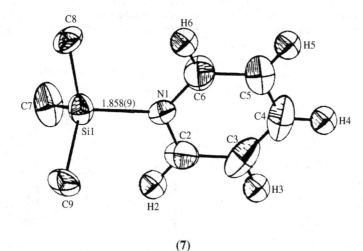

(7)

FIGURE 6. ORTEP drawing of the experimental structure of the trimethylsilyl–pyridine complex (7). Si–N bond length in Å. Reproduced by permission of Wiley-VCH from Reference 69

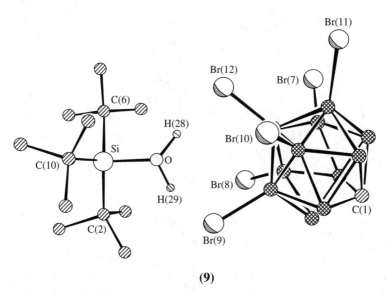

(9)

FIGURE 7. Experimental molecular structure of [t-Bu$_3$Si(OH$_2$)]$^+$ [Br$_6$CB$_{11}$H$_6$]$^-$. Bond distances (Å): Si–O 1.779(9), Si–C(2) 1.897(17), Si–C(6) 1.897(15), Si–C(10) 1.884(12). Bond angles (deg): C(2)–Si–C(6) 116.0(6), C(2)–Si–C(10) 116.1(6), C(6)–Si–C(10) 115.9(7), O–Si–C(2) 100.7(5). Reproduced by permission of the Royal Society of Chemistry from Reference 71

The myth that d-orbitals contribute significantly to silicon bonds has been discredited, e.g. by Schleyer and coworkers[6]. Natural population analysis (NPA)[85] of the bonds to silicon reveals no d-orbital participation in the SiX_3^+, SiX_4, SiX_5^-, SiX_6^{2-} species **1**, **10–12** (X = H), and **13–16** (X = F); see Figure 8. Even the tetrahedral SiH_4 (**10**) and SiF_4 (**14**) do not show the expected sp^3 hybridization, since H and F substituents, which are more electronegative than Si, withdraw more electrons from the higher-lying silicon p-orbitals. The bonds between silicon and first-row elements are both covalent and highly polar[6].

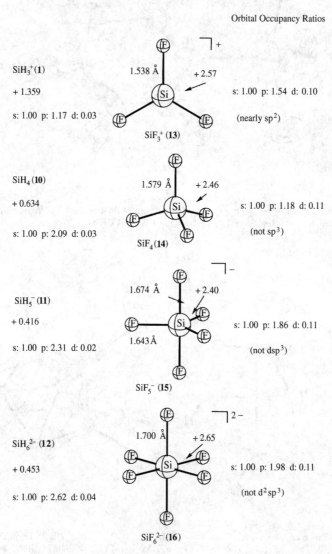

FIGURE 8. Silicon hybridizations and silicon NPA charges in species **1** and **10–16**. Reproduced by permission of Wiley-VCH from Reference 6

10. Silicenium ions: Quantum chemical computations

The complexation energies of silicenium ions with Lewis bases have been studied in detail by several groups[6,83,86]. Schleyer and coworkers reported a decrease in the interaction energies between silicenium ions and Lewis bases in the order: haloalkanes (or haloalkenes)[6] > arenes[73] > alkanes (e.g. methane)[6] > noble gases[6] (Table 3). Both the computed and experimental data show large complexation energies between silicenium ions and arenes, i.e. the silophilicity of arenes even competes with those of n-donors such as fluoroalkanes. It is not expected that such interactions will be suppressed totally in solution. Hence, the triethyl silicenium–toluene aggregate in **2** is best described as a covalent σ-complex with some ionic character[73]. The data in Table 3 show that the more stable the silicenium ion becomes the smaller is its complexation energy with Lewis bases.

TABLE 3. Interaction energies ΔE in binary complexes between silicenium ions and different Lewis bases (kcal mol^{-1}). Coordinating atoms are underlined

R$^+$	Base	PG^i	ΔE^a	Reference
SiH$_3^+$	HN$\underline{C}$	C_{3v}	61.4	83a
SiH$_3^+$	HC$\underline{N}$	C_{3v}	59.9	83a
SiMe$_3^+$	HC$\underline{N}$	C_{3v}	40.1	83a
SiH$_3^+$	$\underline{N}$H$_3$	C_{3v}	76.6	83a
SiMe$_3^+$	$\underline{N}$H$_3$	C_{3v}	54.4	83a
SiH$_3^+$	H$\underline{Cl}$	C_s	26.0	83a
SiMe$_3^+$	H$\underline{Cl}$	C_s	12.2	83a
SiH$_3^+$	H$_2\underline{O}$	C_s	56.4	83a
SiMe$_3^+$	H$_2\underline{O}$	C_s	38.8	83a
SiH$_3^+$	Me–$\underline{O}$–Me	C_s	70.2	83a
SiH$_3^+$	Me$\underline{Cl}$	C_s	40.6	83a
SiH$_3^+$	Me$\underline{F}$	C_s	44.9^b	6
SiMe$_3^+$	Me$\underline{F}$	C_s	30.0^b	6
SiH$_3^+$	Me$\underline{Br}$	C_s	43.1^c	6
SiMe$_3^+$	Me$\underline{Br}$	C_s	28.4^c	6
SiMe$_3^+$	MeC$\underline{N}$	C_{3v}	50.6	83a
SiH$_3^+$	cis-CHF=CH$\underline{F}$	C_1	36.5^b	6
SiMe$_3^+$	cis-CHF=CH$\underline{F}$	C_1	23.7^b	6
SiH$_3^+$	C$_6$H$_6$	C_s	54.8^d	73
SiMe$_3^+$	C$_6$H$_6$	C_s	31.1^e; 23.9^f	73
SiMe$_3^+$	4-Me-C$_6$H$_5$	C_s	34.2^e; 28.4^g	73
Si(NH$_2$)$_3^+$	H$_2\underline{O}$	C_s	32.6^d (28.8)^h	86
Si(NMe$_2$)$_3^+$	H$_2\underline{O}$	C_s	17.3^h	86
SiH$_3^+$	CH$_4$	C_1	16.4^d	6
SiH$_3^+$	He; Ne; Ar	C_{3v}	0.3; 6.8; 7.1	6^d
SiMe$_3^+$	He; Ne; Ar	C_{3v}	0.4; 4.4; 2.6	6^d

^{a}At HF/TZ+P, unless noted otherwise.
^{b}At MP2(fc)/6-31+G*.
^{c}At MP2(fc)/ECP; C,Si/Br: 4/7-ve-MWB-ECP and DZ+P valence basis; H: DZ basis.
^{d}At MP2(fc)/6-31G*.
^{e}At MP2(fc)/6-31G*//HF/6-31G*.
^{f}Experimental gas-phase value taken from A. C. M. Wojtyniak and J. A. Stone, *Int. J. Mass Spectrom. Ion Process.*, **74**, 59 (1986).
^{g}Experimental gas-phase value taken from J. M. Stone and J. A. Stone, *Int. J. Mass spectrom. Ion Process.*, **109**, 247 (1991).
^{h}HF/6-31G*.
^{i}Point-group symmetry.

The comprehensive studies by Cremer and coworkers of silicenium ions and their interactions with Lewis bases and solvent molecules are particularly noteworthy[83]. Nonspecific solvation effects on calculated ^{13}C and ^{29}Si NMR chemical shifts were probed via the PISA continuum model[87] (see discussion below). According to the authors, silicenium ions react in weak or normal nucleophilic media with one or more solvent molecules to form covalently bonded tetra- and penta-coordinate silicon complexes, in which any silicenium ion character is lost[83]. The complexation energies can be as high as 100 kcal mol^{-1} for ternary complexes (see Table 4), e.g. 104.7 kcal mol^{-1} for SiH$_3$(NH$_3$)$_2$$^+$ (at HF/6-311G**)[83]. Similar values have been reported by Schleyer and coworkers[6]. Competitive base coordination to silicenium ion reduces the effectiveness of substituent stabilization; e.g. note the trend in HF/TZ+P computed methyl stabilization energies (in kcal mol^{-1}): Me$_3$Si$^+$ (36.0) > Me$_3$Si$^+$/ClH (22.7) > Me$_3$Si$^+$/OH$_2$ (18.3) > Me$_3$Si$^+$/H$_3$N (13.8)[83a].

Cremer and coworkers have also investigated the effect of increasing the number of 'solvent molecules' in the model complexes SiH$_3$$^+$–(H$_2$O)$_n$ (n = 1, 2, 3, 5)[83a]. The equilibrium Si–O bond lengths in H$_3$Si–OH$_2$$^+$ (**17**) and in Me$_3$Si–OH$_2$$^+$ (**18**), computed at HF/6-31G*, are 1.859 Å and 1.910 Å, respectively[83a], compared to an experimental value of 1.779(9) Å in (t-Bu)$_3$Si–OH$_2$$^+$ (**9**)[71]. One would have expected that the presence of the sterically more demanding alkyl groups in **9** correlates with a longer Si–O bond length. Therefore, we attribute the short experimental Si–O bond distance in **9** to crystal packing forces. In any event, the covalent character of the Si–O bond results in a highly shielded δ ^{29}Si of 99.0 ppm (computed at IGLO/BII[83a]). Based on calculated electron density distributions $\rho(r)$, its associated Laplacian concentration $-\nabla^2\rho(r)$ and the energy density distribution $H(r)$[88], the authors differentiated between covalent and ionic bonding patterns for a large number of silicon compounds and silicenium ions[83a]. Analysis of the electron density in highly hydrated silicenium cations showed that only the axially but not the equatorially positioned water molecules are covalently bonded to silicon. Cremer and coworkers concluded: 'This suggests that only the 1 : 1 and 1 : 2 adducts of water are chemically distinct species, which can be isolated and investigated while additional water molecules do not enlarge the coordination sphere of Si but lead to solvation of the 1 : 2 adduct'[83a]. Thus, one can speak of different 'solvation shells' of the silicenium ion: the first one containing a maximum of two solvent molecules, and a second shell with up to 12 solvent molecules[83b].

Frenking's group investigated the solvation of the silaguanidinium ion, (H$_2$N)$_3$Si$^+$ (**19**), by a water molecule[86]. Compared to the H$_2$O binding energies of the parent H$_3$Si$^+$

TABLE 4. Interaction energies ΔE in ternary complexes between silicenium ions and different Lewis bases (kcal mol^{-1}). Coordinating atoms are underlined

R$^+$	Base (2x)^a	PG^b	ΔE^c	Reference
SiH$_3$$^+$	H$\underline{C}$N	D_{3h}	80.0	83a
SiH$_3$$^+$	$\underline{N}$H$_3$	D_{3h}	105.0; 122.3^d	83a, 6
SiMe$_3$$^+$	$\underline{N}$H$_3$	C_{3h}	65.7	83a
SiMe$_3$$^+$	H$_2$$\underline{O}$	C_s	47.7	83a
SiH$_3$$^+$	H$_2$$\underline{O}$	C_{2v}	79.9	83a
SiH$_3$$^+$	H$_2$$\underline{O}$ (3x)	C_{2v}	87.7	83a
SiH$_3$$^+$	H$_2$$\underline{O}$ (5x)	C_{2v}	97.8	83a

^{a}The number of the base molecules is given in parentheses (2 molecules, unless stated otherwise).
^{b}Point-group symmetry.
^{c}At HF/TZ+P, unless stated otherwise.
^{d}At MP2(fc)/6-31G*.

(57.7 kcal mol^{-1}) and of trialkyl-substituted silicenium ions, e.g. Me$_3$Si$^+$ (40.6 kcal mol^{-1}), the (NH$_2$)$_3$Si$^+$ cation interacts with water to a much smaller extent, i.e. the binding energy is 28.8 kcal mol^{-1}. Hexamethyl substitution on all three nitrogens reduces the water binding energy further to 17.3 kcal mol^{-1}. Thus, the silaguanidinium cation **19** should coordinate to solvent molecules less strongly than alkyl-substituted silicenium ions[86].

Trialkylsilicenium ions and their Lewis base complexes have also been studied by the groups of Olah[77,89], Reed[90,91] and Schleyer[6,73]. Olah and coworkers plotted (see Figure 9) the dependence of the complexation energy (computed at B3LYP/6-31G*//B3LYP/3-21G) in the Me$_3$Si$^+$–benzene complex (**20**) on the Si–C$_{ipso}$ distance[89] (see also Figure 4, with R = Me, R' = H). The authors found an 'anharmonic' energy potential with a computed equilibrium Si–C$_{ipso}$ distance of 2.250 Å[89]. The computed distance in **20** fits nicely the experimental value of 2.18 Å found in **2**[2a], and underlines the reliability of the computational level chosen in this study. The interaction energy between Me$_3$Si$^+$ and the benzene in **20** vanishes only at a distance as large as 4.5 Å[89], illustrating that the 'myth of non-coordinating counterions'[92] or solvent molecules is not valid. Thus, it seems very unlikely that 'free' tricoordinate silicenium ions can exist with Si–C bond distances of ca 2.2 Å to a nearby arene moiety[89].

Reed's group reported the X-ray structures of trialkyl-silicenium ions R$_3$Si$^+$ complexed to low-nucleophilic hexahalo carborane anions[90,91]. The authors varied systematically both the alkyl substituents R (R = Me, Et, i-Pr and t-Bu) and the halogen substitution of the X$_6$CB$_{11}$H$_6^-$ (X = Cl, Br, I) counterions. The degree of silicenium ion character in the

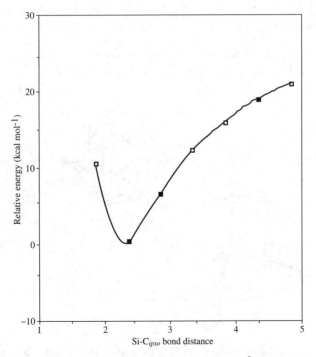

FIGURE 9. Plot of the relative energy vs the Si–C$_{ipso}$ bond distance (Å) in the benzene complex of Me$_3$Si$^+$ (**20**). Reproduced by permission of Elsevier Science from Reference 89

R_3Si^+ moiety was estimated from the sum of the bond angles around silicon, which should reach the ideal 360° value in a free tri-coordinate silicenium ion[90]. Surprisingly, the closest approach to planarity was found for the tri-*iso*-propyl derivatives, rather than for the tri-*tert*-butyl species. The authors also concluded that hyperconjugative stabilization of the cationic silicon centre is not negligible, although it is less pronounced than in carbocations[90]. The hexachloro carborane anion was found to be the least coordinating and, based on structural, energetic and NMR criteria, more than 50% silicenium ion character was assigned to the $i\text{-}Pr_3Si^{\delta+}$ $(Cl_6CB_{11}H_6)^{\delta-}$ complex[91] **21**. The experimental structure of **21** is shown in Figure 10. This viewpoint was challenged by Olah and coworkers, who regard such species as halonium ions[74,89].

The infrared (IR) spectrum of SiH_7^+ (**22**), formally a complex of H_3Si^+ and two molecules of H_2, was reported in 1993 by Okumura and coworkers[93]. The IR spectrum displayed a single band centred around 3866 cm^{-1}, which was red-shifted by 295 cm^{-1} with respect to free molecular hydrogen[93]. The absence of a second band in the 3500 to 4200 cm^{-1} region suggested a symmetric complex with the structure $H_2-SiH_3^+-H_2$, i.e. a planar SiH_3^+ species with two equidistant H_2 ligands bound to opposite faces of the silicenium ion[93]. Moreover, the authors suggested a preferred side-on ligation of the H_2 moieties, since such an arrangement would maximize the electron donation from the σ H−H bonds into the empty 3p-orbital on silicon[93]. Based on an empirical correlation between the lowering of the H−H stretching mode and the binding energy of molecular hydrogen−ion cluster complexes, a crude binding energy estimate of 7−9 kcal mol^{-1} was obtained[93]. The symmetric structure of SiH_7^+ is in sharp contrast to that found for the closely related CH_7^+ complex[94] (**23**), which exhibits an unsymmetric equilibrium geometry. Indeed, **23** has a structure of a protonated methane CH_5^{+}[95] (**24**) complexed with an additional, but weakly bound molecule of hydrogen, attached to one of the electron-deficient hydrogens of the three-centre−two-electron bond. Consequently, two different H−H stretching frequencies are observed in the IR spectrum of CH_7^+ [94].

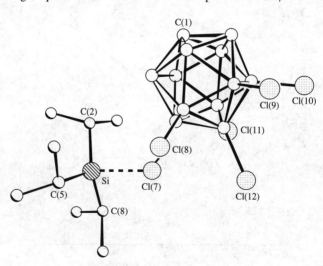

FIGURE 10. Experimental structure of $i\text{-}Pr_3Si^+(Cl_6CB_{11}H_6)^-$ (**21**). The Si−Cl(7) bond length is 2.323(3) Å. Reprinted with permission from Reference 91. Copyright (1996) American Chemical Society

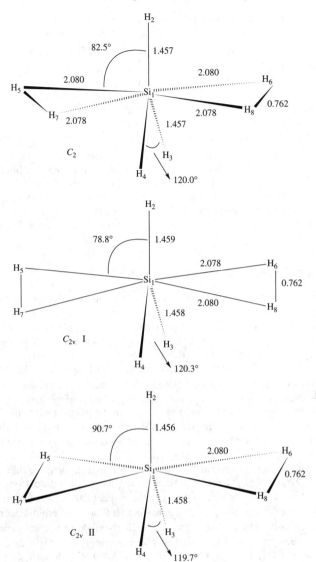

FIGURE 11. Three different structures of SiH$_7^+$ (**22**), optimized at CCSD(T). Bond lengths in Å, bond angles in deg. Reprinted with permission from Reference 98. Copyright (1994) American Chemical Society

Subsequent high level computations on both CH$_7^{+96}$ and SiH$_7^{+97,98}$ confirmed the conclusions drawn from the experimental data. The calculated structures of three different SiH$_7^+$ species are shown in Figure 11. The quite remarkable structural and energetic differences between the CH$_7^+$ and SiH$_7^+$ species could be related to the differences between the CH$_3^+$/SiH$_3^+$ and CH$_5^+$/SiH$_5^+$ pairs. The greater thermodynamic stability of the SiH_3^+ cation (see the previous discussion) requires less stabilization by the ligands, which is reflected in the first complexation energies: CH_3^+ binds molecular hydrogen by 40.0–42.0 kcal mol^{-1}[95] whereas the formation of SiH_5^+ (**25**) from H_3Si^+ and H_2 is much less exothermic (calculated: 10–15 kcal mol^{-1}[99]; experimental: 17.8 kcal mol^{-1}[100]). The calculated equilibrium geometries of SiH_5^+ and of CH_5^+ are also quite different: CH_5^+ is a rather fluxional species in which the electron deficiency is spread over all hydrogens via a non-classical three-centre–two-electron bond[95]. In contrast, SiH_5^+ is best described as a side-on complex between molecular hydrogen and the parent SiH_3^+ cation[99]. The second hydrogen complexation energies, however, show a contrasting trend: they are 4.6 kcal mol^{-1} for SiH_5^+[98,99] but only 1.2 kcal mol^{-1} for CH_5^+[95]. Thus, ligation of a second molecular hydrogen to SiH_3^+ is ca 50% as effective as the first binding, whereas for CH_3^+ second ligation contributes nearly zero. These remarkable differences can be rationalized on both electronic and steric grounds. The more electronegative carbon in CH_3^+ coordinates the first molecular hydrogen more strongly than SiH_3^+, but a hypothetical symmetric $H_2-CH_3^+-H_2$ complex is sterically more crowded than its silicon congener[98].

D. ^{29}Si NMR Chemical Shifts of Silicenium Ions

Nuclear magnetic resonance is very sensitive to structural variations. Thus, ^{29}Si NMR chemical shifts should be one of the best tools for measuring the degree of silicenium ion formation. IGLO NMR computations predict ^{29}Si NMR chemical shifts δ of 264.7 ppm and 346.7 ppm for SiH_3^+ and $SiMe_3^+$ (**26**), respectively[6]. The observed chemical shifts of trialkylsilicenium ions in arene solvents, however, were found at much higher fields (83.6 ppm for $SiMe_3^+$ in d_6-benzene[2b], 81.8 ppm for $SiEt_3^+$ in toluene[2b] and 109.8 ppm for i-Pr_3Si^+ in the solid state[3a]). A summary of calculated and experimental ^{29}Si NMR chemical shifts of silicenium ions and related compounds is given Table 5. Note, for instance, the computed large deshieldings in the isolated silyl-substituted silicenium ions (e.g. $\delta^{29}Si = 925$ ppm for $(Me_3Si)_3Si^+$)[77], compared to the experimental $\delta^{29}Si$ of 111.1 ppm in benzene solution[101]. Hence, the alleged preparation of free $(Me_3Si)_3Si^+$ cations in arene solutions[101] must be questioned.

Lambert and coworkers attributed the 200–250 ppm upfield deviation from the computed gas-phase values of the free silicenium ions to 'distortion from the plane to relieve external sources of strain'[2]. Contrary to these speculations, IGLO computations by Olah's[77] and Cremer's[83] groups predict a *downfield* shift upon pyramidalization, as shown in Figure 12. A similar deshielding of the ^{13}C NMR chemical shift upon pyramidalization of CH_3^+ has been reported by Schindler and Kutzelnigg[32b,c]. The distortion of the HCH bond angle from 120° to 117° results in a downfield shift of 20 ppm. Due to the pyramidalization, the empty p_π-orbital (either on C^+ or on Si^+) mixes with the σ-orbitals gaining partial s-character. The energy of the p_π-orbital is lowered, giving larger paramagnetic contributions since the local $\sigma-\pi$ excitation energies are reduced. Therefore, the central carbon or silicon nuclei in the XH_3^+ species (X = C, Si) are deshielded upon bending[32b,c].

^{29}Si NMR chemical shifts of silicenium ions show a remarkable long-range dependence on the distance to neighbouring molecules (either Lewis bases or solvent molecules). Schleyer and coworkers[6] first noticed that $\delta^{29}Si$ in the H_3Si^+–methane adduct **27** is considerably shielded at C–Si distances below 4.0 Å, e.g. 56.8 ppm at the equilibrium

TABLE 5. Compilation of calculated and measured δ^{29}Si NMR values (ppm) of silicenium ions and of related compounds

Species (PG)	δ^{29}Si, calc.[a]	Reference	δ^{29}Si, exp.
H$_3$Si$^+$ (D_{3h})	270.2 (310.8)	83a	
	264.7[b]	6	
H$_3$Si$^+$/Ng (C_{3v})	Ng = He: 253.7[b]	6	
	Ng = Ne: 188.2[b]		
	Ng = Ar: 112.1[b]		
H$_3$Si$^+$/NH$_3$ (C_{3v})	−28.7	83a	
H$_3$Si$^+$/(NH$_3$)$_2$ (D_{3h})	−127.7	83a	
H$_3$Si$^+$/NCH (C_{3v})	−26.2	83a	
H$_3$Si$^+$/(NCH)$_2$ (D_{3h})	−102.8	83a	
H$_3$Si$^+$/NCMe (C_{3v})	−35.3	83a	
H$_3$Si$^+$/OH$_2$ (C_s)	13.4	83a	
H$_3$Si$^+$/(OH$_2$)$_3$ (C_{2v})	−64.3	83a	
H$_3$Si$^+$/(OH$_2$)$_5$ (C_{2v})	−51.5	83a	
H$_3$Si$^+$/ClH (C_s)	26.7	83a	
MeH$_2$Si$^+$ (C_s)	299.1[b]	119	
Me$_2$HSi$^+$ (C_2)	325.1[b]	119	
Me$_3$Si$^+$ (C_{3h})	355.9	83a	
	346.7	6	
	354.2[c]	89	
Me$_3$Si$^+$/Ng (C_{3v})	Ng = He: 345.0[b]	6	
	Ng = Ne: 311.0[b]		
	Ng = Ar: 275.0[b]		
Me$_3$Si$^+$/C$_6$H$_6$ (C_s)	60[b,d], 80.0[c]	6, 89	83.6 (Ref. 2)
Me$_3$Si$^+$/NH$_3$ (C_{3v})	52.8	83a	
Me$_3$Si$^+$/NCH (C_{3v})	67.0	83a	
Me$_3$Si$^+$/NCMe (C_{3v})	52.2	83a	
Me$_3$Si$^+$/OH$_2$ (C_s)	99.0, 101.9[c]	83a, 89	
Me$_3$Si$^+$/ClH (C_s)	183.5	83a	
Me$_3$Si$^+$/BrH	154.4[c]	89	
Me$_3$Si$^+$/BrMe (C_s)	102.1[b] [−1.9][f], 117.5[c]	6; 89	
Me$_3$Si$^+$/FMe (C_s)	134.4[b] [49.6][f]	6	
Me$_3$Si$^+$/CFH=CFH (C_1)	178.0[b] [74.6][f]	6	
Me$_3$SiBr (C_{3v})	33.9[e]	89	26.4 (Ref. 118)
Et$_3$Si$^+$/(C_{3h})	371.3[c]	89	
Et$_3$Si$^+$/C$_3$H$_6$ (C_1)	70.0[c]	89	
Et$_3$Si$^+$/C$_6$H$_6$ (C_s)	104.6[g]	77	92.3 (Ref. 2)
Et$_3$Si$^+$/C$_6$H$_5$Me (C_s)	79.8[c]	89	81.8 (Ref. 2) 83.2 (Ref. 89)
NH$_2$SiH$_2^+$ (C_{2v})	122.2[h]	86	
(NH$_2$)$_3$Si$^+$ (D_3)	40.0[h]	86	
(NH$_2$)$_3$Si$^+$/OH$_2$ (C_1)	−21.0[h]	86	
Me$_3$SiOH	15.1[c]	89	
(Me$_3$Si)$_2$OH$^+$	67.6[c]	89	
Me$_3$Si$^+$/NCCH$_3$	51.5[c]	89	
(H$_3$Si)$_3$SiH	−139.2[g]	77	

(*continued overleaf*)

TABLE 5. (continued)

Species (PG)	δ^{29}Si, calc.[a]	Reference	δ^{29}Si, exp.
$(H_3Si)_3Si^+$	865.7[g]	77	
$(Me_3Si)_3SiH$	−124.2[g]	77	−117.4 (Ref. 77)
$(Me_3Si)_3Si^+$	925.3[g]	77	
$(H_3Si)_3Si^+/C_6H_6$	55.6[g]	77	

[a] All data at IGLO/BII/HF/6-31G* (values in parentheses at GIAO-MP2/BII), unless noted otherwise.
[b] IGLO/BII//MP2/6-31+G*, except the values of the H_3Si^+ and Me_3Si^+ noble gas (Ng) complexes, which are at IGLO/BII//MP2/6-31G*.
[c] IGLO/BII'//B3LYP/6-31G*.
[d] Using a HF/6-31G* geometry.
[e] IGLO/BII'//B3LYP/6-311+G*.
[f] Data in square brackets refer to the corresponding H_3Si^+ complexes.
[g] IGLO/BII'//HF/6-31G*.
[h] IGLO/BII+sp//MP2/6-31G*.

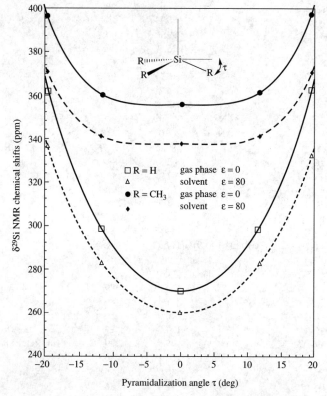

FIGURE 12. Dependence of the calculated δ^{29}Si chemical shifts of SiR_3^+ (R = H, CH_3) on the pyramidalization angle τ at Si. Solid lines denote IGLO/[7s6p2d/5s4p1d/3s1p] (basis II) calculations, dashed lines denote PISA-IGLO/[7s6p2d/5s4p1d/3s1p] (basis II) calculations. Reprinted with permission from Reference 83a. Copyright (1995) American Chemical Society

distance of 2.26 Å [MP2(fc)/6-31G* geometry; see Figure 13]. Thus, the highly deshielded ^{29}Si NMR values, which have been computed for 'naked' silicenium ions and also are anticipated for free silicenium ions in solution, are practically hard to achieve. A similar distance dependence of the ^{29}Si NMR chemical shifts in $H_3Si^+-S_n$ complexes (S = H_2O, NH_3 and HCN; $n = 1, 2$) was also reported by Cremer's group[83a] and their results are shown in Figure 14.

Even non-nucleophilic media, such as nobel gases, were shown to influence the ^{29}Si NMR chemical shifts of silicenium ions to a remarkable extent[6], e.g. in Me_3Si^+-Ng complexes, where Ng = He (**28**), Ne (**29**) and Ar (**30**) (Figure 15), compared to the isolated Me_3Si^+ ion.

A recent achievement by Cremer's group[83] is the direct incorporation of continuum solvent effects, via the PISA self-consistent reaction field (SCRF) model[87], in the computation of chemical shifts. Cremer and his coworkers found that computed solvent effects on ^{29}Si NMR chemical shifts can be as large as 20 ppm *upfield* in highly polar media ($\epsilon = 80$)[83]. The expected ^{29}Si NMR chemical shift range of silicenium cations[83a] as a function of the strength of interaction with nucleophiles, predicted by Cremer and his coworkers, are given in Table 6. The largest δ^{29}Si values, i.e. *ca*. 410 ppm, have been estimated for free alkyl-substituted silicenium ions in the gas phase or in non-coordinating solvents. Weakly interacting solvents, i.e. with donicities D^{84} ranging between 0 and 1,

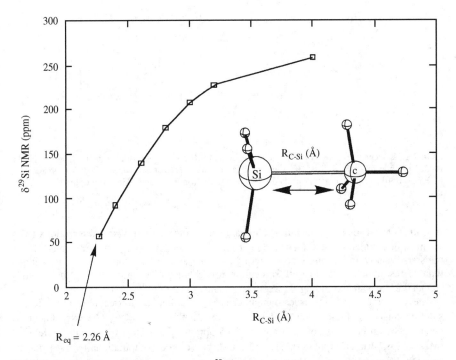

FIGURE 13. Dependence of the computed δ^{29}Si chemical shifts on the Si-C distance in the H_3Si^+-methane complex (**27**); data at IGLO/BII//MP2(fc)/6-31G* versus TMS. Reproduced by permission of Wiley-VCH from Reference 6

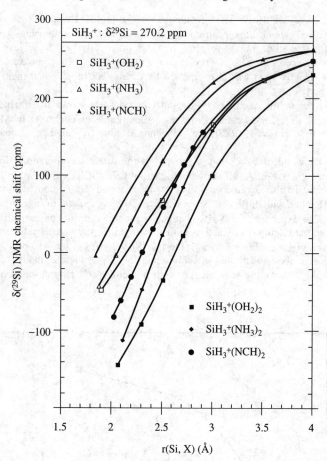

FIGURE 14. Dependence of the δ^{29}Si chemical shifts of SiH$_3$(S)$_n{}^+$ cations, for $n = 1$ (S = HCN, NH$_3$, H$_2$O, HCl) and for $n = 2$ (S = HCN, NH$_3$, H$_2$O), on the Si–S distance, IGLO/[7s6p2d/5s4p1d/3s1p] //HF/6-311G(d,p) calculations. Reprinted with permission from Reference 83a. Copyright (1995) American Chemical Society

lead to a partial loss of silicenium ion character due to van der Waals interactions (with the most deshielded δ^{29}Si values being between 220 and 390 ppm), implying some charge transfer. Weakly and strongly coordinating solvents, i.e. solvents having donicities larger than 1, result in a total loss of silicenium ion character, as indicated by the strongly shielded δ^{29}Si values ranging between -130 and $+110$ ppm[83a].

The preparation of a species which can be described as a free silicenium ion in solution has been finally achieved recently by Lambert's group[4]. The experimentally determined, strongly deshielded δ^{29}Si of 225.5 ppm for the trimesitylsilyl cation Mes$_3$Si$^+$ (Mes = 2,4,6-trimethylphenyl), **31**, could also be verified computationally[5] (see discussion below).

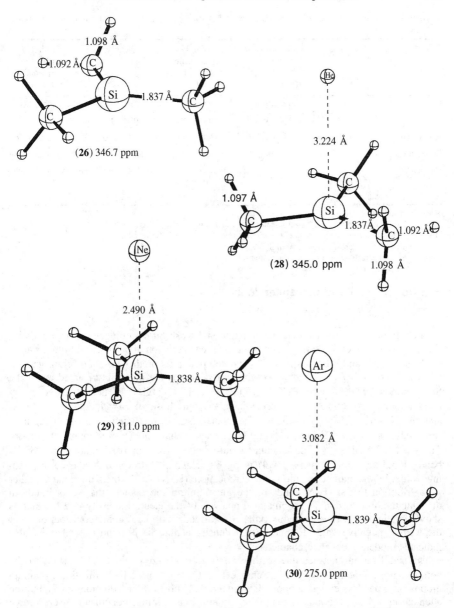

FIGURE 15. Computed δ^{29}Si chemical shifts and geometries of Me$_3$Si$^+$ (**26**) and noble gas complexes, **28–30**. Calculated NMR data at IGLO/BII//MP2(fc)/6-31G* versus TMS. Reproduced by permission of Wiley-VCH from Reference 6

TABLE 6. Estimated ranges of δ^{29}Si chemical shifts (ppm) for silicenium cations in the gas phase and in solution[a,b]

R_3Si^+	Gas phase	Non-coordinating solvent ($0 < \epsilon \leqslant 80$)	Weakly interacting solvents ($0 < D \leqslant 1$)	Weakly coordinating solvents ($0 < D \leqslant 10$)	Strongly coordinating solvents ($D > 10$)
R = H	300[c]	290–300	120–290	10–110	−130 to 10
R = CH$_3$	385[c]	370–385	200–370	90–190	−50 to 90
R = C$_2$H$_5$	410[c]	390–410	220–390	110–210	−30 to 110
Si−S interaction		solvation[d]	van der Waals[e]	weakly bonding	stronger bonding
Ion character		free silicenium ions	partial loss of silicenium ion character	total loss of silicenium ion character	

[a] Data from Reference 83a.
[b] The solvent is characterized by a dielectric constant ϵ and a donicity D (see Reference 84).
[c] Including estimated correlation corrections of 30 ppm (see Reference 83a for details).
[d] Solvation without any charge transfer from S to R_3Si^+.
[e] Van der Waals interactions imply some charge transfer.

E. π-Donor Stabilized Silicenium Ions

1. The silaguanidinium ion

In continuation of their previous computational work on Y-conjugated compounds[102], Frenking and his coworkers presented a detailed study on the silaguanidinium cation $(H_2N)_3Si^+$ (**19**) and related species[86]. The computations predict an equilibrium geometry with an overall D_3 symmetry, i.e. the amino groups are planar but rotated about the Si−N bonds by 19.6° at MP2/6-31G* (12.9° at HF/6-31G*)[86]. The calculated Si−N bond lengths of 1.658 Å are intermediate between those of Si=N double (1.568 Å) and Si−N single (1.748 Å) bonds. The planar D_{3h} form, which has one imaginary frequency, is only 0.7 kcal mol^{-1} higher in energy than the D_3 minimum (at MP4/6-311G**//MP2/6-31G*)[86]. The calculated out-of-plane rotation barrier about the Si−N bond of 5.5 kcal mol^{-1} (same level) is by 6.6 kcal mol^{-1} lower than the corresponding value in the parent guanidinium carbocation $(H_2N)_3C^+$. Thus, there is a higher π-contribution to the C−N bonds of the guanidinium ion than to the Si−N bonds in the silaguanidinium ion. The topological analysis of the electron density in **19**, which is shown in Figure 16, reveals highly polar Si−N bonds with large ionic contributions, but the weakly negative value for the energy density at the Si−N bond critical point also indicates moderate covalent character[86].

The stabilization energies due to the amino groups were evaluated by means of the homodesmotic reactions shown in Scheme 1[86]. The incremental stabilization by amino groups becomes less effective upon higher degree of substitution (equations 5–7), but the attenuation is less pronounced for the silicenium ions. Similar conclusions were drawn earlier by Apeloig[9,10]. The cumulative stabilizing effect of the three amino groups in the silaguanidinium ion, computed via the sum of equations 5–7, namely −77.6 kcal mol^{-1}, amounts to ca 44% of the corresponding value in the analogous guanidinium carbocation (−174.8 kcal mol^{-1}). The total stabilizing effect of triple substitution shows that amino groups (equation 8) are much more effective than methyl groups (equation 9). Thus, the authors concluded that amino-substituted silicenium ions are better targets in the search for stable silicenium ions in solution than alkyl-substituted species[86]. In particular, they

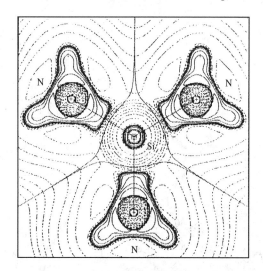

FIGURE 16. Contour line diagrams of the Laplacian distribution $\nabla^2\rho(r)$, at MP2/6-31G(d), of the silaguanidinium ion **19** in the SiN$_3$ plane. Dashed lines indicate charge depletion [$\nabla^2\rho(r) > 0$], solid lines indicate charge concentration [$\nabla^2\rho(r) < 0$]. The solid lines connecting the atomic nuclei are the bond paths; the solid lines separating the atomic nuclei indicate the zero-flux surfaces in the plane. The crossing points of the bond paths and zero-flux surfaces are the bond critical points r_b. Reproduced by permission of Wiley-VCH from Reference 86

suggested that the bis(pyrrolidino)(2,5-di-*tert*-butylpyrrolidono)silicenium ion **32**, whose calculated structure is shown in Figure 17, should benefit from both strong electronic stabilization and steric protection of the cationic centre[86].

Stabilization energies (kcal mol^{-1}) for:			X = S;	X = C	
XH$_3^+$ + H$_3$XNH$_2$	→	XH$_2$NH$_2^+$ + XH$_4$	−37.5	−103.4	(5)
XH$_2$NH$_2^+$ + H$_3$XNH$_2$	→	XH(NH$_2$)$_2^+$ + XH$_4$	−23.3	−44.3	(6)
XH(NH$_2$)$_2^+$ + H$_3$XNH$_2$	→	X(NH$_2$)$_3^+$ + XH$_4$	−16.8	−27.1	(7)
Σ (5) − (7)			−77.6	−174.8	
XH$_3^+$ + HX(NH$_2$)$_3$	→	X(NH$_2$)$_3^+$ + XH$_4$	−63.5	−159.3	(8)
XH$_3^+$ + HX(Me)$_3$	→	X(Me)$_3^+$ + XH$_4$	−36.0	−67.7	(9)

SCHEME 1

The silicon atom in the silaguanidinium ion bears a high positive charge (+2.24) while the nitrogens carry a high negative charge (−1.35)[86]. The large positive charge on silicon is not reflected in a downfield shift of the ^{29}Si NMR chemical shift of the silaguanidinium ion. IGLO calculations for (H$_2$N)$_3$Si$^+$ predict a highly shielded ^{29}Si NMR chemical shift of 40 ppm[86]. The upfield shift of the ^{29}Si NMR resonance, relative to SiH$_3^+$ (270.2 ppm), is due to a significant electron population of the formally empty 3p-orbital on silicon[86]. Recently, the effect of π-back donation on the NMR resonance of the X atom in a series of Me$_n$XCl$_{4-n}$ (X = C, Si, Ti; $n = 0-4$)[103] compounds has been found to vary

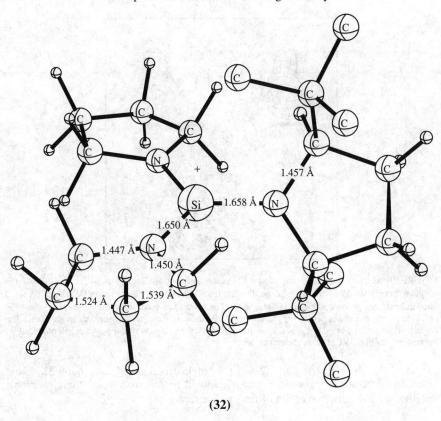

(32)

FIGURE 17. AM1 optimized geometry of the C_2 bis(pyrrolidono)(2,5-di-*tert*-butylpyrrolidino)silicenium ion (**32**), originally reported in C_1 symmetry in Reference 86. For clarity, all hydrogens of the *tert*-butyl groups were omitted from the plot. Bond lengths are given in Å

strongly for different atoms X. In general, there is no simple charge density–chemical shift relationship, although such correlations were found for series of closely related compounds, e.g. for the neutral $H_n SiMe_{4-n}$ compounds[83a].

2. Resonance stabilization by aromatic substituents

The alleged preparation of triphenylsilicenium (sityl) ions in different solvents, e.g. in sulpholane or dichloromethane[79,80], was reported by Lambert and coworkers in the mid-1980s. Based on conductance measurements and $^{35}Cl/^{37}Cl$ NMR spectra of the sityl perchlorate solutions, the authors concluded that the species which they observed were 'free' silicenium ions[80]. Unfortunately, Lambert and his coworkers did not report the ^{29}Si NMR spectra of the silyl compounds. Lambert's claims were challenged by Olah and coworkers[70] and by Eaborn[104] based on experimental data on related compounds. In particular, the reported silicon chemical shift of 3.0 ppm[70] was that of a normal covalently bound, i.e. non-dissociated, triphenylsilyl perchlorate. In the case of $Me_3Si-OClO_3$ ($\delta^{29}Si = 43.4$–47.0 ppm)[105] and $(i\text{-Pr}S)_3Si-OClO_3$ ($\delta^{29}Si = 18$ ppm)[81], similar highly shielded δ^{29}Si NMR chemical shifts have been measured, and these values were confirmed by the subsequent IGLO NMR computations on such silicenium ion–Lewis base complexes (see discussion above).

Nevertheless, aryl rings can provide significant conjugative stabilization to silicenium ions. A single phenyl substituent, for instance, stabilizes $C_6H_5SiH_2^+$ by 31.5 kcal mol^{-1} (HF/6-31G*)[6], compared to hydrogen in the parent SiH_3^+.

'Internally solvated' (chelated) silicenium ions, e.g. **33** shown in Figure 18, have been prepared, both by Corriu's[64,106] and by Belzner's[107] groups, who employed aryl rings with electron-donating amino groups in the *ortho*-positions to stabilize coordinatively the silicenium ion. In essence, the silicenium ion function interacts internally with the *ortho* dimethylamino groups to form a pentacoordinate siliconium ion. The observed δ^{29}Si NMR values in *internally solvated* silicenium ions are highly shielded (ranging from *ca* -10 to -60 ppm)[106,107]. IGLO computations on 2,6-diaminophenyl-substituted silicenium ions confirmed the upfield ^{29}Si NMR chemical shift (-80.5 ppm) in these pentacoordinate species[108].

A detailed theoretical study of phenyl- and anthryl-substituted silicenium ions was reported by Ottosson and Cremer[108]. The silicenium ion **34**, substituted with a 1,8-dimethylanthryl group, whose optimized geometry is shown in Figure 19, has a computed δ^{29}Si of 187 ppm, which is *ca* 150 ppm upfield from the corresponding gas-phase value of $SiHMe_2^+$ (334 ppm)[108]. Thus, the $-SiMe_2^+$ group in **34**, which is positioned between two methyl groups, retains much more silicenium ion character than amino-coordinated

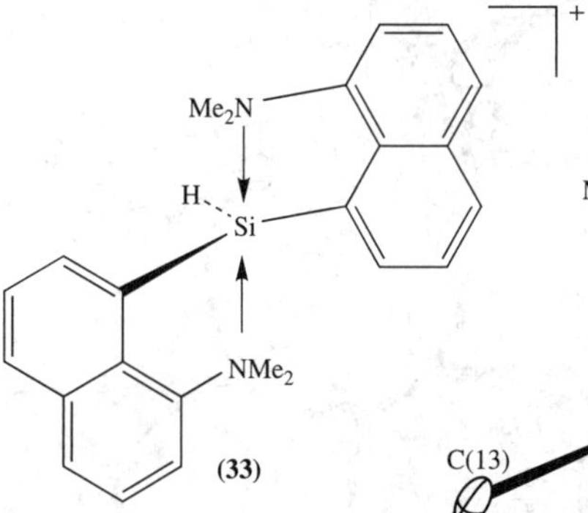

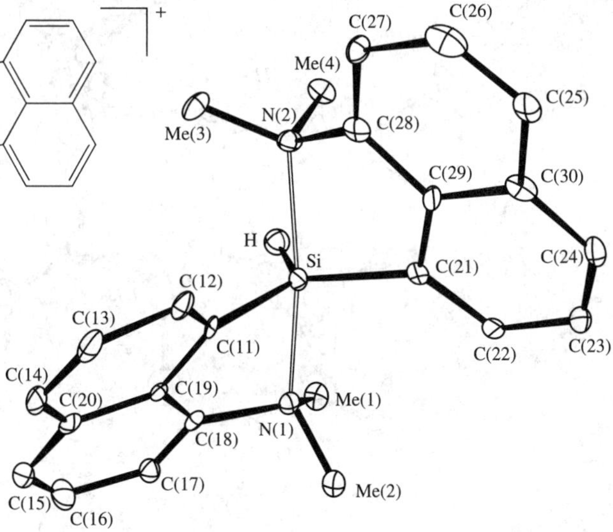

FIGURE 18. ORTEP drawing of the cation **33** showing the molecular conformation and atom numbering. Selected bond lengths (Å) and angles (deg): Si−H 1.73(12), Si−C(11) 1.90(1), Si−C(21) 1.85(1), Si−N(1) 2.08(1), Si−N(2) 2.06(1), N(1)−Si−N(2) 167.8(4), C(11)−Si−C(21) 118.4(6), C(11)−Si−H 126(4), C(21)−Si−H 114(4), Si−C(11)−C(19) 106.5(9), Si−C(21)−C(29) 110.2(9), N(1)−C(18)−C(19) 110.6(1.0), N(2)−C(28)−C(29) 113.4(1). Reproduced by permission of The Royal Society of Chemistry from Reference 106b

silicenium ions (e.g. **33**). Note also that the less nucleophilic methyl groups cannot stabilize the silicenium ion centre via dative lone pair-donation, as is the case with amino groups, although the steric shielding might be similar. Maerker, Kapp and Schleyer showed that even methane can coordinate to silicenium cations[6], and the intramolecular methyl group interactions in **34** are similar.

Very recently, Lambert and Zhao[4] prepared the trimesityl silicenium ion **31** by protonation of allyltrimesityl silane. It had been found earlier that protonation of allylsilanes removes the allyl group (as its corresponding alkene), and this is a good way to generate silicenium ion intermediates[109]. The cleavage of the C—Si bond is rationalized in terms of extensive C—Si hyperconjugation in the initially formed carbocation[6,73]. Unfortunately, Lambert and Zhao were not able to obtain a crystal structure of the trimesityl silicenium tetrakis(pentafluorophenyl)borate salt[4]. However, they could measure the ^{29}Si NMR spectrum of **31** and observed a highly deshielded ^{29}Si NMR absorption at $\delta = 225.5$ ppm[4]. The extent of solvent coordination was further probed by using mixtures of different aromatic solvents, but no significant change in the ^{29}Si chemical shifts was observed[4]. The authors concluded that nucleophilic coordination to the arene moieties is almost completely inhibited sterically by the *ortho*-methyl groups of the mesityl substituents. C_6D_6/CH_3CN (1/3) and C_6D_6/Et_3N (1/1) solvent mixtures, in contrast, result in δ^{29}Si of 37.0 and 47.1 ppm, respectively[4]. Hence, the much smaller alkyl nitriles and amines can bind directly to the silicenium ion centre of **31**.

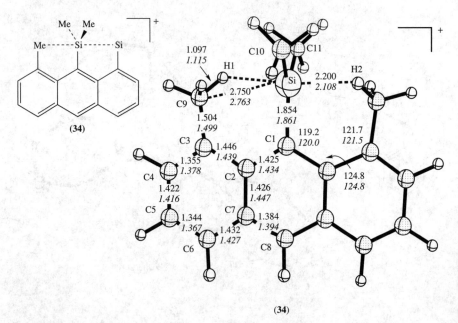

FIGURE 19. HF/6-31G* (normal print) and B3LYP/6-31G* (italics) geometrical parameters of cation **34**. Selected bond lengths (Å), bond and torsion angles (deg): Si—C(10) 1.869 *1.878*, C(10)—Si—C(1) 123.0 *123.1*, H(1)—Si—H(2) 175.5 *175.0*, C(10)—Si—C2—C3 84.3 *82.4*, Si—C(1)—C(2)—C(3) 0.0 *0.3*, C(1)—C(2)—C(3)—C(9) −4.9 *−6.0*, H(1)—C(9)—C(3)—C(2) 50.9 *47.3*, H(1)—C(9)—Si—C(10) 115.3 *110.8*, H(1)—C(9)—Si—C(11) 1.0 *2.9*. Reprinted with permission from Reference 108. Copyright (1996) American Chemical Society

Subsequent *ab initio* computations by Müller and coworkers[5] on triarylsilicenium ions confirmed the formation of a free tricoordinate trimesitylsilicenium ion **31** in solution. The calculated equilibrium geometries of triarylsilicenium ions exhibit considerably twisted aryl rings and propeller-like shapes. The C−H bonds of the *ortho*-methyl groups, however, are too far from the silicon centre to interact (2.990 Å at B3LYP/6-31G*)[5]. Employing B3LYP/6-31G* optimized geometries, the following ^{29}Si NMR chemical shifts were calculated, using GIAO-B3LYP/6-311+G(2df,p) [Si], 6-31G* [C,H]//B3LYP/6-31G*: 205.0 (triphenylsilicenium) and 243.9 ppm (trimesitylsilicenium **31**), respectively[5]. The 18.4 ppm difference between experiment and theory for **31** was attributed to deficiencies in the computational method[5], since GIAO-DFT seems to overestimate the paramagnetic contributions to $\sigma(^{29}Si)$. In the parent triphenylsilicenium ion there is a strong dependence of $\delta^{29}Si$ on the twist angle; the larger the dihedral angle the larger the downfield shift of $\delta^{29}Si$. A $\delta^{29}Si$ of 290.2 ppm is predicted for the D_{3h} symmetric all-perpendicular triphenylsilicenium ion geometry[5]. The authors concluded: 'This computational study strongly corroborates the experimental finding that the trimesitylsilicenium ion (**31**) is the first free trigonal silyl cation lacking any coordination to solvent or counterion'[5].

3. Intramolecular π-stabilization

In 1995, Schleyer and coworkers reported the results of calculations for the silicon congeners of a number of non-classical carbocations: the 2-norbornyl cation, $C_7H_{11}^+$ (**35**), the 7-norbornadienyl cation, $C_7H_7^+$ (**36**) and cyclopropylcarbinyl cation, $C_4H_7^+$ (**37**)[6]. The MP2(fc)/6-31G* calculations predicted that the silicenium ion analogs, **38**–**40** are more stable than the corresponding parent carbocations **35**–**37**, respectively. Thus, the 6-sila-2-norbornyl cation (**38**) is more stable than **32** by 18.1 kcal mol^{-1}; the 7-sila-7-norbornadienyl cation **39** is by 15.1 kcal mol^{-1} more stable than **36** and the cyclopropylsilicenium ion (**40**) is by 12.6 kcal mol^{-1}, more stable than **37**[6]. IGLO/BII'//MP2(fc) /6-31G* chemical shift calculations predicted that the sila-analogs of the two bicyclic carbocations should show unusually shielded ^{29}Si NMR chemical shifts of 1 ppm for **38** and −148.6 ppm for **39**[6]. Thus, the higher coordination number of the silicon atoms in these cations is reflected directly in their ^{29}Si NMR resonances.

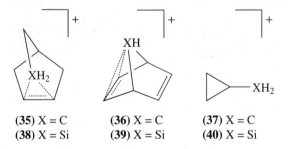

(**35**) X = C (**36**) X = C (**37**) X = C
(**38**) X = Si (**39**) X = Si (**40**) X = Si

More recently, Auner's and Schleyer's groups in a joint effort reported the experimental observation of an alkyl-substituted 6-sila-2-norbornyl cation[110]. The 6,6-Dimethyl-5-neopentyl-6-sila-2-norbornyl cation **41** was prepared by a hydride transfer reaction via the 'π-route'[111] from a suitable cyclopentenyl silane derivative at room temperature (equation 10). The toluene solution of the tetrakis(pentafluorotetraborate) salt of **41** showed a sharp single peak at $\delta(^{29}Si)$ of 87.7 ppm versus TMS[110]. As this value is also in the range of silicenium ion–arene σ-complexes[73], and no X-ray quality crystals could be obtained, the 6-sila-2-norbornyl cation structure of **41** had to be confirmed by quantum chemical computations. Calculations using the hybrid density functional B3LYP/6-31G* method confirmed that **41** indeed has the 'closed' form of a true sila-analog of the non-classical 2-norbornyl cation. Calculations of the effect of interaction with a benzene molecule on the relative stabilities of an 'open' (**42**) and of a 'closed' (**43**) cation show that, although the 'open' arenium ion form benefits much more from solvation by benzene (20.8 kcal mol^{-1}) than the closed form (5.0 kcal mol^{-1}), the latter remains by 6.3 kcal mol^{-1} more stable (see Figure 20)[110].

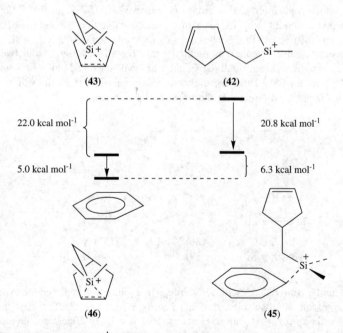

Due to computational savings, the 'open' and 'closed' forms of the 6,6-dimethyl-6-sila-2-norbornyl cation, **42** and **43**, the 'closed' 5,6,6-trimethyl-6-sila-2-norbornyl cation (**44**) and their corresponding benzene complexes **45**, **46** and **47**, respectively, were optimized

FIGURE 20. Energetics (kcal mol^{-1}) of benzene complexation to cations **42** and **43**, forming complexes **45** and **46**, respectively. Reproduced by permission of Wiley-VCH from Reference 110

as reasonable realistic theoretical models[110] of cation **41** in toluene solution. Despite the unsymmetrical substitution pattern, an almost symmetrically bridged 'closed' silicenium ion is predicted to be the most stable structure at B3LYP/6-31G* of **44**, as shown in Figure 21[110]. Benzene coordination has only minor structural consequences, resulting mainly in a slight lengthening (by 0.13 to 0.16 Å) of the Si6−Cl/C2 bonds in **47** compared to those in **44**[110]. The B3LYP/6-31G* optimized geometries of **44, 45** and **47** are shown in Figure 21.

The most important support for the 'closed' structure of **41** was given, however, by comparison of the computed and measured ^{29}Si and ^{13}C NMR chemical shifts and the $^1J_{CH}$ coupling constants. In particular, the deshielding of the 'olefinic' carbons C1 and C2 upon ionization are only reproduced well by the 'closed' structures **43** and **44**. Similarly, the shielded methyl groups are found only in the 'closed' structures. In contrast, the computations predict a strongly deshielded methyl signal for the 'open' silicenium ion isomer **42**[110]. The non-classically bridged bicyclic nature of silicenium ion **41** is further confirmed by the computed $^1J_{Cl/C2-H}$ of 167.4 Hz (in **44**), which is in reasonable agreement with the experimental $^1J_{Cl/C2-H}$ of 174 Hz[110] (in **41**). Relative to the corresponding silane,

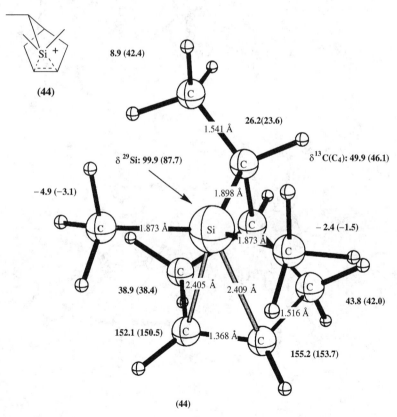

FIGURE 21. B3LYP/6-31G* optimized geometries of species **44, 45** and **47**. Bond lengths in Å; computed ^{13}C and ^{29}Si NMR data in boldface (**44**: IGLO-DFPT/BIII; **45, 47**: IGLO-DFPT/BII), experimental NMR data for species **41** given in parentheses. Reproduced by permission of Wiley-VCH from Reference 110

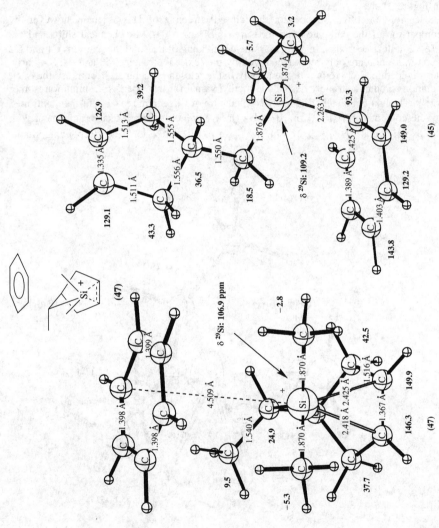

FIGURE 21. (continued)

this is an increase in $^1J_{CH}$ of by 22 Hz. The corresponding difference in $^1J_{CH}$ between the 2-norbornyl cation and cyclopentene is 26.2 Hz (experimental data)[110]. The authors concluded about **41**: '... there is strong evidence for the preparation of the first intramolecularly π-stabilized silanorbornyl cation in solution that is not coordinated to solvent or a counterion'[110]. An alternative description of **41** is: 'a nearly symmetrically bridged β-silyl carbocation with siliconium ion character'[110]. Interestingly, such non-classical pentavalent silicon ('siliranium') ions have been postulated before by Knölker and coworkers in the sila-Wagner–Meerwein rearrangement of organosilicon compounds[112].

F. Stabilization by Organoboryl Groups

It has been shown theoretically by several groups that the tetra- or pentacoordination of trialkylsilicenium ions to counterions or solvent molecules in solution cannot be easily prevented. Thus, the preservation of a major silicenium ion character might be obtained only by combining internal stabilization and steric blocking of the silicon centre. Cremer and coworkers have shown that strong electron-donor groups, such as amino groups (see also Section IV. E.1)[86], are not entirely suitable for this purpose, since π-donation to the formally empty $3p_\pi$ (Si$^+$) orbital is so extensive that the silicenium ion character is essentially lost[108,113]. This is indicated by the strongly shielded ^{29}Si NMR chemical shifts, usually around 40 ppm or lower[86].

Cremer and coworkers have recently explored the stabilization of silicenium ions by electropositive substituents[83,113]. The trimethylsilyl or tris(trimethylsilyl)silyl groups, which have also been used experimentally[2b,101], are thought to stabilize an electron-deficient centre mainly by hyperconjugation and by inductive effects. However, the calculations reveal that the rather long Si–Si$^+$ bonds (ca 2.34–2.40 Å) in (Me$_3$Si)$_3$Si$^+$ result in little hyperconjugative stabilization by the terminal groups. Figure 22 shows the optimized geometry of (Me$_3$Si)$_3$Si$^+$.

In contrast to silyl groups, boron substituents have the following advantages[113]: (a) boron has a lower electronegativity (2.01) than carbon (2.50), which results in smaller distortions of the frontier molecular orbital energies; (b) the Si–B bonds (ca 1.98 to 2.03 Å) are shorter than Si–Si bonds, facilitating interactions between the pseudo-π orbitals of the substituent's alkyl groups and the formally empty $3p_\pi$(Si$^+$) orbital, transmitted via the empty $2p_\pi$(B) orbital.

Following this reasoning, Cremer's group studied computationally several organoboryl-substituted silicenium ions[113]. The tris(dimethylboryl)silicenium ion, (Me$_2$B)$_3$Si$^+$ (**48**), is by 61.9 kcal mol^{-1} more stable than the parent H$_3$Si$^+$ (B3LYP/6-31G*), compared to a stabilization energy of only 43.2 kcal mol^{-1} provided by three methyl groups as in Me$_3$Si$^+$[113]. The (Me$_2$B)$_3$Si$^+$ cation adopts a twisted D_3 equilibrium structure, where the BC bonds are rotated by 86° (relative to a planar structure), rather close to the 90° rotated D_{3h} structure. Presumably, the slight distortion from the ideal orthogonal arrangement reduces the steric repulsions between the organoboryl groups. The barrier to rotation around the Si–B bonds is 6.7 kcal mol^{-1} per each substituent[113]. If one assumes that the rotational barrier is mainly due to the loss of the hyperconjugative stabilization resulting from the interaction of the $3p_\pi$(Si$^+$) orbital and the B–CH$_3$ bond, then 20 kcal mol^{-1} can be attributed to this stabilization while ca 40–42 kcal mol^{-1} of the total 61.9 kcal mol^{-1} stabilization energy of **48** (versus H$_3$Si$^+$) can be attributed to inductive contributions[113].

The tris-(9-borabicyclo[3.3.1]nonyl)silicenium ion, (9-BBN)$_3$Si$^+$ (**49**), adopts a D$_{3h}$ equilibrium geometry, perhaps due to the greater rigidity of the bicyclic cages. Based on non-bonded H–H distances it was estimated that an opening with a diameter of 2.34 Å, through which nucleophiles might interact with the silicenium ion centre, exists in **49**[113].

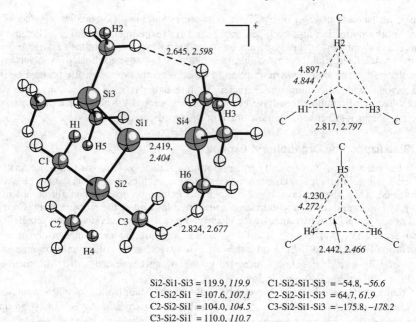

Si2-Si1-Si3 = 119.9, *119.9* C1-Si2-Si1-Si3 = −54.8, *−56.6*
C1-Si2-Si1 = 107.6, *107.1* C2-Si2-Si1-Si3 = 64.7, *61.9*
C2-Si2-Si1 = 104.0, *104.5* C3-Si2-Si1-Si3 = −175.8, *−178.2*
C3-Si2-Si1 = 110.0, *110.7*

FIGURE 22. HF/6-31G* and B3LYP/6-31G* (values in italics) optimized geometries of $(Me_3Si)_3Si^+$. Bond lengths in Å, bond angles in deg. The two inserts on the right side indicate the dimensions of the opening above and below the central Si atom formed by non-bonded hydrogen contacts. Reprinted with permission from Reference 113. Copyright (1997) American Chemical Society

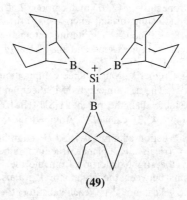

(49)

The complexation of **49** with benzene, forming **50**, has been further examined at HF/6-31+G*[113]. Frequency calculations confirmed that the C_1 structure, which is shown in Figure 23, is a minimum. The shortest Si−$C_{benzene}$ bond length in **50** is 2.57 Å, which is 0.1–0.3 Å longer than silicon−carbon distances in other silicenium ion−arene complexes. At HF/6-31G*, the complexation energy of benzene to **49** is only 4.4 kcal mol^{-1}, which is roughly 20 kcal mol^{-1} lower than the complexation energies of tertiary alkylsilicenium ions to benzene or toluene. The low interaction energy of **49** with benzene was attributed by Cremer and coworkers to the large stabilization of silicenium ion **49** by the three 9-BBN substituents[113]. However, the coordination of **49** to benzene forces the three boryl groups towards each other, increasing the steric interactions within the silicenium cation moiety. Thus, one might speculate whether the energetically unfavourable steric effects in **50** partially compensate the stabilization provided by the interaction of the silicenium ion with benzene. In any event, a low solvent interaction of the silicenium ion with benzene is the outcome. The weak interaction of **49** with benzene was further confirmed by the computed geometries, e.g. the relatively long Si−C distance and the small C−C bond alternation in the benzene moiety (see Figure 23)[113].

Due to computational limitations, the ^{29}Si NMR chemical shift could be computed only for the benzene complex of $(Me_2B)_3Si^+$, whose interaction energy with benzene of 8.9 kcal mol^{-1} is almost two-fold greater than that of **49**[113]. The computed δ^{29}Si NMR chemical shift for the $(Me_2B)_3Si^+$-benzene complex is 116.4 ppm versus TMS, corresponding to ca 28% silicenium ion character[113]. A better model for the (9-BBN)$_3$Si$^+$-benzene complex was constructed by adjusting the $(Me_2B)_3Si^+$-benzene geometry to that of the (9-BBN)$_3$Si$^+$-benzene complex, i.e. the bicyclic cages of the latter complex were 'deleted', and the free valencies were saturated by hydrogens, whose positions were partially optimized. The estimated silicenium ion character for this model of the (9-BBN)$_3$Si$^+$-benzene complex is 41%. This estimation is based on the computed ^{29}Si NMR chemical shift of 189 ppm in $(Me_2B)_3Si^+$-benzene, compared to δ of 571.8 ppm for the free $(Me_2B)_3Si^+$ cation[113]. The reported 41% silicenium-ion character is much higher than estimated for any other R_3Si^+-solvent complex. However, note that even the (9-BBN)$_3$Si$^+$ ion (**49**) is not a 'free' silicenium ion in benzene solution, since it forms a rather weak cation–solvent complex.

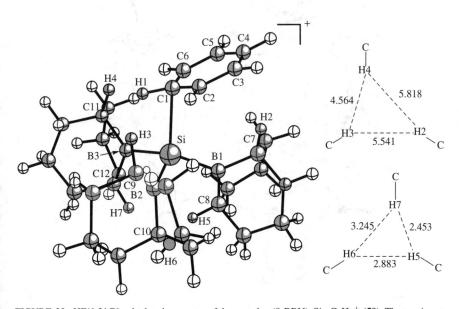

FIGURE 23. HF/6-31G* calculated geometry of the complex (9-BBN)$_3$Si−C$_6$H$_6^+$ (**50**). The two inserts on the right side indicate the dimensions of the opening above the central Si atom formed by atoms H2, H3, H4 (black atoms), and below the central Si atom by atoms H5, H6, H7 (black atoms). Distances in Å. Reprinted with permission from Reference 113. Copyright (1997) American Chemical Society

V. CONCLUDING REMARKS AND OUTLOOK

The experimental search for free stable silicenium ions in solution seems to have finally succeeded, at least for triaryl-substituted species[4,5]. However, computational studies will continue to augment, to verify and to guide the interpretations of experimental findings, as well as to suggest new experiments for preparing other silicenium ions. Schleyer and coworkers, for instance, have recently explored a silicenium ion 'within a cage', namely

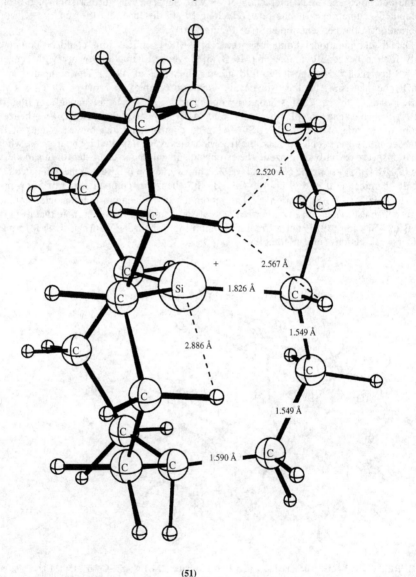

(51)

FIGURE 24. B3LYP/6-31G* optimized geometry of a fully 'caged' silicenium ion **51**. Bond lengths in Å. Non-bonded H–H and H–Si contacts (Å) shown as dashed lines

$C_{17}H_{29}Si^+$ (**51**, C_{3h} symmetry), which corresponds to a local minimum at HF/3-21G*[114]. The B3LYP/6-31G* optimized geometry of **51** is shown in Figure 24. In **51**, the silicenium ion centre is extremely shielded sterically, and the closest non-bonded H—H distances at the peripheral methylene linkages are 2.52–2.57 Å (Figure 24). GIAO-SCF/3-21G//B3LYP/6-31G* chemical shift computations for **51** predict a strongly deshielded $\delta^{29}Si$ of 303 ppm[114]. However, the very efficient steric protection of the deeply buried ionic centre might also hamper the ion's preparation!

Another future direction might be the theoretical exploration of the reactions of silicenium ions with other species, e.g. the application of silicenium ions in cationic polymerization. Work on the heavier analogs of silicenium ions has also progressed significantly during the last three years[115], leading to the isolation and X-ray characterization of the first aromatically stabilized trigermanium ion[116,117]. The fruitful interplay between experiment and theory provides a most promising pathway to new chemical structures and insights, as was demonstrated so clearly for silicenium ions.

VI. REFERENCES

1. S. Patai and Z. Rappoport (Eds.), *The Chemistry of Organic Silicon Compounds*, Wiley, New York, 1989.
2. (a) J. B. Lambert, S. Zhang, C. L. Stern and J. C. Huffman, *Science*, **260**, 1917 (1993),
   (b) J. B. Lambert and S. Zhang, *J. Chem. Soc., Chem. Commun.*, 383 (1993).
3. (a) C. A. Reed, Z. Xie, R. Bau and A. Benesi, *Science*, **262**, 402 (1993),
   (b) Z. Xie, D. J. Liston, T. Jelinek, V. Mitro, R. Bau and C. A. Reed, *J. Chem. Soc., Chem. Commun.*, 384 (1993).
4. J. B. Lambert and Y. Zhao, *Angew. Chem.*, **109**, 389 (1997); *Angew. Chem., Int. Ed. Engl.*, **36**, 400 (1997).
5. T. Müller, Y. Zhao, and J. B. Lambert, *Organometallics*, **17**, 278 (1998).
6. C. Maerker, J. Kapp and P. v. R. Schleyer, in *Organosilicon Chemistry: From Molecules to Materials*, Vol. II (Eds N. Auner and J. Weis), VCH, Weinheim, 1995, pp. 329–359.
7. (a) B. J. DeLeeuw, R. S. Grev and H. F. Schaefer III, *J. Chem. Educ.*, **69**, 441 (1992).
   (b) H. F. Schaefer III, *Acc. Chem. Res.*, **15**, 283 (1982).
8. See also the first chapter by Y. Apeloig and M. Karni in this volume.
9. Y. Apeloig, in *Heteroatom Chemistry* (Ed. E. Block), VCH, Weinheim, 1990, pp. 27–46.
10. Y. Apeloig, in *The Chemistry of Organic Silicon Compounds*, Vol. 1 (Eds. S. Patai and Z. Rappoport), Wiley, New York, 1989, p. 57.
11. M. S. Gordon, in *Molecular Structure and Energetics* (Eds. J. F. Liebman and A. Greenberg), VCH, Deerfield Beach, FL, 1986, p. 101.
12. S. Nagase, *Acc. Chem. Res.*, **28**, 469 (1995).
13. K. Morokuma, in *Organosilicon and Bioorganosilicon Chemistry: Structure, Bonding, Reactivity and Synthetic Applications* (Ed. H. Sakurai), Ellis Horwood, Chichester, 1985, p. 33.
14. Leading texts are:
    (a) J. M. Seminario and P. Politzer (Eds.), *Modern Density Functional Theory: A Tool for Chemistry; Theoretical and Computational Chemistry*, Vol. 2, Elsevier, Amsterdam, 1995.
    (b) D. R. Salahub, R. Fournier, P. Mlynarski, I. Papai, A. St-Amant and J. Ushio, *Density Functional Methods in Chemistry*, Springer-Verlag, Berlin, 1991.
    (c) R. M. Dreizler and E. K. Gross, *Density Functional Theory*, Springer-Verlag, Berlin, 1990.
    (d) R. G. Parr and Y. Wang, *Density Functional Theory of Atoms and Molecules*, Oxford University Press, Oxford, 1989.
15. T. Ziegler, *Chem. Rev.*, **91**, 651 (1991).
16. L. Greengard, *Science*, **265**, 909 (1994).
17. M. C. Strain, G. E. Scuseria and M. J. Frisch, *Science*, **271**, 51 (1996).
18. (a) T. Darden, D. M. York and L. Pedersen, *J. Chem. Phys.*, **98**, 10089 (1993).
    (b) D. M. York, T. Darden and L. Pedersen, *J. Chem. Phys.*, **99**, 8345 (1993).
    (c) D. M. York, A. Wlodawer, L. G. Pedersen and T. A. Darden, *Proc. Natl. Acad. Sci. U.S.A.*, **91**, 8715 (1994).

19. (a) J. Gao, *Acc. Chem. Res.*, **29**, 298 (1996).
    (b) J. Gao, in *Reviews in Computational Chemistry*, Vol. 7 (Eds. K. B. Lipkowitz and D. B. Boyd), Verlag Chemie, Weinheim, 1996, p. 119.
20. R. L. Martino, C. A. Johnson, E. B. Suh, B. L. Trus and T. K. Yap, *Science*, **265**, 902 (1994).
21. T. Clark, *A Handbook of Computational Chemistry. A Practical Guide to Chemical Structure and Energy Calculations*, Wiley, New York, 1985.
22. W. J. Hehre, L. Radom, P. v. R. Schleyer and J. A. Pople, *Ab Initio Molecular Orbital Theory*, Wiley, New York, 1986.
23. K. B. Lipkowitz and D. B. Boyd (Eds.), *Reviews in Computational Chemistry*, Vols. 1–8, Verlag Chemie, Weinheim, 1990–1996.
24. P. v. R. Schleyer, H. F. Schaefer III and N. C. Handy (Eds.), *Encyclopedia of Computational Chemistry*, Wiley, New York, to appear in 1998. Look also at: http://www.ccc.uni-erlangen.de/info/ECC/
25. Browse the following URL: http://hackberry.chem.niu.edu/ECCC4/
26. Browse at http://www.ccc.uni-erlangen.de/info/JMOLMOD/jmolinfo.html
27. J. Gauss, *Ber. Bunsenges. Phys. Chem.*, **99**, 1001 (1996).
28. J. C. Slater, *Phys. Rev.*, **81**, 385 (1951).
29. N. C. Handy, in *Lecture Notes in Quantum Chemistry II*, Vol. 58 (Ed. B. O. Roos), Springer-Verlag, Berlin–Heidelberg, 1994, p. 91.
30. (a) P. C. Hohenberg and W. Kohn, *Phys. Rev.*, **B136**, 864 (1964).
    (b) W. Kohn and L. J. Sham, *Phys. Rev.*, **A140**, 1133 (1965).
    (c) A recent review is: P. C. Hohenberg, W. Kohn and L. J. Sham, *Adv. Quantum Chem.*, **21**, 7 (1990).
31. See, for example, V. G. Malkin, O. L. Malkina, L. A. Eriksson and D. R. Salahub, in *Modern Density Functional Theory: A Tool for Chemistry; Theoretical and Computational Chemistry*, Vol. 2 (Eds. J. M. Seminario and P. Politzer), Elsevier, Amsterdam, 1995, p. 273 and references cited therein.
32. (a) W. Kutzelnigg, C. van Wüllen, U. Fleischer, R. Franke and T. von Mourik, in *Nuclear Magnetic Shieldings and Molecular Structure* (Ed. J. A. Tossell), Kluwer Academic Publishers, Dordrecht, 1993, p. 141.
    (b) W. Kutzelnigg, U. Fleischer and M. Schindler, in *NMR Basic Principles and Progress*, Vol. 23, Springer-Verlag, Berlin–Heidelberg, 1990, p. 165.
    (c) M. Schindler and W. Kutzelnigg, *J. Chem. Phys.*, **76**, 1919 (1982).
    (d) W. Kutzelnigg, *Israel J. Chem.*, **19**, 193 (1980).
33. A. E. Hansen and T. D. Bouman, *J. Chem. Phys.*, **82**, 5035 (1985).
34. R. Ditchfield, *Mol. Phys.*, **27**, 789 (1974).
35. For summaries on the application of the *ab initio*/IGLO/NMR method see:
    (a) For applications on carbocations: P. v. R. Schleyer and C. Maerker, *Pure Appl. Chem.*, **67**, 755 (1995), especially footnote #4.
    (b) For boranes and carboranes: M. Bühl and P. v. R. Schleyer, *J. Am. Chem. Soc.*, **114**, 477 (1992).
    (c) An extensive bibliography has also been given by M. Diaz, J. Jaballas, J. Arias, H. Lee and T. Onak, *J. Am. Chem. Soc.*, **118**, 4405 (1996).
36. R. Ditchfield and D. P. Miller, *J. Am. Chem. Soc.*, **93**, 5287 (1971).
37. K. Wolinski, J. F. Hinton and P. Pulay, *J. Am. Chem. Soc.*, **112**, 8251 (1990).
38. M. Bühl, J. Gauss, M. Hofmann and P. v. R. Schleyer, *J. Am. Chem. Soc.*, **115**, 12385 (1993).
39. (a) S. P. A. Sauer, I. Paidorova and J. Oddershede, *Mol. Phys.*, **81**, 87 (1994).
    (b) S. P. A. Sauer, I. Paidorova and J. Oddershede, *Theor. Chim. Acta*, **88**, 351 (1994).
40. (a) V. Galasso and G. Fronzoni, *J. Chem. Phys.*, **84**, 3215 (1986).
    (b) V. Galasso, *J. Mol. Struct. (THEOCHEM)*, **10**, 201 (1983).
41. C. v. Wüllen and W. Kutzelnigg, *Chem. Phys. Lett.*, **205**, 563 (1993).
42. (a) J. Gauss, *Chem. Phys. Lett.*, **191**, 614 (1992).
    (b) J. Gauss, *J. Chem. Phys.*, **99**, 3629 (1993).
43. (a) D. Sundholm, J. Gauss and A. Schäfer, *J. Chem. Phys.*, **105**, 11051 (1996).
    (b) J. Gauss and J. F. Stanton, *J. Chem. Phys.*, **104**, 2574 (1996).
    (c) J. Gauss and J. F. Stanton, *J. Chem. Phys.*, **103**, 3561 (1995).
    (d) J. Gauss and J. F. Stanton, *J. Chem. Phys.*, **102**, 251 (1995).
    (e) J. Gauss, *Chem. Phys. Lett.*, **229**, 198 (1994).

44. (a) J. R. Cheeseman, G. W. Trucks, T. A. Keith and M. J. Frisch, *J. Chem. Phys.*, **104**, 5497 (1996).
    (b) G. Rauhut, S. Puyear, K. Wolinski and P. Pulay, *J. Phys. Chem.*, **100**, 6310 (1996).
    (c) V. G. Malkin, O. L. Malkina, M. E. Casida and D. R. Salahub, *J. Am. Chem. Soc.*, **116**, 5898 (1994).
    (d) A. M. Lee, N. C. Handy and S. M. Colwell, *J. Chem. Phys.*, **103**, 10095 (1995).
    (e) A. M. Lee, S. M. Colwell and N. C. Handy, *Chem. Phys. Lett.*, **229**, 225 (1994).
45. G. A. Olah and P. v. R. Schleyer (Eds.), *Carbonium Ions*, Vols. 1–5, Wiley, New York, 1968–1976; *Stable Carbocation Chemistry*, P.v.R. Schleyer and G. K. S. Prakash (Eds.), Wiley, New York, 1997.
46. P. Vogel, *Carbocation Chemistry*, Elsevier, Amsterdam, 1985.
47. M. Hanack, D. Lenoir, H.-U. Siehl and L. R. Subramanian, in *Methoden der Organischen Chemie (Houben-Weyl)*, Vol. **E19c** (Ed. M. Hanack), Georg Thieme Verlag, Stuttgart, 1990.
48. N. Muller and R. S. Mulliken, *J. Am. Chem. Soc.*, **80**, 3489 (1958).
49. J. Higuchi, *J. Chem. Phys.*, **31**, 563 (1959).
50. B. Wirsam, *Chem. Phys. Lett.*, **18**, 578 (1973).
51. Y. Apeloig and P. v. R. Schleyer, *Tetrahedron Lett.*, 4647 (1977).
52. A. C. Hopkinson and M. H. Lien, *J. Org. Chem.*, **46**, 998 (1981).
53. S. A. Godleski, D. J. Heacock and J. M. McKelvey, *Tetrahedron Lett.*, **23**, 4453 (1982).
54. Y. Apeloig, S. A. Godleski, D. J. Heacock and J. M. McKelvey, *Tetrahedron Lett.*, **22**, 3297 (1981).
55. W. J. Hehre, R. Ditchfield, L. Radom and J. A. Pople, *J. Am. Chem. Soc.*, **92**, 4796 (1970).
56. J. Berkowitz, G. B. Ellison and D. Gutman, *J. Phys. Chem.*, **98**, 2744 (1994).
57. J. A. Seetula, Y. Feng, D. Gutman, P. W. Seakins and M. J. Pilling, *J. Phys. Chem.*, **95**, 1658 (1991).
58. P. W. Seakins, M. J. Pilling, J. T. Niiranen, D. Gutman and L. N. Krasnoperov, *J. Phys. Chem.*, **96**, 9847 (1992).
59. R. Damrauer, S. R. Kass and C. H. DePuy, *Organometallics*, **7**, 637 (1988).
60. (a) J. Y. Corey, *J. Am. Chem. Soc.*, **97**, 3237 (1975).
    (b) J. Y. Corey, D. Gust and K. Mislow, *J. Organomet. Chem.*, **101**, C7 (1975).
61. See for example, the attempted ionization of $(Me_2N)_3SiCl$ with $AlCl_3$: H. Cowley, M. C. Cushner and P. E. Riley, *J. Am. Chem. Soc.*, **102**, 624 (1980).
62. W. J. Pietro and W. J. Hehre, *J. Am. Chem. Soc.*, **104**, 4329 (1982).
63. J. Kapp, C. Schade, A. M. El-Nahas and P. v. R. Schleyer, *Angew. Chem.*, **108**, 2373 (1996); *Angew. Chem., Int. Ed. Engl.*, **35**, 2236 (1996).
64. C. Chuit, R. J. P. Corriu, C. Reyé and J. C. Young, *Chem. Rev.*, **93**, 1371 (1993).
65. For recent reviews see:
    (a) H. Grützmacher, *Angew. Chem.*, **107**, 323 (1995); *Angew. Chem., Int. Ed. Engl.*, **34**, 295 (1995).
    (b) M. Drieβ and H. Grützmacher, *Angew. Chem.*, **108**, 900 (1996); *Angew. Chem., Int. Ed. Engl.*, **35**, 829 (1996) and references cited therein.
66. A. J. Arduengo III, R. L. Harlow and M. Kline, *J. Am. Chem. Soc.*, **113**, 361 (1991).
67. G. Boche, P. Andrews, K. Harms, M. Marsch, K. S. Rangappa, M. Schimeczek and C. Willeke, *J. Am. Chem. Soc.*, **118**, 4925 (1996).
68. S. K. Shin and J. L. Beauchamp, *J. Am. Chem. Soc.*, **111**, 900 (1989).
69. K. Hensen, T. Zengerly, P. Pickel and G. Klebe, *Angew. Chem.*, **95**, 739 (1983); *Angew. Chem. Int. Ed. Engl.*, **22**, 725 (1983).
70. G. K. S. Prakash, S. Keyaniyan, R. Aniszfeld, L. Heiliger, G. A. Olah, R. C. Stevens, H.-K. Choi and R. Bau, *J. Am. Chem. Soc.*, **109**, 5123 (1987).
71. Z. Xie, R. Bau and C. A. Reed, *J. Chem. Soc., Chem. Commun.*, 2519 (1994).
72. L. Pauling, *Science*, **263**, 983 (1994).
73. P. v. R. Schleyer, P. Buzek, T. Müller, Y. Apeloig and H.-U. Siehl, *Angew. Chem.*, **105**, 1558 (1993); *Angew. Chem., Int. Ed. Engl.*, **32**, 1471 (1993).
74. G. A. Olah, G. Rasul, X.-y. Li, H. A. Buchholz, G. Sandford and G. K. S. Prakash, *Science*, **263**, 983 (1994).
75. J. B. Lambert and S. Zhang, *Science*, **263**, 984 (1994).
76. C. A. Reed and Z. Xie, *Science*, **263**, 985 (1994).
77. G. A. Olah, G. Rasul, H. A. Buchholz, X.-y. Li and G. K. S. Prakash, *Bull. Soc. Chim. Fr.*, **132**, 569 (1995).

78. K. N. Houk, *Chemtracts Org. Chem.*, **6**, 360 (1993).
79. J. B. Lambert, J. A. McConnell and W. J. J. Schulz, *J. Am. Chem. Soc.*, **108**, 2482 (1986).
80. (a) J. B. Lambert, W. J. J. Schulz, J. A. McConnell and W. Schilf, *J. Am. Chem. Soc.*, **110**, 2201 (1988).
    (b) J. B. Lambert and W. Schilf, *J. Am. Chem. Soc.*, **110**, 6364 (1988).
81. J. B. Lambert and W. J. Schulz, *J. Am. Chem. Soc.*, **105**, 1671 (1983).
82. K. B. Wiberg, *Tetrahedron*, **24**, 1083 (1968).
83. (a) L. Olsson, C.-H. Ottosson and D. Cremer, *J. Am. Chem. Soc.*, **117**, 7460 (1995).
    (b) M. Arshadi, D. Johnels, U. Edlund, C.-H. Ottosson and D. Cremer, *J. Am. Chem. Soc.*, **118**, 5120 (1996).
84. V. Gutmann, *Coord. Chem. Rev.*, **18**, 225 (1976).
85. A. E. Reed, L. A. Curtiss and F. Weinhold, *Chem. Rev.*, **88**, 899 (1988).
86. U. Pidun, M. Stahl and G. Frenking, *Chem. Eur. J.*, **2**, 869 (1996).
87. (a) S. Miertus, E. Scrocco and J. Tomasi, *Chem. Phys.*, **55**, 117 (1981).
    (b) R. Bonaccorsi, R. Cimiraglia and J. Tomasi, *J. Comput. Chem.*, **4**, 567 (1983).
    (c) R. Bonaccorsi, P. Pala and J. Tomasi, *J. Am. Chem. Soc.*, **106**, 1945 (1984).
    (d) J. L. Pascual-Ahuir, E. Silla, J. Tomasi and R. Bonaccorsi, *J. Comput. Chem.*, **8**, 778 (1987).
    (e) F. Floris and J. Tomasi, *J. Comput. Chem.*, **10**, 616 (1989).
88. (a) R. F. W. Bader, *Atoms in Molecules*, Oxford University Press, Oxford, New York, 1990.
    (b) R. F. W. Bader, *Acc. Chem. Res.*, **18**, 9 (1985).
89. G. A. Olah, G. Rasul and G. K. S. Prakash, *J. Organomet. Chem.*, **521**, 271 (1996).
90. Z. Xie, R. Bau, A. Benesi and C. A. Reed, *Organometallics*, **14**, 3933 (1995).
91. Z. Xie, J. Manning, R. W. Reed, R. Mathur, P. D. W. Boyd, A. Benesi and C. A. Reed, *J. Am. Chem. Soc.*, **118**, 2922 (1996).
92. M. R. Rosenthal, *J. Chem. Educ.*, **50**, 33 (1973).
93. Y. Cao, J.-H. Choi, B.-M. Hass, M. S. Johnson and M. Okumura, *J. Phys. Chem.*, **97**, 5215 (1993).
94. D. W. Boo and Y. T. Lee, *Chem. Phys. Lett.*, **211**, 358 (1993).
95. (a) P. R. Schreiner, S.-J. Kim, H. F. Schaefer III and P. v. R. Schleyer, *J. Phys. Chem.*, **99**, 3716 (1993).
    (b) P. v. R. Schleyer and J. W. de. M. Carneiro, *J. Comput. Chem.*, **13**, 997 (1992).
    (c) see also W. Klopper and W. Kutzelnigg, *J. Phys. Chem.*, **94**, 5625 (1990).
96. S.-J. Kim, P. R. Schreiner, P. v. R. Schleyer and H. F. Schaefer III, *J. Phys. Chem.*, **97**, 12232 (1993).
97. R. Liu and X. J. Zhou, *J. Phys. Chem.*, **97**, 9555 (1993).
98. (a) C.-H. Hu, P. R. Schreiner, P. v. R. Schleyer and H. F. Schaefer III, *J. Phys. Chem.*, **98**, 5040 (1994).
    (b) P. R. Schreiner, H. F. Schaefer III and P. v. R. Schleyer in *Advances in Gas Phase Ion Chemistry*, N. G. Adama and L. G. Babcock (Eds.), JAI Press, Inc., London, Vol. 2, pp. 125–160.
99. (a) P. v. R. Schleyer, Y. Apeloig, D. Arad, B. T. Luke and J. A. Pople, *Chem. Phys. Lett.*, **95**, 477 (1983).
    (b) C.-H. Hu, M. Shen and H. F. Schaefer, *Chem. Phys. Lett.*, **190**, 543 (1992).
    (c) E. Delrio, M. I. Menendez, R. Lopez and T. L. Sordo, *J. Chem. Soc., Chem. Commun*, 1779 (1997).
100. B. H. Boo and P. B. Armentrout, *J. Am. Chem. Soc.*, **109**, 3549 (1987).
101. J. B. Lambert, S. Zhang and S. M. Ciro, *Organometallics*, **13**, 2430 (1994).
102. (a) A. Gobbi and G. Frenking, *J. Am. Chem. Soc.*, **115**, 2362 (1993).
     (b) A. Gobbi, P. J. MacDougall and G. Frenking, *Angew. Chem.*, **103**, 1023 (1991); *Angew. Chem., Int. Ed. Engl.*, **30**, 1001 (1991).
103. S. Berger, W. Bock, G. Frenking, V. Jonas and F. Müller, *J. Am. Chem. Soc.*, **117**, 3820 (1995).
104. C. Eaborn, *J. Organomet. Chem.*, **405**, 173 (1991).
105. G. A. Olah, G. Rasul, L. Heiliger, J. Bausch and G. K. S. Prakash, *J. Am. Chem. Soc.*, **114**, 7737 (1992).
106. (a) C. Chuit, R. J. P. Corriu, A. Mehdi and C. Reyé, *Angew. Chem.*, **105**, 1372 (1993); *Angew. Chem., Int. Ed. Engl.*, **32**, 1311 (1993).
     (b) C. Brelière, F. Carré, R. Corriu and M. W. C. Man, *J. Chem. Soc., Chem. Commun.*, 2333 (1994).

10. Silicenium ions: Quantum chemical computations

- (c) F. Carré, C. Chuit, R. J. P. Corriu and A. Mehdi, *Angew. Chem.*, **106**, 1152 (1994); *Angew. Chem., Int. Ed. Engl.*, **33**, 1097 (1994).
- (d) M. Chauhan, C. Chuit, R. J. P. Corriu and C. Reyé, *Tetrahedron Lett.*, **37**, 845 (1996).
107. J. Belzner, D. Schär, B. O. Kneisel and R. Herbst-Irmer, *Organometallics*, **14**, 1840 (1995).
108. C.-H. Ottosson and D. Cremer, *Organometallics*, **15**, 5309 (1996).
109. W. Uhlig, in *Organosilicon Chemistry: From Molecules to Materials*, Vol. II (Eds. N. Auner and J. Weis), VCH, Weinheim, 1995, p. 21.
110. H.-U. Steinberger, T. Müller, N. Auner, C. Maerker and P. v. R. Schleyer, *Angew. Chem.*, **109**, 667 (1997); *Angew. Chem., Int. Ed. Engl.*, **36**, 626 (1997).
111. (a) R. G. Lawton, *J. Am. Chem. Soc.*, **83**, 2399 (1961).
 (b) S. Winstein and P. Carter, *J. Am. Chem. Soc.*, **83**, 4485 (1961).
 (c) P. D. Bartlett and S. Bank, *J. Am. Chem. Soc.*, **83**, 2591 (1961).
 (d) P. D. Bartlett, S. Bank, R. J. Crawford and G. H. Schmid, *J. Am. Chem. Soc.*, **87**, 1288 (1965).
112. (a) H.-J. Knölker, G. Baum and R. Graf, *Angew. Chem.*, **106**, 1705 (1994); *Angew. Chem., Int. Ed. Engl.*, **33**, 1612 (1994).
 (b) H.-J. Knölker, N. Foitzik, H. Goesmann and R. Graf, *Angew. Chem.*, **105**, 1104 (1993); *Angew. Chem., Int. Ed. Engl.*, **32**, 1081 (1993).
 (c) H.-J. Knölker, P. G. Jones and J.-B. Pannek, *Synlett*, 429 (1990).
113. C.-H. Ottosson, K. J. Szabó and D. Cremer, *Organometallics*, **16**, 2377 (1997).
114. N. J. R. van Eikema Hommes, C. Maerker and P. v. R. Schleyer, unpublished results.
115. E. D. Jemmis, G. N. Srinivas, J. Leszczynski, J. Kapp, A. A. Korkin and P. v. R. Schleyer, *J. Am. Chem. Soc.*, **117**, 11361 (1995).
116. A. Sekiguchi, M. Tsukamoto and M. Ichinohe, *Science*, **275**, 60 (1997).
117. For recent reviews see:
 (a) P. v. R. Schleyer, *Science*, **275**, 39 (1997).
 (b) J. Belzner, *Angew. Chem.*, **109**, 1331 (1997); *Angew. Chem., Int. Ed. Engl.*, **36**, 1277 (1997).
118. R. K. Harris and B. E. Mann, *NMR and The Periodic Table*, Academic Press, London, 1978, p. 313.
119. H. Jiao and P. v. R. Schleyer, unpublished data.

CHAPTER 11

Silicenium ions — experimental aspects

PAUL D. LICKISS

Department of Chemistry, Imperial College of Science, Technology and Medicine, London SW7 2AY, UK
Fax: +44 (0)171 594 5804; e-mail: p.lickiss@ic.ac.uk

I. INTRODUCTION	557
II. GAS-PHASE STUDIES	559
III. SOLUTION STUDIES	562
A. Conductance Measurements	562
B. Cryoscopic Measurements	565
C. Spectroscopic Measurements	565
1. Electronic spectroscopy	565
2. NMR spectroscopy	565
D. Silicenium Ions as Reaction Intermediates	575
IV. SOLID STATE STUDIES	581
V. CONCLUSIONS	589
VI. ACKNOWLEDGEMENTS	589
VII. REFERENCES	589

I. INTRODUCTION

There have been a large number of comparative studies of the chemistry of analogous carbon and silicon compounds over the last 50 years, many of which were carried out in order to ascertain whether silicon compounds could be demonstrated to exhibit similar chemical properties to those of their more well-known carbon analogues. Such studies have, for example, during the last 15 years led to the successful isolation of compounds containing multiple bonds to silicon and to a much improved knowledge of compounds containing polysilane linkages (see Chapters 15, 16 and 17 by Sakurai, Auner, and Tokitoh and Okazaki respectively for multiple bonding and Chapter 37 by Hengge and Stüger for polysilanes). One of the remaining, and one of the most difficult, practical challenges to silicon chemists has been to demonstrate the existence of, and to isolate, three-coordinate, trivalent cationic species R_3Si^+, analogous to the well known R_3C^+ cations in either

The chemistry of organic silicon compounds, Vol. 2
Edited by Z. Rappoport and Y. Apeloig © 1998 John Wiley & Sons Ltd

solution or the solid state. Even the naming of R_3Si^+ species has been the subject of some discussion; the terms siliconium, silylium, silicenium, silylenium, silyl cation and silico cation have all been used. The current recommendation by IUPAC[1] is silylium, but silicenium will be used here in order to be consistent with other chapters in this volume.

The Pauling scale of electronegativities gives values of 2.5 and 1.8, respectively, for carbon and silicon. The more electropositive nature of silicon would suggest that it should be easier to form R_3Si^+ species than analogous R_3C^+ and this is indeed the case in the gas phase where, for example, R_3Si^+ ions are abundant in the mass spectra of organosilicon compounds[2,3] and ionization potentials for silyl radicals are lower than for alkyl radicals, e.g. 9.84 and 8.01 eV for $H_3C^•$ and $H_3Si^•$ and 6.34 and 5.93 eV for $Me_3C^•$ and $Me_3Si^•$ radicals, respectively (see Table 1 in Reference 4). Silicon is also more polarizable and larger than carbon and might again, therefore, be expected to form more stable R_3Si^+ ions than the analogous carbon species. Why then was there little real progress in preparing silicenium ions in condensed phases until very recently and why do unambiguous examples of such species still remain to be prepared? It is not due to a thermodynamic instability, as gas-phase silicenium ions are readily prepared[2,3] (tables of calculated relative stabilities of carbenium and silicenium ions are given in Reference 4) and a determination by FT ion cyclotron resonance spectroscopy of the hydride affinities for H_2MeSi^+, HMe_2Si^+ and Me_3Si^+ in the gas phase gives values of 245.9, 230.1 and 220.5 kcal mol^{-1}, respectively, which, when compared with the values of 270.5, 251.5 and 233.6 kcal mol^{-1} for the corresponding carbenium ions, shows that the silicenium ions are thermodynamically significantly more stable[5]. The instability is actually kinetic in origin due to the high electrophilicity of R_3Si^+ ions which, in solution or the solid state, leads to such species interacting with a wide variety of both π- and σ-electron donors, such as solvents and other potential nucleophiles such as oxygen- and halogen-containing counter anions. The false notions of non-coordinating solvents and anions are particularly highlighted in this field, the strongly electrophilic silicenium ions being found to interact with many species that in other areas of chemistry are found to be relatively innocent. Unlike carbon which readily forms compounds containing three-coordinate sp^2 carbon and which tends not to increase its coordination number above four, silicon has a tendency to increase its coordination number to five or six which has traditionally been thought to be due to formal hybridization states of sp^{3}d and sp^{3}d^2. Calculations, however, indicate that the orbital participations for five and six-coordinate silicon species are in fact approximately sp^2 with no significant d-orbital participation, and that the bonding to silicon is better regarded as partially ionic[4]. These factors, together with the poorer $\pi-\pi$ overlap between Si and first-row π-donor substituents containing N or O, and the longer Si−R compared to C−R bond lengths which mean that R_3C^+ ions are sterically better protected from anions than R_3Si^+, all militate against the formation of R_3Si^+ ions. The work described below shows how various factors such as solvent, anion, steric and electronic effects of the substituents at silicon, and reaction byproducts can be manipulated in potential preparations of silicenium ions so that the chances of success are maximized. Even with the best current knowledge on the important factors a recent review concludes that 'the prospects for obtaining and observing truly 'free' silyl cations in condensed phases are very poor'[4].

One further problem that arises in the search for a 'free' silicenium ion is what the term 'free' means in this context. In the case of R_3C^+ ions there are crystallographic studies that show that for free ions such as Me_3C^+ there is a separation of about 3 Å or greater between the cationic centre and the closest atoms of its counter anion $Sb_2F_{11}^-$, and that the ions are planar[6] and so it might be expected that a similar or even greater separation might be found between a silicenium ion and its related anion and that R_3Si^+ will be planar. Such separations and planar geometry have yet to be found (see Section IV). The arguments presented in recent years for the existence of silicenium ions have focused on the anion cation separation (ideally greater than 3 Å), the R—Si—R angles in the R_3Si^+ species (ideally 120°), the out-of-plane displacement of the silicon atom (ideally zero) and the ^{29}Si NMR chemical shift values (according to calculations, these should be about 350 ppm or even greater for simple trialkyl silicenium ions). As there may be species isolated that successfully satisfy one of the criteria better than the others, the question of whether a 'free' silicenium ion has been prepared or can be prepared may depend on the definition chosen. Certainly none of the species so far isolated or observed spectroscopically satisfies any of these criteria, and calculations carried out in the last few years have shown that most, if not all, of the species proposed to be silicenium ions are actually better described in other ways. Much discussion has therefore been directed towards the degree or extent of cationic character in cationic silicon-containing species.

The long history of the search for silicenium ions has meant that there have been several reviews or critical comments published on the problem. Early work was reviewed by O'Brien and Hairston[7] and by Corriu and Henner[8] while more recent studies have been reviewed by various authors[3,4,9-19]. This chapter will concentrate on work reported in the last few years, particularly since 1988.

Running in parallel with the numerous experimental attempts to prepare silicenium ions in recent years, there has been a considerable number of computational studies carried out on them. The computational work is described in detail in the chapter by Schleyer and Maerker and this chapter will describe experimental work, concentrating mainly on solution and solid state experimental work with only occasional reference to calculations when necessary.

II. GAS-PHASE STUDIES

The abundance of silicenium ions in the gas phase has allowed many studies to be carried out on their formation and chemistry and much of this work has been reviewed[2,3] (see also the chapter by Goldberg and Schwarz). This chapter will concentrate on studies carried out in condensed phases but some gas-phase studies, particularly where they give information relevant to some of the current work on condensed-phase species, will be described briefly below.

The importance of silicenium ion reactions in the gas phase has become better appreciated in recent years as the use of chemical vapour deposition techniques for the preparation of materials such as silicon carbide[20] and silicon nitride[21-24] has increased. Silicenium ions have been generated in the gas phase by several ionization techniques, often using mass spectrometric methods; see, for example, Reference 25. The methods involved will not be described here but some of the physical characteristics and the chemistry of the derived ions will be discussed. A study of the gas-phase IR spectrum of H_3Si^+ shows the ν_2 and ν_4 fundamentals to be at 838.0669(24) and 938.3969(36) cm^{-1}, respectively[26], while an IR laser absorption spectroscopy shows that, as expected, H_3Si^+ is planar, and gives a value for the ν_2 band centre of 838.0674(7) cm^{-1}, a ground state rotational constant of 5.2153(1) cm^{-1}, and an Si—H bond length of 1.462 Å[27]. The heat of formation of H_3Si^+ generated by the reaction of ground state silicon ions with SiH_4 was found to be 237.1(2) $kcal\,mol^{-1}$, in good agreement with earlier photoionization measurements and calculations[28]. More recent photoionization mass spectrometry studies give heats of formation for H_3Si^+, D_2MeSi^+, HMe_2Si^+ and Me_3Si^+ of 235.3, 204 ± 1, 173.2 and 145.0 $kcal\,mol^{-1}$, respectively[29]. Low energy electron impact dissociation of both SiH_4 and Si_2H_6 gives H_3Si^+ as a primary product[30] and cluster cations containing up to nine silicon atoms in the gas phase are formed in the reactions between disilane and cations

D_nSi^+ ($n = 0$–3) and $D_nSi_2^+$ ($n = 0$–6)[31–33]. Calculations on the clustering reactions of H_3Si^+ with SiH_4 and D_3Si^+ with SiD_4 have also shown that only small clusters seem to be formed by this mechanism and that such clustering is not the route to the hydrogenated silicon dust formed in SiH_4 plasmas[34,35]. However, if a few percent of water is present, then larger clusters leading to the formation of hydrogenated particles of silicon are formed[36]. The reaction rates of H_3Si^+ with D_4Si and D_3Si^+ with H_4Si have been measured and two different mechanisms for the reactions were found, one involving ion–molecule formation, the other 'long-range hydride stripping', which together combine to give a reaction rate that is greater than the collision rate. The overall rate depends significantly on the number of H and D substituents present[37]. Gas-phase reactions in SiH_4/GeH_4 proceed via both H_3Si^+ and H_3Ge^+ ions to give species containing Si–Ge bonds[38] and reactions between H_3Si^+ and PH_3 give a variety of ions containing Si, P and H as determined by mass spectrometry[39]. A study of the association reactions of primary, secondary and tertiary amines with Me_3Si^+ applying high pressure mass spectrometry shows that the Me_3Si^+ affinities (46.5 and 59.5 kcal mol^{-1} for NH_3 and n-$BuNH_2$, respectively) increase in the order NH_3 < $MeNH_2$ < $EtNH_2$ < n-$BuNH_2$, the values increasing linearly with the proton affinity of the amines. The affinities for *sec-* and *tert-*butylamines are lower than for the n-butyl isomer which is attributed to steric effects and the Me_3Si^+ affinities also increase in the order primary < secondary < tertiary amine[40].

An FT ion cyclotron resonance study shows that in the reactions of H_3SiCl with $H_nCl_{3-n}Si^+$ ions ($n = 1$, 2 or 3), H_2ClSi^+ participates in both disproportionation and hydride transfer reactions but H_3Si^+ and HCl_2Si^+ only participate in hydride transfer. The disilylchloronium ion $(H_3Si)_2Cl^+$ is thought to be an intermediate (perhaps analogous to the silylated halonium ions found in condensed phases, see Section IV) in the reaction between H_3Si^+ and H_3SiCl and to have a lifetime at room temperature of several tenths of a second[41]. The values of the chloride affinities for chlorosilyl cations are similar and fall in the order H_2ClSi^+ > HCl_2Si^+ ≈ Cl_3Si^+ while the hydride affinities are in the order Cl_3Si^+ > HCl_2Si^+ > H_2ClSi^+ [42]. Fourier transform mass spectrometric studies show that the predominant mechanism of decomposition for MeH_2Si^+ is dehydrogenation and that of $Me(Cl)HSi^+$ to be a 1,2-elimination of HCl[43]. The dominant reaction between H_3Si^+ and CH_3Cl, CH_2Cl_2 and $CHCl_3$ is chloride abstraction to give SiH_3Cl and H_3C^+, H_2ClC^+ and HCl_2C^+ ions, respectively[44]. Generation of the highly reactive Cl_3Si^+ ion in the gas phase under plasma conditions allows polymeric surfaces such as polyester, polypropylene and polytetrafluoroethylene, that are normally regarded as inert, to be modified by replacement of H, OH or F[45,46].

The silicenium ions derived from the methylsilanes Me_nSiH_{4-n} ($n = 0$–3) have been studied by photoionization mass spectrometric techniques[29] and have been shown to undergo rapid Me/H exchange reactions with propene and with 2-methylpropene[47]. The most stable isomer of formula $SiC_2H_7^+$ has been calculated[48] to be Me_2HSi^+ and this ion has also been found to be thermodynamically more stable than its isomer EtH_2Si^+ using Fourier transform mass spectrometry[49]. Such gas-phase spectrometric techniques have also shown that, after collisional activation, Me_3Si^+ undergoes a rearrangement to $EtMeHSi^+$ followed by extrusion of $H_2C=CH_2$ to give $MeH_2Si^{+\,50}$. The silicenium ions i-$PrMe_2Si^+$ and n-$PrMe_2Si^+$ can be generated in the gas phase from i-$PrMe_2SiCl$ and n-$PrMe_2SiCl$, respectively, by chemical ionization using isobutane. Both ions undergo rearrangements to give isomeric α- and β-silyl substituted carbenium ions which then undergo elimination of C_2H_4 and C_3H_6 to give a mixture of Me_3Si^+ and Me_2HSi^+, respectively, Me_3Si^+ predominating for i-$PrMe_2Si^+$, and Me_2HSi^+ for n-$PrMe_2Si^{+\,51}$. The ions i-$PrMe_2Si^+$ and n-$PrMe_2Si^+$ are also formed in the high pressure (50–760 torr) γ-radiolysis of mixtures of CH_4 and $Me_3SiCH=CH_2$ after initial protonation both α and β to the silicon and subsequent rearrangement of the resulting carbenium ions. Both silicenium ions can be trapped using methanol to give i-$PrMe_2SiOMe$ and n-$PrMe_2SiOMe$[52]. A recent theoretical study of the migratory aptitude of H, Me and Ph groups from silicon to an α carbon atom to give silicenium ions is consistent with the experimental results determined by mass spectrometry in the gas phase[53]. Ionization of $Me_3SiOCH_2CH_3$ in the gas phase followed by alkene elimination leads to $Me_2(OH)Si^+$ formation (equation 1). This cation can then be used to generate the $Me_2(OH)Si^•$ radical by collisional neutralization[54]. The unimolecular reactions of $Me_3SiOSiMe_2{}^+$ and deuterium labelled analogues have been studied using mass-analysed ion kinetic energy spectrometry. Scrambling of the methyl groups occurs initially and this is followed by losses of CH_4 and $Me_2Si=O$. This is in contrast to the carbon analogue $Me_3COCMe_2{}^+$, the most significant reaction of which is loss of $CH_2=CMe_2$[55]. A related study has shown that electron impact induced fragmentation of $Me_2Si(OEt)_2$ leads to formation of $MeSi(OH)_2{}^+$ and $Me_2Si=OH^+$[56]. The differences between the relative stabilities and the decomposition pathways of the products formed in the reactions between Me_3Si^+ and the *cis* and the *trans* isomers of 1,2-cyclopentanediol can be used to distinguish between the two isomers. The *cis* isomer decomposes to give the hydrated trimethylsilicenium ion $Me_3SiOH_2^{+57}$, which is also formed in the reaction between Me_3Si^+ and water[58]. The *t*-Bu analogue of the hydrated ion t-$Bu_3SiOH_2{}^+$ has been prepared in solution and its solid state structure has been determined[59] (Section IV).

$$Me_3SiOCH_2CH_3{}^{+•} \longrightarrow Me_2\overset{+}{Si}OCH_2CH_3 \longrightarrow Me_2\overset{+}{Si}OH + CH_2=CH_2 \qquad (1)$$

Studies carried out on reactions in the gas phase also offer the opportunity to investigate species that interact strongly with solvents etc. in condensed phases. It has thus been possible to study reactions of silicenium ions in the gas phase, solution state analogues of which have become of significant interest. One such type of reaction is that of electrophilic aromatic silylation which was a reaction unknown in solution [but see the discussion of the structure of $Et_3Si(toluene)^+$ in Sections III.C.2 and IV for a recent solution and solid state example] due to the strong influence of both solvent and anion on silicenium ion formation. A mass spectrometric study of the reactions of Me_3Si^+, generated by the reaction between $CH_5{}^+$ and $C_2H_5{}^+$ with Me_4Si, with aromatic compounds, concluded that there was no aromatic substitution with the formation of a Si—C bond to give **1**, but rather that a π-complex of type **2** was formed[60]. A more recent collision-assisted dissociation study of Me_3SiArH^+ (Ar = benzene or toluene) formed by either silylation of the aromatic or by protonation of $ArSiMe_3$ failed to resolve the question of whether **1** or **2** represents the gas-phase structure best[61].

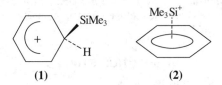

(1)   (2)

Radiolytic studies have now demonstrated that the Wheland intermediate type **1** does in fact form and that earlier work had neglected the possibility that rapid desilylation by bases present in the mass spectrometric study would compete effectively with, and prevent, deprotonation, i.e. that reaction 2 is faster than reaction 3. This would therefore give the appearance that an Si—C bond was not formed[62–65]. If a nitrogen-centred base such as Et_3N is added to the system for the generation of Me_3Si^+, then this does cause deprotonation and this competes effectively with the trace amounts of oxygen containing impurities present that cause desilylation. This then does allow the formation of neutral arylsilane products. If, however, larger amounts of amine are added, then this intercepts the Me_3Si^+ and $C_nH_5^+$ ($n = 1$ or 2) precursor ions thus reducing arylsilane yields[63]. Yields of less than 2% of silylated arenes can be achieved in solution if a similar strategy is used and a hindered base is present to abstract the proton (equation 4). For example, trimethylsilylation of toluene gives an overall yield of 1.0%, the o-, m- and p-isomers being formed in a 0.6 : 34.6 : 64.8 ratio. The very low overall yields are due in part to the disproportionation of the products in the presence of the $AlCl_3$[66].

$$\text{[arenium ion with SiMe}_3\text{ and H]} + B \longrightarrow PhH + BSiMe_3^+ \quad (2)$$
$$\longrightarrow PhSiMe_3 + BH^+ \quad (3)$$

B = oxygenated base, e.g. H_2O, MeOH or CH_2O

$$ArR + R'_3SiCl + i\text{-}Pr_3N \xrightarrow[150\,°C,\,24\,h]{AlCl_3} Ar(R)(SiR'_3) + [i\text{-}Pr_3NH]^+[AlCl_4]^- \quad (4)$$

The selectivity of the gas-phase electrophilic aromatic substitution appears to be dominated by the steric requirements of the silyl substituent, the *meta* : *para* ratio for toluene substitution by Me_3Si^+ is 14 : 86, while mesitylene, in which any substitution is forced to be *ortho* to two methyl groups, is unreactive[62,66]. This substitution pattern is similar to that found in work on the solution or solid state products from the reaction between Et_3Si^+ and toluene in which there is apparently 100% *para* substitution[67] (see Section IV). Once formed, species of type **1** appear to be stable towards 1,2-H shifts as judged by radiolytic techniques[68]. The existence of ions of type **1** has been further confirmed (although not excluding the possibility of ions **2** in equilibrium with **1**) by ion cyclotron resonance mass spectrometry studies. This work shows that transfer of D^+ from $[Et_3SiC_6D_6]^+$ to Et_3N does occur, this type of reaction being unknown for the π-type complex, but being consistent with the σ-structure which can undergo both reactions 2 and 3[69]. The existence of long-lived $[Me_3Si(toluene)]^+$ ions with the σ-bound structure **1** is also confirmed by the observation of a primary kinetic isotope effect in competition reactions of Me_3Si^+ with $C_6H_5CH_3$ and $C_6D_5CD_3$[70].

III. SOLUTION STUDIES

A. Conductance Measurements

Early attempts to demonstrate that silicenium ions were components of conducting solutions relied on methods analogous to those used successfully in carbon chemistry. In solvents, such as liquid SO_2, pyridine or nitrobenzene/$AlBr_3$, in which Ph_3CCl was found to be ionized, there was no significant ionization of Ph_3SiCl[71,72]. There is also no ionization of Ph_3SiOH in liquid HCl, a solvent that does promote ionization of

Ph$_3$COH to give Ph$_3$C^{+}[73]. This and other early work has been reviewed by Corriu and Henner[8]. More recent work by Lambert and coworkers has focused on the possibility that organosilicon perchlorates (prepared from the reaction between silanes R$_3$SiH and Ph$_3$CClO$_4$[74]) might be ionized in suitable solvents. The perchlorates Me$_3$SiClO$_4$[75-77], PhMe$_2$SiClO$_4$[77], Ph$_2$MeSiClO$_4$[77], Ph$_3$SiClO$_4$[75,78], (RS)$_3$SiClO$_4$ (where R = Me[76], Et[76], or i-Pr[76,79]) and (Me$_2$N)$_n$Me$_{3-n}$SiClO$_4$ (where n = 1, 2 or 3)[80] have been studied using CH$_2$Cl$_2$, MeCN, sulpholane and 1,2-dichloroethane solutions. In concentrated solutions in sulpholane or in the solvents of lower ionizing power, CH$_2$Cl$_2$ and 1,2-dichloroethane, the perchlorates R$_3$SiClO$_4$ (where R = Me or Ph) are thought to be covalent R$_3$SiClO$_4$ species or R$_3$Si$^+$ClO$_4^-$ ion pairs in equilibrium with free ions, but in dilute (less than 0.01 M) sulpholane solution conductance measurements suggest that free ions are present. The organosulphur derivative (i-PrS)$_3$SiClO$_4$ was found to give a conducting solution in CH$_2$Cl$_2$[79]. By comparison, the azide Me$_3$SiN$_3$ gave negligible conductance in CH$_2$Cl$_2$, MeCN or sulpholane and the triflate Me$_3$SiOSO$_2$CF$_3$ gave very low conductance in CH$_2$Cl$_2$ but high conductance in MeCN and sulpholane. The interpretations of the results from these conductance studies have been questioned[13,81-83] and they have been proposed to be due not to ionized silyl perchlorates but to hydrolysis, to give free ClO$_4^-$ ions, by residual water in the solvents used, some of which are particularly difficult to dry adequately for studies using very low concentrations of highly moisture sensitive materials. Lambert and coworkers have countered these criticisms[18,77] and have argued that integration of the ^{1}H NMR signals due to hydrolysis products show that the concentration of water present is only about one-tenth of that of the silicon-containing substrate. Hydrolysis is clearly a problem in some of the NMR studies that have been reported (see below) and if free R$_3$Si$^+$ ions are present there seems to be a contradiction between the conductance results and those of ^{29}Si NMR studies on similar solutions which indicate the silicon to be four coordinate[13,18]. Most recent attempts to study silicenium ions have not used conductance as a measure of potential ion formation. The very bulky perchlorate (Me$_3$Si)$_3$CSiMe$_2$ClO$_4$ has been shown to undergo methanolysis relatively slowly with a half-life at 27.5 °C of about 24 min, and little if any solvolysis in CF$_3$CH$_2$OH[12,84]. Although this perchlorate is far bulkier than those used by Lambert, the relief of steric strain on formation of the three-coordinate [(Me$_3$Si)$_3$C]Me$_2$Si$^+$ ion might be expected to aid ionization. It would be expected that in such good ionizing solvents, and at the concentrations used, ionization would occur and there would then be very rapid solvolysis. As rapid solvolysis does not occur, this also seems to cast doubt on the proposal that silyl perchlorates undergo ionization in solution at low concentrations.

A further complication in solution-phase measurements is the possibility that the solvent molecules may play an important role such that even if the silyl perchlorate were to ionize, the silyl moiety may still be four coordinate by interaction with the solvent to give species R$_3$Si–S$^+$ (S = solvent) such as **3** which is both four coordinate and ionic and which might be consistent with both conductance and some NMR measurements.

$$\text{(3)}$$

Several studies[75,78,85] including the use of conductometric titration[18,76,77] showed that for pyridine and N-methylpyrrole in CH$_2$Cl$_2$ complex formation occurs, but for the weaker nucleophiles MeCN and sulpholane solvent complexes were unimportant. Unfortunately, it is likely that in CH_2Cl_2 solution, sulpholane and MeCN cannot displace perchlorate from the substrate due to poor solvation by CH_2Cl_2. It is thus possible that the ions in sulpholane solution are of the form shown in **3**. This interpretation of the conductance measurements is not unreasonable as there are many R_3Si-S^+ (S = solvent) type complexes known (for a review of the work in this area see Reference 3). For example, for R = Me, complexes with $(Me_2N)_3P$[86], Ph_3PO[86], pyridine[87,88] and 3-methyl and 3-(trimethylsilyl)imidazole[87] are all well described. The ions $[R_3Si(MeCN)]^+$ (where R_3 = Me_3, Ph_3 or $PhMe_2$) are also thought to be generated electrochemically from the corresponding disilanes $(R_3Si)_2$ in MeCN solution[89]. A recent study of over sixty combinations of silyl halides, perchlorates, triflates and tetraarylborates in solvents such as CH_2Cl_2, pyridine, DMSO, MeCN, HMPA and sulpholane concluded that in all cases weak covalent Si-solvent bonds were formed, and that no free silicenium ions were generated[90]. Intramolecular stabilitation of a cationic species to give a five-coordinate ion **4** can also be achieved, ^{29}Si NMR spectroscopy indicating that the same five-coordinate species is present both in solution and in the solid state[91]. Multinuclear NMR studies have, however, shown that the cationic species **5**, with no Si—H group present, is in fact an equilibrium mixture of the two four-coordinate silylammonium cations **6** and **7** in CD_3OD, $CDCl_3$ and CD_3CN solutions[92]. More recently it has been shown that the nature of the cations without an Si—H group present is dependent on the anion, the more nucleophilic anions chloride and iodide give four-coordinate silylammonium species **8**, the less nucleophilic tetraarylborate giving a silicon-centred cationic species **9**, and the Ph_4B anion giving a mixture of the four-and the five-coordinate species that slowly converts into the four-coordinate species in CD_3CN solution[93,94]. Calculations have also shown that in aryl-substituted silicenium ions bearing substituents in the *ortho* position, e.g. $[R_2SiC_6H_3(CH_2NMe_2)_2-2,6]^+$ (R = H or Me), intramolecular coordination of the silicon by the amine functions is strong, and that this may prevent intermolecular solvation of the ion by solvent molecules[95].

(4)

(5)

(6)

(7)

(8) X = Cl or I

(9) ⁻B[3,5-(CF$_3$)$_2$C$_6$H$_3$]$_4$

B. Cryoscopic Measurements

Early, unsuccessful work to demonstrate ionization of organosilicon compounds by cryoscopic methods has been reviewed by Corriu and Henner[8]. More recent work has again focused on the possible dissociation of silyl perchlorates, particularly in sulpholane and in CH$_2$Cl$_2$. The perchlorates (i-PrS)$_3$SiClO$_4$, Ph$_n$Me$_{3-n}$SiClO$_4$ (n = 0, 1 or 3) and (Me$_2$N)$_n$Me$_{3-n}$SiClO$_4$ (n = 1, 2 or 3) were all found to give the expected molecular weights, within about 5–10% accuracy, for two-particle systems in sulpholane rather than those expected for either covalent perchlorates or for hydrolysis products such as silanols or siloxanes[18]. The apparent accuracy of these results is surprising as, together with the silyl perchlorate present, there will also be an equivalent number of moles of Ph$_3$CH (formed during the synthesis of the silyl perchlorate), and some R$_3$SiOSiR$_3$ and HClO$_4$. The ionic and covalent forms of the silyl perchlorate are also in equilibrium and, at the concentrations at which the experiments were carried out, only about 45–65% free ions are thought to be present[13,18,75]. It is also possible that the cryoscopic measurements can be explained by the presence of solvent complexes of the type R$_3$Si—S$^+$ClO$_4^-$ (S = solvent) rather than R$_3$Si$^+$ClO$_4^-$ species. The importance of such solvent complexes is discussed in the previous section.

C. Spectroscopic Measurements

1. Electronic spectroscopy

In acidic solution the carbinol [4-(Me$_2$N)C$_6$H$_4$]$_3$COH ionizes to give the deep violet cation [4-(Me$_2$N)C$_6$H$_4$]$_3$C$^+$ but the corresponding silanol [4-(Me$_2$N)C$_6$H$_4$]$_3$SiOH shows very little change on protonation, this being due to protonation of the NMe$_2$ groups rather than to ionization[96]. Similarly, simple perchlorates such as R$_3$SiClO$_4$ (R = Ph, p-tolyl) are colourless whereas the analogous crystalline carbon analogues are highly coloured[97]. Both of these studies thus indicated that silicenium ions did not seem to be present in systems where carbon analogues were well known. Electronic spectroscopy does not seem to have been widely used in silicenium ion studies in more recent years, although for Et$_3$SiB(C$_6$F$_5$)$_4$ in benzene solution a spectrum comprising one broad absorption, λ_{max} = 303 nm, has been reported[67].

2. NMR spectroscopy

NMR spectroscopy has, in the last ten years, become one of the most important tools in the search for silicenium ions in condensed phases. Studies involving ^{1}H, ^{11}B, ^{13}C, ^{19}F, ^{29}Si and ^{35}Cl have all been carried out, with particular recent interest focused on the ^{29}Si nucleus.

a. *¹H and ¹³C NMR spectroscopy.* Proton NMR spectroscopy has been used relatively little in silicenium ion studies, despite its high sensitivity (especially when compared with ^{29}Si), but it is useful in determining the extent of hydrolysis of simple species such as Me_3SiClO_4 for which the hydrolysis product, $Me_3SiOSiMe_3$ (after initial formation of Me_3SiOH and its condensation), has a simple and distinct ^{1}H NMR signal[77]. A series of ^{1}H and ^{13}C chemical shift values for Me_3SiClO_4, $MePh_2SiClO_4$ and $Me_2PhSiClO_4$ in sulpholane containing 10% CD_2Cl_2[77] and ^{13}C shifts for Ph_3SiClO_4 in CD_2Cl_2[76] have been reported, but there is only a small change of about 0.1 ppm in ^{1}H chemical shift on changing the concentration of the silyl perchlorate from 0.27 to 0.009 mol dm^{-3} and about 0.2 ppm ^{13}C chemical shift change on changing the concentration from 0.27 to 0.01 mol dm^{-3}[77]. These changes could be attributed to an increasing ionic content, but such small changes could also readily be attributed to dilution effects.

A study of i-Pr_2MeSi derivatives by ^{1}H and ^{13}C NMR spectroscopy has led to an estimation of the strength of interaction between a silicenium ion and a coordinating solvent[67]. The methyl groups of i-Pr_2MeSiH and other covalent derivatives are diastereotopic and this can be observed by both ^{1}H and ^{13}C NMR spectroscopy, but if the perchlorate were either dissociated into a planar three-coordinate cation, or formed a four-coordinate species that could dissociate to a planar form, then the methyl groups would become homotopic. In C_6D_6 solution both Pr_2MeSiH and i-$Pr_2MeSiClO_4$ show diastereotopic methyls in the NMR spectra, but for i-$Pr_2MeSiClO_4$ in CD_2Cl_2 and i-$Pr_2MeSiB(C_6F_5)_4$ in C_6D_6, CD_3CN or a pyridine/C_6D_6 mixture the methyl groups are observed to be homotopic. If the methyl groups are rendered equivalent by a dissociative exchange of the solvent molecules and formation of an intermediate planar silicenium ion, then the upper limit for the dissociation energy has been calculated to be about 13 kcal mol^{-1}. This value compares with 28.4 kcal mol^{-1} measured by mass spectrometry for the association of Me_3Si^+ with toluene in the gas phase[98] which would be greater than that expected in solution as solvation would help to stabilize free ions. Both the solution-and gas-phase values are significantly less than that for a covalent Si−C bond for which bond energies are about 90 kcal mol^{-1} [99]. It is also possible for the methyl groups in the isopropyl groups to be equivalent if a five-coordinate species with a trigonal bipyramidal structure with two solvent molecules in the axial positions is formed. This possibility has not been disproved for these species, but for other systems involving nitrile exchange there is evidence against a five-coordinate, associative exchange mechanism[100,101].

b. *^{35}Cl NMR spectroscopy.* The ^{35}Cl nucleus is quadrupolar with a spin of 3/2, which means that covalent perchlorates exhibit a ^{35}Cl NMR signal that may be hundreds or thousands of Hz wide. In contrast, symmetrical ionic species such as Cl^- and ClO_4^- have much narrower signals. This difference in linewidths has been used to probe the possible ionization of silyl perchlorates in solution. For a 0.584 mol dm^{-3} solution in sulpholane the ^{35}Cl NMR signal for Me_3SiClO_4 is −28 ppm and has a width at half height of 1960 Hz. As the concentration of the silyl perchlorate is reduced there is a concomitant reduction in linewidth until, at a concentration of 0.0047 mol dm^{-3}, it is only 24 Hz[75]. These results were interpreted as being due to the rapid equilibrium between an associated (covalent or ion pair) species, giving a broad signal and free ions that would give a sharp signal. The degrees of ionization for both Me_3SiClO_4 and Ph_3SiClO_4 were both calculated to be about 50% at a concentration of 0.03 mol dm^{-3} and to be essentially complete below about 0.005 mol dm^{-3} [75]. Unfortunately, the lower concentration samples in this study were made up by taking a fraction of the original sample and diluting with fresh solvent, thus adding new water from incompletely dried solvent. Using such a technique it is clearly possible to introduce sufficient water into the sample to cause complete hydrolysis for dilute solutions[13]. The ^{35}Cl NMR results also do not seem to agree well with those from a ^{29}Si NMR study of Me$_3$SiClO$_4$, which showed that there was little change in the chemical shift over a similar range of concentrations[81].

c. ^{11}B and ^{19}F NMR spectroscopy. Spectra from ^{11}B and ^{19}F nuclei have mainly been used for compounds containing [B(C$_6$F$_5$)$_4$] as the anion. One of the potential problems with anions containing halogens is coordination between the anion and the cation via a halogen atom. If this were the case for B(C$_6$F$_5$)$_4$ containing species, then a shift in a fluorine resonance would be expected; there is, however, no significant difference in ^{19}F chemical shifts for the anion in either the trityl or the Et$_3$Si$^+$ species, suggesting that there is no significant interaction between Si and F. The linewidth and the chemical shift for the boron resonance in the trityl and the Et$_3$Si$^+$ salts is also the same, again indicating little interaction between the anion and the cation[67,102]. The ^{11}B resonance for i-Pr$_3$Si(Br$_5$CB$_9$H$_5$) (Br$_5$CB$_9$H$_5$ = *closo*-6,7,8,9,10-Br$_5$CB$_9$H$_5$$^-$) is essentially the same as that for the Br$_5$CB$_9$H$_5$$^-$ anion in ionic species, suggesting that although X-ray crystallography shows only one Si$\cdots$Br interaction in the solid state (thus rendering the boron atoms inequivalent), that in toluene solution there is rapid rotation of the anion via the four equivalent bromines so that the boron atoms appear equivalent, even down to $-40\,°$C[103].

d. ^{29}Si NMR spectroscopy. The wide availability of high-field FT NMR spectrometers during the last ten years has brought ^{29}Si NMR spectroscopy to the fore in the search for silicenium ions in solution. Clearly a cationic species with the positive charge largely located on a silicon atom should give a distinctive chemical shift, deshielded with respect to related four-coordinate species. More recently, solid state NMR studies have also become important (see Section IV). The availability of reliable computational methods for the calculation of ^{29}Si chemical shifts has also allowed comparisons to be made between experimentally observed shifts and those expected for free ionic species, and there has been considerable discussion about the degree of ionic character indicated by particular chemical shifts. Calculations of chemical shifts are discussed in detail in the chapter by Schleyer and coworkers, but they will be mentioned here where relevant. Chemical shifts for Me$_3$Si$^+$ and Ph$_3$Si$^+$ have been predicted to be 250 ± 25 and 125 ± 25 ppm, respectively, from correlations with carbon analogues[104]. IGLO calculations, however, predict a chemical shift of 355.7 ppm for Me$_3$Si$^+$[83] while an estimate including correlation effects using the GIAO-MP2 method gives a value of 385 ± 20 ppm[105]; other calculations give 354 (GIAO) and 382 (SOS-DFPT) ppm[106]. Similar calculations give values of 371.2 (GIAO) and 415.6 (SOS-DFPT) ppm for Et$_3$Si$^+$ and 342.3 (GIAO) and 371.0 (SOS-DFPT) ppm for i-Pr$_3$Si$^+$[106]. It is unlikely that the highly deshielded values for calculations on free ions in the gas phase will be closely approached for ions in condensed phases, and even at the best levels of theory there is a discrepancy of between 17 and 30 ppm between the calculated and observed values for Me$_3$C$^+$[107].

For the simple silyl perchlorates Me$_3$SiClO$_4$ and Ph$_3$SiClO$_4$ the ^{29}Si chemical shifts are about 47 and 3 ppm, respectively, the values depending a little on the solvent used[18,77]. Reliable chemical shifts for (i-PrS)$_3$SiClO$_4$ and other thioalkyl derivatives have not been obtained because of the poor stability of these ions in solution[18]. The chemical shift of Me$_3$SiClO$_4$ varies little for either neat liquid or sulpholane solutions at concentrations of 0.584, 0.29 and 0.15 mol dm^{-3}, clearly indicating covalency over this range of concentrations[81]. This seems to contradict the results of the ^{35}Cl NMR study described above which indicated that at these three concentrations there were 20 ± 5, 29 ± 5 and

35 ± 7% free ions, respectively[75]. Such degrees of free-ion formation would, according to an IGLO-type calculation value of 355.7 ppm, lead to average chemical shifts of about 108.7, 136.5 and 155.0 ppm, respectively[83]. The discrepancy between these results can be resolved if hydrolysis (which would lead to increasing concentrations of $HClO_4$ being formed as the concentration of the silyl perchlorate was reduced) of the perchlorates is taken into account in the case of the ^{35}Cl NMR measurements.

Table 1 shows ^{29}Si chemical shift values for a variety of silicon species in which there was thought to be a significant degree of positive character to the silicon atom together with some less polar species for comparison. One of the criteria that has been used to argue for the development of cationic character at a silicon centre has been the downfield chemical shift change on going from a silyl hydride R_3SiH, to the analogous cationic species R_3Si^+ after hydride abstraction using a trityl salt. This change is about 101.1, 92.1, 95.5 and 228.5 ppm for methyl, ethyl, isopropyl and trimethylsilyl substituents, respectively. These changes are considerably less than those expected for the Me, Et, and i-Pr derivatives, suggesting that species with predominantly silicenium ion character are not present. However, it has been reported that the large change for the Me_3Si derivative was 'in accord with a substantially free silylium ion'[102]. From comparisons between SiH_4 and H_3Si^+ it has been predicted that a δ ^{29}Si value (for the central silicon) for free $(Me_3Si)_3Si^+$ would be ca 245 ppm and that for the π-complex $(Me_3Si)_3SiC_6H_6^+$, δ ^{29}Si = 175 ppm[108]. The experimentally observed value of δ ^{29}Si = 111 ppm for $(Me_3Si)_3SiC_6H_6^+$ in benzene solution has thus been proposed to indicate 'that a nearly free silylium cation is realized up to 70% by $(Me_3Si)_3SiC_6H_6^+$ $TPFPB^-$,[108]. This optimistic conclusion does not seem to be borne out by more recent calculations which give values of 925.3[109] and 920.4 ppm[110] for $(Me_3Si)_3Si^+$ indicating that a change of over 1000 ppm would be expected on free cation formation, and that in fact the change of 228.5 ppm is indicative of relatively little cationic character to the silicon in the species formed. Calculations also give a value of 205.8 ppm for the central silicon in $(Me_3Si)_3SiC_6H_6^+$ in the gas phase, which is much closer to that of the experimentally determined solution species than the value for $(Me_3Si)_3Si^+$[110]. (Early work by Lambert and Sun on the potential of silyl substitution for the stabilization of silicenium ions derived from silyl perchlorates also concluded that 'Silyl substitution offers no palpable stabilisation of positive charge on trivalent silicon'[111].) The argument that very large chemical shift changes should be associated with a change from R_3SiH to $R_3Si(solvent)^+$ to R_3Si^+ should also be used with caution if the substituents R are not simple alkyl groups. For example, the ^{29}Si chemical shifts for $(Me_2N)_3SiH$, $(Me_2N)_3Si(H_2O)^+$ and $(Me_2N)_3Si^+$ are calculated to be −20.8, −20.5 and 42.1 ppm respectively[112]. Relatively little experimental work has been carried out recently using non-alkyl or aryl substituents and careful comparisons should be made between experimental and calculated ^{29}Si chemical shift values before conclusions can be drawn. It should be noted that Ph_3SiH, $EtMe_2SiH$, $Me_2(Me_3SiCH_2)SiH$ and $Ph_2(Me_3SiCH_2)SiH$ also react with $Ph_3CB(C_6F_5)_4$, but to give, as judged by ^{29}Si NMR spectroscopy, a variety of unidentified products, none of which have a ^{29}Si resonance at lower field than 60 ppm[67]. It is unclear why such relatively closely similar substrates to those that do seem to give single species should undergo apparently significantly different reactions. The reactions of the trityl salt with t-Bu_3SiH and with t-$Bu_2(C_6H_{11}$-$c)SiH$ are very slow and are also reported to give unidentified mixtures of products[67]. It is clear that further work needs to be carried out to ascertain the influence of a range of substituents at silicon and to determine what other products are formed in the reactions with trityl salts.

The ^{29}Si NMR data described above and detailed in Table 1 are derived mainly from solution state spectra for which it is difficult to take into account the effect that the solvent

TABLE 1. Solution ^{29}Si chemical shifts for various compounds with some cationic character containing silicon, together with calculated values and related species for comparison[a]

Compound	Solvent	Chemical shift (δ, ppm from Me$_4$Si)	References
Me$_3$SiH	C$_6$D$_6$	−17.5[b]	102
	CD$_2$Cl$_2$, at −70 °C	−15.5	113
Me$_3$SiClO$_4$	CH$_2$Cl$_2$	44.0	77
	sulpholane	46.5	77
	none	47	81
	C$_5$D$_5$N	42.6	90
	calc. value	40.2	83
	calc. value	58.6	114
Me$_3$SiOTf	none	43.54	115
	CH$_2$Cl$_2$	43.7	90
	CH$_2$Cl$_2$/sulpholane	46.6	90
	C$_5$D$_5$N	40.8	90
Me$_3$SiCl	CH$_2$Cl$_2$	31.1	90
	C$_5$D$_5$N	31.8	90
Me$_3$SiBr	CH$_2$Br$_2$	27.3	104
	benzene	26.41	116
	calc. value	33.9	117
Me$_3$SiBr.AlBr$_3$	CH$_2$Br$_2$	62.7	104
Me$_3$SiBrH$^+$	calc. value	154.4	117
Me$_3$SiBrMe$^+$	calc. value	117.5	117
Me$_3$Si[B(C$_6$F$_5$)$_4$]	C$_6$D$_6$	83.6	67
Me$_3$SiC$_6$H$_6$$^+$	calc. value	77.9	109,117
Me$_3$SiC$_6$H$_5$Me$^+$	calc. value	60.4	109
Me$_3$SiOH	acetone	14.9	118
	calc. value	15.1	117
Me$_3$Si(OH$_2$)$^+$	calc. value	101.9	117
[Me$_3$SiOEt$_2$]$^+$ {B[C$_6$H$_3$(CF$_3$)$_2$-3,5]$_4$}	CD$_2$Cl$_2$, at −70 °C	66.9	113
[(Me$_3$Si)$_2$OEt]$^+$	CD$_2$Cl$_2$	59.0	113
(Me$_3$Si)$_2$OH$^+$ {B[C$_6$H$_3$(CF$_3$)$_2$-3,5]$_4$}	calc. value	67.6	117
(Me$_3$Si)$_3$O[B(C$_6$F$_5$)$_4$]	CD$_2$Cl$_2$	51.1	119
Me$_3$Si(MeCN)$^+$	CD$_2$Cl$_2$, at −10 °C	38.5	120
	calc. value	51.5	117
Me$_3$Si(C$_5$H$_5$NO)$^+$	CD$_2$Cl$_2$	49.4	121
Me$_3$Si(C$_5$H$_5$N)$^+$	CD$_2$Cl$_2$	42.3	121
[Ph$_2$C=O-SiMe$_3$]$^+$ {B[C$_6$H$_3$(CF$_3$)$_2$-3,5]$_4$}	CD$_2$Cl$_2$	52.3	122
Me$_3$Si$^+$	calc. value	354	106
	calc. value	354.2	117
	calc. value	355.7	83
	calc. value	382	106
	calc. value	385 ± 20	105
Et$_3$SiH	C$_6$D$_6$	0.2	102
	MeCN	6.0	102
Et$_3$SiOTf	none	44.46	115
Et$_3$SiOTf.BCl$_3$	none	76.72	115
Et$_3$Si[B(C$_6$F$_5$)$_4$]	C$_6$D$_6$	92.3	67
	C$_6$D$_6$/C$_6$H$_5$Me (3 : 1)	87.1	67
	p-(CD$_3$)$_2$C$_6$H$_4$	85.6	67
	C$_6$D$_5$CD$_3$	81.8	67

(continued overleaf)

TABLE 1. (continued)

Compound	Solvent	Chemical shift (δ, ppm from Me$_4$Si)	References
	sulpholane	58.4	67
	CD$_3$CN	36.7	67
Et$_3$SiC$_6$H$_5$Me$^+$	calc. value	79.7	117
[Et$_3$SiCH$_2$CPh$_2$] [B(C$_6$F$_5$)$_4$]	C$_6$D$_6$	46.2	123
Et$_3$SiC$_3$H$_6^+$	calc. value	70.0	117
[Et$_3$Si(NCPr)]$^+$ {B[C$_6$H$_3$(CF$_3$)$_2$-3,5]$_4$}$^-$	PrCN	37.01	101
Et$_3$Si$^+$	calc. value	371.2	106
	calc. value	371.3	117
	calc. value	376.6	109
	calc. value	415.6	106
Et$_3$Si(C$_6$H$_6$)$^+$	calc. value	104.6	109
Et$_3$Si(C$_6$H$_5$Me)$^+$	calc. value	82.1	109
i-Pr$_3$SiH	C$_6$D$_5$CD$_3$	12.1	103
i-Pr$_3$SiBr	C$_6$D$_5$CD$_3$	45.4	103
i-Pr$_3$SiBr.AlBr$_3$	C$_6$D$_5$CD$_3$	55.8	103
i-Pr$_3$Si(Br$_5$CB$_9$H$_5$)	C$_6$D$_5$CD$_3$	97.9	103
[i-Pr$_3$Si(MeCN)][Br$_5$CB$_9$H$_5$]	CD$_2$Cl$_2$	33.8	103
i-Pr$_3$Si[B(C$_6$F$_5$)$_4$]	C$_6$D$_6$	107.5	67
	C$_6$D$_5$CD$_3$	94.0	103
i-Pr$_3$Si$^+$	calc. value	342.3	106
	calc. value	371.0	106
Bu$_3$SiClO$_4$	CH$_2$Cl$_2$	44.6	90
	CH$_2$Cl$_2$/sulpholane	46.3	90
Bu$_3$Si[B(C$_6$F$_5$)$_4$]	CH$_2$Cl$_2$	33.4^c	90
	CH$_2$Cl$_2$/HMPA-d_{18}	32.8^c	90
i-Bu$_3$Si[B(C$_6$F$_5$)$_4$]	C$_6$D$_6$	99.5	67
i-Pr$_2$MeSi[B(C$_6$F$_5$)$_4$]	C$_6$D$_6$	96.9	67
	CD$_3$CN	41.9	67
	pyridine/C$_6$D$_6$	44.3	67
(C$_6$H$_{13}$)$_3$Si[B(C$_6$F$_5$)$_4$]	C$_6$D$_6$	90.3	67
	mesitylene	65	67
MePh$_2$SiH	CD$_2$Cl$_2$, at $-40°$C	-17.4	113
[MePh$_2$SiOEt$_2$] {B[C$_6$H$_3$(CF$_3$)$_2$-3,5]$_4$}	CD$_2$Cl$_2$	38.0	113
MePh$_2$Si[B(C$_6$F$_5$)$_4$]	C$_6$D$_6$	73.6	67
	CD$_3$CN	4.2	67
[MePh$_2$Si(NCR)]$^+$ {B[C$_6$H$_3$(CF$_3$)$_2$-3,5]$_4$}	MeCN	4.23	103
(mesityl)$_3$Si[B(C$_6$F$_5$)$_4$]	C$_6$D$_6$	225.5	124
	C$_6$D$_6$/toluene (1 : 3)	225.6	124
	C$_6$D$_6$/p-xylene (1 : 1)	225.6	114
	C$_6$D$_6$/MeCN (1 : 3)	37	114
	C$_6$D$_6$/Et$_3$N (1 : 1)	47.1	114
(Me$_3$Si)$_3$SiH	C$_6$D$_6$	-117.4^b	102
	calc. value	-124.2	109
	calc. value	-119.6	110
(Me$_3$Si)$_3$Si$^+$	calc. value	925.3	109
	calc. value	920.4	110
(Me$_3$Si)$_3$Si[B(C$_6$F$_5$)$_4$]	C$_6$D$_6$	111.1	67
	C$_6$H$_5$Me	96.1	67
(Me$_3$Si)$_3$Si(C$_6$H$_6$)$^+$	calc. value	205.8	110
11	e	87.7	125
(Me$_2$N)$_3$Si[B(C$_6$F$_5$)$_4$]	C$_6$D$_6$	-30.8	126
	CH$_2$Cl$_2$	-39.3	126

TABLE 1. (continued)

Compound	Solvent	Chemical shift (δ, ppm from Me$_4$Si)	References
(Me$_2$N)$_3$Si$^+$	calc. value	42.1	112
[Cp$_2$Zr(μ-H)(SiHPh)]$_2^{2+f}$	C$_6$D$_6$	105.9	127
	C$_2$D$_2$Cl$_4$	110.0	127
[Cp$_2$Zr(μ-H)(SiHCH$_2$Ph)]$_2^{2+f}$	C$_6$D$_5$CD$_3$	110.9	127
[CpCp*Zr(μ-H)(SiHPh)]$_2^{2+f}$	C$_6$D$_6$	106.0	127
[(MeCp)$_2$Zr(μ-H)(SiHPh)]$_2^{2+g}$	C$_6$D$_6$	101.5	127
[(p-tolylS)$_2$SiRu(PMe$_3$)$_2$Cp*][BPh$_4$]	CH$_2$Cl$_2$, at -60 °C	259.4	128
[(EtS)$_2$SiRu(PMe$_3$)$_2$Cp*][BPh$_4$]	CH$_2$Cl$_2$, at -60 °C	264.4	128
[(EtS)$_2$Si($trans$-PtH(PCy$_3$)$_2$)][BPh$_4$]	e	308.65	129
[Me$_2$SiRu(PMe$_3$)$_2$Cp*][B(C$_6$F$_5$)$_4$]	CH$_2$Cl$_2$, low temp.	311	130
[Ph$_2$SiRu(PMe$_3$)$_2$Cp*][B(C$_6$F$_5$)$_4$]	CH$_2$Cl$_2$, low temp.	299	130
12	e	−12.1	131

[a] The structural formulae are mostly written as given by the original authors, but, in the light of recent calculations, the species shown as silicenium ions in solution are probably better represented as arenium, bromonium ions etc.; see text for a more detailed description.
[b] The value of ca 17.5 given for Me$_3$SiH in some reports is incorrect.
[c] Reaction with the solvent gives Bu$_3$SiCl.
[d] The value of ca 117.4 given for (Me$_3$Si)$_3$SiH in several reports is incorrect.
[e] Solvent not reported.
[f] Counter anion, [B(Bu)$_n$(C$_6$F$_5$)$_{4-n}$]$_2^{2-}$.
[g] Counter anion, [BH(C$_6$F$_5$)$_3$]$_2^{2-}$.

has on chemical shifts. It is clear from computational work carried out in the last few years that solvents do play a crucial role by forming species in which there is a significant and well defined interaction or bond between the silicon and a solvent molecule, and that these species are calculated to have ^{29}Si NMR chemical shift values close to those found experimentally in solution. (The nature of the Et$_3$Si$^+$ species in aromatic solvents has been particularly controversial and is discussed in more detail in Section IV.) Thus, the ^{29}Si NMR data may better be attributed to species described as R$_3$Si(solvent)$^+$ rather than a silicenium ion salt dissolved in a non-interacting solvent. Details of the many calculations carried out in this area are discussed more fully in the chapter by Schleyer and coworkers. X-ray crystallographic structure determinations have been carried out on some of the species given in Table 1 and they do indeed show interactions between the silicon and either a solvent molecule or the anion; these are described in more detail in Section IV.

The potential for solvent and anion complexation to a cationic silicon centre has led to a move away from the solvents and anions traditionally regarded as potential nucleophiles to both solvents (for example benzene and toluene) and anions (for example [B(C$_6$F$_5$)$_4$]$^-$ and halogenated carboranes) that are less nucleophilic. The reactions between chlorosilanes with Na{[3,5-(CF$_3$)$_2$C$_6$H$_3$]$_4$B} or hydrosilanes with Ph$_3$C{[3,5-(CF$_3$)$_2$C$_6$H$_3$]$_4$B} in CD$_2$Cl$_2$ in the presence of ethers gives oxonium salts, for example [Me$_3$SiOEt$_2$]$^+${[3,5-(CF$_3$)$_2$C$_6$H$_3$]$_4$B}$^-$, which has a ^{29}Si NMR chemical shift of 66.9 ppm (cf 46 ppm for Me$_3$SiClO$_4$), indicative of a significant degree of positive charge buildup on the silicon[113]. Since the silyl perchlorates are not attacked by Et$_2$O alone, these results suggest that it is a silicenium ion that is attacked by the ether. Trisilyloxonium ions, for example (Me$_3$Si)$_3$O$^+$, are formed as relatively long-lived (at -70 °C) products from the reaction between Me$_3$SiH and Ph$_3$CB(C$_6$F$_5$)$_4$ in the presence of a siloxane[119]. Again the use of an anion with low coordinating ability allows the formation of silicenium ions which silylate the siloxane present to give the novel oxonium ions. The ^{29}Si NMR chemical shift for the Me$_3$Si group in (Me$_3$Si)$_3$O$^+$ is 43.4 ppm deshielded compared with (Me$_3$Si)$_2$O, again indicative of the positive charge buildup on the silicon. If the same reaction is carried out with Ph$_3$C{[3,5-(CF$_3$)$_2$C$_6$H$_3$]$_4$B} the more nucleophilic fluorines in the aryl ligand are abstracted by the trisilyloxonium ions to give fluorosilanes[119]. This also occurs in the reaction between Ph$_3$C{[3,5-(CF$_3$)$_2$C$_6$H$_3$]$_4$B} and hydrosilanes in CH$_2$Cl$_2$, but, in the presence of a more strongly coordinating solvent such as butyronitrile, silylnitrilium salts [R$_3$Si(NCC$_3$H$_7$)]$^+$ {[3,5-(CF$_3$)$_2$C$_6$H$_3$]$_4$B}$^-$ are formed (R$_3$ = Et$_3$, Me$_2$Bu-t, t-Bu$_2$Bu-s etc.) which undergo exchange of the nitrile on the NMR timescale indicating a relatively weak Si−N interaction[101]. A crystalline acetonitrilium salt i-Pr$_3$Si(MeCN)]$^+$ [closo-6,7,8,9,10-Br$_5$CB$_9$H$_5$]$^-$ was isolated by Xie and coworkers from a similar reaction[103]. The positive character of the silicon in this case was indicated by both the ^{29}Si NMR chemical shift of 37.2 ppm in CD$_2$Cl$_2$ solution, and the C−Si−C angle in the cation being opened up to an average of 115.8° (see Section IV)[103]. Reaction between R$_3$SiH [R$_3$ = Me$_3$, Ph$_2$Me, (2-thienyl)$_2$Me, or (3,5-t-Bu$_2$C$_6$H$_3$)$_3$] and Ph$_3$C{[3,5-(CF$_3$)$_2$C$_6$H$_3$]$_4$B} in CD$_2$Cl$_2$ solution containing excess MeCN gives [R$_3$Si(MeCN)]$^+$ {[3,5-(CF$_3$)$_2$C$_6$H$_3$]$_4$B}$^-$ which has, in the case for R$_3$ = Me$_3$, a ^{29}Si NMR chemical shift that varies with the amount of MeCN present. If less than two equivalents are present, then the shift is about 38 ppm, but if a large excess is present then the shift is about 28 ppm, which can be attributed to an equilibrium being set up with the five-coordinate species [R$_3$Si(MeCN)$_2$]$^+$ {[3,5-(CF$_3$)$_2$C$_6$H$_3$]$_4$B}$^-$ containing two coordinated MeCN molecules[120]. This effect has also been proposed to explain the difference in chemical shifts (see Table 1) between the spectra of [i-Pr$_3$Si(MeCN)][Br$_5$CB$_9$H$_5$] in CD$_2$Cl$_2$ and MeCN[103]. The use of the weakly coordinating closo-7,8,9,10,11,12-Br$_6$CB$_{11}$H$_6$$^-$ anion has also allowed the first protonated silanol [t-Bu$_3$SiOH$_2$]$^-$ to be prepared[59]. The compounds shown in Table 1 in which there is a halogenated carborane as the anion do not seem to show an interaction between the silicon and a solvent molecule in the solid state but rather a halogen atom in the anion. This may also persist in solution as is indicated by ^{11}B NMR studies as described above. This type of interaction would also explain why the chemical shifts in solution (and in the solid state) are also considerably more upfield than free silicenium ions would be, in a similar manner to the R$_3$Si(solvent)$^+$ species. A more detailed discussion of these compounds is given below in Section IV.

Perhaps the most interesting ^{29}Si NMR data to be reported recently is that on a species thought to be (mesityl)$_3$Si[B(C$_6$F$_5$)$_4$]^{124}. The species seems to be a product from several reactions (Scheme 1), although not from the reaction between (mesityl)$_3$SiH and trityl cation, the cleanest product being obtained from the reaction between (mesityl)$_3$SiCH$_2$CH=CH$_2$ and the β-silylcarbocation (Et$_3$SiCH$_2$)Ph$_2$C$^+$. The product may be obtained as an oil [unfortunately B(C$_6$F$_5$)$_4$-containing species rarely seem to give material suitable for X-ray crystallography in this field] and solutions in various aromatic solvents (see Table 1) give ^{29}Si chemical shifts of ca 225.5 ppm. In mixed solvents containing a good donor such as Et$_3$N, the shift (see Table 1) is what might be expected for an R$_3$Si(solvent)$^+$ species. It is currently unclear exactly what the nature of the species responsible for the strongly deshielded value for the silicon is, but the ^{29}Si NMR chemical shift would seem to be consistent with the presence of a substantially free silicenium ion[132]. Unfortunately, there have been few calculations on triarylsilicon species such as Ar$_3$Si$^+$ or Ar$_3$Si(solvent)$^+$ reported. If the value for Ph$_3$Si$^+$ derived from comparison with Ph$_3$C$^+$[104] similarly underestimates the chemical shift compared with calculation, as does Me$_3$Si$^+$ (250±25 ppm by comparison with the carbon analogue, versus ca 370±15 ppm for more recently calculated values), then a value for Ph$_3$Si$^+$ might tentatively be given as ca 245 ± 25 ppm, quite close to the observed value for the (mesityl)$_3$Si-containing species. Recent preliminary calculations[133] on Ph$_3$Si$^+$ and on (2,6-Me$_2$C$_6$H$_3$)$_3$Si$^+$ show that the ^{29}Si NMR chemical shift depends strongly on the aryl twist angle and that the chemical shifts predicted for such species are indeed close to that found experimentally for (mesityl)$_3$Si$^+$. It should also be noted that there is very little change in the ^{29}Si chemical shift for the species in several different aromatic solvent mixtures (Table 1). This seems to be in contrast to the work reported for Et$_3$Si$^+$ in various solvents where changing the nature of the aromatic solvent has a small but apparently significant effect on the chemical shift (Table 1). The (mesityl)$_3$Si species may thus be interacting relatively little with the solvent, or there may be a strong interaction between the cation and the anion, to the exclusion of the solvent, or a strong interaction between the silicon and C$_6$D$_6$ which is not affected by addition of other aromatic solvents. It would be of interest to record the chemical shift in toluene alone to see if the same shift is obtained in the absence of C$_6$D$_6$. These arguments can only be regarded as speculative at this stage and further computational work needs to be carried out to estimate chemical shifts for Ar$_3$Si$^+$ and Ar$_3$Si(solvent)$^+$ species together with an investigation of the effect of substituents on the aromatic rings. It may be that the steric protection afforded by the *ortho* substituents in a (mesityl)$_3$Si-containing species prevents the close approach of any solvent or the counter anion to the silicon and that a three-coordinate silicon species is indeed present. (Steric effects are clearly important in the gas-phase silylations of aromatic molecules where sites *ortho* to a methyl substituent are strongly shielded from attack.) Pidun, Stahl and Frenking have suggested a similar approach to steric protection of a three-coordinate silicon centre by the use of the 2,5-*t*-Bu$_2$-pyrrolidino group as a substituent[112].

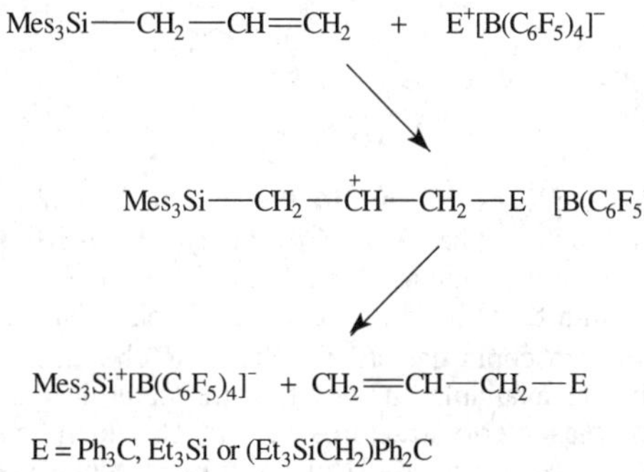

Mes$_3$Si—CH$_2$—CH=CH$_2$ + E$^+$[B(C$_6$F$_5$)$_4$]$^-$

↓

Mes$_3$Si—CH$_2$—$\overset{+}{\text{CH}}$—CH$_2$—E   [B(C$_6$F$_5$)$_4$]$^-$

↙

Mes$_3$Si$^+$[B(C$_6$F$_5$)$_4$]$^-$ + CH$_2$=CH—CH$_2$—E

E = Ph$_3$C, Et$_3$Si or (Et$_3$SiCH$_2$)Ph$_2$C

SCHEME 1

Another recent attempt to generate a silicenium ion has followed from the computational result that the symmetrically bridged 6-sila-2-norbornyl cation is a local minimum on the potential surface and that it is more stable than the 2-norbornyl cation[4]. Auner, Schleyer, Müller and coworkers have reported that treatment of the silicon hydride **10** with a stoichiometric amount of trityl cation affords a species thought to be the bridged cation **11** (equation 5) having a ^{29}Si chemical shift of 87.7 ppm[125]. The calculated value for the fully open, three-coordinate cation is 364.2 ppm but the value for the bridged structure is calculated to be only 101.4 ppm, the sum of the angles at silicon being 350.6°[125]. Although in this case the silicon atom is pseudo-five-coordinate, the relatively good agreement between the experimental and calculated values for the ^{29}Si NMR chemical shift suggests that the species formed in solution does indeed have a structure similar to that shown for the cationic species **11**, and that there is little, if any, specific interaction between the cation and the solvent. Clearly, a solid state structure determination of the norbornyl derviative **11** would be of great interest.

$$\text{(10)} \xrightarrow[\text{toluene}]{\text{Ph}_3\text{C}^+[\text{B}(\text{C}_6\text{F}_5)_4]^-} \text{(11)} \quad [\text{B}(\text{C}_6\text{F}_5)_4]^- + \text{Ph}_3\text{CH} \tag{5}$$

The reaction of decamethylsilicocene with catechol gives the unusual silicon hydride **12**, the structure of which was determined by NMR spectroscopy (equation 6)[131]. The ^{29}Si NMR chemical shift for **12** is 386 ppm deshielded when compared with Cp$_2^*$Si and the Si−H coupling constant is 302 Hz both of which could be regarded as consistent with a silicenium ion. However, although the silicon in **12** is formally only bound to three substituents, its coordination number can be regarded as higher because the Cp* groups are π- rather than σ- bonded.

$$\text{Cp}_2^*\text{Si} + \text{catechol} \longrightarrow [\text{Cp}_2^*\text{SiH}]^+ \quad \text{(12)} \tag{6}$$

Dinuclear zirconium complexes **13** and **14** that are part of a catalytic cycle in the polymerization of silanes such as PhSiH$_3$ by Cp$'_2$MCl$_2$/2BuLi/B(C$_2$F$_5$)$_3$ (where Cp' can be C$_5$H$_5$ or a substituted cyclopentadienyl ligand, and M = Ti, Zr or Hf) can be prepared according to equations 7 and 8. Although no crystallographic data for the complexes are available, the IR and spectroscopic data (the ^{29}Si NMR chemical shifts are all greater than 100 ppm. see Table 1) available suggest that the cationic parts of the complexes exist as a hybrid of two mesomeric structures **15** and **16** which can be thought of as a cationic zirconocene with a α-agostic Si−H or an intramolecularly stabilized silicenium ion, respectively[127,134,135]. Solid state structural data for these complexes would be of obvious interest.

$$\text{Cp}'_2\text{ZrCl}_2 \xrightarrow[\text{3. RSiH}_3]{\text{1. 2BuLi, 2. B(C}_6\text{F}_5)_3} [\text{Cp}'_2\text{Zr}(\mu\text{-H})(\text{SiHR})]_2^{2+}[\text{BBu}_n(\text{C}_6\text{F}_5)_{4-n}]_2^{2-} \tag{7}$$

$$\text{(13)}$$

Cp' = C$_5$H$_5$; R = Ph, PhCH$_2$ etc.

$$(\text{MeCp})_2\text{ZrH}_2 \xrightarrow[\text{2. PhSiH}_3]{\text{1. B(C}_6\text{F}_5)_3} [(\text{MeCp})_2\text{Zr}(\mu\text{-H})(\text{SiHPh})]_2^{2+}[\text{BH}(\text{C}_6\text{F}_5)_3]_2^{2-} \tag{8}$$

$$\text{(14)}$$

A range of transition metal complexes (see Table 1) have been prepared by Tilley and coworkers in recent years that contain three-coordinate silicon[128−130]. The ^{29}Si NMR chemical shifts for these complexes are highly deshielded, possibly indicative of cationic character, and the geometry around the silicon is trigonal planar, also consistent with silicenium ion character. Calculations to estimate the effect of a transition metal substituent on the chemical shifts do not yet seem to have been carried out. There are several possible resonance forms for the complex cations (Scheme 2); the short metal–Si distances found in the solid state, however, are consistent with double bond character between the metal and silicon, and **17** is probably the main contributor. These compounds may therefore best be regarded as silylene complexes even though they fulfil some of the criteria that are often associated with silicenium ions.

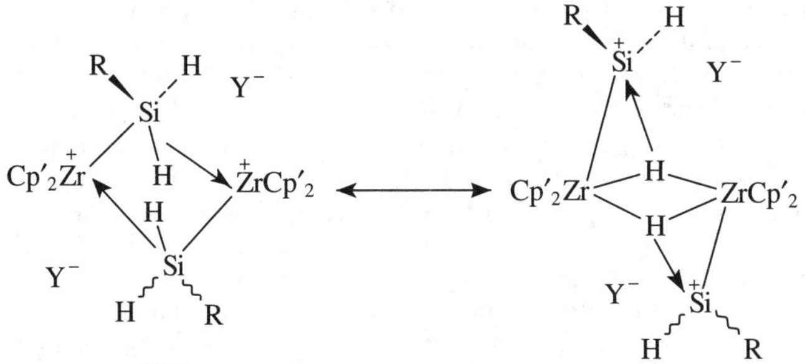

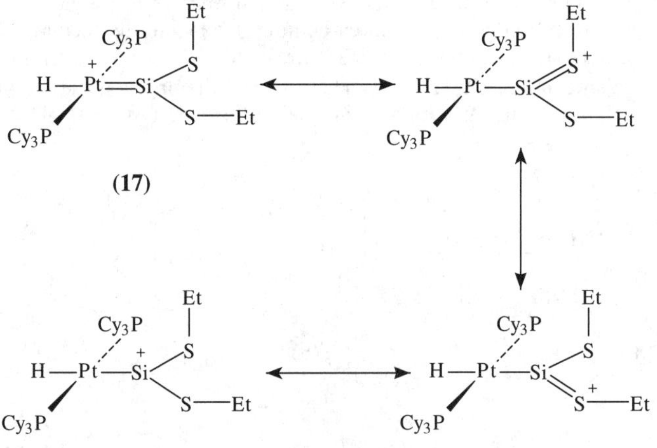

SCHEME 2

D. Silicenium Ions as Reaction Intermediates

The mechanism of the now widely used hydride transfer reaction for the formation of possible silicenium ions has been investigated in some detail. Studies of the reactions between R_3SiH (R = Et, Ph, n-Bu etc.) and trityl salts such as Ph_3CBF_4 and Ph_3CSbF_6 led Chojnowski and coworkers to conclude, on the basis of steric, substituent, kinetic isotope and ring strain effects, that the mechanism involves in the rate-determining step a single electron transfer process which, in a second fast step, yields a silicenium-like species[136,137] (equation 9). This conclusion was, however, questioned by Apeloig[11] who showed by calculations that the experimentally observed substituent effects on the hydride transfer rates are fully consistent with a rate-determining formation of a silicenium-like species. Mayr and coworkers have carried out further detailed studies on the reactions between R_3SiH compounds and diarylcarbenium ions Ar_2HC^+ (Ar = Ph, p-tolyl, p-MeOC$_6$H$_4$, p-PhOC$_6$H$_4$)[138]. They concluded, in contrast to Chojnowski and coworkers, that, based on deuterium labelling studies as well as steric and electronic arguments, the mechanism is not a single electron transfer process but involves formation of a silicenium ion in the rate-determining step (equation 10). It was also shown, using a series of β-substituted silanes such as (Me$_3$MCH$_2$)Me$_2$SiH (M = C, Si, Ge or Sn), that both electronic and steric effects are important in determining the reaction rate, and that hyperconjugation is also effective in stabilizing silicenium ions, but that the effect is much smaller than that for comparable carbenium ions[139]. More recent computational work also shows that the mechanism is not a single electron transfer process but is actually a barrierless synchronous hydride transfer (SHT, equation 10), and that the single electron transfer mechanism is calculated to be highly disfavoured[140].

$$R_3SiH + R'_3C^+X^- \xrightarrow[\text{slow}]{\text{SET}} R_3SiH^{+\bullet}X^-//R'_3C^\bullet \xrightarrow{\text{fast}} R_3Si^+X^- + R'_3CH \qquad (9)$$

$$R_3SiH + R'_3C^+X^- \xrightarrow{\text{SHT}} R_3Si^+X^- + R'_3CH \qquad (10)$$

Work by Apeloig and Stanger on adamantyl substituted silanes has provided another example of a reaction in which formation of a silicenium ion seems to represent the best interpretation of the experimental results. Solvolysis of the adamantyl derivatives **18** in aqueous (CF$_3$)$_2$CHOH proceeds (Scheme 3) to give a mixture of products, one of which, **20**, is a simple substitution product from carbenium ion **19** but the second, **22**, appears to be the result of a 1,2-methyl migration from the silicon to the ring carbon to which it is attached to give silicenium ion **21**[141,142]. The ratios of the products can be interpreted as due to the reactions of solvent separated ion pairs, but it is also possible that a bridged

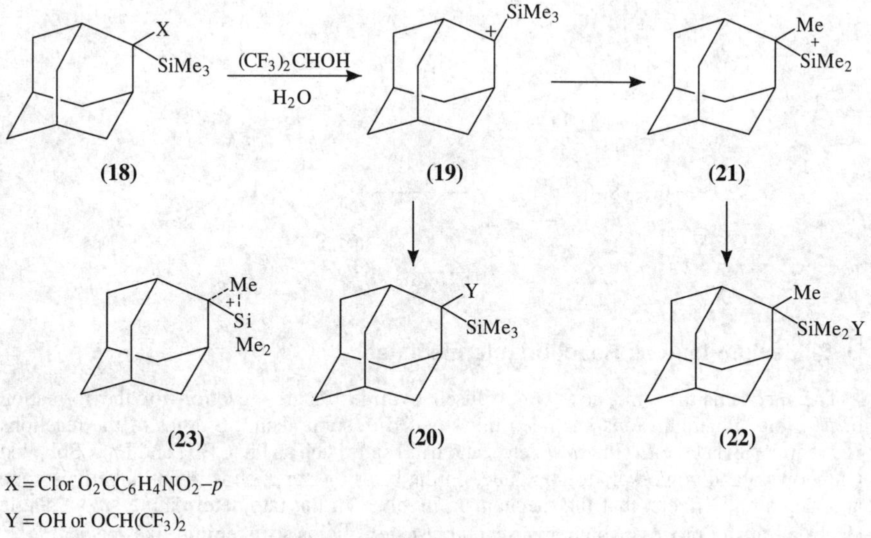

X = Cl or O$_2$CC$_6$H$_4$NO$_2$-p
Y = OH or OCH(CF$_3$)$_2$

SCHEME 3

cation such as **23**, which could be attacked by the solvents at either end of the bridge, could be involved[12,143]. The rearrangement of **19** to **21** is consistent with the relative thermodynamic stabilities for carbenium and silicenium ions found in the gas phase as described in Section II and calculations, for example on $Me_3CSiMe_2^+$ and $Me_3SiCMe_2^+$ which show that the silicenium ion is more stable by 9.7 kcal mol^{-1} [11].

The involvements of 1,3-bridged silicenium ion intermediates, analogous to the 1,2-bridged species **23**, has been proposed for a range of reactions involving silicon compounds bearing the exceptionally bulky $(Me_3Si)_3C$ group. Eaborn and coworkers have found that compounds of the type $(Me_3Si)_3CSiRR'X$ and $(Me_3Si)_2C(SiMe_2X)(SiRR'X)$ (where R = Me, Et, Ph etc. and X = halogen, H etc.) undergo reactions (e.g. equation 11) with some electrophiles including silver salts AgY(Y = $MeCO_2$, CF_3CO_2, ClO_4 etc.), CF_3CO_2H and ICl to give, in the case of $(Me_3Si)_3C$-containing precursors, rearranged products $(Me_3Si)_2C(SiMe_2Y)(SiRR'Me)$ if the substituents R and R' are large (e.g. Ph), or a mixture of the rearranged and the unrearranged products $(Me_3Si)_3CSiRR'Y$ if the R and R' groups are smaller, eg. Et (for a review see Reference 144). Such reactions are thought to proceed via a bridged cationic intermediate of type **24**, which may be attacked by an incoming nucleophile at either end of the bridge, attack at Si(1) leading to unrearranged products and attack at Si(2) leading to rearranged products. (Calculations also reveal that structures of type **24** do correspond to local energy minima[4].) The difference in size between the substituents on the silicon atoms at either end of the bridge would then have a significant effect in determining at which end a nucleophile would attack larger groups, eg. Ph on Si(1), would prevent attack at that silicon leading to exclusive attack at Si(2) to give rearranged products. This work has been extended to include other groups than Me as the bridging or migrating group, for example Ph[145-149], $CH=CH_2$[150], OMe[151,152], Cl[153] and N_3[153,154] have all been found to undergo similar rearrangements. It is now clear, however, that steric effects may not be the overriding factors in determining the degree of rearrangement and that electronic factors may dominate if the electronic environment at each of the two ends of a bridged cation is very different. For example, $(Me_3Si)_3CSi(OMe)_2I$ reacts with ICl to give only the unrearranged product $(Me_3Si)_3CSi(OMe)_2Cl$[155], $(Me_3Si)_3CSiPh(OMe)I$ reacts with AgOCN to give $(Me_3Si)_3CSiPh(OMe)NCO$[156] and $(Me_3Si)_3CSiH_2I$ reacts with $AgBF_4$ in Et_2O solution to give $(Me_3Si)_3CSiH_2F$ [the same reaction in CH_2Cl_2 gives a 1 : 1 ratio of $(Me_3Si)_3CSiH_2F$ and $(Me_3Si)_2C(SiMe_2F)(SiH_2Me)$][157]. The inadequacy of a simple steric argument for rearrangements of this type of bulky silane has also been shown in recent studies using isotopically labelled compounds in which the two ends of the bridged cation are the same size, i.e. by using $(Me_3Si)_3CSi(CD_3)_2I$ as a precursor. In reactions of this bulky iodide with silver salts such as $AgClO_4$, $AgNO_3$ and AgO_2CCF_3, products that are predominantly unrearranged are formed and it is now clear that as more data are accumulated on this type of reaction, more complicated reaction mechanisms involving several intermediates will have to be invoked[158].

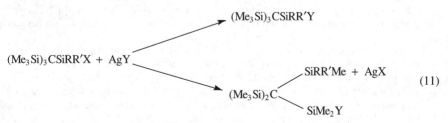

(11)

Y = $MeCO_2$, CF_3CO_2, ClO_4, etc.

$$(Me_3Si)_2C \overset{(2)}{\underset{(1)}{\diagdown}} \overset{RR'}{\underset{Si}{Si}} + Me$$
$$\qquad\qquad Me_2$$

(24)

The reaction between polysilylacylsilanes $(Me_3Si)_3SiCOR$ (R = t-Bu, adamantyl or bicyclo[2.2.2]octyl) with $TiCl_4$ in CH_2Cl_2 solution gives rise, after an aqueous workup, to the 1,3-bis(silanols) 25. The products may result from either of the routes shown in Scheme 4, both of which involve 1,2-migrations of methyl groups to a silicenium ion centre and of silyl groups to a carbocationic centre[159].

The reaction between Ph_3SiH and $Ph_3CB(OTeF_5)_4$ in CH_2Cl_2 solution leads to formation of $Ph_3SiOTeF_5$ and $B(OTeF_5)_3$ due to breakdown of the bulky anion. This may be due to attack by the strongly electrophilic Ph_3Si^+ although further evidence for the existence of the silicenium ion in this reaction is lacking[160]. Silicenium ions are also thought to be short-lived intermediates in the fluorination of silanes R_3SiH (R = Me, Et, t-Bu, $PhCH_2$, Ph etc.) by $NOBF_4$ or NO_2BF_4 (equation 12), although as the reaction is carried out in MeCN solution the formation of $[R_3Si(MeCN)]^+$ species is also likely[161]. The species $[R_3Si(MeCN)]^+$ are thought to be present as short-lived intermediates in the reaction between R_3SiCl (R_3 = $PhMe_2$, Bu_2H, Ph_2Bu-t etc.) and $NaBF_4$ in MeCN solution (equation 13)[162].

$$R_3SiH \xrightarrow{NOBF_4 \text{ or } NO_2BF_4} R_3SiF + BF_3 + HNO \text{ or } HNO_2 \qquad (12)$$

R = Me, Et, t-Bu, Ph, $PhCH_2$ etc.

$$R_3SiCl + NaF \text{ or } NaBF_4 \xrightarrow{MeCN} R_3SiF + NaCl \text{ or } MeCN \bullet BF_3 \qquad (13)$$

R_3 = t-Bu_2H, t-$BuMe_2$, Ph_2Me etc.

The reactions between the silylenes SiX_2 (X = F, Cl, Br or I), generated in the gas phase, and condensed into toluene solutions containing I_2 or ICl at $-90\,°C$, afford mixtures of the *ortho*, *meta* and *para* isomers of $SiX_2I(C_6H_4Me)$ probably due to an electrophilic attack on the aromatic ring by short-lived X_2ISi^+ species[163]. *Ortho*-substituted products are favoured in these reactions, the $o : p : m$ ratio for the Cl_2Si/I_2/toluene system being 10 : 4 : 3, but if ethylbenzene is used in place of toluene the proportion of *ortho*-substituted product is reduced and the ratio is 4 : 2 : 3. This was attributed to steric effects and is also consistent with the formation of only the *para* isomer in the $[Et_3Si(toluene)][B(C_6F_5)_4]$ compound as described above, which results from the silylation of toluene by the relatively bulky Et_3Si^+.

The reaction between $(ArS)_3SiCl$ (Ar = 2,4,6-t-$Bu_3C_6H_2$) and $AgClO_4$ in benzene solution at $50\,°C$ gives several unexpected sulphur-containing compounds such as $(3,5$-t-$Bu_2C_6H_3S)_2$ that are thought to arise from the initial formation of covalent $(ArS)_3SiClO_4$ in equilibrium with its ionic form $(ArS)_3Si^+ClO_4^-$, which then undergoes fragmentation and rearrangement to give the observed products[164]. Attempts to generate silicenium ions from sterically hindered aryloxy and arylthio silanes $(ArO)_2(MeO)SiX$ (X = H or Cl), $(ArS)_3SiH$ or $(Ar'S)_3SiH$ (Ar = 2,4,6-t-$Bu_3C_6H_2$, Ar' = 2,4,6-i-$Pr_3C_6H_2$) have been unsuccessful[165] and no clear evidence was found for the intermediacy of silicenium ions

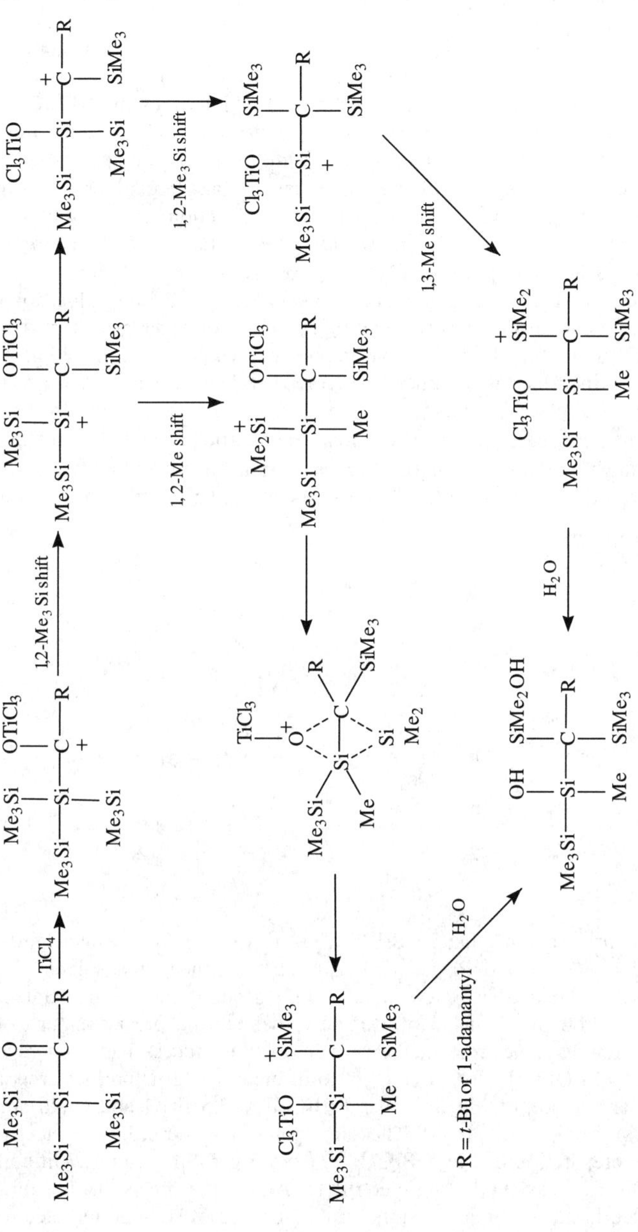

SCHEME 4

in the reaction of Me_3SiX/BX_3 (X = Cl, Br or I) with alcohols[166]. The reaction between glycidyl methacrylate and chlorosilanes is proposed to proceed via an electrophilic attack by R_3Si^+ on an epoxide ring, although in the DMF solution used the ion is undoubtedly solvated and is not a free silicenium ion[167].

A silicenium ion might potentially be stabilized by being coordinated to a $Cr(CO)_3$ unit, but there is no reaction between $(Ph_3SiH)Cr(CO)_3$ and Ph_3CBF_4 and, in contrast to the reaction of $(Ph_3CH)Cr(CO)_3$ with triflic acid which gives the $[(Ph_3C)Cr(CO)_3]^+$ cation, the reactions between $(Ph_3SiH)[Cr(CO)_3]_n$ ($n = 1$, 2 or 3) and triflic acid are thought to give protonated silanols (which are themselves unusual species) as judged by NMR spectroscopy[168]. Extended Hückel molecular orbital calculations also suggest that a silicenium ion can be significantly stabilized by interaction with a suitable transition metal fragment such as $CpMo(CO)_2$ or $Co(CO)_3$, but, so far, no stable complexes of this type containing a silicenium ion seem to have been isolated[169].

A study of the kinetic and thermodynamic aspects of the radiation-induced polymerization of 'super-dry' cyclic siloxanes $(Me_2SiO)_n$ ($n = 3$, 4 or 5) suggests that the reactions proceed via silicenium ions. Initial ionization proceeds via loss of a methyl group to give silicenium ions, which may then either be involved in backbiting reactions in which the silicenium centre is internally solvated by oxygen atoms in the chain **26**, or in propagation in which more reactive unsolvated ions **(27)** participate[170,171]. The formation of $(EtO)_3Si^+$ during the hydrolysis of $(EtO)_4Si$ has been postulated[172,173], but this seems highly unlikely and a mechanism involving five-coordinate intermediates is much more probable[174].

(26)   (27)

A recent patent[175] claims that $Et_3SiB(C_6F_5)_4$ is formed as a white solid from the reaction between Et_3SiH and $Ph_3CB(C_6F_5)_4$ in the absence of a solvent. The silicenium salt can then be used as a co-catalyst with group 4 transition metal complexes in the polymerization of olefins. Unfortunately, neither the precise nature of the silicenium cation nor its exact role in the polymerization process has been reported. An ether adduct $[Me_3Si(OEt_2)]^+[B(C_6F_5)_4]^-$, also of unknown structure, was reported to be formed from the reaction of Me_3SiCl and $LiB(C_6F_5)_4 \cdot 2.5 \ Et_2O$ and could similarly be used as a polymerization co-catalyst. The same patent also describes an electrochemical method for the preparation of $Me_3SiB(C_6F_5)_4$ from $Me_3SiSiMe_3$ in 1,2-difluorobenzene with n-$Bu_4NB(C_6F_5)_4$ as supporting electrolyte. Again, the precise nature of the active species produced is unclear, but reactions in the absence of solvent and electrochemical methods of generation may have some synthetic potential[175]. The 9,10-dicyanoanthracene sensitized photolytic cleavage of aryldisilanes $ArMe_2SiSiMe_3$ (Ar = Ph, 4-MeC_6H_4 or 4-$MeOC_6H_4$) in MeCN gives initially disilane radical cations; subsequent Si−Si bond fission gives radicals ArMe$_2$Si• or Me$_3$Si• and silicenium ions ArMe$_2$Si$^+$ or Me$_3$Si$^+$ (presumably coordinated to the MeCN solvent), which can then undergo reaction with any water present to give silanols and siloxanes[176]. The [Et$_3$Si(C$_6$H$_6$)] [B(C$_6$F$_5$)$_4$] species has recently been used as a useful reagent because it is able to silylate Ph$_2$C=CH$_2$ to give the first example of a stable β-silyl carbocation [Et$_3$SiCH$_2$CPh$_2$][B(C$_6$F$_5$)$_4$][123].

One further method for the preparation of silicenium ions that is independent of the phase in which the reaction occurs and which is not sensitive to solvent, pressure or temperature is to use a nuclear decay to generate a good leaving group from silicon. Russian workers have studied the process of β-decomposition of a tritium nucleus to helium which, if it occurs for tritium bonded to silicon, gives a molecular ion which then loses helium to give a silicenium ion as shown for Me$_2$SiT$_2$ (equation 14). The Me$_2$TSi$^+$ ion formed in this way reacts with MeOH to give Me$_2$TSiOMe, and with (t-BuO)Me$_2$SiH and Me$_3$SiOSiMe$_2$H to give Me$_2$TSiOSiMe$_2$H[9]. The reaction between T$_3$C$^+$ and Me$_3$SiOSiMe$_3$ affords Me$_3$SiOCT$_3$ and Me$_3$Si$^+$, apparently via the intermediate formation of the oxonium ion (Me$_3$Si)$_2$(T$_3$C)O$^{+\,177}$. Although only small amounts of material may be generated and detected in this way, it would be of interest to apply the method to molecules bearing larger and more stabilizing groups than methyl and hydrogen.

$$Me_2SiT_2 \xrightarrow{\beta} [Me_2SiTHe]^+ \longrightarrow Me_2TSi^+ + He^0 \qquad (14)$$

IV. SOLID STATE STUDIES

A variety of compounds containing four-, five- or six-coordinate silyl cations in which the silicon centre is stabilized by N→Si interactions have been prepared and structurally characterized, but they do not fulfil the criteria for description as silicenium ions. The four-coordinate **28** has an Si−N distance of 1.821(2) Å and C−Si−C angles averaging 113.4°, while the five-coordinate **29** is also stabilized intermolecularly, this time by two N-methylimidazole molecules with Si−N distances of 2.034(3) and 2.005(3) Å and a planar geometry at silicon[178]. Intramolecular stabilization of a silicon cation is also possible to give the five-coordinate **30** with Si−N distances of 2.06(1) and 2.08(1) Å and a near-planar geometry at silicon in the equatorial plane[179]. The triflate salt **31** has Si−N distances of 2.072(2) and 2.052(2) Å and has a nearly trigonal bipyramidal geometry at silicon[180]. All of these Si−N distances are significantly longer than would be expected for a covalent bond (1.70−1.76 Å) but are shorter than dative interactions found in neutral species with an intramolecular N→Si interaction, for example 2.291(2) Å in the five-coordinate [2-(Me$_2$NCH$_2$)C$_6$H$_4$]$_2$SiCl$_2$[181].

The first, and probably the most controversial crystallographic study of a compound thought to contain a silicon with substantial silicenium ion character, was that described by Lambert and coworkers who isolated, after crystallization from toluene, the colourless [Et$_3$Si(toluene)]$^+$[B(C$_6$F$_5$)$_4$]$^-$, from the reaction between Et$_3$SiH and Ph$_3$CB(C$_6$F$_5$)$_4^{67,182}$. The original description of the compound was that of 'a silyl cation with no coordination to anion and distant coordination to solvent'[182]. The closest approach between the anion and the cation is an Si···F distance of 4.04 Å, indicative of negligible interaction between them. Figure 1 does, however, show that the geometry at silicon in the Et$_3$Si group is not planar and that the fourth coordination site at silicon is occupied by a toluene molecule (for structural details see Table 2). It is the nature of the interaction between the silicon and the toluene molecule that has been the point of most contention. There are two crystallographically independent molecules in the unit cell, the distance between the silicon and the *para* carbon atom of the toluene molecule averaging 2.18 Å, which

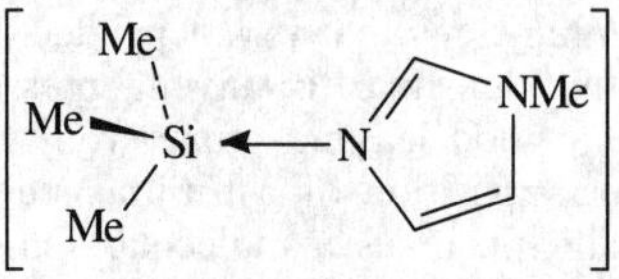

(28)

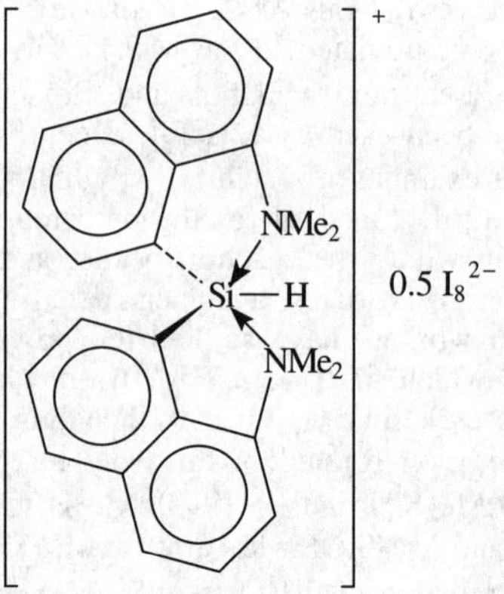

(30)

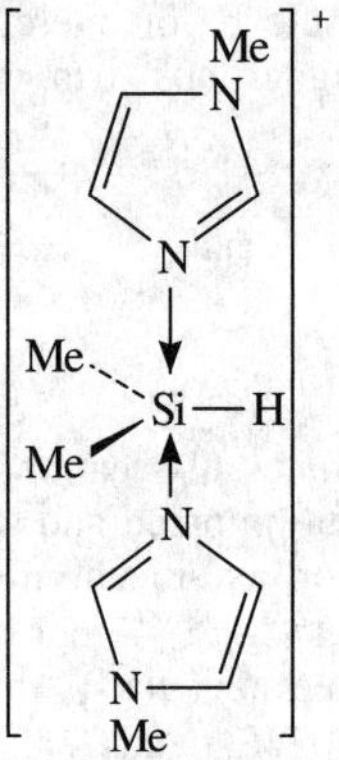

(29)

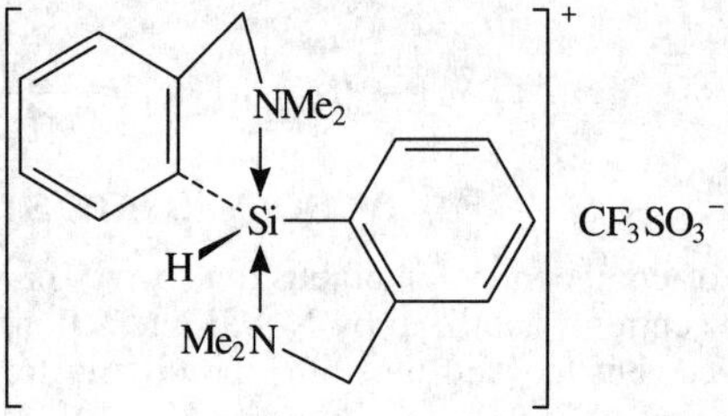

(31)

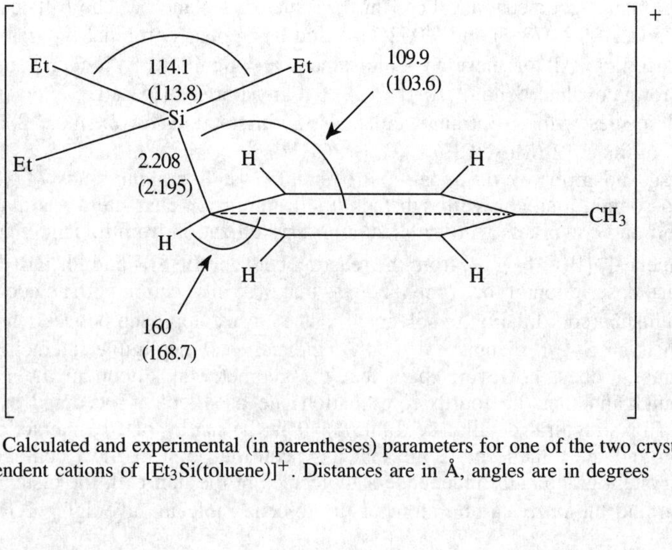

FIGURE 1. Calculated and experimental (in parentheses) parameters for one of the two crystallographically independent cations of $[Et_3Si(toluene)]^+$. Distances are in Å, angles are in degrees

TABLE 2. Structural data for compounds reported to have silicenium ion character[a]

Compound	Si–L distance (Å) obs.	Si–L distance (Å) covalent	Pauling bond order[b]	Σ C–Si–C (°)[c]	Out-of-plane distance (Å)[d]	References
			Compounds of the type [R$_3$Si(solvent)] [anion][e]			
[Et$_3$Si(toluene)] [B(C$_6$F$_5$)$_4$][f]	2.18	1.85	0.28	342	0.462	67,182
[Me$_3$Si(pyridine)] [Br]	1.856	1.73	0.62			88
[Me$_3$Si(pyridine)] [I]	1.858	1.73	0.61		0.485	88
[i-Pr$_3$Si(MeCN)] [Br$_5$CB$_9$H$_5$]	1.82(2)	1.73	0.71	346.7	0.410	103
[i-Bu$_3$Si(H$_2$O)] [Br$_6$CB$_{11}$H$_6$]	1.779(9)	1.64	0.58	348		59
			Compounds of the type [R$_3$Si] [anion][g]			
[Et$_3$Si] [Br$_6$CB$_{11}$H$_6$][h]	2.444(7)	2.24	0.46	345.0(10)	0.419	185
[Et$_3$Si] [Br$_6$CB$_{11}$H$_6$][h]	2.430(6)	2.24	0.48	349.0(9)	0.348	185
[i-Pr$_3$Si] [Cl$_6$CB$_{11}$H$_6$]	2.323(3)	2.08	0.39	351.8(4)	0.307	106
[i-Pr$_3$Si] [I$_6$CB$_{11}$H$_6$]	2.661(6)	2.46	0.46	346.8(9)	0.400	106
[i-Pr$_3$Si] [Br$_6$CB$_{11}$H$_6$]	2.479(9)	2.24	0.40	351.0(13)	0.300	185
[i-Pr$_3$Si] [Br$_5$CB$_9$H$_5$]	2.46(1)	2.24	0.43	347.5	0.40	103
[i-Bu$_2$MeSi] [Br$_6$CB$_{11}$H$_6$]	2.466(12)	2.24	0.42	345.8(21)	0.408	185
[i-Bu$_3$Si] [Br$_6$CB$_{11}$H$_6$]	2.465(5)	2.24	0.42	348.7(7)	0.371	185
Ph$_3$SiClO$_4$	1.74	1.64	0.68	340.9	0.479	82

[a]Table adapted, with additions, from Reference 18; Ph$_3$SiClO$_4$ is included for comparison.
[b]Calculated according to the Pauling equation, $0.60 \log n = D(l) - D(n)$, where $D(l)$ is the distance for a covalent bond order of one, and $D(n)$ is the observed bond distance for a bond order n.
[c]Sum of three C–Si–C angles at silicon, tetrahedral Σ = 328.5°, trigonal planar Σ = 360°.
[d]The distance of the silicon atom from the plane containing the 3α-carbons.
[e]L = carbon for toluene, nitrogen for pyridine and MeCN, and oxygen for water.
[f]Values are average for two crystallographically independent molecules.
[g]L = halogen for carborane anions and oxygen for perchlorate.
[h]Two crystallographically independent molecules.

is significantly longer than the previously longest known Si−C bond of 2.03 Å found in the sterically crowded cyclic $t\text{-Bu}_2\overline{\text{Si}(t\text{-Bu})_2\text{SiOPh}_2\text{C}}$[183] and even more so compared to the sum of the covalent radii, 1.88 Å, of Si and C. Other noteworthy features of the structure are that the C−Si−C angles in the Et$_3$Si groups of the two molecules range from 112.8–114.9°, that the aromatic ring is nearly planar (the largest dihedral angle being 3°), and that the bond lengths within the ring fall in a relatively narrow range. These features are in contrast to those known for Wheland-type complexes containing alkyl rather than silyl substituents, in which severe distortions are found in the aromatic ring[184].

The crystallographic parameters clearly indicate the absence of a 'free' silicenium ion as the geometry at silicon is not planar (indeed the Et−Si−Et angle of 114° is closer to the tetrahedral than to the trigonal planar value), and the fourth coordination site is occupied by a toluene molecule. The solid state ^{29}Si NMR chemical shift is also not indicative of a 'free' ion according to the value expected by calculations. Conversely, it was argued that the relatively long distance between the silicon and the toluene molecule suggested insignificant σ-complexation and that, as the toluene molecule is relatively little perturbed from its normal geometry, there is little positive charge on the carbon atoms[182]. These factors would seem to suggest that there is some significant degree of silicenium ion character to the cation. These arguments have, however, been countered in numerous papers, many of them presenting detailed computational studies that indicate that bonding between the silicon and the toluene molecule is significant and that the species is best described as an arenium ion [Et$_3$Si(toluene)]$^+$ in which the positive charge is largely on the ring with very little silicenium ion character present. A detailed description of this computational work is provided in the chapter by Schleyer; only a brief outline of the arguments will be given here for comparison with the experimental studies. Pauling[186] and Lambert and Zhang[187] have both calculated the bond order n for the Si$\cdots$toluene interaction using the equation $D(n) = D(1) - 0.60\log n$ [where $D(1)$ is the distance for a full single bond and $D(n)$ the distance for a bond of order n] using different values for $D(1)$ to give values of n of 0.35 and 0.28, respectively, i.e. about one-third of a normal covalent bond. Further experimental evidence for a significant σ interaction, analogous to that found in the gas-phase work described in Section II, comes from the reaction between the [Et$_3$Si(toluene)]$^+$ species with the hindered base i-Pr$_2$NH which gives, along with (Et$_3$Si)$_2$O, a 7% yield of p- and m-triethylsilyltoluenes in a 2 : 1 ratio, respectively[109,187]. Thus the *meta*-silylated arenium ion is present in solution even though this isomer is not detected in the solid state. Although the yield of triethylsilyltoluenes is low, their formation does suggest that deprotonation competes with desilylation of a σ complex as is found in the gas phase. Another puzzling aspect of the Et$_3$Si derivatives is that the solid state ^{29}Si NMR chemical shift values for [Et$_3$Si (toluene)]$^+$ and for Et$_3$SiB (C$_6$F$_5$)$_4$ prepared in the absence of solvent are so similar[67] (see Table 3). This could be due to the species prepared in the absence of a solvent interacting with the Ph$_3$CH formed as a byproduct from the reaction to give [Et$_3$Si(Ph$_3$CH)]$^+$, which as a silylated monosubstituted aromatic system might be expected to have a similar chemical shift to that of [Et$_3$Si(toluene)]$^+$.

Calculations by Olah and coworkers[109,117,188], Schleyer and coworkers[4,189] and Olsson and Cremer and coworkers[105,108,114] have, however, shown that the experimentally determined structure is close to that calculated for arenium ions [R$_3$Si(arene)]$^+$ (R = Me or Et, arene = benzene or toluene). The various bond lengths and angles are generally reproduced well as are the ^{29}Si NMR chemical shift data. The unusually long Si−C$_{para}$ bond length can be attributed to β-silicon hyperconjugation as shown in Scheme 5[117,189]. The experimental and calculated structural parameters for [Et$_3$Si (toluene)]$^+$ are shown in Figure 1. Thus, although some of the experimental data do, at first sight, seem to be

TABLE 3. Solid state ^{29}Si chemical shifts for various compounds containing silicon with some cationic character

Compound	Chemical shift (δ, ppm from Me$_4$Si)	References
Me$_3$Si[B(C$_6$F$_5$)$_4$]	84.8	67
t-BuMe$_2$Si(Br$_6$CB$_{11}$H$_6$)	112.8	185
Et$_3$Si[B(C$_6$F$_5$)$_4$]	94.3	67
[Et$_3$Si(toluene)] [B(C$_6$F$_5$)$_4$]	93.5	67
Et$_3$Si(Br$_6$CB$_{11}$H$_6$)	106.2, 111.8	185
i-Pr$_3$Si[B(C$_6$F$_5$)$_4$]	107.6	67
i-Pr$_3$Si(Cl$_6$CB$_{11}$H$_6$)	115	106
i-Pr$_3$Si(Br$_6$CB$_{11}$H$_6$)	110	106
i-Pr$_3$Si(I$_6$CB$_{11}$H$_6$)	97	106
i-Bu$_3$Si[B(C$_6$F$_5$)$_4$]	89.4	67
[t-Bu$_3$Si(OH$_2$)] [(Br$_6$CB$_{11}$H$_6$)]	46.7	59

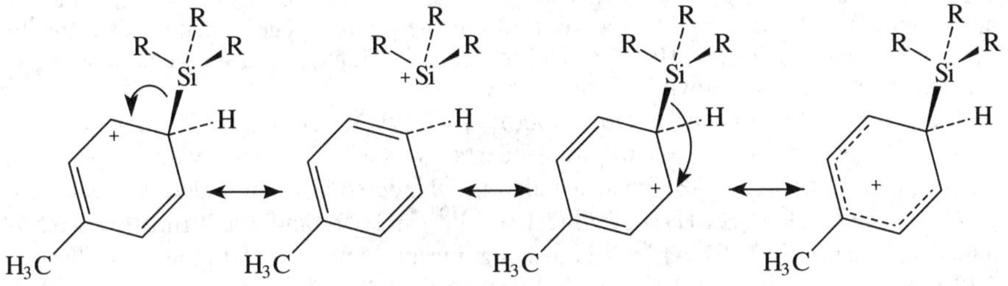

SCHEME 5

consistent with the presence of a significantly silicenium ion like species, the calculations show that the data are, in fact, very similar to those predicted for arenium ions. Lambert and Zhang have, however, countered the theoretical interpretations of their results and point out the deviations between experiment and theory. In particular, there is a pronounced alternation of C—C bond lengths in the arene ring found in the experimentaly derived structure which, although the same trend is reproduced by the calculations, is much less pronounced. Also, the Si—C$_{para}$···C$_{ipso}$ angle (Figure 1) is less well reproduced by theory than other important parameters[187]. A more detailed discussion of the theoretical arguments is given in the chapter by Schleyer. More recent calculations, however, at the correlated level of MP2/6-31G* for [Me$_3$Si(toluene)]$^+$ do agree significantly better than earlier calculations, in particular reproducing the Si—C$_{para}$···C$_{ipso}$ angle in [Et$_3$Si(toluene)]$^+$ well[109]. It would thus seem that fully optimized calculations can reproduce experimental data well for these compounds, and that the [Et$_3$Si(toluene)]$^+$ species is best described as an arenium rather than a silicenium ion in the solid state. Such small discrepancies that do still exist between the experimental and calculated structures may well be inevitable. Other calculations have investigated how the bonding in R$_3$M(arene)$^+$ (M = Si, Ge, Sn and Pb) species varies and these results also suggest that the Et$_3$Si(toluene)$^+$ complex is best described as a σ-bonded arenium complex in which the positive charge is delocalized on the aromatic ring, rather than a π-bonded complex in which the positive charge is mainly found on silicon. This situation changes on descending group 14 with the π-bonded complex structure being favoured for lead[190].

The solid state work on $[Et_3Si(toluene)]^+[B(C_6F_5)_4]^-$ was quickly followed by related work by Reed and coworkers who have used halogenated carboranes as anions of low nucleophilicity in the place of $[B(C_6F_5)_4]^-$. Structural data for the compounds prepared in this work are given in Table 2. The first of these compounds to be reported were i-$Pr_3Si(Br_5CB_9H_5)$ and $[i$-$Pr_3Si(MeCN)][Br_5CB_9H_5]$, prepared in a similar manner to Lambert's compounds, by treating i-Pr_3SiH with the trityl salt of the carborane anion in toluene and in MeCN, respectively[103]. The structural data for the $[i$-$Pr_3Si(MeCN)]^+$ cation show C—Si—C angles that are about half way between those expected for tetrahedral and trigonal planar and an Si—N distance of 1.82(2) Å, which while slightly longer than a typical Si—N bond length which fall in the range 1.70–1.76 Å, is similar to that found in $[Me_3Si(pyridine)]I$ (1.858 Å)[88]. These values together with the ^{29}Si NMR shift of only 33.8 ppm in MeCN solution suggest that the species is best regarded as an N-tri-isopropylacetonitrilium ion. The other compounds containing halogenated carborane anions detailed in Table 2 (apart from the $[t$-$Bu_3SiOH_2]^+$-containing compound which can be regarded as a silaoxonium ion) have been the subject of similar discussion to that for the $[R_3Si(arene)]^+$ species, i.e. do the structural and NMR parameters indicate silicenium or silahalonium ion character? Perhaps the most obvious difference between these compounds and those studied by Lambert and coworkers is the absence of any solvent molecules coordinated to the silicon; even when prepared and crystallized from toluene solution, an ion pair is isolated rather than a silylated toluene product. The structural data for the series of compounds $[R_3Si][Br_6CB_{11}H_6]$ are given in Table 2 and the general structure for such compounds is shown in Figure 2.

The Si—Br bond distances are 2.430(6)–2.479(9) Å, the relatively narrow range suggesting that the effect of the substituents is small[185]. These values are about 0.2 Å greater than the Si—Br bond lengths in unhindered silyl bromides, e.g. Me_3SiBr [2.235(2) Å][191], $(BrH_2SiCH_2)_2$ [2.2362(12) Å][192] and 1-bromo-3,5,7-trimethyl-1,3,5,7-tetrasilaadamantane (2.197 Å)[193], and also significantly longer than in the bulky bromo-9-(9-trimethylsilylfluorenyl)-bis(trimethylsilyl)silane [2.284(5) Å][194]. Although the values

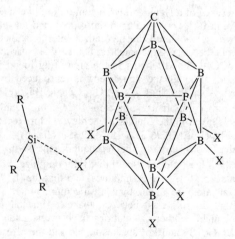

FIGURE 2. A schematic representation of $[R_3Si][X_6CB_{11}H_6]$ compounds where for X = Br, R = Me, Et, i-Pr or i-Bu and for X = Cl or I, R = i-Pr

for the bromocarborane derivatives are significantly longer than the other reported Si—Br distances in four-coordinate compounds, they are much shorter than the value of 3.122 Å found in the five-coordinate 1-[(bromo)dimethylsilylmethyl]-piperid-2-one[195] and well within the value of 3.5 Å estimated for complete charge separation[185]. Another argument for significant silicenium ion character in these compounds is that the halocarborane anion is very little distorted. The B—Br distance for the bromines coordinated to the silicon atoms lie in the range 1.99(2)–2.05(2) Å while those uncoordinated to Si lie in the range 1.92(2)–2.02(3) Å, i.e. there is perhaps a small, but barely significant, increase in B—Br distance on coordination to silicon which perhaps suggests little covalent bonding between the Si and the bromine. The individual C—Si—C angles within a particular R_3Si group may vary by up to 8 or 9°, and so it is probably more instructive to consider the sum of the C—Si—C angles and the deviation of the silicon from the plane of the three carbons to which it is attached (Table 2). There are some significant differences in the parameters for the two independent $Et_3Si(Br_6CB_{11}H_6)$ molecules and these presumably represent some degree of the flexibility of the species under crystal packing forces. Perhaps the most surprising aspects of the structures of the $R_3Si(Br_6CB_{11}H_6)$ compounds are that the sum of the C—Si—C angles and the Si—Br distance are not maximized in the most sterically demanding and electron-releasing t-Bu_3Si derivative. This has been attributed[185] to the flexibility of the Si—Br—B angle which readily opens up to 125° in the t-Bu_3Si derivative compared to only 114.7° in the i-Pr_3Si analogue. The i-Pr_3Si derivative thus seems to be the closest approach to a silicenium ion for a structurally characterized species, as it has the largest sum of C—Si—C angles and is closest to planarity at silicon. It should be noted that the correlation between the C—Si—C angle and the Si displacement from the C_3 plane is not linear, and a distortion from 120° to 117° gives more than half of the full displacement on going from a trigonal planar to a tetrahedral geometry at silicon[15]. The geometry of the substituents R in the $R_3Si(Br_6CB_{11}H_6)$ compounds, particularly for the ethyl and the isopropyl derivatives, is, in the case of at least one of the alkyl substituents on each silicon, highly distorted. This is manifested in large Si—C—C angles, some greater than 120°, and a near-planar geometry at the carbon attached to the silicon. These features have been attributed to C—H bond hyperconjugation, which gives some double bond character to the Si—C bond[185].

A further study by Reed and coworkers has concentrated on the coordinating ability of the carborane anion and the halogenated series i-$Pr_3Si(X_6CB_{11}H_6)$ (X = Cl, Br, I) have been prepared[106]. Structural data are given in Table 2. Again, as in the case of varying the alkyl substituents, there is an unexpected trend and the hexachloro derivative is found to be less coordinating in this case than the more bulky, softer, hexaiodo derivative. This has been explained using the argument that, as the stability of halonium ions is known to follow the trend iodonium > bromonium > chloronium, then the iodo derivative will have the greatest halonium ion character and the chloronium ion the least halonium ion (and consequently the most silicenium ion) character[106]. The average C—Si—C angle, the sum of the C—Si—C angles and the distance the Si is away from the C_3 plane for the chloro and bromo derivatives are very similar, although it does seem that the chloro-containing anion is slightly less strongly coordinated. It has thus been argued that i-$Pr_3Si(Cl_6CB_{11}H_6)$ is an 'ion-like' species and is the closest to a silicenium ion in the condensed phase to be structurally characterized so far[106]. The structural parameters together with the ^{29}Si NMR data have led Reed and coworkers to conclude that i-$Pr_3Si(Cl_6CB_{11}H_6)$ 'may have >50% silylium ion character' and that the term 'silylium ion-like' is most appropriate to describe the structure[106].

Further evidence proposed to support the predominance of silicenium ion character in these compounds is provided by their chemistry, because they react as if they were silicenium ions. For example, they react with nucleophiles such as CH_2Cl_2 and C_6H_5F

to form R_3SiX^{196} and with water to give $R_3Si(OH_2)^{+59}$. In none of these reactions is the B-halogen bond seen to be cleaved. These interpretations of Reed and coworkers' results have, however, as was the case for [Et$_3$Si(toluene)] [B(C$_6$F$_5$)$_4$], been criticized, the alternative explanation being that the compounds are, in fact, halonium ions[117,186]. Olah and coworkers[117] argue that a species such as i-Pr$_3$Si with significant silicenium ion character could not be isolated from toluene solution without reaction occurring, as indeed seems to be the case for Et$_3$Si$^+$ in Lambert's work. It does seem surprising that, if the species are predominantly ionic when generated in the presence of Reed's weakly coordinating halocarborane anions, in the presence of a large excess of toluene silylation of the solvent does not occur. It is unclear why Et$_3$Si$^+$ should not react with solvent and bind to a less coordinating anion than [B(C$_6$F$_5$)$_4$]$^-$ in preference. This might, however, be the case if the product is a halonium ion rather than a silicenium ion.

Another argument against a silicenium ion formulation for the R$_3$Si(X$_6$CB$_{11}$H$_6$) species is that their ^{29}Si NMR chemical shifts, given in Table 3, which generally fall in the range of 85–115 ppm, are well short of the range, 350–400 ppm, calculated for R$_3$Si$^+$ ions (cf Section III.C.2). The calculated values for free gas-phase R$_3$Si$^+$ ions may not be good yardsticks by which to estimate the degree of ionicity in condensed phases. For example, calculations on the Me$_3$Si···Cl system show that approach of the Cl$^-$ ion any closer than the van der Waals distance of 3.3 Å reduces the ^{29}Si NMR chemical shift from a 'free' value of 388 ppm to only about half of this value[106]. Related calculations on the effect of species such as argon and methane, that can be regarded as some of the least nucleophilic potential solvents, also show that the ^{29}Si NMR chemical shift for R$_3$Si$^+$ is significantly reduced in their presence[4]. This would suggest that there is very little chance of observing 'free' silicenium ions in solution or the solid state with chemical shifts greater than 350 ppm unless, perhaps, the silicon bears substituents that sterically hinder the approach of solvent molecules. Unfortunately, large and elaborate substituents may well lead to intramolecular coordination. Such calculations have led Reed and coworkers to suggest a value of about 220 ppm as the upper limit to be expected for R$_3$Si$^+$ ions (R = alkyl) in condensed phases, and hence his halocarborane derivatives would have about 50% ionic character based on this parameter.

Calculations of the chemical shift for the bromonium ion [Me$_3$SiBrMe]$^+$ give a value of 117.5 ppm, which is close to the observed values for Et$_3$Si(Br$_6$CB$_{11}$H$_6$). This again suggests that Reed's compounds may better be regarded as halonium ions[117]. Calculations on the structure of [Me$_3$SiBrMe]$^+$ give values of 2.523 Å and 116.4° for the Si−Br bond distance and C−Si−C angle, respectively, which are close to those (2.444 Å and 116.3°) for the comparable parameters in Et$_3$Si(Br$_6$CB$_{11}$H$_6$), also consistent with the formulation of Et$_3$Si(Br$_6$CB$_{11}$H$_6$) as a bromonium ion[117]. The lack of significant B−Br bond elongation for the bromine involved in an interaction with the silicon compared to the uncoordinated B−Br bonds has also been used to support the proposition that R$_3$Si(X$_6$CB$_{11}$H$_6$) species have significant silicenium ion character. However, the C−Br bond length of 1.996 Å in [Me$_3$SiBrMe]$^+$ is not very much longer than that, 1.939 Å, in free MeBr[197], and can be compared, for example, with the similar lengthening of the B−Br distances for uncoordinated [1.92–1.98(2) Å] to coordinated [1.99(2) Å] in Et$_3$Si(Br$_6$CB$_{11}$H$_6$)[185]. Thus, all of the important structural parameters of Et$_3$Si(Br$_6$CB$_{11}$H$_6$) seem to be reasonably well reproduced in calculations of a bromonium ion structure for [Me$_3$SiBrMe]$^+$ and this has led Olah and coworkers to conclude that Et$_3$Si(Br$_6$CB$_{11}$H$_6$), and by implication other R$_3$Si(X$_6$CB$_{11}$H$_6$) species, 'has very little triethylsilicenium character and is de facto a bromonium zwitterion'[117].

V. CONCLUSIONS

It can be seen from the work described above that there are a whole series of criteria such as $R_3Si \cdots X$ distance, Si—R distance, R—Si—R angle, distance of the Si from the C_3 plane and ^{29}Si NMR chemical shift that have been used as measures of the degree of ionicity in cationic silicon compounds. The degree of silicenium ion character is thus definition-dependent, and in no case does an experimentally observed species satisfy any definition completely. These various experimentally derived parameters have led to a wide range (ca 20–75%) of ionic character being assigned to $R_3Si(X_6CB_{11}H_6)$ species, and this has, in turn, led to conflicts with computational interpretations. The current state of work on these compounds seems to reflect that they best be described as halonium ions but with some silicenium ion character, the degree of which varies with the definition chosen but which probably does not yet exceed 50% in any case. Similarly, the $Et_3Si(toluene)^+[B(C_6F_5)_4]$ compound is probably best regarded as an arenium ion in the light of high level computational studies. Perhaps the most promising candidate for a free silicenium ion is the recently described $(mesityl)_3Si^+$ for which solution ^{29}Si NMR data are in reasonably close agreement with the calculated value (see Section III.C.2). Solid state studies on this ion, perhaps in conjunction with the use of a halogenated carborane as anion, may well lead to convincing evidence that a free silicenium ion can exist in a condensed phase.

It should also be noted here that recent work in Japan has shown that a free germanium cation can be prepared and structurally characterized. Thus, the reaction of tetrakis (tri-*t*-butylsilyl)cyclotrigermene with trityl tetraphenylborate affords $(t\text{-}Bu_3SiGe)_3^+$ BPh_4^- as a yellow crystalline solid. X-ray crystallography shows that there is no close interaction between the anion and the cation. The stability of the compound is presumably due not only to the steric protection afforded by the bulky $t\text{-}Bu_3Si$ substituents, but also to the fact that it contains a 2π-electron aromatic ring.[198]

Finally, it is perhaps well to be reminded of the comment made by Grant Urry[199] more than 25 years ago when describing the chemistry (or lack of it at that time) of polysilanes, and apply it to the field of silicenium ion chemistry: 'It is perhaps appropriate to chide the polysilane (silicenium ion) enthusiast for milking the horse and riding the cow in attempting to adapt the successes of organic chemistry in the study of polysilanes (silicenium ions). A valid argument can be made for the point of view that the most effective chemistry of silicon arises from the differences with the chemistry of carbon compounds rather than the similarities'.

VI. ACKNOWLEDGEMENTS

The author wishes to thank Professor C. Eaborn for the many stimulating discussions concerning silicenium ions over the years, and Professors N. Auner, A. R. Bassindale, J. B. Lambert and P. v. R. Schleyer, and Dr. T. Müller for kindly supplying information prior to publication.

VII. REFERENCES

1. G. J. Leigh (Ed.), *Nomenclature of Inorganic Chemistry*, Blackwell, Oxford, 1990, p. 106.
2. H. Schwarz, in *The Chemistry of Organic Silicon Compounds* (Eds. S. Patai and Z. Rappoport), Part 1, Wiley, Chichester, 1989, pp. 445–510.
3. J. Chojnowski and W. A. Stanczyk, *Adv. Organomet. Chem.*, **30**, 243 (1990).
4. C. Maerker, J. Kapp and P. v R. Schleyer, in *Organosilicon Chemistry II* (Eds. N. Auner and J. Weis), VCH, Weinheim, 1996, pp. 329–359.

5. S. K. Shin and J. L. Beauchamp, *J. Am. Chem. Soc.*, **111,** 900 (1989).
6. S. Hollenstein and T. Laube, *J. Am. Chem. Soc.*, **115,** 7240 (1993).
7. D. H. O'Brien and T. J. Hairston, *Organomet. Chem. Rev. A*, **7,** 95 (1971).
8. R. J. P. Corriu and M. Henner, *J. Organomet. Chem.*, **74,** 1 (1974).
9. V. D. Nefedov, T. A. Kochina and E. N. Sinotova, *Russian Chem. Rev.*, **55,** 426 (1986).
10. J. B. Lambert and W. J. Schulz Jr., in *The Chemistry of Organic Silicon Compounds* (Eds. S. Patai and Z. Rappoport), Part 2, Wiley, Chichester, 1989, pp. 1007–1014.
11. Y. Apeloig, in *Heteroatom Chemistry* (Ed. E. Block), VCH, New York, 1990, pp. 27–46.
12. C. Eaborn, *J. Organomet. Chem.*, **405,** 173 (1991).
13. P. D. Lickiss, *J. Chem. Soc., Dalton Trans.*, 1333 (1992).
14. K. N. Houk, *Chemtracts, Org. Chem.*, **6,** 360 (1993).
15. S. H. Strauss, *Chemtracts, Inorg. Chem.*, **5,** 119 (1993).
16. S. Borman, *Chem. Eng. News*, Nov. 8, 1993, p. 41.
17. J. Chojnowski and W. A. Stanczyk, *Main Group Chem. News*, **2,** 6 (1994).
18. J. B. Lambert, L. Kania and S. Zhang, *Chem. Rev.*, **95,** 1191 (1995).
19. G. Rong, R. Ma, L. Long and C. Zhou, *Huaxue Tongbao*, 19 (1996); *Chem. Abstr.*, **125,** 221889 (1996).
20. A. Tachibana, S. Kawauchi, T. Yano, N. Yoshida and T. Yamabe, *J. Mol. Struct., (THEOCHEM)*, **119,** 121 (1994).
21. L. Operti, R. Rabezzana, G. A. Vaglio and P. Volpe, *J. Organomet. Chem.*, **509,** 151 (1996).
22. J.-F. Gal, R. Grover, P.-C. Maria, L. Operti, R. Rabezzanna, G.-A. Vaglio and P. Volpe, *J. Phys. Chem.*, **98,** 11978 (1994).
23. I. Haller, *J. Phys. Chem.*, **94,** 4135 (1990).
24. A. Tachibana, S. Kawauchi, N. Yoshido, T. Yamabe and K. Fukui, *J. Mol. Struct.*, **300,** 501 (1993).
25. M. Yamamoto, M. Tanaka, Y. Yokota and T. Takeuchi, *J. Mass Spectrom. Soc. Jpn.*, **41,** 277 (1993).
26. P. B. Davies and D. M. Smith, *J. Chem. Phys.*, **100,** 6166 (1994).
27. D. M. Smith, P. M. Martineau and P. B. Davies, *J. Chem. Phys.*, **96,** 1741 (1992).
28. B. H. Boo and P. B. Armentrout, *J. Am. Chem. Soc.*, **109,** 3549 (1987).
29. S. K. Shin, R. R. Corderman and J. L. Beauchamp, *Int. J. Mass Spectrom. Ion Processes*, **101,** 257 (1990).
30. T. Motooka, P. Fons, H. Abe and T. Tokuyama, *Jpn. J. Appl. Phys. Part 2b*, **32,** L879 (1993).
31. W. D. Reents, Jr., M. L. Mandich, and C. R. C. Wang, *J. Chem. Phys.*, **97,** 7226 (1992).
32. M. L. Mandich, W. D. Reents Jr. and K. D. Kolenbrander, *Pure Appl. Chem.*, **62,** 1653 (1990).
33. M. L. Mandich, W. D. Reents Jr. and K. D. Kolenbrander, *J. Chem. Phys.*, **92,** 437 (1990).
34. K. Raghavachari, *Adv. Met. Semicond. Clusters*, **2,** 57 (1994).
35. K. Raghavachari, *J. Chem. Phys.*, **92,** 452 (1990).
36. M. L. Mandich and W. D. Reents Jr., *J. Chem. Phys.*, **96,** 4233 (1992).
37. W. D. Reents Jr. and M. L. Mandich, *J. Chem. Phys.*, **93,** 3270 (1990).
38. L. Operti, M. Splendore, G. A. Vaglio and P. Volpe, *Spectrochim. Acta, Part A*, **49A,** 1213 (1993).
39. P. Antoniotti, L. Oporti, R. Rabezzana, G. A. Vaglio, P. Volpe, J.-F. Gal, R. Grover and P.-C. Maria, *J. Phys. Chem.*, **100,** 155 (1996).
40. X. Li and J. A. Stone, *Int. J. Mass Spectrom. Ion Processes*, **101,** 149 (1990).
41. S. Murthy and J. L. Beauchamp, *J. Phys. Chem.*, **99,** 9118 (1995).
42. S. Murthy and J. L. Beauchamp, *J. Phys. Chem.*, **96,** 1247 (1992).
43. R. Bakhtiar, C. M. Holznagel and D. B. Jacobson, *J. Phys. Chem.*, **97,** 12710 (1993).
44. K. P. Lim and F. W. Lampe, *Int. J. Mass Spectrom. Ion Processes*, **92,** 53 (1989).
45. F. Denes, Z. Q. Hua, C. E. C. A. Hop and R. A. Young, *J. Appl. Polym. Sci.*, **61,** 875, (1996).
46. Z. Q. Hua, F. Denes and R. A. Young, *J. Vac. Sci. Technol.*, **14,** 1339 (1996).
47. K. A. Reuter and D. B. Jacobson, *Organometallics*, **8,** 1126 (1989).
48. I. S. Ignatyev and T. Sundius, *Organometallics*, **15,** 5674 (1996).
49. R. Bakhtiar, C. M. Holznagel and D. B. Jacobson, *Organometallics*, **12,** 621 (1993).
50. R. Bakhtiar and C. M. Holznagel, *Organometallics*, **12,** 880 (1993).
51. Y. Apeloig, M. Karni, A. Stanger, H. Schwarz, T. Drewello and G. Czekay, *J. Chem. Soc., Chem. Commun.*, 989 (1987).
52. G. Angelini, Y. Keheyan, G. Laguzzi and G. Lilla, *Tetrahedron Lett.*, **29,** 4159 (1988).

53. S. G. Cho, *J. Organomet. Chem.*, **510**, 25 (1996).
54. V. Q. Nguyen, S. A. Shaffer, F. Turecek and C. E. C. A. Hop, *J. Phys. Chem.*, **99**, 15454 (1995).
55. S. Tobita, S. Tajima, F. Okada, S. Mori, E. Tabei and M. Umemura, *Org. Mass Spectrom.*, **25**, 39 (1990).
56. S. Tobita, S. Tajima and F. Okada, *Org. Mass. Spectrom.*, **24**, 373 (1989).
57. W. J. Meyerhoffer and M. M. Bursey, *Org. Mass Spectrom.*, **24**, 246 (1989).
58. J. A. Stone, A. C. M. Wojtyniak, *Can. J. Chem.*, **64**, 575 (1986).
59. Z. Xie, R. Bau and C. A. Reed, *J. Chem. Soc., Chem. Commun.*, 2519 (1994).
60. A. C. M. Wojtyniak and J. A. Stone, *Int. J. Mass Spectrom. Ion Processes*, **74**, 59 (1986).
61. X. Li and J. A. Stone, *Can. J. Chem.*, **70**, 2070 (1992).
62. F. Cacace, M. E. Crestoni, S. Fornarini and R. Gabrielli, *Int. J. Mass Spectrom. Ion Processes*, **84**, 17 (1988).
63. S. Fornarini, *J. Org. Chem.*, **53**, 1314 (1988).
64. F. Cacace, *Acc. Chem. Res.*, **21**, 215 (1988).
65. M. Speranza, *Mass Spectrom. Rev.*, **11**, 73 (1992).
66. G. A. Olah, T. Bach and G. K. S. Prakash, *J. Org. Chem.*, **54**, 3770 (1989).
67. J. B. Lambert, S. Zhang and S. M. Ciro, *Organometallics*, **13**, 2430 (1994).
68. F. Cacace, M. E. Crestoni and S. Fornarini, *J. Am. Chem. Soc.*, **114**, 6776 (1992).
69. F. Cacace, M. Attinà and S. Fornarini, *Angew. Chem., Int. Ed. Engl.*, **34**, 654 (1995).
70. M. E. Crestoni and S. Fornarini, *Angew. Chem., Int. Ed. Engl.*, **33**, 1094 (1994).
71. A. B. Thomas and E. G. Rochow, *J. Inorg. Nucl. Chem.*, **4**, 205 (1957).
72. N. N. Lichtin and P. D. Bartlett, *J. Am. Chem. Soc.*, **73**, 5530 (1951).
73. M. E. Peach and T. C. Waddington, *J. Chem. Soc.*, 1238 (1961).
74. J. Y. Corey, *J. Am. Chem. Soc.*, **97**, 3237 (1975).
75. J. B. Lambert and W. Schilf, *J. Am. Chem. Soc.*, **110**, 6364 (1988).
76. J. B. Lambert, W. J. Schulz, J. A. McConnell and W. Schilf, *J. Am. Chem. Soc.*, **110**, 2201 (1988).
77. J. B. Lambert, L. Kania, W. Schilf and J. A. McConnell, *Organometallics*, **10**, 2578 (1991).
78. J. B. Lambert, J. A. McConnell and W. J. Schulz, *J. Am. Chem. Soc.*, **108**, 2482 (1986).
79. J. B. Lambert and W. J. Schulz, *J. Am. Chem. Soc.*, **105**, 1671 (1983).
80. J. B. Lambert, L. Kania, B. Kuhlmann and J. A. McConnell, unpublished results, cited in Reference 18.
81. G. A. Olah, L. Heiliger, X. Y. Li and G. K. S. Prakash, *J. Am. Chem. Soc.*, **112**, 5991 (1990).
82. G. K. S. Prakash, S. Keyaniyan, R. Aniszfeld, L. Heiliger, G. A. Olah, R. C. Stevens, H. K. Choi and R. Bau, *J. Am. Chem. Soc.*, **109**, 5123 (1987).
83. G. A. Olah, G. Rasul, L. Heiliger, J. Bausch and G. K. S. Prakash, *J. Am. Chem. Soc.*, **114**, 7737 (1992).
84. C. Eaborn and F. M. S. Mahmoud, *J. Chem. Soc., Perkin Trans. 2*, 1309 (1981).
85. J. B. Lambert, J. A. McConnell, W. Schilf and W. J. Schulz Jr., *J. Chem. Soc., Chem. Commun.*, 455 (1988).
86. J. Chojnowski, M. Cypryk and J. Michalski, *J. Organomet. Chem.*, **161**, C31 (1978).
87. A. R. Bassindale and T. Stout, *J. Chem. Soc., Perkin Trans. 2*, 221 (1986).
88. K. Hensen, T. Zengerley, P. Pickel and G. Klebe, *Angew. Chem., Int. Ed. Engl.*, **22**, 725 (1983).
89. M. Okano and K. Mochida, *Chem. Lett.*, 819 (1991).
90. M. Arshadi, D. Johnels, U. Edlund, C.-H. Ottosson and D. Cremer, *J. Am. Chem. Soc.*, **118**, 5120 (1996).
91. C. Chuit, R. J. P. Corriu, A. Mehdi and C. Reye, *Angew. Chem., Int. Ed. Engl.*, **32**, 1311 (1993).
92. V. A. Benin, J. C. Martin and M. R. Willcott, *Tetrahedron Lett.*, **35**, 2133 (1994).
93. M. Chauhan, C. Chuit, R. J. P. Corriu and C. Reyé, *Tetrahedron Lett.*, **37**, 845 (1996).
94. M. Chauhan, C. Chuit, R. J. P. Corriu, A. Mehdi and C. Reyé, *Organometallics*, **15**, 4326 (1996).
95. C.-H. Ottosson and D. Cremer, *Organometallics*, **15**, 5309 (1996).
96. H. Gilman and G. E. Dunn, *J. Am. Chem. Soc.*, **72**, 2178 (1950).
97. U. Wannagat and W. Liehr, *Angew. Chem.*, **69**, 783 (1957).
98. J. M. Stone and J. A. Stone, *Int. J. Mass Spectrom. Ion Processes*, **109**, 247 (1991).
99. R. Walsh, in *The Chemistry of Organic Silicon Compounds* (Eds. S. Patai and Z. Rappoport), Part 1, Wiley, Chichester, 1989, pp. 371–391.
100. D. A. Strauss, C. Zhang, G. E. Quimbita, S. D. Grumbine, R. H. Heyn, T. D. Tilley, A. L. Rheingold and S. J. Geib, *J. Am. Chem. Soc.*, **112**, 2673 (1990).

101. S. R. Bahr and P. Boudjouk, *J. Am. Chem. Soc.*, **115**, 4514 (1993).
102. J. B. Lambert and S. Zhang, *J. Chem. Soc., Chem Commun.*, 383 (1993).
103. Z. Xie, D. J. Liston, T. Jelinek, V. Mitro, R. Bau and C. A. Reed, *J. Chem. Soc., Chem. Commun.*, 384 (1993).
104. G. A. Olah and L. Field, *Organometallics*, **1**, 1485 (1982).
105. L. Olsson, C.-H. Ottosson and D. Cremer, *J. Am. Chem. Soc.*, **117**, 7460 (1995).
106. Z. Xie, J. Manning, R. W. Reed, R. Mathur, P. D. W. Boyd, A. Benesi and C. A. Reed, *J. Am. Chem. Soc.*, **118**, 2922 (1996).
107. S. Sieber, P. Buzek, P. v. R. Schleyer, W. Koch, J. W. d. M. Carneiro, *J. Am. Chem. Soc.*, **115**, 259 (1993).
108. L. Olsson and D. Cremer, *Chem. Phys. Lett.*, **215**, 433 (1993).
109. G. A. Olah, G. Rasul, H. A. Buchholz, X.-Y. Li and G. K. S. Prakash, *Bull Chem. Soc. Fr.*, **132**, 569 (1995).
110. C.-H. Ottosson and D. Cremer, *Organometallics*, **15**, 5495 (1996).
111. J. B. Lambert and H.-n. Sun, *J. Am. Chem. Soc.*, **98**, 5611 (1976).
112. U. Pidun, M. Stahl and G. Frenking, *Chem. Eur. J.*, **2**, 869 (1996).
113. M. Kira, T. Hino and H. Sakurai, *J. Am. Chem. Soc.*, **114**, 6697 (1992).
114. D. Cremer, L. Olsson and H. Ottosson, *J. Mol. Struct. (THEOCHEM)*, **313**, 91 (1994).
115. G. A. Olah, K. Laali and O. Farooq, *Organometallics*, **3**, 1337 (1984).
116. E. V. Van Den Berghe and G. P. Van Der Kelen, *J. Organomet. Chem.*, **59**, 175 (1973).
117. G. A. Olah, G. Rasul and G. K. S. Prakash, *J. Organomet. Chem.*, **521**, 271 (1996).
118. A. R. Bassindale, personal communication.
119. G. A. Olah, X.-Y. Li, Q. Wang, G. Rasul and G. K. S. Prakash, *J. Am. Chem. Soc.*, **117**, 8962 (1995).
120. M. Kira, T. Hino and H. Sakurai, *Chem. Lett.*, 153 (1993).
121. A. R. Bassindale and T. Stout, *Tetrahedron Lett.*, **26**, 3403 (1985).
122. M. Kira, T. Hino and H. Sakurai, *Chem. Lett.*, 555 (1992).
123. J. B. Lambert and Y. Zhao, *J. Am. Chem. Soc.*, **118**, 7867 (1996).
124. J. B. Lambert and Y. Zhao, *Angew. Chem., Int. Ed. Engl.*, **36**, 400 (1997) and personal communication
125. N. Auner, paper OB20, XIth International Symposium on Organosilicon Chemistry, Montpellier, September 1996; H.-U. Steinberger, T. Müller, N. Auner, C. Maerker and P. v. R. Schleyer, *Angew. Chem., Int. Ed. Engl.*, **36**, 626 (1997).
126. J. B. Lambert, personal communication to G. Frenking, cited in Reference 107.
127. V. K. Dioumaev and J. F. Harrod, *Organometallics*, **15**, 3859 (1996).
128. D. A. Strauss, S. D. Grumbine and T. D. Tilley, *J. Am. Chem. Soc.*, **112**, 7801 (1990).
129. S. D. Grumbine, T. D. Tilley, F. P. Arnold and A. L. Rheingold, *J. Am. Chem. Soc.*, **115**, 7884 (1993).
130. S. K. Grumbine, T. D. Tilley, F. P. Arnold and A. L. Rheingold, *J. Am. Chem. Soc.*, **116**, 5495 (1994).
131. P. Jutzi and E.-A. Bunte, *Angew. Chem., Int. Ed. Engl.*, **31**, 1605 (1992).
132. P. v. R. Schleyer, *Science*, **275**, 39 (1997).
133. T. Müller, personal communication.
134. V. K. Dioumaev and J. F. Harrod, *J. Organomet. Chem.*, **521**, 133 (1996).
135. V. K. Dioumaev and J. F. Harrod, *Organometallics*, **13**, 1548 (1994).
136. J. Chojnowski, W. Fortuniak and W. A. Stanczyk, *J. Am. Chem. Soc.*, **109**, 7776 (1987).
137. J. Chojnowski, L. Wilczek and W. Fortuniak, *J. Organomet. Chem.*, **135**, 13 (1977).
138. H. Mayr, N. Basso and G. Hagen, *J. Am. Chem. Soc.*, **114**, 3060 (1992).
139. N. Basso, S. Görs, E. Popowski and H. Mayr, *J. Am. Chem. Soc.*, **115**, 6025 (1993).
140. Y. Apeloig, O. Merin-Aharoni, D. Danovich, A. Ioffe and S. Shaik, *Isr. J. Chem.*, **33**, 387 (1993).
141. Y. Apeloig, *Stud. Org. Chem.*, **31**, 33 (1987).
142. Y. Apeloig and A. Stanger, *J. Am. Chem. Soc.*, **109**, 272 (1987).
143. D. N. Kevill, *J. Chem. Res. (S)*, 272 (1987).
144. A. R. Bassindale and P. G. Taylor, in *The Chemistry of Organic Silicon Compounds* (Eds. S. Patai and Z. Rappoport), Part 1, Wiley, Chichester, 1989, pp. 839–892.
145. M. A. M. R. Al-Gurashi, G. A. Ayoko, C. Eaborn and P. D. Lickiss, *Bull. Soc. Chim. Fr.*, **132**, 517 (1995).
146. C. Eaborn, P. D. Lickiss, S. T. Najim and W. A. Stanczyk, *J. Chem. Soc., Chem. Commun.*, 1461 (1987).

11. Silicenium ions — experimental aspects   593

147. C. Eaborn, K. L. Jones and P. D. Lickiss, *J. Chem. Soc., Chem. Commun.*, 595 (1989).
148. C. Eaborn, K. L. Jones and P. D. Lickiss, *J. Chem. Soc., Perkin Trans. 2*, 489 (1992).
149. C. Eaborn, K. L. Jones, P. D. Lickiss and W. A. Stanczyk, *J. Chem. Soc., Perkin Trans. 2*, 59 (1993).
150. A. Ayoko and C. Eaborn, *J. Chem. Soc., Perkin Trans. 2*, 1047 (1987).
151. N. H. Buttrus, C. Eaborn, P. B. Hitchcock, P. D. Lickiss and S. T. Najim, *J. Chem. Soc., Perkin Trans. 2*, 891 (1987).
152. C. Eaborn, P. D. Lickiss, S. T. Najim and M. N. Romamelli, *J. Chem. Soc., Chem. Commun.*, 1754 (1985).
153. C. Eaborn, P. D. Lickiss, S. T. Najim and M. N. Romanelli, *J. Organomet. Chem.*, **315**, C5 (1986).
154. C. Eaborn and M. N. Romanelli, *J. Organomet. Chem.*, **451**, 45 (1993).
155. C. Eaborn and D. E. Reed, *J. Chem. Soc., Perkin Trans. 2*, 1695 (1985).
156. Z. H. Aiube and C. Eaborn, *J. Organomet. Chem.*, **421**, 159 (1991).
157. S. M. Whittaker, Ph. D. Thesis, University of Salford (1993).
158. A. I. Almansour, J. R. Black, C. Eaborn, P. M. Garrity and D. A. R. Happer, *J. Chem. Soc., Chem. Commun.*, 705 (1995).
159. A. G. Brook, M. Hesse, K. M. Baines, R. Kumarathasan and A. J. Lough, *Organometallics*, **12**, 4259 (1993).
160. D. M. Van Seggan, P. K. Hurlburt, M. D. Noirot, O. P. Anderson and S. H. Strauss, *Inorg. Chem.*, **31**, 1423 (1992).
161. G. K. S. Prakash, Q. Wang, X.-y. Li and G. A. Olah, *New J. Chem.*, **14**, 791 (1990).
162. Y. Apeloig and O. Merin-Aharoni, *Croat. Chem. Acta*, **65**, 757 (1992).
163. S. R. Church, C. G. Davies, R. Lümen, P. A. Mounier, G. Saint and P. L. Timms, *J. Chem. Soc., Dalton Trans.*, 227 (1996).
164. N. Tokitoh, T. Imakubo and R. Okazaki, *Tetrahedron Lett.*, **33**, 5819 (1992).
165. A. Schäfer, M. Weidenbruch, S. Pohl and W. Saak, *Z. Naturforsch. Teil B*, **45B**, 363 (1990).
166. M. Labrouillère, C. Le Roux, A. Oussaid, H. Gaspard-Ilhoughmane and J. Dubac, *Bull. Soc. Chem. Fr.*, **132**, 522 (1995).
167. V. A. Kovyazin, I. B. Sokol'skaya, V. M. Kopylov, I. A. Abronin, Yu T. Efimov, O. V. Shtefan and V. Yu Kapustin, *Organomets. in USSR*, **4**, 652 (1991).
168. K. L. Malisza, L. C. F. Chao, J. F. Britten, B. G. Sayer, G. Jaouen, S. Top, A. Decken and M. J. McGlinchy, *Organometallics*, **12**, 2462 (1993).
169. R. Ruffolo, A. Decken, L. Girard, H. K. Gupta, M. A. Brook and M. J. McGlinchey, *Organometallics*, **13**, 4328 (1994).
170. P. Sigwalt and V. Stannett, *Makromol. Chem., Macromol. Symp.*, **32**, 217 (1990).
171. D. M. Naylor, V. T. Stannett, A. Deffieux and P. Sigwalt, *Polymer*, **35**, 1764 (1994).
172. S. S. Jada, *J. Am. Ceram. Soc.*, **70**, C298 (1987).
173. S. S. Jada, *J. Am. Ceram. Soc.*, **71**, C413 (1988).
174. L. D. David, *J. Am. Ceram. Soc.*, **71**, C412 (1988).
175. D. D. Devore, D. R. Neithamer, R. E. Lapointe and R. D. Mussell, World Patent, No. WO 9608519 A2 960321, 1996.
176. K. Mizuno, T. Tamai, I. Hashida and Y. Otsuji, *J. Org. Chem.*, **60**, 2935 (1995).
177. I. S. Ignatyev and T. A. Kochina, *J. Mol. Struct. (THEOCHEM)*, **236**, 249 (1991).
178. K. Hensen, T. Zengerley, T. Müller and P. Pickel, *Z. Anorg. Allg. Chem.*, **558**, 21 (1988).
179. C. Brelière, F. Carré, R. Corriu and M. W. C. Man, *J. Chem. Soc., Chem. Commun.*, 2333 (1994).
180. J. Belzner, D. Schär, O. Kneisel and R. Herbst-Irmer, *Organometallics*, **14**, 1840 (1995).
181. R. Probst, C. Leis, S. Gamper, E. Herdtweck, L. Zybill and N. Auner, *Angew. Chem., Int. Ed. Engl.*, **30**, 1132 (1991).
182. J. B. Lambert, S. Zhang, C. L. Stern and J. C. Huffman, *Science*, **260**, 1917 (1993).
183. A. Schäfer, M. Weidenbruch and S. Pohl, *J. Organomet. Chem.*, **282**, 305 (1985).
184. F. Effenberger, F. Reisinger, K. H. Schöwälder, P. Bäuerle, J. J. Stezowski, K. H. Jogun, K. Schöllkopf and W.-D. Stohrer, *J. Am. Chem. Soc.*, **109**, 882 (1987).
185. Z. Xie, R. Bau, A. Benesi and C. A. Reed, *Organometallics*, **14**, 3933 (1995).
186. L. Pauling, *Science*, **263**, 983 (1994).
187. J. B. Lambert and S. Zhang, *Science*, **263**, 984 (1994).
188. G. A. Olah, G. Rasul, X.-Y. Li, H. A. Buchholz, G. Sandford and G. K. S. Prakash, *Science*, **263**, 983 (1994).

189. P. v. R. Schleyer, P. Buzek, T. Müller, Y. Apeloig and H.-U. Siehl, *Angew. Chem., Int. Ed. Engl.*, **32**, 1471 (1993).
190. H. Basch, *Inorg. Chim. Acta*, **242**, 191 (1996).
191. M. D. Harmony and M. R. Strand, *J. Mol. Spectrosc.*, **81**, 308 (1980).
192. N. W. Mitzel, J. Reide and H. Schmidbaur, *Acta Crystallogr. Sect. C*, **52**, 980 (1996).
193. S. N. Gurkova, A. I. Gusev, V. A. Sharapov, T. K. Gar and N. V. Alexeev, *Zh. Strukt. Khim.*, **20**, 356 (1979); English translation: *J. Struct. Chem.*, **20**, 302 (1979).
194. U. Schubert and C. Steib, *J. Organomet. Chem.*, **238**, C1 (1982).
195. A. A. Macharashvili, Yu. I. Baukov, E. P. Kramarova, G. I. Oleneva, V. A. Shestunovich, Yu. T. Struchkov and V. E. Shklover, *Zh. Strukt. Khim.*, **28**, 107 (1987); English translation: *J. Struct. Chem.*, **28**, 552 (1987).
196. C. A. Reed, Z. Xie, R. Bau and A. Benesi, *Science*, **262**, 402 (1993).
197. S. L. Miller, L. C. Aamodt, G. Dousmanis, C. H. Townes and J. Kraitchman, *J. Chem. Phys.*, **20**, 1112 (1952).
198. A. Sekiguchi, M. Tsukamoto, and M. Ichinoke, *Science*, **175**, 60 (1997).
199. G. Urry, *Acc. Chem. Res.*, **3**, 306 (1970).

CHAPTER 12

Silyl-substituted carbocations

HANS-ULLRICH SIEHL

Abteilung für Organische Chemie I der Universität Ulm, D-86069 Ulm, Germany
Fax: +49-(0)731-502-2800;   e-mail: ullrich.siehl@chemie.uni-ulm.de

and

THOMAS MÜLLER

Fachinstitut für Anorganische und Analytische Chemie der Humboldt Universität
Berlin, D-10115 Berlin, Germany
Fax: +49(0)30-2903-6966;   e-mail: h0443afs@rz.hu-berlin.de

I. INTRODUCTION	596
II. CALCULATIONAL RESULTS	596
III. GAS PHASE STUDIES	601
A. α-Substitution	601
B. β-Substitution	604
IV. SOLVOLYTIC STUDIES	610
A. α-Silicon-substituted Carbocations in Solvolysis	610
B. β-Silicon-substituted Carbocations in Solvolysis	616
1. α-Disilanyl Carbocations in Solvolysis	632
C. γ-Silicon Effect	635
D. δ-Silicon Effect	641
V. STABLE ION STUDIES	645
A. α-Silyl Substitution	645
1. NMR spectroscopic characterization of α-silyl-substituted carbocations	645
a. 1-Phenyl-1-trimethylsilylethyl cation	645
b. Diphenyl(trimethylsilyl)methyl cations	647
2. X-ray structure analysis of the tris(trimethylsilyl)cyclopropenylium cation	648
B. β-Silyl-substituted Carbocations	649
1. sp^2-Hybridized carbocations	649
a. Photochemical generation and UV characterization of transient β-silyl-substituted carbocations	649

The chemistry of organic silicon compounds, Vol. 2
Edited by Z. Rappoport and Y. Apeloig   © 1998 John Wiley & Sons Ltd

            i. 9-(Trimethylsilylmethyl)fluoren-9-yl cations   649
            ii. Silyl-substituted cyclohexadienyl cations   651
        b. X-ray crystallographic study of β-silylcyclohexadienyl
           cations   652
        c. NMR spectroscopic characterization of β-silyl-substituted
           carbocations   655
            i. β-Silyl-substituted secondary p-methoxybenzyl cations   655
               α. 1-p-Anisyl-2-triisopropylsilylethyl cations   655
            ii. Other β-silyl-substituted secondary benzyl cations   659
               α. β-Silyl-substituted tolyl- and phenylmethyl cations   659
               β. 1-Ferrocenyl-2-triisopropylsilylethyl cations   660
            iii. β-Silyl-substituted tertiary benzyl cations   662
               α. 1,1-Diphenyl-2-(triethylsilyl)ethyl cations   662
            iv. β-Silyl-substituted allyl cations   662
               α. 2-[1′-(Trimethylsilyl)ethenyl]adamant-2-yl cations   662
     2. sp-Hybridized carbocations (vinyl cations)   664
        a. β-Silyl-substituted dienyl cations   664
        b. β-Silyl-substituted α-arylvinyl cations   667
            i. α-Ferrocenyl-β-silyl-substituted vinyl cations   667
            ii. α-Mesityl-β-silyl-substituted vinyl cations   669
            iii. α-Tolyl- and α-phenyl-β-silyl-substituted vinyl cations   676
            iv. α-Anisyl-β-silyl-substituted vinyl cations   679
        c. The 1-bis(trimethylsilyl)methyl-2-bis(trimethylsilyl)ethenyl
           cation   685
  C. Silyl-substituted Hypercoordinated Carbocations   688
     1. The α-and-γ-silyl effect in bicyclobutonium ions   688
        a. 1-(Trimethylsilyl)bicyclobutonium ion   690
        b. 3-endo-(Trialkylsilyl)bicyclobutonium ions   693
     2. Silanorbornyl cation   695
        a. The 6,6-dimethyl-5-neopentyl-6-sila-2-norbornyl cation   695
VI. CONCLUSIONS   697
VII. ACKNOWLEDGEMENTS   697
VIII. REFERENCES   697

I. INTRODUCTION

This chapter deals with silyl-substituted carbocations. In Section II results of quantum chemical *ab initio* calculations of energies and structures of silyl-substituted carbocations are summarized[1]. Throughout the whole chapter results of *ab initio* calculations which relate directly to the experimental observation of silyl-substituted carbocations and their reactions are reviewed. Section III reports on gas phase studies and Section IV on solvolytic investigations of reactions which involve silyl-substituted carbocation intermediates and transition states. Section V summarizes the structure elucidation studies on stable silyl-substituted carbocations. It includes ultra-fast optical spectroscopic methods for the detection of transient intermediates in solution, NMR spectroscopic investigations of silyl-substituted carbocations in superacids and non-nucleophilic solvents, concomitant computational studies of model cation and X-ray crystallography of some silyl-substituted carbocations which can be prepared as crystals of salts.

II. CALCULATIONAL RESULTS

In this brief part only the most important results of *ab initio* calculations regarding the energies and structures of silyl-substituted carbocations are summarized[1].

12. Silyl-substituted carbocations

The simplest α-silyl-substituted carbenium ion $H_3SiCH_2^+$ (**1**) is not a minimum on the potential energy surface and it rearranges without a barrier to the more stable (by 49 kcal mol^{-1} at MP4/6-31G(d)//3-21G(d)) silylium ion H_2Si^+Me (**2**)[2]. For α-silylethyl cation (**3**) and α-silylvinyl cation (**4**) calculations at the correlated MP2(fu)/6-31G(d,p) level predict hydrogen bridged structures just as for the parent ethyl and vinyl cations[3,4]. **3** is separated by a barrier of only 5.2 kcal mol^{-1} from the ethylsilylium ion **5**, which is by 25.9 kcal mol^{-1} more stable [at MP2(fu)/6-31G(d,p)]. Similarly, the activation barrier for the 1,2-H shift from the α-silylvinyl cation (**4**) to the isomeric 1-silaallyl cation (**6**) is only 8.9 kcal mol^{-1}[3,4].

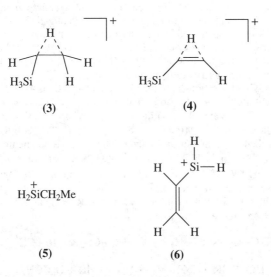

According to *ab initio* calculations by Wierschke, Chandrasekhar and Jorgensen a silyl-substituent stabilizes a directly attached positively charged carbon compared with hydrogen, but it destabilizes it compared with a methyl substituent. Thus, the calculated reaction energies of the isodesmic reaction 1 are 17.8 kcal mol^{-1} (R = SiH$_3$) and 34.0 kcal mol^{-1} (R = Me) [at MP3/6-31G(d)//3-21G(d)][5].

$$H_2\overset{+}{C}-R + CH_4 \longrightarrow CH_3^+ + CH_3-R \qquad (1)$$

$$H_2C=\overset{+}{C}-R + H_2C=CH_2 \longrightarrow H_2C=\overset{+}{C}-H + H_2C=CHR \qquad (2)$$

For dicoordinated cations the situation is slightly different. Apeloig and Stanger showed, using the isodesmic reaction 2, that an α-silyl group, i.e. in **7**, stabilizes the vinyl cation by *ca* 24.9 kcal mol^{-1} (HF/3-21G//3-21G) relative to hydrogen[6]. This value remains nearly unchanged at higher levels of calculations [24.1 kcal mol^{-1} at MP3/6-31G(d)//3-21G(d)][5] and 25 kcal mol^{-1} at QCISD(T)/6-31G(d)//MP2(fu)/6-31G(d)][7]. More interesting is the comparison with the methyl group. It is found that α-silyl and α-methyl groups stabilize the vinyl cation nearly to the same extent: the difference between the calculated reaction energies of equation 2 for R = Me and R = SiH$_3$ is only 2.5 kcal mol^{-1} at QCISD(T)/6-31G(d)//MP2(fu)/6-31G(d)[7], 0.2 kcal mol^{-1} (HF/3-21G//3-21G)[6] and 2.9 kcal mol^{-1} [MP3/6-31G(d)//3-21G(d)][5], with the 2-propenyl cation **8** being the more stable.

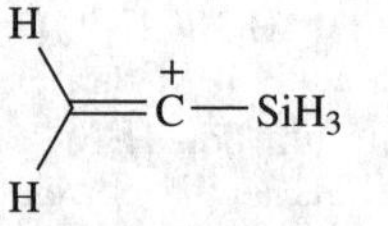

$$\underset{H}{\overset{H}{>}}=\overset{+}{C}-SiH_3 \qquad \underset{H}{\overset{H}{>}}=\overset{+}{C}-CH_3$$

(7) 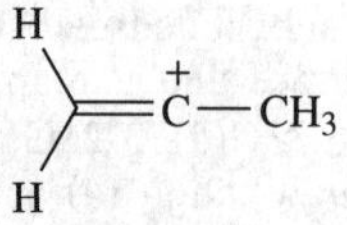   (8)

The destabilizing effect of a silyl group compared with a methyl group in trivalent carbocations was explained by the weaker SiH (relative to CH) hyperconjugation and by electrostatic repulsion between the adjacent positively charged cationic carbon and the electropositive silicon[8]. In the vinyl cation the stronger hyperconjugation with the β-vinylic hydrogens diminishes the demand for hyperconjugative stabilization from the α-substituent, thus reducing the advantage of an α-methyl group. In addition, the vinyl cation is more sensitive to the σ-effects of the α-substituent than an alkyl cation, and the stabilizing σ-donation by the electropositive silyl group is therefore more important for the vinyl cation than for the tricoordinated carbenium ion[6].

In contrast, β-silyl-substitution is predicted by the calculations to be far more stabilizing than β-alkyl substitution. Thus, the isodesmic equation 3 predicts for **9** a stabilization by the β-silyl group of 38 kcal mol^{-1}, while the β-methyl substitution in **10** gives only a stabilization of 28 kcal mol^{-1} [MP3/6-31G(d)//3-21G(d)][5]. The stabilization by the silyl substituent is markedly orientation-dependent. Thus, the perpendicular conformation of the β-silylethyl cation **9p** is higher in energy by 29.6 kcal mol^{-1} compared with the bisected conformation **9** [MP3/6-31G(d)//3-21G(d)][5]. The open β-silyl-substituted vinyl cation **11** is lower in energy by 28.6 kcal mol^{-1} and 20.5 kcal mol^{-1} [MP3/6-31G(d)//3-21G(d)] compared with the vinyl cation (equation 4, R = H) and the 1-propenyl cation (equation 4, R = Me), respectively[5].

$$H_2\overset{+}{C}-CH_2R + C_2H_6 \longrightarrow H_3C-CH_2R + H_3CCH_2^+ \qquad (3)$$

$$H\overset{+}{C}=CHR + C_2H_6 \longrightarrow H_2C=CHR + H_3CCH_2^+ \qquad (4)$$

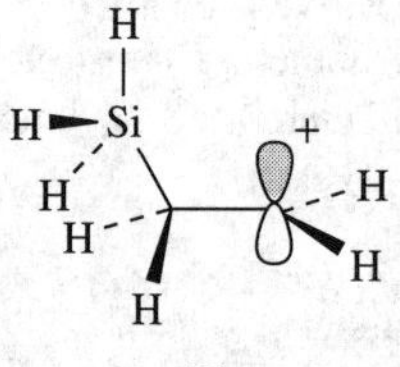

(9) 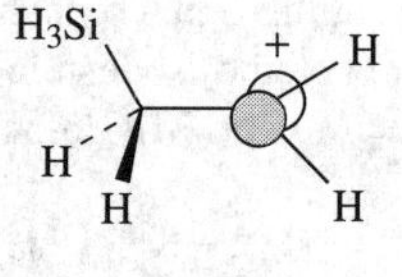   (9p)

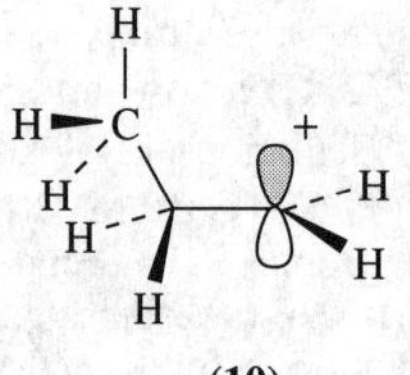

(10) 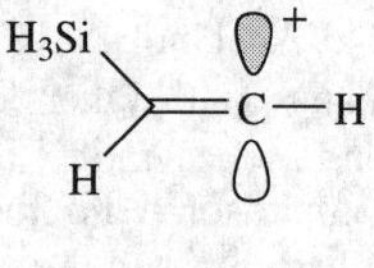   (11)

The β-silyl effect is dependent on the electron demand of the electron-deficient centre and it decreases with the increased stability of the silicon-free cation. Thus, the stabilization energy is reduced from the value for the primary cation **9** (33.2 kcal mol^{-1} compared with the ethyl cation) to the secondary cation **12** (22.1 kcal mol^{-1} compared with the isopropyl cation) and finally attains an additional stabilization of only 15.9 kcal mol^{-1} for the tertiary cation **13** [compared with the *t*-butyl cation; all values calculated at MP2(fc)/6-31G(d)//6-31G(d)]^9. The additional stabilization gained by β-silyl substitution is reduced from 33 kcal mol^{-1} for the primary vinyl cation **11** to only *ca* 10 kcal mol^{-1} for the strongly stabilized α-phenyl-substituted vinyl cation **14** [at MP2(fc)/6-31G(d)//MP2(fc)/6-31G(d)]^7.

(12)   (13)

(14)

The hyperconjugative interaction between the σ-Si—C bond and the formally empty 2p(C$^+$) orbital leads to marked geometrical effects. For example, in the calculated equilibrium structure of the secondary ion **12** (Figure 1) the calculated Si—C bond is distinctly longer than in the perpendicular conformation **12p**, in which the SiCC$^+$ hyperconjugation is switched off [2.070 Å versus 1.938 Å, at HF/6-31G(d)^9]. Furthermore, the calculated C$^+$—C distance shortens from 1.461 Å in the perpendicular **12p** to 1.386 Å in the bisected conformation **12**, which is by 22 kcal mol^{-1} more stable. The series of calculated structures of vinyl cations shown in Figure 2 demonstrates nicely the geometrical consequences of β-C—Si hyperconjugation in vinyl cations. Apeloig, Siehl, Schleyer, and coworkers^{7,10} calculated at the correlated MP2/6-31G(d) level the structures of β-silylvinyl cations **14–20** possessing various α-substituents including methyl, hydroxy, vinyl, isobuten-1-yl, phenyl, cyclopropyl and silyl. The degree of bridging in these β-silyl-substituted vinyl

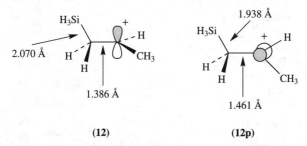

(12)   (12p)

FIGURE 1. Calculated structures of bisected ion **12** and perpendicular ion **12p** (HF/6-31G(d))

cations (taking the SiC$_\beta$C$^+$ bond angle α as a measure) was found to decrease (i.e. increases) with increasing electron-donating ability of the α-substituent. Thus, on going from the strongly electron-donating isobuten-1-yl substituent in **16** to a methyl group in **19**, α decreases from 117.5° to 91.9° and closes even further to 76.3° for an α-silyl group in **20**.

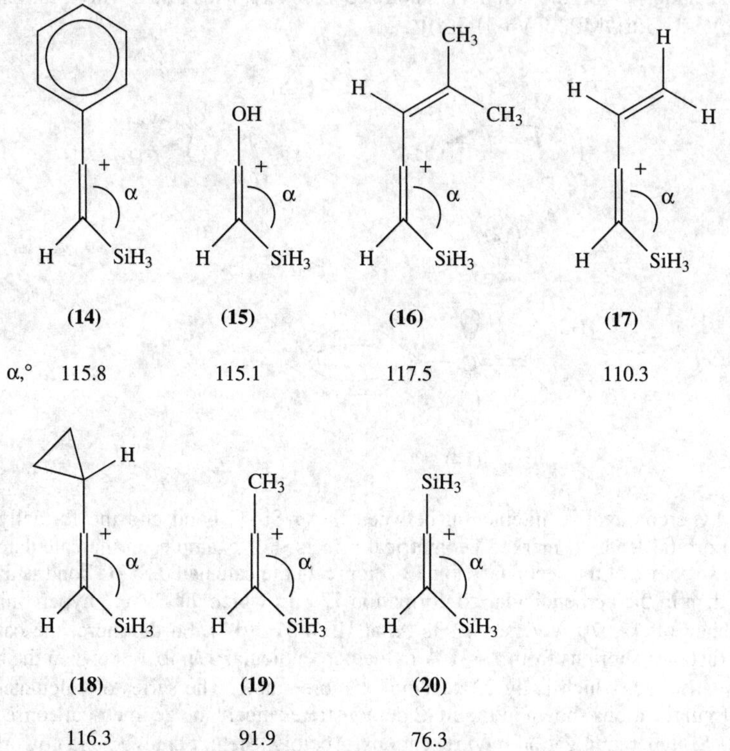

	(14)	(15)	(16)	(17)
α,°	115.8	115.1	117.5	110.3

	(18)	(19)	(20)
	116.3	91.9	76.3

FIGURE 2. Geometrical consequences of β-C−Si hyperconjugation in vinyl cations (MP2/6-31G(d))

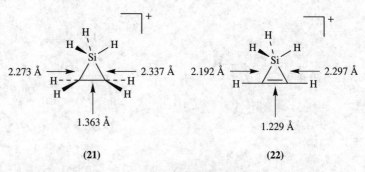

FIGURE 3. Calculated structures of **21** and **22** (MP2/6-31G(d,p))

The 'classical' open β-silylethyl cation **9** and β-silylvinyl cation **11** are no minima at higher level of theory[3,4]. They collapse to the bridged protonated silacyclopropane **21** and silacyclopropene **22**, respectively. On the basis of their calculated structures (Figure 3) both cyclic molecules are best described as π-complexes between a silylium ion and ethene or acetylene, respectively.

Silyl substituents have also distinct effects on the stability of aryl cations. For example, the disilyl (**23**) and trisilyl-substituted (**24**) phenyl cations are calculated to be more stable than the parent phenyl cation **25** by 25 kcal mol^{-1} (HF3-21G)[11] and by 22.4 kcal mol^{-1} (HF6-31G(d))[12], respectively.

(23)   (24)   (25)

III. GAS PHASE STUDIES

A. α-Substitution

In agreement with the calculations, all gas phase studies of α-silyl-substituted carbenium ions show that these intermediates exist only in a very flat potential well. They undergo fast 1,2-H- or 1,2-alkyl shifts, producing the more stable silylium or β-silyl-substituted carbenium ions. For example, Apeloig, Schwarz and coworkers[13] employed elegant gas phase ion techniques combined with *ab initio* MO theory to study the unimolecular dissociation of nascent $C_5H_{13}Si^+$ ions. Their results indicate that the nascent α-silyl-substituted $H_3C-CH^+Si(CH_3)$ ion **26** undergoes a fast 1,2-methyl migration forming the silylium ion **27** with little or no barrier (Figure 4). The 1,2-H shift yielding the β-silyl-substituted carbenium ion **28** is connected with an appreciable barrier of *ca* 10 kcal mol^{-1}, which results from the fact that the 1,2-H shift produces the eclipsed conformation **28a** which lies 37 kcal mol^{-1} higher in energy than the ground-state conformation **28**. This agrees nicely with the results of isotopic labeling experiments which demonstrate that, prior to dissociation to C_2H_4 and Me_3Si^+, the nascent ion **26** rapidly undergoes 1,2-methyl migrations leading to an almost complete exchange of the four methyl groups[13,14]. It is notable that the tertiary silyl-substituted carbenium ion $Me_2C^+SiMe_3$ **29** also exchanges, prior to loss of propene, its five methyl groups in a nearly quantitative way. Thus, the methyl migration to the silylium ion $Me_3CSi^+Me_2$ **30** is still relatively facile while the hydrogen shift to produce the β-silyl-substituted carbenium ion $H_2C^+MeCHSiMe_3$ **31** involves a significant barrier[14].

Ion–molecule reactions were used to identify the structure of the predominant ion resulting from ionization of $MeCHXSiMe_3$ (X = Cl, OR, OH). While Hajdasz and Squires[15] concluded from their afterglow study that ions **27** and **28** are formed in the ionization of the chloride in a 4 : 1 ratio, Jacobson and coworkers[16] clearly identified the exclusively formed $SiC_5H_{13}^+$ ion in their FTMS study as the silylium ion **27**. This disparity was explained by the different reactant gas pressures used in the experiments[16]. The high

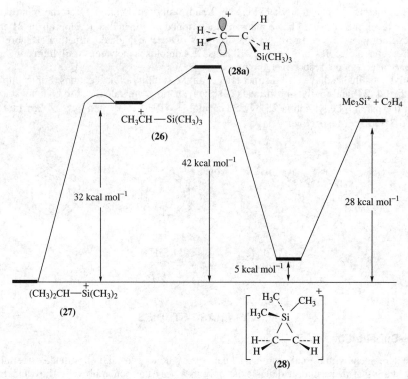

FIGURE 4. Schematic potential energy diagram of some $C_5H_3Si^+$ isomers and their dissociation products. The relative energies are given in parentheses and are based on HF/6-31G(d)//3-21G(d), corrected for substituent effects obtained from MP2/6-31G(d)//6-31G(d) calculations of the smaller $C_2H_7Si^+$ system

pressure used in the afterglow experiment allows the relaxation of the complex ion **28** prior to monomolecular dissociation into C_2H_4 and Me_3Si^+. This relaxation process is not favoured in the low pressure FTMS conditions[16].

In a FTMS study Jacobsen and coworkers[16] produced incipient α-silyl-substituted carbenium ions **32–34** by chloride elimination from the corresponding α-chloroalkylsilanes. They studied the rearrangement of **32–34** to silylium ions by direct 1,2-group migration (H, Me, Ph) from silicon to carbon (equations 5–7). The structures of the resulting silylium ions **35–38** were determined by specific ion/molecule reactions with methanol and isotopically labeled ethene. The product ratio **37/38** (resulting from nascent **34**) and the exclusive formation of **35** from **32** suggested the following relative migratory aptitude: phenyl > hydrogen >> methyl. However, several factors which may contribute to the observed migration distribution, like the relative thermodynamic stability of the products or the orientation of the migrating group in the neutral silane as the elimination of chloride occurs, could not be determined[16]. A theoretical study at the MP2/6-31+G(d)//HF/6-31G(d) level by Cho[17] revealed that all three α-silyl-substituted carbenium ions **32–34** are not minima on the potential energy surface but rearrange spontaneously to **35–38**. According to these calculations the experimentally observed migratory aptitude is determined by the activation energy for the simultaneous C—Cl bond heterolysis of the α-chlorosilanes and the concurrent 1,2-silyl migration[17].

$$(H_3C)_2HSi-\overset{+}{C}H_2 \longrightarrow (H_3C)_2\overset{+}{Si}-CH_3 \quad (5)$$
$$\text{(32)} \qquad\qquad\qquad \text{(35)}$$

$$(H_3C)_3Si-\overset{+}{C}H_2 \longrightarrow (H_3C)_2\overset{+}{Si}-CH_2CH_3 \quad (6)$$
$$\text{(33)} \qquad\qquad\qquad \text{(36)}$$

$$(C_6H_5)(H_3C)HSi-\overset{+}{C}H_2 \longrightarrow (C_6H_5)(H_3C)\overset{+}{Si}-CH_3 + (H_3C)H\overset{+}{Si}-CH_2C_6H_5 \quad (7)$$
$$\text{(34)} \qquad\qquad\qquad \text{(37)} \quad 6\% \qquad \text{(38)} \quad 94\%$$

Gas phase basicities defined by the reaction of alkenes with acids (equation 8) can be used to determine the relative stabilities of the resulting carbenium ions. Tsuno and coworkers found in pulsed ICR gas phase protonation experiments of α-trimethylsilylstyrenes **39** that benzyl cation **40** is exclusively formed. The measured gas phase basicities for **39** are comparable to those of α-alkylstyrenes and they are significantly higher than for styrene[18].

$$Ar-CR=CR'_2 + BH^+ \longrightarrow Ar\overset{+}{C}R-CHR'_2 + B \quad (8)$$

(39)   (40)

Using the isodesmic equations 9a–c Tsuno and coworkers calculated from the experimental gas phase basicities that the α-trimethylsilyl group stabilizes the styryl cation by 4.5 kcal mol^{-1} (equation 9a). Equations 9b and 9c which compare the effect of the α-trimethylsilyl group with those of the *t*-butyl- and the methyl substituent, respectively, are nearly thermoneutral, indicating the order Alkyl = SiMe$_3$ > H of styryl cation stabilization ability[18].

$$PhC(SiMe_3)=CH_2 + Ph\overset{+}{C}HMe \longrightarrow Ph\overset{+}{C}(SiMe_3)Me$$
$$+PhCH=CH_2 \qquad \Delta G° = -4.5 \quad (9a)$$

$$PhC(SiMe_3)=CH_2 + Ph\overset{+}{C}(Me)Bu\text{-}t \longrightarrow Ph\overset{+}{C}(SiMe_3)Me$$
$$+PhC(Bu\text{-}t)=CH_2 \qquad \Delta G° = -0.2 \quad (9b)$$

$$PhC(SiMe_3)=CH_2 + Ph\overset{+}{C}Me_2 \longrightarrow Ph\overset{+}{C}(SiMe_3)Me$$
$$+PhC(Me)=CH_2 \qquad \Delta G° = 0.7 \quad (9c)$$

This seems to be in disagreement with theoretical calculations, which predict a smaller stabilization of carbenium ions by silyl groups compared with alkyl groups[1,5–7,9]. The calculations, however, utilize SiH$_3$ groups instead of trialkylsilyl groups. The lower electron-donating capacity of the silyl group might be the reason for this apparent theoretical–experimental disagreement.

While simple α-silyl-substituted trivalent carbenium ions could be detected in the gas phase only as nascent species and their existence as such is very unlikely, substituted α-silylvinyl cations could be produced in the gas phase and their thermodynamic stability could be measured. MS experiments by McGibbon, Brook and Terlouw showed that $Me_3Si-C^+=CH_2$ **41** formed by ionization of $Me_3SiCl=CH_2$ is stabilized by 29.5 kcal mol^{-1} compared with hydrogen[19]. This experimental finding is in agreement with recent calculations which found that $Me_3Si-C^+=CH_2$ is more stable than the parent (bridged) vinyl cation HC(H)CH$^+$ by 30.1 kcal mol^{-1} at MP2/6-311G(2d,p)//MP2(fu)/6-31G(d) (further corrections for higher order correlation effects and zero-point energy differences lead to an estimate of 33.3 kcal mol^{-1})[7]. On the other hand, according to the same experiments, $Me_3Si-C^+=CH_2$ is by only 2 kcal mol^{-1} more stable than $H_2C=C^+-CH_3$[19], while the calculations predict, using the hydride transfer reaction 10, a value of 11.7 kcal mol^{-1} at MP4SDTQ/6-31G(d)//MP2(fu)/6-31G(d)[7] (15.3 kcal mol^{-1} at HF/3-21G//3-21G[6]). This large experimental–theoretical discrepancy reflects in our view experimental errors rather than the inadequacy of the calculations.

$$H_2C=\overset{+}{C}-SiMe_3 + H_2C=CHMe \longrightarrow H_2C=\overset{+}{C}-Me + H_2C=CHSiMe_3 \qquad (10)$$

B. β-Substitution

Most of the gas phase studies concerning the β-silyl-substituted carbocations were directed towards a quantification of the β-silyl group effect and to assign a structure to the observed ions. We will first report the results of the experiments in which only the energy of these type of cations have been determined and we will then discuss the results of the structure elucidations.

The simplest β-trimethylsilyl-substituted carbenium ion was studied by Hajdasz and Squires[15]. They found that α-protonation of $Me_3SiHC=CH_2$ **42** yields exclusively the β-trimethylsilylethyl cation **43** which undergoes collision-induced dissociation by C_2H_4 loss and reacts with MeOH, C_6H_6 and other Lewis bases by exclusive ethylene displacement. The measured proton affinity (PA) of 199 kcal mol^{-1} ranks **42** among isobutene (PA 196), tetramethylethylene (PA 199) and styrene (PA 202 kcal mol^{-1}) with respect to base strength. Therefore, the total stabilization energy of **43** relative to the bridged ethyl cation is similar to that of tertiary and benzylic carbenium ions.

The authors computed from the measured PA, using the hydride transfer reaction 11, a relative stabilization energy of the β-Me$_3$Si group compared with hydrogen of 39 kcal mol^{-1}. This is in perfect agreement with the results of earlier *ab initio* calculations which predict a stabilization of 38 kcal mol^{-1} [MP3/6-31G(d)//3-21G(d)] for $H_3SiCH_2CH_2^+$, **44**, compared with the open ethyl cation[5]. Note, however, that at higher level **44** is not stable but closes to the bridged silylethyl cation **21**, which is more stable by 30.0 kcal mol^{-1} than the bridged ethyl cation[5,9,10].

$$Me_3SiCH_2CH_3 + H_2C(H)CH_2^+ \longrightarrow Me_3SiCH_2-\overset{+}{C}H_2 + C_2H_6 \qquad (11)$$
$$(43)$$

Theory predicts that the stabilizing effect by a β-silicon group is slightly larger for the vinyl cation than for the ethyl cation. Thus, at MP2(fu)/6-31G(d)//MP2(fu)/6-31G(d) the bridged silylvinyl cation **22** is stabilized by 33.5 kcal mol^{-1} compared with the bridged vinyl cation[7]. Higher level calculations predict a stabilization of 33.1 kcal mol^{-1} for **22** compared with the bridged vinyl cation [QCISD(T)/6-31G(d)//MP2(fu)/6-31G(d)][7]. The stabilizing effect of a trimethylsilyl group is even larger, 51.3 kcal mol^{-1}

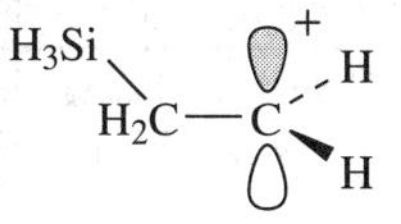

(44) 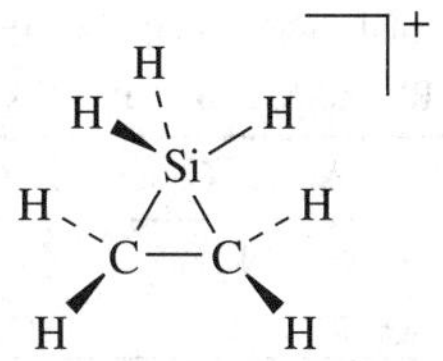 (21)

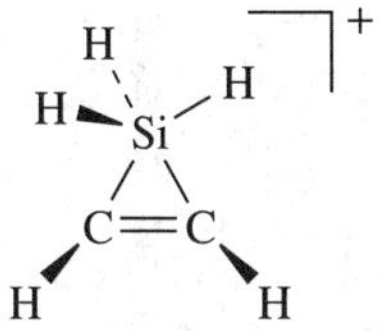

(22) 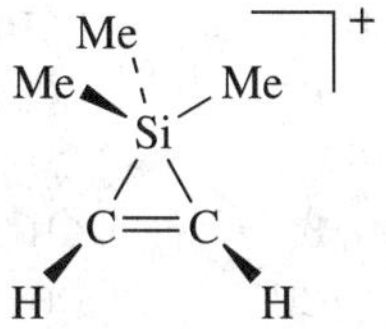 (45)

at MP4(SDTQ)/6-31G(d)//MP2(fu)/6-31G(d), as given by equation 12, for R = Me[7]. Corrections for the restricted basis set size and zero-point energy differences lead to a theoretical value of 49.6 kcal mol^{-1}[7], in agreement with the results of an experimental study by McGibbon, Brook and Terlouw[19] who calculated from their appearance energy measurements of HC(SiMe$_3$)$^+$CH **45** formed by ionization of Me$_3$Si(H)C=CHI **46** a value of 43.5 kcal mol^{-1}.

$$H_2C=CH_2 + HC(SiR_3)^+CH \longrightarrow HCH^+CH + H_2C=CHSiR_3 \quad (12)$$
$$\text{(22 or 45)}$$

Stone and coworkers determined the β-silicon effect in α-alkyl- and aryl-substituted carbenium ions[20] and vinyl cations[21] by measuring in a high-pressure mass spectrometer the thermodynamic data for the association of various alkenes and alkynes with trimethylsilylium ion. From their measured thermodynamic data they calculated, by using equations 13 and 14, the β-silyl stabilization energies listed in Table 1.

$$R^1R^2HC-\overset{+}{C}R^3R^4 + Me_3SiR^1R^2C-CHR^3R^4 \longrightarrow R^1R^2HC-CHR^3R^4$$
$$+ Me_3SiR^1R^2C-\overset{+}{C}R^3R^4 \quad (13)$$

$$Me_3SiR^1C=CHR^2 + R^1HC=\overset{+}{C}R^2 \longrightarrow Me_3SiR^1C=\overset{+}{C}R^2$$
$$+ R^1HC=CHR^2 \quad (14)$$

The measured stabilization energies are consistent with the theoretically predicted large hyperconjugative interaction between the Si–C σ-bond and the formally empty 2p(C) orbital. The results for the trivalent carbenium ions show a consistent decrease of 10 kcal mol^{-1} with each successive methyl group substitution on the carbenium carbon (Table 1, entries 1,4,6). Even the very stable t-butyl cation Me$_3$C$^+$ is stabilized by an additional 24 kcal mol^{-1} by a β-Me$_3$Si substituent (Table 1, entry 6). The stabilization of vinyl cations due to the presence of a β-Me$_3$Si group is found to be smaller than for

TABLE 1. Experimental stabilization energies (kcal mol^{-1}) of β-silyl-substituted carbocations Me$_3$SiCHR^{1}R^2–$\overset{+}{C}$R^{3}R^4 and Me$_3$Si–CR^1=$\overset{+}{C}$R^2 according to equations 13 and 14, respectively

Entry	Cation		β-Silyl-stabilization energy	Reference
Carbenium ions	(equation 13)			
	R^1, R^2	R^3, R^4		
1	H, H	H, H	44	20[a]
2	H, H	H, H	39	16
3	H, H	n-C$_4$H$_9$, H	26	21
4	H, H	CH$_3$, H	34	20[a]
5	CH$_3$, H	CH$_3$, H	34	20[a]
6	H, H	CH$_3$, CH$_3$	24	20[a]
7	CH$_3$, H	CH$_3$, CH$_3$	25	20[a]
8	CH$_3$, CH$_3$	CH$_3$, CH$_3$	22	20[a]
9	H, H	H, C$_6$H$_5$	17	20[a]
Vinyl cations	(equation 14)			
	R^1	R^2		
10	H	n-C$_4$H$_9$	11	21
11	CH$_3$	n-C$_3$H$_7$	12	21
12	H	C$_6$H$_5$	9	21
13	H	H	43.5	19

[a]Original value was corrected by -4 kcal mol^{-1}; for details see Reference 21.

similar carbenium ions (i.e. compare Table 1, entries 3 and 10 or 9 and 12). This seems to be in disagreement with the simple argument that hyperconjugative interactions between the C–Si bond and the formally empty 2p(C) orbital should be more efficient in vinyl cations than in carbenium ions due to the shorter C=C bond and the coplanar alignment of the interacting orbitals in the vinyl cations[22]. Very early calculations of Apeloig, Schleyer and Pople[23] have already suggested that vinyl cations are more sensitive to α-stabilizing effects than trivalent carbenium ions. This is corroborated by the gas phase protonation studies of several alkynes and alkenes by Stone and coworkers[20,21]. These studies show that the same α-alkyl or α-aryl substituent stabilizes a carbenium ion less than a vinyl cation[20]. The total stabilization afforded by both the α-alkyl or α-aryl substituent and a β-Me$_3$Si substituent appears to be approximately the same in both alkyl and vinyl cations and hence the β-silicon effect is considerably smaller for the vinyl cations. Unfortunately, the equilibrium constants for the association between acetylene and trimethylsilicenium ion could not be measured due to the low basicity of acetylene toward Me$_3$Si$^+$. However, Stone and coworkers extrapolated from their study of α-substituent effects on the alkyl cations a lower limit of 20 kcal mol^{-1} on the β-silicon effect for the unsubstituted vinyl cation[21]. This small extrapolated value is in contrast with the most recent high level computations and with the experimental study of Terlouw and coworkers[19] (Table 1, entry 13). Clearly, an additional experimental determination of the stabilization due to the β-silicon in HC(SiMe$_3$)$^+$CH ion (**45**) is desirable.

The structure determination of the observed species is not straightforward since the only direct information from the gas phase MS studies are the m/z values of the species. Further structural details have to be obtained by indirect methods like deuterium labeling experiments and consecutive reactions of the investigated ions. High energy collision activation experiments by Ciommer and Schwarz[24] with ^{13}C- and CD$_2$-labelled C$_5$H$_{13}$Si$^+$ ions formed by dissociative ionization of Me$_3$SiCH$_2$CH$_2$OPh show that the methylene groups in the ethene ligand become equivalent prior to or during collisional activation decomposition (CAD). This finding was originally advanced as proof for the bridged structure **28** of the Me$_3$SiCH$_2$CH$_2$$^+$ ion, but this finding was later interpreted as being consistent with either the open ion **47** or the bridged ion **28**[14]. The bimolecular reactions of the Me$_3$SiCH$_2$CH$_2$$^+$ ion with n- and π-Lewis bases are relatively rapid and result mainly or exclusively in ethene displacement. This behaviour is indicative of an ion structure with an intact but labile ethene ligand. Again, both the open **47** and silicon-bridged **28** structures are consistent with the observed reactivity.

(28)   (47)

The gas phase reactions between the simplest silylium ion H$_3$Si$^+$ and ethene[25], acetylene[26] and benzene[27,28] and the fate of the formed adducts in the high pressure MS (HPMS) were studied by Mayer and Lampe[25]. By using tandem and high pressure mass spectrometry they found that H$_3$Si$^+$ adds to ethene yielding the persistent complex SiC$_2$H$_7$$^+$, which successively adds two more molecules of ethene yielding the ion (C$_2$H$_5$)$_3$Si$^+$. Based on the reactivity of the initially formed SiC$_2$H$_7$$^+$ adduct Mayer and Lampe concluded that it has an alkylsilylium ion C$_2$H$_5$SiH$_2$$^+$ **48** structure and that no silyl-substituted carbenium ion is formed[25]. This is in line with recent *ab initio* calculations by Hopkinson and coworkers[3] (Figure 5) which show that ethylsilylium ion **48** is more stable [at MP2/(fu)/6-31G(d,p)] than the bridged silylethyl cation **21** by 6 kcal mol^{-1}. According to these calculations ion **21** is stabilized by 43.2 kcal mol^{-1} compared with both separated fragments ethene and SiH$_3$$^+$. The activation barrier of 10.2 kcal mol^{-1} for the 1,2-H shift to **48** is small but significant. Thus, the calculations suggest that both ions are formed by the addition of H$_3$Si$^+$ to ethene with the silylium ion **48** being the more abundant species. In the absence of any efficient collisional deactivation **21** will rearrange to the more stable **48**. Formation of the hydrogen-bridged species **49** is very unlikely, considering the unfavourable thermochemistry.

Mayer and Lampe made no hypothesis concerning the structure of the adduct formed in their HPMS experiments with acetylene and H$_3$Si$^+$[26]. More definite answers are given by theory. The initially formed acetylene–H$_3$Si$^+$ complex, e.g. the silyl-bridged vinyl cation **22**, is a minimum at MP2(fu)/6-31G(d,p), but it undergoes a facile two-step isomerization to the more stable (by 22.4 kcal mol^{-1}) 1-silaallyl cation **50**. The overall barrier for this transformation, 17.2 kcal mol^{-1}, is smaller than the energy of its dissociation into H$_3$Si$^+$

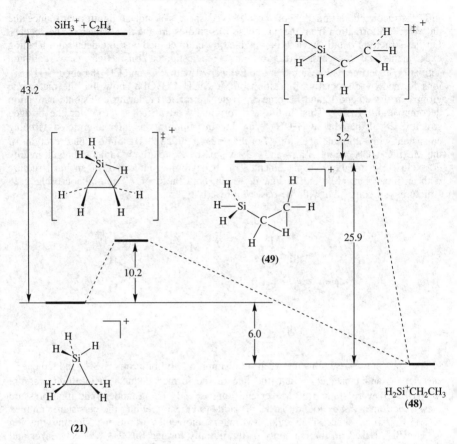

FIGURE 5. Part of the $SiC_2H_7^+$ potential energy surface (MP2/6-31G(d,p), relative energies in kcal mol^{-1})

and acetylene (Figure 6). It is therefore predicted that **22** which is initially formed by the addition of H_3Si^+ to acetylene, has enough excess energy (39.6 kcal mol^{-1}) to equilibrate with **50**, which is the most stable ion on the potential energy surface. This suggests that the species which is formed under high pressure MS conditions[26] is **50** and not the bridged vinyl cations **22** or **51**[4].

The addition of H_3Si^+ to benzene (equation 15) appears to follow a mechanism that is completely analogous to that proposed for the addition of carbenium ions to aromatics[26,27]. The energy-rich complex H_3Si^+/C_6H_6 can be collisionally stabilized and analysed in a tandem-MS. H_3Si^+/C_6H_6 eliminates hydrogen molecule, producing the phenylsilylium ion **52**. The predominant loss of HD from the initially formed complex of D_3Si^+ and benzene (70%) suggests that the formed ion has the structure of a β-silyl-substituted dienyl cation **53**[26,27].

Fornarini and colleagues gave also experimental evidence that in the reaction of trimethylsilicenium ion with arenes, σ-complexes such as **54** are produced[29–31]. The kinetic isotope effect of the gas phase deprotonation reaction and specific deuterium labeling

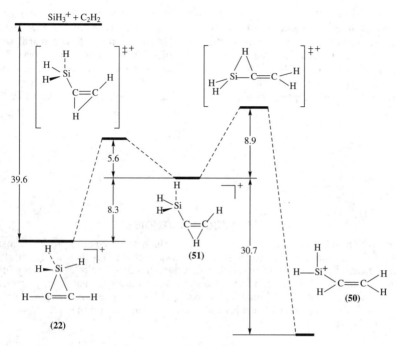

FIGURE 6. Part of the SiC$_2$H$_5$$^+$ potential energy surface (MP2/6-31G(d,p), relative energies in kcal mol^{-1})

experiments discard the possibility of a π-complex like **55**[32] and confirm the dienyl cation-type structure **54** of Me$_3$Si$^+$/C$_6$H$_6$.

Tsuno and coworkers measured the gas phase affinity of substituted styrenes toward trimethylsilylium ions based on Me$_3$Si$^+$ transfer equilibria between different styrenes[33]. The substituent effect on the stability of the Me$_3$Si$^+$/styrene complex has been analysed by means of the LArSR equation and gave a ρ value of -5.76 and an r value of 0.41. These results imply that the long-lived Me$_3$Si$^+$/styrene ions have only little need for stabilization by additional π-donation from the aryl ring. Therefore, the classical open structure **56** for the Me$_3$Si$^+$/styrene complex was discarded in favour of a partially bridged structure **57**[33].

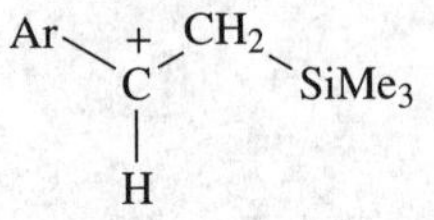

(56) 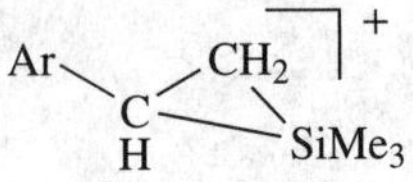   (57)

IV. SOLVOLYTIC STUDIES

In this section, solvolysis reactions are described which are thought to proceed via silyl-substituted carbocations. The reader should be aware of the fact that nearly all effects which are described here are of purely kinetic origin and therefore refer to energy differences between ground states and transition states. Hence they are not strictly applicable to the intermediate silyl-substituted carbocations, although the Hammond postulate suggests a close structural resemblance between the transition state for the ionization and the formed carbocation.

A. α-Silicon-substituted Carbocations in Solvolysis

The first report of a distinct effect of silicon on a neighbouring carbenium ion was published as early as 1946. Sommer, Whitmore and coworkers[34,35] reported that under conditions of basic hydrolysis, (1-chloroethyl)trichlorosilane **58** and (1-chloropropyl)trichlorosilane **59** failed to give cleavage of the α-C−Cl bond, while under the same conditions the analogous 2-chloroethyl **60** and 2-chloropropyl compounds **61** hydrolyse rapidly.

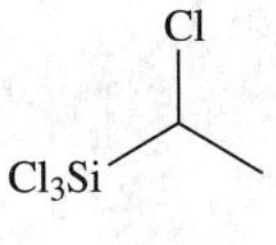

(58) 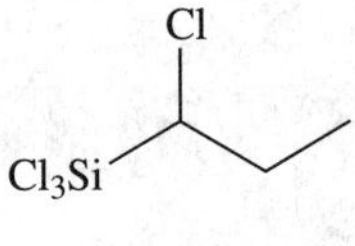   (59)

(60) 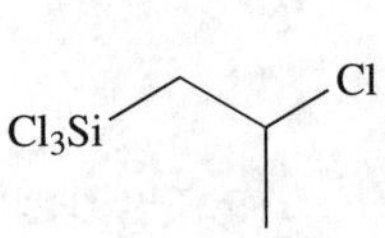   (61)

Since then the kinetic α-silicon effect has been measured in solvolysis of various systems that are considered to involve ionization in the rate-determining step. The results are compiled in Table 2, which also includes for comparison data for the α-silicon effect

TABLE 2. Kinetic α-silicon effects in solvolysis

Compound	$10^5 k$ (s^{-1})	Reference group R	k (compound)/ k (reference compound)	Reference
$S_N 1$ solvolysis				
65				
X = Br	a	t-Bu	2.6×10^{-5}	37
	b	Me	3.3×10^{-4}	40
X = Cl	b	Me	1.3×10^{-3}	40
X = OPBN	b	Me	0.33	40
66				
X = OPBN	3.82^c	Me	0.46	39
		t-Bu	2.8×10^{-6}	39
		H	10^8	39
	72.70^d	Me	1.08	39, 40
X = Cl	a	Me	70	40
73				
X = Br	0.0313^e	Me	6.06×10^{-5}	41
X = OTs	$2.06 \; 10^3$	t-Bu	1.53	41
80				
X = Br	3.17^f	CMe$_2$	2.01×10^{-3}	41
88				
X = OTf	6.34^g	Me	118	45
		i-Pr	0.75	45
93				
X = OTf	0.0282^g	t-Bu	0.021	45
		H	61	45
99				
X = OTf	0.153^g	t-Bu	0.019	45
		H	31	45
Protonation				
83				
	$24,200^h$	t-Bu	0.018	42
		H	1.8	42
	i	t-Bu	0.014	43
		H	15.1	43
$S_N 2$ solvolysis				
62				
X = OTs	0.144^j	t-Bu	1830	8
X = OTf	$21,400^j$	t-Bu	957	8
X = OTs	2.249^k	t-Bu	0.96	8
X = OTf	18.3^l	t-Bu	0.42	8

[a] In 60% EtOH.
[b] In 97% TFE.
[c] In 80% acetone at 125.8 °C.
[d] In 97% TFE at 100.2 °C.
[e] In 97% TFE at 25 °C.
[f] In 60% acetone.
[g] In 50% EtOH at 25 °C.
[h] In 80% acetone at 25 °C.
[i] In H$_2$O at 25 °C.
[j] In 60% EtOH at 25.0 °C.
[k] In 97% TFE at 97 °C.
[l] In 97% TFE at 35 °C.

in S_N2-type substitution experiments. In most cases the α-silicon effect is evaluated by comparing the rate for an α-trimethylsilyl-substituted substrate with that for the silicon-free, the α-methyl or the α-*t*-butyl analogues.

The effect of α-silyl substitution on the stability of a carbenium ion was qualitatively unclear for a long time. Early solvolytic studies by the groups of Eaborn[36] and Cartledge[37] suggest a destabilizing effect of α-silyl substitution compared with alkyl. The measurement and interpretation of the kinetic α-silicon effect in solvolysis reactions is, however, often complicated by the fact that steric and ground state effects may play an important role and that, in addition, the rates of ionization often involve a contribution from nucleophilic solvent assistance.

For example, the experimental finding by Dostrovsky and Hughes[38] that compound **62**, X = Br solvolyses 2600 times more rapidly than the neopentyl analog **63**, X = Br in basic ethanol could be explained by a change of the reaction mechanism. In α-silicon-substituted systems S_N2 reactions are favoured. Stang, Apeloig, Schiavelli and their coworkers[8] found that **62**, X = OTf, OTs ethanolysed 957 and 1830 times faster, respectively, without rearrangement, than the analogous **63** which yield in a k_Δ mechanism (i.e. with neighbouring group participation) almost only rearranged products. The extremely low Grunwald–Winstein *m* values for **62** (0.34 and 0.23 for X = OTf and OTs, respectively) are also consistent with a bimolecular mechanism for the solvolysis of these silicon derivatives. Furthermore, in aqueous trifluoroethanol the relative solvolysis rates are reversed: $k(\mathbf{62})/k(\mathbf{63}) = 0.42$ and 0.96 for X = OTf and OTs, respectively. This serves as strong evidence for a bimolecular mechanism without the intermediacy of an ionic species. The theoretical calculations further corroborate the S_N2-type displacements in neopentyl-like silylmethyl sulphonates. According to the calculations, if an α-trimethylsilylcarbenium ion is formed in a k_C mechanism it should rearrange to the more stable silylium ion. Thus the transient formation of $Me_3SiCH_2^+$ **33** should yield substantial amounts of rearranged products as well. In addition, the calculations also indicate that a Me_3Si group lowers the barrier for S_N2 displacements relative to *t*-Bu by 20 kcal mol^{-1} (at 3-21G//3-21G). This dramatic effect is due to a better σ-electron-accepting ability of the trimethylsilyl group and the comparably long Si−C bonds which alleviates the steric effects that inhibit the S_N2 reaction in neopentyl systems[8].

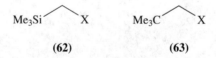

(**62**)   (**63**)

Another complication concerning the kinetic α-silicon effect arises from large ground-state effects in the α-silyl sulphonates or carboxylates used for the kinetic experiments. Cartledge and Jones[37] found that the bromide **64**, X = Br reacted 38000 times faster than bromide **65**, X = Br in 60% aqueous ethanol at 25 °C, in agreement with a strong destabilizing effect of the α-silyl substituent. In contrast, Apeloig and Stanger[39] found comparable reaction rates for Me_3Si and methyl substitution in the k_C solvolysis of tertiary adamantyl systems **66**, X = OPNB and **67**, X = OPNB, i.e. $k(\alpha\text{-Me})/k(\alpha\text{-Me}_3Si) = 2.18$ in 80% aqueous acetone and 0.97 in 97% TFE. This very small solvent effect strongly supports a rate-limiting k_C solvolysis for both compounds. The secondary **68**, X = OPNB solvolyses 10^8 times slower than either tertiary system. The authors argued that the nearly identical solvolysis rates of **66** and **67**, X = OPNB are not due to the similar stabilities of the corresponding carbenium ions but rather reflect that the α-silylcarboxylates are electronically destabilized relative to the corresponding α-alkylcarboxylates. Their calculations corroborate the interpretation of the solvolysis experiments. Thus the α-silylmethanol **69** (which serves as a model for the carboxylate) is destabilized by a geminal interaction between the oxygen and the silyl group relative to the corresponding alkyl alcohols by 6–8 kcal mol^{-1} [at MP2/6-31G(d)//3-21G(d)]. By combining these ground-state energy differences with the similar solvolysis rates of **66** and **67**, X = OPNB the authors concluded that in solution an α-Me$_3$Si substituent destabilizes the 2-adamantyl cation **70** by 6–8 kcal mol^{-1} but that it is by 12.4 kcal mol^{-1} more stabilizing than hydrogen[39].

Me$_3$C—X (**64**)

Me$_3$C—X (**65**)

(**66**) R = SiMe$_3$
(**67**) R = Me
(**68**) R = H

H$_3$SiCH$_2$OH

(**69**)

(**70**)

Further *ab initio* calculations at the MP2/3-21G(d)//3-21G(d) level by the group of Apeloig[40] indicate that the destabilizing geminal interactions between the α-silyl substituent and the leaving group become less important for halogens as the leaving groups. Thus, while the destabilization of **71** compared with **72** due to geminal interactions is 7.8 kcal mol^{-1} for X = formate or 5.6 kcal mol^{-1} for X = sulphonate, it decreases to 4.4 kcal mol^{-1} for X = Cl and becomes very small for X = Br or I (1.9 and 1.0 kcal mol^{-1}, respectively). These theoretical predictions were supported by accompanying solvolysis experiments, which show that the relative rate ratio $k(\mathbf{64})/k(\mathbf{65})$ increases from 3 for X = OPNB to 792 for X = Cl and to 3010 for X = Br[40].

A similar dependence of $k(α\text{-Alkyl})/k(α\text{-SiMe}_3)$ on the leaving group was observed in the solvolysis of benzylic substrates. Thus, Shimizu and coworkers[41] found that the $k(\mathbf{74})/k(\mathbf{73})$ ratio in 97% TFE (25 °C) is 0.66 for X = OTs but the $k(\mathbf{75})/k(\mathbf{73})$ ratio is 16400 for X = Br. The authors argued that the steric repulsion between the *t*-butyl group and the *ortho* hydrogens in the nearly planar incipient benzyl cation **76** is larger than the steric effect in **76**. This actually slows the solvolysis rate of **74**, resulting in a nearly identical solvolysis rate to that of **73**. These destabilizing steric interactions are absent in **78**, thus leading to a faster solvolysis[41]. However, the arguments given by Apeloig and coworkers[39,40] suggest that ground state effects should also be taken into account in explaining the observed solvolysis rates of the benzyl tosylates studied by Shimizu and coworkers[41].

H₃SiCH₂X MeCH₂X

(71) (72)

(73) R = SiMe₃
(74) R = t-Bu
(75) R = Me

(76) R = SiMe₃
(77) R = t-Bu
(78) R = Me

(79) E = C
(80) E = Si

(81) E = C
(82) E = Si

In order to minimize the complications due to steric effects Shimizu and coworkers[41] investigated also the solvolysis reactions of 1-indanyl bromides **79** and **80**. **79** solvolyses via a cationic intermediate **81**. The characteristics of the solvolysis of **80** in aqueous acetone (m = 0.93, no skeletal rearrangement during the solvolysis) are consistent with a k_C mechanism involving the rate-determining formation of the 2-silaindanyl cation **82**. **80** solvolysed 498 times slower than the carbon analogue **79**, i.e., approximately 30-fold slower than the relative solvolysis rate ratio $k(\mathbf{75})/k(\mathbf{73})$ of 16,400 for the open benzylic system. This comparison suggests that the electronic α-silyl effect in benzyl cations leads to a rate retardation of approximately 500-fold, while steric repulsion between the *ortho* hydrogen and the trimethylsilyl group is responsible for another factor of 30. These experiments suggest that in solution an α-silyl group is *ca* 4 kcal mol⁻¹ less effective than an alkyl substituent in stabilizing benzyl cation **78**. The smaller destabilizing effect on **78** compared with that in RC⁺Me₂ probably reflects the reduced electronic demand of the carbenium ion centre in the former cation[41].

The small α-silyl effect found by Soderquist and Hassner[42] and by Kresge and Tobin[43] in the protonation of methoxyvinyl ethers **83–85** [$k(\alpha\text{-H}) : k(\alpha\text{-SiMe}_3) : k(\alpha\text{-}t\text{-Bu}) = 1 : 1.8 : 100$ in aqueous acetone at 33 °C[42] and 1 : 15 : 1025 in H₂O at 25 °C][43] (equation 16) might also be explained by the inherent stability of the formed α-methoxy-stabilized carbocations **86a–c**. Destabilization due to geminal interaction between the SiMe₃ and the MeO group in the α-silyl-substituted vinyl ether **83** (R = SiMe₃) might, however, also contribute to the apparently small α-silyl effect. The relative rate ratio **83** : **87** of 30 : 1 is consistent with smaller stabilization of the α-methoxy-α-triphenylethyl cation by the less electropositive SiPh₃ group[42].

(83) R = SiMe₃ (86a) R = SiMe₃
(84) R = H   (86b) R = H
(85) R = t-Bu (86c) R = t-Bu

(16)

Theory predicts that β-hyperconjugation across the double bond of vinyl cations is more effective than in carbenium ions. In addition, due to the higher electronegativity of the positively charged sp-carbon atom, vinyl cations are more sensitive to σ-hyperconjugative effects than trivalent carbenium ions[22,23,44]. Consequently, it is predicted that α-silyl substitution will stabilize a vinyl cation to the same extent as does a methyl group[44]. Solvolysis studies by Schiavelli, Stang and coworkers[45] revealed that the α-silyl triflate, **88**, which forms an intermediate vinyl cation in the rate-determining step, reacts 118 times faster than its methyl-substituted analogue **89**, but it solvolyses slightly slower than the α-isopropylalkenyl triflate **90** [$k(\mathbf{88})/k(\mathbf{90}) = 0.75$]. Ground-state and steric effects complicate the interpretation of these experiments[6]. The α-silyl-substituted enol $H_2C=C(SiH_3)OH$ (**91**), which is a model for the solvolytically investigated vinyl triflates, is calculated to be destabilivzed by 3.8 kcal mol^{-1} (at HF/3-21G//3-21G) relative to $H_2C=C(CH_3)OH$ (**92**) due to unfavourable α-geminal interactions in **92**. According to these calculations α-silylvinyl triflates are predicted to solvolyse faster than α-alkylvinyl triflates, although the corresponding carbenium ions have nearly the same energy[6]. Therefore, the solvolysis studies by Stang, Schiavelli and coworkers[45] support the calculational results. The deactivating influence of the α-trimethylsilyl substituent in **88** compared with the α-isopropyl group in **90** might result from acceleration of the solvolysis of the α-isopropyl derivative by steric effects. In combination with the calculations[6] it became evident from the solvolysis experiments[45] that α-silyl substitution stabilizes a vinyl cation relative to hydrogen and that it is as effective as an alkyl group.

(**87**)

(**88**) R = SiMe$_3$
(**89**) R = Me
(**90**) R = i-Pr

(**91**) R = SiH$_3$
(**92**) R = Me

Ethanolysis of the α-ethynyl-substituted vinyl triflates **93**–**95** proceeds via the intermediacy of the cumulenic dicoordinated cations **96**–**98** (equation 17) and suggests the following stability order for the carbocations: **98** > **96** > **97**[45]. The same relative reactivity order was found for the triflates **99**–**101**[45].

(17)

(**93**) R = SiMe$_3$
(**94**) R = H
(**95**) R = t-Bu

(**96**) R = SiMe$_3$
(**97**) R = H
(**98**) R = t-Bu

(**99**) R = SiMe$_3$
(**100**) R = H
(**101**) R = t-Bu

Dallaire and Brook[46] studied the protiodemetallation reaction of the bis(silyl) alkynes **102**, E = R$_3$Si to give **104** shown in equation 18. They found small differences in the relative rates for R$_3$E = SiMe$_2$Bu-*t*, SiMe$_3$ and SiPh$_3$ [k(**102**, SiMe$_2$Bu-*t*) : k(**102**, SiMe$_3$) : k(**102**, SiPh$_3$) = 1.2 : 1 : 0.04]. Assuming that the β-stabilization by the trimethylsilyl group in the intermediate vinyl cation **103** is constant for the three different α-silyl groups, these relative rates provide an estimate of the inductive effects, showing the SiPh$_3$ group to be destabilizing compared with trialkylsilyl groups[46].

$$R_3E\text{---}\!\!\equiv\!\!\text{---}SiMe_3 \xrightarrow{H^+} R_3E\text{---}\overset{\pm}{=\!=}\!\!\diagdown\!\overset{SiMe_3}{H}$$

(**102**)   (**103**)

$$R_3E\text{---}\!\!\equiv\!\!\text{---}H$$

(**104**)

(18)

B. β-Silicon-substituted Carbocations in Solvolysis

The dramatic β-effect of silicon in reactions which involve carbocations was noted already in 1937 by Ushakov and Itenberg[47]. During the ensuing sixty years mechanistic, theoretical and synthetic studies have established the β-silicon effect as one of the largest neighbouring group effects which is exceeded only by those of the silicon congeners germanium and tin.

In solvolysis reactions, a silicon at the β-position greatly facilitates the heterolysis of a C−X bond, often affording exclusively elimination products (equation 19). Thus, already Sommer, Whitmore and coworkers[34,35] described a greatly enhanced reactivity of **60** and **61** in their basic hydrolysis in comparison with the corresponding α- and γ-substituted systems.

$$Me_3Si\diagdown C_\beta\text{---}C_\alpha\diagup_X \longrightarrow \diagdown C\!=\!C\diagup + Me_3SiX \qquad (19)$$

It is now commonly accepted that in polar solvents the elimination to the alkene **107** takes place by an E1-like mechanism involving a cationic intermediate[48]. The major issue still in need of clarification is the structure of the intermediate β-silicon carbocation which will be close to the structure of the transition state of the solvolysis reaction. An open carbocation **105** in which the silicon interacts purely by hyperconjugation without significant movement in the transition state (vertical stabilization)[49−52] is as possible as a three-membered siliconium ion **106** formed by neighbouring group participation (non-vertical stabilization). The terms vertical and non-vertical stabilization are referred to here for historical reasons only. They should no longer be used to describe hyperconjugative stabilization which implies no or little reorganization of the atoms, and hypercoordination with bridging, respectively. High level quantum chemical *ab initio* calculations reveal that all electronic stabilization is accompanied by geometrical distortions of the molecule. The stabilization and reorganization of the molecular framework depend on the electronic demand of the positively charged centre[53]. The mechanism via **105** involves a simple rate-determining ionization (k_C mechanism, equation 20), while the latter route via **106** involves a rate-determining nucleophilic attack of the β-silicon substituent at C_α during generation of the incipient carbenium ion (k_Δ mechanism, equation 20).

$$\text{(equation 20)}$$

Both mechanisms are equally compatible with the high E1 character found by Sommer and Baughman for the solvolysis of the primary chloride **108**[54]. The negative Hammett ρ values, found by Vencl and coworkers[55] and Fujiyama and Munechika[56] for the ethanolysis of the primary aryldimethylsilylethyl chloride **109** (-1.2 and -2.3 in 92% EtOH and 80% EtOH, respectively) which indicates a significant build-up of positive charge at the silicon atom in the transition state of the solvolysis, also do not allow one to distinguish between both mechanisms.

(108)   **(109)**

The stereochemistry of the elimination of Me_3SiX from β-trimethylsilyl-substituted precursors has been studied by Jarvie and coworkers[57]. They found that solvolysis of *erythro*-$Me_3SiCHBrCHBrMe$ **110** lead predominantly to *cis*-1-brompropene **111** (equation 21).

$$\text{(equation 21)}$$

The solvent effects found for this reaction excluded an E2 mechanism in favour of an E1 reaction with a rate-determining cleavage of the C—Br bond. Thus, the elimination of bromine from **110** must proceed from an antiperiplanar conformation. Independent hydrolysis experiments by Hudrlik and Peterson[58] gave further support for this mode of the silicodehalogenation. They found that *threo*-**114** gave >90% *cis*-4-octene **115** (equation 22), as expected for an antiperiplanar stereochemistry for the elimination. The stereochemical outcome of the reaction of **110** can be explained most conveniently by the intermediacy of silicon-bridged species **112**. Increasing amounts of *trans*-1-brompropene (*trans*-**111**) were, however, detected in solvent mixtures with increasing ionization power parameter Y [0.3% of *trans*-**111** in pure ethanol ($Y = -2.033$) and 15% in 100% formic acid ($Y = 2.05$)]. This behaviour might be explained by an equilibrium between the bridged **112** and the open ion **113** (equation 21)[57]. However, Lambert[48] argues that a high barrier for the rotation around the C—C bond in **113** resulting from strong β-Si$^-$ hyperconjugation could enforce the antiperiplanar mode for the elimination from **113**. His arguments are supported by the significant rotation barrier of 22 kcal mol^{-1} calculated for ion **116** [at MP2/6-31G(d)//6-31G(d)] in the gas phase[9].

(22)

(**114**) (**115**)

Jarvie and coworkers[59] studied the secondary deuterium isotope effects on the solvolysis rates of bromides **117**, **118** and **119** and concluded from the value of the k_H/k_D of 1.10 for **117** that a C—Br bond cleavage was rate-determining, 'possibly assisted by the β-silicon'[59]. Their arguments were supported by the very small secondary isotope effects found for **118** and **119** ($k_H/k_D = 1.02$ and 0.995, respectively). Interestingly, they found that the reaction of **120** with PBr$_3$ gave equal amounts of **121** and **122** (equation 23), i.e. the CD$_2$ and CH$_2$ groups became equivalent during the reaction. A similar result was obtained by Eaborn and coworkers[60]. Similarly to the non-classical ion debate[61], this can be explained either by the intermediacy of a bridged ion **123** or by a fast equilibrium between two open cation structures **124** and **125** (equation 24) as suggested by Lambert[48]. Eaborn and coworkers favoured the cyclic siliconium ion, based on the fact that **126** solvolyses faster than predicted from its σ_p^+ constant [σ_p^+(Me$_3$SiCH$_2$) $= -0.54$][60].

(**116**) (**117**)

(**118**) (**119**)

$$\text{Me}_3\text{Si}-\text{CH}_2\text{CD}_2-\text{OH} \xrightarrow{\text{PBr}_3} \text{Me}_3\text{Si}-\text{CD}_2-\text{CH}_2-\text{Br}$$

(120)   (121)

$+$ (23)

$$\text{Me}_3\text{Si}-\text{CH}_2-\text{CD}_2-\text{Br}$$

(122)

(123) — bridged cation with SiMe$_3$, H, H, D, D

$$\text{Me}_3\text{Si}-\text{CH}_2-\overset{+}{\text{CD}_2} \rightleftharpoons \overset{+}{\text{CH}_2}-\text{CD}_2-\text{SiMe}_3 \quad (24)$$

(124)   (125)

Experiments by Davis and Jacocks[62] shed a new light on the problem. They found for the deoxysilylation reaction of substituted β-hydroxyalkylsilanes **127–130** a relative rate ratio of $k(\mathbf{127}) : k(\mathbf{128}) : k(\mathbf{129}) : k(\mathbf{130}) = 1 : 10^{3.3} : 10^{5.92} : 10^{6.77}$. The substituent effect on the rate was analysed by considering the sum of the σ^+ values of the substituents on the alcohol carbon ($\sum \sigma^+$). From the linear free energy relationship for this reaction ($\log k_{rel} = \rho^+ \sum \sigma^+$) a ρ^+ value of -11 is obtained. The second CH$_2$SiMe$_3$ group in **129** leads to a rate acceleration of $10^{5.92}$ compared with **127**, resulting in $\sigma^+ = -0.54$ for the second CH$_2$SiMe$_3$ group, identical with that of the first group in **127**. This suggests an additivity of the substituent effects on the relative rates[62] which is inconsistent with an anchimeric effect of the β-trimethylsilyl group on the solvolysis reaction, since only one silyl group can engage in a neighbouring group participation in the transition state of the solvolysis of **129**[62].

Me$_3$Si–CH$_2$CH$_2$–Cl   Me$_3$Si–CH$_2$CH$_2$–OH   Me$_3$Si–CH$_2$–C(CH$_3$)(H)–OH

(126)   (127)   (128)

Me$_3$Si–CH$_2$–CH(OH)–CH$_2$–SiMe$_3$ Me$_3$Si–CH$_2$–C(CH$_3$)$_2$–OH$_2$

(129)   (130)

A good tool to distinguish between the neighbouring group assisted route (k_Δ) with a silicenium ion-like intermediate, and the formation of an open β-silyl carbocation which is only stabilized by hyperconjugation without significant geometrical changes, is the conformational dependence of the rate acceleration by the β-silicon in the solvolysis reactions.

It is well established that hyperconjugation exhibits a cosine-squared dependency on the dihedral angle Θ between the interacting orbitals [for an open cation like **105**, the 2p(C) and the σ(C−Si)][63]. A siliconium ion structure like **106** can only be formed by a backside attack on the carbon attached to the leaving group in an antiperiplanar conformation. Thus, both mechanisms should display a maximum acceleration when Θ is 180°, i.e. at the antiperiplanar arrangement. Smaller angles between the leaving group X and the β-Si−C bond will severely hamper the formation of the three-membered ring and will slow down the solvolysis rate in a k_Δ mechanism. In contrast, for the k_C mechanism, there will be a substantial rate acceleration for all angles Θ, except for Θ = 90° at which the interacting orbitals are orthogonal and no hyperconjugative stabilization of the transition state can occur. A pure inductive β-silicon effect should show no angular dependency. The qualitative relationship between all three modes of stabilization is shown in Figure 7[64].

The experiments by Jarvie and coworkers[59] have established the antiperiplanar mode for the rate-determining ionization. The rate enhancement by the β-trimethylsilyl group compared with a non-substituted reference [$k(\beta$-SiMe$_3)/k(\beta$-H)] was typically about 10^6–10^7 for acyclic substrates. The elimination of the leaving group can, however, also occur from a less favoured conformation and will show a lower rate enhancement. In a series of papers Lambert and coworkers[64−68] examined cyclic systems **131**−**139** with well defined dihedral angles (except **131** and **133**) between the leaving group and the β-trimethylsilyl group and measured the kinetic enhancement of their solvolysis by the latter. The solvolytic results are summarized in Table 3. Solvent effect studies revealed that all the silicon-containing substrates solvolysed by a k_C mechanism

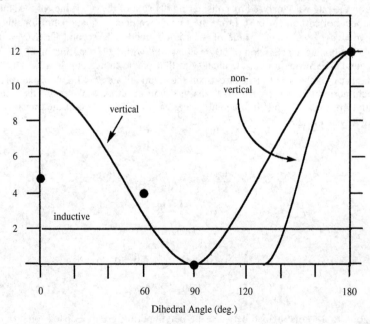

FIGURE 7. Qualitative dihedral dependences for the inductive, vertical and nonvertical modes of the kinetic β-silicon effect. Reprinted with permission from Reference 64. Copyright (1990) American Chemical Society. Experimental data[64,66,67] are indicated with solid dots

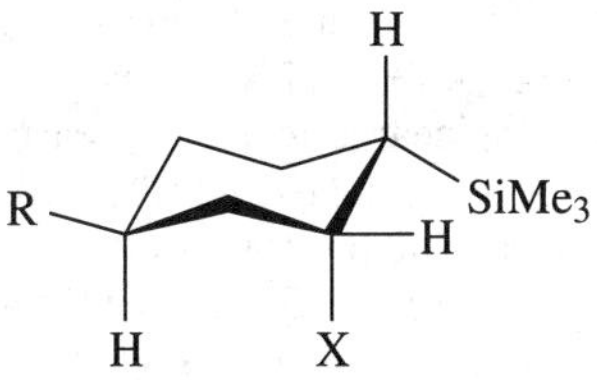

(131) R = H
(132) R = t-Bu

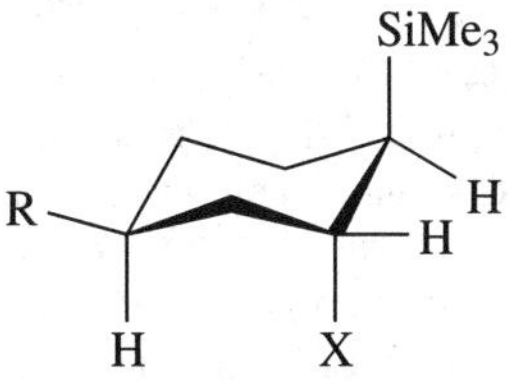

(133) R = H
(134) R = t-Bu

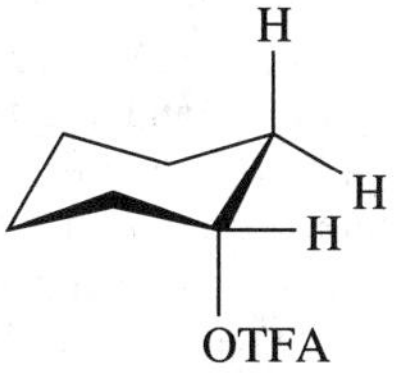

(135)

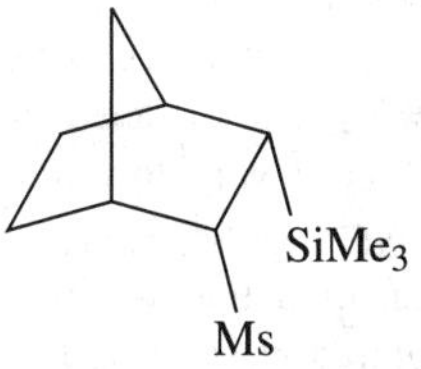

(136) Ms, SiMe3

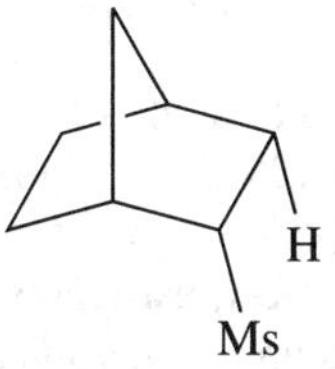

(137) Ms, H

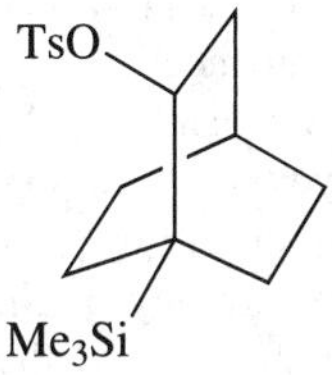

(138) TsO, Me3Si

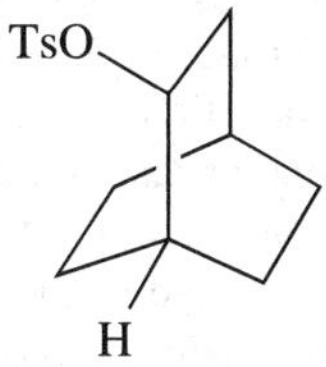

(139) TsO, H

TABLE 3. Kinetic β-silicon effects in solvolysis

Compound	$k(s^{-1})$	Reference group R	k(compound)/ k(reference compound)	Reference
(131) X = OTFA	2.36×10^{-5a}	H	3.3×10^{4b}	65
(132) X = OTFA	2.84×10^{-5a}	H	4.0×10^{4b}	66
(133) X = OTFA	4.0^a	H	5.7×10^{9b}	66
(134) X = OTFA	1.72×10^{3a}	H	2.4×10^{12b}	66
(134) X = ODNB	7.1×10^{-3a}	H	1.0×10^{7b}	66
(136) X = OMs	0.335^a	H	9.4×10^4	64
(138) X = OTs	1.64×10^{-3c}	H	1.3^b	67
(141) X = OTFA	3.68^d	t-Bu	2.99×10^5	72
		H	1.05×10^5	72
(150) X = OTs	6.7×10^2	Me	9.9×10^{10b}	73
(179) X = OTFA	1.4×10^5	H	27	82

[a]In 97% TFE at 25 °C.
[b]The reference solvolyses by a k_S mechanism.
[c]In 97% TFE at 35 °C.
[d]In 30% dioxane at 25 °C.

involving a silyl-substituted carbocation as an intermediate, whereas most of the reference compounds solvolysed by a k_S mechanism[65-67]. Thus, the introduction of even an entirely perpendicular trimethylsilyl group in **138** brings about a change in the mechanism from k_S to k_C[67]. While the solvolytic results of the cyclohexyl trifluoroacetates **131** and **133**, X = OTFA were inconclusive due to the not well defined stereochemical relation between the leaving group and the silyl group[65], the results for the stereochemically biased systems **132** and **134** are highly interesting[66]. **134** (X = OTFA), in which the leaving group and the trimethylsilyl group are in antiperiplanar orientation, reacts in 97% TFE 10^{12} more rapidly than the reference compound cyclohexyl–OTFA **135**[66]. This rate acceleration is equivalent to a relative stabilization of the transition state by the β-silicon compared with β-hydrogen of 16.5 kcal mol^{-1}. Even **131** (X = OTFA)[65] and **132** (X = OTFA)[66] in both of which a syn-clinal arrangement ($\Theta = 60°$) exists between the leaving group and the trimethylsilyl group solvolysed about 4×10^4 times faster than **135** in TFA at 25 °C. Similar results were obtained in five-membered ring systems[68,69]. The β-silicon effect in a syn-periplanar geometry, studied in the endo-norbornane system **136**, is of the same magnitude[64]. The relative rate for the solvolysis of the mesylates in 97% TFE, k [**136** (OMs)]: k [**137** (OMs)], is $9.4 \times 10^4 : 1$. Provided that ground state effects are negligible, this suggest a free energy difference between the transition states of 6.8 kcal mol^{-1}. A dihedral angle of $\Theta = 0°$ was modeled in the 1-trimethylsilyl[2.2.2]bicyclooctane-2-tosylate **138** (X = OTs)[67]. The very low relative rate ratio k (**138**, X = OTs): k(**139**, X = OTs) = 1.34 : 1 ($\Delta\Delta G^{\ddagger} = 0.2$ kcal mol^{-1}) was ascribed to the absence of any β-silicon effect, especially an inductive stabilization of the incipient carbocation by the β-silyl group[67]. This interpretation is hampered by the fact that the reference molecule **139** solvolyses by a k_S mechanism under the applied reaction conditions. The measured rate ratio is therefore only a lower estimate of the β-silicon effect on the transition state stabilization for a k_C solvolysis. Ab initio calculations by Ibrahim and Jorgensen[9], however, do support the experimental finding by Lambert and Liu[67]. According to their MP2/6-31G(d)//6-31G(d) calculations the perpendicular conformation in the β-silyl carbocation **12** is not stabilized compared with the isopropyl cation $(CH_3)_2CH^+$ **140**[9].

(**12**)

The experimental dependency of the β-silyl effect on Θ in solvolysis reactions is sketched in Figure 7[64]. Obviously, it differs from that anticipated for a k_Δ mechanism with rate-determining formation of siliconium ion or from the cosine-squared function expected for the pure hyperconjugative stabilization model. Apparently, the β-silyl effect is operative in the solvolysis of both the syn- and anti-periplanar conformations. The rate acceleration in the latter might be ascribed to a more favourable geometry for the σ-anchimeric assistance.

The α-secondary deuterium isotope effect in the solvolysis of **134** ($k_H/k_D = 1.17$) is similar to these for model reactions proceeding via an open carbocation-like transition state with hyperconjugative stabilization ($k_H/k_D = 1.2$)[70]. The isotope effect is, however, markedly different from effects observed for model reactions proceeding via internal oxygen participation (where ($k_H/k_D) < 1.08$)[71]. Thus, although strong hyperconjugative β-silyl stabilization is involved in the solvolysis reaction, the measured kinetic isotope effect was thought to exclude a rate-determining formation of a bridged siliconium ion-like transition state.

On the other hand, calculations have shown that hyperconjugative interaction of the β-Si—C σ-bond with an empty 2p(C) orbital does have pronounced structural consequences[9]. Thus, in the secondary cation **12** the silyl group is bent towards the positively charged carbon. A *syn* leaving group might prevent the molecule from adopting the optimal geometry for maximum β-silyl stabilization already in the transition state, leading to the observed lower reactivity of **136** compared with **134**[64]. From these arguments it is obvious that a clear distinction between hyperconjugative and bridging stabilization modes is not made easily, since there is always a subtle interplay between the electron demand of the carbocation and the magnitude of the β-silyl effect.

The question arises whether a k_C process can be distinguished kinetically from the k_Δ mechanism for the ionization of β-silicon-containing molecules. Mechanistic studies by Shimizu, Tsuno and coworkers showed that for some systems it is possible to distinguish between the two mechanisms[72,73]. The benzyl trifluoroacetate **141**, X = OTFA solvolyses more rapidly than both the silicon-free substrate **142** and the β-*t*-butyl analogue **143**, X = TFA in 30% dioxane at 25 °C by factors of 1.05×10^5 and 2.99×10^5, respectively, revealing a marked β-silyl effect even in a benzylic system[72]. These $k(\mathbf{141})/k(\mathbf{142})$ and $k(\mathbf{141})/k(\mathbf{143})$ rate ratios amount to β-silyl stabilization energies of 6.9 and 7.5 kcal mol^{-1}, respectively, relative to β-hydrogen or β-*t*-butyl in the solvolysis leading to the cationic benzylic intermediate. The relative small stabilization due to β-silyl-substitution compared with the cyclohexyl system **134** reflects the reduced electronic demand of the α-methylbenzyl cation **144** relative to the secondary cyclohexyl cation **145**[72]. The solvolysis rates of **141** in dioxane–water mixtures are as sensitive to the solvent ionizing power as those for the 1-tolylethyl trifluoroacetate **146**, a reference standard undergoing k_C solvolysis. The solvolysis of **141** and **146** exhibits almost identical secondary α-deuterium isotope effects (k_H/k_D = 1.18–1.19 and 1.178, respectively), in line with a change from sp^3 to sp^2 hybridization at the benzylic carbon atom in the transition state. Substituent effects on the solvolysis of **141** in 90% dioxane indicate that the resonance interaction of the aryl ring with a positive charge at the benzylic position is as effective as in an α-methylbenzyl cation. Thus, all these mechanistic criteria of solvent effects, kinetic isotope effects and substituent effects support a simple k_C mechanism

(**141**) R = SiMe$_3$
(**142**) R = H
(**143**) R = *t*-Bu

(**144**)

(**145**)

(**146**)

via the open β-silyl-substituted carbocation **147** and are inconsistent with a k_Δ mechanism involving a cyclic siliconium ion **148**[72]. It is noteworthy that Tsuno and coworkers did not find in the gas phase any effective π-delocalization of the positive charge into the aryl π-system for a series of 2-trimethylsilyl 1-arylethyl cations having different aryl substituents[33]. This indicates that the thermodynamic stabilities of 2-trimethylsilyl 1-arylethyl cations are only very little influenced by the π-donor ability of the aryl ring. Hence, they concluded that the ground state structure of these cations is not the open structure **147** but a partially bridged structure **149**. The obvious disagreement between gas phase studies[33] and the solvolytic experiments[72] was rationalized by suggesting that in the rate-determining transition state for the bond cleavage step in the solvolysis experiment, no neighbouring group assistance by the β-trimethylsilyl group is operative and that **147** is formed. However, after the bond cleavage takes place a partially bridged ion **149** is formed without any significant barrier [33].

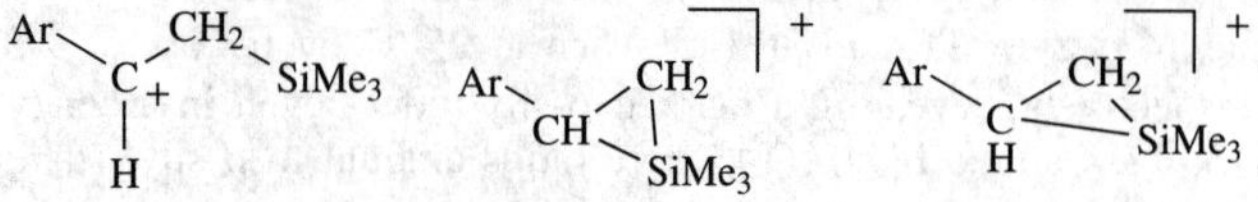

(147)   **(148)**   **(149)**

In contrast, **150** is 2.7×10^4-fold less reactive than **141**, X = OTFA, corresponding to an energy difference of 6.1 kcal mol^{-1} between the transition states provided that the ground state energies are similar. The ethanolysis of **150** gives 1,2-silyl rearranged products in addition to styrene. The substituent effects for the reaction of **150** are remarkably small (i.e. the $k_{p-\text{MeO}}/k_\text{H}$ rate ratio is 2.11, compared with 269 for **141**). These facts are more consistent with the formation of a bridged siliconium ion **151** than of an open β-silyl cation **152**[73].

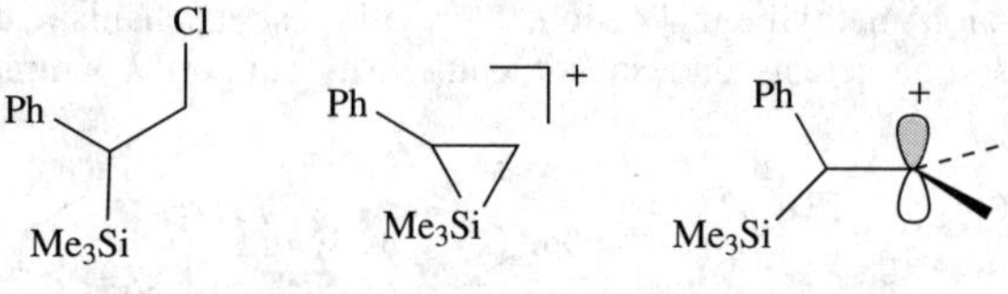

(150)   **(151)**   **(152)**

Competition experiments by Mayr and Pock[74] showed that allylsilane **153** is at −78 °C 30700 times more reactive towards diarylcarbenium ions than propene **154**. Thus, in the

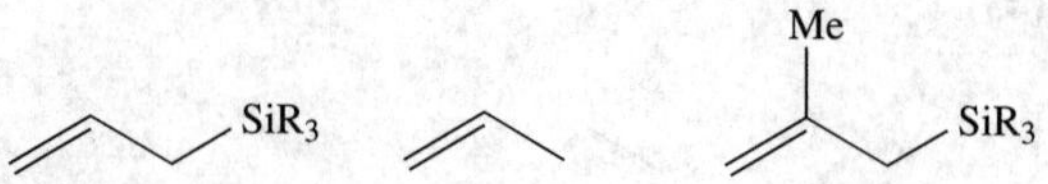

(153) R = Me
(157) R = Ph
(161) R = Et
(162) R = i-Pr
(163) R = n-Bu
(164) R = n-Hex

(154)

(158) R = Me
(159) R = Ph
(160) R = Cl

transition state for the addition of an electrophile to an allylsilane a β-silicon provides only a stabilization of 4.2 kcal mol^{-1}[74]. In subsequent studies Mayr and Hagen[75,76] examined the influence of the substituent on silicon on the rate of the electrophilic addition of *p*-anisylphenylcarbenium ion **155** to allysilanes to give ion **156** (equation 25) and their results are summarized in Table 4.

$$\underset{(\mathbf{155})}{\text{R}^2\text{-CH}_2\text{-SiR}_3^1} + \underset{\text{An}}{\overset{\text{Ph}}{\text{C}}}-\text{H} \xrightarrow[-78°C]{\text{slow} \atop \text{CH}_2\text{Cl}_2} \underset{(\mathbf{156})}{\overset{\text{Ph}}{\underset{\text{An}}{\text{C}}}-\text{CH}_2-\overset{+}{\underset{\text{R}^2}{\text{C}}}-\text{CH}_2-\text{SiR}_3^1} \quad (25)$$

An = *p*-MeOC$_6$H$_4$, R^2 = H, Me

They found a correlation between Taft inductive constants σ_I for the substituents on silicon and the reactivity of the allylsilane towards **155**[76]. Inductively electron-withdrawing substituents at silicon greatly reduce the reaction rates, i.e. $k(\mathbf{158}) : k(\mathbf{159}) : k(\mathbf{160}) = 1 : 0.11 : 1.8 \times 10^{-7}$ at $-78\,°$C in CH$_2$Cl$_2$[76]. Thus, the trichlorosilyl substituent is actually slightly deactivating for the addition of the carbocation to the C=C double bond compared with hydrogen. [$k(\mathbf{160}) : k$(isobutene) $= 355 : 1$)][76]. Replacement of the methyl by larger (branched or unbranched) alkyl groups leads to a slight increase in the reactivity [$k(\mathbf{153}) : k(\mathbf{164}) = 1 : 2.75$, see Table 4]. The stability of the formed carbocation also influences the extent of the kinetic β-silicon effect in this reaction. Thus, while **153**

TABLE 4. Relative rate constants for the reaction of substituted allyl element compounds R$_3^1$ECH$_2$C(R^2)=CH$_2$ with bisarylcarbenium ions in CH$_2$Cl$_2$[74-76] at $-70\,°$C.

Compound	E	R^1	k_{rel}
157[a]	Si	Ph	0.017
153[a]	Si	Me	1.0
161[a]	Si	Et	1.59
162[a]	Si	*i*-Pr	2.23
163[a]	Si	*n*-Bu	2.57
164[a]	Si	*n*-Hex	2.75
170[a]	Ge	Ph	0.096
171[a]	Sn	Ph	27.7
160[b]	Si	Cl	1.8×10^{-7}
158[b]	Si	Me	1
159[b]	Si	Ph	0.113
168[b]	Ge	Ph	0.283
169[b]	Sn	Ph	4.91

[a]Addition of *p*-anisylphenylcarbenium ion; reference: propene $k = 5.0 \times 10^{-6}$.
[b]Addition of bis(*p*-anisyl)carbenium ion; reference: isobutene $k = 6.4 \times 10^{-5}$.

reacts 2×10^5 times faster with **155** than propene, yielding the secondary carbocation **156** (R^2=H) (equation 26) the reaction of **165** with dianisylcarbenium ion **166** yielding the tertiary carbocation **167** (equation 27) is only 15600 times faster than the addition of **155** to isobutene[76,77].

$$\text{(153)} + \text{(155)} \xrightarrow[-78\,°\text{C}]{\text{slow} \atop \text{CH}_2\text{Cl}_2} \text{(156 } R^2 = H\text{)} \tag{26}$$

An = p-C$_6$H$_4$OMe

$$\text{(165)} + \text{(166)} \longrightarrow \text{(167)} \tag{27}$$

The kinetic β-silyl effect for the reactions shown in equations 25 and 26 is smaller than the kinetic effect of β-germanium and β-tin substituents. The relative rate ratio for the addition of **166** is $k(\mathbf{159}) : k(\mathbf{168}) : k(\mathbf{169}) = 1 : 2.5 : 43$. The reactivity differences are, however, larger in the propene series, reflecting the higher electron demand in the transition state of the addition in the secondary system $[k(\mathbf{157}) : k(\mathbf{170}) : k(\mathbf{171}) = 1 : 5.6 : 1614]$.

(168) R = Me   (169) R = Me   (172)
(170) R = H   (171) R = H

A similar dependence of the kinetic β-silicon effect on the electronegativity of the substituents at silicon was found by Brook, Hadi and Neuy[78,79] They analysed the degree of *syn* addition of bromine to a series of (*E*)-β-silylstyrenes (**172**) as a measure of the stabilizing ability of the silyl group (equation 28). While they recovered from the reaction of **172**, $R^1 = R^2 =$ Me with bromine exclusively the *trans*-β-bromostyrene **173**, suggesting a *syn*-addition of the bromine to the C=C double bond, the yield of *trans*-**173** is reduced to only 17% for **172**, $R^1 = R^2 = F$[78]. Furthermore, the same authors found a linear correlation between the group electronegativity of the silyl substituent and the percentage of *syn* addition[79]. This was interpreted as an indication of a reduced β-silyl effect on the intermediate carbocation **174** for electronegative substituents at the silicon.

Kresge and Tobin[80] investigated the β-silicon effect on the hydrolysis of vinyl ethers (equation 29) and found a rate acceleration on the hydrolysis of **175** compared with **176**, and hence a stabilizing effect of the β-silyl group on the intermediate α-ethoxy carbocation **177** compared with **178**. The acceleration is small: the rate factor $k(\mathbf{175}) : k(\mathbf{176})$ of 129 is equivalent to a free energy of activation difference $\Delta\Delta G^{\ddagger}$ of 2.9 kcal mol^{-1}, whereas the solvolytic studies for the generation of a cyclohexyl cation **145** gave $\Delta\Delta G^{\ddagger} = 16.5$ kcal mol^{-1} [76].

$$\text{(172)} \xrightarrow{Br_2} \left[\text{bromonium intermediate} \right]^+ \rightleftharpoons \text{(174)}$$

(28)

anti-addition → cis-**(173)** via anti-elimination (−SiR$_3$Br)

syn-addition → trans-**(173)** via anti-elimination (−SiR$_3$Br)

$$\text{(175) R = SiMe}_3 \quad \text{(176) R = H} \xrightarrow{HA} \text{(177) R = SiMe}_3 \quad \text{(178) R = H} \xrightarrow[-\text{EtOH}]{H_2O} \text{aldehyde} \quad (29)$$

This difference was originally attributed solely to a transition state conformation in the vinyl ether hydrolysis reaction that is unfavourable for hyperconjugative stabilization[80]. The authors argued that the protonation reaction starts at a dihedral angle $\Theta = 90°$ between the β-Si−C bond and the developing empty 2p(C$_\alpha$) orbital that, even in a product-like very late transition state **TS-175** would never decrease to less than $\Theta = 60°$. The cosine-squared dependence of the hyperconjugation interaction would therefore explain the low observed reaction rate[80]. Another important factor which attenuates the β-silyl effect in **177** is the inherent stability of the ethoxy-stabilized carbocation **178**. The interaction with lone pair at the oxygen will lower the energy of the formally empty 2p(C$^+$) orbital in **177** and **178**. Thus, the hyperconjugation interaction with the β-Si−C bond is smaller in **177** than in secondary cyclohexyl cations, which is shown by the smaller additional β-silyl stabilization for **177**. In a subsequent paper Gabelica and Kresge also agreed on the contribution of this effect to an observed small acceleration in the hydrolysis of **175**[81].

The small β-silicon effect found by DeLucca and Paquette[82] in the solvolysis of the cyclopropylmethyl trifluoroacetate **179** (equation 30) might be a result of the inherent stability of the cationic intermediate **180** and its delocalized structure. The 27-fold rate acceleration of **179** compared with cyclopropylmethyl trifluoroacetate is equivalent to a β-silyl stabilization of only 2.2 kcal mol^{-1}.

(30)

Kresge and coworkers[81,83] studied also the β-silyl effect on the rate of protonation of simple alkynes and alkenes. They found for the protonation of phenyl(trimethylsilyl)acetylene **181** in aqueous perchloric acid a rate acceleration compared with **182** of only a factor of 300[83]. They attributed this to a 'surprisingly weak stabilization of a carbocation by a trimethylsilyl group'[83]. They explained this apparently small rate acceleration (which is equivalent to $\Delta\Delta G^{\ddagger} = 3.4$ kcal mol^{-1}) by assuming a transition state geometry for the protonation of **181**, which is not optimal for silyl stabilization of the developing positive charge. In contrast to the transition state (**TS-175**) of the protonation of the vinyl ether **175**, the dihedral angle Θ in the transition state (**TS-181a**) for the protonation of phenyltrimethylsilylacetylene already has the optimum value of $\Theta = 0°$, due to the linear geometry of the alkyne and the incipient vinyl cation. The authors argued, however, that the Me$_3$Si—C=C$^+$ bond angle α begins with an initial value of 180° which will finally reach 120° in the fully formed vinyl cation, but will remain considerably greater than that in the early stages of the protonation of **175**. During the course of the protonation of the alkyne the β-trimethylsilyl group is bent far away, as shown in **TS-181b** (side view), from the developing empty 2p(C$^+$) orbital, leading to a sizable reduction of hyperconjugation interaction between the 2p(C$^+$) orbital and the β-Si—C bond.

(TS-175)   (TS-181a)   (TS-181b)

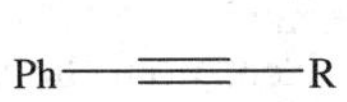

12. Silyl-substituted carbocations

(181) R = SiMe₃ — Ph—≡—R
(182) R = H

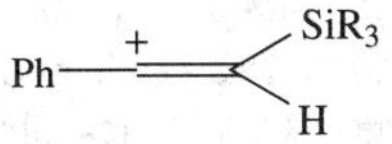

(183) R = Me — Ph—⁺=C(SiR₃)(H)
(184) R = H

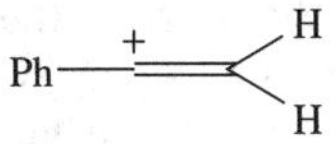

(185) Ph—⁺=C(H)(H)

However, it must be taken into account that the α-phenylvinyl cation **185** is already highly stabilized by the phenyl substituent, leading consequently to a smaller β-silicon effect in the vinyl cation **183**. Ab initio calculations by Buzek predicted for **184** an additional stabilization of 10 kcal mol^{-1} by the silyl group[7]. The thermodynamic stabilization of **183** compared with **185**, experimentally determined by Stone and coworkers in the gas phase, is 9 kcal mol^{-1} [21]. Thus, the kinetically determined stabilization of the transition state is only about 6 kcal mol^{-1} smaller than the β-silyl effect for stabilization of the ground state carbocation.

Similar comparisons between the thermodynamic β-silyl stabilization measured in the gas phase[20,21] and the kinetic β-silicon effect[81,83] found in protonation experiments in solution are possible for the acetylenes **186** and **188** and for the alkene **190**. The data for both solution study and gas phase equilibrium measurements are summarized in Table 5.

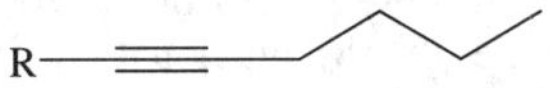

(186) R = SiMe₃
(187) R = H

(188) R = SiMe₃
(189) R = H

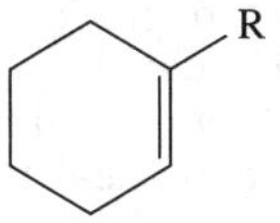

(190) R = SiMe₃
(191) R = H

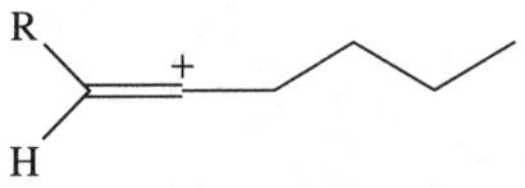

(192) R = SiMe₃
(193) R = H

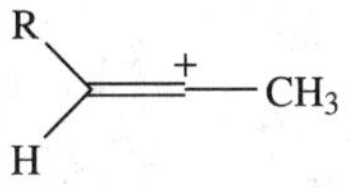

(194) R = SiMe₃
(195) R = H

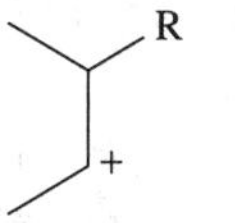

(196) R = SiMe₃
(197) R = H

TABLE 5. Comparison between the kinetic β-silyl effect ($\Delta\Delta G^{\ddagger}$) derived from protonation experiments and thermodynamic β-silyl stabilization ($\Delta H°$) of small carbocations[20,21,81,83]

System	$\Delta\Delta G^{\ddagger}$	$\Delta H°$
181/182	3.5	
183/185		9
186/187	6.5	
188/189	6.5	
192/193		11
190/191	5.7	
196/197		34

In solution the alkynes **186–189** undergo a rate-determining carbon protonation to yield the corresponding vinyl cations **192–195**, as is evident from the measured deuterium isotope effects and the linear Cox–Yates plots[81,83]. The rate accelerations due to the β-silyl substitution of 56,800 and 54,300 measured for **186** and **188**, respectively, reflect stabilization of the transition states for the protonation by $\Delta\Delta G^{\ddagger} = 6.5$ kcal mol^{-1} for both compounds. These effects are considerably greater than those found for protonation of **181**, indicating that the β-silyl effect in **183** is attenuated by the strongly electron-donating phenyl substituent. They are, however, significantly smaller than the $\Delta H° = 11$ kcal mol^{-1} found for **192** in the gas phase, and Gabelica and Kresge[81] attributed the difference between the small $\Delta\Delta G^{\ddagger}$ found in their protonation experiments and the nearly three times larger effect for the kinetic β-effect found for the fixed antiperiplanar stereochemistry in **134** to an unfavourable geometry in the transition state for the protonation, and used arguments similar to those discussed for the protonation of **181**. The question, however, arises as to whether solvent or ground state effects might play an important role in the solution phase protonation.

The large difference between the $\Delta\Delta G^{\ddagger} = 5.7$ kcal mol^{-1} found for the cyclohexene system **190/191**[81] and the $\Delta H° = 34$ kcal mol^{-1} for the similar ions **196/197**[20] point to drastic differences in the mode of stabilization of the transition state for the protonation in solution and the free silyl-substituted carbocation in the gas phase.

Dallaire and Brook[46] studied the protiodemetallation of silyl-, germyl- and stannyl-alkynes **198–200** in order to compare the β-effect of these groups on vinyl cations. The first step of the protiodemetallation (equation 31), i.e. the protonation of the triple bond, was found to be rate-determining for all alkynes. The relative kinetic β-effect arising from the second-order rate constants for EMe$_3$, Me$_3$Sn : Me$_3$Ge : Me$_3$Si (1.5 × 10^8 : 525 : 1), follows the order found for tricoordinated carbenium ions.

$$\text{Me}_3\text{Si}\equiv\text{EMe}_3 \xrightarrow{\text{H}^+} \text{Me}_3\text{Si}-\overset{+}{\underset{}{\text{C}}}=\text{C}\begin{pmatrix}\text{EMe}_3\\ \end{pmatrix}$$

(**198**) E = Si
(**199**) E = Ge (31)
(**200**) E = Sn

$$\downarrow$$

$$\text{Me}_3\text{Si}\equiv\text{H}$$

The first solvolytic generation of an aryl cation by Sonoda and coworkers[84] following theoretical predictions by Apeloig and Arad[11] highlights the exceptional stabilization of a carbocation by a β-silyl substituent. Apeloig and Arad pointed out that a 2,6-disilylphenyl cation **201** is calculated to be stabilized compared with the unsubstituted phenyl cation **202** by 25 kcal mol^{-1} (at 3-21G). Thus, it is as stable as the 2-propenyl cation **195**, which is one of the least stable vinyl cations produced in solvolysis reaction. The solvolytic generation of 2,6-bis(trimethylsilyl)-substituted aryl cation **203** should thus be amenable[11]. This theoretical prediction was verified by Sonoda and coworkers[84] who showed that triflate **204** gives at 100 °C in trifluoroethanol, **205**, quantitatively the trifluoroethyl ether **206**. The reactions followed first-order kinetics and isotopic labeling experiments proved an aryl–oxygen cleavage of **204** during the solvolysis. The kinetic data and the observed low selectivity toward external nucleophiles support a S_N1 mechanism for the solvolysis of **204** proceeding via the aryl cation **203** (equation 32)[84].

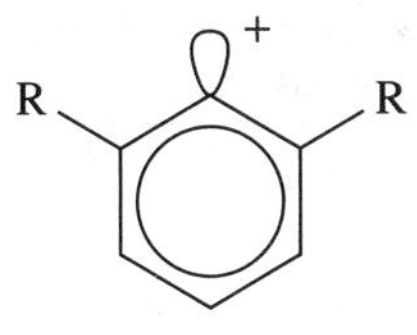

12. Silyl-substituted carbocations

(**201**) R = SiH$_3$
(**202**) R = H
(**203**) R = SiMe$_3$

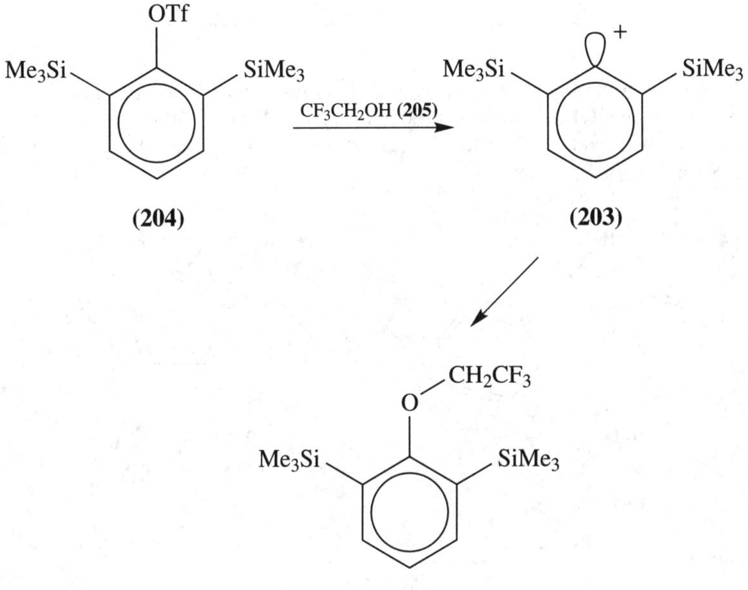

(**204**) → [CF$_3$CH$_2$OH (**205**)] → (**203**)

(32)

(**206**)

Thermodynamic stabilization of carbocations can be accurately determined by calculating the free energies of bond homolysis and bond heterolysis from experimentally measured ionization constants p$K_{(R^+)}$ and standard potentials $E_{NHE}(R^-/R^{\bullet})$ and $E_{NHE}(R^{\bullet}/R^+)$ in a thermodynamic cycle. Bausch and Gong used such a thermochemical cycle to determine in solution the free energies of C—H bond heterolysis in 9-substituted fluorenes[85]. Their data (Table 6) show that the C—H bond is destabilized by an α-methyl group (in **208**) by 8 kcal mol^{-1} compared with the unsubstituted fluorene **207**. An α-trimethylsilyl group (in **210**) weakens the C—H bond by only 2 kcal mol^{-1}. The same small effect on the free energy for the C—H bond was found for the α-t-butylfluorene **209**. These very small α-effects might be due to destabilization of the fluorenyl cations by steric interactions between the bulky α-substituents and the 1- and 8-$peri$ hydrogens of the fluorene ring.

The β-trimethylsilyl group in **212** weakens the C—H bond by 16 kcal mol^{-1} compared with the C—H bond in fluorene **207**. Comparing the data for the tertiary C—H bonds from Table 6, the β-H-substituted fluorenyl cation (from **208**) is less stable by 8 kcal mol^{-1}, and

TABLE 6. Free energy data (in kcal mol^{-1}) for the C—H bond heterolysis of fluorenes[85]

Compound	$\Delta G°$(R—H)
207	105
208	97
209	103
210	103
211	95
212	89

the β-t-butyl-substituted carbocation (from **211**) is less stable by 6 kcal mol^{-1}, compared with the β-trimethylsilyl-substituted carbocation (from **212**). The thermochemical solution data for the β-silyl effect are therefore in qualitative agreement with theoretical predictions and data from kinetic experiments[85].

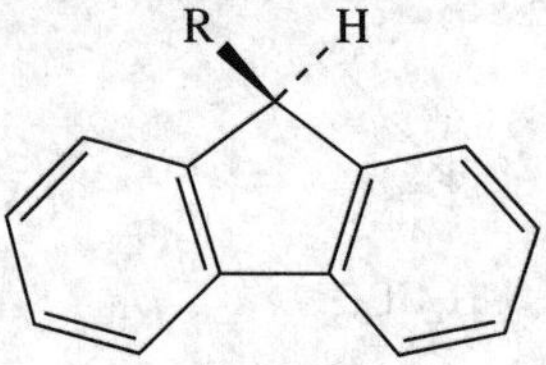

(**207**) R = H
(**208**) R = Me
(**209**) R = t-Bu
(**210**) R = SiMe$_3$

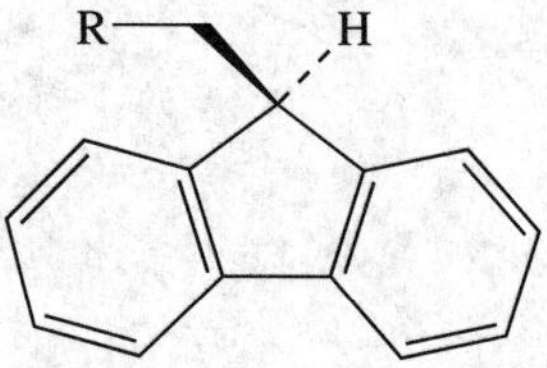

(**211**) R = t-Bu
(**212**) R = SiMe$_3$

1. α-Disilanyl carbocations in solvolysis

The intriguing combination of an α-silyl and a β-silyl substituent in the disilanyl group (—SiR$_2$SiR$_3$) prompted Shimizu, Tsuno and coworkers to study the effect of a β-Si—Si bond on the stability of carbocations[86]. They studied the solvolysis of the benzylic bromides **213–215** and found that the β-SiMe$_3$ substituent in **213** increases the solvolysis rate by a factor of 1.07×10^5 compared with the t-butyl-substituted reference compound **214** (Table 7).

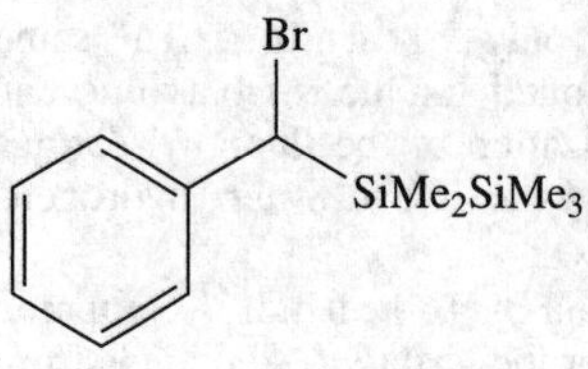

(**213**) PhCH(Br)SiMe$_2$SiMe$_3$
(**214**) PhCH(Br)SiMe$_2$CMe$_3$
(**215**) PhCH(Br)SiMe$_3$

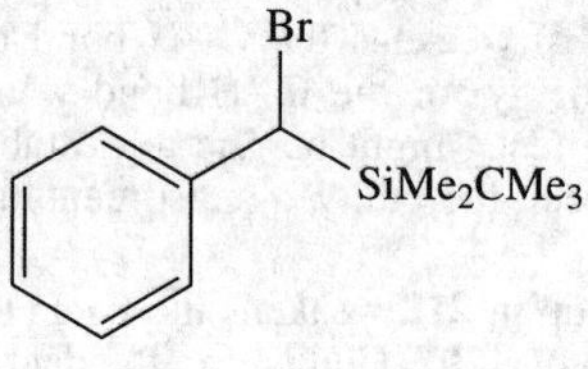

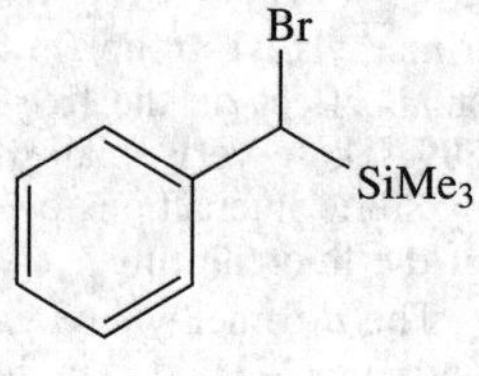

TABLE 7. Kinetic effect of the disilanyl group in solvolysis

Compound	$k(s^{-1})^a$	Reference group R	$k(\text{compound})/k(R)$	Reference
213	6.2×10^{-2}	$-SiMe_2Bu\text{-}t^b$	1.07×10^5	86
		$-SiMe_3^b$	1.98×10^5	86
		$-Me^b$	1.20×10^1	86
220	3.86×10^{-4}	$-SiMe_2CH_2$	0.209	86

[a] In 97% TFE at 25 °C.
[b] R replacing Me_3SiMe_2Si.
[c] R replacing the $SiMe_2SiMe_2$.

Unlike the β-alkyl substituted bromides **214** and **215** which give substitution products without skeletal rearrangements, the solvolysis of the disilanyl compound **213** in various solvents gives exclusively 1,2-trimethylsilyl rearranged products irrespective of the solvent nucleophilicity (equation 33). The Grunwald–Winstein solvent effects for the solvolysis of **213** ($m = 0.91-0.98$) suggest a k_C mechanism with a rate-determining formation of a carbocation. The measured α-deuterium effect (1.16–1.17) and the substituent effects ($\rho = -3.71$, $\sigma = 1.16$) are consistent with the formation of an open sp^2-hybridized benzylic carbocation **216** which undergoes a fast 1,2-trimethylsilyl shift, yielding the rearranged product **217** after capture by the solvent. The experimental findings rule out a k_Δ mechanism via the siliconium ion **218**[86].

(33)

From the relative solvolysis rates of **213** and **215**, Shimizu and coworkers calculated a stabilization by 7 kcal mol^{-1} of the benzyl cation by the β-Si–Si bond compared with a β-C–Si bond. Due to the antagonistic effects of the α- and the β-silyl groups, the net effect of the disilanyl group is relatively small[86]. Thus, **213** is only 12 times as reactive as α-methylbenzyl bromide **219**, X = Br (Table 7; in 30% acetone the relative rate is decreased to 0.66). This might be compared with the rate acceleration of 1.05×10^5 by the CH$_2$SiMe$_3$ group found for the solvolysis of **141** (X = OTf) relative to **142** (X = OTf) (see Table 3). Calculation, however, reveals no stabilizing effect of an α-disilanyl group on the positive charge in carbocations. These theoretical results are supported by solvolysis experiments. 2-Disilanyl-substituted 2-adamantyl sulphonates solvolyse with the same rate as the corresponding 2-trimethylsilyl-2-adamantyl derivatives **66**[87].

(218)   (219)   (141) R = SiMe$_3$
                    (142) R = H

Similar to the β-Si effect of a σ(C−Si) bond, the stabilizing effect of a $\beta - \sigma(Si-Si)$ bond is not operative in an orthogonal arrangement with the 2p(C$^+$) orbital. Thus, the indanyl compound **220** solvolyses via the carbocation intermediate **221** without skeletal rearrangement yielding **222** (equation 34), but **220** is slightly less reactive than its β-alkyl analogue **223** [$k(220)/k(223) = 0.209$]. This rate retardation compared with the large accelerating effect found for **213** provides experimental evidence that the β-silicon effect of a $\beta - \sigma(Si-Si)$ bond in benzylic cations is exclusively hyperconjugative in origin and contributions of inductive effects are insignificant[86].

(220)   (221)   (222)   (34)

(223)

A remarkable change of the mechanism was observed by Shimizu and coworkers[88] in the solvolysis of the disilanyl-substituted cyclopropyl bromide **224c**. While the cyclopropyl bromides **224b** and **224c** give, upon solvolysis, the open allylic ethers **227**, in line with a σ-assisted mechanism (k_Δ)[86] via the allyl cations **228**, **224c** predominantly yielded the cyclopropyl ring-retained product **225** and only smaller amounts of **227c** (product ratio **225**/**227c** = 1.18; equation 35). These findings suggest that the β-silyl effect stabilizes the incipient cyclopropyl cation **226** enough to compete with the σ-assisted ring opening to the allyl cation **228c**[88].

(**224a**) R = SiMe$_3$
(**224b**) R = Me
(**224c**) R = SiMe$_2$SiMe$_3$

(**228a**) R = SiMe$_3$
(**228b**) R = Me
(**228c**) R = SiMe$_2$SiMe$_3$

(**227a**) R = SiMe$_3$
(**227b**) R = Me
(**227c**) R = SiMe$_2$SiMe$_3$

S = CH$_2$CF$_3$

(35)

(**224c**)   (**226**)   (**225**)

The overall kinetic effect of the disilanyl group is, however, small due to the different solvolysis mechanisms. **224** solvolyses only 3.48 times faster than **226** in TFE at 100 °C. For the k_Δ process Shimizu and coworkers calculated the relative rates for **224** : **225** : **226** to be 1.60 : 1.0 : 10.7. They concluded that these small silicon effects reflect the concerted nature of the k_Δ solvolysis involving transition states with a highly delocalized charge distribution[88].

C. γ-Silicon Effect

Sommer, Whitmore and coworkers already recognized the stabilizing effect of a γ-silyl group on a positively charged carbon[34,35]. In the basic hydrolysis they found an enhanced reactivity of **229** compared with the α-substituted chlorosilane **58**, although **229** was found to be less reactive than the β-isomer **59**.

(**58**)   (**59**)   (**229**)

Shiner and coworkers were the first to study quantitatively the γ-silyl effect in solvolysis[90–95]. They found a distinct kinetic acceleration of the solvolysis of cis-3-(trimethylsilyl)cyclohexyl brosylate **230** compared with the alkyl reference **231** in TFA [$k(230)/k(231) = 462$] which is equivalent to a lowering of the ionization barrier by 3.7 kcal mol^{-1} [90,91]. This kinetic γ-silicon effect has striking stereochemical requirements. For example, the *trans* isomer **232** solvolyses by β-H participation similar to the *t*-Bu analogues **231** and **233** as evident from the large β-d$_4$ isotope effects ($k_H/k_D = 2$–3) and **232** is only as reactive as the *t*-Bu substituted cyclohexyl brosylate **233** [$k(232)/k(233) = 1.2$]. In contrast, the β-tetradeuterio analogue of **230** exhibits a small or even inverse secondary isotope effect ($k_H/k_D = 0.972$–1.005). The α-d$_1$ isotope effect is also smaller than expected for the formation of an open sp^2-hybridized carbenium ion. This indicates a participation of the γ-Si–C bond in the rate-determining step of the solvolysis of **230**[90,91].

(230) R = SiMe₃
(231) R = t-Bu

(232) R = SiMe₃
(233) R = t-Bu

Further support was given by the formation of small amounts of bicyclo[3.1.0]hexane **234** and by the retained configuration at C_α in the substitution product **235**. On the basis of these experimental data Shiner and coworkers suggested that the solvolysis of **230** proceeds via a 1,3-bridged carbonium ion **236** as intermediate (equation 36). The preferred conformation of **230** is a 'W' conformation with respect to the orientation of the leaving group and the silyl substituent. The incipient carbocation **236** is stabilized by the silyl group through the so-called 'percaudal' interaction involving overlap of the back lobe of the C_γ–Si bond and the developing 2p(C) orbital on the positively charged carbon atom[90,91].

(36)

92%   8%
(235)   (234)

A theoretical analysis of the γ-silicon effect at the HF/6-31G(d) level of theory by Davidson and Shiner[92] showed that a γ-silyl-substituted cation strongly favoured the *trans* perpendicular structure **237**. In the optimized structure of **237** the γ-carbon approaches the positively charged carbon to within a distance of only 1.75 Å [at HF/6-31G(d)]. This geometry and the γ-stabilizing effect and its orientation dependence were attributed to a significant contribution from the canonical structure **238C**. Consequently, the γ-silyl group stabilizes the carbocation **238** more than a methyl group because SiH_3^+ is more stable than CH_3^+[92].

(237) (238A) (238B) (238C)

R = H₃C, SiH₃

TABLE 8. Kinetic γ-silicon effect in solvolysis

Compound	$k(s^{-1})$	Reference group R^e	k(compound)/ k(reference compound)	Reference
230	176.3^a	t-Bu	462	91
232	1.99^a	t-Bu	1.2	91
239	1100^a	H	138	94
240	84.37^a	H	129	93
241	0.144^a	H	21 800	95
242	707^a	H	4600	95
245	6.15×10^{-4b}	H	8.6	96
247	2.36×10^{-3b}	H	33	97
248 (R = Ms)	0.103^c	H	33 000	98
249 (R = Ms)	1.5×10^{-4c}	H	48	98
251 (R = Bs)	3.8×10^{-6d}	H	2.4	98
252 (R = Bs)	1.7×10^{-8d}	H	0.01	98
257	3.8×10^{-7d}	H	0.24	98
260	7.1×10^{-2c}	H	23 000	98

^{a}In 97% TFE at 25 °C.
^{b}In 80% EtOH at 70 °C.
^{c}In EtOH at 25 °C.
^{d}In 80% EtOH at 25 °C.
^{e}Group R replacing the Me₃Si group in the compound.

Shiner and coworkers also studied optically active acyclic secondary sulphonate esters and found rate accelerations of about 100 in solvolysis in TFA at 25 °C compared with the alkyl reference (Table 8)[93-95]. From the product distribution and the stereochemical outcome of the solvolysis Shiner and coworkers proposed the reaction course outlined in equation 37 for the brosylate **239**[94]. The transition state for the ionization adopts the 'W' conformation and, to a smaller extent, the '*endo*-Sickle' conformation and results in the formation of γ-silicon carbocation which is stabilized by a percaudal interaction. The incoming nucleophile attacks the bridged ion from the front side[94].

A remarkably high rate acceleration by a γ-silyl group was found for the acyclic primary sulphonate **241**, which is 21 800 times more reactive than its silicon-free analogue. In this case a 1,2-migration of Me₃SiCH₂ occurs after the rate-limiting formation of the carbocation **243**, yielding the more stable tertiary carbocation **244** (equation 38). For the secondary sulphonate **242** a smaller kinetic γ-silicon effect was detected[95].

In the adamantyl framework the γ-silicon effect was studied by Grob and coworkers[96,97]. The first-γ-trimethylsilyl substituent (cf **245**) enhances the solvolysis rate of 1-bromoadamantane **246** by a factor of only 8.6 and the second trimethylsilyl group in

247 leads to an overall enhancement of $33^{96,97}$.

(37)

(a) Mechanism for the solvolysis of *threo*-4-(trimethylsilyl)-3-methyl-2-butyl-brosylate
(b) Mechanism for the solvolysis of *erythro*-4-(trimethylsilyl)-3-methyl-2-butyl-brosylate

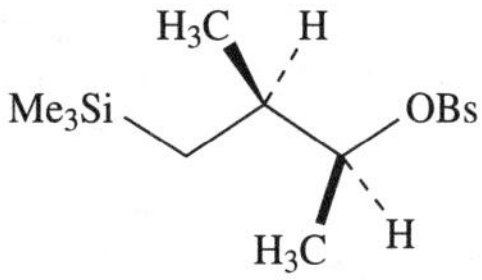

12. Silyl-substituted carbocations

(239)

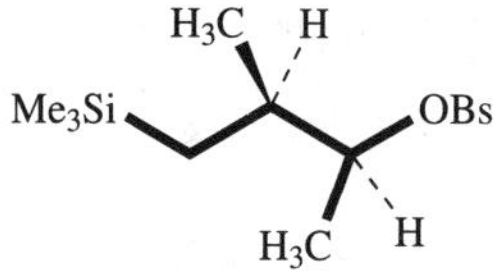

threo-(239)
'W' conformation

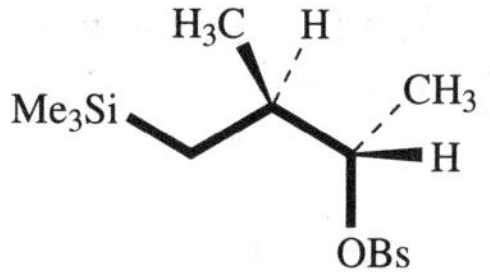

threo-(239)
'*endo*-Sickle' conformation

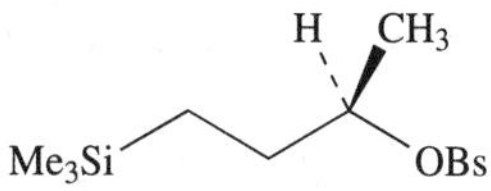

(240)

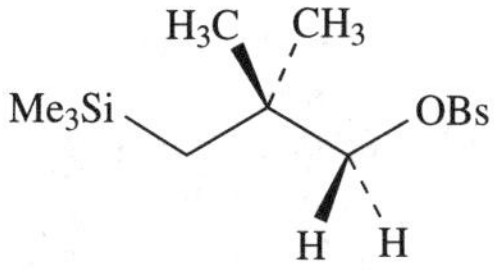

(241)

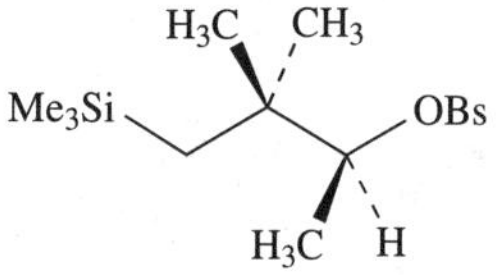

(242)

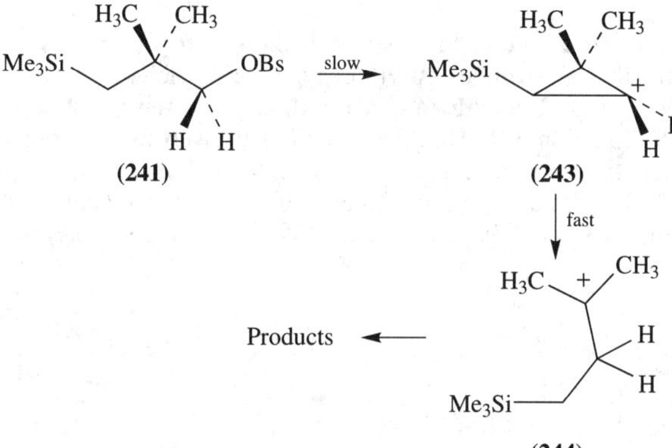

$$(241) \xrightarrow{slow} (243) \xrightarrow{fast} (244) \longrightarrow \text{Products} \tag{38}$$

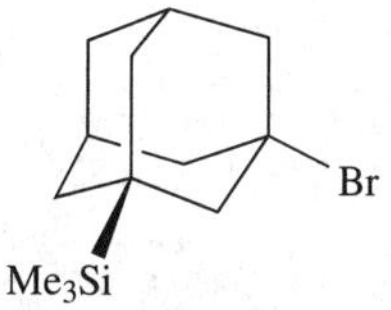

(245)

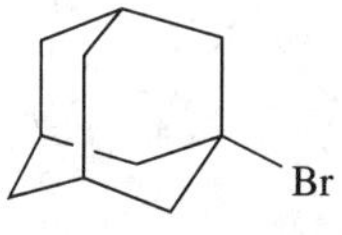

(246)

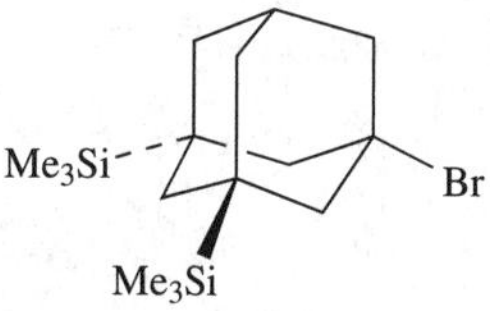

(247)

Further convincing evidence for the strong orientational preference of a 'W'-like geometry for the kinetic γ-silicon effect was given by Bentley, Kirmse and coworkers[98–100]. By using the rigid geometric arrangement in 2,6-disubstituted norbornyl derivatives, they found that for the 2-*exo* substituted compound the 6-*exo* trimethylsilyl substituent in **248** (R = Ms) causes a rate acceleration of 33 000 relative to hydrogen, while the 6-*endo*-Me$_3$Si compound **249** (R = Ms) solvolyses only 48 times faster than the unsubstituted reference compound **250** (R = Ms)[98]. In contrast, in the solvolysis of the corresponding 2-*endo*-mesylate **251** (R = Bs) the 6-*exo*-Me$_3$Si substituent shows only a 2–4-fold rate enhancement and the 6-*endo*-Me$_3$Si substituent in **252** (R = Bs) actually shows a rate retardation. The large kinetic effect of the *exo*-Me$_3$Si substituent in **248** (R = Ms) and the absence of a similar rate enhancement for the *endo* isomer **249** (R = Ms) are consistent with specially favourable interaction of the γ-silicon group with the incipient empty

2p(C^+) orbital in the transition state for the solvolysis. The weak deactivating effect of the 6-*endo*-Me$_3$Si group in **252** (R = Ms) was ascribed to steric hindrance to departure and/or solvation of the leaving group[98,99].

The solvolysis of **252** is one of the rare examples for a norbornyl–norpinyl rearrangement. While the *exo*-trimethylsilyl brosylate **251** yields mainly substitution and elimination products **253–255** with an intact norbornyl framework, **252** gives nearly 86% of norpinene **256** (equation 39). The bis-(trimethylsilyl)substituted compound **257** gives almost exclusively the norpinene derivative **258** (equation 40). While the trimethylsilyl group(s) in **252** and **257** exert no kinetic effect on the reaction rate, the β-effect on the intermediate carbocations **252A** and **259**, respectively, determines the product distribution[99].

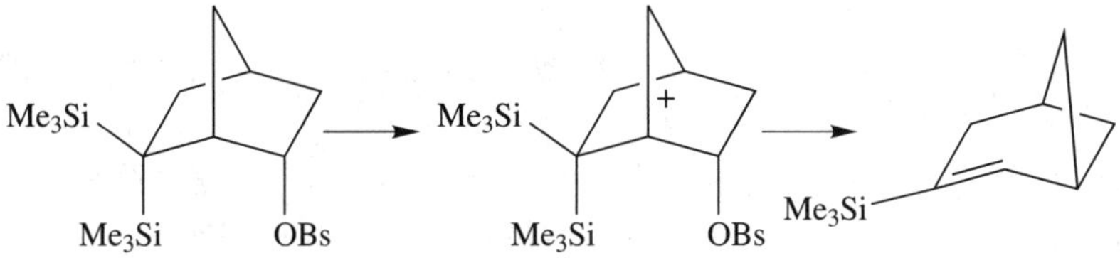

(40)

(257) → **(259)** → **(258)**

In the 2-*exo* isomer **260** the measured rate enhancement of 23 000 compared with **250** indicates a large γ-silicon effect. The intermediate **261** undergoes a fast 2,6 migration and Wagner-Meerwein shifts to give the products via ion **262** (equation 41)[100].

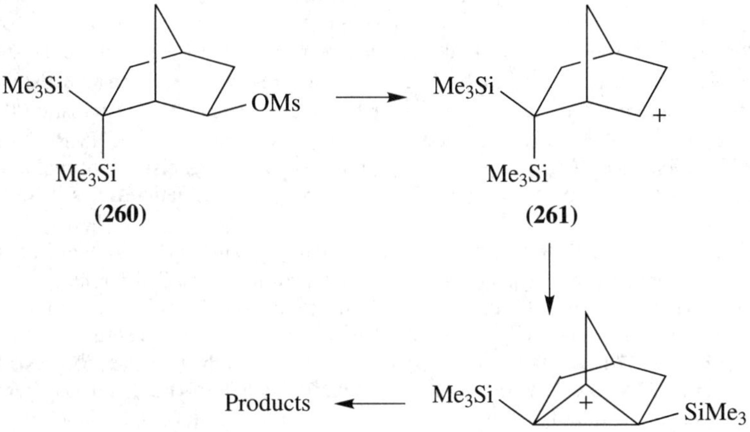

(260) → **(261)** → **(262)** → Products

(41)

D. δ-Silicon Effect

The kinetic influence of a δ-silicon substituent on the solvolysis of cyclohexyl tosylates is insignificant. Fessenden and coworkers[101] found virtually identical rates of ethanolysis of *cis* and *trans* **263** to those of the *t*-butyl reference compounds **264**. However, no evidence for a rate-determining k_c solvolysis was provided.

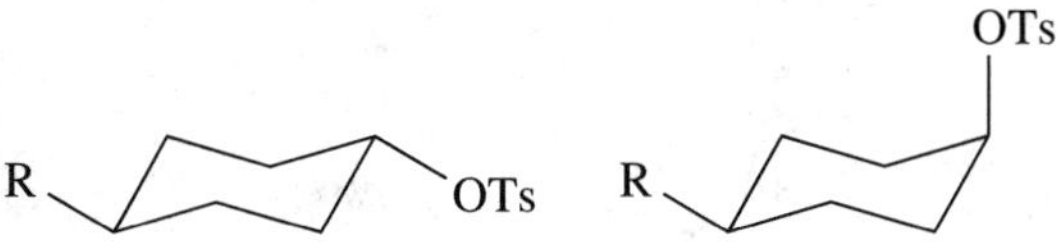

trans-(**263**) Me$_3$Si   *cis*-(**263**) Me$_3$Si
trans-(**264**) *t*-Bu   *cis*-(**264**) *t*-Bu

Adcock, Shiner and coworkers found a significant 48.6-fold rate enhancement for the 4-trimethylsilyl bicyclo[2.2.2]octyl mesylate **265** compared with the silicon-free analogue **266**[102]. The effects of trimethylgermyl and trimethyltin groups in the 4-position (in **267**

(265) Me₃Si ... OMs (266) H ... OMs (267) Me₃Ge ... OMs (268) Me₃Sn ... OMs

and **268**, respectively) are even larger, i.e. the trimethyltin mesylate **268** solvolyses 2841 times faster than **266** (Table 9).

The remarkable δ-effect on the transition state of the solvolysis was ascribed to the occurrence of double hyperconjugation (through bond coupling), i.e. **269A** ↔ **269B** ↔ **269C** in the intermediate cation **269**. The large δ-deuterium effect ($k_H/k_D = 1.05$ in TFE) found for **266-d⁴** indicate a substantial lowering of the C−H force constants and corroborates the idea of through bond coupling in **269**[102]. The alternative explanation in terms of homohyperconjugation[103−105] or back-lobe (percaudal) participation through space (**269A** ↔ **269D**) was discarded on the basis of previous ¹⁹F NMR studies on bicyclo[2.2.2]octanes. According to the authors a pure transmission of polar effects through space could also not account for the relatively large kinetic effects in solvolysis of **265**, **267** and **268**[102].

Similar large δ-effects were found in the adamantyl framework, although the stereochemical arrangement of δ-substituent and leaving group is very different[106,107]. Thus, E-5-trimethylsilyl-2-adamantyl brosylate **270** undergoes ionization 23 times more rapidly than 2-adamantyl brosylate **271**. Again the trimethyltin substituent exerts an even larger effect [$k(272)/k(270) = 9000$] (Table 9)[106]. The major product of the solvolysis of the tin compound **272** was **273**, the result of a five-bond heterolytic Grob fragmentation. Together with the similarity of the kinetic acceleration to the bicyclo[2.2.2]octyl case this was taken as an indication that the effect is transmitted through the bonds rather than

TABLE 9. Kinetic δ-silicon, δ-germanium and δ-tin effects in solvolysis

Compound	$k(s^{-1})$	k(compound)/k(reference compound)[a]	Reference
265	80.7[b]	48.6	102
267	177.9[b]	71.1	102
268	4713[b]	2841	102
270	23.13[b]	51	106
272	3200[b]	7000	106
274	3.69[b]	8	106
275	44.83[b]	98	106
278	1.14×10^{-4}[b]	1319.4	108
280	1.77×10^{-4}[b]	20486	108
286	316[c]	108.9	109

[a]Reference compound is hydrogen instead of the R₃E (E = Si, Ge, Sn) derivative.
[b]In 97% TFE at 25 °C.
[c]In MeOH at 25 °C.

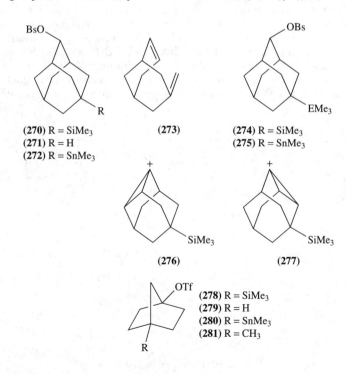

(269A) (269B) (269C)

(269D)

through space[106]. It was postulated[106] that the smaller effect found for the Z-isomers **274** and **275** (Table 9) is due to isomerization of the pyramidal Z-cation, e.g. of **276** to the more stable E-cation **277**[107].

A large kinetic silicon γ,δ-effect was found in the solvolysis of 4-trimethylsilylnorbornane triflate. Adcock and coworkers[108] found a 1319-fold acceleration for **278** compared with **279**. For the tin compound **280** the relative solvolysis rate was even 20486, while a 4-methyl group in **281** causes only a modest acceleration $[k(\mathbf{281})/k(\mathbf{279}) = 7.41]$[108].

(270) R = SiMe$_3$
(271) R = H
(272) R = SnMe$_3$

(273)

(274) R = SiMe$_3$
(275) R = SnMe$_3$

(276)

(277)

(278) R = SiMe$_3$
(279) R = H
(280) R = SnMe$_3$
(281) R = CH$_3$

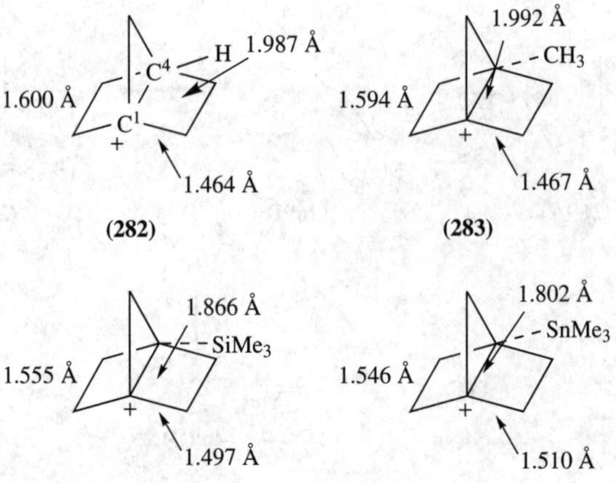

(282) **(283)** **(284)** **(285)**

FIGURE 8. Optimized geometries of cations **282**–**285** (at MP2/6-31G(d)/[DZP])

High level computations[108]. on the cations **282**–**285** at the correlated MP2(6-31G(d))/[DZP] level of theory show that cation **284** is by 11.1 kcal mol^{-1} more stable than its unsubstituted analogue, the 1-norbornyl cation **282**. Furthermore, the calculations reveal that in the norbornyl system the silyl-effect is transmitted by back-lobe interactions (homohyperconjugation), while stabilization by double hyperconjugation is of minor importance. Thus, the optimized structures of **282**–**285** (Figure 8) show smaller C1–C4 distances for the highly stabilized **284** and **285** than for **282**, indicating the importance of C1–C4 interaction as described in the canonical structures **284B**–**D**. In contrast, on the basis of the calculated structures, the occurrence of double hyperconjugation (see canonical structures **284 E**–**F**) was discarded for the 1-norbornyl cation system[108].

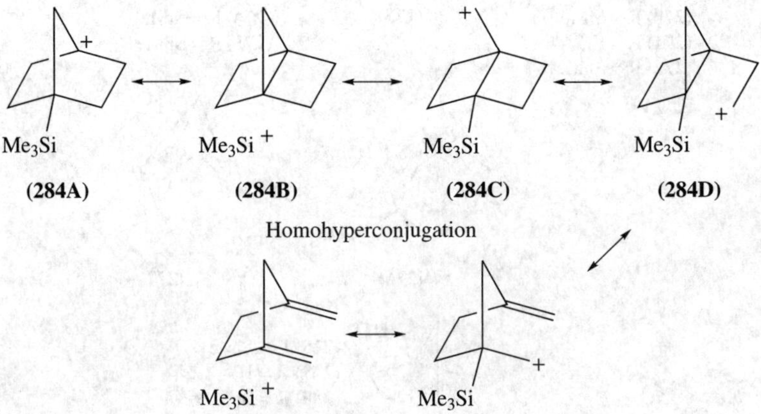

(284A) **(284B)** **(284C)** **(284D)**

Homohyperconjugation

(284F) **(284E)**

Double Hyperconjugation

Eaton and Zhou[109] found in the cubyl system a rate enhancement by a δ-trimethylsilyl group of $k(\mathbf{286})/k(\mathbf{287}) = 109$. For the trimethyltin substituent in **288** a factor of 2800 was extrapolated. This measured δ-effect seems to be too large for the transmission of pure polar effect through the cubyl cage, as suggested by Moriarty, Kevill and coworkers[110,111]. Eaton and Zhou argued that the relatively large δ-effects of silicon and tin are connected with the delocalized structure of the cubyl cation **289** as shown schematically in **289A–289C**, originally suggested by Hrovat and Borden[112].

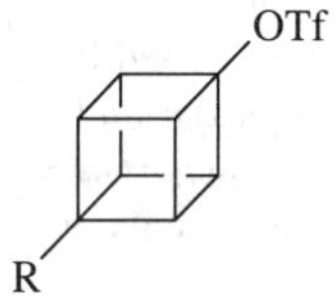

(**286**) R = SiMe$_3$
(**287**) R = H
(**288**) R = SnMe$_3$

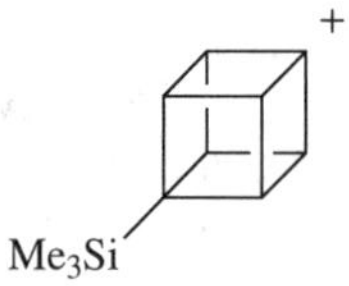

(**289**)

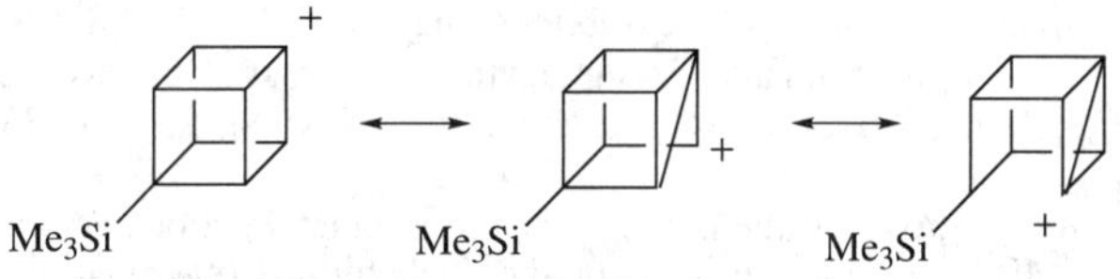

(**289A**)   (**289B**)   (**289C**)

V. STABLE ION STUDIES

This section summarizes the structure elucidation studies on silyl-substituted carbocations. It includes ultra-fast optical spectroscopic methods for the detection of transient intermediates in solution with life-times of about 10^{-7} s. A summary of NMR spectroscopic investigations of silyl-substituted carbocations and concomitant computational studies of model cations is given. A number of reactive silyl-substituted carbocations can be obtained as persistent species in superacids and non-nucleophilic solvents. Some of them have life-times of hours or even longer at low temperatures, and in some cases silyl-substituted carbocations can be prepared which are stable even at room temperature. Some silyl-substituted carbocations form crystals which were investigated by X-ray crystallography at room temperature.

A. α-Silyl Substitution

1. NMR spectroscopic characterization of α-silyl-substituted carbocations

a. 1-Phenyl-1-trimethylsilylethyl cation. Early attempts to generate the 1-phenyl-1-trimethylsilylethyl cation **290** by ionization of the 1-phenyl-1-trimethylsilyl ethanol **291** with SbF$_5$/FSO$_3$H at $-78\,°$C have failed. Only the cumyl cation **292** was observed instead (equation 42)[113].

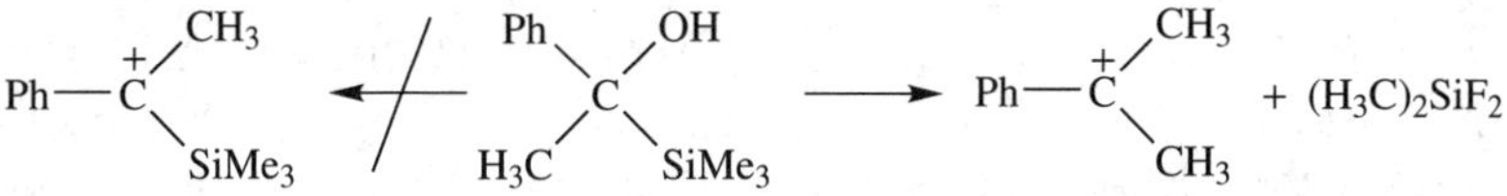

(**290**)   (**291**)   (**292**)   + (H$_3$C)$_2$SiF$_2$

(42)

When 1-phenyl-1-trimethylsilylethyl chloride **293** is reacted with SbF$_5$ under carefully controlled experimental conditions at $-125\,°$C, the 1-phenyl-1-trimethylsilylethyl cation

290 is exclusively generated as indicated by ^{1}H and ^{13}C NMR spectra (equation 43)[114,115].

$$\underset{\textbf{(293)}}{\text{Ph}-\underset{\underset{\text{CH}_3}{|}}{\overset{\overset{\text{SiMe}_3}{|}}{\text{C}}}-\text{Cl}} \xrightarrow[-125\,°\text{C}]{\text{SbF}_5} \underset{\textbf{(290)}}{\text{Ph}-\underset{\underset{\text{CH}_3}{\backslash}}{\overset{\overset{\text{SiMe}_3}{/}}{\text{C}+}}} \quad (43)$$

Due to the benzylic p–π resonance stabilization, the C^+-C_{ipso} bond has partial double bond character and the *ortho, ortho'* and *meta, meta'* CH groups *syn* and *anti* to the silyl group are non-equivalent. The ^{1}H NMR spectrum (Table 10) shows two separate signals for the *ortho* protons at δ = 8.80 and 8.72 ppm and additional signals at 8.63 (*para*-H), 8.02 (*meta*-H), 3.75 ($C^+-\underline{C}H_3$) and 0.65 (Si–Me) ppm.

The ^{13}C NMR spectrum (Table 11) exhibits signals at δ = 292.82 (C^+), 37.27 ($C^+-\underline{C}H_3$) and -0.21 (Si$\underline{C}$H$_3$) ppm and six signals for the aromatic carbons at δ = 158.91 (C_{para}), 148.12 (C_{ipso}), 148.34/138.25 (C_{ortho} *anti/syn* to SiMe$_3$), 133.59/133.07 (C_{meta} *anti/syn* to SiMe$_3$).

The effect of an α-silyl group on the positive charge in benzyl cations can be estimated by comparing the NMR spectroscopic data for the 1-phenyl-1-trimethylsilylethyl cation **290** with those for the phenyl ethyl cation **78** and the cumyl cation **292** prepared under identical conditions from the corresponding chlorides with SbF$_5$. The ^{1}H and ^{13}C NMR data for **78**, **290** and **292**[114] measured under comparable conditions are given in Tables 10 and 11.

TABLE 10. ^{1}H NMR chemical shifts (δ, ppm) for various benzylic carbocations, Ph–C$^+$RR′ in SO$_2$ClF/SO$_2$F$_2$ at $-117\,°$C[a]

Ion	R,R′	R	R′	o	m	p
78	R = Me, R′ = H	3.39	10.1	8.88/8.47 *anti/syn*	8.02	8.73
290	R = Me, R′ = SiMe$_3$	3.75	0.65	8.80/8.72 *anti/syn*	8.02	8.63
292	R = R = Me	3.46		8.77	7.92	8.52

[a] 250 MHz, internal reference δ(NMe$_4^+$) = 3.00 ppm.

TABLE 11. ^{13}C NMR chemical shifts (δ, ppm) for various benzylic carbocations, Ph–C$^+$RR′ in SO$_2$ClF/SO$_2$F$_2$

Ion	R,R′	C^+	C_o	C'_o	C_m	C'_m	C_p	C_{ipso}	R
78	R = Me, R′ = H[a]	231.28	155.36	144.14	134.25	134.01	161.92	141.91	28.05
290	R = Me, R′ = SiMe$_3$[b]	292.82	148.34 *anti*	138.25	133.59 *anti* to Si	133.07 *syn*	158.91	139.43	R(37.27) R′(−0.21)
292	R = R′ = Me[a]	255.85	142.22		132.96		156.14	140.25	33.46
	[c]	256.33	142.79		133.56		156.70	140.91	34.14

[a] 63 MHz at $-117\,°$C in SO$_2$ClF/SO$_2$F$_2$, internal reference δ(NMe$_4^+$) = 55.65 ppm.
[b] 100 MHz at $-124\,°$C, internal reference δ(CD$_2$Cl$_2$) = 53.80 ppm.
[c] 100 MHz at $-110\,°$C.

$$\underset{(78)}{\text{Ph}-\overset{+}{\text{C}}\diagup^{\text{H}}_{\diagdown\text{CH}_3}}$$

The low field shift of the resonance for the C$^+$ carbon of the 1-phenyl-1-trimethylsilylethyl cation **290** is noteworthy. This is the most deshielded shift of a benzyl cation carbon observed so far. NMR chemical shifts are dependent not only on the charge densities but also upon the substituent effect of the neighbouring groups. Consequently, the chemical shift of the C$^+$ carbon is not a good measure of the relative charge in these type of benzyl cations because the substituent effect on chemical shift is different for the substituents R = H, SiMe$_3$ and CH$_3$. The trimethylsilyl substituent has a large deshielding effect on the chemical shift of an adjacent sp^2-hybridized carbon.

The carbonyl carbon signal in trimethylsilylphenyl ketone **294** ($\delta = 238$ ppm) is 42 ppm deshielded compared with that in acetophenone **295** ($\delta = 196$ ppm)[113]

$$\underset{(294)}{\text{Ph}-\text{C}\diagup^{\text{O}}_{\diagdown\text{SiMe}}} \qquad \underset{(295)}{\text{Ph}-\text{C}\diagup^{\text{O}}_{\diagdown\text{CH}_3}} \qquad \underset{(296)}{\text{Ph}-\overset{+}{\text{C}}\diagup^{\text{CH}_3}_{\diagdown\text{SiH}_3}}$$

The *para* carbon in benzyl cations is remote from the substitution site at the benzylic position and has an approximately similar environment in all the three carbocations **78, 290** and **292**. Hence, the *para* carbon chemical shift can be used to monitor the demand for delocalization of the positive charge into the aromatic ring. Semiempirical (AM1) and *ab initio* calculations [B3LYP/6-31G(d)] for the benzyl cation-type structures (Ph−C$^+$ MeR) with R = H, Me, SiH$_3$, SiMe$_3$ show that the aryl π-system and the vacant 2p(π)-orbital in the phenylethyl (**78**), the 1-phenyl-1-silylethyl (**296**), the 1-phenyl-1-trimethylsilylethyl (**290**) and the cumyl (**292**) cations are almost coplanar in all systems, thus p–π-conjugative overlap is not hindered even in **290** which is substituted by the rather bulky SiMe$_3$ group. The better the α-substituent at the C$^+$-carbon stabilizes the positive charge, the less charge needs to be delocalized into the aryl ring, and a less deshielded NMR signal for the *para* carbon is expected. The chemical shift of the *para* carbon in the 1-phenyl-1-trimethylsilylethyl cation **290** ($\delta = 158.9$ ppm) is in between the values of C$_{para}$ in the phenylethyl cation **78** ($\delta = 161.9$ ppm) and the cumyl cation **292** ($\delta = 156.7$ ppm)[116–118]. This shows that the α-trimethylsilyl group is stabilizing compared with α-hydrogen, but destabilizing compared with the α-methyl group in these carbocations. This is confirmed by comparison of the ^{1}H chemical shifts for the *para* hydrogens, which are $\delta = 8.73$, 8.63 and 8.52 ppm in the phenylethyl (**78**), the 1-phenyl-1-trimethylsilylethyl (**290**) and the cumyl (**292**) cation, respectively[114]. These findings are in accord with earlier calculations on model systems, with solvolysis data and also with recent quantum chemical *ab initio* calculations of charge distribution in α-silyl-substituted carbocations[11,13,119]. The ^{29}Si chemical shift of the 1-phenyl-1-trimethylsilylethyl cation **290** ($\delta = 9.61$ ppm in SO$_2$ClF/SO$_2$F$_2$ at $-127\,^\circ$C) is similar to that of the neutral progenitor, 1-phenyl-1-trimethylsilylethyl chloride **293** (δ^{29}Si = 9.08 ppm)[115]. Charge delocalization to the silyl group is not important in **290**, contrary to the case of β-silyl-substituted carbocations (Section V.B).

b. Diphenyl(trimethylsilyl)methyl cations. The diphenyl(trimethylsilyl)methyl cation **297** can be generated from the corresponding alcohol by reaction with FSO$_3$H in SO$_2$ClF[113].

The ^{13}C NMR chemical shift of the C$^+$ carbon is at $\delta = 259$ ppm, 30–60 ppm deshielded from the shift of the C$^+$ carbon in the 1,1-diphenylethyl cation (**298**), $\delta = 222.9$ ppm and the diphenylmethyl cation (**299**), $\delta = 200.6$ ppm, respectively. The *para*-carbons of the aromatic rings show slightly shielded chemical shifts compared with the methyl and hydrogen analogues [$\delta = 147.4$, 148.1 and 150.9 ppm, respectively, for the SiMe$_3$, CH$_3$ and H derivatives]. This was taken as an indication of a smaller positive charge delocalization into the aromatic ring in the trimethylsilyl-substituted analogue (**297**). However, as evident from the results obtained for the 1-phenyl-1-trimethylsilylethyl cation (**290**), this does not imply that the α-trimethylsilyl group is a better stabilizer of the positive charge than the α-methyl or α-hydrogen analogues. The non-planarity of the two phenyl rings does not allow a conclusive comparison for the cations (C$_6$H$_5$)$_2$C$^+$R (**297, 298** and **299**) with R = SiMe$_3$, CH$_3$ and H, respectively

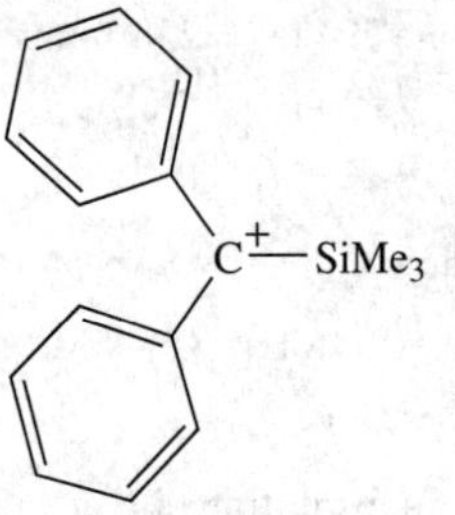

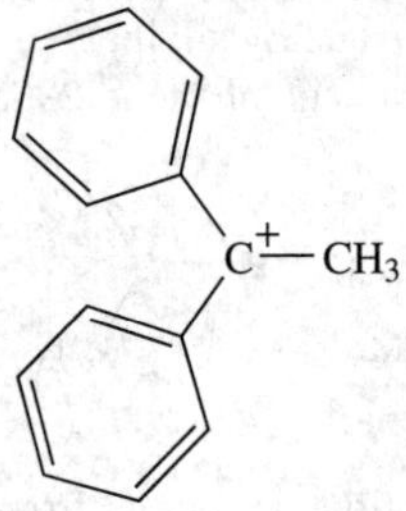

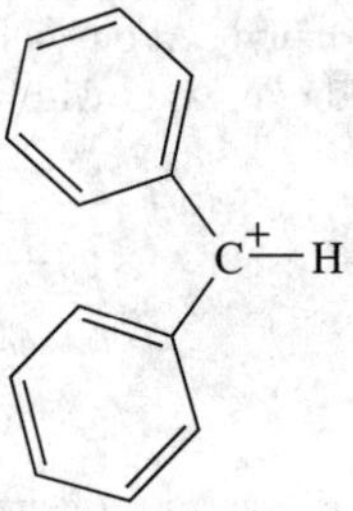

(**297**)   (**298**)   (**299**)

2. X-ray structure analysis of the tris(trimethylsilyl)cyclopropenylium cation

The tris(trimethylsilyl)cyclopropenylium hexachloroantimonate **300**·SbCl$_6^-$ forms crystals which are stable at room temperature[121].

The X-ray structure shows slight distortions of the ring due to the close vicinity of one of the chlorine atoms of the anion. One silyl group is bent out of the ring plane and away from the anion. The tris(silyl)cyclopropenylium cation **301** serves as a close model in quantum chemical *ab initio* calculations for the experimentally investigated trimethylsilyl-substituted carbocation **300**. At both the HF/6-31G(d) and the correlated MP2(fc)/6-31G(d) levels a highly symmetric C_{3h} structure of **301** is a minimum. This is consistent with ^{13}C NMR measurements of the trimethylsilyl-substituted cation **300** in solution which indicate that it has a C_{3h} symmetry[122]. The calculated C—C bond length in the three-membered ring of the SiH$_3$ model **301** (1.388 Å) is in agreement within experimental error with the average value found in the X-ray structure of the SiMe$_3$-substituted cation salt **300**·SbCl$_6^-$ (1.384 Å). The other geometrical data also show good agreement between the experimental and calculated data for the model cation. According to a NBO analysis of the HF/6-31G(d) wave function all the ring carbons in cyclopropenylium cations carry the same charge. The charge decreases with decreasing electronegativity of the substituent [**302** (CH$_3$) +0.2, **303** (H) +0.025 and **301** (SiH$_3$) −0.243]. With silyl substitution the ring carbons are actually negatively charged, although formally they represent the carbocation centres, and the positive charge resides on the silyl groups. As the Me$_3$Si group is probably a better σ-electron donor than H$_3$Si, the total negative charge on the ring carbons is expected to be even higher in the experimentally studied cation **300**. The smaller π-orbital occupation of 2.04 e for SiH$_3$ substitution in **301** as compared with 2.12 e for methyl substitution in **302** indicates a less effective hyperconjugative electron donation from the silyl group across the longer Si—C$^+$ bond in **301**, compared with the C$^+$–CH hyperconjugation in the trimethyl-substituted cation **302**. The calculated ^{13}C NMR chemical shift of $\delta = 218.3$ ppm for the ring carbons, calculated for a geometry of **301** obtained at the MP2(fc)/6-31G(d) level using the IGLO method and basis set II, agrees very well with the value measured experimentally for the SiMe$_3$-substituted cation **300** ($\delta = 214.3$ ppm). The thermodynamic stabilities of tris-substituted cyclopropenylium cations was evaluated from isodesmic reactions. Three silyl groups as in **301** stabilize the parent cyclopropenylium cation **303** by 22.4 kcal mol^{-1} at MP3(fc)/6-311G(d,p)//6-31G(d)+ZPVE. The tris-methyl-substituted carbocation **302** is 31.4 kcal mol^{-1} more stable than the parent cation **303**. The order of substituent stabilizing abilities Me > H$_3$Si > H found for cyclopropenylium ions **302**, **301** and **303** is thus the same as that observed for other carbocations with a corresponding substitution pattern. However, due to the inherent higher stability of the 2π-Hückel aromatic system, the magnitude of the stabilizing effect is smaller than in ordinary alkyl cations. The p$k_{(R+)}$ value of the tris-silyl-substituted cyclopropenylium cation **301** is estimated to be 4 as compared with 7.4 for the tris-methyl-substituted cyclopropenylium cation **302**.

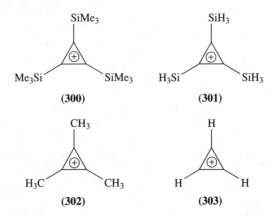

B. β-Silyl-substituted Carbocations

1. sp^2-Hybridized carbocations

a. Photochemical generation and UV characterization of transient β-silyl-substituted carbocations. i. 9-(Trimethylsilylmethyl)fluoren-9-yl cations. When 9-(trimethylsilylmethyl)fluoren-9-ol **304** at ca. 10^{-4} molar concentrations in TFE is submitted to photolysis, both dehydroxylation and desilylation take place (equation 44)^{123}.

The only product observable after short irradiation times is 9-methylenefluorene **305**. At longer times 9-methyl-9-trifluorethyl ether **306** is detected as a secondary product formed by protonation of the exocyclic double bond of 9-methylenefluorene **305** to give the 9-methylfluoren-9-yl-cation **307**, which then reacts with the TFE solvent. The transient absorption spectrum obtained upon laser flash photolysis of 9-(trimethylsilylmethyl)fluoren-9-ol **304** shows three bands with λ_{max} 340, 370 and 425 nm, all decaying with the same rate constant, 2.4 × 10^7 s^{-1} in TFE at 25 °C. The transient spectrum is different from the spectrum of 9-methylfluoren-9-yl cation **307**, which can be obtained as a stable cation in 96% H$_2$SO$_4$ from the corresponding alcohol **308**. Picosecond experiments reveal that the transient is formed on a 25 ps time scale directly from the silyl-substituted precursor **304** and shows the characteristics of a cationic intermediate, being unaffected by oxygen but sensitive to the nucleophilicity of the medium. This leads to the conclusion that the 9-(trimethylsilylmethyl)fluoren-9-yl cation **309** is the observed intermediate of the desilylation. The rate constant decreases on proceeding from TFE to the more weakly nucleophilic 1,1,1,3,3,3-hexafluoro-2-propanol (HFIP). It was concluded that the desilylation is a one-step process proceeding via a transition state **310** with participation of the solvent or other added nucleophiles. The marked different absorptions of the 9-(trimethylsilylmethyl)fluoren-9-yl cation **309** compared with simple 9-alkylfluoren-9-yl cations such as **307** indicate that the interaction of the β-silyl group with the positive charge has a significant perturbing effect. If all the charge would be transferred from the fluorenyl moiety to the silicon, a 9-methylenefluorene/Me$_3$Si$^+$ complex would result. The spectrum of the transient **309** is, however, quite different from that of 9-methylenefluorene **305** and thus the geometry was suggested to be intermediate between a 9-alkylfluoren-9-yl cation and a 9-methylenefluorene-Me$_3$Si$^+$ complex.

(44)

ii. Silyl-substituted cyclohexadienyl cations. 2-Silyl-substituted 1,3-dimethoxybenzenes **311** undergo a desilylation reaction upon photoexcitation in HFIP[124]. The only products observed are 1,3-dimethoxybenzene **312** and the silyl ether $R_3SiOCH(CF_3)_2$ (equation 45). The reaction proceeds via selective protonation at the 2-position.

$$\underset{(311)}{\text{OMe-C}_6H_3(\text{SiR}_3)\text{-OMe}} \xrightarrow[(F_3C)_2CHOH]{h\nu} \underset{(313)}{[\text{cation with H, SiR}_3]^+} \longrightarrow \underset{(312)}{\text{OMe-C}_6H_4\text{-OMe}} \quad (45)$$

When the reaction was conducted in O-deuteriated HFIP, $(CF_3)_2CHOD$, as a solvent, the deuterium was found exclusively at the 2-position of the 1,3-dimethoxybenzene **312**. Laser flash photolysis at 248 nm irradiation carried out with 10^{-3} M solutions of substrates **311** with different silyl groups ($SiMe_3$, $SiMe_2Ph$, $SiMePh_2$, $SiMe_2C_6H_4X$-4: X = OMe, Me, F, I) show absorptions of the transient cations **313** with maxima at 380–390 nm and a weak band at 300 nm. The absorption spectra are similar to that of the parent 2,6-dimethoxybenzenium ion **314** (R = H) (λ_{max} = 410 nm) but show a slight hypsochromic shift.

(314)

The exponential decay is uniform across the entire spectrum, indicating that a single transient species is being produced. The decay rate at 25 °C varies between 8×10^3 for the $SiMe_3$ and 1×10^5 for the 4-$MeC_6H_4(Me)_2Si$ group. The decay is unaffected by oxygen, an efficient quencher of radicals and triplets. Despite its photochemical generation the transient **313** is a ground state species. The relative long life of the transient implies that it cannot be a singlet excited state while the lack of quenching by oxygen mitigates against a triplet. However, the decay is accelerated by added nucleophiles such as alcohols, which is characteristic for cations. The transient is assigned the structure of a β-silyl-substituted cyclohexadienyl cation **313**. The hypsochromic shift compared with the parent cyclohexadienyl cation **314** reflects a perturbation of the cyclohexadienyl π-system induced by the interaction of the silyl substituent with the positive charge. The mechanism suggested initiates by a photoprotonation of the excited 2-silyl-1,3-dimethoxybenzenes **311** at C2 to give the intermediate 1-R_3Si-2,6-dimethoxybenzenium ions **313** which undergo desilylation via transition state **316** to give 1,3-dimethoxybenzene **312** and the protonated silyl ethers **315** (equation 46). Kinetic analysis demonstrates that these cations undergo preferentially desilylation over deprotonation. The desilylation occurs with nucleophilic participation. Quantitative agreement between the kinetics and the product analysis as well as the entropies of activation determined for the reaction with the solvent serve as evidence for a bimolecular reaction.

$$(313) \longrightarrow (316) \longrightarrow (312) + (315) \quad (46)$$

The observation that added alcohol accelerates the decay supports the interpretation that the reaction is associative, with concerted Si−C bond breaking and Si−O bond making. A dissociative pathway with initial cleavage of the C−Si bond to 1,3-dimethoxybenzene and the silylium ion R_3Si^+ is inconsistent with the results. Although a free silyl cation does not form even for the reaction with the less nucleophilic t-BuOH and HFIP, the ρ-values obtained for different R_3Si substituents offer evidence for some silyl cation character in the transition state **316**. The silicon in carbocation **313** bears a partial positive charge due to its interaction with the neighbouring π-system. In an associative transition state **316** this charge can either decrease or increase, depending on the relative amount of C−Si bond breaking and Si−O bond making. For the reaction with methanol, ρ is 0.0, indicating no charge formation at the transition state. The t-BuOH and HFIP reactions, however, have negative ρ values (−0.8 and −1.3, respectively), pointing to an increase in the positive charge at silicon as the reaction proceeds. With these nucleophiles C−Si bond breaking is more advanced than Si−O bond making at the transition state **316** and the silicon has some Si^+ character. This study gives direct evidence that β-silyl-substituted carbocations react with nucleophilic participation in the Si−C_β bond-breaking process. This is in accord with the observation made in superacids that bulky substituents at silicon increase the stability of β-silyl carbocations, i.e. that the rate-determining step of the β-C−Si bond cleavage in superacids is the attack of the nucleophile, which is retarded by the large alkyl groups at silicon.

b. X-ray crystallographic study of β-silylcyclohexadienyl cations. When trityl tetrakis(pentafluorophenyl)borate **317** was reacted with triethylsilane in toluene, a compound with the composition $Et_3Si(C_6F_5)_4B\cdot(PhMe)_2$ was isolated as a solid and investigated by X-ray crystallography[125,126]. The crystal data, in particular the nearly planar geometry of the closer toluene moiety, as well as the relatively long Si−C(ring) distance (2.18 Å) were interpreted as due to a structure of a silylium ion with no coordination to anion and a distant coordination to the solvent. This interpretation was heavily disputed by several authors (see Lickiss' chapter on silylium ions) and the structure was suggested to correspond instead to the p-triethylsilyltoluenium ion **318**[127−131].

The structural data reported from the X-ray analysis are very similar to the *ab initio* calculated structural data for silyl-substituted arenium cations. The calculations of the p-trimethylsilyl (**319**) (Figure 9) and p-triethylsilyl toluenium ions (**318**) [HF/6-31G(d)] and the silylbenzenium ion **53** [MP2(fc)6-31G(d)] all show similar geometries and the characteristic geometric data resemble closely the main features found in the crystal structure.

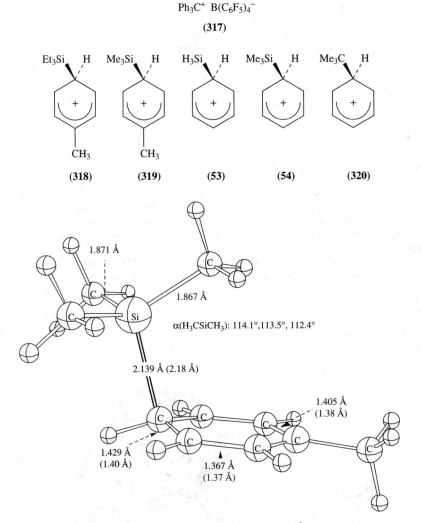

FIGURE 9. Calculated structure of **319** at HF/6-31G* (bond distances in Å, angles in deg), experimental data for **318** in parentheses

The calculated Si—C$_{ipso}$ distances of 2.14, 2.087 and 2.197 Å, the 113.8° average angle at silicon in the silyl groups and the near planarity of the six-membered rings are just like those reported for the X-ray structure.

Furthermore, the ^{29}Si NMR chemical shift ($\delta = 81.8$ ppm) measured in toluene differed very much from what is expected by quantum chemical calculation of model compounds for planar trialkylsilylium ions (Me$_3$Si$^+$, δ^{29}Si $= 355.7$ ppm; Et$_3$Si$^+$, δ^{29}Si $= 354.6$ ppm). For distorted trimethylsilylium ions with C—Si—C angles of 114.0° and 109.5°, even more deshielded ^{29}Si chemical shifts of 368.2 and 397.0 ppm were calculated [IGLO basis II at HF/6-31 G(d) level].

IGLO calculation of the ^{29}Si NMR chemical shift [DZ basis (C,H), basis II (Si)//HF/6-31G(d)] for the *p*-trimethylsilyltoluenium ion **319** and the trimethylsilylbenzenium ion **54** give δ^{29}Si = 60 and 80.9 ppm. These results are in line with the expected trend that more charge will be delocalized towards silicon in the trimethylsilylbenzenium ion **54** compared with the *p*-trimethylsilyltoluenium ion **319**. The chemical shift of δ = 82.1 ppm calculated for the *p*-triethylsilyltoluenium ion **318** agrees very well with the experimental value of δ = 81.8 ppm in toluene.

Contrary to the original interpretation, the structures investigated by X-ray crystallography are regarded as silyl-substituted arenium ions. The characteristic geometric features accompanying the β-silyl stabilization of these carbocations are rationalized as follows. The long Si—C$_{ipso}$ bond is the result of hyperconjugative stabilization of the cyclohexadienyl cation by the β-silyl effect. The silyl group bends over towards the dienyl π-system and the hydrogen at the *ipso*-carbon moves up, thus resulting in a flattening of the ring. The alternation of the ring bond lengths arising from π-resonance stabilization of the positive charge is reduced compared with alkyl-substituted cyclohexadienyl cations. β-Silyl stabilization is more efficient than β-alkyl stabilization and thus the demand for π-resonance stabilization is lower in the β-silyl-substituted carbocations. For the *t*-butyl analogue, i.e. the 6-*t*-butylcyclohexa-2,4-dienyl cation **320**, ring bond lengths with more pronounced alternation are calculated.

A closely related X-ray structure *i*-Pr$_3$Si(Br$_6$CB$_{11}$H$_6$) **321** (Figure 10) also displays similar features[132]. The brominated carborane anion Br$_6$CB$_{11}$H$_6^-$ was considered to be the least nucleophilic anion and the structure was interpreted as having about one third silylium ion character.

Consideration of the X-ray structure (with a Si—Br bond distance of 2.479 Å and average CSiC bond angles of 115.8°) and the relevant ^{29}Si NMR shift of δ^{29}Si = 109.8 ppm indicate, however, that no R$_3$Si$^+$ species was observed, but rather a structure in which

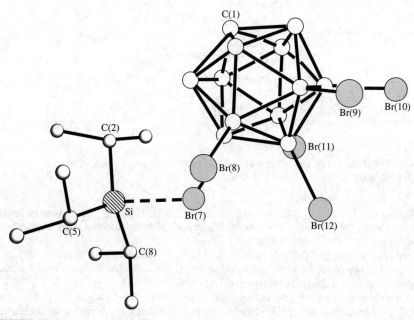

FIGURE 10. X-ray structure of **321**

12. Silyl-substituted carbocations

a single bromine atom was in bonding distance to a silyl group, which resulted in a polarized silylbromonium zwitterion.

c. NMR spectroscopic characterization of β-silyl-substituted carbocations. i. β-Silyl-substituted secondary p-methoxybenzyl cations. α. 1-p-Anisyl-2-triisopropylsilylethyl cations. Protonation of *E*-1-*p*-anisyl-2-triisopropylsilylethene **322** with FSO$_3$H in SO$_2$ClF/SO$_2$F$_2$ yields a mixture of *anti*- and *syn*-**323** 1-*p*-anisyl-2-triisopropylsilylethyl cations **323** (equation 47)[115,133].

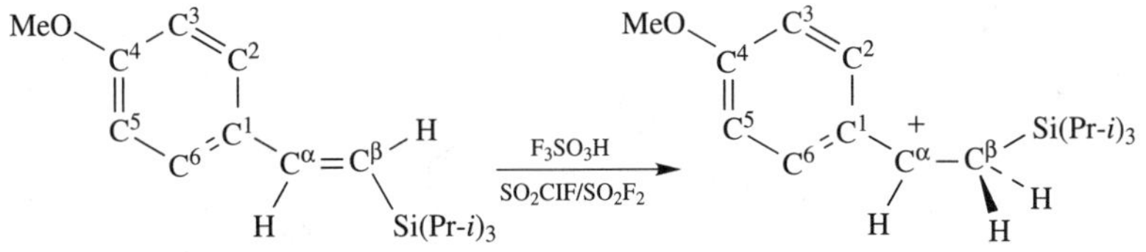

$$\text{(322)} \xrightarrow{\text{F}_3\text{SO}_3\text{H}, \text{SO}_2\text{ClF/SO}_2\text{F}_2} \text{(323)} \tag{47}$$

The *para*-methoxy group is involved in the *p*-π-resonance delocalization of the positive charge as shown in the resonance structure **323B** of the limiting structures **323A–323D**. Hence, the C4–oxygen bond in **323** has a partial double bond character.

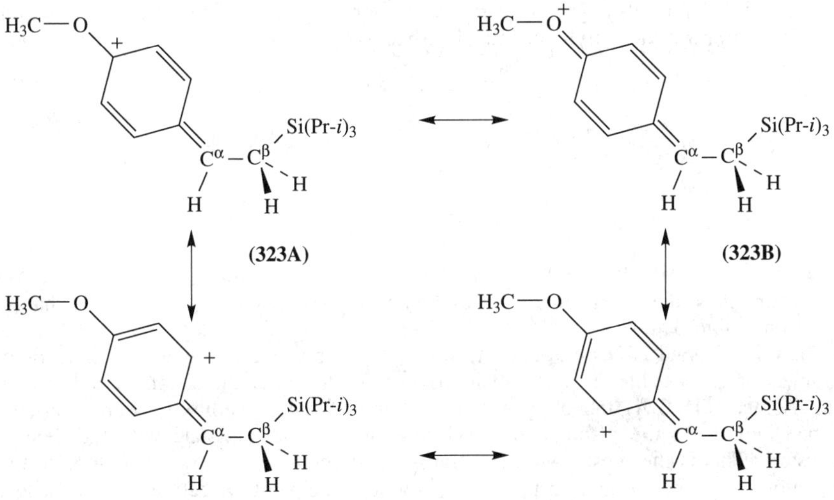

(323A)   (323B)

(323D)   (323C)

Two isomers with the O-methoxy group in the plane of the aryl ring are possible. The O-methoxy group can be in an *anti* or *syn* orientation relative to the —CH$_2$SiR$_3$ group at the benzylic carbon as in *anti*-**323** or *syn*-**323** (equation 48).

Below 120 °C the interconversion of the *syn* and *anti* isomers is slow on the NMR time scale. The *ortho, ortho'* and *meta, meta'* positions in the two isomers are non-equivalent. Two sets of four lines are observed, one for the C2 and C6 *ortho* and C3 and C5 *meta* carbons in the *anti* isomer (*anti*-**323**) and another set with lower intensity for the C2' and C6' *ortho* and C3' and C5' *meta* carbons in the *syn* isomer (*syn*-**323**). Upon warming above the coalescence temperature of about −110 °C the chemical shift for the non-equivalent positions in the two isomers is averaged and only four lines for the aromatic methine carbons are observable. A barrier $\Delta G^{\#}$ (−110 °C) of about 7 −8 kcal mol^{-1} was estimated from a line shape analysis.

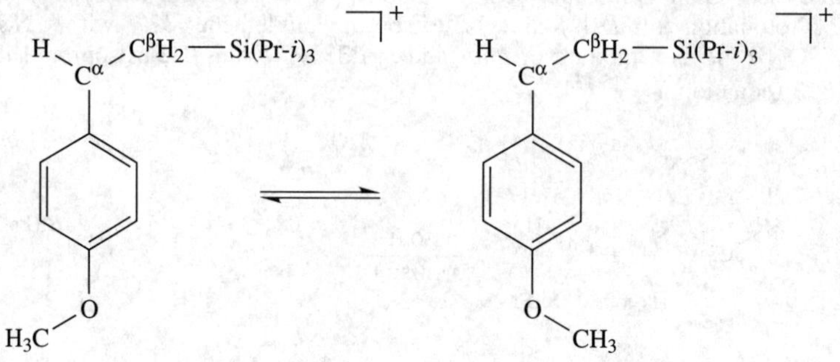
H
C^{α}
$C^{\beta}H_2$—Si(Pr-*i*)$_3$
+
O
H_3C
H
C^{α}
$C^{\beta}H_2$—Si(Pr-*i*)$_3$
+
O
CH_3

anti-**(323)** *syn*-**(323)**

(48)

Ab initio DFT calculations were performed using the B3LYP/6-31G(d) hybrid method for the analogous 1-*p*-anisyl-2-SiH_3-substituted ethyl cation **324**, which is a close model for the experimentally observed cation **323** with alkyl groups at silicon. It has been shown that different alkyl groups at silicon have no significant impact on the electronic stabilization in α-aryl-β-silyl-substituted carbocations.

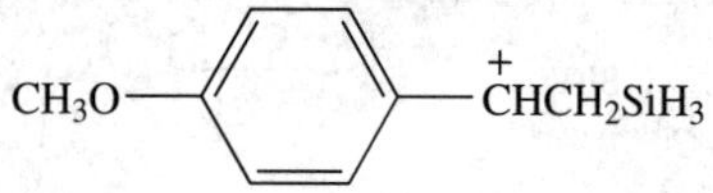
CH_3O
+
$CHCH_2SiH_3$

(324)

The optimized structures for the *anti* and *syn* isomers of **324** (Figure 11) have very similar energies, the *syn* isomer *syn*-**324** being only 0.3 kcal mol^{-1} less stable than the *anti* isomer *anti*-**324**.

The C_{β}−Si bond is perpendicular to the aryl ring plane and thus a maximum β-silyl stabilization is possible. In the transition state for the interconversion of the *syn* and *anti* isomers (TS-**324**) the oxygen−methyl bond of the *p*-methoxy group is oriented perpendicularly to the plane of the aryl ring and cannot conjugate with the aromatic π-system. This is the most favourable transition structure for the *syn/anti* isomerization (equation 48). Rotation around the C1−C^+ bond in benzyl-type cations is known to be higher in energy as compared to rotation around the C4−OCH_3 bond.

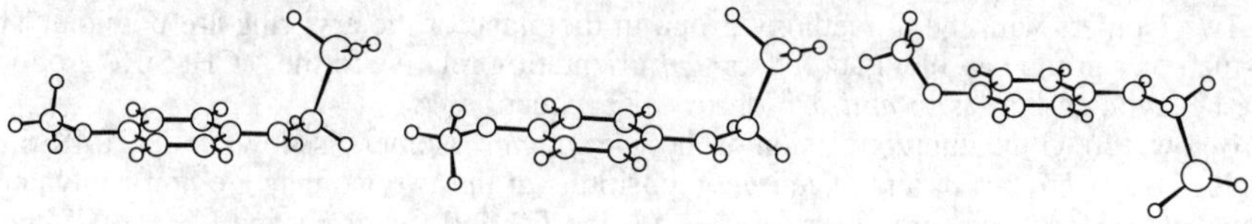

syn-**(324)** *anti*-**(324)** TS-(324)

FIGURE 11. Structures of *syn, anti* and TS-**324**

The energy difference between the transition structure TS-**324** with a perpendicular OCH$_3$ group and the *syn* and *anti* structures of **324** is a measure of the barrier for the methoxy group rotation around the C4–oxygen bond. The transition structure TS-**324** is calculated to be higher in energy by 10.3 kcal mol^{-1} than the *anti* isomer (*anti*-**324**). This is in satisfactory agreement with the experimentally observed energy barrier. The energy barrier for methoxy group rotation is a measure of the electronic demand of the carbocation. The better the β-substituent stabilizes the positive charge, the less delocalization into the aryl ring is necessary and thus the lower the barrier expected for the rotation around the C4–methoxy bond.

The corresponding barriers in the parent anisylmethyl cation **325**, the 1-anisylethyl cation **326**, the anisyl phenylmethyl cation **327** and the anisyl cyclopropylmethyl cation **328** were determined experimentally to be $\Delta G^{\#} = 12$, 10.6, 8.9 and 8 kcal mol^{-1}, respectively. This shows that a β-silyl substituent, here the R$_3$SiCH$_2$ group in **323**, is much more efficient than a methyl group in **326** and about as efficient as an α-phenyl or α-cyclopropyl ring in delocalizing the positive charge in **327** and **328**, respectively. This is in accord with the experimentally observed chemical shift of the *para* carbon C4, which is less deshielded in the R$_3$SiCH$_2$- substituted cation **323** ($\delta = 178.84$ ppm) compared with $\delta = 185.84$ ppm in the 1-*p*-anisylethyl cation **326** and $\delta = 181.93$ ppm in the 1-*p*-anisylpropyl cation **329** (see below).

NMR chemical shift calculations for the B3LYP/6-31G(d) optimized geometries of the *anti* and *syn* SiH_3-substituted model cations *anti*-**324** and *syn*-**324** were performed using various levels of theory (Table 12).

The NMR chemical shifts calculated with the B3LYP/6-31G(d) DFT/SCF hybrid method using the GIAO method implemented in the Gaussian 94 program show deviations for both *syn*-**324** and *anti*-**324** of more than 10 ppm and up to about 20 ppm for the SiH_3-substituted methylene group compared with the corresponding experimental values observed for *syn*-**323** and *anti*-**323**. The deviations from experiment are smaller (<8 ppm) when the 6-311G(d,p) basis set is used. The SOS DFT IGLO III method implemented in the DeMon program which uses the Perdew-Wang91 DFT functional gives small deviations from experiment (<6 ppm, excluding the β-methylene carbon chemical shift)[134]. Computational experience with smaller carbocations, where calculations at higher levels are feasible, have shown that an appropriate level of theory for the optimization of the molecular geometry is very important in order to obtain useful results

TABLE 12. Experimental ^{13}C NMR chemical shifts of 1-*p*-anisyl-2-(trialkylsilyl)ethyl cations (at 100 MHz, $T = -126\,°C$) in SO_2ClF (external capillary TMS, $\delta = 0$ ppm)[a]

NMR method		C1	C_α	C_β	C6	C2	C5	C3	C4	OCH_3
exp (**323**)										
	anti				152.18	138.12	114.98	123.85		
		132.20	205.30	43.17					178.84	58.93
	syn				148.49	141.65	121.95	116.57		
calc (**324**)										
GIAO	*anti*[b]	120.8	185.71	29.0	142.0	127.9	105.5	115.6	167.3	53.5
BLYP/	Δ	11.4	19.6	14.2	10.2	10.2	9.5	8.3	11.6	5.4
6-31G(d)										
GIAO	*anti*[c]	136.9	206.5	37.1	158.8	144.0	118.9	131.5	187.1	61.8
B3LYP/	Δ	−4.7	−1.2	6.0	−6.6	−5.9	−3.9	−7.7	−8.3	−2.9
6-311G(d,p)										
SOS DFT/	*anti*[d]	136.8	198.31	39.9	155.1	140.7	118.3	129.8	182.9	63.4
IGLO III	Δ	−4.6	7	3.3	−2.9	−2.6	−3.3	−6.0	−4.0	−4.5
GIAO	*syn*[b]	120.6	185.73	28.8	138.6	131.0	114.1	107.0	167.2	53.4
BLYP/	Δ	11.6	19.6	14.4	9.9	10.7	7.9	9.6	11.7	5.5
6-31G(d)										
GIAO	*syn*[c]	136.8	206.6	36.9	155.2	147.3	129.8	120.4	187.0	61.7
B3LYP/		−4.6	−1.3	6.3	−6.7	−5.8	−7.9	−3.8	−8.2	−2.8
6-311G(d,p)										
SOS DFT/	*syn*[d]	136.6	198.34	38.8	151.7	143.5	128.1	119.9	183.1	63.2
IGLO III	Δ	−4.4	7	4.4	−3.2	−1.9	−6.2	−3.3	−4.3	−4.3

[a] All geometries calculated at SCF/DFT level using B3LYP Functional and 6-31G(d) basis set. For numbering see equation 47.
[b] Reference TMS $\delta^{13}C$(calc) = 186.01 ppm, geometry (C_1): B3LYP/6-31G(d); NMR calculation: GIAO BLYP/6-31G(d).
[c] Reference TMS $\delta^{13}C$(calc) = 183.9 ppm, geometry (T): B3LYP/6-31G(d); NMR calculation: GIAO B3LYP/6-311G(d,p) ^{13}C.
[d] Reference TMS $\delta^{13}C$(calc) = 183.5 ppm, geometry (C1): B3LYP/6-31G(d); NMR calculation: SOS DFT PW91 (Perdew-Wang 91) IGLO III.

12. Silyl-substituted carbocations

from the calculation of NMR chemical shifts. The general good agreement between the calculated and observed chemical shifts obtained show that DFT-based *ab initio* calculations are practical and reliable methods for the computational investigation of this type of benzyl-substituted carbocations. The calculated differences between the chemical shifts of the two sets of signals for the two isomers allow unequivocal assignment of the experimental signals to the *syn* and *anti* isomers even using the data obtained at the lowest level of the chemical shift calculations.

ii. Other β-silyl-substituted secondary benzyl cations. α. *β-Silyl-substituted tolyl- and phenylmethyl cations.* The protonation of styrenes normally leads to facile oligomerization and polymerization. β-Silyl-substituted styrenes lacking the strong electron-donating *p*-OCH$_3$ group but having bulky alkyl groups at silicon, such as 1-phenyl-2-triisopropylsilylethene **330**, are protonated under strong acidic conditions using FSO$_3$H/SbF$_5$ at low temperatures in SO$_2$ClF/SO$_2$F$_2$ solution to form β-silyl-α-aryl-substituted ethyl cations such as **331** (equation 49)[115,135].

The protonation of silyl-substituted styrenes is regioselective and leads to secondary benzyl cations with a β-silyl substituent (equation 49). The isomeric carbocation **332** is not formed.

Likewise, the 1-*p*-tolyl-2-trisopropylsilylethyl cation **333** is obtained by protonation of 1-*p*-tolyl-2-triisopropylsilylethene **334** (equation 50). Less acidic conditions, i.e. FSO$_3$H without addition of SbF$_5$, are sufficient to generate the carbocation **333** while they are insufficient to generate **331**.

Styrenes with less bulky alkyl groups at silicon, such as 1-phenyl-2-trimethylsilylethene **335**, 1-p-tolyl-2-trimethylsilylethene **337** and even the relatively more bulky substituted 1-p-tolyl-2-dimethylisopropylsilyl ethene **338**, on protonation using either FSO_3H/SbF_5 or FSO_3H, do not lead to solutions of persistent β-trimethylsilyl-substituted carbocations such as **336** from **335** but undergo fast β-silyl fragmentation to silyl-free cations such as **78** (equation 51).

^{29}Si NMR spectroscopy is a suitable tool to monitor the electron demand in β-silyl-substituted carbocations. The ^{29}Si NMR chemical shifts in the α-phenyl-, α-p-tolyl,α-p-anisyl- and α-ferrocenyl-substituted β-silylethyl cations **331**, **333**, **323** and **339** are $\delta = 66.3, 56.9, 38.9$ and 23.5 ppm, respectively, decreasing regularly as expected on increasing the electron-donating capability of the α-substituent.

β. *1-Ferrocenyl-2-triisopropylsilylethyl cations*. The 1-ferrocenyl-2-triisopropylsilylethyl cation **339** can be generated from 1-ferrocenyl-2-triisopropylsilylethene **340** with trifluoroacetic acid in SO_2ClF (equation 52)[115,135]. Cation **339** is the only β-silyl-substituted secondary carbocation which can be generated without β-silyl fragmentation even in trifluoroacetic acid, a less non-nucleophilic solvent than superacids.

The ^{13}C NMR chemical shifts (ppm) of **339** ($-95\,^\circ$C in SO_2ClF/SO_2F_2) are at $\delta = 139.4$ (C$^+$), 98.83 (C1), 76.67/79.29 (C2,5), 90.99/91.64 (C3,4), 80.12 (C1′–5′), 25.29 (CH$_2$), 18.06 (i-Pr–CH$_3$) and 11.12 (i-Pr–CH). The positive charge in α-ferrocenyl-substituted carbocations is highly delocalized into the ferrocenyl moiety and the demand

12. Silyl-substituted carbocations

for hyperconjugative stabilization by the β-silyl-substituent is reduced. This is confirmed by comparing the ^{29}Si NMR chemical shift of $\delta = 23.5$ ppm in the 1-ferrocenyl-2-triisopropylsilylethyl cation **339** with δ^{29}Si $= 66.3$ ppm for the corresponding 1-phenyl-2-triisopropylsilylethyl cation **331**. The β-silyl effect in α-ferrocenyl-substituted carbocations is, however, still verifiable.

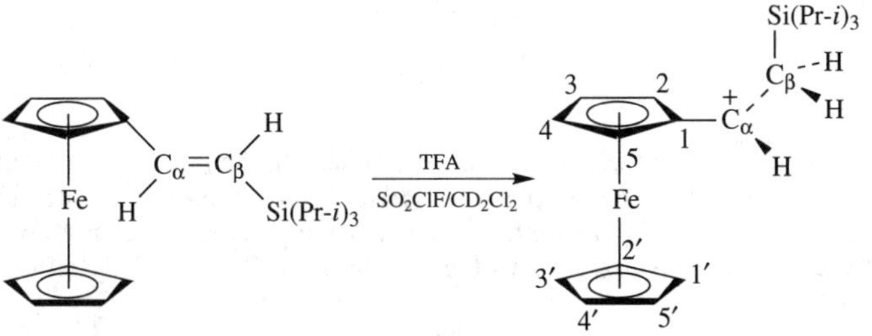

$$\text{(340)} \xrightarrow[\text{SO}_2\text{ClF/CD}_2\text{Cl}_2]{\text{TFA}} \text{(339)} \tag{52}$$

The signals for the C2,5, C3,4 and C1′–5′ positions in the 1-ferrocenyl-2-triisopropylsilylethyl cation **339** are shielded by -7.5, -4.3 and -3.2 ppm, respectively, compared with the parent ferrocenylmethyl cation **341**. The corresponding shieldings for the t-butyl-(**342**) and isopropyl-(**343**) substituted ferrocenylmethyl cations, and the ferrocenylethyl cation **344** are smaller; **342**: -3.7, -0.4, -2.1 ppm; **343**: -4.6, -0.8, -2.0 ppm and **344**: -3.3, -0.35, $+0.1$ ppm, respectively[136]. This shows that even in carbocations with $2p(C^+)$-π conjugative stabilization by a very good electron donor like the α-ferrocenyl group, hyperconjugative stabilization by the β-σ-bonds is operative and the stabilization by $2p(C^+)$-$\sigma(C-Si)$ interaction of a silyl group (as in **339**) is stronger than the $2p(C^+)$-$\sigma(C-H)$ hyperconjugation (as in **344**) or $2p(C^+)$-$\sigma(C-C)$ hyperconjugation (as in **342** and **343**).

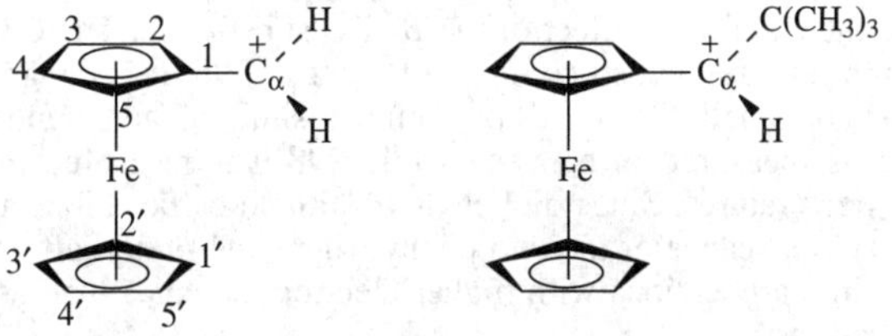

(341)

(342)

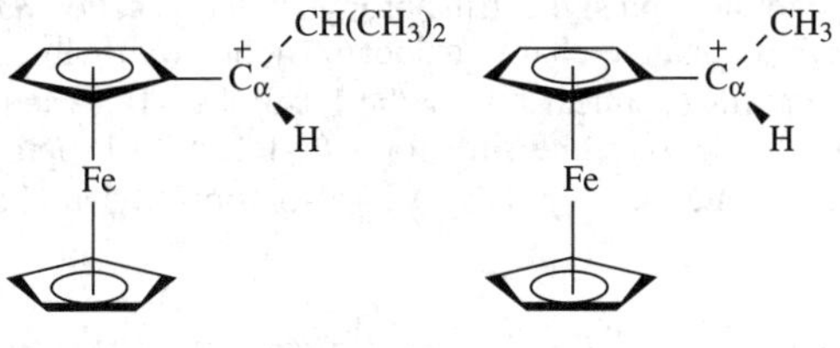

(343)

(344)

iii. β-Silyl-substituted tertiary benzyl cations. Earlier attempts to prepare the 1,1-diphenyl-2-trimethylsilylethyl cation **345** by ionization of 1,1-diphenyl-2-trimethylsilylethanol in FSO_3H/SO_2ClF at both $-78\,°C$ and $-140\,°C$ failed and led only to the formation of 1,1-diphenylethyl cation **298** along with trimethylsilyl fluorosulphate[113].

$$Ph_2C^+CH_2SiMe_3$$

(345)

α. 1,1-Diphenyl-2-(triethylsilyl)ethyl cations. When the benzene complex of triethylsilyl tetrakis(pentafluorphenyl)borate ($[Et_3Si^+(PhH)TPFPB^-]$) **346** was added to a benzene or toluene solution of 1,1-diphenylethene at room temperature, a highly coloured product was formed which was assigned to the 1,1-diphenyl-2-(triethylsilyl)ethyl cation **347** (equation 53)[137].

$$[Et_3Si^+(PhH)TPFPB^-] + H_2C=CPh_2 \longrightarrow Et_3SiCH_2CPh_2^+\, TPFPB^- + PhH \quad (53)$$
$$\textbf{(346)} \qquad\qquad\qquad\qquad\qquad\qquad \textbf{(347)}$$

The room temperature ^{13}C NMR spectrum in benzene-d_6 showed a peak at $\delta = 225.4$ ppm, in addition to signals at $\delta = 130.3, 135.2, 137.6$ and 141.1 ppm which were attributed to carbons in *meta*, *ortho*, *ipso* and *para* positions. The methylene signal was at $\delta^{13}C = 56.2$ ppm and the ethyl groups at silicon at $\delta^{13}C = 5.2$ and 6.3 ppm. A ^{29}Si NMR resonance was reported at $\delta = 46.2$ ppm. UV spectra show a strong absorption maximum at 432 nm and a weaker maximum at 310 nm, resembling those observed for the diphenylmethyl and diphenylethyl cations **299** and **298** at 442 nm and 300 nm, respectively.

In comparison with the 1,1-diphenylethyl cation **298** ($\delta^{13}C = 229.2$, C^+; 131.5, C_{meta}; 141.2, C_{ortho}; 141.5, C_{ipso} and 148.1, C_{para}) the 7 ppm high field shift of the *para* position may indicate the stabilizing effect of the β-silyl substituent. The C^+ carbon in cation **347** is about 4 ppm less deshielded compared to that in **298**. It is dangerous, however, to base interpretations on small differences of chemical shifts obtained under vastly different conditions; **347** was measured in benzene while **298** was measured in FSO_3H/SO_2ClF at very different temperatures. Since diphenyl-substituted carbocations are stabilized very efficiently by π-aryl resonance of the two phenyl rings, the β-silyl effect in **347** is diminished compared with carbocations with higher electron demand, such as cyclohexadienyl and phenylethyl cations.

The ^{29}Si NMR chemical shift for the *p*-triethyltoluenium ion **318** [δ (exp) = 81.8 ppm, δ(calc) = 82.1 ppm] and the 1-phenyl-2-triisopropylsilylethyl cation **331** ($\delta = 66.34$ ppm) indicate a larger hyperconjugative charge delocalization towards the silyl group in these carbocations. The ^{29}Si NMR chemical shift in the 1,1-diphenyl-2-(triethylsilyl)ethyl cation **347** ($\delta = 46.2$ ppm) is in between the shift for 1-(*p*-tolyl)-2-triisopropylsilylethyl cation **333** ($\delta^{29}Si = 56.92$ ppm) and the 1-(*p*-anisyl)-2-triisopropylsilylethyl cation **323** ($\delta^{29}Si = 38.88$ ppm).

iv. β-Silyl-substituted allyl cations. α. 2-[1′-(Trimethylsilyl)ethenyl]adamant-2-yl cations. The 2-[1′-(trimethylsilyl)ethenyl]adamant-2-yl cation **348a** can be obtained by the reaction of [1′-(trimethylsilyl)ethenyl]-2-adamantanol with FSO_3H/SbF_5 at $-130\,°C$[138].

The ^{1}H and ^{13}C NMR data measured at $-100\,°C$ are in accord with the formation of the carbocation structure (equation 54).

The trimethylsilyl substituent in 2-(trimethylsilyl)allyl cations is enforced into a perpendicular conformation and thus β-C−Si hyperconjugative stabilization is essentially not possible. The carbocationic centre C1 (δ^{13}C = 295.0 ppm) and the terminal methylene carbon C3 of the allyl system (δ^{13}C = 166.6 ppm) in **348a** are deshielded by 11.5 and 1.2 ppm, respectively, compared with the silicon-free analogue **348** (δ^{13}C = C1 283.5; C3 165.7 ppm). The deshielding was inferred to arise from a neighbouring silicon-induced anisotropic effect or from the trimethylsilyl group induced twisted π-system. ^{13}C chemical shift calculations [IGLO at the DZ/6-31G(d) level, geometry at 6-31G(d)] of the parent allyl cation $C_3H_5{}^+$ and 2-SiH$_3$- and 2-SiMe$_3$-substituted allyl cations show a deshielding of the same magnitude (10 ppm) for the cationic centres in the silyl-substituted allyl cation model structures (δ^{13}C = C1, C3: 2-SiH$_3$ = 238.6 ppm; 2-SiMe$_3$ = 235.1 ppm) as in the parent allyl cation (δ^{13}C = C1, C3: 228.8 ppm). MM2 and MNDO calculations of the 1-adamantyl-2-trimethylsilylallyl cation **348a** show that the bulky trimethylsilyl substituent severely hinders the coplanarity of the carbocationic carbon and the vinylic π-system. This reduces the allylic stabilization due to a reduced overlap of the π-electrons with the cationic centre compared with the system lacking the silyl group. Thus the interplay of electronic and steric effects produces a slight deshielding of the signal for the C^+ carbon and indicates an overall destabilizing effect of the trimethylsilyl group in this cation. The silicon-substituted central carbon of the 1-adamantyl-2-trimethylsilylallyl cation **348a** (δ^{13}C = 161.2 ppm vs 140.7 ppm for **348b**) is deshielded by 20.5 ppm in accordance with other experimental observations on acylsilanes. For the central carbon of 2-R-substituted allyl cations $C_3H_4R^+$, ^{13}C chemical shifts of δ = 136.9, 144.4 and 151.7 ppm for 2-R = H, SiH$_3$ and SiMe$_3$, respectively, were calculated by IGLO at the DZ/6-31G(d) level using 6-31G(d) geometries.

$$\text{R = Si(CH}_3)_3 \quad \text{R = H} \xrightarrow[-130\,°C]{FSO_3H/SO_2ClF} \textbf{(348)} \quad \text{(a) R = Si(CH}_3)_3 \quad \text{(b) R = H} \tag{54}$$

At $-100\,°C$ the β-C−H carbons of the adamantyl moiety are equivalent, indicating a rapid rotation across the bond between the cationic carbon and the central carbon of the allyl cation. However, in the parent 2-ethenyladamantyl cation **348b** these carbons are non-equivalent, indicating a high barrier for such a rotational process. The low rotational barrier in the silyl-substituted cation can be rationalized by assuming the intermediacy of a perpendicular allyl cation conformation **348ap** in which the β-silyl group is aligned in the plane with the vacant p orbital at the cationic carbon. This conformation lacks stabilization by π-p allyl conjugation but is hyperconjugatively stabilized by the β-silyl effect. The perpendicular conformation of the β-silyl-substituted allyl cation **348ap** is thus lower in energy compared with the perpendicular conformation of the allyl cation **348bp** lacking the β-silyl group (equations 55 and 56). Therefore the barrier for allyl rotation is smaller for the β-silyl-substituted allyl cation **348a** as compared with **348b**, leading at −100 °C to a fast exchange on the ^{13}C NMR time scale and an averaged signal for the β−CH carbons of the adamantyl moiety.

2. sp-Hybridized carbocations (vinyl cations)[139,140]

a. β-Silyl-substituted dienyl cations. The silyl-substituted vinyl cations **349** and **350** are formed by the reaction of the silyl-substituted allenyl alcohols **351** with SbF$_5$ at temperatures below −130 °C[141,142]. At about −110 °C cations **349** and **350** fragment with cleavage of the β-C−Si bond and expulsion of the silyl group, forming the silicon-free cations **352** and **353** (equation 57). The ^{13}C NMR chemical shifts of **349** and **350** (Table 13) are in general similar to those in the other α-vinyl-substituted vinyl cations.

The shielding of the C_α and $C_{\gamma'}$ resonances in **349** and **350** indicates allyl resonance delocalization of the positive charge. The unusual high field resonance of the signals for the sp²-hybridized carbons C_β and $C_{\beta'}$ in accord with those for other vinyl cations, are due to the sp-hybridization of the geminal C_α carbon. The methyl groups at $C_{\gamma'}$ are non-equivalent because of the partial double bond character of the $C_\alpha-C_{\beta'}$ bond.

A closer comparison of the silyl-substituted cations **349** and **350** with the analogous non-silyl-substituted cations **352**, **353** and **355** reveals significant differences. The charge-dispersing effect of the β-trimethylsilyl group is evident when **349** and **350** are compared with their non-silyl-substituted counterparts. The positively charged C_α in **349** is about 33–36 ppm and, in **350**, 54–57 ppm less deshielded than the corresponding carbon in **352**, **353** and **355** (Table 13). The chemical shift of the C_α carbon is, however, not only dependent on the charge density at C_α, but is also influenced by a $\beta-\pi$ substituent effect on the chemical shift from the different substituents at C_β. The composite effect of the charge-density effect and C_β-substituent effect precludes a direct correlation of the chemical shift of the C_α signal with the charge delocalization. The γ'-carbon is, however, remote from the substitution site C_β and the changes of the chemical shift for $C_{\gamma'}$ in dienyl cations are related only to the charge delocalization abilities of the different C_β substituents. The 28 ppm and 25 ppm shielding of $C_{\gamma'}$ in **349** compared with **352** and **353**, respectively, can thus be attributed to the superior electron-donating ability of a β-trimethylsilyl group compared with a β-hydrogen or β-methyl group, respectively. The high field shift for the signals of the γ'-methyl groups in **349** and **350** as compared with those in **352**, **353** and **355** reflects a lower demand for a C–H hyperconjugative stabilization of a positive charge at the γ-carbon by the γ-methyl groups because less charge is localized at this site in **349** and **350**. The signal of the methyl carbons on the silicon group shows no significant shift differences compared with the neutral precursor molecules, and hence the contribution of the alkyl groups at silicon to the delocalization of the positive charge in this type of carbocation is not significant. Both positively charged carbons of the allyl moiety in **349** and **350**

TABLE 13. ¹³C NMR chemical shifts (δ, ppm) for dienyl cations of type **354** in SO₂ClF/SO₂F₂[a]

(354)

	C_α	C_β	$C_{\beta'}$	$C_{\gamma'}$	C_β–R	$C_{\gamma'}$–R	
352[b]	242.50	79.40	114.70	262.10		33.23, 37.22	
353[c]	244.30	90.50	114.51	259.40	9.30	32.53, 36.56	
355	245.39	101.55	113.9	257.64	16.29	32.43, 36.44	
356	202.66	63.66	111.72	228.92	39.63	27.15, 30.62	
349[b]	208.72	80.83	108.94	234.32		28.48, 32.52	−0.4 SiMe₃
350[c]	188.43	86.42	107.85	221.72	11.09	26.84, 30.68	−1.17 SiMe₃

[a] 100.62 MHz, internal reference $\delta(\text{NMe}_4^+) = 55.65$ ppm, $T = -120\,°C$, chemical shifts $\delta(\pm 0.03)$ ppm.
[b] At −137 °C.
[c] At −126 °C.

(355)   **(356)**

are more shielded than the corresponding carbons in **352**, **353** and **355**. This indicates that the positive charge is delocalized away from0 the allylic system **357A** ↔ **357C** towards the β-trimethylsilyl group, as described by the hyperconjugative resonance structure **357B**.

(357B)   **(357A)**   **(357C)**

The electron donating ability of a β-C—Si bond in **349** is nearly as large as the effect of the strained C—C bond of a β-cyclopropylidene substituent in **356**. The difference in the chemical shift of $C_{\gamma'}$ in **356** (228.92 ppm) and in **349** (234.32 ppm) is only 5.4 ppm. The additional methyl substitution at C_β in the dienyl cation **350** causes an unexpectedly large effect on the chemical shifts, and the signals of both positively charged carbons C_α and $C_{\gamma'}$ are shifted upfield compared with **349**. For the dienyl cations **352**, **353** and **355** an additional methyl group leads to a downfield shift for C_α and an upfield shift for $C_{\gamma'}$ of about 1.5 ppm per methyl group (Table 13). This indicates an internal shift of the positive charge in **352**, **353** and **355** between the two terminal carbon atoms of the resonating allylic system from the γ'-site to the α-site in the order **355**, **353** and **352**. However, for the geminal β-methyl and β-trialkylsilyl-substituted dienyl cation **350**, C_α and $C_{\gamma'}$ absorb at 20.3 ppm and 12.6 ppm higher field, respectively, as compared with cation **349**, which has only the trialkylsilyl substituent at C_β. The ^{13}C NMR chemical shifts for cation **350** indicate that the additional methyl group at the β-carbon enhances the hyperconjugative interaction of the β-Si—C bond with the formally empty (2p)π orbital of the vinyl cation C_α carbon. This effect can be rationalized by theoretical calculations. The additional methyl substituent at C_β in the model cation structure **358** decreases the C_α—C_β—Si bond angle as compared with the model structure **359**. This geometrical distortion towards more bridging enhances the hyperconjugative effect of the β-silyl group in cations of type **358**.

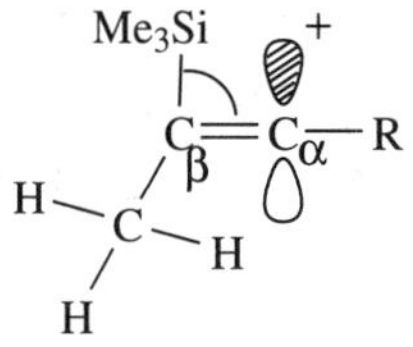

12. Silyl-substituted carbocations

(358) 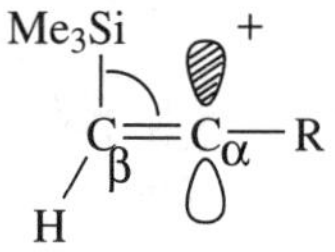   (359)

When the ^{13}C NMR chemical shift of the $C_{\gamma'}$ carbon in dienyl cations **349, 350, 352, 353, 355** and **356** is used as a probe for the ability of β-substituents to hyperconjugatively donate electrons to the formally empty 2p(π) orbital of the vinyl cation C_α, the order in Scheme 35 is obtained;

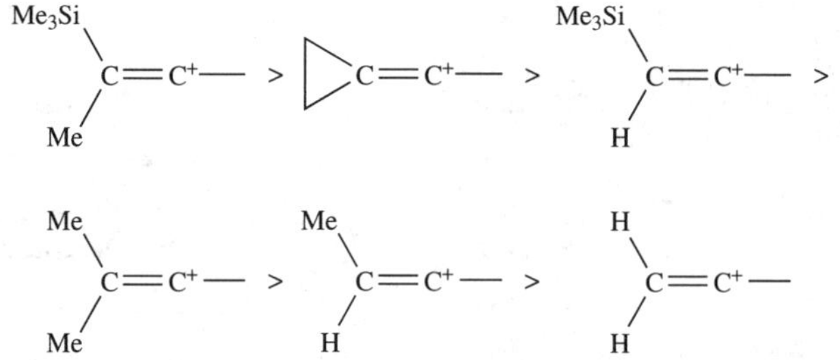

SCHEME 35

b. *β-Silyl-substituted α-arylvinyl cations.* i. *α-Ferrocenyl-β-silyl-substituted vinyl cations.* ^{1}H NMR spectroscopy was used in 1977 to search for β-silyl-substituted ferrocenyl cations[143]. 1-(1′,2-Dimethylferrocenyl)-2-(trimethylsilyl)ethyne **360** was converted in trifluoroacetic acid solution into 1-(1′,2-dimethylferrocenyl)vinyl cation **363**, which quickly captured the solvent to form the 1-(1′,2-dimethylferrocenyl)-1-trifluoracetoxyethyl cation **364**. The reaction was followed by ^{1}H NMR spectroscopy. The first-formed β-silyl-substituted vinyl cation **361**, which was not observed under experimental conditions, presumably fragments to ferrocenylalkyne **362** and trimethylsilyl trifluoroacetate. The alkyne **362** is protonated to give the 1-ferrocenylvinyl cation **363**, which is quickly converted by the acid solvent to cation **364** (equation 58). Related fragmentations of benzyltrimethylsilanes in trifluoroacetic acid are known.

Protonation of 1-ferrocenyl-2-trimethylsilylalkyne **365** by trifluoroacetic acid in SO$_2$ClF at < −80 °C yielded the α-ferrocenyl-β-trimethylsilylvinyl cation **366**. Cation **366** decomposes slowly at −80 °C to the parent α-ferrocenylvinyl cation **367** and trimethylsilyl trifluoroacetate (equation 59)[144].

Steric shielding of the silicon atom increases the stability of the initially formed β-silyl vinyl cations. The 1-ferrocenyl-2-(triisopropylsilyl)vinyl cation **368** and the 1-ferrocenyl-2-(dimethyl-*tert*-hexylsilyl)vinyl cation **369** are formed when the corresponding alkynes are protonated with trifluoroacetic acid in SO$_2$ClF. These cations show no fragmentation at comparable low temperatures[144].

The general ^{13}C NMR spectroscopic data of the β-silyl α-ferrocenylvinyl cations are similar to those of the β-*tert*-butyl-substituted α-ferrocenylvinyl cation **370** (Table 14)[145]

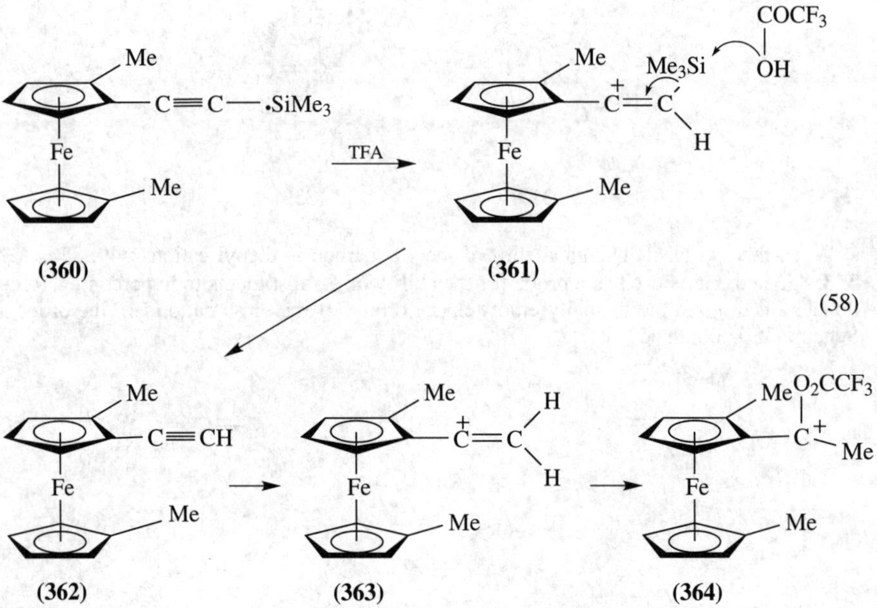

(58)

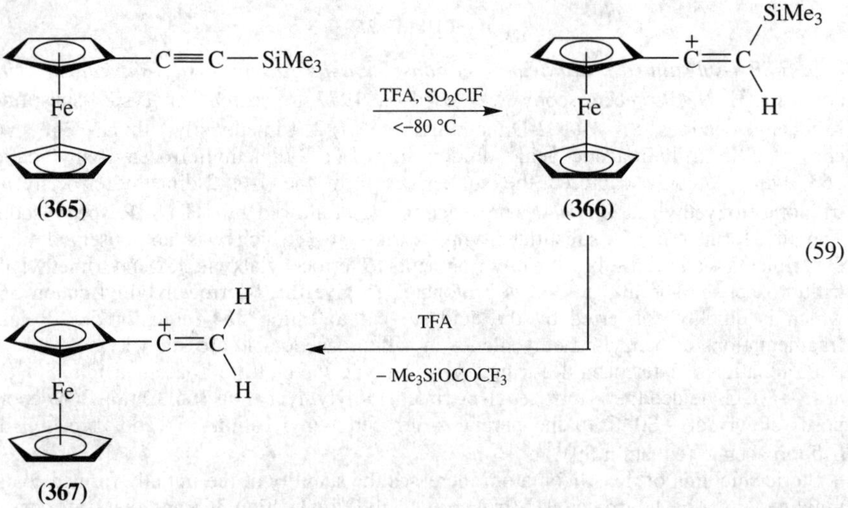

(59)

A detailed comparison of the chemical shifts of the ring carbons in the β-H (**367**), β-*tert*-butyl- (**370**) and β-trialkylsilyl-substituted α-ferrocenylvinyl cations (**366, 368** and **369**) reveals that the β-silyl group is superior to β-alkyl groups in stabilizing the positive charge. In the β-silyl-substituted vinyl cations the C3, C4 signals are about 2.5 ppm shielded as compared with the β-H or β-*tert*-butyl-substituted vinyl cations. The chemical shifts in the silyl-substituted ferrocenylvinyl cations exhibit only minor variations on changing the alkyl groups at silicon, showing that the latter do not contribute significantly to the charge delocalization in these carbocations.

12. Silyl-substituted carbocations

[Structures **(368)**, **(369)**, **(370)** of ferrocenyl-substituted vinyl cations]

The β-silyl effect is attenuated in α-ferrocenyl-substituted vinyl cations because the strong electron-donating effect of the α-ferrocenyl group renders the cation centre less electron-deficient and thus lowers the demand for hyperconjugative stabilization by substituents at the β-carbon[149].

TABLE 14. ^{13}C NMR chemical shifts (δ, ppm) for α-ferrocenyl-substituted vinyl cations **366–370**

	R in Fc-$\overset{+}{C}$ = CHR	C_α	C_β	C3,4	C2,5	C1	C1′–5′
367	H	183.7	99.2	95.9	81.9	72.5	84.9
370	CMe$_3$	177.1	126.6	96.1	83.1	74.7	85.2
366	SiMe$_3$	183.2	105.1	93.4	79.7	63.1	83.1
369	Si(Me)$_2$(Hex-t)	181.9	104.4	93.3	79.6	62.6	83.1
368	Si(Pr-i)$_3$	181.8	99.4	93.3	79.4	62.2	83.1

ii. α-Mesityl-β-silyl-substituted vinyl cations. Steric shielding by bulky alkyl groups at silicon and special experimental precautions are essential for the successful preparation of β-trialkylsilyl-substituted α-arylvinyl cations in solution, which otherwise easily undergo a C$_\beta$—Si bond fragmentation. The α-mesityl-β-silyl-substituted vinyl cations **372–375** are accessible by protonation of the corresponding alkynes with FSO$_3$H/SbF$_5$ in SO$_2$ClF/SO$_2$F$_2$ at −130 °C (equation 60)[114,147]. Figure 12 shows as a typical example the ^{13}C NMR spectrum of the α-mesityl-β-(triisopropyl)silylvinyl cation **375**.

Attempts to generate the β-trimethylsilylvinyl cation **376** by protonation of mesityltrimethylsilyl ethyne **377** (equation 61) were unsuccessful[114,148].

The β-unsubstituted mesitylvinyl cation **378** is not accessible by direct protonation of mesitylethyne **379** (equation 62). All attempts to generate vinyl cations by direct

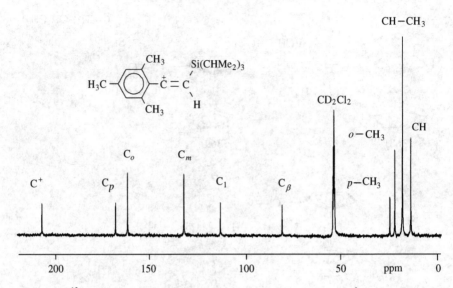

FIGURE 12. ^{13}C NMR of α-mesityl-β-(triisopropyl)silylvinyl cation **375** at $-135\,°$C in SO$_2$ClF/SO$_2$F$_2$, internal reference CD$_2$Cl$_2$.

(371) → (372) R = SiMe$_2$Pr-i
(373) R = SiMe$_2$Bu-t
(374) R = SiMe$_2$Hex-t
(375) R = Si(Pr-i)$_3$

(60)

(377) ↛ (376)

(61)

protonation of monosubstituted alkynes with superacids have failed and led instead only to the formation of rather complex mixtures and partial polymerization.

$$\text{(379)} \quad \xrightarrow{\quad//\quad} \quad \text{(378)} \qquad (62)$$

The mesitylvinyl cation **378** is obtained, however, without apparent formation of any side products, when a solution of α-mesityl-β-dimethylisopropylsilylvinyl cation **372** is warmed up from −130 to −100 °C for 10 minutes. This reaction can be explained as outlined in equation 63.

$$\text{Mes}-\text{C}\equiv\text{C}-\underset{\underset{\text{CH}_3}{|}}{\overset{\overset{\text{CH(CH}_3)_2}{|}}{\text{Si}}}-\text{CH}_3 \xrightarrow[-130\,°\text{C}]{\text{FSO}_3\text{H/SbF}_5} \text{Mes}-\overset{+}{\text{C}}=\text{C}\underset{\text{H}}{\overset{\text{Si(CH(CH}_3)_2)(\text{CH}_3)_2}{\diagup}}$$

(380) (372)

(63)

$$X^- = F^-, FSO_3^- \qquad \text{Mes}-\overset{+}{\text{C}}=\text{C}\underset{\text{H}}{\overset{\text{H}}{\diagup}} \xleftarrow{\text{FSO}_3\text{H}} [\text{Mes}-\text{C}\equiv\text{C}-\text{H}]$$

Mes = Mesityl (378) (379)

with arrow: −Me₂(i-Pr)SiX | X⁻, −100 °C

Protonation of the fairly sterically congested 1-mesityl-2-dimethylisopropylsilyl alkyne **380** at −130 °C leads to a clean formation of the α-mesityl-β-silyl-substituted vinyl cation **372**. The hyperconjugative interaction of the β-C—Si bond with the formally vacant p orbital at C_α increases the partial positive charge at silicon, thus increasing its susceptibility to nucleophilic attack by the anions present in the superacid solution. Fragmentation of the β-silyl group in **372** occurs at somewhat higher temperatures (−100 °C). The nucleophile attacks at silicon likely via an S_N2(Si) transition state, which subsequently leads to fast cleavage of the β-C—Si bond. This cleavage is facile because the vinyl cation moiety is strongly electron-withdrawing, thus acting as a good leaving group to form the trialkylsilane i-PrMe₂SiX (X = OSO₃F or F). The monosubstituted alkyne **379** formed as a transient in the presence of a large excess of superacid is immediately protonated to yield the silicon-free vinyl cation **378**.

An alternative mechanism, proceeding through a double protonation reaction sequence followed by silyl fragmentation, could be envisaged in the highly acidic media (equation 64). The first step at −130 °C leads to C_β protonation of the alkyne **380** with formation of the vinyl cation **381** (= **372**). At the higher temperature (−100 °C) the large excess of superacid permits a second proton to attack the vinyl cation **381**, presumably at the p orbital at C_β which is orthogonal to the vacant p orbital at C_α. This leads to a highly reactive gitonic dication intermediate **382** with two formal positive charges at the C_α carbon. Further reaction to give stable products could occur by concerted attack of the gegen ion X^- at silicon and a C_β–Si bond cleavage to form $XSiR_3$ and the β-unsubstituted mesitylvinyl cation **378**.

$$\text{Mes}-\text{C}\equiv\text{C}-\text{SiR}_3 \xrightarrow{\text{FSO}_3\text{H/SbF}_5} \text{Mes}-\overset{+}{\text{C}}=\text{C}\begin{smallmatrix}\text{SiR}_3\\ \text{H}\end{smallmatrix}$$

(380) (381)

$$\downarrow \text{FSO}_3\text{H/SbF}_5 \qquad (64)$$

$$\text{Mes}-\overset{+}{\text{C}}=\text{C}\begin{smallmatrix}\text{H}\\ \text{H}\end{smallmatrix} \xleftarrow[-\text{XSiR}_3]{X^-} \left[\text{Mes}-\overset{++}{\text{C}}-\text{C}\begin{smallmatrix}\text{SiR}_3\\ \text{H}\\ \text{H}\end{smallmatrix}\right]$$

(378) (382)

The silyl fragmentation in superacids initiated by a controlled temperature increase is a method to generate persistent carbocations, such as the vinyl cation **378**, which are not accessible by direct protonation of unsaturated hydrocarbons because of excessive formation of oligomeric and polymeric products.

For the β-trialkylsilyl-substituted mesitylvinyl cations **373–375** with larger alkyl groups at silicon, fragmentation to the silicon-free vinyl cations is not observed. Cations with larger groups are stable at relatively higher temperatures compared with those substituted by less bulkier alkyl groups at silicon. The β-silyl fragmentation reaction in superacids is controlled, at least in part, by steric crowding caused by the alkyl groups at silicon. However, this is not the only factor controlling the stability of β-silyl-substituted arylvinyl cations towards C_β–Si bond cleavage. Depending on the ability of the α-aryl substituent to delocalize a positive charge, more or less partial positive charge is localized at the silicon centre as a result of the hyperconjugative electron donation from the β-C–Si bond to the formally vacant $2p(\pi)$ orbital at the C^+ carbon. This changes the susceptibility of the silicon towards nucleophilic attack which is followed by cleavage of the β-C–Si bond.

The NMR spectroscopic data of the β-silyl-substituted α-mesitylvinyl cations corroborate the hyperconjugative charge-delocalizing ability of β-silyl groups. Comparison with β-alkyl- and β-H-substituted α-mesitylvinyl cations gives a measure of the magnitude of the β-silyl effect in these type of carbocations. The ^{13}C NMR spectroscopic data of the α-mesitylvinyl cations **372–375** and **378, 383–385** with various β-substituents together with data for the sp^2-hybridized mesityl carbocations **386** and **387** are summarized in Table 15.

The ^{13}C NMR chemical shifts of the positively charged carbon in the vinyl cations in Table 15 cover a range from 239 to 192 ppm. The difference of the chemical shift for

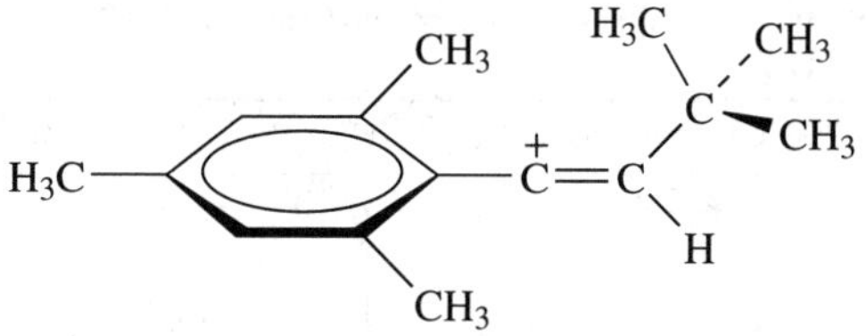

(383)

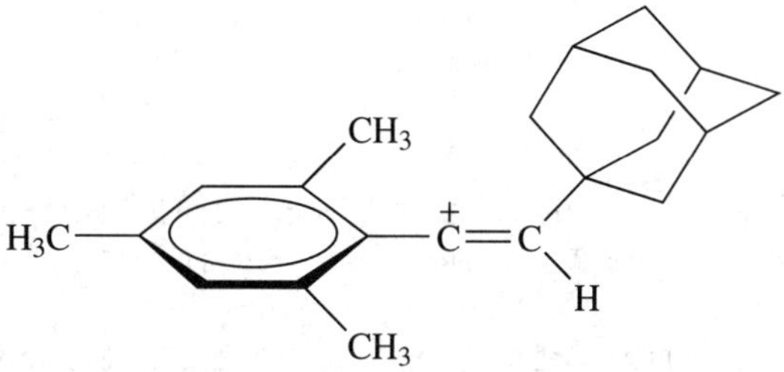

(384)

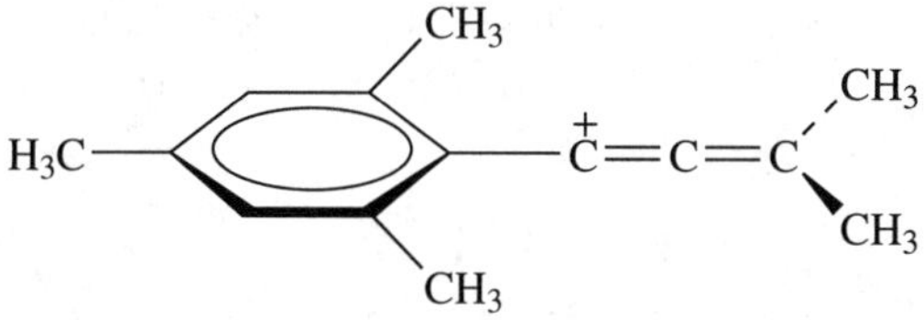

(385)

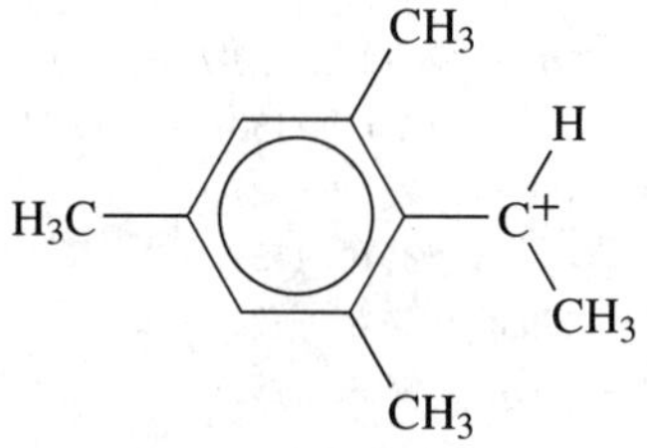

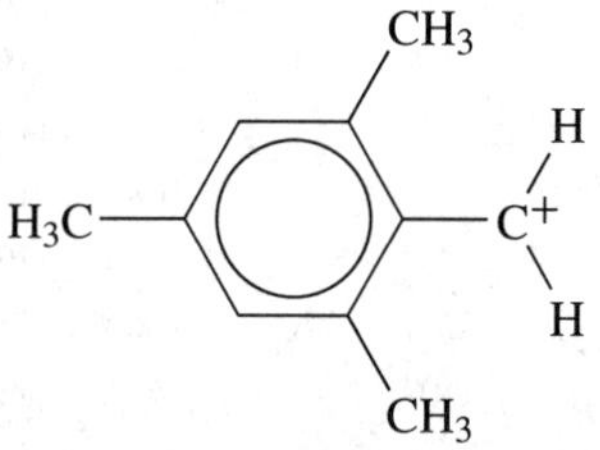

(386)   **(387)**

the C$^+$ carbon in the β-silylvinyl cations **372–375** and in the silicon free analogues **378** and **383–385** is 30–33 ppm. Silyl substituents cause a pronounced shielding effect on the C$_\beta$ carbon shift ($\Delta\delta$: *ca* -24 ppm) as compared with alkyl groups. The aromatic C1 position in the vinyl cations **372–375**, **378** and **383–385** display a large shielding effect of 20–30 ppm relative to the 1-mesityl ethyl (**386**) and the mesitylmethyl cation (**387**). This is due to the different hybridization of the adjacent C$^+$ carbon which is sp-hybridized in the vinyl cations but sp^2-hybridized in the trigonal carbocations **386** and **387**. Analogous high field shift has been observed for the C1 aryl carbon adjacent to the sp-hybridized C$^+$ carbon of the benzoyl cation **388**.

TABLE 15. ^{13}C NMR chemical shifts (δ, ppm) of vinyl cations **372-375**, **378**, **383-385** and cations **386** and **387** in SO$_2$ClF/SO$_2$F$_2$[a,b]

Cation	C$_\alpha$	C$_\beta$	C1	C$_{ortho}$	C$_{meta}$	C$_{para}$	o-Me	p-Me	Other
378	238.5	82.3	116.6	167.8	133.8	180.0	21.7	26.1	
		(177)			(166)		(130)	(131)	
383	237.3	107.1	118.4	166.4	133.4	178.5	21.5	25.8	Cq 40.9, Me 30.0
		(174)			(165)		(130)	(129)	(125)
384	238.7	106.3	118.7	166.1	133.2	177.9	21.5	25.7	C1' 45.0, C2', 8', 9'
		(169)			(166)		(130)	(131)	43.4, C3', 5', 7'
									29.2 (128),
									C6', 4', 10' 35.6
									(125)
372	206.0	83.6	113.5	162.7	132.5	168.5	21.4	24.4	Si-Me −4.2 (122),
		(183)			(166)		(131)	(129)	CH 16.5 (129),
									C-Me 16.7 (129)
373	206.0	83.2	113.2	162.3	132.2	168.2	21.3	24.2	Si-Me −5.4 (122),
		(182)			(165)		(127)	(124)	Cq 19.8,
									C-Me 24.9 (122)
374	207.3	84.3	113.5	162.3	132.3	168.1	21.4	24.3	Si-Me −2.9 (128),
		(184)			(165)		(131)	(129)	Cq 26.6,
									Cq-Me 17.8 (125),
									CH 33.6 (156),
									CH-Me 18.0 (132)
375	207.8	81.1	113.6	162.3	132.6	168.5	21.8	24.4	CH 13.4 (118),
		(175)			(166)		(134)	(132)	Me 17.8 (125)
385	192.2	136.2	119.5	155.7	131.6	165.3	21.0	23.7	C$_\gamma$ 213.9,
					(160)		(129)	(130)	Me 34.9(132)
387	172.0		144.1	168.3	135.3	189.6	20.9	27.3	
	(166)				(170)		(135)	(131)	
386	204.3	27.1	140.5	166.4	136.5	179.6	26.6	21.6	
	(154)	(129)		163.9	134.4		25.3	(131)	
				(165)	(130)				
				(171)	(127)				

[a] 100.6 MHz, $\delta(\pm 0.1$ ppm) at −120 °C; internal reference: CD$_2$Cl$_2$: $\delta = 53.8$ or NMe$_4^+$: $\delta = 55.7$.
[b] $^1J_{CH}$, coupling constants (± 1.8 Hz) in parentheses.

(388)

The chemical shift of the C$^+$ carbon in the vinyl cations **372–375, 378, 383–385** is dependent not only on the charge density, but is also influenced by a substituent effect on the chemical shift from the different substituents at C$_\beta$. Thus the shift of the C$^+$ carbon is not a good probe to evaluate the effect of a β-silyl group on a positive charge in **372–375**. Similar to sp^2-hybridized benzyl cations the shift of the *para* carbon, however, is a suitable probe to monitor the demand for delocalization of the positive charge into the aromatic ring. Figure 13 shows a comparison of the ^{13}C NMR chemical shift of the *para* carbon in the vinyl cations **372–375, 378, 383–385** and the sp^2-hybridized model cations **386** and **387**[149]. The better the positive charge is stabilized by σ-electron donation of the substituent at the β-carbon as in **372–375** and **383–384** or by π-electron donation of the β-vinylidene substituent as in the allenyl cation **385**, the less is the demand for charge delocalization by π-conjugation with the aromatic ring and thus the *para* carbon becomes less deshielded.

The chemical shift for the *para* carbon in the parent mesitylvinyl cation **378** (180.0 ppm) and in the mesitylethyl cation **386** (179.6 ppm) is similar, indicating that the electronic demand is about the same. Both carbocations are stabilized by σ-bond hyperconjugative interaction with the β-hydrogens as compared with the mesitylmethyl cation **387** (189.6 ppm) which lacks a β-substituent. A pronounced upfield shift of the *para* carbon (10–12 ppm) is observed for the β-silylvinyl cations **372–375** relative to the silyl-free vinyl cations **378, 383** and **384**, indicating a decrease in the electron demand when the β-substituent is changed from β-H or β-alkyl to a β-silyl group. This shows that β-C–Si hyperconjugation is more efficient than β-C–H or β-C–C hyperconjugation.

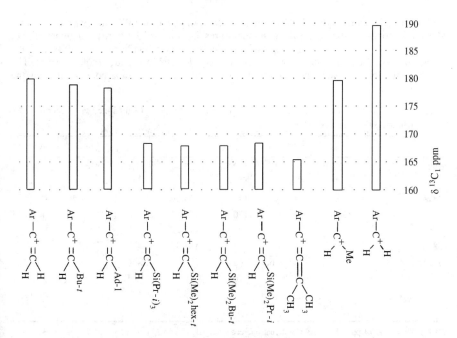

FIGURE 13. Comparison of the *para* carbon NMR chemical shift (ppm) in mesitylvinyl cations **372–375, 378, 383–385** and model cations **386** and **387** (Ar = mesityl).

The close resemblance of the *para* carbon shift of the silyl-substituted cations **372–375** (168–170 ppm) to that of the mesitylallenyl cation **385** (165.9 ppm), which in addition to aryl conjugation is stabilized by β-allyl resonance, shows that hyperconjugative interaction of a β-C—Si σ-bond with the 'vacant' $2p(\pi)$ orbital on C^+ in **372–375** is about as efficient as β-π-conjugation in **385**.

iii. α-Tolyl- and α-phenyl-β-silyl-substituted vinyl cations. Protonation of 1-(*p*-tolyl)-2-triisopropylsilylethyne **388a** leads to formation of the α-(*p*-tolyl)-β-triisopropylsilylvinyl cation **389a** (equation 65)[114,140]. The ^{13}C NMR chemical shifts and $^1J_{CH}$ coupling constants are summarized in Table 16.

Me—⟨⟩—C≡C—Si(CHMe$_2$)$_3$

(**388a**)

(65)

Me—⟨⟩—$\overset{+}{C}$=C$\overset{Si(CHMe_2)_3}{\underset{H}{}}$

(**389a**)

The dimethylisopropylsilyl-substituted 1-*p*-tolylvinyl cation **389b** can be prepared from the corresponding alkyne **388b** at temperatures below $-125\,°$C. At somewhat higher temperature ($-105\,°$C) the cleavage of the β-silyl carbon bond occurs rapidly and the α-(*p*-tolyl)vinyl cation **390** is formed as the only detectable product (equation 66). The β-fragmentation is faster and occurs at lower temperatures compared with that of the

TABLE 16. ^{13}C NMR chemical shifts (δ, ppm) of α-tolyl and α-phenyl-substituted vinyl cations **389a**, **389b**, **390** and **393** in SO$_2$ClF/SO$_2$F$_2^{a,b}$

	C$_\alpha$	C$_\beta$	C1	C$_{ortho}$	C$_{meta}$	C$_{para}$	*p*-Me	Other
390	250.94	77.10	114.24	152.78	135.44	181.11	26.79	
		(177)		(175)	(172)		(134)	
389a	205.84	73.82	111.23	148.11	133.69	167.25	24.45	CH 13.55
		(187)		(170)	(170)		(129)	Me 17.33
					(9.0)			
389b^c	—	—	—	147.89	133.57	—	24.40	
393	205.84	73.82	114.16	148.18	132.51	164.6		CH 14.12
	(4.0)	(192)	(8.4)	(170)	(172)	(6.4)		(116.7)
					(6.3)			Me 17.35
								(126.6)

[a] 100.6 MHz, $\delta(\pm 0.1$ ppm) at $-120\,°$C; internal reference: CD$_2$Cl$_2$ $\delta = 53.8$ or NMe$_4^+$ $\delta = 55.7$.
[b] J_{CH} coupling constants (± 1.8 Hz) in parentheses.
[c] Weak signals.

corresponding α-mesityl-substituted vinyl cation **372**.

$$\text{Me}-\underset{(388b)}{\text{C}_6\text{H}_4}-\text{C}\equiv\text{C}-\underset{\underset{\text{CMe}_3}{|}}{\overset{\overset{\text{CHMe}_2}{|}}{\text{Si}}}-\text{CMe}_3$$

$\downarrow$ FSO$_3$H, −125 °C

$$\text{Me}-\underset{(389b)}{\text{C}_6\text{H}_5^+}-\overset{+}{\text{C}}=\underset{\text{H}}{\overset{\overset{\text{CHMe}_2}{|}}{\text{C}}}\diagdown\underset{\text{CMe}_3}{\overset{\text{Si}-\text{CMe}_3}{}} \qquad (66)$$

$\downarrow$ −FSO$_3$Si(Me$_2$)Pr-i, −105 °C

$$\text{Me}-\underset{(390)}{\text{C}_6\text{H}_5^+}-\overset{+}{\text{C}}=\text{C}\diagdown\underset{\text{H}}{\overset{\text{H}}{}}$$

A comparison of the ^{13}C NMR chemical shift of the *para* carbon in α-(*p*-tolyl)vinyl cations **389a** and **390** and the sp^2 hybridized 1-(*p*-tolyl)ethyl cations with an additional 1-cyclopropyl- (**391**) or 1-methyl substituent (**392**) (Figure 14)[150,151] gives a qualitative measure of the electron demand of various 1-*p*-tolyl- substituted sp^2- and sp-hybridized carbocations.

The demand for π-aryl delocalization of the positive charge is decreasing as the hyperconjugative σ-stabilization of the positive charge by the β-substituents is increasing from β-H in **390** and **392** to a β-silyl group in **389a** and the strained cyclopropane C−C bonds in **391**.

The α-phenyl-β-triisopropylsilylvinyl cation **393** can be prepared by protonation of the corresponding alkyne **394** (equation 67). The ^{13}C NMR spectroscopic data are given in Table 16[114,140].

Attempts to protonate 1-phenyl-2-(trialkylsilyl)alkynes **395** with alkyl groups smaller than isopropyl, in order to prepare 1-phenylvinyl cations **396**, were unsuccessful (equation 68)[114].

The ^{13}C NMR chemical shifts of the *para* carbon for various phenyl-substituted sp^2-and sp-hybridized carbocations (Figure 15) indicate that the demand for π-aryl delocalization of the positive charge for the β-silyl-substituted vinyl cation **393** is lower compared with that in α-phenylethyl (**78**), α-methyl-α-phenylpropyl (**397**) and cumyl (**292**) cations[150,153]. The stabilizing effect of the β-silyl group in **393** is comparable to that of the cyclopropyl substituent in 1-cyclopropylbenzyl cation (**398**). The *para* carbon shift in the β-σ-silyl

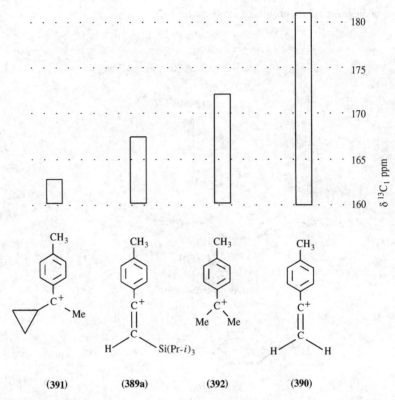

FIGURE 14. Comparison of the *para* carbon NMR chemical shift (ppm) in α-(*p*-tolyl)vinyl cations **389a** and **390** and sp²-hybridized *p*-tolyl model cations **391** and **392**.

stabilized vinyl cation **393** is at about 5 and 4 ppm lower field compared with the β-π-resonance-stabilized phenylallenyl cation **399**, or the sp²-hybridized carbocation **400**, respectively.

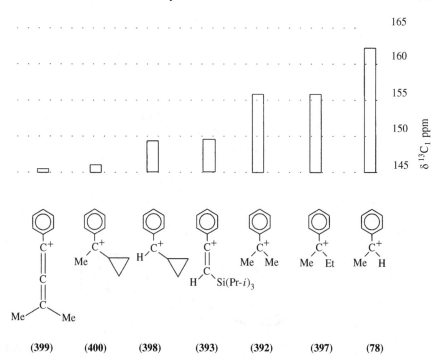

FIGURE 15. Comparison of the ^{13}C NMR chemical shift (ppm) of the *para* carbon in phenylvinyl cations **393** and **399** and selected sp^2-hybridized phenyl cations.

iv. α-Anisyl-β-silyl-substituted vinyl cations. The α-(*p*-anisyl)-β-triisopropylsilylvinyl cation **401** is accessible by protonation of the corresponding alkyne **402** at very low temperatures (equation 69)[154]. The β-silyl group in **401** is readily cleaved, and at temperatures above −115 °C complete conversion to 1-(*p*-anisyl)vinyl cation (**403**) takes place in a few minutes (equation 70). The ^{1}H NMR and ^{13}C NMR spectra of cation **401** are shown in Figures 16 and 17.

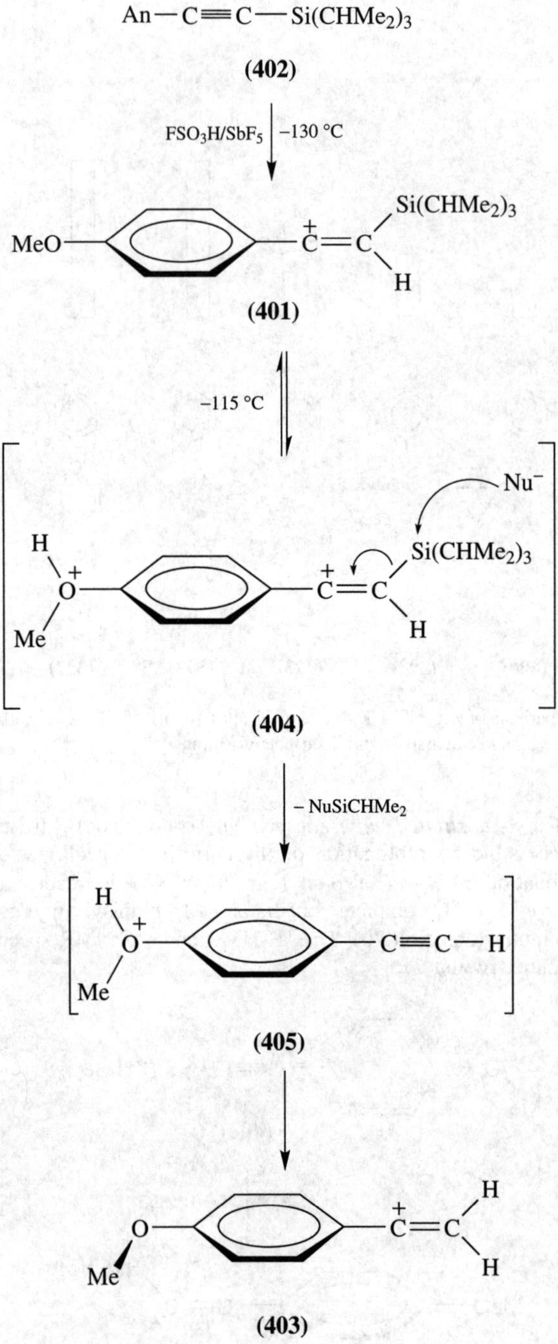

$$\text{An}-\text{C}\equiv\text{C}-\text{Si(CHMe}_2)_3$$

(402)

↓ FSO$_3$H/SbF$_5$ | −130 °C

MeO–C$_6$H$_4$–C$^+$=C(Si(CHMe$_2$)$_3$)H

(401)

⇌ −115 °C

[(H)(Me)O$^+$–C$_6$H$_4$–C$^+$=C(Si(CHMe$_2$)$_3$)H ←Nu$^-$]

(404)

(70)

↓ − NuSiCHMe$_2$

[(H)(Me)O$^+$–C$_6$H$_4$–C≡C–H]

(405)

↓

MeO–C$_6$H$_4$–C$^+$=CH$_2$

(403)

12. Silyl-substituted carbocations

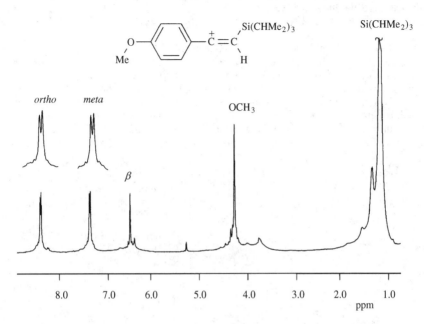

FIGURE 16. 400-MHz ^{1}H NMR spectrum of cation **401** in SO$_2$ClF/SO$_2$F$_2$ at $-120\,^\circ$C

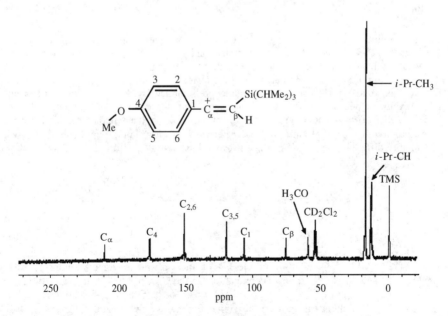

FIGURE 17. 63-MHz ^{13}C NMR spectrum of cation **401** in SO$_2$ClF/SO$_2$F$_2$ at $-124\,^\circ$C.

The C_β–Si bond fragmentation in cation **401** occurs already at temperatures where the corresponding 1-mesityl-, 1-(p-tolyl)- and 1-phenylvinyl-β-(triisopropyl)silyl-substituted vinyl cations **375**, **389a** and **393** are stable. The facile formation of the 1-(p-anisyl)vinyl cation **403** can be ascribed to a second protonation of **401** at the oxygen of the anisyl group.

At $-130\,°C$ the highly sterically hindered alkyne **402** is protonated at C_β to form cation **401**. When the temperature is raised above about $-115\,°C$, the large excess of superacid leads to a reversible protonation at the oxygen of the methoxy group to form a doubly charged carbo-oxonium dication **404**. This dication is not observable, presumably since only a minor fraction of it is in equilibrium with **401** (equation 70). The protonated p-methoxy group in **404** is no longer a good electron-releasing group. Therefore, in the protonated **406** resonance hybrids like **406D** and **406E**, which in **401** contribute significantly to the delocalization of the positive charge, are of minor importance for **404**.

(406A)　(406B)　(406C)　(406D)　(406E)

As the delocalization of the charge at C_α by π-conjugation into the aromatic ring is lower in **404**, the demand for charge delocalization by β-C–Si hyperconjugation (cf **407A** ↔ **407B**) is increased. The formal no-bond hyperconjugative resonance structure **407B** gains more importance and the partial positive charge at silicon is increased. The silicon therefore becomes more susceptible to nucleophilic attack by the gegen ions which led to its expulsion via an $S_N2(Si)$ transition state. The intermediate O-protonated alkyne **405** formed *in situ* is converted in the superacid media to the silicon-free vinyl cation **403** (equation 70). The ^1H and ^{13}C NMR spectra of the 1-(p-anisyl)vinyl cation **403** are remarkably different from those of the corresponding β-silyl-substituted vinyl cation **401**.

(407A)　(407B)

The ^{13}C NMR chemical shift (Table 17) of the positive charged carbons in **401** (209.8 ppm) and **403** (239.7 ppm) is comparable to the C^+ carbon shift in 1-mesitylvinyl cations with analogous β-substituents (207.8 ppm for **375** and 238.5 ppm for **378**, respectively). The chemical shift of the signal for the sp^2-hybridized β-carbon in **401** (76.0 ppm) and **403** (78.4 ppm) is in accord with that in other vinyl cations and with IGLO

TABLE 17. ^{1}H and ^{13}C NMR chemical shifts (δ, ppm) for cations **401** and **403** in SO$_2$ClF/SO$_2$F$_2$[a]

	C$_\alpha$	C$_\beta$	C1	C2, C6		C3, C5		C$_{para}$	OMe
401[b]	209.8	75.9	106.8	151.3		119.8		177.0	59.9
		(179)		(172)		(172)			(152)
		[6.51]		[8.41]		[7.37]			[4.29]
403[c]	239.7	78.4	108.9	156.1	150.9	118.0	124.9	184.4	60.7
		(176)		(174)	(173)	(174)	(173)		(150)
		[6.42]		[8.51]	[8.26]	[7.30]	[7.46]		[4.37]

[a]400-MHz ^{1}H NMR data in square brackets, δ ($\pm$0.01 ppm) at $-120\,^\circ$C, internal reference: CHDCl$_2$: $\delta = 5.32$ ppm.
[b]63-MHz ^{13}C NMR δ ($\pm$0.05 ppm) at $-124\,^\circ$C, $^1J_{CH}$ coupling constants ($\pm$1.8 Hz) are given in parentheses.
[c]100.6-MHz ^{13}C NMR at $-102\,^\circ$C.

calculations of model vinyl cations. A shielding of 4–5 ppm is observed as compared with the corresponding C$_\beta$ signals of 1-mesitylvinyl cations **375** (81.1 ppm) and **378** (82.3 ppm). Due to the sp-hybridization of the C$^+$-carbon in the vinyl cations **401** and **403** the aromatic C1 position shows an upfield shift of about 25–30 ppm compared with 1-(*p*-anisyl)-substituted carbocations with an sp^2-hybridized C$^+$-carbon.

As in other benzylic-type carbocations, the chemical shift of the *para* carbon is a probe to the electronic demand of the carbocation centre and it can be used to evaluate the effect of a β-silyl group on the positive charge.

The signal for the *para* carbon in the β-silyl-substituted cation **401** (177.0 ppm) appears about 7 ppm upfield compared with the signal in the silicon-free cation **403** (184.4 ppm), indicating a lower demand for charge delocalization into the aryl ring when the β-substituent is changed from β-H to a β-silyl group. This shows that the β-silyl stabilization effect is operative even in highly conjugatively stabilized carbocations with α-anisyl substituents. The *para* carbon shift in **401** may be compared with those in 1-phenyl (**408**) or 1-cyclopropyl (**409**) substituted *p*-anisylmethyl (R = H) or 1-(*p*-anisyl)ethyl (R = Me) cations (176–181 ppm)[155]. The similar shifts suggest that β-silyl stabilization in the vinyl cation **401** is about as efficient in dispersing the positive charge as 2p(π)-conjugation of an additional phenyl ring in **408** or a σ-C–C hyperconjugation of an additional cyclopropyl group in **409**.

(408)

(409)

Whereas the NMR spectra for the 1-(*p*-anisyl)vinyl cation **403** show magnetically nonequivalent signals for the C2 and C6 *ortho* and C3 and C5 *meta* carbon and proton signals, these positions remain equivalent in the ^{1}H and ^{13}C NMR spectra of the β-silyl-substituted cation, even at the lowest temperatures.

As in other 1-(*p*-anisyl)-substituted carbocations, in the 1-(*p*-anisyl)vinyl cations **401** and **403** there exists a torsional barrier for a methoxy group rotation around the phenyl–C4–oxygen bond which is due to delocalization of the positive charge to the methoxy group. This is illustrated by the contribution of the valence bond resonance structure **406E**. This rotational barrier is lower than the other barrier around the C1–C$^+$ bond which is due to contributions of resonance structures **406B–406E**.

The demand for delocalizing positive charge into the aryl moiety is higher in **403** than in **401**, as is evident from the *para* carbon chemical shifts (Table 15). In **403** the enhanced participation of the *p*-methoxy group in delocalizing the positive charge results in higher double bond character of the phenyl–C4–oxygen bond as compared with **401**. The free energy of activation ($\Delta G^{\#} = 9.0 \pm 1$ kcal mol^{-1} at $-100\,°$C) for the dynamic process was determined from kinetic line broadening in the ^{13}C NMR spectra. The fact that the rotation around the C4–O bond in **403** is slow, but is fast in **401**, clearly demonstrates that **401** is stabilized by the β-silyl effect and thus needs less π-aryl stabilization compared with **403**.

Quantum chemical *ab initio* calculations were performed to reveal additional details of the structures and charge stabilization of 1-(*p*-anisyl)vinyl cations. The 1-(*p*-anisyl)vinyl cation **403** and the model cation structures **410** and **411** with β-SiH$_3$ and β-CH$_3$ groups, respectively, were calculated along with the corresponding transition state structures **403-TS**, **410-TS** and **411-TS** for the rotation of the *p*-methoxy group around the C4–oxygen bond.

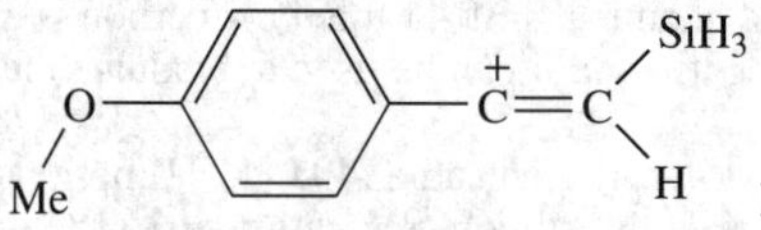

(410) 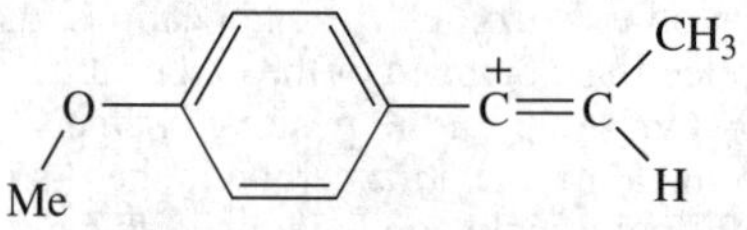   (411)

The geometries were calculated at the HF/6-31G level. Figure 18 shows the structure of vinyl cation **403** and the transition state structure **403-TS**. Computational experience with more electron-deficient, smaller vinyl cation structures lacking π-conjugative stabilizing α-substituents, indicate that at higher level of calculations, which is feasible for the smaller structures, the effects of σ-stabilization by β-σ-C–H, β-σ-C–C and β-σ-C–Si bonds are more pronounced while the relative order for hyperconjugation is not changed. The α-arylvinyl cations **403**, **410** and **411** have an inherently lower demand for β-σ hyperconjugative stabilization than vinyl cations lacking α-π-conjugating substituents, and the

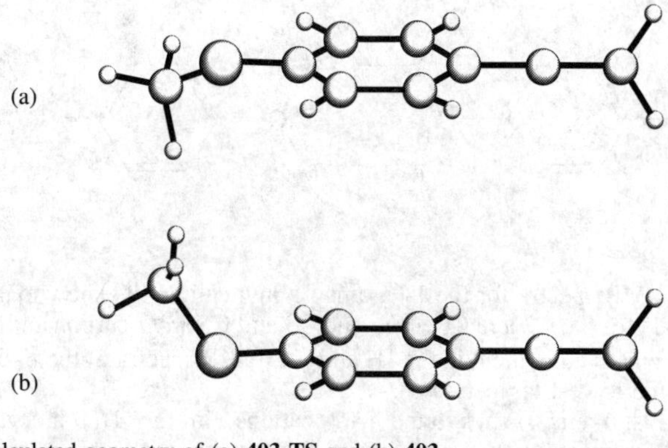

FIGURE 18. Calculated geometry of (a) **403-TS** and (b) **403**

level of the basis set used for the *ab initio* calculation is thus not as crucial. Model calculations for **403** using the 6-31G(d) basis set and for **410** using the 6-31G basis set with inclusion of d orbitals at silicon show that the data obtained at the 6-31G level are qualitatively reliable and thus adequate to allow comparison of a consistent set of calculated data with the experimental NMR results.

In the optimized geometry for the vinyl cations **403**, **410** and **411** the β-C−R bonds are confined to the same plane with the formally vacant $2p(C^+)$ orbital, allowing optimum overlap of the latter with the π-orbitals of the aromatic system as well as with the β-C−R σ-bond.

In the most stable conformation for the 1-(*p*-anisyl)-substituted vinyl cations **403**, **410** and **411** the O−methyl bond of the C4−OCH$_3$ group is calculated to be coplanar with the plane of the aryl ring (Figure 18), as expected due to the contributions of resonance forms **406E**. The calculated geometrical data reveal that charge dispersal to the aryl ring and to the *p*-methoxy group becomes more important as the hyperconjugative interaction of the formally vacant $2p(C^+)$ orbital with the β-σ-bond is decreasing in the order **410** > **411** > **403**, where the β-substituent changes from β-SiH$_3$ to β-CH$_3$ and β-H. The C4−OMe and C1−C$^+$ bond lengths decrease in the same order. Also, the bond length alternation for the aromatic C−C bonds, i.e. shorter C2,C3 and C5,C6 bonds and elongated C1,C2/C1,C6 and C3,C4/C4,C5 bonds, due to increasing importance of quinonoid resonance structures of the type **406B−406E**, becomes more pronounced.

The calculated data confirm the interpretation of the NMR data that, even in highly π-stabilized α-anisyl carbocations, β-σ-bond hyperconjugation contributes to the dispersal of positive charge. The computational data also confirm the relative order of the hyperconjugative electron-donor ability of β-σ bonds to be C−Si > C−C > C−H. The energy differences between the transition state structures and the most stable structures were calculated to be 8.3, 7.9 and 7.4 kcal mol^{-1} for **403**, **411** and **410**, respectively, and they are a measure of the barrier for methoxy group rotation around the C4−oxygen bond. The order obtained is in agreement with the other calculated data, as well as with the observed chemical shifts for **401** and **403**.

c. The 1-bis(trimethylsilyl)methyl-2-bis(trimethylsilyl)ethenyl cation. Protonation of tetrakis(trimethylsilyl)allene **412** with HSO$_3$F/SbF$_5$ gives the 1-[bis(trimethylsilyl)]methyl-2-bis(trimethylsilyl)ethenyl cation **413**[156]. The isomeric allyl cation **414** is not formed (equation 71).

Ph) with HSO_3F/SbF_5, which yields the corresponding allyl cations **416** (equation 72)[157].

$$\underset{\underset{H_3C}{\overset{R}{\diagup}}}{\overset{\overset{R}{\diagdown}}{C}}=C=\underset{\underset{R}{\diagdown}}{\overset{\overset{CH_3}{\diagup}}{C}} \xrightarrow{HSO_3F} \underset{\underset{H_3C}{\overset{R}{\diagup}}}{\overset{\overset{R}{\diagdown}}{C}}\!\!=\!\!\underset{\underset{R}{\diagdown}}{\overset{\overset{H}{\diagup}}{C}}\!\!\overset{+}{\diagdown}C\!-\!CH_3 \qquad (72)$$

(415)   R = H, CH_3, Ph   **(416)**

The different reaction course followed by the silyl- and alkyl-substituted allenes is fully consistent with the hyperconjugation model, which predicts that a β-silyl group stabilizes carbenium ions better than a β-alkyl group, and with previous conclusions that α-silyl substitution in carbocations is destabilizing relative to α-methyl substitution. The larger space requirements of the trimethylsilyl groups compared with methyl groups may also contribute to the kinetic stability of **413**.

At $-100\,°C$ the ^{13}C NMR spectrum of **413** shows five signals (Figure 19, Table 18). The signal of the C^+-carbon C_α in **413** appears at 208.7 ppm. This value is comparable with the chemical shift in vinyl cations stabilized by $2p$-π-conjugation or by hyperconjugation with cyclopropyl substituents. The signal of the C^+-carbon in simple α-alkyl-substituted vinyl cations $H_2C=C^+-CH_3$ and $H_2C=C^+-CH_2CH_3$ have been estimated by IGLO chemical shift calculations to be about 200 ppm more deshielded.

The strong shielding effect for C_α in the tetrasilyl-substituted vinyl cation **413** can be attributed mainly to the hyperconjugative charge dispersal by the silyl groups. At

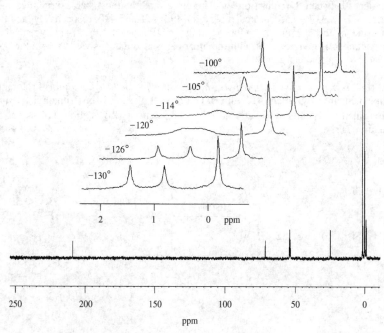

FIGURE 19. 100.6-MHz ^{13}C NMR spectrum of **413** ($-100\,°C$, SO_2ClF/SO_2F_2, reference CD_2Cl_2, $\delta = 53.80$ ppm), and an expanded region (2–0.5 ppm) showing the temperature dependence of the $(CH_3)_3Si$ signals at -130 to $-100\,°C$)

TABLE 18. ^{13}C NMR chemical shifts (δ, ppm) of **413** in SO$_2$ClF/SO$_2$F$_2{}^a$

$T(^\circ C)$	C$_\alpha$	C$_\beta$	CH	β-C–SiMe$_3$	CH–SiMe$_3$
−100	208.73	71.39	25.00	1.59	0.14
−136	208.74	70.58	25.27	1.35/0.70	−0.30

^{a} 100.62 MHz, internal standard CD$_2$Cl$_2$, $\delta = 53.80$ ppm.

temperatures $< -100\,^\circ$C the low field peak of the two signals for the four trimethylsilyl groups broadens and below the coalescence temperature of *ca* $-114\,^\circ$C it splits into two signals at 1.35 and 0.75 ppm. This peak is assigned to the trimethylsilyl groups at the methine carbon which is connected to C$_\alpha$ by a single bond. The observed dynamic process is due to rotation around the C$^+$–CH(SiMe$_3$)$_2$ bond which renders the two SiMe$_3$ groups at the methine carbon non-equivalent. The torsional barrier, $\Delta G^{\#} = 7.7$ kcal mol^{-1} at $-114\,^\circ$C, is determined by the dihedral angle dependence of the hyperconjugative stabilization of the positive charge by the β-CH-silyl groups.

Whereas the SiMe$_3$ groups at the double bonded β-carbon are always in an optimal position for hyperconjugation, the SiMe$_3$ groups at the methine carbon are conformationally mobile. At low temperatures the cation prefers a conformation in which the hyperconjugation between the empty 2p(C$^+$) orbital and the C–Si bond is maximized. The non-equivalence of the two SiMe$_3$ groups at the methine carbon indicate that the conformation in which one silyl group is fully conjugated with the empty orbital and the other has a dihedral angle of about 60° is the most favourable conformation. This conformation would also render the SiMe$_3$ groups at the sp^2 C$_\beta$ carbon non-equivalent. Any ^{13}C NMR chemical shift difference for these groups is likely to be small and was not observed.

Theoretical calculations performed on smaller model cations with SiH$_3$ groups confirm the dihedral dependence of the β-silyl effect. In disilyl-substituted model cation H$_2$C=C$^+$–CH(SiH$_3$)$_2$ **417** the difference between conformation **417b** where each of the C–Si bonds forms a dihedral angle, θ, of *ca* 30° with the empty 2p(C$^+$) orbital, and conformation **417a** with different dihedral angles θ of 0° and 60°, is small, about 3 kcal mol^{-1} [HF/6-31G(d)] in favour of **417b**. A detailed analysis of the computational results leads to the prediction that at higher levels of calculations and with inclusion of electron correlation, conformation **417a** is likely to be the more stable conformation of **417**. The model calculations reveal that the silyl groups involved in hyperconjugation show strong geometric distortions towards bridging. The Si–CH–C$^+$ bond angle is smaller (102.4°) and the C–Si bond is elongated [r(CH–Si) = 2.03 Å] for the SiH$_3$ group in **417a** which is aligned for optimal interaction with the empty 2p orbital, compared with the other silyl group at $\theta = 60°$ for which the Si–CH–C$_\alpha$ bond angle is 117.1° and the r(CH–Si) bond length is 1.96 Å. According to calculations [3-21G(d)] the rotational barrier around the C$_\alpha$–CH-bond of **417** is 5.9 kcal mol^{-1}, in very good agreement with the experimental value of 7.5 kcal mol^{-1}. The theoretical calculations are also consistent with the experimental observation at $-130\,^\circ$C in indicating the appearance of two signals for the CH(SiMe$_3$)$_2$ group. Similar, though less pronounced structural distortions have been observed in tricoordinated carbocations stabilized through C–C hyperconjugation[158].

(**417a**)   (**417b**)

C. Silyl-substituted Hypercoordinated Carbocations

1. The α- and γ-silyl effect in bicyclobutonium ions

The structure and dynamics of the cyclopropylmethyl/cyclobutyl cation system $[C_4H_7]^+$ and its 1-methyl-substituted analogue $[C_4H_6CH_3]^+$ have been a subject of controversy for many years[159,160]. Depending on the substitution pattern, these types of carbocations are prone to undergo facile rearrangements. A three-fold degenerate bicyclobutonium/cyclopropylmethyl cation rearrangement (equation 73) would render the CH_2 carbons equivalent.

(73)

The rate of another dynamic process, a conformational ring inversion, depends in particular on the nature of the substituent R at the α-carbon. The geminal *exo* and *endo* methylene hydrogens are interchanged by opening of the bridging bond and planarization

12. Silyl-substituted carbocations

to a cyclobutyl cation intermediate, followed by closure of the bridging bond to a mirror image bicyclobutonium ion (equation 74).

(74)

Depending on the rates of these processes and related to the NMR time scale, only an averaged signal for the methylene carbons and/or an averaged signal for the geminal *exo/endo* methylene hydrogens are observed in the ^{13}C and ^{1}H NMR spectrum, respectively.

According to contemporary high field NMR spectroscopic measurements, investigations of equilibrium isotope effects on NMR spectra and quantum chemical calculations which include electron correlation, the parent system $[C_4H_7]^+$ is now best described as a degenerate set of rapidly interconverting bicyclobutonium ions **418** with minor contributions from another degenerate set of rapidly equilibrating cyclopropylmethyl cations **419** which are only marginally higher in energy (by <1 kcal mol^{-1}) than **418**[161–165].

For the analogous 1-methyl-substituted cation $[C_4H_6CH_3]^+$, recent experimental investigations and quantum chemical *ab initio* calculations agree that the dynamics of this cation system can be adequately described by considering only one degenerate set of cations, which have the hypercoordinated puckered methylbicyclobutonium ion structure **420** without contributions from a degenerate set of (1′-methylcyclopropyl)methyl cation structures **421**[7,166,167].

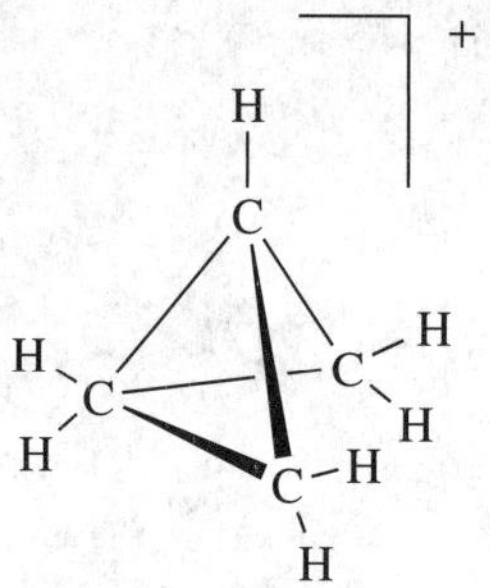

(418)

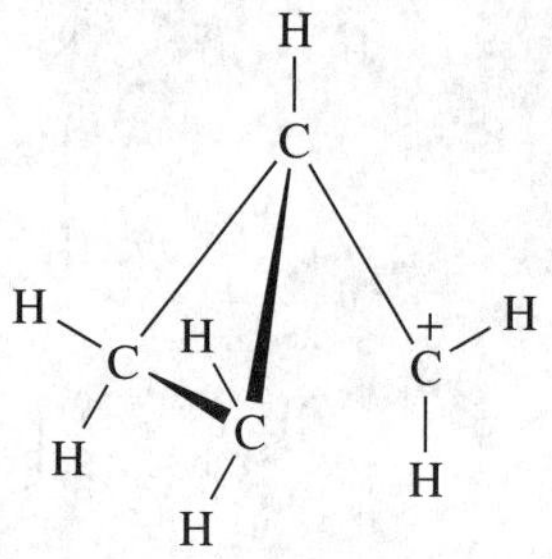

(419)

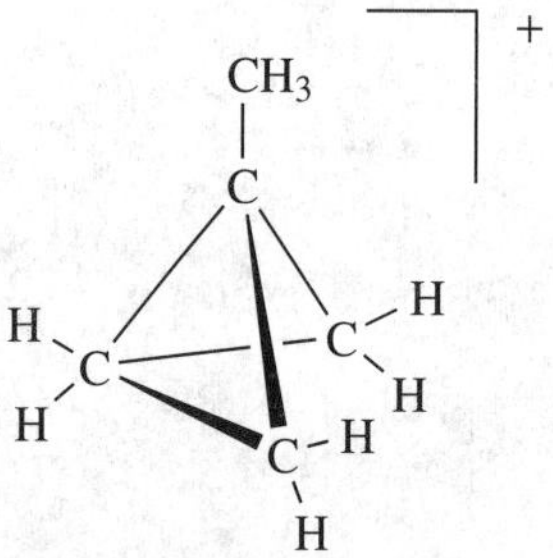

(420)

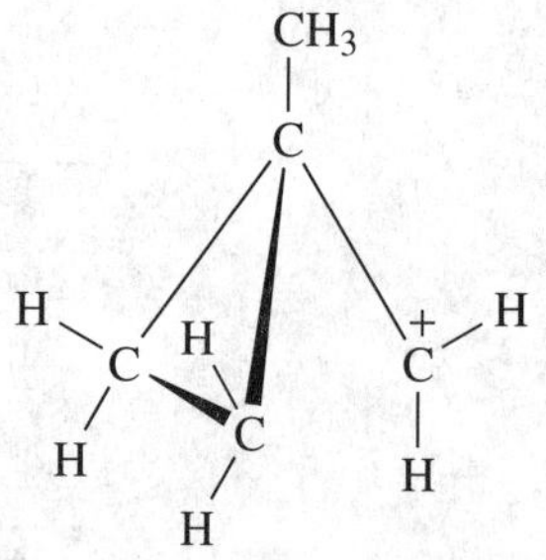

(421)

Recently, trialkylylsilyl-substituted bicyclobutonium ions have been generated in superacid solution and were characterized by NMR spectroscopy.

a. 1-(Trimethylsilyl)bicyclobutonium ion. The 1-(trimethylsilyl)bicyclobutonium ion **422** is obtained from (1′-trimethylsilylcyclopropyl)methanol **423** by reaction with SbF$_5$

(equation 75)[168].

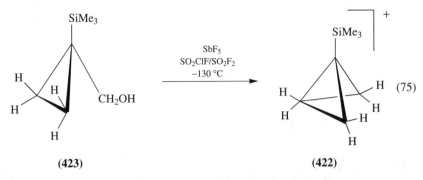

(423)   (422)   (75)

Analogous cations **424** with the alkyl groups R at silicon larger than methyl can be prepared more easily. The NMR spectroscopic data of **422** in SO_2ClF/SO_2F_2 solution at $-128\,°C$ is in accord with a bridged puckered bicyclobutonium structure undergoing a three-fold rapid degenerate rearrangement (equation 76, R = $SiMe_3$) that renders the two β- and one γ-methylene groups equivalent, thus leading to one averaged ^{13}C NMR signal for the CH_2 groups at 48.9 ppm (Figure 20).

(**424**)

(76)

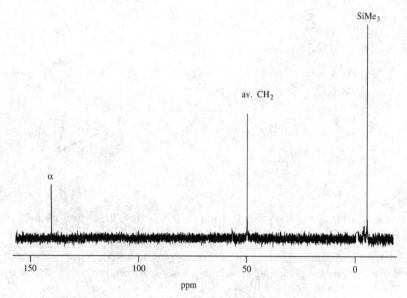

FIGURE 20. ^{13}C NMR spectrum of 1-(trimethylsilyl)bicyclobutonium ion **422** at $-125\,^\circ$C

Conformational ring inversion for cation **422** (equation 74, R = SiMe$_3$) is slow, so that two separate signals for the three averaged *endo*-CH$_2$ (4.04 ppm) and three averaged *exo*-CH$_2$ hydrogens (3.24 ppm) are observed. The deuterium equilibrium isotope effects for *exo*- and *endo*-CHD-labeled cations are different in sign and magnitude and can be explained by assuming different *endo* and *exo* C–H bond force constants at the pentacoordinated carbon. NMR chemical shift calculations for MP2/6-31G(d) optimized geometries of the 1-silylbicyclobutonium ion **425** and the (1′-silylcyclopropyl)methyl cation **426** were performed by the GIAO-SCF and GIAO-MP2 method (Figure 21).

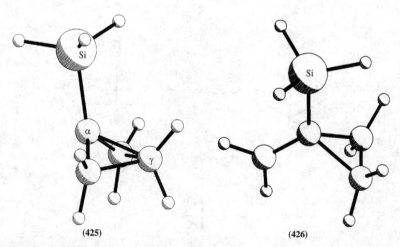

FIGURE 21. Calculated structures of **425** and **426**

The experimental shifts are satisfactorily reproduced by GIAO-MP2//tzp/dz calculated shifts for the 1-silylbicyclobutonium structure **425**. This good agreement between theory and experiment supports a fully degenerate set of interconverting 1-trimethylsilylbicyclobutonium ions for **422** and **424** (R = Me$_2$Bu-t), respectively, and excludes contributions from the other isomers, the 1'-trialkylsilylcyclopropylmethyl cations **427**, to the observed equilibrium process.

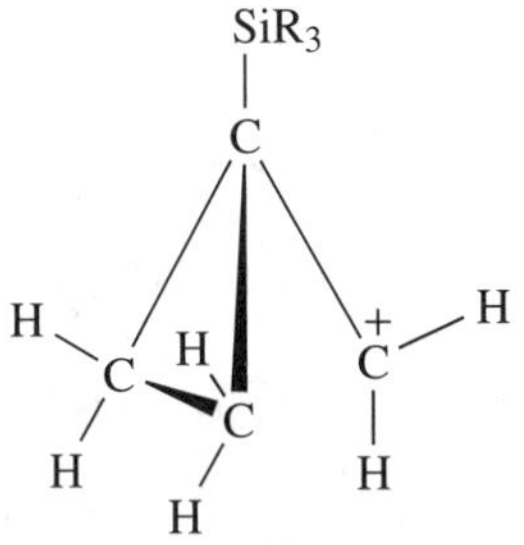

(427)

The geometric and electronic properties of the 1-trialkylsilylbicyclobutonium cations, as concluded from a comparison of the experimental data for **422** and **424** and the calculated data for **425**, are intermediate between that of the parent bicyclobutonium ion **418** and that of the 1-methylbicyclobutonium ion **420**.

b. 3-endo-(Trialkylsilyl)bicyclobutonium ions. The 3-*endo*-(t-butyldimethylsilyl)bicyclobutonium ion **428** is accessible from [1'-(t-butyldimethylsilyl)cyclopropyl]methanol **429** (equation 77)[169].

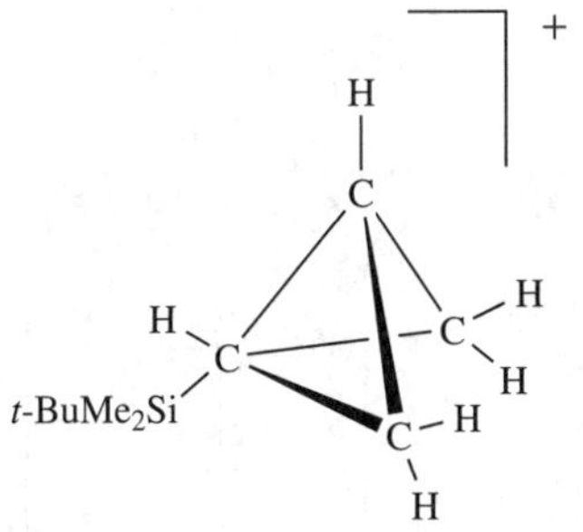

(428)

Reaction of [1'-(t-butyldimethylsilyl)cyclopropyl]methanol **429** with SbF$_5$ at −130 °C initially leads to formation of the 1-(t-butyldimethylsilyl)bicyclobutonium ion **430**. At −115 °C cation **430** is completely converted to the 3-*endo*-(t-butyldimethylsilyl)bicyclobutonium ion **428** in a few minutes.

The 3-*endo*-(trimethylsilyl)bicyclobutonium ion **431** is also obtained directly from a four-membered-ring progenitor, i.e. 3-(trimethylsilyl)cyclobutyl chloride **432**, by reaction with SbF$_5$ at $-130\,°C$ (equation 78).

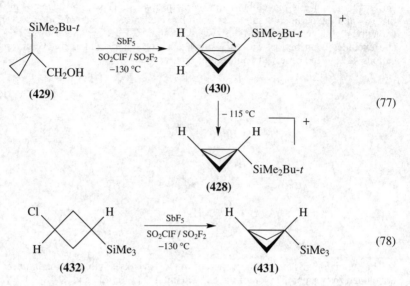

(77)

(78)

The ^{13}C NMR spectrum of the 3-*endo*-(*t*-butyldimethylsilyl)bicyclobutonium ion **428** in SO$_2$ClF/SO$_2$F$_2$/CCl$_3$F solution at $-120\,°C$ (Figure 22) shows three distinct signals for C$_\alpha$ (99.6 ppm), C$_\beta$/C$_{\beta'}$ (66.3 ppm) and C$_\gamma$ (-21.0 ppm).

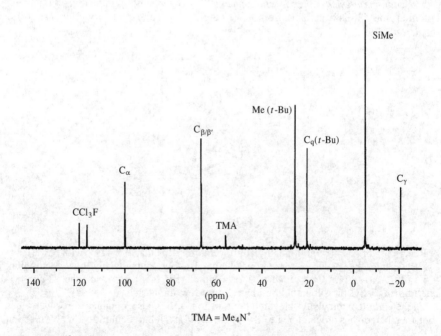

FIGURE 22. ^{13}C NMR spectrum of **428** in SO$_2$ClF/SO$_2$F$_2$/CCl$_3$F at $-120\,°C$

The ^{1}H NMR spectrum shows resonances for H$_\alpha$ (5.95 ppm), H$_\beta$/H$_{\beta'}$ exo (4.01 ppm), H$_\beta$/H$_{\beta'}$ endo (3.12 ppm) and H$_\gamma$ (0.30 ppm). The NMR data are in accord with a bridged, puckered bicyclobutonium ion structure that is static on the NMR time scale. The ^{29}Si NMR chemical shift of 43.1 ppm for ion **428** indicates that the silicon is involved in stabilization of the positive charge. The stabilization occurs by shifting electron density from the C$_\gamma$−Si σ-bond across the bridging bond to the formal carbenium carbon C$_\alpha$. This γ-silyl- type of interaction may be termed silicon homohyperconjugation.

Quantum chemical calculations at MP2/6-31G(d) level for SiH$_3$-substituted model structures predict that only the 3-endo-silylbicyclobutonium cation structure **433** is at a minimum on the potential energy surface, while the 3-exo cation structure exo-**433** has one imaginary frequency and thus corresponds to a transition state. The calculated structures of endo- and exo-**433** are given in Figure 23.

For the model bicyclobutonium ion structure **433** with the SiH$_3$ substituent at the γ-endo position, the NMR chemical shifts calculated at the GIAO-MP2//tzp/dz level and the coupling constants calculated with an SOS-DFT approach are in good agreement with experimental results for **428** and **431**. The $^3J_{HH}$ coupling between H$_\alpha$ and H$_\gamma$, measured experimentally as 5.5 Hz, is calculated to be 5.9 Hz in the endo-SiH$_3$-substituted cation **433** but only 1.2 Hz for the exo-SiH$_3$-substituted cation exo-**433**. The observation of a cross bridge $^3J_{HH}$ coupling is direct proof for the bridged structure and the magnitude of the coupling constant confirms the endo arrangement of the silyl substituent in **428** and **431**.

The calculated C$_\alpha$−C$_\gamma$ bridging distance for the 3-endo ion **433** (1.642 Å) is shorter by 0.012 Å compared with the parent bicyclobutonium ion **418**. This indicates that the homohyperconjugative stabilization of the positive charge by the γ-endo C−Si σ-bond is more efficient than the stabilization by a γ-endo C−H σ-bond. In analogy with a β-silicon substituent, γ-silyl groups with a suitable stereoelectronic arrangement are superior to hydrogen in stabilizing the positive charge in carbocations.

2. Silanorbornyl cation

a. The 6,6-dimethyl-5-neopentyl-6-sila-2-norbornyl cation. This cation, **434**, was prepared as a tetrakis(pentafluorophenyl)borate salt by a hydride transfer reaction of 3-(3-cyclopentenyl)-2,5,5-trimethyl-2-silahexane **435** with trityl tetrakis(pentafluoro-

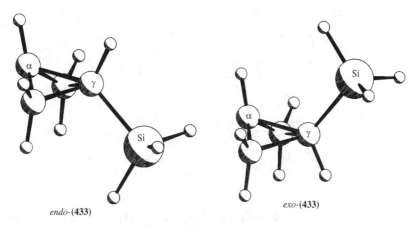

FIGURE 23. Calculated structures [MP2/6-31G(d)] of endo- and exo- **433**.

phenyl)borate in toluene (equation 79)[170].

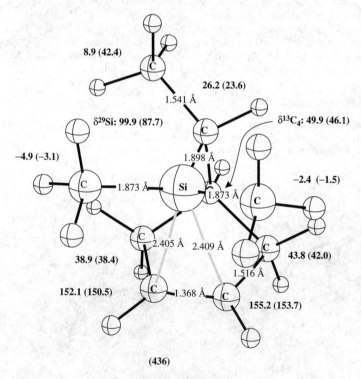

The single peak observed in the ^{29}Si NMR spectrum in toluene at 87.7 ppm is in the ^{29}Si chemical shift range expected for arene–silyl cation complexes (experimental or computed structures). However, the deshielding of the C1 and C2 carbons at 153.74 ($^1J_{CH}$ = 174 Hz) and 150.52 ($^1J_{CH}$ = 174 Hz) ppm compared with the corresponding signals in the starting silane at 130.57 ($^1J_{CH}$ = 156 Hz) and 130.41 ($^1J_{CH}$ = 156 Hz) provides evidence for an intramolecular π-complexation of the positively charged silicon. Quantum chemical calculations on ions **436** and **437** at the B3LYP/6-31G(d) level support the interpretation of the experimental findings. The C_s symmetric model cation structure, 5,6,6-trimethyl-6-sila-2-norbornyl cation **436**, is a local minimum (Figure 24).

FIGURE 24. Calculated structure of ion **436** at B3LYP/6-31G(d). Bond lengths in Å, calculated chemical shifts at IGLO-DFPT/BIII in bold. Experimental chemical shifts for **434** in parenthesis

The deshielding of C1/C2 carbons and the $\delta^{13}C$ values of the methyl groups at silicon are well reproduced by the calculations of NMR chemical shifts in a model structure **437**.

(437)

Further support comes from calculation of $^1J_{CH}$ spin–spin coupling constants at IGLO-DFPT/BIII level. The data were interpreted as giving strong evidence for intramolecularly π-stabilized silanorbornyl cation structure lacking coordination to solvent or counterion. The species can also be regarded as an almost symmetrically bridged β-silyl carbocation with siliconium ion character.

VI. CONCLUSIONS

The data presented in this chapter have resulted from quantum chemical *ab initio* calculations and various experimental studies performed in the gas phase, in nucleophilic solvents, in superacid media and in the soild state. Continued evolution of the techniques to study reactive intermediates and the advance of high-speed computers and sophisticated *ab initio* programs have yield rich rewards. The information gained provides very useful insights into the energetic and structural details of silyl substituted carbocations and the kinetics of their reactions. In particular much has been learned through these studies on the many aspects of σ-π-interactions of silicon with a positive charge in carbocationic intermediates, transition states and stable ions. The principles of the silicon interaction in carbocations are now quite well understood. The stabilizing effect of β- or γ-silyl groups have been used to synthesize hitherto elusive carbocations from silyl substituted precursors. The combined application of experimental and theoretical methods has been especially fruitful in the study of silicon substituted carbocations.

VII. ACKNOWLEDGEMENTS

H.-U. S. gratefully acknowledges financial support by the Deutsche Forschungsgemeinschaft (DFG) and the Fonds der Chemischen Industrie. T. M. thanks the Fonds der Chemischen Industrie for a Liebig scholarship.

VIII. REFERENCES

1. For a recent review see: Y. Apeloig, in *The Chemistry of Organic Silicon Compounds* (Eds. S. Patai and Z. Rappoport), Vol. 1, Chap. 2, Wiley, Chichester, 1989, p. 87.
2. J. A. Pople, Y. Apeloig and P. v. R. Schleyer, *Chem. Phys. Lett.*, **85**, 489 (1982).
3. A. E. Ketvirtis, D. K. Bohme and A. C. Hopkinson, *Organometallics*, **14**, 347 (1995).
4. A. E. Ketvirtis, D. K. Bohme and A. C. Hopkinson, *J. Phys. Chem.*, **98**, 13225 (1994).
5. C. Wierschke, J. Chandrasekhar and W. J. Jorgensen, *J. Am. Chem. Soc.*, **107**, 1496 (1985).
6. Y. Apeloig and A. Stanger, *J. Org. Chem.*, **47**, 1462 (1982); see also addition in *J. Org. Chem.*, **48**, 5413 (1983).
7. P. Buzek, Ph. D. Thesis, Erlangen, 1993.
8. P. J. Stang, M. Ladika, Y. Apeloig, A. Stanger, M. D. Schiavelli and M. R. Hughey, *J. Am. Chem. Soc.*, **104**, 6852 (1982).
9. M. R. Ibrahim and W. J. Jorgensen, *J. Am. Chem. Soc.*, **111**, 819 (1989).

10. T. Müller, Y. Apeloig and H.-U. Siehl, unpublished results.
11. Y. Apeloig and D. Arad, *J. Am. Chem. Soc.*, **107**, 5285 (1985).
12. P. v. R. Schleyer, H. Jiao, M. N. Glukhovtsev, J. Chandresekhar and E. Kraka, *J. Am. Chem. Soc.*, **116**, 10129 (1994).
13. Y. Apeloig, M. Karni, A. Stanger, H. Schwarz, T. Drewello and G. Czekay, *J. Chem. Soc., Chem. Commun.*, 989 (1987).
14. T. Drewello, P. C. Burgers, W. Zummack, Y. Apeloig and H. Schwarz, *Organometallics*, **9**, 1161 (1990).
15. D. Hajdasz and R. Squires *J. Chem. Soc., Chem. Commun.*, 1212 (1988).
16. R. Bakhtiar, C. M. Holznagel and D. B. Jacobson, *J. Am. Chem. Soc.*, **114**, 3227 (1992).
17. S. G. Cho, *J. Organomet. Chem.*, **510**, 25 (1996).
18. M. Mishima, T. Ariga, Y. Tsuno, K. Ikenaga and K. Kikukawa, *Chem. Lett.*, 489 (1992).
19. G. A. McGibbon, M. A. Brook and J. K. Terlouw, *J. Chem. Soc., Chem. Commun.*, 360 (1992).
20. X. Li and J. A. Stone, *J. Am. Chem. Soc.*, **111**, 5586 (1989).
21. W. Zhang, J. Stone, M. A. Brook and G. A. McGibbon, *J. Am. Chem. Soc.*, **118**, 5764 (1996).
22. Y. Apeloig and T. Müller, in *Divalent Carbocations*, (Eds. Z. Rappoport and P. J. Stang), Chap 2, Wiley, Chichester, 1997.
23. Y. Apeloig, P. v. R. Schleyer and J. A. Pople, *J. Am. Chem. Soc.*, **99**, 5901 (1977).
24. B. Ciommer and H. Schwarz, *J. Organomet. Chem.*, **244**, 319 (1987).
25. T. M. Mayer and F. W. Lampe, *J. Phys. Chem.*, **78**, 2433 (1974).
26. T. M. Mayer and F. W. Lampe, *J. Phys. Chem.*, **78**, 2645 (1974).
27. W. N. Allen and F. W. Lampe, *J. Chem. Phys.*, **65**, 3378 (1976).
28. W. N. Allen and F. W. Lampe, *J. Am. Chem. Soc.*, **99**, 2943 (1977).
29. S. Fornarini, *J. Org. Chem.*, **53**, 1314 (1988).
30. F. Cacace, M. E. Crestioni and S. Fornarini, *J. Am. Chem. Soc.*, **114**, 6776 (1992).
31. M. E. Crestioni and S. Fornarini, *Angew. Chem., Int. Ed. Engl.*, **33**, 1094 (1994).
32. A. C. M. Wojtyniak and J. A. Stone, *Int. J. Mass Spectrom. Ion Processes*, **74**, 59 (1986).
33. M. Mishima, C. H. Kang, M. Fujio and Y. Tsuno, *Chem. Lett.*, 2439 (1992).
34. L. H. Sommer, E. Dorfmann, G. M. Goldberg and F. C. Whitmore, *J. Am. Chem. Soc.*, **68**, 488 (1946).
35. L. H. Sommer and F. C. Whitmore, *J. Am. Chem. Soc.*, **68**, 485 (1946).
36. M. A. Cook, C. Eaborn and D. R. M. Walton, *J. Organomet. Chem.*, **29**, 389 (1971).
37. F. K. Cartledge and J. P. Jones, *Tetrahedron Lett.*, 2193 (1971).
38. I. Dostrovsky and E. D. Hughes, *J. Chem. Soc.*, 157 (1946).
39. Y. Apeloig and A. Stanger, *J. Am. Chem. Soc.*, **107**, 2807 (1985).
40. Y. Apeloig, R. Biton and A. Abu-Freih, *J. Am. Chem. Soc.*, **115**, 2522 (1993).
41. N. Shimizu, E. Osajima and Y. Tsuno, *Bull. Chem. Soc. Jpn.*, **64**, 1145 (1991).
42. J. A. Soderquist and A. Hassner, *Tetrahedron Lett.*, **29**, 1899 (1988).
43. A. J. Kresge and J. B. Tobin, *J. Phys. Org. Chem.*, **4**, 587 (1991).
44. Y. Apeloig, P. v. R. Schleyer and J. A. Pople, *J. Am. Chem. Soc.*, **99**, 1291 (1977).
45. M. D. Schiavelli, D. M. Jung, A. K. Vaden, P. J. Stang, T. E. Fisk and D. S. Morrison, *J. Org. Chem.*, **46**, 92 (1981).
46. C. Dallaire and M. A. Brook, *Organometallics*, **12**, 2332 (1993).
47. S. N. Ushakov and A. M. Itenberg, *Zh. Obshch. Khim.*, **7**, 2495 (1937); *Chem. Abstr.*, **32**, 2083[8] (1938).
48. J. B. Lambert, *Tetrahedron*, **46**, 2677 (1990).
49. W. Hanstein and T. G. Traylor, *Tetrahedron Lett.*, 4451 (1967).
50. W. Hanstein, H. J. Herwin and T. G. Traylor, *J. Am. Chem. Soc.*, **92**, 829 (1970).
51. T. G. Traylor, W. Hanstein, H. J. Berwin, N. A. Clinton and R. S. Brown, *J. Am. Chem. Soc.*, **93**, 5715 (1971).
52. T. G. Traylor, H. J. Berwin, J. Jerkunica and M. H. Hall, *Pure Appl. Chem.*, **30**, 599 (1972).
53. P. v. R. Schleyer, D. Lenoir, P. Mison, G. Liang, G. K. S. Prakash and G. A. Olah, *J. Am. Chem. Soc.*, **102**, 683 (1980).
54. L. H. Sommer and G. A. Baughman, *J. Am. Chem. Soc.*, **83**, 3346 (1961).
55. J. Vencl, J. Hetflej, J. Czermak and V. Chvalovsky, *Collect. Czech. Chem. Commun.*, **38**, 1256 (1973).
56. R. Fujiyama and T. Munechika, *Tetrahedron Lett.*, **34**, 5907 (1993).
57. A. W. P. Jarvie, A. Holt and J. Thompson, *J. Chem. Soc. (B)*, 852 (1969).

12. Silyl-substituted carbocations 699

58. P. F. Hudrlik and D. Peterson, *J. Am. Chem. Soc.*, **97**, 1464 (1975).
59. A. W. P. Jarvie, A. Holt and J. Thompson, *J. Chem. Soc. (B)*, 746 (1970).
60. M. A. Cook, C. Eaborn and D. R. M. Walton, *J. Organomet. Chem.*, **24**, 301 (1970).
61. H. C. Brown, *The Non-Classical Ion Problem*, with comments by P. v. R. Schleyer, Plenum, New York, 1977.
62. D. D. Davis and H. M. Jacocks III, *J. Organomet. Chem.*, **206**, 33 (1981).
63. R. Hoffmann, L. Radom, J. A. Pople, P. v. R. Schleyer, W. J. Hehre and L. Salem, *J. Am. Chem. Soc.*, **94**, 6221 (1972).
64. J. B. Lambert and E. C. Chelius, *J. Am. Chem. Soc.*, **112**, 8120 (1990).
65. J. B. Lambert and R. B. Finzel, *J. Am. Chem. Soc.*, **104**, 2020 (1982).
66. J. B. Lambert, G.-T. Wang, R. B. Finzel and D. H. Teramura, *J. Am. Chem. Soc.*, **109**, 7838 (1987).
67. J. B. Lambert and X. Liu, *J. Organomet. Chem.*, **521**, 203 (1996).
68. J. B. Lambert and G.-T. Wang, *J. Phys. Org. Chem.*, **1**, 169 (1988).
69. G. -t. Wang, E. C. Chelius, D. Li and J. B. Lambert, *J. Chem. Soc., Perkin Trans. 2*, 331 (1990).
70. J. B. Lambert, R. W. Emblidge and S. Malany, *J. Am. Chem. Soc.*, **115**, 1317 (1993).
71. R. Eliason, M. Tomic, S. Borcic and D. E. Sunko, *J. Chem. Soc., Chem. Commun.*, 1490 (1968).
72. N. Shimizu, S. Watanabe and Y. Tsuno, *Bull. Chem. Soc. Jpn.*, **64**, 2249 (1991).
73. N. Shimizu, F. Hayakawa, S. Watanabe and Y. Tsuno, *Bull. Chem. Soc. Jpn.*, **66**, 153 (1993).
74. H. Mayr and R. Pock, *Tetrahedron*, **42**, 4214 (1986).
75. H. Mayr and G. Hagen, *J. Chem. Soc., Chem. Commun.*, 91 (1989).
76. G. Hagen and H. Mayr, *J. Am. Chem. Soc.*, **113**, 4954 (1991).
77. H. Mayr and M. Patz, *Angew. Chem., Int. Ed. Engl.*, **33**, 938 (1994).
78. M. A. Brook, M. A. Hadi and A. Neuy, *J. Chem. Soc., Chem. Commun.*, 957 (1989).
79. M. A. Brook and A. Neuy, *J. Org. Chem.*, **55**, 3609 (1990).
80. A. J. Kresge and J. B. Tobin, *J. Phys. Org. Chem.*, **4**, 587 (1991).
81. V. Gabelica and A. J. Kresge, *J. Am. Chem. Soc.*, **118**, 3838 (1996).
82. G. DeLucca and L. A. Paquette, *Tetrahedron Lett.*, **24**, 4931 (1983).
83. A. J. Kresge and J. B. Tobin, *Angew. Chem., Int. Ed. Engl.*, **32**, 721 (1993).
84. Y. Himeshima, H. Kobayashi and T. Sonoda, *J. Am. Chem. Soc.*, **107**, 5286 (1989).
85. M. J. Bausch and Y. Gong, *J. Am. Chem. Soc.*, **116**, 5963 (1994).
86. (a) N. Shimizu, C. Kinoshita, E. Osajima, F. Hayakawa and Y. Tsuno, *Bull. Chem. Soc. Jpn.*, **64**, 3280 (1991).
   (b) N. Shimizu, C. Kinoshita, E. Osajima and Y. Tsuno, *Chem. Lett.*, 1937 (1990).
87. (a) A. Abu-Freih, PhD Thesis, Technion, Haifa, 1997.
   (b) V. Braude, PhD Thesis, Technion, Haifa, 1995.
88. (a) N. Shimizu, C. Kinoshita, E. Osajima, F. Hayakawa and Y. Tsuno, *Bull. Chem. Soc. Jpn.*, **64**, 3280 (1991).
   (b) N. Shimizu, C. Kinoshita, E. Osajima and Y. Tsuno, *Chem. Lett.*, 1937 (1990).
89. (a) P. v. R. Schleyer, W. F. Sliwinski, G. W. Van Dine, U. Schöllkopf, J. Paust and K. J. Fellenberger, *J. Am. Chem. Soc.*, **94**, 125 (1972).
   (b) W. F. Sliwinski, T. M. Su and P. v. R. Schleyer, *J. Am. Chem. Soc.*, **94**, 133 (1972).
   (c) H. C. Brown, C. Gundu Rao and M. Ravindranathan, *J. Am. Chem. Soc.*, **100**, 7946 (1978).
90. V. J. Shiner, Jr., M. W. Ensinger and G. S. Kriz, *J. Am. Chem. Soc.*, **108**, 842 (1986).
91. V. J. Shiner, Jr., M. W. Ensinger, G. S. Kriz and K. A. Halley, *J. Org. Chem.*, **55**, 653 (1990).
92. E. D. Davidson and V. J. Shiner, Jr., *J. Am. Chem. Soc.*, **108**, 3135 (1986).
93. V. J. Shiner, Jr., M. W. Ensinger and R. D. Rutkowske, *J. Am. Chem. Soc.*, **109**, 802 (1987).
94. V. J. Shiner, Jr., M. W. Ensinger and J. C. Huffman, *J. Am. Chem. Soc.*, **111**, 7199 (1989).
95. J. Coope, V. J. Shiner, Jr. and M. W. Ensinger, *J. Am. Chem. Soc.*, **112**, 2834 (1990).
96. C. A. Grob and P. Sawlewicz, *Tetrahedron Lett.*, **28**, 951 (1987).
97. C. A. Grob, M. Gründel and P. Sawlewicz, *Helv. Chim. Acta*, **71**, 1502 (1988).
98. T. W. Bentley, W. Kirmse, G. Llewellyn and F. Söllenböhmer, *J. Org. Chem.*, **55**, 1536 (1990).
99. W. Kirmse and F. Söllenböhmer, *Angew. Chem., Int. Ed. Engl.*, **28**, 1667 (1989).
100. W. Kirmse and F. Söllenböhmer,. *J. Am. Chem. Soc.*, **111**, 4129 (1989).
101. R. J. Fessenden, K. Seeler and M. Dagani, *J. Org. Chem.*, **31**, 2483 (1966).
102. W. Adcock, A. R. Krstic, P. J. Duggan, V. J. Shiner, Jr., J. Coope and M. W. Ensinger, *J. Am. Chem. Soc.*, **112**, 3140 (1990).
103. R. Hoffmann, A. Iwamura and W. J. Hehre, *J. Am. Chem. Soc.*, **90**, 1499 (1968).

104. R. Hoffmann, *Acc. Chem. Res.*, **4**, 1 (1971).
105. D. E. Sunko, S. Hirsl-Starcevic, S. Pollack and W. J. Hehre, *J. Am. Chem. Soc.*, **101**, 6163 (1979).
106. W. Adcock, J. Coope, V. J. Shiner, Jr. and N. A. Trout, *J. Org. Chem.*, **55**, 1411 (1990).
107. M. Xie and W. J. Le Noble, *J. Org. Chem.*, **54**, 3839 (1989).
108. W. Adcock, C. I. Clark and C. H. Schiesser, *J. Am. Chem. Soc.*, **118**, 11541 (1996).
109. P. E. Eaton and J. P. Zhou, *J. Am. Chem. Soc.*, **114**, 3119 (1992).
110. R. M. Moriarty, S. M. Tuladhar, R. Penmasta and A. K. Awasthi, *J. Am. Chem. Soc.*, **112**, 3228 (1990).
111. D. N. Kevill, M. J. D'Souza, R. M. Moriarty, S. M. Tuladhar, R. Penmasta and A. K. Awasthi, *J. Chem. Soc., Chem. Commun.*, 3228 (1990).
112. D. A. Hrovat and W. T. Borden, *J. Am. Chem. Soc.*, **112**, 3227 (1990).
113. G. A. Olah, A. L. Berrier, L. D. Field and G. K. S. Prakash, *J. Am. Chem. Soc.*, **104**, 1349 (1982).
114. F.-P. Kaufmann, Ph. D. Thesis, University of Tübingen, 1992.
115. B. Müller, Ph. D. Thesis, University of Tübingen, 1995.
116. G. A. Olah, R. J. Spear and D. A. Forsyth, *J. Am. Chem. Soc.*, **99**, 2615 (1977).
117. G. A. Olah and P. W. Westerman, *J. Am. Chem. Soc.*, **95**, 3706 (1973).
118. H. C. Brown, M. Periasamy, D. P. Kelly and J. J. Giansiracusa, *J. Org. Chem.*, **47**, 2089 (1982).
119. M. Kavni, Y. Apeloig and H.-U. Siehl, unpublished results.
120. H.-U. Siehl, Y. Apeloig and T. Müller, unpublished results.
121. A. de Meijere, D. Faber, M. Noltemeyer, R. Boese, T. Haumann, T. Müller, M. Bendikov, E. Matzner and Y. Apeloig, *J. Org. Chem.*, **61**, 8564 (1996).
122. G. Maier, D. Volz and J. Neudert, *Synthesis*, 561 (1992).
123. C. S. Q. Lew, R. A. McClelland, L. J. Johnston and N. P. Schepp, *J. Chem. Soc., Perkin Trans. 2*, 395 (1994).
124. C. S. Q. Lew and R. A. McClelland, *J. Am. Chem. Soc.*, **115**, 1156 (1993).
125. J. B. Lambert, S. Zhang, C. L. Stern and J. C. Huffman, *Science*, **260**, 1917 (1993).
126. J. B. Lambert and S. Zhang, *J. Chem. Soc., Chem. Commun.*, 383 (1993).
127. P. v. R. Schleyer, P. Buzek, T. Müller, Y. Apeloig and H.-U. Siehl, *Angew. Chem., Int. Ed. Engl.*, **32**, 1471 (1993).
128. L. Pauling, *Science*, **263**, 983 (1994).
129. G. A. Olah, G. Rasul, X.-Y. Li, H. A. Buchholz, G. Sandford and G. K. S. Prakash, *Science*, **263**, 983 (1994).
130. G. A. Olah, G. Rasul, H. A. Buchholz, X.-Y. Li and G. K. S. Prakash, *Bull. Soc. Chim. Fr.*, **132**, 569 (1995).
131. C. Maerker, J. Kapp and P. v. R. Schleyer, in *Organosilicon Chemistry II* (Eds. N. Auner and J. Weis), VCH, Weinheim, 1996, p. 329.
132. C. A. Reed, Z. Xie, R. Bau and R. Benesi, *Science*, **262**, 402 (1993).
133. H.-U. Siehl, B. Müller and O. Malkina, in *Organosilicon Chemistry III* (Eds. N. Auner and J. Weis) VCH, Weinheim, 1997, p. 25.
134. (a) V. G. Malkin, O. L. Malkina and D. R. Salahub, *Chem. Phys. Lett.*, **261**, 335 (1996).
    (b) V. G. Malkin, O. L. Malkina, M. E. Casida and D. R. Salahub, *J. Am. Chem. Soc.*, **116**, 5898 (1994).
135. B. Müller and H.-U. Siehl, unpublished results.
136. S. Braun, T. S. Abram and W. E. Watts, *J. Organomet. Chem.*, **97**, 429 (1975).
137. J. B. Lambert and Y. Zhao, *J. Am. Chem. Soc.*, **118**, 7867 (1996).
138. G. K. S. Prakash, V. P. Reddy, G. Rasul, J. Casanova and G. A. Olah, *J. Am. Chem. Soc.*, **114**, 3076 (1992).
139. H.-U. Siehl, in *Dicoordinated Carbocations* (Eds. Z. Rappoport and P. Stang), Chap. 5, Wiley, Chichester, 1997.
140. H.-U. Siehl, in *Stable Carbocation Chemistry* (Eds. G. K. S. Prakash and P. v. R. Schleyer), Chap. 5, Wiley, New York, 1997.
141. T. Müller, Ph. D. Thesis, University of Tübingen, 1993.
142. H.-U. Siehl, *Pure Appl. Chem.*, **67**, 769 (1995).
143. T. S. Abram and W. E. Watts, *J. Chem. Soc., Perkin Trans. 1*, 1522 (1977).
144. K. Bertsch, Diplomarbeit, University of Tübingen, 1987.
145. E.-W. Koch, H.-U. Siehl and M. Hanack, *Tetrahedron Lett.*, **26**, 1493 (1985).

146. S. Braun, T. S. Abram and W. E. Watts, *J. Organomet. Chem.*, **97**, 429 (1975).
147. F.-P. Kaufmann and H.-U. Siehl, *J. Am. Chem. Soc.*, **114**, 4937 (1992).
148. F.-P. Kaufmann, Diplomarbeit, University of Tübingen, 1990.
149. H.-U. Siehl, B. Müller, M. Fuss and Y. Tsuji, in *Organosilicon Chemistry II* (Eds. N. Auner and J. Weis) VCH, Weinheim, 1996, p. 361.
150. G. A. Olah and P. W. Westerman, *J. Am. Chem. Soc.*, **95**, 3706 (1973).
151. H. C. Brown. M. Periasamy, D. P. Kelly and J. J. Giansiracusa, *J. Org. Chem.*, **47**, 2089 (1982).
152. H. C. Brown, M. Periasamy and K. T. Liu, *J. Org. Chem.*, **46**, 1646 (1981).
153. G. A. Olah, R. J. Spear and D. A. Forsyth, *J. Am. Chem. Soc.*, **99**, 2615 (1977).
154. H.-U. Siehl, F.-P. Kaufmann and K. Hori, *J. Am. Chem. Soc.*, **114**, 9343 (1992).
155. R. Jost, J. Sommer, C. Engdahl and P. Ahlberg, *J. Am. Chem. Soc.*, **102**, 7663 (1980).
156. H.-U. Siehl, F.-P. Kaufmann, Y. Apeloig, V. Braude, D. Danovich, A. Berndt and N. Stamatis, *Angew. Chem., Int. Ed. Engl.*, **30**, 1479 (1991).
157. C. U. Pittman Jr., *J. Chem. Soc., Chem., Commun.*, 122 (1969).
158. P. v. R. Schleyer, W. de M. Caneiro, W. Koch and D. A. Forsyth, *J. Am. Chem. Soc.*, **113**, 3990 (1991).
159. G. A. Olah, V. P. Reddy and G. K. S. Prakash, *Chem. Rev.*, **92**, 69 (1992).
160. (a) D. Lenoir and H.-U. Siehl, in *Houben-Weyl Methoden der Organischen Chemie*, (Ed. M. Hanack), Vol. E19c, Thieme, Stuttgart, 1990, p. 413.
 (b) G. A. Olah, G. K. S. Prakash and J. Sommer, *Superacids*, Wiley, New York, 1985, p. 143.
161. M. Saunders and H.-U. Siehl, *J. Am. Chem. Soc.*, **102**, 6860 (1980).
162. W. Koch, B. Liu and D. J. DeFrees, *J. Am. Chem. Soc.*, **110**, 7325 (1988).
163. M. L. McKee, *J. Phys. Chem.*, **90**, 4980 (1986).
164. (a) M. Saunders, K. E. Laidig, K. B. Wiberg and P. v. R. Schleyer, *J. Am. Chem. Soc.*, **110**, 7652 (1988).
 (b) C. S. Yannoni, P. C. Myhre and G. G. Webb, *J. Am. Chem. Soc.*, **112**, 8992 (1990).
165. J. S. Staral, I. Yavari, J. D. Roberts, G. K. S. Prakash, D. J. Donovan and G. A. Olah, *J. Am. Chem. Soc.*, **100**, 8016 (1978).
166. (a) H.-U. Siehl, *J. Am. Chem. Soc.*, **107**, 3390 (1985).
 (b) H.-U. Siehl in *Physical Organic Chemistry* 1986 (Ed. M. Kobayashi), Elsevier, Amsterdam, 1987, p. 25.
 (c) J. Gottaut, Diplomarbeit, University of Tübingen, 1990.
 (d) H.-V. Siehl, Habilitationsschvift, University of Tübingen, 1986.
167. M. Saunders and N. Krause, *J. Am. Chem. Soc.*, **110**, 8050 (1988).
168. H.-U. Siehl and M. Fuss, *J. Am. Chem. Soc.*, **117**, 5983 (1995).
169. H.-U. Siehl and M. Fuss, unpublished results.
170. H.-U. Steinberger, T. Müller, N. Auner, C. Maerker and P. v. R. Schleyer, *Angew. Chem., Int. Ed. Engl.*, **36**, 626 (1997).

CHAPTER **13**

Silicon-substituted carbenes

GERHARD MAAS

Abteilung Organische Chemie I, Universität Ulm, Albert-Einstein-Allee 11, D-89081 Ulm, Germany
Fax: (int.)+49(0)731-502-2803; e-mail: gerhard.maas@chemie.uni-ulm.de

```
    I. INTRODUCTION ...................................... 704
   II. GEOMETRIC AND ELECTRONIC STRUCTURE: THEORY AND
       EXPERIMENT ........................................ 704
  III. SILICON-SUBSTITUTED CARBENES IN SYNTHESIS .......... 711
       A. Common Methods of Preparation and General Reactivity Patterns ... 711
       B. The Silylcarbene-to-Silene Rearrangement ................ 712
       C. Intramolecular Carbene Reactions with Substituents Attached to
          Silicon ............................................ 727
          1. C,H insertion .................................... 727
             a. 1,3-C,H insertion ............................. 727
             b. 1,4-C,H insertion ............................. 727
             c. 1,5-C,H insertion ............................. 728
          2. Reactions at C=C bonds .......................... 730
          3. Reactions at C≡C bonds .......................... 732
       D. Carbene Reactions of Bis(diazomethyl)silanes and -polysilanes ..... 732
       E. Silylcarbene Reactions not Involving the Silyl Group .......... 739
          1. Silylcarbenes without a second substituent ............... 739
             a. (Trimethylsilyl)carbene ........................ 739
             b. Other (alkylsilyl)carbenes ...................... 743
             c. (Alkenylsilyl)carbenes ......................... 743
             d. (Trimethoxysilyl)carbene ....................... 743
          2. Alkyl(silyl)carbenes and cycloalkyl(silyl)carbenes ........... 743
          3. Alkenyl(silyl)carbenes and cycloalkenyl(silyl)carbenes ....... 744
             a. 1-Alkenyl(silyl)carbenes ........................ 744
             b. 2-Alkenyl(silyl)carbenes ........................ 746
             c. Cycloalkenyl(silyl)carbenes ..................... 747
          4. Aryl(silyl)carbenes ................................ 748
          5. Acyl(silyl)carbenes ................................ 751
             a. Silyl-substituted ketocarbenes ................... 751
             b. Alkoxycarbonyl(silyl)carbenes ................... 752
             c. Aminocarbonyl(silyl)carbenes ................... 760
```

The chemistry of organic silicon compounds, Vol. 2
Edited by Z. Rappoport and Y. Apeloig © 1998 John Wiley & Sons Ltd

6. Imino(silyl)carbenes 761
7. Phosphino(silyl)carbenes and phosphoryl(silyl)carbenes 761
   a. Phosphino(silyl)carbenes 761
   b. Phosphoryl(silyl)carbenes 765
8. Sulfur-substituted silylcarbenes 766
9. Halo(silyl)carbenes 767
F. Special Methods to Generate Silylcarbenes 767
   1. Silylcarbenes by rearrangement reactions 767
      a. Isomerization of 2-silylfurans 767
      b. Olefin-to-carbene isomerization 769
      c. Ring-opening of 2-lithio-2-silyloxiranes 772
   2. Silylcarbenes by intermolecular reactions 772
      a. Photochemical reaction of (silyl)alkynones with alkenes 772
      b. Insertion of carbon into a Si—H bond 773
IV. CONCLUDING REMARKS 774
V. ACKNOWLEDGMENTS 774
VI. REFERENCES 774

I. INTRODUCTION

Carbenes are species which contain a divalent carbon atom with six electrons in its valence shell. Thus, the general formula of a carbene is CR_2, where R stands for a wide range of substituents, including silyl groups. Typically, carbenes are short-lived intermediates which are generated *in situ* from appropriate precursors and undergo fast intra- or intermolecular reactions. In this manner, silicon-substituted carbenes provide access to a variety of organosilicon compounds, such as silyl-substituted molecular frameworks, silaheterocycles and silaethenes.

Since the first reports in the 1960s, the interest of researchers in the chemistry of silicon-substituted carbenes has not ceased up to the present time. While the now classical books on carbene chemistry by Kirmse[1] and Jones and Moss[2] mentioned silylcarbenes only in passing, detailed reviews on the preparative chemistry of silylcarbenes[3] and on the synthesis of silylcyclopropanes by silylcarbene transfer to alkenes[4] have recently appeared in two volumes of the Houben-Weyl series. The chemistry of (trimethylsilyl)carbene has also been covered in further reviews[5,6].

All silicon-substituted carbenes that have been studied experimentally so far contain a SiR_3 group (as opposed to silicon with other coordination numbers); therefore, the expression 'silylcarbenes' will be used throughout this chapter.

II. GEOMETRIC AND ELECTRONIC STRUCTURE: THEORY AND EXPERIMENT

Depending on whether the two nonbonding electrons at a carbenic carbon atom are spin-paired or not, the carbene can exist either in a singlet or a triplet electronic state. The lowest-energy singlet state (S) has a bent structure with both electrons located in the nonbonding σ orbital, while the lowest triplet state (T) has one electron in each of the two nonbonding orbitals, one with σ- and one with π-character (Figure 1). Typical valence angles of monosubstituted singlet carbenes, according to experimental observations[7] and theoretical calculations[8], are in the range 100–110 deg, whereas those of the corresponding triplet carbenes are larger by at least 20–30 deg.

Methylene itself has a triplet ground state. The best experimental determinations of the singlet–triplet gap have furnished a value of 9.0–9.1 kcal mol^{-1}[9,10], in excellent agreement with recent high-level calculations. Stabilization of the singlet state is achieved by heteroatom substituents such as Hal, OH and NH_2, which can denote electron density

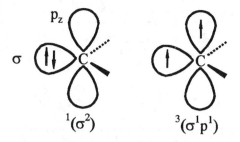

FIGURE 1. Lowest-energy singlet (S) and triplet (T) state of a carbene

from a nonbonding electron pair at the heteroatom to a formally empty p_z orbital of the carbenic carbon atom. On the other hand, triplet states are expected to be stabilized by substituents that are more electropositive than carbon as well as by an increase of the valence bond angle due to bulky substituents. Both substituent effects cause a higher amount of s character in the C—X bond(s) of the carbene. This in turn produces more p character in the nonbonding σ orbital at the carbene center, so that both nonbonding valence orbitals get closer in energy.

Along these lines, it is expected that a silyl substituent will stabilize the triplet ground state of a carbene. No experimental values for the singlet–triplet gap of silyl-substituted carbenes are available, but meaningful comparisons with other carbenes are now possible based on theoretical calculations[8,11] which take care of a balanced treatment of the two spin states. [Earlier calculations suffered from the problem of unbalanced electron correlation and a spin–coupling bias, both in favor of the triplet state, so that a too high value of the singlet–triplet gap was predicted for carbenes with a triplet ground state (e.g. $HCSiH_3$[12-14]) and a too low value for carbenes with a singlet ground state.] Some results so obtained for silylcarbenes and some other archetypical carbenes are given in Tables 1 and 2. These data are in reasonable agreement with available experimental values for CH_2, $CHCl$ and other carbenes (CHF, CF_2) not mentioned in the Tables.

TABLE 1. Geometries of some carbenes (H—C—X—Y) according to theoretical calculations[a,8]

Carbene	Singlet			Triplet		
	C—X (Å)	H—C—X (deg)	H—C—X—Y (deg)	C—X (Å)	H—C—X (deg)	H—C—X—Y (deg)
CH_2	1.109	101.4		1.082	126.1	
HCCl	1.727	102.3		1.702	124.0	
$HCNH_2$	1.342	104.0		1.434	122.9	
$HCSiH_3$	1.950	109.0		1.858	144.8	
$HCSiH_2CH_3$	1.949	105.8	162.0	1.866	138.8	180.0
$HCSiH_2SiH_3$[b]	1.911	106.6	89.0	1.861	139.9	43.6
$HCCH_2SiH_3$	1.508	104.3	91.4	1.526	127.1	49.0
$HCCH_2CH_3$	1.546	102.6	180.0	1.530	126.6	60.2

[a] Double zeta plus polarization basis sets, two-configuration self-consistent field (singlets) or ROHF (triplet) wave functions using the Huzinaga MINI(d,p) basis set were used. For $HCSiH_2SiH_3$, the latter basis set was replaced by an effective core potential (ECP) together with a split valence plus polarization basis set.
[b] Geometry-optimized at CASSCF/ECP(d,p).

TABLE 2. Calculated singlet–triplet energy gaps ($\Delta E_{ST}{}^a$) for several carbenes (kcal mol^{-1})[b]

Carbene	ΔE_{ST}			Carbene	ΔE_{ST}	
	6-31G(d,p)[c]	ECP(d,p)[c]	MRCI[d]		6-31G(d,p)[c]	ECP(d,p)[c]
CH$_2$	8.7	6.8	9.0	HCSiH$_2$CH$_3$	18.1	
HCCl	−5.1	−6.0	−9.3	HCSiH$_2$SiH$_3$		14.7
HCNH$_2$	−32.9	−32.1		HCCH$_2$CH$_3$	3.4	
HCSiH$_3$	17.8	16.7	18.4	HCCH$_2$SiH$_3$	−1.2	−2.3

[a] $\Delta E_{ST} = E_S - E_T$; a positive sign means that the singlet is higher in energy.
[b] Structures were optimized at the theoretical level described in Table 1, footnote a. Final energies were obtained by using multi-reference configuration interaction (MRCI) wave functions including single and double excitations from the reference wave functions.
[c] From Reference 8.
[d] From Reference 11; individual basis sets were used for each atom.

Table 1 presents two remarkable features of silylcarbene geometry. First, the C–Si bond length is dramatically decreased in the triplet carbene with respect to the singlet species; this difference correlates with a higher s character of the C$_{carbene}$–X bonds for the triplet, but more p character for the singlet, as discussed above. The second feature concerns the dihedral angle H–C–X–Y: while (methylsilyl)carbene has an antiperiplanar arrangement in both spin states, (disilanyl)carbene shows torsion angles of 89 (singlet) and 43.6 (triplet) deg. Notably, the conformation of (silylmethyl)carbene is very similar to (disilanyl)carbene, whereas ethylcarbene has an antiperiplanar singlet state conformation. By analogy to the energetic stabilization of a carbenium ion by silyl groups in β-position (see the chapter by Siehl), the β-silicon effect can be invoked to explain the conformational differences. In the perpendicular conformation of (disilanyl)carbene and (methylsilyl)carbene, respectively, overlap of the σ(Si–Si) or σ(C–Si) orbitals with the carbenic π orbital is possible, and it is clear that this interaction is especially important energetically for the singlet carbene where the π orbital is essentially empty (Figure 2). In contrast to HCCH$_2$SiH$_3$, no σ/π-overlap is possible in HCSiH$_2$CH$_3$ (torsion angle, 162 deg). It has been suggested that the magnitude of the β-silicon effect is associated with the amount of overlap between the carbenic π orbital and the attached Si–Si or Si–C (or C–Si) bond[8]. If one takes the distance d between C$_{carbene}$ and the midpoint of X–Y, which is a function of the C–X bond length and the C–X–Y bond angle (Figure 2), as a qualitative measure, it becomes obvious why the β-silicon effect is of minor importance in disilanyl carbene (as suggested by the calculated singlet–triplet gap, see below), in spite of the favorable torsion angle.

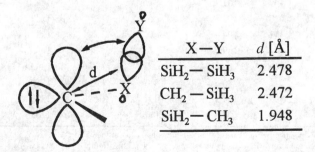

X–Y	d [Å]
SiH$_2$–SiH$_3$	2.478
CH$_2$–SiH$_3$	2.472
SiH$_2$–CH$_3$	1.948

FIGURE 2. The β-silicon effect in a singlet carbene

13. Silicon-substituted carbenes

The calculated singlet–triplet splitting (Table 2) confirms that silylcarbene has a triplet ground state. The gap is about 9 kcal mol^{-1} higher than in methylene, a value which is much lower than in earlier calculations. Replacement of SiH$_3$ by SiH$_2$CH$_3$ does not affect ΔE_{ST} appreciably. However, the gap is already much smaller for (disilanyl)carbene, and in (silylmethyl)carbene the singlet state is so much stabilized with respect to the triplet state that a singlet ground state is predicted. Thus, the importance of the β-silicon effect for stabilization of a singlet state is also reflected in the calculated ΔE_{ST} values. The qualitative correlation between increasing Mulliken population of the carbenic π orbital and increasing stabilization of the singlet state for the carbenes mentioned in Table 2 underlines the importance of electron donation to the formally empty π orbital at the carbene center.

Calculations of relative energies on the CH$_4$Si potential surface established the following order of increasing energy[12]: silaethene (H$_2$C=SiH$_2$), singlet methylsilylene [^{1}(H$_3$CSiH)], triplet methylsilylene [^{3}(H$_3$CSiH)], ^{3}(HCSiH$_3$), ^{1}(HCSiH$_3$).

The geometries, relative energies and net atomic charges of fifteen structural isomers with the chemical formula C$_2$H$_4$Si have been calculated at the (6-31G*//3-21G) SCF level[15]. The four carbenes in this set, which also includes silylenes and Si,C double- and triple-bonded species, have by far the highest relative energy. Two of these carbenes are silicon-substituted carbenes, namely 2-silacyclopropylidene (H$_2\overline{\text{SiCC}}$H$_2$) and (1-silavinyl)carbene [H$_2$C=Si(H)CH].

The silicenium-substituted carbene HCSi$^+$H$_2$ has been studied in the context of the potential energy surface for the unimolecular decomposition of the methylsilicenium ion (H$_3$CSi$^+$H$_2$)[16]. Calculated structures for the singlet and triplet species are shown in Figure 3. In both cases, a remarkably short Si–C distance is found, which points to some degree of π bonding. The linearity of the singlet carbene is remarkable (cf, Table 1) and suggests a description of this species as a silavinyl cation. At the UMP4/6-311++G(2df,2pd) level, triplet HCSi$^+$H$_2$ is 7.7 kcal mol^{-1} lower in energy than the singlet species. Thus, ΔE_{ST} is much lower than for the neutral carbene HCSiH$_3$ and even lower than for CH$_2$. According to the calculations of activation barriers, generation of the carbene by 1,1-dehydrogenation at carbon (H$_3$CSi$^+$H$_2$ $\rightarrow$ HCSi$^+$H$_2$ + H$_2$) plays no role in the experimentally studied[17] thermal decomposition of the methylsilicenium ion, since 1,2-dehydrogenation and cleavage into CH$_4$ and SiH$^+$ occur more easily.

The geometries and relative energies of the singlet carboxycarbene **1** and carboxy(trimethylsilyl)carbene **2** (Figure 4) have been calculated and analyzed by *ab initio* methods[18]. The optimized geometries (RHF/6-31G*) of several conformations of **1** and **2**, i.e., **1-p** or **2-p** (perpendicular, C_1 symmetry), *1-s-cis* and *1-s-trans* and *2-s-cis* and *2-s-trans* (all planar, C_s symmetry), are shown in Figure 4. It was found that the perpendicular conformer represents the global minimum and that *1-s-cis* and *1-s-trans* are transition states for the rotation about the C–C bond. According to MP4/6-31G**//6-31G* calculations, the relative energy differences, $\Delta\Delta H_f$ (*1p* − *1-s-cis*) and $\Delta\Delta H_f$

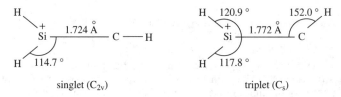

FIGURE 3. Structures for HCSi$^+$H$_2$; singlet species calculated at MP2/6-31G(d,p), triplet species at UMP2/6-31G(d,p))

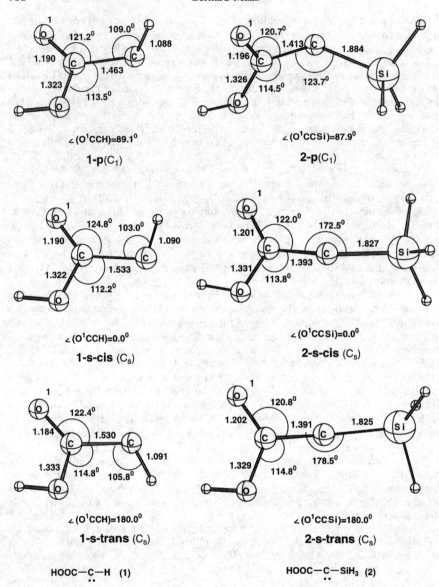

FIGURE 4. Geometry-optimized calculated structures of carbenes **1** and **2**

(**1p** − **1-s-trans**), are about 5–7 kcal mol^{-1} higher in the monosubstituted carbene **1** than in carboxy(silyl)carbene **2**. The geometries of the perpendicular conformation of carbenes **1** and **2** are not very different. In contrast, the planar *s-trans* and *s-cis* conformations of **2** differ from those of **1** as well as from the perpendicular conformation of **2** by having a nearly linear C−C−Si bond angle. The shortening of both the C−C and the C−Si bond lengths in the linear form may also be seen as a consequence of the increased s character of these bonds. Geometries and electronic properties of silylcarbene (HCSiH$_3$), (trimethylsilyl)carbene, and carboxy(silyl)carbene (**2**) had been calculated earlier with the MINDO/3 method[19]. The MINDO/3 optimized geometry of singlet **2** was similar to that of **2-s-cis** or **2-s-trans** which as pointed out above according to the *ab initio* calculations[18] are not local minima.

The matrix-isolation technique is the method of choice for the direct spectroscopic observation of carbenes[7]. However, efforts to generate and to observe silylcarbenes in solid matrices at cryogenic temperatures met with limited success. When (trimethylsilyl)diazomethane, (dimethylsilyl)diazomethane or bis(trimethylsilyl)diazomethane were irradiated in an argon matrix at ⩽10 K, no IR spectra of the corresponding carbenes **3a–c** could be obtained[20,21]. However, weak ESR signals were observed which were typical for a linear carbene with a triplet ground state[21]. These results indicate that at least small amounts of carbenes **3a–c** were present in the matrix.

Me$_3$Si⁀H   Me$_2$HSi⁀H   Me$_3$Si⁀SiMe$_3$ H$_2$MeSi⁀H

(**3a**)   (**3b**)   (**3c**)   (**3d**)

Attempts to trap carbenes **3b** and **3d** with molecular oxygen, a reaction frequently used to characterize triplet carbenes, were successful at high (>5%) concentrations of O$_2$ in the argon matrix[22]. In both cases, the corresponding silyl formate was detected, and it was assumed that a carbonyl oxide and a dioxirane are intermediates in this carbene trapping reaction.

Photolysis of matrix-isolated (trimethoxysilyl)diazomethane at λ ⩾ 305 nm produced carbene **3e** (equation 1), which was characterized by IR and UV-Vis spectroscopy[23]. Under these conditions, no other species besides the carbene could be detected spectroscopically. In an O$_2$-doped (1%) argon matrix, the carbene rapidly reacted with oxygen to give carbonyl oxide **4** which was further photoisomerized to formylsilane **5**. Again, the fast reaction of **3e** with ^{3}O$_2$ points to a triplet ground state of the carbene.

$$(MeO)_3SiCH\!\!=\!\!N_2 \xrightarrow[-N_2]{h\nu\ (>305\ nm)\ Ar,\ 10\ K} (MeO)_3Si\!-\!\ddot{C}\!-\!H\quad (\mathbf{3e})$$

$$\Big\downarrow {+O_2 \atop 40\ K}$$

$$MeO\!-\!\underset{\underset{OMe}{|}}{\overset{\overset{O\diagdown CH_2OH}{|}}{Si}}\!-\!CH\!\!=\!\!O \quad\longleftarrow\quad (MeO)_3Si\!-\!\overset{O\!-\!O^\bullet}{\underset{H}{C}} \qquad (1)$$

(**5**)   (**4**)

Considerable interest has been shown recently in so-called 'stable carbenes' of the type (R$_2$N)$_2$P–C–SiR$_3$[24]. Due to the ability of the phosphino group to share the lone electron pair at phosphorus with the carbenic carbon and to the ability of phosphorus to expand its valence shell, the basic question is whether these species in particular, and phosphinocarbenes in general, are to be considered indeed as carbenes I or better as phosphavinyl ylides II or λ^5-phosphaacetylenes III (Figure 5). The geometries predicted by theoretical calculations for the parent phosphinocarbene H_2P-C-H^{25} and the silyl-substituted phosphinocarbenes $H_2P-C-SiH_3$ and $(H_2N)_2P-C-SiH_3$[26] are given in Figure 5. In all cases, the singlet is predicted to be planar while the triplet is calculated to have a nonplanar, C_s-symmetrical structure featuring a pyramidal phosphorus atom. The P–C distance in the singlet is much shorter than in the triplet and falls in the range of a P≡C triple bond for $(H_2N)_2P-C-SiH_3$, where electron donation by the phosphino group is particularly high. As discussed above, the introduction of a SiH_3 group at the carbene center increases the valence angle at this center both in the singlet and in the triplet, but a decrease of the C–Si bond length in the triplet with respect to the singlet (cf Table 1) is not observed.

In all cases, the calculations identify the singlet species as the more stable one; at the PMP-4 level, the singlet states of $H_2P-C-SiH_3$ and $(H_2N)_2P-C-SiH_3$ are 5.6 and 13.9 kcal mol^{-1}, respectively, below the triplet states. Thus, all computational evidence points to a description of $(R_2N)_2P-C-SiR_3$ as a species with a P,C multiple bond, with an emphasis on the ylide structure II in which both charges are stabilized by the respective neighboring substituents. An X-ray structure determination of one of the isolable [bis(dialkylamino)phosphino](silyl)carbenes is not yet available but, as will be shown in

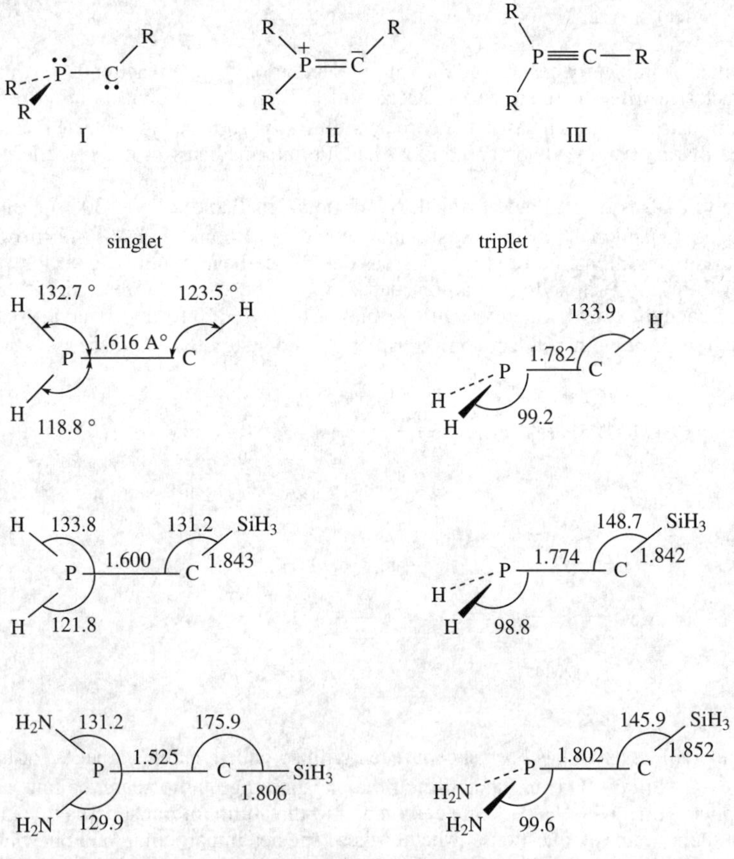

FIGURE 5. Possible structures of phosphinocarbenes and phosphino(silyl)carbenes and the calculated structures for singlet and triplet species; bond lengths (Å) and bond angles (°) are given

Section III.E.7, the chemical behavior has facets of carbene as well as P,C multiple bond chemistry.

III. SILICON-SUBSTITUTED CARBENES IN SYNTHESIS
A. Common Methods of Preparation and General Reactivity Patterns

The most common routes to carbenes[27] are also the major ones to generate silylcarbenes, namely dediazoniation of aliphatic diazo compounds and α-elimination reactions (Scheme 1). The extrusion of N_2 from silyl-substituted diazo compounds can be achieved by UV-irradiation or thermally. The thermal decomposition, however, is of less importance since these diazo compounds are thermally much more stable than their nonsilylated counterparts, so that thermal impact may stimulate noncarbene pathways.

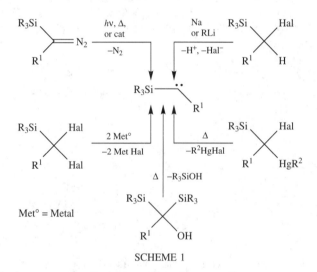

SCHEME 1

Transition-metal catalysis, especially by copper, rhodium, palladium and ruthenium compounds, is another approved method for the decomposition of diazo compounds. It is now generally accepted that short-lived metal–carbene intermediates are or may be involved in many of the associated transformations[28]. Nevertheless, these catalytic carbene transfer reactions will be fully covered in this chapter because of the close similarity in reaction modes of electrophilic carbenes and the presumed electrophilic metal–carbene complexes.

Suitable candidates for α-elimination reactions are silylmethyl halides ($\rightarrow$ base-induced elimination of H-Hal), silylmethyl dihalides ($\rightarrow$ halide/metal exchange followed by elimination of a metal halide) and stable carbenoid-type compounds such as (α-halo-α-silylalkyl)mercury compounds ($\rightarrow$ thermal elimination of mercury(II) halide). Bis(phenylthio)(trimethylsilyl)methyl lithium ($\rightarrow$ elimination of LiSPh) represents a borderline case (see Section III.E.8).

Further methods for the generation of silylcarbenes are available, but their application is usually limited to special types of silylcarbenes. Such methods will be discussed separately in Section III.F.

A specific silylcarbene may be generated from different precursors. Furthermore, these reactive intermediates may undergo different competing transformations. Therefore, this section is not classified by methods of generation nor reaction patterns, but rather according to the different types of silylcarbenes, i.e. to the substituent R^1 in a carbene with the general formula $R_3Si-C-R^1$.

With the exception of some phosphino(silyl)carbenes, which should only formally be considered as carbenes (see Section III.E.7), silylcarbenes are short-lived, electrophilic intermediates. They rapidly undergo intra- or intermolecular reactions which are typical of electrophilic carbenes in general, such as insertion reactions in C—H, X—H and other single bonds, [1 + 2] cycloaddition to alkenes and alkynes, ylide-forming reactions with nonbonding electron pairs of heteroatoms and the formation of (formal) carbene dimers. Furthermore, the possibility to isomerize by a 1,2-migration of a substituent to the carbene center also applies to silylcarbenes. In a disubstituted carbene **6** both R^1 and R^2 can undergo such a migration, which in the case of R^1 generates a silyl-substituted alkene [or, in the Wolff rearrangement of acyl(silyl)carbenes, a silylketene] but a silicon–carbon double bond (silaethene, silene) in the case of R^2 (equation 2). It is obvious that the silylcarbene-to-silene rearrangement is a particularly appealing aspect in silylcarbene chemistry.

$$\text{(2)}$$

The spin state of a reacting carbene has consequences for the product formation[1,2]. Reactions such as 1,2-migration, concerted insertion into C—H and O—H bonds and stereospecific cyclopropanation are considered to be typical for a singlet state, whereas nonstereospecific cyclopropanation and H-abstraction originate from the triplet carbene. It must be taken into account, however, that the spin state of the reacting species is not necessarily the same as that of the electronic ground state. For example, simple UV irradiation of a diazo compound will first generate the singlet carbene, while irradiation in the presence of an appropriate triplet sensitizer populates the triplet state; the further reaction course depends on whether the reaction with spin conservation is faster or not than the (spin-forbidden) intersystem crossing. For silylcarbenes, these issues have not been looked at as closely as for some other carbene classes. In spite of the fact that a silyl substituent stabilizes the triplet state (see Section II), it appears, however, that silylcarbenes often undergo singlet carbene reactions. This is especially true for acyl(silyl)carbenes, which have a high tendency for the Wolff rearrangement (Section III.E.5) and add to 1,2-disubstituted alkenes with nearly complete retention of stereochemistry.

B. The Silylcarbene-to-Silene Rearrangement

Silylcarbenes, especially when generated by photochemical or thermal decomposition of diazo compounds or tosylhydrazone alkali salts or α-hydroxy-α-silylsilanes, tend to

13. Silicon-substituted carbenes

rearrange to silaethenes (silenes) by a 1,2(Si→C) migration of substituents such as H, alkyl, alkenyl, aryl and SiR$_3$ (equation 3). Several studies show the following migratory aptitude: H > Me[22], SiMe$_3$ > Me[29−31], (Me$_3$Si)$_3$Si > Me[32], Ph > Me[31,33]; in all these cases, the first-mentioned substituent migrates exclusively. While the observation Ph > Me was obtained from the photochemical decomposition of methyl (dimethylphenylsilyl)diazoacetate, product studies of the photolysis of diazo compounds Me$_2$(R)SiC(H)N$_2$ in ethanol revealed the competitive migration of substituents in the following, statistically corrected order: Me > Ph > CH$_2$Ph[34]. Methoxy groups do not migrate at all[23].

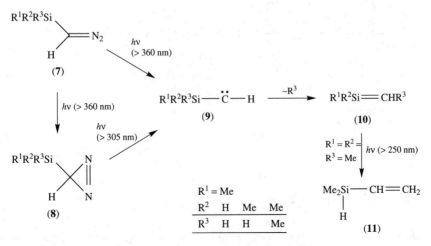

(3)

Met = Metal

All silenes generated so far on the silylcarbene route are reactive intermediates themselves, which were characterized by typical subsequent reactions[35] such as isomerization and dimerization or by trapping reactions (see below). However, photolysis of (silyl)diazo compounds in inert matrices at low temperature allowed the isolation and spectroscopic (IR, UV) characterization of several silenes (Scheme 2, Table 3). Irradiation of (diazomethyl)silanes **7** at λ > 360 nm produced both diazirine **8** and silenes **10**, but at shorter wavelength (λ > 305 nm) the silenes were produced cleanly from both precursors; the

SCHEME 2. Photochemistry of (silyl)diazo compounds in argon matrix

TABLE 3. Matrix isolation of silenes $R^1R^2Si=CR^3R^4$ by photolysis of (silyl)diazo compounds according to Scheme 2.

R^1	R^2	R^{3a}	R^4	Conditions[b]	Reference
Me	H	H	H	A	22
Me	Me	H	H	A	22
Me	Me	Me	H	A,B	20–22,31
Me	Me	Me	D	A	21
Me	Me	SiMe$_3$	H	B	29,31
Me	Me	Me	SiMe$_3$	B	29,31
Me	Me	SiMe$_3$	SiMe$_3$	B	29,31
Me	Me	Ph	COOMe	C	31
Me	Me	SiMe$_3$	COAd[c]	C	31
Me	Me	SiMe$_3$	COOEt	C	31,36

[a] R^3 is the substituent that has undergone the 1,2(Si→C) migration.
[b] A denotes argon matrix, 10 K, λ > 305 nm; B denotes 3-methylpentane matrix, 77 K, λ > 300 nm; C denotes 3-methylpentane matrix, 77 K, λ > 360 nm.
[c] Ad = 1-adamantyl.

carbenes **9** were probably too short-lived to be observed even under these conditions (see Section II)[20–22]. When matrix-isolated 1,1,2-trimethylsilene was irradiated at still shorter wavelengths, isomerization to dimethyl(vinyl)silane (**11**) took place[22].

Several transformations of silenes upon warming of the matrix have been described[20–22]. For example, dimerization of 1,1,2-trimethylsilene (**12**), leading to 1,3-disilacyclobutanes **13**, occurs above 45 K[20–22], addition of methanol before warming to 20 °C yields **14**[22] and warming of an O$_2$-doped matrix (0.5–1%) gives silylhydroperoxide **15** and vinyloxysilanol **16** in temperature-dependent relative yields[22] (Scheme 3).

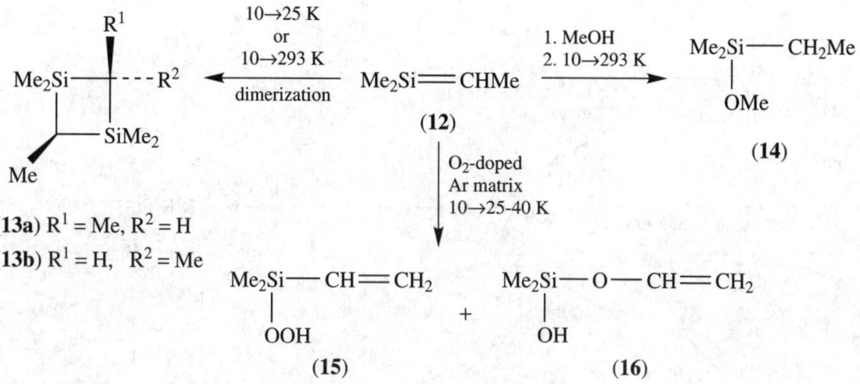

SCHEME 3. Reactions of matrix-isolated silene **12** on warming

Silaacrylate **17** is another example of a matrix-isolated silene[36]. It is formed simultaneously with ketene **18** on irradiation of methyl (pentamethyldisilanyl)diazoacetate

13. Silicon-substituted carbenes

in an argon matrix at 10 K (Scheme 4). At shorter wavelengths, the isomerization of **17** to **18** by a 1,3(C→Si) ethoxy shift takes place. Silene **17** does not react with methanol at 10 K, but above 35 K addition at the Si=C bond takes place and generates ethyl methoxydimethylsilyl–(trimethylsilyl)acetate. Similarly, irradiation of the diazoester in a methanol/tetrahydro-2-methylfuran glass at 77 K gave the (methoxysilyl)acetate (83%), **18** (11%) and methoxydimethylsilyl–(trimethylsilyl)ketene (4%) as the only isolable products.

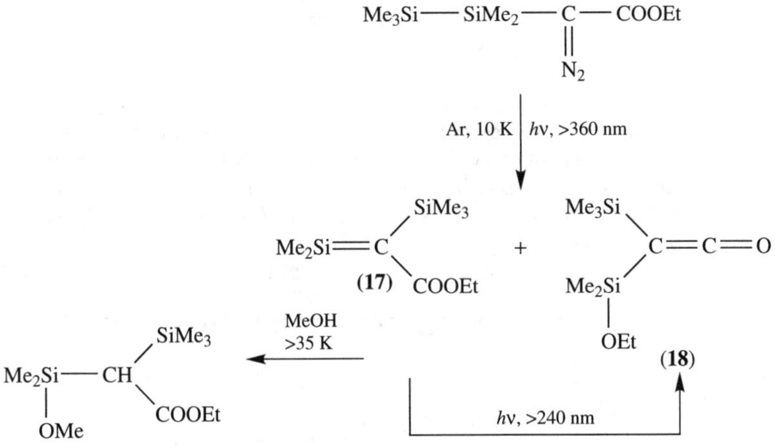

SCHEME 4. Preparation and reactions of silaacrylate **17** in an argon matrix

The following examples in this section illustrate that the silylcarbene-to-silene rearrangement and subsequent silene reactions are common also under nonmatrix conditions. The early research on this topic was reviewed in 1979[37].

When (trimethylsilyl)carbene (**3a**) is generated by photolysis of the diazo precursor **19** in an alcohol solution at room temperature, the rearrangement to the silene is so fast that only the addition products of the latter, but not the O,H insertion product of the carbene, can be isolated (equation 4)[34]. The same holds true for photolysis in diethylamine[38].

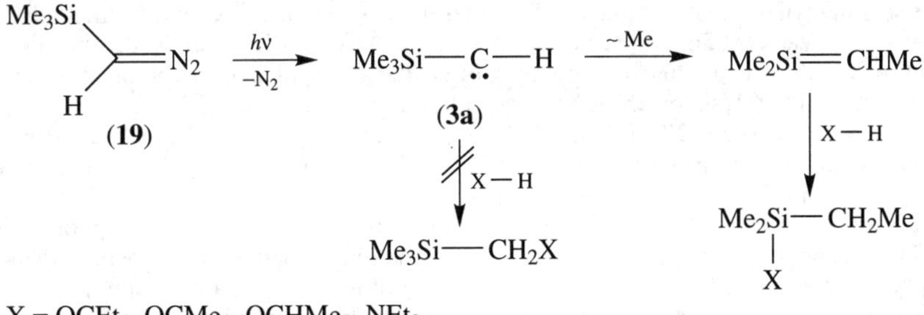

$X = OCEt_3, OCMe_3, OCHMe_2, NEt_2$

Analogously, copyrolysis of **19** with an alcohol at 425 °C also leads to the silene-derived product, but not to the O,H insertion product of the carbene[38]. The formation of 1-phenylpropene upon copyrolysis of **19** with benzaldehyde[38] (equation 5) corresponds to another well-established silene-trapping reaction, namely [2 + 2] cycloaddition between silene and carbonyl group and thermal [2+2] cycloreversion of the resulting silaoxetane.

$$(19) \xrightarrow[-N_2]{425\,°C} Me_2Si{=}CHMe \xrightarrow{PhCHO} \begin{array}{c}\text{silaoxetane}\end{array} \quad (5)$$

(12)

$-[Me_2Si{=}O]$ → PhCH=CHMe, 21%

When phenyl(trimethylsilyl)diazomethane (20) is pyrolyzed in the gas phase, typical reactions of carbene 21 can be observed (see Section III.E.4). However, copyrolysis with alcohols or carbonyl compounds generates again products which are derived from silene 22[39,40] (equation 6). Thus, alkoxysilanes 23 are obtained in the presence of alcohols and alkenes 24 in the presence of an aldehyde or a ketone. 2,3-Dimethylbuta-1,3-diene traps both the carbene (see Section III.E.4) and the silene.

$$Me_3Si(Ph)C{=}N_2 \xrightarrow[-N_2]{500\,°C} Me_3Si{-}\ddot{C}{-}Ph \longrightarrow Me_2Si{=}C(Me)Ph \quad (6)$$

(20)   (21)   (22)

ROH(D) → Me$_2$Si(OR)—C(Me)Ph(H(D)) (23)

R = Me (16%), Et (16%)

(24) 5–28%

| R^1 | Me | (CH$_2$)$_4$ | (CH$_2$)$_3$ | H | H |
| R^2 | Me | | | Ph | Et |

Bis(trimethylsilyl)diazomethane (25) represents an excellent source for silene 26[41]. It appears that carbene 3c, which is expected from the photochemical or thermal decomposition of 25, escapes most trapping efforts due to rapid isomerization to silene 26 (equation 7). Photolysis of 25 in benzene solution yields 27 and 28 in a combined yield of 64% and disilazane 29 (10%); all these products are likely to be derived from 26. Similarly, photolysis in the presence of methanol or D$_2$O traps the silene quantitatively (to give 31 and 32).

Remarkably enough, epoxide 30 was identified among the products of the photolysis of 25 in benzene/benzaldehyde. This seems to be the only reported case where carbene 3c has been intercepted. Pyrolysis of 25 in a nitrogen flow at 400 °C or vacuum pyrolysis at 500 °C led to the same product pattern as the photolysis. Copyrolysis with benzaldehyde or butadiene gave only trapping products of the silene intermediate[41].

Investigations similar to the ones just mentioned have focussed on (trimethylgermyl)(trimethylsilyl)carbene 34[41]. Photolysis of diazo compound 33 in MeOD yielded a mixture of methoxysilane 37 and methoxygermane 38 in a 79 : 21 ratio (equation 8). This result suggests that carbene 34 underwent a 1,2(Si→C) and a 1,2(Ge→C) methyl shift

13. Silicon-substituted carbenes

competingly and that silaethene **35** as well as germaethene **36** were trapped by the alcohol. In contrast, copyrolysis of diazo compound **33** with benzaldehyde allowed one to isolate only compounds that are likely to be derived from the intermediary silaethene **35**[41] (equation 9).

$$33 \xrightarrow[\text{PhCHO}]{\substack{400\,°C \\ N_2\text{ flow}}} 34 \longrightarrow 35 \qquad (9)$$

with 34 formed by ene reaction (PhCHO) giving Me₂Si(OCH₂Ph)–C(=CH₂)GeMe₃, and 35 formed by loss of [Me₂Si=O] giving Me₃Ge–C(Me)=CH–Ph.

Taking into account the better migrating ability of SiMe₃ with respect to methyl, it comes as no surprise that carbenes **36** and **38**, when generated photochemically or thermally from the corresponding diazo compounds, immediately rearrange to silenes **37** and **39** (equation 10). Again, only the silenes are trapped in the presence of alcohols, benzaldehyde or acetone[29,30]. When **37** was generated through flash pyrolysis at 450 °C in the absence of another reagent, it underwent head-to-tail dimerization to the corresponding 1,3-disilacyclobutane[30]. Silylcarbene-to-silene rearrangements by 1,2(Si→C) silyl group migration are also assumed for the photochemical decomposition of 1,3-bis(diazomethyl)trisilanes and 1,4-bis(diazomethyl)tetrasilanes (see Section III.D).

$$\text{Me}_3\text{Si}-\text{SiMe}_2-\ddot{\text{C}}-\text{H} \xrightarrow{\sim \text{Me}_3\text{Si}} \text{Me}_2\text{Si}=\text{CHSiMe}_3$$
$$(36) \hspace{4cm} (37)$$

$$\text{Me}_3\text{Si}-\text{SiR}_2-\ddot{\text{C}}-\text{SiMe}_3 \xrightarrow{\sim \text{Me}_3\text{Si}} \text{R}_2\text{Si}=\text{C}(\text{SiMe}_3)_2 \qquad (10)$$
$$(38) \hspace{4cm} (39)$$

(a) R = Me,   (b) R = SiMe₃

The last three entries of Table 3 tell us that 2-acylsilenes are accessible by irradiation of matrix-isolated (silyl)diazoketones and (silyl)diazoacetates. In fact, the same transformations can be reproduced in solution, but in all reported cases the 2-acylsilenes underwent fast intra- or intermolecular reactions or were trapped by added alcohols or enolizable as well as nonenolizable carbonyl compounds. Short-lived acyl(silyl)carbenes are generally assumed to be the immediate precursors of the 2-acylsilenes, but their intermediacy has never been proven. It appears that a clean silylcarbene-to-silene rearrangement for silyl-ketocarbenes (**40**) requires the presence of a disilanyl substituent (Si–Si) at the carbene, whereas silyl groups having only Si–C bonds favor the silylcarbene-to-silylketene isomerization (**40**→**41a**, Wolff rearrangement) (Scheme 5). Similarly, alkoxycarbonyl(disilanyl)carbenes (**43**, $R^1 = \text{SiR}_3$) rearrange only to 2-acylsilenes **45**, but alkoxycarbonyl(trialkylsilyl)carbenes (**43**, R^1 = alkyl, aryl) can undergo both types of rearrangement. Although acylsilenes **42** and **45** can be trapped (see below), it is also possible that they isomerize to ketenes **41b** and **46** respectively, by a fast 1,3 (C→Si) substituent migration. Thus, acyl(silyl)carbenes **40** and **43** may furnish not only the 'normal' ketenes **41a** and **44** from the Wolff rearrangement, but also the 'doubly rearranged' ketenes **41b** and **46**. In this section, we look only at the acyl(silyl)carbene-to-acylsilene isomerization, and leave the Wolff rearrangement to Section III.E.5.

When acyl(disilanyl)carbenes **48** are generated by photolysis of diazoketones **47** in benzene solution, the acyl(silyl)carbene-to-acylsilene rearrangement appears to be the exclusive consequence. The further fate of the acylsilenes **49** depends on the nature of substituent R (equation 11). When R is bulky (1-adamantyl (1-Ad), *t*-Bu), 1,4-cyclization yields the 1,2-silaoxetanes **50**, which easily decompose to an alkyne and dimethylsilanone at elevated temperature[32,42,43]. Acylsilenes with sterically less demanding substituents undergo a [4+4] cyclodimerization leading to the eight-membered silaheterocycles **51**[32,43].

In some cases, the doubly rearranged (i.e. silene-derived, see Scheme 5) ketenes **52** could be detected as minor by-products[32].

$$Me_3Si—SiMe_2—\underset{\underset{N_2}{\|}}{C}—COR \xrightarrow[-N_2]{h\nu \atop benzene} Me_3Si—SiMe_2—\ddot{C}—COR$$

(47) → (48)

(48) → Me$_2$Si=C(SiMe$_3$)(COR) (49)

(49) → (50) [R = 1-Ad, t-Bu]: 4-membered ring with Me$_2$Si–O–C(R)=C(SiMe$_3$)

(49) → (51) [R = i-Pr, Me, Ph, 2-furyl]: 8-membered dioxa-disila ring with Me$_3$Si, R substituents

(49) → (52) [R = t-Bu, i-Pr, Me]: Me$_3$Si(Me$_2$Si)(R)C=C=O

(11)

When carbonyl compounds were present during the photolysis of diazoketones **47**, the three acylsilene transformations displayed in equation 11 were partly or fully suppressed. (However, this was not the case for **47**, R = 1-Ad, which did not react with carbonyl compounds[42].) Thus, acylsilenes **49** reacted with enolizable ketones such as acetone, acetophenone and acetylacetone in an ene-type reaction forming silyl enol ethers, e.g. **54** (equation 12)[44]. With nonenolizable carbonyl compounds, such as benzophenone, crotonaldehyde and ethyl acetate, the acylsilenes where trapped in a hetero-[4+2] cycloaddition reaction forming 1,3-dioxa-4-sila-5-cyclohexenes **55**[44]. Surprisingly, the doubly rearranged ketenes **52** were no longer found in these reactions; rather, the 'normal' ketenes **53**, corresponding to a Wolff rearrangement, were formed in significant amounts and were characterized by addition of methanol[44]. It is not clear how the added carbonyl compounds partly redirect the reaction pathway from the exclusive silylcarbene-to-acylsilene to the silylcarbene-to-silylketene pathway. It would appear that at least some of these carbonyl compounds act as triplet sensitizers in the photolysis reaction, but it is well established that the Wolff rearrangement (i.e. **47** → **53** or **48** → **53**) occurs either from an excited singlet state of the diazoketone or from the singlet carbene.

In this context, it is worth mentioning that there is only one other, clear-cut example for the simultaneous occurrence of the acyl(silyl)carbene-to-acylsilene and the acylcarbene-to-silylketene rearrangement of an acylcarbene bearing a Si—Si substituent. Carbene **57**, generated by photolysis of diazoketone **56** in benzene, isomerized to both **58** and **59** in about equal amounts[44]. While the acylsilene cyclized to 1,2-silaoxetene **60**, the ketene was isolated and structurally characterized by X-ray diffraction analysis of the derived carboxylic acid **61** (equation 13).

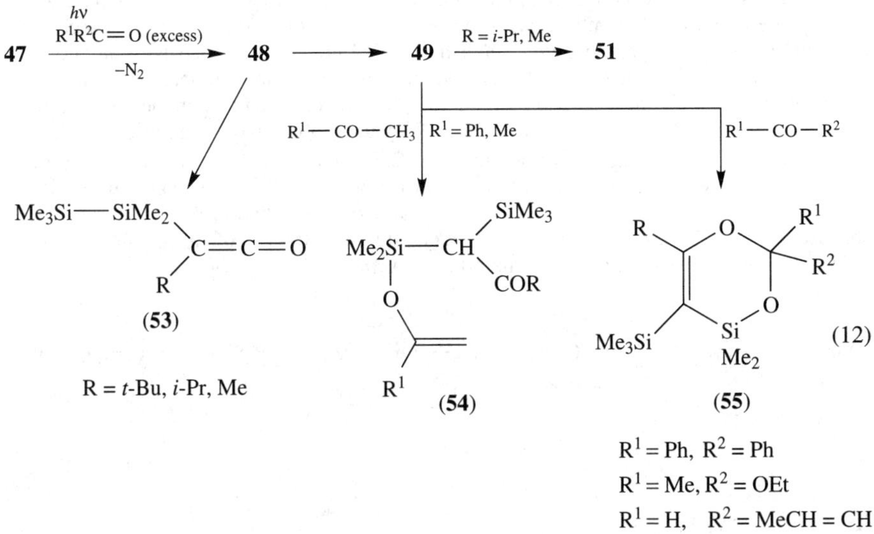

$$
\begin{array}{c}
\textbf{47} \xrightarrow[-N_2]{\substack{h\nu \\ R^1R^2C=O \text{ (excess)}}} \textbf{48} \longrightarrow \textbf{49} \xrightarrow{R = i\text{-Pr, Me}} \textbf{51}
\end{array}
$$

Branches from **48**:

- $R^1-CO-CH_3$, $R^1 = Ph, Me$
- R^1-CO-R^2

Products:

(53) Me$_3$Si—SiMe$_2$\C(R)=C=O; R = t-Bu, i-Pr, Me

(54) Me$_2$Si(SiMe$_3$)—CH(COR)—O—C(=CH$_2$)R^1

(55) Six-membered ring with R, Me$_3$Si, SiMe$_2$, O, O, and CR^{1}R^2 (equation 12)

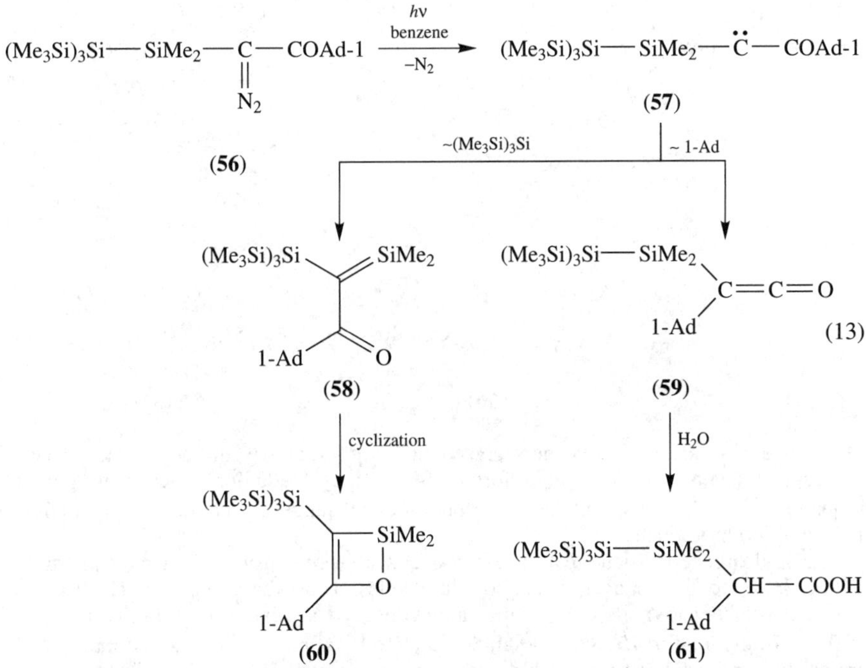

R^1 = Ph, R^2 = Ph
R^1 = Me, R^2 = OEt
R^1 = H, R^2 = MeCH=CH

$$
\text{(Me}_3\text{Si)}_3\text{Si}-\text{SiMe}_2-\underset{\substack{\| \\ N_2}}{C}-\text{COAd-1} \xrightarrow[-N_2]{\substack{h\nu \\ \text{benzene}}} (\text{Me}_3\text{Si})_3\text{Si}-\text{SiMe}_2-\ddot{C}-\text{COAd-1}
$$

(56) → **(57)**

From **(57)**: ~(Me$_3$Si)$_3$Si shift gives **(58)**: (Me$_3$Si)$_3$Si\C(SiMe$_2$)=C(1-Ad)(O)

~1-Ad shift gives **(59)**: (Me$_3$Si)$_3$Si—SiMe$_2$\C(1-Ad)=C=O (13)

(58) → cyclization → **(60)**: four-membered ring with (Me$_3$Si)$_3$Si, SiMe$_2$, O, 1-Ad

(59) → H$_2$O → **(61)**: (Me$_3$Si)$_3$Si—SiMe$_2$\CH(1-Ad)—COOH

As stated above, an alkoxycarbonyl(disilanyl)carbene rearranges to an acylsilene, but not to a silylketene (Scheme 5). For diazoacetate **62**, the result of a matrix-isolation experiment (Table 3) can also be reproduced when the nitrogen extrusion is induced by vacuum pyrolysis at 360 °C, by thermolysis of the neat compound at 180 °C or by photolysis in

THF solution[36,45]. In all cases, the formation of the doubly rearranged ketene **65** in nearly quantitative yield suggests that an isomerization sequence carbene→ silene → ketene has taken place (equation 14). While no direct evidence for the participation of carbene **63** is available, the presumed acylsilene intermediate **64** could be trapped with alcohols, by an ene-reaction with enolizable ketones (cyclohexanone, cyclooctanone, 3-pentanone, 3-heptanone) and by alkene formation with benzophenone or 2-adamantanone. For example, the photolysis of **62** in methanol gave **66** (70% yield), obviously an addition product of methanol to silene **64**, together with ketene **67** (23%) and a small amount of ketene **65**[36]. Since **65** is stable in methanol, **67** cannot stem from direct alkoxy exchange in **65**. It has been suggested that (similar to the migration of alkoxy groups in the Wolff rearrangement, see Section III.E.5.b), the 1,3 (C→Si) alkoxy migration (**64** → **65**) proceeds via an ion pair, which could be susceptible to the alkoxy exchange.

$$(14)$$

When diazoacetate **62** was thermolyzed in 7-norbornanone, the acylsilene **64** could be trapped again, this time in a formal [4 + 2] cycloaddition reaction to give **68** (equation 15)[46,47]. The simultaneous formation of **69** suggests a stepwise, ionic pathway providing both products.

The silylcarbene-to-silene rearrangement has also been considered for the construction of the Si=C bond of silabenzenes and silafulvenes. Thus, the photochemical decomposition of (diazomethyl)silole **70** in methanol produced a mixture of 1,2-dihydrosilin **74** (8%), 1,4-dihydrosilin **75** (8%) and silole **76** (19%)[48,49] (equation 16). Undoubtedly, the primarily formed silylcarbene **71** has undergone a ring expansion to form silabenzene **72**, and a 1,2(Si→C) methyl shift to form silafulvene **73**, and methanol has added to the Si=C bond of both intermediates. When the hydrogen atom in carbene **71** is replaced by a methyl group, an 1,2(C→C)H shift leading to a 1-vinylsilole takes over[49] (see Section III.E.2).

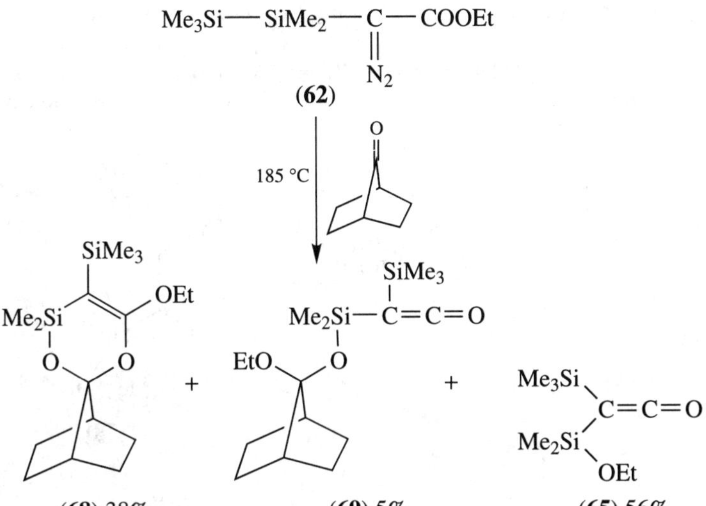

13. Silicon-substituted carbenes

$$\text{Me}_3\text{Si}-\text{SiMe}_2-\underset{\underset{\text{N}_2}{\|}}{\text{C}}-\text{COOEt}$$

(62)

(15)

(68) 38% + (69) 5% + (65) 56%

(16)

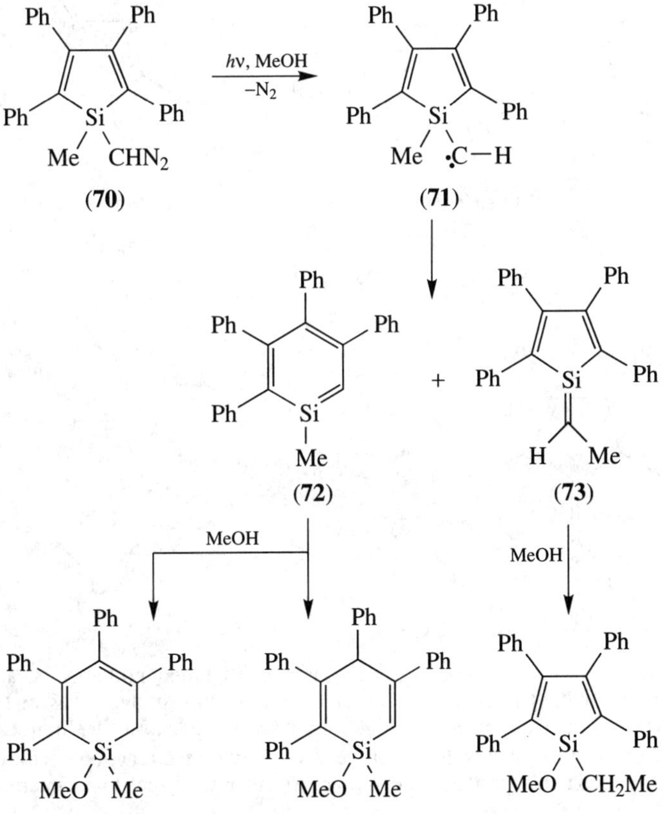

(70) → (71) → (72) + (73)

(72) → (74), (75); (73) → (76)

A silafulvene (**79**) was also formed when the 2-diazo-1,2-dihydrosilin **77** was decomposed under copper catalysis (equation 17)[49,50]. In the presence of *tert*-butanol, **79** was trapped as the (*tert*-butoxy)silane **81**. When **77** was decomposed in the presence of benzophenone or benzaldehyde, fulvenes **80** were obtained in another reaction typical for silenes.

(17)

While formation of a silabenzene from carbene **78** failed probably because of the low tendency of methyl to undergo the 1,2 migration, this strategy was at least partly successful with carbene **83**, where the high migratory aptitude of a Me₃Si group has been exploited[51]. The result depends, however, very strongly on the reaction conditions for the decomposition of **82**; the products are shown, together with presumed intermediates, in equation 18. In a similar manner, silabenzene **84** was generated and could be observed by its IR and NMR spectra at $-100\,°C^{52}$.

Conditions: A, $CuSO_4/Cu_2Cl_2$(cat.), benzene-MeOH, 25 °C, 8 days;
B, $CuSO_4/Cu_2Cl_2$(cat.), benzene-MeOH, >25 °C;
C, $h\nu$, $\lambda \le 385$ nm, MeOH-ether, $-60 \to 25$ °C;
D, $h\nu$, $\lambda \le 435$ nm, ether, 2,3-dimethylbuta-1,3-diene, 25 °C

(18 *continued*)

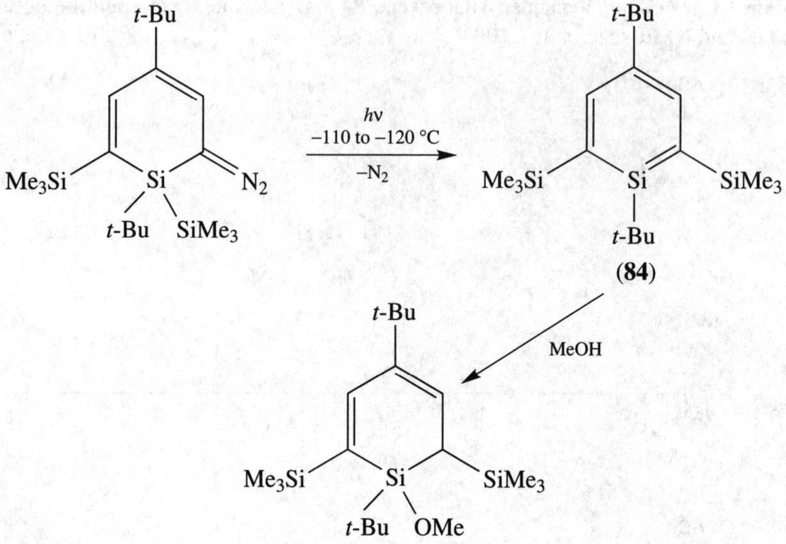

(18)

Photochemical or copper-catalyzed decomposition of diazo compound **85** failed to give a handle on 2-silanaphthalene **87** (equation 19)[49,50]. Instead of the expected 1,2-Ph migration, carbene **86** apparently underwent simply an O,H insertion reaction with methanol in 96% yield.

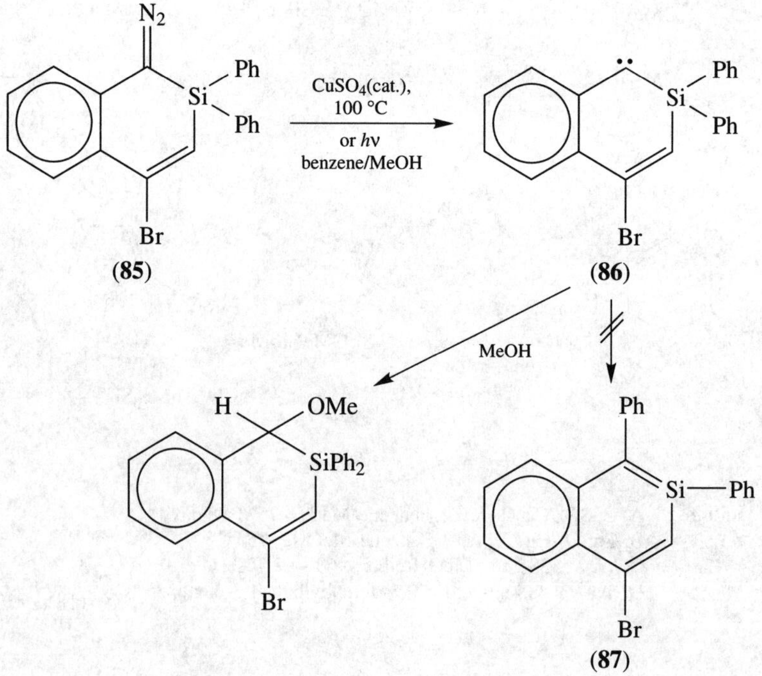

(19)

C. Intramolecular Carbene Reactions with Substituents Attached to Silicon

1. C,H insertion

a. 1,3-C,H insertion. In a silylcarbene, this reaction mode would lead to a silirane. Although silirane formation has been invoked as one of the reaction pathways of (trimethylsilyl)carbene[21,38,53,54], other (trialkylsilyl)carbenes[55,56] and phenyl(trimethylsilyl)carbene[40], a stable silirane has never been isolated from any of these reactions, probably because this strained ring system was not stable under the pyrolytic or nucleophilic reaction conditions. For example, flash vacuum pyrolysis of (trimethylsilyl)diazomethane at 440 °C yields the products **13**, **89** and **90**[21,38] (equation 20). While **13** is obviously the dimer of silene **12** and two mechanistic proposals exist for the formation of **90**[21,54], a deuterium-labeling study[54] [flash-vacuum pyrolysis of $Me_3Si-C(D)N_2$ at 750 °C] supports the proposal that vinylsilane **89** is formed by isomerization partly of silene **12** and partly of silirane **88**. Extrusion of dimethylsilylene from **88** may be the source of ethylene[57], similar to the observation of a small amount of styrene in the pyrolytic decomposition of phenyl(trimethylsilyl)diazomethane[40].

$$Me_3Si-CHN_2 \xrightarrow[-N_2]{440\,°C/\,10\,Torr} Me_3Si-\ddot{C}-H \longrightarrow Me_2Si\triangleleft$$
$$(19) \qquad\qquad (3a) \qquad\qquad (88)$$

$$Me_2Si=CHMe \qquad Me_2Si-CH=CH_2 \qquad H_2C=CH_2$$
$$(12) \qquad\qquad\quad |\atop H \qquad\qquad \sim 2\% \qquad (20)$$
$$\qquad\qquad\qquad (89)\ \ 3-4\%$$

(13) 38%   (90) 4%

The silirane intermediate could also account for the formation of vinylsilane **89** when (trimethylsilyl)carbene is generated from dichloromethyl(trimethyl)silane and Na–K in a gas-phase reaction[53]. The transformations of (trialkylsilyl)carbenes, generated from (α-halomethyl)silanes by α-elimination with a strong base, may involve transient siliranes as well (equation 23 and Section III.E.1.b).

For an eventual 1,3-C,H insertion in the copper-catalyzed decomposition of diazoketone **190**, see Section III.E.5.a.

b. 1,4-C,H insertion. (Trimethoxysilyl)carbene (**3e**), generated in an argon matrix by irradiation (λ > 305 nm) of (trimethoxysilyl)diazomethane (equation 1), undergoes clean 1,4-C,H insertion to form 1,2-silaoxetane **91** (equation 21) upon short-wavelength UV irradiation[23]. A carbenic 1,4-C,H insertion also accounts for the formation of silacyclobutanes **93** upon solution photolysis of diazoesters **92** (equation 22)[58]. Only one diastereomer of **93** was obtained in all cases, but the stereochemistry was not determined. The silacyclobutenes rearrange easily to the 1-oxa-2-silacyclohexenes **94** which in some cases

($X = N_3$, NCS) are already present in the reaction mixture.

$$(Me_3O)_3Si-\ddot{\ddot{C}}-H \xrightarrow[\text{Ar matrix}]{h\nu \; \lambda > 280 \text{ nm}} \text{MeO}-\underset{\underset{O}{|}}{\overset{\overset{OMe}{|}}{Si}}\quad\quad\quad\quad (21)$$

(3e) \qquad\qquad\qquad (91)

$$\underset{(92)}{t\text{-Bu}-\underset{\underset{X}{|}}{\overset{\overset{t\text{-Bu}}{|}}{Si}}-\underset{\underset{N_2}{\|}}{C}-COOEt} \xrightarrow[\substack{-N_2 \\ 57-74\%}]{h\nu \;(>300\text{ nm}) \\ \text{toluene}} \underset{\text{EtOOC}}{X-\overset{t\text{-Bu}}{Si}\square\text{Me}} \;+\; \underset{\text{EtO}}{\text{(94)}} \quad (22)$$

X = Cl, N₃, NCO, NCS (93)

Δ, X = Cl

c. 1,5-C,H insertion. Alkyl (dimethylsilyl)carbenes **95**, generated from an alkyl(chloromethyl)dimethylsilane and sodium, can undergo both 1,3- and 1,5-C,H insertion (equation 23)[56], but 1,4- and 1,6-C,H insertion reactions are not observed. 1,3-C,H insertion can occur at a methyl or a methylene group of **95**. The ratio of the products formed on the silirane route suggests a 1°/2° selectivity of 1 : 2.5, which is to be compared with the ratio 1.0 : 1.3–1.9 reported for 2-butylcarbene undergoing 1,3-C,H insertion[1].

$R^1 = CH_3, C_2H_5, C_3H_7$ further reactions

$R^1 = H, CH_3$

13. Silicon-substituted carbenes

Similarly generated (ω-alkenyl)(dimethylsilyl)carbenes **96** (equation 24) undergo 1,5-C,H insertion reactions giving **97** in competition with intramolecular cyclopropanation to give **98**[59]; however, the yields of all these reactions are low.

Carbenes **100**, generated by UV irradiation of methyl (alkoxysilyl)diazoacetates **99**, furnish 1-oxa-2-silacyclopentanes **101** by 1,5-C,H insertion (equation 25)[60]. In order to suppress competing reaction pathways of the carbene intermediate (e.g. formation of a ketazine with excess diazo compound), the photolysis was carried out at high dilution, but, even then, yields were rather modest. Carbene insertion at CH_2 seems to occur much more easily than at CH_3; a preference of 3 ± 0.6 : 1 for methylene insertion can be calculated from the isolated yields after correction for the number of C—H bonds. It should be noted that for both carbenes, **96** and **100**, C,H insertion occurs only in the 1,5 mode whereas no 1,3-, 1,4- or 1,6-C,H insertion products could be detected.

R^1	H	H	Me
R^2	H	Me	Me

2. Reactions at C=C bonds

[(3-Butenyl)-, (2,3-dimethyl-3-butenyl)- and (4-pentenyl)-dimethylsilyl)]carbene have been generated by treatment of the corresponding chloromethylsilanes with sodium. Intramolecular [1 + 2] cycloaddition of the carbenic carbon atom to the double bond leads to 1-silabicyclo[3.1.0]hexanes and 1-silabicyclo[4.1.0]heptanes, respectively, usually in competition with intramolecular C,H insertion (equation 24)[56]. In contrast, no carbene-derived product could be obtained from (allyldimethylsilyl)carbene. Finally, reaction of chloromethyldimethylvinylsilane with sodium provided, besides the typical products of a Wurtz reaction (**103** and **104**), a small amount of cyclopropane **106** (equation 26)[56]. It has been suggested that (dimethylvinylsilyl)carbene (**102**) isomerizes to silabicyclo[1.1.0]butane **105** by intramolecular cyclopropanation, and nucleophilic ring-opening finally leads to **106**.

$$\text{(26)}$$

In contrast to the failure to observe an intramolecular reaction of (allyldimethylsilyl)carbene[56], photolysis of the (allyldiisopropyl)diazoacetate **107** furnishes 1-silabicyclo[2.1.0]-pentane **109** in good yield[61] (equation 27). Obviously, the intramolecular cyclopropanation of carbene **108** accounts for this result. Altogether, the chemistry of (alkenylsilyl)carbenes (equations 24, 26 and 27) exhibits close similarities to that of the all-carbon analogues[1].

2-Oxa-1-silabicyclo[n.1.0]alkanes (n = 3: **111**; n = 4: **113**) were the only products isolated from the photochemical, thermal or transition-metal catalyzed decomposition of (alkenyloxysilyl)diazoacetates **110** and **112**, respectively (equation 28)[62]. The results indicate that intramolecular cyclopropanation is possible via both a carbene and a carbenoid pathway. The efficiency of this transformation depends on the particular system and on the mode of decomposition, but the copper triflate catalyzed reaction is always more efficient than the photochemical route. For the thermally induced cyclopropanation **112** → **113**, a two-step noncarbene pathway at the high reaction temperature appears as an alternative, namely intramolecular cycloaddition of the diazo dipole to the olefinic bond followed by extrusion of N_2 from the pyrazoline intermediate. A direct hint to this reaction mode is the formation of 3-methoxycarbonyl-4-methyl-1-oxa-2-sila-3-cyclopentenes instead of cyclopropanes **111** in the thermolysis of **110**.

$hv, \lambda > 300$ nm ; Δ, 140 °C ;
cat. CuOTf or $Rh_2(OOCC_3F_7)_4$

Photochemical or thermal decomposition of (vinyloxysilyl)diazoacetates **114** yields the 1-oxa-2-sila-3-cyclopentenes **118** as the major reaction products (equation 29), but much material is lost in the separation steps[61] (isolated yields: 12–24%). It is reasonable to assume that the zwitterion **116** is the direct precursor of **118**. It is open to speculation whether **116** is formed directly by attack of the electrophilic carbene on the electron-rich double bond in **115** or via cyclopropanation (by analogy to **108** → **109**, equation 27) and subsequent ring-opening of the so-formed push-pull-substituted cyclopropane. At least, the first-mentioned reaction mode is more likely when the decomposition of **114** is catalyzed by the highly electrophilic rhodium(II) perfluorobutyrate which leads to **117**, a structural isomer of **118**.

$$(29)$$

3. Reactions at C≡C bonds

The Cu(I)-catalyzed decomposition of (alkynyloxysilyl)diazoacetates **119** furnishes the silaheterocycles **120** and/or **121** (equation 30) in modest yield[63]. In these cases, the photochemical extrusion of nitrogen from **119** does not lead to defined products and the thermal reaction is dominated by the 1,3-dipolar cycloaddition ability of these diazo compounds. In mechanistic terms, carbene **122** or more likely a derived copper carbene complex, is transformed into cyclopropene **123** by an intramolecular [1 + 2] cycloaddition to the triple bond. The strained cyclopropene rearranges to a vinylcarbene either with an exocyclic (**124**) or an endocyclic (**125**) carbene center, and typical carbene reactions then lead to the observed products. Analogous carbene-to-carbene rearrangements are involved in carbenoid transformations of other alkynylcarbenes[64].

D. Carbene Reactions of Bis(diazomethyl)silanes and -polysilanes

When two carbene functions are separated by one or more silicon atoms, one can expect them to enter independently the usual inter- or intramolecular reactions. Among the intramolecular reactions, extensions of those which have been discussed in Sections III.B and III.C are particularly appealing, namely silylcarbene-to-silene rearrangement at one or both carbene centers and intramolecular carbene dimerization to form a C,C double bond and thus an unsaturated silaheterocycle.

$R^1 = R^2 = H$: **(120)** 26% ; **(121)** 25%
$R^1 = H, R^2 = Me$: **(120)** 25%
$R^1 = Me, R^2 = H$: **(121)** 39%

To date, bis(diazomethyl)silanes represent the only precursors to silylene-dicarbenes. It should be noted, however, that the extrusion of N_2 from the two diazo functions can occur more or less simultaneously or successively. In the latter case, a diazocarbene is formed in the first place, which is supposed to undergo a fast carbene reaction before the generation of a carbene from the second diazo function occurs. This should be kept in mind for all transformations which are explained mechanistically with the participation of silylene-dicarbenes. There are even examples where only one of the two diazo groups enters carbene chemistry at all (see equations 35 and 36 below).

Various intramolecular transformations are conceivable for dicarbene **127**, which would result from twofold elimination of molecular nitrogen from bis(diazomethyl)silane **126**. HF energies, calculated by *ab initio* methods, for the possible rearrangement products **128–134** are given in Scheme 6[65].

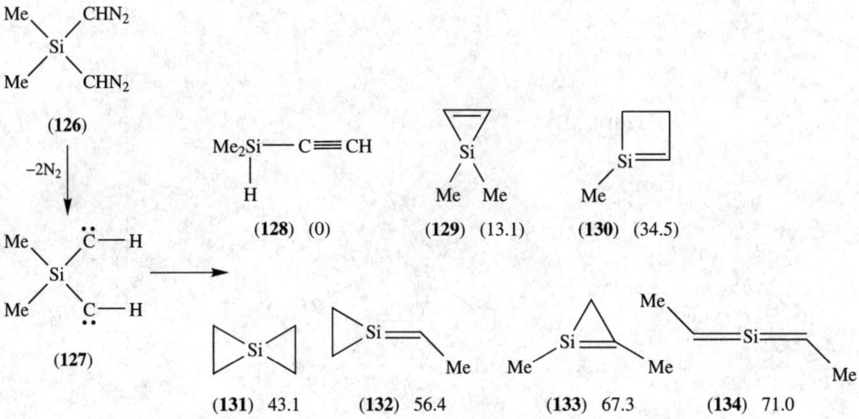

SCHEME 6. Possible intramolecular products from dicarbene **127**; HF energies relative to **128** (kcal mol^{-1}, calculated at the RHF/6-31G(d,p) level) are given in parentheses

Photolysis ($\lambda \geq 305$ nm) of **126** in an argon matrix at 10 K generated a single photoproduct to which the 1*H*-silirene structure **129** was assigned by comparison of the experimental with the calculated IR spectrum[65]. Short-wavelength irradiation ($\lambda >$ 254 nm) of **129** caused its rearrangement to ethynylsilane **128**. The latter was also the main product of a flash vacuum pyrolysis of **126** at 500 °C. When **126** was photolyzed in ethanol (170 and 273 K) or in an EPA glass (77 K), products **135–138** (equation 31) were obtained in temperature-dependent yields. While **135**, **136** and **138** are likely to be derived from 1*H*-silirene **129**, dicarbene **127** and 1-silacyclobutene **130** (formed from the carbene by 1,2-methyl shift followed by 1,4-C,H insertion of the second carbene center), respectively, diethoxysilane **137** could result from the intermediates **131**, **132** or **134**.

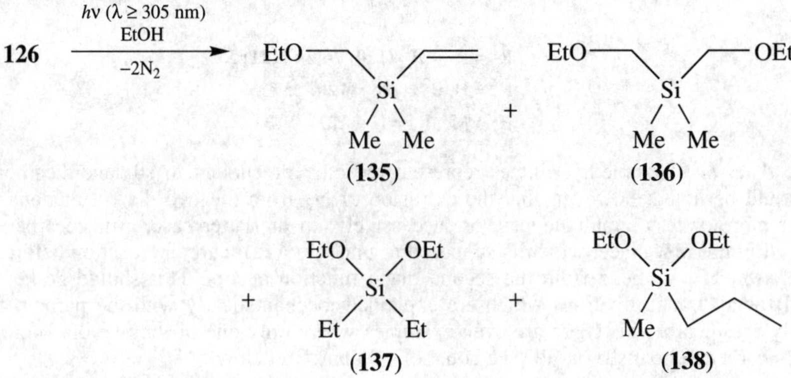

Silirenes (**140**, equation 32) could also be involved in the transition-metal catalyzed decomposition of bis(diazoketones) **139** which provides the electron-rich [4]radialenes **142**[66,67]. While the formation of **142** directly from silirene **140** cannot be excluded *a priori*, it is more reasonable to assume that **140** undergoes twofold ring-expansion to form the cyclic cumulene **141**, which then provides **142** by a cyclodimerization reaction. The intermediacy of **141** is corroborated by the isolation of the Diels–Alder product **143**[66].

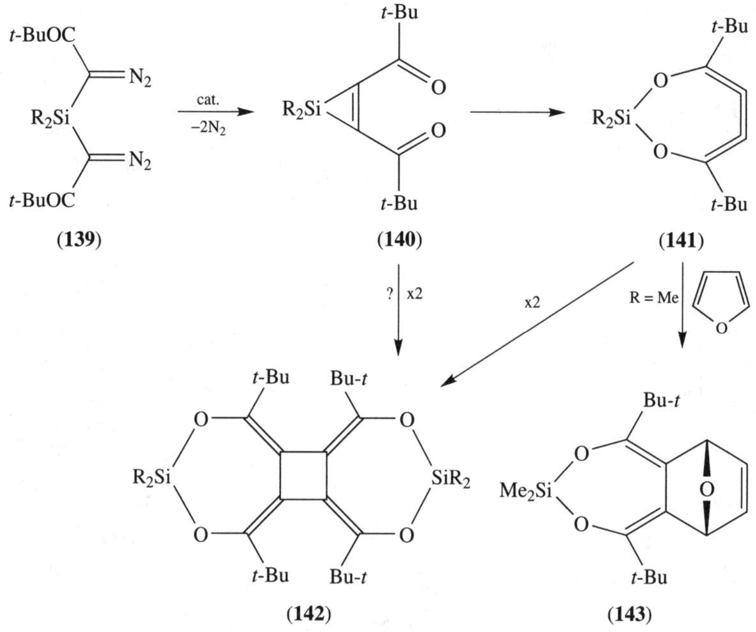

(32)

R = Me: 38-40%
R = *i*-Pr: 5-9%
cat.: CuOTf or Pd(OAc)$_2$

By analogy with the formation of a 1*H*-silirene from a silylene-dicarbene, a disilanediyl-1,2-dicarbene could isomerize to a 1,2-disilacyclobutene. Such a species (**147**) may be involved indeed in the photochemical or metal-catalyzed decomposition of bis(diazoketones) **144** (equation 33), from which silaheterocycles **145** and **146** were obtained in low yield[68]. Electrocyclic ring-opening of **147** and twofold cyclization of bissilene **148** could provide **145**, and twofold ring-expansion of **147** with subsequent hydrogenation of the resulting cyclic [3]cumulene could account for **146**. However, these mechanisms are still a matter of speculation. Apart from the fact that it is not known whether a diazocarbene or a dicarbene is the reacting species, it is also possible that a silylcarbene-to-silene rearrangement is involved in the formation of the products shown.

When the COR groups of bis(diazo) compound **144** are replaced by Ph or SiMe$_3$ substituents, the derived carbene chemistry leads to different results[69] (equation 34). The only identified product obtained from the flow pyrolysis of 1,2-bis(diazobenzyl)-tetramethyldisilane was 1,1,2,2-tetramethyl-3,4-diphenyl-1,2-disilacyclobutene, which rapidly reacts with air under insertion of an oxygen atom into the Si—Si bond. In contrast, the analogous 1,2-disilacyclobutene could not be directly isolated from the vacuum pyrolysis of 1,2-bis[diazo(trimethylsilyl)methyl]-tetramethyldisilane. However, exposure of the pyrolysate to air, N-methyl-1,2,4-triazolinedione or methanol furnished products that point to the presence of this 1,2-disilacyclobutene in the pyrolysate, perhaps in equilibrium with the corresponding bissilene.

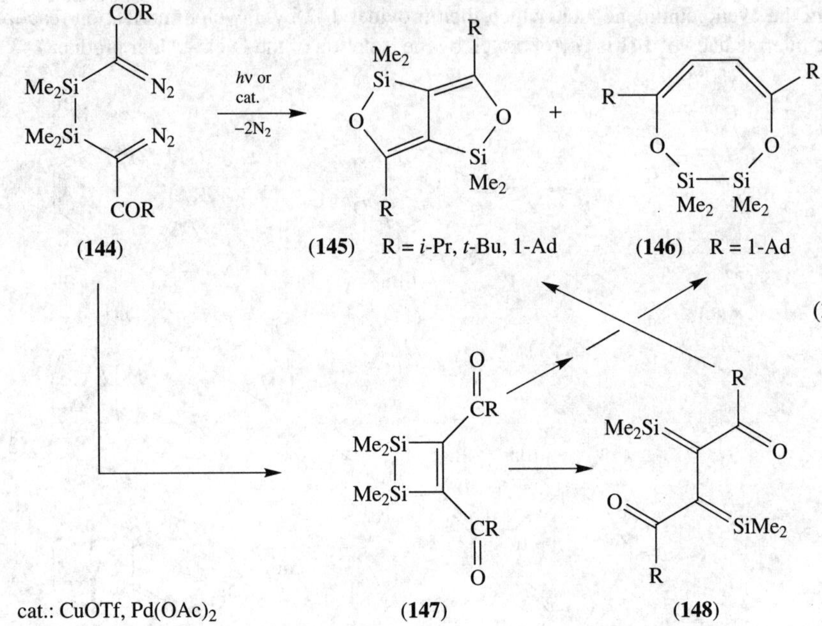

(33)

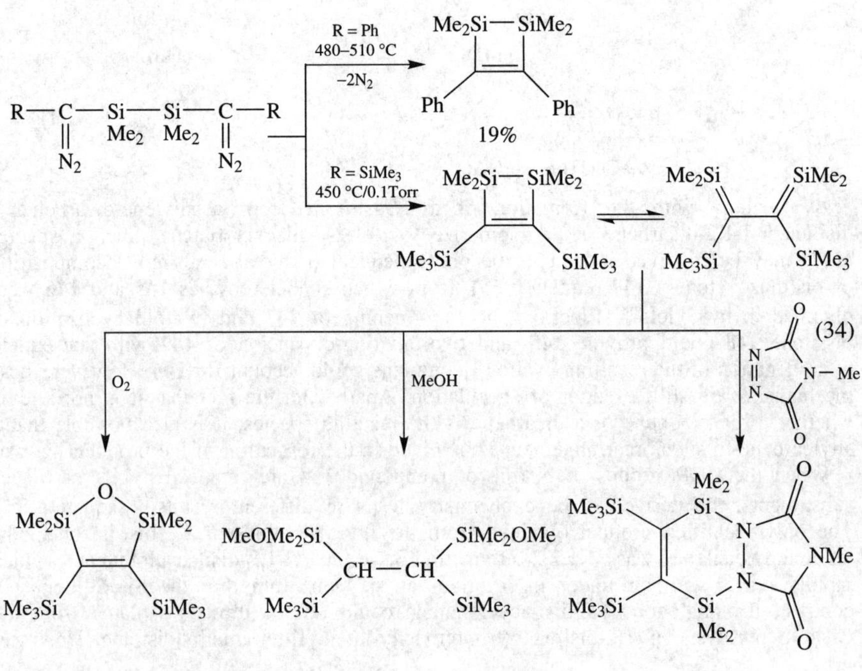

cat.: CuOTf, Pd(OAc)$_2$

(34)

13. Silicon-substituted carbenes

The photochemical decomposition of the same diazo compound in *tert*-butanol/benzene generated five products, in which either one or two alcohol molecules are incorporated and which are obviously derived from diazosilene and bissilene intermediates.

Photochemical or thermal decomposition of 1,3-bis(diazomethyl)trisilanes **149, 152** and of 1,4-bis(diazomethyl)tetrasilanes **155** begins with formation of only one carbene center followed by a silylcarbene-to-silene rearrangement[69–71] (equation 35; suggested intermediates are put in brackets for clarity). Intramolecular [3+2] cycloaddition of the resulting diazosilenes (e.g. **150**) would create bicyclic pyrazolines (e.g. **151, 153**) which can furnish the final isolated products with retention or loss of the azo moiety. Alternatively, the diazosilene intermediate can be transformed into a disilene (e.g. **154**). For the photolysis of **149**, only the pyrazoline route with retention of the azo moiety was observed[69,70]. UV-irradiation of bis(diazo)trisilanes **152** in cyclohexane yields mainly polymeric material, but also small amounts of a highly air-sensitive 2,3,5-trisilabicyclo[2.1.0]pentane and a 2,4,5-trisilabicyclo[1.1.1]pentane[69,70]. While the latter is considered to be the head-to-tail cyclodimer of disilene **154**, the former does not seem to be the corresponding head-to-head dimer but rather a product arising from pyrazoline **153**, by analogy with the pathways suggested for **149** and **155**[69]. Bissilene **154** (R = Me) could be trapped when the photolysis of **152** was carried out in the presence of a teriary amine or in *tert*-butanol[71].

A diazosilene is probably also involved in the photochemical or copper-catalyzed decomposition of bis(diazoacetate) **156** in benzene (equation 36). In both cases, diazoketene **157** was the only identified product[72]. Its formation was explained by the silylcarbene-to-acylsilene-to-silylketene sequence outlined in Scheme 5. Efforts to achieve the N_2 extrusion from the remaining diazo function by thermolysis in boiling toluene or by prolonged photolysis resulted only in unspecific decomposition.

(35 continued)

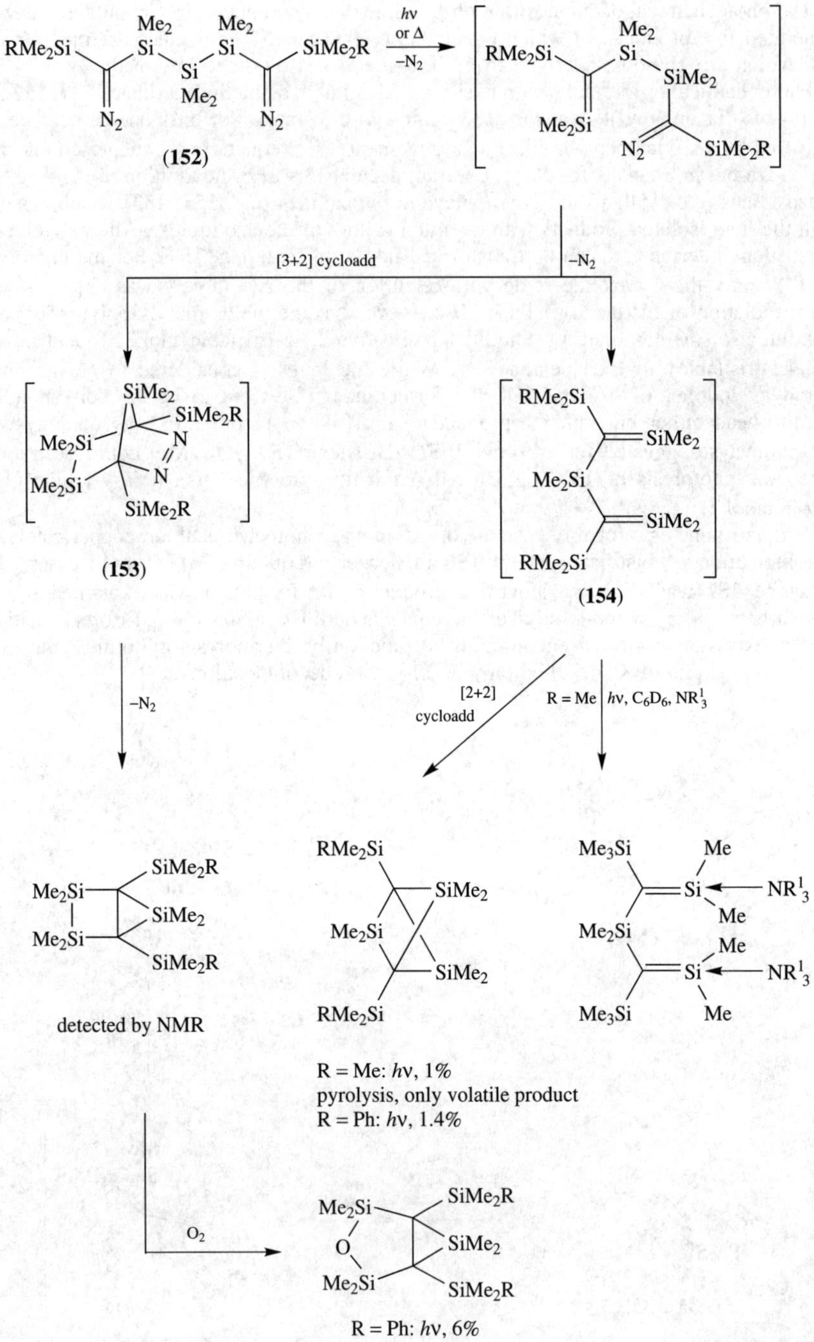
(152)
hν or Δ
−N2
[3+2] cycloadd
−N2
(153)
(154)
−N2
[2+2] cycloadd
R = Me
hν, C6D6, NR¹3
detected by NMR
R = Me: hν, 1%
pyrolysis, only volatile product
R = Ph: hν, 1.4%
O2
R = Ph: hν, 6%

(35 *continued*)

E. Silylcarbene Reactions not Involving the Silyl Group

1. Silylcarbenes without a second substituent

a. (Trimethylsilyl)carbene. This carbene is probably the best investigated silylcarbene. It is assumed to be an intermediate in the reactions of (dichloromethyl)trimethylsilane with Na–K in the gas phase[53] (equation 37, method A) and of (chloromethyl)trimethylsilane with alkali metals[55] (method B) or lithium 2,2,6,6-tetramethylpiperidide[73,74] (method C). Furthermore, the carbene can be generated by thermal decomposition of bis[bromo(trimethylsilyl)methyl]mercury[75] (method D) and by extrusion of molecular nitrogen from diazo(trimethylsilyl)methane[5,6] (method E). It is not clear, however, whether the carbenoid-type species phenylthio(trimethylsilyl)methyl lithium provides a source for this carbene by α-elimination[76] (method F, see also Section III.E.8).

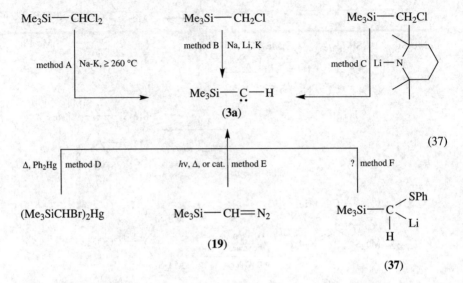

(37)

While method A furnished only dimethyl(vinyl)silane (see Section III.C.1.a), the Wurtz reaction (method B) in an inert solvent gave a mixture of products which stem either from the α-elimination generating **3a** or from the C,C coupling process leading to **158** (equation 38)[55]. It is assumed that **3a** forms 1,1-dimethylsilirane by 1,3-C,H insertion (cf equation 23) and that C,C and Si,C cleavage of this sensitive ring system by (trimethylsilyl)methyl sodium is responsible for the formation of **159–161** (combined yield: 32%). It is obvious that, due to the complexity of this reaction, method B cannot serve as a versatile entry into the chemistry of **3a**.

With respect to the complications of method B, it is remarkable that generation of carbene **3a** according to method C in the presence of *cis*-but-2-ene or various cycloalkenes gave the corresponding (trimethylsilyl)cyclopropanes in yields up to 30%[73,74]. Similarly,

1,2,3-tris(trimethylsilyl)cyclopropene was obtained in 15% yield by carbene transfer to bis(trimethylsilyl)acetylene[77]. The base is crucial for the success of the α-elimination; with sec-butyl lithium as a base, only Me₃SiCH(Li)Cl is formed which is reluctant to the carbene-forming step.

Method D (equation 37) is neither a convenient nor an effective way to generate **3a**. After heating bis[bromo(trimethylsilyl)methyl]mercury for 7 days at 160 °C with cyclohexene and diphenylmercury, 7-trimethylsilylbicyclo[4.1.0]heptane was obtained in a yield of only 9%, accompanied by a trace of the carbene dimers (*cis*- and *trans*-bis(trimethylsilyl)ethene) and a large amount of starting material[75].

By far the best source for **3a** is (trimethylsilyl)diazomethane (**19**). It has already been mentioned that gas-phase pyrolysis of **19** alone[21,38,54] yields products which are derived from intramolecular carbene reactions such as 1,3-C,H insertion and silylcarbene-to-silene rearrangement (see equation 20). Also, copyrolysis of **19** with alcohols or benzaldehyde allowed one to trap the silene but not the carbene **33** (see equation 5). Furthermore, solution photolysis of **19** in the presence of alcohols or amines did not give the X,H insertion products of the carbene but rather trapping products of the silene[33,38]. On the other hand, photochemically generated carbene **3a** did undergo some typical intermolecular carbene reactions such as cyclopropanation of alkenes (ethylene, *trans*-but-2-ene, but not 2,3-dimethylbut-2-ene, tetrafluoroethene and hexafluoropropene), and insertion into Si—H and methyl-C—H bonds[78] (equation 39). The formal carbene dimer, *trans*-1,2-bis(trimethylsilyl)ethene, was a by-product in all photolyses in the presence of alkenes; it is generally assumed that such carbene dimers result from reaction of the carbene with excess diazo compound.

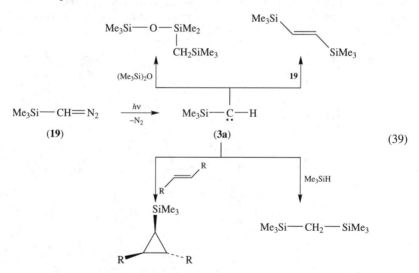

(39)

Transfer of carbene **3a** from **19** to a variety of alkenes and cycloalkenes has been achieved under catalysis by copper(I) chloride[74,79–82]. However, with the exception of cyclohexene[69,70] (72% yield), only moderate yields could be obtained. In all cases, the cyclopropanation was stereospecific with respect to the double bond configuration of the alkene and gave the sterically less crowded cyclopropane diastereomer in excess. As in the photochemical cyclopropanation, the formal carbene dimer *trans*-1,2-bis(trimethylsilyl)ethene is often formed as the major by-product. Cyclopropanation of *trans*-but-2-ene with **19** with copper(II) chloride as catalyst was found to be even less productive than the photochemical carbene transfer reaction[78] (13 vs 23% yield). Not unexpectedly, electron-deficient acyl-substituted alkenes could also be cyclopropanated when $PdCl_2$ was used as catalyst, whereas the copper- or rhodium-catalyzed carbene transfer failed[82] (equation 40).

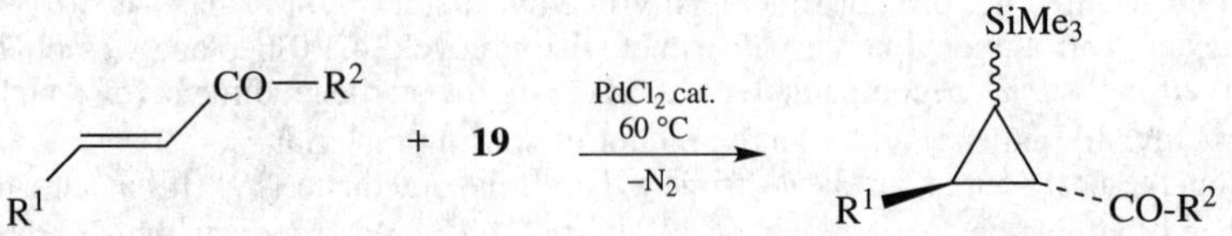

(40)

R^1 =	Ph	Ph	Ph	C_5H_{11}
R^2 =	OEt	Me	Ph	OEt
yield(%)	79	32	21	23

In connection with these catalytic cyclopropanation reactions, it should be mentioned that the isolable ruthenium–carbene complex **162**, which is obtained from **19**, [$RuCl_2$(*p*-cymene)]$_2$ and 2,6-bis(4-isopropyl-1,3-oxazolin-2-yl)pyridine, reacts with styrene at elevated temperature in a carbene transfer reaction[83] (equation 41). Since complex **162** is also catalytically active for (alkoxycarbonyl)carbene transfer to olefins, this reaction represents one of the few connecting links between catalytic and stoichiometric carbene transfer reactions of metal–carbene complexes.

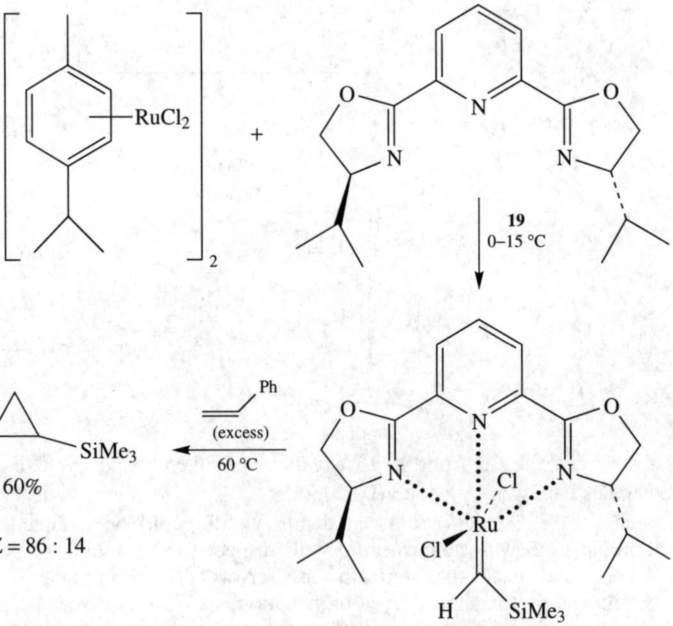

(41)

b. Other (alkylsilyl)carbenes. (Ethyldimethylsilyl)carbene, (propyldimethylsilyl)carbene and (butyldimethylsilyl)carbene have been generated from the corresponding trialkyl(chloromethyl)silane according to method B (equation 37), but no other transformations than their intramolecular C,H insertion reactions (cf equation 23) are known[56]. (Benzyldimethylsilyl)carbene and (dimethylphenylsilyl)carbene, when generated by UV irradiation of the corresponding diazomethylsilane in the presence of ethanol, did not seem to react with the alcohol, probably because of the high rate of the silylcarbene-to-silene rearrangement which led finally to the addition product of the alcohol to the silene[33].

c. (Alkenylsilyl)carbenes. The preparation of these carbenes and intramolecular reactions such as competitive C,H insertion and cyclopropanation have been discussed in Section III.C.2. No intermolecular chemistry has been reported.

d. (Trimethoxysilyl)carbene. Irradiation of (trimethoxysilyl)diazomethane in a CO-doped argon matrix at 35 K produced (trimethoxysilyl)ketene, which is a trapping product of carbene **3e**[21] (equation 42). For the reaction of **3e** with molecular oxygen, see equation 1. No solution chemistry of **3e** has been reported as yet.

$$(MeO)_3Si-CH=N_2 \quad \xrightarrow[\text{3.5\% CO}]{\substack{h\nu\ (\lambda > 360\ nm) \\ 35K,\ argon\ matrix}} \quad (MeO)_3Si-\underset{\cdot\cdot}{C}-H$$

(**3e**)

$$\downarrow CO$$

$$(MeO)_3Si-CH=C=O$$

(42)

2. Alkyl(silyl)carbenes and cycloalkyl(silyl)carbenes

For alkyl(silyl)carbenes where the alkyl contains an α-C–H bond, 1,2(C→C) hydride shift leading to a vinylsilane is the common reaction pathway. Vinylsilane formation has been observed for free (photochemically or thermally generated) carbenes (equations 43[84], 44[48,49,50] and 45[85,86]) but also in carbenoid reactions. In the latter case, the configuration of the alkene could be controlled to a large extent by the choice of the catalyst: The *E*-alkene was formed nearly exclusively with copper(I) chloride as catalyst[87], whereas rhodium(II) pivalate[88] gave mainly the *Z*-alkene (equation 46).

$$Ph_3Si-\underset{\underset{N_2}{\|}}{C}-CH_2R \quad \xrightarrow[-N_2]{h\nu} \quad Ph_3Si-\underset{\cdot\cdot}{C}-CH_2R$$

$$\downarrow$$

$$Ph_3Si-CH=CHR$$

R = H: 55%
R = Ph: 75%; *E/Z* = 2.6

(43)

$$\text{(44)}$$

$$\text{(45)}$$

R = CH$_2$Ph (98%) ; E/Z = 1.0
R = C$_6$H$_{13}$ (93%) ; E/Z = 1.0
R = C$_8$H$_{17}$ (94%) ; E/Z = 0.77

R = PhCH$_2$, C$_9$H$_{19}$, CH$_3$(CH$_2$)$_3$CH(Et),
CH$_2$=CH(CH$_2$)$_8$, CH$_3$(CH$_2$)$_2$C≡C—CH$_2$,

[1,3-dioxan-2-yl]—CH$_2$, Ph, [2-thienyl]—CH$_2$;

Z/E ≤ 6/94

R = (CH$_2$)$_8$Me, CH$_2$Ph, Ph,
1-naphthyl, 2-thienyl ;
Z/E = 78/22 to 97/3

$$\text{(46)}$$

The dramatic influence of a methyl group on the reaction pathway is exemplified by carbene **163** which rearranges to a vinylsilane. In the absence of this methyl group, however, the silylcarbene-to-silene rearrangement (equation 16) occurs[48,49].

Cycloalkyl(silyl)carbenes with an α-C—H bond have not yet been investigated systematically. When cyclohexyl(trimethylsilyl)carbene was generated by thermal α-elimination from cyclohexyl-bis(trimethylsilyl)methanol at 500 °C, only the (1,2) hydride shift took place, whereas cyclopentyl(trimethylsilyl)carbene, generated analogously, gave both the endocyclic alkene and the 1,3-C,H insertion product[85,86] (equation 47).

3. Alkenyl(silyl)carbenes and cycloalkenyl(silyl)carbenes

a. 1-Alkenyl(silyl)carbenes. Acyclic vinylcarbenes tend to rearrange to cyclopropenes. This transformation is also involved in the photochemical conversion of 3*H*-pyrazole **164** into cyclopropene **167**[89] (equation 48). Photochemical ring-opening of **164** generates

13. Silicon-substituted carbenes

diazo compound **165**, which on prolonged irradiation yields the carbene **166**.

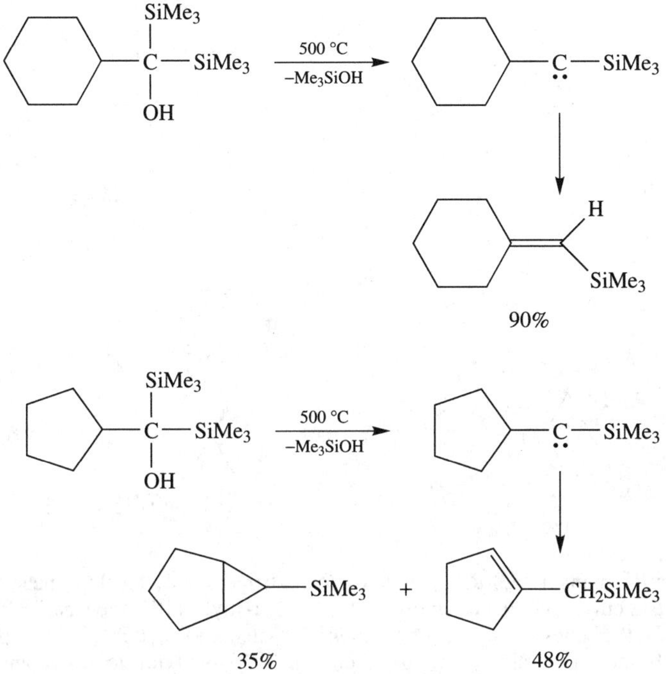

(47)

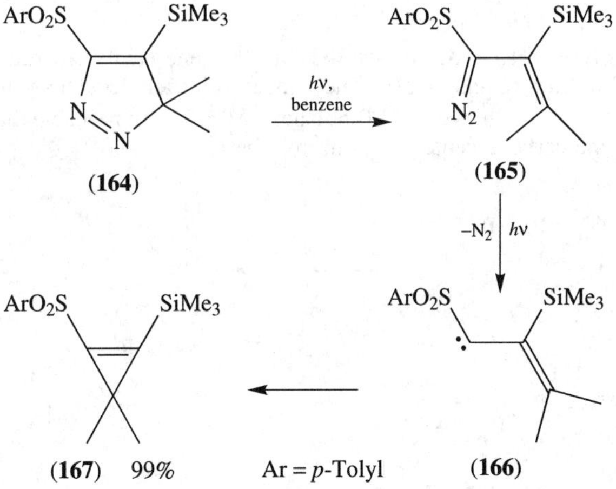

(48)

(**167**) 99%   Ar = *p*-Tolyl   (**166**)

In principle, the vinylcarbene-to-cyclopropene isomerization is reversible. While this has not been reported for **167** (where ring-opening could produce not only alkenyl(sulfonyl)carbene **166**, but also the isomeric 1-alkenyl(silyl)carbene), it was found that 1-trimethylsilylcyclopropenone acetal **168** (equation 49) by thermal ring-opening yields the (trimethylsilyl)vinylcarbene **169** besides traces of the isomeric vinylcarbene **170**. Both carbenes are obviously nucleophilic since they are able to cyclopropanate the electron-poor double bond of acrylonitrile[90a].

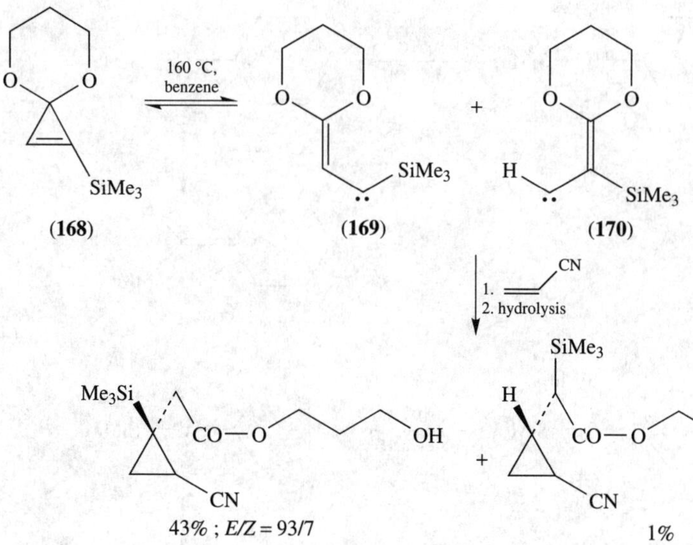

(49)

In a similar manner, 3,3-disubstituted 1,2-bis(trimethylsilyl)cyclopropenes rearrange to 1,1-bis(trimethylsilyl)allenes, most likely by 1,2-silyl shift of primarily formed (1-silylvinyl)silylcarbenes. According to *ab initio* calculations, this reaction pathway is energetically more favorable than those including a 2,2-disilylcyclopropylidene or a 2,3-disilylpropylidene[90b].

b. 2-Alkenyl(silyl)carbenes. Carbenes **172** can be generated from α,α-dibromosilanes **171** and methyl lithium (equation 50). They undergo intramolecular cyclopropanation of the C=C bond to form silylbicyclo[1.1.0]butanes **173**[91]. It is remarkable that an 1,2(C→C) hydride shift to the carbene center plays no role here.

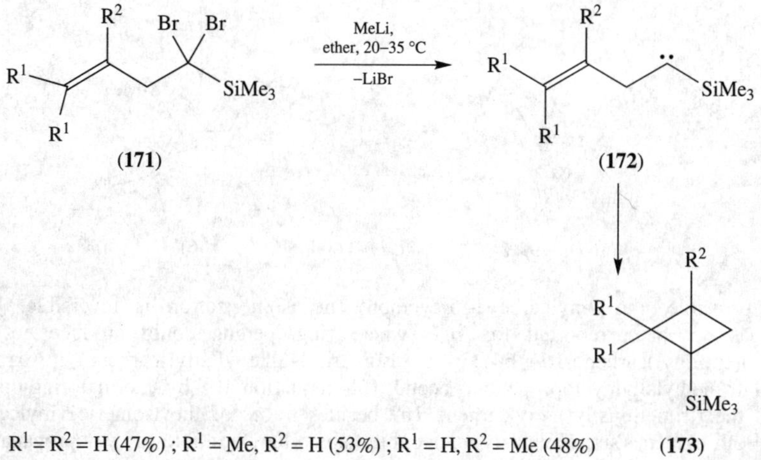

(50)

$R^1 = R^2 = H$ (47%) ; $R^1 = Me, R^2 = H$ (53%) ; $R^1 = H, R^2 = Me$ (48%) (**173**)

Bicyclobutanes are also obtained from the catalytic decomposition of diazo compound **174**[92] (equation 51). Copper(I) iodide was the catalyst of choice, whereas rhodium(II) acetate did not show any activity in this case. When the related diazo compound **175** was decomposed, the product pattern depended in an unusually selective manner on the catalyst[92]. Intramolecular cyclopropanation leading to **176** is obviously less favorable than for carbene **172** and must yield to the 1,2-hydride shift not observed with the former carbene. The configuration of the resulting butadiene **177** can be completely reversed by the choice of the catalyst.

(51)

catalyst	%**176**	% Z-**177**	% E-**177**
$Rh_2(OAc)_4$	–	91	–
$PdCl_2(CH_3CN)_2$	–	–	74
CuI	33	–	49

c. Cycloalkeny(silyl)carbenes. Under UV-irradiation (in solution, in an organic glass at −196 °C or in argon matrix), the carbene generated from various [(2-cyclopropenyl)diazomethyl]silanes **178** (R = Me[93,94], Ph[95], OPr-i[95]) undergoes exclusive fragmentation to give two alkynes (equation 52). By analogy with experiments carried out with the analogue of **178** with COOR instead of $SiMe_2R$[96] it is assumed that this fragmentation is a triplet carbene reaction. As the related *tert*-butylcarbene (**178**, *t*-Bu instead of $SiMe_2R$) undergoes ring-expansion rather than fragmentation, it has been suggested that the silyl group enhances the rate of the intersystem crossing from the singlet to the triplet carbene. While the carbene fragmentation is also one of the pathways of the thermolysis of silylated diazo compounds **178**, the copper(I)-catalyzed decomposition generates the stable cyclobutadienes **179** (equation 52), which can be isolated in pure form after complexation of the copper salt with 1,2-(diphenylphosphino)ethane[93-95]. Obviously, a 1,2-carbon shift bringing about a ring-expansion has taken place at the stage of the intermediate metal carbene.

R = Me, Ph, OPr-*i*

4. Aryl(silyl)carbenes

Phenyl(trimethylsilyl)carbene (**21**) has been generated from phenyl(trimethylsilyl)diazomethane (**20**) by gas-phase pyrolysis[39,40] as well as by thermolysis[97] or photolysis[33,40,98,99] in solution, by flash thermolysis of the tosylhydrazone lithium salt **180**[40], and by pyrolysis of α,α-bis(trimethylsilyl)benzyl alcohol (**181**)[85,86] (equation 53).

$$\text{(53)}$$

The gas-phase pyrolysis of **20**[40] in the absence of other reagents gave the products shown in equation 54. Olefin **182** is probably formed from two carbene molecules or from reaction of the carbene with **20**, and benzyltrimethylsilane is a typical product of a twofold H abstraction attributed to a triplet carbene. A carbene-to-carbene rearrangement, not uncommon for arylcarbenes, should account for the formation of **183**, and styrene is probably a decomposition product of 1,1-dimethyl-2-phenylsilirane, formed by 1,3-C,H insertion at the carbene stage (cf Section III.C.1). The flash thermolysis of **180** led to **183** (20%) and styrene (5%)[40]. Products of the silylcarbene-to-silene rearrangement (**21** → **22**, equation 6) were not detected in these experiments; however, formation of silene **22** under the reaction conditions is suggested by various copyrolyses with alcohols, ketones and aldehydes, or 2,3-dimethyl-buta-1,3-diene, which furnished trapping products of the silene[39,40] (see equation 6). Furthermore, gas-phase pyrolysis of **181** furnished, besides **183** and styrene, pentamethyl-(1-phenylethyl)disiloxane; the latter product appears to result from addition of trimethylsilanol to silene **22**[85,86].

When carbene **21** is generated from the diazo precursor photochemically in solution, it reacts with added alcohols by O,H insertion (equation 55), in contrast to the gas-phase copyrolysis where the silene intermediate is trapped[32,98]. Similarly, photolysis of **20** in the presence of 2,3-dimethylbutadiene gives mainly vinylcyclopropane **184**, while after copyrolysis of **20** and the same diene one finds that most of the vinylcyclopropane rearranged to the cyclopentene, together with the [4+2] cycloaddition product of diene and silene **22**[40]. Furthermore carbene **21**, generated photochemically[99] or thermally[97] (at 117–148 °C) in solution, undergoes [1+2] cycloaddition to alkynes to give cyclopropenes **185**.

$$Me_3Si-\underset{\underset{N_2}{\|}}{C}-Ph \xrightarrow[-N_2]{h\nu} Me_3Si-\underset{..}{C}-Ph \xrightarrow{ROH} Me_3Si-\underset{\underset{OR}{|}}{CH}-Ph \quad (55)$$

(**20**)   (**21**)   R = Me, Et, *t*-Bu

(**185**) R = Me ; R = Ph (78%)

(**184**)

Phenyl(triphenylsilyl)carbene has also been trapped without the interference of a silylcarbene-to-silene rearrangement[84]. It undergoes O,H insertion with alcohols and is oxidized to the ketone by DMSO; the latter reaction is likely to include an S-oxide ylide (equation 56).

$$Ph_3Si-\underset{\underset{N_2}{\|}}{C}-Ph \begin{array}{c} \xrightarrow[-N_2]{h\nu, ROH} Ph_3Si-\underset{\underset{OR}{|}}{CH}-Ph \quad R = Me, Et, \textit{t-Bu} \\ \\ \xrightarrow[-N_2]{DMSO, 150\,°C} Ph_3Si-\underset{\underset{O-SMe_2}{|}}{\bar{C}}-Ph \xrightarrow{-Me_2S} Ph_3Si-\underset{\underset{O}{\|}}{C}-Ph \end{array} \quad (56)$$

5. Acyl(silyl)carbenes

a. Silyl-substituted ketocarbenes. Wolff rearrangement yielding ketenes is a typical result of the photochemically or thermally induced elimination of N_2 from many α-diazocarbonyl compounds[100,101]. It is now generally accepted that this transformation normally proceeds via singlet acylcarbenes as reaction intermediates, but can also occur from an excited singlet state of the diazo compound itself by concerted N_2 loss and (1,2) migration of a substituent. Furthermore, there are many indications that the rearrangement is under conformational control, requiring the *s-cis* conformation at the (O=)C−C(N_2) bond of the diazo compound[100] or a *syn*-periplanar arrangement of the two substituents at the ketocarbene moiety[102]. Photochemical decomposition ($\lambda \geqslant 280$ nm) of (1-diazo-2-oxoalkyl)silanes **186** in benzene produced indeed silyl ketenes **188** as the result of a Wolff rearrangement (equation 57). These ketenes were either isolated or quenched with water or alcohols to give the corresponding acids or esters[103]. When R^1 was a *t*-Bu group, the β-silylcyclobutanones **189** were obtained as by-products; obviously, they result from an intramolecular γ-C,H insertion of carbene **187**.

$$R^1-\underset{\underset{N_2}{\|}}{\underset{O}{\overset{\|}{C}}}-\underset{}{C}-SiR_3 \xrightarrow[-N_2]{h\nu, \text{benzene}} R^1-\underset{\underset{O}{\|}}{C}-\overset{..}{C}-SiR_3 \xrightarrow[39-94\%]{\sim R^1} O{=}C{=}C\underset{R^1}{\overset{SiR_3}{\diagup}}$$

(186)   **(187)**   **(188)**

$R^1 = t$-Bu | 13–25%

(189) with O and SiR_3 substituents

$R_3Si = SiMe_2,Bu$-t, $SiMe(Bu$-$t)_2$, $SiPh_2,Bu$-t, $Si(Pr$-$i)_3$

↓ R^2OH

$$R^1-\underset{SiR_3}{\overset{}{CH}}-COOR^2 \quad (57)$$

R^1 = Me, *t*-Bu, $CHPh_2$, 1-adamantyl, Ph

R^2 = H, Me, Et

$R_3Si = SiEt_3$, $Si(Pr$-$i)_3$, $SiMe_2Bu$-t, $SiMe(Bu$-$t)_2$, $SiPh_2Bu$-t

Due to the bulkiness of the SiR_3 group, diazoketones **186** most likely adopt a *s-trans* conformation, which is prohibitive for a concerted Wolff rearrangement directly from **186**. Therefore, carbenes **187** are the common intermediates in the formation of both **188** and **189**. Furthermore, triplet-sensitized irradiation of **186** [$R^1 = t$-Bu, $SiR_3 = Si(Pr$-$i)_3$] (excess of benzophenone, $\lambda = 320$ nm) gave the same products in similar yields as the direct irradiation. It can be assumed, therefore, that a primarily formed triplet carbene interconverts rapidly with the singlet carbene which then rearranges to the ketene or undergoes intramolecular C,H insertion. Decomposition of diazo compounds **186a** by catalytic copper(I) triflate also gave silyl ketenes **188a** in several cases (equation 58); in contrast to the photochemical reaction, γ-C,H insertion was not observed[103]. A singular result is the formation of 1-oxa-2-sila-4-cyclopentene **191** in the Cu(I)-catalyzed decomposition of diazoketone **190**. It may be assumed that a copper–carbene intermediate undergoes β-C,H insertion to form an acylsilirane, the ring-expansion of which leads to the final product **191**.

$$R^1-\underset{\underset{O}{\|}}{C}-\underset{\underset{N_2}{\|}}{C}-R^2 \xrightarrow[-N_2]{\text{Cu OTf cat.,}\atop\text{benzene, 20 °C}} \underset{R^2}{\overset{R^1}{\diagdown}}C=C=O$$

(186a)   (188a)

R^1	Me	t-Bu	Ph
R^2	SiEt$_3$	SiMe$_2$Bu	Si(Pr-i)$_3$

$$t\text{-Bu}-\underset{\underset{O}{\|}}{C}-\underset{\underset{N_2}{\|}}{C}-\text{Si}(\text{Pr-}i)_3 \xrightarrow[-N_2]{\text{CuOTf, cat.,}\atop\text{benzene, 10 °C}} \tag{58}$$

(190)

$$\left[t\text{-Bu}-\underset{\underset{O}{\|}}{C}-\underset{\underset{[Cu]}{\|}}{C}-\text{Si}(\text{Pr-}i)_3 \longrightarrow t\text{-Bu}-\underset{\underset{O}{\|}}{C}\cdots\overset{i\text{-Pr},\ \text{Pr-}i}{\underset{}{\text{Si}}} \right] \longrightarrow \text{(191)}$$

Remarkably, the thermal decomposition of (1-diazo-2-oxoalkyl)silanes **186** does not generate silylcarbenes. Instead, a 1,3(C→O) silyl shift takes place under rather mild conditions (20–80 °C, depending on the substituents) from which (β-siloxy)diazoalkenes **192** result (equation 59). These short-lived cumulenic diazo compounds rapidly split off molecular nitrogen, and the resulting β-siloxyalkylidene carbenes undergo typical intramolecular reactions, from which siloxyalkynes[104], 1-oxa-2-sila-4-cyclopentenes[103,105] or heptafulvene derivatives[106] result. The presumed diazoalkenes **192** can be trapped by [3 + 2] cycloaddition to electron-deficient alkenes[107], cyclopropenes[108] and phosphaalkenes[109].

$$R^1-\underset{\underset{O}{\|}}{C}-\underset{\underset{N_2}{\|}}{C}-SiR_3 \xrightarrow{\Delta} \underset{R_3SiO}{\overset{R^1}{\diagdown}}C=C=N_2 \xrightarrow{-N_2} \underset{R_3SiO}{\overset{R^1}{\diagdown}}C=C: \tag{59}$$

(186)   (192)

↓ further reactions

A transformation analogous to **190** → **191** was observed when diazoketone MeCO–C(N$_2$)SiEt$_3$ was heated in boiling cyclohexane in the presence of copper powder[110]. In the light of the preceding results, it must be assumed that a purely thermal decomposition via a diazoalkene intermediate is operating here, rather than the ketocarbenoid pathway outlined in equation 58.

b. *Alkoxycarbonyl(silyl)carbenes.* The photochemical[32,33,111–114], thermal[33,115] and transition-metal catalyzed[114] decomposition of readily available silyl-substituted

13. Silicon-substituted carbenes

diazoesters has been studied in much detail. While the first two methods represent a convenient source of carbenes **43** (equation 60), the metal-mediated transformation is likely to include metal–carbene intermediates *(vide infra)* in the product-forming step.

$$R^1-Si-\underset{\underset{N_2}{\|}}{C}-COOR^2 \quad \begin{array}{c} \xrightarrow{ML_n \text{ cat.}} \\ -N_2 \\ \\ \xrightarrow{h\nu \text{ or } \Delta} \\ -N_2 \end{array} \quad \begin{array}{c} R^1-Si-\underset{\underset{ML_n}{\|}}{C}-COOR^2 \longrightarrow \text{products} \\ \\ R^1-Si-\underset{..}{C}-COOR^2 \longrightarrow \text{products} \\ (\mathbf{43}) \end{array} \tag{60}$$

As already discussed in Section III.B, carbenes **43** can rearrange to a silene by migration of R^1 to the carbene center and to a ketene by migration of OR^2. The situation is complicated further by a subsequent $1,3(C \rightarrow Si)$ shift of the OR^2 group in the silene to form a doubly rearranged ketene (see Scheme 5).

Photolysis of ethyl (trimethylsilyl)diazoacetate[112], methyl (triethylsilyl)diazoacetate[114] and methyl (*tert*-butyldiphenylsilyl)diazoacetate[114] in inert solvents did not provide useful information as to this question. More informative was the photolysis of **193** in various alcohols, leading to a similar mixture of products **194–197** in all cases[33,113] (equation 61). Compound **194** is the major product and probably represents the O,H insertion product of carbene **43**. Compound **195** could be derived from a silene intermediate (equation 62). Wolff rearrangement accounts for the alkoxyacetates **196** and **197**. The exchange of both alkoxy residues leading to **197** is rationalized by the assumption that the Wolff rearrangement of alkoxy groups can include an ion pair intermediate which either collapses or exchanges the alkoxy group with an external alcohol molecule.

$$\underset{\underset{N_2}{\|}}{Me_3Si-C-COOEt} \quad \xrightarrow[-N_2]{h\nu, \text{ROH}} \quad Me_3Si\underset{OR}{\overset{OR}{\bigwedge}}COOEt \quad + \quad Me_2Si\underset{OR}{\overset{Me}{\bigwedge}}COOEt$$

(**193**)   (**194**)   (**195**)

$$+ \quad Me_3Si\underset{}{\overset{OEt}{\bigwedge}}COOR \quad + \quad Me_3Si\underset{}{\overset{OR}{\bigwedge}}COOR \tag{61}$$

(**196**)   (**197**)

R = Me, Et, *i*-Pr, *t*-Bu

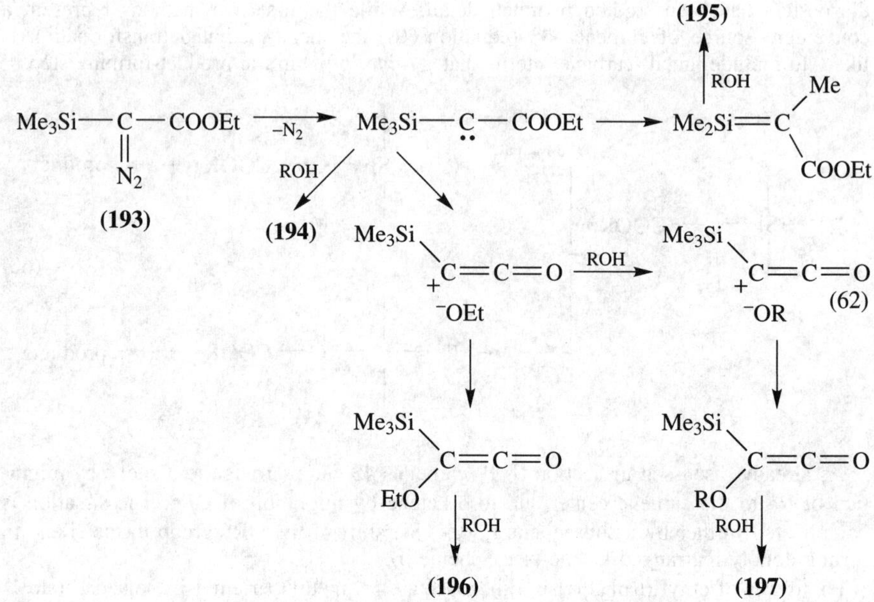

$$\text{Me}_3\text{Si}-\underset{\underset{N_2}{\|}}{C}-\text{COOEt} \xrightarrow{-N_2} \text{Me}_3\text{Si}-\underset{..}{C}-\text{COOEt} \xrightarrow{\text{ROH}} \text{Me}_2\text{Si}=C\underset{\text{COOEt}}{\overset{\text{Me}}{\diagdown}} \quad (195)$$

(193) → (194), with ROH branches leading to:

$$\underset{-\text{OEt}}{\overset{\text{Me}_3\text{Si}}{\diagdown}}\!\!\!+\!\!C\!=\!C\!=\!O \xrightarrow{\text{ROH}} \underset{-\text{OR}}{\overset{\text{Me}_3\text{Si}}{\diagdown}}\!\!\!+\!\!C\!=\!C\!=\!O \quad (62)$$

↓ ↓

$$\underset{\text{EtO}}{\overset{\text{Me}_3\text{Si}}{\diagdown}}C\!=\!C\!=\!O \qquad \underset{\text{RO}}{\overset{\text{Me}_3\text{Si}}{\diagdown}}C\!=\!C\!=\!O$$

↓ ROH ↓ ROH

(196) (197)

When diazoester **198** was subjected to gas-phase pyrolysis and ethanol was added afterwards, a partly different result was obtained[32] (equation 63). Alkoxyacetate **199** is again considered to result from the Wolff rearrangement, but since both ethoxy groups in **200** stem from the added alcohol, a silene cannot be the precursor. In a separate experiment, the doubly rearranged ketene **202** was isolated and shown to react quantitatively with ethanol to give **200**. Also interesting is the formation of a small amount (4%) of β-lactone **201**, which appears to result from 1,4-C,H insertion of the carbene. No product of an intramolecular C,H insertion could be found after photolysis of ethyl (trimethylsilyl)diazoacetate in an inert solvent[112].

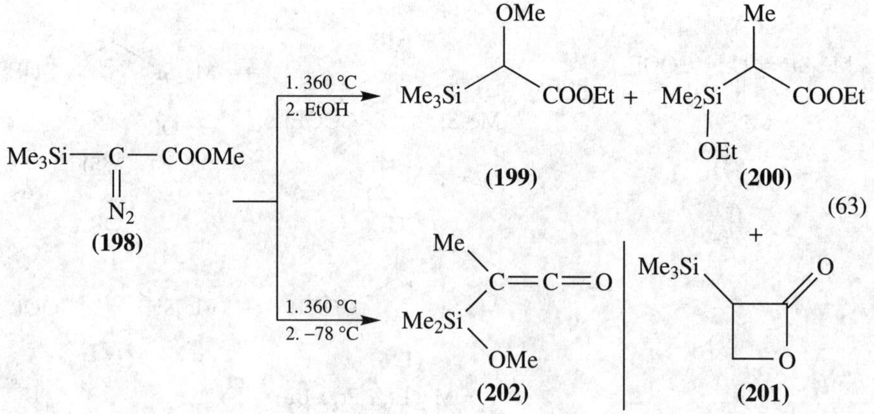

$$\text{Me}_3\text{Si}-\underset{\underset{N_2}{\|}}{C}-\text{COOMe} \quad \begin{array}{c}\xrightarrow[\text{2. EtOH}]{\text{1. 360 °C}}\\ \\ \xrightarrow[\text{2. -78 °C}]{\text{1. 360 °C}}\end{array}$$

(198)

1. 360 °C, 2. EtOH → Me$_3$Si–CH(OMe)–COOEt (**199**) + Me$_2$Si(Me)(OEt)–COOEt (**200**)

1. 360 °C, 2. –78 °C → Me$_2$Si(OMe)–C(Me)=C=O (**202**)   +   Me$_3$Si-β-lactone (**201**)

(63)

In the presence of appropriate substrates, a variety of other typical acylcarbene reactions took place with carbene **43** (equation 64). Photolysis of **193** in hydrocarbons (cyclohexane, butane, 2,3-dimethylbutane) yields C,H insertion products of the carbene in approximately 50% yield[111,112] with the following selectivities: 2°/1° = 3.5; 3°/1° = 1.5. The higher selectivity for insertion into secondary than into tertiary C,H bonds is unusual and may be attributed to unfavorable steric interactions between the disubstituted carbene and the substrate.

(64)

In the presence of alkenes, photolysis of alkyl (silyl)diazoacetates leads mainly to the formation of cyclopropanes as diastereomeric mixtures[4,111,112]. With (Z)- and (E)-but-2-ene, the cyclopropanation is not completely stereospecific with respect to the double bond configuration, but gives a small amount of the 'wrong' isomer; these results point to the participation of a triplet carbene in the cyclopropanation reaction. Allylic C,H insertion products are also formed; their yield increases in the series 1,1-, 1,2-, tri- and tetrasubstituted C=C bond. With 2,3-dimethyl-but-2-ene, the allylic C,H insertion product is formed at the complete expense of the cyclopropane.

Interaction of carbene **43** with thioethers leads first to a sulfonium ylide; a subsequent 1,2-R shift provides the formal C,S insertion product **203** and, if possible, β-elimination generates (alkylthio)acetate **204**[116].

For the transition-metal catalyzed decomposition of silyl-substituted diazoacetates **205** [silyl = SiMe$_3$, SiEt$_3$, SiMe$_2$Bu-t, Si(Pr-i)$_3$SiPh$_2$Bu-t, SiMe$_2$SiMe$_3$], copper(I) triflate and dirhodium tetrakis(perfluorobutyrate) proved to be the best catalysts[114]. While these two catalysts induce the elimination of N$_2$ at 20 °C even with bulky silyl substituents, dirhodium-tetraacetate even at 100 °C decomposes only the trimethylsilyl- and triethylsilyl-diazoacetates. When the decomposition reactions are carried out in tetrachloromethane or toluene solution, the formal carbene dimers **206**, azines **207** (probably formed from diazoester and the carbene or metal–carbene intermediate), furanone **208** and doubly rearranged ketenes **209** can be formed[114] (equation 65). The result depends on the catalyst, the solvent and the particular silyl group. Furanone **208** could result from [3 + 2] cycloaddition of the ketocarbene 1,3-dipole (or a metal complex thereof) to the ketene formed by a Wolff rearrangement.

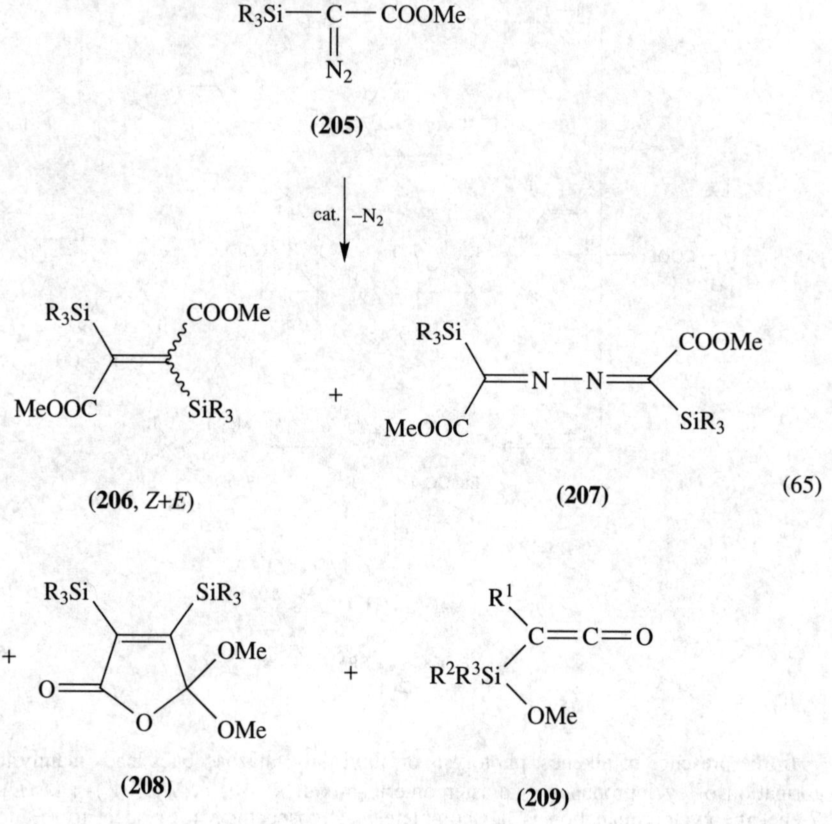

$$R_3Si-\underset{\underset{N_2}{\|}}{C}-COOMe$$

(205)

cat. $|$ $-N_2$

(206, Z+E)   (207)   (65)

(208)   (209)

$R_3Si = Me_3Si, Et_3Si, (i\text{-}Pr)_3Si, t\text{-}BuPh_2Si\ (R^1 = Ph),$
$Me_3SiMe_2Si\ (R^1 = SiMe_3)$

With dirhodium tetrakis(perfluorobutyrate) as catalyst, only ketene **209** was obtained in practically all cases, except for the trimethylsilyl (unseparated product mixture) and triisopropylsilyl cases (no decomposition by this catalyst).

With the same catalysts, carbene transfer from diazoacetates **205** [silyl groups: $SiMe_3$, $SiEt_3$, $Si(Pr\text{-}i)_3$] to 1-hexene and styrene was studied[4,117]. The yields of cyclopropanes were usually higher in the metal-catalyzed than in the photochemical version. Interestingly, the metal-catalyzed procedure gave preferentially *E*-**210**, whereas with the photochemically generated carbene an E/Z ratio <1 was obtained (equation 66). The significance of these results needs further investigation. For the cyclopropanation of cyclohexene with diazoester **205** ($R_3Si = Me_3Si$), the complex $[Ru_2(CO)_4(\mu\text{-OAc})_2]_n$ is the catalyst of choice[118], since only the allylic C,H insertion product of the carbene is obtained with copper(I) triflate and $Rh_2(OOCC_3F_7)_4$ as catalysts.

$$\begin{array}{c} R_3Si-\underset{\underset{N_2}{\|}}{C}-COOMe \\ (\mathbf{205}) \end{array} \xrightarrow[h\nu \text{ or cat.}]{R^1} \begin{array}{c} SiR_3 \\ \triangle \\ R^1 \quad COOMe \\ (E\text{-}\mathbf{210}) \end{array} + \begin{array}{c} COOMe \\ \triangle \\ R^1 \quad SiR_3 \\ (Z\text{-}\mathbf{210}) \end{array}$$

(66)

$R=Me$ | $Me_3Si \equiv SiMe_3$
$h\nu$ or cat. ↓

R^1	SiR_3	E-210 /	Z-210
		$h\nu$	CuOTf cat.
C_4H_9	$SiMe_3$	0.83	2.9
Ph	$SiMe_3$	0.65	3.6
Ph	$Si(Pr\text{-}i)_3$	0.37	4.7

$Me_3Si\diagdown\quad\diagup COOMe$
$\quad\quad\triangle$
$Me_3Si\diagup\quad\diagdown SiMe_3$

[1 + 2] Cycloaddition of the carbene derived from **205** to bis(trimethylsilyl)acetylene yields the expected cyclopropene in low yield both photochemically (20%) and under catalysis by copper(I) triflate at 80 °C (10–13%)[119]. The latter version of the reaction is accompanied by [3 + 2] cycloaddition of the diazo compound to the alkyne, and the photochemical route yields a by-product which obviously comes from carbenic C,H insertion at a $SiMe_3$ group of the alkyne.

Transition-metal mediated carbene transfer from **205** to benzaldehyde generates carbonyl ylides **211** which are transformed into oxiranes **216** by 1,3-cyclization, into tetrahydrofurans **212, 213** or dihydrofurans **214** by [3 + 2] cycloaddition with electron-deficient alkenes or alkynes, and 1,3-dioxolanes **215** by [3 + 2] cycloaddition with excess carbonyl compound[120] (equation 67). Related carbonyl ylide reactions have been performed with crotonaldehyde, acetone and cyclohexanone (equation 68). However, the ylide generated from cyclohexanone could not be trapped with dimethyl fumarate. Rather, the enol ether **217**, probably formed by 1,4-proton shift in the ylide intermediate, was isolated in low yield[120]. In this respect, the carbene transfer reaction with **205** is not different from that with ethyl diazoacetate[121], whereas a close analogy to diazomalonates is observed for the other carbonyl ylide reactions.

Efforts to trap the carbonyl ylide intermediate by intramolecular [3 + 2] cycloaddition to a C=C bond were unsuccessful. Rather, the decomposition of allyl (trimethylsilyl)diazoacetate (**218**) (equation 69) in the presence of aldehydes gave 1,3-dioxolan-4-ones **219**; their formation has been explained by 1,5-cyclization of the carbonyl ylide intermediate followed by a Claisen rearrangement[122]. With acetone as carbonyl component, the reaction proceeds analogously. Clean formation of **219** occurred only with $Rh_2(OOCC_3F_7)_4$ as catalyst, while the copper(I) triflate catalyzed version led to a mixture of **219**, an oxirane and the product of intramolecular carbenoid cyclopropanation. Carbonyl ylide reactions derived from homoallylic ester **220** are shown in equation 70[122].

(67)

R = Me, Et
E = COOMe
cat. = CuOTf, $Rh_2(OOCC_3F_7)_4$, $[Ru_2(CO)_4(\mu\text{-OAc})_2]_n$

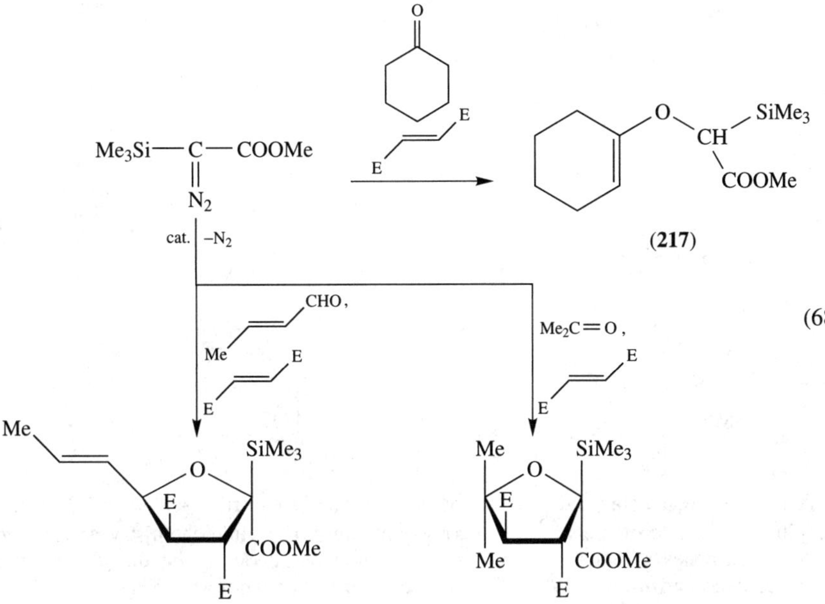

13. Silicon-substituted carbenes

(Scheme 68): Me₃Si—C(=N₂)—COOMe reacts with cyclohexanone (cat., −N₂) to give **(217)**: 1-(cyclohexenyloxy)-CH(SiMe₃)(COOMe). With MeCH=CHCHO and E—CH=CH—E gives the furan product bearing Me, SiMe₃, COOMe and E substituents. With Me₂C=O and E—CH=CH—E gives the analogous dihydrofuran with Me, Me, SiMe₃, COOMe, E.

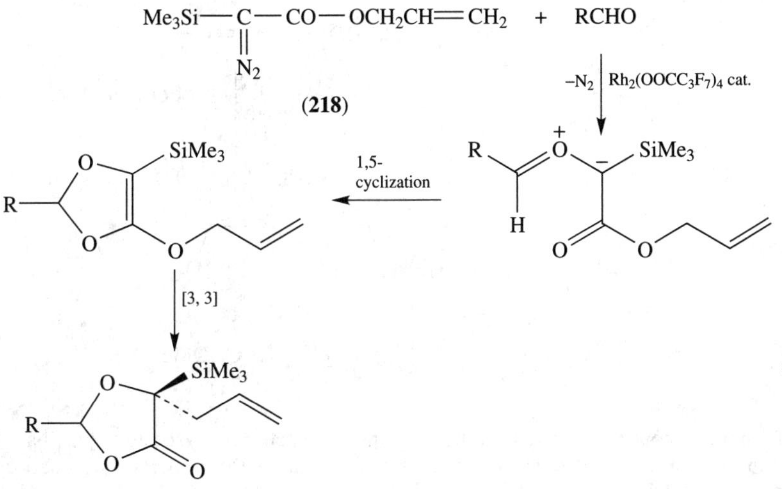

E = COOMe

(Scheme 69): Me₃Si—C(=N₂)—CO—OCH₂CH=CH₂ **(218)** + RCHO, Rh₂(OOCC₃F₇)₄ cat., −N₂ → carbonyl ylide intermediate → 1,5-cyclization → 2-R-4-SiMe₃-5-(allyloxy)-1,3-dioxole → [3,3] → **(219)**.

R = Me, CH=CHMe, Ph, 4-MeOC₆H₄, C₆H₄, 3,4,5-(MeO)₃C₆H₂

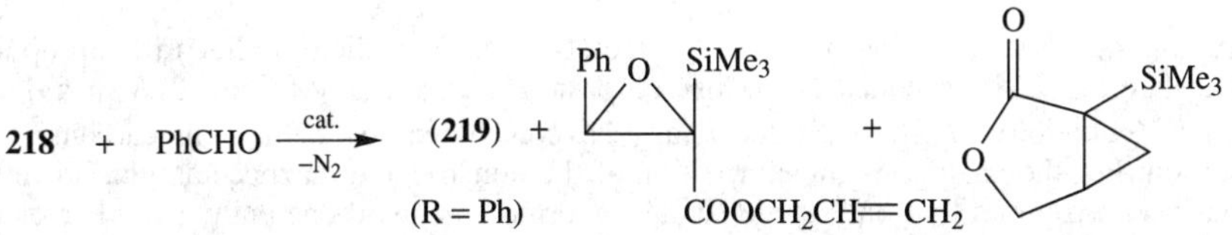

218 + PhCHO $\xrightarrow[-N_2]{\text{cat.}}$ **(219)** (R = Ph) + Ph,SiMe₃-epoxide with COOCH₂CH=CH₂ + bicyclic lactone with SiMe₃

Transition-metal catalyzed transfer of acylcarbenes to nitriles leads to 1,3-oxazoles via nitrile ylide intermediates[123]. The corresponding nitrile ylide chemistry derived from acyl(silyl)carbenes still awaits a closer look, but it has been shown that the rhodium-catalyzed decomposition of **198** in the presence of methyl cyanoformate and benzaldehyde provides 1,3-oxazole **221** (equation 71) exclusively[120]. This implies that the carbene moiety has been transferred only to the nitrile but not to the aldehyde.

In another copper-mediated carbene transfer reaction, diazoester **222** has been decomposed in the presence of bis(triethylsilyl- or -germyl)mercury (equation 72); it was assumed that the obtained ketenes **223** result from the insertion of ethoxycarbonyl(trimethylsilyl)carbene into a Hg-element bond followed by a cyclic fragmentation process[110].

c. *Aminocarbonyl(silyl)carbenes.* UV-irradiation of *N,N*-diethyl diazo(triisopropylsilyl)acetamide (**224**) provides a mixture of β-lactam **225** and γ-lactam **226** in a 1 : 2 ratio[103] (equation 73). Both products may be considered to result from carbenic C,H insertion. For the analogous transformation of the nonsilylated diazoacetamide, however, it has been proposed that only the γ-lactam is formed on a carbene pathway, whereas the β-lactam results directly from an excited singlet state of the diazo compound[124,125].

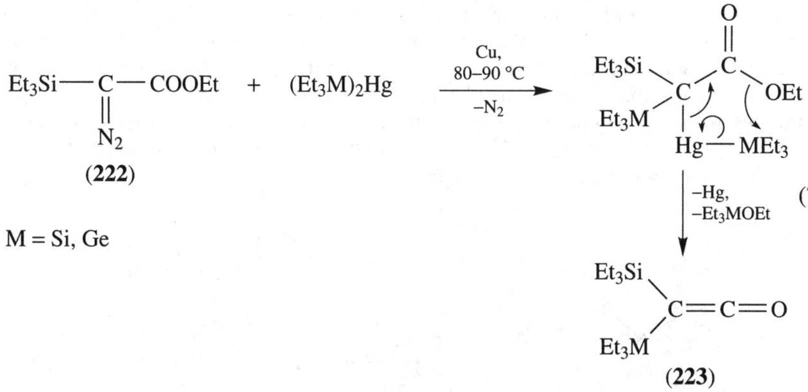

$$Et_3Si—\underset{\underset{(222)}{N_2}}{\overset{\|}{C}}—COOEt + (Et_3M)_2Hg \xrightarrow[-N_2]{Cu, 80-90\,°C}$$

M = Si, Ge

(223)

(72)

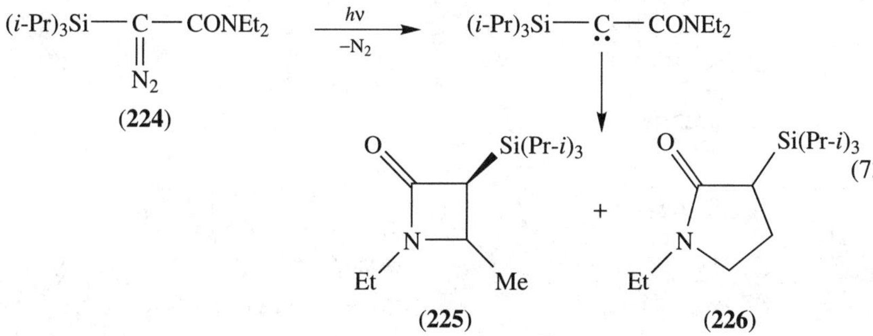

$$(i\text{-Pr})_3Si—\underset{\underset{(224)}{N_2}}{\overset{\|}{C}}—CONEt_2 \xrightarrow[-N_2]{h\nu} (i\text{-Pr})_3Si—\overset{..}{C}—CONEt_2$$

(225) + (226) (73)

6. Imino(silyl)carbenes

Photolysis of 1,2,3-triazole **227** (equation 74) provides the bis(trimethylsilyl)ketenimine **229** in better than 80% yield[126]. As with other triazoles it can be assumed that the light-induced extrusion of nitrogen from **227** generates the iminocarbene **228** which undergoes a fast Wolff rearrangement by 1,2-migration of a trimethylsilyl group. It is remarkable that a 1,5-cyclization of the carbene does not compete with the rearrangement at all, in contrast to the behavior of those N-aryl iminocarbenes where phenyl or alkyl substituents are involved in the 1,2-migration.

7. Phosphino(silyl)carbenes and phosphoryl(silyl)carbenes

a. Phosphino(silyl)carbenes. In 1988, Bertrand and coworkers reported that the thermolysis of [bis(diisopropylamino) phosphino]-(trimethylsilyl)diazomethane (**230a**, equation 75) at 250 °C provides a red oily material which can be distilled at 75–80 °C/0.02 Torr and is stable for several weeks at room temperature in benzene solution[127]. This compound has the constitution **231a** and was therefore called a 'stable carbene'. Since this report, several other carbenes of the same type became known (**231b–e**[128,129], **231f**[130]). They are typically obtained by thermal extrusion of molecular nitrogen from their rather unstable diazo precursors at 25–70 °C. Photolysis of the diazo compounds also generates these carbenes, but a clean preparation was only possible for **231f**. At room temperature, carbenes **231a–d** and **f** are stable for several weeks, while **231e** survives for only a few hours. Together with the salt [(*i*-Pr)$_2$N]$_2$P–C–$^+$PH[N(Pr-*i*)$_2$]$_2$· CF$_3$SO$_3$$^{-131}$, **231a–f** represent the only known examples of this type of 'stable carbene', i.e. compounds in which a phosphino group is attached to a formally divalent carbon atom. By comparison with other substitution patterns, it becomes obvious that one or better two dialkylamino groups in the phosphino groups are important for stabilization of the phosphinocarbenes. Furthermore, the silyl group also contributes to stabilization. By comparison, the analogous stannyl-substituted carbenes [(*i*-Pr)$_2$N]$_2$P–C–SnR$_3$ (R = Me, Ph, *c*-Hex) are only transient intermediates which could be trapped with *tert*-butylisonitrile or methyl acrylate[132].

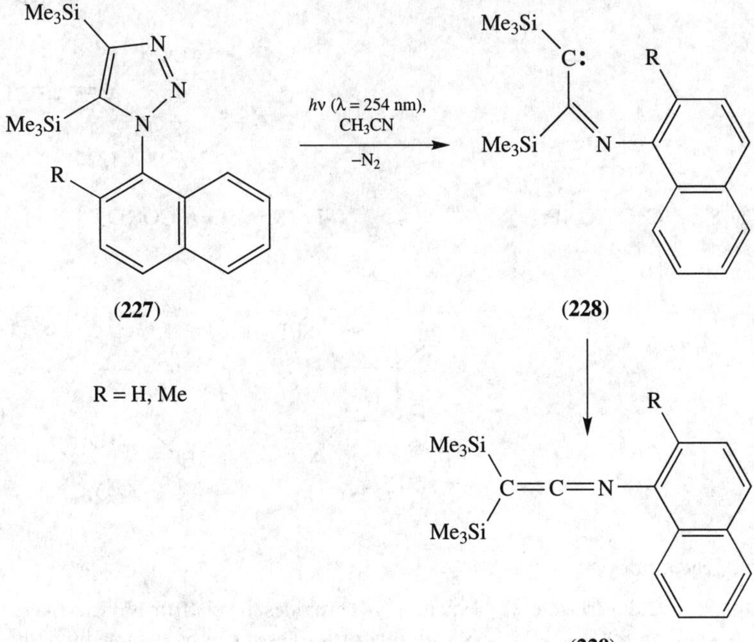

(74)

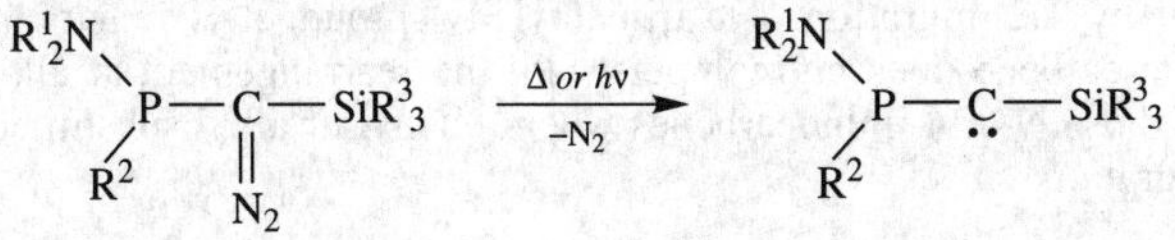

R = H, Me

(75)

230, 231	a	b	c	d	e	f
R$^1_2$N	(*i*-Pr)$_2$N	Tmp	Tmp	Tmp	Tmp	(*c*-Hex)$_2$N
R^2	(*i*-Pr)$_2$N	(*i*-Pr)$_2$N	Me$_2$N	Me$_2$N	Ph	(*c*-Hex)$_2$N
R^3	Me	Me	*i*-Pr	Me	Me	Me

Tmp = 2,2,6,6-tetramethylpiperidide

As carbenes are typically perceived by chemists as short-lived intermediates, the term 'stable carbenes' appears as a *contradictio in se*, and some debate arose about the true chemical nature of Bertrand's stable phosphinocarbenes as well as Arduengo's carbenes of the 1,3-imidazol-2-ylidene type[133]. It has been outlined already in Section II that resonance structures **232A** (carbene), **232B** (phosphavinyl ylide) and **232C** (λ^5-phosphaacetylene) (equation 76) can be used to describe the bond state of singlet phosphinocarbenes and that, according to theoretical calculations, the so-called stable carbenes **231** are best regarded as phosphavinyl ylides (**232B**). For this structure, amino groups obviously stabilize the positive charge at phosphorus, and silyl groups can stabilize the negative charge at carbon by delocalization.

$$\begin{array}{ccc}
R^1 \\
\ddot{P}-\ddot{C}\diagdown R^3 \\
R^2 \\
\text{(232A)}
\end{array}
\quad \longleftrightarrow \quad
\begin{array}{c}
R^1 \\
\overset{+}{P}=\bar{C}\diagdown R^3 \\
R^2 \\
\text{(232B)}
\end{array}
\quad \longleftrightarrow \quad
\begin{array}{c}
R^1 \\
P\equiv C-R^3 \\
R^2 \\
\text{(232C)}
\end{array}$$

(76)

The results of theoretical calculations, spectroscopic data and chemistry of phosphinocarbenes have been reviewed[22,134]. In the context of this chapter, it may suffice to mention that phosphino(silyl)carbenes **231** behave as chemical chameleons which in some of their transformations display a behavior that is thought to be typical for (nucleophilic) carbenes, whereas the reactivity of a species with a P,C multiple bond is perceived in other cases. Examples of carbene-type reactions of **231a** are given in equation 77, namely intramolecular C,H insertion[127], cyclopropanation of electron-deficient alkenes such as methyl acrylate and dimethyl fumarate[135] (dimethyl maleate does not react), stereospecific oxirane formation with aldehydes [135] and ketenimine formation with an isocyanide[135]. Other aminophosphino(silyl)carbenes react analogously[129]. In most cases, the first-formed phosphino-substituted products are rather unstable and were converted to the P-sulfides for isolation. Carbene **231f** undergoes a smooth [1 + 2] cycloaddition to benzonitrile leading to 2*H*-azirene **233**[136]. Under the catalytic action of dichloro(*p*-cymene)ruthenium(II), **233** undergoes ring-expansion to a 1,2(λ^5)-azaphosphete (equation 78).

Photochemically generated carbene **231a** reacts with (*tert*-butylmethylidyne)phosphine to form the stable 1(λ^5),2(λ^3)-diphosphete **234**[137]. It has been proposed that 2*H*-phosphirene **235** is a transient intermediate in this reaction (equation 79); it would be formed by [1 + 2] cycloaddition of **231a** to the P,C triple bond in complete analogy to azirene **233** and would undergo spontaneous ring-expansion to **234**.

It should be mentioned that alkenes **236**, which are often found as (formal) carbene dimers in reactions involving electrophilic carbenes, have never been observed in the context of phosphino(silyl)carbenes. UV-irradiation of a [bis(dialkylamino)phosphino](trimethylsilyl)diazomethane leads to a 1(λ^5),3(λ^5)-diphosphete **237** (equation 80) which can be regarded as the head-to-tail cyclodimer of a phosphavinyl ylide (cf **232B**), whereas irradiation of diphenylphosphino- or dimethoxyphosphino-(trimethylsilyl)diazomethane produces a 1,2(λ^5),4(λ^5),6(λ^5)-azatriphosphorin **238** in a sequence which may also include the corresponding diphosphete **237**[138].

Several other transformations have been reported which highlight either the carbene or the P,C multiple bond nature of compounds **231** or allow a mechanistic interpretation by more than one of the different reactivity patterns[24].

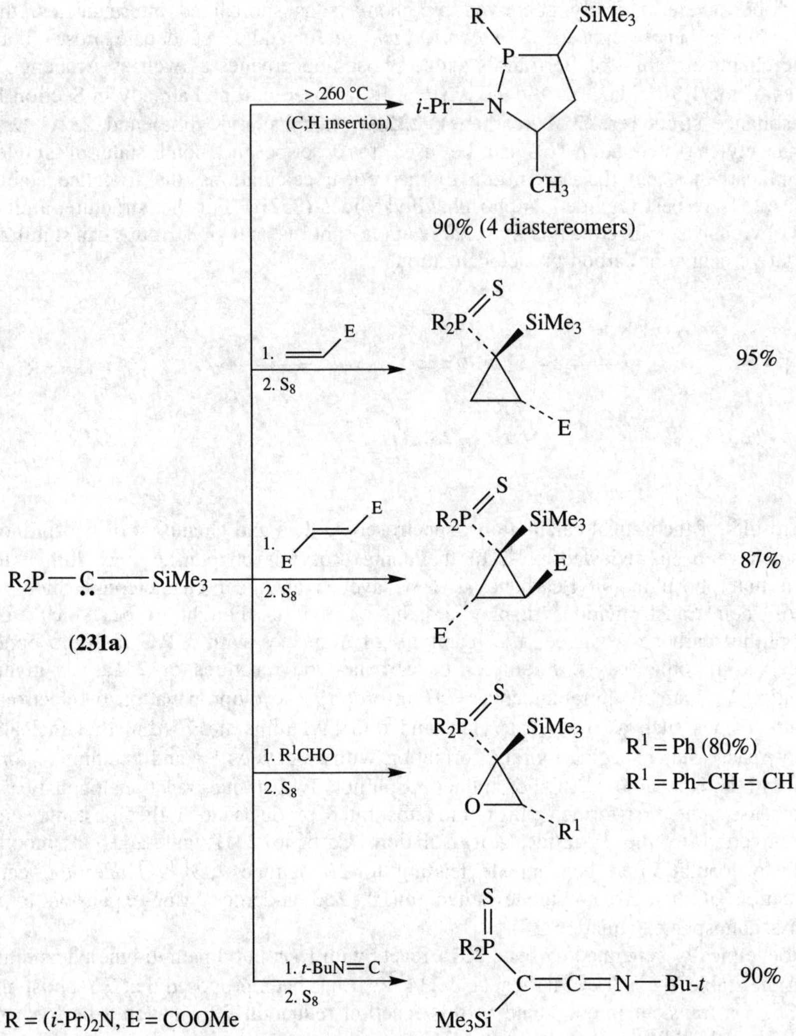

$R_2P-\underset{..}{C}-SiMe_3$

(231a)

$R = (i\text{-Pr})_2N$, E = COOMe

Reactions of (231a):

- $> 260\,°C$ (C,H insertion) → pyrrolidine-type product with i-Pr—N, R, SiMe$_3$, CH$_3$ substituents, 90% (4 diastereomers)

- 1. CH$_2$=CH–E; 2. S$_8$ → cyclopropane with R$_2$P(=S), SiMe$_3$, E — 95%

- 1. E–CH=CH–E; 2. S$_8$ → cyclopropane with R$_2$P(=S), SiMe$_3$, E, E — 87% (77)

- 1. R^{1}CHO; 2. S$_8$ → oxirane with R$_2$P(=S), SiMe$_3$, R^1; R^1 = Ph (80%), R^1 = Ph–CH=CH (82%)

- 1. t-BuN=C; 2. S$_8$ → $R_2P(=S)-C(SiMe_3)=C=N-Bu\text{-}t$, 90%

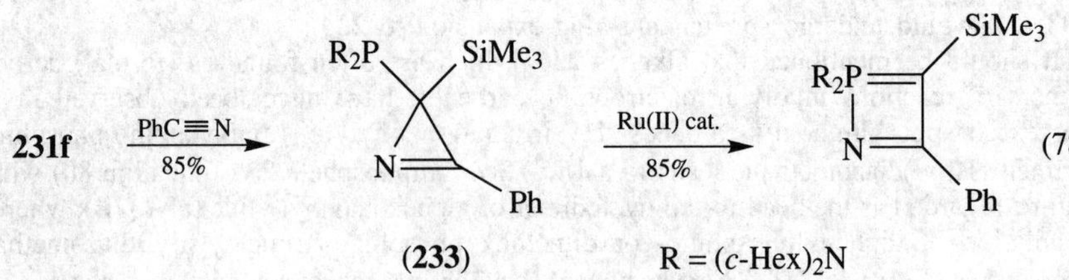

231f $\xrightarrow{PhC\equiv N}_{85\%}$ (233) [azirine with R$_2$P, SiMe$_3$, N, Ph] $\xrightarrow{Ru(II)\ cat.}_{85\%}$ azete with R$_2$P, SiMe$_3$, N, Ph (78)

R = (c-Hex)$_2$N

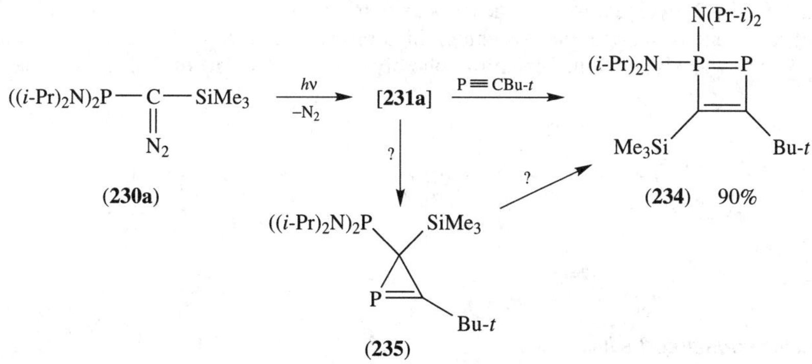

13. Silicon-substituted carbenes

$$((i\text{-Pr})_2N)_2P-\underset{\underset{N_2}{\|}}{C}-SiMe_3 \xrightarrow[-N_2]{h\nu} [231a] \xrightarrow{P\equiv CBu\text{-}t} \underset{Me_3Si}{\overset{(i\text{-Pr})_2N-\overset{N(Pr\text{-}i)_2}{\underset{|}{P}}=P}{\underset{Bu\text{-}t}{}}}$$

(230a)   ?   (234)   90%   (79)

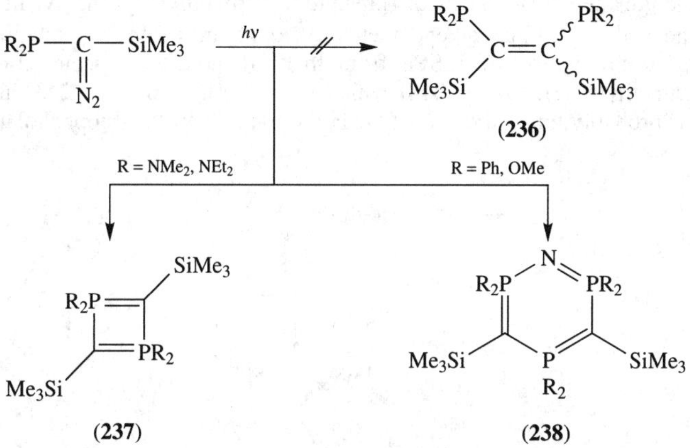

$$((i\text{-Pr})_2N)_2P\underset{(235)}{\overset{SiMe_3}{\underset{Bu\text{-}t}{\triangle}}}$$

$$R_2P-\underset{\underset{N_2}{\|}}{C}-SiMe_3 \xrightarrow{h\nu} \underset{Me_3Si}{\overset{R_2P}{\diagup}}C=C\underset{SiMe_3}{\overset{PR_2}{\diagdown}}$$

(236)

R = NMe₂, NEt₂   R = Ph, OMe   (80)

(237)   (238)

Bis[diazo(trimethylsilyl)methyl] phosphines **239** loose only one equivalent of dinitrogen on warming and are transformed into 1,2,4(λ^3)-diazaphospholes **240**[139] (equation 81). It is obvious to explain this cyclization by intramolecular ketazine formation of a diazocarbene intermediate, but in light of the preceding discussion, speculations about the bond state of this phosphinocarbene (phosphavinyl ylide or phosphaacetylene) are allowed.

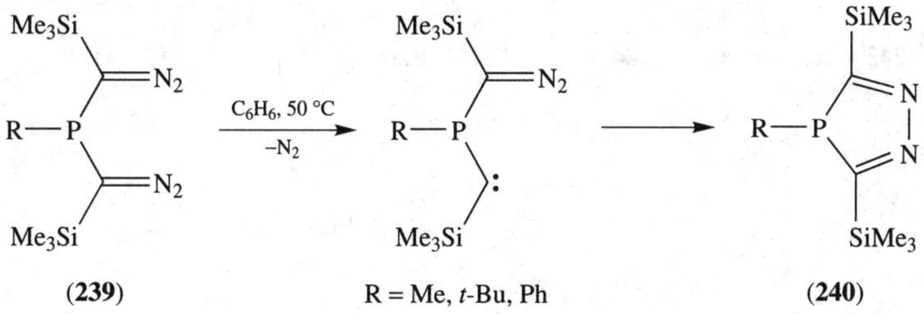

$$\underset{(239)}{\overset{Me_3Si}{\underset{Me_3Si}{R-P}}\diagup\overset{N_2}{\diagdown}N_2} \xrightarrow[-N_2]{C_6H_6, 50\,°C} \underset{Me_3Si}{\overset{Me_3Si}{R-P:}\diagup\overset{N_2}{\diagdown}}} \longrightarrow \underset{(240)}{\overset{SiMe_3}{R-P\diagup\overset{N}{\diagdown}N}\underset{SiMe_3}{}} \quad (81)$$

R = Me, t-Bu, Ph

b. Phosphoryl(silyl)carbenes. Although the chemistry of phosphorylcarbenes is well developed[140], hardly anything is known about phosphoryl(silyl)carbenes. Some experiments in the author's laboratories were rather unproductive[141]. Thus, photolysis of diazo compound **241** in benzene produced an unseparable mixture with at least five major components, one of which is probably a cycloheptatriene according to NMR data, likely to be formed by [1+2] cycloaddition of the carbene to benzene and subsequent norcaradiene-to-cycloheptatriene isomerization. An unidentified product mixture with at least three major components was also obtained from the photolysis ($\lambda = 254$ nm) of **242** in benzene.

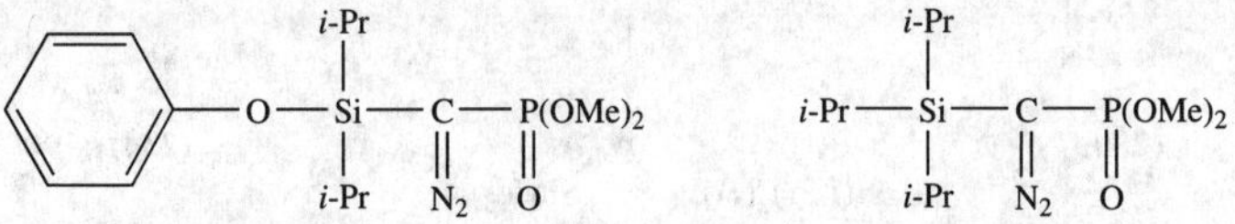

(241)

(242)

8. Sulfur-substituted silylcarbenes

The three-component reaction of bis(phenylthio)-(trimethylsilyl)methyl lithium (**243**), phenyloxirane and a terminal alkene yields cyclopropanes **245**[142] (equation 82). It is assumed that α-elimination of LiSPh from the carbenoid-like species **243** generates phenylthio(trimethylsilyl)carbene (**244**) which is in equilibrium with **243**; although this equilibrium is probably far on the side of the latter, trapping of the thiophenolate ion by the

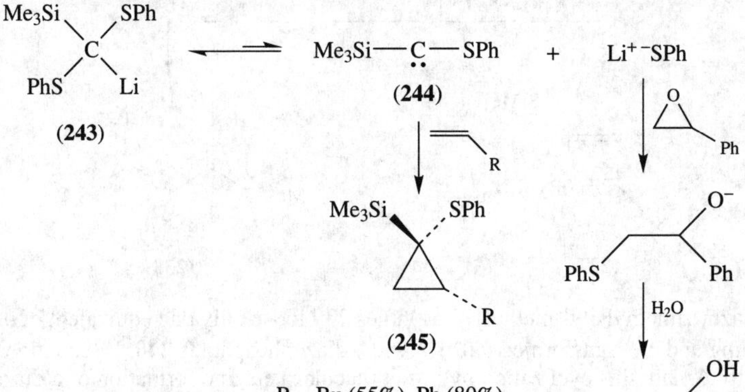

R = Bu (55%), Ph (80%)
OEt (68%), SiMe$_3$ (31%),
SPh (85%)

(82)

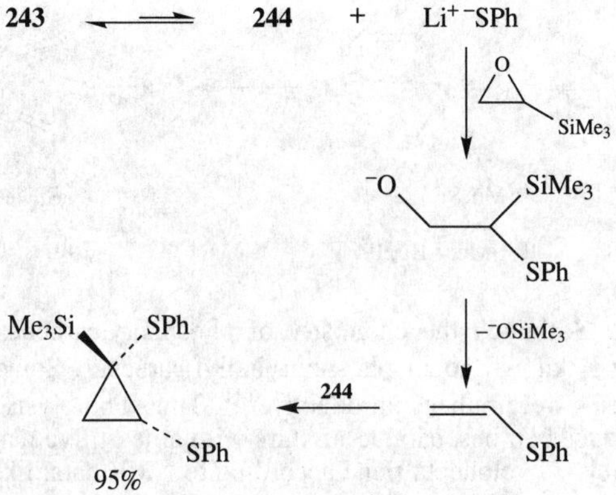

oxirane renders the carbene available for [1 + 2] cycloaddition to the alkene. The chosen alkenes were cyclopropanated diastereospecifically; their successful reaction indicates the electrophilic nature of carbene **244**. Related carbenes (**244**; SiEt$_3$, t-BuMe$_2$Si instead of SiMe$_3$; Tol and 4-ClC$_6$H$_4$ instead of Ph) have been generated analogously[143].

Interestingly, cyclopropane **245** (R = SPh) is also formed when **243** is combined with (trimethylsilyl)oxirane. In this case, the necessary alkene (phenylthioethylene) is provided by silanolate elimination from the alkoxide arising from regiospecific ring-opening of the oxirane by thiophenolate (equation 82)[142].

It should be mentioned that this elegant method to generate a carbene appears to be of limited scope. Lithium salts related to **243**, e.g. (Me$_3$Si)$_2$(PhS)CLi, Me$_3$Si(MeS)$_2$CLi and 2-trimethylsilyl-1,2,3-thiolan-2-yl lithium, do not undergo efficient α-elimination of a lithium thiolate but rather direct nucleophilic addition to the two oxiranes mentioned[142]. These observations are reminiscent of the reaction of phenylthio(trimethylsilyl)methyl lithium with olefins leading to (trimethylsilyl)cyclopropanes[76]; it is not clear whether this transformation really includes the transfer of free (trimethylsilyl)carbene or rather occurs by a Michael addition/ring closure sequence (see also Section III E.1.a).

9. Halo(silyl)carbenes

Carbenes of this type have been generated so far only by thermally induced α-elimination of RHgHal from suitable (α-halo)mercurials (see Section III.A). Thus, chloro(trimethylsilyl)carbene (**248**) can be generated by heating (Me$_3$SiCCl$_2$)$_2$Hg (**246**) at ca 120 °C. It undergoes [1 + 2] cycloaddition with acyclic and cyclic alkenes and inserts into the Si−H bond of triethylsilane[75,144] (equation 83). The disadvantage that only one dichloro(trimethylsilyl) group of **246** is used for carbene formation can be overcome, with the consequence of better yields of the carbene transfer products, by the use of an equimolar mixture of **246** and Ph$_2$Hg. In this case, the mixed organomercury compound **247** is likely to be formed and to serve as the carbene precursor. Similarly, carbene **248** can be generated thermally from an equimolar mixture of Me$_3$SiCCl$_2$HgCl and Ph$_2$Hg[75]. In all cases, by-products were formed which are likely to be derived from the Me$_3$SiCCl$_2$· radical. In fact, when **246** is decomposed at higher temperature (ca 220 °C), the homolytic cleavage of the Hg−C bond, generating this radical, becomes the major pathway.

Analogously to carbene **248**, bromo(trimethylsilyl)carbene **250** was generated by thermal decomposition of an equimolar mixture of **249** and Ph$_2$Hg (equation 84) and transferred to cyclohexene[75]. Although the α-elimination reaction of the carbene precursor occurs faster than in the chloro case, the obtained bromocyclopropane is less stable under the still harsh reaction conditions and is partly decomposed.

F. Special Methods to Generate Silylcarbenes

In this section, some individual transformations are mentioned in which silylcarbene intermediates have been suggested but which do not belong to any of the more familiar and more versatile methods to generate silylcarbenes as shown in Scheme 1 (Section III.A).

1. Silylcarbenes by rearrangement reactions

a. Isomerization of 2-silylfurans. 2-Silyl-substituted furans **251** undergo a remarkably clean photochemical isomerization reaction featuring a ring-opening process[145]. It has been suggested that the latter leads to (1-alkenyl)(trimethylsilyl)carbenes **252** which provide an 1-acyl-3-silylallene by a 1,2(C→C) H shift (equation 85). A similar mechanism has been proposed to explain the formation of a 1,2-oxasilin as one of the products from flash vacuum pyrolysis of **251** ($R^1 = R^2 = H$)[146]; in this case, a 1,2(Si→C) H shift in the intermediate carbene **253** was assumed.

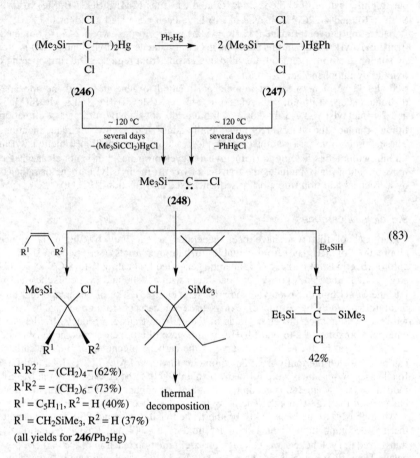

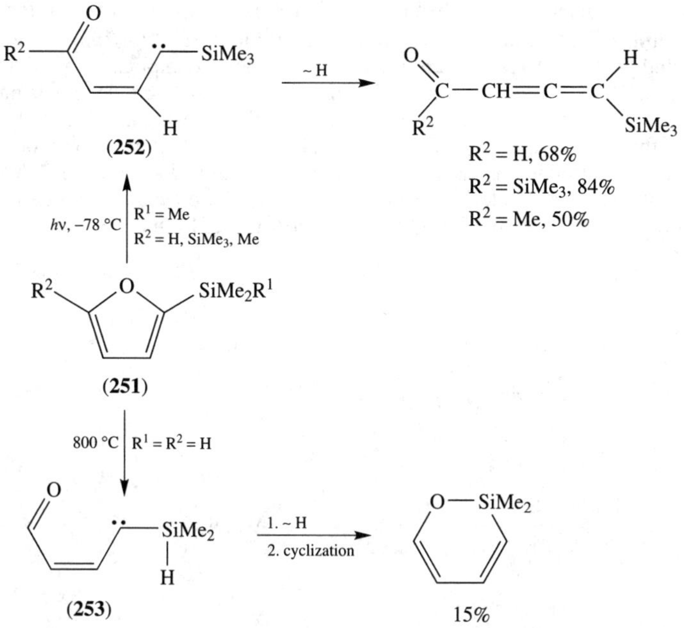

13. Silicon-substituted carbenes

(85)

b. Olefin-to-carbene isomerization. Flash vacuum pyrolysis (FVP) at 700 °C of bis(silyl)ketenes **254** (equation 86) cleanly affords a silyl acetylene and a mixture of permethylcyclosiloxanes[147]. It has been suggested that the decomposition of **254** begins with a 1,2-silyl shift at the C=C bond to produce an acyl(silyl)carbene **255**, which then reacts further as shown in the equation. It should be noted that the proposed isomerization **254** → **255** represents a retro-Wolff rearrangement.

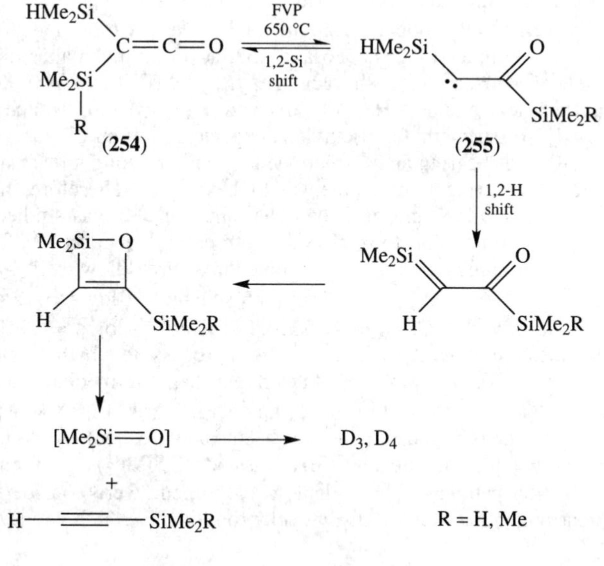

(86)

R = H, Me

An analogous mechanistic scheme (equation 87) has been proposed for the flash vacuum pyrolysis of dimethylsilyl(trimethylsilyl)thioketene[148] (**256**). The pyrolysis of bis(trimethylsilyl)thioketene (**257**) leads to a more complicated product mixture (equation 88). With 47% conversion, a mixture of trimethylsilylacetylene, 1-trimethylsilyl-1-propyne, bis(trimethylsilyl)acetylene, (trimethylsilyl)thioketene, 2,2,4,4-tetramethyl-2,4-disila-1-thietane and 2,2,4,4-tetramethyl-2,4-disila-1,3-dithietane was obtained. All products can be rationalized, however, by the assumption that carbene **258** undergoes not only a silylcarbene-to-silene rearrangement (as in the preceding two cases) but also isomerization to 2-thiirene and insertion into a methyl-C, H bond.

$$\begin{array}{c}\text{HMe}_2\text{Si}\\ \text{Me}_3\text{Si}\end{array}\!\!\!\!\!\!\!\!\!\!\!\!\!\!\!\text{C}\!=\!\text{C}\!=\!\text{S} \quad\xrightleftharpoons[]{\text{FVP}\;700\,°\text{C}}\quad (87)$$

(**256**)

An in-depth experimental and theoretical study on the feasibility of the olefin-to-carbene rearrangement by a thermally induced 1,2-shift of H, Me or SiMe$_3$ has appeared[149]. In contrast to the hetero cumulenes **254**, **256** and **257**, acyclic vinylsilanes do not undergo this rearrangement. For example, flash vacuum pyrolysis (FVP) at 700–800 °C of 1,1-bis(trimethylsilyl)ethylene yields CH$_4$, Me$_3$SiH and trimethyl(vinyl)silane as the major products, but not 1,2-bis(trimethylsilyl)ethylene expected from an olefin-to-carbene rearrangement. In small rings bearing an *exo*-methylene group, the sought rearrangement could be assisted by the relief of ring strain on the way to the carbene. Therefore, the FVP of 2,4-dimethylene-1,3-disiletane **259** and of 2-methylenesiletane **262** was studied. In fact, the 1,2-silyl shift at the olefinic bond took place in both cases (equation 89). The Arrhenius parameters of these rearrangements were determined in a stirred flow reactor system as follows: **259** → **261** at 500–600°C: $E_a = 54.09\pm1.26$ kcal mol^{-1}, $\log A = 12.48\pm0.33$ s^{-1}; **262** → **264** (a+b) at 530–590 °C: $E_a = 47.5\pm0.8$ kcal mol^{-1}, $\log A = 11.3\pm0.2$ s^{-1}. In both cases, the values are in accord with a concerted process, in which alkyl(silyl)carbene **260** and dialkylcarbene **263** are transition states rather than intermediates. For the process involving carbene **260** (SiH$_2$ instead of SiMe$_2$ an energy level of 60.8 kcal mol^{-1} for the transition state, represented by the carbene, has been calculated [MP2/6-31 G(d)//HF-6-31 G(d) + 0.89 ZPE(HF)] for carbene **260** (SiH$_2$ instead of SiMe$_2$). In another theoretical study[90b], isomerization pathways of 1,2-disilylcyclopropene were considered and it was found that the rearrangement to a 2,2-disilylcyclopropylidene, which would a also require a 1,2-silyl shift at the olefinic bond, cannot compete energetically with the ring-opening leading to (1-silylvinyl)silylcarbene.

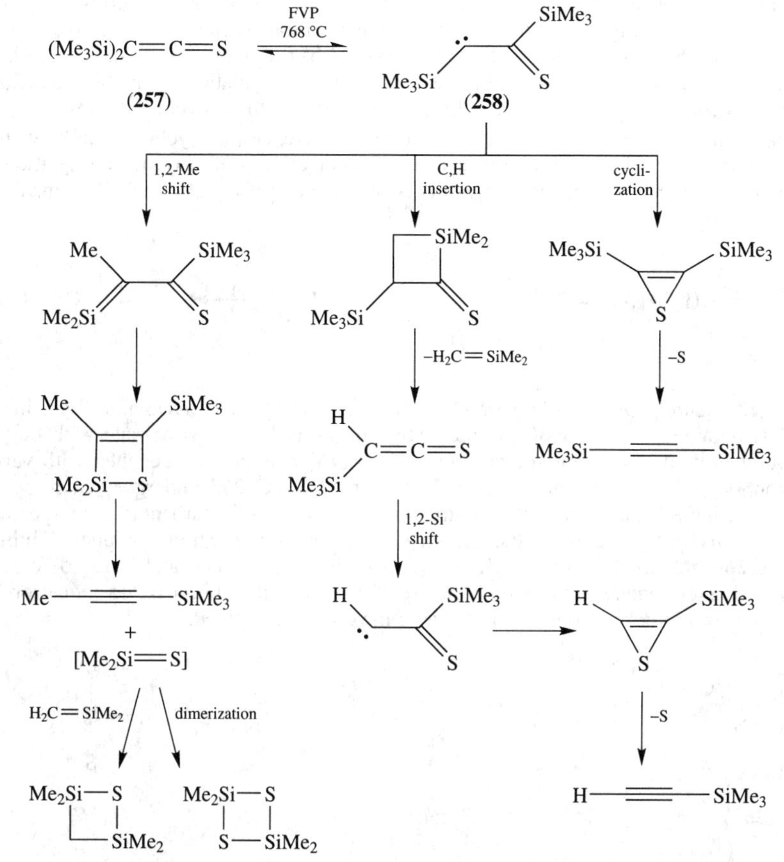

(88)

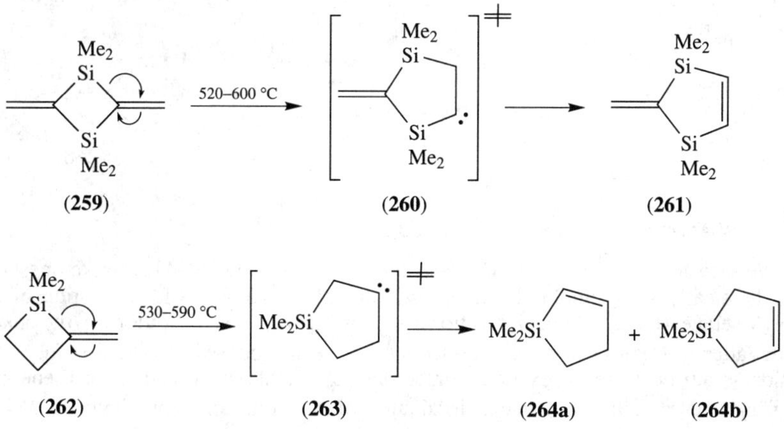

(89)

Notably, the carbon analogue of **259** (CMe$_2$ instead of SiMe$_2$) does not seem to undergo the olefin-to-carbene rearrangement[149]. A comparison of the energy surfaces for the two processes by *ab initio* methods including electron correlation shows that, in contrast to the rearrangement by 1,2-silyl migration (**259** → **260**), the process featuring a 1,2-alkyl migration has an early transition state with a substantial barrier. This fact seems to be prohibitive to the rearrangement, although the relief of ring strain would have been larger than for the silicon case (**259** → **261**) according to the calculations. The thermochemical differences mentioned do not appear to be a property of these cyclic systems. Electron-correlated *ab initio* methods predict an analogous difference for acyclic olefins as well[149], namely barrier-free thermal olefin-to-carbene rearrangements of vinylsilane (equation 90) as opposed to the analogous rearrangements of propene (cf equation 90, CH$_3$ instead of SiH$_3$).

$$\underset{(90)}{\overset{H}{\underset{..}{C}}-CH_2SiH_3 \quad \xleftarrow{\Delta, -SiMe_3} \quad \overset{H}{\underset{H_3Si}{C}}=CH_2 \quad \xrightarrow{\Delta, -H} \quad \overset{H}{\underset{H_3Si}{\underset{..}{C}}}-CH_3}$$

c. Ring-opening of 2-lithio-2-silyloxiranes.[150] 2-(Triphenylsilyl)oxiran-2-yl lithium (**265**) is obtained quantitatively by metalation of 2-(triphenylsilyl)oxirane with butyl or *tert*-butyl lithium in a donor solvent (THF, TMEDA) and can be quenched with various electrophiles, also at low temperature. However, at 25 °C **265** undergoes ring-opening, which can be considered as an α-elimination reaction of this carbenoid-type species to form the silylcarbene **266**. In the presence of excess butyl lithium or phenyl lithium, alkenes **268** are finally obtained. It is assumed that the carbene adds RLi to form the dilithiated intermediates **267**, which can readily eliminate lithium oxide (equation 91). No further trapping reactions of the carbene have been described.

2. Silylcarbenes by intermolecular reactions

a. Photochemical reaction of (silyl)alkynones with alkenes. (1-Alkynyl)ketones react photochemically with simple alkenes in two directions, namely with formation of a 2-alkynyloxetane (Paterno–Büchi reaction) and by [3 + 1] cycloaddition to the alkene; for the latter reaction, a carbene intermediate has been proposed[151]. The first-mentioned reaction is attributed to an excited singlet state of the alkynone and the carbene route to a triplet state[152]. In this context, irradiation of a mixture of (silyl)alkynone **269** and

2,3-dimethylbut-2-ene in benzene (equation 92) furnished a 4-oxaspiro[2,4]heptane **272** (53%), which underwent thermal isomerization to dihydrofuran **273** rather easily, and the expected oxetane (12%)[153]. It was proposed that the photochemical [3 + 2] cycloaddition of **269** to the alkene generates silylcarbene **270**, which isomerizes to **272** via diradical **271** that arises from the triplet state of the carbene by intramolecular hydrogen abstraction. The SiMe$_3$ group has a decisive influence on the further fate of carbene **270**, since the analogous *tert*-butyl-(vinyl)carbene (**270**, *t*-Bu instead of SiMe$_3$) reacts in a different way; the difference may lie in the different spin states (triplet vs singlet) of the reacting carbenes.

b. Insertion of carbon into a Si—H bond. When trimethylsilane and carbon vapor, generated by means of a 16-V electrical arc between graphite electrodes, were codeposited on a surface cooled at 77 K, tetramethylsilane and bis(trimethylsilyl)methane were formed in very low yield together with some other silicon-containing products (equation 93)[154]. It is assumed that the two products mentioned come from (trimethylsilyl)carbene by a twofold H abstraction and by insertion into a Si−H bond of another molecule of Me₃SiH, respectively. By analogy to the generation of alkylcarbenes by insertion of a carbon atom into a C−H bond[155], it is assumed that carbon insertion into the Si−H bond of trimethylsilane yields (trimethylsilyl)carbene.

$$Me_3Si-H + C_{vap} \longrightarrow Me_3Si-\underset{..}{C}-H \xrightarrow{Me_3SiH} Me_3Si-CH_2-SiMe_3$$

$$\downarrow \text{H abstraction}$$

$$Me_3SiMe \tag{93}$$

IV. CONCLUDING REMARKS

This review has shown that a silyl group attached to a carbene center is not simply an innocent bystander. In one way or the other, it is in most cases involved in the further transformation of the carbene. For example, it may influence the electronic properties of a carbene by stabilizing the triplet state. In Bertrand's phosphino(trimethylsilyl)carbenes, the fact that these compounds can be isolated at room temperature is due in part to the SiMe₃ group that seems to stabilize the bond structure of a phosphavinyl ylide by delocalization of a negative charge at the carbon atom. In other silylcarbenes, a substituent or a functional group at the silicon takes part in typical intramolecular carbene reactions; these reaction modes allow the synthesis of a great variety of organosilicon compounds that are often not easily available otherwise, such as silaethenes and silaheterocycles. Finally, in a silylcarbene bearing a different second substituent, the carbene transformation may be determined by the second substituent, but the presence of a bulky silyl group may still exert a more or less pronounced steric effect on product formation. With these principles in mind, silylcarbene pathways should be considered also in future syntheses of organosilicon compounds of new structural type or substitution pattern.

It was mentioned in the Introduction that among the whole class of silicon-substituted carbenes, only those bearing a silyl group (R₃Si) are known experimentally. Substituents containing Si with other coordination numbers [e.g. $R_2C=Si(R)$ and R_2Si^+] are, of course, unusual since silenes and even more silicenium ions are high-energy compounds themselves, and their combination in a corresponding carbene seems rather esoteric. Nevertheless, theoretical calculations on such species already exist (see Section II) and may stimulate the experimental quest.

V. ACKNOWLEDGMENTS

Our own work on silylcarbenes has been supported by the Deutsche Forschungsgemeinschaft and the Volkswagen-Stiftung.

VI. REFERENCES

1. W. Kirmse, *Carbene Chemistry*, 2nd ed., Academic Press, New York, 1971.
2. M. Jones, Jr. and R. A. Moss, *Carbenes*, Vol. I, Wiley, New York, 1973.
3. H. Tomioka, in *Methoden der organischen Chemie (Houben-Weyl)*, Vol. E19b (Ed. M. Regitz), Thieme Verlag, Stuttgart, 1989, pp. 1410–1459.
4. D. Mayer and G. Maas, in *Methoden der organischen Chemie (Houben-Weyl)*, Vol. E17a (Ed. A. de Meijere), Thieme Verlag, Stuttgart, 1997, pp. 811–842.

13. Silicon-substituted carbenes

5. T. Shioiri and T. Aoyama, in *Advances in the Use of Synthons in Organic Chemistry*, Vol. 1 (Ed. A. Dondoni), JAI Press, London, 1993, pp. 51–101; T. Shioiri and T. Aoyama, *J. Synth. Org. Chem. Jpn.*, **44**, 149 (1986).
6. R. Anderson and S. B. Anderson, in *Advances in Silicon Chemistry*, Vol. 1 (Ed. G. L. Larson). JAI Press, London, 1991, pp. 303–325.
7. W. Sander, G. Bucher and S. Wierlacher, *Chem. Rev.*, **93**, 1583 (1993).
8. H. Shimizu and M. S. Gordon, *Organometallics*, **13**, 186 (1994).
9. A. R. W. McKellar, P. R. Bunker, T. J. Sears, K. M. Evenson, R. J. Saykally and S. R. Langhoff, *J. Chem. Phys.*, **79**, 5251 (1983).
10. D. G. Leopold, K. K. Murray and W. C. Lineberger, *J. Chem. Phys.*, **81**, 1048 (1984); D. G. Leopold, K. K. Murray, A. E. S. Miller and W. C. Lineberger, *J. Chem. Phys.*, **83**, 4849 (1985).
11. E. A. Carter and W. A. Goddard III, *J. Chem. Phys.*, **88**, 1752 (1988).
12. B. T. Luke, J. A. Pople, M.-B. Krogh-Jespersen, Y. Apeloig, M. Karni, J. Chandrasekhar and P.v.R. Schleyer, *J. Am. Chem. Soc.*, **108**, 270 (1986).
13. H. J. Köhler and H. Lischka, *J. Am. Chem. Soc.*, **104**, 4085 (1982).
14. J. D. Goddard, Y. Yoshioka and H. F. Schaefer III, *J. Am. Chem. Soc.*, **103**, 7366 (1981).
15. M. S. Gordon and R. D. Koob, *J. Am. Chem. Soc.*, **103**, 2939 (1981).
16. M. S. Gordon, L. A. Pederson, R. Bakhtiar and D. B. Jacobson, *J. Phys. Chem.*, **99**, 148 (1995).
17. S. Tobita, K. Nakajima, S. Tajima and A. Shigihara, *Rapid Commun. Mass Spectrom.*, **4**, 472 (1990).
18. Y. Apeloig and K. Albrecht, personal communication, 1994.
19. R. Nojori, M. Yamakawa and W. Ando, *Bull. Chem. Soc. Jpn.*, **51**, 811 (1978).
20. O. L. Chapman, C.-C. Chang, J. Kolc, M. E. Jung, J. A. Lowe, T. J. Barton and M. L. Tumey, *J. Am. Chem. Soc.*, **98**, 7844 (1976).
21. M. R. Chedekel, M. Skoglund, R. L. Kreeger and H. Shechter, *J. Am. Chem. Soc.*, **98**, 7846 (1976).
22. M. Trommer, W. Sander and A. Paryk, *J. Am. Chem. Soc.*, **115**, 11775 (1993).
23. M. Trommer and W. Sander, *Organometallics*, **15**, 736 (1996).
24. G. Bertrand and R. Reed, *Coord. Chem. Rev.*, **137**, 323 (1994).
25. (a) M. T. Nguyen, M. A. McGinn and A. F. Hegarty, *Inorg. Chem.*, **25**, 2185 (1986),
    (b) M. R. Hoffmann and K. Kuhler, *J. Chem. Phys.*, **94**, 8029 (1991).
26. D. A. Dixon, K. D. Dobbs, A. J. Arduengo III and G. Bertrand, *J. Am. Chem. Soc.*, **113**, 8782 (1991).
27. M. Regitz (Ed.), *Methoden der organischen Chemie (Houben-Weyl)*, Vol. E19b: *Carbene(oide), Carbine*, Thieme Verlag, Stuttgart 1989.
28. (a) M. P. Doyle, *Chem. Rev.*, **86**, 919 (1986).
    (b) G. Maas, *Top. Curr. Chem.*, **137**, 75 (1987).
29. A. Sekiguchi and W. Ando, *Organometallics*, **6**, 1857 (1987).
30. A. Sekiguchi and W. Ando, *Chem. Lett.*, 871 (1983).
31. A. Sekiguchi and W. Ando, *Chem. Lett.*, 2025 (1986).
32. K. Schneider, B. Daucher, A. Fronda and G. Maas, *Chem. Ber.*, **123**, 589 (1990).
33. W. Ando, A. Sekiguchi, T. Hagiwara, T. Migita, V. Chowdhry, F. H. Westheimer, S. L. Kammula, M. Green and M. Jones Jr., *J. Am. Chem. Soc.*, **101**, 6393 (1979).
34. W. Ando, A. Sekiguchi and T. Migita, *Chem. Lett.*, 779 (1976).
35. Recent review on silenes: G. Raabe and J. Michl, in *The Chemistry of Organic Silicon Compounds* (Eds. S. Patai and Z. Rappoport), Chap. 17, Wiley, Chichester, 1989, pp. 1015–1142; A. G. Brook and K. M. Baines, *Adv. Organomet. Chem.*, **25**, 1 (1986).
36. (a) A. Sekiguchi and W. Ando, *Tetrahedron Lett.*, **26**, 2337 (1985).
    (b) A. Sekiguchi, T. Sato and W. Ando, *Organometallics*, **6**, 2337 (1987).
37. L. E. Gusel'nikov and N. S. Nametkin, *Chem. Rev.*, **79**, 529 (1979).
38. R. L. Kreeger and H. Shechter, *Tetrahedron Lett.*, 2061 (1975).
39. W. Ando, A. Sekiguchi, J. Ogiwara and T. Migita, *J. Chem. Soc., Chem. Commun.*, 145 (1975).
40. W. Ando, A. Sekiguchi, A. J. Rothschild, R. R. Gallucci, M. Jones Jr., T. J. Barton and J. A. Kilgour, *J. Am. Chem. Soc.*, **99**, 6995 (1977).
41. T. J. Barton and S. K. Hoekman, *J. Am. Chem. Soc.*, **102**, 1584 (1980).
42. A. Sekiguchi and W. Ando, *J. Am. Chem. Soc.*, **106**, 1486 (1984).
43. G. Maas, K. Schneider and W. Ando, *J. Chem. Soc., Chem. Commun.*, 72 (1988).

44. G. Maas, M. Alt, K. Schneider and A. Fronda, *Chem. Ber.*, **124**, 1295 (1991).
45. W. Ando, A. Sekiguchi and T. Sato, *J. Am. Chem. Soc.*, **103**, 5573 (1981).
46. W. Ando, A. Sekiguchi and T. Sato, *J. Am. Chem. Soc.*, **104**, 6830 (1982).
47. T. J. Barton and G. P. Hussmann, *Organometallics*, **2**, 692 (1983).
48. W. Ando, H. Tanikawa and A. Sekiguchi, *Tetrahedron Lett.*, **24**, 4245 (1983).
49. A. Sekiguchi, H. Tanikawa and W. Ando, *Organometallics*, **4**, 584 (1985).
50. A. Sekiguchi and W. Ando, *J. Am. Chem. Soc.*, **103**, 3579 (1981).
51. G. Märkl, W. Schlosser and W. S. Sheldrick, *Tetrahedron Lett.*, **29**, 467 (1988).
52. G. Märkl and W. Schlosser, *Angew. Chem.*, **100**, 1009 (1988); *Angew. Chem., Int. Ed. Engl.*, **27**, 963 (1988).
53. P. S. Skell and E. J. Goldstein, *J. Am. Chem. Soc.*, **86**, 1442 (1964).
54. M.-H. Yeh, L. Linder, D. K. Hoffman and T. J. Barton, *J. Am. Chem. Soc.*, **108**, 7849 (1986).
55. J. W. Connolly and G. Urry, *J. Org. Chem.*, **29**, 619 (1964).
56. J. W. Connolly, *J. Organomet. Chem.*, **11**, 429 (1968).
57. D. Seyferth and D. C. Annarelli, *J. Am. Chem. Soc.*, **97**, 7162 (1975).
58. S. Bender, Dissertation, University of Kaiserslautern, 1994.
59. J. W. Connolly and P. F. Fryer, *J. Organomet. Chem.*, **30**, 315 (1971).
60. T. Werle, Diploma Thesis, University of Kaiserslautern, 1991.
61. B. Daucher, Dissertation, University of Kaiserslautern, 1996.
62. G. Maas and F. Krebs, University of Kaiserslautern, unpublished work.
63. G. Maas, F. Krebs and V. Gettwert, University of Kaiserslautern, unpublished work.
64. A. Padwa and M. D. Weingarten, *Chem. Rev.*, **96**, 223 (1996).
65. M. Trommer, W. Sander and C. Marquard, *Angew. Chem.*, **103**, 816 (1994); *Angew. Chem., Int. Ed. Engl.*, **33**, 766 (1994).
66. A. Fronda and G. Maas, *Angew. Chem.*, **101**, 1750 (1992); *Angew. Chem., Int. Ed. Engl.*, **28**, 1663 (1992).
67. A. Fronda, F. Krebs, B. Daucher, T. Werle and G. Maas, *J. Organomet. Chem.*, **424**, 253 (1992).
68. G. Maas and A. Fronda, *J. Organomet. Chem.*, **398**, 229 (1990).
69. W. Ando, M. Sugiyama, T. Suzuki, C. Kato, Y. Arakawa and Y. Kabe, *J. Organomet. Chem.*, **499**, 99 (1995).
70. W. Ando, H. Yoshida, K. Kurishima and M. Sugiyama, *J. Am. Chem. Soc.*, **113**, 7790 (1991).
71. W. Ando, *Bull. Chem. Soc. Jpn.*, **69**, 1 (1996).
72. A. Fronda, Dissertation, University of Kaiserslautern, 1991.
73. R. A. Olofson, D. H. Hoskin and K. D. Lotts, *Tetrahedron Lett.*, 1677 (1978).
74. R. G. Daniels and L. A. Paquette, *J. Org. Chem.*, **46**, 2901 (1981).
75. D. Seyferth and E. M. Hanson, *J. Organomet. Chem.*, **27**, 19 (1971).
76. E. Schaumann, C. Friese and S. Spanka, *Synthesis*, 1035 (1986).
77. P. J. Garratt and A. Tsotinis, *J. Org. Chem.*, **55**, 84 (1990).
78. R. N. Haszeldine, D. L. Scott and A. E. Tipping, *J. Chem. Soc., Perkin Trans. 1*, 1440 (1974).
79. D. Seyferth, A. W. Dow, H. Menzel and T. C. Flood, *J. Am. Chem. Soc.*, **90**, 1080 (1968).
80. D. Seyferth, H. Menzel, A. W. Dow and T. C. Flood, *J. Organomet. Chem.*, **44**, 279 (1972).
81. A. J. Ashe III, *J. Am. Chem. Soc.*, **95**, 818 (1973).
82. T. Aoyama, Y. Iwamoto, S. Nishigaki and T. Shioiri, *Chem. Pharm. Bull.*, **37**, 253 (1989).
83. S.-B. Park, H. Nishiyama, Y. Itoh and K. Itoh, *J. Chem. Soc., Chem. Commun.*, 1315 (1994).
84. A. G. Brook and P. F. Jones, *Can. J. Chem.*, **49**, 1841 (1971).
85. A. Sekiguchi and W. Ando, *Tetrahedron Lett.*, 4077 (1979).
86. A. Sekiguchi and W. Ando, *J. Org. Chem.*, **45**, 5286 (1980).
87. T. Aoyama and T. Shioiri, *Tetrahedron Lett.*, **29**, 6295 (1988).
88. T. Aoyama and T. Shioiri, *Chem. Pharm. Bull.*, **37**, 2261 (1989).
89. A. Padwa, M. W. Wannamaker and A. D. Dyszlewski, *J. Org. Chem.*, **52**, 4760 (1987).
90. (a) H. Tokuyama, T. Yamada and E. Nakamura, *Synlett*, 589 (1993).
    (b) Y. Apeloig, T. Müller, A. de Meijere and T. Faber, unpublished results.
91. M. S. Baird, S. R. Buxton and M. Mitra, *Tetrahedron Lett.*, **23**, 2701 (1982).
92. M. S. Baird and H. H. Hussain, *Tetrahedron*, **43**, 215 (1987).
93. G. Maier and D. Born, *Angew. Chem.*, **101**, 1085 (1989); *Angew. Chem., Int. Ed. Engl.*, **28**, 1050 (1989).
94. G. Maier D. Born, I. Bauer, R. Wolf, R. Boese and D. Cremer, *Chem. Ber.*, **127**, 173 (1994).
95. G. Maier, R. Wolf, H.-O. Kalinowski and R. Boese, *Chem. Ber.*, **127**, 191 (1994).

96. M. Regitz and P. Eisenbarth, *Angew. Chem.*, **94**, 935 (1982); *Angew. Chem., Int. Ed. Engl.*, **21**, 913 (1982).
97. B. Coleman, N. D. Conrad, M. W. Baum and M. Jones Jr., *J. Am. Chem. Soc.*, **101**, 7743 (1979).
98. W. Ando, A. Sekiguchi, T. Hagiwara and T. Migita. *J. Chem. Soc., Chem. Commun.*, 372 (1974).
99. H. G. Köser, G. E. Renzoni and W. T. Borden, *J. Am. Chem. Soc.*, **105**, 6359 (1983).
100. H. Meier and K.-P. Zeller, *Angew. Chem.*, **87**, 52 (1975); *Angew. Chem., Int. Ed. Engl.*, **14**, 32 (1975).
101. W. Ando, in *The Chemistry of Diazonium and Diazo Groups* (Ed. S. Patai), Part 1, Wiley, Chichester, 1978, p. 458.
102. M. Torres, J. Ribo, A. Clement and O. P. Strausz, *Can. J. Chem.*, **61**, 996 (1983).
103. R. Brückmann, K. Schneider and G. Maas, *Tetrahedron*, **45**, 5517 (1989).
104. G. Maas and R. Brückmann, *J. Org. Chem.*, **50**, 2801 (1985).
105. R. Brückmann and G. Maas, *Chem. Ber.*, **120**, 635 (1987).
106. R. Brückmann and G. Maas, *J. Chem. Soc., Chem. Commun.*, 1782 (1986).
107. R. Munschauer and G. Maas, *Angew. Chem.*, **103**, 312 (1991); *Angew. Chem., Int. Ed. Engl.*, **30**, 306 (1991).
108. R. Munschauer and G. Maas, *Chem. Ber.*, **125**, 1227 (1992).
109. B. Manz and G. Maas, *Tetrahedron*, **52**, 10053 (1996).
110. O. A. Kruglaya, I. B. Fedot'eva, B. V. Fedot'ev, I. D. Kalikhman, E. I. Brodskaya and N. S. Vyazankin, *J. Organomet. Chem.*, **142**, 155 (1977).
111. U. Schöllkopf and N. Rieber, *Angew. Chem.*, **79**, 906 (1967); *Angew. Chem., Int. Ed. Engl.*, **6**, 884 (1967).
112. U. Schöllkopf, D. Hoppe, N. Rieber and V. Jacobi, *Justus Liebigs Ann. Chem.*, **730**, 1 (1969).
113. W. Ando, T. Hagiwara and T. Migita, *J. Am. Chem. Soc.*, **95**, 7518 (1973).
114. G. Maas, M. Gimmy and M. Alt, *Organometallics*, **11**, 3813 (1992).
115. W. Ando, A. Sekiguchi, T. Migita, S. Kammula, M. Green, and M. Jones Jr., *J. Am. Chem. Soc.*, **97**, 3818 (1975).
116. W. Ando, T. Hagiwara and T. Migita, *Tetrahedron Lett.*, 1425 (1974).
117. M. Alt, Dissertation, University of Kaiserslautern, 1993.
118. G. Maas, T. Werle, M. Alt and D. Mayer, *Tetrahedron*, **49**, 881 (1993).
119. G. Maier, D. Volz and J. Neudert, *Synthesis*, 561 (1992).
120. M. Alt and G. Maas, *Tetrahedron*, **50**, 7435 (1994).
121. H. C. Lottes, J. A. Landgrebe and K. Larsen, *Tetrahedron Lett.*, **30**, 4089 and 4093 (1989).
122. M. Alt and G. Maas, *Chem. Ber.*, **127**, 1537 (1994).
123. R. D. Connell, M. Tebbe, A. R. Gangloff, P. Helquist and B. Åkermark, *Tetrahedron*, **49**, 5445 (1993) and references cited therein.
124. H. Tomioka, H. Kitagawa and Y. Izawa, *J. Org. Chem.*, **44**, 3072 (1979).
125. H. Tomioka, M. Kondo and Y. Izawa, *J. Org. Chem.*, **46**, 1090 (1981).
126. G. Mitchell and C. W. Rees, *J. Chem. Soc., Perkin Trans. 1*, 413 (1987).
127. A. Igau, H. Grützmacher, A. Baceiredo and G. Bertrand, *J. Am. Chem. Soc.*, **110**, 6463 (1988).
128. G. R. Gillette, A. Baceiredo and G. Bertrand, *Angew. Chem.*, **102**, 1486 (1990); *Angew. Chem., Int. Ed. Engl.*, **29**, 1429 (1990).
129. G. R. Gillette, A. Igau, A. Baceiredo and G. Bertrand, *New. J. Chem.*, **15**, 393 (1991).
130. G. Alcaraz, R. Reed, A. Baceiredo and G. Bertrand, *J. Chem. Soc., Chem. Commun.*, 1354 (1993).
131. M. Soleilhavoup, A. Baceiredo, O. Treutler, R. Ahlrichs, M. Nieger and G. Bertrand, *J. Am. Chem. Soc.*, **114**, 10959 (1992).
132. N. Emig, J. Tejeda, B. Réau and G. Bertrand, *Tetrahedron Lett.*, **36**, 4231 (1995).
133. (a) A. Arduengo, H. V. R. Dias, R. L. Harlon and M. Kline, *J. Am. Chem. Soc.*, **114**, 5530 (1992); A. D. Dixon and A. D. Arduengo, *J. Phys. Chem.*, **95**, 4180 (1991).
(b) For a short résumé of recent developments with 'stable carbenes'. see M. Regitz, *Angew. Chem.*, **103**, 691 (1991); *Angew. Chem., Int. Ed. Engl.*, **30**, 674 (1991).
134. (a) G. Bertrand, *Heteroatom Chem.*, **2**, 29 (1991).
(b) G. Bertrand, in *Multiple Bonds and Low Coordination in Phosphorus Chemistry* (Eds. M. Regitz and O. J. Scherer), Thieme Verlag, Stuttgart, 1990, pp. 443–461.
135. A. Igau, A. Baceiredo, G. Trinquier and G. Bertrand, *Angew. Chem.*, **101**, 617 (1989); *Angew. Chem., Int. Ed. Engl.*, **28**, 621 (1989).
136. G. Alcaraz, U. Wecker, A. Baceiredo, F. Dahan and G. Bertrand, *Angew. Chem.*, **107**, 1358 (1995); *Angew. Chem., Int. Ed. Engl.*, **34**, 1246 (1995).

137. R. Armbrust, M. Sanchez, R. Réau, U. Bergsträsser, M. Regitz and G. Bertrand, *J. Am. Chem. Soc.*, **117**, 10785 (1995).
138. H. Keller, Dissertation, University of Kaiserslautern, 1986.
139. H. Keller and M. Regitz, *Tetrahedron Lett.*, **29**, 925 (1988).
140. H. Heydt, M. Regitz and G. Bertrand, in *Methoden der organischen Chemie (Houben-Weyl)*, Vol. E19b (Ed. M. Regitz), Thieme Verlag, Stuttgart, 1989, pp. 1822–1900.
141. G. Maas, K. Schneider and S. Mayer, unpublished work.
142. E. Schaumann and C. Friese, *Tetrahedron Lett.*, **30**, 7033 (1989).
143. C. Friese, in *Methoden der organischen Chemie (Houben-Weyl)*, Vol. E17a (Ed. A. de Meijere), Thieme Verlag, Stuttgart, 1997, pp. 835–842.
144. D. Seyferth and E. M. Hanson, *J. Am. Chem. Soc.*, **90**, 2438 (1968).
145. T. J. Barton and G. P. Hussmann, *J. Am. Chem. Soc.*, **105**, 6316 (1983).
146. T. J. Barton and B. L. Groh, *J. Am. Chem. Soc.*, **107**, 8297 (1985).
147. T. J. Barton and B. L. Groh, *J. Am. Chem. Soc.*, **107**, 7221 (1985).
148. T. J. Barton and G. C. Paul, *J. Am. Chem. Soc.*, **109**, 5293 (1987).
149. T. J. Barton, J. Lin, S. Ijadi-Maghsoodi, M. D. Power, X. Zhang, Z. Ma, H. Shimizu and M. S. Gordon, *J. Am. Chem. Soc.*, **117**, 11695 (1995).
150. J. J. Eisch and J. E. Galle, *J. Organomet. Chem.*, **341**, 293 (1988).
151. S. Hussain and W. C. Agosta, *Tetrahedron*, **37**, 3301 (1981).
152. S. Saba, S. Wolff, C. Schröder, P. Margaretha and W. C. Agosta, *J. Am. Chem. Soc.*, **105**, 6902 (1983).
153. S. Wolff and W. C. Agosta, *J. Am. Chem. Soc.*, **106**, 2363 (1984).
154. P. S. Skell and P. W. Owen, *J. Am. Chem. Soc.*, **94**, 1578 (1972).
155. P. S. Skell and R. R. Engel, *J. Am. Chem. Soc.*, **87**, 4663 (1965); *J. Am. Chem. Soc.*, **88**, 4883 (1966).

CHAPTER **14**

Alkaline and alkaline earth silyl compounds — preparation and structure

JOHANNES BELZNER and UWE DEHNERT

Institut für Organische Chemie der Georg-August-Universität Göttingen, Tammannstr. 2, D-37077 Göttingen, Germany

I. ABBREVIATIONS	780
II. INTRODUCTION	780
III. HYDRIDOSILYL ANIONS	780
IV. ALKYL-SUBSTITUTED SILYL ANIONS	781
A. Methods of Preparation	781
B. Structural Studies	784
V. ARYL-SUBSTITUTED SILYL ANIONS	788
A. Preparation	788
B. Structural Studies	793
VI. SILYL-SUBSTITUTED SILYL ANIONS	794
A. Preparation of Hydrido–Oligosilyl Anions $H_{2n+1}Si_nM$	794
B. Preparation and Properties of Acyclic Oligosilyl Anions	794
1. Linear oligosilyl anions	795
2. Branched oligosilyl anions	799
3. Oligosilyl anions via cleavage of cyclic silanes	801
C. Preparation and Properties of Cyclic Oligosilyl Anions	805
VII. FUNCTIONALIZED SILYL ANIONS	807
A. Lithium Silenolates	807
B. Silyl Anions Bearing a Heteroatom in the α-Position	808
1. Amino-substituted silyl anions	808
2. Alkoxy-substituted silyl anions	810
3. Halogen-substituted silyl anions	812
VIII. MISCELLANEOUS	814
A. Mono- and Dianions of Silacyclopentadienes	814
B. Oligolithiated Monosilanes	819
IX. REFERENCES	821

The chemistry of organic silicon compounds, Vol. 2
Edited by Z. Rappoport and Y. Apeloig © 1998 John Wiley & Sons Ltd

I. ABBREVIATIONS

DME	1,2-dimethoxyethane	Mes	2,4,6-trimethylphenyl
DMI	1,3-dimethylimidazolidine	PMDTA	pentamethyldiethylenetriamine
DMPU	N,N'-dimethylpropyleneurea	TBAF	tetrabutylammonium fluoride
HMPA	hexamethylphosphortriamide	THF	tetrahydrofuran
Tip	2,4,6-tris(i-propyl)phenyl	TMDAP	1,3-bis(dimethylamino)propane
LDMAN	lithium 1-(dimethylamino)naphthalenide	TMEDA	N,N,N',N'-tetramethylethylenediamine

II. INTRODUCTION

This chapter will concentrate mainly on the chemistry of silyllithium compounds, i.e. neutral tetravalent silyl compounds bearing one or more lithium substituents at silicon. In addition, the chemistry of other organosilicon compounds containing alkaline and alkaline earth metals will be outlined in this chapter.

The chemistry of metalated organosilicon compounds has been the subject of several reviews[1], the most recent ones by Lickiss and Smith[1a] and Tamao and Kawachi[1b], which cover the literature up to the year 1994. This chapter will now take into account the developments in the chemistry of metalated silanes up to the middle of 1996; however, for completeness there will be some overlap with former reviews. The emphasis of this review is on the synthesis and structure of these metalated silanes. However, some examples of their utilization for synthetic purposes will also be given where appropriate. For more information about synthetic applications of silyl anions the reader is referred to some leading references in this field[1a,b,h−k].

Another point which needs to be clarified from the start is the nomenclature of metalated silanes: We will frequently use the term 'silyl anion' in this chapter when we talk about metalated silanes. Although the term 'anion' defines, literally taken, an ionic compound, this expression, when used by us, does not necessarily imply that the compound in question is of ionic nature, but covers, as well, in analogy to the use of the term 'carbanion', silicon compounds with a polarized covalent silicon–metal bond.

In order to organize the material to be presented in this chapter, we use a ranking system of substituents at silicon, which starts with hydrogen as the substituent of lowest priority (Section III) and will proceed through alkyl (Section IV), aryl (Section V) and silyl groups (Section VI) as substituents to functional groups (e.g. amino, hydroxy, Section VII) as the highest-ranked substituents; a compound which bears substituents of different priority will be found in the chapter dealing with the substituents of higher priority. The final section will treat metalated siloles and oligolithiated monosilanes.

III. HYDRIDOSILYL ANIONS

The reaction of SiH_4 with potassium in DME or diglyme yields silylpotassium[2]; silylrubidium and silylcesium are obtained analogously[3]. In contrast, the reaction of sodium with SiH_4 yields a mixture of metalated silanes $NaSiH_n(SiH_3)_{3-n}$ ($n = 1-3$)[4]; depending on the reaction conditions, either $NaSiH_3$[4a] or $NaSi(SiH_3)_3$[2f] is formed as the major product.

KSiH$_3$, RbSiH$_3$ and CsSiH$_3$ form a NaCl-type lattice[2b,3]; KSiH$_3$ is reported to crystallize also in an orthorhombic modification at low temperature[2e]. *Ab initio* calculations at the MP4/6-31G*//3-21G* level predict a molecular tetrahedral structure **1** for NaSiH$_3$[5]. An analogous structure was found for LiSiH$_3$. However, an inverted tetrahedron **2** was calculated at the MP4SDTQ/6-31G**//6-31G* level of theory to be by 2.4 kcal mol^{-1} more stable than **1**, because of more favourable electrostatic interactions[6].

$$
\begin{array}{cc}
\text{(1)} & \text{(2)} \quad \text{bond lengths in pm}
\end{array}
$$

Structure (1): H–Si–Na with bond lengths 149.6, 283.7; H bonds at angles. Structure (2): Si with three H and Li, distances 156.8, 238.6, 191.1 pm.

Interestingly, this type of inverted geometry around silicon has been found experimentally in a sodium–oxygen cage compound, [Na$_8$(O$_3$C$_5$H$_{11}$)$_6$(SiH$_3$)$_2$], which contains inverted NaSiH$_3$ moieties. A computational reinvestigation, using the complex [(NaOH)$_3$NaSiH$_3$] as a model, has shown that in this environment the inverted geometry of NaSiH$_3$ is favored by 1.43 kcal mol^{-1} over the 'normal' tetrahedral structure[7].

IV. ALKYL-SUBSTITUTED SILYL ANIONS
A. Methods of Preparation

Three methods are available for the preparation of peralkylated silyl alkaline and alkaline earth compounds.

(1) Cleavage of the Si−Si bond of a disilane with nucleophiles such as organolithium compounds or alkali metal alkoxides or hydrides (equation 1):

$$R_3Si-SiR_3 + MNu \longrightarrow R_3SiM + NuSiR_3 \quad (1)$$

M = Li, Na, K; Nu = Me, OMe, H

(2) Reaction of sodium or potassium hydride with a hydrogen-bearing silane (equation 2):

$$R_3SiH + MH \longrightarrow [R_3SiH_2]^-M^+ \longrightarrow R_3SiM + H_2 \quad (2)$$

M = Na, K

This reaction, which initially was assumed to be a simple deprotonation[8a,b], was shown later by Corriu and coworkers[9] to proceed via a two-step mechanism. The initial step is the reversible formation of a pentacoordinated species, which decomposes irreversibly under formation of molecular hydrogen and the metalated silane.

(3) Transmetalation of silylmercury compounds with lithium (equation 3):

$$R_3SiCl \xrightarrow{Na/Hg} (R_3Si)_2Hg \xrightarrow{Li} 2R_3SiLi + Hg \quad (3)$$

The silylmercury compounds are most conveniently available by reacting a chlorosilane with sodium amalgam. In contrast to the two methods mentioned above, which in most cases require polar, aprotic solvents such as THF, DME or HMPA, transmetalation reactions can be performed in nonpolar solvents and often result in better yields than when performed in ethereal solvents.

Two types of reaction allow the synthesis of silylmagnesium compounds:

(1) Transmetalation of a lithiated silane or of a silylcobalt compound by reaction with a magnesium halide or a Grignard reagent yields the corresponding silyl Grignard compound (equation 4). Silylmagnesium compounds can be also obtained from silylmercury compounds, which undergo metal–metal exchange with magnesium (equation 5).

$$R_3SiM \xrightarrow[\text{or RMgHal}]{\text{MgHal}_2} R_3SiMgHal \quad (4)$$

$$M = Li, Co(CO)_4$$

$$(R_3Si)_2Hg \xrightarrow{Mg} (R_3Si)_2Mg + Hg \quad (5)$$

(2) Silyl Grignard compounds can be obtained from highly reactive magnesium (given as Mg*) and halosilanes (equation 6).

$$R_3SiHal \xrightarrow{Mg^*} R_3SiMgHal \quad (6)$$

The most effective way to obtain Me_3SiLi on a preparative scale is the reaction of $Me_3Si-SiMe_3$ with MeLi in Et_2O with HMPA as cosolvent (equation 7)[10]. The driving force for this conversion is the formation of a Si–C bond, which is stronger than the Si–Si bond of the starting material[10b]. However, the formation of Me_3SiMe_2SiLi as a co-product in this reaction (up to 58%, depending on the reaction conditions[11d]) sometimes limits its usefulness[11] (see Section VI.B.1.) Alternatively, Me_3SiLi may be obtained by the reaction of $(Me_3Si)_2Hg$ with lithium metal[12]. Based on variable-temperature NMR studies, $LiHg(SiMe_3)_3$ and $Li_2Hg(SiMe_3)_4$ were postulated as intermediates in this reaction[12].

$$Me_3Si-SiMe_3 + MeLi \xrightarrow[Et_2O]{HMPA} \begin{array}{l} Me_3SiLi \\ Me_3Si-SiMe_2Li \end{array} \quad (7)$$

MeSiNa and Me_3SiK have been prepared by nucleophilic cleavage of $Me_3Si-SiMe_3$ using different nucleophile/solvent systems such as NaOMe/HMPA[13a], NaOMe/DMI[13b], KOMe/18-crown-6/benzene[13b], KOMe/18-crown-6/THF[13b], KOBu-t/THF[13c] and KOBu-t/DMPU[13c] (equation 8). These reactions are exergonic due to the formation of the strong Si–O bond.

$$Me_3Si-SiMe_3 + MOMe \longrightarrow Me_3SiM + Me_3SiOMe \quad (8)$$
$$M = Na, K$$

The metal hydrides NaH and KH have also been used to cleave $Me_3Si-SiMe_3$ forming Me_3SiM (M = Na, K) and Me_3SiH (Scheme 1)[8a] These reactions are frequently performed in HMPA. Using THF or benzene as solvent requires the presence of 18-crown-6. The initially formed Me_3SiH is transformed by reaction with excess metal hydride into another equivalent of Me_3SiM[8a,b]. On the whole, the reaction of $Me_3Si-SiMe_3$ with two equivalents of NaH or KH converts both silyl groups into the corresponding metalated silanes and thus is, in so far as the yield is concerned, often superior to the cleavage of disilanes with other nucleophiles.

When equivalent amounts of $Me_3Si-SiMe_3$ and tetrabutylammonium fluoride (TBAF) are dissolved in HMPA, an equilibrium between the starting material, a pentacoordinated

$$Me_3Si\text{---}SiMe_3 + MH \longrightarrow Me_3SiM + Me_3SiH$$

$$Me_3SiH + MH \longrightarrow Me_3SiM + H_2$$

$$Me_3Si\text{---}SiMe_3 + 2\,MH \longrightarrow 2\,Me_3SiM + H_2$$

M = Na, K

SCHEME 1

species and a silyl anion with Bu_4N^+ as countercation is established according to ^{1}H NMR studies (equation 9)[14]. This mixture was used by Hiyama and coworkers to transfer the trimethylsilyl group to aldehydes[14], 1,3-dienes[14], and 3,3,3-trifluoropropenes[15].

$$Me_3Si\text{---}SiMe_3 + TBAF \rightleftharpoons [Me_3Si\text{---}SiMe_3F]^-Bu_4N^+ \qquad (9)$$

$$[Me_3Si]^-Bu_4N^+ + Me_3SiF$$

Treatment of the trisilane $Me_3SiSiMe_2SiMe_3$ with TBAF in HMPA, however, does not proceed by formation of $SiMe_3^-$; instead, the more stable disilanyl anion $Me_3SiMe_2Si^-$ is formed in equilibrium, as was concluded from the products which were obtained when the reaction was performed in the presence of aldehydes[16].

Et_3SiLi is formed along with EtLi, when the mercury compound $EtHgSiEt_3$ is reacted with lithium in benzene (equation 10). However, when this reaction is performed in THF, Et_3SiLi is obtained as sole lithiated compound (equation 11)[17].

$$EtHgSiEt_3 \xrightarrow[\text{benzene}]{\text{Li}} Hg + EtLi + Et_3SiLi \qquad (10)$$

$$2\,EtHgSiEt_3 \xrightarrow[\text{THF}]{\text{Li}} 2\,Hg + C_2H_4 + C_2H_6 + 2\,Et_3SiLi \qquad (11)$$

In contrast, Et_3SiK cannot be prepared by mercury–potassium exchange. Reaction of $(Et_3Si)_2Hg$ with potassium metal gives Et_3SiPh in 90% yield (equation 12). The reaction of $(Et_3Si)_2Hg$ with sodium proceeds analogously, albeit with a lower yield of Et_3SiPh[18]. It has been speculated that the initially formed, strongly basic Et_3SiM (M = Na, K) is protonated by benzene thereby forming PhM and Et_3SiH, which subsequently undergo a coupling reaction to yield eventually Et_3SiPh and the corresponding metal hydride[18].

$$(Et_3Si)_2Hg + 2\,K \xrightarrow{\text{benzene}} Hg + 2\,KH + 2\,Et_3SiPh \qquad (12)$$

t-Bu_3SiK and t-Bu_3SiNa are the only peralkylated silylmetal compounds which can be prepared directly by reaction of the appropriately substituted halosilane with the corresponding metal (equation 13)[19]. The coupling reaction of the formed silyl anion with the starting material to give a disilane, which has been observed when other alkylated halosilanes are treated with alkali metals, does not occur in this case due to the severe steric hindrance at the silicon center.

$$t\text{-}Bu_3SiHal + 2\,M \longrightarrow t\text{-}Bu_3SiM + MHal \qquad (13)$$

M = Na, K; Hal = Br, I

A silyl compound with two different alkyl substituents, $(t\text{-Bu})_2\text{MeSiK}$, was prepared by Fürstner and Weidmann in low yield by the cleavage of the Si−Si bond of $(t\text{-Bu})_2\text{MeSi−SiMe}(t\text{-Bu})_2$ with C_8K[20]. This reaction is quite exceptional, as usually the Si−Si bond of peralkylated disilanes is not susceptible to reductive cleavage by alkali metals.

Silylstannanes having at least one bulky substituent can be cleaved with a high-order organocuprate yielding the corresponding silylcuprates (equation 14), which are used *in situ* as silylating agents in organic syntheses[21].

$$\text{RMe}_2\text{Si−SnMe}_3 \xrightarrow[-n\text{-BuSnMe}_3]{+n\text{-Bu}_2\text{Cu(CN)Li}_2} \text{RMe}_2\text{SiBu(Cu)(CN)Li}_2 \quad (14)$$

R = *t*-Bu, 1,1,2-trimethylpropyl

Bis(trimethylsilyl)magnesium has been prepared by treatment of $(\text{Me}_3\text{Si})_2\text{Hg}$ with magnesium metal in either THF[22], DME[22] or 1,3-bis(dimethylamino)propane (TMDAP)/Et_2O[23]. When the reaction is carried out in THF, the initial product is $(\text{Me}_3\text{Si})_2\text{Mg·2THF}$, according to its ^{1}H NMR spectrum, but it readily looses one equivalent of THF to yield $(\text{Me}_3\text{Si})_2\text{Mg·THF}$. The product, which has been obtained in DME as solvent, can be recrystallized from cyclopentane to yield an adduct $(\text{Me}_3\text{Si})_2\text{Mg·DME}$. By treating this ethereal adduct with a 10-fold excess of TMEDA, the $(\text{Me}_3\text{Si})_2\text{Mg·TMEDA}$ complex has been prepared as well (equation 15)[24].

$$(\text{Me}_3\text{Si})_2\text{Hg} + \text{Mg} \xrightarrow{\text{DME}} \text{Hg} + (\text{Me}_3\text{Si})_2\text{Mg·DME}$$
$$\downarrow {-\text{DME} \mid +\text{TMEDA}}$$
$$(\text{Me}_3\text{Si})_2\text{Mg·TMEDA} \quad (15)$$

Silyl Grignard compounds were suggested to be formed in the reaction of silyllithium and silylpotassium compounds with magnesium halides and they were used *in situ* for subsequent reactions[8a,b]. However, an unambiguous characterization of such species or their isolation was missing. The reaction of Me_3SiCl with Riecke magnesium does not lead to Me_3SiMgCl but yields instead the disilane $\text{Me}_3\text{SiSiMe}_3$. The formation of the latter via an initially formed silyl Grignard compound Me_3SiMgCl followed by a subsequent coupling with the starting material seems plausible[8c]. The first unambiguously identified silicon analogues of a Grignard compound were obtained by Ritter and coworkers when trimethylsilyl bromide or trimethylsilyl iodide were treated with highly reactive magnesium metal (synthesized by dehydrogenation of catalytically prepared MgH_2) in the presence of chelating ligands such as TMEDA or pentamethyldiethylenetriamine (PMDTA) (equation 16). In contrast to the classical Grignard synthesis, toluene had to be used as solvent, because ethereal solvents are quantitatively cleaved under the reaction conditions[25].

$$\text{R}_3\text{SiHal} \xrightarrow[L]{\text{Mg}^*} \text{R}_3\text{SiMgHal·L}$$

Hal = Br, I; L = $\text{Me}_2\text{NCH}_2\text{CH}_2\text{NMe}_2$, $\text{MeN(CH}_2\text{CH}_2\text{NMe}_2)_2$ (16)

B. Structural Studies

Studies of the structure of peralkylated silyl anions in solution by spectroscopic methods such as IR, UV or NMR spectroscopy are exceedingly rare, but some information on their structure in the solid is available.

14. Alkaline and alkaline earth silyl compounds — preparation and structure

Trimethylsilyllithium is a hexamer in the crystal (Figure 1)[26]. The Li_6 framework can be described as a six-membered ring, which adopts a 'folded-chair' conformation with an average Li—Li bond length of 270 pm. Alternatively, taking into account the considerably longer 1,3-Li···Li distances within the six-membered ring (average: 327 pm), the Li_6 skeleton may be regarded as a strongly distorted octahedron. Above each Li_3 face a μ^3-bridging trimethylsilyl group is located. The silicon centers are hexacoordinated to three carbon substituents and three lithium centers with Si—Li distances averaging 268 pm. The interaction between silicon and lithium may be regarded as a four-center two-electron bonding. However, in view of more recent results concerning the nature of the Si—Li bond in metalated silanes (Section V.B), a predominately ionic character of the Si—Li interaction may be assumed as well.

When solvent-free $(Me_3SiLi)_6$ is recrystallized from TMEDA, the complex $(Me_3SiLi)_2 \cdot 3TMEDA$ is obtained[27]. Here the hexameric structure of $(Me_3SiLi)_6$ is broken down into a dimeric aggregate which is shown in Figure 2. The lithium centers are coordinated by one Me_3Si substituent and one chelating TMEDA ligand; the fourth coordination site is occupied by the dimethylamino group of another TMEDA molecule, the second amino terminus of which is bridging toward the second $Me_3SiLi \cdot TMEDA$ unit of the dimer.

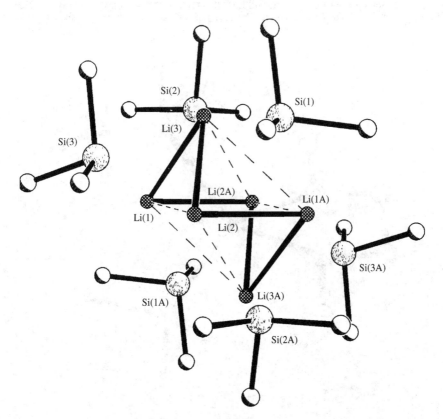

FIGURE 1. Solid state structure of $(SiMe_3Li)_6$ with hydrogen atoms omitted. Reprinted with permission from Reference 26. Copyright 1980 American Chemical Society

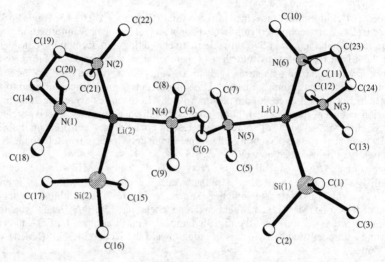

FIGURE 2. Solid state structure of (Me₃SiLi•TMEDA)₂•TMEDA with hydrogen atoms omitted. Reprinted with permission from Reference 27. Copyright 1982 American Chemical Society

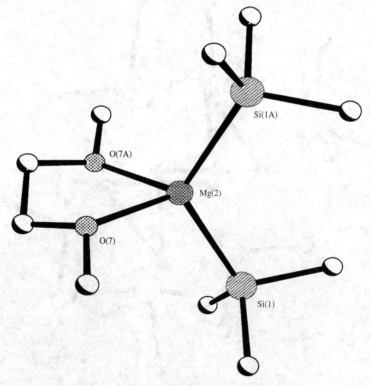

FIGURE 3. Solid state structure of (Me₃Si)₂Mg•DME with hydrogen atoms omitted. Reprinted with permission from Reference 22. Copyright 1977 American Chemical Society

To date, three bis(trimethylsilyl)magnesium compounds have been structurally characterized in the solid state. All of them are monomeric and show quite similar structural characteristics.

The molecular structure of $(Me_3Si)_2Mg\cdot DME$ exhibits C_2 symmetry[22] (Figure 3). The magnesium center is located at the twofold axis and is surrounded by two silicon and two oxygen atoms, thus adopting a distorted tetrahedral coordination sphere. The Mg—Si distance of 263.0 pm is about 9% longer than the sum of the covalent radii of both atoms (241 pm). The steric demand of the trimethylsilyl groups is reflected in the Si—Mg—Si angle of 125.2°, which is appreciably enlarged in comparison with the normal tetrahedral value (109.4°). In contrast, the O—Si—O angle is reduced to 76.3° in order to fit the 'bite angle' of DME. In $(Me_3Si)_2Mg\cdot TMEDA$, the deviation from tetrahedral geometry is smaller. The Si—Mg—Si and N—Mg—N angles are 115.3° and 82.8°, respectively. These changes may be explained by the larger 'bite angle' of TMEDA as well as repulsive interactions between the methyl groups at nitrogen and silicon[24]. The reduced strain of the chelating six-membered ring in $(Me_3Si)_2Mg\cdot TMDAP$ in comparison with the five-membered ring in $(Me_3Si)_2Mg\cdot TMEDA$ is reflected in the enlarged N—Mg—N angle of 93.8°[23].

In contrast to these monomeric structures, $[Me_3SiMgBr\cdot TMEDA]_2$ is reported to form a centrosymmetrical dimer in the solid with two bromine atoms bridging two $Me_3SiMg\cdot TMEDA$ subunits (Figure 4)[25]. The two magnesium–bromine distances differ appreciably in length: The short Mg—Br bond length of 253.4 pm falls into the range of monomeric and dimeric Grignard compounds, whereas the Mg*$\cdots$Br distance of 322.0 pm is longer than the sum of the ionic radii of Mg^{2+} and Br^-, indicating that the interaction between the two $Me_3SiMgBr\cdot TMEDA$ units is quite weak. Substitution of the bidentate TMEDA ligand by the tridentate PMDTA ligand results in deaggregation,

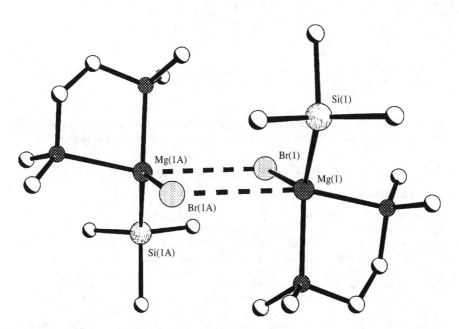

FIGURE 4. Solid state structure of $[Me_3SiMgBr\cdot TMEDA]_2$ with hydrogen atoms omitted. Reproduced from Reference 25 by permission of VCH Verlagsgesellschaft, Weinheim

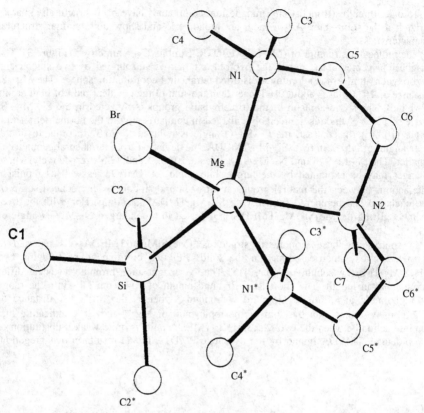

FIGURE 5. Solid state structure of Me₃SiMgBr•PMDTA with hydrogen atoms omitted. Reproduced from Reference 25 by permission of VCH Verlagsgesellschaft, Weinheim

and the monomeric, C_S-symmetric Me₃SiMgBr•PMDTA is formed (Figure 5), in which magnesium is pentacoordinated to bromine, silicon and three nitrogen centers.

V. ARYL-SUBSTITUTED SILYL ANIONS

A. Preparation

The most convenient method to prepare metalated arylsilanes is the reaction of a chlorosilane with lithium or potassium in polar solvents such as THF or DME (Scheme 2).

The first step of this conversion is assumed to be the formation of the silyl anion, which undergoes a subsequent nucleophilic attack on the starting material[28a]. The resulting disilane may be isolated, when stoichiometric amounts of metal are used. However, in contrast to peralkylated disilanes, disilanes which bear at least one aryl substituent at each silicon are susceptible to further reduction. Accordingly, the Si—Si bond of the fully or partially arylated disilane is easily cleaved under the reaction conditions by slow electron transfer from excess metal, eventually transforming both silyl units of the disilane into the desired metalated silane.

14. Alkaline and alkaline earth silyl compounds — preparation and structure

$$Ar_3SiCl + 2M \longrightarrow Ar_3SiM + MCl$$

$$Ar_3SiM + Ar_3SiCl \longrightarrow Ar_3Si-SiAr_3 + MCl$$

$$Ar_3Si-SiAr_3 + 2M \longrightarrow 2Ar_3SiM$$

$$2Ar_3SiCl + 4M \longrightarrow 2Ar_3SiM + 2MCl$$

M = Li, Na, K, Rb, Cs

SCHEME 2

A variety of metalated silanes such as Ph_3SiLi[28a], $(o\text{-Tol})_3SiLi$[28a], Ph_3SiNa[28b], Ph_3SiCs[28a], $(o\text{-Tol})_3SiCs$[28a], Ph_3SiRb[28a], $Ph_2(t\text{-Bu})SiLi$[28c], $Ph_2(o\text{-Tol})SiLi$[28d] and $Ph(o\text{-Tol})_2SiLi$[28e] has been obtained by this method from the corresponding halosilanes. A superior access to Ph_3SiK, Ph_2MeSiK and $PhMe_2SiK$ is the reaction of the corresponding chlorosilane with C_8K in THF. This method yields the silylpotassium compound in a few minutes and in high yield[20].

A disadvantage is that solutions of silyl anions prepared from the halosilanes by one of these methods inevitably contain metal halides as by-product. Salt-free perarylated silyl anions can be prepared by the reaction of a disilane with lithium, sodium or potassium in polar solvents such as THF or DME (Table 1)[28b,29]. Alternative procedures for the preparation of halide-free Ph_3SiK utilize the reductive cleavage of the Si—C bond of $Ph_3Si-CMe_2Ph$ by a sodium–potassium alloy in Et_2O (equation 17)[28f], as well as via the reaction of Ph_3SiH with KH in DME[8a].

$$Ph_3Si-CMe_2Ph \xrightarrow[Et_2O]{Na-K} Ph_3SiK + PhMe_2CK \quad (17)$$

A mixture of the disilane $Me_3Si-SiPh_3$ with TBAF in HMPA has been used by Hiyama and coworkers as a metal-free, synthetic equivalent of the triphenylsilyl anion[15] (equation 18).

$$Me_3Si-SiPh_3 + TBAF \rightleftharpoons Me_3SiF + [Bu_4N]^+[SiPh_3]^- \quad (18)$$

An analogous regioselectivity was observed by Buncel and coworkers[13d], when unsymmetrically substituted disilanes such as $Me_3Si-SiPh_3$, $Me_3Si-SiMePh_2$ or $Me_3Si-SiMe_2Ph$ were treated with potassium t-butoxide in THF or DME. The nucleophilic cleavage of the Si—Si bond proceeded in all three cases to yield exclusively the more stable silyl anion. A salt-free Ph_3SiLi has also been obtained by metal–metal exchange between a silyl–cobalt complex **3** and MeLi (equation 19)[28g,30]. The reaction of **3** with different Grignard reagents has also been used by Corriu and Colomer to generate the corresponding silyl Grignard compound **4** (equation 19)[28g,30].

$$Ph_3Si-Co(CO)_4 + MeLi \begin{cases} \xrightarrow{MeLi} Ph_3SiLi + MeCo(CO)_4 \\ \xrightarrow{RMgBr} Ph_3SiMgBr + RCo(CO)_4 \end{cases} \quad (19)$$

(3)

R = Me, Et, allyl (4)

Analogously, the metal–metal exchange reaction of Ph_3SiK or Ph_3SiLi with $MgBr_2$ was assumed to yield **4**[8b,31]. In contrast, West and Selin found that the reaction of Ph_3SiCl with magnesium is not a practical way to prepare $Ph_3SiMgCl$ (**4**)[32]. The final product of this reaction is $Ph_3Si-SiPh_3$, which most likely arises from the coupling of an initially formed **4** with the starting material Ph_3SiCl. This produced disilane is stable in the presence of excess metal, because magnesium, unlike alkali metals, does not have a sufficiently high reducing potential to effect the cleavage of the Si–Si bond[28a].

Reductive cleavage of disilanes by alkali metals is the most versatile method to prepare metalated arylsilanes, which carry, in addition to the aromatic group, also an aliphatic, benzylic or allylic substituent. Table 1 reports all the metalated arylsilanes which have been prepared by reductive cleavage of disilanes with alkali metals.

Scheme 3 depicts the reaction sequence, which was developed by Sommer and Mason to show that the reductive cleavage of the Si–Si bond of disilanes by lithium proceeds under retention of configuration at a chiral silicon center[39]. The starting silane **5** was converted to the corresponding chlorosilane, which was coupled subsequently with $MePh_2SiLi$ to yield disilane **6**; these reactions are known to proceed with retention and inversion, respectively, i.e. disilane **6** has a configuration at Si* which is inverted in comparison to that of **5**. Cleavage of **6** with lithium metal and subsequent hydrolysis afforded silane *ent*-**5**, which has an inverted configuration of silicon relative to **5**. Because the hydrolysis of the silyllithium compound is assumed to proceed with retention, it has been concluded that the cleavage of the Si–Si bond of **6** by lithium does not invert the configuration at the chiral silicon center[39].

Optically active silyl anions were generated by Corriu and coworkers[28g,30] via the reaction of enantiomerically pure (+)-(α-Naph)PhMeSi*Co(CO)$_4$ with MeLi, which gave after hydrolysis (+)-(α-Naph)PhMeSi*H with 70% retention of configuration. The stereochemistry of the reaction is highly dependent on the nature of the organometallic reagent. With MeMgBr only 55% retention is observed, whereas the reaction with

TABLE 1. Metalated arylsilanes prepared by reductive cleavage of disilanes with alkali metals

Metalated silane	Precursor silane	Solvent	Reference
Ph_3SiLi	$Ph_3Si-SiPh_3$	DME	28b
Ph_3SiNa	$Ph_3Si-SiPh_3$	DME	28b
Ph_3SiK	$Ph_3Si-SiPh_3$	DME, Et_2O	28b, 29
$Ph_2MeSiLi$	$Ph_2MeSi-SiMePh_2$	THF	33
Ph_2MeSiK	$Ph_2MeSi-SiMePh_2$	THF	20[a]
$Ph_2(t\text{-}Bu)SiLi$	$Ph_2(t\text{-}Bu)Si-Si(t\text{-}Bu)Ph_2$	THF	34
$Ph_2(t\text{-}Bu)SiK$	$Ph_2(t\text{-}Bu)Si-Si(t\text{-}Bu)Ph_2$	THF, Et_2O	20[a], 35
$PhMe_2SiLi$	$PhMe_2Si-SiMe_2Ph$	THF	33
$PhMe_2SiK$	$PhMe_2Si-SiMe_2Ph$	THF, Et_2O, DME	20[a], 36
$Ph(i\text{-}Pr)_2SiLi$	$Ph(i\text{-}Pr)_2Si-Si(i\text{-}Pr)_2Ph$	THF	37
$Ph(PhCH_2)_2SiLi$	$Ph(PhCH_2)_2Si-Si(PhCH_2)_2Ph$	THF	37
$Me(H)C=C(Me)CH_2SiPh_2Li$	$[Me(H)C=C(Me)CH_2SiPh_2]_2$	THF	38
$(-)\text{-}neo\text{-}C_5H_{11}PhMeSiLi$	$[(-)\text{-}neo\text{-}C_5H_{11}PhMeSi]_2$	THF	39
Ph_2HSiLi	$Ph_2HSi-SiHPh_2$	THF	40, 45[b]
$[2\text{-}(Me_2NCH_2)C_6H_4]_2HSiLi$	$\{[2\text{-}(Me_2NCH_2)C_6H_4]_2HSi\}_2$	THF	41

[a]C_8K used as the reducing agent.
[b]Lithium 1-(dimethylamino)naphthalenide used as the reducing agent.

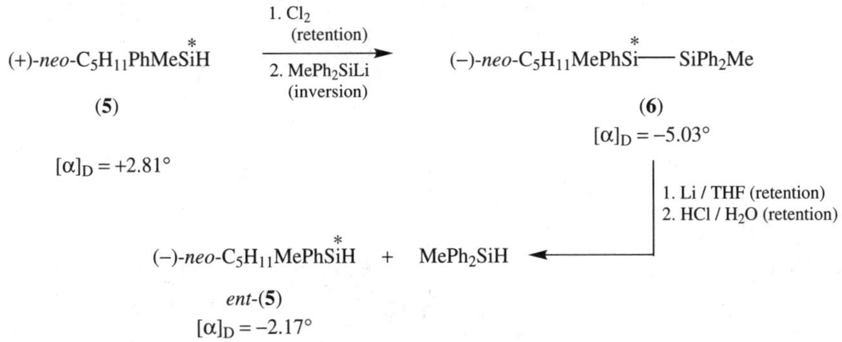

SCHEME 3

(+)-neo-C_5H_{11}PhMeSi*H $\xrightarrow[\text{2. MePh}_2\text{SiLi (inversion)}]{\text{1. Cl}_2 \text{ (retention)}}$ (−)-neo-C_5H_{11}MePhSi*—SiPh$_2$Me

(5) (6)

$[\alpha]_D = +2.81°$ $[\alpha]_D = -5.03°$

1. Li / THF (retention)
2. HCl / H$_2$O (retention)

(−)-neo-C_5H_{11}MePhSi*H + MePh$_2$SiH

ent-(5)

$[\alpha]_D = -2.17°$

n-BuLi proceeds under total loss of chiral information. Similarly, racemic (±)-(α-Naph)PhMeSi*(n-Bu) is obtained when enantiomerically pure (+)-(α-Naph)PhMeSi*H is treated with KH in DME at 50 °C for 24 hours, and the resulting silylpotassium compound is reacted with n-BuBr[11]. The stereochemical course of the metalation is explained by an addition–elimination mechanism via an achiral pentacoordinated silicon compound **7** as the crucial intermediate (equation 20) (see also Section IV.A).

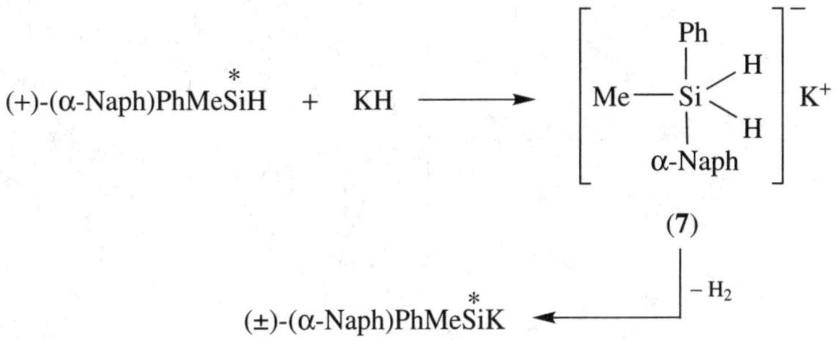

$$(+)\text{-}(\alpha\text{-Naph})\text{PhMeSi*H} + \text{KH} \longrightarrow \begin{bmatrix} \text{Ph} \\ | \\ \text{Me—Si} \begin{smallmatrix} \diagup \text{H} \\ \diagdown \text{H} \end{smallmatrix} \\ | \\ \alpha\text{-Naph} \end{bmatrix}^- \text{K}^+ \quad (20)$$

(7)

$\downarrow -H_2$

(±)-(α-Naph)PhMeSi*K

The cleavage of the Si—Si bond in the 9,10-disilanthracene **8** dimer by lithium metal in THF allows the synthesis of a dimetalated species **9** (Scheme 4)[42]. The reaction can be quenched at this point with electrophiles such as methyl iodide. Extended reaction times in the presence of excess metal produced 9,10-dimetalla-9,10-disilaanthracene **10** (M = Li). The corresponding potassium derivative **10** (M = K) was obtained directly from **8** upon treatment with excess metal. These 1,4-dimetalla species **10** have been used to synthesize a variety of cis-substituted disilaanthracene derivatives[42].

When PhMe$_2$SiLi was reacted with CuCN[43a], CuX (X = Br, I)[43b] or MeCu(CN)Li[43c,d], a variety of silylcopper compounds and silylcuprates was formed: Depending on the stoichiometry used, compounds such as (PhMe$_2$Si)$_2$CuLi·LiX[43b], PhMe$_2$SiCu(CN)Li[43a,c,d], (PhMe$_2$Si)$_2$Cu(CN)Li$_2$[43a,c,d], Me(PhMe$_2$Si)Cu(CN)Li$_2$[43c,d] and (PhMe$_2$Si)$_3$CuLi$_2$[43b−d] were identified by means of multinuclear NMR spectroscopy. These compounds have been used in organic synthesis as silylating agents. They undergo, inter alia, 1,4-addition to enones[43a−c] or may be used for silylcupration of alkynes[43a,c].

The synthesis of silyllithiums containing two aromatic substituents and a hydrogen has been accomplished by the reaction of Ar$_2$SiHCl (Ar = Ph[40], Mes[44]) with lithium metal or lithium 1-(dimethylamino)naphthalenide (LDMAN)[45] in THF. This reaction is assumed

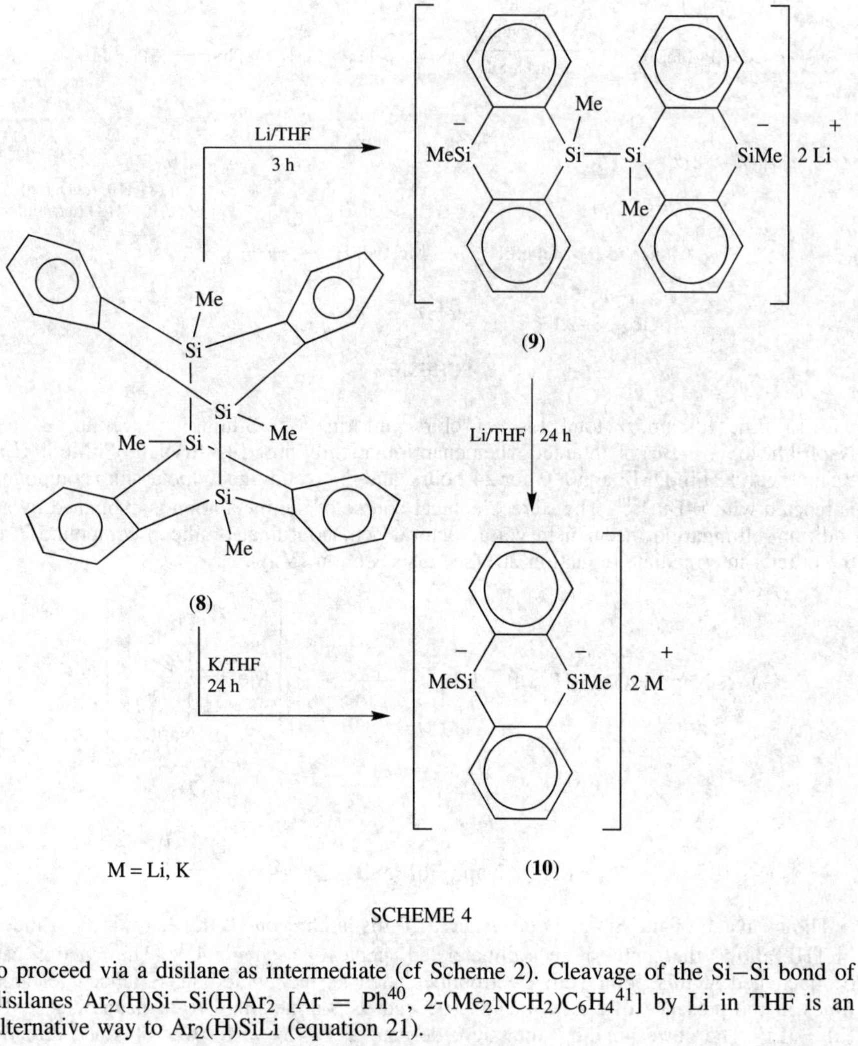

SCHEME 4

to proceed via a disilane as intermediate (cf Scheme 2). Cleavage of the Si—Si bond of disilanes $Ar_2(H)Si-Si(H)Ar_2$ [Ar = Ph[40], 2-$(Me_2NCH_2)C_6H_4$[41]] by Li in THF is an alternative way to $Ar_2(H)SiLi$ (equation 21).

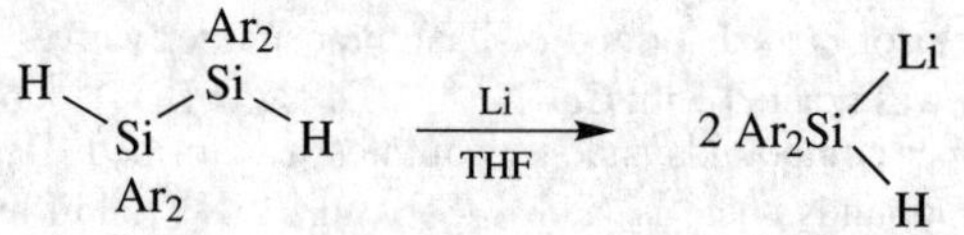

$$H\underset{Ar_2}{\overset{Ar_2}{\underset{Si}{\diagdown}Si\diagup}}H \xrightarrow{\text{Li}}_{\text{THF}} 2\ Ar_2Si\underset{H}{\overset{Li}{\diagdown}} \qquad (21)$$

Ar = Ph, 2-$(Me_2NCH_2)C_6H_4$

In addition, the silylpotassium compounds Ph_2SiHK and $PhSiH_2K$ have been obtained from Ph_2SiH_2 and $PhSiH_3$ by reaction with potassium (the yields were not reported)[46].

B. Structural Studies

IR, UV and, most importantly, NMR spectroscopy have proven to be excellent tools for elucidation of the molecular and electronic structures of silyl anions. Whereas ^{13}C NMR data of metalated alkylsilanes are quite rare, the wealth of ^{13}C NMR data which are available for metalated arylsilanes allows one to draw important conclusions concerning the question of charge delocalization in these compounds (Table 2)[36,47].

Comparing the chemical shift values of chlorosilanes with that of the corresponding silyl alkali metal compounds reveals a strong deshielding of the *ipso* carbons as well as a significant shielding of the *para* carbons; the shift changes of the *ortho* and *meta* carbons on going from the chlorosilane to the silyl anion are less pronounced. This pattern of chemical shift values indicates that charge delocalization into the phenyl ring via mesomeric effects, which would involve the interaction between suitable orbitals at the silicon and the aromatic π-system, is negligible. Instead, the charge distribution in these phenyl-substituted anions is governed by inductive π-polarization effects. Accordingly, the negative charge in phenyl-substituted silyl anions resides almost exclusively at the metalated silicon center in contrast to the corresponding carbon analogues, in which the negative charge density at the α-carbon is significantly reduced by mesomeric delocalization into the aromatic ring system. Similar conclusions have been drawn from UV-spectroscopic investigations, which provide corroborating evidence that there is almost no conjugation between silicon and the phenyl rings[46,48]. The reduced importance of resonance effects in silicon anions relative to carbanions is also reflected in the solid state structures of Ph$_3$SiLi·3THF[49a], as well as of [Ph$_3$SiCu(PMe$_3$)$_2$][49b], in which the silicon center is tetrahedrally surrounded by three carbon and one lithium atoms, whereas in the carbon analogue Ph$_3$CLi·Et$_2$O the carbon center exhibits a planar coordination sphere[50].

Spectroscopic studies have also been devoted to the question of ion-pairing phenomena in silyl anions. It was found that in arylated silyl anions the ^{13}C NMR as well as the ^{7}Li NMR chemical shifts are only slightly influenced by the polarity of the solvent[36,51]. This is in clear contrast to the NMR-spectroscopic behavior of aryl-substituted carbanions, which show a marked solvent dependence. This was interpreted in terms of a significant covalent interaction between silicon and lithium in metalated silanes[36]. Further evidence for a significant covalent nature of the Si—Li bond arises from the observation of a scalar

TABLE 2. ^{13}C NMR chemical shifts (in ppm) of metalated arylsilanes and related species (THF as solvent, cyclohexane (δ 27.7 ppm) as reference)[36]

Compound	ipso	ortho	meta	para	$\Delta\delta^a_{ipso}$	$\Delta\delta^a_{ortho}$	$\Delta\delta^a_{meta}$	$\Delta\delta^a_{para}$
Ph$_3$SiCl	133.8	135.9	128.9	131.5	—	—	—	—
Ph$_2$MeSiCl	134.4	134.0	128.1	130.5	—	—	—	—
PhMe$_2$SiCl	137.0	133.8	128.8	131.0	—	—	—	—
Ph$_3$SiLi	155.9	137.0	126.9	124.6	22.1	1.1	−2.0	−6.9
Ph$_2$MeSiLi	160.1	135.4	126.7	123.9	25.7	1.4	−1.4	−6.6
PhMe$_2$SiLi	166.0	133.8	126.5	122.7	29.0	0.0	−2.3	−8.3
Ph$_3$SiK	158.6	136.9	126.7	123.8	24.8	1.0	−2.2	−7.7
Ph$_2$MeSiK	163.2	135.1	126.5	123.0	28.8	1.1	−1.6	−7.5
PhMe$_2$SiK	170.1	133.6	126.4	121.6	33.1	−0.2	−2.4	−9.4

$^a \Delta\delta = \delta(\text{Ph}_n\text{Me}_{3-n}\text{SiM}) - \delta(\text{Ph}_n\text{Me}_{3-n}\text{SiCl})$, M = Li, K.

^{29}Si–^{6}Li or ^{29}Si–^{7}Li coupling in various arylated silyllithium compounds[52]. For example, the ^{29}Si-NMR signal of Ph$_3$Si^{6}Li is resolved into a 1 : 1 : 1 triplet with a 17 Hz spacing at 173 K in 2-methyltetrahydrofuran[47b]; for Ph$_3$Si^{7}Li a quartet with a ^{29}Si–^{7}Li coupling constant of 45 Hz is observed in d$_8$-toluene at ca $-80\,^\circ$C[49a]. These coupling patterns as well as the magnitude of the coupling constants, which are close to values calculated by $ab\ initio$ methods[53], are in good agreement with a monomeric structure of Ph$_3$SiLi and shows that the Si–Li bond, which was found in the solid state, remains essentially intact in solution at least at low temperatures. However, increase of temperature or changing the solvent to the more polar THF results, due to an acceleration of the intermolecular lithium–lithium exchange, in the disappearance of the ^{29}Si–^{6}Li coupling.

In contrast to these results, which argue in favor of a significant covalent contribution to the Si–Li bonding, other experiments point to a more ionic Si–Li bond. Thus, the fast exchange of lithium in mixtures of Ph$_3$SiLi and Ph$_2$MeSiLi or PhMe$_2$SiLi in THF at temperatures above $-90\,^\circ$C, as observed by ^{7}Li NMR spectroscopy, as well as the observation that the chemical shifts of the lithium nuclei of these compounds are quite similar, has been interpreted in terms of a predominantly ionic Si–Li interaction[51b]. In addition, it was argued that the occurrence of an observable ^{29}Si–^{6}Li or ^{29}Si–^{7}Li coupling speaks for a more ionic bonding, because in case of a covalent Si-Li bond one might expect that a fast quadrupole relaxation would quench the scalar coupling.

The temperature-dependent ^{1}H NMR spectra of Ph(i-Pr)$_2$SiLi in various solvents were shown by Lambert and Urdaneta-Pérez to provide information on the configurational stability of silyllithium compounds[37]. The methyl groups of the isopropyl group are diastereotopic in this compound, and in diglyme they are observed as two separate signals at temperatures up to 185 °C (which is the upper experimental limit). Making the reasonable assumption that the rotation around the Si–C bond is fast on the NMR time scale (and therefore hindered rotation around the C–Si bond cannot account for the chemical inequivalence of the methyl groups) it has been concluded that the configuration of the silicon center is retained up to 185 °C. From this observation, a lower limit of 24 kcal mol^{-1} was calculated for the inversion barrier at silicon. This estimate is in good agreement with the inversion barrier of 26±6 kcal mol^{-1}, which was determined experimentally for SiH$_3^-$ in the gas phase[54], as well as with theoretical calculation[1g].

VI. SILYL-SUBSTITUTED SILYL ANIONS

A. Preparation of Hydrido-Oligosilyl Anions $H_{2n+1}Si_nM$

The reaction of Na with SiH$_4$ in polar solvents such as mono- or diglyme has been reported to yield NaSiH$_3$[4a]; however, under carefully controlled conditions, NaSi(SiH$_3$)$_3$ has been obtained in good yield[4b]. KSi$_2$H$_5$ is formed as the major product when KSiH$_3$ is treated with excess SiH$_4$ in HMPA. This synthesis may be performed as a one-pot procedure, starting from SiH$_4$ and potassium in HMPA: After 1 h, KSiH$_3$ is formed almost exclusively, and if left for longer periods it undergoes subsequent homologization to yield the disilanyl anion. The branched silylpotassium compounds KSiH(SiH$_3$)$_2$ and KSi(SiH$_3$)$_3$ have been prepared by the reaction in HMPA of KSiH$_3$ with Si$_2$H$_6$ or Si$_3$H$_8$[2d,55].

B. Preparation and Properties of Acyclic Oligosilyl Anions

Three methods have been used to synthesize 'organic' metalated oligosilanes:

(a) Cleavage of a Si–Si bond of oligosilanes by nucleophiles such as organolithium or silyllithium compounds (Table 3).

TABLE 3. Metalated oligosilanes prepared by nucleophilic cleavage of a Si—Si bond

Compound	Starting material	Nucleophile	Reference
[Me$_3$SiMe$_2$Si]$^-$[NBu$_4$]$^+$	Me$_3$SiMe$_2$Si—SiMe$_3$	[NBu$_4$]$^+$F$^-$	16
Me$_3$SiMe$_2$SiLi	Me$_3$SiMe$_2$Si—Me	Me$_3$SiLi	10b,11
Me$_3$SiPh$_2$SiLi	Me$_3$SiPh$_2$Si—SiMe$_3$	MeLi	57
Me$_3$SiMes$_2$SiLi	Me$_3$SiMes$_2$Si—SiMe$_3$	MeLi	57
(Me$_3$Si)$_2$MeSiLi	(Me$_3$Si)$_2$MeSi—SiMe$_3$	MeLi	62c
(Me$_3$Si)$_2$(t-Bu)SiLi	(Me$_3$Si)$_2$(t-Bu)Si—SiMe$_3$	MeLi	62c
(Me$_3$Si)$_2$PhSiLi	(Me$_3$Si)$_2$PhSi—SiMe$_3$	MeLi	62b
(Me$_3$Si)$_2$MesSiLi	(Me$_3$Si)$_2$MesSi—SiMe$_3$	MeLi	62a
(Me$_3$Si)$_2$TipSiLi	(Me$_3$Si)$_2$TipSi—SiMe$_3$	MeLi	62b
(Me$_3$Si)$_3$SiLi	(Me$_3$Si)$_3$Si—SiMe$_3$	MeLi	63a,b,c
	(Me$_3$Si)$_3$Si—SiMe$_3$	Ph$_3$SiLi	68
	(Me$_3$Si)$_3$Si—Si(SiMe$_3$)$_3$	MeLi	34, 71a
	(Me$_3$Si)$_3$Si—SiMe$_2$H	MeLi	69b
	(Me$_3$Si)$_3$Si—SiMe(SiMe$_3$)$_2$	MeLi	69b^a
	(Me$_3$Si)$_3$Si—SiMeHSi(SiMe$_3$)$_3$	MeLi	69b
	(Me$_3$Si)$_3$Si—SiMe$_2$SiMe$_3$	MeLi	69b^b
	(Me$_3$Si)$_3$Si—SiMe$_2$Si(SiMe$_3$)$_3$	MeLi	69b
(HMe$_2$Si)$_3$SiLi	(HMe$_2$Si)$_3$Si—SiMe$_2$H	MeLi	68
	(HMe$_2$Si)$_3$Si—SiMe$_2$H	Ph$_3$SiLi	68
t-BuMe$_2$Si(Me$_3$Si)$_2$SiLi	t-BuMe$_2$Si(Me$_3$Si)$_2$Si—SiMe$_3$	MeLi	69a,b
Me$_3$Si(t-BuMe$_2$Si)$_2$SiLi	Me$_3$Si(t-BuMe$_2$Si)$_2$Si—SiMe$_3$	MeLi	69a
Me$_3$SiMe$_2$Si(Me$_3$Si)$_2$SiLi	Me$_3$SiMe$_2$Si(Me$_3$Si)$_2$Si—SiMe$_3$	MeLi	69b^b
t-BuMe$_2$SiSiMe$_2$Si(Me$_3$Si)$_2$SiLi	t-BuMe$_2$SiSiMe$_2$Si(Me$_3$Si)$_2$Si—SiMe$_3$	MeLi	69b
Me(Me$_3$Si)$_2$Si(Me$_3$Si)$_2$SiLi	Me(Me$_3$Si)$_2$Si(Me$_3$Si)$_2$Si—SiMe$_3$	MeLi	69b^a
(Me$_3$Si)$_3$Si(SiMe$_3$)$_2$SiLi	(Me$_3$Si)$_3$Si(SiMe$_3$)$_2$Si—SiMe$_3$	(Me$_3$Si)$_3$SiLi	34
	(Me$_3$Si)$_3$Si(SiMe$_3$)$_2$Si—SiMe$_3$	Ph$_3$SiLi	34,71a
Li(Me$_3$Si)$_2$Si—Si(SiMe$_3$)$_2$Li	Me$_3$Si—Si(SiMe$_3$)$_2$(Me$_3$Si)$_2$Si—SiMe$_3$	(Me$_3$Si)$_3$SiLi	34
	Me$_3$Si—Si(SiMe$_3$)$_2$(Me$_3$Si)$_2$Si—SiMe$_3$	Ph$_3$SiLi	34,71a
Ph$_3$Si(Ph$_2$Si)$_3$Li	(Ph$_2$Si)$_4$	PhLi	79
Ph$_3$Si(Ph$_2$Si)$_4$Li	(Ph$_2$Si)$_4$	Ph$_3$SiLi	79

^{a}(Me$_3$Si)$_3$SiLi and Me(Me$_3$Si)$_2$Si(Me$_3$Si)$_2$SiLi are formed in a ratio 40 : 60.
^{b}(Me$_3$Si)$_3$SiLi and Me$_3$SiMe$_2$Si(Me$_3$Si)$_2$SiLi are formed in a ratio 70 : 30.

(b) Cleavage of a Si—Si bond of oligosilanes by electron transfer from alkali metals (Table 4).

(c) Transmetalation of (oligosilyl)mercury compounds with alkali metal. It appears that only this method allows the use of nonpolar solvents such as toluene or alkanes (Table 5).

1. Linear oligosilyl anions

The formation of Me$_3$SiMe$_2$SiLi as a by-product in the reaction of Me$_3$Si—SiMe$_3$ with MeLi in HMPA/Et$_2$O (which yields mostly Me$_3$SiLi) has been mentioned earlier (Section IV.A). In a large-scale experiment the disilanyl anion has been reported to be

TABLE 4. Metalated oligosilanes prepared by reductive cleavage of a Si—Si bond by means of alkali metals

Compound	Starting material	Solvent	Reference
Me_3SiMes_2SiK	$(Me_3Si)_2Mes_2Si$	THF	57
$Me_3SiAr_2SiLi^a$	$Me_3SiAr_2Si-SiAr_2SiMe_3$	THF	41
$(Me_3Si)_3SiLi$	$(Me_3Si)_3Si-Si(SiMe_3)_3$	THF	71a
$(Me_3Si)_3SiK$	$(Me_3Si)_4Si$	not stated	71b
$Li(Ph_2Si)_4Li$	$(Ph_2Si)_4$	THF	72, 74
$Li(Ph_2Si)_5Li$	$(Ph_2Si)_5$	THF	73
$Li(SiAr_2)_3Li^a$	$(SiAr_2)_3$	1,4-dioxane	75
$Li(SiAr_2)_2Li^a$	$(SiAr_2)_3$	THF	75
$Li[(R_3Si)_2Si]_2Li^b$	$(R_3Si)_2Si=Si(SiR_3)_2$	THF	76
$K[(R_3Si)_2Si]_2K^b$	$(R_3Si)_2Si=Si(SiR_3)_2$	DME	76

^{a} Ar = 2-$(Me_2NCH_2)C_6H_4$.
^{b} R_3Si = $(i\text{-}Pr)_3Si$, $Me(i\text{-}Pr)_2Si$, $Me_2(t\text{-}Bu)Si$.

TABLE 5. Metalated oligosilanes prepared by mercury—metal exchange between silylmercury compounds and alkali metals

Compound	Starting material	Solvent	Reference
Me_3SiMe_2SiLi	$(Me_3SiMe_2Si)_2Hg$	toluene	56
$Me_3SiMe_2SiMe_2SiLi$	$(Me_3SiMe_2SiMe_2Si)_2Hg$	toluene	56
Ph_3SiPh_2SiK	$(Ph_3SiPh_2Si)_2Hg$	THF	83a
$Ph_3SiPh_2SiPh_2SiK$	$(Ph_3SiPh_2SiPh_2Si)_2Hg$	THF	83a
$(Me_3Si)_2MeSiLi$	$[(Me_3Si)_2MeSi]_2Hg$	toluene	56, 61
$(PhMe_2Si)_2MeSiLi$	$[(PhMe_2Si)_2MeSi]_2Hg$	toluene	61
$(Me_3Si)_3SiLi$	$[(Me_3Si)_3Si]_2Hg$	n-heptane	66b
$(Me_3Si)_3SiNa$	$[(Me_3Si)_3Si]_2Hg$	n-heptane	66b
$(Me_3Si)_3SiK$	$[(Me_3Si)_3Si]_2Hg$	n-pentane	66a,b
$(Me_3Si)_3SiRb$	$[(Me_3Si)_3Si]_2Hg$	n-pentane	66a,b
$(Me_3Si)_3SiCs$	$[(Me_4Si)_3Si]_2Hg$	n-pentane	66a,b
$(Me_3SiMe_2Si)_3SiLi$	$[(Me_3SiMe_2Si)_3Si]_2Hg$	THF or n-hexane	69c
$c\text{-}(Me_9Si_5)K$	$[c\text{-}(Me_9Si_5)]_2Hg$	THF	83c
$c\text{-}(Me_9Si_5)Me_2SiK$	$[c\text{-}(Me_9Si_5)Me_2Si]_2Hg$	THF	83b
$c\text{-}(Me_{11}Si_6)K$	$[c\text{-}(Me_{11}Si_6)]_2Hg$	THF	84

formed as the main product when MeLi or n-BuLi is used as the metalating agent[11d]. It has been suggested that Me_3SiMe_2SiLi results from a nucleophilic attack of the initially formed Me_3SiLi on excess disilane under cleavage of a peripheral Si—C bond[10b]; however, an unambiguous demonstration of this hypothesis is still missing.

Crystals of Me_3SiMe_2SiLi, which are suitable for X-ray crystallography, have been obtained by Sekiguchi and coworkers via a lithium–mercury exchange reaction of $(Me_3SiMe_2Si)_2Hg$ with lithium metal in toluene[56]. In contrast to hexameric $(Me_3SiLi)_6$ (Section IV.B), Me_3SiMe_2SiLi is a tetramer in the solid state (Figure 6). The four lithium

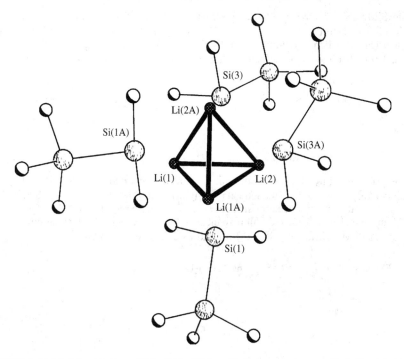

FIGURE 6. Solid state structure of (Me$_3$SiMe$_2$SiLi)$_4$ with hydrogen atoms omitted. Reprinted with permission from Reference 56. Copyright 1995 American Chemical Society

atoms build a central tetrahedron with an average Li—Li distance of 278.0 pm, which is about 20 pm longer than the distances found in tetrameric alkyllithiums; each of the disilyl moieties is located above one of the faces of the Li$_4$ tetrahedron with three nearly equal Si—Li distances averaging 268.3 pm. The broad ^{29}Si NMR signals of isotopically enriched Me$_3$SiMe$_2$Si^{6}Li in toluene-d$_8$ imply the existence of oligomeric aggregates in toluene solution as well. However, in THF-d$_8$ a well resolved 1 : 1 : 1 triplet with 18.8 Hz spacing is observed at 180 K. This signal pattern indicates that the oligomeric aggregates, which exist in the solid as well as in nonpolar solvents, are broken down into a monomeric species by the more polar THF. Moreover, the observation of a ^{29}Si–^{6}Li coupling and its magnitude indicates that the Si—Li bond has a partially covalent character[56].

The aryl-substituted lithiated disilanes, Me$_3$SiPh$_2$SiLi and Me$_3$SiMes$_2$SiLi (Mes = 2, 4, 6-trimethylphenyl), have been prepared by Brook and coworkers from the corresponding trisilanes by nucleophilic cleavage of a Si—Si bond with MeLi in THF/Et$_2$O (equation 22)[57]; an alternative synthesis of Me$_3$SiPh$_2$SiLi utilizes the reaction of Me$_3$SiPh$_2$SiCl with the electron transfer reagent LDMAN[45].

$$R_2Si\begin{matrix}SiMe_3\\SiMe_3\end{matrix} \xrightarrow{MeLi} R_2Si\begin{matrix}SiMe_3\\Li\end{matrix} \qquad (22)$$

R = Ph, 2,4,6-trimethylphenyl(=Mes)

Me$_3$SiMes$_2$SiK has been obtained by treatment of the trisilane (Me$_3$Si)$_2$SiMes$_2$ with Na/K alloy in THF/Et$_2$O/HMPA[57]. The selective rupture of the central Si−Si bond of a symmetrical tetrasilane by lithium in THF allows access to a metalated disilane, bearing the chelating 2-(dimethylaminomethyl)phenyl substituent at the α-silicon (equation 23)[41].

$$\text{Me}_3\text{Si}-\underset{\underset{\text{Ar}_2}{|}}{\overset{\overset{\text{Ar}_2}{|}}{\text{Si}}}-\text{Si}-\text{SiMe}_3 \xrightarrow{\text{Li, THF}} \text{Ar}_2\text{Si}\underset{\text{Li}}{\overset{\text{SiMe}_3}{<}} \tag{23}$$

Ar = 2-(Me$_2$NCH$_2$)C$_6$H$_4$

The reaction of Ph$_3$SiPh$_2$SiCl with lithium metal was reported to produce Ph$_3$SiPh$_2$SiLi and it is likely to proceed via the tetrasilane Ph$_3$Si(Ph$_2$Si)$_2$SiPh$_3$ as an intermediate[58]. The selectivity of this reaction is surprising in view of the more recent report that the reaction of lithium metal with a related chlorosilane, PhMe$_2$SiMe$_2$SiCl, produces an equilibrium mixture of PhMe$_2$SiMe$_2$SiLi, PhMe$_2$SiLi and (Me$_2$Si)$_6$ (equation 24)[59].

$$\text{Ph(Me}_2\text{Si)}_2\text{Cl} \xrightarrow{\text{Li}} \text{Ph(Me}_2\text{Si)}_2\text{Li}$$

$$\updownarrow$$

$$\text{PhMe}_2\text{SiLi} + 1/6\ (\text{Me}_2\text{Si})_6 \tag{24}$$

This reaction has been monitored by means of multinuclear NMR, ESR and UV spectroscopy as well as by HPLC, and four distinct stages have been recognized. In the first step the tetrasilane is formed, as expected. In the second stage, the tetrasilane is further reduced by excess lithium to produce via its radical anion metalated mono-, di- and trisilanes, which subsequently undergo coupling with the starting chlorodisilane to form α,ω-diphenyltri-, tetra- and pentasilanes. In the third stage, (Me$_2$Si)$_6$ is produced by the cleavage of the Si−Si bonds in the diphenylated oligosilanes by silyl anions; linear oligosilanes are produced as well at this point. In the final step, more electrons are transferred from the lithium metal to the system resulting in a complex equilibrium mixture of PhMe$_2$SiMe$_2$SiLi, PhMe$_2$SiLi and (Me$_2$Si)$_6$. It should be mentioned at this point that the third step of the reaction sequence, i.e. the formation of (Me$_2$Si)$_6$ from oligosilanes and silyl anions, has been known also for other α,ω-phenylated permethyloligosilanes, e.g. PhMe$_2$Si(SiMe$_2$)$_n$SiMe$_2$Ph or Ph$_3$Si(SiMe$_2$)$_n$SiPh$_3$ ($n = 1-3$)[60].

2-Lithiotrisilanes, which are used as precursors to polysilane dendrimers, have been obtained by a mercury–lithium exchange reaction (equation 25)[56,61]. The terminally metalated trisilane Me$_3$Si(Me$_2$Si)$_2$Li has been synthesized analogously[56].

$$\underset{\underset{\text{RMe}_2\text{Si}}{|}}{\overset{\overset{\text{RMe}_2\text{Si}}{|}}{\text{Me}-\text{Si}}}-\text{Hg}-\underset{\underset{\text{SiMe}_2\text{R}}{|}}{\overset{\overset{\text{SiMe}_2\text{R}}{|}}{\text{Si}-\text{Me}}} \xrightarrow{\text{Li, toluene}} 2\ \underset{\underset{\text{RMe}_2\text{Si}}{|}}{\overset{\overset{\text{RMe}_2\text{Si}}{|}}{\text{Me}-\text{Si}}}-\text{Li} + \text{Hg} \tag{25}$$

R = Me, Ph

14. Alkaline and alkaline earth silyl compounds — preparation and structure

The reaction of MeLi in Et_2O with alkyl- and aryl-substituted branched tetrasilanes proceeds by cleavage of one Si—Si bond, thus affording various trisilanes, which are lithiated at the central silicon atom (equation 26)[62]. The outcome of these reactions apparently is insensitive as to whether halide-free MeLi or the MeLi·LiBr complex is used[62]. The decrease in the reaction rate in the order: R = Tip > Mes > Ph has been suggested by Fink and coworkers to reflect the relief of steric strain on converting the tetrasilane into the metalated silane[62b].

$$RSi(SiMe_3)_3 \xrightarrow[\text{MeLi·LiBr}]{\text{MeLi or}} RSi(SiMe_3)_2Li + SiMe_4 \quad (26)$$

R = Me, Ph, Mes, Tip

2. Branched oligosilyl anions

The preparation of $(Me_3Si)_3SiM$ and derivatives thereof, which contain the *iso*-tetrasilabutyl subunit, is one of the best investigated topics in the chemistry of metalated oligosilanes.

The Si—Si bond of the easily available pentasilane $(Me_3Si)_4Si$ is smoothly cleaved by organometallic reagents such as MeLi[63] or Ph_3SiLi[63a] yielding $(Me_3Si)_3SiLi$ (**11**). The most convenient access to this silyllithium compound on a large scale appears to be the reaction of $(Me_3Si)_4Si$ with MeLi·LiBr in THF/Et_2O[63b] (equation 27).

$$(Me_3Si)_4Si + MeLi \xrightarrow{THF/Et_2O} (Me_3Si)_3SiLi + SiMe_4 \quad (27)$$
$$\textbf{(11)}$$

The ethereal solution of **11** obtained in this way is of limited stability, and therefore should be used immediately for further reactions. Alternatively, crystalline THF-solvated reagents, **11**·3THF, which can be stored for months without decomposition, may be isolated in a yield of 92% by replacing the ether solvent by *n*-pentane and cooling the resulting solution[63b]. This isolated complex has the further advantage of being soluble in nonpolar solvents such as pentane or toluene, in which, as a rule, subsequent coupling reactions of the anionic species with electrophilic substrates proceed more cleanly and with higher yield than in polar solvents. **11**·3THF is monomeric in the solid state[49a,64]. The Si—Si—Si bond angle is reduced from the ideal tetrahedral value of 109.5° to 102.4°. A similar distortion has been observed for **11**·1.5DME[65] and for the $(Me_3Si)_3SiLi$ unit in the cocrystal **11**·$(Me_3Si)_4Si$[64]. The significant compression of the Si—Si—Si bond angle in these $(Me_3Si)_3SiLi$ compounds was interpreted by Sheldrick and coworkers[64] as an indication of an appreciable ionic character of the Si—Li bond. However, an inherent covalent contribution to the Si—Li interaction is indicated by the $^{29}Si-^{7}Li$ coupling constant of 38.6 Hz measured in toluene-d_8 at room temperature[64]. The quartet pattern of the ^{29}Si NMR signal reveals the momomeric nature of **11** under these conditions.

The entire series of alkali metal compounds $(Me_3Si)_3SiM$ (M = Li, Na, K, Rb, Cs) have been prepared by Klinkhammer and coworkers via a metal–metal exchange reaction from bis[tris(trimethylsilyl)] derivatives of the zinc group, $[(Me_3Si)_3Si]_2M$ (M = Zn, Cd, Hg), and the corresponding alkali metal in *n*-pentane or *n*-heptane (equation 28)[66a,b].

$$M'[Si(SiMe_3)_3]_2 \xrightarrow{M} 2MSi(SiMe_3)_3 + M' \quad (28)$$
$$M' = Zn, Cd, Hg \qquad M = Li, Na, K, Rb, Cs$$

The formation of the cesium compound is complete after a few minutes, when the mercury derivative is used as the starting material. In contrast, the reaction of the zinc compound with potassium metal reaches only 60% conversion in 2 days[66a]. The rubidium and cesium compounds obtained by this method have been characterized in the solid as toluene and THF solvates[66a]. In addition, a cocrystallization product of $(Me_3Si)_3SiCs$ with biphenyl, the origin of which is not stated, is mentioned in Reference 66b. All these complexes are dimeric aggregates having a folded M_2Si_2 ring. The solvate-free lithium, sodium and potassium compounds form cyclic dimers in the solid state as well. The Li_2Si_2 and the K_2Si_2 rings are almost planar, whereas in the sodium compound a folding angle of 170.9° has been found[66b]. In benzene solutions aggregation numbers between 1 and 2 have been determined for all these compounds, which indicates that the dimers found in the solid are partially broken down in benzene solution[66a].

11·3THF has been converted by treatment with CuCl[67a] or CuBr[67b] into $[Li(THF)_4]$ $[Cu_5Cl_4(Si(SiMe_3))_2]$ and $[Cu_2(Si(SiMe_3))_2Li(THF)_3]$, respectively, which are the first lithium silylcuprates to be characterized in the solid state.

The synthesis of metalated iso-tetrasilanes via nucleophilic cleavage of a Si—Si bond by means of organolithium compounds is not restricted to the synthesis of $(Me_3Si)_3SiLi$ itself: $(HMe_2Si)_3SiLi$[68], $(t\text{-}BuMe_2Si)(Me_3Si)_2SiLi$[69a,b], $(t\text{-}BuMe_2Si)_2(Me_3Si)SiLi$[69a] as well as a stannyl derivative, $(Me_3Sn)_3SiLi$[70], have been prepared from the corresponding silane precursors by reaction with MeLi.

$(Me_3SiMe_2Si)_3SiLi·3THF$ and the unsolvated dimer $[(Me_3SiMe_2Si)_3SiLi]_2$ were obtained most recently by Apeloig and coworkers upon treatment of $[(Me_3SiMe_2Si)_3Si]_2Hg$ with lithium metal in THF or hexane[69c]. Both lithiotrisilanes exhibit in the solid a Si—Li distance, which is significantly elongated (i.e. by ca 0.07 Å) in comparison to $(Me_3Si)_3SiLi·3THF$ and $[(Me_3Si)_3SiLi]_2$, respectively. This observation as well as the highfield shift in the ^{29}Si NMR of the metalated silicon center upon changing the Me_3Si group to the Me_3SiMe_2Si group was interpreted in terms of an enhanced ionicity of the silyllithium compounds by β-silyl substitution[69c].

Whereas $(Me_3Si)_4Si$ is not susceptible to electron transfer from lithium (thus allowing its synthesis by lithium-mediated coupling of $SiCl_4$ with Me_3SiCl), hexasilane $(Me_3Si)_3Si-Si(SiMe_3)_3$ (**12**) undergoes smooth reaction with lithium in THF to yield **11** (Scheme 5)[71a]. Alternatively, the central Si—Si bond of **12** may be cleaved by MeLi leading to **11**[71a]. However, the attack of the more bulky Ph_3SiLi or $(Me_3Si)_3SiLi$ on **12** takes a different course. The central Si—Si bond is not broken, and instead one or even two trimethylsilyl groups at the periphery of the molecule are substituted by lithium atoms (Scheme 5)[34,71a].

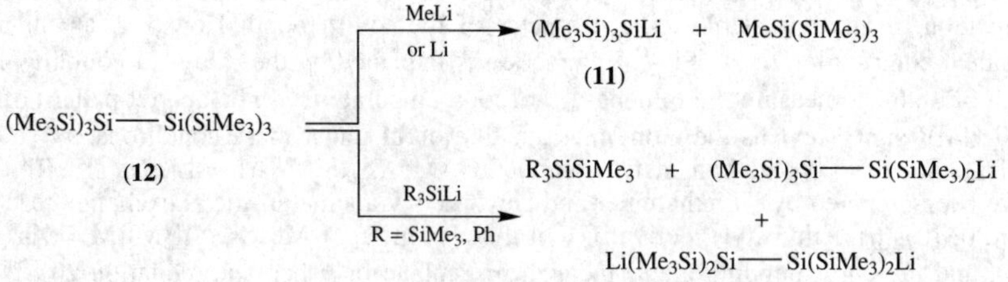

$(Me_3Si)_3Si-Si(SiMe_3)_3$ (**12**)

MeLi or Li → $(Me_3Si)_3SiLi$ + $MeSi(SiMe_3)_3$ (**11**)

R_3SiLi (R = $SiMe_3$, Ph) → $R_3SiSiMe_3$ + $(Me_3Si)_3Si-Si(SiMe_3)_2Li$ + $Li(Me_3Si)_2Si-Si(SiMe_3)_2Li$

SCHEME 5

It may be concluded from the results of Scheme 5 that steric effects are responsible for the removal of the peripheral Me_3Si group. However, the cleavage of unsymmetrical disilanes such as $Me_3Si-SiPh_3$, $Me_3Si-SiMePh_2$ or $Me_3Si-SiMe_2Ph$ by potassium

t-butoxide[13c] or by TBAF[15] appears to be governed by electronic effects, i.e. in these cases the more stable arylated silyl anion is formed.

A more recent study by Apeloig and coworkers of the reaction of MeLi with a variety of oligosilanes $(Me_3Si)_3Si-X$ has shed some more light on the factors which determine the course of the nucleophilic cleavage[69b]. When X is a group having a backbone of one or two silicon centers such as $-SiMe_2H$, $-SiMe_2(t\text{-Bu})$, $-SiMe_2SiMe_3$ or $-SiMe_2SiMe_2(t\text{-Bu})$, MeLi attacks preferentially the sterically less hindered silicon center. Thus $(Me_3Si)_3Si-SiMe_2H$ yields on treatment with MeLi almost exclusively $(Me_3Si)_3SiLi$ (**11**), whereas the reaction of the same nucleophile with $(Me_3Si)_3Si-SiMe_2(t\text{-Bu})$ proceeds to yield exclusively $Li(Me_3Si)_2Si-SiMe_2(t\text{-Bu})$, i.e. the sterically more accessible trimethylsilyl group of the $(Me_3Si)_3Si$ moiety is cleaved by MeLi. In contrast, steric effects appear to be overridden by — not yet fully understood — electronic effects, when X is a branched oligosilyl group such as, *inter alia*, $-SiMe(SiMe_3)_2$ or $-Si(SiMe_3)_3$. In these cases attack of MeLi and cleavage occur preferentially at the silicon center, which is located in the α-position to the $(Me_3Si)_3Si$-moiety, thus yielding **11** as main product.

When the oligosilane $(Me_3Si)_3Si-X$ contains a chloro substituent which is geminal to the $(Me_3Si)_3Si$ group (i.e. $X = -SiR^1R^2Cl$), $Li(Me_3Si)_2SiSiR^1R^2Me$ was obtained, upon reaction with MeLi (Scheme 6). A two-step substitution–lithiation mechanism (a in Scheme 6) was ruled out by demonstrating that the initial substitution product yields a different product in the reaction with MeLi and an elimination–addition mechanism (b in Scheme 6) via a disilene as intermediate was tentatively suggested.

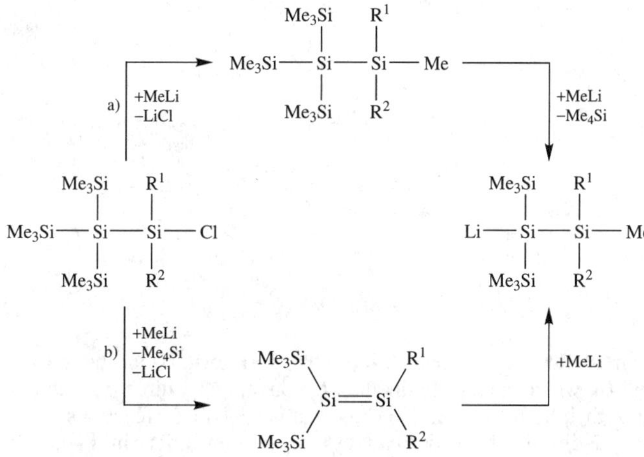

SCHEME 6

3. Oligosilyl anions via cleavage of cyclic silanes

The cleavage of a Si–Si bond in silacyclic compounds by lithium metal is the most versatile method to prepare linear, α,ω-dilithiated oligosilanes. The first dilithio compound obtained by Gilman and coworkers in this way was $Li(Ph_2Si)_4Li$ (equation 29)[72].

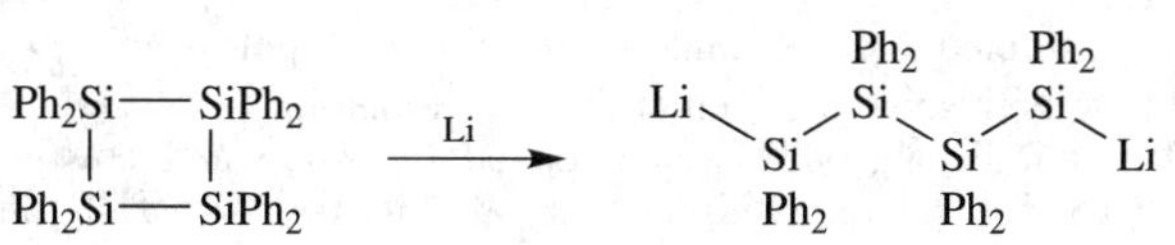

(29)

Analogously, Li(Ph$_2$Si)$_5$Li has been synthesized from the corresponding cyclopentasilane[73]. ^{29}Si NMR and ^{7}Li NMR measurements of Li(Ph$_2$Si)$_4$Li in THF solvents indicate the presence of three different adducts which, at temperatures above 294 K, undergo exchange reactions, which are fast on the NMR time scale[74]. At 173 K, the ^{29}Si NMR signals of the terminal silicon nuclei of each of the adducts show a scalar coupling to one ^{7}Li nucleus (coupling constants: 32.8, 40 and 42 Hz), thus reflecting the partially covalent character of the Si—Li bond. It was shown by X-ray structure determination that Li(SiPh$_2$)$_4$Li is a monomer in the solid state with a planar LiSi$_4$Li backbone with each of the lithium atoms of the centrosymmetrical molecule being bonded to a terminal silicon center in accordance with the solution structures, and adopting a distorted tetrahedral environment due to its coordination to three THF molecules.

1,3-Dilithiotrisilane **13** has been obtained by the reaction of a cyclotrisilane with two equivalents of lithium in 1,4-dioxane (Scheme 7)[75]. **13** is monomeric in the solid state and its structure is shown in Figure 7. Each lithium atom is bonded to a terminal silicon atom and in addition it is coordinated to the two amino groups of the neighboring (Me$_2$NCH$_2$)C$_6$H$_4$ substituents, as well as to an oxygen atom of the dioxane solvent molecule. The LiSi$_3$Li backbone adopts an approximately antiperiplanar conformation.

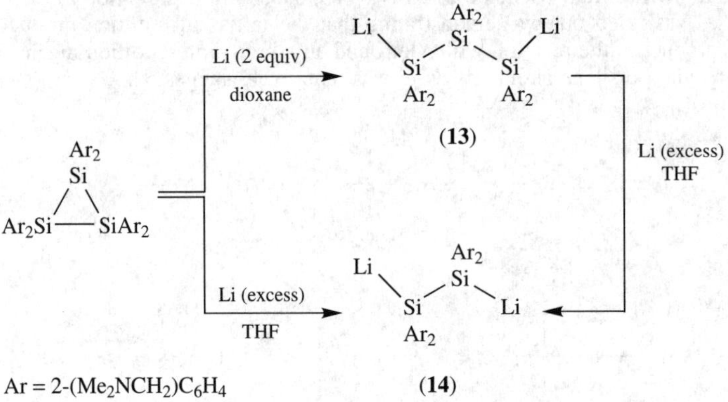

SCHEME 7

The reaction of **13** with excess lithium in THF results in the formation of 1,2-dilithiodisilane **14**, which can be obtained more conveniently directly by the reaction of the cyclotrisilane with lithium in THF. The mechanism of these reactions is still unknown. In contrast to 1,2-dilithioethane derivatives, the lithium atoms in **14** do not undergo a side-on coordination to the Si—Si bond, the length of which is typical for a Si—Si single bond (Figure 8). Instead, each lithium atom is bonded to a silicon atom; Li1 is further coordinated intramolecularly to two Me$_2$NCH$_2$ groups of the geminal aryl substituents and intermolecularly to a THF molecule, while Li2 interacts with only one dimethylamino group and two external THF molecules. **14** is monomeric at room temperature in THF, and the Si—Li bonding is retained in solution, as can be concluded from the ^{29}Si NMR spectrum, which exhibits a 1 : 1 : 1 : 1 quartet at $\delta = -32.8$ with a ^{29}Si-^{7}Li coupling constant of 36 Hz.

Various silyl-substituted disilenes undergo reaction with potassium in DME or with Li in THF to yield the corresponding vicinal dianions (equation 30)[76].

Another 1,2-dilithiodisilane (**16**) was obtained by Ando and coworkers from 1,2-dichlorodisilane **15** by reaction with excess lithium metal in THF under ultrasonic

14. Alkaline and alkaline earth silyl compounds — preparation and structure   803

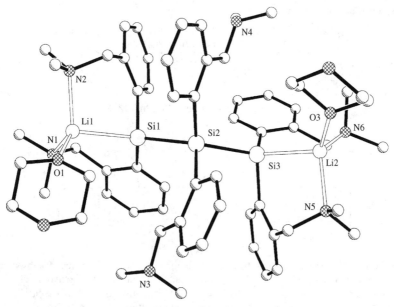

FIGURE 7. Solid state structure of **13** with hydrogen atoms omitted. Reproduced from Reference 75 by permission of VCH Verlagsgesellschaft, Weinheim

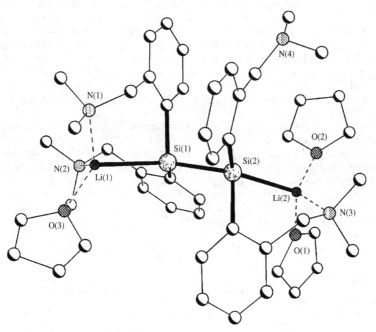

FIGURE 8. Solid state structure of **14** with hydrogen atoms omitted. Reproduced from Reference 75 by permission of VCH Verlagsgesellschaft, Weinheim

activation (equation 31)[77].

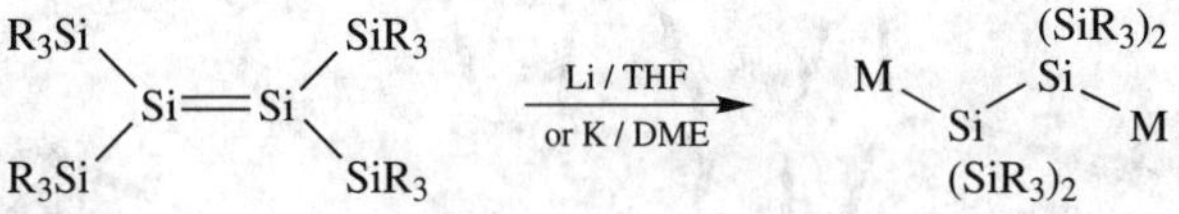

$$R_3Si\!\!-\!\!Si(SiR_3)_2 = Si(SiR_3)_2 \xrightarrow{\text{Li / THF or K / DME}} M-Si(SiR_3)_2-Si(SiR_3)_2-M \qquad (30)$$

$SiR_3 = Si(i\text{-Pr})_3, Si(i\text{-Pr})_2Me, SiMe_2(t\text{-Bu}); M = Li, K$

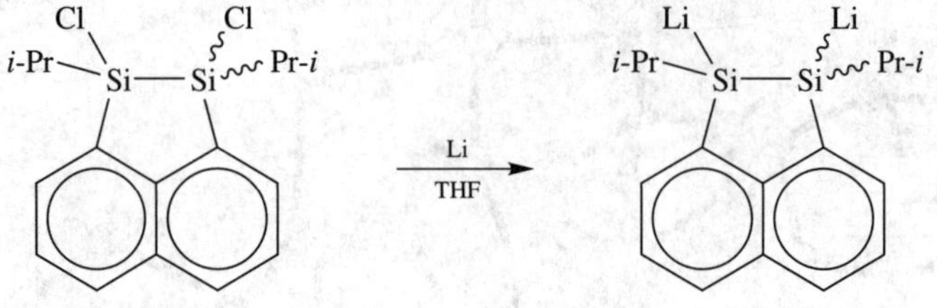

(15) → (16) via Li/THF (equation 31)

One might assume that **16** is formed via a disilene, which is reduced subsequently by the excess metal. However, although the metal-mediated dehalogenation of vicinal dihalodisilanes is known to yield disilenes[78], the formation of such a species in reaction 31 appears unlikely in view of the geometric constraints, which are imposed by the disilaacenaphthene skeleton. Although both the starting material **15** and the products which were obtained on quenching of **16** with D_2O or MeI were shown by NMR spectroscopy to be mixtures of diastereomers, the fact that the NMR spectra of **16** did not show paired signal sets was interpreted in terms of the formation of a single diastereoisomer of **16**. However, the relative stereochemistry at the silicon centers remained undetermined. The ^{29}Si NMR signal of **16** is observed at $\delta = -1.24$ ppm, shifted to lower field compared to other silyl anions. This signal splits into a well-resolved 1 : 1 : 1 : 1 quartet at 173 K in THF-d_8 ($^1J_{^{29}Si-^7Li} = 43.5$ Hz). This finding has been interpreted in terms of a partially covalent character of the Si−Li bond, and is in agreement with calculations at the HF/3-21G* level, which found a minimum structure, in which each lithium interacts with one silicon atom[77].

The behavior of cyclic silanes toward nucleophiles is less uniform. The reactions of $(Ph_2Si)_4$ with PhLi or Ph_3SiLi have been reported to result in mixtures of $(Ph_2Si)_6$ and open-chained oligosilanes; $Ph_3Si(Ph_2Si)_3Li$ and $Ph_3Si(Ph_2Si)_4Li$ have been postulated as reaction intermediates[79]. However, when $(R_2Si)_4$ is reacted with $PhMe_2SiLi$ the initially formed ring-opening product $PhMe_2Si(R_2Si)_4Li$ can be trapped efficiently by $(PhMe_2Si)_2$, yielding the hexasilane $PhMe_2Si(R_2Si)_4SiMe_2Ph$. $PhMe_2SiLi$, which is regenerated in the last step of the sequence, attacks the remaining cyclotetrasilane thus making the overall reaction catalytic in $PhMe_2SiLi$ (Scheme 8)[80].

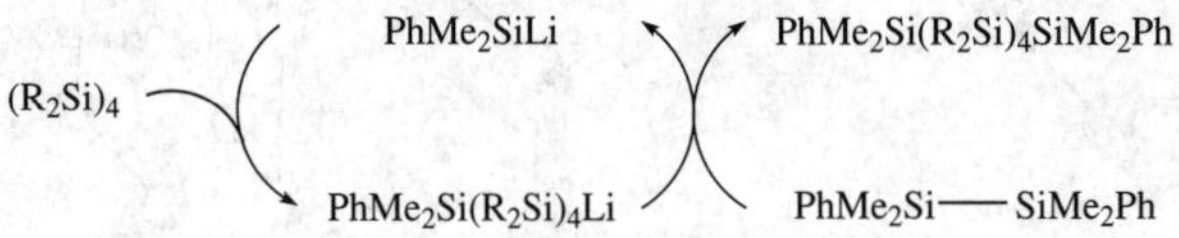

SCHEME 8

When the all-*trans* isomer of $(PhMeSi)_4$ is treated with catalytic amounts of nucleophiles such as *n*-BuLi, Me_3SiK or $(PhMe_2Si)_2CuLi$, a ring-opening polymerization occurs

14. Alkaline and alkaline earth silyl compounds — preparation and structure   805

and polysilanes are formed[81]. Similar reactions were reported for phenylnonamethylcyclopentasilane (17) in polar solvents at low temperatures[82]. In this case the ring-opening of 17 proceeds regioselectively with cleavage of the Si—Si bond to the phenyl-bearing silicon center in each propagation step, eventually forming the ordered polymer 18 (equation 32). It was assumed that the regioselectivity of the Si—Si bond cleaving reaction is due to the formation of the thermodynamically more stable, phenyl-substituted silyl anion[82].

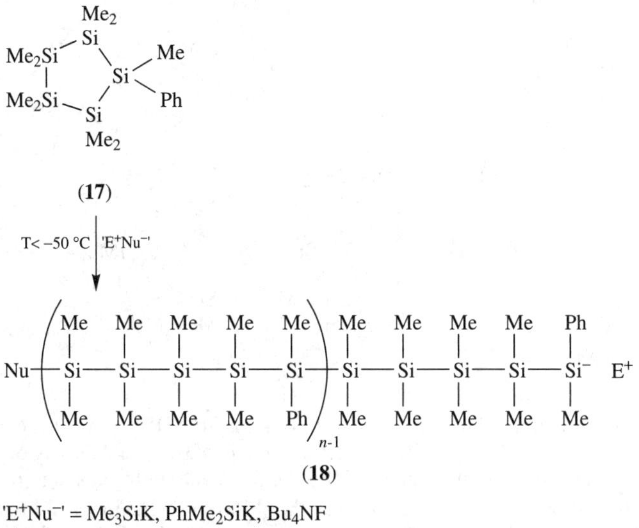

(32)

'E$^+$Nu$^-$' = Me$_3$SiK, PhMe$_2$SiK, Bu$_4$NF

C. Preparation and Properties of Cyclic Oligosilyl Anions

Transmetallation of silylmercury compounds, which are most easily available via the reaction of a silane with $(t\text{-Bu})_2\text{Hg}$, appears to be the most convenient method for the preparation of cyclic potassium oligosilyl compounds: $c\text{-}(Me_9Si_5)K$ (19)[83c] as well as $c\text{-}(Me_{11}Si_6)K$ (20)[84] were obtained by Hengge and coworkers from the corresponding mercury compound by means of a Na/K alloy in THF (equation 33).

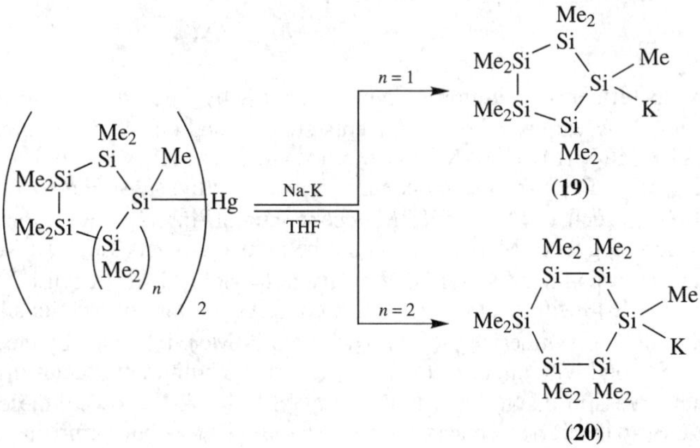

(33)

In addition, **20** is assumed to be formed as the initial product, when the silyl bromide **21** is treated with Na/K alloy in *n*-heptane. The fast coupling of intermediate **20** with bromide **21** results in the formation of dimer **22** in 80% yield (equation 34)[84].

$$\text{(21)} \xrightarrow[n\text{-heptane}]{\text{Na-K}} \text{(20)} \xrightarrow{+21} \text{(22)} \quad (34)$$

The reaction of cyclic silane **23** with lithium metal in THF opens an unexpected way to $c\text{-}(Me_{11}Si_6)Li$ (**24**). Whereas in tri-, tetra- and penta-cyclosilanes a Si−Si bond is cleaved under these conditions (Section VI.B.3), **23** reacts with cleavage of a exocyclic Si−C bond (equation 35)[85]. The reasons for the different behavior of **23** toward lithium remain unclear. It was, however, suggested that the facile formation of **24** is due to its thermodynamic stabilization by electron delocalization into the silicon framework[85].

$$\text{(23)} \xrightarrow[\text{THF}]{\text{Li}} \text{(24)} + \text{MeLi} \quad (35)$$

24 is inert to further substitution of methyl groups by lithium, even in the presence of excess metal[85]. A variety of other reagents such as sodium or potassium metal, RLi (R = *n*-Bu, Ph), MOR (M = Na, K; R = Me, Et, *t*-Bu) and KSiMe₃ have been shown to metalate $(Me_2Si)_6$ in Et_2O/HMPA, inducing a loss of a methyl substituent. However, when using nucleophiles such as PhLi or KSiMe₃ concomitant ring cleavage is observed[85].

The lower homologue, $c\text{-}(Me_9Si_5)Li$, cannot be prepared from $(Me_2Si)_5$: Reaction with lithium metal proceeds under Si−Si bond cleavage to yield $Li(Me_{10}Si_5)Li$[73], whereas on treatment of $(Me_2Si)_5$ with MeLi in HMPA a complex mixture of acyclic silyl anions is formed[85]. Attempts to synthesize $c\text{-}(Me_9Si_5)Li$ from $c\text{-}(Me_9Si_5)SiMe_3$ by cleavage of the exocyclic Si−Si bond with alkali metal alkoxides or organolithium species in various solvents were unsuccessful due to the stability of $c\text{-}(Me_9Si_5)SiMe_3$ toward these reagents[86]. However, KOBu-*t* in diglyme reacts with $c\text{-}(Me_9Si_5)SiMe_3$, but surprisingly a mixture of c-(Me$_9$Si$_5$)K (**19**) and c-(Me$_{11}$Si$_6$)K (**20**) was formed[86]. A similar product mixture, albeit in low yield, is formed when c-(Me$_2$Si)$_5$ is treated either with NaOBu-t or KOBu-t in THF. It was shown, that c-(Me$_2$Si)$_5$ as well as c-((Me$_3$Si)Me$_9$Si$_5$) rearrange under the influence of alkoxides to c-(Me$_2$Si)$_6$, which subsequently undergoes metalation with cleavage of a Si—C bond to yield an equilibrium mixture of metalated cyclohexa- and cyclopenta silanes (Scheme 9)[86]. This equilibrium mixture reacts further with alkoxide. Eventually, an almost quantitative yield of potassium compound **19** was accomplished by the reaction of c-(Me$_2$Si)$_6$ with one equivalent of KOBu-t in DME or diglyme, and this is now the best route to prepare synthetically useful amounts of this silylpotassium compound[86].

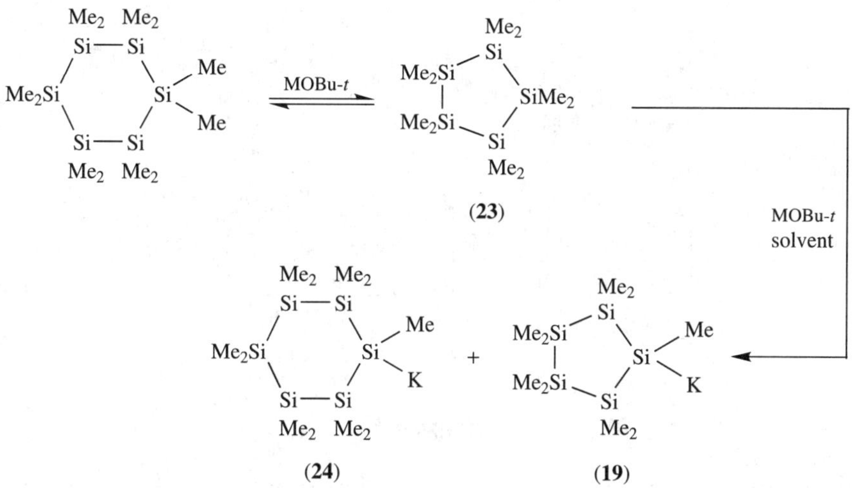

SCHEME 9

VII. FUNCTIONALIZED SILYL ANIONS

A. Lithium Silenolates

Silenolates **26**, i.e. silicon analogues of enolates, are formed, as shown by Ishikawa and coworkers, when sterically congested tris(trimethylsilyl)acylsilanes **25** are treated with silyllithium compounds (Scheme 10)[87a,b,c]: The silyllithium reagent does not add to the C=O bond but exclusively cleaves a Si—Si bond, yielding the corresponding silenolate **26**. According to *ab initio* calculations at the CISD/6-31G*//6-31G* level the model reaction given in equation 36 is exothermic by -30.0 kcal mol^{-1}[87c].

The NMR chemical shift of the central silicon atom of the mesityl-substituted **26**, which is a stable compound at room temperature in THF, is shifted by 15.4 ppm to lower field in comparison to the starting material which resonates at $\delta = -59.9$ ppm. It was suggested that this lowfield shift reflects the sp^2 character of the central silicon atom in **26**[87c]. According to the ^{1}H and ^{29}Si NMR spectra, the Me$_3$Si groups are magnetically nonequivalent at $-40\,°$C due to hindered rotation around the Si—C bond. From variable-temperature NMR

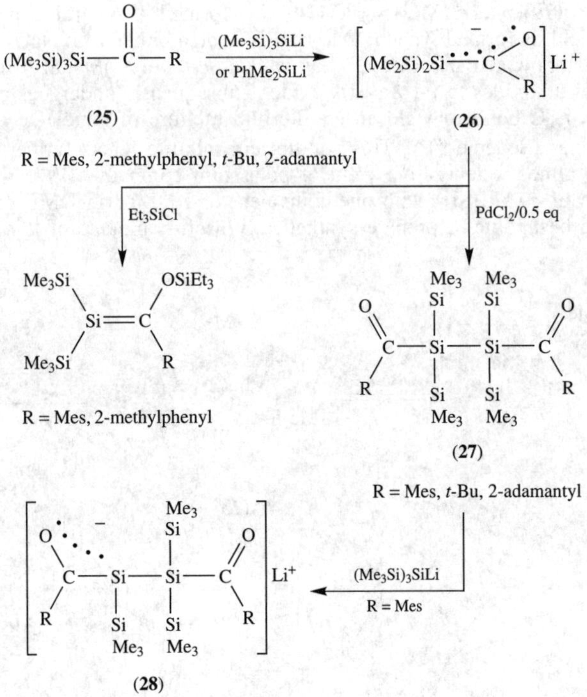

SCHEME 10

experiments, a rotation barrier of 14.3 kcal mol^{-1} has been determined[87c]. The silenolates **26** (R = Mes, 2-methylphenyl) can be converted to silenes by silylation of the anionic oxygen center with Et$_3$SiCl[87a,c]. On treatment of the silyl enolates with 0.5 equivalents of PdCl$_2$, a Si—Si coupling reaction occurs, and the corresponding bis(acyl)-substituted hexasilanes **27** are obtained[87b]. When R = Mes, addition of 1 equivalent of (Me$_3$Si)$_3$SiLi (**11**) converts **27** (R = Mes) into a new silenolate **28** (Scheme 10). However, the adamantyl- and *t*-butyl-substituted **27** did not react with **11** even at room temperature[87b]. Attempts to prepare silenolates with sterically less demanding substituents such as Ph or Me at the carbonyl center were unsuccessful. The phenyl-substituted **26** was formed upon reaction of the corresponding acylsilane **25** with **11**, but it underwent rapid dimerization and subsequent reactions, whereas the methyl-substituted **25** was deprotonated at the methyl group by **11** yielding the corresponding carbon enolate[87c].

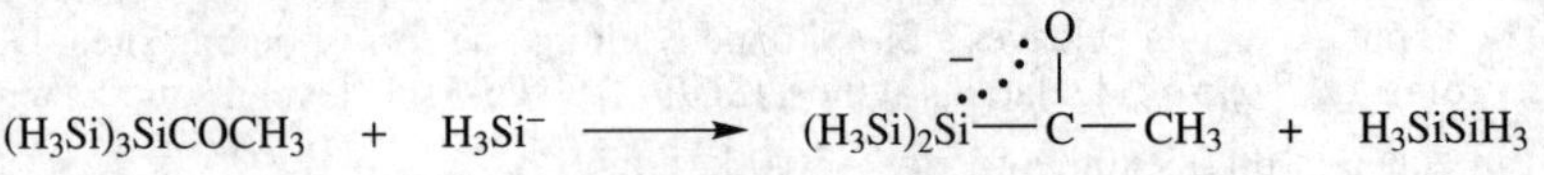

$$(H_3Si)_3SiCOCH_3 + H_3Si^- \longrightarrow (H_3Si)_2Si\text{---}C\text{---}CH_3 + H_3SiSiH_3 \qquad (36)$$

B. Silyl Anions Bearing a Heteroatom in the α-Position

1. Amino-substituted silyl anions

Amino-substituted silyllithium compounds have been prepared by Tamao and coworkers from the corresponding chlorosilanes by treatment with either a dispersion of lithium metal in THF[88] (equation 37 and 38) or LDMAN[45] (equation 39).

$$(Et_2N)_nPh_{3-n}SiCl \xrightarrow[THF]{Li} (Et_2N)_nPh_{3-n}SiLi \quad (37)$$
$$n = 1, 2 \qquad\qquad 97\text{-}98\%$$

$$(Et_2N)PhMeSiCl \xrightarrow[THF]{Li} (Et_2N)PhMeSiLi \quad (38)$$
$$80\%$$

$$(Et_2N)Ph_2SiCl \xrightarrow[THF]{[\text{naphthalenide-NMe}_2]^{-\bullet} Li^+} (Et_2N)Ph_2SiLi \quad (39)$$
$$82\%$$

These α-functionalized lithiosilanes are stable for up to 6 days at 0 °C, when at least one substituent at silicon is an aryl group. Alkylated mono- or diaminochlorosilanes, however, undergo with lithium in THF homocoupling reactions to yield the corresponding symmetrical di- and tetraaminodisilanes (equation 40)[89].

$$(Et_2N)_nR_{3-n}SiCl \xrightarrow[THF]{Li} (Et_2N)_nR_{3-n}SiSiR_{3-n}(NEt_2)_n \quad (40)$$
$$R = \text{alkyl}, n = 1, 2$$

(Amino)silyllithium compounds can be obtained alternatively by transmetalation of stannylated aminosilanes with n-BuLi or t-BuLi (equation 41)[90].

$$(Et_2N)_nPh_{3-n}SiSnMe_3 \xrightarrow[-RSnMe_3]{RLi} (Et_2N)_nPh_{3-n}SiLi \quad (41)$$
$$n = 1, R = n\text{-Bu} \qquad\qquad 90\%$$
$$n = 2, R = t\text{-Bu} \qquad\qquad 77\%$$

In sharp contrast to the aforementioned lithiated (amino)arylsilanes, 1,3-dilithio compounds **30**, which were prepared from the corresponding chlorosilanes **29** by treatment with lithium naphthalenide (equation 42), decompose completely after 3 h at 0 °C[91].

[Structure of (29) with Cl, Ph, Si, NR, Si, Cl, Ph] → [naphthalenide Li+] → [Structure of (30) with Li, Ph, Si, NR, Si, Li, Ph] (42)

(**29**) (**30**)
R = n-Bu, t-Bu

The ^{29}Si NMR signal of $(Et_2N)Ph_2SiLi$ is observed at $\delta = 19.34$ ppm. The ^{13}C NMR shifts of the aromatic carbon atoms resemble that of $MePh_2SiLi$, thus indicating that the charge distribution in both lithiosilanes is quite similar despite the fact that a methyl group was substituted by the electronically very different Et_2N group. $(Et_2N)Ph_2SiLi$ as well as $(Et_2N)Ph_2SiMgMe$ and $(Et_2N)Ph_2SiCu(CN)Li$, which are accessible from the corresponding lithium compound by reaction with MeMgBr or copper cyanide, respectively, have been used in reactions with allylsilanes, β-hydroxysilanes and vinylsilanes as synthetic equivalent of the hydroxy anion[92].

2. Alkoxy-substituted silyl anions

The first evidence for the existence of short-lived (alkoxy)silyl anions (e.g. **32**) was obtained by Watanabe and coworkers in the reaction of NaOMe with the corresponding disilanes (**31**), which yielded in the presence of alkyl halides the expected trapping product of the intermediate (alkoxy)silyl sodium compounds **32** (equation 43)[93].

$$(MeO)_nMe_{3-n}Si\text{---}SiMe_{3-n}(MeO)_n \xrightarrow{NaOMe} (MeO)_nMe_{3-n}SiNa$$
$$(31) \hspace{4cm} (32)$$

$$\downarrow RX \hspace{3cm} (43)$$

$$n = 1, 2 \hspace{3cm} (MeO)_nMe_{3-n}Si\text{---}R$$

When the stannylated allyloxysilane **33** was treated with n-BuLi in THF at −78 °C, a rearranged product **36** was isolated after work-up with Me_3SiCl (equation 44). This reaction was assumed to proceed via the initial formation of the allyloxy-substituted silyllithium **34**, which subsequently — similar to a [2,3] Wittig rearrangement — undergoes a [2,3] sigmatropic rearrangement to yield the corresponding silanolate **35**[94].

(t-BuO)Ph$_2$SiLi (**37**) has been prepared by stannyl–lithium exchange from the corresponding silylstannane at low temperature (equation 45)[95]. In contrast to the analogous amino-substituted silyllithium $(Et_2N)Ph_2SiLi$, **37** is stable in THF at −78 °C only for

14. Alkaline and alkaline earth silyl compounds — preparation and structure

several hours.

$$(t\text{-BuO})\text{Ph}_2\text{Si}-\text{SnMe}_3 \xrightarrow[\text{THF}/-78\,°\text{C}]{n\text{-BuLi}} (t\text{-BuO})\text{Ph}_2\text{SiLi} \quad (45)$$
$$\text{(37)}$$

Other alkoxy-substituted silyllithium compounds such as $(i\text{-PrO})\text{Ph}_2\text{SiLi}$ and $(\text{MeO})\text{Ph}_2\text{SiLi}$ have been obtained by reduction of the corresponding chlorosilanes with LDMAN at $-78\,°\text{C}$ (equation 46)[45].

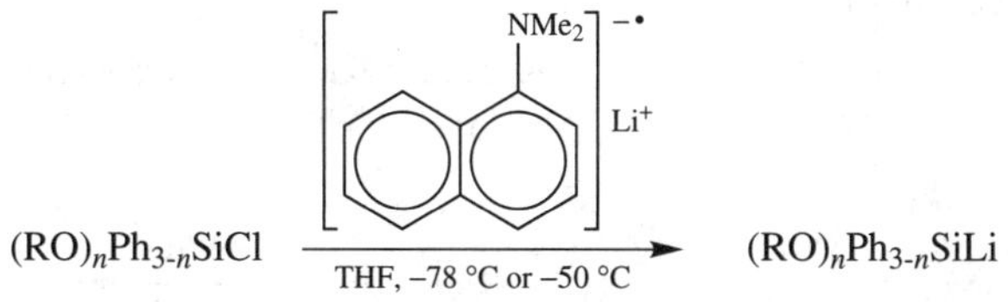

$$(\text{RO})_n\text{Ph}_{3-n}\text{SiCl} \xrightarrow[\text{THF, }-78\,°\text{C or }-50\,°\text{C}]{\text{[LDMAN]}} (\text{RO})_n\text{Ph}_{3-n}\text{SiLi} \quad (46)$$

$R = t\text{-Bu}\ (n = 1, 2),\ i\text{-Pr}\ (n = 1),\ \text{Me}\ (n = 1)$

The use of LDMAN as the electron transfer reagent also allows (though a higher temperature of $-50\,°\text{C}$ is required) the preparation of a di(alkoxy)lithiosilane, $(t\text{-BuO})_2\text{PhSiLi}$, from the corresponding chlorosilane[45]. Alternatively, this compound may be obtained from the chlorosilane by reaction with lithium metal at $0\,°\text{C}$[96].

The ^{29}Si nucleus of $(t\text{-BuO})\text{Ph}_2\text{SiLi}$ (37) resonates at $\delta = 11.3$ ppm, i.e. the ^{29}Si NMR signal is strongly lowfield shifted in comparison with $(t\text{-BuO})\text{Ph}_2\text{SiCl}$ ($\delta = -23.3$ ppm). Tamao and Kawachi[95] pointed out that the deshielding of the ^{29}Si nucleus of 37 is similar to the deshielding observed in the ^{13}C NMR spectra of carbenoids[97]. Moreover, the chemical properties of these alkoxy-substituted silyllithium compounds argue in favor of a silylenoid character of 37 (Scheme 11)[95]. Thus, when 37 is warmed to $0\,°\text{C}$, it undergoes a homocoupling reaction to yield, after work-up with Me_3SiCl, the trisilane 38; i.e. one molecule of 37 acts as a nucleophile, whereas a second molecule takes the role of the electrophile. Upon treatment with excess of n-BuLi in the presence of TMEDA, BuPh_2SiLi, the result of a nucleophilic attack of n-BuLi on 37, is obtained. However, the ambiphilic character of 37 is completely suppressed in the presence of 12-crown-6, which promotes the formation of a solvent-separated ion pair under these conditions. Self-condensation does not occur, and only reactions with electrophiles, such

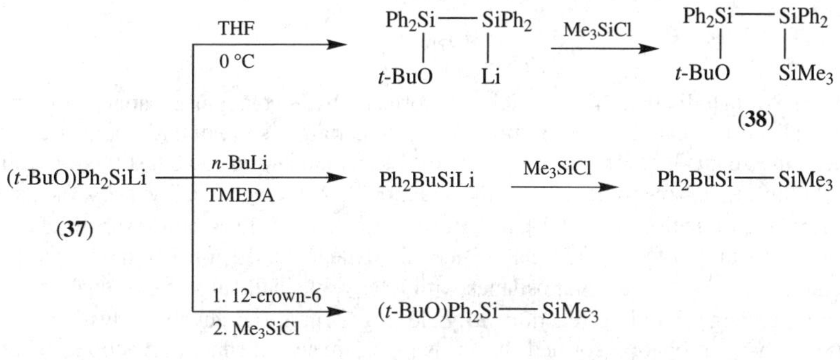

SCHEME 11

as Me$_3$SiCl, are observed. A similar ambiphilic reactivity was reported for $(i$-PrO)Ph$_2$SiLi and (MeO)Ph$_2$SiLi[45]. The di(alkoxy)lithiosilane $(t$-BuO)$_2$PhSiLi is thermally more stable than the lithiated monoalkoxy compounds: It can be kept in a THF solution at room temperature for longer periods without formation of even traces of the self-condensation product[45]. In addition, no nucleophilic substitution occurs with n-BuLi at this temperature, thus revealing the decreased electrophilicity of $(t$-BuO)$_2$PhSiLi relative to monoalkoxy silyllithium compounds.

3. Halogen-substituted silyl anions

Metalated halosilanes were postulated as intermediates in the reductive coupling reactions of dihalosilanes **39** to yield linear and cyclic oligo- and polysilanes (Scheme 12).

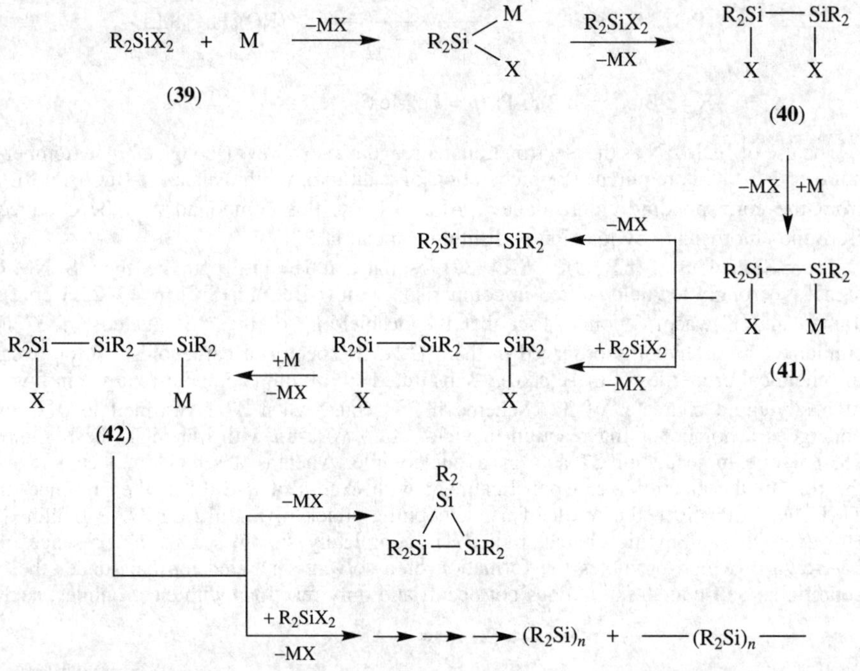

SCHEME 12

The 1,2-dihalodisilane **40**, which is formed by nucleophilic attack of the α-functionalized metalated silane on the starting material, is subsequently converted to the 1-halo-2-metalladisilane **41**. At this stage, the steric demand of the substituents at silicon strongly influences the outcome of the reaction[98]. With very bulky substituents (e.g. R = Tip[78]), elimination of metal halide forms a disilene[99]. Less bulky substituents such as t-butyl or i-propyl favor, via 1-halo-3-metallatrisilane **42**, the formation of cyclotri- and tetrasilanes[99a,100], whereas dihalosilanes with relatively small substituents such as methyl or phenyl undergo coupling reactions to yield larger rings and acyclic polysilanes[101].

LiSiCl$_3$ was probably formed by a halogen–metal exchange reaction, when an attempt was made to couple the highly hindered aryllithium compound **43** with BrSiCl$_3$

14. Alkaline and alkaline earth silyl compounds — preparation and structure

(equation 47)[102].

$$\text{(43)} \quad \text{Ar*Li} + \text{BrSiCl}_3 \longrightarrow \text{Ar*Br} + \text{LiSiCl}_3 \quad (47)$$

The formation of $(i\text{-Pr})_3\text{NH}^+$ upon treatment of HSiCl_3 with $(i\text{-Pr})_3\text{N}$ in refluxing CH_3CN indicates that HSiCl_3 is deprotonated by this and also by other amines such as $(n\text{-Bu})_3\text{N}$ or TMEDA[103,104]. According to ^1H NMR studies this proton transfer is reversible (equation 48).

$$\text{HSiCl}_3 + (i\text{-Pr})_3\text{N} \rightleftharpoons \text{Cl}_3\text{Si}^- + (i\text{-Pr})_3\text{NH}^+ \quad (48)$$

Mixtures of tertiary amines with HSiCl_3 have been used to introduce the trichlorosilyl group, *inter alia*, into olefins[105a] (equation 49) or to convert aromatic carboxylic acids to the corresponding toluenes[105b] (equation 50).

$$\text{Ph-CH=CH}_2 \xrightarrow[(n\text{-Bu})_3\text{N or pyridine}]{\text{HSiCl}_3} \text{Ph-CH}_2\text{-CH}_2\text{-SiCl}_3 \quad (49)$$

$$\text{ArCO}_2\text{H} \xrightarrow[\text{CH}_3\text{CN}]{\Delta, \text{HSiCl}_3} \xrightarrow[\Delta]{(n\text{-Pr})_3\text{N}} \xrightarrow[\text{MeOH/H}_2\text{O}]{\Delta, \text{KOH}} \text{ArCH}_3 \quad (50)$$

Similarly, the mixture of TMEDA with RCl_2SiH (R = Me, Ph) was reported to act in the presence of copper(0), copper(I) or copper(II) catalysts as a useful synthetic equivalent of the corresponding RCl_2Si^- anion (equation 51)[104].

$$\text{RCl}_2\text{SiH} + \text{H}_2\text{C=CH-C(=O)OR}' \xrightarrow[\text{TMEDA}]{[\text{Cu}]} \text{H}_2\text{C(Cl}_2\text{SiR})\text{-CH}_2\text{-C(=O)OR}' \quad (51)$$

R = Me, Ph, Cl ; R' = Me, Et

VIII. MISCELLANEOUS

A. Mono- and Dianions of Silacyclopentadienes

Remarkable advancements have been made in the chemistry of metalated silacyclopentadienes (siloles) during the last few years.

The interest in silole monoanions centers mainly on the question of aromaticity in these compounds. According to early 3-21G/STO-2G calculations by Gordon and coworkers, $[SiC_4H_5]^-$ is about 25% as aromatic as $[C_5H_5]^{-106}$. The optimized structure of the silole anion has C_s symmetry with a pyramidal silicon center. The calculated energy barrier for the inversion at silicon through a planar transition state of C_{2v} symmetry is 16.2 kcal mol^{-1} at HF/6-31G*//6-31G*^{107}, significantly lower than the value of 26 kcal mol^{-1} calculated for $[SiH_3]^-$, and it was proposed that this reflects the small, but not negligible, stabilization of the transition state by electron delocalization[107]. More recent calculations by Schleyer and Goldfuss using the considerably better RMP2/6-31+G*//RMP2/6-31+G* level reveal a silicon center of $[SiC_4H_5]^-$, which is still pyramidal, but strongly flattened (angle sum around Si is 321.6°). The calculated inversion barrier is reduced at this computational level to only 3.8 kcal mol^{-1}[108a]. These results, which imply a certain degree of aromaticity in $[SiC_4H_5]^-$, are in agreement with calculations of the magnetic properties of this anion such as the diamagnetic susceptibility exaltation, which is about 50% of that found for $[C_5H_5]^{-108b}$. Calculations on $[SiC_4H_5]$Li at the RMP2/6-31+G*//RMP2/6-31+G* level found a C_s structure, which features a η^5 coordinated lithium atom. In C_4H_5SiLi, planarization at the silicon center has increased (angle sum around Si is 340.2°) and the ring C—C bond lengths are nearly equalized, as expected for an aromatic system[108a]. Moreover, the aromatic stabilization energy (ASE), which has been calculated by using appropriate isodesmic reactions, increases on going from the silole anion to the lithium compound: $[SiC_4H_5]^-$ exhibits 55% of the ASE of $[C_5H_5]^-$, whereas the ASE of $[SiC_4H_5]$Li amounts to 80% of that computed for $[C_5H_5]$Li. In summary, according to these calculations, the silole anion $[SiC_4H_5]^-$ 'exhibits at least moderate aromaticity, which is enhanced strongly by η^5 coordination of Li$^+$'[108a].

The reaction of silole **44** with KH in THF or DME yields, after work-up with D$_2$O, quantitatively the corresponding deuterated silole (equation 52); metalated siloles have been suggested as intermediates[8b]. In contrast, the analogous treatment of silole **45** with KH yields a mixture of three NMR spectroscopically characterized potassium compounds (equation 53). The main product of this reaction is the pentavalent silicate **46**, which results from the nucleophilic attack of a hydride at the silicon center[109]. The other two products result from hydride addition to one of the ring carbons.

The first metalated silole, **48**, which was characterized unambiguously by means of NMR spectroscopy, has been obtained by Boudjouk and coworkers via reductive cleavage of the Si—Si bond of disilane **47** with lithium or sodium under ultrasonic activation (equation 54)[110a].

Most significant is the appreciable deshielding of the silicon nucleus upon metalation: The ^{29}Si NMR signal is shifted to low-field on going from the disilane to the metalated silole [$\Delta\delta = 21.48$ (Li) and 22.50 (Na) ppm, respectively], which sharply contrasts with the upfield shift, which is characteristic for the conversion of aryl-substituted disilanes into the corresponding silyl anions (Section V.B). The deshielding of the silicon nucleus as well as the shielding of the C_α and C_β nuclei of the five-membered ring indicates appreciable charge transfer from the silicon center into the butadiene moiety. These trends

14. Alkaline and alkaline earth silyl compounds — preparation and structure 815

are nicely reproduced by the calculated (IGLO) chemical shifts for $[SiC_4H_5]Li$[108a].

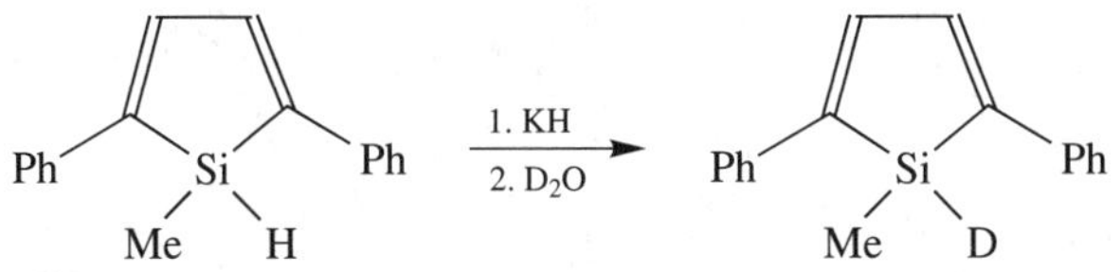

(52)

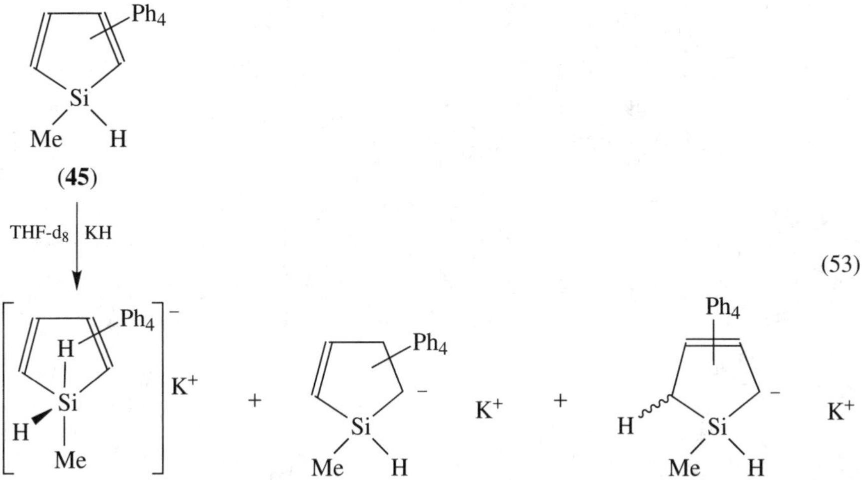

(53)

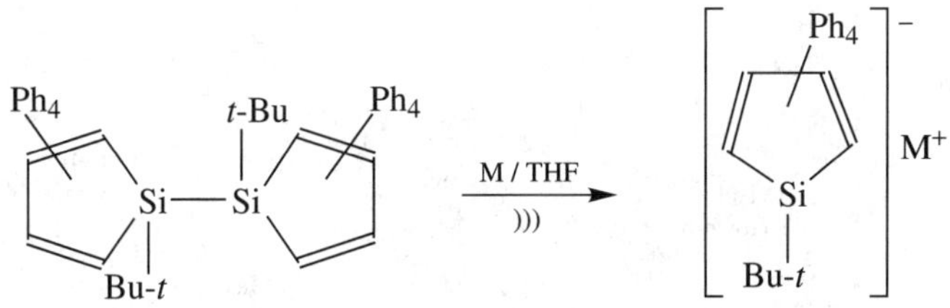

(54)

In contrast to these experimental and computational results, lithiated 1-silafluorenide **50**, which was prepared by the reductive cleavage of the central Si—Si bond of disilane **49** with lithium under ultrasonic activation, provides some evidence for the existence of localized metalated siloles (equation 55)[111]. Thus, upon metalation of **49** to form **50**, a highfield shift of the ^{29}Si nucleus ($\Delta\delta = -47.9$ ppm) is observed. In addition, the chemical shifts of the phenyl carbons indicate that there is no accumulation of π electron density, which would be expected for a delocalized lithium silafluorenide[111].

The reductive dehalogenation of permethylated 1,1-dibromosilole **51** and the subsequent reactions were used by Tilley and coworkers to synthesize the metalated siloles

52–54 (Scheme 13)[112a].

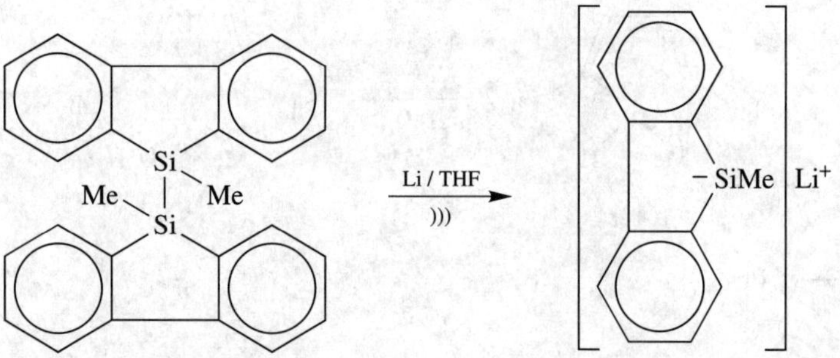

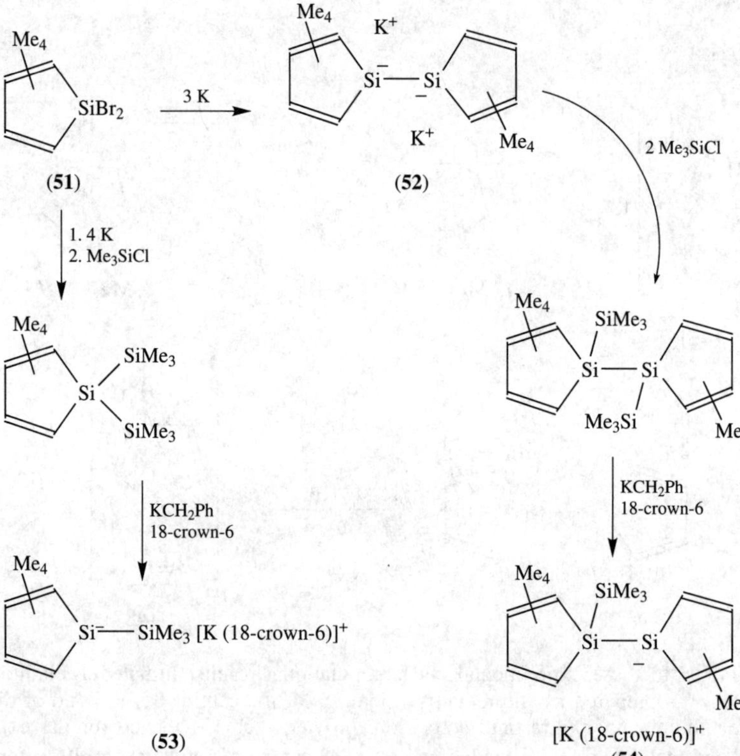

SCHEME 13

The highfield shifted ^{29}Si NMR signal of the 18-crown-6 complex of [C$_4$Me$_4$(SiMe$_3$)Si]K (**53**) in solution as well as the strongly pyramidalized silicon center (angle sum around Si is 279.3°) and the bond localization, which were found in the solid state (Figure 9), support a nonaromatic nature with fixed double bonds for the silole

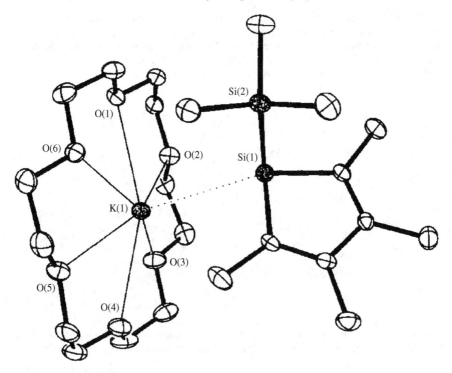

FIGURE 9. Solid state structure of **53** with hydrogen atoms omitted. Reprinted with permission from Reference 112b. Copyright 1996 American Chemical Society

ring[112b]. Variable-temperature NMR spectroscopy of the tetraethyl-substituted siloles, with potassium or the 18-crown-6 complex of potassium as countercations, allows one to estimate the inversion barrier of these compounds as being lower than 8 kcal mol^{-1}. These exceptionally low inversion barriers were attributed to stabilization of the planar transition state by delocalization of the π-electron density[112b]. In conclusion, when comparing the experimental results for siloles such as **48**, **50** and **53**, it appears that the electronic structure of metalated siloles is highly sensitive about the nature of the substituent at silicon and at the ring carbon atoms.

According to theoretical results, silole dianions and dimetalated siloles are aromatic compounds. The geometries of $[H_4C_4Si]^{2-}$, $[H_4C_4Si]Li^-$, $[H_4C_4Si]Li_2$, $[H_4C_4Si]Na_2$ and $[H_4C_4Si]K_2$ have been calculated by Schleyer and coworkers at the RMP2/6-31+G* level, and in all these compounds the five-membered C_4Si ring is planar[113]. The dimetalated siloles prefer an inverted C_{2v} sandwich-type structure, in which the metal centers are located above and below the C_4Si ring, thus adopting an η^5-coordination. The equalized ring bond lengths as well as the calculated aromatic stabilization energies point to a substantial degree of aromaticity in these compounds, which approaches the aromaticity of the $[C_5H_5]^-$ anion. The diamagnetic susceptibility exaltations, which may be used as a gauge for the aromaticity of a system, indicate large aromatic ring currents; these are responsible for the strong shielding of the ^7Li nucleus, which results in a calculated chemical shift (IGLO) of $\delta = -7.7$ ppm.

The first experimental evidence for the existence of a dimetalated silole **56** was obtained by Joo and coworkers from treatment of 1,1-dichlorosilole **55** with sodium in dioxane[114]. A red solid was isolated from this reaction, which yielded upon addition of electrophiles such as Me$_2$HSiCl or Me$_3$SiCl the corresponding **57**. Upon treatment with *t*-BuCl or Me$_3$SnCl, coupling products **58** were obtained (Scheme 14).

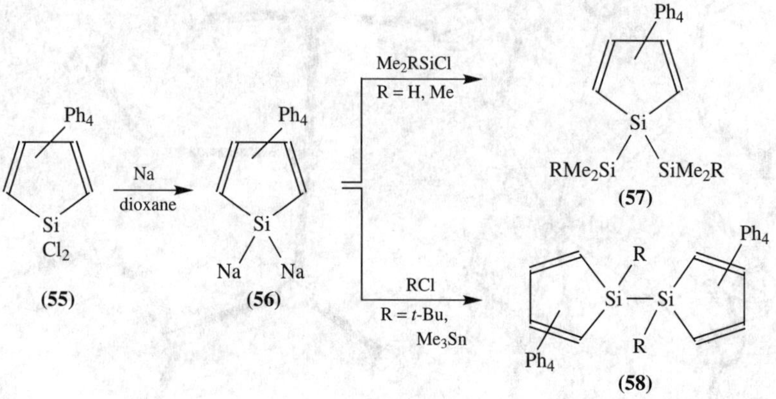

SCHEME 14

A mixture of analogous products was formed when **55** was treated with lithium metal in THF, and the resulting solution was worked up with Me$_3$SiCl (equation 56)[115].

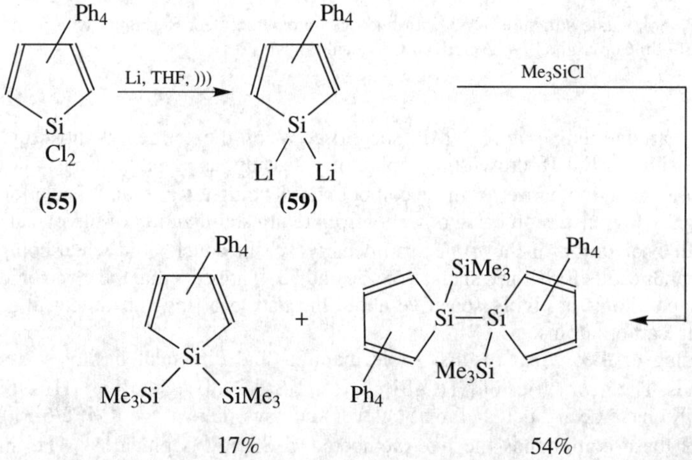

(56)

NMR studies showed that the only anionic intermediate which is present in solution is dilithiosilole **59**[115]. The marked downfield shift in the ^{29}Si NMR spectrum (δ = +68.4 ppm) of **59** as well as the shielding of the C$_\alpha$ and C$_\beta$ nuclei indicate, in agreement with calculations[113], an aromatic character of this compound. This conclusion is supported by the solid state structure of **59** (Figure 10)[116]. The silole ring in **59** is planar and exhibits nearly equal C—C distances, as one would expect for a delocalized system. However, in contrast to the computational results[113,116], the lithium atoms in **59** are not

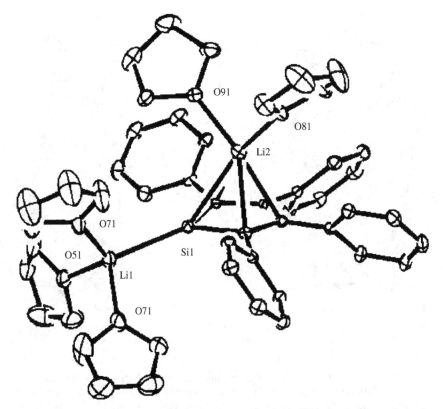

FIGURE 10. Solid state structure of **59** with hydrogen atoms omitted. Reprinted with permission from Reference 116. Copyright 1995 American Chemical Society

equivalent: one of them is η^5-coordinated to the planar silole ring, whereas the other is η^1-bonded to the silicon center. This is in clear contradiction to the calculated minimum structure of the parent bismetalated silole[113,116] for which a η^1-η^5 structure has been calculated to be 21 kcal mol^{-1} less stable than the C_{2v} symmetrical η^5-η^5 structure[116]. However, as pointed out by West and coworkers, solvation of the cationic centers by THF as well as crystal packing effects could reverse this stability order[116].

In contrast to these experimental findings, the 18-crown-6 complex of [Me$_4$C$_4$Si]K$_2$, which was prepared by the reaction of the corresponding dibromosilole with potassium metal in the presence of 18-crown-6 (2 equivalents), adopts in the solid state, in agreement with calculations[113,116], an inverted sandwich structure (Figure 11)[112], thus supporting the theoretical results[113] that the C_{2v} structure is more stable than the η^1-η^5 structure.

B. Oligolithiated Monosilanes

When lithium vapor was reacted with SiCl$_4$ in a Knudsen cell, and the reaction product was treated with excess methyl chloride, Me$_4$Si was obtained in 5–10% yield. This result was interpreted in terms of the formation of SiLi$_4$[117]. Interestingly, tetrahedral SiLi$_4$ is a saddle point of 3rd order at the 3-21G* level of theory[118]. The minimum structure of

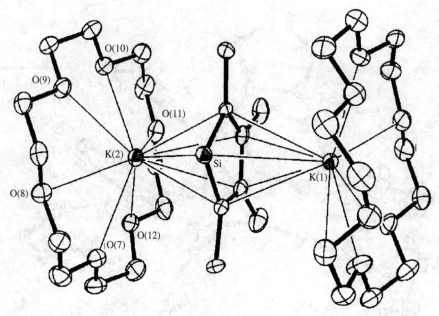

FIGURE 11. Solid state structure of [(18-crown-6)K]$_2$[SiC$_4$Me$_4$] with hydrogen atoms omitted. Reproduced from Reference 112a by permission of VCH Verlagsgesellschaft, Weinheim

SiLi$_4$ has C_{2v} symmetry and it exhibits an inverted tetrahedron **60**; the equatorial and axial lithium atoms exchange easily via a C_{4v} symmetrical transition state, which is only 3.2 kcal mol^{-1} higher in energy than **60** (bond lengths in pm)[118].

$$\text{Si} \overset{\overset{\text{Li}_{ax}}{\diagdown\;\text{Li}_{eq}}}{\underset{\underset{\text{Li}_{ax}}{242.9\diagup\;\text{Li}_{eq}}}{\;251.4}}$$

(**60**) C_{2v}

Thermolysis of (Me$_3$Si)$_3$SiLi at 140–150 °C *in vacuo* for 4 h appears to result in the formation of 1,1-dilithiosilane (Me$_3$Si)$_2$SiLi$_2$, which was quenched with D$_2$O to yield 35% of (Me$_3$Si)$_2$SiD$_2$ (equation 57)[119].

$$(Me_3Si)_3SiLi \xrightarrow[4\,h]{140\text{-}150\,°C} \xrightarrow{D_2O} (Me_3Si)_3SiD_2 + Me_3SiSiMe_3$$

35%   not stated

(57)

$$+\;(Me_3Si)_3SiD\;+\;(Me_3Si)_4Si\;+\;(Me_3Si)_3SiSi(SiMe_3)_3$$

22%   10%   27% (**12**)

14. Alkaline and alkaline earth silyl compounds — preparation and structure   821

The major by-product in this reaction is the hexasilane **12**, which is formed at the expense of the dilithio compound, when the reaction time is extended. In the flash vaporization mass spectrum of solid $(Me_3Si)_2SiLi_2$ the parent ions of the monomer and dimer of the dilithiosilane have been observed. According to *ab initio* calculations for H_2SiLi_2 at the MP4SDTQ/6-31G**//6-31G** level, a structure **61** (bond lengths in pm) in which the lithium atoms are placed unsymmetrically with respect to the mirror plane of the SiH_2 moiety is 6.3 kcal mol^{-1} lower in energy than the conventional tetrahedral geometry[120].

$$\begin{array}{c} 155.5 \quad H \\ Si \\ 256.4 \; / \quad \diagdown H \\ Li \quad\quad Li \\ 241.2 \end{array}$$

(**61**) C_s

IX. REFERENCES

1. Recent reviews on silyl anions:
   (a) P. D. Lickiss and C. M. Smith, *Coord. Chem. Rev.*, **145**, 75 (1995).
   (b) K. Tamao and A. Kawachi, *Adv. Organomet. Chem.*, **38**, 1 (1995).
   (c) M. Fujita and T. J. Hiyama, *Yuki Gosei Kagaku Kyokaishi*, **42**, 293 (1984); *Chem. Abstr.*, **101**, 55130b (1984).
   Structural studies of silyl anions:
   (d) W. S. Sheldrick, in *The Chemistry of Organic Silicon Compounds* (Eds. S. Patai and Z. Rappoport), Wiley, Chichester, 1989, pp. 268–272.
   (e) E. Lukevics, O. Pudova and R. Strukovich, *Molecular Structure of Organosilicon Compounds*, Ellis Horwood, Chichester, 1989, pp. 76–84.
   (f) J. B. Lambert and W. J. Schultz, Jr., in *The Chemistry of Organic Silicon Compounds* (Eds. S. Patai and Z. Rappoport), Wiley, Chichester, 1989, pp. 1007–1010.
   Theoretical studies of silyl anions:
   (g) Y. Apeloig, in *The Chemistry of Organic Silicon Compounds* (Eds. S. Patai and Z. Rappoport), Wiley, Chichester, 1989, pp. 80, 201–204.
   Synthetic applications of silyl anions:
   (h) D. D. Davis and C. E. Gray, *Organomet. Chem. Rev.*, **6**, 283 (1970).
   (i) I. Fleming, in *Comprehensive Organic Chemistry*, Vol. 3 (Eds. D. Barton and W. D. Ollis), Pergamon, Oxford, 1979, pp. 541–669.
   (j) P. D. Magnus, T. Sarkar and S. Djuric, in *Comprehensive Organometallic Chemistry*, Vol. 3 (Eds. G. Wilkinson, F. G. A. Stone and E. W. Abel), Pergamon, Oxford, 1982, pp. 608–614.
   (k) E. W. Colvin, *Silicon in Organic Synthesis*, Academic Press, London, 1988, pp. 51–55.
2. (a) M. A. Ring and D. M. Ritter, *J. Am. Chem. Soc.*, **83**, 802 (1961).
   (b) M. A. Ring and D. M. Ritter, *J. Phys. Chem.*, **65**, 182 (1961).
   (c) E. Amberger and E. Muehlhofer, *J. Organomet. Chem.*, **12**, 55 (1968).
   (d) H. Bürger, R. Eujen and H. C. Marsmann, *Z. Naturforsch., Teil B*, **29B**, 149 (1974).
   (e) O. Mundt, G. Becker, H.-M. Hartmann and W. Schwarz, *Z. Anorg. Allg. Chem*, **572**, 75 (1989).
   (f) T. Lobreyer, J. Oeler and W. Sundermeyer, *Chem. Ber.*, **124**, 2405 (1991).
3. E. Weiss, G. Hencken and H. Kuehr, *Chem. Ber.*, **103**, 2868 (1970).
4. (a) F. Fehér, M. Krancher and M. Fehér, *Z. Anorg. Allg. Chem*, **606**, 7 (1991).
   (b) T. Lobreyer, J. Oeler, W. Sundermeyer and H. Oberhammer, *Chem. Ber.*, **126**, 665 (1993).
5. B. T. Luke, J. A. Pople, M.-B. Krogh-Jespersen, Y. Apeloig, J. Chandrasekhar and P. v. R. Schleyer, *J. Am. Chem. Soc.*, **108**, 260 (1986).
6. P. v. R. Schleyer and T. Clark, *J. Chem. Soc., Chem. Commun.*, 1371 (1986).
7. W. Sundermeyer, T. Lobreyer, H. Pritzkow, N. J. R. v. E. Hommes and P. v. R. Schleyer, *Angew. Chem.*, **106**, 221 (1994); *Angew. Chem., Int. Ed. Engl.*, **33**, 216 (1994).

8. (a) R. J. P. Corriu and C. Guérin, *J. Chem. Soc., Chem. Commun.*, 168 (1980).
   (b) R. J. P. Corriu, C. Guérin and B. Kolani, *Bull. Soc. Chim. Fr.*, 973 (1985).
   (c) L. Rösch, W. Erb and H. Müller, *Z. Naturforsch., Teil B*, **31B**, 281 (1976).
9. R. J. P. Corriu, C. Guérin, J. L. Brefort and B. Henner, *J. Organomet. Chem.*, **370**, 9 (1989).
10. (a) W. C. Still, *J. Org. Chem.*, **41**, 3063 (1976).
    (b) E. B. Nadler and Z. Rappoport, *Tetrahedron Lett.*, **31**, 555 (1990).
11. (a) P. F. Hudrlik, M. A. Waugh and A. M. Hudrlik, *J. Organomet. Chem.*, **271**, 69 (1984).
    (b) P. F. Hudrlik, A. M. Hudrlik, T. Yimenu, M. A. Waugh and G. Nagendrappa, *Tetrahedron*, **44**, 3791 (1988).
    (c) L. Gong, R. Leung-Toung and T. T. Tidwell, *J. Org. Chem.*, **55**, 3634 (1990).
    (d) K. Krohn and K. Khanbabaee, *Angew. Chem.* **106**, 100 (1994); *Angew. Chem., Int. Ed. Engl.*, **33**, 99 (1994).
12. (a) T. F. Schaaf and J. P. Oliver, *J. Am. Chem. Soc.*, **91**, 4327 (1969).
    (b) E. Hengge and N. Holtschmidt, *J. Organomet. Chem.*, **12**, P5 (1968).
13. (a) H. Sakurai, A. Okada, M. Kira and K. Yonezawa, *Tetrahedron Lett.*, 1511 (1971).
    (b) H. Sakurai and F. Kondo, *J. Organomet. Chem.*, **92**, C46 (1975).
    (c) M. A. Shippey and P. B. Dervan, *J. Org. Chem.*, **42**, 2654 (1977).
    (d) E. Buncel, T. K. Venkatachalam and U. Edlund, *J. Organomet. Chem.*, **437**, 85 (1992).
14. T. Hiyama, M. Obayashi, I. Mori and H. Nozaki, *J. Org. Chem.*, **48**, 912 (1983).
15. T. Hiyama, M. Obayashi and M. Sawahata, *Tetrahedron Lett.*, **24**, 4113 (1983).
16. T. Hiyama and M. Obayashi, *Tetrahedron Lett.*, **24**, 4109 (1983).
17. N. S. Vyazankin, G. A. Razuvaev, E. N. Gladyshev and S. P. Korneva, *J. Organomet. Chem.*, **7**, 353 (1967).
18. E. N. Gladyshev, E. A. Fedorova, L. O. Yuntila, G. A. Razuvaev and N. S. Vyazankin, *J. Organomet. Chem.*, **96**, 169 (1975).
19. (a) N. Wiberg and K. Schurz, *J. Organomet. Chem.*, **341**, 145 (1988).
    (b) N. Wiberg, H. Schuster, A. Simon and K. Peters, *Angew. Chem.*, **98**, 100 (1986); *Angew. Chem., Int. Ed. Engl.*, **25**, 79 (1986).
20. A. Fürstner and H. Weidmann, *J. Organomet. Chem.*, **354**, 15 (1988).
21. B. H. Lipshutz, D. C. Reuter and E. L. Ellsworth, *J. Org. Chem.*, **54**, 4975 (1989).
22. A. R. Claggett, W. H. Ilsley, T. J. Anderson, M. D. Glick and J. P. Oliver, *J. Am. Chem. Soc.*, **99**, 1797 (1977).
23. L. Rösch, J. Pickardt, S. Imme and U. Börner, *Z. Naturforsch., Teil B*, **41B**, 1523 (1986).
24. J. P. Oliver, D. W. Goebel, Jr. and J. L. Hencher, *Organometallics*, **2**, 746 (1983).
25. R. Goddard, C. Krüger, N. A. Ramadan and A. Ritter, *Angew. Chem.*, **107**, 1107 (1995); *Angew. Chem., Int. Ed. Engl.*, **34**, 1030 (1995).
26. W. H. Ilsey, T. F. Schaaf, M. D. Glick and J. P. Oliver, *J. Am. Chem. Soc.*, **102**, 3769 (1980).
27. B. Teclé, W. H. Ilsley and J. P. Oliver, *Organometallics*, **1**, 875 (1982).
28. (a) M. V. George, D. J. Peterson and H. J. Gilman, *J. Am. Chem. Soc.*, **82**, 403 (1960).
    (b) A. G. Brook and H. J. Gilman, *J. Am. Chem. Soc.*, **76**, 278 (1954).
    (c) B. K. Campion, R. H. Heyn and T. D. Tilley, *Organometallics*, **12**, 2584 (1993).
    (d) H. Wagner and U. Schubert, *Chem. Ber.*, **123**, 2101 (1990).
    (e) J. Meyer, J. Willnecker and U. Schubert, *Chem. Ber.*, **122**, 223 (1989).
    (f) R. A. Benkeser and R. G. Severson, *J. Am. Chem. Soc.*, **73**, 1424 (1951).
    (g) E. Colomer and R. J. P. Corriu, *J. Chem. Soc., Chem. Commun.*, 176 (1976).
29. H. Gilman and T. C. Wu, *J. Am. Chem. Soc.*, **73**, 4031 (1951).
30. E. Colomer and R. J. P. Corriu, *J. Organomet. Chem.*, **133**, 159 (1977).
31. D. L. Comins and M. O. Killpack, *J. Am. Chem. Soc.*, **114**, 10972 (1992).
32. T. G. Selin and R. West, *Tetrahedron*, **5**, 97 (1959).
33. H. Gilman and G. D. Lichtenwalter, *J. Am. Chem. Soc.*, **80**, 608 (1958).
34. C. M. Smith and P. D. Lickiss, in *Xth International Symposium on Organosilicon Chemistry*, Poznań, Poland, 1993, P-142.
35. B. Reiter and K. Hassler, *J. Organomet. Chem.*, **467**, 21 (1994).
36. E. Buncel, T. K. Venkatachalam, B. Eliasson and U. Edlund, *J. Am. Chem. Soc.*, **107**, 303 (1985).
37. J. B. Lambert and M. J. Urdaneta-Pérez, *J. Am. Chem. Soc.*, **100**, 157 (1978).
38. I. Fleming and S. B. D. Winter, *Tetrahedron Lett.*, **34**, 7287 (1993).
39. L. H. Sommer and R. Mason, *J. Am. Chem. Soc.*, **87**, 1619 (1965).

14. Alkaline and alkaline earth silyl compounds — preparation and structure   823

40. H. Gilman and W. Steudel, *Chem. Ind. (London)*, 1094 (1959).
41. J. Belzner and U. Dehnert, unpublished results.
42. W. Ando, K. Hatano and R. Urisaka, *Organometallics*, **14**, 3625 (1995).
43. (a) S. Sharma and A. C. Oehlschlager, *Tetrahedron*, **45**, 557 (1989).
    (b) S. Sharma and A. C. Oehlschlager, *J. Org. Chem.*, **54**, 5383 (1989).
    (c) S. Sharma and A. C. Oehlschlager, *J. Org. Chem.*, **56**, 770 (1991).
    (d) R. D. Singer and A. C. Oehlschlager, *J. Org. Chem.*, **56**, 3510 (1991).
44. (a) M. Weidenbruch, K. Kramer, K. Peters and H. G. v. Schnering, *Z. Naturforsch., Teil B*, **40B**, 601 (1985).
    (b) D. M. Roddick, R. H. Heyn and T. D. Tilley, *Organometallics*, **8**, 324 (1989).
45. K. Tamao and A. Kawachi, *Organometallics*, **14**, 3108 (1995).
46. E. Buncel, R. D. Gordon and T. K. Venkatachalam, *J. Organomet. Chem.*, **507**, 81 (1996).
47. (a) G. A. Olah and R. J. Hunadi, *J. Am. Chem. Soc.*, **102**, 6989 (1980).
    (b) U. Edlund, T. Lejon, T. K. Venkatachalam and E. Buncel, *J. Am. Chem. Soc.*, **107**, 6408 (1985).
48. A. G. Evans, M. A. Hamid and N. H. Rees, *J. Chem. Soc. B*, 1110 (1971).
49. (a) H. V. R. Dias, M. M. Olmstead, K. Ruhlandt-Senge and P. P. Power, *J. Organomet. Chem.*, **462**, 1 (1993).
    (b) A. H. Cowley, T. M. Elkins, R. A. Jones and C. M. Nunn, *Angew. Chem.*, **100**, 1396 (1988); *Angew. Chem., Int. Ed. Engl.*, **27**, 1349 (1988).
50. R. A. Bartlett, H. V. R. Dias and P. P. Power, *J. Organomet. Chem.*, **341**, 1 (1988).
51. (a) U. Edlund, T. Lejon, P. Pyykkö, T. K. Venkatachalam and E. Buncel, *J. Am. Chem. Soc.*, **109**, 5982 (1987).
    (b) E. Buncel, T. K. Venkatachalam and U. Edlund, *Can. J. Chem.*, **64**, 1674 (1986).
52. For a comprehensive table, see: K. Tamao and A. Kawachi, *Adv. Organomet. Chem.*, **38**, 48–49 (1995).
53. T. Koizumi, K. Morihashi and O. Kikuchi, *Organometallics*, **14**, 4018 (1995).
54. M. R. Nimlos and G. B. Ellison, *J. Am. Chem. Soc.*, **108**, 6522 (1986).
55. H. Bürger and R. Eujen, *Z. Naturforsch., Teil B*, **29B**, 647 (1974).
56. A. Sekiguchi, M. Nanjo, C. Kabuto and H. Sakurai, *Organometallics*, **14**, 2630 (1995).
57. A. G. Brook, A. Baumegger and A. J. Lough, *Organometallics*, **11**, 310 (1992).
58. O. W. Steward, G. L. Heider and J. S. Johnson, *J. Organomet. Chem.*, **168**, 33 (1979).
59. K. E. Ruehl, M. E. Davis and K. Matyjaszewski, *Organometallics*, **11**, 788 (1992).
60. (a) M. Kumada, S. Sakamoto and M. Ishikawa, *J. Organomet. Chem.*, **17**, 231 (1969).
    (b) M. Kumada, S. Sakamoto, M. Ishikawa and S. Maeda, *J. Organomet. Chem.*, **17**, 223 (1969).
61. A. Sekiguchi, M. Nanjo, C. Kabuto and H. Sakurai, *J. Am. Chem. Soc.*, **117**, 4195 (1995).
62. (a) M. J. Fink and D. B. Puranik, *Organometallics*, **6**, 1809 (1987).
    (b) D. B. Puranik, M. P. Johnson and M. J. Fink, *Organometallics*, **8**, 770 (1989).
    (c) K. M. Baines, A. G. Brook, R. R. Ford, P. D. Lickiss, A. K. Saxena, W. J. Chatterton, J. F. Sawyer and B. A. Behnam, *Organometallics*, **8**, 693 (1989).
63. (a) H. Gilman and C. L. Smith, *J. Organomet. Chem.*, **14**, 91 (1968).
    (b) G. Gutekunst and A. G. Brook, *J. Organomet. Chem.*, **225**, 1 (1982).
    (c) W. Biffar and H. Nöth, *Z. Naturforsch., Teil B*, **36B**, 1509 (1981).
64. A. Heine, R. Herbst-Irmer, G. M. Sheldrick and D. Stalke, *Inorg. Chem.*, **32**, 2694 (1993).
65. G. Becker, H. M. Hartmann, A. Münch and H. Riffel, *Z. Anorg. Allg. Chem.*, **530**, 29 (1985).
66. (a) K. W. Klinkhammer and W. Schwarz, *Z. Anorg. Allg. Chem.*, **619**, 1777 (1993).
    (b) K. W. Klinkhammer, G. Becker and W. Schwarz, in *Organosilicon Chemistry II* (Eds. N. Auner and J. Weis), VCH, Weinheim, 1996, pp. 493–498.
67. (a) A. Heine and D. Stalke, *Angew. Chem.*, **105**, 90 (1993); *Angew. Chem., Int. Ed. Engl.*, **32**, 121 (1993).
    (b) A. Heine, R. Herbst-Irmer and D. Stalke, *J. Chem. Soc., Chem. Commun.*, 1729 (1993).
68. H. Gilman, J. M. Holmes and C. L. Smith, *Chem. Ind. (London)*, 848 (1965).
69. (a) Y. Apeloig, M. Bendikov, M. Yuzefovich, M. Nakash and D. Bravo-Zhivotovskii, *J. Am. Chem. Soc.*, **118**, 12228 (1996).
    (b) Y. Apeloig, M. Yuzefovich, M. Bendikov, D. Bravo-Zhivotovskii and K. Klinkhammer, *Organometallics*, **16**, 1265 (1997).
    (c) Y. Apeloig, personal communication.

70. (a) W. Biffar, T. Gasparis-Ebeling, H. Nöth, W. Storch and B. Wrackmeyer, *J. Magn. Reson.*, **44**, 54 (1981).
    (b) R. H. Heyn and T. D. Tilley, *Inorg. Chem.*, **29**, 4051 (1990).
71. (a) H. Gilman and R. L. Harrell, Jr., *J. Organomet. Chem.*, **9**, 67 (1967).
    (b) H. Gilman and C. L. Smith, *J. Organomet. Chem.*, **8**, 245 (1967).
72. H. Gilman, D. J. Peterson, A. W. Jarvie and H. S. J. Winkler, *J. Am. Chem. Soc.*, **82**, 2076 (1996).
73. H. Gilman and G. L. Schwebke, *J. Am. Chem. Soc.*, **85**, 1016 (1963).
74. G. Becker, H.-M. Hartmann, E. Hengge and F. Schrank, *Z. Anorg. Allg. Chem.*, **572**, 63 (1989).
75. J. Belzner, U. Dehnert and D. Stalke, *Angew. Chem.*, **106**, 2580 (1994); *Angew. Chem., Int. Ed. Engl.*, **33**, 2450 (1994).
76. M. Kira, T. Maruyama, C. Kabuto, K. Ebata and H. Sakurai, in *Xth International Symposium on Organosilicon Chemistry*, Poznań, Poland, 1993, P-66.
77. W. Ando, T. Wakahara, T. Akasaka and S. Nagase, *Organometallics*, **13**, 4683 (1994).
78. See e.g.: M. Weidenbruch, A. Pellmann, Y. Pan, S. Pohl, W. Saak and H. Marsmann, *J. Organomet. Chem.*, **450**, 67 (1993).
79. A. W. Jarvie and H. Gilman, *J. Org. Chem.*, **26**, 1999 (1961).
80. Y. Hatanaka and T. Hiyama, in *40th Symposium on Organometallic Chemistry*, Sapporo, Japan, 1993, p. 136
81. M. Cypryk, Y. Gupta and K. Matyjaszewski, *J. Am. Chem. Soc.*, **113**, 1046 (1991).
82. M. Suzuki, J. Kotani, S. Gyobu, T. Kaneko and J. Saegusa, *Macromolecules*, **27**, 2360 (1994).
83. (a) E. Hengge and F. K. Mitter, *Monatsh. Chem.*, **117**, 721 (1986).
    (b) E. Hengge, P. K. Jenkner, P. Gspaltl and A. Spielberger, *Z. Anorg. Allg. Chem.*, **560**, 27 (1988).
    (c) E. Hengge, A. Spielberger and P. Gspaltl, in *Xth International Symposium on Organosilicon Chemistry*, Poznań, Poland, 1993, P-145.
84. E. Hengge and P. K. Jenkner, *J. Organomet. Chem.*, **314**, 1 (1986).
85. A. L. Allred, R. T. Smart and D. A. V. Beek, Jr., *Organometallics*, **11**, 4225 (1992).
86. F. Uhlig, P. Gspaltl, M. Trabi and E. Hengge, *J. Organomet. Chem.*, **493**, 33 (1995).
87. (a) J. Oshita, Y. Masaoka, S. Masaoka, M. Ishikawa, A. Tachibana, T. Yano and T. Yamabe, *J. Organomet. Chem.*, **473**, 15 (1994).
    (b) J. Oshita, S. Masaoka and M. Ishikawa, *Organometallics*, **15**, 2198 (1996).
    (c) J. Oshita, S. Masaoka, Y. Masaoka, H. Hasebe, M. Ishikawa, A. Tachibana, T. Yano and T. Yamabe, *Organometallics*, **15**, 3136 (1996).
88. K. Tamao, A. Kawachi and Y. Ito, *J. Am. Chem. Soc.*, **114**, 3989 (1992).
89. K. Tamao, Y. Ito and A. Kawachi, *Organometallics*, **12**, 580 (1993).
90. K. Tamao, A. Kawachi, and Y. Ito, in *Xth International Symposium on Organosilicon Chemistry*, Poznań, Poland, 1993, I-9.
91. K. Tamao, G.-R. Sun and A. Kawachi, *J. Chem. Soc., Chem. Commun.*, 2070 (1995).
92. K. Tamao, A. Kawachi, Y. Tanaka, H. Othani and Y. Ito, *Tetrahedron*, **52**, 5765 (1996).
93. (a) H. Watanabe, K. Higuchi, M. Kobayashi, M. Hara, Y. Koike, T. Kitahara and Y. Nagai, *J. Chem. Soc., Chem. Commun.*, 534 (1977).
    (b) H. Watanabe, K. Higuchi, T. Goto, T. Muraoka, J. Inose, M. Kageyama, Y. Izuka, M. Nozaki and Y. Nagai, *J. Organomet. Chem.*, **218**, 27 (1981).
94. A. Kawachi, N. Doi and K. Tamao, *J. Am. Chem. Soc.*, **119**, 233 (1997).
95. K. Tamao and A. Kawachi, *Angew. Chem.*, **107**, 886 (1995); *Angew. Chem., Int. Ed. Engl.*, **34**, 818 (1995).
96. K. Tamao and A. Kawachi, *Organometallics*, **15**, 4653 (1996).
97. See e.g.:
    (a) D. Seebach, R. Hässig and J. Gabriel, *Helv. Chim. Acta*, **66**, 308 (1983).
    (b) G. Boche, F. Bosold, J. C. W. Lohrenz, A. Opel and P. Zulauf, *Chem. Ber.*, **126**, 1873 (1993).
98. H. Watanabe, T. Muraoka, M. Kageyama, K. Yoshizumi and Y. Nagai, *Organometallics*, **3**, 141 (1984).
99. (a) T. Tsumuraya, S. A. Batcheller and S. Masamune, *Angew. Chem.*, **103**, 916 (1991); *Angew. Chem., Int. Ed. Engl.*, **30**, 902 (1991) and references cited therein.
    (b) N. Tokitoh, H. Suzuki, R. Okazaki and K. Ogawa, *J. Am. Chem. Soc.*, **115**, 10248 (1993).

14. Alkaline and alkaline earth silyl compounds — preparation and structure

100. (a) R. West, *Angew. Chem.*, **99**, 1231 (1987); *Angew. Chem., Int. Ed. Engl.*, **26**, 1201 (1987) and references cited therein.
 (b) M. Weidenbruch, *Chem. Rev.*, **95**, 1479 (1995) and references cited therein.
101. (a) R. West, in *The Chemistry of Organic Silicon Compounds* (Eds. S. Patai and Z. Rappoport), Wiley, Chichester, 1989, pp. 1210–1217.
 (b) K. Matyjazewski, D. Greszta, J. S. Hrkach and H. K. Kim, *Macromolecules*, **28**, 59 (1995).
102. H. Öhme and H. Weis, *J. Organomet. Chem.*, **319**, C16 (1987).
103. R. A. Benkeser, *Acc. Chem. Res.*, **4**, 94 (1971) and references cited therein.
104. P. Boudjouk, S. Kloos and A. B. Rajkumari, *J. Organomet. Chem.*, **443**, C41 (1993).
105. (a) R. A. Pike, *J. Org. Chem.*, **27**, 2186 (1962).
 (b) R. A. Benkeser, K. M. Foley, J. M. Gaul and G. S. Li, *J. Am. Chem. Soc.*, **92**, 3232 (1970).
106. M. S. Gordon, P. Boudjouk and F. Anwari, *J. Am. Chem. Soc.*, **105**, 4972 (1983).
107. J. R. Damewood, Jr., *J. Org. Chem.*, **51**, 5028 (1986).
108. (a) B. Goldfuss and P. v. R. Schleyer, *Organometallics*, **14**, 1553 (1995).
 (b) P. v. R. Schleyer, P. K. Freeman, H. J. Jiao and B. Goldfuss, *Angew. Chem.*, **107**, 332 (1995); *Angew. Chem., Int. Ed. Engl.*, **34**, 337 (1995).
 (c) P. v. R. Schleyer and H. J. Jiao, *Pure Appl. Chem.*, **68**, 209 (1996).
109. J.-H. Hong and P. Boudjouk, *Organometallics*, **14**, 574 (1995).
110. (a) J.-H. Hong and P. Boudjouk, *J. Am. Chem. Soc.*, **115**, 5883 (1993).
 See also:
 (b) H. Gilman and R. D. Gorsich, *J. Am. Chem. Soc.*, **80**, 3243 (1958).
 (c) M. Ishikawa, T. Tabohashi, H. Ohashi, M. Kumada and J. Iyoda, *Organometallics*, **2**, 351 (1983).
 (d) P. Jutzi and A. Karl, *J. Organomet. Chem.*, **214**, 289 (1981).
111. J.-H. Hong, P. Boudjouk and I. Stoenescu, *Organometallics*, **15**, 2179 (1996).
112. (a) W. P. Freeman, T. D. Tilley, G. P. A. Yap and A. L. Rheingold, *Angew. Chem.*, **108**, 960 (1996); *Angew. Chem., Int. Ed. Engl.*, **35**, 882 (1996).
 (b) W. P. Freeman, T. D. Tilley, L. M. Liable-Sands and A. L. Rheingold, *J. Am. Chem. Soc.*, **118**, 10457 (1996).
113. B. Goldfuss, P. v. R. Schleyer and F. Hampel, *Organometallics*, **15**, 1755 (1996).
114. W.-C. Joo, J.-H. Hong, S.-B. Choi and H.-E. Son, *J. Organomet. Chem.*, **391**, 27 (1990).
115. J.-H. Hong, P. Boudjouk and S. Castellino, *Organometallics*, **13**, 3387 (1994).
116. R. West, H. Sohn, U. Bankwitz, J. Calabrese, Y. Apeloig and T. Müller, *J. Am. Chem. Soc.*, **117**, 11608 (1995).
117. J. A. Morrison and R. J. Lagow, *Inorg. Chem.*, **16**, 2972 (1977).
118. P. v. R. Schleyer and A. E. Reed, *J. Am. Chem. Soc.*, **110**, 4453 (1988).
119. S. K. Mehrotra, H. Kawa, J. R. Baran, Jr., M. M. Ludvig and R. J. Lagow, *J. Am. Chem. Soc.*, **112**, 9003 (1990).
120. A. Rajca, P. Wang, A. Streitwieser and P. v. R. Schleyer, *Inorg. Chem.*, **28**, 3064 (1989).

CHAPTER 15

Mechanism and structures in alcohol addition reactions of disilenes and silenes

HIDEKI SAKURAI

Department of Industrial Chemistry, Faculty of Science and Technology, Science University of Tokyo, Noda, Chiba 278, Japan

I. INTRODUCTION	827
II. DISILENES	828
A. $E-Z$ Isomerization of Disilenes	828
B. Generation of Phenyl-substituted Disilenes. Spectra and Kinetics of Addition of Alcohols	829
C. Generation of Alkoxy- and Amino-substituted Disilenes	831
D. Regiochemistry of Addition of Alcohols to Disilenes	835
E. Diastereochemistry of Addition of Alcohols to Disilenes	839
F. Competition Between Nucleophilic and Electrophilic Mechanisms	842
III. SILENES	844
A. Generation of Silenes	844
B. Nucleophilic Addition to Silenes	844
C. Diastereochemistry of Addition of Alcohol to Silene	845
IV. CONCLUDING REMARKS	853
V. ACKNOWLEDGMENTS	853
VI. REFERENCES	853

I. INTRODUCTION

Organosilicon chemistry has expanded its scope considerably in the last two decades. One of the most remarkable achievements is the progress which was made in the elucidation of the mechanisms in silicon chemistry, which now become comparable to those in carbon chemistry. The behavior of reactive intermediates such as silylenes[1], silyl radicals[2] and silyl anions[3,4] are well explored, although the chemistry of silyl cations is still controversial[5]. Doubly-bonded silicon species are now well understood[6–12] but triply-bonded silicon is still elusive.

The chemistry of organic silicon compounds, Vol. 2
Edited by Z. Rappoport and Y. Apeloig © 1998 John Wiley & Sons Ltd

The mechanism of substitution reactions at saturated silicon centers is well studied, regarding both kinetics and stereochemistry[13,14]. In contrast, addition reactions to unsaturated silicon centers, such as to disilenes and silenes, are relatively unexplored. The reason is clear: suitable substrates for investigations of regio- and stereochemistry and reaction kinetics are not readily available due to inherent kinetic instability of disilenes and silenes. Kinetically stabilized disilenes and silenes are now available, but these are not always convenient for studying the precise mechanism of addition reactions. For example, stable disilenes are usually prepared by the dimerization of silylenes with bulky substituents. Therefore, it is extremely difficult to prepare unsymmetrically substituted disilenes necessary for regio- and/or stereochemical studies.

Nevertheless, mechanistic investigations on the addition reaction to disilenes and silenes advanced considerably in recent years. In this chapter, the author tries to survey the progress achieved recently. The author will not try to review all aspects of the chemistry of silicon unsaturated species since many extensive reviews on this topic are already available[6–12].

II. DISILENES

A. E–Z Isomerization of Disilenes

After Roark and Peddle generated in 1972 tetramethyldisilene, $Me_2Si=SiMe_2$, as the first silicon–silicon unsaturated species[15], it was of major interest to examine the double-bond character of these reactive species. In 1979, Sakurai and coworkers generated independently the (E)- and the (Z)-1,2-dimethyl-1,2-diphenyldisilene and demonstrated

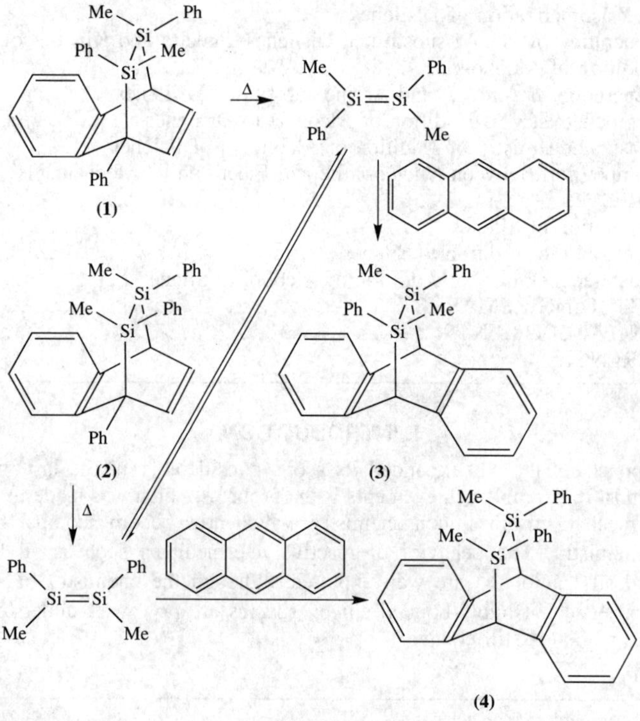

SCHEME 1

that these isomeric disilenes underwent only slow $E-Z$ isomerization compared with trapping reactions by dienes as shown in Scheme 1[16]. The thermal decomposition of **1** at 300 °C in the presence of anthracene afforded a mixture of **3** (96%) and **4** (4%) in 81% yield. A similar thermolysis of the isomeric **2** gave **4** (94%) and **3** (6%) in 88% yield. Since 1,2-dimethyl-1,2-diphenyldisilene generated by the dimerization of PhMeSi: gave a 1 : 1 mixture of **3** and **4** in the reaction with anthracene, [17], these results are remarkable and demonstrate clearly that the strength of the Si=Si double bond in 1,2-dimethyl-1,2-diphenyldisilene is sufficient to prevent the geometrical isomerism from occurring prior to the trapping reactions.

Upon increasing the reaction temperature to 350 °C, thermolysis gave slightly less stereospecific results. Thus **1** gave **3** (94%) and **4** (6%) in 84% yield, and **2** gave **3** (10%) and **4** (90%) in 85% yield, respectively. Later Olbrich, Walsh and coworkers estimated the $E-Z$ isomerization energy of 1,2-dimethyl-1,2-diphenyldisilene to be 25.8 ± 5 kcal mol^{-1} on the basis of these and other data[18].

Thermal $E-Z$ isomerization was also observed for kinetically stabilized disilenes **5–9** (equation 1)[19–22]. The π bond strength was estimated to range from 24.7 to 30.6 kcal mol^{-1}. These data are in good agreement with those of 1,2-dimethyl-1,2-diphenyldisilene and those predicted by *ab initio* calculations for $H_2Si=SiH_2$ (22–28 kcal mol^{-1})[23,24].

$$\begin{array}{c} R^2 \\ \diagdown \\ R^1 \end{array} Si = Si \begin{array}{c} R^1 \\ \diagup \\ R^2 \end{array} \underset{}{\overset{\Delta}{\rightleftharpoons}} \begin{array}{c} R^2 \\ \diagdown \\ R^1 \end{array} Si = Si \begin{array}{c} R^2 \\ \diagup \\ R^1 \end{array}$$

(5) $R^1 = t$-Bu; R^2 = Mes
(6) R^1 = Mes; R^2 = Ad
(7) R^1 = Mes; $R^2 = N(SiMe_3)_2$
(8) R^1 = Mes; R^2 = Dep   (1)
(9) R^1 = Mes; R^2 = Dip

Ad = 1-adamantyl; Dep = 2,6-diethylphenyl; Dip = 2,6-diisopropylphenyl;
Mes = 2,4,6-trimethylphenyl

These $E-Z$ isomerization studies of disilenes indicate that the π overlap between two 3p orbitals of silicon is sufficiently effective to retain the configuration around the double bond, although the π bonding of disilenes is significantly weaker than that of the C=C double bond. Therefore, it is expected that if appropriately substituted disilenes can be generated, regiochemistry as well as diastereochemistry of addition reactions to disilenes can be investigated even with transient reactive disilenes.

B. Generation of Phenyl-substituted Disilenes. Spectra and Kinetics of Addition of Alcohols

Several methods are available for generating disilenes[8] but photolysis of masked disilenes[25,26] is most convenient for mechanistic studies. 7,8-Disilabicyclo[2.2.2]octa-2,5-dienes, the formal adducts of the addition of disilenes to benzene, naphthalene, anthracene and biphenyl, are well established to generate the corresponding reactive disilenes by either thermolysis or photolysis. The parent 7,7,8,8-tetramethyl-7,8-disilabicyclo[2.2.2]octa-2,5-diene (**10**) generates tetramethyldisilene (**11**) in an argon matrix by photolysis at 10 K[27]. Tetramethyldisilene (**11**) was also observed at 344 nm in UV spectra by 3-methylpentane (3-MP) and EPA (ether: isopentane: ethanol = 5 : 5 : 2) matrices (equations 2 and 3, respectively)[27]. On annealing the EPA matrix, a product of addition of ethanol to **11**,

1-ethoxy-2-hydrotetramethyldisilane (**12**), was obtained (equation 3).

$$\text{(10)} \xrightarrow{h\nu\ (254\ \text{nm})}_{\text{3-MP, 77 K}} \underbrace{Me_2Si=SiMe_2}_{\textbf{(11)}} + \bigcirc \quad (2)$$

λ_{max} 344 nm

$$\text{(10)} \xrightarrow{h\nu\ (254\ \text{nm})}_{\text{EPA, 77 K}} Me_2Si=SiMe_2 + \bigcirc$$
(11)

$$\xrightarrow{\text{anneal}} \begin{array}{c} Me_2Si-SiMe_2 \\ |\quad\quad | \\ H\quad\ OEt \end{array} \quad (3)$$
(12)

Phenyltrimethyldisilene (**15**) and (*E*)- and (*Z*)-1,2-dimethyl-1,2-diphenyldisilene (**16**) were also generated (equation 4) by the photolysis of precursors **13**, (*E*)-**14** and (*Z*)-**14**, respectively[28]. Absorption maxima of **15**, (*E*)-**16** and (*Z*)-**16** were observed at 386, 417 and 423 nm, respectively. These absorption bands are also observed in an organic glass (3-MP, EPA), the ethanol addition products being detected after annealing of the EPA matrix, similar to the case of **11**.

$$(4)$$

(**13**) $R^1 = R^2 = Me$ (**15**) $R^1 = R^2 = Me$
(*E*)-(**14**) $R^1 = Me, R^2 = Ph$ (*E*)-(**16**) $R^1 = Me, R^2 = Ph$
(*Z*)-(**14**) $R^1 = Ph, R^2 = Me$ (*Z*)-(**16**) $R^1 = Ph, R^2 = Me$

Phenyltrimethyldisilene (**15**) and (*E*)- and (*Z*)-1,2-dimethyl-1,2-diphenyldisilene (**16**) were also observed as transient absorption spectra by laser flash photolysis of the precursors in methylcyclohexanes[28]. The absorption band at 380 nm, assigned to the disilene **15**, reached maximum intensity at *ca* 10 ns after the excitation and then started to decrease. The half-life assigned to **15** was 700 ns. The logarithm of the decay profile of the transient absorption at 380 nm versus time shows a very good linear relationship, indicating that the decay of the transient absorption fits first-order kinetics. This result shows that intramolecular isomerization or proton abstraction from the solvent is the origin for the decay of the disilene **15**, which survives in solution only for several nanoseconds.

Photochemical generations of transient (*E*)- and (*Z*)-1,2-dimethyl-1,2-diphenyldisilenes (*E*)-**16** and (*Z*)-**16** in solution is also confirmed by laser flash photolysis of the precursors, (*E*)-**14** and (*Z*)-**14**, respectively, in degassed methylcyclohexane at 293 K[28]. The absorption bands at 415 and 417 nm are assigned to the disilenes (*E*)-**16** and (*Z*)-**16**, respectively.

TABLE 1. Rate constants for quenching of phenyl-substituted disilenes by alcohols[28]

Trapping agent	$k_2/M^{-1} \cdot s^{-1}$		
	(E)-PhMeSi=SiMePh	(Z)-PhMeSi=SiMePh	PhMeSi=SiMe$_2$
EtOH	1.7×10^8	1.9×10^8	1.9×10^8
EtOD	1.7×10^8	1.7×10^8	1.8×10^8
i-PrOH	1.2×10^8	1.2×10^8	1.3×10^8
t-BuOH	0.9×10^7	1.1×10^7	1.6×10^7

The half lives of (E)- and (Z)-**14** obtained from the decay profiles at 415 nm are 800 and 710 ns, respectively. These half-lives are almost the same, but both are slightly larger than that of **15**. The logarithms of decay profiles of the transient absorptions versus time also show very good linear lines, indicating a first-order kinetics for decay.

The transient absorption of **15** and of (E)- and (Z)-**16** are successfully quenched by addition of alcohols. The quenching rate constants for these disilenes, determined from the change in the half-lives in the presence of alcohols, give the second-order rate constants k_2, which are summarized in Table 1.

The second-order rate constants k_2 for the reactions of phenyl-substituted disilenes with various alcohols indicate several interesting points. First, the addition reaction of alcohols to the disilenes proceeds so rapidly that it is probable that these phenyl-substituted disilenes are quenched prior to $E-Z$ isomerization. The order of **15** > (E)-**16** ~ (Z)-**16** is found for the relative reactivity of these disilenes. Rate constants of phenyltrimethyldisilene **15** are slightly larger than those of dimethyldiphenyldisilenes (E)-**16** and (Z)-**16**, probably due to steric factors. However, there is little difference in the reactivity among the (E)- and (Z)-isomers, the rate constants of these disilenes also being almost independent of the number of the attached phenyl group(s). Second, the relative rates decrease in the order of EtOH > i-PrOH ≫ t-BuOH, consistent with the increased steric bulkiness of the alcohols. In particular, the reactivity of t-BuOH toward the disilene is lowest, being only about one-tenth that of other alcohols. Third, the magnitude of the kinetic isotope effect (k_H/k_D) observed for the reaction with ethanol and ethanol-d is only 1.0 – 1.1, indicating that there is no significant deuterium isotope effect in these reactions. This fact strongly indicates that the rate-determining step of the addition reaction must be the nucleophilic attack of the alcoholic oxygen on the coordinatively unsaturated silicon center.

C. Generation of Alkoxy- and Amino-substituted Disilenes

Alkoxy- and amino-substituted disilenes (**17**, **18** and **19**) are produced by photolysis of the corresponding masked disilenes (**20**, **21** and **22**)[29]. Thus, when a degassed 3-methylpentane (3-MP) matrix of **20** was irradiated at 77 K, the corresponding alkoxy-substituted disilene **17** was produced (equation 5) as indicated by its broad band at 373 nm.

(5)

After similar irradiation of a mixture of **20** and ethanol in a mixed matrix of 3-MP and isopentane (4 : 1) at 77 K, GC-MS analysis of the solution showed the formation of the expected ethanol adduct of the disilene. The regiochemistry of the product will be discussed in the next section.

The formation of amino-substituted disilenes **18** and **19** was also confirmed in similar experiments using **21** and **22** (equations 6 and 7, respectively). Broad bands growing at 395 and 408 nm, respectively, were observed. Absorption maxima of several transient disilenes are summarized in Table 2, in which the absorption spectral data of the recently prepared (trimethylsilylmethyl)trimethyldisilene[30] is also included. The electronic spectral data of stable disilenes are well documented[8].

$$\text{(21)} \xrightarrow{h\nu\ (254\ \text{nm}),\ 3\text{-MP, 77 K}} \text{(18)} \tag{6}$$

$$\text{(22)} \xrightarrow{h\nu\ (254\ \text{nm}),\ 3\text{-MP, 77 K}} \text{(19)} \tag{7}$$

It is clear that introduction of the functional groups to the silicon–silicon double bond causes a large red shift of the absorption maxima of disilenes. Similar substituent effects have been known for a variety of alkenes[31]. Qualitatively, the reason for the red shift should arise from destabilization of the HOMO of the disilene by an appreciable interaction between the n-orbital of the heteroatoms and the π-orbital of the disilene, although the effect of their distorted structure should also be considered as discussed later.

TABLE 2. Absorption maxima of some reactive disilenes

Disilene	Medium	λ_{max} (nm)	$\Delta \nu$ (cm^{-1})[a]	Reference
Me$_2$Si=SiMe$_2$	Argon matrix	344		27
PhMeSi=SiMe$_2$	Argon matrix	386	3160	28
(E)-PhMeSi=SiPhMe	Argon matrix	417	5090	28
(Z)-PhMeSi=SiPhMe	Argon matrix	423	5430	28
(t-BuO)MeSi=SiMe$_2$	3-MP matrix	373	2260	29
(Me$_2$N)MeSi=SiMe$_2$	3-MP matrix	395	3750	29
(Me$_2$N)MeSi=SiMe(NMe$_2$)	3-MP matrix	408	4560	29
(Me$_3$SiCH$_2$)MeSi=SiMe$_2$	3-MP matrix	360	1290	30

[a] Difference (in wave numbers) from the absorption of tetramethyldisilene.

15. Mechanism and structures in alcohol addition reactions

There is rather little information about disilenes with substituents other than alkyl and aromatic groups. An example is the bis(trimethylsilylamino)-substituted disilene **23** which was isolated as a stable disilene by West and coworkers[19]. More recently, the dialkoxydisilenes **24** were generated by the dimerization of methoxy-substituted silylenes in matrices[32]. However, only the change of UV spectra through an annealing of the matrices containing alkoxysilylenes was observed; no trapping adducts, except for cyclotrisilane, were obtained. Other attempts to generate amino- or fluoro-substituted disilenes were unsuccessful[33].

The formation of an intermediate difluorodisilene **25** (equation 8) was proposed by Jutzi and coworkers[34] in the reaction of decamethylsilicocene with tetrafluoroboric acid. The disilene which was characterized by the ^{29}Si NMR spectrum, then formed the isolable cyclotetrasilane by [2 + 2] cycloaddition.

Later, Maxka and Apeloig calculated the geometry of $Si_2H_2F_2$ by an *ab initio* (6-31G*) method and suggested a nonclassical bridged structure for the fluorosilylene dimer[35]. Thus, even in this case, no clear evidence for generation of the disilene has been indicated. More recently, the existence of the nonclassical bridged structure of the tetraaminodisilene has been suggested in the dimerization reaction of diaminosilylene[36], where scrambling of amino groups occurs between differently substituted diaminosilylenes (see Scheme 2).

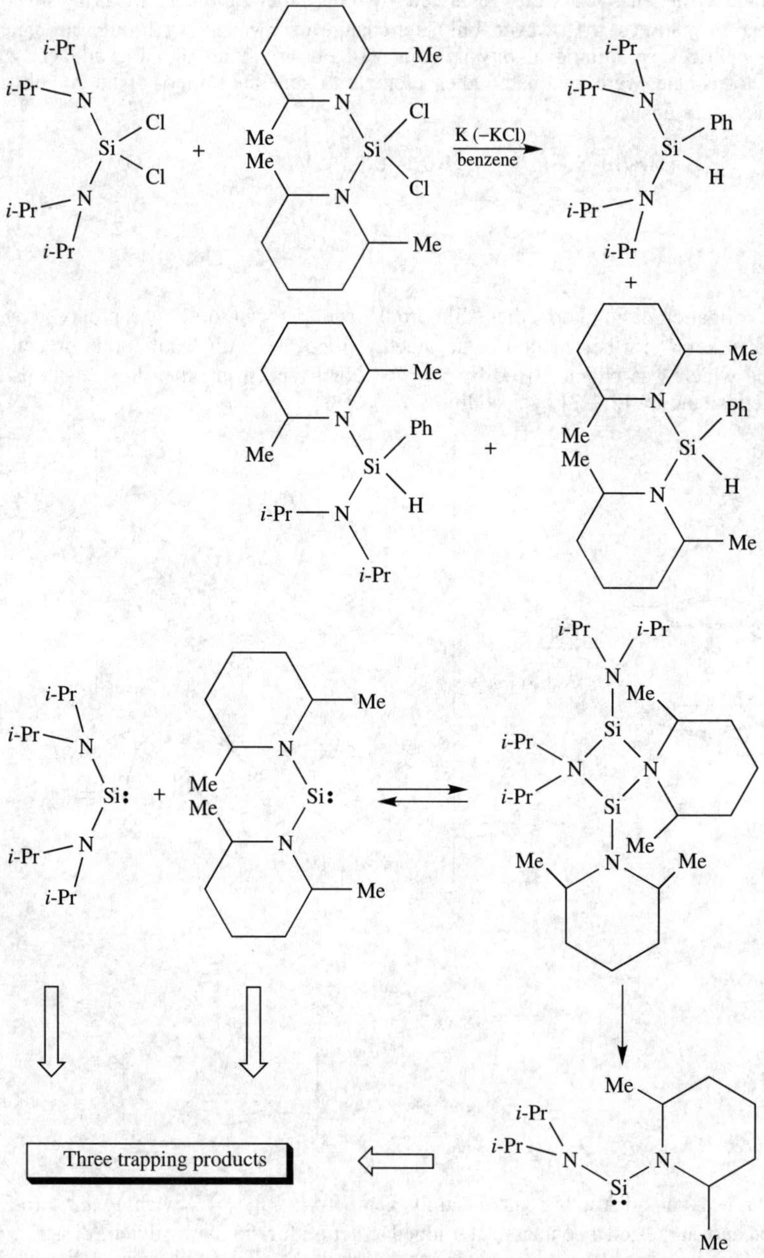

SCHEME 2

Diaminosilylenes generated by the reaction of the corresponding diaminodichlorosilane with potassium react with benzene to give products derived by insertion into the Ph—H bond. Interestingly, the reaction of simultaneously generated bis(diisopropylamino)- and bis(*cis*-2,6-dimethylpiperidino) silylenes gave three products due to scrambling of the amino substituents on silicon atoms.

In relation to the structure of disilene, several *ab initio* MO calculations of the geometry and energies of various disilenes have been reported[37–43]. Several possibilities regarding the conformations around the Si=Si double bond (planar, *trans*-bent, twist and bridged etc.) have to be considered. According to these calculations the most stable geometry depends significantly on the substituents. The parent $H_2Si=SiH_2$ adopts a *trans*-bent conformation, having a bent angle of 12.9°, whereas electropositive substituents such as Li, BH_2 and SiH_3 are predicted to give disilenes with a preferred planar geometry. In contrast, the electronegative and π-donating substituents such as NH_2, F and OH induce large distortions from planarity[35,43].

D. Regiochemistry of Addition of Alcohols to Disilenes

Generation of various phenyl-substituted disilenes by the photolysis of the masked disilenes, 7,8-disilabicyclo[2.2.2]octadiene derivatives, is quite useful, especially for unsymmetrically substituted disilenes. Investigation of the regiochemistry as well as the diastereochemistry of alcohol addition to phenyltrimethyldisilene was made possible for the first time by using this method[27].

Phenyltrimethyldisilene **15**, produced by irradiation of the precursor **13** ($\lambda > 280$ nm) in the presence of several alcohols, gives rise to the formation of 1-alkoxy-2-hydrido-1,1,2-trimethyl-2-phenyldisilane (**26**) as the major product along with a small amount of the isomeric 1-alkoxy-2-hydrido-1,2,2-trimethyl-1-phenyldisilane (**27**) (see Scheme 3). As shown in Table 3, very high regioselectivity was observed. This is the first example demonstrating a regioselective addition reaction to the unsymmetrically substituted disilenes.

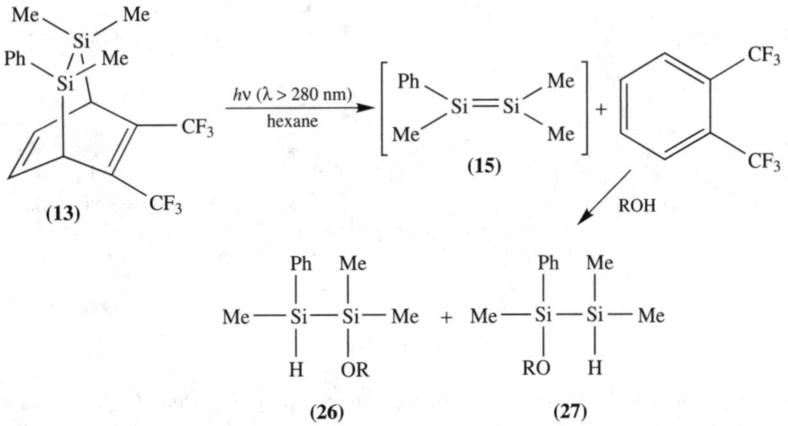

26a and 27a R = Me
26b and 27b R = Et
26c and 27c R = *i*-Pr
26d and 27d R = *t*-Bu

SCHEME 3

TABLE 3. Product ratio in the photolysis of **13** in the presence of various alcohols[28]

ROH	**26/27**
MeOH	92/8
EtOH	95/5
i-PrOH	97/3
t-BuOH	>99/<1

The results in Table 3 were explained as shown in Scheme 4. From the fact that no kinetic isotope effect was observed in the reaction of phenyl-substituted disilenes with alcohols (Table 1), it is assumed that the addition reactions of alcohols to phenyltrimethyldisilene proceed by an initial attack of the alcoholic oxygen on silicon (nucleophilic attack at silicon), followed by fast proton transfer via a four-membered transition state. As shown in Scheme 4, the regioselectivity is explained in terms of the four-membered intermediate, where stabilization of the incipient silyl anion by the phenyl group is the major factor favoring the formation of **26** over **27**. It is well known that a silyl anion is stabilized by aryl group(s)[44a]. Thus, the product **26** predominates over **27**. However, it should be mentioned that steric effects also favor attack at the less hindered SiMe$_2$ end of the disilene, thus leading to **26**.

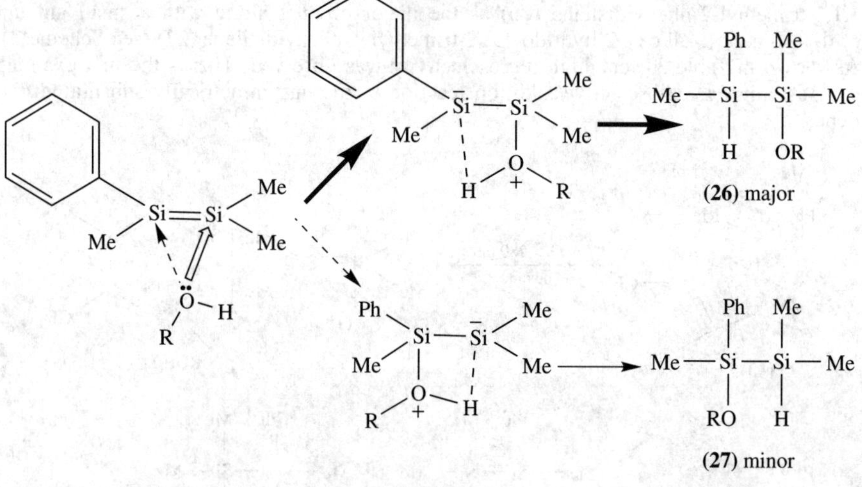

SCHEME 4

Methoxy- and amino-substituted disilenes behave differently from **15**[29]. Irradiation of a hexane solution of **20** in the presence of various alcohols at room temperature afforded 1,1-dialkoxyhydrodisilanes **28a-32a** together with a small amount of the regioisomers **28b-32b** (Scheme 5). Thus alkoxy groups direct the alcohol addition to the alkoxy-substituted silicon atom. The ratios of regioisomers (**a/b**) were 100/0 (EtOH, **28**), 96/4 (i-PrOH, **29**) and 93/7 (t-BuOH, **30**). Steric bulkiness is not the only factor that determines the regioselectivity, since bulky but acidic alcohols, such as 2,6-dimethylphenol

15. Mechanism and structures in alcohol addition reactions

and 1,1,1,3,3,3-hexafluoro-2-methyl-2-propanol, give products in higher regioselectivities (98/2, **31** and >99/<1, **32**, respectively) than *t*-BuOH (Table 4).

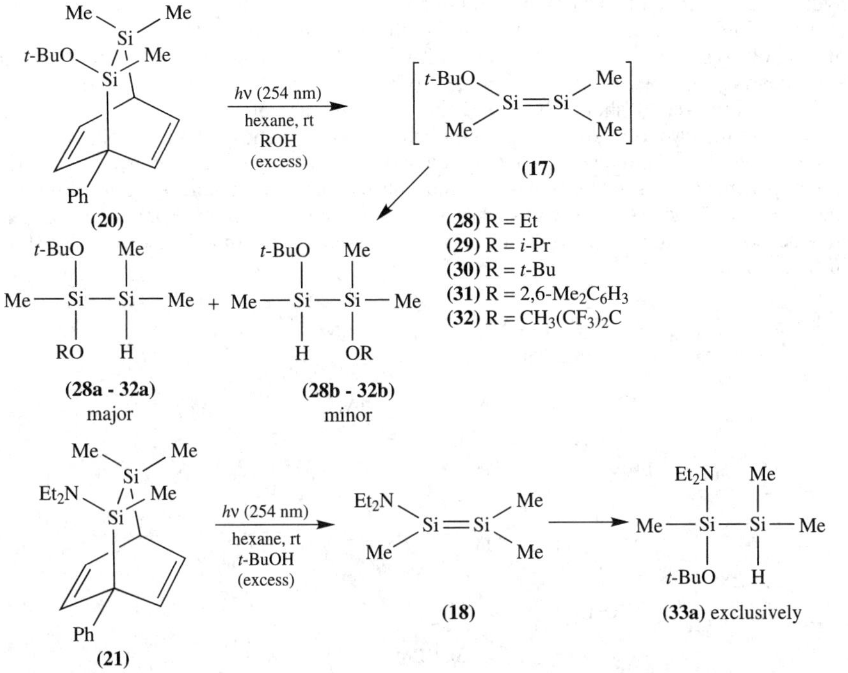

SCHEME 5

TABLE 4. Addition reaction of alcohols to silicon-substituted disilenes XMeSi=SiMe$_2$[29]

X	ROH	Product	a/b[a]
t-BuO	EtOH	28	100/0
	i-PrOH	29	96/4
	t-BuOH	30	93/7
	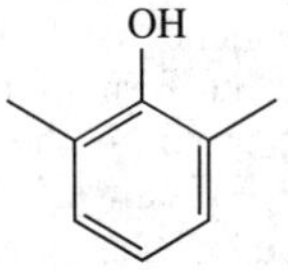 2,6-Me$_2$C$_6$H$_3$OH	31	98/2
	(CF$_3$)$_2$(CH$_3$)COH	32	>99/<1
Et$_2$N	*t*-BuOH	33a	100/0

[a] Determined by ^{1}H NMR.

Addition of an alcohol to the amino-substituted disilene **18** proceeds with an even higher regioselectivity than that to the methoxy-substituted disilene **17**. Thus, the addition of *t*-BuOH to **18** is completely regioselective, where only 1-diethylamino-1-*t*-butoxy-2-hydro-1,2,2-trimethyldisilane **33a** is obtained.

A very high regioselectivity is indicated in the previous section in the addition reactions of alcohols to phenyl-substituted disilene **15**, where the alcoholic proton adds to the silicon atom bearing the phenyl group. Both electronic and steric factors favor this direction of addition. However, it should be noted that the direction of the addition is completely opposite in the case of alkoxy- and amino-substituted disilenes, where the alkoxy groups of the alcohols add to the hetero-substituted and the sterically more hindered silicon atom. The experimental observations indicate that the mechanisms should be quite different in the two cases. The regioselectivity observed for alkoxy- and amino-substituted disilenes may be explained in terms of a strong π-donation effect of the functional groups as shown in Scheme 6.

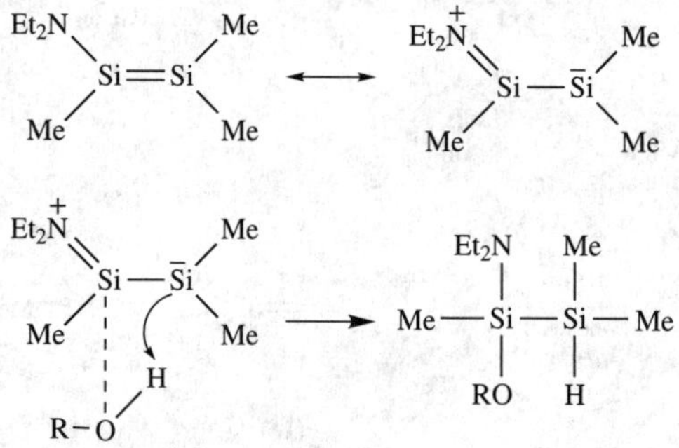

SCHEME 6

Donation of π-electrons from the alkoxy and amino groups makes the Si=Si bond very polar. Although it is uncertain whether nucleophilic interaction between the alcoholic oxygen and the silicon still operates in this mechanism, proton transfer (electrophilic process) is probably the rate-determining step. This suggestion has to be supported by kinetic isotope effect studies similar to those carried out in the case of phenyl-substituted disilenes. Unfortunately, at this point data on the kinetic isotope effect are not available in this system, but the fact that the more acidic but sterically similar 1,1,1,3,3,3-hexafluoro-2-methyl-2-propanol gives higher selectivity (>99/<1) than *t*-BuOH (93/7) supports the rate-determining proton transfer. To some extent the mechanism may be similar to that of the reaction of silyl enol ethers and enamines with electrophiles, since strongly π-donating groups destabilize olefinic systems favoring electrophilic attack[44b]. Therefore, alcohol addition to alkoxy- and amino-substituted disilenes is governed by electrophilic factors. This, however, does not necessarily exclude the formation of a four-centered transition state, even though electrophilic attack by the proton is the rate-determining step.

The trimethylsilylmethyl group (Me_3SiCH_2-) is a weaker electron-donating group than the amino and methoxy groups. The σ^+ value of the trimethylsilylmethyl group on the Brown–Okamoto scale was reported to be -0.66[45], which is smaller (in absolute value) than those of the amino (-1.7) and methoxy (-0.78) groups. Therefore, it was interesting to investigate the regioselectivity of the alcohol addition to the trimethylsilylmethyl-substituted disilene, $(Me_3SiCH_2)MeSi=SiMe_2$ (**34**). Indeed **34**, generated from the corresponding masked disilene, reacted with ethanol to give a 1 : 1 mixture of **35** and **36** (Scheme 7)[30]. In this case, probably both electronic and steric factors compete in determining the regiochemistry.

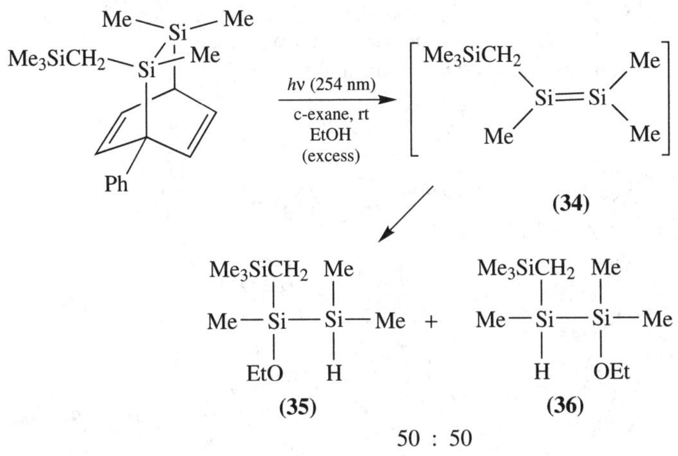

15. Mechanism and structures in alcohol addition reactions

[Scheme 7 depicts the photolysis of a norbornadiene-type precursor (with Me₃SiCH₂, Me, Me, Me substituents on two Si atoms and a Ph group) under hν (254 nm) in c-hexane, rt, EtOH (excess), giving a disilene intermediate **(34)** (Me₃SiCH₂(Me)Si=Si(Me)Me), which reacts with EtOH to give the two regioisomers **(35)** and **(36)** in a 50:50 ratio.]

SCHEME 7

E. Diastereochemistry of Addition of Alcohols to Disilenes

Although the chemistry of disilenes has been developed considerably after the isolation of stable disilenes by West, Finle and Michl[46], rather little has been known about the mechanism, especially on the stereochemistry for the addition reactions to Si=Si bonds of nucleophiles or electrophiles. West and coworkers reported that the addition reaction of ethanol to (E)-1,2-di-t-butyl-1,2-dimesityldisilene in THF gave a 1 : 1 mixture of two diastereomers of alkoxysilanes (equation 9), suggesting a stepwise mechanism[47]. However, the bulky substituents necessary to stabilize disilenes sometimes complicate the stereochemistry. In fact, for the addition reaction of water to the parent disilene (H₂Si=SiH₂), theoretical calculation predicted a concerted-type four-center-like transition state, leading to a *syn*-addition product[48].

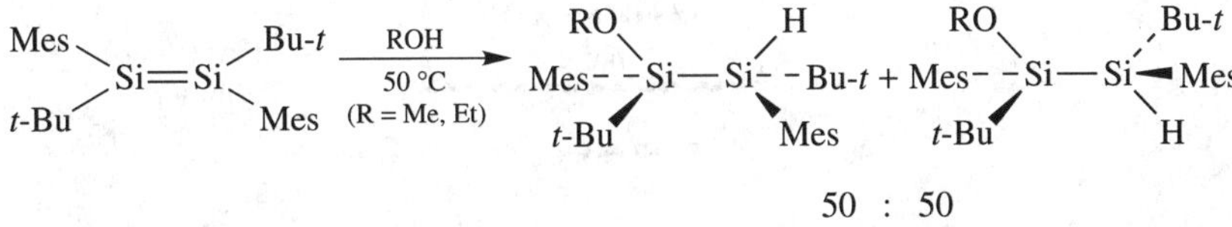

$$\text{Mes}(t\text{-Bu})\text{Si}=\text{Si}(\text{Bu-}t)(\text{Mes}) \xrightarrow[50\ °C]{ROH\ (R=Me,\ Et)} \text{Mes}(t\text{-Bu})(RO)\text{Si-Si}(H)(\text{Bu-}t)(\text{Mes}) + \text{Mes}(t\text{-Bu})(RO)\text{Si-Si}(\text{Mes})(H)(\text{Bu-}t) \quad (9)$$

50 : 50

As stated before, there is little knowledge on the unsymmetrically substituted stable disilenes because stable disilenes are usually prepared by dimerization of silylenes, thus leading to symmetrical disilenes. Unsymmetrically substituted disilenes are produced mostly as transient species (see the preceding section), and it was found that (E)- and (Z)-1,2-dimethyl-1,2-diphenyldisilenes undergo the addition reaction with alcohols very rapidly ($k_2 = 10^7 - 10^8$ M^{-1} s^{-1}). The rates are only 1 to 2 orders of magnitude smaller than the diffusion rates and this guarantees that these disilenes can react with alcohols prior to rotation around the Si–Si bond or prior to $E-Z$ isomerization[28]. This makes it possible to investigate the diastereoselectivity in the addition reaction of alcohols to these disilenes.

Irradiation ($\lambda > 280$ nm) of a hexane/isopropyl alcohol solution of the disilene precursor (E)-**14** (Scheme 8) in a quartz tube at room temperature produced *threo*-1-isopropoxy-1,2-dimethyl-1,2-diphenyldisilane **37a**, which is the product of *syn* addition of isopropyl alcohol to (E)-**16** together with a small amount of the *erythro*-isomer **38a** (**37a/38a** => 99/1, 62% yield). High *syn*-addition diastereoselectivity was also found for (Z)-**16** (**38a**/**37a** => 99/1, 49% yield). The diastereoselectivity in the addition of isopropyl alcohol depends on the concentration of the alcohol (Table 5). At lower concentrations a higher selectivity is indicated[28].

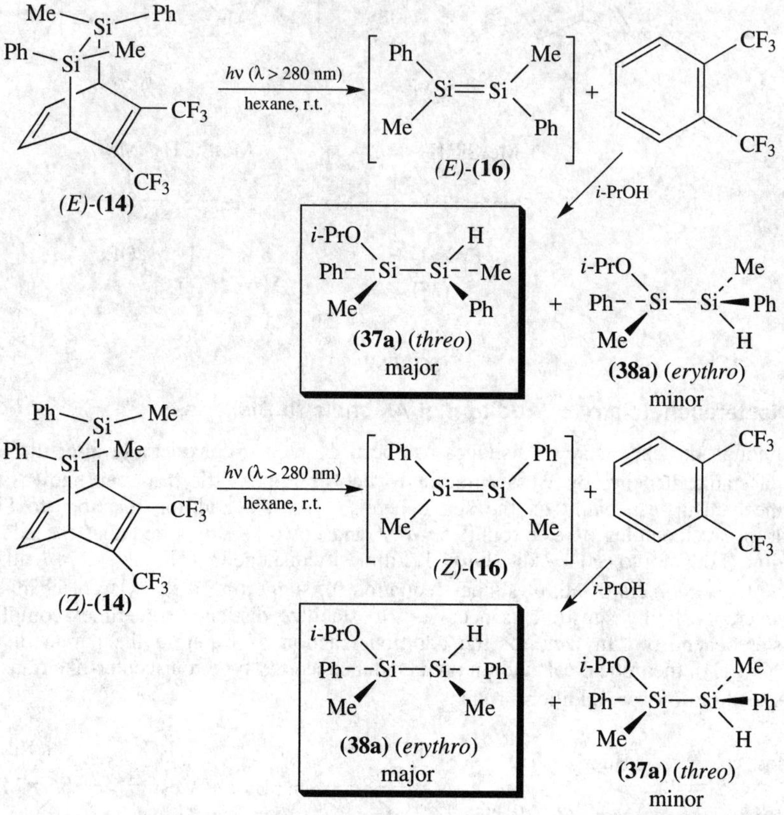

SCHEME 8

TABLE 5. Product ratio in the photolysis of (E)- and (Z)-**16** in the presence of alcohols[28]

ROH	Alcohol concentration (M)	Product ratio, **37/38**	
		(E)-**16**	(Z)-**16**
EtOH	0.85	92/8	8/92
EtOH	1.26	82/18	25/75
EtOH	1.69	78/22	27/73
EtOH	2.48	70/30	44/56
EtOH	5.65	51/49	49/51
i-PrOH	1.31	>99/<1	<1/>99
i-PrOH	4.33	89/11	9/91
t-BuOH	3.51	94/6	5/95

15. Mechanism and structures in alcohol addition reactions

Similar high diastereoselectivities are observed for the addition reaction of *t*-butyl alcohol to (*E*)- and (*Z*)-**16**[28]. High diastereoselectivities are also observed for the reaction with ethanol at low concentration, but the selectivity decreases rather sharply by increasing the ethanol concentration. The diastereoselectivities for the reaction of ethanol, i.e. the [*syn*]/[*anti*] product ratios, are linearly and inversely correlated with the concentration of the alcohol:

$$[syn]/[anti] = (k_a/k_b)/[\text{EtOH}]$$

Where k_a and k_b are the rate constants for intramolecular and intermolecular proton transfer reactions, respectively.

The high diastereoselectivity in the addition of *i*-PrOH, *t*-BuOH and EtOH (at low concentration) suggests that $E \rightleftarrows Z$ photoisomerization of (*E*)- or (*Z*)-**16** does not occur in solution at room temperature or that the trapping of (*E*)- or (*Z*)-**16** by alcohols proceeds faster than the $E \rightleftarrows Z$ isomerization. In addition, the results show that proton transfer in the intermediate adduct formed by the disilenes and alcohols occurs much faster than rotation around the Si—Si bond. However, in the reaction with ethanol, an appreciable amount of the *anti* addition product was formed. Thus, the diastereoselectivity remarkably depended on the concentration of ethanol.

A mechanism involving coordination of the alcoholic oxygen to a silicon atom, followed by a fast intramolecular proton transfer explains the predominant *syn* addition of alcohol to disilene. However, intermolecular proton transfer leading to the *anti*-product competes with intramolecular proton transfer at high alcohol concentrations. The proposed mechanism is presented in Scheme 9.

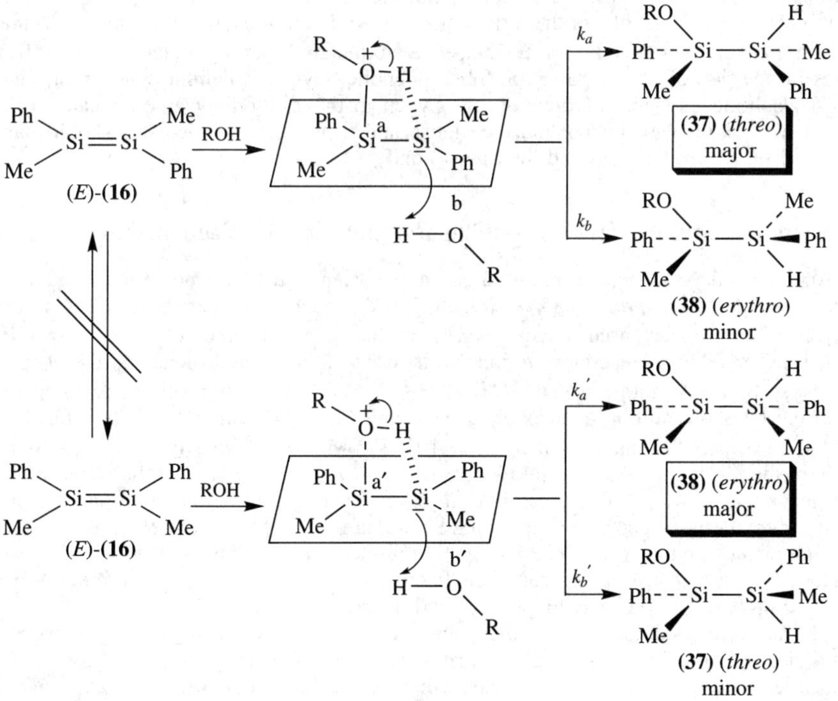

SCHEME 9

The difference in the selectivities as a function of the alcohol used is explained in terms of the differences in the acidities of the silyl alkoxy oxonium cations (protonated alkoxysilanes) formed as intermediate[28]. The rate of intramolecular proton transfer in the intermediate adduct is expected to increase by increasing the acidity of protonated alkoxysilanes. No data are available for acidities of protonated alkoxysilanes, but the pK_a values of protonated alcohols may be used instead. These are in the following order: EtO^+H_2 (−2.4) >> $i\text{-PrO}^+H_2$ (−3.2) > $t\text{-BuO}^+H_2$ (−3.8)[49]. Addition of t-BuOH gives rise to the most acidic intermediate (among those examined) and therefore gives the highest diastereoselectivity. Ethanol forms the least acidic intermediate, leading to *anti*-addition product competing with the intramolecular *syn* adduct, especially at high concentrations of the alcohol.

Recently, West and coworkers have reported similar diatereoselectivities in the reaction of stable disilenes, (*E*)-**5** and (*E*)- and (*Z*)-1,2-di-*t*-butyl-1,2-bis(2,4,6-triisopropyl phenyl) disilene, with ROH (R = Me, Et and *i*-Pr)[50]. In these studies, alkoxy oxygen-coordinated complexes have been suggested as key intermediates (Scheme 9). Computational studies on the reaction of $RSiH=SiH_2$ with water also suggest the importance of water coordinated disilene molecules as intermediate[51]. However, it should be noted that in the X-ray crystallographic structure of $Mes_2Si=SiMes_2 \cdot THF$ complex, the THF molecule is not coordinated to silicon[52].

More recently, Apeloig and Nakash have studied diastereoselectivity in the reaction of (*E*)-**5** with *p*-methoxyphenol[53]. In both benzene and THF, the stereochemistry of the products was independent of the phenol concentration. The *syn/anti* ratios of the addition products were 90 : 10 in benzene and 20 : 80 in THF. They have suggested that intramolecular proton transfer after rotation of the Si−Si bond of the phenol-coordinated intermediate is responsible for the formation of the *anti*-addition rather than intermolecular proton transfer. This must be a special case due to much slower (by a factor of 10^9-10^{12}) rates of addition of phenol to (*E*)-**5**. Since phenolic oxygen is definitely less basic than alkyl alcoholic oxygen, coordination of oxygen in the zwitterionic intermediate in the reaction of (*E*)-**5** with phenol must be loose and hence the intermediates should have much chance of rotation around the Si−Si bond.

F. Competition Between Nucleophilic and Electrophilic Mechanisms

Apeloig and Nakash have reported recently a Hammett-type study for the addition reactions of seven *para*- and *meta*-substituted phenols to tetramesityldisilene **39** (equation 10)[54]. They used a large excess of the phenol to enforce pseudo-first-order kinetics. The addition reactions are indeed firstorder in both the disilene and the phenol.

The second-order rate constants ($k = 10^{-4}-10^{-2}$ M^{-1} s^{-1}) are much lower compared with the rates of addition of alcohols to (*E*)- and (*Z*)-1,2-dimethyl-1,2-diphenyldisilene and 1,2,2-trimethyl-1-phenyldisilene ($k = 10^7-10^8$ M^{-1} s^{-1} (Table 1). The larger steric bulk of the mesityl substituents in **39** may account for this large reactivity difference, but the magnitude of the difference is indeed extremely large. The resulting Hammett plot (see Figure 1) has a concave shape with a minimum for the parent phenol. This means that all phenols, either with electron-donating or electron-withdrawing substituents, react faster than the parent unsubstituted phenol. The concave Hammett plot is formed by two intersecting straight lines having positive and negative ρ values.

The negative ρ value (−1.77) in the reaction of the tetramesityldisilene **39** with phenols with electron-donating substituents indicates that a positive charge is developing on the phenolic oxygen in the transition state, whereas the positive ρ value (1.72) observed

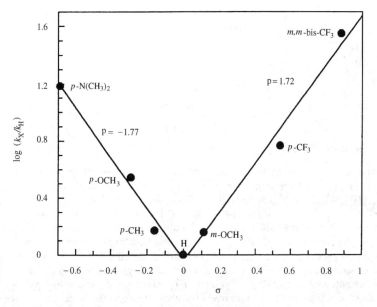

FIGURE 1. A Hammett plot for the addition of ArOH to $Mes_2Si=SiMes_2$

with electron-withdrawing substituents means that a negative charge develops on the phenolic oxygen in the transition state of the rate-determining step. Thus these opposite Hammett slopes indicate a change in mechanism of the reaction of **39** with phenols, from a rate-determining nucleophilic step for electron-rich phenols, to a rate-determining electrophilic step for electron-poor phenols (Scheme 10)[54].

Kinetic isotope effect measurements support this interpretation. A small kinetic isotope effect ($k_H/k_D = 0.71$) is observed for *p*-methoxyphenol in agreement with a rate-determining nucleophilic attack, while a large kinetic isotope effect ($k_H/k_D = 5.27$), observed for *p*-trifluoromethylphenol, strongly supports a mechanism in which a phenolic H (or D) is transferred to **39** in the rate-determining step. Unfortunately, **39** is a symmetric disilene so that diastereoselectivity could not be determined. It will be interesting to examine whether the diastereoselectivity will be effected by the change in the addition mechanism from electrophilic to nucleophilic.

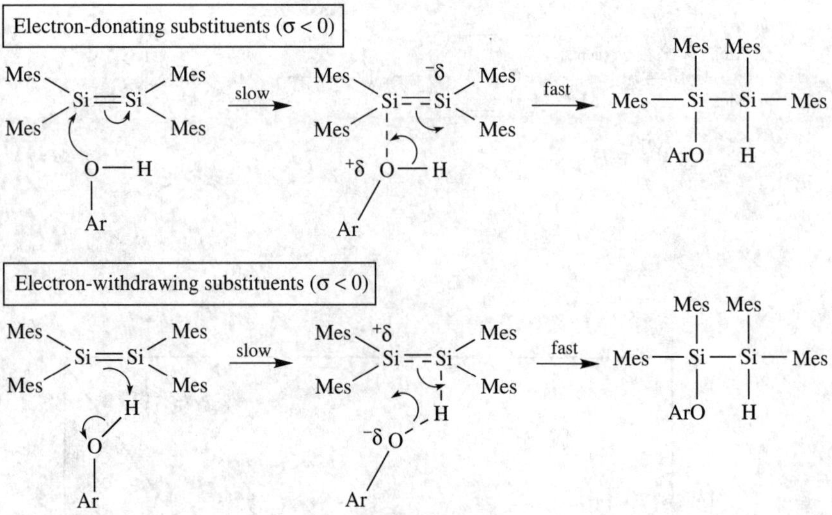

SCHEME 10

III. SILENES

A. Generation of Silenes

After the first recognition of generating silenes in vapor phase by thermolysis of silacyclobutane[55], a large number of reports appeared on the reactions of both reactive and stable silenes[6]. Sterically protected silenes are prepared by [1,3]-sigmatropic shift of a silyl group from silicon to the oxygen of the acylsilanes. Brook has isolated the first kinetically stabilized silenes by the photolysis of certain acylsilanes which lead to a [1,3]-sigmatropic shift[56]. Thermal 1,2-elimination of lithium fluoride from α-lithiated fluorosilanes also provides a variety of sterically hindered silenes[57–62]. Reactive silenes can be generated by 1,2-shift, 1,3-shift, electrocyclic ring-opening, [2+2] cycloreversion, [2+4] cycloreversion and disproportionation of silyl radicals. Raabe and Michl summarized these reactions in their extensive reviews[7,8]. Generation of geometric isomers of stable or of reactive silenes is difficult and therefore mechanistic studies on the reaction of silenes are still quite limited.

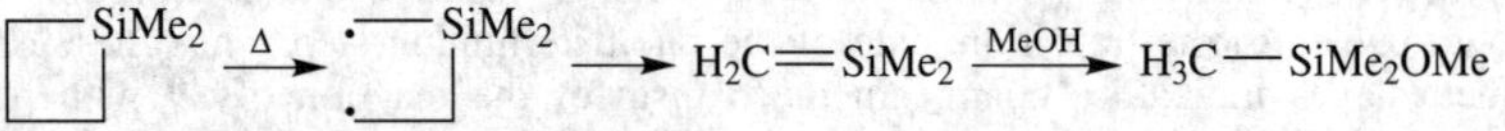

B. Nucleophilic Addition to Silenes

Unlike disilenes, silenes are dipolar species with the silicon and the carbon atoms charged positively and negatively, respectively. Hence, the silicon centers of silenes are strong Lewis acids which can form donor–acceptor complexes with Lewis bases. 1,1-Dimethyl-2,2-bis(trimethylsilyl)silene, for example, forms an adduct with trimethylamine (equation 12)[62]. Other Lewis bases can form adducts with silenes and it is possible to replace a weaker donor by a stronger donor. The order of donor strength thus determined is: $F^- >$ NMe$_3$ $>$ NEt$_3$ $>$ Br $>$ THF[62].

Silenes also react efficiently with alcohols to give addition products. Indeed, addition is the most characteristic reaction of silenes and has been used for trapping silenes. Alcohols react regiospecifically to form alkoxysilanes.

$$\text{Me}_2\text{Si}=\text{C}(\text{SiMe}_3)_2 + \text{NMe}_3 \rightleftharpoons \text{Me}_2\text{Si}(\text{NMe}_3)-\text{C}(\text{SiMe}_3)_2 \quad (12)$$

This is generally true, but it should be noted that silicon–carbon double bonds with inverse electron demand behave differently. Silatriafulvene, for example, did not react with alcohol but underwent 1,2-silyl migration followed by ring expansion reaction to silacyclobutadiene[63]. Silacalicene also adds alcohol in a reverse fashion[64].

C. Diastereochemistry of Addition of Alcohol to Silene

In the first stereochemical study, Brook has observed nonstereospecific addition of methanol to certain isolable silenes[65,66]. Although the precise stereochemistry of the products was not established, a 1/3 mixture of *syn/anti* or *anti/syn* isomers was obtained (equation 13). The results indicate a nonconcerted process for the addition of alcohol to silenes. Wiberg has proposed a two-step mechanism involving an initial formation of a silene–alcohol complex, in accord with the formation of nucleophilic adducts, followed by proton migration from the alcohol to the carbon of the silene (equation 14)[59,61]. This mechanism may be compatible with the results obtained by Brook and coworkers, if rotation around the silene's Si—C bond occurs faster than the proton migration.

$$(\text{Me}_3\text{Si})_2\text{Si}(t\text{-Bu})\text{-Ad-1}(\text{C=O}) \xrightarrow{h\nu\ (>360\ \text{nm})} \underset{\text{Me}_3\text{Si}}{\overset{t\text{-Bu}}{>}}\text{Si}=\text{C}\underset{\text{Ad-1}}{\overset{\text{OSiMe}_3}{<}} \xrightarrow{\text{MeOH}}$$

$$\underset{\text{Me}_3\text{Si}}{\overset{t\text{-Bu}}{>}}\text{Si}-\text{C}\underset{\text{H}}{\overset{\text{OSiMe}_3,\text{Ad-1}}{<}}\ (\textit{anti}) + \underset{\text{Me}_3\text{Si}}{\overset{t\text{-Bu}}{>}}\text{Si}-\text{C}\underset{\text{Ad-1}}{\overset{\text{MeO},\text{H}}{<}}\text{OSiMe}_3\ (\textit{syn}) \quad (13)$$

syn/anti or *anti/syn* = 1/3

$$\underset{\text{Me}}{\overset{\text{Me}}{>}}\text{Si}=\text{C}\underset{\text{SiMe}_3}{\overset{\text{SiMe}_3}{<}} \cdots \text{O}-\text{H}-\text{R} \longrightarrow \underset{\text{Me}}{\overset{\text{Me}}{>}}\overset{-\delta}{\text{Si}}=\text{C}\underset{\text{SiMe}_3}{\overset{\text{SiMe}_3}{<}}\cdots\overset{+\delta}{\text{O}}-\text{H}-\text{R} \longrightarrow \text{Me}-\underset{\text{RO}}{\overset{\text{Me}}{\text{Si}}}-\underset{\text{H}}{\overset{\text{SiMe}_3}{\text{C}}}-\text{SiMe}_3 \quad (14)$$

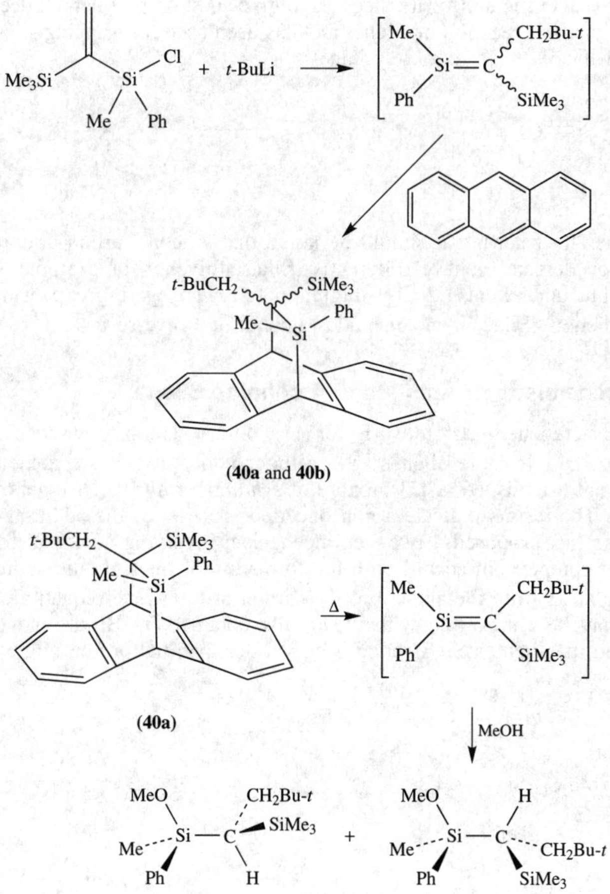

(40a and 40b)

(40a)

(41b) anti

(41a) syn

syn:anti = 100:0

SCHEME 11

Jones and coworkers developed a new method of generating silenes based on the addition–elimination reaction. Addition of t-BuLi to an appropriately substituted chloro(vinyl)silane produces a neopentyl-substituted silene[67,68]. Among many reactions, it has been shown that the transient silene adds to anthracene to afford stereoisomers, **40a** and **40b**, as isolable compounds (Scheme 11). Fractional crystallization of the adduct **40** from hexane gave pure **40a**, leaving a 69/31 mixture of **40a** and **40b**.

The adducts can produce silenes by thermolysis at 190 °C. Thermolysis of **40a** in the presence of a 9-fold excess of methanol gave only **41a**. Similar thermolysis by using a 69/31 mixture of **40a** and **40b** afforded a 69/31 mixture of **41a** and **41b**. Thus methanol adds to the silene stereospecifically[69]. A similar stereospecific addition of alkoxysilane to silene was reported by the same group[70]. The stereospecific *syn* addition can be explained

15. Mechanism and structures in alcohol addition reactions 847

to result from a concerted process involving a [4 + 2] ($\sigma + n + \pi$) 6-electron[71], but Jones and coworkers have concluded that the reaction must be a stepwise process similar to that proposed by Wiberg, where the intermediate adduct must have a substantial Si=C double-bond character. This is expected from the fact that the silene–THF adduct has a relatively short Si–C bond length as determined by X-ray crystallographic analysis[59].

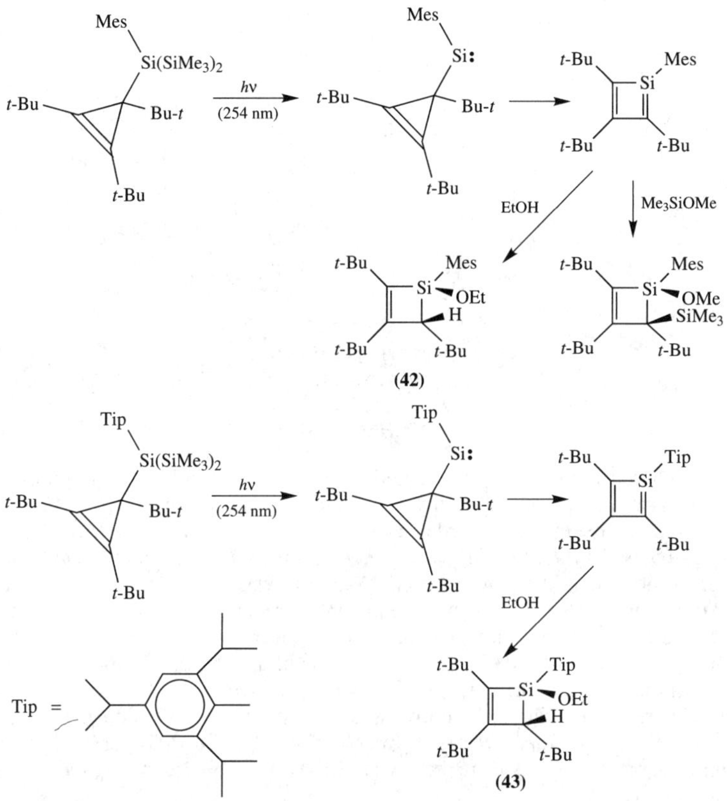

SCHEME 12

Stereospecific *syn* addition of methoxysilane to silacyclobutadiene was reported by Fink and coworkers (Scheme 12)[72]. Ethanol also adds the silacyclobutadiene stereospecifically in a *syn* fashion to give **42** (Scheme 12), but the ethanol adduct undergoes photoisomerization to a photostationary mixture of **42** and its stereoisomer. A similar system, but with the more bulky 2,4,6-triisopropylphenyl (Tip) groups, also gave the *syn* ethanol adduct **43**[73]. The rather complicated photochemical–thermal isomerization process was discussed by Fink[74].

Sakurai and coworkers[75] generated the five-membered silene **44** by a photochemical 1,3-silyl shift in the cyclic divinyldisilane **45** (Scheme 13). Since the silene **44** is constrained to be planar, no bond rotation is possible during the reaction. Contrary to the previous observations i.e. a simple two-step or a concerted four-centered mechanism, alcohols add to **44** nonstereospecifically, although in the cyclic silene bond rotation is prohibited.

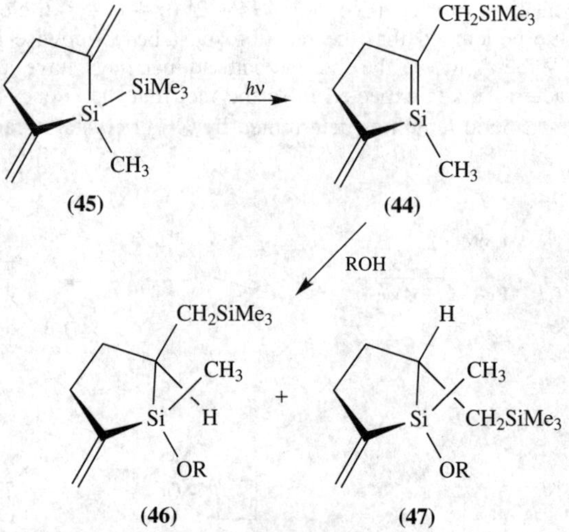

SCHEME 13

(a) R = CH₃; (b) R = CH₂CH₂CH₃;
(c) R = CH(CH₃)₂; (d) R = C(CH₃)₃

The stereochemical outcome for the addition of various alcohols to **44** depends markedly on the concentration and on the acidity of the alcohols used. The *syn/anti* **46/47** product ratio increased in the following order as a function of the alcohol: MeOH (27/73) < n-PrOH (35/65) < i-PrOH (46/54) << t-BuOH (100/0). t-Butyl alcohol gave only a *syn* adduct. Since no interconversion between **46** and **47** was observed under irradiation, the stereochemical outcome must reflect the inherent stereochemistry of the reaction.

Further interesting findings are the dependence of the **46/47** product ratio on the concentration of alcohol. Plots of **46/47** versus the inverse of alcohol concentration gave straight lines at the initial stage of the reaction. This was observed when **45** was irradiated in the presence of various amounts of alcohols in acetonitrile. The slope depended remarkably on the alcohol, i.e. the slopes were 4.6 for MeOH, 9.2 for n-PrOH and 32 for i-PrOH. An infinitely large slope was estimated for t-BuOH, because t-BuOH gave only the *syn* compound.

Based on these results, Sakurai and coworkers proposed for the addition of alcohols to the silene **44** the mechanism shown in Scheme 14. In the first stage the silene forms an alcohol–silene complex **48** as suggested by Wiberg[61], and this stage is followed by an intramolecular proton migration in **48** (the first-order rate constant, k_1) which competes with the intermolecular proton transfer from an additional external alcohol (the second-order rate constant, k_2). These two processes give the *syn* and *anti* addition products, respectively.

This mechanism is fully compatible with the observed linear relationship between the **46/47** product ratio and the reciprocal concentration of the alcohol, since the initial product ratio should be represented by the following equation:

$$d[46]/d[47] = (k_1/k_2)/[ROH] \qquad (15)$$

The relative rate constants (k_1/k_2) reflect the relative rates for intra- and intermolecular proton transfer, respectively. As discussed for disilene, the Brønsted catalysis law must be applied to determine the relative rates, where k_2 and k_1 are expected to increase with the

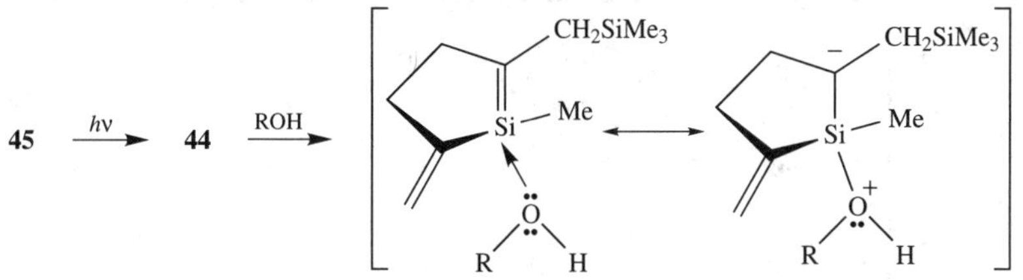

15. Mechanism and structures in alcohol addition reactions

SCHEME 14

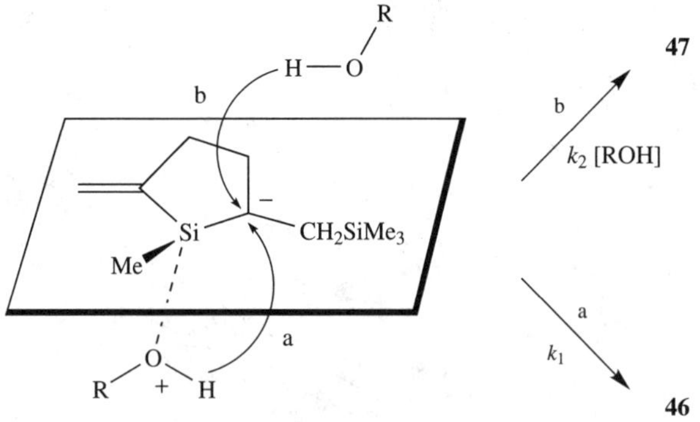

a, Intramolecular H$^+$ transfer
b, Intermolecular H$^+$ transfer

increasing acidity of ROH and of the protonated alcohol, respectively. The pK_a values of alcohols increase in the following order (in DMSO)[76]: MeOH (29.0) < n-PrOH (29.8)[77] < i-PrOH (30.25) < t-BuOH (32.2). The inverse order is known for the pK_a values of the corresponding protonated alcohols, RO$^+$H$_2$: t-BuO$^+$H$_2$ > i-PrO$^+$H$_2$ > n-PrO$^+$H$_2$ > MeO$^+$H$_2$[76]. The less acidic the alcohol is, the more acidic the corresponding protonated alcohol. Thus k_1/k_2 is expected to increase in the following order: MeOH < n-PrOH < i-PrOH < t-BuOH, as observed[75].

The solvent effect on the diastereoselectivity of the reaction is noteworthy[79]. Figure 2 shows the relationship between the **46/47** product ratio and the reciprocal concentration of methanol, in various solvents. The k_1/k_2 values are: CH$_3$CN (4.8), 9 : 1 hexane/Et$_2$O (5.9) and C$_6$H$_6$ (11). The largest value was obtained in benzene, the most nonpolar solvent among those examined. This implies that the relative importance of intramolecular proton transfer (Scheme 14, process a) increases relative to the intermolecular bimolecular process (Scheme 14, process b) as the solvent polarity decreases. Since process a is a reaction starting from a dipolar reactant to give a nonpolar product, polar solvents should retard the rate. Interestingly, the **46/47** product ratio does not follow a linear relationship but shows a concave curve in ether. Ether can coordinate to the silene to give **49** (Scheme 15). The lower the concentration of methanol, the more important the role of the ether–silene complex becomes. Then intermolecular proton transfer gives adduct **50**. The complex **50** undergoes S$_N$2-like substitution by MeO$^-$ or MeOH to give the *syn* adduct **46**.

By using nanosecond laser flash photolysis of **51**, Leigh and coworkers have studied alcohol addition reactions of the 1,3,5-(1-sila)hexatriene derivative **52** (Scheme 16)[80,81] and of 1,1-diphenylsilene[82–84].

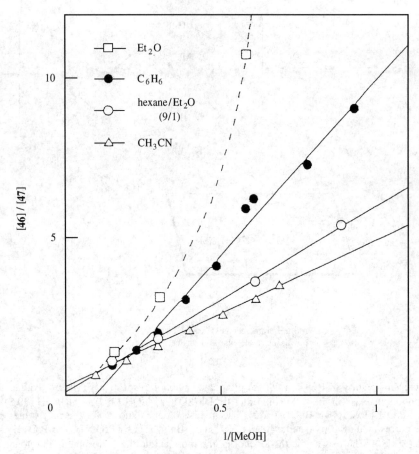

FIGURE 2. Relationship between the **[46]/[47]** ratio and 1/[MeOH] in various solvents

Although the products do not allow one to distinguish between intra- and intermolecular processes, they found that the plots of k_{decay} versus quencher (alcohol) concentration are nonlinear. They have analyzed the data according to the quadratic expression shown in equation 16, where $k_q^{(2)}$ is the third-order rate constant corresponding to transient quenching by two molecules of alcohol.

$$k_{decay} = k_o + k_q[Q] + k_q^{(2)}[Q]^2 \tag{16}$$

From these spectroscopic and kinetic analyses, the authors proposed a mechanism of alcohol addition to **52** which is similar to one proposed earlier[75].

Rate constants for the reaction of silatrienes **52a–c** with alcohols are listed in Table 6. According to the proposed mechanism, k_q and $k_q^{(2)}$ correspond to k_1 and k_2 of equation 15, respectively. It is interesting to compare the very low value of $k_q/k_q^{(2)}$ (0.015–0.047) in the reaction of **52a–52c** with methanol with the corresponding k_1/k_2 value of 4.6 for the silene **44**. The latter is about two orders of magnitude greater than the former. Apparently, silatrienes derived from aryldisilanes behave very differently from **44**.

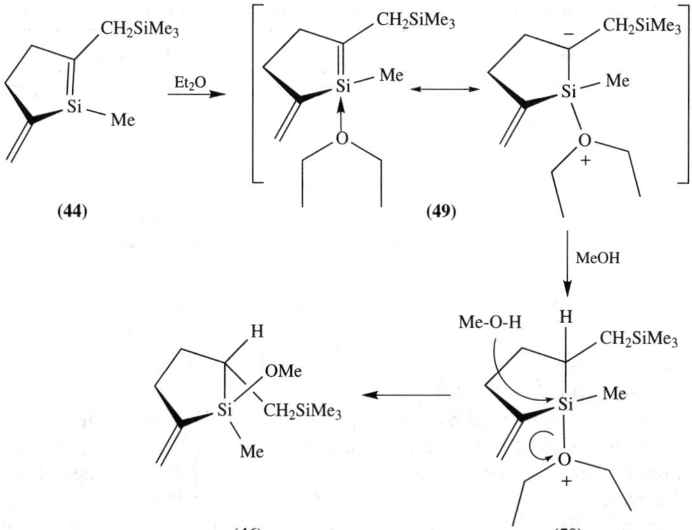

15. Mechanism and structures in alcohol addition reactions

SCHEME 15

SCHEME 16

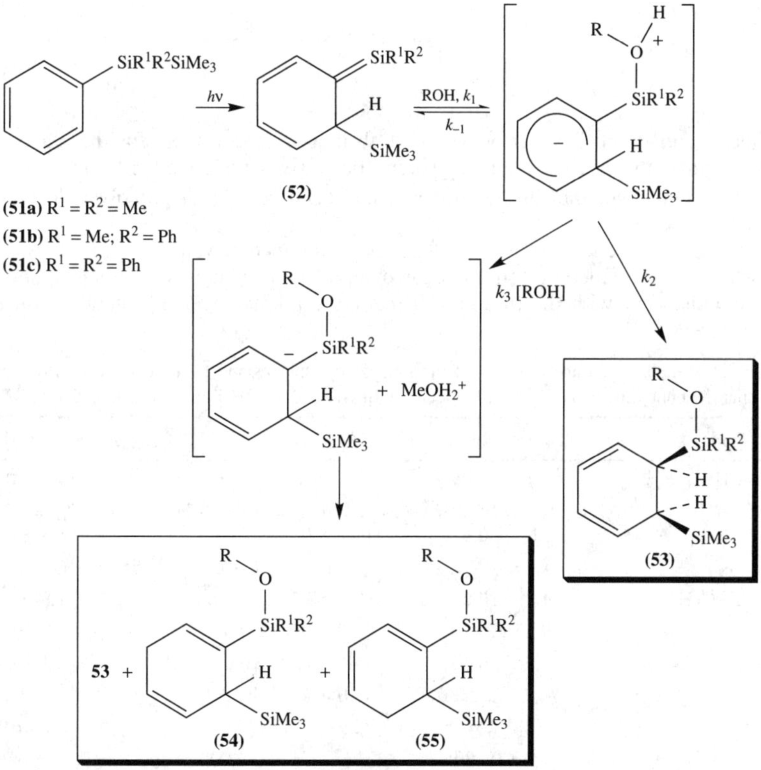

(51a) $R^1 = R^2 = Me$
(51b) $R^1 = Me$; $R^2 = Ph$
(51c) $R^1 = R^2 = Ph$

TABLE 6. Rate constants for reaction of silatrienes **51a, b, c** with alcohols in acetonitrile[84]

Reagent	Rate constant	52a	52b	52c
MeOH	$10^{-8}k_q$	2.3 ± 0.7	0.60 ± 0.25	0.17 ± 0.10
	$10^{-8}k_q{}^{[2]}$	49 ± 10	28 ± 5	11 ± 4
MeOD	$10^{-8}k_q$	1.2 ± 0.2	0.34 ± 0.08	0.09 ± 0.04
	$10^{-8}k_q{}^{[2]}$	51 ± 4	23 ± 2	7.9 ± 0.7
CF_3CH_2OH	$10^{-8}k_q$	0.236 ± 0.004	0.056 ± 0.001	0.017 ± 0.001

In the photolysis of pentamethyldisilane **51a** three methanol addition products are obtained. Compound **53a**, which has not been observed by previous workers[85], is attributed to be the product of intramolecular proton transfer. Photochemical reactions of aryldisilanes have been recently reviewed[86].

1,1-Diphenylsilene, generated by similar photolysis of silacyclobutane[82–84], also reacts very rapidly with water, alcohols and acetic acid (equation 17). Rate constants are only one order of magnitude slower than the diffusion-controlled limit and depend only slightly on the nucleophilicity or acidity of the quencher. Although it is not easy to distinguish kinetically unimolecular and bimolecular processes, a similar mechanism to that shown in Scheme 16 was suggested also for the addition of alcohols to 1,1-diphenylsilene.

$$\text{Ph}-\underset{\text{Ph}}{\overset{\text{Ph}}{\text{Si}}}\underset{}{\overset{R}{\diagup}} \xrightarrow[-RCH=CH_2]{h\nu} Ph_2Si=CH_2 \xrightarrow{R'OH} Ph_2Si-CH_2 \quad (17)$$
$$\qquad\qquad\qquad\qquad\qquad\qquad\qquad\qquad\qquad\quad | \\ \qquad\qquad\qquad\qquad\qquad\qquad\qquad\qquad\qquad OR'$$

Bradaric and Leigh have also measured absolute rate constants for the reaction of a series of ring-substituted 1,1-diphenylsilene derivatives with methanol, *t*-butyl alcohol and acetic acid in acetonitrile by similar nanosecond laser flash photolysis techniques[87] (Table 7).

All the three reactions show small positive Hammett ρ-values at 23 °C. According to Scheme 17, the reaction constants for complexation of alcohols to silene should be positive, consistent with the mechanism involving initial, reversible nucleophilic attack

TABLE 7. Bimolecular rate constants, deuterium isotope effects and Hammett ρ-values for reactions of 1,1-diarylsilenes (**56**; R′ = H)[87] with MeOH, *t*-BuOH and AcOH in acetonitrile at 23 °C

58 (R^1 = H)	$k_{MeOH}/(10^9\ M^{-1}\ s^{-1})$	$k_{t-BuOH}/(10^9 M^{-1}\ s^{-1})$	$k_{AcOH}/(10^9 M^{-1}\ s^{-1})$
Ar = 4-CH$_3$C$_6$H$_4$	1.12 ± 0.06	0.130 ± 0.006	1.41 ± 0.05
	($k_H/k_D = 1.9 \pm 0.1$)	($k_H/k_D = 1.9 \pm 0.2$)	($k_H/k_D = 1.2 \pm 0.2$)
Ar = C$_6$H$_5$	1.5 ± 0.1	0.22 ± 0.02	1.5 ± 0.2
	($k_H/k_D = 1.5 \pm 0.1$)	($k_H/k_D = 1.6 \pm 0.1$)	($k_H/k_D = 1.1 \pm 0.1$)
Ar = 4-F–C$_6$H$_4$	1.89 ± 0.08	0.33 ± 0.022	1.8 ± 0.2
Ar = 4-Cl–C$_6$H$_4$	2.13 ± 0.10	0.39 ± 0.02	2.0 ± 0.2
Ar = 4-CF$_3$–C$_6$H$_4$	2.99 ± 0.16	0.75 ± 0.04	2.3 ± 0.4
	($k_H/k_D = 1.0 \pm 0.1$)	($k_H/k_D = 1.7 \pm 0.1$)	($k_H/k_D = 1.1 \pm 0.1$)
ρ	$+0.13 \pm 0.05$	$+0.55 \pm 0.084$	$+0.17 \pm 0.02$
	($r^2 = 0.980$)	($r^2 = 0.985$)	($r^2 = 0.987$)

at silicon to form a σ-complex. Proton transfer within the complex is the rate-limiting process since relatively large kinetic isotope effects are observed. However, the polar nature of the proton transfer should be small and overall polar character of the reaction is governed by the initial complexation steps. They also found that proton transfer within the complex was entropy-controlled, resulting in negative activation energies for the reaction. However, the reaction of $Ph_2Si=CH_2$ ($E_a = +1.9 \pm 0.3$ kcal mol^{-1}) and more reactive (p-$CF_3C_6H_4)_2Si=CH_2$ ($E_a = +3.6 \pm 0.5$ kcal mol^{-1}) with acetic acid gave positive activation energies. They have suggested that acetic acid adds by a stepwise mechanism but with formation of the complex being rate-determining.

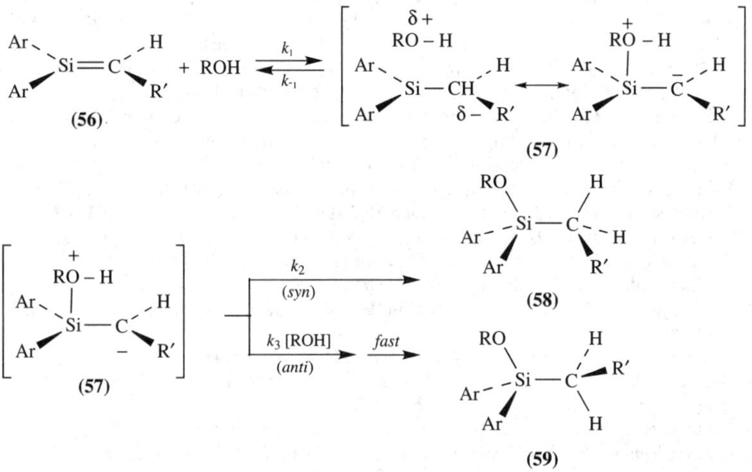

SCHEME 17

IV. CONCLUDING REMARKS

It is evident from the above discussion that much progress has been made in the past decade in deepening our understanding of the chemistry of silicon–silicon and silicon–carbon double bonds. However, the field is still in its infancy and many additional studies should be carried out in order to broaden our knowledge. Proper design of substrates combined with transient spectroscopy is needed. Theoretical studies will also assist the better understanding of the chemistry of multiple bonds to silicon. In this respect, the author did not discuss the theoretical aspects of silicon-containing double bonds, because an excellent review on this topic is available elsewhere[88].

V. ACKNOWLEDGMENTS

The author wishes to thank his colleagues for their indispensable contributions to the studies on the chemistry of silicon–silicon and silicon–carbon double bonds. Their names are cited in the references. The author also wishes to thank the Ministry of Education, Science, and Culture (Specially Promoted Research 02102004) and the Japan Society for the promotion of Sciences (RFTF95P00303) for financial support.

VI. REFERENCES

1. P. D. Gaspar, in *Reactive Intermediates* (Eds. M. Jones, Jr. and R. A. Moss), Vol. 1, Chap. 7 (1978); Vol. 2, Chap. 9 (1981); Vol. 3, Chap. 9 (1985), Wiley, New York.

2. (a) H. Sakurai, in *Free Radicals* (Ed. J. K. Kochi), Chap. 24, Wiley-Interscience, New York, 1975.
   (b) C. Chatgilialoglu, *Chem. Rev.*, **95**, 1229 (1995).
3. I. Fleming, in *Comprehensive Organic Chemistry* (Eds. D. Barton and W. D. Ollis), Vol. 3. (Ed. D. Neville Jones), Pergamon Press, Oxford, 1979.
4. K. Tamao and A. Kawachi, *Adv. Organomet. Chem.*, **38**, 1 (1995).
5. See chapters by P. v. R. Schleyer (Chapter 10) and P. D. Lickiss (Chapter 11) in this volume.
6. A. G. Brook and K. M. Bains, *Adv. Organomet. Chem.*, **25**, 1 (1985).
7. G. Raabe and J. Michl, in *The Chemistry of Organic Silicon Compounds*, Vol. 1, Part 2 (Eds. S. Patai and Z. Rappoport), Chap. 17, Wiley, New York, 1989.
8. G. Raabe and J. Michl, *Chem. Rev.*, **85**, 419 (1985).
9. R. West, *Pure Appl. Chem.*, **56**, 163 (1984).
10. R. West, *Angew. Chem., Int. Ed. Engl.*, **26**, 1201 (1987).
11. T. Tsumuraya, S. A. Batcheller and S. Masamune, *Angew. Chem., Int. Ed. Engl.*, **30**, 902 (1991).
12. R. Okazaki and R. West, *Adv. Organomet. Chem.*, **39**, 231 (1996).
13. A. R. Bassindale and P. G. Taylor, in *The Chemistry of Organic Silicon Compounds*, Vol. 1 Part 2 (Eds. S. Patai and Z. Rappoport), Chap. 17, Wiley, New York, 1989.
14. R. J. P. Corriu, C. Guerin and J. J. E. Moreau, in *The Chemistry of Organic Silicon Compounds*, Vol. 1 Part 2 (Eds. S. Patai and Z. Rappoport), Chap. 4, Wiley, New York, 1989.
15. D. N. Roark and G. J. Peddle, *J. Am. Chem. Soc.*, **94**, 5837 (1972).
16. H. Sakurai, Y. Nakadaira and T. Kobayashi, *J. Am. Chem. Soc.*, **101**, 487 (1979).
17. Y. Nakadaira, T. Kobayashi, T. Otsuka and H. Sakurai, *J. Am. Chem. Soc.*, **101**, 486 (1979).
18. G. Olbrich, P. Potzinger, B. Reinmann and R. Walsh, *Organometallics*, **3**, 1267 (1984).
19. M. J. Michalczyk, R. West and J. Michl, *Organometallics*, **4**, 826 (1985).
20. B. D. Shepherd, D. R. Powell and R. West, *Organometallics*, **8**, 2664 (1989).
21. S. A. Batcheller, T. Tsumuraya, O. Tempkin, W. Dacis and S. Masamune, *J. Am. Chem. Soc.*, **112**, 9394 (1990).
22. H. Jacobsen and T. Ziegler, *J. Am. Chem. Soc.*, **116**, 3667 (1994).
23. M. W. Schmidt, P. N. Truong and M. S. Gordon, *J. Am. Chem. Soc.*, **109**, 5217 (1987).
24. W. Kutzelnigg, *Angew. Chem., Int. Ed. Engl.*, **23**, 272 (1984).
25. H. Sakurai, K. Sakamoto, Y. Funada and M. Yoshidain, in *Inorganic and Organometallic Polymers II, Advanced Materials and Intermediates*, Chap. 2, ACS Symposium Series 572 (Eds. P. Wisian-Neilson, H. R. Allcock and K. J. Wynne), 1994.
26. K. Sakamoto, K. Obata, H. Hirata, M. Nakajima and H. Sakurai, *J. Am. Chem. Soc.*, **111**, 7641 (1989).
27. A. Sekiguchi, I. Maruki, K. Ebata, C. Kabuto and H. Sakurai, *J. Chem. Soc., Chem. Commun.*, 341 (1991).
28. A. Sekiguchi, I. Maruki and H. Sakurai, *J. Am. Chem. Soc.*, **115**, 11460 (1993).
29. H. Sakurai, K. Sakamoto and M. Yoshida, submitted for publication.
30. T. Hoshi, T. Shimada, T. Sanji and H. Sakurai, *Chem. Lett.*, submitted.
31. H. Suzuki, *Bull. Chem. Soc. Jpn.*, **33**, 619 (1960).
32. G. R. Gillette, G. Noren and R. West, *Organometallics*, **9**, 2925 (1990).
33. R. S. Archibald, Y. v. d. Winkel, A. J. Millevolte, J. M. Desper and R. West, *Organometallics*, **11**, 3276 (1992).
34. P. Jutzi, U. Holtmann, H. Bögge and A. Müller, *J. Chem. Soc., Chem. Commun.*, 305 (1988).
35. J. Maxka and Y. Apeloig, *J. Chem. Soc., Chem. Commun.*, 737 (1990).
36. K. Sakamoto, S. Tsutsui, H. Sakurai and M. Kira, *Bull. Chem. Soc. Jpn.*, **70**, 253 (1997).
37. B. T. Luke, J. A. Pople, M.-B. Krogh-Jesperson, Y. Apeloig, M. Karni, J. Chandrasekhar and P. v. R. Schleyer, *J. Am. Chem. Soc.*, **108**, 270 (1986).
38. H. Teramae, *J. Am. Chem. Soc.*, **109**, 4140 (1987).
39. J.-P. Malrieu and G. Trinquier, *J. Am. Chem. Soc.*, **111**, 5916 (1989).
40. G. Trinquier and J.-P. Malrieu, *J. Phys. Chem.*, **94**, 6184 (1990).
41. C. Liang and L. C. Allen, *J. Am. Chem. Soc.*, **112**, 1039 (1990).
42. R. S. Grev, *Adv. Organomet. Chem.*, **33**, 125 (1991).
43. M. Karni and Y. Apeloig, *J. Am. Chem. Soc.*, **112**, 8589 (1990).
44. (a) M. V. George, D. J. Peterson and H. Gilman, *J. Am. Chem. Soc.*, **82**, 403 (1960).
    (b) H. O. House, *Modern Synthetic Reactions*, 2nd ed., Benjamin, Menlo Park, CA, 1977, p. 568.
45. W. Hanstein, H. J. Berwin and T. G. Traylor, *J. Am. Chem. Soc.*, **92**, 829 (1970).
46. R. West, M. J. Fink and J. Michl, *Science (Washington, D.C.)*, **214**, 1343 (1981).

15. Mechanism and structures in alcohol addition reactions

47. D. J. De Young, M. J. Fink, R. West and J. Michl, *Main Group Met. Chem.*, **10**, 19 (1987).
48. S. Nagase, T. Kudo and K. Ito, in *Applied Quantum Chemistry* (Eds. V. H. Smith Jr., H. F. Schaefer III and K. Morokuma), Reidel, Dordrecht, 1986.
49. E. M. Arnett, *Prog. Phys. Org. Chem.*, **1**, 223 (1963).
50. J. Budaraju, D. R. Powell and R. West, *Main Group Met. Chem.*, **19**, 531 (1996).
51. S. Tsutsui, K. Sakamoto, H. Sakurai, M. Kira, T. Veszprèmi, 72 Annual Meeting of the Chemical Society of Japan, March 27–30 (1997); Abstracts I, 2E313.
52. M. Wind, D. R. Powell and R. West, *Organometallics*, **15**, 5772 (1996).
53. Y. Apeloig and M. Nakash, *Organometallics*, in press.
54. Y. Apeloig and M. Nakash, *J. Am. Chem. Soc.*, **118**, 9798 (1996).
55. L. E. Gusel'nikov, N. S. Nametkin and V. M. Vdovin, *Acc. Chem. Res.*, **8**, 18 (1975).
56. A. G. Brook, S. C. Nyburg, W. F. Reynolds, Y. C. Poon, Y.-M. Chan, J.-S. Lee and J.-P. Picard, *J. Am. Chem. Soc.*, **101**, 6750 (1979).
57. N. Wiberg, G. Wagner and G. Muller, *Angew. Chem., Int. Ed. Engl.*, **24**, 229 (1985).
58. N. Wiberg and G. Wagner, *Angew. Chem., Int. Ed. Engl.*, **22**, 1005 (1983).
59. N. Wiberg, G. Wagner, G. Muller and J. Riede, *J. Organomet. Chem.*, **271**, 381 (1984).
60. N. Wiberg and G. Wagner, *Chem. Ber.*, **119**, 1467 (1986).
61. N. Wiberg, *J. Organomet. Chem.*, **273**, 141 (1984).
62. N. Wiberg and H. Köpf, *J. Organomet. Chem.*, **315**, 9 (1986).
63. K. Sakamoto, J. Ogasawara, H. Sakurai and M. Kira, *J. Am. Chem. Soc.*, **119**, 3405 (1997).
64. H. Sohn, J. Merritt and R. West, 30th Organosilicon Symposium, May 30–31, 1997, London, Ontario, Canada, Abstracts, A19.
65. A. G. Brook, K. D. Safa, P. D. Lickiss and K. M. Baines, *J. Am. Chem. Soc.*, **107**, 4338 (1985).
66. K. M. Baines, A. G. Brook, R. R. Ford, P. D. Lickiss, A. K. Saxena, W. J. Chatterton, J. F. Sawyer and B. A. Behnam, *Organometallics*, **8**, 693 (1989).
67. P. R. Jones and M. E. Lee, *J. Am. Chem. Soc.*, **105**, 6725 (1983).
68. P. R. Jones, M. E. Lee and L. T. Lin, *Organometallics*, **2**, 1039 (1983).
69. P. R. Jones and T. F. Bates, *J. Am. Chem. Soc.*, **109**, 913 (1987).
70. P. R. Jones, T. F. Bates, A. F. Cowley and A. M. Arif, *J. Am. Chem. Soc.*, **108**, 3122 (1986).
71. G. Bertrand, J. Dubac, P. Mazerolles and J. Ancelle, *Nouv. J. Chim.*, **6**, 381 (1982).
72. M. J. Fink, D. B. Puranik and M. P. Johnson, *J. Am. Chem. Soc.*, **110**, 1315 (1988).
73. D. B. Puranik and M. J. Fink, *J. Am. Chem. Soc.*, **111**, 5951 (1989).
74. M. J. Fink, in *Frontiers of Organosilicon Chemistry* (Eds. A. R. Bassindale and P. P. Gaspar), Royal Society of Chemistry, Cambridge, 1991.
75. M. Kira, T. Maruyama and H. Sakurai, *J. Am. Chem. Soc.*, **113**, 3986 (1991).
76. F. G. Bordwell, *Acc. Chem. Res.*, **21**, 456 (1988).
77. pK_a value of propanol in DMSO from Reference 78.
78. W. N. Olmsted, Z. Margolin and F. G. Bordwell, *J. Org. Chem.*, **45**, 3295 (1980).
79. T. Maruyama, M. Sc. Thesis, Tohoku University (1989).
80. G. W. Sluggett and W. J. Leigh, *J. Am. Chem. Soc.*, **114**, 1195 (1992).
81. W. J. Leigh and G. W. Sluggett, *J. Am. Chem. Soc.*, **116**, 10468 (1994).
82. W. J. Leigh, C. J. Bradaric and G. W. Sluggett, *J. Am. Chem. Soc.*, **115**, 5332 (1993).
83. C. J. Bradaric and W. J. Leigh, *J. Am. Chem. Soc.*, **118**, 8971 (1996).
84. W. J. Leigh, C. J. Bradaric, C. Kerst and J.-A. H. Banisch, *Organometallics*, **15**, 2246 (1996).
85. M. Ishikawa, T. Fuchikami and M. Kumada, *J. Organomet. Chem.*, **118**, 155 (1976).
86. M. G. Steinmetz, *Chem. Rev.*, **95**, 1527 (1995).
87. C. J. Bradaric and W. J. Leigh, *Can. J. Chem.*, **75**, 1393 (1997).
88. Y. Apeloig, in *The Chemistry of Organic Silicon Compounds*, Vol. 1 Part 1 (Eds. S. Patai and Z. Rappoport), Chap. 2, Wiley, New York, 1989.